Molecular Neurobiology of the Mammalian Brain

SECOND EDITION

Molecular Neurobiology of the Mammalian Brain

SECOND EDITION

Patrick L. McGeer

The University of British Columbia
Vancouver, British Columbia, Canada

Sir John C. Eccles

Abteilung Neurobiologie
Max-Planck-Institut für Biophysikalische Chemie
Göttingen, West Germany

and

Edith G. McGeer

The University of British Columbia
Vancouver, British Columbia, Canada

PLENUM PRESS • NEW YORK AND LONDON

Library of Congress Cataloging in Publication Data

McGeer, Patrick L.
  Molecular neurobiology of the mammalian brain.

  Bibliography. p.
  Includes index.
  1. Brain. 2. Molecular biology. 3. Neurobiology. 4. Mammals—Physiology. I. Eccles, John C. (John Carew), Sir, 1903–   . II. McGeer, Edith G. III. Title.
[DNLM: 1. Brain. 2. Molecular Biology. WL 300 M478m]
QP376.M14   1986   599′.01′88   86-25333
ISBN 0-306-42329-4
ISBN 0-306-42511-4 (pbk.)

© 1987 Plenum Press, New York
A Division of Plenum Publishing Corporation
233 Spring Street, New York, N.Y. 10013

Printed in the United States of America

Preface

The human brain is the inner universe through which all external events are perceived. That fact alone should ensure that neuroscience will eventually receive top priority in the list of human endeavors. The brain represents the pinnacle of sophistication in the realm of living systems. Yet it is an imperfect organ, whose failures in disease processes lead to the occupation of more than half of all hospital beds and whose variable performance in the healthy state contributes in undetermined degree to the world's social problems. Every significant advance in or understanding of the brain has yielded enormous practical dividends. There is every reason to believe the future holds even greater promise.

In the preface to our first edition, we drew attention to the establishment of graduate programs in dozens of universities around the world and the emergence of numerous international journals devoted to interdisciplinary work on the brain. The discoveries that have flowed from this activity have required extensive updating of the details of this book, which is a testimony to the fruitfulness of neuroscience research.

Yet the basics remain the same. It is more important than ever that the neuroscientist be presented with the fundamental subdisciplines that together make up the total of brain research in an integrated manner.

The intention is not to present in detail any aspect of neuroscience. It is not, like many textbooks, a compendium of discoveries. Rather, certain basic information is selected from each of the subdisciplines to build a picture of brain function. The book is especially intended for graduate students and postdoctoral fellows who wish to be exposed to a broad view of the brain in its entirety. We anticipate as well that the book will have value to the specialist, not so much in his own particular realm but in those areas with which he may not be familiar. Thus, it is our hope that the neurophysiologist and neuroanatomist will gain greater insight into neurochemistry and neuropharmacology, while the neuropharmacologist and neurochemist will gain greater insight into neurophysiology. Particular attention has been devoted to the neuroscience underlying many human disease processes and the drugs used to treat them. Neurologists, neurosurgeons, and psychiatrists who wish a strong grounding in those aspects of basic science important to their fields will find the book especially valuable.

Every neuroscientist must select some aspect of the brain in which to work in detail. It is important that the investigator never lose sight of the broader role his work plays in developing an understanding of the magnificent creation that is the human brain. Such a vision stirs the imagination for research and confers wisdom in the approach to it. It provides the necessary background for the linking of new ideas and data into a broader conceptual framework. The book moves from the fine detail of molecular genetics to the broad phenomenon of human consciousness. The chapters are divided into three parts. Part I commences by describing the basic molecular components and subcellular organelles that make up the neuron. It goes on to describe the properties of electrical excitability that are unique to neurons and the peripheral muscles they innervate.

Part II begins by defining neurotransmitters and the ways in which they interact with their receptors. It then defines in a series of chapters the chemistry, anatomy, physiology, and pharmacology of each of the known neurotransmitters.

Part III deals with functional systems, integrating the physiological and biochemical information provided in the first two parts. It commences with a chapter on the building of the brain and its eventual deterioration with age. It then provides functional chapters on movement, behavior, learning, and finally consciousness.

To satisfy the needs of economy of space and clarity of presentation, we have avoided detailed documentation and have not attempted to credit every investigator who has made an outstanding contribution to his specialty. However, there is an extensive reference list, including numerous review articles, which should permit the reader to gain ready access to the details of most facets of brain research. While these may not present a proper balance of prominent authors, they should provide a satisfactory guide to the reader who wishes to pursue any subject in more depth.

ACKNOWLEDGMENTS.   A book of this nature is dependent on good illustrations, and we are grateful to the publishers and scientists who generously granted permission to publish figures, many of which are from articles or books by the following authors: K. Akert, G. I. Allen, J. Altman, P. Andersen, H. Asanuma, R. P. Barber, T. Beach, T. Berger, C. Blakemore, P. I. Bliss, R. Bowker, D. Calne, P. Constantinides, F. Conti, C. Cotman, M. Cowan, R. De Lorenzo, R. M. Eccles, C. Ekerot, E. Evarts, P. Fatt, C. A. Fox, K. Fuxe, R. Granit, E. G. Gray, P. Greengard, J. Hamori, H. K. Hartline, T. Hattori, E. Henneman, A. L. Hodgkin, T. Hökfelt, J. Iles, M. Ito, J. Jack, M. Jacobson, J. K. S. Jansen, H. Kamo, B. Katz, S. Kawaguchi, R. D. Keynes, H. Kimura, H. H. Kornhuber, S. Landgreň, B. Libet, T. Lømo, A. Lundberg, G. S. Lynch, C. Marsden, R. Marshbanks, P. B. C. Matthews, V. B. Mountcastle, N. H. Neff, R. A. Nicoll, S. Ochs, T. Oshima, D. W. Pfaff, R. Poritsky, R. Porter, G. Raisman, P. Rakic, C. E. Ribak, E. Roberts, P. Roland, A. Rose, B. Sakmann, K. Sasaki, P. Scheid, P. H. Schiller, G. E. Schneider, W. Schultz, T. A. Sears, R. W. Sperry, J. Szentágothai, H. Tago, C. A. Terzuolo, W. T. Thach, C. B. Trevarthen, N. Tsukahara, U. Ungerstedt, S. Vincent, T. Watanabe, and H. Yamamura.

We wish as well to thank the late Professor Rolf Hassler and Professor Wolf Singer and Manfred Klee of the Max-Planck-Institut für Hirnforschung (Frankfurt). Many of the illustrations are the work of Mr. Bruce Stewart and the staff of the Biomedical Communications Department of the University of British Columbia. We wish to thank Dr. Helena Táboříková Eccles and Miss Anna C. Marnik for assistance in preparing the chapters, Mrs. Joane Suzuki and Miss H. Thomas for assisting with the artwork, and Mrs. Elizabeth Lyson for helping with the typing of the manuscript.

Contents

Chapter 11

Other Heterocyclic Putative Neurotransmitters: Histamine and Purines 349

List of Figures

Chapter 2
Signaling in the Nervous System

Chapter 3
Chemical Synaptic Transmission at Peripheral Synapses

Chapter 4

Synaptic Transmission in the Central Nervous System

Part II

Specific Neuronal Participants and Their Physiological Actions

Chapter 5

Principles of Synaptic Biochemistry

Chapter 6
Putative Excitatory Neurons: Glutamate and Aspartate

Chapter 7
Inhibitory Amino Acid Neurotransmitters

Chapter 8
Cholinergic Neurons

Chapter 9

Catecholamine Neurons

Chapter 10

Serotonin and Other Brain Indoles

Chapter 11

Other Heterocyclic Putative Neurotransmitters: Histamine and Purines

Chapter 12

The Prominent Peptides

Part III

The Integrative Aspects of Brain Function

Chapter 13

The Building of the Brain and Its Adaptive Capacity

Chapter 14

Control of Movement by the Brain

Chapter 15

Basic Behavioral Patterns

Chapter 16
Neuronal Mechanisms Involved in Learning and Memory

Chapter 17

Perception, Speech, and Consciousness

I

Architecture and Operation of the Nervous System

It is essential that all investigators of the nervous system understand the principles of operation of the neural machinery that they are investigating. Usually, this research is being carried out on some small component that is isolated for study by highly sophisticated techniques in, for example, neurochemistry, neuroimmunology, electron microscopy, or molecular genetics. The study may be at an even more elemental level, with such techniques as homogenization of some brain constituents such as cerebral nuclei and isolation from fractions of the homogenates of interesting molecules such as enzymes or other proteins and putative transmitters or related molecules. The aim of this kind of investigation is to construct metabolic systems that are vitally concerned in chemical synaptic transmission. This important field of inquiry is dealt with extensively in Part II of this book.

To achieve neurobiological meaning, however, the detailed accounts of neurochemical and neuropharmacological studies must be related to the structure and operation of the neural machinery of the brain. In Part I, this information is presented in four chapters. It is not a bare factual treatment. Rather, the attempt is made to outline the important historical events and to show the growth of understanding that occurs with a progressively changing problem situation. Too often, it is erroneously believed that science consists of the discovery and presentation of facts. On the contrary, it is concerned with the attempt to understand the phenomena of nature by the development of hypotheses and their rigorous testing by experiments. We hope that this book is an exemplar of this philosophy of science, which makes science an exciting adventure both in creative imagination and in experimental investigation, two aspects of science in vital interaction.

Chapter 1 gives an account of the structure of the components of the nervous system: neurons, synapses, nerve fibers, glia. Special treatment is given to organelles, both in the perinuclear region and in the presynaptic terminals, because of their significance in relation to the theme of Part II. Chapter 2 concentrates on the beautiful and efficient signaling system of nerve impulses. In Chapter 3, synaptic transmission is studied in two locations of peripheral nervous systems because at these favorable sites the principles of operation have been investigated with great precision. When the principles were established in this way, it was possible to show that similar principles obtained with transmission at some synapses of the central nervous system that were amenable to comparable investigations. For example, the quantal composition demonstrated for the endplate potential could be recognized with central synapses [cf. Figure 4.3 (Section 4.2)]. Chapter 4 comprises a wide range of investigations on excitatory and inhibitory synapses and finishes with a description of some simpler neuronal pathways. This description will serve as an introduction to the much more complex situations in Parts II and III.

1

The Fine Structure of the Mammalian Brain

1.1. Introduction

The brains of six mammals drawn to the same scale (Figure 1.1) illustrate the evolutionary development of the brain from the primitive marsupial up to man, in whom the cerebrum and cerebellum are dominant. But size alone does not account for the preeminent performance of the human brain. This organ, weighing about 1.5 kg, is the most highly organized matter known in the universe. At our present level of understanding, the brains of higher mammals are not greatly inferior in microstructure or in most aspects of operational performance. The brains of elephants and whales are many times larger—up to five times for a whale—and even the brain of the dolphin may be a little larger than the human brain. Yet there is something very special about the human brain. Its performance in relationship to culture, to consciousness, to language, and to memory uniquely distinguishes it from even the most highly developed brains of other animals. We shall discuss this problem in Chapter 17, in which we shall see that it is beyond our comprehension how these subtle properties of the conscious self came to be associated with a material structure, the human brain, that owes its origin to the biological process of evolution. We can state with complete assurance that each of our brains forms the material basis for our experiences and memories, our imaginations, our dreams. Furthermore, it is through our brain that each of us can plan and carry out actions and so achieve expression in the world. We are able to do this because our streams of conceptual thinking can somehow or other activate neuronal changes that eventually result in all the complex movements that give expression. For each of us, our brain is the material basis of our personal identity—distinguished by our selfhood and our character. In summary, it gives for each of us the essential "me." Yet when all this is said, we are still only at the

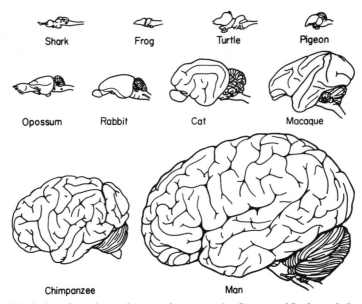

Figure 1.1. Brains of vertebrates drawn to the same scale. Courtesy of Professor J. Jansen.

beginning of comprehending the mystery of what we are. This fundamental philosophical problem is still far beyond our understanding, though in Chapter 17 we will suggest that remarkable progress is being made in this, the greatest of all problems that confront us. It is essentially the problem formulated long, long ago and defined in the question of Plotinus: "What am I?" Meanwhile, we will be engaged in the task of trying to understand the nervous system as neurobiologists. It is convenient at this stage to consider the brain as a machine, but it is a special kind of machine, of a far higher order of complexity in performance than any machine designed by man, even the most complex of computers.

It will be our task in this chapter to describe the essential structure of the brain—the components from which is is built and how they are related to each other as well as some of the fundamental techniques for examining it in detail. In Chapters 2, 3, and 4, we will give an account of the mode of operation of the simplest components. These two modes of investigation, the morphological or structural and the physiological, are complementary, and together they form the basis of a further phase of our inquiry that concerns the linkage of individual components into the simplest levels of organization. We are beginning to understand the simpler patterns of neuronal organization and the way they work. We are still at a very early stage of our attempt to understand the brain, which may well be the last of all the frontiers that man may attempt to penetrate and encompass. We predict the search will extend hundreds of years into the future. We will never run out of problems on this greatest of all problems confronting man because the problems multiply far faster than we solve them! Vigorous and exciting new disciplines emerge in such fields as neurochemistry, molecular neurobiology, neurogenetics, neuropharmacology, neuromathematics, and neurocommunications.

The cellular components of the brain fall into two major categories: neurons, or nerve cells, and neuroglia. The neurons subserve all the specific properties of the brain such as

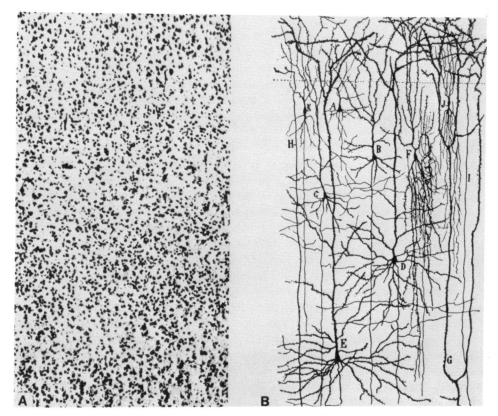

Figure 1.2. Magnified sections of cerebral cortex. (A) All cell bodies are stained. From Sholl (1956). (B) Composite drawing of a Golgi preparation by Ramón y Cajal.

impulse transmission, which operates on the input side in perception and on the output side in movement, and which in all of its patterned diversity underlies the analytical and synthetic properties of the brain. Neuroglial cells are ubiquitous throughout the brain. The word "glia" is derived from "glue" because neuroglia were originally thought of as merely the supporting structures of the brain, but now they are known to have more important and subtle functions, as described in Section 1.3.

1.2. The Neuron

1.2.1. General Morphology

Figure 1.2A is a low magnification of a histological section of a small segment of the human visual cortex (Sholl, 1956). Each of these densely packed dots is a neuron or a glial cell, and these are the individual components of which the whole brain is built. You have to imagine that the human cerebrum, as drawn in Figure 1.1, has about 10 billion of these individual neurons; in Figure 1.2A, they are shown tightly packed together, but actually the density of the packing is much greater. In Figure 1.2A, only the bodies of the neurons are stained. You do not see all the interlacing branching structures stemming from each one of these cell bodies, as shown in Figure 1.2B for a few nerve cells. In this Golgi-stained section of the human cerebral cortex (Ramón y Cajal, 1911) at about 2 times higher magnification, you see one of these neurons or nerve cells (D) displayed right in the center. Growing out upward and sideways from the cell body are several branching "dendrites," as they are called, and projecting straight downward is the fine axon, a nerve fiber. It gives a few branches before leaving the cortex, as do also the axons of cells B, C, and E. Other nerve cells or neurons, A, F, J, and K, in various shapes and sizes, can be seen with their dendrites extending in different directions and their axons branching and terminating in that section of the cerebral cortex.

The clarity of Figure 1.2B depends on the fact that, in this thin section, only about 2% of the cells were stained by this special Golgi technique. But for this lucky accident of random selection by the stain, which is still not understood, the composite of nerve cells seen in Figure 1.2A with their dendrites and axons would be solid black, impenetrable to discrimination.

The packing shown in Figure 1.2 is not only dense, but also extremely intricate, as demonstrated by the electron micrograph in Figure 1.3. The extrasomatic space, which seems plentiful in Figure 1.2A and vast in Figure 1.2B, is actually jammed with processes of neurons and glia. Figure 1.3 also shows many synapses, as well as such subcellular organelles as the nucleus and nucleolus, Golgi apparatus, mitochondria, Nissl bodies, microtubules, and other structures that are described in later sections. We shall see that despite the congestion, the neuronal components of the cortex [see Figure 17.2 (Section 17.1)] or other regions of the brain are selectively connected to form organized patterns. The white matter of brain is, of course, mainly neuronal interconnections, but even in gray matter most of the space is taken up by neuronal and glial processes.

We owe particularly to Ramón y Cajal (1909, 1934), the great Spanish neuro-anatomist, the concept that the nervous system is made up of neurons that are isolated cells, not joined together in some syncytium, but each one independently living its own biological life. This concept is called the *neuron theory*. It is already evident from Figure 1.2B that neurons characteristically are complex, branched structures with a central body or soma from which there rise the branching dendrites and the single axon. These form what we may call, respectively, the receptive pole and the transmitting pole of the neuron, but this terminology must not be taken too literally because the soma also participates in the receptivity of the neuron and usually in the transmissive function as well. In fact, the dendrites can be regarded simply as extensions of the soma, and we shall see that they have a similar cytoplasmic composition. It is otherwise with the axon, right from the site of its origin from the soma.

Many figures in later chapters show special morphological features of neurons. In the mammalian brain, the largest neurons may have a soma 70 μm across, whereas the smallest would be less than 5 μm. Dendrites usually arise as multiple branches from the soma [cf. Figures 1.2B and 1.4A (Section 1.2.2)], often with a dominant *apical dendrite* as in cells B, C, D, and E of Figure 1.2B and in the two cells of Figure 1.3. But the most elaborately branched dendrite, that of the Purkinje cell, arises unipolarly [cf. Figure 14.14 (Section 14.4.2)]. With pyramidal cells of the cerebral cortex [see Figures 1.2B and 1.4B and Figure 17.2 (Section 17.1)], the apical dendrite extends up to the surface, giving off

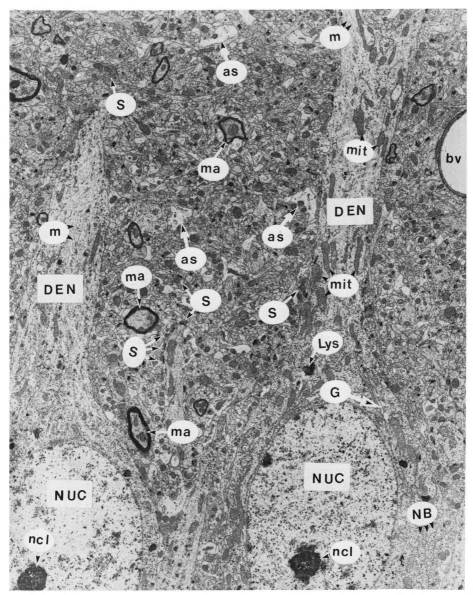

Figure 1.3. Electron micrograph of two hippocampal pyramidal cells, showing proximal segments of their apical dendrites (DEN). Processes are identifiable as dendrites because of the absence of myelin sheaths and the presence of synapses along the surfaces. Nuclei (NUC) are at the bottom of the photomicrography, with both showing a prominent nucleolus (ncl). The Nissl bodies (NB) are represented by highly ordered rough endoplasmic reticulum with associated ribosomes attached to long and shallow cisterns. Lysosomes (Lys) are seen throughout the cytoplasm. One representative of a Golgi apparatus (G) can be easily identified. Mitochondria (mit) are spread throughout the cells. They tend to be elongated, with their long axes oriented in the direction of the dendrites. Microtubules (m) can be seen as faint structures along dendrites. Some appreciation of the incredible complexity of the neuropil can be gained by examining the multitudinous processes among the cell bodies. Several myelinated axons (ma), astrocytic processes (as), and synapses (S) can be identified as well as a blood vessel (bv). Courtesy of A. Rose (University of British Columbia).

several branches en route (Ramón y Cajal, 1911; Szentágothai, 1969). In most species of neurons, the axon arises from a conical projection of the soma, the initial segment [see Figure 1.5A (Section 1.2.2)]. It may continue for a distance of several millimeters to a meter as a transmission line from the soma, being then termed a *nerve fiber*. Usually, it gives off several branches soon after its origin, the *axon collaterals*. The most numerous neurons are those with short axons. Usually, as in Figure 1.2B, cells A, F, and J, the axons of these neurons terminate in profuse branching soon after their origin (the so-called ''Golgi cells''). In other cases, the axon may extend for a few millimeters with very restricted branching. A notable example is the granule cell of the cerebellum [cf. Figure 14.14 (Section 14.4.2)], in which the axon dichotomizes to form a parallel fiber with a trajectory of several millimeters (Ramón y Cajal, 1911; Fox, 1962; Eccles *et al.*, 1967).

An important generalization is that all neurons are completely ensheathed by a surface membrane that is about 7 nm thick and is composed of a bimolecular leaflet of phospholipid molecules with the hydrophilic polar groups pointing both outward and inward [see Figure 1.14 (Section 1.2.3)]. The detailed structure with the associated proteins is treated in Section 1.2.3.

The complexity of neuronal organization is so great that only vague estimates are so far available as to the numbers of neurons, various types of glia, and neuronal synapses in any but the simplest of brains. An exception is the monkey striate cortex, of which O'Kusky and Colonnier (1982a,b) have completed a landmark study. Their results show the following totals per cubic millimeter of adult monkey cortex: 119,000 neurons, 37,700 astrocytes, 16,700 oligodendroglia, 3800 microglia, and 276,000,000 synapses. The total number of neurons for all the visual cortex in one cerebral hemisphere of the monkey was estimated to be 161,000,000.

1.2.2. Synapse

Since a neuron is a biological unit completely enclosed by the surface membrane, we are confronted by the problem of how neurons communicate with each other. This communication is the essence of all brain action, from the simplest to the unimaginably complex. It happens by means of the fine branches of axons of other neurons that make contact with its surface and end in little knobs scattered all over its soma and dendrites, as indicated in Figure 1.4A and B. It was the concept of Sherrington (1906) that these contact areas are specialized sites of communication, which he labeled *synapses* from the Greek word ''synapto,'' which means ''to clasp tightly'' [for a historical review, with confirmation of the neuron theory, reference can be made to Chapter 1 of Eccles (1964b)].

Diagrammatically, the neuronal connectivity is indicated in the model of a neuron suitably enlarged in Figure 1.4A. From the soma or cell body, there are shown just the stumps of the dendrites that would be extending great distances in all directions, and projecting downward from this soma is the single axon that eventually ends in many branches. The arrows indicate that messages or impulses come to the neuron by the many fine nerve fibers that make contacts on the surface of the soma and dendrites. The sole output line is along the axon. Figure 1.4B shows the varieties of contacts made by nerve fibers on the dendrites and soma of a pyramidal cell of a special part of the cerebral cortex, which is similar to the pyramidal cells of Figure 1.2B. Many contacts are made on small

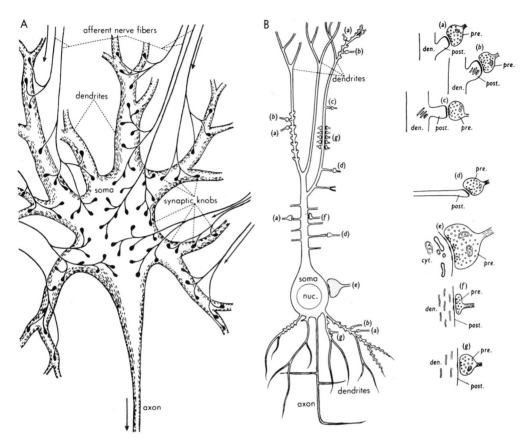

Figure 1.4. Synaptic endings on neurons. (A) General diagram. (B) Drawing of a hippocampal pyramidal cell to illustrate the diversity of synaptic endings on the different zones of the apical and basal dendrites, and the inhibitory synaptic endings on the soma. (*a–g*) Various types of synapses shown at higher magnification. From Hamlyn (1962).

spines that branch out from the dendrites. Numerous small spines can be seen projecting from the dendrites of the cells in Figure 1.2B. The total number of synapses of all kinds on central human cortical cells has been estimated to be about 40,000 per neuron (Cragg, 1975), but in the monkey striate cortex, a count showed 2300 per neuron (O'Kusky and Collonier, 1982a,b).

Electron micrographs show that central synapses resemble neuromuscular synapses [see Figure 3.2 (Section 3.2.2)] in their essential features. The presynaptic nerve fiber expands to the terminal bulb, or it may traverse several synaptic zones, transiently expanding in each, forming the "boutons de passage." In the presynaptic fiber, there are synaptic vesicles that are concentrated toward the synaptic cleft, often in a discrete cluster (see Figure 1.5B). The numerous mitochondria are more remotely placed. The synaptic cleft has a remarkably uniform width of about 20 nm. The presynaptic and postsynaptic membranes that bound the cleft tend to be darkly stained. On the dendrites of some neurons, e.g., pyramidal cells of the cerebral cortex (Figure 1.4B), there are numerous

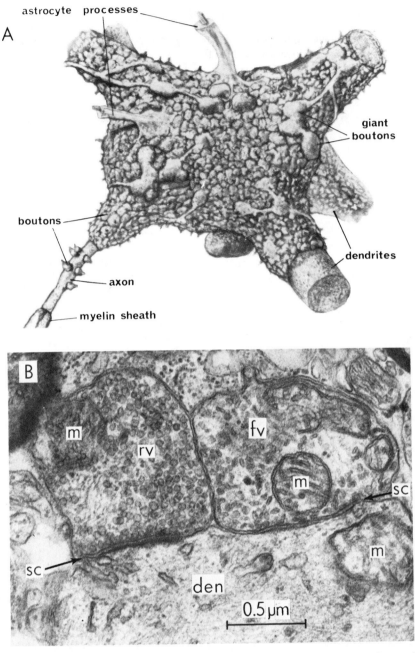

Figure 1.5. Synapses on the surface of motoneurons. (A) Reconstruction from serial electron micrographs of a motoneuron with its tightly packed surface scale of synaptic knobs (boutons), some being large. From Poritsky (1969). (B) Electron micrograph of two synaptic knobs on a motoneuron in a fish spinal cord. That to the left is excitatory, with spherical vesicles (rv) and dense staining on each side of the synaptic cleft (SC); that to the right is inhibitory, with ellipsoid vesicles (fv) and a much lighter staining of the membranes on each side of the synaptic cleft. (m) Mitochondrion; (den) dendrite. From Gray (1970).

spines, each of which is the postsynaptic element of a synapse (Gray, 1959, 1983). These spine synapses can be extremely numerous—up to 100,000 on a Purkinje cell of the human cerebellum [see Figure 14.14 (Section 14.4.2)].

The synaptic knobs cover the surface far more densely than is shown in Figure 1.4A. For example, in the reconstruction of Figure 1.5A, a synaptic scale provides an almost complete coverage of the surfaces of the soma, the dendritic stumps, and even the axonal origin. On the axon, there is shown the beginning of the myelin sheath that is discussed in Section 2.5 [cf. Figures 2.12 and 1.9 (Section 1.2.3)]. When the synaptic knobs are highly magnified by electron microscopy [see Figures 1.4B (Section 1.2.1) and 1.5B], their location at the ends of fine nerve fibers (Figure 1.4) is not often shown because the fine nerve fiber is usually not in the section. The most important finding is that the synaptic knobs can be seen to be completely separated from the surface of the dendrite by an extremely narrow space, the synaptic cleft. Synapses such as those of Figure 1.5B are the structural basis of the chemical transmission whereby the messages that come down the fine fibers to synaptic knobs on the recipient cell act on that cell by means of the secretion of minute amounts of specific chemical substances stored in the synaptic vesicles. The mode of operation of synapses is described in Chapters 3 and 4.

In Figures 1.4B and 1.5A, synapses are seen to vary in size, which is in agreement with the electron microscopic observations of Conradi (1969), who described five types of synapses on motoneurons. The very large synapses (cf. Figure 1.5A) are presumably the endings from afferents of annulospiral endings that give the monosynaptic excitation that is diagrammed in Figure 4.1A (Section 4.1). Figure 1.5A gives some feeling for the busy life that a nerve cell has, with incessant bombardment by impulses activating its dense coverage by synapses, which are even on the axon at its origin. We shall see, however, that the neuron has some relief because, as first recognized by Sherrington (1906), many of these synapses, the inhibitory synapses, are specialized to prevent the neuron from firing impulses in response to the excitatory synapses. The neuron is in the hands, as it were, of the two opposing operations of excitation and inhibition, which are the theme of much of Chapter 4.

Two broad categories of synapses, Type 1 and Type 2, have been defined (Whittaker and Gray, 1962) which have classically been thought to be excitatory and inhibitory, respectively. As shown diagrammatically in Figure 1.6A, Type 1 is distinguished from Type 2 by three criteria: (1) The postsynaptic membrane thickening is denser than the presynaptic; hence, Type 1 synapses are referred to as asymmetric and Type 2 as symmetric, (2) The synaptic cleft of Type 1 is wider and includes a band of dense material. (3) With Type 1, the specialized contact area is large and continuous, whereas with Type 2, the specialized dense areas tend to be small and fragmented. With Types 1 and 2, there is an accumulation of synaptic vesicles in close relation to the dense areas, and the presynaptic dense structures can be seen in favorable preparations. When fixed in glutaraldehyde, Type 2 vesicles become ellipsoid in shape (Uchizono, 1967).

A third specialized contact area between neurons resembles the desmosomal contacts between epithelial cells. These desmosomes are characterized by symmetrical membrane thickening as with Type 2 synapses, but there are no associated synaptic vesicles; hence, these structures are assumed to function as attachment sites and not as channels of communication.

Spine synapses such as those labeled (a)–(d) in Figure 1.4B are almost invariably

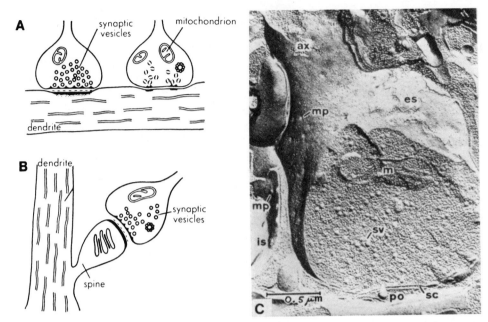

Figure 1.6. Synaptic structure. (A) Diagrams of two synapses on a dendrite, the left and right being, respectively, Type 1 with spherical vesicles and Type 2 with ellipsoid vesicles and other characteristic features of inhibitory synapses. (B) A synapse on a spine of a pyramidal cell (cf. Figure 1.4B) or of a cerebellar Purkinje cell, with all the features of an excitatory synapse. (A, B) from Whittaker and Gray (1962). (C) Photograph of a synapse as revealed by the freeze–etch technique. (ax) Axon; (es) external surface; (is) internal surface; (m) mitochondrion; (mp) micropinocytosis; (po) postsynaptic element; (sc) synaptic cleft; (sv) synaptic vesicle. From Akert *et al.* (1969).

Type 1, as shown in Figure 1.6B, and have spherical vesicles; hence, an excitatory identification is indicated. Synapse (e) of Figure 1.4B is of Type 2. The vesicles in Figure 1.4B are all shown spherical, as they would be because glutaraldehyde fixation was not employed. It is now recognized that the synapse illustrated by (e) is made by basket cells and is inhibitory with ellipsoid vesicles. Synapse (f) may also be inhibitory, but (g) is excitatory from the Schaffer collaterals [see Figure 16.3A (Section 16.3)].

The small round vesicles that occur at peripheral cholinergic synapses, such as at the neuromuscular synapses, also appear to characterize central synapses mediated by acetylcholine. On the other hand, the vesicles of γ-aminobutyric acid (GABA)ergic nerve endings in the deep cerebellar nuclei [see Figure 7.8 (Section 7.2.1)] and the substantia nigra are flattened. It is possible that the multitudes of synapses that operate by as yet unidentified excitatory transmitters also have small round vesicles and that glycine and other inhibitory transmitters have flattened ones. Pleomorphic morphology is also seen, for example, in vesicles of dopaminergic and serotonergic nerve endings. In terminals of neurosecretory cells, in chromaffin cells, and in peripheral noradrenergic varicosities, vesicles of different sizes exist that often contain a dense core (Cooper *et al.*, 1974). Other varieties of synaptic vesicles have been recognized, e.g., coated vesicles and complex vesicles (Gray, 1964).

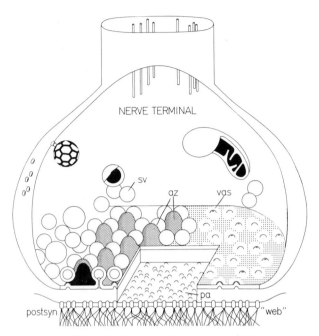

Figure 1.7. Schema of the mammalian central synapse. The active zone (az) is more complex and allows far more vesicle attachment sites (vas) per square unit of surface than the motor endplate. The postsynaptic aggregation of intramembranous particles is restricted to the area facing the active zone. The connection between the particles and the "web" of De Robertis is hypothetical. (sv) Synaptic vesicle; (pa) particle aggregations on postsynaptic membrane (postsyn.). From Akert *et al.* (1975).

A remarkable technique for displaying the ultrastructure of synapses is the freeze–etch technique illustrated in Figure 1.6C (Akert *et al.,* 1969). Synaptic vesicles and a mitochondrion are in the cut surface of the synaptic knob, which also has its external surface and its axon displayed. Below is the narrow synaptic cleft that separates the knob from the postsynaptic element.

By means of this freeze–etch technique, Akert *et al.* (1975) have examined the inner and outer aspects of the surface membrane surrounding a synaptic knob. The idealized cutaway perspective drawing of Figure 1.7 shows that the synaptic vesicles tend to be in a hexagonal array separated by dense structures (cf. Figure 1.6A and B) (Gray, 1964). To the right, the vesicles and dense structures have been removed, disclosing the underlying small protuberances, the vesicle attachment sites, in hexagonal array. Below is a cutaway area of the knob revealing the outer face of the postsynaptic membrane encrusted with fine particles that may be the postsynaptic receptor sites [cf. Figure 3.11A (Section 3.2.9)]. This illustration gives some indication of the cellular machinery associated with the movement and discharge of synaptic vesicles and with the postsynaptic action of the transmitter, but the mode of operation of this machinery is as yet largely unknown.

It is now recognized that the plane of cleavage in Figures 1.6C and 1.7 is not the synaptic cleft, but is along the middle of the bimolecular leaflets that form the cell membrane [see Figure 1.14 (Section 1.2.3)] (Akert and Peper, 1975). This reinterpretation of the cleavage planes, however, does not materially affect the general description given here. Similar fine structures occur at the neuromuscular junction, but there the attachment sites of the synaptic vesicles are arranged in longitudinal bands that are congruent with the folds on the postsynaptic membrane [see Figure 3.9A (Section 3.2.7)].

Only brief reference need be made to electrical synapses, where transmission is by electric currents flowing from the presynaptic to the postsynaptic component. Structurally, such synapses are identified by the very narrow cleft, 2–4 nm, and consequently they are called "gap junctions" (Pappas, 1975). Electrical synapses are important special sites in invertebrate and lower vertebrate nervous systems. Gap junctions rarely occur in the mammalian brain and appear to have little or no functional significance; hence, they will not be further considered in this book. The examples cited above make up the most common forms of synapses, but by no means do they cover the full repertoire of interneuronal connections in the brain. Other recorded types of synapses include the following: axoaxonic (Somogyi *et al.*, 1982a); reciprocal dendrodendritic, i.e., excitatory in one direction and inhibitory in the other (Rall *et al.*, 1966); somatodendritic (Lieberman, 1971); dendrosomatic (Pinching, 1970); somatosomatic (Pinching, 1970); and even dendroaxonic (Pinching and Powell, 1971). In addition, a number of neurotransmitters seem to be released in significant amounts from axonal swellings (varicosities) along their length, comparable to "boutons de passage" except that they make no synaptic contact, and therefore must act at a distance (see Chapters 7–11), while others are quite clearly released at dendrites, where no particular presynaptic dendritic structure can be seen by electron microscopy. It is well to keep in mind, therefore, that the classical descriptions presented here fall far short of telling the complete story of how neurons communicate by chemical messengers.

1.2.3. Components of the Neuron

In common with all eukaryotic cells, the neuron possesses a nucleus and a series of cytoplasmic organelles to carry out its biological functions (Figure 1.8). The overall chemical hierarchy that operates within these physical structures is now well established. It is controlled by DNA, the master molecule, which resides in the nucleus. DNA is the master molecule because it is the only one that replicates itself. It is made up of only four bases—adenine (A), cytosine (C), thymine (T), and guanine (G)—each of which is linked to 2-deoxyribose. The chain is formed by phosphate bridges between the 5′ position of one deoxyribose and the 3′ position of the next (the prime distinguishes between ribose and base ring numbers). The order of bases forms the code. It is conventional to read the DNA code in the direction from the 5′ position toward the 3′ position and to abbreviate it by the first letter of the base, e.g., ACTGACTG etc. The secret of DNA is that it always consists of two identical or complementary chains tightly bound to one another (see Figures 1.10 and 1.11). Complementarity is important because it ensures that two identical daughter strands can be produced from one mother strand. This is possible because in the double-helix form, hydrogen bonds are formed between A of one chain and T in the other, and between G and C (see Figure 1.11). The direction of the two chains is reversed, since the 5′ position of one chain is equal to the 3′ position of the other.

In addition to making copies of itself at the time of cell division, DNA acts as a

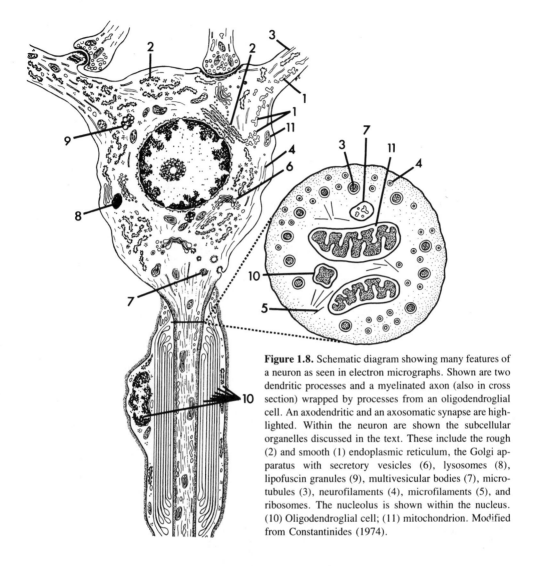

Figure 1.8. Schematic diagram showing many features of a neuron as seen in electron micrographs. Shown are two dendritic processes and a myelinated axon (also in cross section) wrapped by processes from an oligodendroglial cell. An axodendritic and an axosomatic synapse are highlighted. Within the neuron are shown the subcellular organelles discussed in the text. These include the rough (2) and smooth (1) endoplasmic reticulum, the Golgi apparatus with secretory vesicles (6), lysosomes (8), lipofuscin granules (9), multivesicular bodies (7), microtubules (3), neurofilaments (4), microfilaments (5), and ribosomes. The nucleolus is shown within the nucleus. (10) Oligodendroglial cell; (11) mitochondrion. Modified from Constantinides (1974).

template between cell divisions. Portions of its code are copied or *transcribed* into various RNA molecules, which are synthesized in the 5' to 3' direction. These RNA molecules are expendable, but in their limited lifetime are directors of cellular operations. RNA is also made up of only four bases—adenine (A), cytosine (C), guanine (G), and uracil (U)—each of which is attached to ribose and similarly joined from the 5' to 3' positions by phosphate bridges. The RNA produced by the DNA is identical except that U replaces T and ribose replaces deoxyribose.

In the cytoplasm, RNA directs protein synthesis in a process called *translation* (see Figure 1.12). Three letters of the RNA code called a "codon" translate into one amino acid in the protein chain. Thus, GCU translates into alanine, UUU into phenylalanine, and so on. Actually, there are 61 codons for the 20 amino acids. Multiple codons exist for all amino acids except tryptophan and methionine. Proteins are synthesized starting from the amino end and finishing at the carboxyl end, corresponding to the 5'-to-3' direction of the RNA.

Proteins, as enzymes, govern the remaining chemical operations of the cell. These are extremely sophisticated and many require their own carefully constructed factories known as organelles. We shall give brief functional and morphological descriptions of these organelles, it being understood that knowledge about them is still in a primitive state.

Nucleus.   The flow of information commences in the nucleus (Figures 1.9 and 1.10). The nucleus of every cell contains all the DNA in the mammalian genome. In the case of humans, this is thought to consist of roughly 3.3×10^9 bases in each of the two complementary chains. If strung out, this would consist of a double helix about 200 cm long! Only a tiny proportion of this DNA will ever be transcribed by any given neuron. The overwhelming majority represents DNA required by precursor cells or other types of cells in the body and not expressed in the neuron. Some DNA will be transcribed at a modest rate, but to maintain the functions of the cell, special portions must be transcribed at an extremely brisk rate. Obviously, a clever packaging system is required. Most of the vast DNA helix must be permanently filed away in the limited space of the nucleus, but

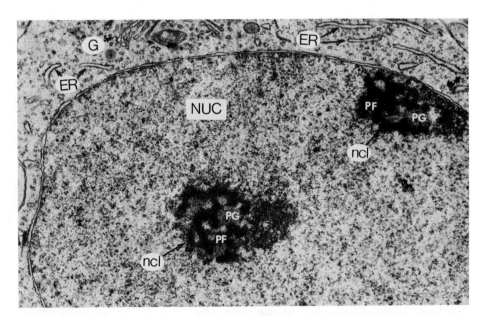

Figure 1.9. Nucleus (NUC) of immature neuron of neonatal rat cerebral cortex, showing primarily the double-layered nuclear membrane. Note the many gaps in the membrane and the apparent attachment of ribosomes to the external surface. The gaps represent holes in the spherical membrane. Ribosomes on the nuclear surface are thought to be newly formed. Two nucleoli (ncl) can be seen showing the pars granulosa (PG) and pars fibrosa (PF) segments. They participate in the formation of ribosomes. Many ribosomes are seen in the cytoplasm, including some in association with rough endoplasmic reticulum (ER). A Golgi apparatus (G) is also present. Courtesy of T. Hattori (University of Toronto).

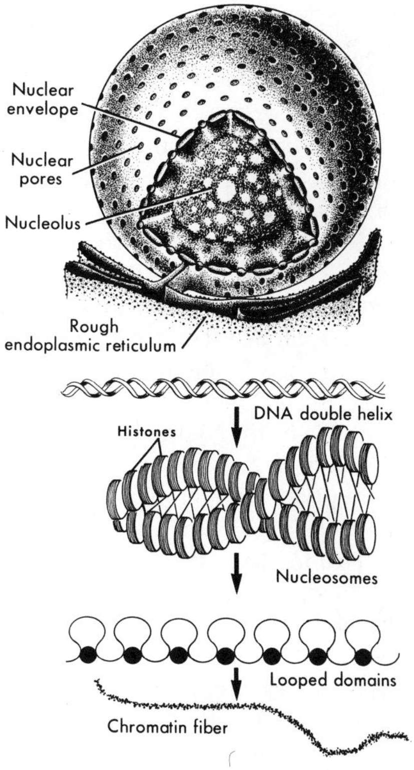

Figure 1.10. Structure of the nucleus and packing of DNA. The physical structure of the nucleus is shown at the top of the figure. It consists of a double membrane. The outer membrane is continuous with the rough endoplasmic reticulum as shown in the figure. The most prominent body inside the nucleus is the nucleolus. The nuclear envelope contains many pores. Inside the nucleus is contained all the genomic DNA, the arrangement of which is shown in the lower part of the figure. Progressively smaller scales are shown to illustrate how its 200-cm length can be coiled into a nucleus about 10 μm in diameter. First, the two chains are twisted to form a double helix. The double helix is then wound around histones to form nucleosomes. Those parts of the DNA that need to be readily available for transcription may be organized into looped domains, but finally the nucleosomes are organized into chromatin fibers. See the text for details.

small sections must be readily available for transcription. The method is illustrated in Figure 1.10. The first step is the formation of nucleosomes, by winding the DNA helix around a spool-shaped body made up of histones. Two copies each of four histones known as H2A, H2B, H3, and H4 combine to form the spool-shaped particle, which is about 11 nm in diameter. The spool is an extremely early invention, because these four highly basic histone proteins are virtually identical throughout the biological kingdom, whether the species is a pea or a cow. Each spool binds the acidic functions of 146 base pairs of DNA as shown in Figure 1.10. Linker DNA of about 60 base pairs joins one spool to the next. The spool plus the linker DNA make up the nucleosome. The human genome thus contains about 1.5×10^7 nucleosomes. The next step is to bind the nucleosomes together into a chromatin fiber. Another histone, H1, in concert with various proteins, forms this ropelike material, which can be detected under favorable circumstances by electron microscopy. The final step is packing the chromatin fibers with still more protein to form the typical granular material known as "nuclear chromatin" (Figure 1.9). Proteins may bind the fibers in such a way that accessible areas, known as "looped domains," are easily available for transcription (Figure 1.10). Perhaps specific stretches of DNA are not tightly protected by histones, but are reversibly bound to proteins specific for activating or repressing a gene.

Neurons cannot divide, so the neuronal DNA cannot reproduce itself. It does, however, have a remarkable capacity for self-repair. A missing or damaged base, for example, can be recognized by a repair nuclease that cuts out the damaged segment and uses the good chain as a template for repair.

The nucleus is concerned with directing the life of the cell by transcribing the three fundamental types of RNA: messenger (mRNA), ribosomal (rRNA), and transfer (tRNA). With the help of specific RNA polymerases, sections of the double helix part briefly, and the genetic information is read off one of the strands. An RNA polymerase initiates the process by binding to a DNA sequence known as the "promoter." It then unwinds the DNA helix, starting transcription at the initiation site and building up the RNA by adding bases complementary to the DNA one at a time, from the 5′ to the 3′ end, until a terminator signal is reached. At this point, the single-stranded complementary RNA chain is released, and the double-helix DNA is reconstituted.

The most rapid rate of RNA synthesis is in the nucleolus, which is the most prominent structure within the nucleus (Figure 1.9). The nucleolus is usually single and darkly stained with the conventional dyes used for light microscopy and by the heavy metals used for electron microscopy. It is so conspicuous on the background of the poorly stained karyoplasm that it provides an identifying feature of neurons. There may be more than one nucleolus, as Figure 1.9 shows. The nucleolus is an assembly of duplicated genes constituted for the purpose of synthesizing ribosomes. A loop of DNA is contributed by each of ten separate chromosomes, which join together for the heavy transcriptional duties that lie ahead. The nucleolus has no limiting membrane and is held together by a method of binding that is not really understood. The size of the nucleolus seems to be a reflection of its degree of activity. Its duplicated genes continuously transcribe rRNA at a furious rate, accounting for 80–90% of all RNA produced by the nucleus. This newly synthesized rRNA, produced by RNA polymerase I, is immediately bound with proteins imported from the cytoplasm. A complex series of steps involving cutting and splicing the RNA and joining with various proteins takes place. The large and small subunits of the ribosome are finally transported separately in a still immature state through pores in the nuclear membrane for finishing in the cytoplasm. The process is partly visible under the electron microscope, as shown in Figure 1.9. The granular compartment of the nucleolus contains 15-nm ribosomal precursor particles in later stages of assembly, while the fibrillar compartment contains many 5-nm ribonucleoprotein fibers representing the early RNA transcripts (Sommerville, 1985).

While the nucleolus has a specialty function for ribosomes, the main part of the nucleus is where the many thousands of different mRNAs that will be translated into the specific proteins of the cell are produced. The human genome is estimated to contain about 200,000 structural gene loci, of which about 80,000 are expressed in the brain (McKusick, 1982), many of which will be rare copies expressed only in subsets of neurons.

The initial transcripts are referred to as "heterogeneous nuclear RNA" (hnRNA). The initiator and terminator sites are different from those for rRNA, and the enzyme required is RNA polymerase II, but the process is similar. Studies on isolated nuclei show that typical hnRNAs are 8000–20,000 nucleotides long, which is much longer than the 1000–2000 nucleotides required to code for the average protein. The explanation is that further processing of hnRNA is required before the RNA is ready for shipment to the cytoplasm as mature mRNA. The process is illustrated in Figure 1.11. First, a cap of 7-methylguanosine triphosphate is placed on the 5′ end to permit subsequent binding to a

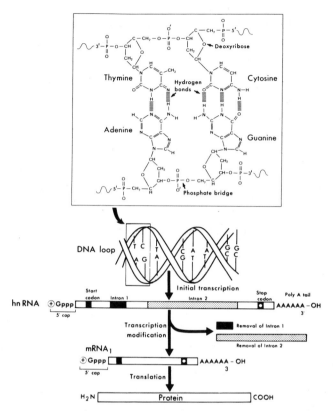

Figure 1.11. Process of transcription of DNA to rare copy RNA in the nucleus. Molecules from two chains are seen joined together with thymine and cytosine in the top chain, forming hydrogen bonds with adenine and guanine in the bottom chain. The chains are formed by phosphate bridges that go in opposite, or complementary, directions. At the top, the linkage is from the 3′ deoxyribose position on the left to the 5′ on the right, while the reverse is true for the bottom chain. A DNA loop parts to permit initial transcription of a length of DNA giving heterogeneous nuclear RNA (hnRNA). A 5′ cap and a poly (A) tail are added. The hnRNA is shown containing two intron sequences that are excised by transcription modification. Messenger RNA₁ is formed by the removal of intron 1, while MRNA₂ is formed by the removal of intron 2. These different mRNAs can leave the nucleus and later be translated into two different proteins by the process shown in Figure 1.12. These two proteins may represent different genes from the same transcribed segment of DNA. See the text for details.

ribosome, and then comes transcription followed typically, but not always, by the addition of 100–200 adenylic residues to the 3′ end as a polyadenylate [poly(A)] tail.

The hnRNA is then cut, with large interior sections known as "introns" being removed. The cut ends, referred to as "exons," are then spliced back together to produce the final mRNA. The mRNA leaving the nucleus will have a 5′ cap followed by sequences of unknown function prior to an AUG triplet codon for methionine to initiate protein translation. This will be followed by the translation sequence ending in a UGA, UAA, or UAG codon to signal termination. Apparently, the process of cutting out introns adds gene variability, because the sequences removed from the same hnRNA transcript can vary. Different extrons are then spliced together to produce a series of mRNAs from a single hnRNA. In addition, introns are evidently required for export from the nucleus, because isolated DNA transcripts without introns, when inserted into the nucleus, remain there. In total, only about 5% of the RNA produced in the nucleus ever leaves for the cytoplasm, so the majority of synthesized RNA must be in the form of intron sequences.

The poly(A) tail added to the 3' end is of great assistance to biochemists in sorting out the less plentiful mRNAs of the cell from the larger, relatively homogeneous populations of rRNA and tRNA. The number of copies of any given mRNA in a cell can be extremely small, depending on how often a given gene needs to be expressed.

As with rRNA, tRNA must be produced in large quantities by the cell. Thus, there exist multiple DNAs for tRNA. It is synthesized in the extranucleolar region by RNA polymerase III, which also makes one of the small components of the ribosome. Less is known about tRNA than about the other forms, but rapid progress in our understanding can be anticipated. The final product, the result of the normal intranuclear splicing procedures, is a series of folded L-shaped nucleotides of about 70–80 bases, at least one for each amino acid. They are constructed in a similar way. The specific amino acid attaches to a CCA terminus at one end of the L. An anticodon of three bases recognizes its codon on mRNA at the time of protein synthesis. Each tRNA is named for the amino acid it transfers.

Ribosomes and Protein Synthesis.   The function of the tiny ribosome granule is to receive genetic instruction from mRNA produced in the nucleus and to translate it into protein in the cell. The ribosome, no larger than 250 Å in its longest dimension, is about one tenth the size of a wavelength of visible light and therefore cannot be detected by the optical microscope. It has the equivalent weight of about 100–150 protein molecules and thus is too large and cumbersome for structural identification by X-ray diffraction. Its size is reasonably suited to detection by electron microscopy, and the main features of its rather complex shape have been worked out by this method (Lake, 1981). The eukaryotic ribosome consists of two subunits: a small subunit of sedimentation rate 40 S with a head, a base, and a platform; and a large subunit of sedimentation rate 60 S with a central protuberance flanked by a ridge and a stalk. The complete ribosome has a sedimentation coefficient of about 80 S. Ribosomes are produced in prodigious quantities by the nucleolus, because ribosomes make possible the final amplification stage in which a single mRNA is translated into many protein copies. Ribosomes can be observed attached to the outer surface of endoplasmic reticulum, in which case the reticulum is designated as "rough endoplasmic reticulum" (see Figures 1.8, 1.9, and 1.10), or in the cytosol, often in rosettelike clusters of five or six. The clusters are probably evidence of the translation amplification process, in which several ribosomes in convoy are simultaneously engaged in synthesizing protein from a single mRNA molecule as shown in Figure 1.12.

The protein synthesis process is started by a special initiator, *N*-formyl-methionyl-tRNA. It differs from other tRNAs in having a blocked amino group to protect the end of a growing protein chain. This initiator attaches to a complex formed between the specific mRNA to be read and the small 40 S subunit of the ribosome with the help of initiator proteins. The large subunit then joins to form a complete ribosome. The *N*-formylmethionyl-tRNA occupies what is called the "P" site on the ribosome, which also has the enzyme peptidyl-tRNA. Elongation then involves the following three steps: codon recognition, peptide-bond formation, and translocation. Complex interactions requiring protein-elongation factors and GTP are involved in these three steps, which are shown in

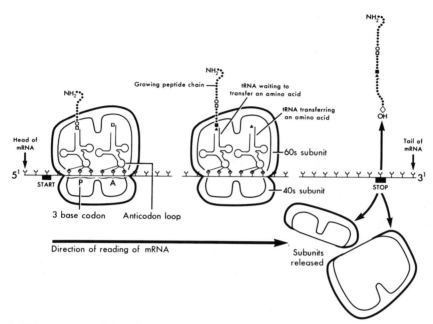

Figure 1.12. Process of translation of RNA into protein in the cytoplasm. Two ribosomes in a ribosomal array are shown reading the same length of mRNA that is being threaded through from the 5′ end to the 3′ end. Each ribosome has two sites for tRNA molecules on its surface, one at the A site and one at the P site (see the text for details). When the stop codon is reached, the protein chain is released and the ribosome parts into its two subunits.

Figure 1.12. Each codon, which consists of three bases on the mRNA, is read at the A site. This site is then occupied by the tRNA that has the appropriate anticodon and carries the next amino acid to be added to the chain. Peptidyl transferase now forms the bond between the amino acids that occupy the P and the A sites. Translocation involves the spent tRNA leaving the P site, a shift of the amino acid tRNA that now carries the growing chain from the A to the P site, and the reading of the next codon at the A site. The elongation process continues until a stop codon, recognized by protein-releasing factors, terminates the process. The specificity of peptidyl transferase is altered so that water rather than an amino group is the acceptor of the activated peptidyl moiety. The protein chain is then released, and the ribosome dissociates into its two separate subunits.

The synthesized proteins have an inbuilt address code as well as a function code. The proteins designed for the internal milieu of the cell are synthesized by free ribosomes in the cytosol. Proteins that we describe here as "sequestered proteins" have a different set of address codes, which require them to be synthesized by ribosomes attached to endoplasmic reticulum. As they are synthesized on the outer side of the rough endoplasmic reticulum membrane, they are threaded through on the inside of the membrane to the separated milieu.

Endoplasmic Reticulum. Endoplasmic reticulum occurs in all neurons. It comes in two varieties, rough and smooth, according to whether or not ribosomes are studded along the outside surface. Electron micrographs show that the two can be continuous (A. Peters *et al.*, 1976). Rough endoplasmic reticulum corresponds to the Nissl bodies that are part of classic neurohistology. The electron-microscopic appearance can be seen in Figures 1.3 and 1.9 and the diagrammatic appearance in Figures 1.8 and 1.18 (Section 1.3). The endoplasmic reticulum is highly convoluted. It is thought to comprise a single membrane sac that is separated from the cytosol but is continuous with the outer nuclear membrane. It is also separated from the Golgi apparatus, which is described below. The endoplasmic reticulum extends out into the axons and the dendrites. The ribosomes on the reticulum also extend out along the dendrites [see Figure 1.3 (Section 1.2.1)], often having a characteristic longitudinal orientation, but not along the axon. Structurally, the larger dendrites resemble the soma, but there is a progressive diminution of the organelles with distance from the nucleus. The origin of the axon, often from an axon hillock [cf. Figure 1.4A (Section 1.2.2)], can be distinguished from the dendrites by the sparsity of rough endoplasmic reticulum as though some process arrested the attachment of ribosomes.

As Figure 1.8 shows, the rough endoplasmic reticulum is organized in flattened cisterns and joins to the more extended, pipelike arrangement of smooth endoplasmic reticulum. The reticulum can be imagined as a stack of communicating hot-water bottles attached at one end (rough endoplasmic reticulum) to the outer nuclear membrane and at the other to a jumbled series of pipes constructed by a mad plumber (smooth endoplasmic reticulum) extending out into the processes of the neuron. Ribosomes are found only on the outside of the rough endoplasmic reticulum. The mRNAs for sequestered proteins exist in the cytosol and are thought to be distinguished by a special signaling code on the 5' end. Contact with a ribosome is made in the cytosol, and synthesis of the protein chain starts with this signaling sequence. The emerging peptide sequence is recognized by a signal-recognition particle in the cytosol; this particle, in turn, binds the complex to a docking protein on the outer surface of the endoplasmic reticulum (D. L. Meyer *et al.*, 1982). Any ribosome starting to synthesize such a protein in the cytosol will have the process arrested until the emerging peptide sequence attaches itself to the rough endoplasmic reticulum so it can be threaded through the membrane as it is synthesized. The signal sequence is removed inside the reticulum, and carbohydrates or other moieties are added. Those complex molecules, particularly glycoproteins, that are designed for the plasma membrane all contain proteins synthesized in association with the endoplasmic reticulum and have an inbuilt inside–outside membrane directionality.

Smooth endoplasmic reticulum is particularly prominent in cells, such as liver hepatocytes, that specialize in lipid metabolism. In this case, it can be shown that phospholipids and cholesterol, the principal building blocks of all bilipid membranes, are synthesized in the smooth endoplasmic reticulum. The endoplasmic reticulum is itself a membrane factory, continually replenishing its surface.

A characteristic of the membrane of the endoplasmic reticulum is that it can seal itself off. It is possible to convert the membranes into microsomes by homogenization and to separate the microsomes by density-gradient centrifugation. It is even possible to separate rough from smooth endoplasmic reticulum. The membrane is asymmetrical, and thus the surfaces exposed to the cytosol and the internal milieu of the reticulum are always preserved. This ability of the membrane to seal itself off has reinforced the belief that vesicles can bud off from the endoplasmic reticulum, complete with their contents, and be transported to the Golgi apparatus or other structures, thus providing a mechanism for moving materials from place to place within the cell without exposing the contents to the cytosol.

The endoplasmic reticulum does not appear to be totally walled off, however, since microtubules and neurofilaments can pass through openings in the reticulum (A. Peters *et al.*, 1976). It is obvious that much remains to be learned about the nature of this large and complex organelle that is so prominently distributed in the cell cytoplasm. Like the Golgi apparatus, it is a specialized chemical factory, probably with specific functions being carried out in its various sections.

Golgi Apparatus. The Golgi apparatus, which is present in all mammalian cells, was described before the turn of the century. Its precise function is still not known, but evidence is accumulating that it modifies, sorts, and exports protein complexes provided by the endoplasmic reticulum. Like the endoplasmic reticulum, it consists of a stack of half a dozen or more flattened cisternae about 1 μm wide, bound by a single membrane (see Figures 1.3, 1.8, 1.9, and 1.18). It has two distinct faces: a *cis*, or forming, face, and a *trans*, or maturing, face. The *cis* face is closest to the rough endoplasmic reticulum, while the *trans* face is closest to the plasma membrane (Figure 1.8). The outer edges of the cisternae tend to bulge, and secretory vesicles are given off, particularly from the *trans* face. A working hypothesis as to the way in which the Golgi apparatus works has been offered by J. E. Rothman (1981), who proposes that it functions somewhat in the fashion of a petroleum distillation plant. Crude products including proteins, glycoproteins, and other complex materials are shipped in vesicles to the Golgi apparatus after production in the endoplasmic reticulum. In the *cis* section of the Golgi apparatus, the molecules are refined by trimming or addition of certain sugars or other molecules in an orderly fashion. In general, O-linked oligosaccharides are added to proteins and their N-linked oligosaccharides are modified. They proceed as they are being refined through to the *trans* section of the Golgi apparatus. They are then systematically sorted and prepared for export via secretory particles that bud off (Dantry-Yarsat and Lodish, 1983). Golgi-processed compounds include components of the plasma membrane, various proteins for secretion, the digestive enzymes of lysosomes, and other essential cellular materials.

Much is currently being learned about the details of the chemical refinery that constitutes the Golgi apparatus. It can be anticipated that a greatly improved understanding of its function will be available in the not too distant future.

Mitochondria. The mitochondria are the power plants of the cell. They have dimensions of 1000–3000 nm, in comparison with the 20-nm diameter of a ribosome. A single mitochondrion may contain as many as 15,000 complexes of respiratory-chain enzymes, each of which contains a dozen or more separate enzymes (Figure 1.13). The inner mitochondrial membrane is not merely an inner skin, but a sheet of multienzyme systems. These are specialized for stripping carbon chains and extracting their energy. The energy is banked in the form of high-energy phosphate bonds, particularly ATP. Carbon dioxide and water are formed by this Krebs cycle oxidative process [see Figure 6.1 (Section 6.2)], which is coupled to phosphorylation. The mitochondria contain other enzymes. Monoamine oxidase is one that is considered in several chapters that deal with neurotransmitter metabolism (see Chapters 9, 10, and 11).

Neuronal mitochondria are generally similar to their counterparts in other cells. In

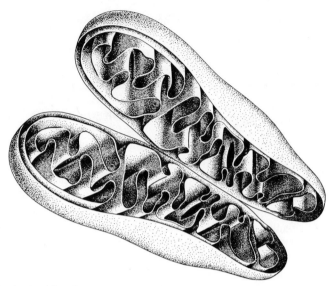

Figure 1.13. A mitochondrion sliced in two. It consists of an external membrane and an internal membrane. The latter is intensively folded and covered with enzyme systems in a highly organized sequence for the purpose of forming high-energy phosphate bonds. The mitochondrion is a plastic structure, able to assume a variety of shapes. Mitochondria lead independent lives within the cell. They have the necessary nucleic acids to reproduce themselves, and spent mitochondria are destroyed by the cell.

Figures 1.3 (Section 1.2.1) and 1.8, mitochondria can be seen as numerous small ovoid organelles with the characteristic internal structure. They are often clustered in the interstices between the endoplasmic reticulum and the Golgi apparatus. In Figure 1.3, the mitochondria are seen to be linearly arranged along the dendrite, and this orientation becomes progressively more evident further along the dendrites. It is also very evident along the axon. Mitochondria are concentrated in axonal terminals, where there is the intense metabolic activity associated with the synthesis, storage, release, and reuptake of neurotransmitters [cf. Figures 1.5B and 1.6A (Section 1.2.2)].

Lysosomes, Lipofuscin Granules, and Multivesicular Bodies.   These cytoplasmic bodies are all thought to be involved in scavenging operations of the cell [see Figures 1.3, 1.8, and 1.17C (Section 1.2.5)]. The best-characterized is the lysosome, a round or spherical particle, typically about 0.3–0.5 μm in diameter. It is thought to bud off from the Golgi apparatus. It is bounded by a single membrane and is filled with a fine granular material. It contains acid phosphatase and other hydrolytic enzymes. With aging, lysosomes become more numerous, and many seem to develop into lipofuscin granules. Lipofuscin accumulates with age and is generally considered to be an undigested residue of neuronal wear and tear. It is yellow to brown in unstained material and fluoresces with an orange-green color, making it an unmistakable marker of older brain tissue. Lipofuscin itself is confined to the lysosome-like particles. Multivesicular bodies are spherical and of similar size, but their interior is filled with a variable number of small spherical and ellipsoid vesicles. Horseradish peroxidase, when injected extraneuronally, will appear within alveolate vesicles and will then be transferred to multivesicular bodies (Figure 1.17C). This may then represent a variant of the hydrolytic process of destroying foreign or damaged proteins.

Cytoskeletal Structures: Microfilaments, Intermediate Filaments, and Microtubules. These three structures exist in all cells and in all elements of the neuron, although they are especially prominent in the processes. They are named according to size, being roughly 4 nm, 10 nm, and 20–25 nm in outside diameter. The latter two structures are referred to in brain as *neurofilaments* and *neurotubules*. They are depicted diagrammatically in Figure 1.8 as they would appear in an axon.

Brain microfilaments, like those in muscle, are made up of actin polymers. They are hard to visualize in brain, although actin is an abundant brain protein. Actin filaments are solid. In muscle, they consist of two strands of globular molecules about 4 nm in diameter, twisted into a helix of 13.5 molecules per turn.

Neurofilaments, unlike microfilaments and neurotubules, vary in structure among species and are not readily disassembled. In mammalian neurons, they are composed of a fundamental triplet of three fibrous proteins having molecular weights of 70,000, 140,000, and 210,000. They are tubular and of indefinite length. The outside diameter is 10 nm, with a clear center 7 nm in diameter. Short spokelike processes radiate from the wall, but they are unbranched. They are seen sparsely in the nucleus, moderately in dendrites, and abundantly in axons. They run down to the synaptic knobs, often forming a loop. They are readily stained by silver and correspond with the classic neurofibrils of light microscopy. In certain pathological conditions, especially Alzheimer's disease, they have a curious tendency to pair and twist about each other, making a complete turn every 160 nm. These paired helical filaments accumulate as a jumbled mass in the cell soma, forming what are known as *neurofibrillary tangles,* which can quite literally fill the cell body (Katzman, 1978).

Like neurofilaments, microtubules are also tubular structures, but have a larger diameter of 20–40 nm. In contrast to the neurofilaments, they are found most prominently in axons of small diameter. They are also found in all dendrites [cf. Figure 1.6A and B (Section 1.2.2)]. In immature animals, nerve fibers contain many microtubules but few neurofilaments. With increasing age, there is a reversal in some axons.

The microtubule is an assembly of subunits composed of a dimer of α- and β-tubulin, which have molecular weights of 56,000 and 53,000, respectively. Tubulin is almost identical in composition throughout a large variety of species. In laboratory preparations, the subunits will neatly assemble themselves first in strands and then in circular tubules familiar in electron-microscopic photomicrographs.

Microtubules are apparently in equilibrium with soluble tubulin and can be readily assembled and disassembled *in vivo*. Microtubules are able to change their shape rapidly in such primitive organs as cilia, apparently by interacting with contractile proteins in the cells. They should be regarded, therefore, as a scaffolding and not as a permanent and rigid structural element (Dustin, 1980).

Specificity of microtubular function may come about as a result of proteins attached to them. High-molecular-weight proteins, for example, have been found to be associated with dendritic, but not axonal, tubules (Matus *et al.*, 1981), and it has been suggested that they are important in distinguishing axonal from dendritic export routes and thus the polarity of the neuron.

Colchicine specifically binds to selected sites on tubulin, throwing into disarray the orderly arrangement of tubules. It is renowned as a specific antidote for gout, but it has achieved biological fame as a fundamental tool for investigating many functions associated with tubules, including axoplasmic flow, which it arrests.

The so-called "cytoskeletal" components—microfilaments, neurofilaments, and neurotubules—should not be regarded as either rigid or independent structures. Rather, they are dynamic elements, working in concert with a heirarchy of proteins to orchestrate in an orderly fashion the busy to-and-fro traffic of the cell.

Plasma Membrane and Its Receptors.   The surface membranes of all nerve fibers and nerve cells have essentially the same structure and properties throughout the whole of the invertebrate and vertebrate kingdoms. The cell membrane was a basic innovation at a very early stage of evolution, and it was such a good discovery that it can be seen to be essentially the same over a wide range of invertebrates and in all vertebrates. Furthermore, the nerve impulse itself was also a very early innovation, and in its essentials it has been preserved right up to the vertebrates and the mammals. No engineer could design or imagine anything so beautiful, efficient, and effective as a nerve impulse and the communication that it gives along nerve fibers, and so from one neuron to others.

The main components of the 7-nm membrane are arranged as a bimolecular leaflet of phospholipid molecules, with hydrophilic polar groups pointing both outward and inward and being about 4 nm apart (Figure 1.14). The fatty acid tails are both saturated and unsaturated. If all were saturated, the membrane would be rigid. The larger the proportion of unsaturated fatty acids, the more fluid is the membrane and the less orderly is the packing. Such membranes can be made artifically, and in many of their properties the artificial membranes resemble natural cell membranes.

The lipid bilayer represents a relatively impermeable barrier to the flow of water-soluble materials but a highly plastic structure for functional modification. Protein complexes are "dissolved" in this bilayer as shown in Figure 1.14, modifying its function in a variety of ways. The bilayer is asymmetrical, presenting one face to the extracellular milieu and another to the internal milieu. This is reflected in both the lipid composition of the bilayer and the orientation of proteins dispersed within it. As mentioned previously, physical disruption as in freeze–fracturing for electron microscopy will cleave the lipid bilayer down its middle, where the bonding is weakest, and reveal microstructural variations involving receptors [see Figure 1.6C (Section 1.2.2)].

A typical lipid bilayer unit is shown diagrammatically in Figure 1.14. The example shows two fatty acids, one saturated and straight and the other unsaturated and bent, both being attached to a single choline–phosphate–glycerol hydrophilic head in a hairpin-like arrangement. Among other hydrophilic heads are phosphatidylethanolamine and phosphatidylserine, with many lipid combinations being included in the hydrophobic regions. For example, about 6% of the lipids in the outer monolayer are made up of glycolipids, particularly gangliosides, which contain one or more sialic acid residues, which presumably are the cell adhesive molecules (CAMs) described by Edelman (1984) (see Section 13.1.1).

Protein molecules are distributed throughout the bilayer, as shown in Figure 1.14,

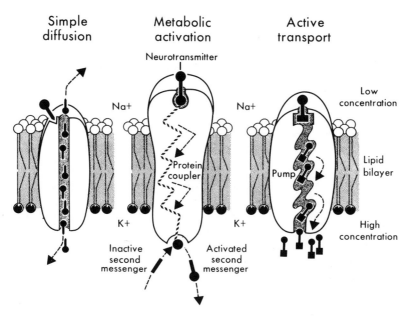

Figure 1.14. Organization of the surface membrane. The basic structure is a bimolecular leaflet of phospholipid molecules with hydrophilic structures on the external surface. One set is oriented to the extracellular fluid while the other is directed toward the intracellular fluid (black spheres). In freeze–fracturing, the membrane parts in the middle, where its structural strength is least. Structural proteins are applied to both surfaces to provide stability. Diagrammatic depictions of three types of receptors: (1) an ionotropic receptor that opens ion channels on activation by a neurotransmitter; (2) a metabotropic receptor that triggers a transmembrane reaction to a second messenger on activation by a neurotransmitter; and (3) a high-affinity neurotransmitter uptake system that "pumps" the neurotransmitter from a low concentration extracellularly to a high concentration in a nerve ending. Many additional types of receptors may exist. See the text for details.

and act as various types of receptors. They may extend to both sides of the bilayer or to only one. They may rotate or diffuse laterally, but their orientation to the internal or external milieu is predetermined and they do not flip-flop.

Figure 1.14 depicts a number of classes of receptors diagrammatically. Some open ion channels (ionotropic receptors as in 1), others trigger second messenger activities (metabotropic receptors as in 2), and still others stimulate transport against a concentration gradient (uptake receptors as in 3). There are many more categories that could be mentioned, all directed at conferring special properties on the neuron. Much is said in later chapters about receptor properties at the molecular level. The receptor concept had its earliest beginnings in pharmacology at the time of Langley (1905), long before the model shown in Figure 1.14 could be imagined. Recognition of the types of molecules that act as receptors has come about much more recently as a result of binding studies between various ligands and protein-containing membrane elements. Closure between pharmacology, membrane biochemistry, and morphology is now possible as working models of the type shown in Figure 1.14 have emerged. Obviously, developing greater knowledge of the detailed structure of receptors and the way in which they interact with their specific messengers will be fruitful areas for future research.

Plasma membranes are continuously produced by the cell in at least two locations, the endoplasmic reticulum and the Golgi apparatus. The membranes can close on themselves to form vesicles or fuse with membranes in different locations. The mechanisms for budding and fusing are completely unknown, but a basic principle seems to be preservation of the orientation of external and internal faces.

With respect to the bioelectrical properties of the membrane, it will be realized that the bimolecular phospholipid leaflet is very highly resistant to electric current and has the considerable electrical capacity of about 1 $\mu F/cm^2$, a value that accords with a dielectric coefficient of 5 and a plate separation of 5 nm. The surface membrane of a neuron or nerve fiber may be likened to an outer wrapping of a leaky condenser, with a resistançe of about 1000 Ω-cm^2, because ions diffuse through these many channels across the membranes that are formed by proteins (cf. Figure 1.14).

The mammalian motoneuron is used in Chapter 4 for the exposition of the electrical and ionic properties of the surface membrane because the theme of this book is the mammalian brain, especially the human brain, and the motoneuron has been the most-studied neuron. As shown in Figure 1.15, the surface membrane of a neuron separates two aqueous solutions that have very different ionic compositions. The external concentrations are approximately the same as for a protein-free filtrate of blood plasma. The internal concentrations are derived more indirectly from investigations of the equilibrium potentials for some physiological processes that are specifically produced by one or two ion species. Within the cell, sodium and chloride ions are at a lower concentration than on the outside, the ratios being about 10:1 and 14:1. In contrast, with potassium there is an even greater disparity—almost 30-fold—but in the reverse direction. Under resting conditions, potassium and chloride ions move through the membrane much more readily than sodium. Of necessity, the electrical potential across the membrane influences the rates of diffusion of charged particles between the inside and outside of the cell. The potential across the surface membrane is normally about -70 mV, the minus sign signifying inside negativity, as demonstrated in Figure 2.4 (Section 2.2).

The equilibrium potentials in Figure 1.15A are derived from the concentrations by the Nernst equation. The equation gives the electrical potential that just balances a concentration difference of charged particles such as ions, so that there is equality in their inward and outward diffusion rates across the membrane.

In Figure 1.15C, there is shown across the membranes a large electrochemical potential difference for sodium ions (130 mV) that is derived from 70 mV for the membrane potential plus 60 mV from the concentration differences as calculated by the Nernst equation. This large electrochemical gradient causes the inward diffusion of sodium ions to be more than 100 times faster than the outward diffusion. Fortunately, the resting membrane is much more impermeable to sodium ions than to potassium and chloride ions, or else the cell would be rapidly swamped with sodium ions. But, of course, there must be some other factor concerned in balancing sodium transport across the membrane even at an unbalanced slow diffusion rate. This is accomplished by a kind of pump that uses metabolic energy to force sodium ions uphill (up the electrochemical gradient) and so outward through the cell membrane, as shown diagrammatically in Figure 1.15C. This diagram further shows that there is an excess of diffusion outward of potassium down the electrochemical gradient of about 20 mV for potassium, and again the

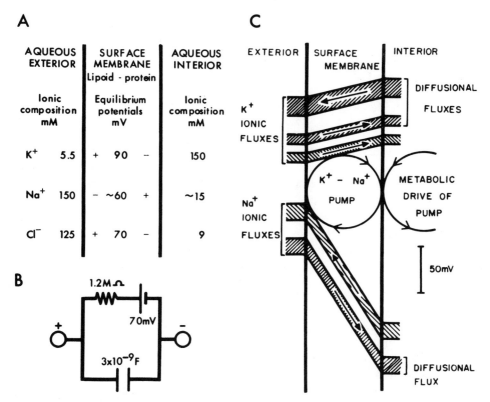

Figure 1.15. Various conditions across the membrane of a cat motoneuron. (A) Approximate ionic compositions inside and outside, and the respective equilibrium potentials. (B) Formal electrical model for an average motoneuron as measured by a microelectrode in the soma. (C) Ionic fluxes through the membrane. The sizes of the electrochemical gradients are roughly indicated by the 50-mV scale bar at the right. The fluxes are in part diffusional down the electrochemical gradients and in part due to specific ion pumps driven by metabolism. The fluxes due to diffusion and the operation of the pump are distinguished by different hatching, and the magnitudes are indicated by the respective widths of the channels. From Eccles (1957).

transport of potassium ions is balanced by an inward pump. In fact, as shown, the sodium and potassium pumps are loosely coupled and are driven by the same metabolic process. These pumps are not triggered by transmitters, differing in this respect from that in Figure 1.14.

With the squid axon, all ionic concentrations are several times larger because they are isotonic with the seawater in which the animal lives; nevertheless, the ratios are similar, being, for inside/outside, 20:1 for potassium, 1:9 for sodium, and 1:14 for chloride. Thus, the diagrammatic representations in Figure 1.15C may be assumed also for the squid axon, both for electrochemical gradients and for ionic fluxes.

As shown in the formal electrical diagram of Figure 1.15B, under resting conditions, the surface membrane of the nerve cell and its axon resembles a leaky condenser charged at a potential of about -70 mV (inside to outside). If this charge is suddenly diminished by about 20 mV, i.e., to −50 mV, there is a sudden further change in the membrane potential—a nerve impulse has been generated, as is described in Chapter 2.

1.2.4. Axoplasmic Flow

Axoplasmic flow is a fundamental property of neurons, just as secretion is of gland cells. It is the method by which materials produced in the cell soma reach the axon and dendrite tips, as well as the method by which materials from these processes are returned to the perikarya. Virtually all dynamic components found in nerve processes are moved by this energy-consuming action, rather than by diffusion or Brownian movement. It was first described by P. Weiss and Hiscoe (1948), who observed the damming of materials in a peripheral nerve following application of a ligature. However, despite extensive investigation since then, the underlying mechanisms are only partly understood (cf. J. H. Schwartz, 1979; Ochs, 1981). The following description of the principal findings will preface a consideration of explanatory hypotheses.

1. Axoplasmic transport can be *anterograde* from the soma down the axon or *retrograde* from the nerve terminals up to the soma.
2. The velocity of anterograde axoplasmic transport varies widely for different particles and substances. In the classic experiments of Ochs (1972, 1982), many labeled proteins travel at the surprisingly high rate of about 400 mm/day, as revealed in the serial radiotracer investigations illustrated in Figure 1.16A. However, different organelles and substances exhibit a wide range in velocities below this maximum down to a very slow transport of 0.5–2 mm/day. All transitions have been observed, and there does not appear to be a special category of slow transport (J. H. Schwartz, 1979; Ochs, 1981).
3. The velocity of retrograde transport is usually fast, 200–300 mm/day. There is evidence that proteins and polypeptides such as horseradish peroxidase (HRP) are loaded onto lysosomes that travel retrogradely by such fast transport to the soma (J. H. Schwartz, 1979).
4. Mitochondria and other large organelles can be seen by Nomarski optics and time-lapse cinematography to move in an intermittent manner; there are brief episodes of proceeding at a high rate, alternating with rest periods, producing a consequent overall slow rate.
5. Except for possibly the slowest transport (J. H. Schwartz, 1979), the continual expenditure of energy along the axon is necessary for transport, Ca^{2+}-Mg^{2+}-ATPase releasing the energy of ATP (Ochs and Iqbal, 1982).

The remarkable constancy of the fast flow rate, regardless of species or type of nerve preparation, suggests that a fundamental process exists that is preserved in detail throughout much of the animal kingdom (Ochs, 1981). Ochs pioneered the method of injecting tritiated leucine or tritiated lysine into a dorsal root ganglion of the spinal cord (Figure 1.16A), showing that the ganglion cells picked up the amino acid and built it into proteins and polypeptides. These tritiated macromolecules were followed by radiotracer techniques and shown to move along the nerve fibers in both directions from the ganglion, out to the periphery and centrally up the spinal cord. By sacrificing animals at different times

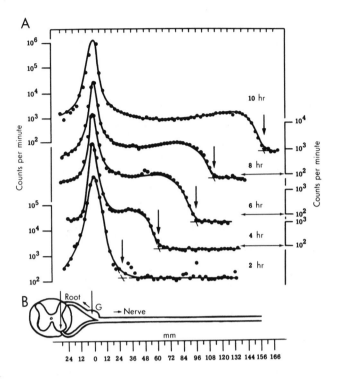

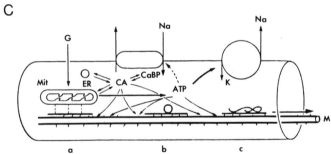

Figure 1.16. Fast axon transport down sensory fibers. (A) A lumbar dorsal root ganglion (G) of the cat was injected with tritiated leucine, and the radioactivity was measured in separate experiments at 2, 4, 6, 8, and 10 hr later at various distances along the sciatic nerve (B). The radioactivity is plotted on a logarithmic scale in counts per minute, with separate scales as shown for each of the curves. From Ochs (1972). (C) Transport filament model. Glucose (G) enters the fiber, and after glycolysis followed by oxidative phosphorylation in the mitochondrion (Mit), ATP is generated. The sum of ATP and creatine phosphate supplies energy to the sodium pump that controls the level of Na^+ and K^+ in the fiber and to the side arms, where, with calmodulin (CaBP), Ca^{2+}-Mg^{2+}-ATPase utilizes ATP to produce a movement of transport filaments along microtubules (M) or neurofilaments or both. Transport filaments are shown as black bars to which various components transported are bound and carried down on the fiber by the side-arm activity. Components transported include mitochondria (a), attaching temporarily as indicated by dashed lines; soluble protein (b), shown in folded or globular configuration; polypeptides and small particulates (c). Thus, a wide range of various components are transported at the same fast rate of transport. From Ochs (1983).

and cutting the peripheral nerve into sections, Ochs was able to show the movement of a wavefront of activity down the nerve as a function of time and thus to record the rate of flow in a dynamic fashion (Figure 1.16A). Radioactivity was observed to travel at the surprisingly fast but constant velocity of about 400 mm/day in a wide range of mammalian species.

These experiments have been repeated many times using many different nerve preparations. The same rate of transport was observed for motor nerve axons, with injections into the motoneuron pool of the spinal cord (inset in Figure 1.16A). Also with the long dorsal roots of the monkey lumbar cord, the rate was the same for central transport from the injected spinal ganglion, though the amount of transport was much less, about one third (Ochs, 1981).

The most complete and challenging hypothesis of the mechanism of axonal transport has been developed by Ochs (1982, 1983). It is illustrated in Figure 1.16C and is called the "transport filament model." Since it applies to rates of axoplasmic transport from the fastest to the slowest, it is also called by Ochs a "unitary hypothesis."

The essential feature is that a microtubule of a nerve fiber has short side arms that move like cilia and have special relationships to transport filaments, which are the short structures shown above the microtubule. The organelles and other components are transported when attached to these transport filaments as indicated in Figure 1.16C. When the attachment is secure, there is a maximum transport rate of the system, 400 mm/day (Figure 1.16A). However, structures such as a mitochondrion are too large for the transport filament and are attached only intermittently to move at the maximum rate, as already stated. The hypothesis thus accounts for the whole range of slow transport rates. Figure 1.16C also indicates the essential ionic and energy source for this active transport. Ca^{2+} is essential at a low level, where it combines with calmodulin to activate Ca^{2+}-Mg^{2+}-ATPase in releasing energy from the ATP that somehow results in the lashing activity of the side arms that moves the transport filaments along the outer surface of the microtubule. Ochs (1982, 1983) gives details of the experimental basis for his hypothesis. A novel ATPase that binds to microtubules and has the properties expected for the fast axonal transport motor has been reported (Brady, 1985). As would be expected, anoxia blocks the transport. So does trifluoperazine, which prevents activation of calmodulin by Ca^{2+}. Colchicine and vinblastin destroy the microtubules and so are also blockers.

A disturbing recent finding is that the microtubules are discontinuous. In fact, they are in very short lengths of several to several hundred micrometers staggered axially along the length of the nerve fiber (Ochs, 1983). How the transport filaments are assembled for each discontinuous length of microtubule is not clear. Some of the unloaded proteins such as tubulin can be used locally for microtubule reassembly. Off-loaded proteins may even be loaded on transport filaments for retrograde transport. It is remarkable that an apparently disordered mechanism can give on the average the discrete progressing waveform of Figure 1.16A.

The hypothesis of Droz (1975) may be regarded as a possible alternative or addition to the transport-filament hypothesis. It relates particularly to the origin of the anterogradely transported material. On the basis of electron microscopic autoradiography, Droz showed that newly synthesized proteins appeared in the axon after passage through the Golgi apparatus. More recent experiments have established that monensin, a specific poison of the Golgi apparatus, inhibits fast transport of all macromolecules, that removal of Ca^{2+} inhibits movement of protein to the Golgi apparatus and subsequent movement of glycoproteins to the axon, and that inhibition of protein synthesis depresses equally both lipid and protein transport (Hammerschlag and Stone, 1982).

The implication of these results is that fast axoplasmic transport requires an appropriate preassembly of materials starting with proteins from the endoplasmic reticulum, which are further processed by the addition of saccharides from the Golgi apparatus and, possibly, lipids from the endoplasmic reticulum. The assembled product is accepted by the transporting mechanism in the axon. Rambourg and Droz (1980) propose that rapidly transported material moves in the anterograde direction in smooth endoplasmic reticulum. Their electron microscopic evidence suggests that a continuous tubular sac with oblique branches extends to the tips of the processes, providing a suitable vehicle for the processes.

Tubulin and neurofilament proteins are transported at the slow rate of 0.5–2.0 mm/day. Many soluble proteins generated in the cytosol and designed for the internal milieu move at the same speed, suggesting a close association with these elements. Some myosin-like proteins, actins, calmodulin, and creatine kinase move at rates of 20 mm/day (D. G. Weiss, 1982), suggesting a primary association with still other elements, possibly even microfilaments.

The subject of axoplasmic flow should not be left without mention being made of the practical use being made of the phenomenon in neuroanatomy. The application of anterograde and retrograde flow to chart anatomical pathways has quite literally revolutionized the field of neuroanatomy. Numerous new tracts, never suspected on the basis of classic neuroanatomical techniques, have been traced with relative ease using a variety of easily detected markers transported by axoplasmic flow. For example, tritiated amino acids, injected in a tightly confined space, are synthesized into proteins and detected by means of anterograde flow in distant nerve endings by autoradiography. Various foreign proteins, particularly horseradish peroxidase, have been used to trace anterograde and retrograde pathways (Figure 1.17). Avidin–biotin, tetanus toxin, and such dyes as granular blue, nuclear yellow, propidium iodide, and others (Heimer and Robards, 1981) have been used as markers. By injecting different markers at separate terminal sites and noting the labels in a single cell body, neurons with branching axons can be detected. Much is said in subsequent chapters about the use of these methods for practical neuroanatomy.

1.2.5. Pathological Changes in the Neuron

Since such essential metabolic processes as protein (and enzyme) synthesis are concentrated in the perikaryal region, it is to be expected that any parts of the neuron severed from this vital region will rapidly die, as indeed they do. This forms the basis of histological techniques (Nauta and Fink–Heimer) for determining the distribution of the axonal branches of neurons (Guillery, 1970). The degenerative changes in the synaptic knob are treated in Chapter 13 [see Figure 13.11 (Section 13.2.2)]. The region central to the section also shows changes (Cragg, 1970; Lieberman, 1971). There is disorganization and degeneration of the Nissl bodies, the rough endoplasmic reticulum being stripped of polyribosomes, which are dispersed in the cytosol (J. P. Johnson *et al.*, 1985). There may be a partial fragmentation of the Golgi apparatus and an eccentric position of the nucleus. These changes have been termed *chromatolysis* or the *axon reaction* (Section 13.3.2) and

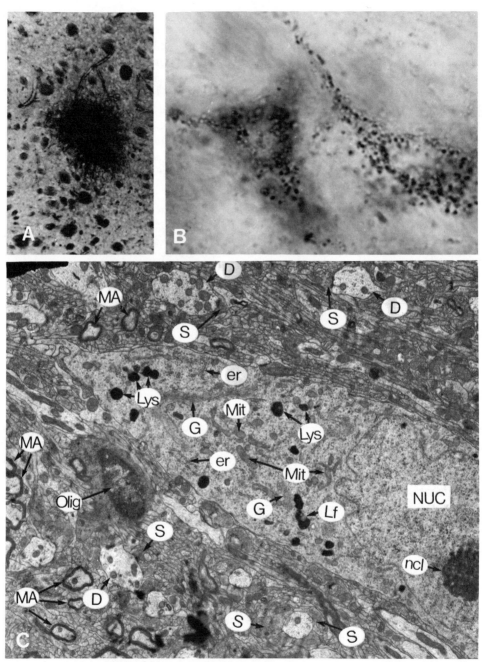

Figure 1.17. (A) Injection site of horseradish peroxidase in the rat caudate–putamen. Material is picked up by nerve endings and transported by retrograde axoplasmic flow to the cell bodies. This nigrostriatal dopaminergic pathway is discussed in detail in Chapters 9 and 14. Staining for peroxidase is by a diaminobenzidine oxidation reaction. (B) High-power photomicrograph of two zona compacta cells. Note the distinctly granular appearance of the peroxidase stain caused by its concentration in lysosomes. (C) Electron micrograph of a cell similar to those in (B). Many of the features shown in Figure 1.3 can also be seen in this electron micrograph. Particularly prominent in the cytoplasm are many dark lysosomes (Lys) containing horseradish peroxidase that are the

are the basis of the Gudden method for studying the axonal path from a neuron. There is a great variation in the susceptibility of neurons to axotomy, neuronal death being common in young animals. The usual outcome, however, is that the neuron recovers over some weeks, the stump meanwhile sprouting and regenerating connections if it is in a peripheral nerve. Besides the characteristic somatic changes, there are also changes in the surface membrane of the soma, with glial displacement of synaptic knobs from their postsynaptic attachment sites (Blinzinger and Kreutzberg, 1968). Section 13.3.3 presents a hypothesis that chromatolysis results from the deprivation of an essential neuronotropic factor that is retrogradely transported from the axonal terminals. This hypothesis is corroborated experimentally, nerve growth factor causing a recovery from the chromatolysis of sympathetic ganglion cells (Section 13.3.2). As described in Section 13.4, there is an intensive search for neuronotropic factors for other classes of neurons.

When neurons are deprived of a considerable fraction of their synapses, atrophic changes set in, which result in diminution of their size or in death, a process called *anterograde transneuronal degeneration* (Cowan, 1970). It is fully treated in Section 13.3.5.

1.3. Neuroglial Cells

There are three types of glia: astrocytes or astroglia, oligodendroglia, and microglia. Astroglia are thought to participate in the following functions: formation of the blood–brain barrier (BBB) and the exchange of materials between capillaries and neurons, biochemical modulation of synaptic activity, guiding of neuronal processes during development and possibly during repair, reservoir activity for potassium and water, some phagocytic activity, and structural support for nervous tissue. Oligodendroglia provide the myelin sheath for axons and are the central counterpart of peripheral Schwann cells. Microglia are traditionally regarded as the potential phagocytes of the central nervous system (CNS). The overall numbers of glia in brain are unknown, although for the gray matter of monkey striate cortex (see Section 1.2.1), the ratio of glia to neurons is roughly 1:2. The proportions in this area are: astroglia 65%, oligodendroglia 28.5%, and microglia 6.5% (O'Kusky and Colonier, 1982a). Oligodendroglia are the most populous glial species overall, since they predominate in white matter. Most estimates suggest that glia outnumber neurons for the total brain (Sears, 1982), although more careful work of the type done in monkey striate cortex is required.

The two major classes of neuroglia, oligodendroglia and astroglia, have the same ectodermal derivation as neurons. The microglia are mesodermal in origin and can act as phagocytes.

The oligodendroglia have rounded nuclei and a few short processes that are concerned with the formation and maintenance of the myelin sheaths. In the adult, it is difficult to demonstrate any connection between the oligodendroglial cytoplasm and the

---

counterparts of the granules shown in (B). Note also the indented membrane of the nucleus (NUC), the nucleolus (ncl), many mitochondria (Mit), endoplasmic reticulum (er), Golgi apparatus (G), and lipofuscin (Lf). The dense neuropil shows many dendrites (D) in cross section with easily detected synapses (S). Numerous myelinated axons (MA) are also visible, as well as oligodendroglia (Olig). Courtesy of T. Hattori (University of Toronto).

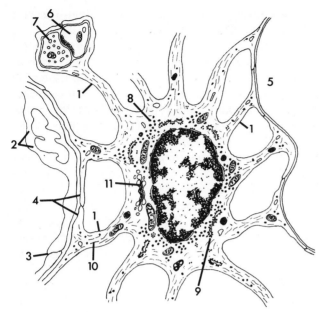

Figure 1.18. Glial cell. An astrocyte is shown demonstrating its many thin processes (1). At the left, they flatten and form a tight junction around a capillary; at the upper left, they surround a synapse; at the right, they are apposed to the surface of a neuron. Within the watery cytoplasm, note the mitochondria, numerous fibrils (10), lysosomes, and glycogen granules (8) in addition to the rough endoplasmic reticulum (9) and Golgi apparatus (11). (2) Capillary endothelial cell; (3) capillary endothelial junctions; (4) pericapillary end feet; (5) neuronal cell body; (6) dendritic spine; (7) nerve ending. Modified from Constantinides (1974).

myelin lamellae in a manner comparable with that shown in Figure 1.8 (Section 1.2.3) and observed for Schwann cells and peripheral axons [cf. Figure 2.12 (Section 2.5)]. However, the role of oligodendroglia in myelination is clearly shown in electron micrographs of the developing brain (Bunge *et al.*, 1961), confirming the relationship depicted in Figure 1.8.

As already mentioned, astroglia have a much wider range of functions than oligodendroglia. Some of their anatomical features and physical relationships are shown in Figure 1.18. The relatively small perikaryal region gives rise to many widely branching processes. The nuclei are large, with finely distributed chromatin. The cytoplasm is pale, with very few organelles, although some cells are distinguished by the extensive bundles of fibrils in the cytoplasm. They have numerous glycogen granules. Doubtless the interlacing astroglial processes provide a structural support to the neurons as was originally surmised.

Capillaries are always separated from neurons by the astroglial sheath surrounding the capillary (Figure 1.18). The astroglial processes extend and flatten to produce pericapillary end feet that form tight junctions with each other. This creates the BBB, which is the subject of the next section. Astroglia act as intermediaries in providing neurons with nutrient substances from the capillaries by being interposed between them (Figure 1.18).

The thin astroglial processes break up the neuropil into a mosaic of small regions, each containing a synaptic field, one of which is shown in Figure 1.18. This intimate relationship between synapses and astroglial processes implies not only that astroglia may make an important contribution to synaptic function, but also that biochemical subtypes may exist. For example, glia have been identified that have high-affinity binding sites for dopamine, noradrenaline, and other neurotransmitters (Hertz, 1982). More efficient uptake of glycine is found with glia from spinal cord, where glycine is a neurotransmitter (see Chapter 7), than with glia from cerebral cortex (Henn, 1976). Mesencephalic glia produce a different morphogenetic effect on dopamine neurons than do striatal glia (Denis-Donini *et al.*, 1984). Astroglia from different brain regions also produce different antigens (Schachner, 1982). Some glia have uptake systems for glutamate or GABA to terminate neurotransmitter action. They possess the enzyme glutamine synthetase to recycle these transmitters back to the neuron (see Chapter 7), reinforcing the concept of biochemical specialization of glia. It has even been suggested that under certain conditions, glia may release transmitters. As discussed in Chapter 13, astroglia play an important role in guiding the growth of nerve fibers to their correct destination during development (Rakic, 1971, 1972). They express N-CAM-like peptides *in vitro* (Noble *et al.*, 1985), and antibodies to N-CAM prevent neuron–glia adhesion *in vitro* (Keilhauer *et al.*, 1985). Possibly they also function during regeneration in the mammalian brain by triggering the sprouting of nerve fibers and guiding the sprouts to the vacant synaptic sites (Raisman and Field, 1973; Eccles, 1976a).

Astroglia serve as a potassium reservoir in the brain. During physiological activation of neurons, seizures, or the spreading depression of anoxia, progressively larger amounts of potassium are released from neurons. In each case, the excess is rapidly taken up by glia for later restoration to neurons (Hertz, 1982).

The astroglia can also act as a water reservoir. The large bulk of their "watery" cytoplasm excellently fits them for this role. Cerebral edema may be specifically due to excess uptake of water by astroglia.

Finally, astroglia function as phagocytes in removing degenerated fragments of neurons. All metabolic waste production from neurons must pass through astroglia to be carried away by the bloodstream. They can proliferate following injury to form an astroglia scar [see Figure 13.20 (Section 13.2.2)].

Microglia are the smallest cells in the CNS. They are present in only minor quantities unless there is some insult to the brain. Under these circumstances, they swell and proliferate (see Section 1.2.5), phagocytosing debris. There is at present some confusion regarding the origin of the large, lipid-filled scavenger cells observed after injury, since monocytes from the blood can assume the same form.

1.4. Blood–Brain Barrier

It has long been known that the relationship of the brain to the blood circulating through it is radically different from that of any other organ. For a wide range of substances, there is a barrier to diffusion from blood capillary to neurons. Full accounts of investigations of the blood–brain barrier (BBB), as it is called, are available in a number of reviews (Davson, 1976; Oldendorf, 1975; Brightman and Reese, 1975; Fenstermacher, 1975; Meisenberg and Simmons, 1983; Heistad, 1984). Quantitative studies have been made on a wide range of substances, and correlations have been made with predicted penetrations through cell membranes having a phospholipid structure. First, the endothelial cells of capillaries differ from those elsewhere in that they are sealed together in overlapping fashion. The oblique tight junction probably blocks the passage out of the capillary of all substances that do not penetrate the capillary endothelial cells (Oldendorf, 1975). Second, beyond the permeable basement membrane, there is an enclosure of the capillary by processes of astroglia that are sealed tightly to one another, the intercellular gap being extremely narrow. The problem now is to determine the blocking properties of these two "tight" junctions. There is experimental evidence (reviewed by Oldendorf,

1975) that the tight sealing of the capillary endothelium is the principal factor concerned in the BBB. On the other hand, an important function can be given to the astroglia that insert themselves between the capillary endothelia and the neurons (Figure 1.18). In contrast to a negative blocking function, they can operate positively in regulating the supply to the neurons of such essential metabolic substances as sugars, amino acids, salts, and peptides (Fenstermacher, 1975; Davson, 1976; Meisenberg and Simmons, 1983). Also, they can aid in the transport of excretory substances to the capillaries. There is evidence, however, that the capillary endothelium may be principally concerned in active transport requiring metabolic energy (Oldendorf, 1975). It must not be assumed that the BBB has properties that specifically protect the brain from harmful organic materials. The penetration that is not carrier-mediated is related to lipoid solubility and not to potential toxicity. It could be predicted that such substances as ethanol, caffeine, opioids, and nicotine score well in lipoid solubility, so that they can penetrate rapidly through the capillary endothelial cells without any impediment by the BBB.

The "tight" junction astroglial system of the BBB does not apply to several brain structures, known as the circumventricular organs, located in the walls of the ventricles. These include the median eminence, the neural lobe of the pituitary gland, the organum vasculosum of the lamina terminalis, the subfornical organ, the subcommissural organ, the area postrema, and the pineal body. At these sites, a low-resistance barrier is formed by ependymal cells that make contact with the capillaries, rather than astroglia. Ependymal cells line the ventricles. They are characterized by luminal cilia, numerous microvilli, and the basal infoldings that contact capillaries. Similar histological features are encountered in the choroid plexus, the major site of cerebrospinal fluid (CSF) formation. The surface area of the blood–CSF barrier has been estimated to be about 1/5000th that of the BBB (Meisenberg and Simmons, 1983). These differing structural features, visible by microscopy, are made more complex by invisible chemical features that affect the regional penetration of materials into various brain nuclei through specific receptors and uptake systems.

1.5. Molecular Genetics Applied to the Nervous System

1.5.1. DNA Amplification and Probe Preparation

DNA is a highly personal molecule. Only in monozygotic twins is there exact duplication of a given code. The DNA in every one of your own neurons has locked within it your evolutionary history and your physiological potential. The RNA in the cytoplasm of one of your neurons at any given moment tells a different story. It is a snapshot in time of the genes being expressed by that particular cell. Since DNA molecules are relatively stable, the ability to read your fortune by studying your nucleic acids may be possible in the future.

We now consider the methods that have so far been developed to read and decipher the DNA code of the neuron and the RNA of the cytoplasm. These remarkable new techniques are producing a biological revolution. They will eventually provide insights into the deepest secrets of living systems. This deciphering of the genetic code represents one of the greatest scientific undertakings of all time. Whatever the ultimate significance, the work has now been commenced, and even the early findings carry much of the drama of a detective novel.

It is beyond the scope of this book to do more than describe the broad principles on which this new science is based and provide an introduction to the new terminologies that are emerging. It can be anticipated that rapid developments will take place, and the reader is advised to consult textbooks (B. Lewin, 1985; Schmitt *et al.*, 1982) and reviews as they become available (W. F. Anderson, 1984; McClintock, 1984; Hayden and Nichols, 1984) as a method of extending the preliminary information provided here.

In Section 1.2.3, we discussed the methods by which the genetic code in the nucleus is read and transcribed into RNA. We also described the processes by which the various forms of RNA cooperate to produce the proteins that then direct the chemical machinery of the cell. The information is applicable to those cells that have a nucleus, i.e., eukaryotic cells, especially the mammalian neuron. To understand the techniques of molecular genetics, it is necessary for us to broaden the information somewhat to include the functioning of bacterial, or prokaryotic, cells, as well as cloning vectors such as plasmids and bacteriophages that grow within them.

Experimental approaches to molecular genetics are based on the following principles:

1. Mammalian DNA can be extracted and then severed by restriction endonucleases into lengths suitable for more detailed study. Restriction endonucleases are usually bacterially derived enzymes that cleave DNA wherever a specific sequence of several bases occurs. The DNA fragments produced are sometimes called *restriction-length fragments*. They vary in length according to the genetic inheritance of the individual. Differences occur even within families because gene variants produce base sequences that are recognized by the restriction endonucleases in some members but not in others.

2. These small stretches of mammalian DNA can be inserted into the DNA of vectors such as plasmids or bacteriophages. The new DNA is recombinant DNA.

3. The modified subcellular vector is then inserted into bacterial cells such as *Escherichia coli*. Propagation of the bacterial cells that contain the vector will produce multiple copies of the particular recombinant DNA. The process is known as *molecular cloning*.

4. The DNA so amplified can be excised from the vector DNA. It can be labeled using such radioactive materials as tritium or ^{32}P, or by molecules such as biotin which will react with an enzyme-linked marker such as avidin–horseradish peroxidase. These labeled lengths of DNA can be used as probes.

5. Probing may involve hybridization of a single strand of the cloned, labeled DNA with a single strand of unknown DNA or RNA from a source that is being investigated. The probe, for example, might be derived from the DNA of a normal individual, while the unknown DNA could come from an individual with a genetic disease such as Huntington's chorea.

6. Messenger RNA can be used as the starting material, rather than a length of genomic DNA, to create a probe. In this case, the enzyme reverse transcriptase is used to create from the mRNA a length of DNA that is complementary to it. This complementary DNA (cDNA) does not necessarily represent the same sequence as in the genome because of the extensive splicing that takes place in the nucleus to create the mRNA [see Figure 1.12 (Section 1.2.3)]. Nonetheless, these probes have proved to be extremely valuable for both DNA and mRNA hybridization experiments.

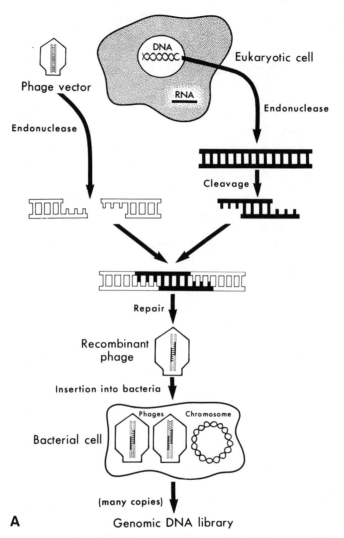

Figure 1.19. Formation of DNA libraries. (A) Genomic DNA from cell nuclei is treated with an endonuclease. DNA from a bacteriophage is cleaved by the same endonuclease. The "sticky ends" recombine with a length of mammalian DNA now incorporated into the phage DNA. The phage then infects a bacterial cell, where it is replicated many times during bacterial growth. The DNA is harvested to give many copies of the inserted DNA. If different fragments of DNA are inserted at the same time into phage vectors, then propagation of bacteria containing the vectors results in a mixed bacterial colony containing a library of genomic DNA. Subsequent cloning is required to separate out from the mixed colony bacteria containing any desired genomic DNA insert. (See the text for details.) (B) Messenger RNA from the cytoplasm of the cell is treated with the enzyme reverse transcriptase. A double stranded DNA is formed, to which polythymidylate tails are added. A plasmid vector is cleaved, and a polyadenylate tail is added to the free ends. When the two are mixed together, the adenylate tails attach to the thymidylate tails, reconstituting a plasmid containing the length of DNA complementary to the mRNA. This is known as a *chimeric plasmid*. The plasmid is now inserted into a bacterial cell, where it is allowed to replicate. Lysis of the bacterial cell will produce many copies of the cDNA. If many mRNAs are used simultaneously to create many cDNAs that are then inserted by recombinant methods into plasmid vectors, then propagation of bacteria containing the vectors results in a mixed bacterial colony containing a library of cDNAs that match the original mRNAs. Again, cloning is required to separate particular bacteria containing a desired cDNA insert. (See the text for details.)

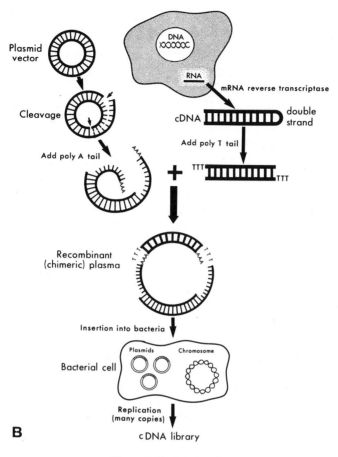

Figure 1.19. (*continued*)

7. Probes can also be created by synthesizing oligonucleotides from appropriate bases. A meaningful length of code, containing at least 16 complementary bases, is considered to be required for stable hybridization to occur with a length of endogenous DNA or RNA.

8. Probes are often used to try to supply "titles" for individual components in DNA "libraries." Libraries are created by simultaneously inserting unidentified mixtures of DNA or RNA into vectors to provide colonies from which clones containing individual inserts may be derived.

The various methods for obtaining appropriate investigative probes or creating libraries are shown in Figure 1.19. If the starting material is nuclear, or genomic, DNA, it must first be divided into workable lengths for study. A class of bacterial enzymes known as *restriction endonucleases* makes this possible. Each can digest DNA, but only where a specific sequence of bases that it recognizes occurs. Usually, the recognition code is 4–6 bases long. If each of the 4 bases, A, C, G, T, were distributed randomly, which they are not, the chances of any given 6-base sequence occurring would be once every 4^6, or 4096, bases in the DNA chain. Fragments of DNA from digestion with any given endonuclease can vary enormously, but typical lengths are a few thousand bases. The fragments can be sized with moderate accuracy by gel electrophoresis and, if not too long to be accommodated by the limited DNA of the vector being used, can be inserted by splicing.

The manner of inserting sequences is illustrated in Figure 1.19 for a plasmid and for a bacteriophage. A plasmid is an independent, extrachromosomal, autonomously replicating, circular double-stranded DNA (dsDNA) body that is readily incorporated into many bacteria. A bacteriophage is a parasitic bacterial virus consisting of linear dsDNA that is injected into the bacterium and directs the bacterial DNA machinery. Growth of bacterial cells results in the production of many copies of the recombinant DNA.

One method, illustrated in Figure 1.19B using a plasmid vector, is to cleave the DNA, converting it from a circular to a linear molecule. A polyadenylate nucleotide tail is then added with the help of suitable enzymes to the $3'$ ends of the plasmid DNA. Similarly, a polythymidylate nucleotide tail is added to the $3'$ ends of the mammalian DNA. When the two are mixed together, adenylate nucleotide tails of the plasmid pair with the thymidylate nucleotide tails of the mammalian DNA insert. The plasmid reconstitutes itself into a circular length of DNA, now called a *chimeric plasmic* because of the mammalian insert. The newly formed chimeric plasmid will be taken up by bacteria, such as *E. coli,* in which it is copiously reproduced to yield many copies of the inserted length of mammalian DNA.

Another method, illustrated for bacteriophage in Figure 1.19A, is to employ a restriction enzyme that makes staggered cuts in the dsDNA of both the phage and the DNA source. This results in single-stranded "sticky ends" for both the phage and the mammalian DNA sequence. Because they have been cut by the same endonuclease, the sticky ends have a complementarity that allows them to join together. The enzyme DNA ligase is then used to sew in place the final bonds joining the ends. The method can also be used, of course, for plasmids. A disadvantage is that the sticky ends of the vector can rejoin without including the desired mammalian DNA sequence. However, cloning techniques permit the selection of the recombinant phage to be made. Because of the rapid rates at which bacterial cells grow and divide, plus the fact that many phage or plasmid molecules may exist per bacterial cell, an almost unlimited supply of the inserted stretch of mammalian DNA can be obtained.

These techniques can be used to construct genomic DNA libraries. To form a genomic DNA library, nuclear DNA is digested with one or more restriction endonucleases with all possible fragments being incorporated into vectors. The vectors are simultaneously cloned so that a resulting mixture is produced from which a desired recombinant vector can later be selected. This is achieved by hybridizing to a more defined length of DNA or probe, which can select out a single clone that possesses a DNA segment for detailed characterization.

The methodology for making a library from mRNAs employs a different strategy. First, the mRNAs from brain or other tissue are extracted and the rare copies concentrated. These usually have a polyadenylated tail so that they can be separated by a column to which oligodeoxythymidylate has been coupled. The mRNAs are then added to a system containing reverse transcriptase so that a corresponding set of DNAs is produced. This set of cDNAs is then inserted into vectors and "shotgun" cloned to produce the second type of library, the cDNA library.

If each gene could be thought of as a complete book, then these DNA libraries would consist of shelves and shelves of neatly bound volumes, the titles of which have been erased. All the library books are there, but locating any one is a formidable task.

The cDNA library should contain highly relevant information about the state of the tissue from which the mRNA has been extracted. All the genes that have been translated to the detectable number of copies, whether normal or abnormal, should be represented.

Starting material for either the genomic or the cDNA library can be chosen from a variety of sources. In theory, genomic DNA libraries should be similar whether brain, fibroblast, or other body tissue is selected. Even cells obtained from amniocentesis can be used. In practice, lymphocytes are the preferred source. On the other hand, the cDNA libraries will vary according to the particular cell type chosen because different genes are expressed in different cells. Particularly diverse cDNA libraries should come from brain tissue because of its complex function.

Sutcliffe *et al.* (1983) made a rat brain cDNA library and then hybridized the cDNA clones back to rat brain mRNA and mRNA from other tissues. They divided the cDNA clones into four categories: those expressed equally in brain and other body tissues, those expressed differentially in brain and other tissues, those expressed exclusively in brain, and those failing to back-hybridize. The final two categories represented over 50% of the clones and were interpreted as representing mRNAs expressed only in brain. The category that did not hybridize was considered to represent very rare brain mRNA species.

All that is required to convert a cloned section of DNA into a probe, as Sutcliffe *et al.* (1983) did, is to add a label. However, to be efficient, the probe needs to be a unique sequence of DNA. The labeling can be done during cloning by the addition of precursor nucleotides labeled with ^{3}H or ^{32}P. A widely used and more efficient method is nick translation, which employs the same technique that DNA uses to repair itself. The double-stranded clone is nicked in one strand. A short length of the strand is removed by DNA polymerase I and replaced by a newly synthesized strand using labeled nucleotide precursors and the remaining strand as the template. A promising advance is the use of biotin-linked nucleotides as precursors (Brigati *et al.*, 1983). Biotin will strongly bond to the protein avidin, which can have an enzyme such as horseradish peroxidase linked to it. Thus, detection can be with an enzyme reaction rather than the much more time-consuming autoradiography.

1.5.2. DNA Hybridization and Gene Tracing

How is it possible to find a DNA sequence that corresponds to any given gene or portion of a gene? That is the overriding question of molecular genetic detective work. The following discussion indicates approaches that are being used. The best established methods of seeking an answer are the Southern and Northern blotting techniques. For the Southern blotting technique (named for the originator), genomic DNA is digested with a restriction enzyme, and the fragments are loaded onto an agarose gel for electrophoresis. The dsDNA is spread out following electrophoresis according to molecular weight and charge. It is denatured to give single-stranded fragments, which are then transferred from the agarose gel to a nitrocellulose filter on which they become immobilized. The transfer process is somewhat akin to blotting. A labeled single-stranded DNA (ssDNA) probe is then applied to the paper. Only those fragments on the nitrocellulose filter that have a minimum number of bases complementary to those on the probe will hybridize with it. Hybridized areas are detected by autoradiography if the label is radioactive, or by enzyme reaction if a biotin–avidin system is used.

In Northern blotting (Sutcliffe *et al.*, 1983), a mixture of RNAs is applied to the agarose gel for sizing by electrophoresis, and a somewhat comparable procedure is followed to transfer ssRNA onto nitrocellulose paper. Again, detection of appropriate RNA molecules is by hybridization with a labeled ssDNA probe.

At present, the basic defect in most inherited neurological disorders is unknown. For these diseases, recombinant DNA technology offers an indirect approach that can be used to infer the presence of the abnormal gene. The difficult task of mapping the human genome was begun in the 1930s using classic Mendelian techniques and population studies. This approach included classic markers such as red blood cell antigens and attempted to determine whether the gene of interest was inherited together with one of these markers. The major limiting factor was that there were not enough available genetic markers. Base changes are believed to be very common in the DNA of normal persons; they have been estimated to occur in approximately 1 of every 200 base pairs. Most of these changes occur in noncoding regions and therefore do not cause disease. However, the DNA polymorphism that results from such variations provides a tool for the study of genetic diseases. Some base changes occur in sites attacked by restriction enzymes. When the base change causes a site to disappear, a larger restriction fragment results. If this occurred only on one chromosome, two sequences could be distinguished by the sizes of the gene fragments obtained after digestion and hybridization. These restriction-fragment-length polymorphisms are inherited as simple Mendelian codominant markers and provide an unlimited number of markers for the study of genetic disease.

Houseman and Gusella (1982) illustrated how the Southern blotting technique could be used to approach the problem of Huntington's disease, an autosomal dominant disorder of the brain, for which no prior knowledge existed concerning either the chromosomal location of the gene or the gene product. The technique was to digest DNA from affected and unaffected individuals with an endonuclease and then hybridize the DNA fragments with various standard probes.

At the outset, it might have seemed that the chances of selecting an endonuclease digest of DNA at random, and attempting to hybridize fragments to a labeled probe, also selected at random, to reveal an unknown genetic defect would be somewhat less than those of hitting the jackpot at a gambling casino. However, Houseman and Gusella (1982) point out that it should take only about 150–600 markers to cover the total human genome, if they each hybridized with a different, appropriately spaced territory of the DNA. This is because positive linkage of a gene can usually be readily detected within a distance of about 10 centimorgans (10^7 base pairs of DNA) because markers linked this closely are inherited in concert 90% of the time.

Gusella *et al.* (1983) digested genomic DNA from Huntington's families with the restriction endonuclease *Hind*III. The digests were put through the Southern blotting procedure, and the ssDNA was tested with a series of labeled DNA probes, the criterion for which was simply their availability and their ability to detect DNA polymorphisms. Of these potential markers only one, designated G8, showed linkages to different restriction fragments in affected compared with unaffected individuals of Huntington's families. The

G8 marker was shown to be a stretch of human DNA 17.6 kilobases long and containing no repetitive DNA sequences.

When tested against DNA obtained from a series of cultured human–mouse cell hybrids, G8 was found to hybridize only when human chromosome 4 was present in the nuclear material. Thus, G8 represents part of the DNA sequence from human chromosome 4, and the differences in affected and unaffected individuals from Huntington's families derive from different restriction-length fragments obtained from this chromosome. A large family from Venezuela showed a definite pattern, as did an American family, with Huntington's cases and family members free of the disease giving distinctly different restriction-length fragments. The authors estimate that the G8 probe is within 0–10^7 base pairs of the actual gene. The probe attaches to the terminal band of the short arm of chromosome 4, as determined from studies of patients suffering from another genetic defect involving this chromosome, the Wolf–Hirschhorn syndrome (Gusella *et al.*, 1985). With the knowledge that the Huntington's gene is localized to a very restricted portion of chromosome 4, it should be possible to obtain many further clones from that region and thus eventually to pinpoint the exact site of the aberrant DNA. This, in turn, should provide information on the fault in the gene and thus the cause of neuronal death in the disease. By analyzing amniotic fluid, it should be possible to detect antenatally individuals who will be at risk for the disease some four decades later. Depending on the nature of the genetic defect, compensatory mechanisms might be devised that would constitute effective therapy for Huntington's disease cases.

While much work remains to be done in this particular disease, the extraordinary power and generality of the technique have been revealed.

Obviously, many laboratories will be following the lead established by Houseman, Gusella, and colleagues in developing the probes that will be required initially to scan the human genome in all manner of hereditary disorders. In addition, more detailed and specific probes will be fashioned to explore in depth the precise locus of a genetic fault, as is currently being done for Huntington's disease and Duchenne muscular dystrophy (J. M. Murray *et al.*, 1982). Duchenne muscular dystrophy is sex-linked and therefore associated with the X chromosome. In this case, an X-chromosome-derived DNA probe consisting of a 6100-base-pair sequence and designated λRC8 hybridizes to a distinct polymorphic segment of DNA in 90% of Duchenne muscular dystrophy cases, meaning that the faulty gene and the λRC8 probe are in the same general region of the X chromosome. Worton *et al.* (1984) have narrowed the location to the middle region of the short arm. A genetic linkage map of the human X chromosome with many DNA marker loci is now well advanced (Drayna and White, 1985).

A totally different approach to reading the genetic code is to isolate the mRNA produced by the gene from the cellular cytoplasm. Several methods have been used to concentrate specific mRNAs from the total mRNA population of the cell and thus make specific cDNAs. If a given mRNA, such as that coding for insulin in pancreatic cells, makes up more than 1% of total mRNA, then it can be concentrated by sucrose density-gradient fractionation, with the activity of fractions being monitored by *in vitro* translation to a definite product such as insulin. The cDNA clone can finally be prepared from the appropriate RNA concentrate.

If antibodies to the protein translation product exist, then another approach is to add these to an *in vitro* translational system. The effect is to precipitate polysomes as they are synthesizing the protein *in vitro* [see Figure 1.12 (Section 1.2.3)]. The mRNA becomes trapped in the precipitated ribosomal convoy. The mRNA can later be isolated for reverse transcriptase production of the corresponding cDNA, which can then be cloned (Krause and Rosenberg, 1982).

For certain peptides, in which some of the amino acid sequences are known, a hybridization probe can be synthesized of short lengths (10–20 base pairs) of the presumed cDNA sequence using the genetic code appropriate for the amino acid sequence. This can then be used to hybridize with cDNA clones from a library. This approach has been used to clone for cDNAs corresponding to low-abundance mRNAs with brain levels as low as 0.03–0.01% (Sutcliffe *et al.*, 1983). Complementary DNA clones now exist for many peptides, peptide precursors, enzymes, and enzyme receptors. They include those for the sodium-channel protein (Noda *et al.*, 1984) (Chapter 2), the acetylcholine receptor (Takai *et al.*, 1985) (Chapter 8), and enzymes such as tyrosine hydroxylase (Lamouroux *et al.*, 1982), dopa decarboxylase (Albert and Joh, 1985) (Chapter 9), glutamic acid decarboxylase (D. L. Kaufman *et al.*, 1985) (Chapter 7), and choline acetyltransferase (Strauss *et al.*, 1985) (Chapter 8).

1.5.3. Hybridomas and Monoclonal Antibodies

An astonishing property of cells maintained in culture is their ability to fuse and thus to form novel hybrid cells. We have briefly discussed the existence of these somatic hybrid cells in connection with establishing the chromosomal location of certain labeled DNA probes.

The discovery that two different cells grown in culture could fuse into a single cell, retaining some chromosomes from each and being able to reproduce itself as a new hybrid cell line, goes back to 1960 (see Ringertz and Savage, 1976). The technique is widely used in genetics, but one of its most important applications is in the production of monoclonal antibodies. This Nobel-prize-winning application required many years of effort by Milstein and colleagues (G. Köhler and Milstein, 1975; Milstein, 1981). Their strategy was to select a mouse myeloma cell line as a neoplastic antibody-producing type that could be grown indefinitely and hybridize it with normal endogenous lymphocytes that could produce specific antibodies in response to antigenic stimulation, but lacked the ability to propagate in culture. Their classic experiment was to immunize mice with sheep erythrocyte antigen to induce appropriate antibody-producing lymphocytes. These were harvested from the spleen and fused with myeloma cells by standard techniques. The resulting hybrid cells were cultured and cloned by selecting for further culture only those cells that expressed the desired anti-sheep erythrocyte antibodies, thus producing a new and valuable cell line that could be preserved forever.

This classic experiment has been reproduced countless times using a wide variety of antigens. It is now a general technique in biology, with many applications in neuroscience (Valentino *et al.*, 1985). The technique is illustrated in Figure 1.20. Any antigen may be used to immunize a mouse. One advantage is that it does not have to be pure, as is the case if serum antibodies are being sought. The immunized animal is sacrificed, and spleen cells are fused with mouse myeloma cells. The particular myeloma cell line chosen is one that will grow in the presence of the DNA blockers azaguanine and bromodeoxyuridine, but

Figure 1.20. Antibody production showing the hybridoma technique. An antigen with three recognition sites is shown. On the left side of the figure, a mouse is injected with the antigen, and a suspension is made of the spleen cells. These are fused with myeloma cells to create many different hybrids. The resulting mixture of cells is cultured in HAT medium so that only newly formed hybridomas will grow. These are then separated into different wells, with only those wells that show antibody production being recultured. The cloning process continues until a cell line that produces antibodies against a single recognition site is obtained. The cell line now produces a monoclonal antibody. In contrast, on the right side of the figure, the same antigen is used to immunize a series of rabbits, producing many different antisera. These antisera contain a mixture of antibodies, some against each of the recognition sites. These polyclonal antibodies are not sustained, and the rabbits must be repeatedly immunized with a purified antigen.

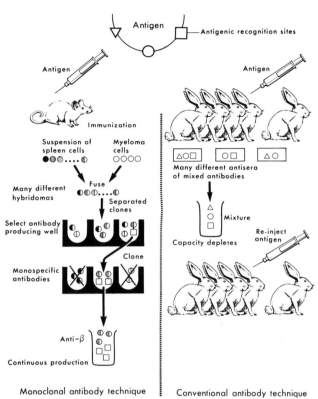

will die in the presence of aminopterin, which blocks endogenous DNA synthesis. Normal cells can overcome an aminopterin block if hypoxanthine and thymidine are added, because they have salvage pathways that are absent in the parent myeloma line. After fusion, therefore, only those cells that successfully hybridize will grow if hypoxanthine–aminopterin–thymidine (HAT) medium is used. A few cultured hybridized cells that show good growth in many wells in a large culture plate are tested for the presence of antibodies by reacting with small amounts of antigen in a different set of wells. Those wells that have specific antibody-secreting cells are recultured. The remainder are discarded. This process of cloning is continued until a single population of cells that secrete the antibody is produced. Then an immortalized cell line for a given monoclonal antibody has been cloned. It is directed against a single recognition site on this antigen. In theory, every type of specific antibody that can possibly be produced by the immune system can be produced in quantity in pure form and, through storage of appropriate hybridoma cells in liquid nitrogen, maintained indefinitely.

The results differ from what takes place *in vivo,* also demonstrated in Figure 1.20. The body simultaneously produces a variety of cell lines, each directing an antibody against a different recognition site. Thus, the serum of the animal will be polyclonal, containing different antibodies that react at different sites on the same antigen.

Each type of antibody production has specific applications, as we shall see in later chapters. Numerous commercial companies have sprung up whose sole business is the production of various monoclonal antibodies. The antibodies have been used for cancer chemotherapy, for immunosuppression after marrow transplants, for characterization of blood groups, for identifying infective agents, for gene mapping of cell-surface antigens, and for a host of immunohistological identifications, as is described in other chapters (for reviews, see McMichael and Fabre, 1982; G. Köhler, 1986).

In summary, the few experiments that have been undertaken to date have opened up impressive horizons. Young scientists are bound to be attracted to the field of molecular genetics, where opportunities appear so great. More sophisticated techniques will speed results, and the development of a larger array of standard tools will accelerate progress. Here, then, is a rich potential resource for solving the many problems of neuroscience, as will become evident in subsequent chapters.

2

Signaling in the Nervous System

2.1. Signaling by Nerve Impulses

Chapter 1 was devoted to giving an account of what a neuron is in itself. Now we come to consider how neurons are concerned in receiving and in giving signals. In Figure 2.1A, the diagram of the spinal cord is partly transverse and partly longitudinal, and below is a muscle that has a stretch receptor, or annulospiral ending [cf. AS in Figure 4.1A (Section 4.1)]. When you give a brief pull to a muscle, as when you make a knee jerk, impulses run up the primary afferent fibers from the stretch receptors and excite nerve cells in the spinal cord (the so-called ''motoneurons'') which fire impulses out to the muscles that are thus made to contract, the reflex circuit being indicated by arrows. That is the knee jerk. It is the very simplest central reflex pathway, and we will be dealing with it in Chapter 4.

To the right in Figure 2.1B, there is shown diagrammatically a cutaneous receptor that is excited by touch or pressure or irritation on the skin (Figure 2.2A). The impulses run up the primary afferent fiber into the spinal cord, then via an interneuron to motoneurons and so to the flexor muscles that bend joints; three such muscles are shown. The resulting flexor reflex withdraws the limb from the irritating stimulus.

On the surface of the hairy skin of a cat, there are very sensitive tactile receptors that give responses like those shown diagrammatically in Figure 2.1B. In Figure 2.2A, steady indentations of the indicated amounts are put upon such a tactile receptor, causing impulses to be discharged along its nerve fiber, which are seen as brief potentials. With 88 μm of indentation the receptor fires fast at the start, then slows but keeps going during the indentation. This same fast–slow sequence can be seen with the larger indentations—154, 374, and 706 μm. The receptor fires more frequently the stronger the indentation. This is a typical signal system. All our touch sensations come like this from special cutaneous receptors firing along the primary afferent nerve fibers. In Figure 2.2A, it can be seen that all the impulses are of the same amplitude. With stronger indentation, there is a faster

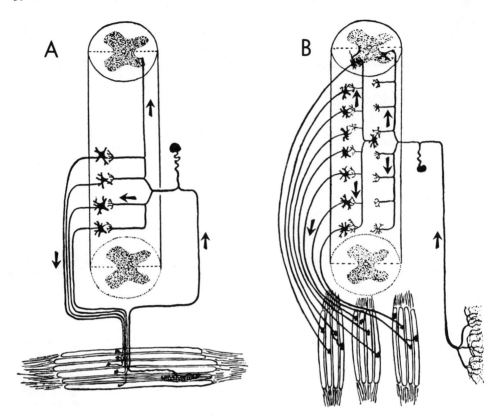

Figure 2.1. Reflex pathways for monosynaptic reflex arc. (A) With afferent fiber from annulospiral ending. (B) For polysynaptic flexor reflex with cutaneous afferent fiber. The three muscles represent flexors of hip, knee, and ankle. Modified from Ramón y Cajal (1909).

train of impulses. It is a universal property of the nervous system that signaling is by coded information like this. Trains of similarly sized impulses signal intensity by frequency. It is like a Morse code with dots only.

Figure 2.2B shows impulses discharged from a quite different receptor, a light receptor in the eye of an invertebrate. When the intensities of light increase by factors of 10, that single unit fires more and more frequently. Again, intensity is coded by frequency, and the impulses are always of the same amplitude.

In Figure 2.3A, a pull on a muscle excites a stretch receptor [cf. AS in Figure 4.1A (Section 4.1)] that discharges during the whole duration of the stretch. Some receptors, called phasic, fire only at the onset of the stimulus. Others, as in Figures 2.2 and 2.3A, discharge throughout the whole duration of an applied stimulus and are called "tonic." There are all varieties in between.

The same general principle of signaling occurs on the output side of the nervous system to muscles. For example, in Figure 2.3B, the firing of a motoneuron to an eye muscle is associated with a downward movement of the eye, which is signaled by the lower trace. A wide range of frequencies of impulse discharge is related to the amount of

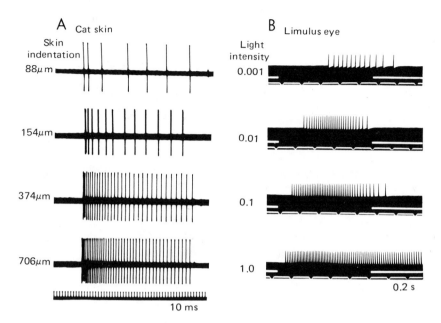

Figure 2.2. Impulse discharges from receptor organs. (A) Tactile receptor of cat furry skin discharging into a single afferent fiber in response to the indicated skin indentations. From Mountcastle (1966). (B) Photoreceptor of limulus eye discharging into a single afferent fiber in response to a flash of light signaled by the gap in the white bar and of the indicated intensity. From Hartline (1934).

downward eye movement. With strong upward movement, the motoneuron is silent. Note in Figure 2.3B(b) that the discharge begins at about position 0 and increases progressively with further downward movement.

Throughout the whole nervous system—with the complexities of organization suggested in Figures 1.2 and 1.3 (Section 1.2.1)—the signaling is by coded information of uniform-size impulses. Intensity is signaled by frequency as in Figures 2.2 and 2.3. By studying the firing patterns of single nerve cells that have this coded language in all their responses to controlled sensory inputs, we learn to understand the mode of operation of some parts of the brain. In fact, nerve impulses are the only language used in the brain for communication at a distance.

All peripheral receptor organs generate the firing of impulses along the nerve fiber that projects from them by reducing the electrical potential across the surface membrane of the terminals of that afferent nerve fiber—a process called "depolarization." This has been established by most thorough investigations on many species of receptor organs, though the retina is an exception, but their consideration is beyond the scope of a book that concentrates on the brain.

2.2. The Nerve Impulse

The reader is referred to Hodgkin (1964) for a general reference.

We now come to the question: What is this impulse that is the basis of all signaling?

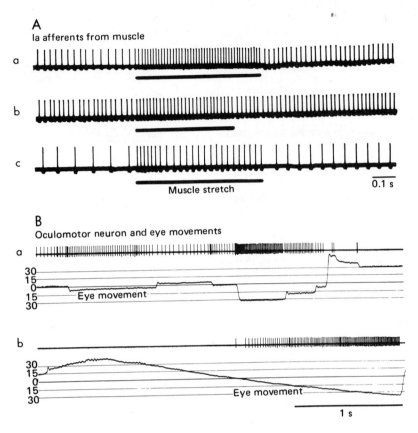

Figure 2.3. Impulse discharges from and to muscle. (A) Discharges of impulses from an annulospiral ending of cat soleus muscle in response to a slowly augmenting stretch (2.9 mm/sec) during the period denoted by the solid horizontal bars. The muscle spindle was under strong influence from a gamma motoneuron discharge [cf. Figure 14.9 (Section 14.3.5)] in (a), less in (b), and zero in (c). From Jansen and Matthews (1962). (B) Shown in the upper traces of (a) and (b) are discharges from an oculomotor neuron of a cat and in the lower traces the simultaneously recorded eye movements given in degrees of angle up or down. From Schiller (1970).

We have regarded it so far as an extremely brief message that runs along the nerve fibers. The frequency of firing may be higher or lower, but the impulse is always full-size. It is the universal currency of the nervous system. It is the only currency that the nervous system knows for any action at a distance. Signals from one nerve cell to another are conveyed by impulses. A nerve cell not firing impulses is mute. It is not communicating. There are a few minor exceptions to this generalization with action at short distances; they are referred to in Chapter 4 [Figures 4.13 and 4.14 (Section 4.6)].

In experimental efforts to study the nature of the nerve impulse, the giant axon of the squid has been of inestimable value. It has been known since the 1930s and has been utilized enormously, most fruitfully by Hodgkin (1964) and Huxley (1959), for which they received the Nobel Prize in 1963. It turns out that this giant axon has essentially all the properties of mammalian nerve fibers, and because of its enormous size, about 0.5

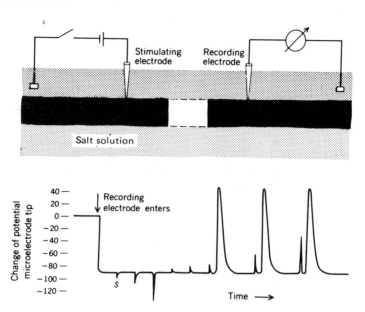

Figure 2.4. Method for stimulating and recording with a single giant fiber. From Katz (1966).

mm diameter, it can be investigated much more effectively. The giant fiber axon is an elongated cylinder with a uniform core surrounded by a thin membrane only 7 nm thick. The content of the axon has the consistency of a jelly, but it is possible to squeeze out this axoplasm through a cut end by a kind of microroller, leaving a collapsed axon that can be filled by an appropriate salt solution, an isotonic potassium solution, thereby restoring impulse transmission for hours (P. F. Baker *et al.*, 1962). In subsequent investigations on the ionic permeability of the surface membrane, a wide variety of solutions were utilized both in the core and in the surround (cf. Keynes, 1983).

Figure 2.4 illustrates a simple yet fundamental experiment on a single giant axon immersed in a salt solution resembling the blood of the squid. The axon (black) is shown in two segments to indicate a long distance between the stimulating and recording electrodes. As mentioned earlier, stimulation by an electric current is effective when it takes charge off the membrane, depolarizing it and so setting up impulses. In Figure 2.4, a brief current was applied through the stimulating electrode just on the surface of the axon. As shown in the trace below the diagram, when the recording microelectrode was outside, there was zero potential against the indifferent electrode; then, suddenly, as the recording electrode was advanced to penetrate the membrane, there appeared a membrane potential of −80 mV, which is the voltage across the surface membrane from inside to outside [cf. Figure 1.15 (Section 1.2.3)]. This immediate change was quite dramatic. Since the microelectrode has a tip diameter of less than 1 μm, it must be either outside or inside. The membrane itself is only 7 nm across, which is only about 1% of the tip diameter.

In Figure 2.4, after intracellular recording had been established, three brief currents (S) of increasing intensity were applied in the direction to increase the membrane potential. Even quite a large current had no effect except for the brief downward artifacts that are seen to be proportional to the intensity of the current. In the reverse direction, the two weakest currents were also ineffective, but at a certain intensity of the brief depolarizing current, a full-size impulse was recorded—nothing less. With two further increases in the applied currents, there was no increase in the response, only in the artifact. This is a good illustration of the all-or-nothing nature of the response; the explanation will be given later. Already we have seen examples of all-or-nothing impulses in Figures 2.2 and 2.3 (Section 2.1). In fact, all that can be propagated along an axon is a response that is full size for the condition of that axon. The alternative is nothing. And this sharp antithesis holds whether the axon is fired by receptors (Figures 2.2 and 2.3A), by nerve cells (Figure 2.3B), or by stimulating electrodes (Figure 2.4).

A further important discovery is illustrated in Figure 2.4. During the impulse, the membrane potential does not just go up to zero. It attains 30 mV or more in reverse. This discovery in 1939 by Cole and its elaboration by Hodgkin and Huxley (1952) proved false all the theories up to that time about the nature of the impulse, which was thought to arise as a brief removal of the membrane charge. The observed reversal gave rise to most interesting investigations. The impulse became a more complicated phenomenon than had been thought. The elegant biological mechanism that was disclosed in the subsequent investigations can be understood in terms of the nature of the surface membrane of the nerve fiber and of the ionic concentration differences across it, which were treated in Chapter 1 [see Figure 1.15 (Section 1.2.3)].

2.3. Ionic Mechanism of the Nerve Impulse

The essentials of the ionic mechanism of the impulse are shown in Figure 2.5, in which the time scale is in milliseconds. To the right is the membrane potential scale showing that the resting potential is about -65 mV. The time course of the membrane potential change during the impulse is given by the dashed line (V) rising to a reversal potential of almost $+30$ mV and then rapidly coming down again. The explanation of this sequence of potential changes is provided by the time courses of the changes in the sodium and potassium conductances (see left ordinate) across the membrane (g_{Na} and g_K).

This explanation is based on investigations by Hodgkin, Huxley, Katz, and Keynes on the squid giant axon using such refined techniques as studies of currents flowing during voltage clamping of the membrane and measurements of ionic fluxes by radiotracers. These investigations have been carried out under conditions in which the internal and external ionic concentrations have been widely varied. For a full description of these elegant investigations, the reader should refer to Hodgkin (1964).

In Figure 2.5, the sodium conductance (g_{Na}) begins to increase when the membrane depolarization (V trace) reaches about 30 mV. The influx of sodium ions causes an extremely steep rise of the membrane potential (V), which approaches toward the sodium equilibrium potential (V_{Na}) at about $+50$ mV. Quite quickly, however, the potential ceases to rise and then falls toward the initial resting level. This occurs for two reasons. First, even if the membrane potential is clamped at a depolarized level (cf. Figure 2.7A), the sodium conductance is quite transient, declining rapidly to the very low level of the resting state. This is the important inactivation response. Also important is the increase in potassium conductance (g_K) that begins a little later than the increase in g_{Na} and runs a

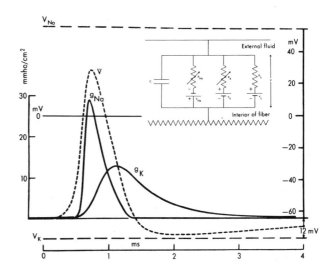

Figure 2.5. Theoretical action potential (V) and membrane conductance changes (g_{Na} and g_K) obtained by solving the equations derived by Hodgkin and Huxley for the giant axon of the squid at 18.5°C. V_{Na} and V_K are the equilibrium potentials for sodium and potassium across the membrane. The inset shows an element of the excitable membrane of a nerve fiber. Note the constant capacity, the channel for K^+, the channel for Na^+, and the leakage channel. Modified from Hodgkin and Huxley (1952).

slower time course. The large membrane potential change during the impulse ($\dot V$) moves the membrane far from the equilibrium potential for potassium ions (V_K at about -75 mV in Figure 2.5), establishing a high gradient for the outward flux of potassium ions across the membrane and counteracting the effect of sodium influx. At first it slows, then halts, the rise of the potential ($\dot V$), and then causes it to fall to the initial level and even reverses it for some milliseconds. Meanwhile, this restoration of the resting membrane potential has accomplished the turning down of both g_{Na} and g_K to their resting levels. The depolarization of the impulse ($\dot V$) occupies only about 1 msec in Figure 2.5, but for a mammalian nerve cell or fiber, the changes are still faster, an impulse having a total duration of only 0.5 msec [cf. Figure 2.10 (Section 2.4)].

The inset in Figure 2.5 shows an electrical model of the sodium and potassium conductances across the membrane, with the capacity also shown as in Figure 1.15B (Section 1.2.3). The contributions of ions passing through channels that do not change with activity are represented by a leakage resistance R_L and a battery V_L; this leakage current is rather small with the squid giant axon and is disregarded, but it is important in myelinated fibers (see Section 2.5). The essential mechanism concerned in the rising phase of the impulse is a kind of autocatalyzing reaction. An initial depolarization leads to an increase in sodium conductance, which in turn leads to entry of sodium and that to more depolarization, and so on up to the peak of the impulse. This explains why the impulse is all or nothing. It is the product of a self-regenerative or explosive reaction. Once it starts, it rises to its full effect, the energy being provided by the ions running down their electrochemical gradients.

Figure 2.6 is a diagram of an impulse showing the sodium and potassium channels through the membrane with their control by gates. The membrane is shown with separate channels for the sodium and potassium ions because these channels and their gates have quite distinctive properties. To give an example, not only do the gates open at different times in response to a depolarization (Figure 2.6), but also the sodium channels are selectively blocked by the poison tetrodotoxin (TTX) when it is applied on the outside,

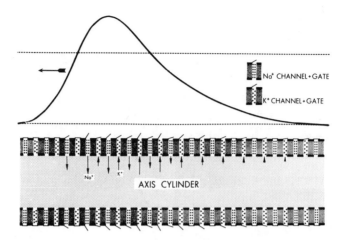

Figure 2.6. Opening of sodium and potassium gates at an instant during the propagation of a nerve impulse. In the upper part of the diagram is a plot of the membrane potential along the fiber, showing reversal as in Figure 2.5. In the lower part, the opening of the two species of gates is symbolically represented by the angles of the respective gates. The thicknesses of the membranes on either side are greatly exaggerated with respect to the axis cylinder.

while the potassium channels are selectively blocked by tetraethylammonium (TEA) injected on the inside. There are various other distinctions between them, so they are depicted as separate channels with their own gates. Presumably, each species of channel is formed by some specific protein and enzyme structure across the membrane resembling that shown in Figure 1.14 (Section 1.2.3) intercepting the bimolecular leaflet, as is indicated for the sodium channel in Figure 2.8.

Figure 2.6 shows that, at a certain stage of membrane depolarization, the sodium gates open wide in the autocatalyzed reaction already described and then rapidly close. The potassium gates begin to open a little after the sodium gates are fully open during the falling phase of the impulse. Their full closure is seen to be delayed until after the end of the impulse, as in Figure 2.5.

It was predicted by Hodgkin and Huxley (1952) that the opening of the sodium channels across the membrane would be associated with a minute current, but it was then too small for detection. Recent work of Armstrong, Keynes, and their associates on the squid axon has detected this current and has provided evidence on the molecular mechanisms that control the Na^+ channels (Bezanilla and Armstrong, 1975; Rojas and Keynes, 1975; Hille, 1976). Their work revealed that the opening of the sodium gates, *per se*, is associated with an outward current across the membrane that has been called the "gating current."

A greatly improved method of recording the gating current is illustrated in Figure 2.7A–C (Keynes, 1983). Figure 2.7A is the time course of current when the voltage of the membrane was suddenly clamped at zero. The capacity transient was canceled out electronically on the assumption that it was symmetrical with the capacity transient for the injection of an equal but opposite clamping current. Complications by potassium current were eliminated by replacing the internal potassium with cesium. The inward sodium

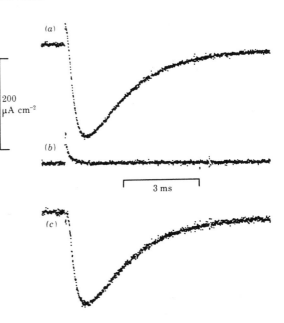

Figure 2.7. Method of separating the sodium and gating currents that flow during a voltage-clamp pulse taking the membrane potential to zero from a holding level of -100 mV. Potassium current was eliminated by perfusion with 350 mM CsF, and the external bathing solution was K-free artificial seawater containing 86 mM NaCl. The temperature was 8°C. (a) $I_{Na} + I_g$ recorded after removal of the symmetrical capacity transient by the P/3 divided-pulse procedure of C. M. Armstrong and Bezanilla (1974). (b) I_g alone after adding 2 μM TTX to the bathing solution to block the sodium channels. (c) I_{Na} alone, obtained by subtracting (b) from (a). From Keynes (1983).

current has a time course resembling that of sodium conductance (g_{Na}) in Figure 2.5, when allowance is made for the much lower temperature and for the steady voltage in contrast to the changing voltage ($\dot{V}$) in Figure 2.5. Of special interest is the brief initial outward current that precedes the inward sodium current in Figure 2.7A. This is the gating current. When the sodium current was eliminated by adding TTX to the external bathing solution, the full time course of the gating current was revealed in Figure 2.7B. Subtraction of B from A gives in Figure 2.7C the time course of the uncontaminated sodium current that is generated by the voltage clamp.

It was originally conjectured that the sodium channels are controlled by rotating dipoles (Keynes, 1975), but experimental testing has caused this conjecture to be replaced by the so-called "parallel activation model" illustrated in Figure 2.8 (Keynes, 1983). There are two quite distinct components of the sodium channel, an outer selectivity filter drawn like a funnel and the channel through the bimolecular phospholipid membrane. The actions of TTX and also saxitoxin provide conclusive evidence for this design. When applied externally, saxitoxin and TTX completely block the sodium channel without affecting the voltage-gating mechanism. TTX has no action when applied internally. It is assumed that TTX and saxitoxin block the selectivity filter (Hille, 1975). In the resting state, the sodium channel across the membrane is closed to the passage of Na$^+$ ions by double-gating protrusions as well as by positive charges in the wall of the channel. In response to a depolarizing current, there is movement of the positive charges outward, which gives the gating current, and also the retraction of the double-gating protrusions by a protein conformational change.

Finally, in the inactivation phase, there is outward movement of the inner set of positive charges and the blocking of the sodium channel by another protrusion. When the membrane potential is restored, as in Figure 2.5, there is a recovery in a few milliseconds

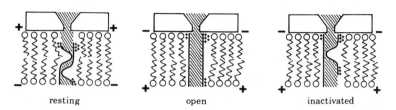

resting   open   inactivated

Figure 2.8. Diagrammatic representation of the structure of the sodium channel according to the parallel activation model. From Keynes (1983).

from the inactivation phase to the resting state. The channel opening is an all-or-nothing phenomenon with a conductance of 8×10^{-12} mhos.

The isolated sodium channel from rat sarcolemma is a protein complex with an estimated molecular weight of 314,000. It consists of a large glycoprotein subunit and several smaller subunits (Barchi, 1983). Figure 2.8 depicts a model (Keynes, 1983) in which the channel changes conformation to go from a resting state to an open and then an inactivated state. It remains a challenge to protein chemists and physiologists to work out the structural details and mechanism of action of the channel. Noda *et al.* (1984) have made a great advance by cloning the cDNA of the protein that constitutes the sodium channel of the electric eel, using the techniques described in Section 1.5.2. They isolated the protein, prepared antibodies to it, and used the antibodies to isolate appropriate cDNA clones from a cDNA library prepared from the electroplax organ mRNA. They also partially sequenced the protein and from these sequences prepared synthetic nucleotides as probes. One clone hybridized with a synthetic probe, and this was used to work out the total nucleotide sequence of the cDNA that corresponds to the full mRNA. From this, the protein structure was deduced. It corresponds to a 1820-amino-acid protein with a molecular weight of 208,000. There are four repeating segments, each of which has hydrophobic and hydrophilic elements. Noda and co-workers postulate that each unit spans the membrane and that the four units surround the channel. The structure with its charge distribution provides much insight into how the model of Figure 2.8 might work, but a further property that must be explained is related to the specificity of the channel. Some univalent cations, such as Na^+ and Li^+, pass through the channel, while others, such as K^+, Rb^+, and Cs^+, do not. This property is not explicable by ion size.

It is possible to estimate that the density of the sodium gates on the membrane is almost $500/\mu m^2$ for the squid axon. This figure is in very good agreement with the density determined for these gates by measuring the radioactivity when the gates are blocked by combination with tritiated TTX. There is as yet no evidence that there is a related process for potassium gates (Keynes, 1983). Its existence is theoretically predictable, but it has not been demonstrated for two reasons. First, it is not possible to block the K^+ currents with TEA as effectively as it is to block Na^+ currents with TTX. Second, from other evidence, there is reason to believe that the density of the K^+ gates is smaller by an order of magnitude.

Observations comparable to those described for the squid giant axons have been made by Nonner *et al.* (1975) on vertebrate myelinated fibers and are illustrated in Figures 2.11–2.13 (Section 2.5). There is the important difference that, except for the nodes of Ranvier, the cylindrical fiber membrane is inactive, being covered with a myelin sheath. There is the quantitative difference that in the nodal membrane, the density of the sodium channels ($5000/\mu m^2$) is 10 times higher than in the squid axon. A remarkable difference was disclosed by a voltage clamping study of single fibers isolated from a rabbit sciatic nerve (Chiu et al., 1979). Depolarization caused a brief inward Na^+ current that resembled that in Figure 2.5, but there was no detectable outward K^+ current corresponding to the potassium conductance (g_K) shown in the theoretical diagram (Figure 2.5). The quick recovery from the depolarization of the impulse was accomplished solely by leakage currents (G_L) across the membrane, i.e., through the $R_L V_L$ shown in the inset of Figure 2.5. The peak g_{Na} was about 7 times the g_L, but that is still sufficient for rapidly repolarizing the node.

In the diagram for the mammalian nodes of Ranvier, there would be only the $R_{Na} V_{Na}$ and $R_L V_L$ channels. In the frog node of Ranvier, there is still the $R_K V_K$ channel. As would be expected, the K^+ channel blockers, 4-aminopyridine and TEA, had no effect on the time course of the action potential of the mammalian node of Ranvier (Bostock et al., 1981). However, with mammalian unmyelinated fibers, there were still effective K^+ channels.

Ritchie (1979) discusses the density of the Na^+ channels at the mammalian node of Ranvier that is given as $10,000/\mu m^2$, which is considerably higher than in the frog. Is it the case that evolutionary pressure for maximum density of Na^+ channels crowded out the K^+ channels? As Ritchie points out, even with a conservative estimate for the size of a Na^+ channel, 20% of the nodal membrane would be occupied, leaving little room for Na^+ pumps and leakage channels. The reader should refer to Ritchie (1979) for a most illuminating discussion of the size of the node and of its channel densities.

As described by Hille (1977), the local anesthetic procaine has effective action from the axoplasmic side, when the sodium channel is in the open state (Figure 2.8). It then binds with the proteins of the channel, producing a conformational change to the inactivation state (Figure 2.8). Keynes (1983) describes experiments with a lidocaine derivative that support this mode of action of local anesthetics. The reader should refer to Ritchie (1979) for a systematic survey of the action of local anesthetics on sodium channels.

Calcium channels across the neuronal membrane are of great importance because the internal calcium concentration of neurons is essentially concerned in neuronal activity and in plasticity, which will become evident in Chapter 16. There is a most comprehensive review by Hagiwara and Byerly (1981) on voltage-dependent calcium channels in a wide variety of organisms and tissues, but necessarily our attention must be concentrated on a very limited field. The squid giant axon and synapse is important as a model, but slice preparations of the mammalian brain now allow precise studies.

The internal Ca^{2+} concentration is very low, 10^{-7}–10^{-8} M, for mammalian neurons (Cheung, 1980), which would give an electrochemical gradient of about -200 mV, inside to outside. Thus, for even a small opening of Ca^{2+} gates, there would be a considerable influx of Ca^{2+} ions. As Hagiwara and Byerly (1981) insist, it is important to distinguish between Na^+ and Ca^{2+} voltage-dependent channels. Nevertheless, with aequorin testing of squid giant fibers, there is evidence (P. F. Baker et al., 1971) for a small influx of Ca^{2+} ions, about 1% of the Na^+ influx, through typical Na^+ channels. This influx is blocked by TTX just as is that of Na^+ [cf. Figure 2.7(b)]. Meeves (1975) finds a 10-fold higher ratio for Ca^{2+} to Na^+ influx through Na^+ channels of squid axons, but points out that this discrepancy may be due to the abnormal conditions of these axons. Depolarization of the presynaptic terminals of the squid giant synapse treated with TTX reveals additional Ca^{2+} channels that are not blocked by TTX (Katz and Miledi, 1969a), but are blocked by Mn^{2+} or Mg^{2+}. These Ca^{2+} channels are also open to Sr^{2+} and Ba^{2+}. A rather large membrane depolarization (about 40 mV) is required to open the Ca^{2+} channels in the presynaptic fiber, as can be seen in Figure 3.12 (Section 3.3) by using the criterion of neurotransmitter release as an index of Ca^{2+} influx (Katz and Miledi, 1966).

Slice preparations of the mammalian hippocampus were used to investigate Ca^{2+} channels in pyramidal cells, CA1 by Schwartzkroin and Slawsky (1977) and CA3 by R. K. S. Wong and Prince (1978). In response to depolarizing pulses, intracellular recording showed two types of spike response: a fast spike eliminated by TTX and a slower, TTX-resistant spike superimposed on a long depolarization, both these latter being increased by increase in external Ca^{2+} and eliminated by Mn^{2+}. There is thus a clear identification. The former is the typical spike generated by opening of Na^+ channels (see Figures 2.5–2.8); the latter are due to the opening of Ca^{2+} channels. The Ca^{2+} spike required a threshold depolarization of about 17 mV and had a duration of about 6 msec (R. K. S. Wong and Prince, 1978). Schwartzkroin and Slawsky (1977) give a much longer duration for the Ca^{2+} spike.

In mammalian cerebellar slices, depolarization of Purkinje cells produces typical fast Na^+ spikes. Later spike responses superimposed on a slow depolarization of the dendrites are revealed as due to opening of Ca^{2+} channels by their resistance to TTX, by their blockage through removal of external Ca^{2+}, and by the action of Cd^{2+}, Co^{2+}, and Mn^{2+} (Llinás and Sugimori, 1980). The dendritic spikes have a characteristic slow time course and a threshold level of about 10 mV dendritic depolarization. The fast Na^+ spikes do not invade the Purkinje cells dendrites, which is in contrast to hippocampal granule cells, in which the dendritic spike response is mostly due to Na^+ and is eliminated by TTX (Jefferys, 1979). Activation of Purkinje cells by a climbing fiber impulse results in an initial Na^+-generated repetitive discharge that blends into a Ca^{2+} response of slower spikes superimposed on a depolarization of variable duration (Llinás and Sugimori, 1980). In the *in vivo* preparation Ekerot and Oscarsson (1981) have recorded similarly a long depolarization of up to 1 sec after the initial spike responses evoked by a climbing fiber impulse, which they attribute to opening of Ca^{2+} channels in the dendrites.

In all cases, the Ca^{2+}-generated spike potentials have a much slower time course of rise and decline than the Na^+ spikes. Their all-or-nothing character is attributable to the opening of voltage-dependent Ca^{2+} channels, much as with the Na^+ spikes. The slower rise indicates a relative sparseness of the Ca^{2+} channels, and the slower decline is due to the delayed inactivation response. Inactivation seems to be in abeyance during the bump-like depolarizations on which Ca^{2+} spikes are superimposed and which are also sensitive to Mn^{2+}, Mg^{2+}, and Co^{2+} ions and to lowering of external Ca^{2+} (Schwartzkroin and Slawsky, 1977; R. K. S. Wong and Prince, 1978; Llinás and Sugimori, 1980). It is presumed that Ca^{2+} channels are in specific structures just as for the Na^+ channels (Figure 2.8), but without the TTX-sensitivity filter. However, no details are known. The significance of the Ca^{2+} channels is considered in Chapter 16.

2.4. Conduction of the Nerve Impulse

After this detailed study of the events at one site on the nerve fiber, we are in a position to answer the question: How does the impulse travel along the fiber? So far, we

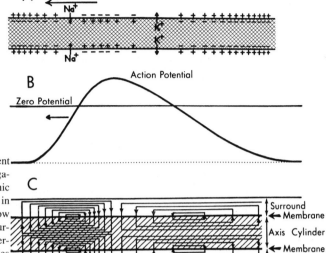

Figure 2.9. Ionic fluxes and current flow at an instant during the propagation of a nerve impulse. (A) Ionic fluxes. (B) Membrane potential, as in Figure 2.6. (C) Lines of current flow between the axis cylinder and the surround, illustrating the cable properties of the nerve fiber. From Eccles (1953).

have considered in detail the happenings in one segment of the nerve fiber as the nerve impulse passes along it. The scale of Figure 2.5 (Section 2.3) is in milliseconds. Figures 2.6 (Section 2.3) and 2.9A are diagrams with a scale in length, not time. In Figure 2.9A, the axon is shown in longitudinal section with the charge across the membrane (negativity inward) ahead of the impulse and with the reversal of charge during the impulse brought about by the influx of sodium. Later, the efflux of potassium restores the charge.

When we come to consider the propagation of this impulse along the fiber in the direction of the arrow, we have to introduce the concept that, in addition to the ionic mechanisms of the surface membrane, the fiber as a whole is acting like a kind of cable. This cable property of the nerve, however, is terribly poor by any engineering standards. In electrical transmission along a submarine cable, there is a conducting core and an insulating sheath. In the nerve fiber, the specific conductance of the core is about 10^8 times worse than that of the copper core that an electrical engineer would use. Moreover, the sheath is about 10^6 times leakier than that of a good cable. Nevertheless, in evolutionary design, this very discouraging performance of the biological cable was circumvented by boosters, a device also used in cable transmissions over long distances, where attenuation becomes serious. Boosters are inserted at intervals to strengthen the attenuated signals.

In the unmyelinated nerve fiber, a booster mechanism is built in all the way along the surface. In Figure 2.9B, the impulse is drawn as though it were frozen in its propagation along the fiber, from right to left. It is shown in Figure 2.9C that ahead of the impulse, there is a passive cable-like spread, with current leaving the outer surface of the fiber and circling back toward the impulse zone. Thus, charge is taken off the membrane by this quite limited cable-like spread ahead of the impulse, with the result that this zone becomes depolarized sufficiently to open the sodium gates (cf. Figure 2.5) and thus to turn on the self-regenerative process of sodium conductance that leads to the full-size potential change of the impulse. All that the cable-like property of the nerve fiber has to do is to transmit the depolarization for a minute distance along the surface. Then, at a critical level of depolarization, the booster mechanism of the membrane, i.e., the self-regenerative sodium conductance, takes over and builds the full impulse. The process goes on seriatim, giving the indefinite propagation of a full-size impulse at a uniform conduction velocity.

Figure 2.9 thus gives in diagrammatic form the essential features of the ionic mechanism of nerve transmission that occurs for invertebrates and vertebrates. It is the explanation of the universal currency of the nervous system, and undoubtedly it is the most efficient biological mechanism that could have been developed for the fast transmission of messages over the relatively long distances required in large animals.

We have now explained how the impulse has moved along the fiber and how recovery occurs in its wake. The whole process has been accomplished with the loss of some potassium and the gain of some sodium. This ionic exchange has been measured by radiotracer techniques and is in good agreement with prediction. There is the further requirement that ionic recovery has to be brought about by ionic pumps such as those diagrammed in Figure 1.15C (Section 1.2.3). It has been found that the more sodium there is inside, the more strongly the pumps work, so there is an automatic recovery mechanism governed by feedback control. But there is no urgency in this ionic restoration because nerve fibers contain enormously more ions than they need for one impulse. A squid giant axon loses only about one one-millionth of its potassium per impulse, but the thinnest nerve fibers, with a much larger surface/volume ratio, may lose as much as one one-thousandth. Nevertheless, there is an ionic reserve for hundreds of impulses, and meanwhile the ionic composition is being continuously restored by metabolically operated ion pumps. As would be expected, conduction of nerve impulses can continue for hundreds of impulses in the anaerobic state.

Figure 2.9C illustrates a remarkable property of impulse transmission. The cable-like spread ahead of the impulse is accomplished by electric currents that flow outward from the membrane and thence through the surrounding medium to enter the membrane at the zone of the impulse and finally complete the circuit by returning up the conducting core of the nerve fiber. In accord with predictions, alterations in the conductivity of the "surround" change the effectiveness of the cable-like spread, the conduction velocity being faster with an increase in conductance and slower with a decrease. It might be thought that this use of the surround for return current flow is an undesirable feature of design because an impulse in one fiber might spread to adjacent fibers (cross-talk) under the usual conditions of close packing. Reference to Figure 2.9C shows, however, that the privacy of the fibers is safeguarded by virtue of the directions of current flow into adjacent fibers of the surround. There is an initial strong inward current into them before the outward current associated with the peak of action potential, and finally there is a terminal inward current. Thus, as an impulse travels along a nerve fiber, the adjacent fibers are subjected to the depressant action of anodal current before and after the cathodal current; hence, cross-talk does not occur, even with the closest packing of normal nerve fibers. Injured regions of nerve fibers, however, can be zones of cross-talk, and this can result in sensory disturbances, as in causalgia and neuralgia.

It has long been known that a nerve impulse is followed by a period during which it is impossible to set up a second impulse (the absolute refractory period) and a further period during which a stronger stimulus is required to set up an impulse (the relative refractory period), which in addition is smaller in size. The usual illustrations from the older

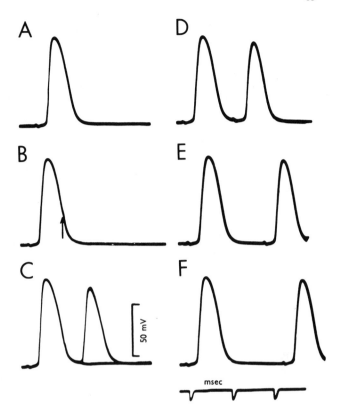

Figure 2.10. Refractoriness of nerve fiber: Responses to double stimulation of a motor axon. Intracellular recording from a motor axon in the cat spinal cord that was stimulated about 20 mm away, there being double stimulation in (B) [note the small stimulus artifacts such as marked by the arrow in (B)]. From Coombs *et al.* (1957a).

literature give a confused picture because they are recorded from nerve trunks that often contain thousands of fibers. In contrast, Figure 2.10 is an intracellular recording from a motor nerve fiber in a cat spinal cord of action potentials evoked by double stimulation of that fiber after it has left the spinal cord [cf. Figure 2.1 (Section 2.1)]. At the shortest stimulus interval [Figure 2.10B (0.7 msec)], the second stimulus was ineffective; at Figure 2.10C (0.8 msec), it was sometimes effective, the second action potential being depressed. As the stimulus interval was progressively lengthened to 2.3 msec (Figure 2.10D–F), the action potential recovered almost to full size.

This rapid recovery is typical of the nerve fibers in the brain. As a consequence, nerve fibers can carry high frequencies of discharge. In Figure 2.3B (Section 2.1), the highest frequencies of the motoneuronal discharge were over 200 per second. Neuronal discharge frequencies also exceeded 200 per second in Figure 2.17A and 2.17B (Section 2.9). But more remarkable are the extremely high frequencies often attained by the discharges of many species of interneurons in the central nervous system. For example, in Figure 4.15B (Section 4.7.2), the frequency of the Renshaw-cell discharge was initially at 1600 per second; yet its axon carried this high frequency of impulses, though with a diminished size, and its synapses operated effectively for each successive impulse (Figure 4.15D). The absolute refractory period for mammalian nerve fibers is usually about 0.5 msec. Evidently, nerve fibers have a functional design that gives a potentiality for frequency response adequate for the most extreme demands made by intensely discharging neurons or receptor organs.

2.5. Conduction Velocity and the Myelinated Fibers

The reader is referred to Hodgkin (1964) for general references.

Now we come to a disadvantage that is inherent in the initial invertebrate design of the nerve fiber as a conducting device. If the impulse is to glide along the fiber in the manner of successive invasion by cable transmission and supplementary boosting as in Figure 2.9 (Section 2.4), the progress is slow. For example, with a large crab axon 30 μm in diameter, the conduction velocity is only 5 m/sec. Cable transmission is more expeditious if the fiber diameter is larger, but, if other factors are unchanged, it can be shown theoretically that the reward in speed is disappointing, for it is proportional to the square root of the diameter. Thus, in accordance with expectation, increasing the diameter by 16 times, as from crab axon to squid axon (from 30 to 500 μm), gives only a 4-fold increase in velocity (i.e., from 5 to 20 m/sec).

The situation was transformed by the brilliant innovation of coating the nerve fibers with a thick insultation that was interrupted at intervals. Actually, this innovation was foreshadowed by an invertebrate, the prawn, that achieved a conduction velocity equal to that of the squid giant axon with a fiber diameter of only 20 μm. But the vertebrates perfected the design. Figure 2.11B illustrates the essential features of impulse conduction in a nerve fiber coated by the insulating myelin sheath, in contrast to the unmyelinated fiber in Figure 2.11A. In Figure 2.11A, the impulse glides along the fiber by smooth progressive cable invasions as in Figure 2.9 (Section 2.4). In Figure 2.11B, only the spaced nodal interruptions of the myelin sheath are active in the propagation, and cable

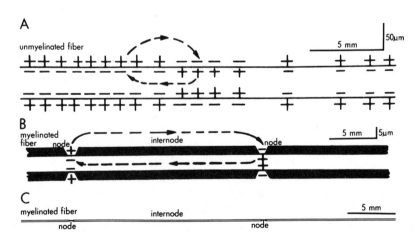

Figure 2.11. Propagation of impulse along a nerve fiber. (A) There is propagation as in Figure 2.9C, but with only one line of current flow drawn in. (B) The propagation is in a myelinated fiber, with the current flow restricted to the nodes. The dimensions in (B) are transversely exaggerated as shown by the scale, but are correctly shown in (C). Modified from Hodgkin (1964).

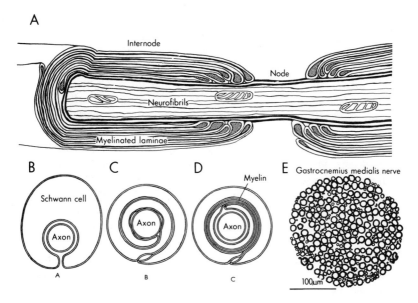

Figure 2.12. Myelinated nerve fibers. (A) A node shown diagrammatically in a longitudinal section. The terminals of the laminated myelin wrapping are shown applied to the surface of the nerve fiber adjacent to the node. Also shown are the neurofibrils and three mitochondria. (B–D). Sequences in the formation of the myelin sheath by rotary migration of the Schwann cell. From Robertson (1958). (E) Tight packing of myelinated fibers of a wide range of sizes in a muscle nerve of the cat. From Eccles and Sherrington (1930).

transmission is by currents flowing from the inactive node at the left into the active node with its reverse membrane potential. There is thus activation in succession of node to node without any active contribution from the long internodal zones. Actually, the myelinated fiber in Figure 2.11B is drawn distorted by the large transverse magnification shown by the scale. In the undistorted drawing of Figure 2.11C, the relative lengths of nodes and internodes in an ordinary large vertebrate nerve fiber can be appreciated.

Extraordinary efficiency is achieved by having the impulse hop from node to node (so-called "saltatory transmission"), with the long internodal section passive. It is often mistakenly thought that the myelin sheath achieves this remarkable result simply because it is an insulator that prevents the flow of current across the nerve membrane of the internode. At least as important a contribution by the myelin sheath, however, is given by the enormous reduction, even to less than 1%, in the electrical capacity between the axis cylinder and the surround. This reduction simply results from the increased plate separation of the myelinated cylindrical condenser enveloping the fiber in the internode. It is for this reason that the brief currents shown in Figure 2.11B spread so effectively from node to node, even with the long distances shown in Figure 2.11C. No more than half the inward current at an active node is lost by leakage through the resistance and capacity of the myelinated membrane of the internode.

Figure 2.12A shows diagrammatically the myelin wrapping around the internode and the manner of its interruption at the node, where the bare nerve membrane has channels for Na^+ ions controlled by gates, just as for the surface of an unmyelinated fiber [see

Figure 2.6 (Section 2.3)], but at a considerably greater density, about 10 times. Since the booster operations by ionic fluxes during activity are restricted to less than 1% of the surface area of the nerve fiber, the total ionic fluxes are greatly reduced; hence, there is an enormous metabolic advantage in myelination.

Figure 2.12B–D illustrate the remarkable manner in which the spiral wrapping of myelin is applied to the nerve fiber. There is one enveloping cell called a Schwann cell around each internode (Figure 2.12B). The Schwann cell then starts a spiral migration around the fiber, wrapping its surface membrane around in layer after layer. All the Schwann cell cytoplasm is eventually squeezed out as the wrapping continues (Figure 2.12C and D), so that just the myelin sheath is left, with layer after layer of two opposed Schwann cell membranes, each about 8.5 nm thick. Around a large nerve fiber, there may be more than 100 of the double membranes. Thus, the myelin wrapping of each internode is the creation of a single Schwann cell. It is remarkable that, with respect to the direction of the spiral wrapping, there is no collusion between Schwann cells of adjacent internodes.

Figure 2.12E is a transverse section of a mammalian nerve showing the dark myelin sheath around the fibers, the wide variation in fiber size, and the close packing of the fibers. Yet, as we have seen, though its transmission is by electric currents, there is no risk of cross-talk. The conduction velocity is approximately proportional to the fiber diameter, the ratio being about 6:1 for mammalian nerves. Thus, a large nerve fiber 20 μm in diameter (including the myelin) would conduct impulses at about 120 m/sec. It should be mentioned that all large nerve fibers in the brain also have a myelin sheath. Impulse conduction is comparable to that in peripheral nerve. The only difference is that the myelin is made by the oligodendroglia, which are a special variety of glial cells that were discussed in Section 1.3.

Figure 2.13 gives an elegant demonstration of the way in which a nerve impulse hops from node to node, as was first discovered by Tasaki (1939) and then fully investigated by Huxley and Stampfli (1949). Figure 2.13B shows a nerve fiber with two nodes in a rat dorsal rootlet. Recording is by two electrodes on the nerve fiber at the positions indicated by the two stems of the "Y," about 350 μm apart, and these are moved progressively along the fiber as shown. The impulse propagates downward (↓). In the four upper records of Figure 2.13A, there is no change in the time of the electrical response, then a brief delay as the electrodes record from the next internode, then again a constant time until crossing the third internode. Plotting latency against distance (Figure 2.13C) shows this stepwise delay in the impulse from node to node. The average time between the successive nodal invasions is very brief (25 μsec). Since the distance between nodes was 1.25 mm, the conduction velocity would be about 50 m/sec.

This technique is very valuable in disclosing local defects in the myelin sheath, which can be produced experimentally by diphtheria toxin. With mild damage, there is delay at each node, so that the internodal time may increase from 25 μsec to as long as 300 μsec under conditions just critical for blockage.

It was further found that after several days of demyelination, saltatory conduction with these long delays was replaced by continuous conduction at the slow velocity of 1.1–2.3 m/sec, which is about 1/20th to 1/40th of the normal saltatory velocity (Bostock and Sears, 1978). The continuous conduction in the internodes is associated with electrical excitability, so it can be assumed that the demyelination results in the development of

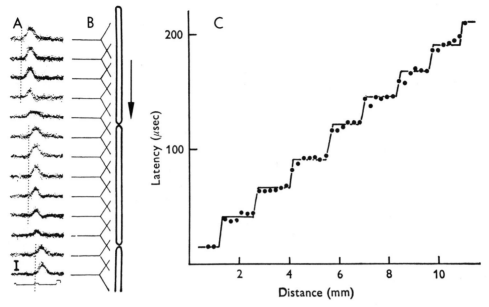

Figure 2.13. Saltatory conduction along a myelinated fiber. (B) The fiber is shown with two nodes, the arrow giving the direction of propagation. The fiber is in a very fine filament of a spinal cord root, and recordings of the impulse are made by two fine electrodes about 350 μm apart that are applied at various positions along the nerve fiber as shown by the two forks of the "Y"'s, the stems of which point to the recorded potentials (A). (A) The vertical dotted lines define three sets of records with simultaneous onsets and with steps of 25 μsec between. (C) Measurements are plotted showing eight steps in another fiber. From Rasminsky and Sears (1971).

Na$^+$ channels along the internodal membrane. Furthermore, potassium channels have developed, just as with mammalian unmyelinated fibers (Bostock *et al.*, 1981). Diphtheria toxin damage provides an experimental model for the state of the pathways to and from the brain in multiple sclerosis and also in optic neuritis. For an as yet unknown reason, the myelin of the nerve fibers disintegrates, hence the failure of impulse transmission and the resultant severe incapacity. The nerve fibers are not killed, so there is hope that the disease could be cured if their myelin wrapping could be restored. The prospects for success are much brighter than for diseases in which the nerve cells and fibers have died.

A remarkable feature of mammalian nerves is the extremely wide range of fiber diameters (cf. Figure 2.12E). This wide variation also occurs in the brain and in the pathways to and from the brain. This finding has to be considered in relation to impulse-conduction velocity, the velocity in meters per second being approximately equal to 6 times the diameter in microns. Thus, the range in velocities for the normal variation of myelinated fiber diameter, 2–20 μm, would be 12–120 m/sec. In addition, there are a great number of unmyelinated fibers from 0.2 to 1.0 μm in diameter that have much slower velocities, 0.2–2 m/sec. There is much evidence for the attractive postulate that the fiber velocity is related to the urgency of the information it is called on to transmit. The fastest fibers are concerned in the control of movement, both the afferent input from the muscles [cf. Figure 2.3A (Section 2.1)] and up to the brain and the descending pathways from brain to muscles. Some of the pathways of cutaneous sense [cf. Figure

2.2A (Section 2.1)] are also fast. At the other extreme, pathways carrying visceral information and neuron-modulator function are not urgent and are slow conducting; they are even unmyelinated over much of their course. In general, it can be postulated that in the evolutionary design of the nervous system, nerve fiber size is related to the urgency of the information it carries. More simply put: Nerve fibers are no larger than they should be!

2.6. Metabolic Considerations

It is useful now to provide a general survey of the movements of Na^+ and K^+ across the nerve membrane that were diagrammed in Figure 1.15C (Section 1.2.3). It is important to recognize that there is a very clear distinction between the passive (diffusional) and active (metabolically driven) transports across the membrane, because they travel by entirely distinct pathways. Table 2.1 summarizes these differences (Keynes, 1975). It is to be noted that the passive properties are for both Na^+ and K^+ channels, whereas the active properties (pump) are for Na^+ only, though as indicated in Figure 1.15C (Section 1.2.3), the K^+ pump is coupled with the Na^+ pump and both depend on energy provided by ATP.

The propagation of nerve impulses must be associated with the production of heat,

TABLE 2.1
Evidence of the Independence of the Sodium Pump from the Na^+ and K^+ Excitability Channels[a]

Property	Na^+ and K^+ channels	Na^+ pump
Direction of ion movements	Down the electrochemical gradient	Against the electrochemical gradient
Source of energy	Preexisting concentration gradient	ATP
Voltage dependence	Regenerative link between potential and Na^+ conductance	Independent of membrane potential
Blocking agents	TTX blocks Na^+ at 10^{-9} M; TEA blocks K^+ at 10^{-3} M. Ouabain has no effect.	TTX and TEA have no effect. Ouabain blocks at 10^{-7} M.
External calcium	Increase in Ca^{2+} raises threshold for excitation; decrease lowers threshold.	No effect
Selectivity	Li^+ not distinguished from Na^+	Li^+ pumped much more slowly than Na^+
Number of channels or pump sites	Rabbit vagus has 27 TTX-binding sites/μm^2; squid axon has 500/μm^2.	Rabbit vagus has 750 ouabain-binding sites/μm^2; squid axon has 4000/μm^2.
Maximum rate of movement of Na^+	10^{-8} mole/cm²-sec during rising phase of spike	6×10^{-11} mole/cm²-sec at room temperature
Metabolic inhibitors	No effect: Electrical activity is normal in axon perfused with pure salt solution.	CN^- (1 mM) and DNP (0.2 mM) block as soon as ATP is exhausted.

[a]From Keynes (1975). (DNP) Dinitrophenol.

initially due to the flow of the electricity (action currents) and probably to other events immediately associated with the ionic mechanisms during the impulse. Then, later, there must be heat production associated with the operation of the ion (Na^+ and K^+) pumps that restore the original ionic composition of the nerve fiber. The detection of this heat had to await the very sensitive techniques developed by Downing, Gerard, and Hill in 1926. For several decades, progressive improvement in techniques by A. V. Hill (1960) and his associates has revealed complications that are still not fully understood. At least the initial output of heat is recognized for all varieties of nerves, unmyelinated and myelinated, but with the slower processes that occur in the unmyelinated nerves, the initial output (positive) is followed by a prolonged negative heat. Approximate values for the initial heat are 14×10^{-6} cal/g per impulse for ummyelinated and 0.8×10^{-6} cal/g per impulse for myelinated. Thus, an impulse in a myelinated fiber results in a temperature rise of no more than 10^{-6}°C. These differences reflect the increased efficiency of transmission in myelinated fibers [cf. Figures 2.11 and 2.12 (Section 2.5)]. No figures are available, however, for the heat produced by the slow recovery processes associated with the pumping of ions.

It may now be asked: How far does the heat production associated with impulses account for the heat production of the brain? Kety (1961) reports that the oxygen consumption of the normal human brain is 3.3 ml/100 g per min, or 46 ml/min for the whole brain. This gives an energy consumption of almost 20 watts. The heat production would be 2.75×10^{-3} cal/g brain per sec. Actually, this figure is too low for the gray matter of the cerebral cortex, which has about 2–5 times the metabolic rate of the white matter of brain (L. Sokoloff, 1981). It is further of interest that Kety's measurements show no significant fall in oxygen consumption of the brain during sleep. This matches the finding that even neuronal discharges are not much depressed in sleep (Evarts, 1961; Eccles, 1980). Some are depressed and other aroused. On the other hand, in surgical anesthesia and in coma of various kinds, there is a reduction of the oxygen consumption of the brain, even down to half.

There has been in recent years an amazing technical development in the study of the regional metabolic activity of the human brain under a wide variety of activities and disorders (Cunningham and Cremer, 1985). There are two main techniques: The most direct have been those that study the regional cerebral metabolic rate (rCMR) employing short-lived isotopes, particularly $^{15}O_2$ with a half-life of 2 min (Frackowiak et al., 1980a), [^{11}C]deoxyglucose (L. Sokoloff, 1981), or [^{18}F]fluorodeoxyglucose (FDG) (Phelps and Mazziotta, 1985), and using positron emission tomography (PET). There is, however, the simpler alternative of studying the regional cerebral blood flow (rCBF) with ^{133}Xe or ^{77}Kr (Lassen et al., 1978; Roland et al., 1980a, 1982; Ingvar, 1983). Investigations reveal that changes in rCBF generally indicate changes in cerebral metabolism. The mean resting blood flow is 80 ml/100 g-min for gray matter (Ingvar, 1983; Roland et al., 1982) vs. only 20 ml/100 g-min for white matter.

With the use of FDG to determine the rCMR for glucose by PET, values for normal humans have been reported in the range of 25–40 micromoles per 100 grams of tissue per minute for various areas of the cerebral cortex (N. L. Foster et al., 1984; Kuhl et al., 1982; Phelps and Mazziotta, 1985). Values are substantially increased by appropriate stimulation. For example, visual stimuli increase the glucose metabolic rate in the visual cortex, memory tasks increase the rate in the medial temporal cortex, and auditory stimuli increase it in Heschl's gyri and the planum temporale (Phelps and Mazziotta, 1985). In organic dementia, particularly Alzheimer's disease, there is a grave depression of metabolism in the parietal, temporal, and frontal lobes (for a review, see P. L. McGeer, 1986). In appropriate sections of this book, there is reference to the highly specific regional activities of the brain as revealed by these technical procedures.

2.7. Impulse Propagation in Neurons: Somata and Dendrites

Hitherto, attention has been restricted to nerve fibers, which are geometrically the simplest components of neurons, being long uniform cylinders in the cases investigated [see Figures 2.6, 2.9, 2.11, and 2.12 (Sections 2.3–2.5)]. Figure 1.2B (Section 1.2.1), however, reveals that many neurons have short axons that branch profusely. It is assumed that impulses also propagate along these branches to their terminals, and this has been established in special cases that give favorable opportunity for investigation. However, there are notable exceptions [cf. Figures 4.13 and 4.14 (Section 4.6)]. More complicated situations arise in the propagation of impulses over the neuronal somata and the dendrites that branch therefrom (Eccles, 1957, 1964a).

Some of the principal features of the spike potentials generated by a neuron are illustrated in Figure 2.14. The impulse is initiated by stimulation of the axon of this motoneuron in the ventral root, and it propagates antidromically as indicated by the arrow in Figure 2.14B, where the three specifically responding zones are labeled M, IS, and SD. Recording is from the intracellular microelectrode, through which steady current can be passed to vary the membrane potential, as indicated to the left of the series of responses in Figure 2.14A. This motoneuron has a high membrane potential (-80 mV), and it is displaced over the range -87 to -60 mV, as indicated. The antidromic spike potential is observed to have three distinct components, each all-or-nothing in character. There is first the very small spike (about 5 mV) that is seen alone, sometimes at -82 mV and always at -87 mV, and that is shown by threshold differentiation (lowest record of Figure 2.14A) to be generated by an impulse in the motor axon of the impaled motoneuron. Second, there is the larger spike of about 40 mV (which is always set up at membrane potentials of -80 mV or less, and also sometimes at -82 mV). Finally, the full-size spike is superimposed at membrane potentials of -77 mV or less, and rarely at -78 mV. Figure 2.14B shows the antidromic pathway, together with the regions of the motoneuron (M, IS, and SD) in which the three components of the spike potentials are believed to be generated.

This identification has been firmly established by many investigations, particularly the elegant analysis by Terzuolo and Araki (1961) using simultaneous recording from a motoneuron by intracellular and extracellular microelectrodes. Furthermore, they could record simultaneously with one microelectrode inserted into a dendrite and the other into the soma of the same motoneuron (Figure 2.14C and D). The dendritic potential shows a summit delayed by about 0.3 msec in Figure 2.14C and 0.5 msec in Figure 2.14D, and it has a slower time course. It is full-size in Figure 2.14C, but in the region more remote from the soma (Figure 2.14D), the size was reduced and the time course much lengthened. In Figure 2.14B, the arrows indicate the zones of blockage between M and IS, between the IS and the SD, and between the large dendrites and their terminal branches. The location of this last blocking site is believed to be much more variable than the others.

Figure 2.14. Intracellular responses evoked by an antidromic impulse, indicating stages of blockage of the antidromic spike in relation to the initial level of membrane potential. (A) The initial membrane potential (indicated to the left of each record) was controlled by the application of intracellularly applied currents. Resting potential was at -80 mV. The bottom record was taken after the amplification had been increased 4.5 times and the stimulus had been decreased until it was just at threshold for exciting the axon of the motoneuron. From Coombs *et al.* (1955a). (B) Schematic drawing of a motoneuron showing dendrites (only one is drawn with terminal branches), the soma, the initial segment of the axon (IS), and the

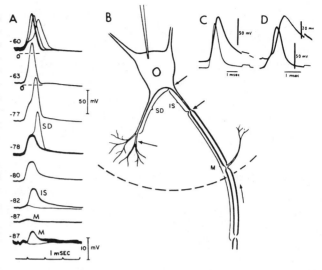

medullated axon (M) with two nodes, at one of which there is an axon collateral. The three arrows indicate the regions in which delay or blockage of an antidromic impulse is likely to occur. The regions that produce the M, IS, and somatodendritic (SD) spikes are indicated approximately by the labeled brackets. From Eccles (1957). (C, D). Simultaneous intracellular traces of dendritic and soma spike potentials from the same motoneuron. From Terzuolo and Araki (1961).

Three methods can be employed to elicit the motoneuronal spike potential: antidromic [see Figures 2.14A–D and 2.16B (Section 2.8)], synaptic excitation [see Figure 4.5A–D (Section 4.3)], and direct stimulation by an applied current through the recording microelectrode (see Figure 2.15A). The normal sequence of events for synaptic excitation of these species of neurons (Figure 4.5A–D) is, first, the synaptic depolarization of the soma and dendrites; then, by the current flow, the IS membrane is depolarized. When an impulse is generated there, the current flow reverses and adds greatly to the synaptic depolarization of the SD membrane, so that the much higher threshold is attained and the SD spike is generated.

With cat motoneurons, the average threshold depolarizations for IS and SD spikes are, respectively, 10 and 25 mV. An important functional consequence is that with the generation of impulse discharge in the IS region, there is the best possible arrangement for integration of the whole synaptic excitatory and inhibitory action on that neuron. All these influences are electrotonically transmitted to the initial segment and there algebraically summed, as is described in Chapter 4. Many species of neurons exhibit a low threshold zone in the initial segment comparable to that in the motoneurons, e.g., the neurons in the spinal cord projecting up to the brain and also the motor pyramidal neurons of the cerebral cortex (Oshima, 1969). However, there seems to be no such threshold discrimination with short-axon neurons in the brain or spinal cord (Eccles, 1964a).

There is a long history relating to impulse generation and propagation in dendrites. The extensive earlier work was reviewed by Eccles [1957 (pp. 223–227)]. It was recognized that synaptic excitation on remote dendrites would be relatively ineffective in exciting impulse discharges from the neuron at the axonal origin unless there were strategic sites on the dendrites boosting by local impulse generation (G. H. Bishop, 1958). Dendritic spikes in hippocampal pyramidal cells were observed to propagate very slowly (0.4 m/sec) along the dendrites. At that time, Fatt (1957) had studied the depth profile of the antidromically generated extracellular potential in a single motoneuron and had shown that the antidromic impulse propagated from the soma for 0.2–0.3 mm along the dendrites at a slow velocity of 0.7–1.0 m/sec. Such a dendritic spike potential was displayed by intracellular recording (Terzuolo and Araki, 1961) from motoneuronal soma and dendrite (Figure 2.14 C and D).

A few years later, W. A. Spencer and Kandel (1961) made a detailed study of fast prepotentials observed on the rising phase of the spike when recording intracellularly from hippocampal pyramidal cells excited synaptically. It was proposed that there is a strategic branching site of the apical dendrite, a ''bifurcation trigger zone'' where impulses ''act as a booster zone for otherwise ineffectual distal dendritic regions.''

This conjecture has been confirmed by a depth profile analysis of CA1 pyramidal cells (Andersen and Lømo, 1966; Andersen et al., 1966). The Schaffer collaterals have a restricted synaptic zone on the apical dendrites and trigger a dendritic spike that propagates to the soma at about 0.4 m/sec. The slow conduction velocity for dendritic spikes is attributable to the high threshold for opening Na^+ gates, about 30 mV, and to the profuse dendritic branching (Eccles, 1960). Antidromic invasion of hippocampal granule cells propagates for 200 μm up dendrites at the very slow rate of 0.08–0.12 m/sec (Jeffreys, 1979).

As documented above by intracellular recording from dendrites of cerebellar Purkinje cells, there was demonstration of the synaptic generation of dendritic spike potentials, which were mostly Ca^{2+} spikes (Llinás and Sugimori, 1980; Ekerot and Oscarsson, 1981), but no attempt was made to measure a conduction velocity.

In contrast to normal mammalian motoneurons, an active role of dendrites is exhibited by motoneurons that are suffering from chromatolysis (see Sections 1.2.5 and 13.3.2) on account of severance of their axons some weeks previously (Figure 2.15). Synaptically induced depolarizations [cf. Figure 4.5A–D (Section 4.3)] evoke partial spike responses of localized regions of the soma or dendrites (Figure 2.15B and C). The additional depolarization may then cause further partial spikes (Figure 2.15F–J), and eventually a full propagated spike may arise either at the initial segment of the axon or elsewhere, in the soma or up some dendrite (Figure 2.15F, H, and J). In Figure 2.15A–D, there is, in addition, recording from the motoneuron axon in the ventral root, the arrows indicating an impulse discharge just after the large intracellular spike (Figure 2.15A, and D). Figure 2.15 gives an excellent illustration of the postulated booster function of dendritic spike potentials. Similar observations were reported by Kuno and Llinás (1970). In alligator Purkinje cells, Llinás and Nicholson (1971) made a systematic intracellular study of cumulative dendritic spikes and likewise suggested trigger zones on dendrites.

There has been much speculation about the opportunities for local integrative actions of excitatory and inhibitory synapses and a large neuron can be assumed to function as a conglomerate of many such interacting zones. It has to be recognized, however, that a neuron is a unit by virtue of the fact that it has only one axon along which it can fire. In any case, we have the overwhelming challenge of trying to understand the operation of the

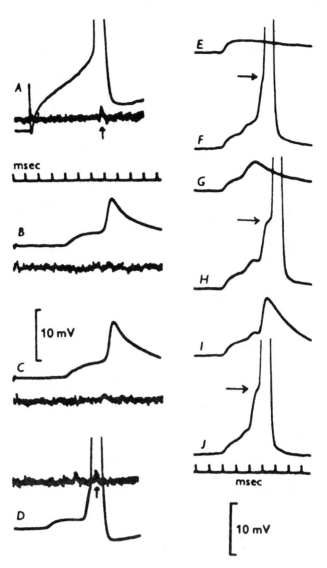

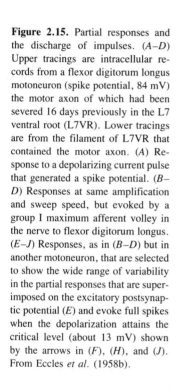

Figure 2.15. Partial responses and the discharge of impulses. (A–D) Upper tracings are intracellular records from a flexor digitorum longus motoneuron (spike potential, 84 mV) the motor axon of which had been severed 16 days previously in the L7 ventral root (L7VR). Lower tracings are from the filament of L7VR that contained the motor axon. (A) Response to a depolarizing current pulse that generated a spike potential. (B–D) Responses at same amplification and sweep speed, but evoked by a group I maximum afferent volley in the nerve to flexor digitorum longus. (E–J) Responses, as in (B–D) but in another motoneuron, that are selected to show the wide range of variability in the partial responses that are superimposed on the excitatory postsynaptic potential (E) and evoke full spikes when the depolarization attains the critical level (about 13 mV) shown by the arrows in (F), (H), and (J). From Eccles *et al.* (1958b).

immensely complicated circuits of neuronal action, as will be seen in Chapters 4 and 14–17, which are based on the neuron operating as a unit. Attempts to theorize on a possible fragmentation of this unity are certainly out of place at this time.

2.8. Afterpotentials

The spike potential of the squid giant axon [see Figure 2.5 (Section 2.3)] was followed by a phase of hyperpolarization that is attributable to the continuance of a relatively high level of K^+ conductance across the membrane. Similarly, the spike

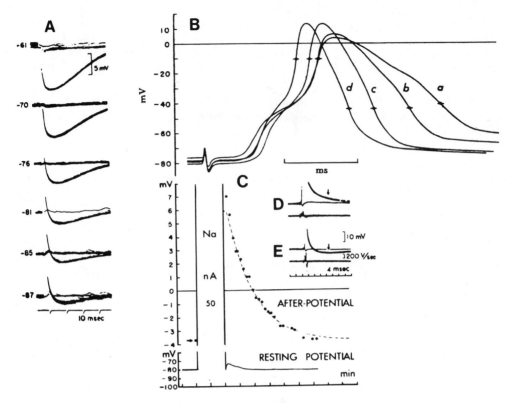

Figure 2.16. (A) AHPs of a motoneuron, occurring at various levels of membrane potential as controlled by intracellularly applied currents. For each record, the stimulus applied to the ventral root was adjusted to the critical strength at which the axon of the particular motoneuron was sometimes excited and sometimes not. The membrane potentials in millivolts at which the action potentials were evoked are given alongside each record. The resting potential varied from −76 to −79 mV. The spike component of the antidromic action potential does not appear in these records, the amplification being too high and the sweep too slow to display it. From Coombs *et al.* (1955a). (B) Superposition of antidromic spike potentials recorded at 17 (*a*), 52 (*b*), 151 (*c*), and 348 sec (*d*) after a sodium injection into an anterior biceps motoneuron by passing a current of 50 nA for 60 sec. (C) Changes in the AHP following the indicated sodium injection from an NaCl-filled single microelectrode inserted into a gastrocnemius–soleus motoneuron. The measurements were made at the points indicated by the arrows in (D) and (E). The dashed line through the plotted points is an exponential curve with a time constant of 104 sec. (D, E) Specimen records of the membrane potential (upper traces) obtained at 36 (D) and 342 sec (E) after a sodium injection. (B–E) From M. Ito and Oshima (1964).

potential of a motoneuron is followed by an afterhyperpolarization (AHP) that usually reaches a maximum of 5–10 mV at about 10 msec and continues for some 100 msec (Figures 2.16A and E). Its duration is longer—up to 150 msec—for motoneurons that innervate slow muscle fibers than for those that innervate fast fibers, usually 60–80 msec (Eccles *et al.*, 1958a). There is evidence for two phases of K^+ conductance—the initial phase associated with the decline of the spike potential, analogous to that in Figure 2.5, and a later phase that is partly separated, which gives the AHP (Eccles, 1957; Gustafsson, 1974). The two phases of K^+ conduction following a depolarization may be related to the two phases of potassium current investigated in frog ganglion cells, I_K and I_C (P. R.

Adams *et al.*, 1982). I_C is triggered by the entry of Ca^{2+} ions and is sensitive to external TEA.

In Figure 2.16A, varying the membrane potential by a steady background current as in Figure 2.14A (Section 2.7) had a remarkable effect on the AHP following an antidromic spike potential. Because of blocked antidromic transmission, it was not possible in this experiment to investigate membrane potentials larger than -87 mV, but extrapolation of the linear curve relating size of AHP to membrane potential gave an equilibrium potential of -90 mV. When Na^+ was injected electrophoretically into a motoneuron, thus reducing its K^+ concentration, the AHP was temporarily inverted as in Figure 2.16C and D. In Figure 2.16C, recovery to the normal level occurred exponentially with a time constant of about 100 sec. Reference to Figure 1.15B (Section 2.7) gives the explanation, namely, that the linked Na^+–K^+ pump is restoring the normal ionic concentration and, in particular, building up the K^+ concentration to its normal level with an equilibrium potential of -90 mV. Thus, it can be postulated that the AHP of mammalian spinal motoneurons is due to an increased K^+ conductance of the neuronal membrane. This hypothesis of Coombs and colleagues (Coombs *et al.*, 1955a; Eccles, 1957) was called into question, but a systematic study (Gustafsson, 1974) decisively corroborated the original hypothesis.

Immediately following the spike potential, there is a slower decline of depolarization that, after some milliseconds, leads to the AHP. This phase is called the "afterdepolarization" and is probably due to a residuum of raised Na^+ permeability, particularly that persisting in the dendrites, as indicated by the slow dendritic spike potentials (cf. Figure 2.14C and D). There is much variability in this afterdepolarization, depending particularly on the amount of injury inflicted on the motoneuron by the impalement. When a motoneuron is in excellent condition, as in Figure 2.16E, there is a large afterdepolarization, the reversal to the AHP taking several milliseconds.

The electrophoretic injection of Na^+ into a motoneuron, with the raising of internal Na^+ and depletion of internal K^+, changes the spike potential (Figure 2.16B), the rise and decline being greatly slowed. This is exactly in accord with the hypothesis that the spike potential of neurons is produced by the linked Na^+–K^+ permeability increase as illustrated in Figures 2.5 and 2.6 (Section 2.3) for the squid axon. Recovery is complete after about 6 min, which corresponds well with the recovery of the AHP (Figure 2.16C). This correspondence is in accord with the hypothesis that both recoveries are due to a restoration of the normal ionic composition by the operation of the linked Na^+–K^+ pump.

Gustafsson (1974) has shown that the cells of origin of the dorsal spinocerebellar tract (DSCT) in the spinal cord differ from motoneurons in that the AHP is smaller and reaches an earlier maximum than occurs with motoneurons. Otherwise, the AHP had similar properties, e.g., an associated conductance increase to K^+ ions and summation with repetitive stimulation. There is evidence that all neurons that have been sufficiently investigated tend to be grouped into three classes. First, there are the large alpha motoneurons as above, the large pyramidal tract cells of the motor cortex, the trochlear and hypoglossal motoneurons. Second, there are the DSCT neurons and other tract neurons, the rubrospinal and vestibulospinal. To these rather similar two classes, there should be added a third, a wide-ranging class comprising interneurons of various types. These are characterized by very high frequencies of firing—up to 1000 sec or even higher. It is evident that a large AHP is not conducive to such high frequencies. Further reference to these classes of neurons is made when considering repetitive responses.

2.9. Repetitive Neuronal Firing

In the normal operation of brain synapses, there is rarely a sharp single action by a synchronized volley of impulses, but instead there are prolonged repetitive discharges [cf. Figure 2.3B (Section 2.1)]. Hence, the question now arises: How effective is a long, continued membrane depolarization in causing a prolonged impulse discharge? This question can be answered most directly by investigating the effects of steady depolarizing currents, instead of using synapses to fire a neuron. In Figure 2.17A, a microelectrode was inserted into a motoneuron and steady currents were passed to take charge off the surface membrane, while at the same time the responses of the neuron were recorded. It is seen that a current of 7 nA just caused the neuron to fire, but the firing was slow and irregular. As the steady current was progressively increased, it produced more and more depolarization of the surface and faster and faster firing of the cell. Finally, with 4 times the threshold current, the neuron fired initially at the very high rate of at least 200/sec. In all cases, the initial frequency slowed during the steady current application; nevertheless, this experiment on prolonged impulse discharge reveals the immense range in the responses of neurons. Continuous synaptic bombardment is active in the same way. There are many species of neurons in the central nervous system that can be driven by intense synaptic bombardment to fire at 500–1000/sec. This is evidence of real synaptic power, and these neurons correspondingly exert strong synaptic power at their synapses.

In Figure 2.17B, a cell in the cerebral cortex gave similar responses. The weakest

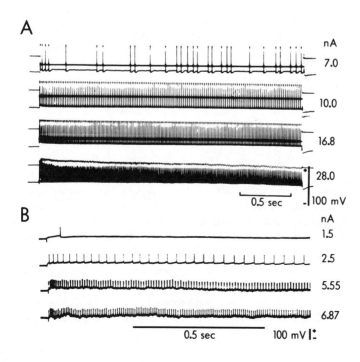

Figure 2.17. Generation of impulse discharges in neurons by prolonged depolarizing currents. (A) Intracellular responses of a rat motoneuron to prolonged (2.5 sec) rectangular depolarizing currents applied through the same electrode. Strengths of currents are indicated to the right in nanoamperes. From Granit *et al.* (1963). (B) Intracellular responses of a pyramidal tract cell of the motor cortex to currents indicated in nanoamperes for each trace. From Oshima (1969).

current caused only one discharge. Again, frequency of response increased sharply with intensity, and with a current just over 4 times threshold, the initial frequency was over 200/sec.

It is immaterial whether depolarization is due to a steady synaptic bombardment or to an applied current. There is essentially the same response, depolarization generating impulse discharge in a manner illustrating that intensity is coded by frequency, just as demonstrated with receptor organs in Figures 2.2 and 2.3A (Section 2.1). The synapses and the neurons of the central nervous system are well adapted metabolically to respond continuously at frequencies as high as 100/sec and with peak responses as high as 500/sec. This is a performance far superior to that of neuromuscular synapses. It used to be thought that one of the distinguishing properties of synapses was fatigue (Sherrington, 1906), but this is belied by the performance of synaptic mechanisms in the normal functioning of the brain.

The AHP following the spike potential [see Figure 2.16 (Section 2.8)] exerts an important contributing influence on the frequency of neuronal firing in response to a steady depolarizing current. After each spike potential, there is a phase of increased membrane potential, the AHP, the next discharge occurring when the depolarizing influence has overcome this hyperpolarization, the membrane depolarization again reaching threshold (cf. Eccles, 1957; Schwindt and Calvin, 1972; Calvin and Schwindt, 1972; Gustafsson, 1974). In the upper trace of Figure 2.17B, the AHP can be seen after the single spike. The second trace also illustrates the origin of each successive spike response on the rising phase of depolarization following the AHP. With stronger depolarizing currents in Figure 2.17B, the firing frequency is so high that this relationship is obscured, but it has been studied in detail by Schwindt and Calvin (1972). It has already been mentioned that the motoneurons that supply slow tonic muscles have a longer AHP than those that supply fast muscles. This is a good physiological arrangement, for it ensures that slow muscles are activated at a slow frequency. The motoneurons are thus tuned to the frequencies appropriate for the muscles, slow discharge rates for slow muscles and fast rates for fast muscles.

When considering the mechanism of impulse generation by a motoneuron, it is important to recognize the two distinct zones of the neuron [see Figure 2.14B (Section 2.7)]. There is, first, the soma–dendritic zone (SD) that is subjected to the influences of synapses, the excitatory depolarizing and the inhibitory hyperpolarizing, as described in Chapter 4. It also generates the AHP after each SD spike discharge. Second, there is the initial segment (IS), which is the trigger zone for firing impulses, but in which the firing of an impulse produces a negligible AHP (cf. Eccles, 1957). The normal sequence is: (1) synaptic depolarization of the SD and IS membranes to the threshold level for initiating the IS spike, (2) generation of an SD spike by the IS spike, (3) generation of the AHP by the SD zone. Thus, there is a division of function between the IS and the SD zone of the neuron. The IS zone is simply a firing zone. The SD zone is a controller of this firing by virtue of the potential changes it transmits electrotonically to the IS zone, namely, the algebraically summed, synaptically generated depolarizations and hyperpolarizations and the AHP autogenously initiated by the SD spike response. With repetitive discharge, there is summation of the AHPs, sometimes with decrement. Baldissera and Gustafsson (1974) have developed computer models on the basis of these concepts and have shown that they account very satisfactorily for all the details of the effects produced by steady depolarizing currents in producing repetitive discharge as illustrated in Figure 2.17.

In Figure 2.17, it can be seen that the intracellular spike potentials were reduced by the depolarizing currents. The more the depolarization, the greater the reduction. This effect is more obvious in Figure 2.14A, where there was a large reduction of spike height when the membrane potential was depolarized to -60 mV. Similar reduction of spike potential by depolarization occurs with impulses in nerve fibers [cf. Figure 3.13 (Section 3.3)] and is attributable to some interaction with the gating mechanism. It is of general pharmacological interest because spike reduction is observed when various putative synaptic excitants are tested. When there is a powerful, pharmacologically induced depolarization, there may even be suppression of the spike discharge, which can be a source of confusion.

3

Chemical Synaptic Transmission at Peripheral Synapses

3.1. Discovery of Chemical Synaptic Transmission

In Chapter 2, it was shown how transmission along a nerve fiber occurs by brief all-or-nothing impulses and that this impulse propagation depends on two main factors: (1) the cable-like properties of the nerve fiber, the conducting core being ensheathed by a relatively resistant membrane; (2) the large and momentary increase in sodium conductance of this membrane when its resting potential is suddenly diminished. As a consequence, the extremely poor cable-like transmission along the nerve fiber is amplified at each segment in a self-regenerative manner. The all-or-nothing character of the propagation derives from the amplification that each segment of the nerve gives to the attenuated signal transmitted in the manner of a cable from the adjacent active region.

In contrast to this impulse propagation along nerve fibers in a decrementless manner, there has long been the idea that some special mechanism is responsible for transmission from one conducting element to the next, e.g., from nerve fiber to muscle fiber. The concept of the neuron was developed by the neuroanatomists His, Forel, and Ramón y Cajal in the latter part of the 19th century and led to the concept of the synapse by Sherrington, namely, specialized transmitting structures by which one neuron affects another [cf. Figures 1.4–1.7 (Section 1.2.2)]. Much earlier, evidence had been presented by Claude Bernard that the junctional region between nerve and muscle is blocked by curare, and in 1877, Du Bois Reymond is reputed to have suggested that neuromuscular transmission is in part chemically mediated. Unfortunately, he based this suggestion on the mistaken notion that there is a protoplasmic continuum (K. Krnjevic, 1974). At the turn of the century, there were only the first beginnings of the hormonal story, in which so-called "humoral" transmission occurs by "chemical messengers, or hormones," which are specific chemical substances that are carried by the bloodstream and act slowly for prolonged periods.

In the first three decades of this century, it was shown that transmission across the synapse at the neuromuscular junction is very fast, with a delay of at most 1 msec in transmission from nerve impulse to muscle impulse. This contrast between the slowness of the chemical transmission by hormones and the extremely fast and brief transmission at neuromuscular junctions and synapses made it appear that chemical transmission is too impossibly slow. Hence, the concept arose that these latter transmissions were electrically mediated, i.e., that the electrical currents responsible for transmission along the nerve fiber were themselves responsible for exciting the nerve cell or muscle fiber across the synaptic junction. A great difficulty confronted this hypothesis because of the extraordinary mismatch in the electrical properties between the nerve fiber and the muscle fibers. This was largely overlooked, however, because of the inadequacy of the evidence on the respective electrical parameters.

The first suggestion of chemical transmission was put forward in 1904 by T. R. Elliott, a Cambridge University student. He was struck by the similarity in effects of injection of adrenaline and stimulation of the sympathetic nervous system. It was an Oxford student, W. E. Dixon (1906), who followed up on the concept with regard to the parasympathetic nervous system. Dixon's experiments drew attention to the comparable action of injected muscarine and stimulation of parasympathetic fibers. The idea of chemical transmission was not well received, and Dixon unfortunately became discouraged by his inability to extract an active principle.

Some years later, Dale (1914) broke open the field of autonomic pharmacology. He recognized the attractiveness of the speculation for adrenaline mediation of sympathetic nerve impulses and for acetylcholine (ACh) mediation of parasympathetic impulses. The great potency of ACh demonstrated by R. Hunt and Taveau (1906) and its blockade by atropine greatly impressed Dale. In 1914, Dale identified a whole series of agents that mimicked electrical stimulation of autonomic fibers which he termed either "parasympathomimetic" or "sympathomimetic." He was on the verge of substantiating the hypothesis of chemical transmission in the autonomic nervous system. Not only that, he was working with the transmitters themselves. Nevertheless, this crowning step eluded him, just as it had Elliott and Dixon.

In retrospect, Dale (1938) spoke in his characteristically vivid style of this preliminary stage of the chemical transmitter hypothesis:

> Such was the position in 1914. Two substances were known, with actions very suggestively reproducing those of the two main divisions of the autonomic system; both for different reasons were very unstable in the body, and their actions were as a consequence of a fleeting character; one of them was already known to occur as a natural hormone. These properties would fit them very precisely to act as mediators of the effects of autonomic impulses to effector cells, if there were any acceptable evidence of the liberation at nerve endings. The actors were named, the parts allotted, and almost forgotten; but only direct and unequivocal evidence could ring up the curtain and this was not to come until 1921.

Dale was referring to the experiment of Otto Loewi that established the first satisfactory proof of chemical transmission. Loewi's elegantly simple experiment with two frog hearts set up in series will go down as one of the classic experiments of all time. A dramatic flourish to the occasion has been given by Loewi's (1960) own recollection of the event almost 40 years later:

> The night before Easter Sunday of that year (1921) I awoke, turned on the light, and jotted down a few notes on a tiny slip of thin paper. Then I fell asleep again. It occurred to me at 6:00 in the morning that during the night I had written down something most important, but I was unable to decipher the scrawl. The next night, at 3:00 a.m., the idea returned. It was the design of an experiment to determine whether or not the hypothesis of chemical transmission that I had uttered seventeen years ago was correct. I got up immediately, went to the laboratory, and performed a simple experiment on a frog heart according to the nocturnal design. . . . the hearts of two frogs were isolated, the first with its nerves, the second without, both hearts were attached to Straub cannulas filled with a little Ringer solution. The vagus nerve of the first heart was stimulated for a few minutes. Then the Ringer solution that had been in the first heart during stimulation of the vagus was transferred to the second heart. . . . if carefully considered in the daytime, I would undoubtedly have rejected the kind of experiment I performed. . . . it was good fortune that at the moment of the hunch I did not think but acted immediately.

Loewi described the active material released by stimulation of the vagus nerve as "Vagusstoff." In 1926, he and his co-worker Navratil presented evidence that it was identical with ACh. The year before, Loewi had also reported on "Acceleranzstoff," obtained from stimulation of accelerator fibers to the heart. This experiment pointed to adrenaline as the neurotransmitter for sympathetic nerves.

Loewi's choice of the frog heart was providential. The cholinesterase content in this organ is small compared to that in mammalian hearts, the low temperature favors stability of ACh, and the saline medium avoids the cholinesterase in erythrocytes and plasma. Thus Loewi was successful, whereas Elliott, working on the same logical principle but under less favorable circumstances, had been unsuccessful. Loewi was also fortunate in choosing the frog to demonstrate his "Acceleranzstoff" because in the spring, the season of Loewi's accelerator experiments, the adrenaline content is very high. Loewi and others could not know that the frog is an exception in that it liberates adrenaline from sympathetic nerves, whereas most other species release noradrenaline.

The beautifully simple experiments of Loewi "rang up the curtain." For the first time, it was shown experimentally that nerve impulses act by chemical transmission at synapses, albeit synapses of a very specialized character, from nerves to heart muscle.

In the following decade, a stream of outstanding experiments was performed by the physiologists of the time, all experts on the peripheral autonomic nervous system. Feldberg and Krayer (1933) confirmed Loewi's experiment by vagal stimulation of the dog heart, using the recently developed leech-muscle bioassay to demonstrate the active principle. Dale and Dudley (1929) isolated ACh and identified it chemically from extracts of horse spleen. Feldberg and Gaddum (1934) established that peripheral ganglia used ACh as their neurotransmitter and demonstrated that adding ACh to the perfusing fluid mimics the effects of preganglionic stimulation. Dale proposed in 1933 that those fibers that liberate ACh be called "cholinergic" and those that liberate adrenaline or allied substances be called "adrenergic."

The techniques developed by Dale, Feldberg, Gaddum, Brown, and Vogt were next applied with positive results to the neuromuscular junction (Dale et al., 1936). The chemical transmissions in the autonomic system were very slow, with time of action measured in tenths of seconds; hence, for many years, there was a controversy with respect to Dale's hypothesis that the very fast neuromuscular transmission is mediated by ACh. There is a recent account of this controversy by Eccles (1982d).

Originally, the hypothesis was proposed by Dale and his colleagues (cf. Dale, 1935, 1938) on the basis of experiments that showed that when motor nerve fibers were stimulated, minute quantities of ACh were liberated into fluid perfused through the muscle, and that intraarterial injection of very small amounts of ACh evoked impulses in muscle fibers and hence caused their contraction. In addition, there was evidence that curare-like substances blocked neuromuscular transmission by depressing the response of muscle to injected ACh, and not by depressing the liberation of ACh. The action of inhibitors of cholinesterase provided further supporting evidence, for it was predicted that under such conditions, the ACh effect would be prolonged and intensified, and this was observed. The excitatory effect of a single nerve volley was prolonged to give a brief waning tetanus of the muscle, and in curarized muscles, there was restoration of neuromuscular transmission. This pioneer work demonstrated that impulses in nerve fibers excite muscle by the liberation of ACh. Stemming from this hypothesis are the amazingly elegant investigations during the last three decades by Katz, Kuffler, and their associates that are described below. It has turned out that chemical transmission is a much more complicated biological process than Dale had supposed.

In the peripheral synapses, the essential physiological problem that was solved in their functional design was the transmission of impulses across the synaptic gap with the minimum of delay and distortion, and also with a safety factor high enough to give reliability. These chemically transmitting synapses were designed to compensate for electrical mismatch between the presynaptic and postsynaptic components of the synapse, e.g., the very small nerve terminal and the large area of the muscle fiber membrane with its high capacity. Though these same functions are fulfilled at the synapses of the central nervous system, a quite different functional design is called for in relation to the general problem of connectivity. Major differences occur because of the integrative function of the nervous system, as is discussed in Chapter 4.

3.2. Neuromuscular Transmission

3.2.1. Introduction

Before embarking on the attempt to study neuronal mechanisms in the brain, it is important to study the transmission of peripheral synapses because the investigations on some of these provide the basis for our understanding of synapses in the brain. Essentially the same biological processes go on in the peripheral and central synapses, but those in the brain are more elusive when it comes to the precise investigations that are discussed in Chapter 4. Of the many peripheral synapses that have been studied, there are two of particular value in our attempts to study brain synapses: neuromuscular synapses, especially in the frog and snake; and the giant synapses in the stellate ganglion of the squid. These two exemplars of peripheral synapses form an adequate experimental base for the understanding of the essential features of central synapses.

Figure 3.1A shows diagrammatically a motoneuron in the spinal cord with the course of its axon down to the muscle fibers that it innervates. Limb muscles of the cat, for

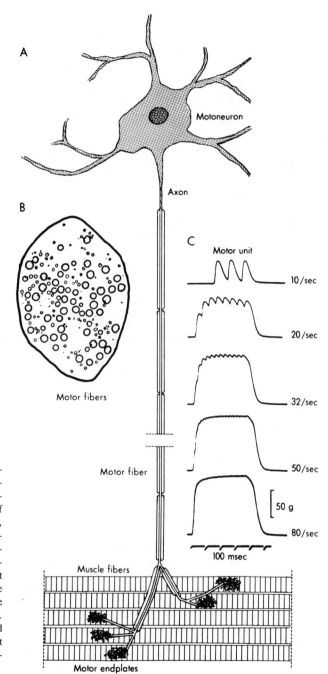

Figure 3.1. Motor unit. (A) A motoneuron with its axon passing as a myelinated nerve fiber to innervate muscle fibers. (B) Transverse section of motor fibers supplying a cat muscle, all afferent fibers having been degenerated. (A, B) From Eccles and Sherrington (1930). (C) Isometric mechanical responses of a single motor unit of the cat gastrocnemius muscle. The responses were evoked by repetitive stimulation of the motoneuron [cf. (A)] by pulses of current applied through an intracellular electrode at the indicated frequencies. From Devanandan *et al.* (1965).

example, are innervated by some hundreds of motoneurons and the large motor fibers (cf. Figure 3.1B) stemming therefrom. Usually the motor axon branches to innervate some hundreds of muscle fibers in a particular muscle, not the five shown here, and usually a muscle fiber receives only one such innervation. Thus, a motoneuron makes its contribution to movement by discharging impulses down its axon and so exciting its own group of muscle fibers to contract. The whole executive ensemble is called a "motor unit." It is a true unitary component of movement because the discharged impulse is usually effective in setting up impulses and the ensuing contraction in all the innervated muscle fibers. The total number of motoneurons with their dependent motor units is about 200,000 for the human spinal cord. That number is responsible for the contractions of all the muscles of the limbs, body, and neck, i.e., for our total muscular performance except for the head.

Figure 3.1C shows the isometric contraction tensions produced by tetanization of a motor unit over a wide range of frequencies. In the cat, the contraction tension produced by rapid repetitive activation of a single motor unit ranges from 50 to 100 g for the large extensor muscles and from 10 to 20 g for smaller muscles (Devanandan *et al.*, 1965). Within limits, the higher the frequency of the nerve cell discharge, the larger the resulting contraction. There are three essentially distinct twitches of the muscle when the frequency of discharge from the single motoneuron is 10/sec. At 20/sec, the contractions tend to fuse, and this tendency increases at 32 and at 50, until at 80/sec there is a smooth, strong contraction. It is perhaps somewhat surprising that a single cat motoneuron can command a contraction of 100 g in one of the leg muscles. In man, the unit contraction would be stronger, but it is not accurately known. Stronger actions of the muscles are secured by the faster firing of its motoneurons. But it is more effective to bring into action many more motoneurons. Any large muscle is innervated by many hundreds of motoneurons. In Figure 3.1B, there are about 60 large motor axons in the transverse section of a nerve branch to a cat muscle.

3.2.2. Structural Features of the Neuromuscular Synapse

In the mammal, the nerve terminal is a tightly clustered structure on the surface of the muscle fiber, as indicated in Figure 3.1A and as drawn in microscopic section in Figure 3.2A, in which the axon makes three small contacts on the surface of the muscle fiber but does not fuse with it. There is the separation that is characteristic of all chemical synapses and that is displayed in brain synapses in Figures 1.4–1.7 (Section 1.2.2). The special structural enlargement of the sarcoplasm of the muscle fiber at the junctional region is called a "motor endplate." The problem is that the nerve impulse travels down to the nerve ending and, after only 1 msec of transmission time, a new impulse starts up in the muscle fiber at the region of the endplate and runs along it in both directions, thereby setting up the contraction.

Light microscopy was not adequate to reveal the essential structures of the nerve–muscle synapse. It was only with the advent of electron microscopy that the structural bases of chemical transmission were revealed. The frog neuromuscular synapse is much less compact than that in the mammal. Fine nerve fibers run for hundreds of microns in longitudinal grooves on the surface of the muscle fiber, as is well illustrated in the drawings from a photomicrograph in Figure 3.4A, in which two such fibers are shown. Figure 3.2B is a drawing made from an electron micrograph of a longitudinal section of a

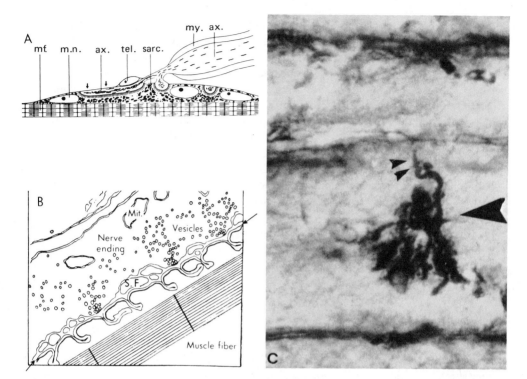

Figure 3.2. Microscopic structure of neuromuscular synapses. (A) Schematic drawing of a motor endplate. (ax.) Axoplasm with its mitochondria; (my.) myelin sheath; (tel.) teloglia (terminal Schwann cell); (sarc.) sarcoplasm of muscle fiber; (m.n.) muscle nuclei; (mf.) muscle fiber. The three (one separate) terminal nerve branches lie in ''troughs.'' From Couteaux (1958). (B) Drawing of electron micrograph of part of a neuromuscular junction from frog sartorius. Longitudinal section of the muscle. (Mit.) Mitochondria; (S.F.) small projection from enveloping Schwann cell. From Birks *et al.* (1960). (C) Motor endplate of guinea pig diaphragm stained for choline acetyltransferase by immunohistochemistry. The axon is shown (double arrowhead), as well as the neuromuscular junction (single arrowhead). From H. Kimura *et al.* (1980).

segment of one such long nerve fiber in the frog. The muscle can be recognized by the longitudinal striations with two transverse striations. The nerve terminal can be recognized by the dense aggregation of vesicles, with mitochondria lying further back from the muscle. The surface of the muscle fiber is shown by the heavy line with frequent foldings. Opposite three of the foldings are clusters of synaptic vesicles, but in several places, small projections from the enveloping Schwann cell intrude between the nerve and muscle surfaces. Elsewhere, there is separation by a cleft about 40 nm across, as marked by the arrows at upper right and lower left. The numerous spherical vesicles are about 50 nm in diameter and are characteristic of all chemically transmitting synapses [cf. Figures 1.3B (Section 1.2.1) and 1.4B, 1.5, and 1.6 (Section 1.2.2)]; hence, they are appropriately called synaptic vesicles. The folds run transversely to the elongated nerve terminal (Figure 3.2B), and vesicle attachment sites [cf. v.a.s. in Figure 1.7 (Section 1.2.2)] are revealed by the freeze–etch technique to be congruent with the folds (Akert and Peper, 1975).

Figure 3.2C shows a motor endplate of the guinea pig diaphragm stained for choline acetyltransferase (ChAT), the specific synthesizing enzyme for ACh (Chapter 8). It shows the axon expanding into the terminal region, which is embedded in the muscle fiber. This high concentration of ChAT in the terminal region is consistent with the necessity of synthesizing large quantities of ACh to fill the synaptic vesicles as shown in Figures 3.10 (Section 3.2.8) and Figure 3.11 (Section 3.2.9).

There are two essential structural features of a chemical synapse, as illustrated diagrammatically in Figure 3.2B: (1) the synaptic vesicles and (2) the synaptic cleft, the space (note arrows) across which transmission has to occur. As illustrated in Figures 4.8A and B (Section 4.4), this synaptic cleft provides the essential space for the flow of currents that are generated by the action of the transmitter substance on the membrane across the synapse, which is the postsynaptic membrane—in this case, the membrane of the motor endplate.

The essential mechanism of the synapse is that the nerve impulse causes some of the synaptic vesicles to liberate their contents into the synaptic cleft. The vesicles at the neuromuscular junction contain prepackaged ACh, about 5000 molecules of ACh in each, according to latest estimates (see Section 3.8), and they squirt their contents into the synaptic cleft. By diffusion across the cleft, the ACh acts on the membrane of the motor endplate. The transmitter opens the ion gates and ions stream through, causing a change in membrane potential of the muscle fiber in the direction of a depolarization. When this change reaches a critical level, a muscle impulse is generated that propagates along the muscle fiber and sets off the complex sequence of events responsible for a muscle contraction. We now turn to consider in some detail the experiments by which this synaptic mechanism has been revealed. Special reference should be made to two books by Katz (1966, 1969).

3.2.3. Physiological Features of the Neuromuscular Synapse

In Figure 3.3A, there is a nerve fiber terminating on the motor endplate with the fine nerve terminals running along the muscle fiber as in Figure 3.4A (Section 3.2.4). The nerve fiber can be stimulated by a brief electrical pulse through the applied electrode and so send an impulse down to the nerve terminal, as described in Chapter 2. As shown in Figure 3.3A, recording from the muscle fiber is by an intracellular microelectrode either right at the motor endplate or 1, 2, or 3 mm away (Fatt and Katz, 1951).

At all sites, the potential across the muscle membrane is about 90 mV, with the inside negative; this is the normal value for a frog muscle fiber. Just as with the giant nerve fiber in Figure 2.4 (Section 2.2), when the microelectrode crosses the membrane, a membrane potential of -90 mV is immediately registered. In Figure 3.3B, with recording at the motor endplate (0 mm), after a brief latency, a muscle impulse is evoked, a complex potential change with humps on the rising and falling phases (explained below). At 1 and 2 mm away, the response looks simpler. It rises rapidly and smoothly and starts to go down just like an ordinary nerve impulse [Figures 2.5 and 2.6 (Section 2.3)], but is greatly slowed in its latter part. Two changes have occurred in moving away from the motor endplate. One is that the impulse is later. That is because the muscle impulse started at the endplate and moved along, just as with an impulse in a nerve fiber, except that it is much slower, only about 1 m/sec. The other is that the humps have disappeared, and the

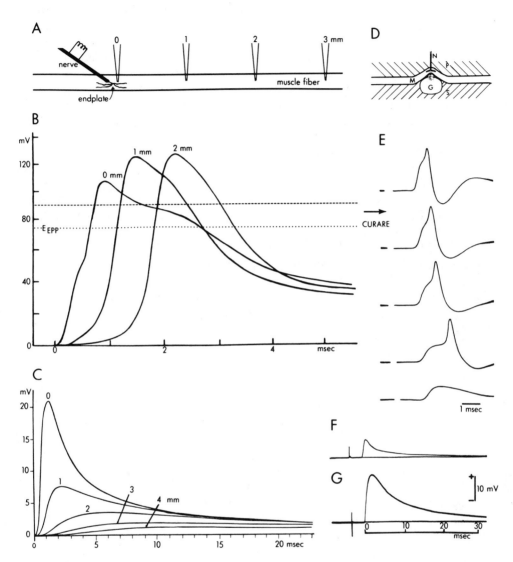

Figure 3.3. Endplate potentials (EPPs) and muscle action potentials. (A) Nerve fiber with its terminal branches innervating a frog muscle fiber. (B) Action potentials evoked by a single nerve impulse and recorded as in (A) at 0, 1, and 2 mm from the motor endplate. (- - -) Level of zero membrane potential, the resting membrane potential being −90 mV; (·····) equilibrium potential for the EPP (E_{EPP}). (C) Action potentials evoked and recorded as in (B), but after blockage of neuromuscular transmission by curare. (A–C) From Fatt and Katz (1951). (D) Isolated single nerve (N)–muscle (M) preparation that was mounted on a recording electrode. (P) Paraffin; (S) saline; (G) glass rod. (E) Series of potentials recorded as in (D) showing progressive action of curare. (D, E) From Kuffler (1942). (F, G) EPPs of curarized muscle recorded at endplate region as in (C), before and after addition of 10^{-6} M prostigmine. From Fatt and Katz (1951).

summit is higher. How can this be explained? What is happening at the endplate zone that reduces the muscle action potential and adds the humps before and after? The answer is that a special operation due to the chemical transmitter causes these changes at the endplate.

The initial hump is in fact due to depolarization by the transmitter, and at a critical level of this depolarization (about 50 mV), it can be seen to fire the impulse. But this impulse is much smaller than at 1 or 2 mm from the endplate. Evidently, the transmitter action on the endplate membrane prevents the impulse from reaching as high a reversal point as elsewhere. The transmitter depolarization has an equilibrium potential at about −15 mV, so it effectively pulls the muscle impulse potential down toward this value. The ionic mechanism of the impulse would tend to give a membrane potential of at least +40 mV, as has been seen with the nerve impulse in Figures 2.5 (Section 2.3) and 2.9 (Section 2.4). The net result is the compromise at about +15 mV for the summit of the action potential at the endplate (0 mm). Finally, the depolarization by the residual transmitter action satisfactorily explains the hump on the declining phase of the impulse at the endplate. These endplate effects are very slight at 1 mm and virtually absent at 2 mm (Figure 3.3B).

If the size of the endplate potential (EPP) could be reduced so that it failed to fire an impulse, it would be much simpler to investigate the events at the neuromuscular synapse. This can be done by poisoning with curare. In recent years, curare and related substances have become of great clinical value because with them the anesthetist can greatly depress neuromuscular transmission so that the surgeon can operate with the advantage of having the patient with perfect muscular relaxation.

Figure 3.3D and E gives a good illustration of progressive curarization. In the early 1940s, about 10 years before intracellular recording, Kuffler (1942) dissected out a single nerve–muscle fiber (Figure 3.3D) and held it between paraffin and saline while recording by a fine platinum wire supported by a curved glass rod shown in transverse section. It was the first example of an investigation on a single nerve–muscle preparation in isolation. When the nerve is stimulated and the recording is from the endplate zone, there is a local potential followed by an action potential (Figure 3.3E), just as in Figure 3.3B. After curare is added to the saline, there is a progressive change, as shown in the subsequent records of Figure 3.3E. The initial potential was progressively reduced and fired the impulse progressively later, eventually failing. The muscle was then paralyzed. Until that happened, the muscle action potential was full-size. It is a beautiful example of the all-or-nothing. The nerve impulse either fires a full-size muscle action potential or it does not fire at all.

As described below, curare works simply by preventing the ACh from effectively depolarizing the muscle at the endplate. When the muscle impulse fails, there remains the EPP, as in the bottom trace of Figure 3.3E. In Figure 3.3C, there are superimposed traces of such EPPs recorded intracellularly as in Figure 3.3A at the endplate zone (0) and at 1, 2, 3, and 4 mm distal thereto. The EPP is decremented progressively with distance and becomes slower in time course. It spreads simply by the cable properties of the muscle fiber, approximately halving every millimeter, so it becomes very much depleted from 0 to 1 to 2 to 3 to 4 mm. Cable transmission is even poorer in muscle fibers than in nerve fibers.

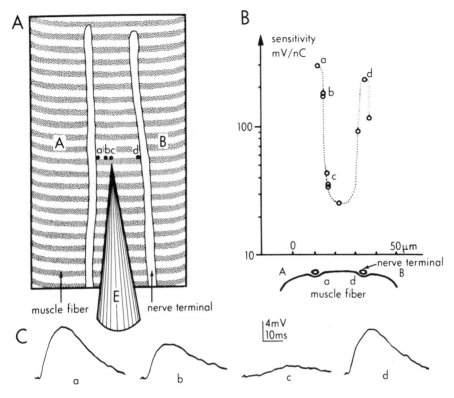

Figure 3.4. Distribution of ACh receptors in the vicinity of nerve terminals on skeletal muscle of the frog. See the text for a full description. From Peper and McMahan (1972).

3.2.4. Pharmacological Features of the Neuromuscular Synapse

We have already discussed how it was established that the neuromuscular transmitter is ACh. Though the work of Dale and his colleagues seems pretty crude by present standards, it was a remarkable achievement at that time.

A great improvement came with the electrophoretic injection of ACh with a fine micropipette that could be manipulated into very close proximity to a motor endplate. Recording was by an intracellular electrode in a preparation that was curarized so that only an EPP was evoked, much as in Figure 3.3C (Section 3.2.3) at 0. With the best location, K. Krnjevic and Miledi (1958) found that electrophoretic injection of 1.5×10^{-14} g ACh just outside the motor endplate gave a potential very similar to that produced by a nerve impulse that was estimated to liberate 1.5×10^{-15} g at the endplate. This discrepancy of a factor of 10 is as good as can be expected when it is remembered that the injection could be at only one end of the cleft, whereas the ACh liberated by the nerve impulse would be directly onto the endplace surface (cf. Figure 3.2) and would be acting over the whole of this surface.

A still higher level of technical excellence has been achieved by McMahan, Kuffler, and their associates (Peper and McMahan, 1972). The drawing in Figure 3.4A is from a photomicrograph of a living muscle fiber identified by its cross striations, and on its surface can be seen two terminal nerve fibers running longitudinally. This beautiful display is due to a special technique, Nomarski interference microscopy, that allows the living tissue to be studied at very high magnification and with sharp optical sections. It was a particularly good arrangement for the experimenter that the two nerve fibers are so nicely parallel and terminate together. Below the graph (Figure 3.4B) can be seen the muscle fiber in section with the two nerve terminals in little grooves (a, d). The muscle fiber has a recording intracellular electrode that is out of the picture in (A), but at a distance that is quite small relative to the length constant of the muscle fiber, so that it would record with little distortion the potential changes across the muscle membrane.

Electrophoretic injections of ACh were made by the micropipette [Figure 3.4A (E)] that is shown accurately applied at a point (c) on the muscle membrane between the two nerve terminals. The injecting current was 32 nA for 1 msec, and the applied points are shown as black dots along the line between the two nerve terminals. When the point of application [(a) in Figure 3.4A] was close to the visualized nerve terminal, it produced in Figures 3.4B and C the large depolarization labeled (a). The same injection only about 5 μm away [(b) in Figure 3.4A] gave the smaller depolarization (b); when injection was between the two nerve terminals [(c) in Figure 3.4A], the ACh hardly affected the membrane at all (c). Finally, when the injection was near [(d) in Figure 3.4A] the other nerve terminal, there was the large response (d) again. The results of several additional injections are also plotted on the graph (Figure 3.4B) and show how subtle and delicate this performance is, the two nerve terminals being shown below the graph on the same scale.

Now what does this show? It shows that only the muscle membrane under the nerve fibers is sensitive to ACh. The remainder is less sensitive by about two orders of magnitude. The effects at other sites are largely due to diffusion of the injected ACh onto the sensitive sites at the nerve contacts. The same decrement of response as from (a) to (c) is observed if the tip of the electrode is moved by the same distance vertically away from the muscle surface. It can be concluded that the sensitivity to ACh is exactly at the actual groove, just where the nerve contact is.

A still more exquisite experiment was carried out by Kuffler and Yoshikami (1975a,b) on the synapses on snake muscle fibers. Pretreatment by collagenase to remove fibrous tissue allows the nerve terminal with the whole presynaptic apparatus to be lifted gently off the postsynaptic sites, which remain as "craters" directly accessible to the electrophoretic application of ACh. High sensitivity to ACh is found to be restricted precisely to the "craters." Quantitative details are discussed in Section 3.2.8.

3.2.5. Quantal Liberation of Acetylcholine

We now come to an investigation that related the liberation of ACh to the vesicles that are seen in the nerve terminal in Figure 3.2B (Section 3.2.2). It was based largely on the classic experimental work of Katz, Fatt, Miledi, and their associates. In fact, at least two years before the vesicles were recognized by electron microscopy, there was evidence that ACh is liberated from nerve terminals in packages (Fatt and Katz, 1952).

In Figure 3.5A, the microelectrode was inserted into the muscle fiber quite close to the motor endplate. In 1952, when Fatt and Katz first detected the sequence of irregularly

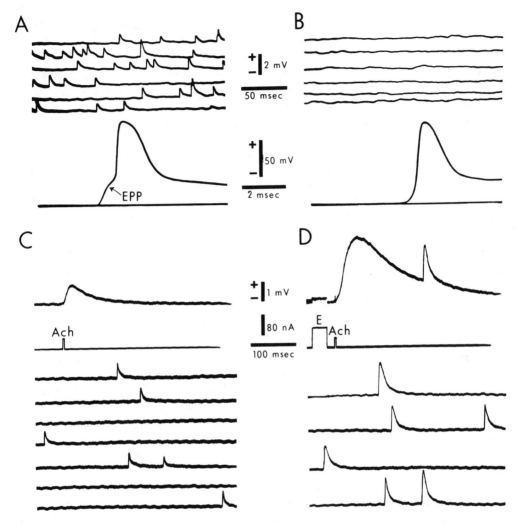

Figure 3.5. Spontaneous miniature EPPs (minEPPs). (A) Intracellular recording at an endplate. (B) Recording from 2 mm away in the same muscle fiber. The upper traces were recorded at low speed and high amplification; they show the localized spontaneous activity at the endplate region. The lower records show the electrical response to a nerve impulse, taken at high speed and lower gain (cf. Figure 3.3B). (A, B) From Fatt and Katz (1952). (C, D) The upper traces show the potentiating effect of physostigmine [applied by pulse (E)] on endplate depolarization produced by a brief pulse of ACh. In the lower traces, it can be seen that the spontaneous minEPPs in (C) are greatly increased and prolonged in (D) during steady electrophoretic application of physostigmine. From Katz (1969).

recurring small potentials, as shown in the upper traces of Figure 3.5A, they did not realize that this "biological noise" was a major discovery! The accurate location of the recording electrode at the endplate is shown by the action potential with an initial EPP set up by a nerve impulse in the lowest trace of Figure 3.5A, which resembles that in Figure 3.3B (Section 3.2.3) at 0 mm. When the recording electrode was 2 mm away, as in Figure

3.5B, there was the simple action potential (cf. Figure 3.3B) and almost no "biological noise."

How can it be shown that this "biological noise" in Figure 3.5A is due to ACh? First, there is the action of curare, which depresses the EPP and the "biological noise" in the same way. Figure 3.5C shows another test. An electrophoretic injection of ACh at the endplate region caused a brief depolarization as in Figure 3.4 [(upper trace) Section 3.2.4]. The lower traces of Figure 3.5C show "biological noise" as in Figure 3.5A, but at a much slower frequency. In Figure 3.5D, the electrophoretic injection of ACh is preceded by a substance, physostigmine, that very rapidly inactivates acetylcholinesterase (AChE), the enzyme that rapidly destroys ACh. Physostigmine works so rapidly that when injected electrophoretically at E just before the ACh, the depolarization produced by the ACh jet was increased and prolonged because the ACh was not being destroyed by the AChE. It can be seen in the lower traces of Figure 3.5D that, at the same time, the units of the biological noise were increased and prolonged. This is good evidence that the "biological noise" is due to brief jets of ACh that are acting on the endplate membrane of the muscle fiber. In every respect, these miniature potentials have the properties of the EPP, hence the justification of their designation as miniature EPPs (minEPPs).

Synaptic vesicles [cf. Fig. 3.2B (Section 3.2.2)] were recognized by electron microscopy of nerve terminals just after the minEPPs were discovered. There was such an obvious and attractive correlation that suspicions became fashionable and still were as late as 1985, but the review of Silinsky (1985) should now be regarded as definitive. After over three decades of investigation, it can be taken as established that the synaptic vesicles are preformed quantal packages of the transmitter substance and that under resting conditions, vesicles discharge their contents into the synaptic cleft at relatively infrequent intervals, thus producing the minEPPs. A further remarkable property, discovered by Katz and his colleagues (Katz, 1966), is that the spontaneous release of the quanta is timed in a strictly random manner.

If the transmitter is packaged in the synaptic vesicles, it would be expected that the release of neurotransmitter by a nerve impulse would also have a quantal composition. Normally, however, the EPP is of a size that indicates that it is due to 100–300 quanta, so a "quantal grain" could not be detected. By reducing the calcium in the bathing solution and adding some magnesium, a reduction of the EPP can be effected even to the point where it resembles the minEPPs. For example, in the 18 traces of Figure 3.6A, the nerve impulse evoked a very small EPP only four times, at the time of the second dotted line. The two spontaneous minEPPs seen late in two traces resemble three of these EPPs, the fourth clearly being a double. Figure 3.6B shows another series with a less severe depression of neuromuscular transmission (Liley, 1956). In two traces, there was no EPP; in two, it had the size of the minEPPs; and in three, it was double that size. Evidently, these observations show that when the size of the EPP is sufficiently reduced, the EPP exhibits a quantal grain corresponding to that of the minEPPs.

The series of double traces of Figure 3.6C were recorded by an extracellular electrode in close proximity to the endplate region, the quantal composition of the EPP being reduced by low calcium (Katz and Miledi, 1965b). It can be seen that there is a wide range of latencies of the quantal EPPs. In contrast, there is no latency variation in the initial diphasic spike produced by the nerve impulse. The latency variation of the EPP is shown in the histogram of Figure 3.6D for a large number of observations, a few of which are

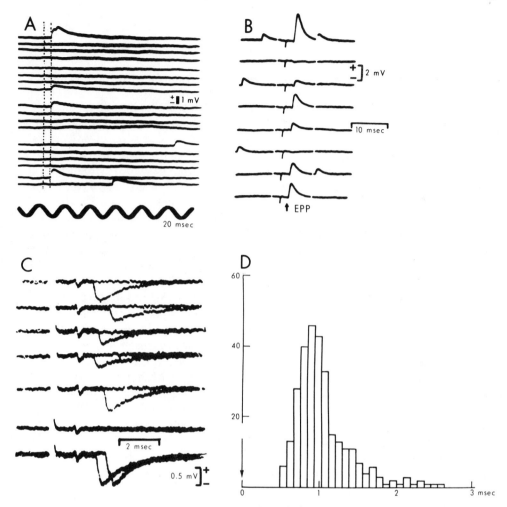

Figure 3.6. Quantal components of the EPP. (A) Intracellular recording from a frog neuromuscular synapse in calcium-deficient and magnesium-rich medium. The times of nerve stimulus and of EPP onset are indicated by the pair of vertical dotted lines. From del Castillo and Katz (1954). (B) Intracellular records from a muscle fiber of the rat diaphragm with stimulation of the phrenic nerve (note artifact), also in a calcium-deficient and magnesium-rich medium. From Liley (1956). (C) Extracellularly recorded EPPs of frog muscle fiber with the calcium level adjusted so that only about half the nerve impulses evoked EPPs. (D) Histogram for latencies of EPPs for the experiment partly illustrated in (C). (C, D) From Katz and Miledi (1965b).

illustrated in Figure 3.6C. It can be concluded that the quantal liberation of transmitter by a nerve impulse has a considerable range in latency. An explanation of this variance is given in Section 3.2.7.

Statistical analysis using Poisson's theorem (Katz, 1958, 1962) gives a precise confirmation of this quantal composition of the EPPs and shows that the quanta are identical to those randomly released to give the minEPPs. Indeed, detailed studies by Katz and associates have shown that the quantal emission of ACh conforms with predictions from Poisson's theorem for the independent release of small numbers of quanta (Katz, 1966, 1969).

3.2.6. Factors That Control Quantal Emission from the Nerve Terminal

An immense range of experimental data conforms with the hypothesis that at the mammalian neuromuscular synapse, a nerve impulse causes the liberation of a large number of quanta of ACh. The synaptic vesicles empty their contents of ACh into the synaptic cleft by a process called "exocytosis." Two quantitative questions now arise: (1) How many quanta are released at an endplate in a normal calcium environment? The estimates range from 100 to 300 (Hubbard, 1973). (2) How many ACh molecules are there in one quantum? There was a large range in the estimates (cf. Hubbard, 1973), but as described in Section 3.2.8, Kuffler and Yoshikami (1975b) have experimentally determined that the upper limit is 10,000 molecules, which would give a moderately hypertonic solution in a vesicle. It was originally thought that the quantal packages were of very uniform size, but it now seems that the largest may be twice the size of the smallest.

Katz and Miledi (1965a) have shown that the impulse normally travels along the fine terminal nerve fibers of the frog neuromuscular junction [see Figure 3.4A (Section 3.2.4)] and that this propagation is essential for its effectiveness. With present techniques, it is possible to demonstrate only that nerve impulse transmission to within 50 μm of the release sites is necessary for effective action. However, the impulse *per se* has no effective action. All that is required is a sufficient depolarization of the nerve terminals. After block of impulse transmission by tetrodotoxin (TTX), EPPs can still be generated by electrotonic transmission of strong depolarizing currents applied close to presynaptic terminals (Katz and Miledi, 1967b). With this technique, neurotransmitter release occurs very effectively even in a sodium-free medium. On the other hand, there is a wealth of experimental evidence that increase in intracellular Ca^{2+} ions is essential for neurotransmitter release at all types of chemical synapses. For example, in Figure 3.6, the great depression of the EPP was due to the calcium-deficient, magnesium-rich medium.

3.2.7. Essential Role of Calcium in Quantal Release

The experimental arrangements for Figure 3.7A are shown below the three series of superimposed traces evoked by stimulation of the nerve (Katz and Miledi, 1965c). The extracellularly recording micropipette was filled with 0.5 M $CaCl_2$, but outward diffusion of Ca^{2+} ions was prevented by a restraining negative voltage in the upper trace of Figure 3.7A. The nerve impulse was completely ineffective because there were the same highly unfavorable conditions for release of transmitter, namely, a Ca^{2+}-free bathing solution containing 0.84 mM Mg^{2+}. In the second and third traces, the negative bias on the micropipette was reduced in two steps so that some outward diffusion of Ca^{2+} could occur. The Ca^{2+} level in the vicinity of the recording site was raised sufficiently to allow some quantal release by the nerve impulse. This effectiveness was lost as soon as the original restraining voltage was restored, at which point there was a return to the zero response illustrated in the first trace. Figure 3.7B presents the averages of 600 traces of the corresponding records in Figure 3.7A. It is interesting to note that the triphasic nerve impulse potential remains unaltered by the increase in calcium; it is only the release of neurotransmitter that is changed.

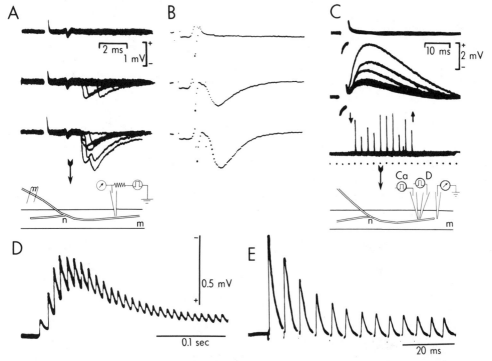

Figure 3.7. Extracellular calcium and neuromuscular transmission. (A, B) The experimental arrangement is indicated by the arrow below the traces of (A). The extracellular recording was by a micropipette filled with 0.5 M CaCl₂. From Katz and Miledi (1965c). (C) The experimental arrangement is indicated by the arrow below the traces. From Katz and Miledi (1967a). D and E are described in the text.

The essential role of extracellular calcium in neurotransmitter release can also be demonstrated in the absence of nerve impulses (Katz and Miledi, 1967a). The experimental arrangements are indicated below Figure 3.7C. Elimination of nerve impulses was secured by TTX, and two micropipettes were located close together on one of the nerve terminals on the muscle fiber from which the intracellular recording was being taken as shown. The bathing solution was Ca^{2+}-free with 1.7 mM Mg^{2+}. Brief applications of depolarizing currents by the D electrode caused no transmitter release, as shown in the upper trace of Figure 3.7C. Then the restraining negative voltage on the Ca^{2+} electrode was reduced, allowing Ca^{2+} to diffuse around the nerve terminal. The consequence was that the same depolarizing pulse then released transmitter, as shown by the large EPPs of the second trace. The third trace shows the same phenomenon at a much slower sweep speed. The depolarizing pulses are indicated by the dots at 0.5-sec intervals below the trace, and the successive EPPs appear as vertical lines. The bias on the Ca^{2+} pipette was released between the two upper arrows. It is seen that immediately there was a production of EPPs by each pulse, though there was a considerable variation in their quantal content. On restoration of the restraining voltage, there was again a complete failure of EPP production.

In Figure 3.8A, a vesicle is shown being filled with transmitter molecules and then moving through the presynaptic terminal to reach a dense protrusion [d.p. in Figure 3.9A]

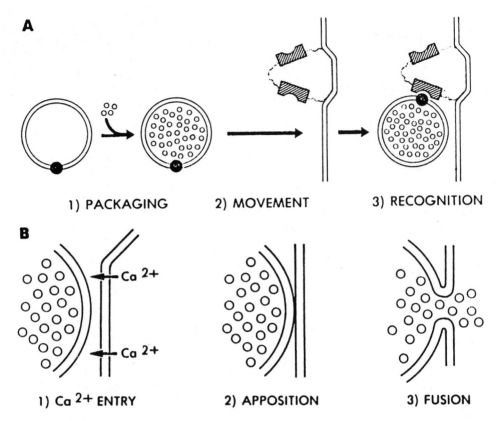

A

1) PACKAGING   2) MOVEMENT   3) RECOGNITION

B

1) Ca $^{2+}$ ENTRY   2) APPOSITION   3) FUSION

Figure 3.8. Partial reactions during exocytosis. (A) Steps involved in bringing a filled vesicle to the active site. (B) Three steps involved in release of neurotransmitter. From R. B. Kelly *et al.* (1979).

on the presynaptic membrane confronting the cleft, where it achieves a specific attachment. Then, in Figure 3.8B, depolarization of the nerve terminal by an impulse causes Ca^{2+} ions to enter from the synaptic cleft as demonstrated by the aequorin technique (Baker *et al.*, 1971). Not shown is a further development of the exocytosis hypothesis (De Lorenzo, 1981), that the incoming Ca^{2+} ions combine with the protein calmodulin that is associated with the vesicle sheath. Calmodulin is a small protein (molecular weight 16,700) that has 4 binding sites for Ca^{2+} ions (Cheung, 1980, 1982; Means *et al.*, 1982), and, when activated by binding of 4 Ca^{2+} ions, it becomes a powerful excitant of phosphorylating enzymes. It has long been known from a precise analysis of the relationship of Ca^{2+} concentration to size of EPP (Dodge and Rahaminoff, 1967) that each binding site must be occupied by 4 Ca^{2+} ions for the exocytosis of one synaptic vesicle. This remarkable stoichimetric reaction can now be accounted for by the necessity for the binding of 4 Ca^{2+} ions in the activation of calmodulin, which leads to the last stages of exocytosis in Figure 3.8B. De Lorenzo (1981) has developed this calcium–calmodulin hypothesis of exocytosis on the basis of systematic investigations on the calmodulin content of synaptic vesicles and the calcium–calmodulin action on a synaptic tubulin–kinase system that could cause the exocytosis [Figure 5.5 (Section 5.4.2)]. Important support for this hypothesis is provided by a study of the effects of trifluoperazine, phenytoin, and diazepam, which depress Ca^{2+}–calmodulin activation of phosphorylating enzymes and also the release of transmitter from intact synaptosomes.

By a special albumin technique, Gray (1983) has displayed microtubules in the presynaptic terminals of both neuromuscular and central synapses (Figure 3.9). These microtubules extend down to the presynaptic active sites and have synaptic vesicles attached along their length. Gray postulates that the microtubules function to transport the synaptic vesicles from their sites of generation (cf. Hubbard, 1973) down to the active presynaptic sites [see Figure 3.2B (Section 3.2.2)]. In the case of muscle, these sites are opposite the transverse postsynaptic folds. It can be further assumed that the calcium–calmodulin action would accelerate this transport and as well (De Lorenzo, 1981) cause the fusion of the synaptic vesicle to the presynaptic membrane and its rupture (see Figure 3.8B).

It is assumed that these chemical processes are fast enough to accomplish the exocytosis in a fraction of a millisecond, but this assumption has been called into question by M. B. Kennedy (1983). Nevertheless, reaction velocities in a test tube may be orders of magnitude slower than in biological microsites. When the EPP produced by a nerve impulse is reduced to single miniature EPPs, as in Figures 3.6C and D (Section 3.2.5) and 3.7A, it can be seen that there is a scatter of over 1 msec in the latencies of the individual exocytotic events. The most attractive explanation is that this is the range of time involved in assembling the 4 Ca^{2+} ions onto the calmodulin molecule so that it is activated to bring about the exocytosis. Related to this requirement, there is an immense excess of entering calcium ions, even 10,000 times more than the 4 Ca^{2+} ions required for each synaptic vesicle.

So far, we have concentrated on the action of single impulses at the neuromuscular synapse. Movements are brought about by discharges of repetitive impulses to muscles as in Figure 2.3B (Section 2.1). This rapid repetitive activation of synapses raises two new problems. First, at fairly high frequencies, there is a progressive increase in the successive EPPs (Figure 3.7D), which can be attributed to the buildup of residual Ca^{2+} in the nerve terminal (Katz and Miledi, 1968) and possibly of the undissociated calcium–calmodulin complex. This complex rapidly dissociates as the Ca^{2+} ion concentration falls by extrusion or absorption by intracellular organelles such as mitochondria. Second, there is an eventual decline in the EPP (Figure 3.7D) that is attributable to depletion of the immediately available vesicles. Depression is a dominant feature of mammalian EPPs (Figure 3.7E). These two features of potentiation and depression are exhibited by repetitive stimulation of synapses in the central nervous system and are considered in Chapters 4 and 16.

3.2.8. Molecular Action of Acetylcholine

Until a little more than a decade ago, the action of ACh was investigated by the membrane depolarization produced either by the electrophoretic application of quite large numbers of ACh molecules—about 20 million in Figure 3.4 (Section 3.2.4)—or by the quantal emission of some thousands of molecules, as in Figures 3.5 and 3.6 (Section 3.2.5). Then Katz and Miledi (1972, 1973) achieved a great advance in their identification of the size and time course of the depolarizations produced by single ACh molecules. This identification was accomplished by spectral frequency analysis of the electrical noise that

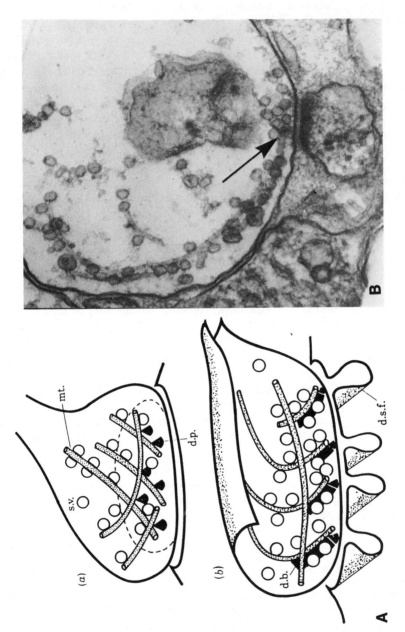

Figure 3.9. (A) Diagrams of a central nervous system synapse (*a*) and a motor endplate (*b*), showing the relationship between microtubules (mt.) and dense projections (d.p.) or bars (d.b.). (s.v.) Synaptic vesicles; (d.s.f.) dips in subjunctional folds. (B) Albumin–treated brain synapse showing a microtubule with attached vesicles apparently anchored to a dense projection (→). From Gray *et al.* (1982).

the application of ACh evokes across the postsynaptic membrane. Computer analysis of the noise revealed that it was made up of a random assemblage of extremely minute elements, each about 0.3 μV and of very brief duration, about 1 msec. It will be recognized that this noise is at least 1000 times less than the "biological noise" that was discovered by Fatt and Katz (1952) and that turned out to be miniature EPPs (see Figure 3.5). Rigorous testing procedures and controls established that these minute potentials were produced by single ACh molecules that momentarily (about 1 msec) attach to receptor sites on the postsynaptic membrane and open ion channels, presumably by a brief conformational change in the protein structure of the channel [cf. Figure 1.14 (Section 1.2.3) and Figure 2.8 (Section 2.3)]. Since the average minEPP of frog muscle is about 1500 times larger, it could be concluded that about 1500 ACh molecules participate in generating a minEPP and hence that there are about 1500 molecules of ACh in a quantum (or synaptic vesicle). This would be a minimum number, however, because an appreciable fraction—at least 30%—of the quantal population is hydrolyzed by cholinesterase before it can reach a receptor site, as indicated by the ACh–AChE interaction in Figure 3.11B (Section 3.2.9).

This very interesting and important feature, quantal number, has been studied in a quite different manner by Kuffler and Yoshikami (1975a,b). The bared "craters" of snake neuromuscular junctions were subjected to electrophoretically applied jets (1-msec duration) of ACh. With extremely close apposition, the injection produced a depolarization with a remarkable similarity to the minEPP. These workers utilized a refined biological assay for the estimation of the numbers of ACh molecules actually injected electrophoretically and arrived at a quantal number of 10,000. They regard this as an overestimate, however, because the injection site over the "crater" of necessity could not be in such a snug relationship as the nerve terminal. A possible value of 5000 molecules for the snake is not so far removed from the minimum of 1500 molecules derived from Katz and Miledi's results for the frog. It is interesting that 5000 ACh molecules in a spherical vesicle 45 nm in diameter give an approximately isomolar concentration (Hubbard, 1973).

The noise analysis technique has now been superseded by the technique of employing a glass micropipette with a tip about 1 μm tip in diameter to lead selectively from a very small patch of membrane to which it has been closely applied. This is the patch-clamp technique pioneered by Neher and Sakmann (1976). In a further refinement, Neher found in 1979 that the clean polished tip of the micropipette could be made, by gentle suction, to seal very tightly to the clean surface membrane to which it was applied. This is called the "giga-seal" technique because the resistance of the seal is many giga-ohms, often above 10×10^9 ohms (Corey, 1983). So the responses of a very small area of the membrane, a few square micrometers, are being selectively led off by the micropipette.

When the physiological solution in the micropipette contains an extremely dilute solution of ACh, the activation of the ACh receptor sites on the external surface of the patch opens ion channels across the membrane. As shown in Figure 3.10A, opening of the channels by 100 nM ACh results in inward current pulses that are of widely different durations but of the same magnitude, there being a very small variance from the mean at 2.7 pA. When a long run of channel openings is analyzed, there is found to be an excess of brief openings of no more than 1 msec. These brief openings are interpreted as being effected by a single molecule of ACh, the longer being effected when two molecules of

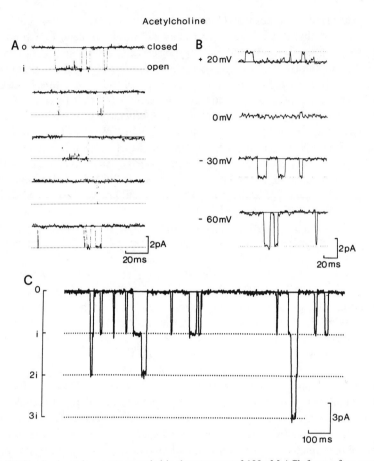

Figure 3.10. (A) Single channel currents recorded in the presence of 100 nM ACh from a frog muscle fiber at the resting membrane potential of −92 mV. The temperature was 10°C. Downward deflection indicates inward current. The distribution of current pulse amplitudes has a peak at 2.7 pA. (B) Voltage dependence of single channel currents activated by 2 μM ACh in rat embryonic muscle. The temperature was 23°C. Downward deflection indicates inward current; upward deflection, outward current. (C) Single channel currents recorded in the presence of 100 nM ACh from a frog muscle fiber at −130 mV membrane potential. The temperature was 10°C. Currents of three individual channels superimpose to form regularly spaced amplitude levels of −3.9 pA. From Colquhoun and Sakmann (1981).

ACh become attached to the channel receptor. It can also be seen in Figure 3.10A that there are brief partial reductions of the current (Colquhoun and Sakmann, 1981).

The current flowing through the channel is due to ions diffusing down the electrochemical gradient across the membrane. In Figure 3.10B, this gradient is varied from +20 mV to −60 mV by an applied current, and the sizes of the current pulses vary according to the expected linear relationship.

When there are several channels across the patch that is being recorded from, there is the expected summation of currents, as illustrated in Figure 3.10C for the current pulses of one, two, or three channels, each of 3.9 pA.

The patch-clamp technique will be of the greatest value in studying the receptors and ion channels. With variants of the technique as described by Corey (1983), in addition to the on-cell patch of Figure 3.10, the patch attached to the micropipette can be removed from the cell and studied either with the inner surface outward—the inside-out patch—or with the outer surface outward—the outside-out patch. This gives the advantage of being able to control the solutions on both sides of the membrane, with consequent analysis of the ion permeability of the channel, as is described for the inhibitory patches on nerve cells in Chapter 4.

The molecular structure of the ACh channel has been subjected to intensive investigation, as reviewed by Karlin (1983). The protein composition has been identified, and the structural relationships of the proteins are being investigated (Stevens, 1985). Patch-clamp techniques will be of the greatest value in these investigations and also in investigations of the mode of action of desensitizing processes and of receptor blocking agents.

3.2.9. Diagrammatic Representation of Neuromuscular Synapses

Figure 3.11 summarizes the essential elements of the hypothesis of transmission at the neuromuscular synapse (cf. Hubbard, 1973). In the nerve terminal (Figure 3.11A), there can be seen synaptic vesicles inside which are tightly packed ACh molecules, about 5000 per vesicle. One vesicle is liberating its transmitter into the synaptic cleft—a quantal emission. As stated in relation to Figure 3.8 (Section 3.2.7), this quantal emission is brought about by a calmodulin molecule combined with 4 Ca^{2+} ions by means of the activation of protein kinases [De Lorenzo, 1981; Figure 5.5 (Section 5.4.2)]. The vesicle itself is not ejected into the synaptic cleft—only its contents—and then it is assumed to be refilled with ACh and so to be available for reuse. To the left of the vesicle being emptied is shown a vesicle that has been previously squeezed out and is now being recharged with ACh. An approximate estimate for neuromuscular synapses is that the reserve of synaptic vesicles is enough for 2000–5000 impulses, which is only a few minutes' supply with ordinary muscular activity. Evidently, efficient and fast replenishment is essential. The local new formation of synaptic vesicles or their transport from manufacturing sites in the motoneuronal soma would be much too slow.

Also shown on the postsynaptic membrane of Figure 3.11A are two types of receptor sites that have different molecular configurations. One type is related to transmembrane channels for ions, and these are shown opened up when an ACh molecule is attached. The other type is an AChE site that destroys the attached ACh.

Figure 3.11B illustrates other features of the neuromuscular synapse. There is, first, the vesicle with stored ACh, and arrows indicate its release and subsequent fate. Part goes to AChE and is very rapidly hydrolyzed into choline and acetic acid, and part goes on to the receptor sites on the postsynaptic membrane, staying there momentarily to open the ion channels and so produce the EPP. Within 1–2 msec, it leaves these sites and is rapidly destroyed by the AChE. The effectiveness of this destruction is illustrated in Figures 3.3F and G (Section 3.2.3), in which inactivation of the AChE resulted in the large increase of the curarized EPP. It is also shown in Figure 3.5D (Section 3.2.5), in which the fast-acting anticholinesterase physostigmine caused a large increase in the depolarization produced by both injected ACh and quantally liberated ACh. Figure 3.11B further indicates the operational sequence: ACh on receptor sites $\rightarrow$ EPP $\rightarrow$ generation of a muscle

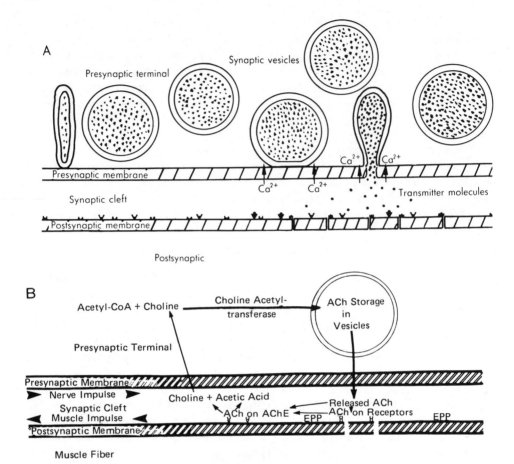

Figure 3.11. Essential elements in neuromuscular transmission. (A) Portion of synaptic cleft with synaptic vesicles in close proximity to the presynaptic membrane, one actually discharging transmitter molecules (ACh) into the synaptic cleft. Five ACh molecules are shown combined with receptor sites on the subsynaptic part of the postsynaptic membrane with the consequent opening up of channels through that membrane. Arrows labeled Ca^{2+} show movement of Ca^{2+} through the presynaptic membrane when it is depolarized. Also shown by different symbols are the molecules of the enzyme AChE attached to the cleft side of the postsynaptic membrane. (B) Schematic representation of the sequence of events in neuromuscular transmission (see text).

impulse → muscle contraction. The diagram, of course, shows only the surface membrane on one side of the muscle fiber.

A nerve impulse releases about 10^6 molecules of ACh into the synaptic cleft of a single motor endplate. On the postsynaptic membrane, there are about 2×10^7 receptor sites for ACh in the rat or mouse junction and also 2×10^7 active centers of AChE. There is thus a large excess of both receptor and destroying sites relative to the quantal liberation of ACh (Hubbard, 1973).

Also shown in Figure 3.11B is the recycling of choline after it has been produced by hydrolysis of ACh. This is discussed in more detail in Chapter 8.

Figure 3.12. Synaptic transmission in the squid stellate ganglion. Presynaptic depolarization and transmitter release with impulses eliminated by TTX. (H) Drawing of the whole synaptic structure of the stellate ganglion of the squid, with experimental arrangments: (Pre) presynaptic terminal; (Post) postsynaptic giant axon. (G) In the upper diagram are the special arrangements for this experiment: Length of synaptic contact, 0.8 mm. (a) Current-passing electrode; (b) prerecording electrode; (c) postrecording electrode. (A–F) Sample recordings in six frames. Upper trace: applied current pulse; middle and lower traces: presynaptic and postsynaptic potentials recorded through electrodes (b) and (c), respectively. (G) Input–output relationship obtained with 1-msec current pulses. Abscissa: peak of presynaptic depolarization; ordinate: size of postsynaptic response. Inset: initial part of curve in greater detail. The temperature in these experiments was about 10°C; the concentration of TTX was 2×10^{-7} g/ml. From Katz and Miledi (1966).

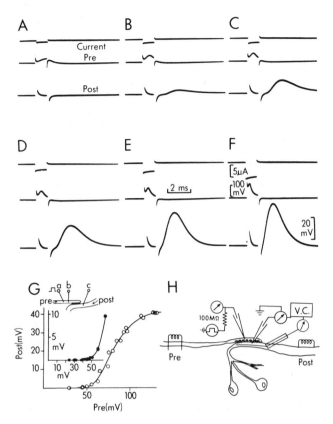

3.3. Transmission across the Giant Synapse of the Squid Stellate Ganglion

Further insights into the physiological events that occur in chemical synaptic transmission have been given by investigations on the giant synapse in the squid stellate ganglion. The synapse is truly of gigantic dimensions, there being a contact up to 1 mm long between a presynaptic and a postsynaptic fiber, each of which is several hundred micrometers in diameter. There is not, however, a continuous synaptic contact over this whole area of approximation, but a multitude of small contacts, as indicated in Figure 3.12H. This synapse gives unique opportunities for investigation. For example, electrodes can be inserted into the presynaptic fiber for direct recording of its membrane potential and its action potential, or for passing currents to alter the membrane potential, a facility that is not possible with the neuromuscular synapse because the presynaptic nerve terminal is far too small.

That it is a chemically transmitting synapse is established by much evidence: There is the characteristic synaptic delay between the presynaptic and postsynaptic responses; no direct electrical transmission of an electrotonic potential can be detected in either direction (Hagiwara and Tasaki, 1958); Ca^{2+} and Mg^{2+} ions play mutually antagonistic roles, just as with neuromuscular transmission (A. Takeuchi and N. Takeuchi, 1962); and there are miniature synaptic potentials analogous to the minEPPs, but much smaller (Miledi, 1967).

Miledi (1969) considered critically the possibility that the transmitter could be glutamate. There is now very convincing evidence that it is. Synaptic transmission is irreversibly blocked by a spider toxin (N. Kawai *et al.*, 1983), yet presynaptic and postsynaptic impulse transmission is unaffected. Furthermore, the glutamate depolarization of the postsynaptic membrane is also blocked by the toxin. In the lobster neuromuscular preparation, transmission is abolished by this toxin that specifically blocks the glutamate receptors.

Figure 3.12 illustrates a remarkable experiment by Katz and Miledi, which shows the way in which the large dimensions of this giant synapse can be utilized to disclose special features of chemical synaptic transmission. Impulse transmission was blocked by TTX, and presynaptic depolarization was effected by intracellular application of a brief current pulse [(a) in inset of Figure 3.12G], as shown in the upper traces of the series of Figures 3.12A–F. Below are shown the presynaptic and postsynaptic membrane potentials. In Figure 3.12A, the presynaptic depolarization had no postsynaptic effect, there being only brief artifacts when the current was turned on and off. In Figure 3.12B, there was a slight depolarizing action, and in Figure 3.12C–F, it became increasingly greater. The plotted points in Figure 3.12G show that the more the terminal is depolarized (plotted as abscissae), the greater the amount of neurotransmitter liberated, as indicated by the postsynaptic depolarization (ordinates). Before any transmitter was liberated, however, there had to be almost 40 mV presynaptic depolarization (note the more sensitive plotting in the inset of Figure 3.12G).

Figure 3.13 illustrates essentially the same results as in Figure 3.12 under conditions in which presynaptic currents were employed to change the size of the presynaptic impulse in a preparation not poisoned by TTX. In the frame labeled 0, no current was passed and the large presynaptic action potential (upper trace) evoked a later postsynaptic potential like an EPP. It is called a "synaptic potential" or, in accordance with usual terminology, an "excitatory postsynaptic potential" (EPSP). When the presynaptic impulse was increased by making the membrane potential larger, the output of transmitter was increased, as indicated by the EPSPs in the frames labeled +3 to +8. If, on the other hand, the presynaptic impulse was decreased by reducing the membrane potential (frames −3 to −9), the EPSPs revealed that the liberation of transmitter was reduced, even to almost zero.

It is a remarkable finding in Figures 3.12 and 3.13 that a threshold depolarization of about 40 mV is necessary for transmitter liberation from the presynaptic terminal. This can be regarded as the threshold for opening the calcium channels on the presynaptic terminal so that Ca^{2+} can enter the presynaptic terminal (cf. the Ca_D channels in Figure 3.14B). Just as with the neuromuscular synapses, there is direct experimental evidence that the influx of Ca^{2+} ions is essential for the quantal liberation of the transmitter, as is diagrammed in Figures 3.8 (Section 3.2.7), 3.11A (Section 3.2.9), and 3.14B. When the Ca^{2+} of the bathing fluid was greatly reduced, synaptic transmission was blocked, just as with the neuromuscular synapse [see Figures 3.7A–C (Section 3.2.7)]. It is restored by injection of Ca^{2+} ions close to the external surface of the synapse even at a rate as low as 10^{-14}–10^{-15} mole of Ca^{2+} ions in 1 sec. The role of Ca^{2+} ions in causing the emission of neurotransmitter has been elegantly displayed by Miledi (1973), who has injected

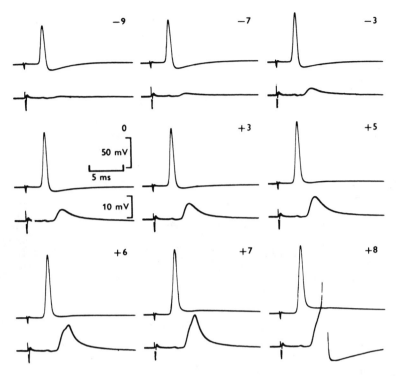

Figure 3.13. Presynaptic depolarization and transmitter release. The experimental arrangements are shown in Figure 3.12H. Prolonged currents of the intensities indicated for each frame in relative units were passed through the electrode inserted into the presynaptic element. This electrode was also used for recording the presynaptic action potentials (upper traces of each frame) evoked by a presynaptic stimulus (note the stimulus artifact). The lower traces are the simultaneously recorded postsynaptic responses. From Miledi and Slater (1966).

Ca^{2+} ions electrophoretically into the presynaptic terminal in a squid giant synapse pretreated with TTX. The Ca^{2+} injection resulted in an enormous output of neurotransmitter, which was in the form of quantal release.

There has been a convincing test of the hypothesis that presynaptic depolarization opens Ca^{2+} gates and Ca^{2+} ions run down their steep electrochemical gradient into the presynaptic terminals and so cause the vesicles to discharge their contents. If in the experiment of Figure 3.12 a prolonged presynaptic depolarization was increased up to the equilibrium potential for Ca^{2+} ions, then no Ca^{2+} should enter and no transmitter liberation or EPSP should occur. This prediction was precisely observed (Figure 10 in Katz and Miledi, 1967a), and at the end of the pulse, there was a large EPSP signaling the output of neurotransmitter as the membrane potential passed through levels at which Ca^{2+} could enter. After supression of Na^+ and K^+ channels, it was possible to carry out this crucial test of the Ca^{2+} hypothesis on neuromuscular synapses of frog and rat with similar results (Katz and Miledi, 1977).

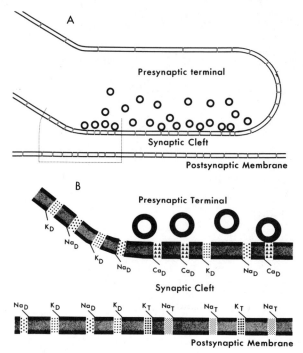

Figure 3.14. Chemically transmitting excitatory synapse. (A) Presynaptic terminal separated by the synaptic cleft from the postsynaptic membrane and containing many spherical synaptic vesicles, some close to the membrane fronting the synaptic cleft. Note the various channels through the membrane. (B) Segment outlined by the dotted line in (A) shown enlarged to give fine detail. Note that all channels have a gate control and that symbols in the channels indicate five different kinds of chemical constitution for the carrier molecules: Na_D, K_D, Ca_D, K_T, and Na_T. The subscripts D and T indicate opening by depolarization and by the transmitter substance, respectively.

3.4. Ion Channels across the Presynaptic and Postsynaptic Membranes of a Chemically Transmitting Synapse

Figure 3.14 displays diagrammatically the various ion channels across the postsynaptic membrane of the neuromuscular junction and of the squid giant synapse, there being separate Na_T and K_T channels for the postsynaptic currents. Also shown in the presynaptic nerve fiber are the Na^+ and K^+ channels that are opened by depolarization and that are responsible for the rise and fall of the action potential [see Figure 2.5 (Section 2.3)]. They are labeled Na_D and K_D to signify their relationship to depolarization. Though selectively permeable to Na^+ ions, the Na_D channel is also effective for Li^+ transport (Hodgkin, 1964) and is even slightly effective for K^+, Rb^+, and Cs^+ (Chandler and Meeves, 1965). Ammonium and some substituted ammonium ions can pass through the Na_T channel [references in Eccles, 1964b (p. 52)]. The Ca_D channel is also permeable to Sr^{2+} and Ba^{2+}, but less effectively (Dodge *et al.*, 1969). It is important that TTX blocks the Na_D channel selectively, not affecting the K_D and Ca_D channels (Katz and Miledi, 1969a), while intracellular tetraethylammonium (TEA) blocks only the K_D channel. The Ca_D channel is distinctive in that it is very effectively blocked by external Mn^{2+} and Mg^{2+} (Katz and Miledi, 1969b). When the Na_D and K_D channels are eliminated by TTX and TEA, respectively, the Ca_D channel becomes so effective that it even exhibits a regenerative depolarizing response (Katz and Miledi, 1969b). By an extremely elegant fluorescence method for detecting calcium, P. F. Baker *et al.* (1970) have shown that membrane depolarization is associated with a two-phase influx of Ca^{2+}, one with about the same time course as Na^+ and the other slower in onset and continuing during the depolarization. The former is blocked by TTX and is plausibly attributed to a slight Ca^{2+} transport through the Na_D gates as mentioned above. The other is TTX-resistant and is blocked by Mn^{2+} and Mg^{2+}, corresponding precisely to the Ca_D channel of Figure 3.14.

Figure 3.14 illustrates two features of the five types of ionic channels displayed. (1) There are the relative specificities of the channels themselves, as symbolized by the various ions principally concerned under normal conditions. Presumably, these specificities have chemical bases, and the discovery of these chemical mechanisms would be a great scientific achievement. Because of their different properties, it is postulated that each type has a unique chemical composition. (2) There are the gates that control these channels, which are opened by depolarization for the Na_D, K_D, and Ca_D channels and by a chemical neurotransmitter for the Na_T and K_T channels—ACh at the neuromuscular synapse and glutamate at the squid synapse. A further difference is that the Na_D and K_D channels are blocked by extracellular TTX and intracellular TEA, respectively, whereas the Na_T and K_T channels, opened by transmitter action at the neuromuscular and squid synapses, are unaffected. For this reason, these channels are shown in Figure 3.14 filled by different chemical mechanisms, as indicated by the different symbolic shadings and by the labels Na_T and K_T.

Figure 3.14 further illustrates a special feature of the presynaptic membrane, namely the presence of specific calcium channels operated by membrane depolarization (Ca_D). The influx of Ca^{2+} ions along these channels is essentially concerned in the discharge of transmitter into the synaptic cleft [see Figure 3.7 (Section 3.2.7)]. In accordance with the evidence of Katz and Miledi (1965a), the Na_D and K_D channels are shown along the nerve terminal in Figure 3.14B, but these channels are not shown in the subsynaptic component of the postsynaptic membrane, where there are exclusively the Na_T and K_T channels. It is not known how far the Na_D and K_D channels are interspersed with the Na_T and K_T channels, but at least they are in close proximity. It is of interest that the Ca_D channels are opened at a rather higher level of depolarization [see Figures 3.12 and 3.13 (Section 3.3)] than are the Na_D and K_D channels associated with the action potential. The Ca_D channels shown in Figure 3.14 are restricted to the presynaptic membrane and are not blocked by TTX (Katz and Miledi, 1969a), which is in contrast to the findings on Ca_D channels of some excitable membranes (P. F. Baker *et al.*, 1971).

The neuromuscular synapse and the squid giant synapse exemplify very well the mechanism of chemical transmission and are seen to be very similar in essentials, though they differ in respect to neurotransmitter substances and dimensional relationships. They also differ in the ionic conductance ratios for the Na_T and K_T channels of the postsynaptic membrane. A ratio of 1.3 was determined by Takeuchi and Takeuchi (1960) for the neuromuscular synapse, but with the squid synapse, the ratio is probably in excess of 5 (P. W. Gage and Moore, 1969). In the diagram of Figure 3.14B, this would be symbolized by the relative number of Na_T and K_T channels. In place of the ratio of about 1.3 for the neuromuscular synapse, there would be, for example, a ratio of 5 for the squid synapse.

3.5. Concluding Remarks

We have finished the story of the investigations on peripheral synapses. In the principal example, a nerve impulse propagates to the nerve terminals of a synapse and there acts with extraordinary efficiency and speed to inject ACh into the exceedingly narrow cleft between the nerve fiber and the muscle fiber. The electrical activity of the nerve impulse has been transformed into a chemical injection. Then, in turn, the chemical transmitter momentarily opens ion gates on the postsynaptic membrane and so reconstitutes an electrical process—an endplate potential leading on to a muscle impulse—on the far side of the synapse. It may be asked: What is the point of this double transformation? Could not the electrical energy of the nerve impulse be utilized to excite the muscle fiber directly? The answer is easy and obvious. There is a large electrical mismatch between the very fine nerve fiber and the large muscle fiber [see Figure 3.4A (Section 3.2.4)], which means that the currents responsible for the propagation of the nerve impulse [cf. Figure 2.9 (Section 2.4)] would be far too small to excite the muscle fiber no matter how efficiently the nerve–muscle junction was arranged. The chemical transmitting mechanism overcomes this extreme disability by introducing an amplification of an order larger than 100-fold. In fact, there is a safety factor of about 5 in transmission from nerve impulses to muscle fibers. This is a very important safeguard for our muscular performance; even under extreme conditions when, for example, the supply of neurotransmitter may be depleted, muscle remains a reliable servant in carrying out the commands of the nervous system, as illustrated in Figure 3.1C (Section 3.2.1).

In Chapter 4, there is an account of chemical synapses in the brain, where we shall see that they are essential for the operation of complex nervous systems. The alternative synaptic mechanism, by electrical transmission, apparently is used only at special sites in the brains of lower animals where accurate timing is essential, but not to any significant extent in the mammalian brain. The two types of peripheral synapses, the neuromuscular and the squid giant, are important as models for the chemically transmitting synapses in the brain, where such fine analytical control is now becoming possible.

4

Synaptic Transmission in the Central Nervous System

4.1. Introduction

In Chapter 3, there was an account of transmission across the neuromuscular synapse whereby a nerve impulse generates a muscle impulse and thus a muscular contraction. In this chapter on synaptic transmission in the central nervous system (CNS), it is convenient to start with some simple concepts of the mode of action of the neuron that operates the motor unit, i.e., the motoneuron, because this study links with the classic reflex work and because the motoneuron has been studied very extensively by modern analytical techniques.

At the beginning of this century, Sherrington (1906) made integration the basic theme in his classic book, *The Integrative Action of the Nervous System.* He developed the concept of central integration as a consequence of a study of reflexes selected to give insight into one or another aspect of integration. Specific sensory inputs were employed, and observations were made on the muscle contractions produced thereby. It had already been established that the muscle is a reliable servant of the nervous system, the observed contractions faithfully reporting the output signals from the CNS, which of course are the impulse discharges from motoneurons. In this manner, Sherrington was able to define the performance of simple arrangements of neurons [as, for example, in Figure 2.1 (Section 2.1)], the so-called "reflex pathways," and to relate it specifically to the properties of the synapses between neurons. Since these synaptic contact areas were very small, Sherrington surmised that many had to be activated at about the same time to cause neurons to discharge an impulse, hence the concepts of convergence and summation. A complementary property derives from the profuse branching of nerve fibers with the consequent extensive dispersion of activity entering the nervous system by some afferent pathway.

The "functional principle of divergence," as it was called, ensures that any particular input has at least the potentiality for a widespread influence on reflex discharges. The most revolutionary concept introduced by Sherrington was that some of the synapses on a membrane have an inhibitory action, preventing the discharge that otherwise would be evoked by the excitatory synaptic action.

In a later publication, *Reflex Activity in the Spinal Cord,* by Creed *et al.* (1932), there was an account of the later analytical study of reflexes by the Sherrington school of neurophysiology at Oxford. Already, the concept of the motor unit had been developed and quantitatively studied [shown diagrammatically in Figure 3.1 (Section 3.2.1)]. A motor unit is defined as the axon of a motoneuron and the muscle fibers innervated thereby. By extensive axonal branching, this number of muscle fibers is usually in excess of 100, and usually each is exclusively innervated by a single motoneuron. The 1932 book was essentially an attempt to describe a variety of reflexes of the spinal cord in terms of their motor-unit composition. Reflex contractions were quantitatively explained by the number and temporal arrangements of the individually activated motor units, and in fact this statement can be extended to all behavioral reactions involving the contractions of striated muscle.

After the Sherrington era, there were great transformations in our detailed understanding of the CNS, but his work provided the foundation for all this new building. Great advances have been made possible by technical developments that sharpened the focus with which the brain could be studied in its basic structure and activity. Intracellular recording from neurons gave important insights into synaptic mechanisms. Examples are the mode of action of excitatory and inhibitory synapses and their interaction, and also the special synaptic mechanisms of presynaptic inhibition and of reciprocally acting synapses. Electron microscopy provided wonderful insights into the structural detail of synapses and neurons, as illustrated in Chapter 1. Radiotracer techniques and improved histological methods have given new insights into the communication systems of the brain and spinal cord. Most recently, horseradish peroxidase (HRP) has been of great value in these respects, as also have electrophysiological studies. And overarching all these achievements is the immense range of neurochemical and neuropharmacological studies that are the theme of many chapters of this book. In each section, there is reference to the historical aspects of the studies (cf. Eccles, 1982d).

Figure 4.1A shows the simplest excitatory and inhibitory pathways through the CNS from muscles that act antagonistically at the knee joint. Annulospiral stretch receptors of the knee extensor muscle are connected to the spinal cord by large fibers (group Ia) that exert a powerful synaptic excitatory action directly on the motoneurons of that muscle and related muscles. If there is a strong enough pull on the muscle, the annulospiral endings will produce, by their discharges along their fibers, an excitation of motoneurons powerful enough to cause the discharge of impulses along their axons, so that the muscle contracts. This is the familiar tendon jerk or stretch reflex of neurophysiology and the classic exemplar of monosynaptic reflexes. The brief pull produced by a tap on the tendon gives a tendon reflex such as the knee jerk or ankle jerk. A prolonged steady pull evokes the prolonged contraction of the stretch reflex (Creed *et al.,* 1932). Figure 4.1C gives an example of temporal facilitation in this monosynaptic pathway. Double stimulation of the Ia fibers by the SE electrode (Figure 4.1A) evokes the discharge of an impulse from many motoneurons (the spike response), whereas either stimulus alone is ineffective.

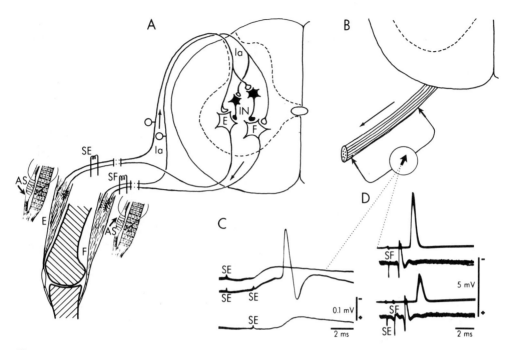

Figure 4.1. Simple reflex pathways and responses. (A) Diagrammatic representation of the pathways from and to the extensor (E) and flexor (F) muscles of the knee joint. The small insets show the details of the origin of the Ia afferent fibers from the annulospiral endings (AS) of muscle spindles. Thence, the afferents run via the dorsal roots to enter the spinal cord and form excitatory synapses on motoneurons (E, F) that supply the muscle of origin and on interneurons (IN) that inhibit synaptically the motoneurons of the antagonistic muscle. The pathway from motoneuron to muscle is as for the motor unit in Figure 3.1 (Section 3.2.1), the endplates being shown enlarged in the insets. The interruptions in the nerves beyond the electrodes (SE, SF) indicate a discontinuity in the diagram. The ventral roots were sectioned in experiments on afferent inputs to the spinal cord, and the dorsal roots were sectioned in antidromic experiments such as in Figure 2.14 (Section 2.7). (B) A ventral root by which motor axons leave the spinal cord. On it are recording electrodes, one close to the spinal cord, the other on the cut end. (C) Double stimulation of the extensor Ia fibers [SE electrode in (A)] evoked an impulse discharge, whereas single stimulation failed and produced only a slow depolarizing potential that was electronically propagated from the motoneurons. (D) This is the opposite action. A single stimulation of flexor Ia fibers [SF electrode in (A)] evoked the monosynaptic reflex discharge in the upper trace. A preceding stimulation of extensor Ia fibers [SE electrode in (A)] inhibited this discharge by about half. The lower traces show nerve volleys recorded from the dorsal root.

In Figure 4.1A, there is another pathway to the E motoneuron, that from the annulospiral endings of the antagonistic knee flexor muscle. Each of these afferent fibers gives off collaterals in the spinal cord that then form excitatory synapses on small nerve cells (interneurons) that in turn form inhibitory synapses on the extensor motoneurons, either preventing their reflex discharge or reducing the probability of that discharge. This reduction appears as a diminution in the monosynaptic reflex response (Figure 4.1D) recorded (Figure 4.1B) as a "population spike" from the whole assemblage of motor axons that supply the muscle. The spike response to SF stimulation (Figure 4.1A) is reduced to about half by a preceding SE stimulus. Figure 4.1A diagrams the simplest pathway for an inhibitory action from the periphery and further shows the reciprocal relationship of the pathways from the antagonistic muscles, the extensors and flexors of the knee joint. The afferent fibers from the annulospiral endings are excitatory at all their synapses, whether on motoneurons or on other neurons. To have an inhibitory action, there is interpolated a special class of interneurons that are exclusively inhibitory at their synapses. It was first found by Lloyd (1946) that the afferent fibers from a muscle have an inhibitory action on the motoneurons of antagonistic muscles, as indicated by a reduction of the population spike recorded from the motor axons in the ventral root that is shown in Figure 4.1D. Only much later was it established that there was interpolation of the inhibitory neuron, as diagrammed in Figure 4.1A (cf. Eccles, 1957, 1969).

It must be understood that the diagram of Figure 4.1A is greatly simplified both in the monosynaptic excitatory path to a single motoneuron and in the inhibitory path through a single interneuron. For example, A. G. Brown (1981) has presented a most detailed quantitative study for a large cat muscle (triceps surae) that is innervated by 450–500 motoneurons and from which about 150 Ia afferent fibers project to the spinal cord to make monosynaptic contacts with these motoneurons. Each afferent fiber branches profusely to give a total of 1000–2000 synapses to some hundreds of motoneurons, which is a measure of the *divergence*, and on each motoneuron there is *convergence* of the collateral branches from over 100 of the Ia afferent fibers to make a total of 200–450 synapses. There would be a corresponding divergence and convergence at each stage of the inhibitory pathway in Figure 4.1A, which likewise has to be conceived as having many lines in parallel. Figure 4.1A indicates correctly, however, that despite this massive in-parallel arrangement, there are only single interneurons in series in the inhibitory pathway; hence, this inhibitory pathway is disynaptic.

4.2. Excitatory Synaptic Action

The reader is referred to Eccles (1964b) for a general reference.

The experimental investigation of synaptic transmission in the simplest pathways of the spinal cord [cf. Figure 4.1 (Section 4.1)] was transformed in 1951 by the technique of recording electrically from the interior of nerve cells, using for this purpose the intracellular glass microelectrode and a cathode-follower amplifier (Brock *et al.*, 1952), as had been used a year or so previously by Fatt and Katz (1951) in investigating neuromuscular transmission. With the motoneuron, there was the added difficulty that it had to be located and penetrated without visual control, and, of course, the necessary rigid fixation was harder to secure. Nevertheless, even the earliest experiments with primitive techniques were surprisingly successful, and soon new insight into the nature of excitatory and inhibitory synaptic action on motoneurons had been gained (cf. Eccles, 1957). After a few minutes, many motoneurons become stabilized to give, even for hours, responses that appear to be unaffected by the impalement. The most satisfactory electrodes are those with a very gradual taper and with a tip diameter of about 0.5 μm, which gives them a resistance of 10–15 MΩ when filled with 3 M KC1. Potassium methyl sulfate (2 M KCH_3SO_4) electrodes are to be recommended.

Figure 4.2 shows schematically the stimulating electrode on the many large fibers from the annulospiral endings of muscle (cf. Figure 4.1A) that converge on and give excitatory synapses to a motoneuron of that muscle. In Figure 4.2B–J, a progressively

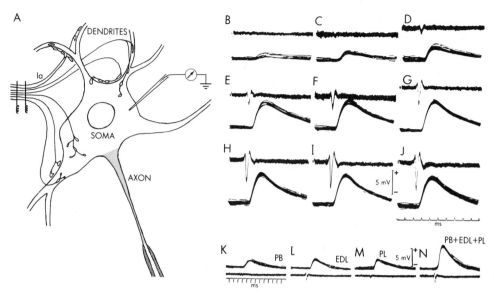

Figure 4.2. Monosynaptic excitation of motoneurons by the group Ia afferent pathway. (A) Drawing of a motoneuron showing the central dendritic regions, the soma, the initial segment of the axonal origin [cf. IS in Figure 2.14B (Section 2.7)], and the beginning of the axonal medullation. Shown on the dendrites and soma are the excitatory synaptic endings of seven group Ia afferent fibers that have an applied stimulating electrode (actually on the peripheral muscle nerve). The intracellular recording microelectrode is shown diagrammatically. (B–J). The upper traces give the size of the afferent volley as it enters the spinal cord, and the lower, the simultaneously recorded EPSPs. All records are formed by the superimposition of about 25 faint traces. (K–M) EPSPs recorded in another motoneuron (peroneus longus) in response to maximum group Ia volleys in the nerves to three muscles: peroneus brevis (PB), extensor digitorum longus (EDL), and peroneus longus (PL). (N) All three muscles combined. From Eccles *et al.* (1957).

stronger stimulation was applied through the stimulating electrode, and the number of fibers so excited was monitored by the recording electrode placed so as to record the impulses as they entered the spinal cord (upper traces in Figure 4.2B–J). The lower traces are the intracellularly recorded potentials produced by the synaptic excitatory action. They are seen to resemble the endplate potentials (EPPs) described in Chapter 3 [Figure 3.3C and F (Section 3.2.3)] in being in the depolarizing direction, there being a temporary diminution of the resting membrane potential of about -70 mV. The size, but not the time course, is changed by an increase in the number of activated fibers, as indicated by the spike potentials in the upper traces. This is a good illustration of the way in which convergence of afferent fibers onto a neuron gives summation of the excitatory synaptic potentials that are recorded across the postsynaptic membrane and hence are named, "excitatory postsynaptic potentials" (EPSPs). After Figure 4.2G, further increase in the stimulus did not result in any increase in the EPSP. In Figure 4.2G, there was stimulation of all the group Ia fibers, the increase in afferent spikes in Figure 4.2H–J being due to the stimulation of group Ib fibers that do not have a monosynaptic excitatory action on motoneurons (Eccles *et al.*, 1957; A. G. Brown, 1981). Usually, there is a linear relationship between the amount of monosynaptic excitation and the size of the EPSP. This is illustrated in Figure 4.2K–N, in which EPSPs were produced in a motoneuron by the convergence of Ia afferent fibers in the nerves to three different muscles (Figure 4.2K–M), and, when synchronized (Figure 4.2N), the EPSP was exactly the sum of the three component EPSPs. According to the formal electrical diagram of Figure 4.4C, an appreciable deviation from the linearity would be expected only when the EPSPs were large. Thus, it may be assumed that the observed EPSPs arise by a simple summation of the EPSPs produced by each synapse.

The EPSPs evoked in a motoneuron monosynaptically by a group Ia afferent volley (Figure 4.2B–N) represent the sum of several unitary EPSPs, each produced by the synaptic knobs stemming from a single Ia fiber (cf. Figure 4.2A). The unitary EPSP produced by single presynaptic impulses can easily be recognized when there is a random bombardment of the motoneuron that gives what we may term "synaptic noise" (cf. Brock et al., 1952; R. E. Burke, 1967). Each of the EPSPs of this noise is produced by a single impulse in a Ia fiber.

By averaging of EPSPs, it has been shown that a single Ia fiber is distributed very widely to the motoneurons of its muscle of origin. For example, in Figure 4.3A, the same group Ia fiber produced the EPSPs in six different motoneurons (Mendell and Henneman, 1971). Some of the extremely small unitary EPSPs of Figure 4.3A would certainly have been missed before the introduction of the averaging techniques. Amplitudes of EPSPs varied from 17 to 600 μV, with an average of 102 μV. The wide variation in time course is attributable to the location of the synapses on the soma or dendrites. The more remote they are, the greater is the electrotonic slowing of the EPSP recorded in the soma (Rall, 1970). With this very sensitive technique, it was found that a single Ia fiber gives excitatory synapses to nearly all the 300 motoneurons supplying the muscle (cat gastrocnemius medialis) from which it arises. A large Ia fiber may give as many as 2000 synapses to motoneurons, and it also gives synapses to other neurons in the spinal cord. This widely distributed synaptic potency is certainly impressive.

By a double HRP technique, it has been possible to identify all the synapses made by a single Ia fiber on a motoneuron (R. E. Burke et al., 1979; A. G. Brown, 1981). For example, the diagram of Figure 4.3B represents the locations of synaptic endings by a single Ia afferent from the medial gastrocnemius muscle on a medial gastrocnemius motoneuron. The five synaptic endings are on three different dendrites, two being rather close to the soma, the others more distant. Wide ranges of distribution are found, but there is a tendency toward clustering. This accounts for the range in time course of EPSPs produced by a single Ia fiber in Figure 4.3A. On some motoneurons, there is clustering of its boutons close to the soma, on others more dispersal, and on the second lowermost trace, there would be a synaptic clustering far out on dendrites. It will be recognized that the Ia synapses are in general much more peripheral than shown in Figure 4.2A. From a survey of the available data in the literature as well as from his own finding, A. G. Brown (1981) states that only about 10–15% of the boutons of Ia fibers make contact with the soma and the proximal 60 μm of the dendrites. However, most are within 600 μm of the soma.

A still more refined level of inquiry concerns the EPSP produced by a single bouton, which is a component of the several bouton endings of a single Ia fiber (Figure 4.3B). By a technique of fluctuation analysis (Redman, 1981; Hirst et al., 1981; Jack et al., 1981a,b), it has been possible to distinguish among the EPSPs generated by each bouton

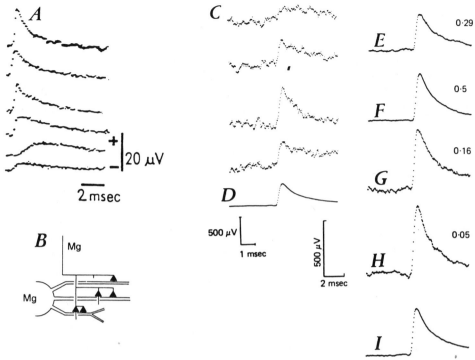

Figure 4.3. (*A*) Averaged recordings of EPSPs produced by impulses in the same Ia fiber terminating on six different motoneurons. From Mendell and Henneman (1971). (*B*) Summary diagram of the location of Ia synapses from a single medial gastrocnemius Ia fiber onto a medial gastrocnemius motoneuron at five sites on three different dendrites as indicated. From A. G. Brown (1981). (*C*) Four EPSPs evoked by a single Ia impulse, selected from a population of 800 responses. (*D*) Average of all the 800 responses. (*E–H*) Components 1–4 of the EPSP derived from fluctuation analysis—the probabilities of the occurrence of these components are indicated to the right of each. (*I*) Reconstructed EPSP obtained by adding the weighted values of *E–H*: 0.29 (*E*) + 0.5 (*F*) + 0.16 (*G*) + 0.05 (*H*). From Jack *et al.* (1981a).

on a motoneuron when activated by a single Ia impulse. For example, Figure 4.3*C* illustrates the wide range of fluctuating EPSPs produced by a single Ia impulse; in Figure 4.3*D*, there is summation of 800 such responses to give a typical EPSP, such as those of Figure 4.3*A*. By the fluctuation analysis, this EPSP is shown to be composed of elements, each generated by a single bouton. It is rare that an activated bouton liberates more than one vesicle. Figure 4.3*E–H* shows four EPSPs derived by the fluctuation analysis of Figure 4.3*C*, each arising from a single bouton. The probability of quantal emission from a bouton of a single synaptic vesicle ranged from 0.5 to 0.05; Figure 4.3*E–H* shows the time course of the EPSPs generated by the synaptic emission from each bouton. In sequence for *E–H*, the sizes of the quantal EPSPs are 302, 406, 505, and 607 μV and, when combined with allowance for probability, yield an accurate reconstruction of the EPSP produced by a single fiber, Figure 4.3*I* being identical to Figure 4.3*D* after allowance for the different scaling. It can be assumed that the four derived EPSPs of Figure 4.3*E–H* are each produced by a single bouton at various distances from the soma, *H* being closest and *E* most remote, as in Figure 4.3*B*. Actually, Redman (1981) has illustrated a triple EPSP composition for an Ia fiber that has been shown by HRP to make three boutons with that motoneuron at the expected distances from the soma.

From this remarkable analysis (Jack *et al.*, 1981a) are derived four conclusions concerning the synaptic functioning of a single Ia fiber on a motoneuron:

1. There is a wide gradation of intermittency, e.g., 0.5–0.05 in Figure 4.3*E–H*. Some boutons may even approach a probability of 1, but above 1 is very rare, i.e., for two vesicles ever to be emitted from a single bouton in response to one impulse.
2. Usually, a Ia fiber gives 3–5 boutons to a motoneuron (cf. Figure 4.3*B*), but the observed range is 1–10.
3. It is proposed that the neurotransmitter emitted from one vesicle is sufficient to saturate all the available subsynaptic receptor sites. In posttetanic potentiation, the synapses with 100% probability are not potentiated, the potentiation depending on diminished intermittency of other synapses (Hirst *et al.*, 1981). There is a similar conclusion for the EPSP potentiation by 4-aminopyridine (Jack *et al.*, 1981b).
4. It is remarkable that the size of the EPSP recorded in the soma was approximately the same for synapses remote on the dendrites as for those close to the soma (Iansek and Redman, 1973; Jack *et al.*, 1981a), despite the electrotonic decrement that would be expected to effect a reduction of at least 5 times. Various suggestions have been made, such as clustering of distal boutons (denied by A. G. Brown, 1981), higher probability of neurotransmitter emission, or increased postsynaptic receptor density. This most valuable physiological arrangement remains an enigma. The last explanation is contrary to the hypothesis of receptor saturation [conclusion (3) above]. It will be seen in Chapter 15 that the explanation of long-term potentiation invokes a more efficient use by receptor sites of the neurotransmitter emitted from a bouton, more receptor sites becoming available.

Korn and associates (Korn *et al.*, 1982; Korn and Faber, 1986) have studied a very different synapse, the inhibitory synapse on the Mauthner cell in the fish spinal cord. It was possible to carry out a fluctuation analysis of the inhibitory postsynaptic potentials produced by a single presynaptic inhibitory fiber. They employed a technique different from that used by Redman and associates, that of binomial analysis, which showed a composition of quanta in good agreement with the histologically determined number of boutons. There was agreement with Redman and associates that the probability of release of quanta from a bouton was less than 1—usually about 0.4. Nelson *et al.* (1983) and Neale *et al.* (1983) studied fetal mouse spinal cord neurons in tissue culture and found comparable results. A presynaptic impulse caused quantal release from a bouton with a probability below unity.

It can be assumed that, like the EPP, the EPSP is generated by a chemical neurotransmitter, the identification of which is as yet unsure but is probably glutamate or aspartate (see Chapter 6) (Curtis and Johnston, 1974). There are, for example, several distinguishing characteristics: the synaptic delay seen in Figure 4.2 between the synaptic impulse and the EPSP; the miniature EPSPs of quantal character and the time course of the monosynaptic EPSP (Figure 4.2), which is explicable (Figure 4.4*B*) as being due to a brief

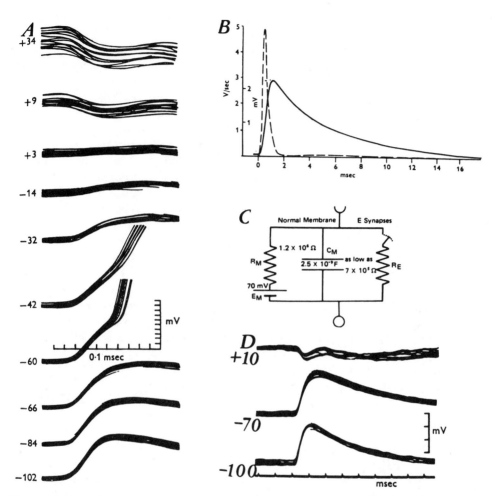

Figure 4.4. (*A*) Monosynaptic EPSPs set up in a cat bicep–semitendinosus motoneuron at various levels of membrane potential (in mV). Each record is formed by the superimposition of about 20 faint traces. The membrane potential was shifted to the indicated values from its resting value of −66 mV by steady currents through the other barrel of the double microelectrode. From Coombs *et al.* (1955c). (*B*) Mean curve (———) of an EPSP recorded in a motoneuron as in Figure 4.2 and analyzed on the membrane time constant to determine the time course (- - -) of the postsynaptic currents that generate the potential change. (*C*) Formal electrical diagram of a motoneuronal postsynaptic membrane [cf. Figure 1.15B (Section 1.2.3)] with R_M, E_M, and C_M values for the motoneuron plus dendrites as "seen" by an intracellular electrode. At right is the representation of the monosynaptic excitatory (E) synapses showing the variable resistance (R_E) and an equilibrium potential of zero in approximate accord with experimental determination as in Figure 4*A*. (*D*) As in (*A*), monosynaptic EPSPs in a gastrocnemius motoneuron with membrane potential changed from the resting potential of −70 mV to +10 mV and −100 mV by the passage of steady current through one barrel of the double microelectrode. (B–D) From Coombs *et al.* (1955c).

initial depolarizing current of 1–2 msec and a subsequent return to the resting potential by membrane currents given by the components at left in Figure 4.4C. The approximately exponential decay can thus be attributed to the effective membrane resistance and capacitance [R_M and C_M in Figure 4.4C]. The time constant of motoneurons has been directly measured (Eccles, 1961) and corresponds to the 3 msec calculated for the exponential decline from the summit of the EPSP in Figure 4.4B. With other species of neurons, there are many variants from this simple diagram.

The chemical transmitter hypothesis of generation of the EPSP had been crucially tested at the time of constructing a formal diagram such as that of Figure 4.4C. The close analogy to the EPP was realized, and it was proposed that the chemical transmitter opened channels across the postsynaptic membrane, as shown for the R_E element in Figure 4.4C. Unfortunately, there are grave technical problems in applying the conventional voltage-clamp technique to neurons deep in the spinal cord. However, the EPSP hypothesis can be crucially tested by investigation of the effects on the EPSP of changes in the level of the membrane potential. The transmitter is assumed to close the switch on the R_E channel (Figure 4.4C) and allow the flow of ions that act to depolarize the membrane, the brief initial inward current flow in Figure 4.4B. On the model of Figure 4.4C, it would be predicted that, if the membrane potential is changed beyond a critical level, the synaptic current flow would be reversed. This level is set by the integrated electrochemical potential of the ions traversing the channel, just as with the EPP. The initial investigation (Coombs et al., 1955c) encountered severe difficulties with the new technique of intracellular recording and current passing by double-barrel microelectrodes. Large currents of up to 200 nA had to be passed through one barrel in order to reverse the EPSP. However, in specially favorable cases, this was accomplished, as illustrated in Figure 4.4A, in which the reversal was at a membrane potential of about 0 mV. Reversed EPSPs are also illustrated in Figure 4.4D with a reversed membrane potential of +10 mV. The triphasic time course of Figure 4.4D suggests that some synapses were on rather distal dendrites (cf. Figure 4.3A and B) and so were not reversed at a membrane potential of +10 mV in the soma.

In 1955, then, it was optimistically assumed that the synaptic model of Figure 4.4C would be accepted as a simple case of chemical transmission at excitatory synapses in the CNS, the EPSP being closely analogous to the EPP. But it was not to be! It was found impossible to repeat the crucial test of Figure 4.4A and D with EPSPs reversed beyond a critical membrane depolarization. Thus, doubt was cast on the original experiments, and it was even proposed that the initial part of the EPSP is set up by electrical synaptic transmission. The dramatic story of this controversy has been told by Engberg and Marshall (1979), who employed double microelectrodes and confirmed precisely the findings of 24 years earlier (Figure 4.4A and D), giving the reversal potential for the EPSP as 0 to +10 mV. Most investigators had used only single microelectrodes for passing current and recording and so had no information on the membrane potential. Large depolarizing currents set up anomalous rectification with high membrane conductance to K^+ ions, so that extremely large currents may still fail to reverse the membrane potential and the EPSP; this was apparently the case with the otherwise excellent study of Edwards et al. (1976) using unitary Ia EPSPs, as in Figure 4.3A. Finkel and Redman (1983) have recently developed a very sophisticated technique for voltage-clamping a motoneuron in the spinal cord. In response to a single Ia impulse into boutons close to the soma, an EPSP of 100 μV was generated by a peak excitatory postsynaptic current of 330 pA. The relationship of current to voltage was rather similar to that shown in the original figure (Figure 4.4A). With change in the membrane potential, there was reversal of the current at +4.6 ± 2 mV, which is very close to the original findings of Figure 4.4A and D. Reference should also be made to the tissue culture studies of Macdonald et al. (1983).

By electrophoretic injection through an intracellular electrode, large changes can be made in the ionic composition of motoneurons. These injections of anions or cations may cause a considerable decrease in the membrane potential, and the EPSP is then diminished correspondingly, but is not otherwise changed (cf. Eccles, 1964b). In light of the recent evidence that anions are not appreciably concerned in generation of the EPP, further investigation is desirable, particularly with isolated preparations, with which it should be possible to change the extracellular ions in the way that was done in investigations by A. Takeuchi and N. Takeuchi (1960) on the EPP. Presumably, the principal ions that generate the EPSP are Na^+ and K^+ as in the endplate membrane, but Ca^{2+} is probably also involved, as is proposed in Chapter 16. Since the equilibrium potential is at about 0 mV membrane potential, sodium conductance should be larger relative to potassium conductance than at the motor endplate. The sodium ion conductance, however, is less dominant than has been observed for the giant synapse of the squid (P. W. Gage and Moore, 1969).

The synapses made by Ia afferent fibers on spinal motoneurons have been given concentrated attention because they have been by far the most investigated mammalian synaptic mechanism using in vivo techniques. The mechanism serves as a model of a simple synaptic system. The new in vitro techniques have already made significant contributions to our understanding of synaptic excitatory mechanisms, e.g., the investigations on cultures of selected cells from the brain or spinal cord. Under suitable treatment, these cultures arrange themselves as a unicellular layer in which individual cells and fibers can be identified in the living preparation and cells and synapses can be investigated selectively (Neale et al., 1983; Nelson et al., 1983; Macdonald et al., 1983; Gähwiler, 1984). Voltage-clamping techniques are possible, and already patch-clamping has been used to study inhibitory synapses in cultured mammalian hippocampal tissue, as described in Section 4.4 (Borman et al., 1983). But there is as yet no study of voltage-clamping or patch-clamping on excitatory synapses in the cultured mammalian CNS. Despite the great technical advantages of these new in vitro techniques, the results have to be evaluated relative to the in vivo investigations described above. This precaution is much more necessary than for the neuromuscular synapse, in which in vitro and in vivo preparations are virtually equivalent.

4.3. Impulse Generation by Synaptic Action

As shown in Figure 4.5A–D, if the EPSP attains a critical threshold level (Figure 4.5B), it causes the neuron to discharge an impulse, the latency being briefer the larger the EPSP (Figure 4.5B–D). In Figure 4.5B–D, this was brought about by increasing the size of the presynaptic volley, i.e., by increasing the number of activated synapses on the motoneurons [cf. Figure 4.2A (Section 4.2)]. As would be expected, a subliminal EPSP can also be made to generate an impulse by procedures that change the membrane potential toward the critical threshold level of depolarization. For example, in Figure

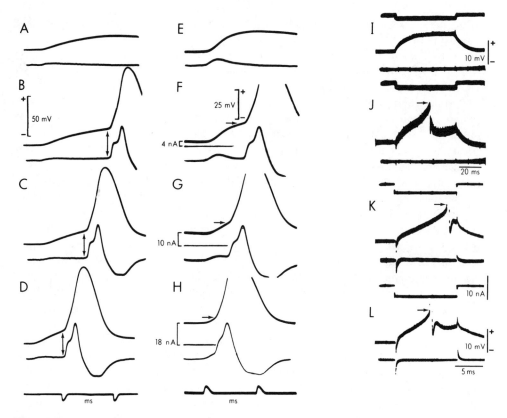

Figure 4.5. Impulse generation by membrane depolarization. (A–D) Intracellularly recorded potentials of a cat gastrocnemius motoneuron (resting membrane potential: −70 mV) evoked by monosynaptic activation that was progressively increased from (A) to (D). The lower traces are the electrically differentiated records, the double-headed arrows indicating the onsets of the IS spikes in (B)–(D) [cf. Figure 2.14A (Section 2.7)]. (E–H) Intracellular records evoked by monosynaptic activation that was applied at 12 msec after the onset of a depolarizing pulse (in nA). A pulse of 20 nA was just below threshold for generating a spike. (E) Control EPSP in the absence of a depolarizing pulse. The lower traces give electrically differentiated records. Note that spikes are truncated. (E–H) From Coombs *et al.* (1957b). (I–L) A pyramidal tract cell of the motor cortex was subjected to a depolarizing current of an intensity and duration signaled by the uppermost traces. The middle traces are the intracellular potential changes and the lowermost, the extracellular control records. Peaks of spike potentials are off the record, but the voltages at the spike origins are indicated by horizontal arrows. Note that (K) and (L) were taken with a much faster sweep. From Koike *et al.* (1968).

4.5F–H, the same EPSP as in Figure 4.5E was made effective by the operation of a background depolarizing current that was commenced 12 msec before. The shift in membrane potential is indicated to the left. The impulse arose at the arrows when the total level of depolarization was about 18 mV, which was made up in varying proportions by a conditioning depolarization and the superimposed EPSP. The threshold level of depolarization may also be attained by superimposing the EPSP on the depolarization produced by a preceding EPSP, which provides a sufficient explanation of the reflex phenomenon known as "temporal facilitation"—a single afferent volley fails to evoke a reflex discharge, but two such volleys in a short interval are effective, as in Figure 4.1C. Comparable observations have been made with impulse generation by the EPSPs of a wide variety of nerve cells.

In Figure 4.5I–L, the critical level of depolarization for generation of an impulse is displayed by a cortical pyramidal cell that was subjected to depolarizing pulses of increasing size, as shown in the upper traces. In Figure 4.5I, the depolarization did not reach threshold. In Figure 4.5J–L, the impulse discharge was generated progressively earlier, but, as shown by the arrows, at approximately the same level of depolarization of about 13 mV.

There is now general agreement that the synaptic excitatory action is effective in generating the discharge of an impulse solely by producing an EPSP that attains the threshold level of depolarization.

4.4. Inhibitory Synaptic Action

The reader is referred to Eccles (1964b) for general references.

There is a second class of synapses that, because they oppose excitation and tend to prevent the generation of impulses by excitatory synapses, are called "inhibitory synapses." There is general agreement that these two basic modes of synaptic action govern the generation of impulses by nerve cells. Figure 4.1A (Section 4.1) shows that the simplest inhibitory pathway in the spinal cord goes via a type of neuron that is specialized to exert an inhibitory synaptic action and nothing else. In 1946, Lloyd (1946) discovered that the Ia afferent fibers of a muscle inhibit motoneurons of the antagonist muscle. But the mechanism of this action was not known, nor was the inhibitory pathway diagrammed as in Figure 4.1A. There had been developed an electrical theory of inhibitory synaptic action, and eventually, in 1951, the theory was tested by inserting a microelectrode inside the motoneuron [see Figure 4.2A (Section 4.2)] to see what inhibition did to the membrane potential. The electrical theory was proved wrong by the discovery that the inhibitory synapses produced a generalized increase of charge on the neuronal membrane, a hyperpolarization (Brock et al., 1952).

In Figure 4.6A–C, there was graded stimulation of the group Ia fibers of the nerve to the antagonistic muscle (cf. Figure 4.1A); the upper traces show the action potential of these fibers just as in Figure 4.2B–J. There was an associated gradation of the inhibitory hyperpolarization of the membrane in the lower traces. This hyperpolarization was termed the "inhibitory postsynaptic potential" (IPSP). Its size is dependent on the number of converging inhibitory synapses, just as with the excitatory synapses and the EPSPs in Figure 4.2B–G. Figure 4.6D and E shows that the EPSP and IPSP of the same motoneuron are nearly mirror images. In both cases, there is a brief flow of synaptic current for about 1 msec that effects the membrane potential change. Subsequently, the membrane potential is restored to the initial level by ionic fluxes across the motoneuronal membrane in general [cf. Figure 4.6F and Figure 1.15B (Section 1.2.3)]. This explains the similarity in time courses of the EPSP and the IPSP. It is also a regular feature that the IPSP has a central latency about 1 msec longer than that of the EPSP, which is the time for traversing the inhibitory interneuron that is interpolated in the pathway (cf. Figure 4.1A).

The inhibitory action of Ia impulses from a muscle on the motoneuron of an antag-

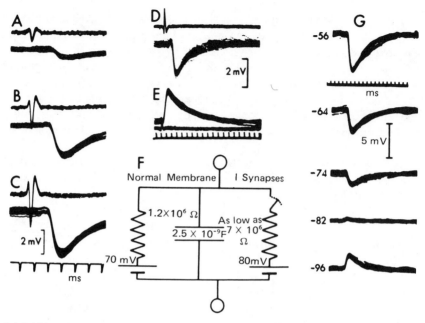

Figure 4.6. Inhibitory postsynaptic potentials (IPSPs) of a motoneuron. (A–C) Graded responses as described in the text. (D, E) Comparison of EPSP and IPSP [time below (E) in milliseconds]. (A–E) From Curtis and Eccles (1959). (F) At left is the formal electrical diagram of the membrane of a motoneuron (cf. Figure 4.4C); at right, the mode of operation of inhibitory synapses is symbolized by the closing of the switch. (G) This resembles Figure 4.4A in showing the effect of varying membrane potential on the IPSP. (F, G) From Coombs *et al.* (1955b).

onist muscle was the first example of an inhibitory mechanism studied by intracellular recording in the CNS. It has turned out that the same general principles obtain for almost all inhibitory actions in the brain, with the exception of presynaptic inhibition [discussed in Section 4.8 (cf. Figure 4.18)]. The techniques used in the spinal cord have been applied at all levels of the CNS. In many laboratories, there is now a great ongoing enterprise of identifying the excitatory and inhibitory neurons in all regions of the brain and determining the manner of their interaction in complex circuits. We have to discover at all these levels, not only the neuronal pathways, but also the mechanisms at all synaptic relays, in particular the synaptic neurotransmitter substances at these relays and the ionic channels that cause the different postsynaptic actions. It is remarkable that we are more advanced in answering these questions with respect to inhibition than to excitation. With inhibition, two neurotransmitters, γ-aminobutyric acid (GABA) and glycine, have been investigated (see Chapter 7).

Figure 4.7 gives a kind of overall view of excitatory and inhibitory synaptic action as we now understand it. In Figure 4.7C, the neuron is shown with an intracellular microelectrode. The motoneuronal excitatory synapses tend to be out on the dendrites as drawn in this figure [cf. Figure 1.4B (Section 1.2.2)], while the inhibitory synapses tend to be concentrated on the soma. Moreover, there are structural differences, as illustrated in the two synapses of Figures 1.5B and 1.6A (Section 1.2.2), in which, with glutaraldehyde

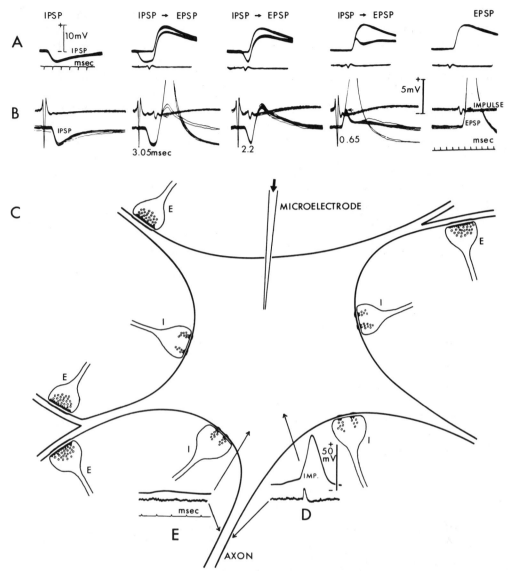

Figure 4.7. Summarizing diagram of excitatory and inhibitory synaptic action. See the text for a full description. (C) The excitatory (E) and inhibitory (I) synapses are shown with characteristic histological features [cf. Figure 1.5B (Section 1.2.2)].

fixation, the synaptic vesicles of the inhibitory synapse at the right have an ellipsoid configuration, while in the excitatory, at the left, they are spherical. In essential structural features, however, the two kinds of synapses are similar.

In Figure 4.7A, there are records of pure IPSPs and EPSPs. In the former, charge is put on the membrane, increasing the membrane potential from −70 to about −73 mV. The EPSP has an antagonistic action, decreasing the charge across the membrane. If the

IPSP is superimposed on the EPSP, as in the fourth frame, the IPSP is seen to be very powerful in counteracting the EPSP, which is also shown alone in the same frame to aid comparison. In the third frame of Figure 4.7B, the IPSP very effectively suppresses the impulse discharge generated by the EPSP alone, which is illustrated in the last frame.

Figure 4.7D is a good illustration of the EPSP generating an impulse discharge (cf. Figure 4.5B–D) that travels down the axon to be recorded as the brief spike in the lower trace. In Figure 4.7E, an initial inhibitory action reduces the EPSP so that it no longer generates an impulse, and the spike (D) in the axon is also suppressed. Evidently, inhibitory synapses achieve their effectiveness by generating the IPSP that directly counteracts the depolarizing action of the EPSPs. Thus, we have the question of what inhibitory synapses are doing solved at that level. There are, however, many more questions to be answered.

Figure 4.8B shows diagrammatically the flow of current under an activated inhibitory synapse—outward through the subsynaptic membrane, along the synaptic cleft, and circling back to hyperpolarize the postsynaptic membrane by inward flow over its whole surface, which is the reverse of that for an excitatory synapse (Figure 4.8A). The outwardly directed current across the inhibitory subsynaptic membrane could be due to the outward movement of a cation such as potassium, to the inward movement of an anion

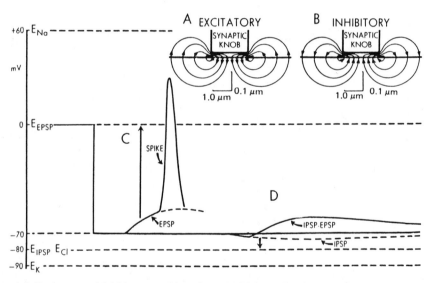

Figure 4.8. Excitatory and inhibitory synaptic action. (A) Diagram showing an activated excitatory synaptic knob. As indicated below the figure, the synaptic cleft is shown at 10 times the scale for width as against length. The current is seen to pass inward along the cleft and in across the activated subsynaptic membrane. Elsewhere, as shown, it passes outward across the membrane, thus generating the depolarization of the EPSP. (B) Similar diagram showing the reverse direction of current flow for an activated inhibitory synaptic knob. (C, D) Diagram for cat motoneuron showing the equilibrium potentials for sodium (E_{Na}), potassium (E_K), and chloride (E_{Cl}) ions [cf. Figure 1.15 (Section 1.2.3)], together with the equilibrium potential for postsynaptic inhibition (E_{IPSP}). The equilibrium potential for the EPSP (E_{EPSP}) is shown at zero. (C) An EPSP is seen generating a spike potential at a depolarization of about 18 mV (see Figure 4.5B–D). (D) An IPSP is shown alone (- - -) and then interacting with an EPSP (——). As a consequence of the depressant influence of the IPSP, the EPSP that alone generated a spike (C) is no longer able to attain the threshold level of depolarization (IPSP-EPSP); i.e., the inhibition has been effective.

such as chloride, or even to a combination of anionic and cationic movements such that there is a net outward flow of current driven by a battery of about -80 mV in series with a fairly low resistance (see Figure 4.6F).

Figure 4.8C and D serves to illustrate the simplest findings on the EPSP and the IPSP and their interaction. The approximate equilibrium potentials for sodium, chloride, and potassium ions are shown by the horizontal dashed lines, the equilibrium potential for chloride ions (E_{Cl^-}) being assumed to be identical to the E_{IPSP} (cf. Figures 4.6G and 4.11E). In Figure 4.8C, the EPSP is seen to be large enough to generate a spike potential, the course of the EPSP in the absence of a spike being shown by the dashed line. In Figure 4.8D, there is an initial IPSP that is seen to diminish the depolarization produced by the same synaptic excitation so that it is no longer adequate to generate a spike. We can now ask: What is the ionic mechanism responsible for the outward current across the subsynaptic membrane of the inhibitory synapse? A first step is to determine the equilibrium potential in the manner illustrated in Figure 4.6G (see also Figure 4.11E). The IPSP observed at the resting membrane potential (-74 mV) was increased by depolarizing the membrane (-64 and -56 mV) and was inverted by hyperpolarizing the membrane to -82 mV and still more at -96 mV. In this way, one can determine the point at which there is zero ionic flux during the IPSP. This is the equilibrium potential E_{IPSP}, which is about -80 mV (Figure 4.8C and D). Under these conditions, when the ionic channels are open, the ionic fluxes carry a zero net charge.

Experimental investigations on the ionic mechanisms of inhibitory synaptic action involve altering the concentration gradient across the postsynaptic membrane for one or another species of ion normally present and, in addition, employing a wide variety of other ions to test the ionic permeability of the subsynaptic membrane. Investigations of the inhibitory synapses on invertebrate nerve and muscle cells are usually performed on isolated preparations.

Changes in relative ionic concentration across the postsynaptic membrane are readily effected by altering the ionic composition of the external medium, as for the EPPs by A. Takeuchi and N. Takeuchi (1960). This method is not suitable for mammalian neurons *in vivo* because they have to be observed with the blood circulating, and the cardiac contraction would be greatly disturbed or suppressed by the necessary large changes in the ionic composition of the blood. Instead, the procedure of electrophoretic injection of ions with the impaling microelectrode has been employed to alter the ionic composition within the postsynaptic cell. For example, the species of anions that can pass through the inhibitory membrane have been identified by injecting one or another species into a nerve cell and seeing whether the increase in intracellular concentration effects a change in the IPSP. These injections are accomplished by filling microelectrodes with salts containing the anions under investigation. When the microelectrode is inserted into a nerve cell, a given amount of the anion can be injected electrophoretically into the cell by passing an appropriate current through the microelectrode for a chosen time, usually 60 sec.

The IPSP in Figure 4.9A was changed to a depolarizing potential (Figure 4.9B) by the addition of about 5 picoeq of chloride ions to the cell, which would more than triple the concentration. After more than twice this injection of sulfate ions into another cell, the IPSP was unchanged (Figure 4.9E and F). This simple test establishes that under the action of the inhibitory transmitter, the subsynaptic membrane becomes momentarily permeable to chloride ions, but not to sulfate. In Figure 4.9I and J, it can be seen that with

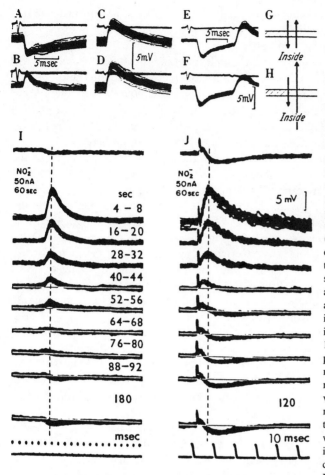

Figure 4.9. Ionic injection into a motoneuron. IPSPs (A) and EPSPs (C) were generated in a biceps–semitendinosus motoneuron by afferent volleys as in Figure 4.6D and E, respectively. After these were recorded, a hyperpolarizing current of 20 nA was passed through the microelectrode, which had been filled with 3 M KCl. Note that the injection of chloride ions converted the IPSP from a hyperpolarizing (A) to a depolarizing response (B), while the EPSP (C) was not appreciably changed (D). (E, F) Passing a much stronger hyperpolarizing current (40 nA for 90 sec) through a microelectrode filled with 0.6 M K_2SO_4 caused no significant changes (E to F) in either the IPSP or the later EPSP. (G, H) These represent the assumed fluxes of Cl^- ions across the membrane before (G) and after (H) the injection of chloride ions, which is shown greatly increasing the efflux of chloride (Eccles, 1964b). (I, J) Effects of electrophoretic injection of NO_2^- ions into motoneurons. (I) An IPSP in a motoneuron is evoked by quadriceps Ia volley. (J) A Renshaw IPSP in a motoneuron [cf. Figure 4.15C (Section 4.7.2)], the innervation of which was not identified, induced by a maximal L7 ventral root stimulation. Records in the top row show control IPSPs evoked before the injection. Records from the second to the ninth rows illustrate IPSPs at the indicated time [identical in (I) and (J)] after the injection of NO_2^- ions by the passage of a current of 50 nA for 60 sec. The bottom records are IPSPs at the end of recovery. Note the different time scales of (I) and (J). All records were formed by the superimposition of about 20 faint traces. From Araki *et al.* (1961).

two types of inhibitory synaptic action on motoneurons in the mammalian spinal cord, the inhibitory membrane was permeable to nitrite ions and recovery from the effect of the ionic injection was complete in 2–3 min.

It is essential to recognize that Figure 4.9I and J exemplifies two quite distinct processes of ion exchange: (1) The *specialized subsynaptic areas,* under the influence of the inhibitory transmitter, momentarily develop a specific ionic permeability of a high order, thus giving the greatly increased conductivity of the resistance shown at right in Figure 4.6F. This increased ionic conductance for 1–2 msec is responsible for the ionic fluxes that carry the inhibitory subsynaptic currents shown diagrammatically in Figure 4.8B. (2) The ionic permeability of the *whole* postsynaptic membrane (at left in Figure 4.6F) controls the intracellular ionic composition of the neuron and is responsible for its

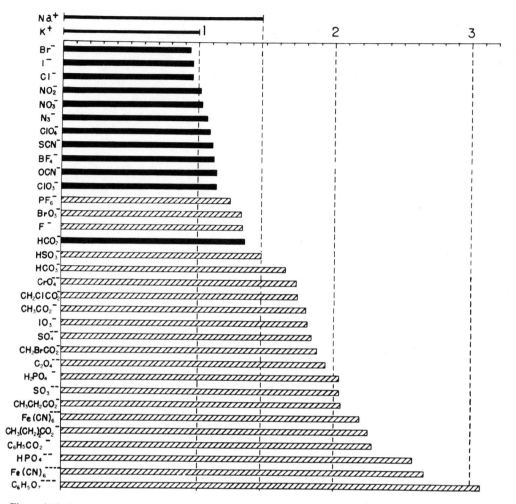

Figure 4.10. Correlation between ion sizes in aqueous solution and the effects of their injection on the IPSP. The lengths of the bands indicate ion size in aqueous solutions as calculated from the limiting conductance in water. (■) Anions effective in converting the IPSP into the depolarizing direction (as in Figure 4.9I and J): (▨) anions not effective (as in Figure 4.9E and F). The sizes of the hydrated K^+ and Na^+ ions are shown above the length scale, the former being taken as the unit for representing the sizes of other ions. From M. Ito *et al.* (1962).

restoration to normal after it has been disturbed by the ionic injection. This loss by outward diffusion of the injected ion occupies rather more than 2 min in Figure 4.9I and J.

Experimental tests such as those illustrated in Figure 4.9 have disclosed that 12 anions can pass through the gates opened by the inhibitory neurotransmitter. In Figure 4.10, these permeable anions can be seen to have small diameters in the hydrated state, whereas the impermeable anions are larger. The formate ion is the only exception to this generalization; otherwise, the activated inhibitory membrane is permeable to all anions that, in the hydrated state, have a diameter not more than 1.14 times that of the K^+ ion, i.e., not more than 0.29 nm, which is the size of the hydrated ClO_3^- ion. The ion diameters in Figure 4.10 are derived from limiting ion conductances by Stokes's law, on the assumption that the hydrated ions are spherical. Possibly the hydrated formate ion may have an ellipsoid shape and hence be able to negotiate membrane pores that block smaller spherical ions. Similar series of permeable and impermeable ions, even to the anomalous formate permeability, have been observed in comparable investigations on inhibitory synapses in fish, toads, and snails. It will certainly be remarkable if the ionic mechanism of central inhibition proves to be almost the same throughout the whole animal kingdom (cf. Eccles, 1964b). Eccles *et al.* (1977) found that the very large IPSPs of hippocampal pyramidal cells [see Figure 4.16 (Section 4.7.3)] display a discrimination between anions comparable to that of Figure 4.10.

The patch-clamp technique [cf. Figure 3.10 (Section 3.2.8)] has now been applied to the ion channels of inhibitory synapses on neurons of the mammalian CNS. The surfaces of the nerve cells become accessible to the micropipette application in cultures of either the spinal cord (Sakmann *et al.*, 1983; Hamill *et al.*, 1983) or the hippocampus (Borman *et al.*, 1983).

The two inhibitory neurotransmitters open ionic channels with different properties, as shown in Figure 4.11. Since the membrane potential was depolarized to 0 mV, far from the chloride equilibrium potential, the ionic currents induced by glycine (A, B) and GABA (C, D) were outward. They displayed the same variability of duration as did the acetylcholine (ACh) currents in Figure 3.10 (Section 3.2.8). Characteristically, the currents generated by glycine application were much larger than the GABA-generated currents, the amplitudes being 1.4 and 0.8 pA, respectively.

When the membrane potential was varied, the ionic currents generated by application of 5 μM GABA were linearly related to the membrane potential with zero at the membrane potential of -60 mV (Figure 4.11E), the equilibrium potential for chloride ions. The patch-clamp technique lends itself to a searching investigation of the anions that can traverse the channels opened by the inhibitory transmitter, a matter which was originally studied *in vivo*, as illustrated in Figures 4.9 and 4.10. Hamill *et al.* (1983) found that the inhibitory channels opened by glycine and GABA are identical. The channels are also permeable to other halide ions, the ratios for the series Br^-, I^-, Cl^-, and F^- being 1.8:1.5:1.0:0.4 for glycine receptor channels and 1.7:1.4:1.0:0.4 for GABA receptor channels. The low F^- permeability was not observed *in vivo* (see Figure 4.10), but this failure may be attributable to the enzyme poisoning produced by intracellular injection of F^- ions (Araki *et al.*, 1961). It was confirmed that the inhibition channels are not permeable to acetate or sulfate ions (cf. Figure 4.10).

Because of the similarity of the ionic permeabilities, it is suggested by Hamill *et al.* (1983) that the two inhibitory neurotransmitters may actually operate the same ionic channels from their respective receptor sites. The distinction of the receptor sites is of course demonstrated by the sharp differentiation in blocking agents, strychnine for glycine receptors and bicuculline and picrotoxin for GABA receptors.

Figure 4.12 gives idealized models of inhibitory synaptic action. Figure 4.12A and B shows two ionic channels through the bimolecular leaflet of the surface membrane as illustrated in Figure 1.14 (Section 1.2.3). One gate (Figure 4.12A) is closed and one (Figure 4.12B) is opened by the steric actions of the transmitter molecule that has become attached to the receptor site. These transmitters are packaged in the synaptic vesicles in inhibitory synaptic knobs [cf. Figure 1.5B (Section 1.2.2)] and liberated into the synaptic

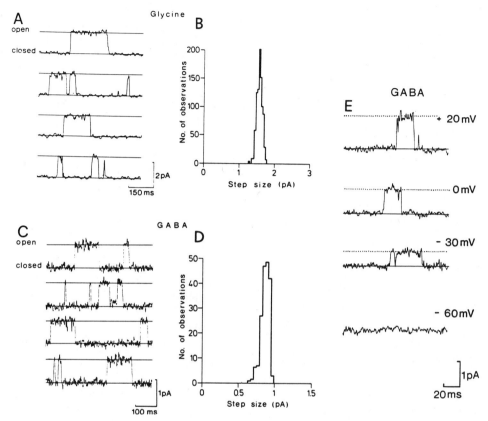

Figure 4.11. (A–D) Single channel outward currents induced by 10 μM glycine (A) and GABA (C) in mouse spinal neurons at 0 mV membrane potential. The temperature was 23°C. The step-size distributions for glycine- and GABA-activated currents yield mean amplitudes of 1.4 pA (B) and 0.8 pA (D), respectively. (A–D) From Sakmann *et al.* (1983). (E) Voltage dependence of single channel currents activated by 5 μM GABA in a rat hippocampal neuron. The temperature was 23°C. Downward deflection indicates inward current; upper deflection, outward current. From Sakmann (personal communication, 1984).

cleft in a quantal manner. By diffusion, a transmitter molecule would find a steric site as in Figure 4.12B and so be able to effect a conformational change [cf. Figure 1.14 (Section 1.2.3)], which effectively opens the channel for about 1 msec. According to the original theory (Coombs *et al.*, 1955b), potassium and chloride ions move by diffusion through the ion channel so opened, producing the subsynaptic current that generates the IPSP (Figure 4.12C and D).

Lux *et al.* (1970) and Lux (1971) criticized this model on the basis of evidence that there is an outward chloride pump across the neuronal membrane. The principal evidence was that, during intravenous infusion of ammonium acetate, the inhibitory synapses were affected, the equilibrium potential of the IPSP approaching the membrane potential. At the same time, there was a slowing in the rate of decline of internal chloride after a test injection. They postulated that there is normally an outward chloride pump across the membrane; that the equilibrium potential for chloride is, as a consequence, about −80 mV

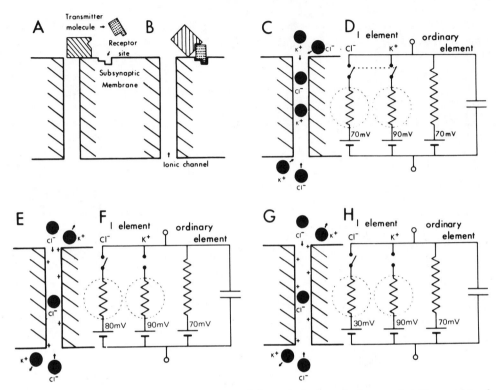

Figure 4.12. Diagrams summarizing the hypotheses relating to the ionic mechanisms employed by a variety of inhibitory synapses in producing postsynaptic membrane potentials. See the text for a full description.

(as is indicated in Figure 4.8); and that this pump is reversibly inactivated by ammonium ions. Lux was thus able to explain the hyperpolarization of the IPSP as due solely to the opening of Cl^- gates, K^+ ions playing no significant role.

The ionic mechanism for generation of motoneuronal IPSPs according to the Lux hypothesis is illustrated in Figure 4.12E and F, in which the battery on the Cl^- channel in Figure 4.12F is shown at -80 mV in accord with the value postulated for E_{Cl} as a consequence of the outward Cl^- pump. The ion channel is shown as being impermeable to K^+. Not shown in Figure 4.12 are many invertebrate inhibitory synapses in which the opening of the ion gates results in no change in the normal resting membrane potential. In the absence of a Cl^--pumping mechanism, there is Cl^- equilibrium, but the inhibition is still effected because the depolarization of excitatory synapses is counteracted by the inhibitory ionic mechanism that acts to hold the membrane potential at the resting level. It has been suggested that, if there are fixed positive charges on the channel, the anions will go through and the cations will be repelled, as shown in Figure 4.12E and G.

Further investigation of the Lux hypothesis has disclosed unexpected difficulties. The large IPSPs of hippocampal neurons [see Figure 4.16 (Section 4.7.3)] were found to be unaffected by intravenous ammonium acetate at much higher levels than were effective with the IPSPs of the spinal cord, trochlear nucleus, or cerebral cortex (G. I. Allen *et al.*,

1977a). Moreover, there was no evidence for an effective outward Cl^- pump. Alger and Nicoll (1983) have repeated these experiments with hippocampal slices and confirmed that the IPSPs set up by synaptic stimulation or electrophoretic GABA were not depressed by 2 mM ammonium chloride, which was fully effective in the Lux *in vivo* experiments. With higher levels of ammonium (4–8 mM), IPSPs were depressed, but there were complications from presynaptic blockage and depression of EPSPs. Nevertheless, it is concluded that Cl^- ions are the major contributor to hippocampal IPSPs, but the outward Cl^- pump postulated by Lux and associates is relatively insensitive to ammonium. The question arises whether the outward Cl^- pump may be different at higher levels of the CNS from that in the motoneurons of the spinal cord and brainstem and in the crayfish stretch receptor. Strangely, there is only one report of an apparent effect of ammonium on IPSPs of the cerebral cortex (Raabe and Gumnit, 1975). In view of the hippocampal evidence for a much less sensitive Cl^- ion outward pump and other complications produced by high ammonium, the action of ammonium on IPSPs of the neocortex needs reinvestigation by more convincing techniques.

Aickin *et al.* (1982) have employed intracellular Cl^- electrodes on the stretch receptor cell of the crayfish, in which ammonium acetate also reduces the IPSP by its depressant action on the outward Cl^- pump (Meyer and Lux, 1974). This study of the outward Cl^- pump disclosed that the outward pumping is for $K^+ + Cl^-$ and that, if internal K^+ is low, the pump is slowed; hence an explanation is provided for the observation that an excess of internal Cl^- ions declines more slowly when K^+ is depleted (Eccles, 1964b; G. I. Allen *et al.*, 1977a). It is concluded that the ion channels for the IPSP are exclusively anionic, as in Figure 4.12E and F.

However, the K^+ ion hypothesis for the IPSP was revived by the admirable studies by Kehoe (1972) on the pleural ganglion cells of *Aplysia*. The IPSPs, generated either by synaptic action or by application of the transmitter, ACh, display a double time course. The fast component is due to an increased Cl^- conductance and the slow to an increased K^+ conductance, the respective equilibrium potentials being -60 and -80 mV, rather as modeled in Figure 4.12C and D. In this elegant investigation, there was complete separation between these two actions of the same transmitter, and each ionic conductance was in good accord with predictions for the ionic composition by the Nernst equation [see Figure 1.15A (Section 1.2.3)]. In a related investigation, Gerschenfeld and Paupardin-Tritsch (1974) have shown that, on other mollusk neurons in *Helix* and *Aplysia*, the inhibitory transmitter is serotonin (cf. Chapter 10), and it likewise generates fast and slow IPSPs due to increases in conductance for, respectively, Cl^- and K^+.

4.5. General Features of Transmission by Postsynaptic Ionotropic Action in the Brain

Let us now return to the general diagram of chemical transmitting synapses in Figure 3.14 (Section 3.4). The wonderful biological process of evolution was not so uncompromisingly innovative. When good solutions were developed, they were retained by natural selection. They had survival value, and certainly chemical synaptic transmission had good survival value. It was invented long ago in relatively primitive invertebrates. And even the transmitter substances at mammalian synapses mostly have an invertebrate lineage. The detailed features of Figure 3.14A apply to both excitatory and inhibitory synapses of the CNS [cf. Figures 1.4B, 1.5B, and 1.6A (Section 1.2.2)]. There are the vesicles in the presynaptic terminals, a synaptic cleft of similar dimension, and a postsynaptic membrane with sensitivity to the transmitter substance that opens ion gates, as indicated in the enlargement in Figure 3.14B for excitatory synapses.

Essentially the same events occur at the excitatory synapses in the CNS as occur at the neuromuscular junction. The all-or-nothing impulse propagates to the terminal with the sodium Na_D and potassium K_D gates opening at a critical level of depolarization [see Figure 2.5 (Section 2.3)]. At the terminal there is, in addition, opening of the Ca_D gate so that Ca^{2+} ions enter and participate in the process whereby the vesicles discharge their quantal content of transmitter into the synaptic cleft. This transmitter then diffuses to the subsynaptic membrane as in Figure 3.11A (Section 3.2.9) and lodges momentarily on steric receptor sites [cf. Figure 1.15 (Section 1.2.3)] that allow it to open the sodium (Na_T) and potassium (K_T) channels shown in Figure 3.14B.

The ionic fluxes through the Na_T and K_T channels depolarize the postsynaptic membrane enough to open the sodium and potassium gates concerned in impulse transmission, i.e., the Na_D and K_D gates shown open in Figure 3.14B. The Na_D gates for impulse transmission in the brain, as in the periphery, are blocked by tetrodotoxin, and it is presumed that the potassium gates (K_D) are similarly blocked by intracellular tetraethylammonium. All evidence indicates that impulse transmission in the CNS is essentially the same as in peripheral nerves. Likewise, it is assumed that the central depressant actions of manganese and magnesium are due to their effects in preventing Ca^{2+} ions from moving in through the Ca_D channels.

With central inhibitory synapses, the only differences are that the transmitters are the amino acids glycine or GABA (see Chapter 7) and that these transmitters open up channels across the subsynaptic membrane [see Figure 4.11 (Section 4.4)] for Cl^- or for Cl^- and K^+, and strictly not for Na^+. Nevertheless, the essence of the story is the same. The presynaptic impulse opens the calcium gates at a certain level of depolarization, and the inward movement of Ca^{2+} ions liberates the transmitter quantally. The transmitter diffuses across the synaptic cleft and opens ion gates, with the consequent ionic fluxes that increase the charge across the subsynaptic membrane, causing the flow of current [see Figure 4.8B (Section 4.4)] that hyperpolarizes the postsynaptic membrane.

4.6. Inhibition by Reciprocal Synapses

The reader is referred to Shepherd (1974) for general references.

In the schematic diagrams of Figure 4.17 (Section 4.7.4), it is assumed that synaptic excitation of the inhibitory neurons generates impulse discharges [cf. Figure 4.15B (Section 4.7.2)] and that the inhibitory action is consequent on the passage of these impulses to the inhibitory synapses and the liberation of neurotransmitter therefrom, in the manner that is described in detail in Chapter 3. It is shown there, however, that a presynaptic impulse is not essential for the liberation of neurotransmitter from a presynaptic terminal. It is sufficient merely to produce an equivalent depolarization of the terminal by an applied current so that the Ca_D gates are opened and the quanta of neurotransmitter are

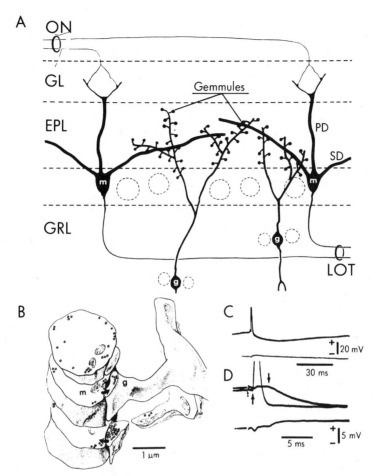

Figure 4.13. (A) Layers and connections in the olfactory bulb. (GL) Glomerular layer; (EPL) external plexiform layer; (GRL) granular layer; (ON) olfactory nerve; (LOT) lateral olfactory tract; (g) granule cell; (m) mitral cell; (PD) primary mitral dendrite; (SD) secondary mitral dendrite. Adapted from Ramón y Cajal (1911). (B) Graphic reconstruction (exploded) of a granule synaptic ending (g) on a mitral secondary dendrite (m). The granule ending is shaped like a gemmule and arises from a granule dendrite lying approximately perpendicular to the mitral dendrite. Within a single ending are two synaptic contacts with opposite polarities (indicated by arrows). From Rall *et al.* (1966). (C, D) Antidromic responses of mitral cells on stimulation of the LOT. The traces are at low and high amplification and slow and fast speeds, as indicated. The lower traces give control extracellular potentials. In (D), the stimulus was just at threshold for the axon of the mitral cell that was excited on one trace and not on the other. From Nicoll (1969).

released. It has now been shown that, in certain inhibitory neurons, the liberation of neurotransmitter is similarly produced by an electrotonically spreading depolarization, not by the depolarization of an impulse.

Rall *et al.* (1966) and Nicoll (1969) have provided convincing evidence, both physiological and histological, that, in the olfactory bulb (Figure 4.13A), the granule cells inhibit mitral cells by a unique synaptic arrangement. By electron microscopy, it has been shown that the secondary mitral cell dendrites have synapses on the gemmules that bud off from the branches of the granule cells in the external plexiform layer; these same gemmules in turn give synapses to mitral cell dendrites, as illustrated in the graphic reconstruction (Figure 4.13B). When the mitral cell dendrite contains vesicles, it is presynaptic to the gemmule; the consequent synaptic transmission is in the direction of the upper arrow. The lower synapse is in the reciprocal direction. Analysis of field potentials recorded in depth through the laminated structure of the olfactory bulb [cf. the technique illustrated in Figure 4.16D and E (Section 4.7.3)] reveals that the former synapses are excitatory; i.e., the mitral cell dendrites directly excite the gemmules of the granule cells. Analysis of field potentials and also intracellular recording from the mitral cells (Figure 4.13C and D) (Nicoll, 1969) show that, after a delay of about 3 msec, mitral cell excitation results in an IPSP of mitral cells, even of those not initially excited. This inhibition apparently occurs in the absence of impulse discharge in granule cells.

Thus, the proposed inhibitory pathway from mitral cell to mitral cell is by a very short linkage through the inhibitory granule cells, even being within a single gemmule. It is pointed out that this reciprocal synaptic action may occur in the absence of impulses. First, impulses in the mitral cell somata depolarize the secondary dendrites by an electrotonic spread of depolarization, there being a consequent liberation of excitatory transmitter with depolarization of the gemmules (upper arrow in Figure 4.13B). Second, this depolarization of the gemmule-studded branches of the granule cell causes the liberation of inhibitory neurotransmitter at their inhibitory synapses (lower arrow in Figure 4.13B), with consequent hyperpolarization of the secondary mitral dendrites. Moreover, it is assumed that the electrotonic spread of depolarization in the granule cell dendrite is adequate for the depolarization of one gemmule to spread to others that in turn inhibit secondary dendrites from many mitral cells, as indicated in Figure 4.13A. The physiological significance of this reciprocal synaptic mechanism can be understood when it is recognized that a dense meshwork is made by the gemmule-studded branches of the granule cells and by the secondary dendrites of mitral cells. It is thus possible for mitral cells to inhibit each other very effectively via the reciprocal synapses. GABA is the inhibitory neurotransmitter of the granule cells (Ribak et al., 1977a).

This remarkable development illustrates an additional manner in which inhibitory interneurons can be effective in producing postsynaptic inhibition. Furthermore, similar reciprocal synapses are observed with horizontal and amacrine cells in the retina (Dowling and Boycott, 1966), which likewise are inhibitory.

These fundamental discoveries concerning the synaptic mechanisms of the olfactory bulb and the retina have opened the way to an understanding of many perplexing observations on the synaptic structures in relay nuclei on afferent pathways to the cerebral cortex. In these nuclei, there are short-axoned interneurons (Golgi type II neurons) that, in general, resemble the granule cells of Figure 4.13 in that they act as a local inhibitory mechanism, their dendrites forming inhibitory synapses on the dendrites of the relay neurons, as indicated in Figure 4.14 for the lateral geniculate nucleus (Hámori et al., 1974). There are three differences in detail: (1) the Golgi cells are synaptically excited in the usual manner by impulses in the retinal afferent fibers; (2) there is often a triadic synaptic arrangement as shown in the electron micrograph of Figure 4.14 and encircled in the diagram below; (3) the axon of the Golgi cell gives inhibitory synapses to the dendrites of the relay cell. The principal mode of operation is that an afferent impulse in the retinal

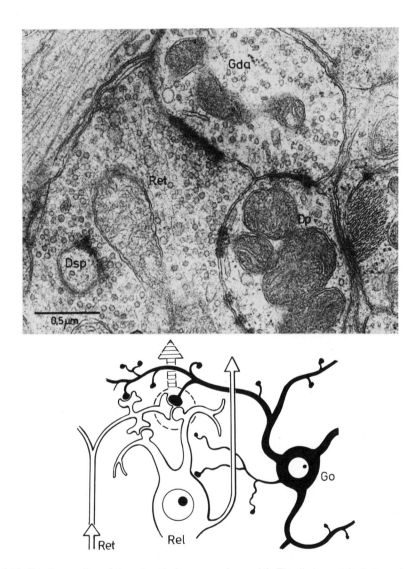

Figure 4.14. Triadic coupling of the subcortical sensory relay nuclei. The diagram at the bottom shows the neuronal arrangement in the lateral geniculate body, in which a retinal (Ret) afferent terminal contacts both a dendritic protrusion of a geniculate cortical relay (Rel) neuron and the large dendritic appendage of a Golgi (Go) type II neuron. The electron micrograph at the top shows the ultrastructural arrangement of the three neuronal constituents. The retinal (Ret) afferent terminal is presynaptic (by excitatory contacts with round vesicles and asymmetrical membrane attachments) to dendritic protrusions (Dp) of the relay cells and to the dendritic appendages (Gda) of Golgi type II neurons. The latter are presynaptic (by inhibitory contacts with flattened vesicles and symmetrical membranes) to the same relay cell dendritic protrusion. (Dsp) An invaginated dendritic spine. From Hámori *et al.* (1974).

afferent fiber excites both the relay cell dendrite and the protrusion from the Golgi cell dendrite, and this latter in turn inhibits the relay cell dendrite. The purpose of this arrangement is presumably to give a very effective depolarization of the Golgi cell protrusion with a consequent very effective inhibitory action by it. But the Golgi cell dendrite is excited at other sites besides the triadic. It can be assumed that, as with the mitral cells of the olfactory bulb, electrotonic transmission along the Golgi cell dendrite is adequate for pooling the depolarizations in order to inhibit many relay cells. And there is also a third inhibitory mechanism, as in the conventional manner, by the axodendritic synapses shown from the Golgi cell to the soma and dendrites of the relay cell.

Besides the lateral geniculate nucleus of the monkey (Hámori *et al.*, 1974), essentially similar inhibitory mechanisms have been identified in the cat in the medial geniculate nucleus (Morest, 1971), in both the medial and lateral geniculate (Jones and Powell, 1969), in the ventrobasal thalamic nucleus (Ralston and Herman, 1969), and in the superior colliculus (Lund, 1969). In addition, Sloper (1971) has reported that dendrodendritic synapses occur, though rarely, in the deeper laminae of the primate motor cortex. Unfortunately, there have been no complementary electrophysiological studies that give a quantitative evaluation of the inhibitory effectiveness of dendrodendritic synapses relative to the standard synapses made by axons of the inhibitory neurons (cf. Figure 4.14). In the olfactory bulb there is no such complication, and the primary discovery was electrophysiological. It has been authoritatively reviewed by Shepherd (1974), who also gives a comprehensive review of the retinal and thalamic dendrodendritic synapses.

4.7. Simple Neuronal Pathways in the Brain

We now come to the neuronal pathways in the brain. There are many questions to be answered: How are the properties of individual synaptic actions effective in the linking of neuronal assemblages into some meaningful performance? What principles of neuronal organization can we define? How is information from receptor organs transmitted up to higher centers of the brain? Chapter 14 treats the complementary problem of how the higher centers of the brain act on lower centers eventually to bring about motoneuronal discharge and thus movement.

4.7.1. Pathways for Ia Impulses

Figure 4.1A (Section 4.1) represents a very simple pathway, in which an Ia afferent fiber from a muscle enters the spinal cord and branches so that the discharges of this annulospinal receptor organ can exert both an excitatory and an inhibitory action. The inhibitory pathway has to be relayed through an interneuron because the afferent neuron itself can act only in an excitatory manner, presumably because it can make only the excitatory neurotransmitter substance. Thus, the transformation comes via an interneuron that has the biochemical competence to make a different neurotransmitter which has an inhibitory synaptic action. One can assume that this inhibitory interneuron has a genome specified for this purpose and that this dates back to its origin in a differentiating mitosis, as is discussed in Chapter 13. So far as we know, all afferent fibers entering the spinal cord are excitatory in their action. For central inhibitory action to occur, there must be the transformation via interneurons specialized for inhibition, as shown for the Ia input in Figure 4.1A. This requirement explains many of the design features in the brain. Figure 4.1A is an example of the simplest pathway producing excitation and reciprocal inhibition via a feed-forward inhibitory pathway.

Since the discovery of the Ia inhibitory circuit, there has been an immense amount of scientific investigation, particularly by the Göteborg school of Lundberg (A. Lundberg, 1979a; Jankowska, 1979; Hultborn, 1976). The simple diagram of Figure 4.1A has become complicated by the discovery that, converging on the Ia inhibitory interneurons, there are pathways from a wide range of afferents at the segmental level and also from higher centers. Figure 4.15 (Section 4.7.2) shows an important inhibitory line from a Renshaw cell onto the Ia inhibitory interneuron. The demonstrated connectivities are far beyond the level of our understanding of the control of movement by the brain as described in Chapter 14.

4.7.2. Renshaw Cell Pathway

In Figure 4.15, there is a simple example of a feedback inhibitory pathway. A motoneuron is shown with its axon going all the way out to the muscle, as illustrated in Figure 3.1A (Section 3.2.1). It acts there by the transmitter ACh. The dashed lines of the

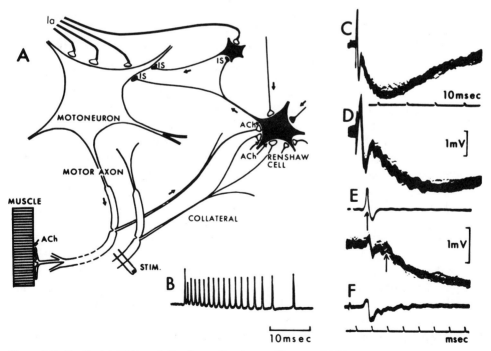

Figure 4.15. Feedback inhibition via Renshaw cell pathways. Shown are inhibitory synapses (IS) and excitatory synapses (ACh) operating with ACh as neurotransmitter. See the text for a full description. From Eccles *et al.* (1954).

axon signify that it extends many centimeters after it leaves the spinal cord before it reaches the muscle. But, in the spinal cord, this motoneuronal axon sends off collateral branches that make synapses on special neurons called "Renshaw cells," after Birdsey Renshaw, a distinguished American neurobiologist who died from polio when he was very young. He discovered these cells and recorded their unique responses, but unfortunately did not have time to work out their mode of operation. This was done much later (Eccles *et al.*, 1954). Both the identification and the location of these neurons have been established beyond all doubt by the beautiful technique of injecting the dye procion yellow into neurons recognized as Renshaw cells by their characteristic intracellular responses (Jankowska and Lindstrom, 1971). There is complete agreement between these anatomical results and the original findings with electrophysiological techniques.

When, by electrical stimulation, impulses are set up in the axons of motoneurons, the Renshaw cells fire with a long burst, often at extremely high frequency, as in Figure 4.15B. All kinds of pharmacological tests have established that the excitatory transmitter is ACh (cf. Chapter 8), which is the same as that at the neuromuscular synapses that are made by these same motor axons. This is in precise accord with Dale's principle, which is defined by stating that a neuron uses the same transmitter mechanism at all of its synapses. For example, all synapses made by a motoneuron have ACh as the transmitter despite great differences in the location (central or peripheral), in the mode of termination, and in the target cells. The Renshaw cell sends its axon to form inhibitory synapses on motoneurons. Figure 4.15C and D shows the large IPSP (at two different sweep speeds) produced in the motoneuron by the rapid Renshaw cell discharge (Figure 4.15B). The latent period for the IPSP is shown by comparing the extracellular record of the motoneuron (Figure 4.15F) with the intracellular record (lower trace in Figure 4.15E). The IPSP begins at the arrow, which is only 1.1 msec after the volley in the motor axons indicated by the arrow in the upper trace in Figure 4.15E (note the faster time scale for Figure 4.15E and F). This latency shows that there is just time for the pathway via Renshaw cells to the motoneuron that is drawn in Figure 4.15A.

Furthermore, the Renshaw cell is also seen in Figure 4.15A to give inhibitory synapses to an inhibitory interneuron. It so happens that, in many cases, this is the same inhibitory interneuron that is on the Ia inhibitory pathway to motoneurons (cf. Figure 4.1A). The action of the Renshaw cell on this inhibitory interneuron is to effect an inhibition of its inhibitory action, i.e., to diminish its inhibitory action, which is the equivalent of excitation. Removal of inhibition thus causes excitation by a process that is called "disinhibition" (V. J. Wilson and Burgess, 1962). Hultborn (1976) and the Göteborg group have carried out a precise analysis of the manner in which inhibition by the Renshaw cell pathway is utilized in motor control (Hultborn and Pierrot-Deseilligny, 1979). Each motoneuron activates many Renshaw cells via its axon collaterals, and each of these Renshaw cells inhibits many motoneurons in a selective manner and also the Ia inhibitory interneurons. It is suggested that the depression of Renshaw cells during voluntary activity could play an important role by favoring reciprocal Ia inhibition.

4.7.3. Hippocampal Basket Cell Pathway

Figure 4.16 is another example of a feedback pathway, this one to a pyramidal cell in the hippocampus, which is a primitive part of the cerebral cortex. The neuron labeled "p"

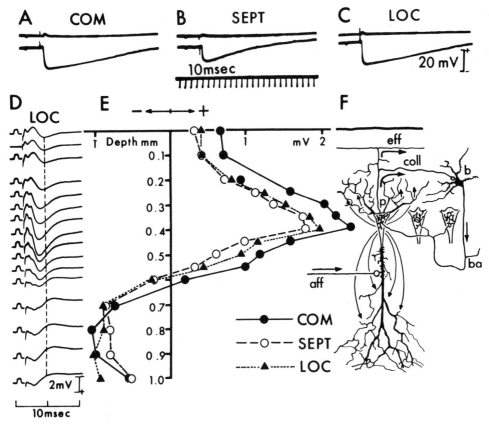

Figure 4.16. Feedback inhibition via basket cells in the hippocampus. See the text for a full description. From Andersen *et al.* (1964).

in Figure 4.16F is the same kind of neuron as that in Figure 1.4B (Section 1.2.2), but it is shown inversely, with its apical dendrite pointing downward and its axon upward. From the axon, there is a collateral just like that of the motoneuron in Figure 4.15A, and it makes an excitatory synapse on a neuron (b) that we can liken to a Renshaw cell because its axon (ba) branches profusely to give inhibitory synapses on the somata of this and adjacent pyramidal cells. This is another example of feedback inhibition. The more this pyramidal cell fires, the more it activates the feedback inhibition.

These inhibitory neurons (b) were called "basket cells" by Ramón y Cajal (1911) because of the basket-like embracement of the somata of the pyramidal cells. He thought they were excitatory, but following investigations by Kandel *et al.* (1961), it was established that they are inhibitory by a series of tests that are partly illustrated in Figure 4.16 (Andersen *et al.*, 1964).

The very large IPSPs produced in a hippocampal pyramidal cell by the basket cell synapses can be seen in the lower traces of Figure 4.16A–C, the upper traces being the controls just outside the cell. Recordings COM, SEPT, and LOC signify three different locations of the stimulation for exciting the basket cells. Extracellular recording at the indicated depths gives the field potential profile in Figure 4.16D and E. The plottings in Figure 4.16E show the field potentials recorded (Figure 4.16D) at the depths corresponding to the drawing of the pyramidal cell in Figure 4.16F and measured at the time of the vertical dashed line in Figure 4.16D. The maximum positivities occur at the depth of the somata in Figure 4.16F, which indicates that the inhibitory synapses are located there. Reference to Figure 4.8B (Section 4.4) reveals that the inhibitory site is the source of current flow to the rest of the cell, as shown by the arrows in Figure 4.16F.

In this way, it was established that the basket cells of Ramón y Cajal were inhibitory, and hence the location of inhibitory synapses was defined for the first time (Figure 4.16F). Their clustered position on the soma of the hippocampal pyramidal cell is strategically an excellent place because their function is to stop the cell firing. This location on the soma provided the opportunity to identify inhibitory synapses in electron micrographs and to define their morphology, as shown in Figure 1.4B(e) (Section 1.2.2).

It has also been shown by Andersen et al. (1966) that the numerous spine synapses on the dendrites [types (a)–(d) in Figure 1.4B] are excitatory. These synapses are very powerful and can generate the discharge of impulses that propagate down the apical dendrite, over the soma, and down the axon. The inhibitory synapses are also very powerful, however, as can be seen in the large IPSPs of Figure 4.16A–C. Their location on the soma at the axonal origin gives them the final say as to whether the impulse is to be allowed to propagate down the axon. It is a good strategic design to have the inhibitory synapses located at the optimal site. If they were located elsewhere, e.g., on the dendrites, impulses generated nearer the soma would be able to fire down the axon without effective inhibitory constraint. Inhibitory synapses of neurons are usually located on or close to the somata.

4.7.4. Operative Features of Inhibitory Pathways

Figure 4.17A is a general diagram of the feedback pathways the operations of which are illustrated in Figures 4.15 and 4.16. Figure 4.17B diagrams the feed-forward pathway that has been illustrated for the Ia afferent fibers from muscle receptors [see Figure 4.1A (Section 4.1)], in which collateral branches of the excitatory afferent fibers excite inhibitory interneurons that inhibit neurons in the forward direction. Both these inhibitory pathways are now known to have more than a general value in keeping down the level of excitation and thus suppressing discharges from all weakly excited neurons. In addition, they participate very effectively in neuronal integration, molding and modifying the patterns of neuronal responses. Inhibition can be thought of as a sculpturing process. The inhibition, as it were, chisels away at the diffuse and rather amorphous mass of excitatory action and gives a more specific form to the neuronal performance at every stage of synaptic relay. This suppressing action of inhibition can be recognized very clearly at higher levels of the brain—particularly in the cerebellum, as we shall see in Chapter 14.

The other point that we wish to emphasize is that inhibitory cells are uniquely specified neurons, as has already been stated. They act via unique neurotransmitter substances, which is almost always glycine in the spinal cord and almost always GABA at supraspinal levels (see Chapter 7). These two neurotransmitters are very similar in their action, opening the same ion channels [see Figure 4.11A–D (Section 4.4)]. Substances that interfere with inhibitory synaptic action would be expected to cause unfettered excit-

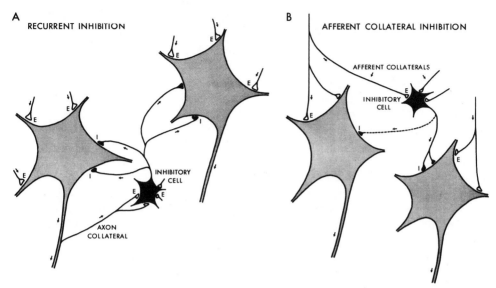

Figure 4.17. The two types of inhibitory pathways. (A) Feedback; (B) feed-forward.

atory action of neuron onto neuron and so lead to convulsions, and this is indeed the case. In fact, as shown by Curtis and his associates, two important classes of convulsants are: (1) drugs such as strychnine that selectively block glycine transmission, and (2) drugs such as bicuculline and picrotoxin that selectively block GABA transmission (Curtis and Johnston, 1974).

The unfolding of the central inhibitory story of the brain has been a great success, and it is now better understood than the central excitatory story. Numerous supraspinal inhibitory pathways are now known to work by GABA, and in many spinal pathways the transmitter is known to be glycine. As opposed to this impressive list of purely inhibitory pathways in the mammalian CNS (Eccles, 1969a), there is no known example of a nerve cell in the mammalian brain that is ambivalent, exercising excitatory action by some of its axonal branches and inhibitory action by others. Originally, it was envisaged that the inhibitory interneurons were interpolated merely to effect a transformation in the synaptic neurotransmitter substance, being just "a commutator-like device," to use Granit's phrase, and that they were very localized in their action, with short axons restricted to the gray matter. This was true of the examples then known in the spinal cord. But now very extensive investigations, particularly at higher levels of the nervous system and, par excellence, the cerebellum (Eccles *et al.*, 1967), have disclosed examples of complex integrating functions of inhibitory neurons (cf. Chapter 14). Some inhibitory neurons have axons that are measured in centimeters in the descending direction (V. J. Wilson *et al.*, 1970; M. Ito *et al.*, 1970) or in the ascending direction (Ekerot and Oscarsson, 1975) (cf. Chapter 7). A further remarkable discovery was that inhibitory neurons can inhibit inhibitory neurons, thereby effecting, by disinhibition, an apparent excitatory action. This was first established by V. J. Wilson and Burgess (1962) for the Renshaw cell system, in which activation of some Renshaw cells can cause a depolarization of motoneurons by inhibiting the background discharge of other inhibitory neurons [cf. Figure 4.15A (Section 4.7.2)]. Several other disinhibitory pathways have since been recognized, particularly in the cerebellum, where basket cells and stellate cells inhibit Purkinje cells that are also inhibitory (Eccles *et al.*, 1967).

In attempts to unravel the neuronal pathways in the brain, it is a great help to recognize that the first big step is to identify the excitatory and inhibitory neurons. This criterion of a sharp classification is of essential importance in trying to discover the mode of operation of cell assemblages in the nuclear and cortical structures of the brain.

4.8. Presynaptic Inhibition

The reader is referred to Nicoll and Alger (1979) for a general reference.

The other method of inhibitory control in the CNS was first recognized by Frank and Fuortes (1957), but several years passed before there was an understanding of the mechanism of presynaptic inhibition and an appreciation of its preeminent role in negative feedback control of the sensory pathways (Eccles, 1964c; R. F. Schmidt, 1971). The inventiveness of the evolutionary process is well illustrated by this quite different neuronal mechanism for depressing synaptic excitatory action.

The mode of operation of presynaptic inhibition is illustrated above the neuron in Figure 4.18. The presynaptic inhibitory fiber is shown making a synapse on the excitatory fiber, and many such synapses have been seen in electron micrographs (Gray, 1962; Conradi, 1969; T. Kojima *et al.*, 1975). Presynaptic inhibition is explained by the following sequence of events: By a chemical transmitter action, the excitatory synaptic knob is depolarized (Figure 4.18C); consequently, a spike potential in this knob is diminished and the output of excitatory neurotransmitter substance is depressed, as has been shown for peripheral synapses [cf. Figure 3.13 (Section 3.3)]. It is thus postulated that inhibitory action is exerted on excitatory presynaptic terminals and not at all on the postsynaptic membrane, hence it is called presynaptic inhibition. Presynaptic inhibition is exerted on the central synaptic terminals of all varieties of large afferent fibers that have been examined so far. At the first synaptic relay, presynaptic inhibition is usually of greater potency than postsynaptic inhibition. All large afferent fibers exert a presynaptic inhibitory action on the central terminals of afferent fibers, but to a considerable extent, the distribution is dependent on the respective modalities of the afferent fibers (Eccles, 1964c; R. F. Schmidt, 1971).

Presynaptic inhibition can be demonstrated by various experimental procedures. A direct test is to record from the excitatory fiber by an intracellular electrode and observe the depolarization directly, as in the records of Figure 4.18C, in which the upper traces are the observed intrafiber potentials produced by one, two, or four stimuli to the conditioning afferent nerve. Another method is to record intracellularly from a neuron that is being synaptically excited and show that the EPSP is reduced because the output of excitatory transmitter is depressed, as can be seen in Figure 4.18A and B. In Figure 4.18A, the EPSP depression results in the delay or prevention of the generation of the impulse discharge that occurred at the arrow in the control (Con). In Figure 4.18B, after a conditioning tetanus, the EPSP is seen to be greatly reduced relative to the control, but not altered in its time course. A very convenient way of evaluating presynaptic inhibition is provided by

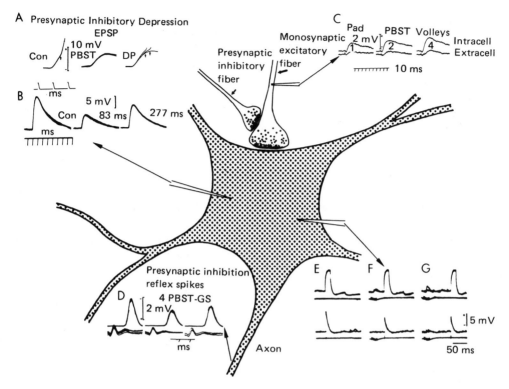

Figure 4.18. Specimen records of various types of responses associated with presynaptic inhibition. (C) An excitatory synapse is shown on the neuron, with a presynaptic inhibitory ending on it, and by intracellular recording (note the electrode), the excitatory fiber is shown to be depolarized—this being primary afferent depolarization (Pad). Note the difference between the intracellular and extracellular traces for one, two, and four volleys in the presynaptic inhibitory pathway. (A, B) Effects of presynaptic depolarization in reducing the excitatory synaptic action (EPSP), recorded intracellularly. (B) Note the great diminution of the EPSPs with respect to the control (Con) when tested at 83 and 277 msec after a brief tetanic stimulation of the presynaptic inhibitory pathway. (A) Presynaptic inhibition from two different muscle nerves depresses the EPSP so that it often fails to generate a spike, as it regularly does in the control (Con) at the arrow. (D) The resultant diminution of reflex spike discharge is recorded by the population spike (cf. Figure 4.1B) in the ventral root, the first being the control response (Eccles, 1964c). The lower trace in (E) gives the control monosynaptic EPSP and, in (F) and (G), its diminution by the presynaptic inhibitory action of a conditioning train of volleys (6 at 250/sec) beginning, respectively, 45 and 95 msec earlier. The upper traces are the potentials produced by an intracellular current pulse of 15-msec duration under conditions identical with the EPSPs of the corresponding lower traces. From Eide *et al.* (1968).

the diminution in the discharge of impulses evoked by a test excitatory input, as shown in Figure 4.18D, the first response being the control population spike. This test is similar to that for postsynaptic inhibition in Figure 4.1D (Section 4.1) and is not discriminative. All these tests have been applied systematically. Another way of demonstrating presynaptic inhibition (technically the simplest) is to test the excitability of the excitatory fibers subjected to the presynaptic inhibition, i.e., of the primary afferent fibers such as the excitatory presynaptic fiber in Figure 4.18. When depolarized by the presynaptic inhibitory action as in Figure 4.18C, these become more excitable. This method alone is unreliable, however, because the excitability may be enhanced by other depolarizing mechanisms, e.g., by an increase in extracellular potassium.

In the initial report on presynaptic inhibition, Frank and Fuortes (1957) stated that the postsynaptic neuron exhibited no change in its electrical properties such as membrane potential or excitability. A very important confirmation by Eide *et al.* (1968) is illustrated in Figure 4.18E–G. The lower traces show the control monosynaptic EPSP (Figure 4.18E) and its diminution (Figure 4.18F and G) by presynaptic inhibition at two test intervals. The upper traces show that there was no accompanying postsynaptic change in conductance when tested by the response produced by a brief pulse applied intracellularly. Furthermore, no change was detectable in the time course of the EPSP during presynaptic inhibition (cf. Figure 4.18B) (Eccles, 1964c; Eide *et al.*, 1968); hence, all precise experimental testing is in agreement with the hypothesis that presynaptic inhibition is indeed exclusively presynaptic (Eccles, 1964b,c; Eide *et al.*, 1968; R. F. Schmidt, 1971). There is no need to consider further the alternative suggestion that presynaptic inhibition is due to inhibitory synapses acting so remotely on the dendrites of the postsynaptic neuron that no effect can be detected with intracellular recording from the soma. Another source of confusion arises from attempts to account for the presynaptic inhibitory effect by the extracellular accumulation of potassium, which of course does occur with intense neuronal activity. Nicoll (1975), however, has shown that presynaptic inhibition has a time course quite different from that of the potassium accumulation, which in any case exerts a very much smaller depression.

There is now good evidence that the presynaptic inhibitory transmitter is GABA (Curtis, 1978; Curtis and Lodge, 1982; Nicoll and Alger, 1979). GABA resembles presynaptic inhibitory synapses in producing a depolarization of the presynaptic terminals of primary afferent fibers. Both actions (of GABA and of the presynaptic synapses) are selectively depressed by bicuculline and picrotoxin. GABA effects the depolarization of the presynaptic terminals or dorsal root ganglion cells largely by opening Cl^- gates (Nicoll, 1975; Nishi *et al.*, 1974; J. P. Gallagher *et al.*, 1978), as indicated in Figure 4.12G and H (Section 4.4). The equilibrium potential for this ionic mechanism is about -30 mV, which indicates that there is an inward Cl^- pump in the presynaptic fibers that causes the E_{Cl} to be about 40 mV in the depolarizing direction from the resting membrane potential. The intracellular Cl^- concentration must be as high as 50 mM. In the frog, this E_{Cl} value obtains even as far peripherally as the dorsal root ganglion cells. This is consistent with the fact that GABA opens Cl^- gates in its action in both postsynaptic and presynaptic inhibition. Finally, presynaptic terminals have been shown in some instances by immunohistochemistry to contain the GABA-synthesizing enzyme glutamic acid decarboxylase (see Section 7.2).

One generalization about presynaptic inhibition is that it has no patterned topography. It is widely dispersed over the afferents of a limb, with but little tendency to focal application. For example, presynaptic inhibitory action by the afferent fibers of a muscle is effective on the afferents from all muscles of that limb, regardless of function. It is not selective on one class of muscle or on the muscles acting at any joint of a limb. This widespread, nonspecific character is exactly what would be expected for the general suppressor influence of negative feedback. Nevertheless, there is organization or pattern in the distribution of presynaptic inhibition; this pattern depends on the class or modality of the afferent fiber on which the presynaptic inhibition falls, the cutaneous afferent fibers being the most strongly affected (cf. Eccles, 1964c; R. F. Schmidt, 1971).

Another generalization is that presynaptic inhibition is much more effective at the primary afferent level than at the higher levels of the brain. However, it has been shown to exercise an important inhibitory influence on pathways through the thalamus and lateral geniculate body, and it is utilized by descending pathways from the cerebrum for inhibitory action on synapses made by primary afferent fibers either in the spinal cord or in the dorsal column nuclei (R. F. Schmidt, 1971). So far, there has been no direct physiological evidence for presynaptic inhibition at the highest levels of the mammalian brain, i.e., the cerebellar cortex and the cerebral cortices—both the neocortex and the hippocampus. However, a question is raised by the widespread existence of $GABA_B$ binding sites at higher levels of the brain [see Table 5.3 (Section 5.3.4) and Section 7.4]. These are thought to be mainly presynaptic. At these sites, GABA may act to reduce the outflow of other neurotransmitters (Bowery et al., 1980).

4.9. Principles of Neuronal Operation

What can we say now as a result of this study of individual nerve cells in the brain and of some of the simplest organizations of them? There will be further examples in the chapters of Part III. Can we define principles of operation? Our understanding of the brain has now advanced so far that we can enunciate quite dogmatically a number of general principles. We now proceed to do this seriatim because it will clear the way for advancing in our further attempts to understand the brain in the following chapters.

First, in the brain, transmission at a distance is by the propagation of nerve impulses. These are the all-or-nothing messages that travel along fibers and that were the main theme of Chapter 2. There is a minor reservation with respect to synaptic actions at close range of the type illustrated in Figures 4.13 and 4.14 (Section 4.6), where the transmission can be cable-like. It is to be understood that this all-or-nothing transmission does not apply to the chemical transport of macromolecules along nerve fibers that was considered in Chapter 1.

Second is the principle of divergence, as already defined. There are numerous branchings of all axons with a correspondingly great opportunity for wide dispersal because the impulses discharged by a neuron travel along all its axon branches to activate the synapses thereon. The divergence may be as low as ten, but values may be in the hundreds or even thousands, as is illustrated in Chapter 14.

Third is the complementary principle of convergence. All neurons receive synapses from many neurons, usually of several different species, as illustrated for the hippocampal pyramidal cell in Figure 1.4B (Section 1.2.2); they also receive both excitatory and inhibitory synapses. It is doubtful whether any neuron in the brain receives only excitation, and certainly there is no example of a purely inhibitory reception. The numbers of synapses on individual neurons are usually measured in hundreds or thousands, the highest recorded being around 100,000 (see Chapter 14).

Fourth is the successive transmutation from electrical to chemical and back to electrical in each synaptic transfer. From this process, there arise the integrational properties of the nervous system. There is the necessity for convergence of many synaptic excitations before a neuronal discharge is evoked, and there is the further opportunity for synaptic inhibitory action to prevent this discharge. In Chapter 3, we concentrated on the principles of chemical synaptic transmission: the presynaptic impulse, the chemical neurotransmitter substances, their quantal liberation by impulses, and their action on the postsynaptic membrane, opening ion gates with the consequent transmutation to an electrical event. This is the basic principle of communication in the brain. We are stressing it because it is fashionable for neuroscientists these days to propose that there is another kind of communication by what they call the "slow potential microstructure." They endow these potential fields with integrative or synthetic properties in their own right, whereas we can with assurance ascribe their production to the flow of currents generated by synaptic action in the ordinary process of synaptic transmission, as diagrammed in Figure 4.8A and B (Section 4.4) and revealed in the depth profile in the hippocampal cortex [see Figure 4.16D and E (Section 4.7.3)]. There is no neurobiological evidence that these currents are anything more than a spinoff from chemical synaptic transmission. Doubtless, if electrical fields were large enough, they would slightly modify the production of impulse discharge by synaptic action. But this would be diffuse and carry no significant information. Neurophysiological investigations strongly emphasize the almost exclusive role of impulses in coding information transmission both in the peripheral nervous system and in the brain, intensity being coded by frequency of firing. Figures 4.13 and 4.14 illustrate the rare exceptions. This principle can be extended indefinitely for sequence after sequence of synaptic transmission.

Fifth, in the brain, there is almost always background firing of neurons. If there is coding by frequency of firing, then it has to be remembered that this coding is superimposed on a background of incessant, irregular discharge. Even when one is asleep, the neurons of one's cerebral cortex are firing impulses. Some fire even faster when one is asleep than when one is awake. The problem is to extract reliable performance from the nervous system, considering that it has so much background noise. This is done by having many convergent lines in parallel, all carrying much the same signals. The modular arrangement of input areas in the cerebrum (see Chapter 17) does just this, and there are other examples now being discovered in which neurons of similar connectivities are arranged in clusters, as, for example, in the cerebellar cortex and its nuclei (see Chapter 14). The same kinds of neurons are organized together, receiving the same kinds of messages on the whole and transmitting the same kind of coded output to other clusters of neurons. Because of the incessant background noise, the responses of one neuron are ineffective. The neurons have to "shout together," as it were, to get the message across and so make a reliable signal above all the background noise. This is one of the problems considered in Chapters 14, 16, and 17—namely, to see how signals are lifted out of noisy backgrounds by collusion of many cells arranged together in clusters and working together in parallel.

Sixth is the whole question of inhibition coming in and sharpening signals and controlling neuronal discharges by the feed-forward, the feedback, and the presynaptic inhibitory circuits that have been described in this chapter.

Seventh, we would stress that the activity in the brain has a basis in neuronal events and can be measured in terms of signals that fire neurons by synaptic operation, and so on in most complex organizational patterns. Chapter 14 is an introduction to some attempts to illuminate some simpler levels that are concerned with the control of movement. We are attempting to understand some simpler levels—some of the simpler jigsaw components of the immense ensemble of complexity—and so gradually to understand more and more complex levels of brain action. Finally, in Chapter 17, we will consider the most important and challenging problem of all—the events in the cerebral cortex, the highest level of the brain, which again have to be understood as the weaving by impulses of the complex spatiotemporal patterns that have been likened by Sherrington to the weaving of an enchanted loom. But again, all we will postulate at the neuronal level is impulse or electrotonic transmission and the excitatory and inhibitory synaptic mechanisms responsible for controlling the impulse discharges.

These basic principles of neuronal operation form a background to Part II, in which the concentration will be on neurotransmitters, both the classical ones, which have already been introduced in Part I, and the immensely complicated development of putative neurotransmitters in the last decade, there now being more than 50 when the peptides are included.

II

Specific Neuronal Participants and Their Physiological Actions

As we learned in the first series of chapters, each neuron manufactures and stores one chemical transmitter. This becomes its biochemical fingerprint and its principal method of communicating with other neurons.

In this next section we describe the known neurotransmitters, the types of cells that use them, and the physiological actions that can be observed.

We commence in Chapter 5 by introducing some new principles of neurotransmission not covered in the first section. We first classify the neurotransmitters into three categories: amino acid (category I), amine (category II), and peptide (category III). They differ widely in their concentration, their methods of synthesis and release, and their action at postsynaptic receptors. We then describe the properties of receptors and divide their interaction with transmitters into two types: ionotropic and metabotropic. They result in very different types of neurotransmission. In the ionotropic type, ionic channels are opened by the interaction of the transmitter and its receptor. The effects have already been described in detail in Part I. In the metabotropic type, the interaction of the transmitter with its receptor produces a transmembrane effect that activates an intracellular second messenger such as cyclic AMP, cyclic GMP, polyinositolphosphide, or calmodulin. Conductance changes are not observed but many intracellular responses are generated. With this background information we discuss the phenomenon of coexistence of neurotransmitters in the same neuron. This does not represent a violation of Dale's principle.

We then proceed to a discussion of the various families of transmitters, describing briefly how they came to the attention of neuroscientists, followed by a description of their chemistry, anatomy, physiology, pharmacology, and pathology.

Chapter 6 is a brief chapter on the ionotropic excitatory amino acids glutamate and aspartate. The evidence that these compounds are neurotransmitters is less well developed than that for other neurotransmitters, but they may be the "workhorses" that operate the great majority of the excitatory synapses.

Chapter 7 deals with the classical inhibitory amino acids GABA and glycine. Their ionotropic mechanism of action is reasonably well understood.

Chapter 8 is on acetylcholine, the compound that "rang up the curtain" on chemical transmission in the nervous system and which has been the focus of much discussion in Chapters 2–4. In the brain its principal action is metabotropic rather than ionotropic and we thus encounter a different physiology surrounding this neurotransmitter.

Chapter 9 deals with the catecholamine family. There is an immense amount of literature on these interesting compounds. Despite the relatively tiny population of catecholamine neurons in brain, their effects are widespread, and their pharmacology is an important area of neuroscience.

Chapter 10 is on serotonin neurons. They have a different role to play in brain function from the catecholamines, and yet there is much overlap in their chemistry and pharmacology.

Chapter 11 is on histamine. There is a much smaller histamine neuronal population than exists for other amine transmitters but there is a distinctive and important physiology.

Chapter 12 is on the peptides. They represent a distinctive class of neurotransmitters with numerous representatives. Their study is one of the most rapidly growing fields of neuroscience.

5

Principles of Synaptic Biochemistry

Classic neuroscience holds that each neuron manufactures one neurotransmitter that is released in a calcium-dependent fashion on axonal stimulation. This is the biochemical fingerprint of the neuron and the specific molecule stored in its synaptic vesicles. Unfortunately, the transmitters are not known for the vast majority of the neurons in the central nervous system (CNS). Moreover, it now appears that some neurons may release a cotransmitter in addition to a principal transmitter.

Such a situation is not a violation of Dale's principle (Dale, 1935) as developed by Eccles (1986a). It merely indicates that the characteristic signaling mechanism of that particular neuron requires more than one molecule to achieve its postsynaptic effect.

In these chapters of Part II, we shall be classifying neurons according to their principal transmitters. This approach is different from the usual one of considering neurons according to their anatomical distribution or their participation in a given physiological system such as sight or hearing. The purpose is to provide greater insight into the action of drugs and the nature of many disease processes. A large proportion of drugs that affect the nervous system do so by modifying transmitter activity. Many diseases of the nervous system result from selective destruction of neurons that can be identified chemically by their transmitter fingerprint. In this chapter, we will consider some general properties of transmitters and their receptors.

We must start by broadening our concepts of synaptic biochemistry to include the following principles:

1. Neurotransmitters may stimulate metabolic rather than electrical postsynaptic activity. We refer to this as a "metabotropic" function as opposed to the classic "ionotropic" function whereby ion channels are opened and the membrane potential is altered.
2. The postsynaptic action of a given transmitter depends on the nature of the

receptor; multiple receptor types, which are pharmacologically distinguishable, are known for each transmitter.

3. In addition to its principal transmitter, a neuron may also possess a cotransmitter.

5.1. Criteria for Neurotransmitters

The first step is to determine whether a compound is a neurotransmitter at all. This is an important area for research, since the transmitters for the vast majority of neurons are still unknown. Many criteria have been suggested as being essential for neurotransmitter status. Some may apply well in certain situations, but be of little value in others. All that really can be said is that multiple tests should be applied to determine whether a material is a specific molecule stored in vesicles and released from the terminals of a given neuron. If there is reasonable doubt, the compound is described only as a "putative transmitter." If the information is speculative, the compound is usually referred to as a "transmitter candidate."

The tests described in Section 5.1.1–5.1.4 are the more important ones.

5.1.1. Anatomical

1. The substance must be present in the nervous system in adequate concentration. This is an obvious but necessary condition. There are many compounds that can be isolated from the nervous systems of submammalian species—or from other tissues of mammalian species—that have biological activity. They do not qualify as neurotransmitters if they are absent from brain or occur only in minuscule concentrations.

2. The substance should be concentrated in the nerve ending fraction of subcellular homogenates. Since a transmitter must be stored in nerve terminals for there to be an adequate physiological supply, failure to find a preferential concentration in this compartment is regarded as important negative evidence.

3. The substance should probably be distributed unevenly in brain, this being the anticipated consequence of an association with particular neurons. Such uneven distribution was a key point in the identification of dopamine (DA), glycine, and several other transmitters.

4. The substance should drop in concentration following lesions of known or suspected brain pathways in which it is involved. Since degeneration always takes place distally to a lesion, and also proximally in the absence of sustaining axon collaterals (chromatolysis), such a concentration change is strong evidence of association of a substance with a pathway. The technique has been widely used in establishing the biochemical nature of long-axoned pathways, but is of less value for short-axoned neurons.

5. In those brain regions where it is functionally involved, the substance should increase in concentration postnatally, concurrent with axonal growth and synaptic development.

6. The substance or its specific synthetic enzyme may be localizable to specific neurons and pathways by histochemical or immunohistochemical studies at the cellular level. This is extremely strong evidence. Fluorescent histochemistry was the crucial technique for the establishment of serotonin [5-hydroxytryptamine (5-HT)] and the cate-

cholamines as transmitters. Unfortunately, such histochemical methods exist for very few compounds in the CNS. Immunohistochemical procedures offer far more flexibility and scope and, despite considerable technical difficulty, are now being widely used.

5.1.2. Chemical

1. The material should be released from suitable tissue preparations *in vitro* by a process that is Ca^{2+}-dependent and K^+-stimulated. These conditions simulate the *in vivo* situation. K^+ is a nonspecific neuronal depolarizer, while Ca^{2+} mobilization in the nerve ending is necessary for release of all transmitters.

2. The material should be bound to presynaptic vesicles. Since the binding is reversible, being only for temporary storage purposes, it is sometimes hard to demonstrate.

3. Synthetic enzymes for the transmitter should be transported from the neuronal soma by the process of axoplasmic flow; the neurotransmitter itself may also be transported because of the necessary transport of vesicular binding proteins. Although techniques now exist for demonstrating such flow in central as well as peripheral pathways, they are not among the more definitive methods. Retrograde transport of the neurotransmitter has also been found in some cases.

4. A sodium-dependent, high-affinity uptake system may exist to pump the material or its precursor into nerve endings against a concentration gradient. This process may be important for removal of the neurotransmitter from the synapse as well as in the maintenance of an adequate supply in the nerve ending. Since pumping is a selective, rather than an absolutely specific process, nontransmitters may also show high-affinity uptake. No high-affinity uptake has been convincingly demonstrated for peptide neurotransmitters (see Chapter 12).

5. The presence of enzymes for synthesis and destruction should be demonstrable, particularly for the nonpeptide neurotransmitters (cf. Section 5.2); a method must exist for supplying large quantities of the neurotransmitter to the nerve ending and then rapidly disposing of it after release. The synthetic enzymes are generally more specific than the destroying enzymes and are localized mainly in the terminal fraction. The destroying enzymes may be in dendrites or glial cells or on cellular membranes. Any synthetic enzyme that is concentrated in synaptosomes should decrease along with its neurotransmitter following axotomy. A synthetic system for building most peptides from amino acids does not exist in the nerve endings. The peptide neurotransmitters are usually produced by cleavage of a precursor polypeptide synthesized in the rough endoplasmic reticulum of the soma, but the cleavage enzymes are found in axons and nerve endings (see Chapter 12).

6. The neurotransmitter should bind with high affinity to postsynaptic membranes that contain its receptor.

5.1.3. Physiological

1. Since the neurotransmitter must have action at its receptor sites, it should show definite physiological action when administered by various techniques. Ideally, it should exactly duplicate the effects of stimulation of the neuronal pathway for which it is believed to be the transmitter. Classical tests of neurotransmitter status in peripheral

systems require such duplication, but it is very difficult to achieve centrally. Iontophoretic application is the most widely used technique, and while quantitative comparisons are seldom possible, excellent correlation can often be found in postsynaptic neurons between stimulation of a pathway and the iontophoretic applications of the presumed transmitter. The iontophoretic technique has been instrumental in the recognition of amino acid neurotransmitters and in suggesting many transmitter candidates.

2. Neurotransmitter release should be demonstrable on nerve stimulation. Again, such release is relatively easy to achieve peripherally by perfusing isolated organs or ganglia and collecting the released agent. Centrally, it is much more difficult. Cups applied to the surface of the brain, however, or push–pull cannulas embedded in tissue, have been successfully applied to collecting increased quantities of transmitters such as glutamate, acetylcholine (ACh), DA, or 5-HT following appropriate stimulation.

5.1.4. Pharmacological

1. It should be possible to find pharmacological agents that interfere with the neurotransmitter at any of the stages of synthesis, storage, release, or action at receptor sites. Specificity is critical, and yet extremely hard to prove. Selective blockers were extremely valuable in confirming ACh and the amino acids GABA and glycine as transmitters.

2. It should also be possible to find pharmacological agents that mimic the action of the proposed neurotransmitter at receptor sites or indirectly enhance its action in other ways. This could be by stimulating its synthesis or release, by inhibiting its reuptake or destruction, or, in some cases, by action at a coupled receptor, as exemplified by the enhancement of GABAergic effects by benzodiazepines.

5.2. Classification of Neurotransmitters

Three broad categories of neurotransmitters can now be defined: amino acids (Type I), amines (Type II), and peptides (Type III). Neurotransmitters in these categories differ in concentration, structure, and principle mode of action (Table 5.1). The amino acid transmitters of Type I (glutamate, aspartate, GABA, and glycine) occur in a concentration range of micromoles per gram of tissue. Their action is *ionotropic*, which means that their receptors function to open ion gates on the postsynaptic membrane and thus to induce large conductance changes and consequent rapid excitatory or inhibitory actions. They share this characteristic with ACh at the neuromuscular junction and Renshaw cells in the spinal cord (see Chapters 4 and 8). Glycine in the spinal cord and brainstem (see Chapters 4 and 7) and GABA at these and higher levels in the CNS (see Chapters 4 and 7) exert inhibitory activity, while glutamate and aspartate (see Chapter 6) exert excitatory activity. Together, the Type I neurotransmitters account for the vast majority of all central neurons. Cell somata that use these Type I neurotransmitters are widely scattered throughout the brain, GABA and glycine occurring in many short-axoned inhibitory neurons as well as in some long-axoned tracts. Glutamate and aspartate systems are not yet well defined, but it appears that many of the corticofugal or cortical commissural tracts use one of these excitatory amino acids.

The amine neurotransmitters of Type II occur in a concentration range of nanomoles

TABLE 5.1
Approximate Levels of Various Proposed Neurotransmitters in Rat Brain[a]

Type I		Type II	
Range 1–15	μmoles/g	Range 1–25	nmoles/g
Glutamate	14[b]	ACh	25
Aspartate	4[b]	DA	6.5
GABA	2.5	NA	2.5
Glycine	2[b]	5-HT	2.5
		Histamine	1[b]

Type III			
Range 1–500	pmoles/g	Range <1	pmole/g
CCK	470	β-End	<1
met-Enk	350	ACTH	<1
SP	100	α-MSH	<1
VIP	40	Bradykinin	<1
SOM	30	TRH	<0.3
LHRH	7	Bombesin	<0.2
Dynorphin	4		
Vasopressin	2.5		
Neurotensin	1.5		

[a]Neurotransmitters: (ACh) acetylcholine; (ACTH) adrenocorticotropic hormone; (CCK) cholecystokinin; (DA) dopamine; (β-End) β-endorphin; (GABA) γ-aminobutyric acid; (5-HT) serotonin (5-hydroxytryptamine); (LHRH) luteinizing-hormone-releasing hormone; (met-Enk) methionine-enkephalin; (α-MSH) α-melanocyte-stimulating hormone; (NA) noradrenaline; (SP) substance P; (SOM) somatostatin (somatotropin-release-inhibiting factor); (TRH) thyrotropin-releasing hormone; (VIP) vasoactive intestinal polypeptide.
[b]Also in nontransmitter pools.

per gram of tissue, roughly 1000-fold lower than Type I neurotransmitters. Their principal mode of action in the CNS is *metabotropic*. Their action at receptors involves a different kind of mechanism that is typically slower in onset. It does not involve the large conductance changes characteristic of rapid movement of ions across membrane boundaries. Instead, the metabotropic neurotransmitter appears to initiate a series of chemical changes in the postsynaptic membrane (or other target membrane) that may have short-term effects, altering the membrane potential, or longer-term, bringing about trophic and plastic changes (Eccles and McGeer, 1979).

Information on the physiology of metabotropic neurotransmitters is scanty compared with that on the ionotropic ones, as we shall see in succeeding chapters. There is evidence suggesting that the metabotropic substances, notably noradrenaline (NA), may be released not only at synaptic junctions but also diffusely from axons and dendrites to influence the metabolism and function of surrounding glia and capillaries as well as neurons. The typical action of metabotropic neurotransmitters may be through second messenger systems, as outlined in Section 5.4. Three of these systems are discussed, all of which operate through the common mechanism of activating proteins through phosphorylation.

TABLE 5.2
Suggested Occurrences of Two Neurotransmitters or Putative Neurotransmitters in Single Mammalian Neurons or Secretory Cells[a]

Neurotransmitters			Regions and ref. nos.[b]

Type I	and	Type II	
GABA		5-HT	Raphe neurons[1]
GABA		(probably DA)	Olfactory bulb[2]

	Two Type II		
5-HT		Adrenaline	Adrenal medulla[3]

Type I	and	Type III	
GABA		Motilin	Cerebellar Purkinje cells[4]
GABA		Enkephalin	Caudate[6,7] and amygdala[7]
GABA		β-End	Caudate and amygdala[7]
GABA		SOM	Cat thalamic reticular n.[5]; cat, rat, and monkey cortex[8,9] and hippocampus[8]
GABA		CCK	Cat cortex and hippocampus[8]
GABA		SP	Cat dorsal column nuclei[10]

Type II	and	Type III	
ACh		SP	Ascending tracts from pons/medulla[11]
ACh		VIP	Autonomic ganglia; some postganglionic sympathetic fibers[12]
ACh		Enkephalin	Sympathetic preganglionic neurons[13]
ACh		CGRP	Cranial nerve nuclei[14]
ACh		Galanin	Septohippocampal tract[15]
DA		CCK	Rat and human A10 projections[16,17,c]
DA		Enkephalin	Carotid body[59]
DA		Neurotensin	Hypothalamus[59]
NA		SOM	Sympathetic ganglia,[18,d] SIF cells,[18] adrenal medulla[18] (many species[59]), guinea pig intestines[19]
NA		Enkephalin	Superior ganglion,[20] SIF cells,[20] splenic nerve[20,e] (many species[59])
NA		Neurotensin	Locus coeruleus, adrenal medulla[59]
NA		Neuropeptide Y	Sympathetic nerves—many species[21]
NA		Vasopressin	Locus coeruleus and subcoeruleus[22]
NA or adrenaline		Neuropeptide Y	Human[25] or rat[26,f] medullary neurons
NA or adrenaline		Enkephalin	Cat locus coeruleus and subcoeruleus[23]
Adrenaline		Enkephalin	Adrenal medulla[24]
Adrenaline		SP	Medulla to spinal cord, vasomotor[27]
5-HT		SP	Raphe[28,29] to spinal cord,[30,g] enterochromaffin cells[29]
5-HT		SP and TRH	Medulla to spinal cord[30–32]
5-HT		CCK	Medulla to spinal cord[33]
5-HT		Enkephalin	Raphe[24] to spinal cord[30]
5-HT		Motilin	Enterochromaffin cells[59]

	Two Type III		
CCK		VIP	Thalamus[34]
CCK		Neurotensin	Thalamus[34]

TABLE 5.2 (*Continued*)

Neurotransmitters		Regions and ref. nos.[b]
CCK	Oxytocin	Hypothalamus[16]
CCK	SP	Periaqueductal gray to spinal cord,[31,35] dorsal root ganglia[36]
CCK/gastrin	SOM	Gastrointestinal tract[59]
CRF	Oxytocin	Hypothalamus[37]
CRF	Vasopressin	Rat[38,39] and guinea pig[40] hypothalamus (increased after adrenalectomy)
CRF	Dynorphin$_{1-8}$	Hypothalamus[41]
CGRP	SP	Trigeminal ganglion,[42] sensory neurons[43]
Neurotensin	VIP	Thalamus[34]
Neurotensin	β-End	Pituitary[44]
Neuropeptide Y	SOM	Human or rat cortex,[45,46,f] striatum[47,f]
Neuropeptide Y	met-Enk	Sacryl sympathetic system[26,f]
Oxytocin	met-Enk or β-End	Hypothalamus[48]
Oxytocin	Renin-like	Hypothalamus[49]
Vasopressin	Angiotensin II	Hypothalamus[50]
Vasopressin	Dynorphin$_{1-8}$	Hypothalamopituitary[48,51]
Vasopressin	leu-Enk	Hypothalamus[52]
SRIF	Enkephalin	Rat or guinea pig median eminence[53]
SRIF	met-Enk	Superior ganglion[54]
SP	met-Enk and/or SOM	Superior ganglion[54]
SP	GRP-BO	Pons/medulla and hypothalamus,[38] dorsal-root ganglia[57]
LHRH	β-Endorphin	Human fetal hypothalamus[55]
LHRH	ACTH$_{17-39}$	Guinea pig median eminence[53,56]
VIP	His-iso-Leu (PHI)	Cat salivary gland[58,d]

[a]Evidence is from the rat unless otherwise specified and in many cases is only double immunohistochemical staining. Neurotransmitters: (ACh) acetylcholine; (ACTH) adrenocorticotropic hormone; (BO) bombesin; (CCK) cholecystokinin; (CGRP) calcitonin-gene-related peptide; (CRF) corticotropin-releasing factor; (DA) dopamine; (β-End) β-endorphin; (GABA) γ-aminobutyric acid; (GRP) gastrin-releasing peptide; (5-HT) serotonin (5-hydroxytryptamine); (leu-Enk) leucine-enkephalin; (LHRH) luteinizing-hormone-releasing hormone; (met-Enk) methionine-enkephalin; (NA) noradrenaline; (PHI) porcine intestinal peptide; (SOM) somatostatin (somatotropin-release-inhibiting factor); (SP) substance P; (T-OH) tyrosine hydroxylase; (TRH) thyrotropin-releasing hormone; (VIP) vasoactive intestinal polypeptide.

[b](SIF cells) Small, intensely fluorescent cells. References: (1) Belin *et al.* (1983); (2) Kosaka *et al.* (1985); (3) Verhofstad and Jonsson (1983); Jaim-Etcheverry and Zieher (1982); (4) Chan-Palay *et al.* (1981); (5) Oertel *et al.* (1983a); (6) Aronin *et al.* (1984); (7) Oertel *et al.* (1983b); (8) Somoygi *et al.* (1984); (9) Schmechel *et al.* (1984); (10) Westman *et al.* (1984); (11) Vincent *et al.* (1983c); (12) J. M. Lundberg *et al.* (1979); (13) H. Kondo *et al.* (1985); (14) Takami *et al.* (1985); (15) Melander *et al.* (1985); (16) Hökfelt *et al.* (1980a); (17) Fallon *et al.* (1983); R. G. Williams *et al.* (1981); Gilles *et al.* (1983); (18) H. Saito *et al.* (1984); (19) Costa and Furness (1984); (20) S. P. Wilson *et al.* (1980); (21) J. M. Lundberg *et al.* (1983); J. L. Morris *et al.* (1985); (22) Caffe *et al.* (1985); (23) Charnay *et al.* (1985); (24) Chaminade *et al.* (1983); (25) Hökfelt *et al.* (1983); (26) Hunt *et al.* (1981a); (27) Lorenz *et al.* (1985); (28) Björklund *et al.* (1979); (29) Lovick and Hunt (1983); (30) Bowker *et al.* (1983); (31) Hökfelt *et al.* (1982); (32) Johansson *et al.* (1981); Marsden *et al.* (1982); (33) Mantyh and Hunt (1984); (34) Sugimoto *et al.* (1985); (35) Skirboll *et al.* (1983); Conrath-Verrier *et al.* (1984); (36) Dalsgaard *et al.* (1982); (37) Burlet *et al.* (1983); (38) Roth *et al.* (1982); (39) Wolfson *et al.* (1985); (40) Tramu *et al.* (1985); (41) Roth *et al.* (1983); (42) Y. Lee *et al.* (1985); (43) Gibbins *et al.* (1985); Wiesenfeld-Hallin *et al.* (1984); (44) Miyoshi *et al.* (1985); (45) Chronwall *et al.* (1984); (46) Vincent *et al.* 1982d,e); (47) Vincent and Johansson (1983); (48) R. Martin *et al.* (1983); (49) Fuxe *et al.* (1982); (50) Kilcoyne *et al.* (1980); Hoffman *et al.* (1982); (51) Whitnall *et al.* (1983); Molineaux and Cox (1982); (52) R. Martin and Voigt (1982); (53) Tramu and Voigt (1982); (54) Ariano and Kenny (1985); (55) Leonardelli and Tramu (1979); (56) Beauvillain *et al.* (1981); (57) Fuxe *et al.* (1983); (58) J. M. Lundberg *et al.* (1984); (59) Hökfelt *et al.* (1980b).

[c]Crawley *et al.* (1985) present physiological and behavioral evidence that CCK potentiates dopamine effects in n. accumbens (which receives a projection from A10), but not in caudate.

[d]Co-release demonstrated.

[e]S. P. Wilson *et al.* (1980) suggest that the 1 : 60 opioid peptides/NA ratio in splenic nerve is the highest yet demonstrated for coexistence.

[f]Neuropeptide Y wrongly identified as avian pancreatic polypeptide (see Chapter 12).

[g]Agnati *et al.* (1983) report that SP reduces K_D and increases B_{max} for 5-HT binding in spinal cord.

Type II neurotransmitters have somata in brain stem nuclei—and, in the case of ACh (see Chapter 8), the medial basal forebrain—that send diffuse projections to many areas of brain. The chemistry of this group of neurotransmitters is reasonably well understood.

Type III comprises the peptide neurotransmitters. This is by far the largest category, including at this stage at least two dozen peptides, of which 15 are listed in Table 5.1. However, they exist at much lower concentrations than neurotransmitters of Types I and II, usually in the range of picomoles per gram of tissue. Their functions and modes of action are as yet largely unknown.

The peptides probably have special metabotropic actions and, in some cases, have been shown to modulate the actions of amine or amino acid neurotransmitters. Thus, they may act as cotransmitters. However, much more evidence as to their mode of action is needed. Some of the peptides have cell somata only in the hypothalamus, while others have scattered cells in many brain areas (see Chapter 12).

Numerous cases have now been documented in which more than one neurotransmitter has been found within a given neuronal type (Table 5.2). The list is expanding rapidly, and it can be anticipated that many more examples will be found now that effective cytochemical techniques are available. The significance of cotransmitters has not yet been assessed. It should prove a fruitful area for future neurophysiological research. In most cases, Type III (peptide) neurotransmitters are involved. But in the case of dorsal raphe neurons, GABA, a Type I neurotransmitter, exists with 5-HT, a Type II neurotransmitter (see Chapters 7 and 10). As yet, no convincing evidence has appeared suggesting selective release of one neurotransmitter over another at a subset of terminals of a given neuron. Selective axonal transport would, of course, make this possible, but it would represent an exception to Dale's principle.

Neurotransmitter action can be modified in many ways. Synthesis, storage, or release may be impaired. Receptor sites may be blocked or activated by competitors. Synthesis may be stimulated or destroying enzymes impaired. Reuptake mechanisms may be influenced. Postsynaptic action may be modified by other neurotransmitters or related compounds. Therefore, knowledge of the chemistry and pharmacology of neurotransmitters can be a powerful tool in the hands of the neuroscientist or physician.

5.3. Properties of Receptors

Neurotransmitters, like hormones, act on receptors. They are highly specialized and highly sensitive glycoproteins that are present in postsynaptic membranes at concentrations considerably below those of the presynaptic neurotransmitters that act on them. We have already discussed in a preliminary way the manner in which receptor components are produced in the cytoplasm, transported to appropriate locations, and "dissolved" in the lipid bilayer of the cell membrane [see Figure 1.14 (Section 1.2.3)]. Each neuron has a multiplicity of receptors on its membranes [see Figure 1.5 (Section 1.2.2)], and it is through the repertoire of receptors that neuronal versatility is developed. We shall now consider the nature of these receptors in more detail, recognizing that appropriate techniques for their study are only now emerging and that our concepts are still in a primitive stage.

The low concentration and high selectivity of receptors explain why they are the locus of action of many drugs. Theories regarding their structure and function must explain how some drugs are agonists, mimicking neurotransmitter activity; others are antagonists, blocking activity; and still others have a mixed agonist–antagonist action. Metabotropic reception involves the initiation of second messenger chemical reactions. A typical receptor complex might be thought of as consisting of three components: a binding site, a coupler, and an effector. The binding site attracts and holds the neurotransmitter or drug; the effector carries out the intended action. The coupler, like an automobile trans- mission, mediates the interaction between the two. The coupler must be tied closely to the binding site, and may even be a part of it, as many simple models depict. But it must nevertheless be separable from the binding process to account for antagonist or partial agonist action of many drugs.

Most receptors are initially studied by their interaction with a ligand. A ligand is a low-molecular-weight compound that possesses a prosthetic group capable of interacting with the high-molecular-weight receptor.

Binding is a physical, reversible mass-action effect between the ligand and the receptor. It usually obeys the laws of first-order kinetics and does not involve chemical bonding of the ligand to the receptor. If chemical bonding occurs, there is irreversible binding that inactivates the receptor.

The coupler and effector actions may merely involve a conformational change in which channels are opened to accommodate ions of suitable charge and diameter. These are ionotropic actions. The nicotinic ACh receptor that opens channels to Na^+, K^+, and ions of similar size is the best-known example. The GABA receptor that opens channels to Cl^- ions is another well-studied system (see Chapter 4).

Alternatively, the coupler and effector may trigger chemical reactions that initiate further events within the cell. In this case, the receptor is metabotropic. The Types II and III neurotransmitters that have been identified to date appear to exert most of their central actions via various types of metabotropic receptors.

At least one specific receptor must exist for each neurotransmitter. However, it would appear as though multiple receptors exist for each of the known transmitters, indicating that the possible modes of neurotransmitter action far exceed the number of neurotransmitters. The criteria that have been used to judge the existence of a particular receptor type are: (1) kinetic, (2) pharmacological, (3) anatomical, and (4) chemical.

5.3.1. Kinetic Criteria

1. There must be high-affinity binding of an appropriate ligand to membranes pre- sumed to contain the receptor. Binding is an extremely common phenomenon in nature, and this is therefore a necessary but far from sufficient criterion. For example, ster- eospecific high-affinity opiate binding to glass filters has been found (Synder, 1984).

2. The binding must be easily saturable. Receptors exist only in low concentration, often in the picomole per gram tissue range or lower. A binding site that occurs in high concentration is likely to be nonspecific.

3. The binding should follow first-order kinetics. At equilibrium, it should obey the Scatchard equation:

$$\frac{B}{F} = \frac{-B}{K_D} + \frac{B_{max}}{K_D}$$

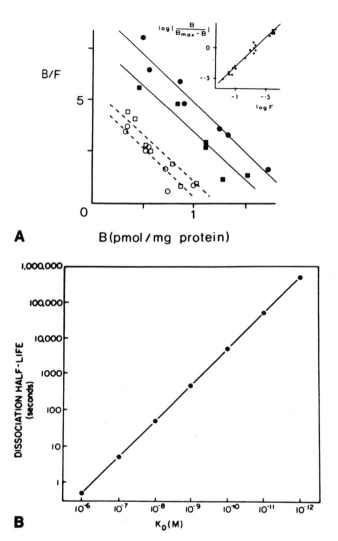

Figure 5.1. (A) Typical Scatchard and Hill (insert) plots of [3H]-QNB binding in the pyriform cortex ○, ● and amygdala □, ■ of rat brain membranes. Secondary lesions were produced in these areas following intracerebral injections of folic acid in the substantia innominata area. (●, ■) Data from control (unlesioned) brains; (○, □) data from lesioned brains. All four Scatchard plots have the same slope, corresponding to a K_D of 0.21 nM. The x intercepts give B_{max}'s of 2.0 and 1.1 pmoles/mg protein for the pyriform cortex and 1.7 and 1.2 for the amygdala. The Hill plot showed the four sets of Scatchard plot data converging to form a single line with a Hill coefficient (slope) of 0.97. From P. T.-H. Wong *et al.* (1983). (B) Theoretical relationship between the equilibrium dissociation constant K_D and the half-life of dissociation of a ligand–receptor complex formed during a binding experiment. It is assumed that first-order kinetics are obeyed and that the association rate is 10^6 $M^{-1} \times sec^{-1}$. After Bennett and Yamamura (1985).

where B is the concentration of bound ligand, F is the concentration of free ligand, B_{max} is the total number of binding sites, and K_D is the dissociation binding constant.

It is important to establish that, in the ligand concentration range being investigated, there is a linear relationship between binding and tissue concentration. On a Scatchard plot (Figure 5.1A), a straight line should be obtained. K_D is obtained from the slope and B_{max} from the x intercept. A curve indicates nonuniform kinetics and, frequently, the presence of several binding sites with different affinities. If, for example, the data fit two separate straight lines, there may be both a high- and a low-affinity binding site.

4. The binding of the ligand to its receptor, and its release from the receptor, should obey classic mass-action principles. For example, there should be no cooperation with other specifically bound ligand molecules and no interference by extraneous molecules binding to the same sites. The membranes should not be heterogeneous in terms of receptor affinity. A test for satisfaction of these conditions is the existence of a slope of unity ($n = 1$) on a Hill plot. The Hill equation, usually expressed in logarithmic terms for convenience in constructing graphs, is:

$$\log [B/(B_{max}-B)] = n\log F - \log K_D$$

If n is greater than 1, cooperative binding exists; if it is less than 1, there is competition for, or interaction with, normal binding.

The manner in which these principles are applied in practice is illustrated in Figure 5.1A. This shows results of binding experiments using the muscarinic cholinergic ligand quinuclidinylbenzilate ($[^3H]$-QNB) (P. T.-H. Wong et al., 1983). Membranes were used from two different regions of brain—pyriform cortex and amygdala—from control rats and from rats with brain lesions that affected the number but not the nature of binding sites. Scatchard plots showed the same K_D of 0.21 nM in each case, although the B_{max}, which reflects the number of binding sites, differed in the two regions and was reduced in both regions in the lesioned animals. The Hill plot (insert), combining the data from the four Scatchard plots, had a slope of 0.97, within experimental error of unity. Thus, reliable kinetics were being followed, and neither positive nor negative cooperative binding was being measured.

For reasonable data to be obtained, it is important that the ligand bind tightly to the receptor so that significant dissociation does not take place during the time course of the experiment wherein separation of bound from free ligand must be achieved. This requires that the K_D be in the range of 10^{-6}–10^{-12} M.

Figure 5.1B is a representative graph that shows what the dissociation half-life would be for various equilibrium dissociation binding constants (K_D) for hypothetical receptor–ligand complexes obeying first-order kinetics, assuming a rate of association of 10^6 M^{-1} sec^{-1} (Bennett and Yamamura, 1985). The latter figure is reasonable considering the in vivo rate at which drugs or neurotransmitters typically initiate a response. The graph shows that the dissociation half-life can be extremely short for ligands with only modest affinity for a receptor. This is probably the case for most neurotransmitters and their receptors in vivo, since their action is typically short-lived. However, it greatly complicates in vivo binding experiments.

The higher the K_D and therefore the lower the affinity, the more rapidly must the binding experiment be carried out. Indeed, for K_D's of the order of 10^{-6}–10^{-8} M, allowable times for separation of bound ligand from nonspecifically sequestered material may be in the range of 0.01–10 sec. The requirement for such rapid separation was the major problem encountered in pioneering experiments aimed at isolating the opiate receptor (Snyder, 1984). Antagonists, on the other hand, usually have a high affinity for receptors, since they would be ineffective if displaced by endogenous neurotransmitters. The higher affinity and lower K_D result in a longer half-life of the bound complex, longer *in vivo* drug action, and easier investigation of binding kinetics *in vitro*.

In practice, the concentration of a radioactive ligand can be held constant in a series of binding experiments with the ligand concentration being increased by adding unlabeled ligand. The labeled and unlabeled ligand can be regarded from a kinetic point of view as, respectively, agonist and antagonist competing for the binding site. Assuming they behave similarly at the binding site, the concentration of unlabeled ligand at which the maximum binding of labeled ligand is displaced by 50% (IC_{50}) will be a reasonable first approximation of K_D. The data can be used to compute the inhibitor $K_i = IC_{50}/(1 + L/K_D)$, where K_D is the dissociation constant of the labeled ligand and L is its concentration. It is beyond the scope of this book to derive the equations involved or to provide more than the barest outlines of the standard techniques that are used. The reader is referred to two monographs (Bylund, 1980; Yamamura *et al.*, 1985) and an authoritative review (Snyder, 1984) for appropriate details.

5.3.2. Pharmacological Criteria

1. The ligand that shows high-affinity binding to the presumed receptor site should be pharmacologically active *in vivo*. Thus, ligands presumed to be binding to the nicotinic ACh receptor in *in vitro* experiments should have either agonist or antagonist pharmacological action at these receptor sites *in vivo*.

2. The time course of the binding should correlate appropriately with the rate of action of the drug *in vivo*. Similarly, the dissociation rate of the receptor–ligand complex should be at least as fast as the washout of biological effect *in vivo*. Such considerations do not apply to irreversible ligands, such as α-bungarotoxin at the nicotinic cholinergic receptor, or materials with similar effects at other receptors.

3. Stereospecificity *in vivo* should be paralleled by stereospecificity *in vitro*. Many receptor sites distinguish between stereoisomers of drugs because their endogenous ligands are themselves stereoisomeric. An inactive isomer of a drug *in vivo* should not show high-affinity binding *in vitro*. Such an isomer, however, is often an excellent tool for assessing nonspecific binding.

5.3.3. Anatomical Criteria

1. The membranes that show high-affinity binding of the ligand must be from tissue known to possess the presumed receptor. Organs or tissues that show no biological response to the neurotransmitter or hormone should not bind the ligand.

2. In brain, the concentration of binding sites should be appropriate to the area of brain known to contain high or low concentrations of the neurotransmitter and its receptor. Glycine, for example, is known to be a neurotransmitter in the spinal cord and brainstem,

TABLE 5.3
Examples of Multiple Receptor Types for Some Neurotransmitters

Subtype	Properties
Glutamate/aspartate types (Chapter 6)	
A-1 (NMDA)	Binds [^{3}H]-2-amino-5-phosphonovaleric acid (AP5), [^{3}H]glutamate, and [^{3}H]aspartate independently of Cl$^-$, Ca^{2+}, and Na$^+$; activated strongly by N-methyl-D-aspartate (NMDA) and weakly by quisqualic acid (QA) and kainic acid (KA); mediates some excitatory responses; binding enhanced by freezing.
A-2 (QA)	Binds [^{3}H]glutamate and [^{3}H]-α-amino-3-hydroxy-5-methyl-4-isoproprionic acid (AMPA); binding unaffected by freezing; very sensitive to QA; synaptic role is unclear.
A-3 (KA)	Binds [^{3}H]-KA strongly, glutamate and QA somewhat; binding unaffected by freezing; sensitive to KA; modulates synaptic action.
A-4 (AP4)	Binds [^{3}H]-L-aminophosphonobutyric acid (AP4), [^{3}H]-L-glutamate, and [^{3}H]-L-aspartate in Cl$^-$-dependent fashion; sensitive to freezing; modulates synaptic action.
γ-Aminobutyric acid type (Chapter 7)	
GABA$_A$	Muscimol-selective; postsynaptic to GABA neurons; antagonized by bicuculline and picrotoxin; binding enhanced by detergent.
GABA$_B$	Baclofen-sensitive, bicuculline-resistant; on nonGABAergic nerve terminals; binding Ca^{2+}-sensitive, inhibited by guanyl nucleotides, reduced by detergent.
Muscarinic cholinergic type (Chapter 8)	
M$_1$	Muscarine is the classical agonist; scopolamine is an antagonist; binds [^{3}H]quinuclidinylbenzilate (QNB); pirenzipine is a selective antagonist; closes K$^-$ channels; largely postsynaptic.
M$_2$	Muscarine is the classical agonist; scopolamine is an antagonist; binds QNB and oxotremorine, but low affinity for pirenzipine; regulated by GTP; inhibits adenylate cyclase; may be on cholinergic neurons.
N	Nicotine is the classical agonist; atropine is the classical antagonist; may be subtypes in brain distinguished by binding or nonbinding of α-bungarotoxin.
Dopamine type (Chapter 9)	
D$_1$	Enhances adenylate cyclase; labeled by [^{3}H]thioxanthenes; absent in pituitary; present in parathyroid.
D$_2$	Lowers adenylate cyclase; labeled by [^{3}H]butyrophenones; present in anterior pituitary; responsible for antipsychotic and extrapyramidal actions.
α-Adrenergic type (Chapter 9)	
α$_1$	Postsynaptic in sympathetic; prazosin- and indoramin-selective; acts through Ca^{2+} channels; little affected by guanine nucleotides.
α$_2$	Located on sympathetic nerve terminals to regulate noradrenaline release, but also postsynaptic, especially in brain; clonidine is a selective agonist; yohimbine and piperoxan are selective antagonists; lowers adenylate cyclase.
β-Adrenergic type (Chapter 9)	
β$_1$	Adrenaline and noradrenaline are equally potent agonists; practalol is a selective antagonist; more in heart than in lung; regional variations in brain; neuronal localization.
β$_2$	Adrenaline more potent than noradrenaline; terbutaline- and salbutamol-selective agonists; more in lung than in heart; few regional variations in brain; more on glia than on neurons.

(continued)

TABLE 5.3 (*Continued*)

Subtype	Properties
Serotonin type (Chapter 10)	
5-HT$_{1A}$	Labeled with [^{3}H]-5-HT at nanomolar concentrations; high affinity for spiperone and 8-hydroxy-2-(di-*n*-propyl)aminotetralin (PAT); regulated by guanine nucleotides, probably linked to adenylate cyclase; mediates contraction of dog basilar artery.
5-HT$_{1B}$	High affinity for 5-HT, but low affinity for spiperone and little binding of PAT; high in basal ganglia, where 5-HT$_{1A}$ sites are low; may be linked to adenylate cyclase.
5-HT$_2$	Labeled with [^{3}H]spiperone and [^{3}H]ketanserin, but low affinity (micromolar) for 5-HT and little binding of PAT; less affected by guanine nucleotides; may be linked to a phosphoinositide system; high in cortex and hippocampus.
Histamine type (Chapter 11)	
H$_1$	Site of action of classic antihistamines such as mypyramine and diphenhydramine; may act through a phosphoinositide system.
H$_2$	Antagonized by metiamide, cimetidine, etc.; unaffected by classic antihistamines; probably linked to an adenylate cyclase.
Adenosine type (Chapter 11)	
A$_1$	Extracellular; labeled by [^{3}H]cyclohexyladenosine, [^{3}H]phenylisopropyl-adenosine (PIA), 2-[^{3}H]chloroadenosine; lowers adenylate cyclase; adenosine analogues potent at nanomolar concentrations; stereospecific for PIA.
A$_2$	Extracellular; stimulates adenylate cyclase; adenosine analogues potent at micromolar concentrations; little stereoselectivity for PIA; labeled with 5-*N*-[^{3}H]methylcarboxamidoadenosine.
P	Intracellular site labeled by [^{3}H]-2,5-dideoxyadenosine; adenosine active at 10^{-5} M concentrations; lowers adenylate cyclase.
Opiate type (Chapter 12)	
μ	Morphine-selective; localized in pain-modulating brain regions.
μ$_1$	Identified by very high-affinity binding of numerous opiates; blocked selectively by naloxonazine; meptazinol a specific agonist; implicated in analgesia, but not in respiratory depression.
δ	Enkephalin-selective; localized in limbic brain regions.
κ	Mediates sedating, less addicting analgesia; localized to deep layers of cerebral cortex; dynorphin has high affinity; mediates rabbit vas deferens contractions.
σ	Naloxone-insensitive; mediates psychotomimetic opiate effects; concentrated in hippocampus.
ε	β-Endorphin-selective; mediates rat vas deferens contractions.

but not in the cerebral cortex. Therefore, receptors for glycine might appropriately be found in the first two areas, but not in the cortex.

3. Receptors for neurotransmitters should be concentrated in the synaptosomal fraction, since postsynaptic membranes adhere to presynaptic elements following subcellular fractionation.

5.3.4. Chemical Criteria

1. It should be possible to isolate and characterize the receptor chemically. Methods of solubilizing receptors are known, as are methods for subsequently concentrating them. However, this is a major task. It has been done for the nicotinic cholinergic receptor (Heidmann and Changeux, 1978; Takai *et al.*, 1985) (see Chapter 8) and the β-adrenergic receptor (Cerione *et al.*, 1983) (see Chapter 9), and there have been reports on the solubilization and concentration of a number of other receptor types (for a review, see Venter *et al.*, 1983.)

2. It should be possible to generate antibodies against isolated, purified receptors. Such antibodies should interact with the receptors *in vitro* and *in vivo*.

3. It should be possible to isolate mRNAs for receptors from cells and to prepare corresponding cDNAs by cloning (Takai *et al.*, 1985). This has been achieved only for the nicotinic cholinergic receptor, which consists of four different proteins, only one of which ACh (Mishina *et al.*, 1985) (see Chapter 8).

Table 5.3 is a partial list of receptors for some of the neurotransmitters listed in Table 5.1. This list is far from complete, but it gives examples of receptors that have been reasonably characterized.

5.4. Coupler and Effector Systems

Neurotransmitters and their receptors are specific for neuronal pathways in brain, but they appear to be coupled to general systems for achieving their transmembrane effects. The membrane barriers that must be crossed are not restricted to the external surfaces of the cells. They also exist intracellularly between the cytosol and various organelles. Of particular interest presynaptically are those that involve vesicles and the smooth endoplasmic reticulum, because these organelles are specialized for the compartmentation of active molecules in neuronal processes. Postsynaptically, the components of receptor action beyond the specific binding site should be discussed. Three coupler and effector systems will be discussed—those that involve (1) the cyclic nucleotides (cAMP and cGMP), (2) Ca^{2+}–calmodulin, and (3) phosphatidylinositol. The proposed individual systems are all complicated, involving a cascade of reactions that often include more than one of the second messenger systems. However, in each case, receptor activation leads to the formation of high-energy phosphate bonds. These bonds are used to phosphorylate a specific protein or class of proteins. The activated proteins then catalyze the appropriate effector reactions. Finally, phosphatases cleave the activated phosphate groups, thus terminating the second messenger action and paving the way for regenerating the system. The simple principles outlined in Figure 5.2 form an underlying theme that can be traced through the series of chemical reactions in each system, the details of which are still incomplete and uncertain.

5.4.1. Cyclic Nucleotides as Second Messengers

The two nucleotides that have so far been identified as second messenger compounds are cAMP and cGMP. Their structures are shown in Figure 5.3. They are synthesized by their own cyclases and are destroyed by a common phosphodiesterase. The essential

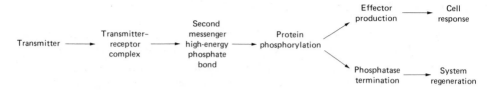

Figure 5.2. General scheme for second messenger action.

feature of these compounds is that they contain a labile, high-energy phosphate bond that is utilized to phosphorylate proteins that then initiate a variety of intracellular physiological events. The enzymes necessary for their formation are localized primarily in the cell membrane, with the active cyclic nucleotide being produced at or near the inner membrane surface, where it can exert its effects on other compounds in the cell membrane or intracellularly.

Cyclic AMP. Cyclic AMP is formed by adenylate cyclase according to the following equation:

$$ATP \rightarrow cAMP + PP_i + H^+$$

It appears that cAMP is involved in at least three distinct processes in the nervous system: regulation of biosynthesis of certain neurotransmitters, functioning of microtubules, and as a second messenger for neurotransmitters. It exerts its actions by phosphorylating cAMP-dependent protein kinases that fall into two broad categories: soluble and membrane-bound. These cAMP-dependent kinases contain a single catalytic subunit complexed to one of two classes of regulatory subunits designated R_1 and R_2.

The most studied adenylate cyclase system is that linked to postsynaptic receptor

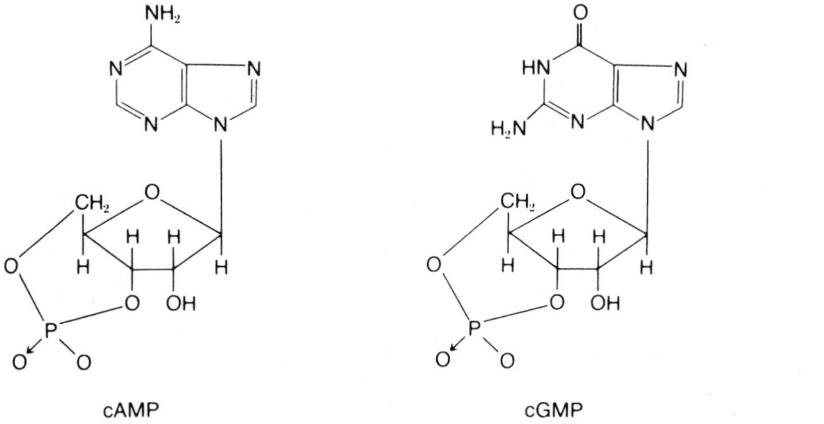

cAMP   cGMP

Figure 5.3. Structures of cyclic 3′,5′-adenosine monophosphate (cAMP) and cyclic 3′,5′-guanosine monophosphate (cGMP).

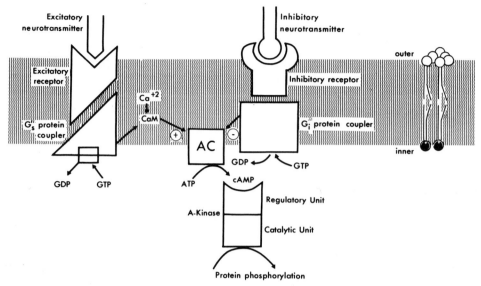

Figure 5.4. Proposed mechanism of action of the cAMP second messenger system. An excitatory or an inhibitory neurotransmitter interacts with its own receptor, which then couples to its own stimulatory (G_s) or inhibitory (G_i) protein. The consequence is to displace an inactive GDP molecule with an active GTP molecule. In the case of the G_s protein, there is activation of a Ca^{2+}–calmodulin complex (CaM), which in turn stimulates adenylate cyclase (AC) to produce cAMP. Cyclic AMP binds to the regulatory unit of A-kinase, releasing the catalytic unit to phosphorylate proteins. On the other hand, when the G_i protein is activated by GTP, it inhibits AC.

action. It is membrane-bound and involves a GTP–GDP system (for reviews, see Schramm and Selinger, 1984; Berridge, 1985). A variety of neurotransmitters or putative neurotransmitters are able to stimulate this system. They include DA, NA, 5-HT, histamine, adenosine, and various peptides. Formation is also stimulated by some prostaglandins (Hildebrandt *et al.*, 1982; Koski *et al.*, 1982).

The actual receptor–coupler–effector system, which is rather complicated, is shown diagramatically in Figure 5.4. It appears to involve the following steps:

1. Binding of the neurotransmitter to its receptor, which may be excitatory or inhibitory, on the outer surface of the postsynaptic membrane.
2. Displacement of inactive GDP from a G protein coupler on the inner surface of the membrane. A stimulatory G protein (G_s) couples to an excitatory receptor site, while an inhibitory G protein (G_i) couples to an inhibitory receptor.
3. Attachment of GTP to the vacated site on the G protein.
4. In the case of stimulatory effects, interaction of the G protein with Ca^{2+}–calmodulin (CaM) to activate adenylate cyclase (AC).
5. Formation of cAMP from ATP by activated AC.
6. Coupling of cAMP to the regulatory subunit of A-kinase.
7. Protein phosphorylation by the catalytic unit of A-kinase.
8. Intracellular actions by the phosphorylated protein.

9. In the case of inhibitory action, coupling between the G_i protein and AC to inhibit cAMP formation (with subsequent inhibition rather than stimulation of steps 4–8).
10. Termination of G protein action by hydrolysis of GTP to GDP.
11. Termination of cAMP action by phosphodiesterase.

Cyclic GMP. The chemistry of cGMP is parallel to that of cAMP in a number of ways. It is formed by guanyl cyclase according to the following equation:

$$GTP \rightarrow cGMP + PP_i + H^+$$

Guanyl cyclase is localized primarily to postsynaptic components of neurons and not in the presynaptic terminals. It is also found in some glia. Cyclic GMP is hydrolyzed by a phosphodiesterase in the same fashion as cAMP, and in analogy to cAMP, there is a cGMP-dependent protein kinase.

Cyclic GMP is present in concentrations far below those of cAMP except in the cerebellum, where the concentrations are approximately equal. The location in the cerebellum has not been determined with certainty. It is greatly decreased in mutant mice lacking Purkinje cells, and its kinase system is highly localized to Purkinje cells. This suggests that it occurs primarily in Purkinje cells (Uhler and Greengard, 1982). On the other hand, in an immunohistochemical study of the localization of cGMP, it was found in cerebellar glial cells, white matter, deep nuclei, and some stellate and basket cells (Chan Palay and Palay, 1979).

Cyclic GMP may be involved in the action of ACh. Cholinergic agents increase cGMP, while atropine blocks such increases.

5.4.2. Calcium and Its Binding Proteins

We have already learned that calcium plays an essential role in the budding and fusion processes of membranes. We have also shown that neurotransmitter release is dependent on calcium through the fusion process of vesicles and presynaptic membranes (see Section 3.2.7). Calcium is also a vital component of many postsynaptic neurotransmitter actions. Therefore, the ion that is specialized for facilitating transmembrane functions is Ca^{2+}. Ca^{2+} accomplishes this by having its own ion channel through membranes, ATP-linked pump, specialized binding proteins, and specialized protein kinases. At one time, it was believed that Ca^{2+} channels differ markedly in their properties according to the tissue, but patch-clamping has revealed that the Ca^{2+} channel may be a fundamental biological device (Reuter et al., 1982) (see Sections 2.3 and 3.4). The calcium-sensitive dye Arsenazo III permits identification of Ca^{2+} channels, since entry of Ca^{2+} into a compartment containing the dye will cause a detectable increase in light absorption owing to the formation of Ca^{2+}–Arsenazo III complexes (Miledi and Parker, 1981). In the squid giant synapse, changes in light absorption can be seen only in nerve terminals and not in preterminal axon regions.

Once Ca^{2+} enters the nerve ending, it is sequestered by virtually every intracellular organelle. Appropriate levels of Ca^{2+} must be reestablished by the existence of specific Ca^{2+}-transporter pumps. At least two of these have been identified that are directly dependent on ATP, and there may be others dependent on ion gradients. Neuronal ATP-dependent Ca^{2+} translocators have been purified (Reichardt and Kelly, 1983).

Ca^{2+} is thought to exert most of its effects by binding to specific proteins that are then transformed from an inactive to an active state. Calmodulin is the most important of these proteins. Calmodulin can be easily separated from other CNS proteins because of its almost unique ability to withstand degeneration by heat. It is a 148-amino-acid peptide containing four Ca^{2+}-binding sites. It consists of two globular lobes—each of which binds two Ca^{2+} ions—connected by a long, exposed α-helix (Babu et al., 1985). It is present in nerve terminals and also in the postsynaptic membrane.

The schematic model of Figure 5.5 summarizes the functions that are thought to be activated by Ca^{2+} and calmodulin. Following the polarization of the nerve terminal, Ca^{2+} in the vicinity of the synaptic cleft diffuses through the Ca^{2+} channel from the extracellular fluid into the nerve ending; there it contacts calmodulin and becomes bound. When 4 Ca^{2+} ions have bound to a single molecule of calmodulin, the calmodulin is transformed from the inactive to the active state. The necessity of binding 4 Ca^{2+} ions could explain the dependence of transmitter release on the fourth power of Ca^{2+} concentration. Activated calmodulin can then initiate a number of reactions, as depicted in Figure 5.5. The most important is the fusing of synaptic vesicles to the presynaptic membrane to permit exocytosis of the neurotransmitter. Anticalmodulin antibodies prevent Ca^{2+}-dependent fusion of granules to the plasmid membrane of sea urchins, an inhibition that can be reversed by excess calmodulin (Steinhardt and Alderton, 1982). The Ca^{2+}-stimulated release of NA, DA, and ACh from isolated vesicles is blocked when calmodulin is removed (De Lorenzo, 1982).

As Figure 5.5 indicates, Ca^{2+} is involved in a number of other intracellular processes. For example, it stimulates glycogenolysis, mitochondrial respiration, endocytosis, and neurotransmitter synthesis. Some of these changes are homeostatic in nature, restoring depleted levels of ATP, neurotransmitters, and synaptic vesicles (Reichardt and Kelly, 1983). Others, such as the Ca^{2+}–calmodulin activation of the tubulin kinase system (B. E. Burke and De Lorenzo, 1982), are probably associated with membrane activities including participation in vesicle fusion. Ca^{2+} is proposed to play a key role in the long-term potentiation of synaptic input that may underlie some mechanisms associated with memory (see Chapter 16).

Ca^{2+} has been shown to stimulate the endogenous protein phosphorylation of a number of brain proteins of various molecular weights in a system that is distinct from the cAMP protein kinases. Thus, both Ca^{2+} itself and the calcium–calmodulin complex can activate protein kinases, some of which are cytosolic and some membrane-bound. There is evidently no shortage of candidate proteins that could mediate exocytosis or other intracellular activities in which Ca^{2+} plays a role.

Calmodulin and Ca^{2+}-dependent kinases exist postsynaptically as well as presynaptically. The major postsynaptic density protein is an endogenous substrate for calmodulin-dependent phosphorylation (Goldenring et al., 1984).

Calcineurin is another Ca^{2+}–binding protein in brain that is particularly concentrated in dendrites and postsynaptic densities but that apparently counteracts Ca^{2+}–calmodulin activation. It also binds 4 moles of Ca^{2+} with an even higher affinity than calmodulin. It can thus act as an extremely fast off switch for Ca^{2+}-dependent biological processes (Klee and Haiech, 1980; Klee et al., 1980).

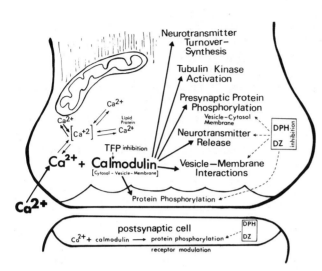

Figure 5.5. Schematic model summarizing present eivdence that supports a role for calmodulin in modulating some of the effects of Ca^{2+} at the synapse. Presynaptic and postsynaptic calmodulin serves as a Ca^{2+} receptor that activates several enzyme processes and synaptic events. The inhibition of calmodulin by trifluoperazine (TFP) is shown, and the model indicates that this drug would be expected to affect all calmodulin-dependent processes. The Ca^{2+}-kinase inhibitors [phenytoin (DPH) and diazepam (DZ)] have been shown to inhibit not only kinase activity but also vesicle–membrane interactions and neurotransmitter release, suggesting that synaptic protein phosphorylation may play a role in these processes. From De Lorenzo (1982).

In summary, Ca^{2+}, its binding proteins, and its kinases are ubiquitous in tissues of all kinds. Fundamental mechanisms that have developed in an evolutionary way have been incorporated into the delicate and specialized requirements of neuronal functions. So far, only general outlines of the way in which these mechanisms are coupled to pre- and postsynaptic effector systems are available.

5.4.3. Polyphosphoinositides

There is rapidly accumulating evidence that polyphosphoinositides are yet another class of second messengers (for brief reviews, see Cooper and Mayer, 1984; Nishizuka, 1984; Berridge, 1985; Marx, 1985a). Catabolism of this membrane-bound material is stimulated when a cell surface receptor is activated. A complex sequence of events is then set in motion. The net result is to release phosphate groups for activating protein kinases and to mobilize intracellular Ca^{2+}. The structure of the central compound, phosphoinositol 4,5-biphosphate (PIP_2), is shown in Figure 5.6. The dotted line shows where this membrane-bound starting material is cleaved by a phospholipase C following activation by an appropriate neurotransmitter or other signaling agent contacting its receptor. As with the cAMP system [see Figure 5.4 (Section 5.4.1)], a G protein is involved as the coupling agent. In this case, it acts between the receptor and phospholipase C. Diacylglycerol (DG) and inositol triphosphate (IP_3) are the products of this cleavage.

A proposed cycle for what may subsequently take place is shown in Figure 5.7. It must be remembered that knowledge of this system is as yet primitive, so that the model may be subject to major revisions in the future.

The DG and IP_3 products of the PIP_2 cleavage are both compounds that possess labile phosphate groups. DG is thought to activate protein kinase C, which is then available for promoting further intracellular reactions, presumably by activating proteins.

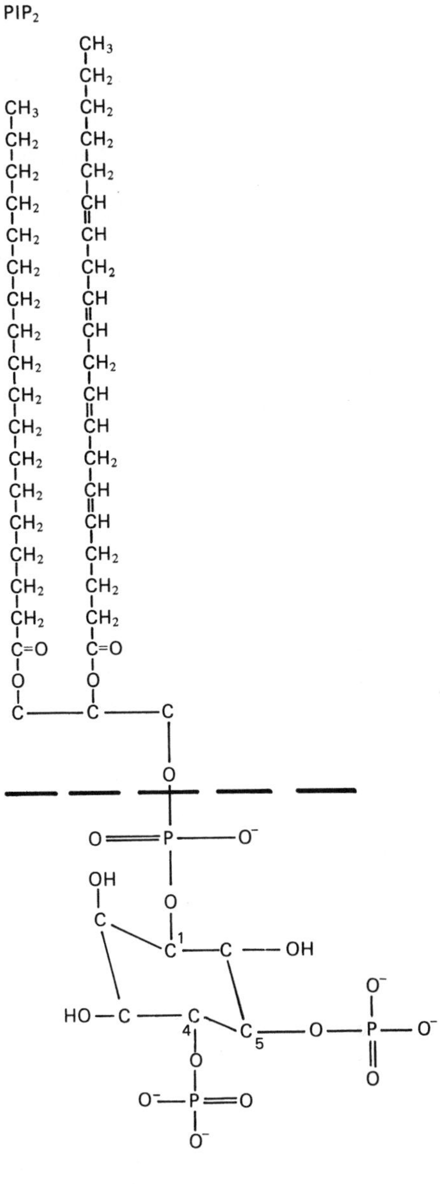

PIP$_2$

Figure 5.6. Structure of phosphatidylinositol 4,5-bi-phosphate (PIP$_2$). The two lipophilic tails, which are long-chain fatty acids making up the acyl groups of diacylglycerol (DG), are buried in the neuronal membrane. The hydrophilic phosphorylated inositol head projects into the neuron from the inner surface of the membrane. (- - -) Point at which the molecule is cleaved to form DG and inositol triphosphate (IP$_3$). The numbers indicate the positions on the inositol ring. (4, 5) Positions phosphorylated in PIP$_2$. After cleavage at the phosphate bridge, the 1, 4, and 5 positions are phosphorylated in IP$_3$. Each ring C also carries an H.

IP$_3$ has the property of releasing intracellular stores of calcium. The mobilized Ca^{2+} can then initiate its own series of reactions through Ca^{2+}–calmodulin activation, possibly in cooperation with protein kinase C. As yet, the spectrum of action of these products has not been determined, but they range from second messenger neurotransmitter action to the target system for tumor promoters and oncogenes (Marx, 1985a). Among the neurotransmitters that have been reported to affect this phosphoinositide system at some receptor types are glutamate/aspartate (see Chapter 6), ACh (see Chapter 8), DA and NA (see

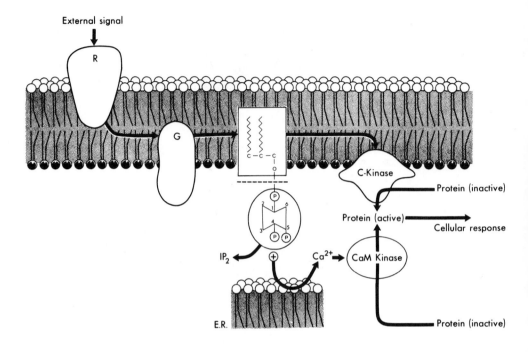

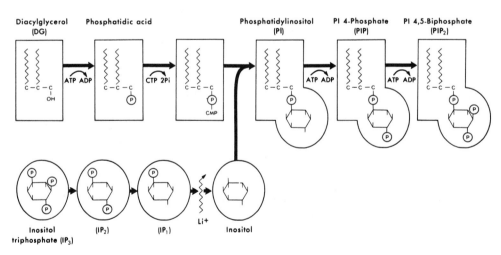

Figure 5.7. Proposed model for the role of the polyphosphoinositides in receptor activation. Binding of a hormone, neurotransmitter, or growth factor to its receptor (R) activates a G protein (G) that requires GTP; it then activates phospholipase C, which splits the polyphosphoinositide (PIP_2) to inositol triphosphate (IP_3) and diacylglycerol (DG). IP_3 releases Ca^{2+} ions from internal membrane stores; the Ca^{2+} ions then combine with calmodulin (CaM) to activate proteins or participate in other intracellular events. DG activates protein kinase C (C-kinase), which can also activate proteins to produce further intracellular events. Resynthesis of PIP_2 involves the conjunction of two cycles that are shown as enlargements of the DG area (rectangles) and the IP_3 area (circles). In the circle series, there is stepwise removal of the three phosphate groups to form first inositol biphosphate (IP_2), then inositol phosphate (IP_1), and finally inositol. In the rectangle series, DG is converted to a phosphatidic acid by the addition of a phosphate group and then, by the addition of cytosine triphosphate, to the cytosine nucleotide derivative (CDP-DG). The products of the two cycles, inositol and CDP-DG, then combine to produce phosphatidylinositol (PI), which is phosphorylated to PIP. Phosphorylation of PIP regenerates PIP_2.

Chapter 9), 5-HT (see Chapter 10), histamine (see Chapter 11), and various peptides (see Chapter 12).

Regeneration of the system is complex. DG is converted to a phosphatidic acid by the addition of a phosphate group and then to the cytosine nucleotide derivative, CDP-DG, by reaction with cytosine triphosphate. Meanwhile, IP_3 is stripped of its phosphate groups, going through the intermediates IP_2 and IP to inositol. Inositol and CDP-DG then combine to produce phosphatidylinositol (PI). Finally, PI is phosphorylated to phosphatidyl-4-phosphate (PIP), which regenerates PIP_2.

One theory of lithium action is that it interferes with the regeneration at the stage where inositol combines with CDP-DG. This drug has been used for some years on an empirical basis to treat manias. Lithium causes a large decrease in inositol concentration in neurons. There is a concomitant increase in the concentration of inositol-1-phosphate. Lithium would thus decrease activation of cells depending on polyphosphoinositides as second messengers by interrupting this resynthesis cycle.

This proposed second messenger system, like the others previously described, exists in cells of many types. While its true significance will not be known for many years, it does fit the established pattern of second messengers in the nervous system in that it developed from fundamental biological mechanisms that have application in many tissues and are therefore of historic evolutionary origin.

5.5. Summary

We have discussed in this chapter the anatomical, chemical, physiological, and pharmacological criteria that can be used to identify neurotransmitters. We have divided them into three categories: amino acids, amines, and peptides. We have defined two general types of action based on the nature of the neurotransmitter receptors: ionotropic and metabotropic. Amino acid neurotransmitters are generally ionotropic, opening up ion channels in postsynaptic membranes. Amines and peptides are generally metabotropic, acting through second messenger systems, three of which are described. The rapidly expanding list of existing cotransmitters would suggest many examples of simultaneous but differing postsynaptic effects from the same presynaptic excitation. We have not discussed herein a third type of transmission: genotropic transmission. It involves action at intracellular receptors that are then translocated to the nucleus to initiate transcriptional events. This concept is discussed in Chapter 15, although it is implicit in many of the events described in Chapter 13.

In Chapters 6–12, we discuss the specifics of neuronal systems by considering individual neurotransmitter types.

6

Putative Excitatory Neurons: Glutamate and Aspartate

6.1. Introduction

Glutamate and aspartate meet many of the criteria for neurotransmitter status listed in Chapter 5. They are present in appropriate concentration in brain. They are released in a calcium-dependent fashion on electrical stimulation *in vitro*. They have powerful excitatory effects on neurons when iontophoresed *in vivo*. They have high-affinity uptake systems in nerve endings of a number of neuronal pathways. They have selective binding sites that can be demonstrated radiochemically *in vitro* and selective receptor sites that can be identified pharmacologically *in vivo*.

While the evidence for neurotransmitter status is not as thorough as it is for some other neurotransmitters that are discussed in later chapters, glutamate and aspartate may be the ionotropic transmitters for most excitatory neurons in brain and therefore could belong to some of the most populous neuronal types.

The excitatory effects of glutamate and aspartate on cerebral cortical cells were first demonstrated after direct application of intracarotid injections by Japanese workers studying epileptic phenomena (Okamoto, 1951; T. Hayashi, 1952).

T. Hayashi actually anticipated the iontophoretic method with his technique of injecting small volumes of glutamate and other substances directly into the cortex through a small metal tube, but he did not anticipate that glutamate [or γ-aminobutyric acid (GABA), which he was also investigating] would later be considered neurotransmitters. Subsequent iontophoretic work by Curtis and Watkins (1960, 1963) confirmed the excitatory potency of glutamate as well as the structural requirements for amino acid excitatory activity, but glutamate and aspartate were still not regarded as serious neurotransmitter candidates because they did not have all the properties that were anticipated for such materials. It was not until the systematic studies of K. Krnjevic and Phillis (1963) on

TABLE 6.1
**Content of Some Amino Acids in Whole Rat
Brain**

Amino acid	Content (μmoles/g wet wt.)
Glutamate	13.6 ± 0.4
Taurine	4.8 ± 0.3
Glutamine	4.4 ± 0.2
Aspartate	3.7 ± 0.2
GABA	2.3 ± 0.1
Glycine	1.7 ± 0.1
Serine	1.4 ± 0.1
Alanine	1.1 ± 0.1
Lysine	0.4 ± 0.0

cortical cells that it was realized that most of the appropriate microphysiological criteria for neurotransmitter action were being met by glutamate and aspartate.

Chemical evidence for a neurotransmitter role for these excitatory amino acids was missing in the early stages because there was no means of differentiating a neurotransmitter amino acid pool from other metabolic pools. A major breakthrough on the chemical side came when Wofsey *et al.* (1971) demonstrated high-affinity uptake of glutamic and aspartic acids into a unique population of synaptosomes from rat brain and spinal tissue. This provided a method for identifying nerve terminals that concentrate these amino acids and, through lesions, a means of determining the cell bodies of origin. Nevertheless, it must be recognized that truly definitive markers that can be applied at the cellular level do not exist for glutamate and aspartate as they do for several other neurotransmitters. Therefore, evidence for neuronal identification and for pathways involving these amino acids must in all cases be considered as tentative.

Glutamate and aspartate are present in abundance in the central nervous system (CNS) (Table 6.1), where they serve many roles. Glutamic acid is incorporated into proteins and peptides, is involved in fatty acid synthesis, contributes (along with glutamine) to the regulation of ammonia levels and the control of osmotic or anionic balance, serves as precursor for GABA and for various Krebs cycle intermediates, and is a constituent of important cofactors such as glutathione and folic acid. In view of these many roles, it is not surprising that L-glutamic acid should be the most plentiful amino acid in the adult CNS (Table 6.1). Brain has 3–4 times as much glutamate as it has taurine, glutamine, or aspartate, the three amino acids that are next most abundant.

6.2. Chemistry and Metabolism of Glutamate and Aspartate

Glutamate and aspartate are nonessential amino acids that do not cross the blood–brain barrier (BBB); therefore, they are not supplied to the brain by the blood. Instead they are synthesized from glucose and other precursors by several routes. Glutamate is produced from α-ketoglutarate by transamination, from glutamine by glutaminase action, from α-ketoglutarate by glutamic acid dehydrogenase action, from ornithine by ornithine aminotransferase (Orn-T) via glutamate semialdehyde, and from proline by proline ox-idase with subsequent oxidation of the intermediate Δ^1-pyrroline-5-carboxylic acid (P5C). The precise role of glutamate formed by each of these routes is unknown.

Metabolic studies have revealed the existence of several chemical compartments or pools for both glutamate and aspartate (for a review, see Fonnum, 1984). Unfortunately, it has not yet been possible to associate definitively any of these metabolic pools with anatomical structures such as neurons or glia or with distinct glutamate or aspartate neuronal systems.

Glutamate and aspartate are different from most of the neurotransmitters we shall discuss in later chapters where the production of the particular chemical is totally dedi-cated to the service of its neuronal type. Glutamate particularly may be involved in nontransmitter functions in many different neuronal and glial types. Nevertheless, it can be anticipated that some specialized glutamate and aspartate chemistry will be found in those neurons where they are neurotransmitters. Our discussion will therefore focus on the aspects of glutamate and aspartate biochemistry that may apply to nerve terminal neuro-transmitter function, recognizing that it may be some time before the results of metabolic studies and those of cellular localization studies can be reconciled.

Figure 6.1 shows the tricarboxylic acid cycle in which α-ketoglutarate and oxaloace-tate, the precusor keto acids to glutamate and aspartate, are formed. Mitochondria in all tissues carry out oxidative phosphorylation via this cycle. Transamination to L-glutamate and L-aspartate can take place from these parent keto acids with the help of the enzyme aspartate aminotransferase [(Asp-T) EC 2.6.1.1] according to the reversible conversion shown in the figure. The enzyme is also known as glutamic-oxaloacetic transaminase (GOT) and glutamic-aspartic transaminase. In this book, we refer to the enzyme as Asp-T.

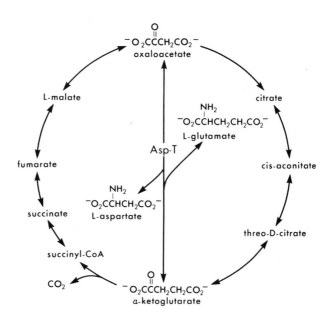

Figure 6.1. Intermediates of the Krebs cycle showing the formation of L-glutamate and L-aspartate by aspar-tate transaminase (Asp-T).

The enzyme is widely distributed in the body and exists in at least two forms in brain (Fonnum, 1968; Heydorn *et al.*, 1985), one of which has been shown by immunocytochemistry to be localized to some nerve terminals. If Asp-T were to be mainly responsible for the synthesis of neurotransmitter pools of glutamate and aspartate, it might be possible for some neurons to secrete both amino acids, since they might be freely interconvertible in the nerve terminals. Such a possibility would explain the frequent conflicts in the literature as to whether a particular path uses aspartate or glutamate. Co-release of aspartate and glutamate has been reported from granule cells (Flint *et al.*, 1981).

Evidence in favor of this enzyme being associated with nerve terminal production of neurotransmitter glutamate is its reported specific histochemical localization in the granule cell area of the rat olfactory bulb, in the nuclear layer of the retina (Recasens and Delaunoy, 1981; R. A. Altschuler *et al.*, 1982), and in the fiber endings of basket cells in the chicken cerebellum (Martinez-Rodriguez *et al.*, 1974). Asp-T has also been localized in terminals of the auditory nerve (Altschuler *et al.*, 1981; Fex *et al.*, 1982), which may utilize either glutamate or aspartate as the transmitter (Wenthold and Gulley, 1978; Wenthold, 1979; M. R. Martin and Adams, 1979); this nerve contains 2–5 times the Asp-T found in other nerves. Furthermore, Asp-T decreases in the ventral cochlear nucleus after the auditory nerve is lesioned (Wenthold, 1980). Asp-T could therefore regulate the relative levels of glutamate and aspartate in these nerve terminals.

If this enzyme were the principal source of glutamate in all glutamate pathways, it should be present in the corticostriatal tract, since this tract is believed to be a fairly massive glutamate pathway (see Section 6.3). No specific histochemical localization of Asp-T occurs in nerve endings in the striatum or hippocampus (Altschuler *et al.*, 1985). Furthermore, P. T.-H. Wong *et al.* (1982a) and Sandberg *et al.* (1985) were unable to find a decrease in this enzyme after lesioning the corticostriatal tract.

Another method for producing neurotransmitter glutamate might be from glutamine, which is present in brain at about one third the level of glutamate. The cycle would need to involve glial cells, which have an uptake system for glutamate (as discussed in Chapter 7). They also have an active synthetic system for converting glutamate into glutamine by the following reaction:

$$\underset{\text{Glutamic acid}}{HOOCCH_2CH_2\overset{NH_2}{CHCOOH}} + NH_3 + ATP \xrightarrow[\text{synthetase}]{\text{Glutamine}}$$

$$\underset{\text{Glutamine}}{H_2NOCCH_2CH_2\overset{NH_2}{CHCOOH}} + ADP + P_i \quad (1)$$

The key glial enzyme is glutamine synthetase, which is not present in neurons. The glutamine formed by this process can be transferred to neurons and might be a reservoir for neuronal glutamate through the action of glutaminase (equation 2). Glutamine has been reported to be a satisfactory precursor for a presumed neurotransmitter glutamate pool (Hamberger *et al.*, 1979; Reubi, 1980).

E. G. McGeer and P. L. McGeer (1979) lesioned the corticostriatal pathway and

$$H_2NOCCH_2CH_2\overset{\underset{\displaystyle NH_2}{|}}{C}HCOOH + H_2O \xrightarrow{\text{Glutaminase}}$$
$$\text{Glutamine}$$

$$HOOCCH_2CH_2\overset{\underset{\displaystyle NH_2}{|}}{C}HCOOH + NH_3 \qquad (2)$$
$$\text{Glutamic acid}$$

found no decrease in striatal glutaminase, suggesting that glutaminase is not associated with glutamate terminals of the pathway originating in the cortex. Kainic acid (KA) lesions of the striatum, which destroy intrinsic neurons but not nerve terminals of remote neurons projecting into the striatum, produced a decrease in glutaminase paralleling the drop of glutamic acid decarboxylase. These data suggest that glutaminase is associated with intrinsic striatal GABA cells rather than projecting cortical glutamate cells. However, Ward *et al.* (1982), using a different assay method, found a 20% reduction in striatal glutaminase, and Sandberg *et al.* (1985) found a transient reduction of about 10% following such lesions—findings that may indicate some association with a corticostriatal tract.

R. A. Altschuler *et al.* (1984, 1985) used an antiserum to glutaminase to localize glutaminase-like immunoreactivity to terminals of the auditory nerve and to granule cells of the dentate gyrus and pyramidal cells of regio superior and inferior and in layer II cells of the entorhinal cortex, all of which are believed to use glutamate as their neurotransmitter.

Glutamate can be formed from α-ketoglutarate by nitrogen fixation using the enzyme glutamic acid dehydrogenase:

$$HOOCCH_2CH_2COCOOH + NH_4{}^+ + DPNH \xrightarrow[\text{dehydrogenase}]{\text{Glutamic acid}}$$
$$\text{α-Ketoglutaric acid}$$

$$HOOCCH_2CH_2\overset{\underset{\displaystyle NH_2}{|}}{C}HCOOH + DPN^+ + H_2O \qquad (3)$$
$$\text{Glutamic acid}$$

So far, no direct evidence exists that this enzyme is concentrated in any presumed glutamate pathway. It seems to be localized to mitochondria (Reijnierse *et al.*, 1975). Humans with a partial deficiency of this enzyme in leukocytes show chronic progressive extrapyramidal rigidity, suggesting a possible involvement of the enzyme with extrapyramidal neuronal pathways, although the disease has been linked to olivopontocerebellar atrophy (Duvoisin *et al.*, 1983).

Another possible route to neurotransmitter glutamate might be from ornithine through transamination and subsequent oxidation of the intermediate P5C (equation 4) (for a review, see E. Roberts, 1981). A parallelism exists between the regional distributions of Orn-T activity and glutamate uptake. About 85% of the enzyme is in the synaptosomal fraction (P. T.-H. Wong *et al.*, 1981; P. T.-H. Wong and McGeer, 1981). Activity in the cortex and striatum of patients with Huntington's disease is decreased (P. T.-H. Wong *et al.*, 1982b). However, only small decreases of Orn-T activity occurred after lesions of the corticostriatal tract. Somewhat larger decreases were observed if intrinsic neurons in

$$\underset{\text{Ornithine}}{H_2NCH_2CH_2CH_2\overset{\overset{\displaystyle NH_2}{|}}{C}HCOOH} \xrightarrow{\text{(Orn-T)}} \underset{\underset{\text{P5C}}{\underset{\displaystyle N}{HC\diagdown\diagup CHCOOH}}}{\overset{\displaystyle H_2C\text{---}CH_2}{\underset{|}{|}}} \xrightarrow{\text{(P5C dehydrogenase)}} \underset{\underset{\text{Glutamic semialdehyde}}{O\quad NH_2}}{\overset{\displaystyle H_2C\text{---}CH_2}{\underset{\displaystyle HC\quad CHCOOH}{|\qquad|}}} \quad (4)$$

Proline (with Proline oxidase arrow), Glutamate

the striatum were destroyed by KA. The majority of activity was unaffected by either treatment, indicating the possibility of a significant glial pool (E. G. McGeer *et al.*, 1983). Thus, Orn-T cannot be exclusively involved in the synthesis of transmitter glutamate.

The possibility that proline or its oxidation product P5C [see equation (4)] could be the precursor would seem to be eliminated by histochemical studies (E. G. McGeer *et al.*, 1983). Neurons failed to stain for either proline oxidase or P5C dehydrogenase, although there was staining for both enzymes in Bergmann glia of the cerebellum. Neither of these enzymes showed a regional distribution that paralleled that of high-affinity glutamate uptake (S. G. Thompson *et al.*, 1985).

In summary, no definite source of neurotransmitter glutamate has been identified, although several possible sources exist and evidence for some of them has been proposed. It has been suggested that a combination of routes may exist for glutamate production, with the critical factor being the existence of specialized synaptic vesicles that could concentrate the available glutamate.

As far as the aspartate transmitter pool is concerned, glutamate (Canzek and Reubi, 1980), glutamine (Bradford and Ward, 1976), and asparagine (Reubi *et al.*, 1980) have all been suggested as precursor sources. Lesioning Glu/Asp afferents to the hippocampus (Nadler *et al.*, 1978) did not cause a decrease in Asp-T, which argues against this being an exclusive or even primary source for the transmitter pool of aspartate. Evidence for synthesis from glutamine or asparagine is equally slim.

6.3. Anatomical Distribution of Glutamate and Aspartate

The distribution of glutamate and aspartate in various brain areas is shown in Table 6.2 for human postmortem tissue. In the areas measured, there is little more than a 2-fold variation in either substance. This is weak evidence for specific pathways, but would nevertheless be anticipated if the amino acids were serving generally as excitatory transmitters as well as filling other metabolic roles.

The following types of evidence have been used to define presumptive neuronal pathways for glutamate or aspartate (Glu/Asp):

1. Reduction in high-affinity uptake following lesions.
2. Reduction in endogenous concentration following lesions.

TABLE 6.2
Glutamine, Glutamate, and Aspartate Concentrations in Human Brain[a]

Brain region	Concentration (μmoles/g wet wt.)		
	Glutamate	Glutamine	Aspartate
Frontal cortex	8.93	3.84	1.92
Temporal cortex	10.78	4.41	1.89
Occipital cortex	8.64	3.75	2.56
Cerebellum	9.64	5.17	2.29
Corpus callosum	5.54	2.56	1.16
Thalamus	8.37	3.70	2.48
Putamen/globus pallidus	10.54	3.72	1.84
Caudate	10.48	3.18	1.30
Amygdala	9.36	4.65	1.73
Substantia nigra	5.08	2.65	2.59
Red nucleus	4.95	2.38	1.96
Hypothalamus	5.77	3.23	2.26

[a]From T. L. Perry *et al.* (1971).

3. Release of Glu/Asp on stimulation.
4. Retrograde transport of [^{3}H]-D-aspartate.
5. Immunohistochemical localization of glutamate, glutamine, and Asp-T.

Evidence for Glu/Asp pathways is based largely on changes in uptake following lesions of known pathways. Additional evidence has been obtained from release of glutamate following stimulation of slices *in vitro*; some data have also been obtained *in vivo*. Early immunohistochemical results have lent support to the uptake and release data.

The proposed pathways for Glu/Asp are listed in Table 6.3 and presented diagram-

TABLE 6.3
Some Pathways Proposed to Use Acidic Amino Acid Transmitters in the Mammalian Brain: Summary of Evidence

Pathway	Evidence and ref. nos.[a]	
	Presynaptic criteria	Postsynaptic criteria
Cerebellar		
Granule cells	U,[7] C,[8] R[6,10]	P[9]
Deep cerebellar n. to red nucleus and thalamus	I[62]	—
Cerebral cortex to:		
Amygdala	U,[11] P[57]	—
Cuneate nucleus	U[12,22]	P[13]
Dentate gyrus	U,[14,17] R[14,16,18]	P[15,30]
Lateral geniculate nucleus	U[19]	—

(*continued*)

TABLE 6.3 (*Continued*)

Pathway	Evidence and ref. nos.[a]	
	Presynaptic criteria	Postsynaptic criteria
Nucleus accumbens	U[20,58]	—
Olfactory tubercle	U[11]	—
Pontine nuclei	U,[22,49,54] R[23]	—
Red nucleus	U[22],[53]	—
Spinal cord	U[22,49,53]	—
Striatum	U,[20,21,24,49,53] C,[26] R[27]	P[25]
Substantia nigra	U,[11] C[55]	—
Ventral tegmental area	U,[56] C[56]	—
Thalamus	U[11,21,22,28,49,53]	P[29]
Hippocampal		
Commissural/associational	U,[14,17] R[14]	P[30]
Schaffer collaterals	R[10,31]	P,[33] A[32]
Hippocampus to:		
Bed nucleus of stria terminalis	U/C[34]	—
Hypothalamus	U,[34,35] C[34]	—
Lateral septum	U,[34–36,39] C,[34,37] R[38]	—
Mammillary body	U/C[34]	—
Nucleus accumbens	U/C[34]	—
Nucleus of the diagonal band	U,[34,39] C[34]	—
Lateral olfactory tract	C,[40] R[41]	P[42]
Brainstem		
Auditory nerve afferents	C,[1] R,[2] I[59]	P[3]
Inferior olive to cerebellum	U,[4] C,[5] R,[4,6] I[61]	—
Spinal trigeminal n. to thalamus	I[60]	—
Pontine n. to cerebellum	I[59]	—
Spinal cord		
Descending tracts	U,[22] C,[43] R[44,50]	—
Interneurons	C[45,47]	P[46]
Primary afferents	C,[47,48] R[50]	P,[52] A[51]

[a]The evidence listed is derived from studies involving the presynaptic criteria of uptake (U), content (C), evoked release (R), and immunohistochemistry (I) and the postsynaptic criteria of identity of action (A) and pharmacological antagonism (P). References: (1) Wenthold and Gulley (1978); (2) Canzek and Reubi (1980); Wenthold (1979); (3) M. R. Martin (1980); (4) Wiklund et al. (1982); (5) Rea et al. (1980); (6) Flint et al. (1981); (7) Rohde et al. (1979); A. B. Young et al. (1974); (8) McBride et al. (1976a,b); (9) Stone (1979b); (10) Skrede and Malthe-Sørenssen (1981); (11) Fonnum et al. (1981b); (12) Rustioni and Cuenod (1982); (13) Stone (1979a); (14) Nadler et al. (1978); (15) Koerner and Cotman (1981); Wheal and Miller (1980); (16) Hamberger et al. (1979); (17) Storm-Mathisen (1977); (18) W. F. White et al. (1977); (19) Baughman and Gilbert (1981); Lund-Karlsen and Fonnum (1978); (20) Walaas (1981); (21) Fonnum et al. (1981a); (22) A. B. Young et al. (1981); (23) Thangnipon and Storm-Mathisen (1981); (24) Divac et al. (1977); P. L. McGeer et al. (1977a); Streit (1980); (25) H. J. Spencer (1976); (26) Hassler et al. (1982); J. S. Kim et al. (1977); (27) Cuenod et al. (1981); Druce et al. (1982); Godukhin et al. (1980); Reubi and Cuenod (1979); Rowlands and Roberts (1980); (28) Rustioni et al. (1982); (29) Haldeman et al. (1972); (30) Hicks and McLennan (1979); J. F. Krnjevic and Cotman (1981); (31) Malthe-Sørenssen et al. (1979); Wieraszko and Lynch (1978); (32) Hablitz and Langmoen (1982); (33) M. Segal (1981); W. F. White et al. (1979); (34) Walaas and Fonnum (1980); (35) Storm-Mathisen and Woxen-Opsahl (1978); (36) Fonnum and Walaas (1978); (37) Nitsch et al. (1979); (38) Malthe-Sørenssen et al. (1980b); (39) Zaczek et al. (1979); (40) Collins (1979); Harvey et al. (1975); (41) Bradford and Richards (1976); Collins and Probett (1981); Yamamoto and Matsui (1976); (42) Collins (1982); Hori et al. (1981); (43) Rizzoli (1968); (44) Fagg et al. (1978); (45) Davidoff et al. (1967); Homma et al. (1979); (46) Biscoe et al. (1977); J. Davies et al. (1980); J. Davies and Watkins (1979); R. H. Evans et al. (1979); Francis et al. (1980); Lodge et al. (1978); J. C. Watkins and Evans (1981); (47) Graham et al. (1967); G. A. R. Johnston (1968); (48) A. W. Duggan and Johnston (1970); J. L. Johnson and Aprison (1970); I. M. Jones et al. (1974); P. J. Roberts and Keen (1974); T. Takahashi and Otsuka (1975); (49) A. B. Young et al. (1983); (50) P. J. Roberts and Mitchell (1972); (51) Curtis (1965); Engberg and Marshall (1979); Zieglgansberger and Puil (1973); (52) J. Davies and Watkins (1982); R. H. Evans et al. (1982); (53) Nieoullon and Dusticier (1983); (54) Thangnipon et al. (1983); (55) J. Kornhuber et al. (1984); (56) Christie et al. (1985a); (57) Brothers and Finch (1985); (58) Christie et al. (1985b); (59) R. A. Altschuler et al. (1985); (60) Magnusson et al. (1985); (61) Beitz et al. (1985); (62) Monaghan et al. (1985).

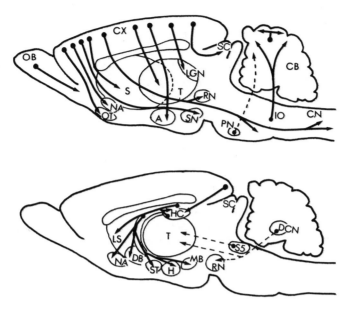

Figure 6.2. Some pathways proposed to use acidic amino acid neurotransmitters in the mammalian brain. The evidence available for each pathway is summarized in Table 6.3 and described in the text. (A) Amygdala; (CB) cerebellum; (CN) cuneate nucleus; (CX) cerebral cortex; (DB) nucleus of the diagonal band; (DCN) deep cerebellar nuclei; (H) hypothalamus; (HC) hippocampus; (IO) inferior olive; (LGN) lateral geniculate nucleus; (LS) lateral septum; (MB) mammillary body; (NA) nucleus accumbens; (OB) olfactory bulb; (OT) olfactory tubercle; (PN) pontine nuclei; (RN) red nucleus; (S) striatum; (SC) superior colliculus; (SN) substantia nigra; (ST) bed nucleus of the stria terminalis; (S5) spinal nucleus of nerve 5; (T) thalamus.

matically in Figure 6.2. The evidence supporting them is also given. It is important to recognize that present techniques are limited in their ability to distinguish between glutamate and aspartate because the uptake systems are so highly similar. It should also be emphasized that these pathways are highly tentative, although they serve as a base on which future data may be built.

Nearly all the pathways that have so far been delineated are corticofugal pathways from the neocortex and hippocampus. Evidence favors glutamate over aspartate as the transmitter for most of these pathways. Other glutamate systems that have been suggested are cerebellar granule cells and afferents of the deep cerebellar nuclei, spinal trigeminal nucleus, and pontine nuclei. For aspartate, evidence exists for hippocampal commissural fibers, spinal interneurons, and climbing fibers of the cerebellum. Other possible Glu/Asp pathways include the lateral olfactory tract, Schaffer collaterals in the hippocampus, and primary sensory afferent pathways, including the auditory and optic nerves.

P. L. McGeer *et al.* (1977a) and Divac *et al.* (1977) lesioned the massive, excitatory corticostriatal pathway and found a drop of 40–50% in high-affinity glutamate uptake into the synaptosomal fraction of the ipsilateral striatum. The uptake of GABA, which has intrinsic and projecting striatal neurons (see Chapter 7), and of dopamine, which has striatal nerve endings from nigral projections (see Chapter 9), was not affected. Since the concentration of glutamate is about 6 times as high as that of aspartate in the striatum (see

Table 6.2), as opposed to only 3.7 for the whole brain [see Table 6.1 (Section 6.1)], it was proposed that the pathway probably uses glutamate, a conclusion that has been confirmed by the selective loss of endogenous glutamate after such lesions (Fonnum *et al.*, 1981a; Hassler *et al.*, 1982). Lesions of other corticofugal pathways have produced losses in high-affinity glutamate uptake in a number of other subcortical structures as listed in Table 6.3 (A. B. Young *et al.*, 1983; Lund-Karlsen and Fonnum, 1978; J. E. Walker and Fonnum, 1983; Walaas and Fonnum, 1979c; Fonnum *et al.*, 1981a,b). These include the nucleus accumbens and amygdala as well as some thalamic, midbrain, pontine, and spinal cord nuclei that receive pyramidal cell projections from the cerebral cortex. Another apparent glutamate cortical pathway is from the entorhinal cortex to the dentate area of the hippocampus. Lesioning this pathway produced a 52% loss of glutamate uptake in the terminal area (Storm-Mathisen, 1977). Bilateral lesions led to a decrease of Ca^{2+}-dependent, K^+-evoked release of glutamate, but not aspartate (Nadler *et al.*, 1978).

The pyramidal cells of the hippocampus, which project to the limbic forebrain and diencephalon, also appear to use glutamate (see Figure 6.2 and Table 6.3). Lesioning of the efferent fibers of these cells in the fornix–fimbria area decreased glutamate uptake by 40–70% and endogenous glutamate by 15–34% in all the terminal areas (Walaas and Fonnum, 1979c).

Retrograde transport of [³H]-D-aspartate has also been taken as evidence of Glu/Asp pathways, since the uptake and possibly the transport system should be specific for this class of neurons. Thus, Rustioni and Cuenod (1982) noted retrograde labeling of layer V pyramidal cells in the sensory motor cortex following injection of [³H]-D-aspartate into the terminal field area in the cuneate nucleus. Labeling was also observed in dorsal root ganglion cells, providing suggestive evidence that these primary afferent neurons, which also project to the cuneate nucleus, are of the Glu/Asp type. Using a similar technique, Matute and Martinez-Millan (1985) detected [³H]-D-aspartate in the visual cortex and superior colliculus following injections into the lateral geniculate nucleus.

An association of glutamate with cerebellar granule cells has been suggested by the fact that "staggerer" and "weaver" mutant mice that have granule cell dysgenesis show lowered glutamate levels (McBride *et al.*, 1976a; Hudson *et al.*, 1976). Destruction of granule cells by viral infections or X-irradiation brings about reduced endogenous glutamate as well as reduced high-affinity glutamate uptake (A. B. Young *et al.*, 1974; Rohde *et al.*, 1979).

Auditory nerve fibers from the spiral ganglion in the cochlea projecting to the cochlear nucleus in the brainstem are thought to use glutamate on the basis of release of the amino acid following stimulation (Canzek and Reubi, 1980). Asp-T is localized to terminals in the cochlear nucleus, which may be supporting evidence.

Immunohistochemistry for glutamate itself is perhaps the most promising approach to definitive localization of glutamate neurons and pathways. It depends on developing antibodies to a glutamate–protein complex that has been created by chemically coupling glutamate to a protein. If the glutamate–protein complex is sufficiently like the endogenous system that is created when neurotransmitter glutamate becomes bound to tissue through a fixation process, then the antibody should be capable of correctly identifying the bound neurotransmitter material. This approach has been used with great success in the histochemical localization of serotonin (see Chapter 10) and histamine (see Chapter 11) and with some success in the localization of GABA (see Chapter 7). Potential difficulties

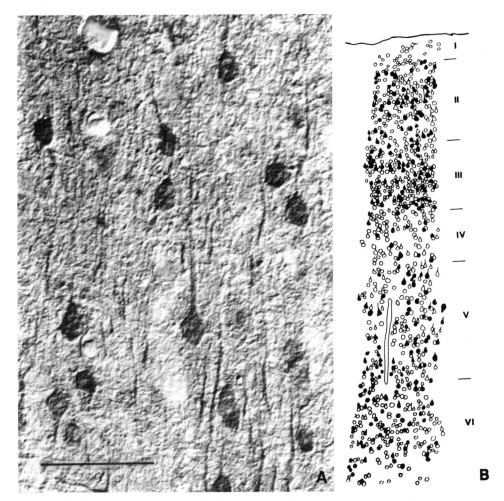

Figure 6.3. (A) Immunohistochemical staining for glutamate-like immunoreactivity using a rabbit antibody to hemocyanin-coupled glutamate. Notice the darkly staining pyramidal cells in this 25-μm vibratome section of rat somatosensory cortex using Nomarski optics. Scale bar: 100 μm. (B) Camera lucida drawing of rat somatosensory cortex showing the distribution of glutamate-positive neurons in the various layers. Courtesy of F. Conti, A. Rustione, and P. Petrusz (University of North Carolina).

are in cross-reactivity of the antibody with other small endogenous molecules that become bound or with glutamate that is not associated with a glutamate neurotransmitter pool. Despite these potential difficulties, preliminary success has been obtained by Storm-Mathisen *et al.* (1983) and by Conti *et al.* (1985).

Conti *et al.* (1985) have obtained excellent staining of neurons of the rat somatosensory cortex as shown in Figure 6.3A for layer V pyramidal cells. Antibodies were developed in rabbits to glutamate conjugated to hemocyanin by glutaraldehyde. There was no staining of GABA cortical neurons (Conti, personal communication). The distribution of positive neurons in the somatosensory cortex is shown in Figure 6.3B, indicating the extensive distribution in all layers except layer I.

The evidence for aspartate pathways is much slimmer than for glutamate. The commissural connection between the two hippocampi is, however, one well-documented exception. Disruption of this connection results in a reduced release of aspartate but not glutamate from stimulated hippocampal slices (Nadler *et al.*, 1978).

Two lines of evidence suggest that climbing fibers from the inferior olive to the cerebellum are of the aspartate type. Wiklund *et al.* (1982) showed retrograde labeling of the inferior olive following injection of $[^3H]$-D-aspartate into the climbing fiber terminal areas in the cerebellar cortex or its deep nuclei, and Rea *et al.* (1980) reported that 3-acetylpyridine-induced destruction of the inferior olive brought about a 16% decrease in aspartate release from the cerebellum.

In some cases, the evidence is controversial or uncertain with respect to whether glutamate or aspartate may be the transmitter. The lateral olfactory tract is a case in point. Harvey *et al.* (1975) reported a reduction in glutamate and aspartate in the guinea pig olfactory cortex following bulbectomy. Bradford and Richards (1976) provided supporting evidence by showing that stimulation of the tract enhanced the release of glutamate but not aspartate. But Collins and Probett (1981) found that lesioning of the tract decreased the release of aspartate but not glutamate from slices of olfactory cortex in the rat. Ffrench-Mullen *et al.* (1985) have presented some chemical and physiological evidence that the true transmitter is *N*-acetylaspartylglutamate (NAAG). NAAG has a diverse regional distribution in brain and was previously suggested as a possible neurotransmitter in the corticostriatal and mammiloseptal tracts, as well as in striatal, hippocampal, and spinal cord neurons, on the basis of lesion-induced declines in concentration (Koller *et al.*, 1984). Its excitatory properties have, however, been denied (Riveros and Orrego, 1984; Luini *et al.*, 1984).

While the anatomical data at this stage must still be regarded as highly tentative, it can be said that glutamate and aspartate meet many of the generally accepted anatomical criteria for neurotransmitter status. It is possible that they constitute the main source of excitatory projecting neurons from the cortex, as well as primary incoming afferents from the periphery. It may be anticipated that many more pathways will be identified in the future. Better methods will be required, however, to distinguish neurotransmitter from metabolic pools. These methods will need to be applied at the cellular level before truly convincing anatomical data will emerge (for reviews, see Fagg and Foster, 1983; Fonnum, 1984).

6.4. Physiology of Glutamate and Aspartate

Highly convincing evidence that L-glutamate and L-aspartate should be neurotransmitters comes from their iontophoretic actions. Both these dicarboxylic amino acids powerfully excite virtually all neurons with which they come in contact. The characteristics of glutamate excitation are its extraordinary sensitivity (under optimal conditions, cells can be excited with as little as 10^{-14} mole), its instantaneous onset, and the rapid termination of its action. This is accompanied by a marked fall in membrane resistance and an increase in sodium and other ion permeability comparable to that observed when acetylcholine is applied near the muscle endplate [see Figure 3.3 (Section 3.2.3)] (K. Krnjevic, 1974). The action is not blocked by tetrodotoxin, so that it is a specific excitation of receptor sites [see Figure 3.12 (Section 3.3)] and not a stimulation of ionic conductance as in Figure 2.6 (Section 2.3). Therefore, it cannot be mediated by the sodium channels responsible for the propagated action potential (Zieglgansberger and Puil, 1973). In most respects, L-aspartate has an iontophoretic action comparable to that of glutamate, although it is generally less potent.

The authenticity of glutamate as a neurotransmitter at the locust neuromuscular junction (Usherwood, 1981) and the squid ganglion giant synapse (N. Kawai *et al.*, 1983) (see Section 3.3) seems well established. But, as mentioned in Chapter 4, significant changes in the internal concentration of cations cannot be produced by electrophoretic injection of any material. Thus, it has not been possible to test for the ionic mechanisms of excitatory synaptic transmission of putative synaptic excitants such as glutamate in the way that was done by A. Takeuchi and N. Takeuchi (1962) for neuromuscular transmission. Nevertheless, considerable physiological evidence has been accumulated for glutamate and aspartate as excitatory neurotransmitters (for a review, see Puil, 1981), including the identification of specific receptors and specific binding sites in brain tissue.

An exciting aspect of the physiology of glutamate systems is their apparent long-term potentiation in the hippocampus, a phenomenon that is probably involved in memory processes (see Chapter 16).

6.4.1. Receptor Sites

Glutamate and aspartate excite a multiplicity of receptors in brain when applied iontophoretically, but it has so far not been possible to associate these receptors with any particular glutamate or aspartate pathway. A variety of agonists and antagonists for these iontophoretically defined receptors have been defined through detailed studies led by Watkins and his colleagues (for reviews, see J. C. Watkins and Evans, 1981; A. C. Foster and Fagg, 1984).

Some of the many excitatory and blocking agents that have been discovered are shown in Table 6.4. They reveal four main receptor subtypes—N-methyl-D-aspartate (NMDA), quisqualic acid (QA), kainic acid (KA), and L-aminophosphonobutyric acid (AP4) receptors.

The NMDA receptor has a mild preference for aspartate over glutamate. On this basis, it has tentatively been assigned by some as the physiological aspartate receptor. However, NMDA antagonists block the excitation of corticofugal fibers to the cuneate nucleus and caudate nucleus (Stone, 1979a), which are considered to be of the glutamate type. The receptor is thought to exist in synapses in the spinal cord, cerebral cortex, cuneate nucleus, caudate nucleus, lateral geniculate body, hippocampus, cochlear nucleus, olfactory cortex, and cerebellum. This distribution could apply to both glutamate and aspartate pathways (for a review, see A. C. Foster and Fagg, 1984). As to its physiology, the receptor is activated strongly by NMDA and only weakly by QA and KA. It is strongly antagonized by the dipeptides β-D-aspartylaminomethylphosphonic acid (Asp-Amp) and γ-D-glutamylaminomethylphosphonic acid (Glu-Amp) (A. W. Jones *et al.*, 1984). These dipeptides have shown some promise as anticonvulsant agents. Thus, this physiological approach may lead to better treatments for epilepsy. The receptor is also strongly antagonized by 2-amino-7-phosphonoheptanoic acid (AP7) and 2-amino-5--phosphonovaleric acid (AP5). Moderate antagonism is shown by D-isomers of the long-chain diaminocarboxylic acids, α-aminoadipate and α-aminosuberate.

The QA receptor is activated by QA and has a high affinity for L-glutamate and α-

TABLE 6.4
Acid Amino Acid Receptor Classes

Active compounds	Receptor class[a]			
	NMDA	QA	KA	AP4
Agonists with strong to moderate potency	NMDA	QA	KA	L-AP4[b]
	NMLA	Willardiine	Domoic acid	(L-Glutamate)?[b]
	Ibotenate	AMPA		
	ADCP	L-Glutamate		
	L-Homocysteate			
	D-Glutamate			
Agonists with weak potency	QA	NMDA		
	L-Glutamate	NMLA		
	KA	Ibotenate		
Inactive compounds	L-AP7	D-α-Aminoadipate		
Antagonists	D-AP7	GDEE		
	D-AP5	GAMS		
	Asp-Amp			
	Glu-Amp			
	D-α-Aminoadipate			
	D-α-Aminosuberate			

[a] Abbreviations: (ADCP) *cis*-1-amino-1,3-dicarboxycyclopentane; (AMPA) α-amino-3-hydroxy-5-methyl-4-isoxazolepropionic acid; (AP4) 2-amino-4-phosphonobutyric acid; (AP5) 2-amino-5-phosphonovaleric acid; (AP7) 2-amino-7-phosphonoheptanoic acid; (Asp-Amp) β-D-aspartylaminomethylphosphonic acid; (GAMS) γ-D-glutamylaminoethyl sulfonate; (GDEE) diethyl glutamate; (Glu-Amp) γ-D-glutamylaminomethylphosphonic acid; (NMDA) N-methyl-D-aspartic acid; (NMLA) N-methyl-L-aspartic acid.

[b] It is not known whether the compound is agonist or antagonist at this receptor.

amino-3-hydroxy-5-methyl-4-isoxazolepropionic acid (AMPA). It is weakly blocked by diethyl glutamate (GDEE) (H. J. Spencer, 1976) and by γ-D-glutamylaminoethyl sulfonate (GAMS) (J. Davies and Watkins, 1985). Synaptic responses blocked by GDEE, which presumably correspond to QA receptors, have been described in the cerebral cortex, thalamus, striatum, and hippocampus.

The KA receptor is activated by KA and domoic acid and is weakly blocked by GAMS and γ-D-glutamylglycine. Although it is not one of the major glutamate receptors, it seems to be particularly important for excitotoxicity (see Section 6.4.3). Since effective antagonists of KA are not known, it has not been possible to evaluate the endogenous KA receptor types iontophoretically.

Only limited studies have been conducted to determine the nature of AP4-sensitive receptors. It is not yet known whether this agent is an agonist or antagonist at the receptor (A. C. Foster and Fagg, 1984). However, L-AP4 is a potent antagonist of excitatory synaptic responses at such sites as the spinal cord and perforant path synapses, which presumably use Glu/Asp as their neurotransmitter agent. NMDA antagonists are not effective at these sites. KA and QA responses at their own receptors are unaffected by concentrations of AP4 that maximally depress these AP4-sensitive synapses. Thus, AP4 receptors represent a separate class, with a physiological significance that is yet to be determined.

6.4.2. Binding Sites

As described in Chapter 5, the existence of selective binding sites for neurotransmitter candidates is powerful evidence for an authentic neurotransmitter role. Such sites have been identified for glutamate and aspartate, but there is only a very tentative correspondence between the pharmacological receptors identified by iontophoresis and the chemical receptors identified by binding. Neither class of receptor has been associated with definite glutamate or aspartate pathways.

The problem of correlating binding sites with genuine amino acid receptors is made much more difficult by the high sensitivity of the identified binding sites to the conditions of assay. The tissues used, buffers employed, preparation of tissues, and presence of ions have all been found to have major effects on the equilibrium dissociation constants (K_D) and maximum numbers (B_{max}) of binding sites [see Figure 5.1 (Section 5.3.1)].

A. C. Foster and Fagg (1984) have attempted to rationalize the literature by dividing the amino acid binding sites into four classes, A-1–A-4 (Table 6.5), and correlating them with iontophoretic receptors.

The A-1 amino acid binding site is enhanced by freezing and thawing and is not influenced by ion concentration. Thus, it is independent of the concentration of sodium, calcium, and chloride ions. Glutamate and aspartate are highly effective ligands for this site, which is more sensitive to NMDA than to QA or KA. Presumably, therefore, this major binding site corresponds to the NMDA iontophoretic receptor.

The A-2 binding site is also enhanced by freeze–thawing, but has glutamate and AMPA as preferred ligands. Since binding is displaced by QA and glutamate, the A-2 site is postulated to correspond to the QA iontophoretic receptor group.

The A-3 site is unaffected by freeze–thawing and strongly binds KA. It is displaced by domoic acid, QA, and glutamate. It presumably corresponds to the KA iontophoretic receptor.

The A-4 binding site is present in high density in fresh tissue, but is abolished by freeze–thawing. It is unmasked by the inclusion of calcium and chloride in the incubation medium. Glutamate and AP4 are the ligands of choice, and the binding is displaced by QA, ibotenate, and serine-O-sulfate. On subcellular fractionation, this type of binding was found to be highly concentrated in the synaptic membrane fraction of rat brain (Butcher *et al.*, 1985). This binding site has been assigned to the AP4 iontophoretic category of receptors.

In addition to these four binding sites, there is also a high-affinity uptake site which is presumably for the purpose of removing the amino acids from the synapse and concentrating them in nerve endings. This uptake site is sodium-dependent, with cysteinesulfinate being an effective competitor of L-glutamate and L-aspartate.

6.4.3. Excitotoxicity

A physiological consequence of the fact that glutamate, aspartate, and a number of structurally related amino acids powerfully excite neurons in the CNS is that they will bring about neuronal destruction if administered in sufficient excess. This neurotoxic consequence of excitatory activity, unnoticed for many years, is being exploited by neuroscientists as a means of providing much new information about the operation of neuronal systems.

TABLE 6.5
Classification of Acidic Amino Acid Binding Sites

Binding site	Radioligands	Distinguishing properties	Receptor and function
A-1	D-AP5 L-Glutamate L-Aspartate	Binding is independent of Cl^-, Ca^{2+}, and Na^+; enhanced by freeze–thawing; displaced by micromolar AP5 and AP7; moderately sensitive to NMDA, less sensitive to AP4, QA, and KA	NMDA Mediates some synaptic excitatory responses
A-2	AMPA L-Glutamate	Binding is enhanced by freeze–thawing; displaced by micromolar QA and L-glutamate; less sensitive to aspartate and ibotenate	QA Mediates excitation; synaptic role unclear
A-3	KA	Binding is unaffected by freeze–thawing; displaced by nanomolar to micromolar domoic acid, QA, and glutamate; less sensitive to cysteate, NMDA, and ibotenate; isopropylene side chain essential for activity of KA	KA Mediates excitation and neurotoxicity; synaptic role unclear
A-4	(D)L-AP4 L-Glutamate	Binding sites present at high density; dependent on Cl^-, stimulated by Ca^{2+}, inhibited by Na^+; abolished by freeze–thawing; displaced by nM QA, ibotenate, and serine-O-sulfate and by µM AP4; less sensitive to NMDA and KA	AP4 Modulates synaptic excitation; precise role and location unclear
Transport site	L-Glutamate L-Aspartate D-Aspartate	Binding sites present at high density; dependent on >40 mM Na^+; displaced by µM L-glutamate, L-aspartate, and cysteine sulfinate; less sensitive to ibotenate, QA, and KA	Na^+-dependent transport; removal of "transmitter" amino acids from synapse

The excitotoxic hypothesis was first articulated by Olney and his colleagues in 1974 (Olney, 1978). It proposes that a depolarization mechanism underlies the neurotoxic effects of a variety of excitatory amino acids. During excitation, ion channels are opened wide, permitting an exchange of intracellular and extracellular ionic species. This causes energy-dependent homeostatic mechanisms to draw heavily on the cell's energy stores in an effort to restore the ionic balance. If these energy sources become irreversibly depleted, or if the ionic exchange reaches a state in which cell membrane pumps can no longer function, then cell death will occur. Following intracerebral injection of KA or one of the other neurotoxic excitatory amino acids, evidence of damage to the dendrites and to the soma appears in a few minutes (Olney, 1978). Large changes in high-energy phosphate and glucose levels are observed, consistent with this hypothesis (Retz and Coyle,

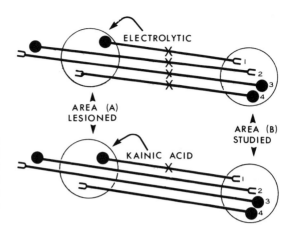

Figure 6.4. Presumed selectivity of the lesion caused by local injection of KA as compared to an electrolytic (or surgical) lesion. All four neurons shown as ending in area B would be destroyed by an electrolytic or surgical lesion of area A, but only neuron 1 would be destroyed by the injection of KA into area A.

1982). Since amino acid receptors are absent, or present only in low concentration in axons and their terminals, these processes are spared. They only become involved if other parts of the neuron are first damaged.

Figure 6.4 is an illustration of why the local effects of intracerebrally injected excitotoxins make them useful tools for neuroscientists. Two brain areas, A and B, are illustrated, with area A being lesioned. If the lesion in area A is electrolytic as shown in the top drawing, it will interrupt all structures that reside in, pass through, or terminate in the area. Contrast this with the bottom drawing, which illustrates the effect of an excitotoxic lesion in the same area. Only neuron 1 is destroyed. Neurons 2–4 are preserved because they terminate in or pass through the lesioned area and their processes do not possess amino acid receptors. Thus, precise information can be elicited about pathways that otherwise would be difficult to obtain. It must be remembered, however, that Figure 6.4 indicates a general but not an absolute result. Close to the injection site, all structures may be damaged. It is only outside that very limited area where selectivity is seen.

While the excitotoxic concept is a general one, it can be anticipated that individual neurotoxic agents will have different effects on specific regions of the nervous system according to the precise amino acid receptors they activate.

The excitotoxins include those compounds that powerfully stimulate the receptors discussed in Section 6.4.1 (see Table 6.4). Figure 6.5 shows the structures of some of the more commonly used compounds. They all have two acidic groups separated by 3–4 atoms, with one acidic group being part of an α-amino acid structure or its equivalent. These compounds can be viewed as analogues of L-glutamate and L-aspartate. Cysteinesulfinic acid and cysteic acid (not shown) are minor amino acids in brain that are concerned with cysteine metabolism, while homocysteinesulfinic acid (not shown) and homocysteic acid are concerned with methionine metabolism. Folic acid is a vitamin required by all tissues. Ibotenic acid is obtained from the mushroom *Amanita strobiliformis*. KA comes from the Japanese seaweed *Digenea simplex,* as do its less active analogues L-allo-KA (not shown) and dihydro-KA (not shown). Quinolinic acid may be of particular interest because this metabolite of tryptophan, an essential amino acid, is reported to occur in mammalian brain in nanomole/gram concentrations. It has a diverse regional distribution (Stone and Connick, 1985). Its catabolic enzyme, quinolinic acid phos-

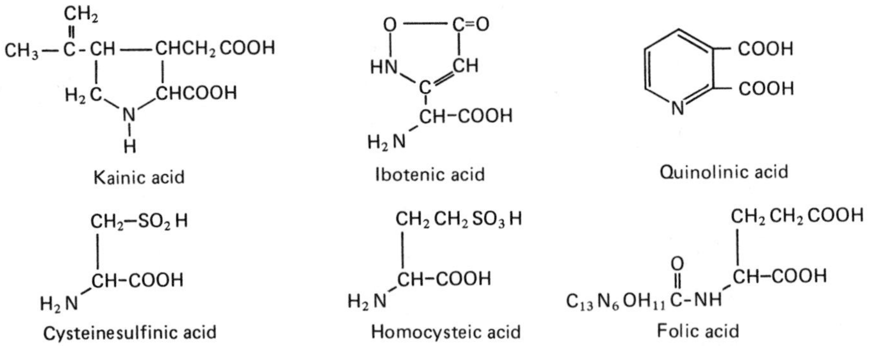

Figure 6.5. Structures of some excitatory and neurotoxic amino acids.

phoribosyltransferase [(QPRT) EC 2.4.2.19], is preferentially localized in the synaptosomal fraction of rat brain homogenates (A. C. Foster *et al.*, 1985a) and is present in particularly high concentrations in human caudate (A. C. Foster *et al.*, 1985b).

Three factors play a role in the extent of neuronal damage produced by a given neurotoxin: (1) the nature and concentration of amino acid receptors affected by that neurotoxin, (2) the synergistic effect of endogenous Glu/Asp input to the injected area, and (3) transsynaptic spread of damage due to induced epileptiform activity.

The types of amino acid receptors that exist have already been referred to in Table 6.4. The potential interplay between an excitotoxin and endogenous Glu/Asp input to the injected area is illustrated by the effect that the corticostriatal pathway has on KA neurotoxicity in the striatum. Injections of 5–10 nmoles KA into the neostriatum will normally cause degeneration of most neuronal cell bodies in a large portion of that nucleus. However, when the corticostriatal pathway has been sectioned, the neurotoxicity of KA is reduced by at least two orders of magnitude (E. G. McGeer *et al.*, 1978). The neurotoxicity of KA is restored if glutamate is injected along with the KA (Biziere and Coyle, 1978). KA, however, also often causes damage in areas remote from the injection site, and such remote damage seems to depend on the severe epileptiform activity that KA often induces, depending on the conditions of injection (P. L. McGeer and E. G. McGeer, 1981a).

Ibotenic and quinolinic acids produce local neuronal lesions with the same type of selectivity as found with KA, but have little or no tendency to produce epileptiform activity and remote damage (Guldin and Markowitsch, 1981; Schwarcz *et al.*, 1979, 1983). They clearly differ from KA in that the local neurotoxicity appears to depend on some direct action on postsynaptic glutamate receptors, and the effect persists even after lesioning of glutamate afferents. Different cell groups in brain show markedly different sensitivities to KA and ibotenate (C. Kohler and Schwarcz, 1983). Moreover, the α-amino-ω-phosphonocarboxylate antagonists, such as AP7, block ibotenate but not KA neurotoxicity in rat hippocampus (Schwarcz *et al.*, 1982). Ibotenate is far more toxic than KA in developing rat striatum or hippocampus, and even dopaminergic nerve endings are markedly affected in the developing striatum (H. X. Steiner *et al.*, 1984).

Quinolinic acid probably acts at the subclass of excitatory receptors that are more sensitive to NMDA than to either glutamate or KA. Quinolinic acid seems to have more selective excitatory actions than glutamate or KA, and pronounced regional differences have been reported for both its excitatory (Perkins and Stone, 1983) and its neurotoxic effects. Thus, on local infusion, quinolinic acid is effective in killing neurons of rat striatum or hippocampus, but not those of several other brain regions (Schwarcz et al., 1983, 1984; Schwarcz and Kohler, 1983).

Folic acid is still different in that it seems to cause little local damage, but does elicit convulsions and cause remote damage (P. L. McGeer et al., 1983; Olney et al., 1981). The kind of receptor involved in the action of folic acid is not yet established.

The distant effects of KA and folic acid are largely blocked by pretreatment of the animal with large doses of an anticonvulsant such as diazepam (P. L. McGeer et al., 1983), which does not affect the local neurotoxicity of KA, ibotenic acid, or quinolinic acid. The distant damage presumably depends on overexcitation of excitatory pathways, but these do not appear to be necessarily neuronal tracts served by excitatory amino acid neurotransmitters. There is some evidence, for example, that overstimulation of the cholinergic neurons of the substantia innominata or pontine reticular formation (see Chapter 8) by either folic acid or KA can cause extensive damage in projection areas (P. L. McGeer et al., 1983; E. G. McGeer and P. L. McGeer, 1984).

The local damage caused by intracerebral injections of excitotoxins (see Figure 6.4) is widely used not only in neuroanatomical and neurophysiological work, but also to produce animal models that mimic some of the pathology of Huntington's disease (see Section 6.5), Alzheimer's disease, and various other pathological conditions (E. G. McGeer and P. L. McGeer, 1985). For these purposes, the tendency of KA to cause distant as well as local damage is disadvantageous but one that can be overcome if the animal is adequately pretreated with an anticonvulsant. The tendency of KA and folic acid to cause distant damage has also proven valuable in the production of models of diseases such as epilepsy (see Section 6.5) and interstitial myocarditis (E. G. McGeer and P. L. McGeer, 1985).

Most work has been done with intracerebral injections of excitotoxins because the BBB excludes most excitatory amino acids from brain. However, the exclusion is not absolute; systematically administered excitatory amino acids can gain partial access to the brain and do selective damage to the CNS. This is particularly the case for infant animals, in which the BBB is not fully developed. At all ages, the so-called "circumventricular organs" (CVOs) are more exposed because in these regions the BBB is absent or less well developed. The CVOs include the median eminence, organum vasculosum of the lamina terminalis, subfornical organ, subcommissural organ, and the area postrema. Two adjacent areas in the hypothalamus, the medial preoptic region (which is continuous with the lamina terminalis) and the arcuate nucleus (which is continuous with the median eminence), are also vulnerable. If minimal toxic doses of glutamate are given, the only areas affected in addition to the arcuate nucleus of the hypothalamus are the inner retina and the subfornical organ. However, high doses of glutamate will, in addition, produce damage to such areas as the medial preoptic region, rostral medial dentate, hippocampal gyrus, medial habenula, and superior colliculi. A long-term consequence is a severe hypoplasia of the anterior pituitary in later life. This constitutes a model of one type of neuroendocrine disorder (Olney and Price, 1978).

There is legitimate room for concern as to whether excessive glutamate and aspartate would be toxic in humans, but there are no firm data to support this supposition. Caution should be exercised, however, with infants, in whom the BBB might not be fully developed. At one time, glutamate was prescribed as an adjuvant in the control of petit mal and psychomotor epilepsy (J. C. Price *et al.*, 1943) and, as with aspartic acid (W. O. Evans, 1968), was supposed to increase performance efficiency and a sense of well-being in normal individuals. For a while, glutamic acid was used in the treatment of mental retardation (F. T. Zimmerman *et al.*, 1946), and there was at least one report that it significantly increased maze-learning in rats (F. T. Zimmerman and Ross, 1944). After a number of years, however, resistance to the dosage of 25–60 g/day and the advent of control studies put an end to the popularity of glutamate as a treatment for mental deficiency (Lombard *et al.*, 1955). No report of neural toxicity from this glutamate therapy appeared, but Kwok (1968) described the strange sensations he experienced after eating in Chinese restaurants, which commonly use monosodium glutamate (MSG) as a flavor enhancer. This "Chinese restaurant syndrome" involves the production of a burning sensation, facial pressure, and chest pains in persons who are particularly sensitive to glutamate ingestion. The effects are probably at least partially peripheral.

6.5. Pharmacology and Pathology of Glutamate and Aspartate

So far there are few, if any, clinically useful pharmacological agents that are known to interact specifically with glutamate and aspartate systems. Baclofen, which is widely used in spasticity and is reported to show some beneficial effects in chorea, is believed to be effective because it inhibits the release of glutamate, but since its action is at a special type of GABA receptor, this drug is discussed in Chapter 7. A number of NMDA antagonists, such as AP7 (see Section 6.4.1), have shown more powerful anticonvulsant activity in many animal models of epilepsy than the generally used GABA agonists or benzodiazepines (see Chapter 7) (D. W. Peterson *et al.*, 1983, 1984; Meldrum *et al.*, 1983). A problem is to design antagonists capable of crossing the BBB at reasonable peripheral doses. When this is accomplished, such compounds may well prove useful in the treatment of epilepsy and possibly other conditions as well.

Mean glutamate levels have been reported to be significantly higher in epileptogenic foci than in control cortex. Levels of glutathione, GABA, and glycine were also slightly elevated, while there was no difference in aspartate or taurine (T. L. Perry and Hansen, 1981). As discussed in Section 6.4.3, convulsions and brain damage similar to that seen in temporal lobe epilepsy are readily induced by intracerebral or, in young animals, peripheral injections of excitotoxins. For example, Ben-Ari *et al.* (1979a,b) used KA injected directly into the amygdaloid nucleus unilaterally to produce status epilepticus. Termination of the episode, and thus prevention of death, was brought about by intraperitoneal doses of 20 mg/kg of diazepam. As expected, neuronal lesions occur in the amygdala. These, however, extend to the hippocampal field, particularly to the highly vulnerable CA3 field. This secondary hippocampal damage is similar to that observed following status epilepticus in man (Blackwood and Corselis, 1976). Similar seizure activity and pathological consequences were seen in baboons given amygdala or temporal pole injections of KA (Menini *et al.*, 1980). Comparable results can also be induced in young rats by peripheral administration of KA, and even fairly low doses of glutamate and related amino acids can activate epileptic foci in mice on intraperitoneal injection (Bradford and Dodd, 1977).

Other diseases in which Glu/Asp neurons might play a role include Huntington's disease and inherited cerebellar disorders. Huntington's disease is an autosomal dominant hereditary disorder (see Section 1.5.2). Typically, it manifests itself in the late 30s or early 40s, which is just beyond the normal childbearing age. The time of onset accounts for its genetic preservation despite its devastating physical and social characteristics. The histological, biochemical, and behavioral changes seen in animals following intrastriatal injections of KA, ibotenic acid, or quinolinic acid have been repeatedly found to resemble those seen in Huntington's disease (E. G. McGeer and P. L. McGeer, 1985).

The most striking pathological changes in Huntington's disease occur in the basal ganglia, particularly the caudate, where there is marked atrophy and severe loss of neurons. The putamen is also heavily involved, and damage to the globus pallidus is usually prominent. Negligible histological changes are observed in the substantia nigra, although severe biochemical losses are noted. The pattern is consistent with destruction of striatal glutamate-receptive neurons by overactivity of the corticostriatal tract. In agreement with such a hypothesis, Greenmyre et al. (1985) found a decrease in glutamate binding in the striatum in Huntington's disease with no change in the cortex or other areas examined. It has been proposed (E. G. McGeer and P. L. McGeer, 1976; Coyle and Schwarcz, 1976) that the mechanism of cell death in Huntington's disease might be an excitotoxic one. This hypothesis does not require the formation in Huntington's disease of a unique KA-like neurotoxin, and indeed there is evidence against such a possibility (Beutler et al., 1981). Rather, the hypothesis suggests either excessive production of an endogenous neurotoxic amino acid or some abnormality in the postsynaptic membrane that would render the neurons more sensitive. Candidates for excessive production are quinolinic acid (Schwarcz et al., 1983, 1984) and glutamate itself, which causes striatal cell loss in rats on chronic infusion (McBean and Roberts, 1984). Preliminary studies indicate no excessive excretion of quinolinic acid (Heyes et al., 1985) or deficiency of the catabolic enzyme, QPRT (A. C. Foster and Schwarcz, 1985), in the blood in Huntington's disease. This does not preclude a local excess of quinolinic acid in the striatum. QPRT activity is somewhat elevated in the striatum in both Huntington's disease and in rats given intrastriatal injections of quinolinic acid (A. C. Foster et al., 1985b).

Glutamate or aspartate or both are clearly important neurotransmitters in the cerebellum (see Section 6.3), and there have been some interesting reports of low glutamate levels in the cerebellum and spinal cord in Friedreich's ataxia (Huxtable et al., 1979; Butterworth and Giguere, 1982) and of low cerebellar aspartate concentrations in persons with dominantly inherited cerebellar disorders (T. L. Perry et al., 1978). The decreased blood levels of glutamic acid dehydrogenase reported in persons with olivopontocerebellar atrophy were mentioned in Section 6.2.

The probable importance of Glu/Asp systems in the brain suggests that they may be involved in many pathological conditions and that pharmacological agents that affect these systems may have diverse clinical uses. Exploration of these possibilities, however, is only beginning and is handicapped by the lack of good indices of Glu/Asp systems applicable to postmortem tissue and of pharmacological agents that readily cross the BBB.

6.6. Summary

Glutamate and aspartate are amino acids that meet many of the anatomical, chemical, physiological, and pharmacological criteria for ionotropic neurotransmitter status. They open sodium and potassium ion channels and bring about a rapid excitatory response. Glutamate is the more abundant and powerful of the two dicarboxylic acids. They are both present in all brain cells because of their involvement in many metabolic roles. A high-affinity uptake system exists for them in nerve endings, and this uptake has been helpful in defining a number of probable Glu/Asp neuronal systems.

These pathways include the main projecting pathways from the neocortex and hippocampus to subcortical structures, as well as cerebral granule cells and climbing fibers. Retrograde transport of [³H]-D-aspartate and release of glutamate/aspartate on stimulation have also been used as criteria to help define excitatory amino acid pathways. Methods for distinguishing glutamate from aspartate pathways are not definitive, although selective release or changes in concentration following lesions have provided suggestive evidence in some cases.

Four classes of Glu/Asp receptors have been defined by electrophysiological studies–the N-methyl-D-aspartate, quisqualic acid, kainic acid, and L-aminophosphonobutyric acid types. These have been tentatively correlated with four distinct binding sites, A-1, -2, -3, and -4, respectively.

Excessive doses of excitatory amino acids and their analogues produce toxic effects on neurons that possess a sufficient receptor concentration. These excitotoxins have become major tools for neuroscientists working in many different areas and have been exploited to produce much new information about neuronal systems, as well as animal models of several human diseases.

Glu/Asp systems may well be involved in the pathology of a number of neurological diseases; their probable involvement in epilepsy is supported by both postmortem and pharmacological studies.

7

Inhibitory Amino Acid Neurotransmitters

7.1. Introduction

In Chapter 6, we discussed the excitatory ionotropic neurotransmitters. We must now consider what neurotransmitters represent the inhibitory ionotropic neurons, because these are probably as plentiful as the excitatory ones. Inhibitory properties are possessed by amino acids with a terminal amino group separated by 2–5 atoms from an acid group. Two compounds of this class, γ-aminobutyric acid (GABA) and glycine, are neurotransmitters in mammalian brain; others, such as taurine, proline, serine, and β-alanine, have physiological activity and have been suggested as neurotransmitters.

GABA and, to a lesser extent, glycine probably account for the majority of inhibitory action in the central nervous system (CNS), although both were ignored as potential neurotransmitters long after their presence in nervous tissue had been established. E. Roberts and Frankel (1950) reported the presence of GABA in brain tissue, having discovered it while working with the then new technique of paper chromatography. In an entirely separate line of investigation, Florey (1953) discovered that extracts of mammalian brain had an inhibitory action on the slowly adapting neuron of the abdominal stretch receptor organ of the crayfish. Bazemore, Elliott and Florey (1957) concluded that GABA was the material mainly responsible for the activity of Florey's Factor I. Meanwhile, Kuffler and Edwards (1958) produced evidence that GABA duplicated the action of the physiological transmitter in one of the crayfish cord synapses, providing the first sound evidence that GABA might be an inhibitory neurotransmitter. And Roberts and others soon discovered that the formation and destruction of GABA fit in with a shunt of the well-known, energy-producing Krebs cycle. K. Krnjevic and Schwartz (1966), using advanced microiontophoretic techniques, presented evidence that GABA mimics the activity of inhibitory neurons of the mammalian cerebral cortex. In 1967, Obata *et al.*

(1967) first demonstrated that GABA applied iontophoretically mimicked the inhibitory action of Purkinje cells and that it was released into the fourth ventricle when Purkinje cells were stimulated. This constituted the first really convincing proof that GABA was an inhibitory neurotransmitter in mammalian brain. Elegant final proof of GABA's role in Purkinje cells was developed by the Roberts team in 1974 when they showed by the immunohistochemical method that the specific synthetic enzyme is localized to Purkinje cell terminals.

The discovery of glycine followed a different route—one that illustrates the advantages to be gained from close collaboration between neurochemists and neurophysiologists. Aprison, a neurochemist, and Werman, a neurophysiologist, seeking ways of collaboration, decided that a method for identifying potential new transmitters in the central nervous system was to determine the relative distribution of various amino acids in different areas. About 1965, they and their colleagues discovered the uneven distribution of GABA, aspartate, and glutamate within the spinal cord of the cat (Aprison and Werman, 1965). They then showed that, after temporary aortic occlusion, there was a significant loss of glycine and aspartate, but not of glutamate or GABA, in the gray matter of the spinal cord. A rigorous neurophysiological comparison showed that glycine iontophoresed onto motoneurons duplicated the action of the inhibitory transmitter released on stimulation of the spinal interneurons. Strychnine was shown to have no effect on the action of GABA, but to block the inhibition of glycine on the motoneurons. Confirmatory chemical evidence came when it was shown that there was a high-affinity uptake system for glycine in slices and homogenates of cat or rat spinal cord and that glycine was preferentially localized to synaptosomes on subcellular fractionation studies.

A more thorough account of the key events that led to the recognition of the neurotransmitter roles of GABA, glycine, aspartate, and glutamate is given in the first edition of this book.

7.2. Chemistry and Metabolism of γ-Aminobutyric Acid

The discovery of GABA in 1950 coincided with the introduction of radiolabeling techniques, making it one of the first major brain materials to be investigated by this powerful methodology. It was quickly established that glutamic acid is the immediate precursor of GABA and that glutamic acid can be formed from either glutamine or α-ketoglutarate. But incubation of brain with various other radiolabeled starting materials revealed a whole complex of metabolic interrelationships involving GABA that are not completely understood to this day.

Glucose, the main energy source of the brain, is a particularly efficient precursor of GABA and is probably the principal *in vivo* source. Pyruvate and several other amino acids can also serve by entering the Krebs cycle. Eventually, the more remote precursors must be converted via α-ketoglutarate to glutamate or glutamine, which together constitute 60% of the free α-amino nitrogen in brain tissue. It is thought that they serve as a reservoir system for producing GABA.

Central to the GABA system is the so-called "*GABA shunt*" (Figure 7.1). The GABA shunt is actually a closed loop that acts in such a way as to conserve the supply of GABA by producing a precursor molecule of glutamate for every molecule of GABA consumed. The first step in the shunt is the transamination of α-ketoglutarate, an inter-

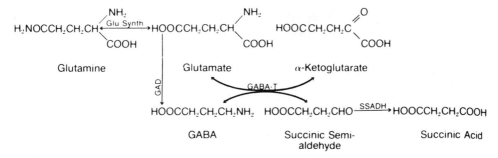

Figure 7.1. Schematic diagram showing the reactions of the GABA shunt.

mediate in the Krebs cycle, to glutamic acid. Glutamic acid is then decarboxylated to form GABA. The next step is the key. GABA is transaminated to form succinic semialdehyde, but the transamination can only take place if α-ketoglutarate is the acceptor of the amine group. This transforms α-ketoglutarate into the GABA precursor glutamate, thereby guaranteeing a continuity of supply. Thus, a molecule of GABA can be destroyed metabolically only if a molecule of precursor is formed to take its place. The succinic semialdehyde formed from GABA is rapidly oxidized to succinic acid, which is a normal constituent of the Krebs cycle.

Three enzymes are involved in the shunt: glutamic acid decarboxylase [(GAD) EC 4.1.1.15], which converts glutamic acid to GABA; GABA-α-oxoglutarate transaminase [(GABA-T) EC 2.6.1.19], which converts GABA to succinic semialdehyde; and succinic semialdehyde dehydrogenase [(SSADH) EC 1.2.1.24], which returns the metabolic remnant to the Krebs cycle. From an energy point of view, the GABA shunt yields 3 ATP molecules. It is therefore less efficient than that portion of the Krebs cycle that converts α-ketoglutarate through the usual pathway to succinic acid. This direct route, used principally by the brain as well as exclusively by all other tissues, yields 3 molecules of ATP and 1 of GTP.

Ancillary to the GABA shunt is a *glutamine loop:* Released GABA can be picked up by glial cells, in which transamination by GABA-T can take place. The glutamate formed in the transamination cannot, however, be converted to GABA in the glia, since they lack GAD. Instead, it is transformed by glutamine synthetase (GluSynth) into glutamine, after which it can be returned to the nerve ending. In the nerve ending, the enzyme glutaminase converts the glutamine back to glutamate, thus completing the loop and conserving the supply of GABA precursor.

Figure 7.2 is a schematic diagram showing the intracellular localization of the components of the GABA shunt and glutamine loop in the region of the synapse. Such tiny bodies as the nerve ending depicted in Figure 7.2 must be capable of functioning in many ways independently of the nucleus, almost like a minuscule cell. Their major job is to synthesize, store, and release a particular neurotransmitter, in this case GABA. To accomplish this, they utilize glucose and oxygen and are capable of a net synthesis of ATP and phosphocreatine. In the case of the GABA nerve ending shown in Figure 7.2, GABA-T and SSADH are attached to mitochondria. Glutaminase and GAD are free in the intraneuronal cytoplasm. GluSynth is limited to glia and occurs in the glial cytoplasm. The shunt itself operates entirely within the nerve ending. A molecule of α-ketoglutarate

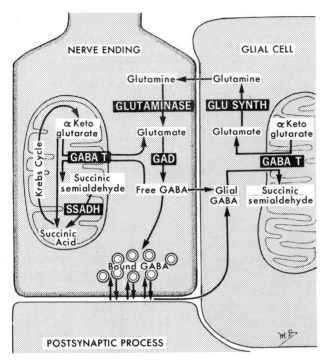

Figure 7.2. Schematic diagram of the relationship of a GABA nerve ending, a postsynaptic process, and a glial cell. Enzymes are printed white on black, endogenous intermediates black on white.

leaves the Krebs cycle (which is confined to the mitochondrion) to form a molecule of glutamate. The reaction is catalyzed by GABA-T, and the amine group required is obtained from GABA. Replacement GABA is formed by the decarboxylation of glutamate, a reaction catalyzed by GAD and occurring exclusively in GABA nerve endings. Released GABA can be taken back into the nerve ending or picked up by glial cells by high-affinity transport processes; in either location, transamination by GABA-T can take place. The glutamate formed in the glial cells is synthesized into glutamine before being returned to the nerve ending. The glutamine loop is completed in the nerve ending by the enzyme glutaminase converting the glutamine back to glutamate.

The regional distribution of glutamine synthetase correlates significantly with that of GAD, but not with that of glial fibrillary acidic protein, a marker of astrocytes (Patel *et al.,* 1985). This is another indication of biochemical specialization of glia (see Section 1.3).

GAD, the key enzyme in the formation of GABA, has been purified, initially from mouse brain (cf. J. Y. Wu, 1976). It has a molecular weight of 85,000 and requires pyridoxal phosphate as a cofactor. The K_m, or concentration of substrate required to half-saturate the enzyme, is 0.7 mM for glutamate and 0.05 mM for pyridoxal phosphate. *In vivo* tissue levels of these materials are so high that the enzyme should normally be fully saturated with its substrate and cofactor. The degree of saturation of GAD with cofactor is said, however, to vary considerably from region to region in brain (M. Itoh and Uchimura, 1981).

GAD is highly localized to the synaptosomal fraction of subcellular homogenates of brain. This indicates that production of GABA is concentrated in the nerve endings rather than in the cell bodies or dendrites of neurons or in the glial cells. This localization is in accord with results from immunohistochemical studies (see Figure 7.3). Within the synaptosomes, it is in a soluble form. Thus, it is not attached to mitochondria of nerve endings, as are the Krebs cycle enzymes (J. Y. Wu, 1976).

It is still not known how the synthesis of GABA is governed and what controls the physiological pool. It has been suggested that Zn^{2+} (M. Itoh et al., 1983) and Ca^{2+} (Gold and Huger, 1982) may play a regulatory role, with Zn^{2+} being inhibitory and Ca^{2+} stimulatory. The concentration of ATP in the nerve ending may also be important, since micromolar concentrations of ATP strongly inhibit GAD by an allosteric effect that shifts the balance between the inactive apoenzyme and the active form. The inhibitory effects of ATP are opposed by both pyridoxal phosphate and inorganic phosphate.

The complementary DNA corresponding to GAD messenger RNA (mRNA) has been cloned from brain (D. L. Kaufman et al., 1985) and pheochromocytoma cell lines (Tillakaratne et al., 1985) by techniques similar to those described in Section 1.5.1 (Figure 1.19). The mRNA is approximately 3.7 kilobases in size.

GABA is destroyed by the enzyme GABA-T. The reaction involves a reversible transamination of GABA with α-ketoglutarate. No transamination of GABA takes place with any other α-keto acid, and of the common amino acids, only β-alanine can substitute for GABA as a substrate.

As with GAD, GABA-T requires pyridoxal phosphate and has two nonequivalent binding sites for this cofactor (Churchich and Moses, 1981). The enzyme from mouse, rat, and rabbit brain has been purified to homogeneity, and antibodies to it have been prepared (K. Saito et al., 1974b; Edwardson et al., 1985). The concentration of GABA required to half-saturate the enzyme is variously reported as 1.1–2 mM (Schousboe et al., 1973; L. J. Fowler et al., 1983). The availability of α-ketoglutarate plays an important role in the destruction of GABA. For example, when respiration ceases at death, the level of α-ketoglutarate rapidly declines. GABA cannot then be destroyed, although it can still be formed from glutamate. There is therefore a rapid increase in brain GABA levels postmortem accompanied by a rapid decline in glutamate. GABA-T is attached to free mitochondria in GABA-containing synaptosomes and in glial cells. In the first location, they participate in maintaining appropriate presynaptic concentrations of GABA, while in both they participate in its destruction after release and reuptake (Figure 7.2).

The final enzyme in the GABA shunt, SSADH, is closely coupled to GABA-T and is similarly distributed in brain. The high tissue content required to saturate the enzyme relative to that for GABA-T probably explains why succinic semialdehyde has never been detected as an endogenous metabolite in neural tissue. That which is formed from GABA is very rapidly converted to succinic acid, which then enters the Krebs cycle.

Most of the metabolic studies on the formation and destruction of GABA have involved the exogenous addition of various radiolabeled precursors. Depending on the precursor used and on the conditions of the experiment, the amounts of labeled glutamate and GABA produced relative to the endogenous tissue levels vary widely. This has given rise to the concept of "pools" or "compartments" of metabolites. It is important to understand that this is a chemical definition and not a morphological one. A compound, such as glutamate or GABA, has a distinct metabolic rate in one chemical compartment independent of other compartments or pools. Although such kinetic data are obtained without any physiological concept of where such compartments might be, it is obvious that there must be some cellular or subcellular basis of chemical compartmentation.

In the case of GABA, evidence for compartmentation comes, in part, from attempts to estimate the half-life. Labeled GABA has been injected directly into the cerebral ventricles and its rate of disappearance measured. The data suggest in each region one rate of disappearance of GABA with a half-life between 0.5 and 1hr and a second with a half-life exceeding 1 hr; regional differences in turnover rates generally parallel GAD but not GABA-T activities (Collins, 1972). Similar conclusions have been reached by inhibiting GABA-T and measuring the rate of increase of GABA.

Several physical possibilities for the GABA compartments are suggested by these varying chemical kinetics. Much GABA is obviously held in nerve endings, where it can exist either free or bound to synaptic vesicles. But it is also taken up by glial cells and, in between, must spend a finite time in the extracellular space. Each of these could represent a chemical pool, and there might easily be many more. For example, other routes have even been described for the formation of GABA in brain. It has been shown, for example, that GABA can be formed from putrescine in a series of steps, apparently by the action of histaminase (see Chapter 11) (Tsuji et al., 1978). Moreover, GABA may be a precursor of other significant brain metabolites. It is known that homocarnosine, the dipeptide derivative of GABA and histidine, exists in brain. Similarly, the interconversion of GABA and γ-hydroxybutyrate has been shown. γ-Hydroxybutyrate is also inhibitory, and its half-life of 0.44 hr in brain and binding to brain membranes have been said to suggest some physiological role (Benavides et al., 1982; Maitre et al., 1983; Rumigny et al., 1981). Other possible conversions are to γ-butyrylbetaine, γ-guanidinobutyric acid, and several other derivatives. None of these alternate routes, however, has been established as being of major importance either metabolically or physiologically (Baxter, 1976).

In summary, metabolic studies have established the probable route for the formation and destruction of GABA. Large quantities of GABA are synthesized from glutamate in nerve endings; part of this may be metabolized immediately for energy production. The neurotransmitter reserve is probably bound in vesicles until it is released by nerve stimulation. The sites for destruction of GABA are localized in mitochondria, apparently in glial cells and in the GABAergic nerve endings. The destruction of GABA causes a new molecule of glutamate to form from α-ketoglutarate. The newly synthesized glutamate in the glia may be converted to glutamine so that it can reenter the nerve ending.

7.3. Anatomical Distribution of γ-Aminobutyric Acid Pathways in Brain

The brain content of GABA is 200- to 1000-fold greater than that of the group II amine neurotransmitters such as dopamine (DA) and noradrenaline (NA) (see Chapter 9), acetylcholine (ACh) (see Chapter 8), and serotonin [(5-HT) 5-hydroxytryptamine] (see Chapter 10), suggesting that it serves a much larger neuronal population. Indices of GABA function—including GABA itself, its synthetic enzyme GAD, and its high-affinity binding sites—are widely and relatively evenly distributed in brain. Such a distribution might be anticipated, since inhibitory interneurons are found in almost all regions of the CNS. Uptake studies with labeled GABA suggest that 25–45% of nerve endings, depending on the brain area, may contain this neurotransmitter (Belin et al., 1980).

Where are these GABA neurons located? While this question cannot be completely answered as yet, the following techniques have provided a great deal of information regarding the general anatomy of GABA neurons:

1. Immunohistochemical localization of GAD.
2. Immunohistochemical localization of GABA itself.

3. Histochemical localization of the GABA-destroying enzyme GABA-T.
4. Uptake of labeled GABA followed by autoradiography.

In theory, the most reliable method should be the immunohistochemical localization of GAD. In practice, the enzyme has been difficult to detect in neuronal cell bodies. Eugene Roberts and his colleagues were first to apply the technique and to provide the initial evidence of GAD-containing neurons in brain. They purified the enzyme, developed monospecific antibodies to it, and carried out the standard immunohistochemical staining procedure (Figure 7.3) (K. Saito *et al.*, 1974a; McLaughlin *et al.*, 1974). Antibodies to GAD have now been produced in a number of laboratories and used for immunohistochemical investigation. GAD-positive fibers and nerve terminals are particularly well visualized with these antisera, as exemplified in Figure 7.3, and with some antisera, there is neuronal staining as well (Oertel *et al.*, 1981). Cell bodies have not always stained readily (Barber *et al.*, 1978), suggesting that newly synthesized GAD does not linger in the cell body but is instead rapidly transported down the axon to the nerve endings. To overcome this problem, colchicine can be administered to arrest axonal transport. Under these conditions, GAD-positive cell bodies can be demonstrated in many regions of brain (Table 7.1). However, colchicine is a highly toxic material that must be administered locally. As a result, only a regional and often a highly variable effect is produced in any given animal. This may account in part for the limited mapping of GABA neurons by the GAD immunohistochemical method despite many years of effort.

Another promising immunohistochemical method that may circumvent this difficulty is to produce antibodies to GABA itself (Storm-Mathisen *et al.*, 1983; Kimura, personal communication). Since GABA is too small a molecule to be antigenic, it must first be chemically linked to a protein such as bovine serum albumin before being injected into a rabbit or other animal to generate specific antibodies. So far, only initial immunohistochemical results have been obtained, but the specificity of the technique has been established. A similar approach has been used for glycine (Section 7.8), glutamate (see Chapter 6), ACh (see Chapter 8), 5-HT (see Chapter 10), and histamine (see Chapter 11).

An alternative, less specific, but more flexible technique depends on the histochemical localization of the enzyme GABA-T. Although it is present in glia, it is more highly concentrated in GABA neurons. Differentiation between the two locations is achieved by the pharmacohistochemical method of pretreating animals with an irreversible inhibitor of GABA-T several hours before sacrifice. Under these conditions, preexisting GABA-T is inactivated, but newly synthesized enzyme can be visualized. The new enzyme appears first in neurons presumed to be of the GABA type, since they would be the ones requiring high concentrations of the enzyme. Glial cells regenerate GABA-T much more slowly.

Figure 7.4 shows staining of rat cerebral cortical cells by the three techniques. GABA and GABA-T show a highly concordant pattern of staining, providing convincing evidence of the nature and distribution of cerebral cortical GABA cells. The GAD method (Figure 7.4A) shows relatively light staining of cells, but intense staining of fibers and nerve endings. The GABA method (Figure 7.4B) shows staining of both cells and fibers, while the GABA-T method (Figure 7.4C) shows more intense staining of cells. The latter result would be anticipated, since GABA-T must reach the fibers and nerve endings by axoplasmic transport (see Section 1.2.4). GABA cells are observed in all layers of the cortex, but their morphology and distribution contrast with those of glutamate cells [see Figure 6.3 (Section 6.3)]. The former show stellate and fusiform morphology, as

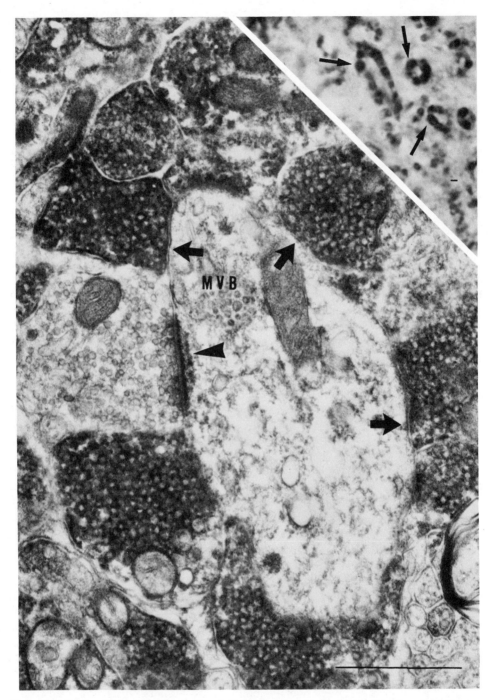

Figure 7.3. GAD-containing terminals in the substantia nigra from specimens incubated in anti-GAD serum. *Inset:* Semithin (1-μm) section of the pars reticulata with obliquely and transversely sectioned dendrites that are encircled by punctate structures containing GAD-positive reaction product (→). Scale bar: 1 μm. The accompanying electron micrograph shows an obliquely sectioned dendrite in the substantia nigra that is surrounded by many axon terminals filled with GAD-positive reaction product that are equivalent to the puncta seen in the inset. Some of these terminals form symmetrical synapses (←), whereas the unstained terminal contains round synaptic vesicles and forms an asymmetrical synapse (◄) with this dendritic shaft. (MVB) Multivesicular body. Scale bar: 1 μm. From Ribak *et al.* (1977b).

TABLE 7.1

Brain Regions in Which GABAergic Neurons Have Been Identified by GAD Immunohistochemistry and/or Tentatively Identified by GABA Immunohistochemistry, GABA-T Pharmacohistochemistry, or Radioactive GABA Uptake[a]

Region and cell type	Immunohistochemistry		GABA-T[b]	GABA uptake
	GAD	GABA		
Olfactory bulb				
Granule	58	—	√	15, 16
Perigranular	—	—	√	32
Cerebral cortex	57	74	√	4, 17, 19, 22, 26, 73, 82, 83
Hippocampus	14, 35, 60, 84	74	—	—
Basket cells	—	—	√	31
Stellate cells	—	—	√	31
Hilar projection neurons	71	—	—	—
Islands of Calleja	65	—	√	—
Bed n. of stria terminalis	—	—	√	—
Olfactory tubercle (granule)	—	—	√	36
Septal n. and diagonal band	9, 34	—	√	—
Substantia innominata	—	—	√	—
Striatum, globus pallidus and entopeduncular n., medium cells	53, 61, 63, 81	—	√	5, 25
Amygdala	—	42	√	—
Hypothalamus, large cells	39, 54, 77, 79	—	√	76
Thalamus				
Reticular n.	25	—	√	—
Other nuclei	—	44	√	—
Medial geniculate	—	78	√	—
Lateral geniculate, small	11, 21	12, 44, 80	√	75
Substantia nigra, small to medium	53, 59, 62	—	√	—
Oculomotor nucleus	—	—	—	38
Superior colliculus	—	—	√	49
Central gray	—	—	√	—
N. of Darkschewitsch	—	—	√	—
Interstitial n. of Cajal	—	—	√	—
Dorsal raphe	3, 50	—	√	3, 13, 56
N. parabrachialis	—	—	√	—
N. cuneiformis	—	—	√	—
Reticular formation	6, 68	—	√	—
Cochlear n., especially dorsal division	72	78	√	70
Medial vestibular n.	—	—	√	—
Lateral vestibular n.	67, 68	—	√	—
N. tractus spinalis nervi trigemini	—	—	√	—
Dorsal motor n. of the vagus	—	—	√	—
Tractus solitarius	29, 48	45	—	—
Inferior colliculus	66	78	√	—
Superior olivary complex	—	78	—	—
Inferior olive	—	—	√	—
N. ambiguus	—	—	√	—
Cerebellum	43, 50, 52, 60, 69, 84	—	—	24, 33

(continued)

TABLE 7.1 (*Continued*)

Region and cell type	Immunohistochemistry			
	GAD	GABA	GABA-T[b]	GABA uptake
Purkinje cells	—	—	√	—
Golgi cells	—	—	√	—
Basket cells	—	—	√	—
Stellate cells	—	—	√	—
Fasciculus cuneatus	—	—	√	—
Spinal interneurons	1, 28, 84	—	—	30, 64
Retina	4, 7, 8, 37, 40, 51, 84	51	—	2, 7, 10, 18, 31, 40, 46, 47, 55, 85

[a]References: (1) Barber *et al.* (1978); (2) Beale and Osborne (1983); (3) Belin *et al.* (1979); (4) F. E. Bloom and Iversen (1971); (5) Bolam *et al.* (1983); (6) Border and Mihailoff (1985); (7) Brandon *et al.* (1979); (8) Brandon (1985); (9) Cheney *et al.* (1982); (10) Ehinger (1970); (11) Fitzpatrick *et al.* (1984); (12) Gabbott *et al.* (1985); (13) Gamrani *et al.* (1979); (14) Goldowitz *et al.* (1982); (15) Halász *et al.* (1979); (16) Halász *et al.* (1981); (17) Hamos *et al.* (1983); (18) Hampton and Redburn (1983); (19) Harandi *et al.* (1981); (20) Hattori *et al.* (1973); (21) Hendrickson *et al.* (1983); (22) S. H. C. Hendry and Jones (1981); (23) S. H. Hendry *et al.* (1983); (24) Hökfelt and Ljungdahl (1972); (25) Houser *et al.* (1980); (26) Houser *et al.* (1983a); (27) Houser *et al.* (1983b); (28) S. P. Hunt *et al.* (1981b); (29) Hwang and Wu (1984); (30) Iversen and Bloom (1972); (31) Iversen and Schon (1973); (32) Jaffe *et al.* (1983); (33) J. S. Kelly *et al.* (1975); (34) C. Kohler and Chan-Palay (1983a); (35) C. Kohler and Chan-Palay (1983b); (36) N. R. Krieger *et al.* (1982); (37) Lam *et al.* (1979); (38) Lanoir *et al.* (1982); (39) Leranth *et al.* (1985); (40) Lin *et al.* (1983); (41) Lin *et al.* (1985); (42) McDonald (1985a); (43) McLaughlin *et al.* (1974); (44) Madarasz *et al.* (1985); (45) Maley and Newton (1985); (46) Marc *et al.* (1978); (47) Marshall and Voaden (1974); (48) Meeley *et al.* (1985); (49) Mize *et al.* (1981); (50) Nanopoulos *et al.* (1982); (51) Nishimura *et al.* (1985); (52) Oertel *et al.* (1981); (53) Oertel *et al.* (1982a); (54) Oertel *et al.* (1982b); (55) Pourcho (1980); (56) Pujol *et al.* (1981); (57) Ribak (1978); (58) Ribak *et al.* (1977a); (59) Ribak *et al.* (1977b); (60) Ribak *et al.* (1978); (61) Ribak *et al.* (1981); (62) Ribak *et al.* (1980); (63) Ribak *et al.* (1981); (64) Ribeiro and da Silva Coimbra (1980); (65) Richman *et al.* (1980); (66) R. C. Roberts *et al.* (1985); (67) Robson and Martin-Elkins (1985); (68) Ruggiero *et al.* (1985); (69) Saito *et al.* (1974a); (70) I. R. Schwartz (1981); (71) Seress and Ribak (1983); (72) Shiraishi *et al.* (1985); (73) Somogyi *et al.* (1981); (74) Somogyi *et al.* (1985); (75) Sterling and Davis (1980); (76) Tappaz *et al.* (1980); (77) Tappaz *et al.* (1983); (78) G. C. Thompson *et al.* (1985); (79) Vincent *et al.* (1982c); (80) Wadhwa *et al.* (1985); (81) Weinberg *et al.* (1985); (82) W. F. White *et al.* (1980); (83) J. K. Wolff and Chronwall (1982); (84) J. G. Wood *et al.* (1976); (85) J. Y. Wu *et al.* (1981).

[b]Complete maps of GABA-T staining may be found in Nagai *et al.* (1983a) for rat forebrain and in Nagai *et al.* (1985) for rat hindbrain.

opposed to the pyramidal morphology of glutamate cells, and are found in layer I and deep layer VI, in addition to the other layers.

Only the GABA-T pharmacohistochemical method has been used so far for complete brain mapping (Figure 7.5) (Nagai *et al.*, 1983a, 1985). This method gives excellent staining of cell bodies and proximally located nerve fibers, as exemplified by Figure 7.6 in the septal area, if appropriate intervals (8–16 hr) between administration of the GABA-T inhibitor and sacrifice of the animal are chosen. Without the inhibitor, or at longer survival times (24–48 hr), glial cell staining predominates, indicating that glial cells are a major reservoir for GABA-T. Stained cells include almost all neuronal groups previously reported to be GABAergic on the basis of GAD/colchicine immunocytochemistry and other methods (Table 7.1). Groups known to be nonGABAergic do not stain, even though they may be strongly GABAceptive. It may even be that such GABAceptive neurons are precluded from possessing GABA-T. The product of the reaction (glutamate), being excitatory to neuronal membranes, might counteract the intended inhibitory effect of GABA. The mapping not only agrees in almost all respects with previously published evidence of GABA cell identification, but suggests previously unreported groups that may also prove to use GABA as their neurotransmitter (Table 7.1).

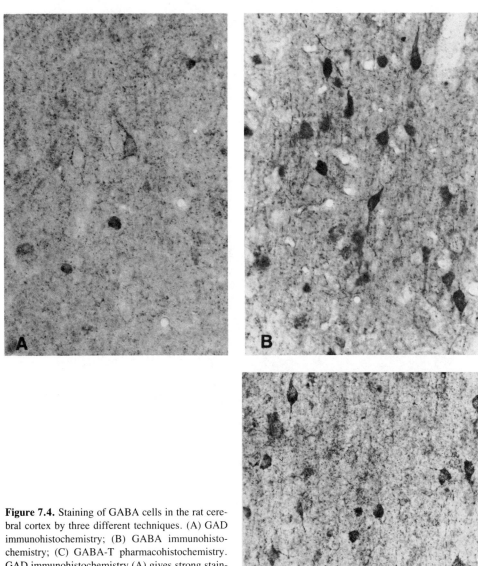

Figure 7.4. Staining of GABA cells in the rat cerebral cortex by three different techniques. (A) GAD immunohistochemistry; (B) GABA immunohistochemistry; (C) GABA-T pharmacohistochemistry. GAD immunohistochemistry (A) gives strong staining of processes but weak staining of cell bodies, while GABA immunohistochemistry (B) and GABA-T pharmachohistochemistry (C) give strong staining of cell bodies. The positively staining cells are distributed through all layers and are mostly of stellate and fusiform morphology. Photomicrographs are from the layer III–IV transition zone. Courtesy of Dr. H. Tago (University of British Columbia). Antisera kindly supplied by Dr. H. Oertel *et al.* (1981) (A) and by Dr. H. Kimura (Shiga University, Japan) (B).

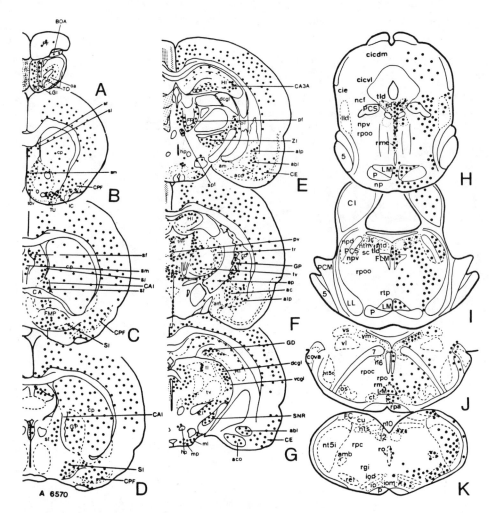

Figure 7.5. Coronal maps of rat brain showing the location of GABA-T-positive cells in rostral-to-caudal order (A–K). Abbreviations: (5, 7, 12) Cranial nerve tracts; (abl) nucleus amygdaloideus basalis, pars lateralis; (ac) nucleus amygdaloideus centralis; (aco) nucleus amygdaloideus corticalis; (alp) nucleus amygdaloideus lateralis, pars posterior; (am) nucleus amygdaloideus medialis; (amb) nucleus ambiguus; (BOA) bulbus olfactorius accessorius; (CA) commissura anterior; (CA3A) region of the hippocampus; (CAI) capsula interna; (CE) cortex entorhinalis; (CI) colliculus inferior; (cicdm) nucleus centralis colliculi inferious, pars dorsomedialis; (cicvl) nucleus centralis colliculi inferious, pars ventrolateralis; (cie) nucleus externus colliculi inferioris; (cova) nucleus cochlearis ventralis, pars anterior; (cp) nucleus caudatus putamen; (CPF) cortex piriformis; (ct) nucleus corporis trapezoidei; (cu) nucleus cuneatus; (d) nucleus Darkschewitsch; (dcgl) nucleus dorsalis corporis geniculati lateralis; (ep) nucleus entopenduncularis; (FC) fasciculus cuneatus; (FLM) fasciculus longitudinalis medialis; (FMP) fasciculus medialis prosencephali; (FR) fasciculus retroflexus; (GD) gyrus dentatus (hippocampus); (GP) globus pallidus; (HI) hippocampus; (hp) nucleus periventricularis hypothalamicus; (IC) inferior colliculus; (io) nucleus olivaris inferior; (iod) inferior olive, dorsalis region; (iom) inferior olive, medialis region; (lc) locus coeruleus; (LG) lamina glomerulosa bulbi olfactorii; (LGI) lamina granularis interna bulbi olfactorii; (LL) lemniscus lateralis; (lld) nucleus lemnisci lateralis dorsalis; (LM) lemniscus medialis; (ml) nucleus mamillaries lateralis; (mp) nucleus mamillaries posterior; (n6) nucleus originis nervi abducentis; (n10) nucleus originis nervi vagi; (ncf) nucleus cuneiformis; (np) nucleus parabrachialis ventralis; (npd) nucleus parabrachialis dorsalis; (npv) nucleus paraventricularis hypothalami; (ntd) nucleus tegmenti dorsalis (Gudden); (ntm) nucleus tractus mesencephali; (nts) nucleus tractus solitarius; (nt5i) nucleus tractus spinalis nervi trigemini pars interpositus; (nt5o) nucleus tractus spinalis nervi trigemini pars oralis; (oa) nucleus olfactorius anterior; (os) nucleus olivarius superior; (P) tractus corticospinalis; (PCM) pedunculus cerebellaris medius; (PCS) pedunculus cerebellaris

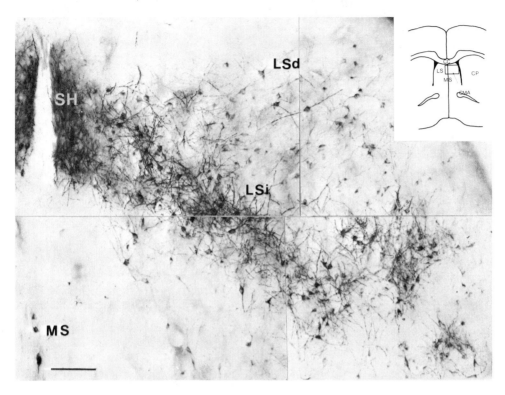

Figure 7.6. GABA-T staining of the rat septal area by the pharmacohistochemical method. Note the long varicose neuronal processes, especially in the lateral part. *Inset:* Area from which picture was taken. (CC) Corpus callosum; (CMA) commissura anterior; (CP) caudate–putamen; (LS) lateral septal nucleus; (LSi) intermediate part; (LSd) dorsal part; (MS) medial septal nucleus; (SH) septohippocampal nucleus. Scale bar: 100 μm. From Nagai *et al.* (1983a).

Another method that has been used to identify presumptive GABAergic neurons is the uptake of radioactive GABA. When brain slices are incubated with [³H]-GABA, the labeled material is pumped into cellular elements by a high-affinity uptake system. The tissue can then be fixed, sliced, and coated with a photographic film for autoradiography. Label can be detected by either light or electron microscopy. GABA is a particularly good material to work with in this regard because it apparently reacts with the fixative to make

superior; (pf) nucleus parafascicularis; (PF) polus frontalis; (pv) nucleus periventricularis thalami; (rel) nucleus reticularis lateralis; (rgi) nucleus reticularis gigantocellularis; (rm) nucleus raphe magnus; (rme) nucleus raphe medialis; (ro) nucleus raphe obscurus; (rpa) nucleus raphe pallidus; (rpc) nucleus reticularis parvocellularis; (rpo) nucleus raphe pontis; (rpoc) nucleus reticularis pontis caudalis; (rpoo) nucleus reticularis pontis oralis; (rtp) nucleus reticularis tegmenti; (sc) superior colliculis; (sf) nucleus septalis fimbrialis; (SI) nucleus septi intermedium; (sl) nucleus septi lateralis; (sm) nucleus septi medialis; (SNR) substantia nigra, pars reticularis; (spf) nucleus subparafascicularis; (st) nucleus triangularis septi; (ST) stria terminalis; (tdv) nucleus tractus diagonalis (of Broca) vertical limb; (tl) nucleus lateralis thalami; (tld) nucleus tegmental lateralis dorsalis; (tm) nucleus medialis thalami; (TO) tractus olfactorius; (tr) nucleus reticularis thalami; (TU) tuberculum olfactorium; (tv) nucleus ventralis thalami; (vcgl) nucleus ventralis corporis geniculati lateralis; (vl) nucleus vestibularis lateralis; (vm) nucleus vestibularis lateralis; (vs) nucleus vestibularis superior; (ZI) zona incerta. From Nagai *et al.* (1983a, 1985).

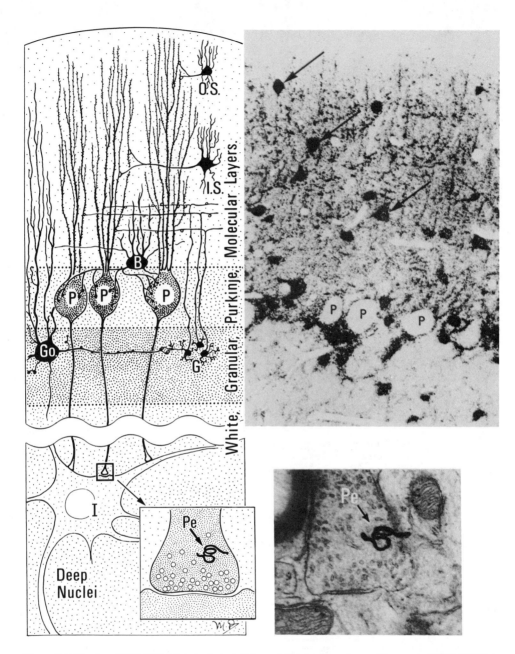

Figure 7.7. Uptake of [³H]-GABA into basket, stellate, and Golgi cells and axonal transport of [³H]-GABA to Purkinje cell terminals in the interpositus nucleus. (A) Diagrammatic representation of the cellular organization of the cerebellar cortex with molecular, Purkinje, and granular layers and white matter delineated (·····). Cells: (O.S.) outer stellate; (I.S.) inner stellate; (B) basket; (P) Purkinje; (Go) Golgi; (G) granule. Granule cells and their processes do not accumulate GABA. (B) Continuation of (A) to include neurons of the nucleus interpositus. *Inset:* A 100-fold enlargement of the insert area showing the Purkinje cell nerve ending (Pe). From P. L. McGeer *et al.* (1975). (C) Light microscopic autoradiograph of area comparable to that shown in (A) following the injection of [³H]-GABA. (←) Some cell bodies with high activity. Note the correspondence of cell bodies that accumulate radioactivity with areas occupied by inner and outer stellate, basket, and Golgi cells. Note also the dot and fiber-like accumulations along tracks of their processes. Purkinje (P) cell bodies have low activity even though they use GABA as their neurotransmitter. ×500. From Hökfelt and Ljungdahl (1972). (D) Electron microscopic autoradiograph of a Purkinje cell nerve ending (Pe) in the nucleus interpositus following injection of [³H]-GABA into the cerebellar cortex. Transport of GABA to the nerve ending was by axoplasmic flow *in vivo.* (→) Silver grain of exposed film. ×54,000. From P. L. McGeer *et al.* (1975).

an insoluble derivative that remains at the original site. Most of the activity is found in nerve terminals, although some uptake into cells that are presumed to be GABAergic also occurs (Figure 7.7 and Table 7.1).

While these methods may be sufficient to demonstrate GABA cells and terminals, other techniques are required to determine projections. These have included decreases in GAD or GABA after various lesions, axoplasmic flow of radioactive GABA, release of GABA on electrical stimulation of the pathway, and demonstration that the tract exerts an inhibitory effect on target cells that can be blocked by GABA antagonists. The GABA-T pharmacohistochemical method has been combined with retrograde tracing techniques in dual-labeling experiments (Araki *et al.*, 1984a,b, 1985) to allow precise localization of presumptive GABAergic tracts. The evidence collected with these various methods indicates a number of long-axoned GABAergic tracts as well as GABA interneurons in many regions of brain (Table 7.2). The pertinent literature on some of these pathways is summarized in Sections 7.3.1–7.3.9.

TABLE 7.2
Proposed Neuronal Pathways Involving GABA[a]

Anatomical system	Evidence and ref. nos.[b]
Long-axoned tracts	
Cortex to substantia nigra	L[97]
Corticorubral	P[2]
Hippocampal commissural	P[8]
Diagonal band to med. habenula	L[13]
Septum to habenula	L[13]
Septum to interpeduncular n.	L[13]
Amygdala to bed nucleus of the stria terminalis	L[48]
N. accumbens to globus pallidus	P[19,40,79] L[95]
N. accumbens to substantia innominata (ventral pallidum)	P[89] L[95]
N. accumbens to entopeduncular	P[72]
N. accumbens to hypothalamus	L[59]
N. accumbens to substantia nigra	(p[89,19*]), (L[96,19*])
N. accumbens to A10	P[99,100] (L[94,96,25*,53*])
Striatopallidal and striatoentopeduncular	P[71] L[26,35;47,65,84]
Striatonigral neurons	P[14,31,71] R[45,90] (L[c]), A[54,88] D[7,98]
Pallidosubthalamic	P[76] (L[27,91*])
Pallidonigral neurons	R[45] (L[c]), A[54,88] D[7]
Entopeduncular to lat. habenula	P[40] L[33,41,54,65] D[6]
Entopedunculothalamic	L[73]
Hypothalamus to cortex	D[93]
Hypothalamus to lat. habenula	D[6]
Hypothalamus to central gray	L[78]
Nigrothalamic, nigrotectal	P[4,10,15,56,68,101] R[85] L[16,42,85,92] A[92] D[5]
Nigrotegmental	P[68] (L[11,43*]), A[11]
Subthalamus to entopeduncular n.	P[77]
Zona incerta to superior colliculus	D[5]
Purkinje cells	P[38] R[70] A[9,52] L[24,75]

(*continued*)

TABLE 7.2 (*Continued*)

Anatomical system	Evidence and ref. nos.[b]
Interneurons in:	
Cortex	P[46] L[30,39] A[82]
Olfactory bulb	P[67,74] L[36,55]
Olfactory tubercle	L[29]
Hippocampus	P[3] L[17,23,80,81,86] A[44]
Septal area	L[58]
Amygdala	L[48]
N. accumbens	L[94–96]
Neostriatum	P[83] L[51,66]
Thalamus	P[64]
Hypothalamus	L[57]
Inferior colliculus	L[1]
Raphe	L[50]
Cerebellum	P[38] R[22] L[28] D[98]
Cochlear nucleus	P[21] L[18]
Rostral ventrolateral medulla	L[60]
Spinal cord	P[20,62] L[37,61]
Retina	P[69,87] R[12] L[32,34,49,63]

[a]See Table 7.1 for immunohistochemical staining for GAD and GABA, pharmacohistochemical staining for GABA-T, and uptake of radioactive GABA.

[b]Types of evidence: (P) inhibitory action; (R) GABA release; (L) lesion effect; (A) axoplasmic flow; (D) dual labeling. (Parentheses) signify that the literature is controversial; (*) denotes a negative report. References: (1) J. C. Adams and Wenthold (1979); (2) H. Altman *et al*. (1976); (3) Andersen *et al*. (1964); (4) M. Anderson and Yoshida (1977); (5) Araki *et al*. (1984a); (6) Araki *et al*. (1984b); (7) Araki *et al*. (1985); (8) Buzsaki and Czeh (1981); (9) Chan-Palay *et al*. (1979); (10) Chevalier *et al*. (1981); (11) Childs and Gale (1983); (12) Collins *et al*. (1979); (13) Contestible and Fonnum (1983); (14) Crossman *et al*. (1973); (15) Deniau *et al*. (1978a,b); (16) Di Chiara *et al*. (1979); (17) Di Lauro *et al*. (1981); (18) W. E. Davis (1977); (19) Dray and Oakley (1978); (20) Eccles *et al*. (1963); (21) E. F. Evans and Nelson (1973); (22) Flint *et al*. (1981); (23) Fonnum and Walaas (1978); (24) Fonnum *et al*. (1970); (25) Fonnum *et al*. (1977); (26) Fonnum *et al*. (1978a); (27) Fonnum *et al*. (1978b); (28) A. C. Foster and Roberts (1980); (29) Gilad and Reis (1979); (30) Godfrey *et al*. (1980); (31) Goswell and Sedgwell (1971); (32) Goto *et al*. (1981); (33) Gottesfeld and Jacobowitz (1978a); (34) Guarneri *et al*. (1981); (35) Hattori *et al*. (1973); (36) Hirsch and Margolis (1980); (37) Homma *et al*. (1979); (38) M. Ito and Yoshida (1964); (39) M. V. Johnston and Coyle (1980); (40) D. L. Jones and Mogenson (1980); (41) Kataoka *et al*. (1977); (42) Kilpatrick *et al*. (1980); (43) Kilpatrick and Starr (1981); (44) C. Kohler and Chan-Palay (1983b); (45) Kondo and Iwatsubo (1978); (46) K. Krnjevic and Schwartz (1966); (47) Kurihara *et al*. (1980); (48) Le Gal La Salle *et al*. (1978); (49) Lund-Karlsen (1978); (50) E. G. McGeer *et al*. (1979); (51) P. L. McGeer and E. G. McGeer (1975); (52) P. L. McGeer *et al*. (1975); (53) P. L. McGeer *et al*. (1977b); (54) P. L. McGeer *et al*. (1978b); (55) P. L. McGeer *et al*. (unpublished data); (56) Macleod *et al*. (1980); (57) Makara *et al*. (1975); (58) Malthe-Sørenssen *et al*. (1980a); (59) Meyers *et al*. (1980); (60) Meeley *et al*. (1985); (61) Y. Miyata and Otsuka (1975); (62) Miyata and Otsuuka (1972); (63) I. G. Morgan and Ingram (1981); (64) Mushiake *et al*. (1984); (65) Nagy *et al*. (1978); (66) Nagy and Fibiger (1980); (67) Nicoll (1969); (68) Niijima and Yoshida (1982); (69) Noell (1959); (70) Obata and Takeda (1969); (71) Obata and Yoshida (1973); (72) Pacitti *et al*. (1982); (73) Penney and Young (1981); (74) Rall *et al*. (1966); (75) Roffler-Tarlov *et al*. (1979); (76) Rouzaire-Dubois *et al*. (1980); (77) Rouzaire-Dubois *et al*. (1983); (78) Sandner *et al*. (1981); (79) Scarnati *et al*. (1983); (80) Schmid *et al*. (1980); (81) Schwarcz *et al*. (1978a); (82) Somogyi *et al*. (1983); (83) Spehlmann *et al*. (1977); (84) Staines *et al*. (1980); (85) Starr and Kilpatrick (1981a); (86) Storm-Mathisen and Fonnum (1971); (87) Straschell and Perwein (1969); (88) Streit *et al*. (1979); (89) Strahlendorf and Barnes (1983); (90) van der Heyden *et al*. (1979); (91) van der Kooy *et al*. (1981a); (92) Vincent *et al*. (1978); (93) Vincent *et al*. (1983a); (94) Waddington and Cross (1978); (95) Walaas and Fonnum (1979a); (96) Walaas and Fonnum (1979b); (97) Walaas and Fonnum (1980); (98) Weinberg *et al*. (1985); (99) Wolf *et al*. (1978); (100) C. Y. Yim and Mogensen (1980); (101) Yoshida and Omata (1979).

[c]There are many references with conflicting results [see the text and E. G. McGeer *et al*. (1982)].

7.3.1. Cerebellum

The cerebellar Purkinje cell is the classical example of a cell that has been proved to exert an inhibitory action by release of GABA. A review of the evidence gives a good idea of the various techniques used to identify presumptive GABAergic systems as well as the postsynaptic actions of GABA:

1. Stimulation of Purkinje cells results in hyperpolarization of postsynaptic cells in the deep cerebellar nuclei and Deiters' nuclei. This hyperpolarization is reversed when the resting membrane potential of the postsynaptic cell is increased beyond the equilibrium potential for the inhibitory postsynaptic potential.
2. The postsynaptic membrane is permeable to chloride and other anions of comparably small size during transmitter action.
3. Picrotoxin and bicuculline, which are classic GABA antagonists, block the effect of stimulation, while strychnine, which is not a GABA blocker, is ineffective.
4. Iontophoretic application of GABA on single Deiters' neurons mimics Purkinje cell stimulation. The effect is blocked by picrotoxin and bicuculline, but not by strychnine (M. Ito and Yoshida, 1964). Similar results are reported for interpositus neurons (Kawaguchi and Ono, 1973).
5. There is a large difference in GABA concentration in different neuronal groups in Deiters' nucleus. Those in the dorsal part, which receive nerve endings from Purkinje cell neurons, have much higher GABA concentrations than cells in the ventral part of the nucleus.
6. When the cerebellar cortex is removed by suction or the axons are undercut, there is a decrease in GABA and GAD content in the dorsal part, but not in the ventral part, of Deiters' nucleus (Fonnum et al., 1970).
7. Stimulation of Purkinje cells results in the release of detectable amounts of GABA into the perfusion fluid of the fourth ventricle or into the output fluid of a push–pull cannula inserted into the area of the deep cerebellar nuclei (Obata and Takeda, 1969).
8. GAD has been detected by immunohistochemistry in Purkinje cell terminals in the deep cerebellar nuclei (McLaughlin et al., 1974; K. Saito et al., 1974a) and Purkinje cell bodies (Ribak et al., 1978; Oertel et al., 1981).
9. Purkinje cells show intense staining for GABA-T a few hours after treatment with an irreversible GABA-T inhibitor (Vincent et al., 1980; Nagai et al., 1983a).
10. Radioactive GABA is conveyed by specific axonal transport processes from the Purkinje cell bodies to nerve terminals in Deiters' nucleus (P. L. McGeer et al., 1975) (Figure 7.7).

Other cells in the cerebellum also appear to use GABA as the neurotransmitter. The most obvious of these are the basket cells, which lie close to the Purkinje cell layer [see Figure 14.12 (Section 14.4.2)]. They are excited by granule fibers just as are Purkinje cells. But the basket cells send their axons at right angles to the Purkinje cells and act to inhibit rows of Purkinje cells that lie parallel to the row being primarily excited [see Figure 14.14 (Section 14.4.2)]. The basket cells accomplish this action by clustering their terminals in a pouch-like fashion around the axon hillocks of the Purkinje cells.

Uptake studies in the cerebellum using radioactive GABA clearly show heavy labeling in the basket, stellate, and Golgi cell regions (Figure 7.7). The uptake in basket cell terminals seems to be so concentrated that the Purkinje cells themselves, which would normally be expected to concentrate GABA, appear relatively clear. GAD immunohistochemistry and GABA-T pharmacohistochemistry (see Table 7.1) also indicate that all these cell types are GABAergic. Thus, in the cerebellum, only the granule cells are nonGABAergic, and these cells appear to use glutamate (see Chapter 6).

7.3.2. Hippocampus

The hippocampus is a layered structure that in some respects parallels the ·cerebellum. The most prominent feature is a layer of large pyramidal cells lying in a sheet approximately 0.5 mm deep to the surface [see Figure 16.3 (Section 16.3)]. Stimulation of the known inputs to the hippocampus, i.e., from the hippocampal commissure and the septum, as well as local stimulation, produce identical inhibitory effects [see Figure 4.16A–C (Section 4.7.3)]. Such a physiological action could take place only if special inhibitory interneurons were excited by all three inputs, with these interneurons ramifying broadly to the pyramidal cells. Basket cells, lying just superficial to the pyramidal cells (i.e., in the region of the basal dendrites), fulfill these criteria. They are activated by all the inputs to the hippocampus and discharge just prior to the onset of the inhibition of pyramidal cells. GABA itself inhibits pyramidal cells, an effect that is blocked by bicuculline (Andersen et al., 1964). The basket cells are also driven by axon collaterals of pyramidal cells, and, in turn, inhibit surrounding pyramidal cells. The wide distribution of the ramifying axons of basket cells suggests that each one may inhibit hundreds of pyramidal cells.

Both immunohistochemistry for GAD (Ribak et al., 1978) and pharmacohistochemistry for GABA-T (Nagai et al., 1983a) (see Table 7.1) have demonstrated that these inhibitory basket cells of the hippocampus, as well as some of the stellate cells, use GABA as their neurotransmitter agent.

Storm-Mathisen and Fonnum (1971) had earlier studied the GAD activity in the various layers of the hippocampus and found it to be highest in the area of basket cells and their terminals on pyramidal cells. This GAD was not altered by lesioning of any of the known hippocampal afferents (Storm-Mathisen, 1975), but was markedly reduced by local injections of kainic acid (KA) (Schwarcz et al., 1978a), indicating that it was associated with basket cells or other interneurons.

Seress and Ribak (1983) have found that about 60% of the neurons in the hilar region are GAD-positive, and since it is known that about 80% of hilar neurons give rise to commissural projections, this suggests that GABA, as well as glutamate or aspartate, may play a role in the commissural pathway. This is consistent with some physiological data (Buzsaki and Czeh, 1981) that suggest that stimulation of the commissural pathway directly inhibits granule cells in the contralateral hippocampus.

7.3.3. Basal Ganglia

The basal ganglia have a rich supply of GABA-containing structures. The caudate, putamen, and globus pallidus have all been shown to contain medium-size GABA neurons as evidenced by their content of GAD (Ribak et al., 1979) and GABA-T (Nagai et al.,

1983a). In addition, there is an intense GABA-T-positive neuropil in the neostriatum, possibly reflecting the presence of many GABAergic interneurons that have been demonstrated by lesion studies (P. L. McGeer and E. G. McGeer, 1975).

The basal ganglia are also the source of long GABA pathways that have been demonstrated both by lesions and by axoplasmic transport. They project from the caudate, putamen, globus pallidus, and substantia nigra. The best established of these are from the neostriatum to the globus pallidus and substantia nigra, from the globus pallidus to the thalamus and substantia nigra, and from the substantia nigra to the thalamus and optic tectum.

Electrolytic lesions of the globus pallidus, hemitransections of the brain at the level of the subthalamus, or destruction of the striatum by suction produces significant decreases in GABA and GAD in the substantia nigra. Nerve endings in the substantia nigra have been shown by immunohistochemistry to contain GAD (Ribak et al., 1977b) (see Figure 7.3). GABA is released in the substantia nigra following stimulation of the striatum (Y. Kondo and Iwatsubo, 1978; van der Heyden et al., 1979), and the inhibitory effect of striatal stimulation is blocked by GABA antagonists (Crossman et al., 1973; Goswell and Sedgwell, 1971; Obata and Yoshida, 1973).

Retrograde transport studies with horseradish peroxidase in rats that were subsequently treated with an irreversible GABA-T inhibitor showed that most of the GABA-T-intensive (and hence presumably GABAergic) neurons projecting to the area of the substantia nigra injected with HRP were in the lateral margin of the globus pallidus and, to a lesser extent, the lateral margin of the neostriatum. Most of the caudatonigral projection would appear to arise from non-GABA-T-intensive neurons that presumably use other neurotransmitters such as substance P or dynorphin (see Chapter 12) (Araki et al., 1985).

The globus pallidus and the substantia nigra pars reticulata (SNR) have dense populations of GAD-positive and GABA-T-intensive neurons, and GABA is the only neurotransmitter so far definitively identified in their efferents (E. G. McGeer et al., 1984) (Table 7.2). Thus, for example, the SNR sends GABAergic projections (partly by collaterals) to the superior colliculus (tectum), thalamus, and tegmentum; double labeling experiments indicate that the projection from the SNR (and the nearby zona incerta) to the superior colliculus is wholly GABAergic (Araki et al., 1984a).

The projection from the rostral entopeduncular nucleus (and the nearby lateral hypothalamus) to the lateral habenula has been shown by similar double labeling to contain both GABAergic and nonGABAergic fibers (Araki et al., 1984b).

It is clear from the many pathways already defined in the basal ganglia (Table 7.2) that GABA must play a major role in the functions of the extrapyramidal system [see Chapter 14 and E. G. McGeer et al. (1984)].

7.3.4. Raphe System

The raphe system, long known as the principal source of 5-HT neurons in the brain, also has neurons that stain positively for GAD and GABA-T, particularly in the dorsal raphe. Some of these can be doubly stained for 5-HT and GAD (Belin et al., 1983). As well, some of these serotonergic neurons take up radioactive GABA (Nanopoulos et al., 1982).

Physiological studies of inhibition in the raphe following stimulation of the habenula were initially interpreted as indicating a direct GABAergic tract, but were later thought to involve excitation of inhibitory interneurons in the raphe. Lesioning of the fasciculus retroflexus does not cause decreases of GAD in either the interpeduncular nucleus or the raphe (Belin *et al.*, 1979; Gottesfeld *et al.*, 1978; E. G. McGeer *et al.*, 1979; Mata *et al.*, 1977), a finding that argues against descending GABAergic fibers.

7.3.5. Diencephalon

The thalamus and hypothalamus both contain GABA neurons, as revealed by GAD immunohistochemistry and GABA-T histochemistry. The thalamus has a much more complex system. In addition to extensive GABA input from the various nuclei of the basal ganglia to both the lateral and medial nuclei, there are many positive cell groups. GABA-T-positive cells are arranged roughly in two concentric shells with a dense neuropil between them (see Figure 7.5). The outer shell consists mainly of the reticular nucleus, which has heavy neuronal staining for both GAD and GABA-T. The inner shell, consisting of intralaminar cell groups, has neurons that stain positively for GABA-T, as do the midline nuclei.

The hypothalamus has a few positive cells in the lateral aspect and an interesting group in the postmammillary caudal magnocellular region (Figure 7.5). These same neurons appear to stain positively for GAD (Leranth *et al.*, 1985), GABA-T (Nagai *et al.*, 1983a), and histamine (Panula *et al.*, 1984) [see Figure 11.2 (Section 11.1.2)]. This again, as in the cells of the raphe, raises the interesting possibility of neurons that contain GABA plus an amine, or class II, transmitter. Many fundamental neuroscience questions are posed by such coexistence of neurotransmitters.

7.3.6. Spinal Cord

Presynaptic inhibition in the spinal cord and its role in negative feedback control of sensory pathways are discussed in Chapter 4 [see Figure 4.18 (Section 4.8)]. The spinal cord interneurons responsible for this effect are GABAergic. Figure 7.8 shows GAD-containing nerve endings surrounding a primary afferent terminal. The dorsal root has been sectioned, permitting the afferent terminal to be identified because of its degeneration. The degenerating terminal is postsynaptic to GABAergic terminals and presynaptic to surrounding dendrites.

7.3.7. Olfactory Bulb

The reciprocal dendritic inhibitory synapses of granule cell interneurons in the olfactory bulb are described in Chapter 4 [see Figure 4.13 (Section 4.6)]. GAD has been identified in the granule cells and in the gemmules of their dendrites, establishing that these unusual inhibitory interneurons are GABAergic (Ribak *et al.*, 1977a).

7.3.8. Retina

The retina is a layered structure, with the content of the GABA system varying widely in different layers. By far the highest levels are found in areas near ganglion cells (Graham, 1972). GABA inhibits the firing of retinal ganglion cells (Noell, 1959; Straschill and Perwein, 1969). Amacrine cells are located in this area, have an inhibitory

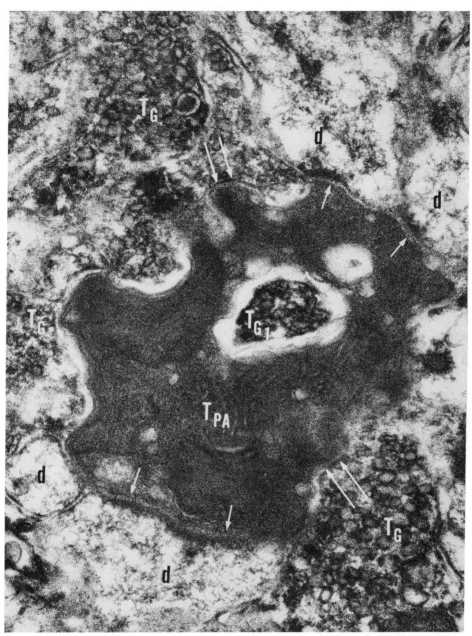

Figure 7.8. Immunocytochemical localization of the GABA-synthesizing enzyme GAD in laminae II–III of rat lumbar spinal cord. Ipsilateral dorsal root lesions were performed 24 hr prior to tissue preparation for immunocytochemistry. A dense, degenerating primary afferent terminal (T_{PA}) is surrounded by GAD-positive axon terminals (T_G) and by dendritic profiles (d). One of the GAD-positive axon terminals (T_{G1}) invaginates into the primary afferent terminal from a tissue level not included in this sectioning plane. Two of the GAD-positive terminals appear to be presynaptic (double arrows) to the degenerating primary afferent terminal. These axoaxonal synaptic relationships are consistent with the evidence that implicates GABA as one of the neurotransmitters that mediate presynaptic inhibition in the dorsal spinal cord. In addition to being postsynaptic to GAD-positive terminals, the primary afferent terminal is also presynaptic ($\rightarrow$) to some of the surrounding dendritic profiles. ×79,000. Courtesy of James E. Vaughn (City of Hope National Medical Center, Duarte, California). (Barber *et al.*, 1978).

physiological action, and form multiple synapses with ganglion and bipolar cells. Uptake and immunohistochemical studies (see Table 7.1), as well as lesion work with neurotoxins (see Table 7.2), suggest that one population of amacrine cells is probably GABA-containing; other populations may use glycine, DA, or one of various peptides as the transmitter. There is some indication that in some species, some of the horizontal or interplexiform cells or both may also use GABA (Lam *et al.*, 1980; S. M. Wu and Dowling, 1980; Brandon *et al.*, 1985).

7.3.9. Peripheral γ-Aminobutyric Acid

The GABAergic system seems largely confined to the CNS, including the spinal cord, but GABA, like almost every known central neurotransmitter, also appears to serve some neurons of the intestinal tract, since the presence of GAD and GABA, high-affinity uptake of GABA, and release on electrical stimulation have all been demonstrated (Taniyama *et al.*, 1982a,b; Saffrey *et al.*, 1983; Jessen *et al.*, 1979; Jessen, 1981).

GABA and GAD also occur in some nonneuronal tissues, especially the pancreas, ovary, and fallopian tubes, in which the concentrations are about one third of those in brain (Erdo *et al.*, 1982; Martin del Rio and Caballero, 1980). In the pancreas, GAD and GABA-T occur only in beta cells and not in neuronal elements (Vincent *et al.*, 1983b), but the GABA in the fallopian tubes appears to be neuronal.

7.4. Physiology of γ-Aminobutyric Acid

Much of the physiological action of GABA is discussed in Chapter 4. As detailed there, GABA exerts its inhibitory action on postsynaptic membranes by acting at specific receptor sites to activate a Cl^- channel. However, this is not the only action. Both physiological and binding studies indicate the existence of more than one type of GABA receptor [see Table 5.3 (Section 5.3.4)], indicating more than one type of response. D. R. Hill and Bowery (1981) described two GABA binding sites in rat brain:

GABA$_A$: Sensitive to bicuculline (antagonist), picrotoxin (antagonist), muscimol (agonist), 3-aminopropanesulfonic acid (agonist), and isoguvacine (agonist); insensitive to baclofen; binding insensitive to ions, but enhanced by Triton X-100 treatment.

GABA$_B$: Insensitive to bicuculline, 3-aminopropanesulfonic acid, and isoguvacine; only weakly sensitive to muscimol, but stereospecifically sensitive to baclofen (agonist); binding inhibited by guanyl nucleotides, dependent on Ca^{2+} or Mg^{2+}, and reduced by Triton X-100 treatment (D. R. Hill *et al.*, 1984; Corda and Guidotti, 1983).

GABA$_A$ sites are thought to be largely postsynaptic and concentrated on central neurons and peripheral sympathetic neurons, while GABA$_B$ sites are on presynaptic autonomic and central nerve terminals. Culture studies indicate that there is no binding of GABA to glial cells (Hosli *et al.*, 1980).

Activation of GABA$_A$ receptors is responsible for the classic postsynaptic inhibitory action (see Section 4.4). At these sites, GABA can produce depolarization of the ganglion

Figure 7.9. Theoretical model of the GABA receptor–ionophore complex, including recognition sites for convulsant and anticonvulsant drugs as viewed from outside the postsynaptic membrane looking down the Cl⁻ channel. The complex contains three binding sites: the GABA recognition site, the picrotoxinin recognition site, and the benzodiazepine recognition site. The ion channel may be a separate component or part of one of the ligand binding proteins. The model proposes that GABA function (opening of the Cl⁻ channel) can be modulated by endogenous ligands or drugs that bind to the picrotoxinin or benzodiazepine recognition sites or both. The effect can be either positive (depressants) or negative (excitants) modulation of the GABA–chloride ionophore. Modified from Olsen and Leeb-Lundberg (1981).

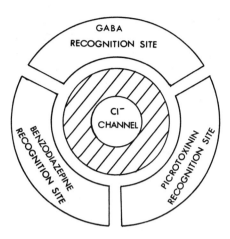

cell body or mediate hyperpolarization within the CNS by increasing membrane conductance to chloride ions. At the $GABA_B$ receptor, GABA (or baclofen) acts to reduce the outflow of other neurotransmitters (Bowery *et al.*, 1980), such as NA, glutamate, DA, or 5-HT. However, its own autoreceptors, where it inhibits release of further GABA, are of the $GABA_A$ type (Brennan *et al.*, 1981). The $GABA_B$ receptor may work by decreasing the inward flux of Ca^{2+} into the presynaptic nerve terminal. It is possible that an adenylate cyclase may be involved, since guanyl nucleotides modulate the affinity at these sites (see Section 5.4.2).

Some of the trophic effects of GABA on neuronal development, in particular those on the ultrastructure of cultured cerebellar granule cells, appear to be mediated by bicuculline-sensitive $GABA_A$ receptors (Meier *et al.*, 1985).

Much remains to be learned about these GABA receptors. Some, but not all, appear to be coupled to modulatory sites. At these modulatory sites, endogenous ligands, or drugs such as the benzodiazepines and picrotoxinin (the active form of picrotoxin), can act to modify the effect of GABA on the Cl⁻ channel (Figure 7.9). The picrotoxinin site is believed to be distinct from the benzodiazepine site. The benzodiazepines affect GABA binding, while picrotoxinin does not. Instead, picrotoxinin inhibits the increase in Cl⁻ flux triggered by the binding of GABA to its receptor. As far as the GABA–benzodiazepine receptor site combination is concerned, purification studies suggest that it is on the same macromolecule (Asano *et al.*, 1983). The endogenous ligands for the benzodiazepine and picrotoxinin sites have not been definitively identified. However, some purines and some peptides, such as cholecystokinin (Bradwejn and de Montigny, 1984) and thyrotropin-releasing hormone (TRH) (Sharif *et al.*, 1983), compete for some benzodiazepine binding sites, suggesting that these putative neurotransmitters (see Chapters 11 and 12) may be candidates. Another possibility is a brain peptide with 104 amino acid residues that, at micromolar concentrations, is capable of displacing benzodiazepines from binding sites and of antagonizing the "anticonflict" action of benzodiazepines in rats on intraventricular injection. This peptide, named "diazepam-binding inhibitor" (DBI) and called by some the "anxiety peptide" (Marx, 1985b), has been localized by immunohistochemistry in nonGABAergic neurons in rat cortex, but may be colocalized with GABA in some cerebellar and hippocampal neurons (Alho *et al.*, 1985). DBI contains at least two copies of an active 18-amino-acid fragment that is even more potent than the parent peptide (Ferrero *et al.*, 1984). Other, uncharacterized, brain peptides have also been reported to affect GABA binding (e.g., Vaccarino *et al.*, 1985).

7.5. Pharmacology of γ-Aminobutyric Acid

Pharmacological agents exist that are capable of interacting with GABA in all the classic areas for manipulation of neurotransmitter systems, i.e., at sites of synthesis, storage, extraneuronal release, presynaptic reuptake, postsynaptic destruction, and postsynaptic activation. In addition, many drugs appear to modulate GABA binding or the coupling of the GABA receptor to the Cl^- channel (see Section 7.4 and Figure 7.9). It is these modulators that comprise the most clinically effective drugs.

The better-known agents are listed in Table 7.3. Some interact prominently with other systems, so that their specificity is limited. Although inconsistencies appear in a number of cases between the presumed mechanisms of action and the overall physiological effects, drugs that diminish GABA activity generally cause excitation leading to convulsions, while those that enhance its activity cause sedation leading to depression. Such global effects would be anticipated for a general inhibitory neurotransmitter.

7.5.1. Receptor Site Modulators

The benzodiazepine class of tranquilizers, of which diazepam (Valium) represents the prototype compound, are among the most widely used drugs in clinical practice. They are extremely effective in alleviating anxiety and also have uses as anticonvulsants and muscle relaxants. Considerable evidence now indicates that the benzodiazepines exert many of their effects by action at a site coupled to a postsynaptic GABA receptor (see Section 7.4 and Figure 7.9) in such a way as to increase GABA binding, augmenting GABA's physiological action and thus acting as anticonvulsants.

However, not all benzodiazepine sites are coupled to GABA sites and not all GABA sites are coupled to benzodiazepine sites, indicating that there are multiple benzodiazepine receptors. At least two major classes have been distinguished, they being referred to as Types I and II, or as BZ_1 and BZ_2. Type I is defined as a GABA-independent site at which the benzodiazepines have their anxiolytic effect, while Type II is defined as a GABA-dependent site at which the benzodiazepines have their anticonvulsive and sedative hypnotic effects. Type II is more readily solubilized than Type I (Lo *et al.*, 1983). The BZ_1 and BZ_2 sites are defined as those that have, respectively, high and low affinity for the antagonist propyl β-carboline-3-carboxylate. Type II is thought to be equivalent to BZ_1, but the equivalence of Type I with BZ_2 has not been fully established (K. W. Gee *et al.*, 1983).

To make matters more complicated, benzodiazepines bind specifically to a pharmacologically distinct site that is called a ''peripheral-type'' receptor because it was first found in kidney. This type is widespread in brain, being most highly concentrated in the olfactory bulb, median eminence, choroid plexus, and ependyma (D. W. Gallager *et al.*, 1981; Schoemaker *et al.*, 1982), but does not appear to be involved in the tranquilizing actions of benzodiazepines, since there is no correlation between clinical potency and the affinity for such sites. The binding is insensitive to picrotoxinin, barbiturates, and GABA.

TABLE 7.3
Some Drugs That Affect GABA Action[a]

Drug	Presumed action mechanism	Physiological effect
Receptor site modulators		
Diazepam (benzodiazepines)	Facilitate GABA binding	Anticonvulsants, tranquilizers
Barbiturates	Facilitate GABA binding	Anticonvulsants
Valproate	Facilitates GABA binding	Anticonvulsants
Pyrazolopyridines	Facilitate GABA binding	Anxiolytics
Tracazolate	Facilitates GABA binding	Anxiolytic
Ivermectin[b]	Facilitates GABA binding	Potent anthelminthic
Ethyl β-carboline-3-carboxylate	Decreases GABA binding	Convulsant
Picrotoxinin	Uncouples GABA site from Cl$^-$	Convulsant
Synthesis		
Allylglycine	GAD inhibitor	Convulsant
β-N-γ-Glutamyldiaminopropionic acid	GAD inhibitor	Convulsant and neuritic
3-Mercaptopropionic acid[c]	GAD inhibitor	Convulsant
High-pressure oxygen	GAD inhibitor	Convulsant
Isonicotinylhydrazide	B_6 antagonist	Convulsant at high doses
Thiosemicarbazide	B_6 antagonist	Convulsant at high doses
Methionine sulfoximine	Glutamine synthetase inhibitor	Convulsant
Release		
Tetanus toxin	Inhibitor of GABA and glycine release	Convulsant
Pump		
cis-3-Aminocyclohexanecarboxylic acid	GABA neuronal pump inhibitor	—
Nipecotic acid	GABA neuronal pump inhibitor	—
β-Alanine	GABA glial pump inhibitor	—
Haloperidol	Monoamine blocker and weak neuronal pump inhibitor	Antipsychotic
Chlorpromazine	Monoamine blocker and weak neuronal pump inhibitor	Antipsychotic
Imipramine	Monoamine and GABA pump inhibitor	Antidepressant
Destruction		
Aminoxyacetic acid	B_6 antagonist for GABA-T	Sedative
n-Dipropylacetate	GABA-T inhibitor	Anticonvulsant
Hydrazinopropionic acid	GABA-T inhibitor	Sedative
Gabaculine	GABA-T inhibitor	Anticonvulsant
γ-AcetylenicGABA	GABA-T inhibitor	—
γ-VinylGABA	GABA-T inhibitor	—
(5)-4-Amino-5-fluoropropionic acid	GABA-T inhibitor	—
3-Amino-4,4-difluorobutyric acid	GABA-T inhibitor	—
Antagonist		
Bicuculline	GABA$_A$ antagonist	Convulsant

(continued)

TABLE 7.3 (*Continued*)

Drug	Presumed action mechanism	Physiological effect
Agonist		
Muscimol	$GABA_A$ agonist	Psychotomimetic
Imidazoleacetic acid	Weak GABA agonist	Weak sedative
Progabide	Metabolized to GABA agonist	Anticonvulsant
Baclofen	$GABA_B$ agonist	Muscle relaxant
γ-Hydroxybutyric acid	Possible GABA agonist	Sedative

[a]For references, see the text and P. L. McGeer *et al.* [1978a (pp. 220–225)].
[b]Also increases GABA release.
[c]Also decreases GABA release.

Work with cultures suggests that some of the binding is on glia (Zlobina *et al.*, 1984). Studies in different species indicate a very late evolutionary appearance of this type of binding site in both periphery and brain, implying that the site may have a highly significant function or functions (Bolger *et al.*, 1985). Activation of the receptor seems to cause an increase in membrane fluidity by means of sequential methylation of phosphatidylethanolamine to phosphatidylcholine (Strittmatter *et al.*, 1979). A specific ligand for this site is 1-(2-chlorophenyl)-*N*-methyl-*N*-(1-methylpropyl)-3-isoquinolinecarboxamide.

There have been suggestions that barbiturates act either at a benzodiazepine site to facilitate GABA binding or at the same site as picrotoxinin (Olsen, 1981). One action of the anticonvulsant valproate (sodium dipropylacetate) is believed to be at the picrotoxinin binding site to facilitate the coupling of the GABA site with the Cl^- channel.

The anxiolytic pyrazolopyridines have also been reported to act both at a benzodiazepine site and at the picrotoxinin recognition site, while the related anxiolytic tracazolate is believed to act only at the picrotoxinin site.

Avermectin B_1, a paralyzing anthelminthic, is an irreversible activator of the GABA–benzodiazepine–chloride ionophore receptor complex; like GABA and Cl^-, it markedly increases benzodiazepine binding in rat cortex, but unlike GABA and Cl^-, the effect cannot be reversed by extensive washing of the membranes. Avermectin increases synaptosomal release of GABA at least 3-fold *in vitro*, without affecting that of glutamate. A dihydro derivative of avermectin B_1, called "ivermectin," has achieved commercial importance as a potent antiparasitic for veterinary medicine. Its action is presumed to be the same as that of avermectin B_1 and to involve primarily the stimulation of GABA-mediated Cl^- ion conductance. This could result from stimulating GABA release, from potentiating GABA binding, or from both. Ivermectin itself does not react with the GABA recognition site (Campbell *et al.*, 1983).

7.5.2. Synthesis Promotion or Inhibition

So far, no way is known for increasing GABA levels by administering either the material itself or substrates for its synthetic enzyme. Since neither glutamate nor GABA crosses the blood–brain barrier (BBB) with ease, administering these materials is without effect beyond the neonatal period. The administration of either glutamine or pyridoxal phosphate is also ineffective, even though they do cross the BBB. Apparently, precursor availability is not critical to GAD activity, and normal tissue levels of both precursor and cofactor are sufficient to saturate the enzyme.

On the other hand, GABA levels can be lowered by a variety of agents that inhibit GABA synthesis. The so-called "carbonyl trapping agents," which act against the cofactor pyridoxal phosphate, form the largest family. Classic members are such hydrazides as isoniazid, semicarbazide, and thiosemicarbazide, which will inhibit GAD in the concentration range of 10^{-4}–10^{-3} M *in vitro* and *in vivo*. Carbonyl trapping agents affect both GAD and GABA-T, however, since both enzymes depend on pyridoxal phosphate. Thus, if GABA-T is more severely inhibited than GAD by a pyridoxal phosphate antagonist, GABA levels may increase rather than decrease. Some agents, such as nicotinyl hydrazide, will either raise or lower GABA levels, depending on the time after administration of the drug and the dose. The carbonyl trapping agents that appear to inhibit GAD more specifically than GABA-T have the greatest tendency to produce seizures in animals.

There is a question as to whether the convulsant action of the hydrazides is really due to inhibition of GAD. It has been shown that at the time when convulsions first commence, GABA levels are not sharply reduced and GAD is not completely inhibited. There is much evidence, nevertheless, that lowered GABA levels do result in convulsions. 3-Mercaptopropionic acid inhibits GAD almost totally at a concentration of 10^{-3} M and also markedly inhibits K^+-induced release of GABA (Fan *et al.*, 1981). Following administration of this agent to rats, convulsions occur in about 7 min. At that time, GABA levels are reduced by about one third (Karlsson *et al.*, 1974). Allylglycine has a similar effect (Alberici *et al.*, 1969). Furthermore, high-pressure oxygen, which also inhibits GAD activity and reduces GABA levels, produces the same effect (Wood, 1975).

The importance of glutamine as a precursor of GABA is suggested by the convulsant activity of methionine sulfoximine, an irreversible inhibitor of glutamine synthetase (Ronzio *et al.*, 1969).

7.5.3. Storage and Release Mechanisms

While it is presumed that GABA is stored in presynaptic vesicles in the same fashion as the catecholamines and 5-HT, direct evidence has been hard to accumulate. Subcellular fractionation studies have failed to show GABA in synaptic vesicles, although the reason for this might be the failure of GABA to remain bound during the lengthy centrifugation procedures. Drugs that will specifically alter the storage of GABA have not yet been found.

Tetanus toxin, a convulsant protein, acts on the CNS apparently by reducing the synaptic release of the inhibitory neurotransmitters GABA and glycine (Curtis *et al.*, 1973). When injected directly into the cerebellum, it suppresses basket cell inhibition or cerebellar Purkinje cells. It does not influence the postsynaptic inhibitory effect of GABA when administered iontophoretically, however, which indicates that it has no effect on receptor sites and must therefore be affecting the presynaptic release of GABA. In studies *in vitro*, tetanus toxin has been shown to affect GABA release from slices of the hippocampus, substantia nigra, and striatum (Collingridge *et al.*, 1981).

7.5.4. Pump Inhibition

Uptake inhibitors enhance and prolong the inhibitory action of GABA, but the physiological effects of the compounds so far investigated vary according to their actions on other systems. GABA pump inhibitors include such GABA analogues as 2-hydroxy-2-chloroGABA and 4-methylGABA, along with agents, such as chlorpromazine, imipramine, and haloperidol, that have their major effects on aromatic amine systems.

A GABA pump system exists for glia as well as synaptosomes. The two processes are not identical because they are affected by different inhibitors. β-Alanine and proline, for example, inhibit glial uptake hundreds of times more powerfully than they do neuronal uptake. They have therefore been used to differentiate glial from neuronal uptake. *cis*-3-Aminocyclohexanecarboxylic acid, (RS)-3-hydroxy-5-aminovaleric acid, and (3RS,4SR,-5SR)-4-hydroxy-5-methyl-nipecotic acid are selective neuronal uptake inhibitors. Two bicyclic isoxazoles that inhibit glial uptake have been suggested as anticonvulsants (Schousboe *et al.*, 1981).

7.5.5. Inhibition of Metabolism

Levels of GABA in brain can be increased by inhibiting the metabolizing enzyme, GABA-T. Many of the inhibitors are relatively nonspecific. Some are highly toxic. Others do not cross the BBB. Thus, they are of limited clinical usefulness. In animals, they tend to produce sedation.

The most useful GABA-T inhibitor so far developed is sodium *n*-dipropylacetate (valproate), which, however, may owe its anticonvulsant activity primarily to an interaction with the GABA receptor–ionophore complex (see Section 7.5.1). It may also have some interaction with excitatory amino acid systems (see Chapter 6). Clinically, it has proven effective in treating certain forms of epilepsy.

Other GABA-T inhibitors are γ-acetylenicGABA, γ-vinylGABA, ethanolamine-*O*-sulfate, gabaculine (5-amino-1,3-cyclohexadienecarboxylic acid), (5)-4-amino-5-fluoropropionic acid, and 3-amino-4,4-difluorobutyric acid, all of which have a long-lasting effect *in vivo,* but are not completely specific. The last three are the most potent on peripheral administration. They owe their long-lasting action to the fact that they irreversibly destroy the GABA-T present at the time of administration; hence, metabolism is slowed until new enzyme can be synthesized. Their use in a pharmacohistochemical method for GABA-T was described in Section 7.3.

7.5.6. Receptor Agonists and Antagonists

Antagonists. Picrotoxin and bicuculline have been regarded as potent and specific blockers of GABA. Both cause convulsions. As mentioned in Section 7.4, picrotoxin probably does not act directly at the GABA recognition site, but modifies the effect of GABA at that site. Bicuculline, however, appears to block the GABA$_A$ receptor and is thus a true antagonist. Bicuculline has a mild anticholinesterase action, but is nevertheless regarded as being highly specific (Curtis *et al.*, 1970). It will not, for example, antagonize the inhibitory effects of glycine (see Section 7.9).

Agonists.   It has been proposed that anticonvulsants might be made by introducing into GABA-like molecules hydrophobic N-protecting (Galzigna *et al.*, 1978) and C-protecting (Frey and Loscher, 1980) groups to yield substances that would cross the BBB and be converted in brain into GABA agonists. A number of such so-called "GABA pro-drugs" have been made and shown to have anticonvulsant activity in animals; a recent example is *N*-pivaloyl-leucyl-γ-aminobutyric acid (Galzigna *et al.*, 1984). The drug of this type that has so far been the most widely studied is "Progabide" (SL 76002) {4-[2'-hydroxy-4'-α-(4-chlorophenyl)benzenemethanimino]butyramide}, which readily enters the brain following peripheral administration and is metabolized to a GABA agonist. Favorable clinical results have been reported in L-dopa-induced dyskinesias, tardive dyskinesia (Morselli *et al.*, 1985), hemiballismus (Gonce *et al.*, 1983), and spasticity (Mondrup and Pedersen, 1984a,b), while both positive and negative results have been reported in epilepsy (Loiseau, 1983; Dam *et al.*, 1983).

Muscimol, a psychotomimetic isoxazole isolated from the mushroom *Amanita muscaria*, is a potent agonist at $GABA_A$ receptors. The reason for its seemingly unrelated psychotomimetic pharmacological action is unknown. Binding of labeled muscimol has been employed as a method for mapping $GABA_A$ receptor sites (Lester and Peck, 1979; de Feudis, 1978).

β-(*p*-Chlorophenyl)GABA, also termed "baclofen" or "lioresal," has become the drug of choice in treatment of spasticity (Terence *et al.*, 1983) and is reported effective in some forms of chorea (Trauner, 1985). It is believed to have its clinical effects because it acts on $GABA_B$ receptors (see Section 7.4).

γ-Hydroxybutyric acid (γ-OH) is another GABA-like agent in clinical use as a basal anesthetic or, at lower doses, as a mild sedative. It readily crosses the BBB. It has inhibitory effects on iontophoresis, which, however, are not antagonized by bicuculline. The most clear-cut response to administered γ-OH is an increase in the level of DA in the brain that has been ascribed to a decreased firing rate of DA neurons. The mechanism of action is unclear.

4,5,6,7-Tetrahydroisoxazolo(5,4-c)pyridin-3-ol (THIP) is a specific GABA agonist with potent analgesic properties; it has been reported to bind to a subpopulation of GABA receptors that are not coupled to a benzodiazepine recognition site.

δ-Aminolevulinic acid, a compound that accumulates in porphyria, acts as a GABA agonist in iontophoretic and binding experiments (Müller and Snyder, 1977); it also reduces the release of GABA from synaptosomal preparations (Brennan and Cantrill, 1979). It has been suggested that these actions might be responsible for some of the symptoms of porphyria.

7.6. Pathology of γ-Aminobutyric Acid

A succinic semialdehyde dehydrogenase deficiency is the first documented inborn error of GABA metabolism in man; it leads to retardation of both mental and motor development, ataxia, and muscular hypertonia. The activity of the enzyme in lymphocytes is only about 10% of that found in normal individuals (K. M. Gibson, 1983).

There is good evidence for abnormalities in GABA systems in parkinsonism and chorea, and they may also play a role in drug-induced dyskinesias (see Chapter 14). There is a considerable body of pharmacological data indicating the importance of GABA in epileptic phenomena. It is established beyond question that induced shortages of GABA in animals cause convulsions, and most of the anticonvulsants in present clinical use are drugs that increase GABAergic activity in the brain. Despite all the pharmacological data, however, there is no convincing evidence of a GABA deficiency in human epilepsy. The basic problem may turn out to be a deficiency in a GABA–benzodiazepine–ionophore system, but it could alternatively involve an excess of glutamate/aspartate activity (see Chapter 6) or even some other neurotransmitter abnormality.

There have also been numerous reports that GABA abnormalities may be involved in psychoses (see Chapter 15), but the evidence is so far inconclusive.

7.7. Chemistry and Metabolism of Glycine

The other demonstrated ionotropic inhibitory neurotransmitter is glycine. In mammals, glycine is a nonessential amino acid that makes up 1–5% of typical dietary proteins. It crosses the BBB with ease and thus may be transported to brain and spinal cord from blood. It is incorporated into peptides, proteins, nucleotides, and nucleic acids; its fragments participate in other metabolic sequences. It can be synthesized from glucose and other substrates. Thus, it is one of the more versatile compounds in brain.

The immediate precursor of glycine in the body is serine. Studies using radioactive precursors suggest that some of the glycine in brain is derived from *de novo* synthesis from glucose through serine (Figure 7.10) and not from transport. But it is not known whether this is the route to the neurotransmitter pool.

The enzyme serine hydroxymethyltransferase (SHMT) is responsible for converting

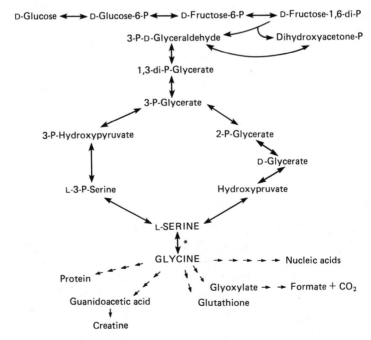

Figure 7.10. Possible synthetic pathways and metabolic routes for glycine in nervous tissue.

serine to glycine. It requires tetrahydrofolic acid, pyridoxal phosphate, and manganese ions and is strongly inhibited by such antipyridoxal compounds as aminooxyacetic acid. Unfortunately, the activity of SHMT cannot be used as an index of the presence of glycinergic neurons. The rate of formation from serine appears constant from area to area, and there is no distinct correlation with the presumed presence of glycinergic cell bodies or nerve endings.

Another suggested route is by transamination of glyoxylate. Glyoxylate aminotransferase activity with L-glutamate, L-glutamine. and L-alanine as amino donors has been found in rat, guinea pig, and goat retinae, suggesting that glyoxylate may serve as a precursor of glycine in vertebrate retina (Dasgupta and Narayanaswami, 1981).

The metabolic disposal of glycine is also unclear. It has been shown *in vitro* that glycine can be converted to glutathione, guanidinoacetic acid, glyoxylate, or even back to serine through a cleavage mechanism (Figure 7.10). Again, however, there is no correspondence between the activity of this cleavage system or other disposal mechanisms and the probable occurrence of glycine neurons.

Obviously, much remains to be discovered about the chemical reactions important for glycine synthesis and disposal and the control of glycine levels.

7.8. Anatomical Distribution of Glycine

As a dietary amino acid, glycine is found in all tissues, and its presence *per se* in an area does not indicate a neurotransmitter role in that area. Nevertheless, distribution studies do offer valuable clues. Thus, the glycine concentrations found in spinal cord, medulla, and pons are much higher than those in other CNS regions (Figure 7.11). This is in accord with studies showing that strychnine binding sites are concentrated in these regions, as is the Na^+-dependent, high-affinity ($K_m \simeq 10^{-5}$ M) uptake system for glycine into synaptosomal fractions of brain (Figure 7.11). Lesion and radioautographic uptake studies have suggested some specific neuronal localizations and pathways (Table 7.4), and an immunohistochemical study based on antibodies raised against a glycine–albumin conjugate was said to show glycine-positive neurons in the retina and brain stem (Pourcho and Goebel, 1986).

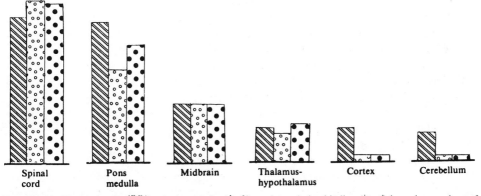

Figure 7.11. Glycine levels (▨), glycine uptake (⬚), and strychnine binding (●) in various regions of rat brain relative to values in midbrain.

TABLE 7.4
Some Pathways That Are Suggested to Use
Glycine[a]

Spinal cord interneurons
Spinal afferents from the raphe and reticular formation
Cortical projection to the hypothalamus
Brainstem afferents to the substantia nigra
Cerebellar Golgi cells
Glossopharyngeal nerve
Retinal amacrine cells

[a] In most cases, the only evidence is physiological, but see the text for details and references.

Within the spinal cord, glycine shows an uneven distribution, with the highest concentrations being found in the ventral and dorsal gray (Aprison et al., 1970; S. J. Berger et al., 1977). This is again in accord with uptake studies using tritiated glycine. These studies have shown that preferential uptake into synaptosomes of the ventral horn gray matter takes place (Ljungdahl and Hökfelt, 1973), with particular accumulations occurring around the perikarya of motor neurons (Matus and Dennison, 1972). The synaptosomes that accumulate glycine contain "flat" synaptic vesicles (Matus and Dennison, 1972; Ljungdahl and Hökfelt, 1973) that make axodendritic or axosomatic contacts (Ljungdahl and Hökfelt, 1973; D. L. Price et al., 1976). These contacts presumably act to inhibit spinal motor neurons (D. L. Price et al., 1976). The synaptosomes that accumulate glycine are different from those that accumulate GABA and constitute about 25% of all terminals in the spinal cord (Iversen and Bloom, 1972). The concentration of glycinergic endings in spinal cord is therefore believed comparable to the concentration of GABAergic endings in the same tissue and is considerably higher than the 10–15% estimates for DA and ACh terminals in areas where they are rich.

There is also some uptake of glycine into nerve cell bodies; cells that accumulate glycine are concentrated in lamina III of the cervical spinal cord of rats, while those that accumulate GABA are principally in lamina I (Ribeiro and da Silva Coimbra, 1980).

Glycine occurs in higher concentrations in the cervical and lumbar than in the thoracic cord. The cervical and lumbar enlargements are the areas of major motor outflow to the extremities, and the marked localization of glycine in these areas correlates with the large number of synaptic junctions and interneurons. Fish, which have no limbs, have no differences in glycine concentration in various levels of the spinal cord.

The existence of K^+-stimulated, Ca^{2+}-dependent release of glycine in the spinal cord (Fagg et al., 1978; Kerwin and Pycock, 1979a; Lopez-Colome et al., 1978b) lends further support to a neurotransmitter role for glycine in that area. In amyotrophic lateral sclerosis, in which there is a preferential loss of anterior horn cells, there is also a preferential loss of glycinergic receptors as evidenced by a large decrease in [^{3}H]-strychnine binding (H. Hayashi et al., 1981).

Physiological evidence suggests that glycine may also be a transmitter in some of the descending fibers of supraspinal origin that inhibit the firing of Renshaw cells (Biscoe and Curtis, 1966), in reticulospinal neurons in the medullary reticular formation (Trebecis,

1973), and in some projections to the spinal cord from the inferior central nucleus of the raphe (Belcher et al., 1978).

The exact location of glycine in the medulla and pons is unknown, but it has been reported (Guertzenstein and Silver, 1974) that the glycine-sensitive area in the medulla is restricted to a 1.5-mm-wide strip situated 1–2.5 mm caudal to the trapezoid bodies.

The existence of glycinergic afferents to the dopaminergic neurons of the substantia nigra has been suggested on the basis of high strychnine binding and marked high-affinity uptake (V_{max} = 787 nmoles/g-min) in the substantia nigra, with no change in uptake following local KA injections (James and Starr, 1979). It is consistent with this hypothesis that strychnine binding is decreased in the substantia nigra of parkinsonian patients who have lost many of the dopaminergic neurons (de Montes et al., 1982), and that local injections of either glycine or strychinine initiate rotation (James and Starr, 1979) and affect dopamine release (Leviel et al., 1979; Kerwin and Pycock, 1979b).

Glycinergic innervation of the dopaminergic neurons of the ventral tegmental area has also been suggested on the basis of the high level of strychnine binding in the ventral tegmental area (Gundlach and Beart, 1981) and uptake and release of glycine in rat ventral tegmental slices (Gundlach and Beart, 1982).

Electron microscopic studies of glycine and GABA uptake in the cerebellum indicate the possibility of at least two separate populations of Golgi cells—one using glycine as the neurotransmitter and the other using GABA (Wilkin et al., 1981). Some such role for glycine in the cerebellum is consistent with the increased local levels of glycine seen in ataxic mouse strains that have reduced numbers of granule cells (Muramoto et al., 1981).

Accumulation of radioactive glycine into a population of amacrine cells lying in the outer two thirds of the inner plexiform layer of rat retina (Pourcho, 1980; J. Marshall and Voaden, 1974; Beale and Osborne, 1983) and particularly high concentrations of endogenous glycine in the inner nuclear and inner reticular layers of the monkey (S. J. Berger et al., 1977) and mouse (A. I. Cohen et al., 1973) have been demonstrated. Moreover, destruction of retinal interneurons and ganglion cells by administration of sodium glutamate to newborn rats led to a 60–70% reduction in glycine uptake (Lund-Karlsen, 1978). Treatment of chick retina with KA also caused a marked reduction in glycine levels (Pasantes-Morales et al., 1981). Furthermore, some of the inhibitory cells in the rabbit retina appear to be blocked by strychnine (H. J. Wyatt and Daw, 1976). Saturable, specific [³H]strychnine binding (K_D = 60 nM; B_{max} = 15 pmoles/mg protein) has been reported in bovine retina (Borbe et al., 1981), and K^+-stimulated, Ca^{2+}-dependent release of glycine has been demonstrated in chick and rat retina (Lopez-Colome et al., 1978a; L. F. P. Smith and Pycock, 1982). Light-evoked release of glycine from the retina of rats (Coull and Cutler, 1978) and cats (Ehinger and Lindberg, 1974) can be demonstrated both in vivo and in vitro, but photic stimulation does not alter release of glycine from dark-adapted rabbit retina (Neal and Massey, 1980; Collins et al., 1979).

The suggestion from physiological evidence (Kita, 1978; Ohta and Oomura, 1979; Kita and Oomura, 1982) that the direct inhibitory pathway from the frontal cortex to the lateral hypothalamus in the rat might be glycinergic received some support from data indicating falls in glycine uptake in the hypothalamus following frontal cortical lesions (E. G. McGeer and Singh, 1980).

On the basis of neurophysiological studies using iontophoresed glycine or the blocking action of strychnine against inhibitory neurotransmission or both, it has been suggested that glycine may also be a neurotransmitter in the inhibition of hypoglossal motoneurons by the glossopharyngeal nerve (A. W. Duggan *et al.*, 1973), in the commissural inhibition of vestibular neurons (Precht *et al.*, 1973), in Deiters' nucleus (Obata *et al.*, 1970), and in the cuneate nucleus (Galindo *et al.*, 1967).

7.9. Physiology, Pharmacology, and Pathology of Glycine

Glycine fulfills the physiological criteria expected of the neurotransmitter released by inhibitory interneurons activated by Ia muscle afferents (see Chapter 4). Its iontophoretic action duplicates the effects of synaptic activation. It hyperpolarizes the membrane, brings about a large fall in membrane resistance, and increases Cl^- permeability. Although these actions are also produced by GABA, glycine is blocked by strychnine and GABA by bicuculline. The action of spinal inhibitory interneurons is prevented by strychnine but not by bicuculline, confirming that glycine is the true neurotransmitter (Werman *et al.*, 1968).

Relatively few specific pharmacological agents have been found that interfere with glycine neuronal systems. This is probably due to a lack of information regarding the factors that control the synthesis and metabolism of glycine in these neurons. The only types of specific agents that have been definitively identified are receptor blockers and uptake inhibitors.

Strychnine is the best known of the receptor blockers. The distribution of strychnine binding sites has been studied by both filtration binding assays (Figure 7.11) and light microscopic autoradiography which permits a very detailed analysis of localization. The density of binding sites is greatest in the gray matter of the spinal cord and decreases progressively in regions more rostral in the neuroaxis. There is a striking correlation between areas of high density of strychnine binding sites and areas in which glycine has been found to be electrophysiologically active. It has been suggested that the anatomical localizations of strychnine binding sites may help explain many of the signs and symptoms of strychnine poisoning (Zarbin *et al.*, 1981).

There are several other substances that are known to block the action of glycine but not of GABA. These include brucine, thebaine, 4-phenyl-4-formyl-*N*-methyl-piperidine, *N,N*-dimethylmuscimol, and *N*-methyl-THIP. In general, these compounds are much less potent than strychnine (Curtis *et al.*, 1967; Krogsgaard-Larsen *et al.*, 1982).

Some compounds have been identified that inhibit glycine uptake and, if analogies can be drawn with other neurotransmitters, might therefore enhance glycine action. *p*-Chloromercuriphenylsulfonate is the most potent inhibitor so far found, but imipramine, chlorpromazine, and hydrazinoacetic acid all have some inhibitory action (Aprison *et al.*, 1970). Synaptosomal glycine uptake is inhibited by methylmalonate at concentrations similar to those found in methylmalonic acidemia, a genetically linked deficiency of methylmalonyl-CoA mutase. This inhibition could be responsible, at least in part, for the neurological spinal cord damage characteristic of this disease (Lopez-Lahoya *et al.*, 1981).

As far back as 1932, glycine was used in the treatment of muscular dystrophy. Apparently the treatment was ineffective, although in some cases up to 60 g of glycine was reported to be ingested in a day without apparent neurotoxicity. It must therefore be considered that glycine is not a particularly potent pharmacological compound. At low doses, it protects dogs from the effects of several emetic drugs, and on this basis, a specific action on the medullary or vomiting center has been proposed.

Nonketotic hyperglycemia, which is due to a primary defect of the glycine cleavage system, is usually fatal in early infancy. Attempts to treat this condition with strychnine in three infants were unsuccessful (von Wendt *et al.*, 1980), so that excessive stimulation of glycine receptors is unlikely to be the critical lesion. On postmortem examination, the brains showed a spongy myelinopathy that suggests that structural alterations in myelin may be more important (Agamanolis *et al.*, 1982).

Glycine concentrations in brain appear to be unaffected by hypoxia, insulin-induced hyperglycemia, deprivation of protein, or infusion of ammonia.

It is clearly evident that our understanding of glycine neurons is extremely limited and that almost nothing is known about how to modify their action with pharmacological agents.

7.10. Other Suggested Inhibitory Amino Acid Transmitters

7.10.1. Taurine

For a review, see *Adv. Exp. Med. Biol.* **139**:1–551 (1981).

A decade after the discovery of high concentrations of taurine in brain (Awapara *et al.*, 1950; E. Roberts and Frankel, 1950), Curtis and Watkins (1960) demonstrated that it had a strong inhibitory action on spinal neurons. Jasper and Koyama (1969) then showed an increased release from the cerebral cortex of cats during arousal. There has since been speculation that taurine might play a neurotransmitter role in brain, but it is now generally believed that in both the central and the peripheral nervous system, it modulates membrane excitability by decreasing the concentration of intracellular free Ca^{2+} (cf. Pasantes-Morales and Gamboa, 1980; Pasantes-Morales *et al.*, 1979; Remtulla *et al.*, 1979). Such an action could result in inhibiting the release of other neurotransmitters (Kuriyama, 1980). The effects of taurine on Ca^{2+} binding to microsomes depend on the presence of K^+. Other amino acids suggested as putative inhibitory neurotransmitters do not show this property (Izumi *et al.*, 1977). There may be a specific role for taurine as a modulator of the GABA–benzodiazepine receptor complex; the enhancement of benzodiazepine binding by GABA is prevented by taurine (Medina and De Robertis, 1984; Iwata *et al.*, 1984).

Taurine is believed to be formed in brain by a route starting from cysteine (Figure 7.12). Cysteine is converted to cysteinesulfinic acid, hypotaurine, and finally to taurine. The key enzyme is believed to be the decarboxylase that converts cysteinesulfinic acid to hypotaurine. Alternative routes through cysteic acid or cysteamine are generally considered less probable. Most brain taurine seems to be taken up from the periphery rather than being formed *in situ*, especially in neonatal animals (Pasantes-Morales *et al.*, 1980).

Taurine is mostly excreted unchanged or conjugated with bile acids. Turnover of total brain taurine is far slower than for any of the established neurotransmitters (9–16 hr for fast decay and 40–238 hr for slow decay).

Although the regional distribution of tuarine is rather even (Table 7.5), higher than average levels are found in the pituitary and pineal glands and the retina, cerebellum, olfactory bulb, and striatum. Particular taurine systems proposed therefore include cerebellar stellate cells, retinotectal neurons, and interneurons of the retina, olfactory bulb, and striatum. Convincing evidence of a neurotransmitter role in these locations is lacking (P. L. McGeer and E. G. McGeer, 1981b).

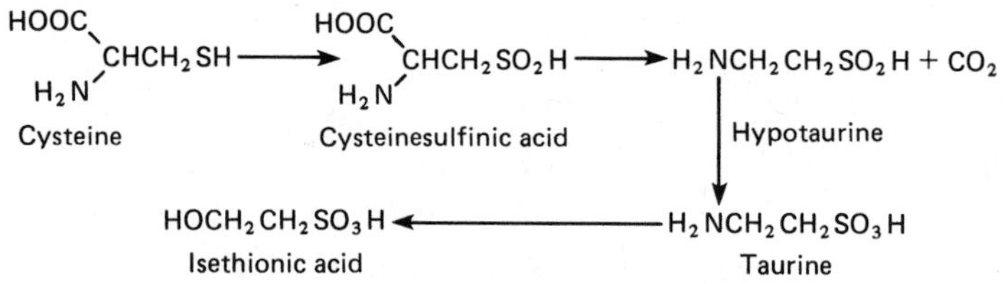

Figure 7.12. Synthesis and metabolism of taurine.

TABLE 7.5
Taurine Concentrations in
Various Areas of Cat Brain
(in μmoles/g)[a]

Pituitary	8.08
Pineal	5.20
Cortex	2.10
Caudate	2.90
Thalamus	2.31
Lateral geniculate	4.30
Medial geniculate	1.70
Hypothalamus	1.81
Cerebellum	2.71
Pons	2.10
Medulla	1.70
Spinal cord	1.70
Heart (rabbit)	10.8

[a]From Guidotti et al. (1972).

Taurine nevertheless seems clearly to be involved in maintaining the structural integrity of the retina. The synthetic enzyme, cysteinesulfinic acid decarboxylase, has been localized by immunohistochemistry in a variety of retinal cells (Lin et al., 1985). A defect in taurine transport or storage, or both, may be related to retinitis pigmentosa, a degenerative disease in humans (Barnett, 1980). Experimental taurine deficiency in animals leads to retinopathy, and persons suffering from retinitis pigmentosa show defective uptake of taurine even into blood platelets (Airaksinen et al., 1979). An association of taurine deficiency with epilepsy has also been suggested, and a mild anticonvulsive activity for taurine has been reported in humans and in some, but not all, experimental models of epilepsy (Huxtable and Barbeau, 1976).

7.10.2. Proline

Although proline is present in relatively low concentrations in mammalian brain (0.2–0.6 μm/g), very high concentrations have been reported in the nervous systems of crabs and lobsters. Significantly higher concentrations occur in dorsal than in ventral roots in the spinal cord of mammals, and proline has a weak glycine-like action on cat spinal neurons. Proline also appears to be taken up into synaptosomes by a high-affinity, sodium-dependent system. Some believe that it may be taken up as a precursor for peptides such as TRH (see Chapter 12). Others say the ability to take up and retrogradely transport [^{3}H]proline appears to be a common property of central neurons and therefore unrelated to its possible status as a neurotransmitter (LeVay and Sherk, 1983) or as a precursor of TRH. Giacobini (1983) concluded that a case could be made for proline (and pipecolic acid) as inhibitory neuromodulators in the mammalian CNS. Inborn errors of metabolism in which high blood levels of proline or pipecolic acid occur are generally characterized by mental retardation, suggesting a possible role in brain development (Giacobini, 1983).

7.10.3. Pipecolic Acid (Piperidine-2-carboxylic Acid)

This material is the major metabolite of lysine in mammalian brain and has been suggested as a putative inhibitory neurotransmitter on the grounds that it (1) depresses the firing of cortical and hippocampal neurons on iontophoretic application (Kase *et al.*, 1980; Takahama *et al.*, 1982), (2) has a variable regional distribution in brain (highest level in the cerebellum is of the order of 15 nm/g), (3) has a largely synaptosomal localization (J. S. Kim and Giacobini, 1984), (4) shows high-affinity uptake into synaptosomes (Nomura *et al.*, 1980), (5) is released by K^+ in a Ca^{2+}-dependent manner from rat brain slices (Nomura *et al.*, 1979), and (6) shows Na^+/Cl^--dependent high-affinity binding to membrane fragments. On intracerebral injection, pipecolic acid produces sedation in rats (T. Miyata *et al.*, 1973). There are arguments, however, against a neurotransmitter role as such. The physiological action is bicuculline-sensitive and may therefore be at GABA receptors. As is suspected for proline, pipecolic acid may be taken up by the GABA transport systems and act as a false transmitter in GABAergic neurons. The uptake of pipecolic acid, like that of proline, may be a relatively common property of central neurons; this seems especially likely in view of the report that L-proline is equally efficacious as L-pipecolic acid in displacing the Na^+-dependent binding of [^{3}H]pipecolic acid to membrane fragments (Gutierrez and Giacobini, 1985).

Piperidine, the decarboxylation product of pipecolic acid, mimics the nicotinic action of ACh on some synaptic sites and has also been proposed as a possible neurotransmitter (P. L. McGeer *et al.*, 1978a). Decarboxylation does not occur readily in brain, which argues against a neurotransmitter role for the imine (Nomura *et al.*, 1983).

7.10.4. Other Amino Acids with Inhibitory Action

Serine, which is interconvertible with glycine in the CNS, and α-alanine have weak glycine-like actions in iontophoretic studies. Levels in brain are low. There is little regional variation, a lack of high-affinity uptake, and no specific binding to brain tissue. Thus, evidence for neurotransmitter action is very weak.

β-Alanine also has depressant effects on iontophoresis and may be taken up into synaptosomes by a high-affinity system (G. A. Johnston, 1977). However, the uptake appears to be into GABAergic structures, where β-alanine may act as a false transmitter (B. Bauer and Ehinger, 1977). Very low levels (0.8 nm/g) of β-alanine occur in the CNS, and there is little reason to believe that it is a neurotransmitter.

7.11. Summary

GABA has been identified as one of the two classic ionotropic inhibitory neurotransmitters in the mammalian CNS. Glycine is the other. GABA, released from nerve endings, hyperpolarizes the postsynaptic neuron by opening membraneous pores or gates to permit Cl^- ions to flow from the extracellular fluid into the postsynaptic cell, thus increasing the negative intracellular potential [see Figure 4.12 (Section 4.4)]. GABA and its synthesizing enzyme, GAD, are widely distributed throughout the CNS in high concentrations, suggesting that a sizeable proportion of all central neurons use GABA as their transmitter. This view is supported by uptake studies that indicate that 25–40% of all nerve endings accumulate GABA. Many techniques have been used to identify GABA neuronal systems. The present list, which is far from complete, includes many long-axoned projections of the substantia nigra and globus pallidus, cerebellar Purkinje cells, and interneurons in most brain nuclei as well as in the retina and spinal cord.

GABA is synthesized in nerve endings by GAD and destroyed presynaptically and in glial cells by GABA-T. Despite much detailed knowledge of the metabolism of GABA, little is known about the control of synthesis and release of physiologically active molecules. A number of pharmacological agents have been discovered that interact with GABA at most of the known sites of neurotransmitter regulation. Those agents that impair GABA functioning to a major degree generally seem to produce convulsions, while those that enhance its action tend to decrease excitability of nerve cells. The clinically most important are the benzodiazepine anxiolytic and anticonvulsive agents; these act by modulating the binding of GABA at postsynaptic receptors.

The action of glycine cannot be distinguished from that of GABA except by the use of specific blocking agents. GABA is blocked by bicuculline and glycine by strychnine. As might be expected, both blockers are convulsant agents.

Glycine is established as the neurotransmitter for inhibitory spinal interneurons activated by Ia muscle afferents. It is thought to be the neurotransmitter for inhibitory fibers of the glossopharyngeal nerve to hypoglossal motoneurons, for reticulospinal neurons in the medullary reticular formation, and for commissural inhibitory fibers of vestibular neurons in Deiters' nucleus, in the cuneate nucleus, and in the retina. The concentration of glycine is highest in the spinal cord, particularly in the ventral gray area, and the medulla. In these areas, a high-affinity uptake system is present for glycine, and it has been shown by autoradiography to accumulate in nerve endings.

The metabolic routes by which glycine is formed and destroyed in nerve endings are uncertain. Apart from specific blockers, which have no therapeutic application, appropriate pharmacological agents that act specifically with glycine systems have not been found.

A number of other amino acids have been shown to have GABA- or glycine-like actions on iontophoresis, but there is little firm evidence that any of them serve as neurotransmitters in mammalian brain. The most important, taurine, seems to act as a general modulator of membrane excitability and, possibly, of neurotransmitter release by regulating the concentration of intracellular free Ca^{2+}.

8

Cholinergic Neurons

8.1. Introduction

In Chapters 3 and 4, we described how acetylcholine (ACh) was the first neurotransmitter to be identified, thus ushering in the concept of the transposition of electrical transmission within neurons to chemical transmission between neurons. We discussed how this was achieved, with the quantal release of ACh at nerve endings of the neuromuscular junction [see Figure 3.11 (Section 3.2.9)] serving as the model. We detailed the effect ACh has in opening ion channels across the postsynaptic membrane [see Figure 3.14 (Section 3.4)] and the way it induces an excitatory postsynaptic potential [see Figure 3.10 (Section 3.2.8)]. These we have defined as ionotropic neurotransmitter functions. We say in Chapter 4 that the classic ionotropic actions of ACh were observed only with nicotinic receptors. The picture with muscarinic receptors was much less clear, opening the possibility of metabotropic action (see Chapter 5).

We must now explore some further concepts that have evolved as a result of studying the cholinergic neuron, including varying responses due to different cholinergic receptors on cholinoceptive cells. At least 95% of all cholinergic receptors in brain are of the muscarinic type, indicating that this mode of ACh transmission is clearly of major importance in the central nervous system (CNS).

Although postsynaptic cholinergic effects may be variable, the chemistry of the cholinergic neuron itself seems to be constant. Quastel *et al.* (1936) were the first to demonstrate the capability of brain tissue to synthesize ACh. They obtained a material they referred to as "choline ester" from brain slices treated with eserine and incubated with glucose and oxygen. The material was indistinguishable from ACh by bioassay. Some three years later, Stedman and Stedman (1939) described the isolation of ACh from beef brain, and in subsequent papers, Quastel and associates attributed the activity they observed to ACh. These early papers established the principal requirements for ACh synthesis, and for almost two decades little new information was added.

It had been recognized that ACh exists in tissue in both free and bound forms, but the nature of the binding was obscure. At first, it was attributed to mitochondrial binding, because early differential centrifugation studies showed that bound ACh appeared in a crude mitochondrial fraction. But this view began to be modified after De Robertis and Bennett (1955) discovered vesicles in nerve endings. They speculated that these synaptic vesicles could be the site of storage and synthesis of transmitter substances.

Shortly thereafter, Del Castillo and Katz (1956) developed the concept of a quantized release of ACh at the neuromuscular junction (see Section 3.2.5). The notion that such quantized release might be related to vesicles led to attempts to correlate electron microscopic and biochemical experiments. Working independently, De Robertis and his colleagues in Argentina and Whittaker and his team in England achieved fundamental breakthroughs using subcellular fractionation techniques.

Homogenized brain tissue was subjected to centrifugation at high speeds (up to $100,000g$) through solutions of different density to separate some of the subcellular organelles discussed in Chapter 1 [see Figure 1.8 (Section 1.2.3)]. One fraction, which initially sedimented with the mitochondria but could be separated off by further centrifugation on sucrose gradients, was of intense interest. This contained pinched-off nerve endings also called "synaptosomes." The homogenization procedure, while it broke down the cell bodies and sheared off axons and dendrites, preserved the nerve endings and even seemed to seal off their axonal stumps. These synaptosomes frequently contained a portion of the postsynaptic membrane, as though the presynaptic terminal had been tightly glued to the postsynaptic dendritic spine (De Robertis et al., 1962, 1963) (Figure 8.1A).

Within the nerve endings could be seen synaptic vesicles. These vesicles were concentrated by exploding the synaptosomes with hypotonic medium followed by centrifugation of the released vesicles. Electron micrographs of the pellets showed a high concentration of intact synaptic vesicles (Figure 8.1B). It was found that both choline acetyltransferase (ChAT), the enzyme that synthesizes ACh, and bound ACh were concentrated in the synaptosomal fraction, while acetylcholinesterase (AChE), the enzyme that metabolizes ACh, was associated with membrane fragments. When the synaptosomes were disrupted, bound ACh was associated primarily with the sedimented vesicles.

This pioneering work not only verified previous theories regarding the association of ACh with synaptic vesicles, but also led to a general understanding of the nature of nerve endings of all neurotransmitter types. Such subcellular work is now a standard test for neurotransmitter candidacy (see Chapter 5).

Much of the very early work on ACh was done using delicate bioassay techniques, including the leech dorsal muscle, the frog rectus abdominus muscle, the guinea pig ileum, and cat blood pressure. While these methods were reliable and sensitive, they were painstaking and foreign to the biochemist's background. Simple chemical methods of ACh assay are now used, particularly high-performance liquid chromatography (HPLC) (Damsma et al., 1985; Bymaster et al., 1985; Potter et al., 1984). In all studies on ACh levels, however, care must be taken in the method of sacrifice, since ACh is rapidly destroyed unless it is sequestered from the metabolizing enzyme, AChE. The most reliable method involves sacrifice by microwave irradiation (Jenden, 1977).

Of at least equal importance to an understanding of central cholinergic function was the development of an immunohistochemical method that allowed detailed mapping of cholinergic structures in the brain at the cellular level (H. Kimura et al., 1981; for a review, see P. L. McGeer et al., 1984a). This method depended on the purification of ChAT and preparation of monospecific antibodies to the purified protein. Interest in

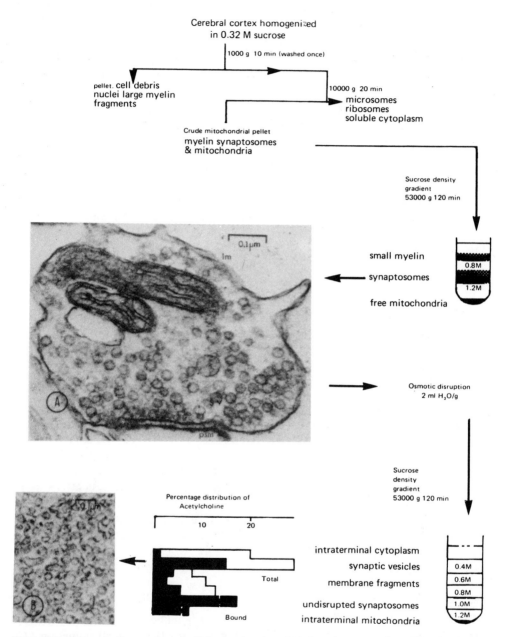

Figure 8.1. Schematic diagram for subcellular fractionation of cholinergic nerve endings following the procedure of Whittaker *et al.* (1964). (A) An isolated nerve terminal. (lm) Limiting membrane, apparently fully sealed; (m) mitochondria; (sv) synaptic vesicles; (psm) an adhering piece of postsynaptic membrane. (B) Appearance of the synaptic vesicles after isolation. Total ACh in the density gradient is that found in the presence of an AChE inhibitor, whereas bound is that which does not require esterase inhibition for its conservation. Free ACh is regarded as the difference between total and bound ACh. Diagram from Marchbanks (1975).

central cholinergic systems has been intensified by evidence that they are important to memory. Loss of cholinergic neurons may be an important contributor to the symptoms of senile dementia of the Alzheimer type (SDAT), a major health problem of growing importance. It can be anticipated that the next decade will see even more attention paid to the precise, detailed anatomy of central cholinergic neurons, as well as to their physiology and pharmacology.

8.2. Chemistry of Acetylcholine

8.2.1. Synthesis of Acetylcholine

The synthesis of ACh involves the single step shown in equation (1), in which the reaction of acetyl-Co A and choline is catalyzed by ChAT (EC 2.3.1.6):

$$\begin{array}{c} O \\ \parallel \\ CH_3C \\ \diagdown \\ SCoA \end{array} + HOCH_2CH_2\overset{+}{N}(CH_3)_3 \overset{ChAT}{\longrightarrow} \begin{array}{c} O \\ \parallel \\ CH_3C \\ \diagdown \\ OCH_2CH_2\overset{+}{N}(CH_3)_3 \end{array} + CoA\text{-}SH \qquad (1)$$

It is this enzyme that characterizes cholinergic neurons. The enzyme is synthesized in cholinergic cell bodies and is transported by axoplasmic flow to nerve endings. It also appears in dendrites (Hattori et al., 1979).

The synthesis of ACh in the nerve ending was discussed and shown diagrammatically in Figure 3.11 (Section 3.2.9) and is presented in slightly more detail in Figure 8.2 (Section 8.2.3). The intrasynaptosomal localization of ChAT is not certain. About 50% of the brain enzyme can be easily solubilized from tissue by high salt concentrations; a less easily solubilized and apparently membrane-bound form also exists (Peng et al., 1986). The easily soluble form may be associated with the synaptosomal sap, while the less easily solubilized form could be associated with the surface of storage vesicles, as seems to have been implied by electron microscopic localization (Hattori et al., 1976, 1977).

The substrates for ACh synthesis are widely available and are not exclusive to the cholinergic neuron. Choline is produced in the liver. It follows two routes in brain; a minor, high-priority route, where it is used directly for ACh synthesis, and a major, low-priority route, where it is phosphorylated by choline kinase and then forms membrane lipid derivatives such as sphingomyelin. Only a small percentage of the choline that reaches the brain is used for ACh synthesis (Jope, 1979). Choline is concentrated in cholinergic nerve endings by a pumping or high-affinity, sodium-dependent uptake system with a K_m in the micromolar range (see Figure 8.2).

Acetyl-CoA is also a general substrate used in many metabolic reactions and thus is common to a variety of cells. Cholinergic cells do generate their own acetyl-CoA, but the process is somewhat complicated and acetate is not a precursor, as is so often stated in textbooks. The originating source of acetyl-CoA for ACh is glucose, as was shown by the earliest experiments on ACh synthesis (Quastel et al., 1936). The more immediate source is the glucose breakdown product, pyruvate. A complex pyruvate dehydrogenase system exists in mitochondria which ultimately produces acetyl-CoA. The system has been extensively studied in *Escherichia coli* bacteria, pigeon breast, beef kidney, and other tissues, as well as brain. Details of the system can be obtained by reference to standard textbooks or reviews (Stryer, 1975; Jope, 1979). The pyruvate dehydrogenase system requires thiamine pyrophosphate. G. E. Gibson and Blass (1976) have proposed that deficiencies in glucose, or in the pyruvate dehydrogenase system, differentially affect cholinergic neurons because of their high acetyl-CoA requirements. Pantothenic acid is a precursor of CoA, and hence pantothenic acid deficiency may also impair cholinergic activity.

ChAT has now been purified from a wide range of sources. The material in most cases has been brain tissue from various species such as rat, pig, beef, human, chicken, and *Drosophila*. However, squid head ganglia, squid optical lobe, snail esophagal ganglia, *Torpedo* electric organ, vertebrate retina, and placenta have also been used as rich sources of ChAT (for reviews, see Chao, 1980; P. L. McGeer *et al.*, 1984a). Sepharose affinity columns coupled to CoA organomercury, styrylpyridinium, naphthylvinylpyridine, blue dextran, and hydrophobic ligand glycidyl ether have been utilized for purifying the protein. Immunoaffinity columns have also been used now that there is wide availability of monoclonal and monospecific polyclonal antibodies (Levey *et al.*, 1982; Peng *et al.*, 1983; Bruce *et al.*, 1985).

A remarkably consistent molecular weight of about 68,000 has been obtained for the native enzyme using such diverse sources as *Drosophila* head, chicken brain, human placenta, beef or human striatum, and pig whole brain. This suggests the preservation of a common gene through many species.

The complementary DNA (cDNA) for *Drosophila* head ChAT has been cloned (Itoh *et al.*, 1986). It has also been tentatively cloned for human basal ganglia (Strauss *et al.*, 1985). Determination of the nucleic acid structure of the messenger RNA (mRNA) of human ChAT and comparison with ChAT mRNAs from other sources should verify the commonality of the gene. Partial characterization of ChAT by proteolytic digestion has indicated that the human and placental enzymes are probably identical (Bruce *et al.*, 1985). The apparent isoelectric pH is 8.1.

The enzyme has a much higher affinity for acetyl-CoA than for choline. The K_m's are roughly 5–10 μM and 250–1000 μM, respectively. A complex can be isolated by column chromatography after the enzyme comes in contact with acetyl-CoA, but not after contact with choline. This suggests that, during *in vivo* synthesis, the enzyme first forms a complex with acetyl-CoA. Choline, after occupying an adjacent site on the enzyme, could then be transferred to the acetyl-CoA with the newly formed ACh leaving the protein. To be active, ChAT seems to require free thiol and imidazole groups, because agents such as cupric ions and photooxidized methylene blue, which interact with these prosthetic groups, inhibit the reaction.

8.2.2. Destruction of Acetylcholine

ACh is disposed of in a single fashion. It is destroyed by the action of cholinesterases, which hydrolyze the ester bond, as follows:

$$CH_3C \overset{O}{\underset{OCH_2CH_2\overset{+}{N}(CH_3)_3}{\big\|}} + H_2O \xrightarrow{AChE} CH_3COOH + HOCH_2CH_2\overset{+}{N}(CH_3)_3 \quad (2)$$

The products are choline and acetate.

AChE (EC 3.1.1.7) (for a review, see Brimijoin, 1983) is contained in substantial concentration in all cholinergic neurons. This preferential localization has been utilized as a principle in identifying cholinergic neurons (L. L. Butcher *et al.*, 1975). However it also occurs in many other neurons and in glia, so that the cholinergic localization is not absolute. In addition, there is a wider series of enzymes known as the "pseudocholine-sterases" that can be found in most organs of the body and that also carry out the reaction. They can be distinguished from AChE by their preference for substrates such as butyrylcholine and benzoylcholine and by their differing sensitivities to various inhibitors (see Section 8.5.7). ACh was once thought to be the sole natural substrate for AChE, but the purified enzyme hydrolyzes a number of peptides, including the physiologically occurring enkephalins (see Chapter 12) (Chubb *et al.*, 1983). This may help to explain the wide distribution of AChE as compared with ACh.

There must be a considerable biological advantage to destroying ACh, because the AChE and pseudocholinesterases are among the most active enzymes in the body. Their blockade by so-called "nerve gases" produces rapid paralysis and death. Purified AChE has been found capable of hydrolyzing the amazing total of 150 g of ACh per milligram of protein per hour. It is also incredibly stable, retaining some catalytic activity after the harsh treatment of formaldehyde fixation. There is little wonder that ACh is absent from the urine and bloodstream and only trace amounts are found in the cerebrospinal fluid.

The mechanism of hydrolysis involves the quaternary choline group being attracted to an anionic site on the enzyme with the acetate attaching to a serine hydroxyl moiety in an adjoining histidine–serine esteratic site. The ester bond is cleaved and the hydrolyzed acetate group released. Irreversible cholinesterase inhibitors, such as diisopropyl fluo-rophosphate (DFP), form a strong bond with the serine hydroxyl moiety, while reversible ones, such as physostigmine, form a weaker carbamylate bond at the same locus. Pyridine-2-aldoxime methiodide (2PAM) and other substituted oximes have the valuable property of breaking the strong bond formed at the esteratic site by irreversible, phos-phorylating AChE inhibitors by forming a chemical bond with them. The AChE is thus regenerated [see Table 8.3 (Section 8.5)].

AChE has been isolated, but the structure is still uncertain. The most common theory is that it is comprised of a monomeric glycoprotein unit of about 65 kilodaltons that can be built up to a tetrameric unit. These have been designated the G_1–G_4 forms (Brimijoin, 1983). Most tissues contain at least two forms of AChE that sediment at 10–11 S (G_4) and 4–5 S (G_1). They have the same catalytic activity and the same K_m for ACh (about 10^{-4} M) (Zakut *et al.*, 1985) and appear to differ only in molecular weight. However, various asymmetric species with a number of monomers attached to a collagen tail apparently exist. The proportions of the various forms, which can be separated by differential centrifugation, vary from region to region in brain and with development (Sung and Ruff, 1983).

Like other proteins, AChE is transported by axoplasmic flow from the cell bodies, where it is synthesized, to the nerve endings. It has the distinction of being the first enzyme to be studied by axoplasmic flow (C. O. Hebb and Waites, 1956). Subcellular distribution studies indicate that the G_1 and G_2 forms are intracellular, while the G_4 form is almost entirely membrane-bound with the catalytic activity pointing outward (Brimi-join, 1983).

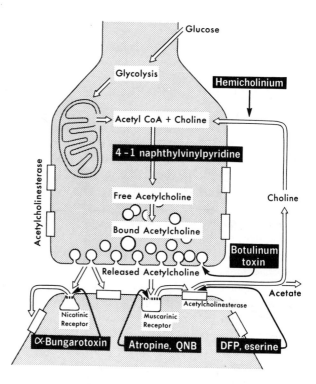

Figure 8.2. Schematic diagram of hypothesized cholinergic synapse. Outline rectangles and arrows show the sites of action of endogenous cholinergic elements; black rectangles and arrows show the sites of action of some common drugs that interact with the cholinergic system [see Table 8.3 (Section 8.5)]. On the postsynaptic side, both nicotinic and muscarinic receptors are shown, although this would not usually occur on the same dendritic spine. (QNB) Quinuclidinylbenzilate; (DFP) diisopropyl fluorophosphate.

8.2.3. Storage, Release, and Turnover of Acetylcholine

The special biochemistry of the cholinergic neurons is reflected in the actions of the cholinergic synapse depicted diagrammatically in Figure 8.2. Acetyl-CoA is formed from glucose and choline is picked up from the extracellular space as starting materials for ACh synthesis. ACh is synthesized by ChAT and then sequestered from AChE by vesicular binding. It is released in quantal fashion from vesicles into the synaptic cleft by the process of exocytosis [see Figure 3.11 (Section 3.2.9)]. The release process requires Ca^{2+} and seems to involve inhibition of an $Na^{+}-K^{+}$ ATPase (E. M. Meyer and Cooper, 1981). The ACh in the cleft reaches a receptor site and can be destroyed either directly or subsequently by AChE. The postsynaptic receptors on the cholinoceptive cell may be nicotinic, muscarinic, or both. Figure 8.2 also shows the sites of action of representatives of various classes of drugs. These are discussed further in Section 8.5.

Receptors of both the muscarinic and the nicotinic type may be located at sites other than on a postsynaptic dendrite or cell soma as depicted in Figure 8.2. For example, both release and lesion evidence indicates the existence of ACh receptors on various types of nerve endings. Here the ACh acts to modulate the release of other neurotransmitters, as is the case for dopaminergic nerve endings of the striatum. In many instances, the released ACh may also act on autoreceptors on its own nerve endings to inhibit the release of further ACh. Evidence for such autoinhibition has been provided, for example, in the neostriatum, nucleus accumbens, and cortex. In the latter case, the autoreceptors appear to be mainly muscarinic in nature (de Belleroche and Gardiner, 1982).

ACh receptors have also been reported on cerebral blood vessels, where the ACh, like other neurotransmitters such as noradrenaline, probably acts to control blood flow (Shimohama *et al.*, 1985; A. Saito *et al.*, 1985).

It is not known how the synaptic machinery actually controls the supply of ACh to the receptor sites, although such information would be extremely valuable for developing methods of modifying the action of cholinergic pathways. As with all neurotransmitters, there must be a system of regulating the supply to equal the demand. Since this can obviously vary widely according to the rate of neuronal stimulation, the capacity to synthesize ChAT must considerably exceed the average requirements. Thus, a generous oversupply of enzyme must be present, along with a regulating mechanism, like the throttle on a car, to modulate the supply of transmitter.

ACh levels can be reduced by limiting the availability of either choline or acetyl-CoA (Jope, 1979). The rate at which ACh is utilized by brain is still uncertain, partly because of multiple pools and partly because of its high lability. Measurements of ACh levels postmortem are highly dependent on the method of sacrifice. So are the levels of free choline. In one set of experiments, for example, ACh levels of whole rat brain were 24.8 nmoles/g tissue with rapid sacrifice by microwave heating, but dropped to 69% of that value if there was a 5-min delay between decapitation and homogenization of the tissue. Choline levels increased 496% under these conditions, from 26.3 to 156.7 nmoles/g. These data suggest that choline phospholipids are in an even more labile pool than ACh itself.

In most turnover studies, a radioactive precursor such as [^{14}C]choline, phosphoryl-Me-[^{14}C]-choline, or [^{14}C]glucose has been used, and the ratio of labeled to unlabeled ACh has been measured at various intervals after the administration of the precursor. Reported turnover rates for ACh *in vivo* have ranged from 0.9 to 104 nmoles/g tissue per min depending on species, area of brain, and method of sacrifice. Potter *et al.* (1984), using a microwave irradiation method, determined the ACh turnover time in mouse brain to be 7 nmoles/g per min for the cortex, 46.6 for the striatum, and 3.9 for the hippocampus.

The amount of free choline measured in brain is only 25–45 nmoles/g (25–45 μM), depending on the species and methods of measurement. This is far less than is apparently required to satisfy ChAT. The concentration to half-saturate the enzyme in the test tube (K_m) was only 5 μM for acetyl-CoA but 250 μM for choline in one set of experiments using the human enzyme (Singh and McGeer, 1977). However, the cholinergic nerve ending possesses a pumping mechanism (see Figure 8.3) that allows it to pick up choline from the extracellular fluid even when the extracellular concentration is very low [cf. Figure 1.14B (Section 1.2.3)]. Choline is transported against a concentration gradient into the synaptosomal sap. As a consequence, the choline concentration in a cholinergic nerve ending could be far higher than the average brain concentration and adequate to saturate the enzyme. The availability of choline has been suggested as the means by which ACh production is controlled, but this hypothesis remains very speculative.

If labeled choline is added to brain tissue in the micromolar concentration range, almost all the label appears as ACh. Choline is phosphorylated and used for phospholipid synthesis only when the concentration is high enough so that the priority requirement of pumping into cholinergic nerve endings for ACh synthesis has been satisfied. This uptake into nerve endings for ACh synthesis is referred to as the "high-affinity uptake system," while that for other, more general purposes in both cholinergic and noncholinergic structures is referred to as the "low-affinity uptake system" (Yamamura and Snyder, 1973).

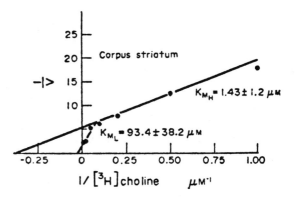

Figure 8.3. Double-reciprocal plot of the uptake of [^{3}H]choline into homogenates of rat corpus striatum. The data can be resolved into two linear components representing high-affinity uptake (K_m 1.43 ± 1.2 μM) and low-affinity uptake (K_m 93.4 ± 38.2 μM). Velocity (V) is expressed as micromoles of [^{3}H]choline accumulated per gram of protein in 4 min. From Yamamura and Snyder (1973).

The different features of the two affinities are illustrated in Figure 8.3, in which the reciprocal of the velocity of [^{3}H]choline uptake is plotted against the reciprocal of the choline concentration. Such double-reciprocal (Lineweaver–Burke) kinetic plots are routinely used in enzyme and drug studies. With simple kinetics, a single straight line is obtained, which can be used to calculate K_m, the concentration of substrate required to half-saturate the active site. In Figure 8.3, which is a Lineweaver–Burke plot using a synaptosomal fraction of rat corpus striatum, two straight lines with sharply differing slopes are obtained. These correspond to the K_m's of the high- and low-affinity uptake systems and have values of 1.43 and 93.4 μM, respectively.

The high-affinity uptake system requires Na$^+$ and is sensitive to various metabolic inhibitors. The low-affinity system is not associated with rapid ACh synthesis, is not Na$^+$-dependent, and is not affected by the same inhibitors.

The activity of this high-affinity uptake system generally parallels the distribution of ChAT and ACh and is considered to be a useful index of the activity of cholinergic terminals (Table 8.1). For example, lesioning the known cholinergic pathway from the septum to the hippocampus causes a sharp loss in this high-affinity uptake, but no change in the low-affinity uptake (Kuhar et al., 1973). This is consistent with degeneration of cholinergic nerve terminals.

It has been suggested that changes in high-affinity uptake of choline by synaptosomes are the most sensitive in vitro measure of the physiological capacity of an area to synthesize ACh in vivo. This may differ from the number of cholinergic synaptosomes. Levels of ChAT would presumably reflect the concentration of cholinergic terminals, but not necessarily the physiological turnover of ACh. Use is therefore sometimes made of a cholinergic uptake index, which is the ratio of high-affinity choline uptake to ChAT activity. This index is increased in the hippocampus and cortex in mice treated with atropine (Eckernas et al., 1980). The homeostatic implications are that the choline pumping system has become enhanced to provide additional ACh to the blocked receptors and that it is the pump, rather than numbers of ChAT synthetic enzyme molecules, that responds to increased physiological demand.

A further complication in determining how ACh synthesis is controlled is that not all ACh is handled equivalently by the nerve ending. There are different "pools" or "compartments." The original concept of compartmentalization was suggested by the now classic studies of Birks and McIntosh (1961). Using the isolated perfused superior cervical

TABLE 8.1

Regional Distribution of Specific [³H]-QNB Binding, [³H]Choline Uptake, and ChAT and AChE Activities in Rhesus Monkey Brain Regions[a]

Region	[³H]-QNB binding	[³H]Choline uptake	ChAT	AChE
Cerebral hemispheres				
Frontal pole	439	1.72	4.4	15.7
Occipital pole	578	1.16	3.28	14.8
Precentral gyrus	483	1.73	6.47	18.6
Postcentral gyrus	516	2.46	7.95	30.5
Cingulate gyrus	546	2.48	6.35	18.5
Pyriform cortex	474	5.97	18.6	29.3
White matter areas				
Corpus callosum	107	0.94	1.4	11.7
Corona radiata	87	0.55	3.49	18.4
Optic chiasma	34	—	3.60	32.9
Cerebral peduncles	140	1.2	3.2	44.2
Limbic cortex				
Amygdala	496	—	26.2	122.0
Hippocampus	502	6.25	13.2	45.2
Hypothalamus	241	9.67	13.9	57.4
Thalamus				
Anterior	285	7.29	14.6	59.1
Medial	369	7.20	23.9	73.1
Lateral	360	3.96	12.7	59.8
Extrapyramidal areas				
Caudate nucleus				
Head	976	28.1	72.8	281.0
Body	1061	56.1	90.0	172.0
Putamen	1126	35.0	111.0	354.0
Globus pallidus	168	2.4	9.3	72.0
Midbrain				
Superior colliculi	381	10.5	35.2	121.0
Inferior colliculi	279	4.58	10.5	46.1
Raphe area	149	4.54	36.0	61.1
Cerebellum–lower brainstem				
Pons	212	2.70	9.2	28.5
Cerebellar cortex	125	1.45	0.8	54.3
Floor of fourth ventricles	96.5	5.05	21.4	45.3
Medulla oblongata	114	4.29	10.2	36.5
Inferior olivary nucleus	47.6	3.85	2.4	68.8
Pontine tegmentum	146	—	12.6	63.8
Spinal cord				
Cervical cord	47.6	3.19	8.1	22.2

[a]From Yamamura *et al.* (1974). Specific [³H]quinuclidinylbenzilate ([³H]-QNB) binding values are given in pmoles/g protein [see also Table 8.2 (Section 8:4.1)]. Velocities of choline uptake are given in nmoles/g protein per 4 min. ChAT activities are given in nmoles ACh synthesized/mg protein per hr. AChE activities are given in nmoles ACh hydrolyzed/mg protein per min.

ganglion, they identified a readily releasable pool of ACh that comprised some 20% of the total releasable amount. A larger component, termed the "reserve pool," was then released on further stimulation. They identified a third pool—which they termed "surplus" ACh—that may have been artifactual due to the use of anticholinesterases in their perfusate. The stimulated release is initially from vesicles as illustrated in Figure 3.11 (Section 3.2.9) and Figure 8.2, but on sustained activity, much of the released ACh may come from the cytoplasm (Carroll and Aspry, 1980).

Many investigators have undertaken experiments since then using labeled choline or acetyl-CoA to label a pool of ACh. The newly synthesized labeled ACh is then compared with the unlabeled pool already present in the tissues. All investigations have come to the same conclusion: Newly synthesized ACh is released in preference to old ACh (Jope, 1981; von Schwarzenfeld, 1979). This, in fact, seems to be true for most neurotransmitters so far studied.

A deficiency of choline in the diet can lead to reduced brain levels of ACh, which would be consistent with failure to provide sufficient substrate for ChAT. Similarly, conditions in which synthesis of acetyl-CoA is reduced while choline uptake remains normal also lead to reduced ACh levels (Jope, 1979). There is no evidence that choline loads will increase ACh levels above normal.

8.3. Anatomy of Cholinergic Neurons: Cholinergic Cell Groups and Pathways

For a review of this subject, see P. L. McGeer *et al.* (1984a).

The first comprehensive map of cholinergic structures in brain was that produced by H. Kimura *et al.* (1981) for the cat using the ChAT immunohistochemical method. Since then, investigators have published other relatively complete or partial maps of the brain of the rat (H. Kimura *et al.*, 1984; Sofroniew *et al.*, 1982; Houser *et al.*, 1983c; Mesulam *et al.*, 1983b; D. M. Armstrong *et al.*, 1983), monkey (Mesulam *et al.*, 1984; Hedreen *et al.*, 1983), baboon (Satoh and Fibiger, 1985), and human (Nagai *et al.*, 1983b,c; P. L. McGeer *et al.*, 1984b; German *et al.*, 1985; Mizukawa *et al.*, 1986). Unlike the situation with respect to glutamic acid decarboxylase (GAD), where cell bodies stain only with difficulty [see Figure 7.5 (Section 7.3)], a number of neuronal groups stain strongly for ChAT, as indicated in Figure 8.4A. Based on the data available to date, five major cholinergic systems in the brain are known. There are, in addition, a number of minor systems.

The major and minor cholinergic systems in brain are described below.

8.3.1. Medial Forebrain Complex

This is a more or less continuous sheet of giant cholinergic cells that starts on the medial surface of the cortex just anterior to the anterior commissure and extends in a caudolateral direction, always maintaining its position close to the medial and ventral surfaces of the brain, terminating toward the caudal aspect of the lentiform nucleus. The names usually given to the various subregions of this complex from rostral to caudal order are (cf. Figure 8.4A): medial septal nucleus (S), nucleus of the vertical limb of the diagonal band of Broca (V), nucleus of the horizontal limb of the diagonal band of Broca

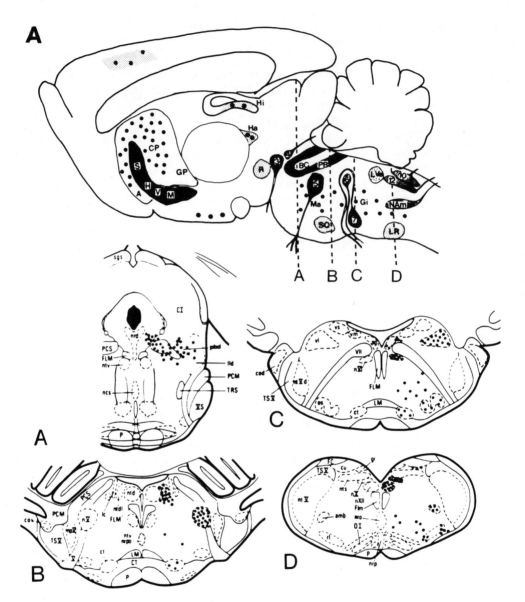

Figure 8.4A. Sagittal diagram of rat brain showing ChAT-containing neurons with cross-sectional diagrams at four brain stem levels (A, B, C, D). Major neuronal systems are indicated by black dots or heavy stippling, minor ones by light stippling. Abbreviations: (A) Nucleus accumbens; (Am) amygdala; (BC) brachium conjunctivum; (CP) caudate–putamen; (Gi) gigantocellular division of the reticular formation; (GP) globus pallidus; (H) nucleus of horizontal limb of diagonal band of Broca; (Ha) habenula; (Hi) hippocampus; (IC) inferior colliculus; (IP) interpeduncular nucleus; (LR, LRe) lateral reticular nucleus; (LVe) lateral vestibular nucleus; (M) nucleus basalis of Meynert; (Ma) magnocellular division of the reticular formation; (NAm) nucleus ambiguus; (PB) parabrachial complex; (R) red nucleus; (S) medial septal nucleus; (SN) substantia nigra; (SO) superior olive; (V) nucleus of vertical limb of diagonal band of Broca.

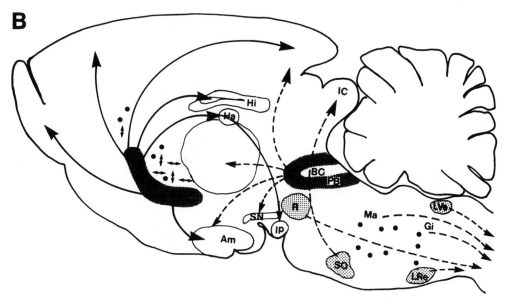

Figure 8.4B. Some proposed cholinergic pathways. Well established routes are indicated by (→); suspected routes by (--→); interneurons (↕ ↔).

(H), and the nucleus basalis of Meynert (M). This cell complex, originally established by H. Kimura *et al.* (1980) as cholinergic by immunohistochemistry in the rat using poly-clonal, monospecific rabbit antisera, has been extensively documented by ChAT immu-nohistochemistry in every species studied.

The complex projects to the neo-, paleo-, and archicortex. The projection of the most rostromedial part of the complex from the medial septum and vertical limb of the diagonal band (S and V in Figure 8.4A) to the hippocampus had been identified on the basis of lesion studies more than a decade earlier (E. G. McGeer *et al.*, 1969). Projections from the caudolateral areas to the neocortex were revealed later (Jones *et al.*, 1976; Divac, 1975; Kievit and Kuypers, 1975), and their relationship to the cholinergic system was again established by lesion studies (M. V. Johnston *et al.*, 1979). Since then, the distribu-tion has been extensively mapped in the monkey with retrograde tracing methods (Jones *et al.*, 1976; R. C. Pearson, 1983; Hedreen *et al.*, 1983) and with retrograde tracing combined with ChAT immunohistochemistry and with AChE–DFP histochemistry (Mes-ulam *et al.*, 1983a, 1984) or with only AChE–DFP or AChE histochemistry (E. G. Jones *et al.*, 1976; Lehmann *et al.*, 1980; Bigl *et al.*, 1982; Kristt *et al.*, 1985). The projections are more or less topographically arranged, but with considerable overlap of the various cortical fields. The rostromedial cells (S and V in Figure 8.4A) project primarily to the hippocampus, but also to the cingulate gyrus and dorsomedial frontal lobe. The more caudal cells of the horizontal nucleus of the diagonal band and the nucleus basalis of Meynert (H and M in Figure 8.4A) project primarily to the remainder of the frontal and the parietal and occipital lobes, while the most caudal and lateral cells (M) project primarily to the amygdala (Nagai *et al.*, 1982) and temporal lobe. The ChAT-positive terminals and cholinoceptive neurons seen in various regions of the cat cortex have also been mapped in detail (H. Kimura *et al.*, 1983).

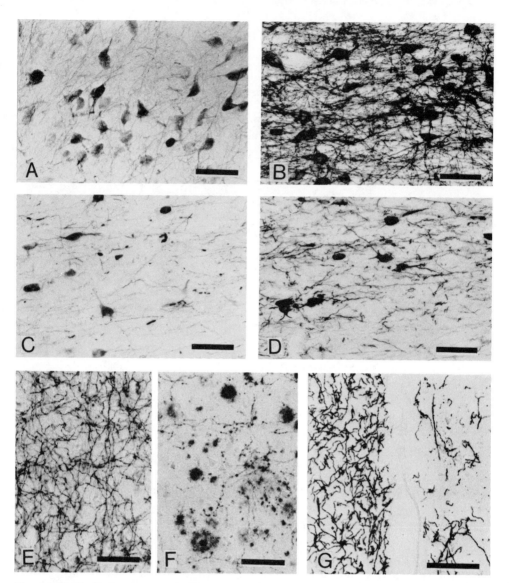

Figure 8.5A–G. Cholinergic cells of the nucleus basalis of Meynert and their presumed termination in the neocortex. (A) High-power photomicrograph of normal human nucleus basalis Meynert cells stained by immunohistochemistry with a monoclonal antibody to human neostriatal ChAT. Notice the larger multipolar cells and network of stained processes. (B) Cells from the same area of a different human brain stained by the ultrasensitive AChE histochemical method of Tago. Notice that the cells of the same size and morphology are overshadowed by the extremely dense neuropil. (C,D) Cells from the same area in a case of Alzheimer's disease stained for ChAT (C) and AChE (D). Only a few cells are left, most of the positively staining neuropil having vanished. (E–G) Fiber staining of neocortex by the Tago AChE histochemical method. Notice the dense fiber network and varicose appearance of the fibers in the normal human cortex (E) as compared with the staining of the same area in a case of Alzheimer's disease (F) where the fiber network is drastically reduced, with the appearance of many AChE-positive senile plaques. (A–F) Scale bars: 10 μm. (G) Rat neocortex stained for AChE following a right-sided electrolytic lesion in the nucleus basalis of Meynert. Notice the greatly reduced fiber staining on the right side. Scale bar: 50 μm.

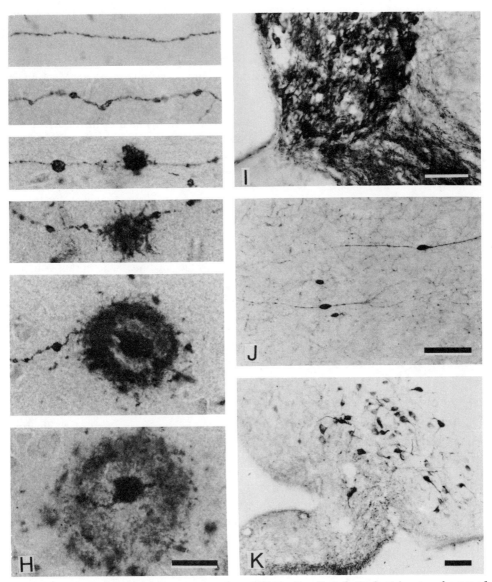

Figure 8.5H–J. (H) Senile plaques developing along AChE-positive fibers in the frontal cortex of a case of Alzheimer's disease. Panels from top down: normal fiber; mildly swollen fiber; two significant enlargements along a single fiber; small plaque with irregular surface; fiber with one small swelling and a distinct plaque with an AChE-positive core and developing halo; fiber with a large plaque with a strong AChE-positive core and a diffuse AChE-positive halo. Scale bar: 20 μm. (I–K) ChAT immunohistochemical staining in the rat with a polyclonal antibody to human placental ChAT. (I) Small neurons in the medial habenula. (J) Small bipolar neurons in the neocortex. (K) Small neurons in the arcuate nucleus with processes extending toward the infundibulum. (I–K) Scale bars: 50 μm. Photomicrographs kindly supplied by H. Tago, University of British Columbia, and polyclonal ChAT antibody by Drs. L. Hersh and G. Bruce of the University of Texas Medical School, Dallas, Texas.

Technical advances in both ChAT and AChE staining have greatly improved the ability to detect both cells and fibers of this complex. Figure 8.5A shows ChAT-positive staining of the nucleus basalis area of normal human brain. It reveals the large, multipolar cholinergic cells that have been seen in every mammalian species studied. Figure 8.5B shows the same area stained by the AChE histochemical method of Tago. Much more intense staining of cholinergic fibers is possible with this technique in comparison with ChAT immunohistochemistry. While it has been possible to stain cortical cholinergic fibers in well-fixed specimens with ChAT antibodies (H. Kimura *et al.*, 1981, 1983; Houser *et al.*, 1983c), it has so far not been possible to do this with human tissue where prompt fixation is not possible. However, the Tago method will permit AChE-positive fibers in human cortex to be detected (Figure 8.5E, F). There is a remarkable loss of cholinergic cells in the basal forebrain of Alzheimer's cases (cf. Figure 8.5A, B, C, and D), which is reflected in a loss of AChE-positive fibers in the cortex (cf. Figure 8.5E and F), similar to that seen following lesions to the basal forebrain in rats (Figure 8.5G). Senile plaques can be seen developing along the AChE-positive fibers (Figure 8.5H).

One of the most intense cholinergic projections in the rat is to the interpeduncular nucleus. One source for this projection comes from the mediobasal forebrain complex (Contestible and Fonnum, 1983; Gottesfeld and Jacobowitz, 1978b, 1979). Lesions of this area, however, do not totally eliminate ChAT from the interpeduncular nucleus, indicating a second source. This must traverse the fasciculus retroflexus, because lesions of that tract almost totally eliminate ChAT in the interpeduncular nucleus (Kataoka *et al.*, 1973). ChAT-positive cells have been described in the medial habenula (Houser *et al.*, 1983c; Hattori *et al.*, 1977; Keller *et al.*, 1984) (Figure 8.5I).

8.3.2. Striatal Interneurons

It has long been known from lesion studies that an internal system of cholinergic neurons exists in the striatum (P. L. McGeer *et al.*, 1971). A group of large cells within the striatum and nucleus accumbens has been identified as being cholinergic, both by ChAT immunohistochemistry (H. Kimura *et al.*, 1980, 1981) (Figure 8.6) and by AChE–DFP histochemistry (Butcher *et al.*, 1975; Lehmann and Fibiger, 1979; Vincent *et al.*, 1983d). These large cells have also been stained by other groups using ChAT immunohistochemistry (Houser *et al.*, 1983c; Sofroniew *et al.*, 1982; D. M. Armstrong *et al.*, 1982; Satoh *et al.*, 1983; Mesulam *et al.*, 1984). They are not known to project beyond the neostriatum. The ChAT and AChE levels of the caudate, putamen, and accumbens are extremely high, and these large neurons probably represent no more than 1% of the neuronal population (Kemp and Powell, 1971). Smaller cholinergic cells have also been reported in the striatum (D. M. Armstrong *et al.*, 1983; Hattori *et al.*, 1976), and while the giant cells are certainly the most prominent ones, they may not represent the entire cholinergic population.

8.3.3. Motor Nuclei for Peripheral Nerves

Large cells, equivalent to anterior horn cells, stain positively for ChAT in cranial nerve nuclei III–VI and IX–XII as the source for efferent fibers to skeletal muscle and autonomic ganglia (D. M. Armstrong *et al.*, 1983; H. Kimura *et al.*, 1981, 1984; Mufson *et al.*, 1982). The counterparts in the spinal cord in the anterior and lateral horns also stain positively (P. L. McGeer *et al.*, 1974a) (see Figure 8.4).

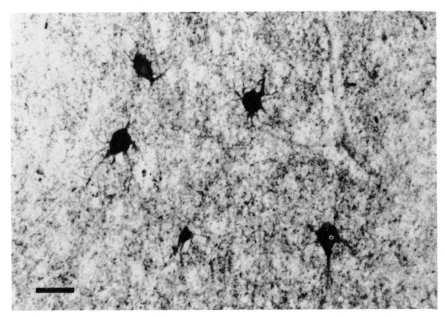

Figure 8.6. Photomicrograph of the human putamen stained by the same monoclonal antibody to human neostriatal ChAT as in Figure 8.5A and C. The neurons are the typical large multipolar neurons that stain heavily for ChAT. The smaller, positively staining profile is a neuron cut in the plane of the section. Scale bar: 50 μM. Courtesy of Dr. H. Tago (University of British Columbia).

8.3.4. Parabrachial Complex

This system (PB in Figure 8.4) is the most intense and concentrated cholinergic cell group in the brainstem (H. Kimura *et al.*, 1981). It surrounds the brachium conjunctivum (superior cerebellar peduncles), commencing in the most rostral aspect of the pons, and follows the direction of the brachium conjunctivum in a caudodorsal direction. Various subnuclei have ChAT-positive neurons in this particular region. The most commonly described nucleus is the pedunculopontine tegmental nucleus in the lateral aspect of the most rostral portion of the complex. Caudal to this nucleus are three adjoining nuclei, the dorsal parabrachial, the Kolliker–Fuse, and the ventral parabrachial nuclei. They surround the cerebellar peduncles in the dorsal, lateral, and ventral regions, respectively. Several groups have reported various subnuclei in this complex to be ChAT-positive (D. M. Armstrong *et al.*, 1983; Satoh *et al.*, 1983; Mesulam *et al.*, 1984; Satoh and Fibiger, 1985).

The parabrachial–pedunculopontine system is the major cholinergic cell group requiring intensive elucidation. The rostral aspect, the pedunculopontine nucleus, projects to the medial thalamic nuclei (Sugimoto and Hattori, 1983), substantia nigra (Beckstead *et al.*, 1979), and subthalamus (Sugimoto and Hattori, 1983), and these pathways appear to be partly cholinergic on the basis of retrograde transport of labeled choline (Sugimoto and Hattori, 1983). Lesion evidence also supports a cholinergic component of the projection to the substantia nigra (E. G. McGeer and P. L. McGeer, 1985). Projections have been described from this area to the cortex (D. M. Armstrong *et al.*, 1982; Saper and Loewy, 1980, 1982), which is also innervated by some fibers from the Kolliker–Fuse nucleus

(Saper and Loewy, 1982). Efferent projections have also been described from these nuclei into the midline, medial, and ventral basal thalamic nuclei and the hypothalamus and amygdala (Saper and Loewy, 1980). It would therefore appear that this region is a major supplier of afferents to all parts of the diencephalon and the limbic system, and possibly even to the cerebral cortex. Some of the rostrally projecting cholinergic neurons have been reported to contain substance P (Vincent *et al.*, 1983c).

8.3.5. Reticular Formation

A scattered collection of very large ChAT- and AChE-positive cells extends throughout the gigantocellular and magnocellular tegmental fields of the reticular formation (Gi and Ma in Figure 8.4). Caudally, the giganto- and magnocellular ChAT-containing neurons are gradually aggregated medially toward the granular layer of the raphe and ventrally to the area near the inferior olivary nucleus. Thus, these cells extend continuously from the pons into the medulla as a longitudinally oriented cluster (H. Kimura *et al.*, 1981, 1984; Mizukawa *et al.*, 1986).

8.3.6. Minor Cholinergic Systems

Minor groups that are probably cholinergic on the basis of ChAT immunohistochemistry include neurons in the lateral reticular nucleus, superior olivary complex, lateral vestibular nucleus, magnocellular red nucleus, and central canal region in the spinal cord (H. Kimura *et al.*, 1981, 1984). In addition, small interneurons that stain for ChAT have been described in the cerebral cortex (Houser *et al.*, 1983c; Eckenstein and Thoenen, 1983) (Figure 8.5J), hypothalamus (Figure 8.5K), and hippocampus (D. A. Matthews *et al.*, 1983). Retinal cholinergic amacrine cells have also been identified (M. Schmidt *et al.*, 1985; Millar *et al.*, 1985).

It should be anticipated that there will be some differences in immunohistochemical results among laboratories, since particular antibodies may recognize different sites on ChAT or because the conformation of ChAT may vary according to the method of fixation or its location within the CNS. It is also possible that some cholinergic cell groups may possess low levels of ChAT in their somata and thus be difficult to stain, as is the case for GAD in many γ-aminobutyric acid cells [see Figure 7.5 (Section 7.3)]. Thus, modifications and additions may be anticipated as techniques for detection improve.

These techniques have so far not been applied to any extent in the periphery, but it is part of classic neuroanatomy that in the spinal cord, all anterior horn cells that supply preganglionic nerves to autonomic ganglia are cholinergic. It is also part of classic neuroanatomy that ganglion cells that give rise to postganglionic parasympathetic fibers are nearly all cholinergic. In addition, cranial nerves that contain voluntary motor fibers (III–VI, IX, XI, and XII) and those that contain preganglionic parasympathetic fibers (III, VII, IX, and X) are cholinergic and are derived from cholinergic cell bodies. The reader is referred to standard textbooks of neuroanatomy for the details of the tracts and to standard pharmacology texts for the classic evidence on which these cholinergic assignments have been made. The sympathetic postganglionic fibers to sweat glands are also cholinergic, and there is suggestive evidence, which needs to be pursued, of an even more widespread distribution of such postganglionic cholinergic fibers in the sympathetic system. More sophisticated information on peripheral cholinergic anatomy can be anticipated now that specific methods for tracing cholinergic cells and fibers are available.

8.4. Physiological Actions of Acetylcholine and the Cholinergic Receptors

8.4.1. Types of Cholinergic Receptors

Two types of cholinergic receptors have been distinguished for years. The classical descriptive terms for these two types, *muscarinic* and *nicotinic,* derive from the fact that the physiological effects of ACh at certain sites can be mimicked by nicotine, an alkaloid from *Nicotiana tobacum,* while others can be simulated by muscarine, a drug obtained from the fungus *Amanita muscaria.* As described in Chapters 4 and 5, nicotinic receptors mediate ionotropic actions of ACh, while muscarinic receptors mediate metabotropic actions.

The nicotinic receptor occurs in ganglia of both the sympathetic and the parasympathetic system in cells innervated by the preganglionic cholinergic fibers. The neuromuscular junction, innervated by cholinergic fibers from anterior horn cells, also has a nicotinic-type receptor. The muscarinic receptor occurs in smooth muscle innervated by postganglionic fibers of the parasympathetic system. This also includes cardiac muscle and certain exocrine glands. Atropine is the classic blocking agent.

Central synapses have both muscarinic- and nicotinic-like receptors. The ligands that have been most commonly used for binding to these receptors are [^{3}H]quinuclidinylbenzilate (QNB) for muscarinic sites and [^{3}H]-α-bungarotoxin ([^{3}H]-α-BTX) for nicotinic sites. Both types of binding show considerable regional variation (Table 8.2). It is clear from Table 8.2 that the distributions are not parallel. In every major brain region, the absolute number of QNB binding sites is much greater than the number of α-BTX sites; in two hypothalamic nuclei (the suprachiasmatic and dorsomedial nuclei), however, Block and Billier (1981) report twice as many nicotinic as muscarinic sites. It is known that some cholinoceptive cells, such as the noradrenergic cells of the superior cervical ganglion, possess both nicotinic and muscarinic receptors, so that a mixed action is possible on the cholinoceptive cell. It is not known how commonly this may occur.

Considerable variability in the absolute values for cholinergic ligand binding, and

TABLE 8.2
Some Representative Data on Binding of Labeled α-BTX and QNB in Brain

Area	Binding as a ratio of that in striatum (absolute, fmoles/mg prot.)					
	α-BTX		QNB			
	Rat[a]	Rat[b]	Rat[c]	Rat[d]	Man[e]	Man[f]
Striatum	100%	100%	100%	100%	100%	100%
	(16)	(4)	(603)	(2120)	(740)	(600)
Cerebellum	50%	25%	8%	12%	14%	5%
Pons	—	360%	41%	29%	27%	—
Medulla	—	250%	26%	29%	—	—
Hypothalamus	469%	650%	28%	35%	31%	—
Midbrain	—	200%[g]	45%	44%	22%	50%[g]
Hippocampus	312%	616%	84%	73%	61%	60%
Cortex	325%	350%	86%	81%	60%	67%

[a-f]References: [a]B. J. Morley *et al.* (1977); [b]M. Segal *et al.* (1978); [c]Kobayashi *et al.* (1978); [d]R. M. Dawson and Jarrott (1980); [e]P. Davies and Verth (1977); [f]Luabeya *et al.* (1984).
[g]Thalamus.

thus the presumed concentration of cholinergic receptors, is found by different investigators. The reason may be that binding is greatly affected by factors such as conformational changes, the presence or absence of ions or GTP or both, and the possibility of cooperative interactions [see Figure 5.1 (Section 5.3.1)] among several cholinergic binding sites or between a cholinergic binding site and a recognition site for some other molecule as, for example, vasoactive intestinal polypeptide or vasopressin (for a review, see Sokolovsky, 1984).

The situation is further complicated by studies indicating the existence of multiple subtypes of both muscarinic and nicotinic receptors in brain [see Table 5.3 (Section 5.3.4)]. Two muscarinic subtypes (M_1 and M_2) have been differentiated in both rat (R. Hammer *et al.*, 1980; M. Watson *et al.*, 1982) and human brain (Garvey *et al.*, 1984). They differ in their affinities for pirenzepine, a compound that is clinically useful in the treatment of peptic ulcer, and oxotremorine. M_1 sites have high affinity for pirenzepine and are probably largely postsynaptic. M_2 sites have low affinity for pirenzepine, but high affinity for oxotremorine (Wamsley *et al.*, 1980). M_2 sites may be largely on cholinergic cells (I. T. Potter *et al.*, 1984). They undergo anterograde transport in the cholinergic splenic, vagus, and sciatic nerves (Kuhar and Zarbin, 1984; Wamsley *et al.*, 1981; Laduron, 1980).

The two nicotinic sites in brain are distinguished by preferential binding of α-BTX and nicotine. The one that preferentially binds α-BTX resembles the neuromuscular junction. The other, which preferentially binds nicotine [or ACh in the presence of atropine (R. D. Schwartz *et al.*, 1982)], resembles sites on sympathetic ganglia. The autoradiographic binding pattern for [^{3}H]nicotine in brain is strikingly different from that for [^{3}H]-α-BTX (Clarke *et al.*, 1984), which supports the dual nicotinic receptor hypothesis.

It is clear from the literature on subtypes of cholinergic receptors that there can be multiple responses to ACh, even on the same cell. For more details, we need to consider the iontophoretic activities of ACh.

8.4.2. Iontophoretic Effects of Acetylcholine on Central Neurons

The iontophoretic action of ACh on brain cells varies widely. It may cause an excitation with rapid onset (nicotinic), an excitation with slow onset (muscarinic), or an inhibition with slow onset (muscarinic).

The classical example of nicotinic action of ACh on CNS cells is that of motor axon collaterals on Renshaw cells [see Figure 4.15 (Section 4.7.2)]. The Renshaw cells respond with an extremely short latency to stimulation by ventral roots. This can be duplicated by iontophoretic application of ACh or nicotine. The activity is blocked by the nicotinic antagonists, dihydro-β-erythroidine and hexamethonium. Other areas in which somewhat similar effects can be observed are the medulla, areas of the thalamus and hypothalamus, and the cerebellum. Occasional cortical cells also respond similarly (for reviews, see K. Krnjevic, 1974; Phillis, 1970).

A much more commonly observed response to iontophoretically applied ACh on CNS cells is a slow and prolonged excitatory action that is blocked by atropine. With extracellular application of ACh and intracellular recording of the response, it can be quite clearly demonstrated that this depolarizing effect is not associated with the decrease in membrane resistance that would accompany the opening of typical sodium and other ion channels [see Figure 2.8 (Section 2.3) and Section 8.4.3]. Instead, the membrane resistance increases, with the depolarizing action having a reversal level close to -100 mV. This result can best be explained by a decrease in the conductance of either Cl^- or K^+, which have equilibrium potentials in this vicinity. Since intracellular injections of Cl^- cause large positive shifts in the Cl^- equilibrium potential but do not change the character of response to ACh, it must be concluded that a reduction in K^+ conductance (G_K) is responsible for the effect. Such slow, excitatory effects, blocked by atropine, are seen with cells in many areas of brain. Even Renshaw cells have some of these muscarinic receptors (K. Krnjevic, 1974).

ACh can also depress the action of a wide variety of CNS cells (Phillis, 1970; K. Krnjevic, 1974). In most areas, ACh is usually excitatory, but it is exclusively inhibitory on the majority of neurons in the nucleus reticularis of the thalamus (Ben-Ari et al., 1976). This inhibitory action is also clearly muscarinic, because it is slow in onset and can easily be blocked by atropine. In this case, however, the membrane changes are thought to be due to an increase, rather than a decrease, in potassium conductance (Cole and Shinnick-Gallagher, 1984).

As discussed in Chapter 5 for metabotropic neurotransmitters in general, the actions of ACh at muscarinic sites are believed to be mediated by second messengers. At some, ACh appears to inhibit cyclic AMP (cAMP) formation [see Figure 5.4 (Section 5.4.1)], while at others it stimulates hydrolysis of phosphoinositides [see Figure 5.7 (Section 5.4.3)] (J. H. Brown and S. L. Brown, 1984; Torda, 1973). Muscarinic agonists have been classified as class A (e.g., carbachol or oxotremorine-M) or class B (e.g., beth-anechol or oxotremorine-1), depending on whether or not they stimulate phosphoinosi-tide turnover (S. K. Fisher et al., 1983). Whether different subtypes of muscarinic receptors or different conformations of the same subtype are involved remains to be demonstrated.

8.4.3. Separation of the Acetylcholine Receptor

Efforts have been made to define the nature of the macromolecules that constitute cholinergic receptors. Remarkable progress has been made in the case of the nicotinic receptor from the electric organ of *Torpedo*. Certain snake venoms, such as ^{125}I-labeled α-BTX or α-cobra toxin, which bind very strongly and specifically to this receptor, were mixed with electric organ membranes, and separation of the receptor was achieved by following the label through various purification procedures. The glycoprotein that con-stitutes the receptor has been isolated and completely purified, the mRNA has been isolated, and cDNAs corresponding to the RNA have been cloned by the techniques shown in Figure 1.19 (Section 1.5.1). The purified glycoprotein has a pentameric struc-ture of four subunits (2α, β, γ, and δ). Each subunit is encoded in a different gene.

On the basis of the cDNA sequences, it is calculated that the two α-subunits have a molecular weight of 53.6 kilodaltons (kd) each, the β-subunit 56 kd, the γ-subunit 58 kd, and the δ-subunit 59.8 kd. The calculated molecular weight is below the experimentally measured value of about 295 kd because the receptor is glycosylated at least eight times.

The subunits are highly similar in overall structure, having stretches of almost identical amino acid sequence (Noda et al., 1984; Claudio et al., 1983; Devillers-Thierry et al., 1983; E. F. Young et al., 1985).

An ingenious method has been developed for determining which parts of the mole-cule are on the outside of the membrane and which are on the inside. Antibodies were prepared to peptides that correspond to the amino terminal, the carboxy terminal, and

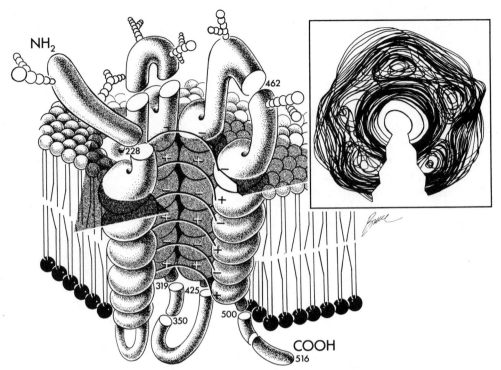

Figure 8.7. Model of the nicotinic cholinergic receptor. The bimolecular leaflet has been opened to expose the glycoprotein that comprises the nicotinic receptor. The protein structure is glycosylated at least eight times, as shown by the indicated glycosyl groups. The NH_2 terminus of the protein is located outside the membrane. The protein has five stretches of amino acids, wound in an α-helix, that traverse the membrane and presumbaly correspond to short stretches of the 2 α, 1 β, 1 γ, and 1 δ subunits of the receptor. Cuts indicate where stretches of the protein are not shown. The locations of identified sequences in the protein that are inside or outside the membrane are indicated by the amino acid number in the chain, which is 516 amino acids long. Simulated charges along the channel are indicated on the basis of presumed amino acid sequences in the α-helix. Adapted from E. F. Young *et al.* (1985). *Inset:* Three-dimensional contour map of a receptor molecule based on electron microscope images as seen from the synapse looking down the channel. Adapted from Brisson and Unwin (1985).

some intermediate sequences. The binding of these antibodies to the receptor-containing membranes was then observed by electron microscopy (E. F. Young *et al.*, 1985). It was determined that the carboxy terminal was on the inside of the membrane and the amino terminal on the outside, with sequences between being identified as existing inside or outside the cell. A model that fits this information is depicted in Figure 8.7, in which the protein is shown traversing the membrane five times. The traversing peptide sequences, which are presumed to be identical, are wound in an α-helix but make up only a minority of the total protein. The α-helix portion is mainly hydrophobic, to be compatible with the lipid bilayer of the membrane, but polar units are oriented toward the pore, creating an ion-conducting channel. Electron microscopic evidence suggests a hollow cylindrical structure, 7–9 nm in diameter. Opening of the channel takes place on binding of ACh or other agonist to the α-subunits. A conformational change takes place, perhaps in the fashion depicted in Figure 2.8 (Section 2.3) for the Na^+ channel. In any event, activation of the ionotropic channel, i.e., the nicotinic cholinergic channel, results in an estimated 10^4 Na$^+$ ions/msec passing through the membrane (E. F. Young *et al.*, 1985).

In addition to the four subunits of the *Torpedo* receptor, the comparable α-, β-, and γ-subunits of the mammalian nicotinic receptor have been elucidated by cloning and sequencing cDNAs or genomic DNA that encodes these polypeptides. The cDNA that encodes the α-subunit has been artificially altered and then used to prepare model receptors with the α-subunits altered at specific sites. In this way, it has been possible to define the regions of the molecule that are involved in ACh binding and in forming a transmembrane ion channel (Mishina *et al.*, 1985). Furthermore, a receptor aggregating factor has been identified in the synaptic cleft region of the neuromuscular junction using monoclonal antibodies (Fallon *et al.*, 1985).

8.5. Pharmacology of Acetylcholine

ACh differs from the amino acid transmitters discussed in Chapters 6 and 7 in that it plays a dominant role in the operation of the peripheral nervous system. It is the principal neurotransmitter for the neuromuscular system and autonomic ganglia, as well as for the postganglionic parasympathetic systems. A broad spectrum of pharmacological agents exists, but the effects are largely peripheral. Many are highly polar and do not cross the blood–brain barrier (BBB); of those that do, the global effects are usually a mixture of central and peripheral actions. Since the focus of this book is on the mammalian brain, only a few of the representative cholinergic drugs will be mentioned (Table 8.3). Readers

TABLE 8.3
Some Representative Agents That Interact with Cholinergic Systems

Receptor interaction
  Muscarinic receptors
    Agonists: carbachol (class A), oxotremorine (class A), bethanechol (class B), muscarine, pilocarpine, arecoline, methacholine
    Antagonists: atropine, scopolamine, methylscopolamine, artane
    Binding agents: pirenzepine (M$_1$), oxotremorine (M$_2$), quinuclidinyl benzilate (QNB)
  Nicotinic receptors
    Agonists: nicotine, carbachol, methacholine
    Antagonists: dihydro-β-erythroidine (Renshaw cells), hexamethonium (Renshaw cells, autonomic ganglia), D-tubocurarine, decamethonium (receptor depolarizer)
    Binding agents: α-bungarotoxin, nicotine
Acetylcholinesterase inhibitors
  Reversible: physostigmine
  Irreversible: diisopropyl fluorophosphate (DFP)
  Regenerator: pyridine-2-aldoxime methiodide (2PAM)
Acetylcholine synthesis
  Choline acetyltransferase inhibitor: 4-(1-naphthylvinyl)pyridine
  Choline uptake inhibitors: hemicholinium-3 (HC-3)
  Substrate replenishers: choline, lecithin
Acetylcholine release
  Inhibitor: botulinus toxin
  Promoter: black widow spider venom

are referred to such excellent standard textbooks as that by A. G. Gilman *et al.* (1985) or to various reviews (e.g., Karczmar, 1981) for more comprehensive discussions.

8.5.1. Muscarinic Agonists

Muscarine, the active agent in the *Amanita* genus of mushrooms, has well-known physiological effects. They include salivation, sweating, miosis, dyspnea, abdominal pains, watery diarrhea, vertigo, confusion, weakness, coma, and, at sufficient doses, death. Pilocarpine and arecoline are two cholinomimetics with primarily muscarinic action. Arecoline does not cross the BBB.

Methacholine and carbachol are two ACh-like agents that are hydrolyzed more slowly by AChE than is ACh. Their actions are also weighted toward muscarinic action, but they do have significant nicotinic action. They do not cross the BBB. Bethanechol, a relatively pure muscarinic agonist that does not cross the BBB, has been administered by intracranial infusion with a peripherally implanted minipump to Alzheimer's cases, with some reported benefit (Harbaugh *et al.*, 1984).

8.5.2. Muscarinic Antagonists

In contrast to most nicotinic antagonists, standard muscarinic blockers often have central as well as peripheral actions. This is consistent with the predominance of muscarinic receptors in the CNS. Atropine and scopolamine are the classic agents. They cause dilation of the pupil (mydriasis), decreased salivation, decreased secretions of the pharynx and respiratory tract, increased heart rate, decreased secretory activity, decreased motility of the gut, and decreased bladder tone. They also cause loss of short-term memory, drowsiness, euphoria, fatigue, and dreamless sleep. Higher doses result in excitement, restlessness, hallucinations, and delirium. They have a mild antiparkinsonian effect (see Chapter 14) and reverse the tremor caused by the central muscarinic agonist tremorine. High doses of artane, a classic agent for the treatment of Parkinson's disease, have proven useful in many cases of childhood-onset dystonia (Fahn, 1983). Methylscopolamine does not cross the BBB and thus gives only the peripheral signs of muscarinic blockade.

8.5.3. Nicotinic Agonists

Nicotine is, of course, the principal nicotinic agonist. While it has no therapeutic application, its presence in tobacco testifies to its physiological activity. Its action in the periphery is biphasic. Stimulation rapidly turns to blockade and paralysis of the neuromuscular junction as the dose increases. It causes tachycardia and vasoconstriction. Centrally, it produces tremors followed by convulsions. Methacholine and carbachol also have peripheral nicotinic effects.

8.5.4. Nicotinic Antagonists

Most nicotinic antagonists are neuromuscular blockers and therefore have a devastating physiological effect. D-Tubocurarine, the chemical warfare agent discovered by primitive South American Indian tribes, is the classic agent. It stabilizes the receptor by combining with cholinoceptive sites at the postjunctional membrane, thereby blocking the neurotransmitter action of ACh. The muscles will no longer respond to applied ACh or to nerve stimulation. Muscle fibers will still respond, however, to the depolarizing action of K^+ ions or to direct electrical stimulation, clearly indicating that it is the cholinergic receptor channel and not the Na^+ channel that is being blocked.

Decamethonium has effects similar to those of D-tubocurarine, but a slightly different mechanism of action. It combines with the receptor in such a way as to leave it depolarized for a prolonged period. The action can be partially antagonized by D-tubocurarine, but not by cholinesterase inhibitors. Furthermore, K^+ ions cannot antagonize the block because the muscle fiber itself is already depolarized.

Hexamethonium represents yet another class of nicotinic blockers that act preferentially on ganglionic receptors. Although most ganglionic blocking agents penetrate the BBB poorly, those that do also have little central effect. This again emphasizes the lack of dominance in the CNS of nicotinic receptors.

8.5.5. Release Inhibitors and Promoters

Botulinus toxin, the causative agent in certain types of food poisoning, inhibits the release of ACh from all types of nerve fibers. The mechanism is unknown. Death is by respiratory paralysis. Injections of small doses of the toxin into the eye muscle in cases of blepharospasm or into the neck muscles in spasmodic torticollis are reported to give substantial relief of the spasms in these forms of dystonia for periods averaging about 8 weeks (Frueh et al., 1984; Tsui et al., 1985).

Black widow spider venom promotes the release of ACh. It is not a specific agent, however, and it induces morphological changes in synaptic vesicles.

8.5.6. Anticholinesterases

Many cholinesterase inhibitors have a broad spectrum of action on both AChE and the pseudocholinesterases. These inhibitors are divided into two classes: reversible and irreversible. The outstanding representative of the reversible group is physostigmine (eserine). The best known of the irreversible AChE inhibitors is DFP. This is one of a class of organophosphorus compounds that were discovered shortly before World War II. They were used initially as agricultural insecticides, but later were developed as chemical warfare agents. Organophosphorus compounds react with AChE, phosphorylating the enzyme on the serine OH group at the esteratic site, releasing HF. The enzyme is thereby rendered inactive, unless the catalytic site is regenerated. This can be achieved through the use of 2PAM (see Section 8.2.2).

The effects of AChE inhibitors are manifold. They cause marked miosis, watery nasal discharge, wheezing respiration due to bronchiolar secretions, nausea and vomiting, cramps and diarrhea, involuntary urination, excessive sweating, involuntary twitchings, and weakness. Centrally, the effects are slurred speech and confusion, followed by loss of reflexes, convulsions, coma, and death. The effects can be countered by huge doses of atropine and, in the case of DFP poisoning, by 2PAM. Therapeutically, anti-AChE compounds have three principal uses, myasthenia gravis, bladder atony, and glaucoma (for reviews, see Main, 1976; Koelle, 1970), although they have been tried in Huntington's disease and have some transient beneficial effects on memory in senile dementia.

Some inhibitors that are selective for either AChE or the pseudocholinesterases are known and are useful in *in vitro* studies on these enzymes. Two examples are tetraisopropylpyrophosphoramide, which is selective for the pseudocholinesterases, and 1,5-*bis*-(4-allyldimethylammoniumphenyl)pentan-3-one (BW284C51), which is selective for AChE.

8.5.7. Synthesis Stimulators and Inhibitors

No good method has been found to stimulate ACh synthesis. There is some literature suggesting that choline supplements will raise brain ACh levels, but the majority of investigators find that such supplements have no effect in animals on a normal diet. There have been numerous trials of choline or its derivative, lecithin, in senile dementia of the Alzheimer type, without significant clinical improvement (for a review, see Hollister, 1985).

Hemicholinium-3 (HC-3) limits synthesis and hence lowers brain ACh levels by interfering with choline uptake. Autoradiographic studies of [^{3}H]-HC-3 binding sites in brain slices have been used as a method of mapping the distribution of cholinergic nerve terminals (Vickroy *et al.*, 1985).

The azidium mustard derivative AF64A is an irreversible inhibitor of choline uptake, and intraventricular administration has been said to selectively decrease cholinergic activity in brain without, however, causing a disappearance of cholinergic neurons (T. J. Walsh *et al.*, 1985; Rylett and Colhoun, 1984). However, when injected directly into brain tissue, it causes nonselective damage.

These compounds do not cross the BBB and hence must be administered intracerebrally for any central effect.

Most searches for specific inhibitors of ChAT have been disappointing. Compounds such as 4-(1-naphthylvinyl)pyridine (Haubrich and Pflueger, 1979; Goldberg *et al.*, 1971) and 3′,4′-dichloro-4-stilbazole (B. R. Baker and Gibson, 1971) offer some promise. They are much more effective *in vitro* than *in vivo*, suggesting that ample enzyme exists in tissues and that choline uptake is a more sensitive locus for chemical attack.

8.6. Pathology of Cholinergic Systems

8.6.1. Myasthenia Gravis

Fundamental work on the nicotinic receptor of the neuromuscular junction proved the key to an understanding of myasthenia gravis. It is now accepted that myasthenia gravis is an autoimmune disease in which the attack is directed specifically against the nicotinic receptor, with the circulating antibody both blocking the receptor and causing its accelerated degradation. The use of thymectomy and adrenal corticosteroids in the treatment of myasthenia gravis is predicated on their apparent interference with the autoimmune reaction (Drachman, 1978). Myasthenia gravis is characterized by weakness and fatiguability of muscle. The similarity between the symptoms of the disorder and those of curare poisoning, coupled with the remarkable response of many patients to AChE inhibitors, first pointed to the neuromuscular junction as the site of pathology. However, it was not known whether the problem was presynaptic or postsynaptic until [^{125}I]-α-BTX binding studies were done which demonstrated a 70–89% reduction in the number of ACh receptors in muscle biopsies from myasthenic patients. There remained the problem of what caused this loss. Injection of purified nicotinic receptor glycoprotein into rabbits, in an attempt to raise antibodies, led to the development in the animals of marked muscular weakness and respiratory insufficiency. Further studies indicated many other similarities between this animal model of experimental allergic myasthenia gravis and the human disease. These findings led to the detection of antireceptor antibody in the majority of myasthenic patients, who nevertheless showed normal synthesis of the cholinergic receptor protein.

8.6.2. Senile Dementia of the Alzheimer Type

The initial clue to possible cholinergic involvement in SDAT came from three groups in the United Kingdom who reported sharply decreased levels of ChAT in the neocortex, particularly in the hippocampus and the temporal, and frontal lobes (D. M. Bowen et al., 1976; P. Davies and Maloney, 1976; E. K. Perry et al., 1977a,b). Since then, numerous groups have confirmed these findings (cf. Bartus et al., 1982). These decreases are over and above those attributable to the aging process (P. L. McGeer and E. G. McGeer, 1976; E. K. Perry et al., 1977c; P. Davies, 1979; Rossor et al., 1982a; S. J. Allen et al., 1983).

D. L. Price et al. (1982) (cf. Whitehouse et al., 1981, 1982) first reported loss of large cells from the nucleus basalis of Meynert area, which innervates the neocortex [see Figures 8.4 and 8.5 (Section 8.3.1)]. This finding has now been corroborated in a number of studies (R. H. Perry et al., 1982; Nagai et al., 1983b; Tagliavini et al., 1984; D. Mann et al., 1984a; Rogers et al., 1985; Saper et al., 1985; Arendt et al., 1985) [see Figure 8.5 (Section 8.3.1)]. The giant cells of the medial basal forebrain are a single population of cholinergic cells. There is also a loss of these cells during the normal aging process. Thus, the substantia innominata of a young person will contain in excess of 450,000 cholinergic neurons. This declines to about 150,000 in the elderly, while Alzheimer's cases have fewer than 100,000 such cells (P. L. McGeer et al., 1984b; D. M. Mann et al., 1984b). The dropout of cells during aging helps to explain the increasing incidence of Alzheimer's disease with age (Kay, 1986; A. S. Henderson, 1986), as well as the increasing effects of cholinergic blockade on memory.

The cholinergic hypothesis of Alzheimer's disease holds that the main physiological deficit that produces the symptoms of the disease is the loss of these cells (Bartus et al., 1982; E. K. Perry, 1986). It has long been known that scopolamine, a muscarinic cholinergic blocker, induces deficits in learning and memory (Ostfeld and Araguette, 1962; Dundee and Pandit, 1972; Crow and Grove-White, 1973; Drachman and Leavitt, 1974; Hrbek et al., 1974) that are highly similar to the memory deficits seen without drugs in elderly subjects (Drachman, 1977). These deficits can be reversed by physostigmine, but not by amphetamine. Furthermore, cholinergic agonists such as arecoline will improve the ability of young subjects to acquire new information (Sitaram et al., 1978; K. L. Davis et al., 1979). These findings in humans have also been reproduced in nonhuman primates (Bartus and Johnson, 1976; Bartus, 1979) and more recently in other species (Lerer and Friedman, 1982; Flicker et al., 1983).

Deficits in the acquisition of long-term memory were noted in rats with bilateral nucleus basalis lesions even though the septal–hippocampal cholinergic pathway was intact (C. L. Murray and Fibiger, 1985). Dubois et al. (1985) found a disorganization of both spontaneous and learned behavior in similarly treated rats, emphasizing the importance of this cholinergic pathway to the memory process. Unfortunately, the mechanism has not been approached at the synaptic level. It does not yet form part of the detailed theories of Sections 16.3–16.5, but it should be a fruitful field for future research.

Several studies have shown that muscarinic binding sites are comparable in Alzheimer's disease patients and age-matched controls (E. K. Perry *et al.*, 1977c; P. White *et al.*, 1977; P. Davies and Verth, 1977; Reisine *et al.*, 1978; D. J. Bowen *et al.*, 1979; Lang and Henke, 1983). However, Mash *et al.* (1985) have reported that there is a selective loss of the M_2 receptor, which is thought to be presynaptic. These data suggest that the physiological deficit is due to presynaptic cholinergic losses.

Cell losses in Alzheimer's disease are not restricted to the medial basal forebrain cholinergic complex (Mountjoy, 1986; Emson and Lindvall, 1986), although some other cholinergic cell groups, such as those in the basal ganglia, seem to be spared (Nagai *et al.*, 1983b). The locus coeruleus, for example, shows severe neuronal dropout in SDAT (D. M. A. Mann *et al.*, 1980; J. J. Mann *et al.*, 1981; Bondareff *et al.*, 1982), and there is substantial cortical cell loss as well (Mountjoy *et al.*, 1983). All these could contribute to the severe glucose metabolic deficits noted in the cerebral cortex of Alzheimer's cases *in vivo* using position emission tomography (see Foster *et al.*, 1983a; P. L. McGeer *et al.*, 1986) (for a review, see P. L. McGeer, 1986).

8.6.3. Diseases That Affect Anterior Horn Cells

Poliomyelitis and amyotrophic lateral sclerosis (ALS) are diseases in which CNS neurons are attacked but there is a preferential loss of anterior horn cells. Poliomyelitis is caused by a retrovirus. The bulbar form, in which cells in the brainstem are attacked, is frequently fatal. The cells involved are primarily cholinergic cells from cranial nerve nuclei in the brainstem as well as the reticular formation (Bodian, 1972). Respiratory complications are among the most severe of the deficits that develop, possibly due to the involvement of cells in the respiratory centers in the parabrachial and reticular areas as well as in the direct cholinergic motor systems.

ALS is a progressive disease of unknown etiology that is universally fatal. It is hypothesized to be of viral origin.

Cholinergic systems are involved in other disease entities, such as Huntington's (see Chapter 14). They may also play some role in mental illness (see Chapter 15) and epilepsy, but they are not believed to be the primary source of pathology. (For a review of cholinergic systems in human pathology, see P. L. McGeer and McGeer, 1984).

8.7. Summary

ACh was the first compound proven to be a neurotransmitter and thus was responsible for establishing the concept of chemical transmission in the nervous system. Through elegant physiological and pharmacological experiments, now part of classic neurobiology, it was established that cholinergic neurons supply all preganglionic fibers of the autonomic nervous system, all postganglionic fibers of the parasympathetic system, all fibers of the musculoskeletal system, and a few fibers of the postganglionic sympathetic system.

All components of the cholinergic system are widely distributed within the brain.

Five principal cholinergic neuronal systems have now been identified: the medial basal forebrain complex, which innervates the neo-, archi-, and paleocortex; the extrapyramidal system, which provides interneurons to the caudate, putamen, and accumbens; the parabrachial, which has neurons in several nuclei surrounding the brachium conjunctivum (superior cerebellar peduncles); the reticular, with neurons in the magno- and gigantocellular fields of the reticular formation and in the lateral reticular nucleus; and the cranial motor nerves. In addition, there are several minor cholinergic groups.

Synthesis of ACh is a relatively simple process involving acetyl-CoA, produced in brain mitochondria, and choline, produced in liver. The synthetic enzyme ChAT has been completely purified, antibodies to it have been produced in rabbits, and an immunohistochemical method for its localization in brain has been developed.

Regulation of ACh production is not well understood. There is an excess of enzyme for normal rates of *in vivo* synthesis and sufficient substrates under normal circumstances. Administration of choline to animals on a normal diet does not increase ACh levels in brain.

Chemically, ACh is found in both bound and free states. These pools probably correspond to the cytoplasmic and vesicular fractions identified in synaptosomes. Newly synthesized ACh is preferentially released on stimulation.

ACh is destroyed by the extraordinarily active enzyme AChE. This enzyme is found not only in cholinergic neurons, but also in cholinoceptive neurons such as the dopaminergic neurons of the substantia nigra. Thus, AChE is an unreliable marker of cholinergic neurons. Pseudocholinesterase, which exists in many organs of the body, including brain, can also destroy ACh. Various neuropeptides, as well as ACh, may be endogenous substrates for AChE.

Two broad classes of cholinergic receptors exist: nicotinic and muscarinic. Subtypes of both appear to exist. Nicotinic receptors are excitatory and operate by opening ion channels. They are therefore ionotropic. Muscarinic receptors appear to be metabotropic. They probably involve cAMP, cGMP, and phosphoinositides as second messengers, with the membrane alterations being interpreted in relation to K^+ conductance changes. They are slow in onset and may be excitatory or inhibitory. Other neurotransmitters, such as noradrenaline, substance P, thyrotropin, vasoactive intestinal polypeptide, and arginine vasopressin may modulate the muscarinic effects of ACh.

Nicotinic-type receptors include all ganglionic receptors, all receptors of the musculoskeletal system, and a few central receptors. Muscarinic receptors include all smooth muscle and exocrine gland receptors of postganglionic parasympathetic fibers and the overwhelming majority of brain receptors.

A broad spectrum of agents that interact with the cholinergic system are available. Many of these are part of classic cholinergic pharmacology with clinical application to peripheral cholinergic systems. Included are agents that stimulate or block nicotinic and muscarinic receptors and agents that specifically bind to subtypes of these receptors. Choline uptake inhibitors exist, but success has been limited with inhibitors of ChAT. Numerous AChE inhibitors exist, of both the reversible and the irreversible type.

Loss of the cholinergic receptors of the neuromuscular junction occurs in the autoimmune disease myasthenia gravis. Preferential cholinergic neuronal loss occurs in the medial basal forebrain complex in Alzheimer's disease, and preferential loss of anterior horn cells occurs in poliomyelitis and ALS.

9

Catecholamine Neurons

9.1. Introduction

Our discussion now shifts to the catecholamines. An enormous literature exists on these transmitters. The reader might infer from this that catecholamine neurons are quantitatively important in the brain. On the contrary, they constitute only a minute fraction of the total, numbering only about 50,000 in the rat brain and probably no more than 2 million in the human brain. As mentioned in Chapter 1, there are probably more than 10 billion neurons in the human neocortex alone.

The importance of catecholamine neurons, then, cannot be attributed to their numbers; it must be found elsewhere. One possible reason may be their remarkable axonal divergence. For example, each dopaminergic neuron in the rat substantia nigra is estimated to have about 500,000 boutons in the neostriatum, and in the human it may be as high as 5 million. Another reason may be the nature of their postsynaptic effects. The overall catecholamine action is much slower than that of many other cells, and their action at some receptors seems linked to cyclic nucleotides. Thus, they may modulate the reaction of cells to more specific, ionotropic neurotransmitters, or they may initiate other types of postsynaptic activity.

The anatomy of catecholamine neurons is now known in great detail, but the complete story of the catecholamines cannot be told because so little is known of their detailed postsynaptic actions. As a starting point in providing perspective on this important group of compounds, we shall give an account of the evolution of our present state of knowledge.

The term *catecholamine* is derived from the structure, which couples an amine side chain with a dihydroxyphenyl (catechol) ring (see Figure 9.1). The catecholamine family is comprised of dopamine (DA), noradrenaline (NA), and adrenaline. They are formed from the dietary amino acid tyrosine by the following sequence of reactions:

265

L-Tyrosine → L-3,4-dihydroxyphenylalanine → dopamine →
                                      → L-noradrenaline → L-adrenaline

The only compound in the sequence which is always a transient intermediate is L-dihydroxyphenylalanine (L-dopa). As will be described later, however, it can be employed as a pharmacological agent and thus has a special importance in its own right.

The catecholamines were discovered in reverse order of their position in the biosynthetic sequence. This is due to the location of their biological prominence. DA is the dominant catecholamine in brain, NA in sympathetic neurons, and adrenaline in the adrenal gland. Adrenaline was the first body hormone to be isolated in crystalline form. Oliver and Schafer (1895) used the cardiovascular actions of extract of adrenal medulla as the assay system. This led to its isolation and finally to its synthesis by Stolz (1904). As described in Chapter 3, a Cambridge University student, T. R. Elliott, initiated the idea of chemical transmission by proposing that adrenaline was the effector agent released from sympathetic nerve endings on stimulation. As with the proposal made shortly after by Dixon for acetylcholine (ACh), this was not taken seriously until the experiments of Dale a few years later. He and Barger lent credence to the hypothesis with their classic examination of a series of synthetic amines that they termed "sympathomimetic." Their observation that some of these sympathomimetic derivatives were superior to adrenaline in mimicking the action of sympathetic stimulation was the first clue that NA and not adrenaline might be the true transmitter.

Otto Loewi, whose most important contribution was described in Chapter 4 on ACh, carried the work to the next stage by obtaining "Acceleranzstoff" from the frog's heart (the frog is unique in releasing adrenaline rather than NA as the transmitter agent), showing that stimulation of sympathetic nerves also led to the release of neurotransmitters.

The next chapter in the unfolding story of the catecholamines was written by Cannon. He had found evidence of a "special and unknown factor," different from adrenaline, that was released following hepatic nerve stimulation (Cannon and Uridil, 1921). More thorough experimentation a dozen years later led him to propose that his "sympathin" obtained by stimulation of autonomic nerves was really composed of sympathin E and sympathin I. Bacq (1934), working in Cannon's laboratory, suggested that the inhibitory sympathin I was identical with adrenaline, while the excitatory sympathin E was identical with NA. The insight of Bacq was not to be proved until a dozen years later, when von Euler showed that extracts of sympathetic nerves did indeed contain NA.

For years, there had been speculation on the biosynthetic origin of adrenaline. Funk (1911) synthesized its precursor, L-dopa, believing that it might reveal the route of synthesis. But the critical experiment linking L-dopa to the catecholamines was not to be performed until 1939. It was then that Holtz (1939) incubated an extract from guinea pig kidney, which contained L-dopa decarboxylase, with its substrate. When the incubate was injected intravenously, a strong rise in blood pressure was obtained. It was DA that they isolated and presumed was the active principal. Thus, the third active member of the catecholamine family was biologically identified. In subsequent experiments, Holtz et al. (1947) proved their hypothesis by injecting themselves with L-dopa intravenously and isolating DA from their urine.

Blaschko (1939), noting that L-dopa but not N-methyldopa was decarboxylated by mammalian tissue, reasoned that DA and NA must be the precursors of adrenaline. He correctly predicted the complete biosynthetic sequence starting from tyrosine. It was not until 1951, however, that Goodall (1951) finally demonstrated that DA actually exists in mammalian tissues.

Thus, the three catecholamines had all been identified in peripheral tissue, but their presence in brain was still not suspected. Furthermore, there was much confusion about whether DA and NA were physiologically active in their own right or merely precursors of adrenaline.

Marthe Vogt (1954) first measured the "sympathin" content of brain, noting that the distribution could not be explained on the basis of brain vascularity. For example, very high concentrations were found in the midbrain and hypothalamus.

The rauwolfia alkaloid reserpine had been introduced into clinical medicine with no physiological explanation at that time as to why this ancient Indian remedy possessed antihypertensive and tranquilizing properties. Moreover, there was no anticipation of an unexpected side effect, that of a parkinsonian-like rigidity, which was noted by several clinical groups using the drug.

Carlsson and his group developed a chemical assay for DA and showed extraordinarily high concentrations in the corpus striatum. Like the other catecholamines and serotonin [(5-HT) 5-hydroxytryptamine], it was depleted following the administration of reserpine. This correlation among high levels of DA in the corpus striatum, its depletion by reserpine, and the accompanying parkinsonian-like side effects led Carlsson (1959) to propose that dopamine is involved in extrapyramidal function.

About the same time, Brodie made his proposal that NA is involved in ergotropic functions. This concept helped to explain the tranquilizing action of reserpine, which depleted the amines, as well as that of the phenothiazines, which blocked their action (Brodie et al., 1959).

Almost simultaneously, Kline and co-workers discovered that iproniazid, a monoamine oxidase inhibitor, is a mood elevator. It took little time to make the connection between enhanced levels of catecholamines and 5-HT and the psychological improvement (Loomer et al., 1957).

Birkmayer and Hornykiewicz (1961) followed up on Carlsson's hypothesis by measuring DA levels in the autopsy brains of a series of parkinsonian patients. They found that there were sharply decreased levels. Since Carlsson had shown some years previously that the administration of L-dopa would overcome the effects of DA depletion, it followed naturally that clinicians should administer L-dopa to parkinsonian patients.

The first striking clinical results were reported by Cotzias and co-workers, who titrated their patients with very large doses of L-dopa over periods of several weeks. This has since become the method of choice for treating parkinsonian patients (Cotzias et al., 1969).

By the early 1960s, the importance of DA and NA in brain had been firmly established, but the biological source of L-dopa was still not clear. E. G. McGeer et al. (1963) injected labeled tyrosine directly into the neostriatum and showed that it was converted to labeled catecholamines. Thus, brain was capable of converting the dietary amino acid tyrosine to L-dopa and catecholamines. Udenfriend (1966) and colleagues isolated and purified tyrosine hydroxylase from adrenal gland and showed that it was the rate-limiting step in the biosynthetic sequence.

The climactic event in the catecholamine story came with the extension by Falck et al. (1962) of the histofluorescence technique of Eranko (1955) into a method for detecting catecholamines and 5-HT in nerve tissue. The Falck technique was rapidly exploited by Dahlstrom, Fuxe, and many others to lay out the pathways of DA and NA neurons in brain (Dahlstrom and Fuxe, 1964a,b).

The position of adrenaline in brain was less certain. Vogt and others had demonstrated the presence of adrenaline by bioassay, but chemical techniques made the finding seem less certain. P. L. McGeer and E. G. McGeer (1964) demonstrated the conversion of labeled NA to adrenaline by the hypothalamus, but there was again no proof that there are distinct adrenaline-containing cells. That demonstration was to wait another decade until the elegant immunohistochemical procedures of Hökfelt and his colleagues were developed. Goldstein purified phenylethanolamine-N-methyltransferase from adrenal tissue and prepared antibodies to it. These were then used to establish by immunohistochemistry the presence of distinct adrenaline-containing neurons in the brainstem (Hökfelt *et al.*, 1974).

9.2. Chemistry of Catecholamine Neurons

The catecholamine neurons will be looked at as a group in terms of their chemistry because they have so much in common.

9.2.1. Synthesis of Catecholamines

The catecholamines are a family, with each member differing in the number of enzymes in the catecholamine synthetic chain it possesses. The metabolic sequence is shown in Figure 9.1. Joh *et al.* (1984) have postulated that the enzymes tyrosine hydroxylase (TH), dopamine-β-hydroxylase (DBH), and phenylethanolamine-N-methyltransferase (PNMT) are coded for by a single gene or linked genes, suggesting that they may have a common ancestral origin.

Tyrosine Hydroxylase.   TH (EC 1.14.16.1) is believed to be the rate-controlling step for catecholamine biosynthesis *in vivo* for the following reasons:

1. Free tyrosine, but not L-dopa, is detected in tissues. This implies that all tyrosine that is converted to L-dopa is rapidly carried through the remaining steps.
2. Catecholamine levels are not influenced by changing dietary levels of tyrosine or by parenteral administration of large doses of the amino acid. On the other hand, catecholamine levels are dramatically changed by oral or parenteral administration of L-dopa.
3. Catecholamine levels in tissue are sharply reduced by the administration of TH inhibitors, but only slightly by dopa decarboxylase (DDC) inhibitors.

In keeping with this concept, TH protein seems to be present in tissue in much lower concentrations than DDC and DBH, which is probably the reason that this initial step was the last in the biosynthetic sequence to be demonstrated experimentally. TH can be studied in material from DA-, NA-, or adrenaline-containing tissues. Two commonly employed sources are the corpus striatum, which is rich in dopaminergic nerve endings, and the adrenal medulla, which is largely composed of adrenaline- and NA-secreting

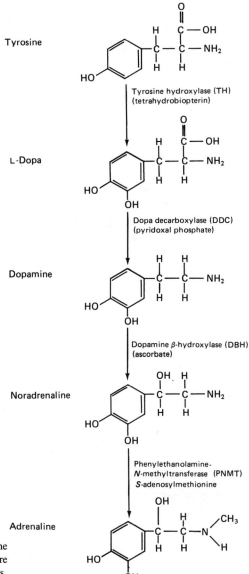

Figure 9.1. Schematic presentation of catecholamine synthesis from tyrosine to adrenaline. Enzymes are indicated with their essential cofactors in parentheses.

cells. Pheochromocytoma-derived cell lines are convenient for tissue culture study.

TH has been completely purified from adrenal medulla by several laboratories, and antibodies to it have been prepared (Nagatsu *et al.*, 1964; Okuno and Fujisawa, 1982; Oka *et al.*, 1983). In common with many enzymes, it appears to exist in a number of molecular weight forms. The TH in bovine adrenal has a molecular weight of 280,000; in NA neurons, approximately 200,000; and in dopaminergic neurons, only 65,000. The higher-molecular-weight form in the adrenal medulla is thought to be made up of four identical subunits (Oka *et al.*, 1983). A form of molecular weight approximately 34,000 can be produced by treating the enzyme *in vitro* with trypsin. For catalytic activity, TH requires molecular oxygen, and tetrahydrobiopterin (BH_4) as a cofactor (Nagatsu *et al.*, 1964). The reaction sequence for the conversion of tyrosine to L-dopa is presumed to be as follows:

$$TH\text{-}H_2 + tyrosine \rightarrow TH\text{-}H_2 \cdot tyrosine$$
$$TH\text{-}H_2 \cdot tyrosine + \tfrac{1}{2} O_2 \rightarrow TH + dopa + H_2O$$
$$BH_4 + TH \rightarrow TH\text{-}H_2 + Q\text{-}BH_2$$
$$Q\text{-}BH_2 + 2NADH \xrightarrow{\text{(DHPR)}} BH_4 + 2NAD$$

where $Q\text{-}BH_2$ is quinonoid dihydropteridine and DHPR is dihydropteridine reductase (EC 1.6.99.7), an enzyme linked with the NAD reducing system.

During the hydroxylation of tyrosine to L-dopa, the $TH\text{-}H_2$ protein becomes oxidized to the inactive TH form. It is reactivated by a BH_4 molecule that donates its hydrogen atoms. The K_m of TH for tyrosine is in the micromolar range and hence it is virtually saturated by the high tissue concentrations of endogenous tyrosine.

DHPR is required to maintain the pteridine cofactor in its active tetrahydro form. $Q\text{-}BH_2$ is very unstable and can rearrange to $7,8\text{-}BH_2$. $7,8\text{-}BH_2$ is not a substrate for DHPR (Nichol *et al.*, 1983), but is a substrate for dihydrofolate reductase.

The molecular genetic techniques described in Section 1.5 have been applied to solving the structure of TH. The appropriate messenger RNA (mRNA) was isolated and the complete coding sequence for TH protein was worked out. Moreover, through DNA hybridization experiments, the TH gene was localized in humans to chromosome 11. A few of the details of the experiments will be described because they provide an excellent illustration of how molecular genetics can open new vistas for neurochemistry.

The first step was to prepare an appropriate complementary DNA (cDNA) probe for TH (Lamouroux *et al.*, 1982). This was accomplished by cloning bacterial colonies that contained recombinant plasmids carrying DNA sequences complementary to rat TH mRNA [see Figure 1.19 (Section 1.5.1)]. The PC12 cell line was used as the source of the mRNA, since this cell line vigorously expresses the TH gene. A method known as "differential colony hybridization" was used to weed out unwanted bacterial colonies from the initial mixtures of bacteria that contained recombinant DNA sequences complementary to the initial mRNA isolates. Comparable preparations from rat liver cells were used for this purpose, since the TH gene is not expressed in them. The few colonies that contained cDNAs that were expressing genes in the PC12 clones but not in the rat liver clones were further screened to locate those that were associated with TH cDNA. This was done by growing the clones, isolating their mRNA, and translating it into protein using a rabbit reticulocyte system. The clones that yielded protein that reacted with TH antibody were identified as the ones that contained portions of the TH gene. One, designated pTH-1, was chosen as a probe to be used in a search for others. It contained only a portion of the nucleic acid sequence that codes for TH, but could be used to identify other bacterial colonies that contained more complete sequences.

A cDNA that contained the full sequence was finally obtained by further screening.

Using reverse transcriptase (see Figure 1.19), cDNAs were prepared from a mixture of polyadenylated mRNAs isolated from the PC12 cell line (Grima *et al.*, 1985). These cDNAs were then inserted into plasmids by oligo (dG·dC) tailing as shown in Figure 1.19 so they could then be inserted into bacteria and amplified into a cDNA library. Approximately 50,000 different recombinant bacterial clones were produced in the initial mixture or library. Separate colonies were then grown by plating the bacteria thinly enough in a culture dish so that each resulting colony represented a single bacterial clone. From this, the correct clones expressing the TH gene were selected by the technique of *in situ* hybridization using the pTH-1 TH probe. This involved transferring a sample of the colonies to filter paper and treating them to expose the DNA. The double-stranded DNA of the plasmid was denatured as described in Section 1.5.2 to create single strands. The labeled probe was then added so it could cross-hybridize with the DNA strand from the plasmid carrying the TH gene. By this technique, it was possible to select the appropriate clones, which could then be grown in quantity. The cDNA inserts in the plasmids that coded for TH were then separated off by endonuclease digestion. Finally, the nucleotide sequence was determined by the method of Sanger *et al.* (1977). One clone, designated pTH-51, contained 1758 base pairs, including the complete coding sequence as well as the 3′ untranslated region. From this, the exact order of amino acids was deduced. The sequence gave a molecular weight of 55,900 for TH, which is slightly lower than the apparent molecular weight of 62,000 obtained following enzyme purification.

Knowledge of the DNA and mRNA coding sequences can provide great insight into the charge distribution of a protein, the nature of bridges that form after synthesis to provide a three-dimensional structure, the location of catalytic sites, the location of sites for activation of the enzyme, as, for example, by phosphorylation, and the location of binding sites that can later attach the enzyme to intracellular organelles. The techniques of computer-aided design and computer-aided manufacturing can even be applied to the analysis of three-dimensional protein structure to reveal how it can exert its various biological actions.

Using somewhat different techniques, J. F. Powell *et al.* (1984) assigned the human TH gene to chromosome 11. A marker for human TH cDNA was first produced. The previously available rat TH cDNA probe (Lamouroux *et al.*, 1982) was used to isolate the appropriate human mRNA by Northern blot (DNA–RNA hybridization) analysis of RNA mixtures. The rat TH cDNA probe hybridized with a single mRNA species identical in length to the TH mRNA previously found for rat. Since the rat probe proved satisfactory for initial screening of human mixtures, it was used to isolate the appropriate human cDNA from a library consisting of 6000 colonies made from a human mRNA mixture. The human cDNA was then converted into a probe by labeling. A panel of previously characterized human–mouse hybridoma cell lines that collectively had all, or nearly all, human chromosomes were then screened for cross-hybridization with the human TH cDNA probe. The appropriate human–mouse hybrid cell lines expressing the TH gene were then detected with this human TH cDNA probe by Southern blot analysis (DNA–DNA hybridization) (see Section 1.5.2). The cells that contained fragments of chromosome 11 were the ones that expressed the TH gene. This somewhat complicated methodology illustrates the bootstrap approach that is necessary in these early days of molecular genetics. Nevertheless, assignment of many genes to particular chromosomes has already been achieved by comparable approaches.

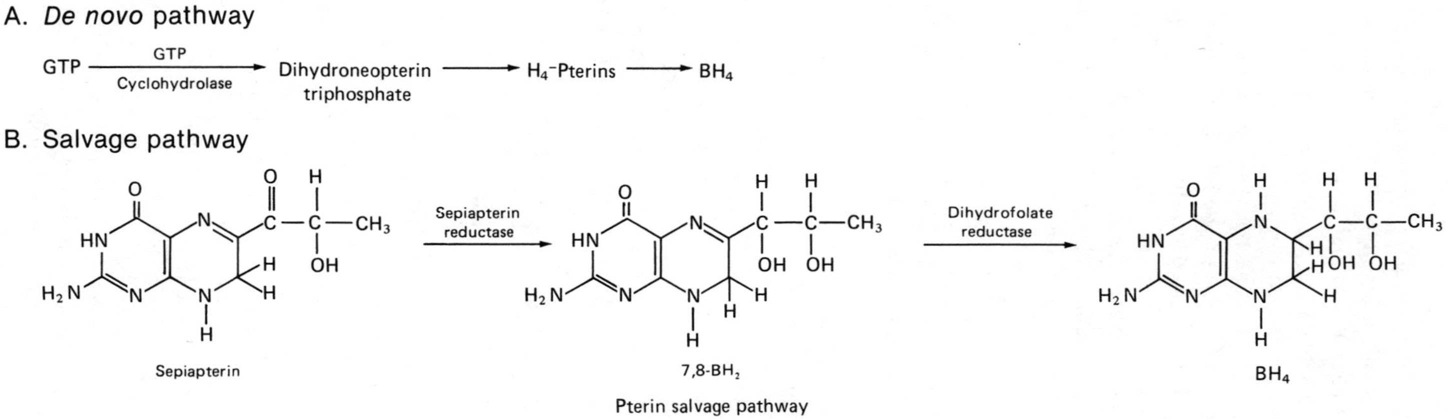
A. De novo pathway
GTP
GTP
Cyclohydrolase
Dihydroneopterin triphosphate
H4-Pterins
BH4
B. Salvage pathway
Sepiapterin
Sepiapterin reductase
7,8-BH2
Dihydrofolate reductase
BH4
Pterin salvage pathway

Figure 9.2. Schematic presentation of the *de novo* and salvage pathways for BH_4 synthesis.

Tetrahydrobiopterin. The cofactor for tyrosine hydroxylation is BH_4, which donates the necessary hydrogen atoms to the enzyme. It is not exclusive to this hydroxylation. As already mentioned, it participates in a number of comparable reactions, such as the hydroxylation of phenylalanine and tryptophan. It is also involved in the cleavage of plasmalogens, so it is widely distributed in the periphery. The synthesis of BH_4 is somewhat complicated since two pathways exist: the *de novo* pathway and the salvage pathway (Figure 9.2). Many body tissues including brain can readily synthesize BH_4 by the *de novo* pathway starting from GTP (Milstien and Kaufman, 1983; Nichol *et al.*, 1983). Cyclohydrolase converts GTP to dihydroneopterin triphosphate, which is converted to H_4-pterin intermediates on the way to BH_4 (G. K. Smith and Nichol, 1984). It is the principal route for producing BH_4. It is blocked by N-acetylserotonin (Milstien and Kaufman, 1983). The salvage pathway proceeds from sepiapterin through 7,8-BH_2 to BH_4.

If the salvage pathway involving sepiapterin and 7,8-BH_2 is the principal route for BH_4 synthesis, then a deficiency of BH_4 should be brought about by treating animals with the dihydrofolate reductase inhibitor, methotrexate, or with its lipophilic relative, metoprine. No such deficiency can be produced (Nichol *et al.*, 1983). GTP cyclohydrolase activity, on the other hand, is induced by reserpine administration, which increases BH_4 demand (Viveros *et al.*, 1981).

Congenital failure of BH_4 synthesis leads to a malignant form of phenylketonuria. About 2% of babies with hyperphenylanemia have this defect (Nixon *et al.*, 1980; Matalon, 1984; Dhondt, 1984) rather than the usual form of the disease, which results from a failure of liver phenylalanine hydroxylase. Infants with the BH_4-deficient form do not respond to a phenylalanine-low diet, presumably because hydroxylation of tyrosine and tryptophan and other reactions involving BH_4 are affected. There are at least two variants of BH_4 deficiency; one is caused by a deficiency in DHPR, the other by a "biopterin synthetase deficiency." Therapy includes supplements of BH_4 as well as L-dopa and 5-hydroxytryptophan (5-HTP). Reduced BH_4 levels have also been reported in a variety of neurological diseases, including Parkinson's disease, Alzheimer's disease, and Huntington's chorea, but there is no evidence that this is the primary lesion (Leeming *et al.*, 1981).

Dopa Decarboxylase. DDC (EC 4.1.1.28) is the second enzyme in the catecholamine synthetic chain (see Figure 9.1). It is the terminal synthetic enzyme in dopaminergic neurons. The proper designation is L-aromatic amino acid decarboxylase. It is a pyridoxal-dependent enzyme, and since it catalyzes the conversion of L-dopa to DA, it is also present in all catecholamine neurons. But it also has many other functions. It is responsible for the decarboxylation of 5-HTP to 5-HT (see Section 10.2.1) and is therefore present in 5-HT neurons. Furthermore, it apparently occurs in neurons that are neither catecholaminergic nor serotonergic. These have been described as D1–D14 (see Section 9.4.1). These neurons may produce neurotransmitters directly from amino acid substrates such as tyrosine, phenylalanine, or tryptophan (Jaeger *et al.*, 1984). In the periphery, DDC is found in widely disbursed endocrine cells that synthesize and secrete small polypeptide hormones (Pearse and Takor Takor, 1979), and it occurs in a virulent small cell carcinoma of the lung.

DDC was originally purified from pig kidney, but has now been purified from species as diverse as *Drosophila* (W. C. Clark *et al.*, 1978) and human (Maneckjee and

Baylin, 1983). The structure of subunits from these sources is virtually identical, suggesting a high degree of conservation of the molecule across a wide range of species. Maneckjee and Baylin (1983) used the enzyme-activated, irreversible inhibitor monofluoromethyldopa as a method of detecting which subunit of the enzyme was catalytically active. They were able to show that this inhibitor binds exclusively to a subunit of molecular weight 50,000. The native human enzyme behaves as a dimer of this monomer subunit. This is a size similar to that previously proposed for DDC from both *Drosophila* (W. C. Clark *et al.*, 1978) and hog kidney (Christenson *et al.*, 1972).

Many antibody preparations to the purified enzyme have been reported, and the immunohistochemical results with these preparations are described in Section 9.4.

Dopamine-β-hydroxylase (Dopamine-β-monooxygenase). Noradrenergic neurons are characterized by the presence of the copper-containing enzyme DBH (EC 1.14.17.1), which hydroxylates the ethylamine side chain of DA (see Figure 9.1). It is also present in adrenaline neurons. There are two copper atoms per catalytic subunit (Klinman *et al.*, 1984). Valence changes in copper evidently play an essential part in the activity of the enzyme, because copper-chelating agents inhibit its activity. One of the problems in measuring DBH activity in brain and other tissues is the presence of endogenous inhibitors that have this copper-chelating property. Thus, to assay the enzyme, it is necessary either to purify it partially or to add sulfhydryl agents that neutralize the endogenous inhibitors. Copper chelators such as diethyl dithiocarbamate are powerful inhibitors.

During the oxidation, ascorbate reduces the cupric copper of the enzyme, which is then oxidized when the reduced enzyme reacts with DA and oxygen. Thus, the enzyme-bound copper undergoes cyclic reduction and oxidation during the reaction.

The enzyme also catalyzes the side-chain hydroxylation of several amines that are structurally related to dopamine, such as α-methyl-DA, *N*-methyl-*p*-tyramine, α-methyl-*p*-tyramine, α-methyl-*m*-tyramine, *p*-hydroxy-*N*-methylamphetamine, and mescaline.

DBH has been purified to homogeneity by several methods. S. Friedman and Kaufman (1965) originally estimated the molecular weight of their purified beef adrenal DBH to be 290,000. It has been determined more recently that rat adrenal DBH is a tetrameric glycoprotein, each unit of which has a molecular weight of 88,000 (Okuno and Fujisawa, 1984). The specific activity of this preparation is 1500 nmoles/min per mg protein. Monoclonal antibodies have been prepared and coupled to Sepharose to form a purification column.

A clone that codes for a rat cDNA copy of a portion of the DBH mRNA has been isolated (O'Malley *et al.*, 1983).

One of the striking features of DBH is its association with storage vesicles. This means that, in noradrenergic neurons, the final step in the synthesis takes place in the vesicles or on their surface. In subcellular fractionation studies, DBH has proved to be a reliable vesicle marker, and one of the consequences of NA release in the periphery is a simultaneous release of DBH from the vesicles into the extracellular compartment. From there, it is cleared to the serum.

Phenylethanolamine-*N*-methyl-transferase. PNMT (EC 2.1.1.28), the enzyme that synthesizes adrenaline from its precursor NA, is a soluble enzyme that requires *S*-adenosylmethionine to carry out the reaction:

NA + *S*-adenosylmethionine → adrenaline + *S*-adenosylhomocysteine

It occurs only in adrenaline neurons and in adrenal medullary chromaffin cells. Unlike its precursor enzyme, DBH, it is not associated with vesicles. The enzyme has been purified to homogeneity (Connett and Kirshner, 1970; Joh and Goldstein, 1973). Its molecular weight is 40,000, but it also exists in two higher-molecular-weight forms of approximately 80,000 and 160,000. It will methylate only phenylethylamines with a hydroxyl group. L-NA is the most active substrate with the lowest K_m, indicating that it is much preferred over most other substrates such as normetanephrine, phenylethanolamine, and octopamine. Inhibition by phenethylamines and amines related to tranylcypromine has been demonstrated. Antibodies have been prepared against the purified enzyme and have been employed in immunohistochemistry as a definitive marker for adrenaline neurons [see Figure 9.9 (Section 9.4.3)].

9.2.2. Catabolism of Catecholamines

We have described the synthetic sequence by which catecholamines are formed. We now consider the methods by which they are catabolized. Two enzymes are involved: monoamine oxidase (MAO) and catechol-*O*-methyltransferase (COMT). MAO is a mitochondrial enzyme that acts both pre- and postsynaptically. COMT is attached to external cell membranes and acts only postsynaptically. Numerous metabolites are produced by the action of these enzymes on the three catecholamines, as shown in Figure 9.3. COMT action alone produces 3-methoxytyramine from DA, normetanephrine from NA, and metanephrine from adrenaline. MAO action alone produces dihydroxyphenylacetic acid (oxidation) from DA, and dihydroxymandelic acid (oxidation) or dihydroxyphenylethylene glycol (DHPG) (reduction) from NA and adrenaline. Action of MAO and COMT, apparently regardless of their order, produces 3-methoxy-4-hydroxyphenylacetic acid [(HVA) homovanillic acid] from DA and both 3-methoxy-4-hydroxymandelic acid [(VMA) vanillylmandelic acid] (MAO oxidation) and 3-methoxy-4-hydroxyphenylethylene glycol (MHPG) (MAO reduction) from NA and adrenaline.

Monoamine Oxidase. The enzyme was first described by Hare (1928) and was called "amine oxidase" by Richter (1937). The term "monoamine oxidase" was adopted later to distinguish the enzyme from others that attack diamines, such as histamine and spermine. Mitochondrial MAO is exclusively located on the surface of mitochondria. It is particularly abundant in the liver, kidney, and intestine, in addition to the central nervous system (CNS). It oxidizes a wide variety of aromatic amines, including the catecholamines, 5-HT, tyramine, phenylethylamine, and others.

Mitochondrial MAO should not be confused with plasma amine oxidase, which catalyzes the deamination of benzylamine, kynuramine, spermine, and spermidine. This latter type of amine oxidase has also been found in connective tissues, such as the skin, bone, aorta, and dental pulp, and is supposed to participate in the cross-linking reactions of collagen.

Despite extensive study, mitochondrial MAO remains an enigmatic enzyme. J. P. Johnston (1968) originally reported two forms of MAO in homogenates of rat brain. He designated them A and B, with enzyme A being inhibited by clorgyline. Inhibitors of enzyme B were subsequently found, the most effective of which is deprenyl.

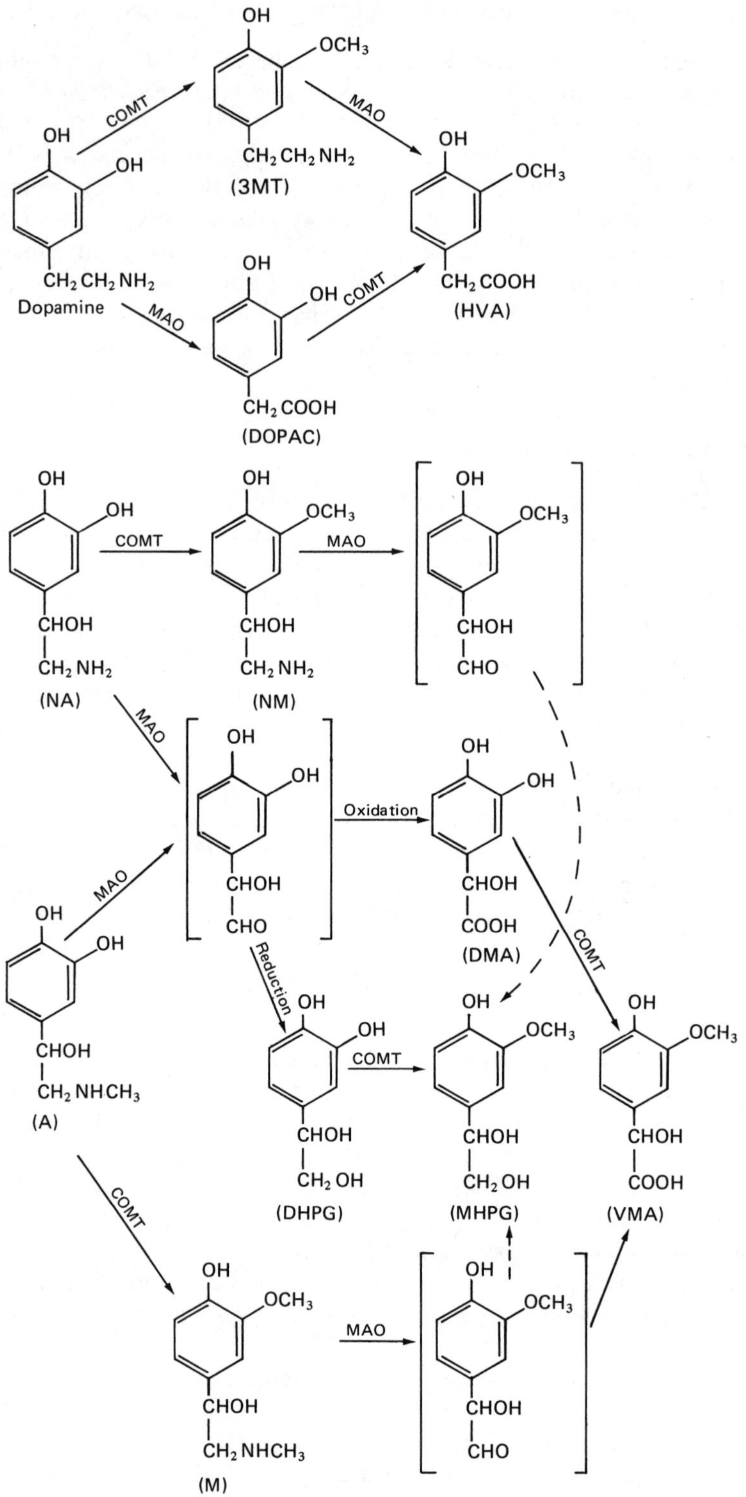
COMT
MAO
Dopamine
(3MT)
(HVA)
(DOPAC)
(NA)
(NM)
Oxidation
Reduction
(DMA)
(A)
(DHPG)
(MHPG)
(VMA)
(M)

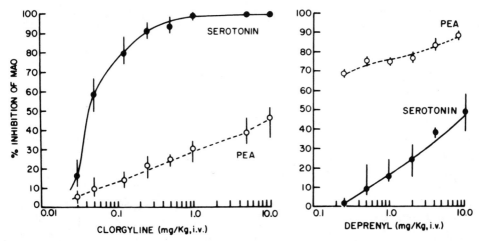

Figure 9.4. Blockade of rat brain MAO activity by increasing doses of clorgyline or deprenyl. Animals were killed 2 hr after injection of the drugs, and enzyme activity was assayed with 5-HT or β-phenylethylamine (PEA) as substrate. 5-HT is a substrate for type A MAO and is preferentially blocked by clorgyline, while PEA is a substrate for type B MAO and is preferentially blocked by deprenyl. From Neff *et al.* (1974).

Figure 9.4 indicates the manner in which MAO is inhibited by clorgyline or deprenyl using 5-HT (type A substrate) and phenylethylamine (type B substrate) as indicators of enzyme activity. Two different curves are seen in the figure, indicating the presence of two different forms of the enzyme. The B form is concentrated in erythrocytes and the A form in placenta.

There are three theories as to the origin of these two forms: (1) They are derived from a single gene with a different lipid environment of the mitochondrial membrane acting as a modifier; (2) they are derived from a single gene that is modified by glycosylation or proteolytic cleavage in the cytosol; or (3) they are derived from two different genes (Powell, 1984).

Cawthon *et al.* (1981) isolated the A form from human placental trophoblasts, the B form from human erythrocytes, and both forms from human cultured skin fibroblasts. An apparent molecular weight of 63,000 was obtained for MAO-A and 60,000 for MAO-B. Following limited proteolysis, different peptide patterns were obtained, establishing that these two forms are different molecular types.

Denney *et al.* (1982) developed a monoclonal antibody to type B MAO from human placenta. They used the antibody to make an immunoaffinity column and then isolated type B MAO from a mixture of A and B obtained from human liver. MAO-B had a molecular weight of 59,000. They also developed a monoclonal antibody to type A from human erythrocytes and used the two monoclonal antibodies to screen monkey brain for distinct neuronal populations using either the A or the B form.

←

Figure 9.3. Schematic presentation of the breakdown of catecholamines by the actions of MAO and COMT. Compounds: (A) adrenaline; (DHPG) dihydroxyphenylethylene glycol; (DMA) dihydroxymandelic acid; (DOPAC) dihydroxyphenylacetic acid; (HVA) homovanillic acid; (M) metanephrine; (MHPG) 3-methoxy-4-hydroxyphenylethylene glycol; (3MT) 3-methoxytyramine; (NA) noradrenaline; (NM) normetanephrine; (VMA) vanillylmandelic acid (3-methoxy-4-hydroxymandelic acid).

In the monkey, MAO-A-positive neurons were primarily those of the catecholamine type (see Section 9.4), while MAO-B-positive neurons were primarily those of the 5-HT type (see Section 10.4) (Westlund *et al.*, 1985). Approximately 90% of the MAO in the superior cervical ganglion, where noradrenergic neurons predominate, is type A, while in the pineal gland, where melatonin-producing neurons predominate, it is mostly type B.

The existence of such preferential presynaptic localization does not mean that MAO-A is concerned exclusively, or even primarily, with catecholamine metabolism and MAO-B with 5-HT metabolism. MAO also exists in postsynaptic neurons and in glia. The B form predominates in glia. The physiological effects of released monoamines may be monitored more closely by MAO at these other sites.

MAO does not convert its various substrate amines all the way to the corresponding acids, although such conversion appears to be the usual end result of its action in the periphery. The initial oxidation goes only as far as the corresponding aldehyde. The aldehyde can then be oxidized further to the acid by aldehyde oxidase or reduced to the alcohol by alcohol dehydrogenase. In brain, the aldehyde formed from DA appears to be almost exclusively oxidized to the acid, but that formed from NA seems to be predominantly reduced. Thus, major metabolites of brain NA are DHPG and MHPG. In the primate brain, MHPG remains in the free state, but it is sulfated in many species. Thus, MHPG and MHPG·SO$_4$ are useful indices of central NA turnover, although some may come from peripheral sources (Li *et al.*, 1984).

Many compounds have been described as MAO inhibitors. Most inhibit both the A and B forms and are therefore broad-spectrum. Hydrazine-type MAO inhibitors such as iproniazid, and acetylenic-type agents such as pargyline, clorgyline, and deprenyl, are noncompetitive and irreversible. Other types, such as the cyclopropylamine tranylcypromine, are competitive and reversible. The various types of MAO inhibitors are discussed further in Sections 9.5.2 and 15.5.2.

Catechol-*O*-methyltransferase.   COMT (EC 2.1.1.6) was discovered and described by Axelrod and Tomchek (1958) following the finding of M. D. Armstrong *et al.* (1957) of VMA in urine. COMT is the enzyme that methylates catecholamines (see Figure 9.3), as well as their corresponding alcohols and acids. The enzyme catalyzes the transfer of the methyl group of *S*-adenosylmethionine to the *m*-hydroxy group of catechols. It is distributed widely in many tissues, including erythrocytes. It is particularly abundant in liver and kidney and in astrocytomal cells. This latter localization suggests that the enzyme is not predominantly associated with presynaptic elements. It is generally believed that the COMT near synapses is either on glial cells or in or on the surface of dendritic elements of postsynaptic neurons. Soluble and membrane-bound forms of the enzyme exist (Jeffery and Roth, 1985), which would be compatible with either or both locations.

O-Methylation is the major metabolic pathway for all extracellular catecholamines. It is assumed, therefore, that much released DA is first converted to 3-methoxytyramine before being attacked by MAO to form HVA (see Figure 9.3). HVA, 3-methoxytyramine, and O-methyl derivatives of DA and other catecholamines have only a tiny fraction of the physiological activity of the catechols themselves. Evidently, the 3-methoxy group greatly reduces the capability of binding to receptor sites, which accounts for the deactivating capability of the enzyme.

It is not possible to determine from measuring the levels of HVA, the dominant metabolite of DA in tissue and urine, whether MAO or COMT acts first on the neurotransmitter. It is of some importance to determine this, because DA metabolized presynaptically by MAO would have no physiological action.

Some COMT inhibitors are known, but they do not have the physiological effects of MAO inhibitors (see Section 9.5.2).

Membrane-bound and soluble COMT display comparable kinetics and other properties and are therefore thought to have a highly similar, if not identical, molecular structure. The molecular weight of human brain COMT is 27,500, with a pI value near pH 5.0. The dead-end inhibitor tropolone shows competitive inhibition with dopamine as substrate, but noncompetitive inhibition with S-adenosylmethionine. These data suggest that the reaction is an ordered one, with S-adenosylmethionine being the leading substrate. The K_m for DA is 290 μM and that for S-adenosylmethionine is 9 μM.

9.3. Storage, Release, and Turnover of Catecholamines

9.3.1. Storage of Catecholamines

The discovery that catecholamines and other neurotransmitters are stored within highly specialized granular particles in nerve endings represented a great conceptual advance in the 1950s (see Section 8.1). These granules can be separated by subcellular fractionation studies, and their makeup can then be studied by a variety of techniques. Catecholamine and 5-HT vesicles differ from many others in that reserpine treatment makes them incapable of storing their amines.

We shall first consider the storage and release of NA in peripheral sympathetic neurons because they serve as a useful and well-studied model for what may take place centrally.

The concentration of catecholamines in the specialized granules has been estimated to be 0.6 M—more than twice the osmolarity of mammalian body fluid and near the solubility limits. The means by which this storage is achieved is therefore of some considerable interest. Granules have been found to contain adenine nucleotides, chiefly as ATP, in an approximate ratio of 4 molecules of catecholamine to 1 of the adenine nucleotide. The composition of the granules is roughly as follows: water, 68.5%, catecholamine, 6.7%; adenine nucleotides, 4.5%; protein, 11.5%; lipids and other molecules, 7.8%. The principal protein in granules from adrenal chromaffin cells and splenic nerve is chromogranin (Somogyi et al., 1984), which is also believed to be involved in the storage process. The granules possess an outer limiting membrane that must play a role in taking up and storing NA, which is the principal catecholamine in these structures. Obviously, NA cannot be in free solution inside the vesicle, or the vesicle would swell and burst due to hyperosmotic conditions. Therefore, some kind of a salt linkage between the anionic phosphate groups of ATP and the amine group of NA must exist. The granules contain an Mg^{2+}- and Ca^{2+}-dependent ATPase.

It is generally believed that the vesicles or components thereof are formed in the neuronal cell body and are subsequently transported to the nerve terminal region by the process of axoplasmic flow [see Figure 1.16 (Section 1.2.4)]. Alternatively, it has been suggested that vesicles are formed from the outer membrane of the nerve terminal by the process of pinocytosis, but electron microscopic studies on the sciatic nerve just proximal to a ligated area reveal a substantial increase of granular elements, which favors the concept of transport. Additional evidence is that, following reserpine administration, NA begins to reappear in cell bodies, but not in nerve endings, after a few hours. This implies a restoration of binding capacity through the synthesis of new proteins associated with vesicles. This is consistent with the observation of decreased transport of DA in the nigrostriatal tract immediately following reserpine treatment, but greatly enhanced transport 24 hr later, when new binding material has been synthesized.

Noradrenergic storage vesicles differ from dopaminergic ones in possessing DBH. Most of the enzyme appears to be associated with the limiting membrane, because, when the vesicles are subjected to membrane-disruptive procedures such as osmotic shock and then recentrifuged, only 10–20% of the DBH is solubilized. This soluble component can evidently also be released on stimulation. DBH is normally present in small quantities in the bloodstream, and the amount of DBH has been found to increase with nerve stimulation (Axelrod, 1972). This is regarded not only as evidence of the association of DBH with NA storage vesicles, but also as evidence for the concept of exocytotic release of neurotransmitters. If NA were released from the synaptosmal cytoplasm, with storage vesicles merely acting to maintain an appropriate intrasynaptic concentration of NA, then DBH would not be released on nerve stimulation unless it too were coming from the cytoplasm. But if this were the case, then TH, DDC, and other soluble enzymes should also appear in the perfusates—an appearance that has never been observed.

9.3.2. Release of Catecholamines

The release mechanisms for NA appear to be comparable to those for DA, but they have been more carefully studied. Isolated organs such as the spleen and intestine have been extensively investigated through preloading of the nerve terminals by prior intravenous administration of labeled NA. Following stimulation or treatment with drugs, labeled material is released into the perfusate, allowing quantitative estimation of the effects. The most striking aspect of sympathetic nerve stimulation is the relatively slow rate of excitation that is required to produce a significant increase in neurotransmitter release. Most tissues respond to changes in the frequency of stimulation in the range of 1–3 per sec, and maximum responses can usually be obtained with stimulus frequencies of less than 20 per sec. Using cat spleen, the highest amounts of NA release per stimulus were found at frequencies of less than 20 per sec. In the same preparation, the highest amounts of NA release per stimulus were found at frequencies of the order of 10 per sec and declined as the stimulation frequency increased to 30 per sec. With higher frequencies of stimulation, there was an overflow of NA into the venous circulation, but with low frequencies as much as 90% was inactivated by local processes. Nevertheless, the quantities of VMA and other metabolites of NA excreted in the urine suggest that the equivalent of 5–10 mg NA is consumed each day in humans. The majority is derived from sympathetic nerve endings, since the excretion of metabolites does not change appreciably after adrenalectomy (Iversen, 1967). Studies of central catecholamine neurons using implanted electrodes have repeatedly shown this same phenomenon of relatively slow neuronal discharge rates (see Sections 9.5.1 and 9.6.1).

It is clear that Ca^{2+} plays a highly significant role in release, because the amount of NA discharged on sympathetic nerve stimulation is considerably reduced if the Ca^{2+} content of the perfusing medium is lowered. It is analogous to the release of ACh from cholinergic nerve endings, where depolarization of the presynaptic terminal is accompanied by the attachment of vesicles to the inner surface of the presynaptic fibers [see Figure 3.11 (Section 3.2.9) and Figure 5.5 (Section 5.4.2)]. The neurotransmitter packets attached to these sites discharge their contents to the exterior of the synaptic cleft, with Ca^{2+} operating as the essential cofactor. The impulse arriving at the nerve ending produces the necessary alteration in permeability to Ca^{2+}, and the influx is the main stimulus responsible for catecholamine mobilization and secretion by the process of exocytosis. ATP, chromogranin, and a little DBH are also extruded into the synaptic cleft.

Inactivation is primarily by reuptake of the NA into the nerve endings. Whether or not the protein that is also released by this process is recaptured again by the nerve ending is undetermined. If it is not recaptured, then the process of axoplasmic flow must be extremely vigorous to replace the lost material.

While it can be presumed that similar processes take place in the CNS, it must be remembered that duplicating the kinds of stimulation and perfusion experiments that can be easily performed in the periphery is extremely difficult with brain tissue. It has been possible, however, to perform some crude experiments using perfusates, cortical cups, brain slices, and push–pull cannulae, and some interesting data of a comparable nature have been obtained.

We can now turn to the rate at which central catecholamines are actually turned over.

9.3.3. Reuptake of Synaptically Released Catecholamines

The principal method for inactivating released catecholamines is reuptake into the nerve endings. Data indicate that as much as 80% of the NA released by stimulation may be inactivated by the process and recaptured for future use. For example, inhibition of uptake with cocaine leads to an approximate 4-fold increase in the outflow of NA and its metabolites from isolated tissues following sympathetic nerve stimulation (Iversen, 1973). This is in sharp contrast to ACh, the activity of which is terminated by acetylcholinesterase breakdown of the neurotransmitter. Of course, MAO and COMT both exist in postsynaptic neurons and glial cells, so that catecholamine molecules that diffuse away from the presynaptic uptake pump will be destroyed. This reuptake process was first described by Axelrod (1971) and his colleagues, who discovered that, when labeled NA was injected intravenously, it was accumulated in synaptic vesicles of sympathetic nerve endings. The details of the reuptake process were soon worked out in the periphery and in the brain, through both *in vivo* studies and *in vitro* experiments using synaptosomal fractions.

This phenomenon, which was referred to briefly in Section 5.1.2, has been best studied in the catecholamines, and the general properties that have been found may have application to some other Type I (amino acid) and Type II (amine) neurotransmitters, although Type III (peptide) neurotransmitters appear to be excluded. The pump, or uptake mechanism, carries the catecholamine against a substantial concentration gradient from the outside to the inside of the catecholamine cell surface membrane. The pumping process obeys Michaelis–Menten kinetics and appears to be specifically coupled to certain energy-dependent mechanisms. The process is easily saturated and has a K_m in the range of 0.2–1 μM in most rat tissues. It is inhibited competitively by a number of pharmacological agents (see Sections 9.5.2 and 15.5.2), as well as noncompetitively by agents such as ouabain that inhibit membrane $[Na^+–K^+]$-ATPase. Separate pumps exist for DA and NA, even though the general process is the same. For example, desipramine is 1000 times less potent at inhibiting DA uptake than NA uptake. On the other hand, DA uptake, but not NA uptake, is strongly inhibited by benztropine. The mechanism obviously has an extremely important bearing on the apparent turnover rates of catecholamines, since there is such substantial re-use of released neurotransmitter.

9.3.4. Turnover of Catecholamines

Dopamine Turnover. Several methods have been used to determine the rate at which DA is turned over within a given tissue. The turnover rate, of course, is considerably below the maximum synthetic rate of a tissue because excess capacity is always present. The turnover rate also does not necessarily reflect the speed at which some of the molecules are metabolized. It is a well-established principle of the nervous system that such rates can vary tremendously and that the most recently synthesized transmitter is usually the most rapidly metabolized. Different methods of measurement of turnover rates do not always yield the same answers, perhaps reflecting the fact that different molecules of the same transmitter are metabolized at different rates. Nevertheless, turnover rates do give some indication of the average length of survival of a neurotransmitter molecule and also establish the effects of various drugs on neurotransmitter utilization. The following methods have been used:

1. Administration of a labeled pulse of DA intraventricularly or directly into tissues followed by measurement of the rate of decline of DA. This method suggests for DA a half-life in rat striatum of approximately 1.5 hr. The weakness of this method is that the DA cannot be entirely limited to dopaminergic neurons, since some of it will be taken up by noradrenergic neurons. Furthermore, the labeled DA may not enter into an "average" pool, but may preferentially find its way into a storage pool different from the "newly synthesized" pool.
2. Measurement of the rate of appearance or disappearance of labeled DA after administration of a pulse of labeled tyrosine or L-dopa. If the rate of appearance following tyrosine is to be used, a decarboxylase inhibitor is often administered and only L-dopa formation is measured. This, of course, does not distinguish between DA and NA except by selection of dopaminergic- or noradrenergic-rich areas of brain. Both synthesis and disappearance methods suggest for striatal DA a half-life of 2.5–4 hr.
3. Measurement of the disappearance of endogenous DA following the administration of α-methyl-p-tyrosine or other potent inhibitors of TH. As with method (2), there may not be an adequate distinction between DA and NA neurons, and the drug used may influence the metabolism of either amine. Nevertheless, the half-life figures obtained by this method are generally comparable to those obtained by other methods, being approximately 4 hr for striatal DA.
4. Measurement of the tissue or fluid levels of DA metabolites. The appearance of labeled HVA or labeled 3,4-dihydroxyphenylacetic acid (DOPAC), or of both, following a pulse of labeled DA should be a reflection of the amount of DA consumed. This method generally indicates turnover times of about 2 hr in the striatum. Diffusion of the metabolites is a factor to consider, as is possible entrance of the pulse of DA into a special pool.

In all methods utilizing radioactive material, the cold pool sizes should be measured. This is sometimes neglected, which may account for some of the variability in estimates of DA half-life obtained by various laboratories. Nevertheless, all the estimates are of the same order of magnitude (1.5–4 hr), and there is general agreement on the probability that DA turnover is somewhat faster than NA turnover (E. Costa, 1970; Iversen and Glowinski, 1966; Persson, 1970).

Noradrenaline Turnover. The same techniques used for measuring the turnover of DA have also been used for NA. The data are more extensive for NA and are generally considered to be more precise. Using these techniques, Iversen (1967) has reported the mean half-life of NA in selected regions of brain to be 3.7 hr for the hypothalamus, 3.6 hr for the medulla, 3.0 hr for the hippocampus, 2.7 hr for the cortex, and 2.2 hr for the cerebellum. This compares with turnover times reported by E. Costa (1970) of 14.3 hr for NA in heart, 8.3 hr for brain, 5.3 hr for hypothalamus, and 2.1 hr for the superior cervical ganglion.

Adrenaline Turnover. Little is yet known about central adrenaline turnover rates. It is a minor catecholamine in mammalian brain.

9.3.5. *In Vivo* Control of Catecholamine Synthesis

The method by which control of catecholamine synthesis is regulated is unknown, but control occurs at the tyrosine hydroxylation step. Spano and Neff (1972) estimated the rate of formation of DA from tyrosine in the guinea pig caudate nucleus *in vivo* to be roughly 32 nmoles/g per hr. By contrast, the maximum velocity (V_{max}) levels of TH *in vitro* for this same area have been reported to be as high as 1200 (Cicero *et al.*, 1972), depending on the assay conditions used.

Obviously, some mechanism other than the amount of TH protein must govern the minute-to-minute conversion, even though general control, at least for DA, is exerted at that level. At one time, it was believed that feedback inhibition by the catecholamines themselves was responsible, since they are inhibitory in *in vitro* homogenates. But relatively high concentrations are required to achieve effective inhibition. On the other hand, relatively low concentrations of the DA agonist apomorphine will inhibit TH activity in striatal slices. This is powerful evidence for the presence of DA autoreceptors on nerve endings, since the nigral cell bodies have been eliminated in the procedure. Further evidence is that the inhibition is partially reversed by compounds that block presynaptic dopaminergic receptors. This evidence suggests that control could be exerted by a feedback inhibitory action of released catecholamines on autoreceptors, located on the external surface of nerve endings, while activation could be achieved by nerve stimulation. How could such a system operate? One way would be through stimulation or inhibition of kinases that activate TH by phosphorylation (M. Goldstein, 1984; Vulliet *et al.*, 1984).

Using DEAE–cellulose chromatography, Sze *et al.* (1983) separated from the striatum two forms of TH that were interpreted as being the phosphorylated and nonphosphorylated forms. The percentage of the phosphorylated form is increased in rats sacrificed 90 min after treatment with haloperidol.

Although the quantity of TH protein is normally far in excess of short-term requirements, it nevertheless can be increased by nerve stimulation (Zigmond *et al.*, 1980) or receptor blockade. The time course is measured in days rather than minutes because TH protein must be synthesized in the cytosol, where there are polysomes, and transported by axoplasmic flow to nerve endings, where most of the catecholamine synthesis takes place. Drugs that block catecholaminergic receptor sites, such as chlorpromazine, or that deplete the nerve endings, such as reserpine, induce increased TH protein (Nyback, 1972; D. S. Segal *et al.*, 1971).

These data bear only indirectly on the more precise question of exactly how the regulatory mechanisms work to maintain the levels of each catecholamine at a constant level while yielding the necessary amount for minute-to-minute activation of postsynaptic receptors. Such details will have to be supplied by future research.

9.4. Anatomy of Catecholamine Neurons

9.4.1. Methods and General Anatomical Considerations

Information concerning the anatomical distribution of the catecholamine neurons and their pathways has been accumulated using a variety of techniques. The complementary nature of these techniques has provided extensive confirmation of basic data and has extended the range of information considerably beyond the limits of any single method. The techniques are:

1. Formaldehyde-induced fluorescence histochemistry using the Falck–Hillarp technique. This classic method detects dopamine (DA) and noradrenaline (NA) cell bodies, but does not distinguish between the two. It is not sufficiently sensitive to pick up adrenaline neurons or even some of the less intense DA groups such as the A15.
2. Glyoxylic acid histochemistry. This technique (Lindvall *et al.*, 1973) is considerably more sensitive than the Falck–Hillarp technique, although it also fails to pick up all cell groups and is unable to distinguish among the various catecholamines.
3. Tyrosine hydroxylase (TH) immunohistochemistry. This technique stains all catecholamine cell groups, but cannot distinguish among the various types. A powerful antiserum (Hökfelt *et al.*, 1984) has permitted more detailed identification of the various catecholamine cell groups than has been possible with histofluorescence techniques.
4. Aromatic L-amino acid decarboxylase (AADC) immunohistochemistry. This method picks up both catecholamine and serotonin (5-HT) cell bodies. The nonspecificity makes it useful only as a confirmatory technique. Moreover, Jaeger *et al.* (1984) have defined 14 AADC-positive cell groups, which they describe as D1–D14, that contain neither catecholamines nor 5-HT.
5. Dopamine-β-hydroxylase (DBH) immunohistochemistry. This method is capable of picking up both NA and adrenaline cell bodies. It has proved to be invaluable in distinguishing between DA and NA cell groups.
6. Phenylethanolamine-*N*-methyltransferase (PNMT). This method is definitive for adrenaline cells, as opposed to DA and NA cells. It has, however, picked up cells that do not stain positively for TH and DBH. This may reflect a relatively high sensitivity of the antibody preparation or in some cases a false-positive result (Hökfelt *et al.*, 1984).

TABLE 9.1
Catecholamine Cell Groups and Fiber Tracts

Alphanumeric designation	Anatomical location	Projection system
Noradrenaline		
A1	Ventrolateral reticular medulla	Ascending and descending fibers
A2	Dorsomedial medulla	Ascending and descending fibers
A3	Lateral medullary tegmentum	Ascending and descending fibers
A4	Locus coeruleus	Ascending and descending fibers
A5	Subcoeruleus	Ascending and descending fibers
A6	Locus coeruleus	Ascending and descending fibers
A7	Pontine lateral tegmentum	Ascending and descending fibers
Dopamine		
A8	Retrorubral	Nigrostriatal
A9	Substantia nigra	Nigrostriatal
A10	Ventral tegmentum	Mesolimbic–mesocortical
A11	Periventricular gray	Diencephalon and spinal cord
A12	Arcuate nucleus	Tuberoinfundibular
A13	Dorsal hypothalamus	Incertohypothalamic
A14	Rostral periventricular	Incertohypothalamic
A15	Dorsal preoptic	Local diencephalic
A16	Periglomerular	Olfactory bulb interneurons
Adrenaline		
C1	Ventrolateral reticular medulla	Ascending and descending fibers
C2	Dorsomedial medulla	Medullary fibers

Fiber tracts

1. Medial forebrain bundle
2. Nigrostriatal tract
3. Dorsal catecholamine bundle
4. Ventral tegmental tract
5. Medullary catecholamine bundle

Lesion, axoplasmic transport, and pharmacological manipulation have all been used in conjunction with these six basic techniques to help define catecholamine cell bodies and their pathways and terminals.

These methods have revealed at least 18 distinct catecholamine cell groups in brain. They are listed in Table 9.1 by type of catecholamine, anatomical location of the cell group, and alphanumeric designation. Of these 18 groups, 9 are dopaminergic (A8–A16), 7 are noradrenergic (A1–A7), and 2 are adrenergic (C1–C2). In addition to these groups, there are retinal DA and adrenaline interneurons and some DA cells in the periphery, as well as the well-known peripheral noradrenergic neurons of sympathetic ganglia and the noradrenergic and adrenergic cells of the adrenal medulla.

The alphanumeric designation system is of historical origin. In their original cataloguing of aminergic neurons in brain using the Falck–Hillarp formaldehyde-induced fluorescence technique, Dahlstrom and Fuxe (1964a,b) distinguished 12 catecholamine cell groups, which they designated A1–A12 moving from caudal to rostral location in the brain stem. They also described 9 indoleamine cell groups that they designated B1–B9 (see Section 10.4). Further DA groups were later identified that were more rostrally

located; they were designated A13–A16. Finally, through immunohistochemical techniques, adrenaline cell groups were identified. These, however, were in the caudal brain
stem in the same general area as the A1 and A2 cell groups. To avoid confusion, they
were designated C1 and C2.

The projections of the cell groups intermingle and so are listed separately in Table
9.1. They are assembled into the following five major tracts, all of which are largely
uncrossed:

1. Medial forebrain bundle: carries major dopaminergic and noradrenergic fibers
   into the telencephalon.
2. Nigrostriatal tract: carries dopaminergic nigral efferents through the globus pallidus to the neostriatum.
3. Dorsal catecholamine bundle: carries noradrenergic projections of the locus coeruleus into the diencephalic area, where the tract turns ventrally to join the
   medial forebrain bundle.
4. Ventral tegmental tract: carries NA and adrenaline fibers from the medulla and
   pontine areas into the forebrain.
5. Medullary catecholamine bundle: a caudal extension of the ventral tegmental
   tract.

The actual numbers of catecholamine cells in most of the groups listed in Table 9.1
are quite modest. For rat brain, the total noradrenergic count (Groups A1–A7) has been
estimated to be about 5000 per side (Swanson and Hartman, 1975), all of which are
located in the pons and medulla. The largest single group is in the locus coeruleus (A4 and
A6). The locus coeruleus has about 1500 neurons on each side in the rat (Swanson, 1976).
Although some DA cells may be present in the A2 complex of the medulla, the vast
majority are in the three mesencephalic groups designated A8–A10. There are about
15,000–20,000 neurons on each side in this complex in the rat, about half of which are in
the substantia nigra (Guyenet and Crane, 1981; Hedreen and Chalmers, 1972; Swanson,
1982). For human brain, the catecholamine neuronal count is about 20 times higher in
young people and decreases with age. There are initially about 450,000 DA neurons in the
substantia nigra (P. L. McGeer *et al.,* 1976) and 30,000 neurons in the locus coeruleus on
each side. The entire diencephalic dopaminergic neuron population amounts to less than
one tenth that of the mesencephalon (Björklund and Lindvall, 1984). Similarly, the
adrenaline neuronal population forms only a small component of the medullary catecholamine population.

In summary, the total catecholamine neuronal population of rat brain is about 25,000
neurons on each side, of which roughly 5% are in the medulla, 15% in the pons, 70–75%
in the mesencephalon, and 5–10% in the diencephalon. About 80% of the catecholamine
cells are dopaminergic, less than 20% are noradrenergic, and only a very small percentage
are adrenergic.

An approximate idea of the relative density of nerve endings of each of the catecholamines in various brain areas can be formed from Table 9.2. This table gives the
values for DA, NA, adrenaline, and 5-HT in a variety of cortical, limbic, diencephalic,
midbrain, and hindbrain areas for adult dog (Mefford *et al.,* 1982). The amine levels
range from 1.6 to 11,930 ng/g tissue, depending on the monoamine and area of brain. In
general, the highest values are found for DA in the extrapyramidal and limbic systems.

TABLE 9.2
Catecholamine and Serotonin Levels in Different Areas of Dog Brain[a]

Brain area	Level (ng/g tissue)			
	Dopamine	Noradrenaline	Adrenaline	Serotonin
Cortex				
Gyrus rectus	97	313	4	443
Gyrus presylvius	21	138	2	158
Gyrus frontalis	49	111	4	152
Basal ganglia				
N. caudatus med.	11,551	68	—	621
Putamen	8,122	121	—	653
N. accumbens septi	7,631	498	56	1,803
Globus pallidus	1,552	178	26	1,578
Limbic system				
Gyrus cingulate	89	219	4	154
Area septalis	925	335	44	796
Hippocampus	19	249	2	301
Dorsal amygdala	738	468	19	557
Ventral amygdala	194	129	5	723
Hypothalamus				
Area preoptica	393	1,569	367	1,384
N. hypothal. dorsomed.	133	1,416	204	575
N. hypothal. ventromed.	249	2,451	447	663
N. post. hypothal.	101	836	178	665
N. lat. hypothal.	305	1,479	193	693
Corpus mammilare	25	248	30	129
Thalamus				
N. anterovent. thal.	45	138	13	197
N. med. dors. thal.	75	257	15	380
N. vent. lat. thal.	100	455	15	140
N. pulvinaris thal.	18	202	5	75
Subst. gris. cent. thal.	128	406	32	409
Corp. geniculati lat.	13	180	6	218
Corp. geniculati med.	28	328	3	103
N. habenulae	121	422	39	510
Brainstem				
Substantia nigra	833	338	11	1,391
N. retic. pontis oralis	27	545	28	255
N. retic. pontis caud.	74	1,069	74	310
N. retic. ventralis	53	448	27	218
N. centralis superior	351	1,035	93	1,995
N. dorsalis raphe	69	1,032	43	1,268
N. raphe	44	768	46	469
N. abducentis	118	432	13	301
N. praepositus	52	334	22	474
N. n. hypoglossi	80	497	16	543
N. olivaris inf.	50	587	55	152
N. n. oculomotorii	232	2,463	126	1,419
N. ruber	307	803	202	568
N. vestibularis sup.	50	375	27	153
Colliculus inferior	28	307	9	179
Colliculus superior	72	578	11	866
Subst. grisea cent.	127	1,045	117	795
N. tractus solitari	97	887	49	511
Cerebellum	6.5	43	3	47

[a]From Mefford et al. (1982).

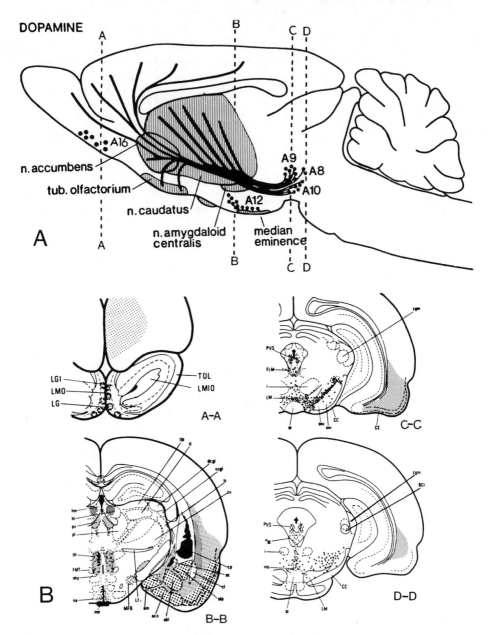

Figure 9.5. DA neurons and pathways in brain. (A) Sagittal section of rat brain showing the principal DA neuronal groups and major rostrally directed pathways. For arrangement of pathways relative to other amine groups, see Figure 10.4B (Section 10.4.1). (B) Four cross sections of brain at levels A–A, B–B, C–C and D–D as indicated in the sagittal diagram. Dots indicate the relative numbers of cell bodies. Abbreviations: (abl) nucleus amygdaloideus, basalis, pars lateralis; (ac) nucleus amygdaloideus centralis; (aco) nucleus amygdaloideus corticalis; (alp) nucleus amygdaloideus lateralis, pars posterior; (am) nucleus amygdaloideus medialis; (BCI) brachium colliculi inferioris; (CC) crus cerebri; (CE) cortex entorhinalis; (cgm) nucleus centralis corporis geniculati medialis; (cl) claustrum; (cp) nucleus caudatus putamen; (dcgl) nucleus dorsalis corporis geniculati lateralis; (FLM) fasciculus longitudinalis medialis; (FMT) fasciculus mammillothalamicus; (hi) hippocampus; (hm) nucleus habenulae medialis; (IH) incertohypothalamic DA terminal system; (ip) nucleus interpeduncularis;

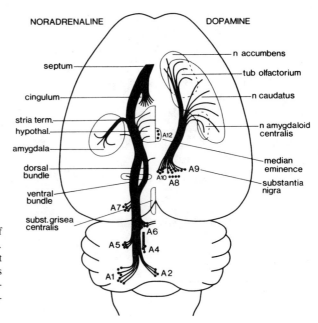

Figure 9.6. Horizontal projections of the ascending DA and NA pathways. Terminal fields in the cortex are not shown. For arrangement of pathways relative to other amine groups, see Figure 10.4B (Section 10.4.1). From Ungerstedt (1971).

The next highest values are found for NA in the hypothalamus and certain brain stem areas. Values for adrenaline are generally quite low except in some hypothalamic areas.

We will now consider the detailed anatomy of each of the catecholamine cell types.

9.4.2. Anatomical Distribution of Dopamine

The distribution of the major dopaminergic groups in rat brain is shown in sagittal section in Figure 9.5A and as a series of cross sections in Figure 9.5B; their projections are also shown in horizontal section in Figure 9.6 [for general reviews of the anatomy, see Björklund and Lindvall (1984) and Hökfelt *et al.* (1984)].

Mesencephalic Dopamine Neurons. The main mass of DA neurons is in the mesencephalon in the retrorubral (A8), substantia nigra (A9), and ventral tegmental (A10) groups. They are not really distinct entities, but rather form a continuum of cells. The A8 cells can be thought of as a caudal extension of the substantia nigra. The A9 group is the pars compacta of the substantia nigra, which forms a cap over the pars reticulata. Some DA-positive cells are also found in the pars reticulata and pars lateralis of that nucleus.

(LG) lamina glomerulosa bulbi olfactorii; (LGI) lamina granularis interna bulbi olfactorii; (LM) lemniscus medialis; (LMIO) lamina medullaris interna bulbi olfactorii; (LMO) lamina molecularis bulbi olfactorii; (me) median eminence; (MFB) fasciculus medialis prosencephali (medial forebrain bundle); (n$_x$) nucleus originis nervi vagi; (na) nucleus arcuatus; (nhp) nucleus hypothalamicus posterior; (nlo) nucleus linearis oralis; (pf) nucleus parafascicularis; (pv) nucleus periventricularis thalami; (PVS) periventricular DA fiber system; (r) nucleus ruber; (snc) substantia nigra, pars compacta; (snt) sensory nucleus of trigeminal nerve; (tl) nucleus lateralis thalami; (tlp) nucleus lateralis thalami, pars posterior; (TOL) tractus olfactorius lateralis; (tr) nucleus reticularis thalami; (tv) nucleus ventralis thalami; (vcgl) nucleus ventralis corporis geniculati lateralis. After Ungerstedt (1971) and Björklund and Lindvall (1984).

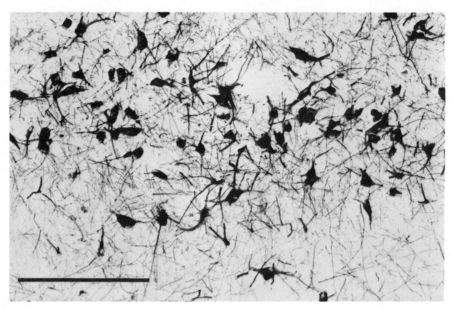

Figure 9.7. Immunohistochemical staining of human substantia nigra dopaminergic cells using a polyclonal antibody to TH. Scale bar: 0.5 mm. Photomicrograph courtesy of Dr. Hisaki Kamo (University of British Columbia).

The A10 cells are a medial extension of this group into the ventral tegmentum. They spread to form a cap over the interpeduncular nucleus.

The localization of dopaminergic cells in the human substantia nigra is well illustrated in Figure 9.7. The figure shows immunohistochemical staining using an antibody to TH. The same cell group is shown in Figure 1.17 (Section 1.2.4) for the rat where horseradish peroxidase has been retrogradely transported from the caudate–putamen.

Diencephalic Dopamine Neurons. These neurons are arranged more or less in two horizontal columns running through the rostrocaudal extent of the hypothalamus, one near the dorsal surface of the hypothalamus and the other near the ventral surface. They are joined by two vertical columns, one medial in the periventricular region and the other a smaller, irregular lateral column. These are partially illustrated in cross section in Figure 9.5B.

The A11 cells are the most caudally located of the diencephalic neurons. They are in the periventricular gray matter of the posterior dorsal region of the hypothalamus and adjacent thalamus and are therefore just rostral to the A10 cells. The largest diencephalic group is the A12 complex located in the arcuate nucleus of the hypothalamus and adjacent dorsal area. It innervates the median eminence and neurohypophysis, as well as the pars intermedia. The A13 cells are mostly in the medial part of the zona incerta. They, and the A14 group in the rostral periventricular area, form the origin of the incertohypothalamic complex.

The A15 cell group starts at the same level as the A14 rostrally, but is more laterally placed. The group extends laterally into the paraventricular nucleus.

Other Dopaminergic Groups.   The A16 DA cells are those periglomerular cells of the olfactory bulb that are TH-positive but DBH-negative (see Figure 9.5). They are interneurons. In the caudal brain stem, some cells of the A2 group appear to be dopaminergic because they are also TH-positive but DBH-negative. DA cells are found in the amacrine cell population of the retina. Some DA cells may also exist in the spinal cord. In the periphery, the small, intensely fluorescent (SIF) cells of the superior cervical ganglion, as well as cells in the carotid body, are positive. In addition, DA is found in large quantities in the liver, duodenum, heart, and lungs of certain ruminant animals, where it seems to be correlated with a special type of mast cell (Bertler *et al.*, 1959).

Dopamine Fiber Bundles.   The substantia nigra cells (A8 and A9) give off fibers mainly to the nigrostriatal bundle, which originates in the medial aspect of the complex, travels rostrally with the medial forebrain bundle, and then turns laterally to traverse the globus pallidus en route to its major terminals in the neostriatum. Some fibers travel further to provide sparse innervation to some amygdala and cortical areas. Some A9 fibers also go to the nucleus accumbens. The ventrotegmental cells (A10) give rise to the more medially situated mesolimbic–mesocortical fiber tract, which also ascends with the medial forebrain bundle. The mesolimbic fibers innervate mainly the prefrontal cortex, nucleus accumbens, amygdala, olfactory structures, septal area, and nucleus of the stria terminalis (see Figures 9.5 and 9.6).

Moore and Bloom (1978) proposed the following general principles for topographic arrangement of telencephalic DA terminals with their cell bodies of origin in the brain stem: (1) Dorsal cells project to ventral terminals and ventral cells to dorsal terminals; (2) medial cells project to medial terminals and lateral cells to lateral terminals; and (3) anterior cells project to anterior terminals and posterior cells to posterior terminals. The topography in the substantia nigra, however, is oblique, with anterior cells innervating dorsomedial structures and posterior cells innervating ventrolateral structures. Whether there is dopaminergic innervation of the hippocampus is still uncertain, although there is some innervation of the pyriform cortex and entorhinal cortex. In the prefrontal cortex, the dopaminergic innervation is concentrated in layers 5 and 6, while in the cingulate gyrus, it is concentrated in layers 2 and 3. The innervation is mostly ipsilateral. While the major dopaminergic innervation is in the frontal cortex, there is also dense innervation in the temporal cortex in primates. The posterior cortex apparently is not innervated.

The tuberohypophyseal system from the A12 neurons provides short axons to innervate the median eminence, the stalk, and the pars intermedia and pars posterior of the pituitary. The A11 group provides a descending pathway in the spinal cord. The fibers are in the intermediolateral cell column and in the areas surrounding the central canal. Terminals occur in these areas and throughout the dorsal horn (Skagerberg *et al.*, 1982). The remaining dopaminergic neuronal groups provide intrinsic innervation to their areas of origin. The dorsal periventricular diencephalic cells innervate thalamic and hypothalamic nuclei. The incertohypothalamic system is intradiencephalic.

9.4.3. Anatomical Distribution of Noradrenaline

Noradrenergic neurons are classified on the basis of being positive for catecholamine histochemistry by the Falck–Hillarp and glyoxalic acid techniques and positive by immunohistochemistry for TH, AADC, and DBH, but negative for PNMT. The noradren-

NORADRENALINE

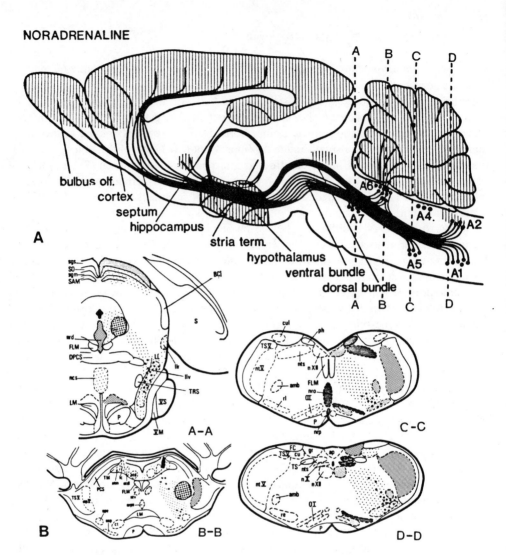

Figure 9.8. Noradrenaline neurons and pathways. (A) Sagittal section of rat brain showing the principal noradrenaline neuronal groups and major rostrally projecting pathways. (B) Four cross sections of brain at levels A–A, B–B, C–C, and D–D as indicated in the sagittal diagram. Dots indicate the relative numbers of cell bodies. Abbreviations: (amb) nucleus ambiguus; (ap) area postrema; (BCI) brachium colliculi inferioris; (ct) nucleus corporis trapezoidei; (cu) nucleus cuneatus; (cul) nucleus cuneatus lateralis; (DPCS) decussatio pedunculi cerebellarium superiorum; (FC) fasciculus cuneatus; (FLM) fasciculus longitudinalis medialis; (gr) nucleus gracilis; (lc) locus coeruleus; (LL) lemniscus lateralis; (llr) nucleus lemnisci lateralis rostralis; (llv) nucleus lemnisci lateralis ventralis; (LM) lemniscus medialis; (ncs) nucleus centralis superior; (npV) nucleus sensorius principalis nervi trigemini; (nrd) nucleus reticularis medullae oblongatae, pars dorsalis; (nro) nucleus raphe obscurus; (nrp) nucleus reticularis paramedianus; (nrpo) nucleus raphe pontis; (ntd) nucleus tegmenti dorsalis (Gudden); (ntdl) nucleus tegmenti dorsalis lateralis; (ntm) nucleus tractus mesencephali; (nts) nucleus tractus solitarius; (ntv) nucleus tegmenti ventralis (Gudden); (ntV) nucleus tractus spinalis nervi trigemini; (nX) nucleus originis nervi vagi; (nXII) nucleus originis nervi hypoglossi; (OI) oliva inferior; (osp) nucleus parolivarius superior; (P) tractus corticospinalis; (PCS) pedunculus cerebellaris superior; (ph) nucleus prepositus hypoglossi; (re) nucleus reuniens; (rl) nucleus reticularis lateralis; (S) subiculum; (SAM) stratum album mediale colliculi superioris; (sgm) stratum griseum mediale colliculi superioris; (sgs) stratum griseum superficiale colliculi superioris; (SO) stratum opticum colliculi superioris; (TM) tractus mesencephalicus nervi trigemini; (TRS) tractus rubrospinalis; (TS) tractus solitarius; (TSV) tractus spinalis nervi trigemini; (VM) nervus trigeminus, radix motoria; (VS) nervus trigeminus, radix sensoria. After Ungerstedt (1971) and Moore and Card (1984).

ergic nature of the A1–A7 groups had been established by pharmacological and fluorescent histochemical methods so that later histochemical and immunohistochemical studies were largely confirmatory. The location of the cell groups is shown in cross and sagittal sections in Figure 9.8 and in horizontal section in Figure 9.6. The cell groups combine their rostrally directed efferents into two major systems: the dorsal noradrenergic bundle and the lateral tegmental bundle. Noradrenergic cell bodies also project caudally. The dorsal noradrenergic system is made up mainly of rostrally projecting fibers from the locus coeruleus (A4–A6) neurons. The remainder (A1–A3, A5, and A7) contribute to the ventral tegmental system. The caudally projecting fibers descend through the reticular formation and the ventral and lateral funiculi of the spinal cord. They give off terminals at all levels to the dorsal and ventral gray matter and, in the thoracic cord, to the intermediolateral cell column.

Using axonal transport plus DBH immunocytochemistry, Westlund et al. (1982) showed that spinal cord projections arise from NA cells of the ventral locus coeruleus, the nucleus subcoeruleus, the medial and lateral parabrachial nuclei, the nucleus of Kolliker–Fuse, and the region around the superior olivary nuclei. Medullary NA cell groups do not contribute projections to the spinal cord. Of all NA cells that project to the spinal cord, 79% are located in the nucleus subcoeruleus and locus coeruleus (Westlund et al., 1982, 1983). The existence of single catecholamine neurons in the locus coeruleus that project to both the cerebellum and the spinal cord or to both the cerebellum and the frontal cortex has been demonstrated (Nagai et al., 1981).

9.4.4. Anatomical Distribution of Adrenaline

Adrenergic neurons contain the four catecholamine-synthesizing enzymes. They are therefore positive by immunohistochemistry for TH, AADC, DBH, and PNMT. They have not been detected by histochemical methods because of the weak fluorescence of adrenaline derivatives compared with those of the other catecholamines. Immunohistochemistry, particularly for PNMT, has been the critical method.

The cell bodies and presumed pathways are shown in Figure 9.9. The cell groups are in the medulla close to, but only mildly overlapping, the NA medullary groups A1 and A2. The ventral C1 group is just rostral to the A1 group, while the dorsal C2 group is just rostral to the A2 group. There are also some slightly more rostral dorsomedial neurons.

In addition to these groups, there are some amacrine cells in the retina (Hökfelt et al., 1984) and possibly some in the hypothalamus. PNMT-positive cell bodies in the hypothalamus have been reported by Ross et al. (1984), Ruggiero et al. (1983), and Hökfelt et al. (1984). However, since these cells have not been positive for other catecholamine-synthesizing enzymes by immunohistochemistry, their authenticity has not been unequivocally confirmed. Adrenergic fibers are both ascending and descending and originate largely from the C1 cell group. The C2 group produces a minor, dorsally located projection that innervates mostly periventricular structures in the medulla.

Fibers ascend in the dorsal aspect of the ventromedial part of the reticular formation with fibers intermingling with the noradrenergic ascending projection. Nerve endings are considerably more restricted in their distribution than DA and NA nerve endings. A few are seen in the limbic forebrain, olfactory structures, and nucleus accumbens. Most are in the hypothalamus, especially in the periventricular area. The intensity of innervation of the midbrain, pons, and medulla generally follows the levels of adrenaline shown in Table

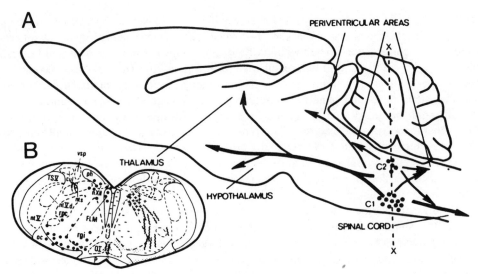

Figure 9.9. Adrenaline neurons and pathways. (A) Sagittal section of rat brain showing the principal adrenaline neuronal groups and the major ascending and descending pathways. (B) Cross section at level X–X in A. Cell bodies are indicated by dots on the left and axonal pathways by + and − on the right. Abbreviations: (cul) nucleus cuneatus lateralis; (FLM) fasciculus longitudinalis medialis; (nts) nucleus tractus solitarius; (ntV) nucleus tractus spinalis nervi trigemini; (ntVd) nucleus tractus spinalis nervi trigemini, pars dorsomedialis; (nXII) nucleus originis nervi hypoglossi; (oc) tractus olivocerebellaris; (OI) oliva inferior; (P) tractus corticospinalis; (ph) nucleus prepositus hypoglossi; (rgi) nucleus reticularis gigantocellularis; (rpc) nucleus reticularis parvocellularis; (TSV) tractus spinalis nervi trigemini; (vsp) nucleus vestibularis spinalis. After Hökfelt *et al.* (1974, 1984).

9.2. Only a few terminals can be found in the cerebellum, and in the cord only the sympathetic lateral column has significant innervation.

9.5. Physiology and Pharmacology of Dopamine

In previous sections, we have discussed the catecholamines as a family, stressing the many common features of their behavior. We learned in Section 9.4, however, that they have separate and distinct neuronal localizations. They also have distinctive pathways, but there is considerable overlap in the terminal fields. Therefore, we must now emphasize the differences among the catecholamines to understand their separate physiological actions. It is necessary to integrate information concerning their pathways with information concerning the multiple types of postsynaptic receptors that they activate [see Table 5.2 (Section 5.2)].

9.5.1. Physiology of Dopamine

The three major dopaminergic tracts that have been identified are the nigrostriatal, mesolimbic–mesocortical, and tuberoinfundibular. The major receptor types are D_1 and D_2.

The nigrostriatal tract is concerned with initiation and execution of movement. Loss of function produces the well-known akinesia and rigidity of Parkinson's disease. The greatest difficulty parkinsonian patients encounter is that of initiating a movement. Once initiated, movements proceed in a grudging fashion, from which it is difficult to make modifications. The parkinsonian launched in a festinating gait toward a wall may have extreme difficulty in arresting the movement or changing direction. This phenomenon is discussed further in Sections 9.5.2 and 14.5.4.

The functions of the mesolimbic–mesocortical system are not well understood in physiological terms. Presumably, there is an emotional component as well as one of thought organization. The system has been implicated in a number of behaviors, including stereotypy, self-stimulation, and consummatory behaviors such as eating and drinking. It is also the one implicated in DA theories of schizophrenia (see Section 15.6.1).

The tuberoinfundibular system plays a role in coupling the hypothalamus to the pituitary. Basal hypothalamic DA neurons project to all parts of the median eminence, the stalk, the neural lobe, and the pars intermedia of the pituitary (Björklund and Lindvall, 1984). The terminals are close to the capillaries of infundibular portal vessels. It is presumed that DA exerts its main effects on the hypophysis by means of this portal circulation, much as the adrenal corticosteroids do on adrenal medullary cells (see Section 9.7). Dopaminergic stimulation inhibits prolactin (PRL) secretion, stimulates growth hormone (GH) secretion, inhibits ovulation, inhibits melanocyte-stimulating hormone (MSH) release, and may have an inhibitory effect on gonadotropin release. Both PRL and MSH exert feedback control on the activity of tuberal DA neurons (Lichtensteiger, 1979).

There are other diencephalic DA neurons. Some of these may act on the para-ventricular and supraoptic nuclei (Moos and Richard, 1982) to facilitate the reflex release of oxytocin and to modulate the release of vasopressin.

The demonstration of descending DA axons from the A11 diencephalic group to the dorsal horn and preganglionic sympathetic neurons suggests a possible involvement in the regulation of nociception and sympathetic outflow. To date, such a role has not been established.

The behavior of DA cells *in vivo* is illustrated in Figure 9.10. This figure illustrates the resting rate of discharge of dopaminergic neurons of the monkey substantia nigra in relation to three stages of movement: prior to a behavioral trigger, between the behavioral trigger and the onset of movement, and during the movement itself (Schultz, 1986a,b). The spontaneous discharge of these neurons is slow, between 0.1 and 8 impulses/sec, in keeping with all catecholamine neurons in the brain and in the periphery that have been studied. After the behavioral trigger, which was the opening of a door to a food-containing box, there was a latency of about 100 msec followed by a transient increase in the firing rate. This subsided prior to the onset of movement 200–300 msec later, but most DA cells continued to show a modest increase over the background level during the performance of the task itself. Following chemical destruction of substantia nigra neurons in these monkeys, the time of onset of movement was lengthened following the behavioral trigger, and the time to complete the movement was lengthened even more. The implication of these results is that the discharge burst prior to movement was required to mobilize sufficient DA to permit appropriate signals to pass through to other motor circuits [for a fuller description, see Figure 14.26 (Section 14.5.5)]. In the case of the lesioned substantia nigra, not enough DA could be mobilized, and thus the movement was delayed in onset and delayed even more in execution.

Other studies have substantiated the slow background discharge rates of nigral and

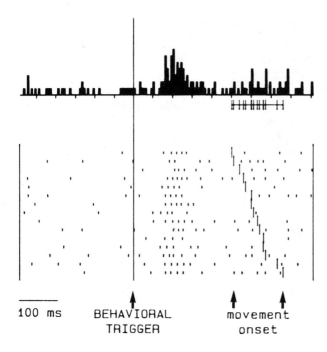

Figure 9.10. Extracellular discharge activity of a dopamine neuron in monkey substantia nigra. The neuron responds to the behavioral trigger (door opening of a food-containing box) before arm movement. The response is time-locked to the behavioral trigger (central solid line) and not to the later occurring movement (small lines below the histogram and in the right half of the dot display). Data are shown in the form of a dot display in which each dot represents the moment of a neuronal discharge, and each horizontal line of dots represents one performance in the task. The sequence of dotted lines is rearranged according to increasing intervals between the behavioral trigger and movement onset (reaction time). The peristimulus time histogram shown above is a sum of the discharges in the dot display. Bin width is 5 msec. Courtesy of Schultz (1986).

100 ms   BEHAVIORAL   movement
              TRIGGER   onset

midbrain DA neurons and the phenomenon of bursting under stimulus (Bannon and Roth, 1983; Chiodo and Benjamin, 1983; Walters *et al.*, 1975).

There are currently two methods for following what is taking place in terms of DA release as a result of these neuronal discharges. One is the push–pull cannula technique as developed by Glowinski and colleagues (e.g., Cheramy *et al.*, 1981) and the other is voltammetry as developed by R. N. Adams and his colleagues (see R. N. Adams and Marsden, 1982). Voltammetry is the more promising technique because of its versatility and speed of response (for a review, see Stamford, 1985). Following stimulation of the nigrostriatal tract, DA release can be detected in the striatum by voltammetry (Ewing *et al.*, 1983). The release is of short duration, being detectable in the extracellular fluid for only about 15 sec. An ingenious method of establishing that the voltammetric signals correspond to DA was obtained by placing microaluminate probes next to the voltammetric recording electrode; the aluminate probes absorbed the DA that had been released. There was a correspondence between the voltammetric values and the amount of DA subsequently obtained by high-performance liquid chromatography following elution from the alumina (Rice *et al.*, 1985).

Voltammetry has been used to determine the turnover rate constant for DA (0.046/min) and DOPAC (0.053/min) in the striatum following electrical stimulation of the nigrostriatal pathway (Michael *et al.*, 1985). The DA voltammetric signal in the striatum increased in an exponential manner following an increase in the stimulation rate of the nigrostriatal tract (frequency 0–25 Hz) (Gonon and Buda, 1985). Stimulation pulses applied in bursts were more potent than an equivalent number of pulses regularly spaced. Interruption of the dopaminergic impulses either by an electrolytic lesion or by low doses of apomorphine caused an immediate decrease of the DA current. These experiments provide confirmation of many less sophisticated studies relating DA release to the activity of dopaminergic pathways. They also illustrate the elegant promise of the voltammetric technique.

An aspect of the physiology of nigral dopaminergic neurons that deserves extensive evaluation is that of dendritic release. Such release can have a variety of effects on signal transmission within the substantia nigra (Cheramy *et al.*, 1981; Ruffieux and Schultz, 1980; Waszczak and Walters, 1983). For example, DA released from the dendrites of substantia nigra pars compacta cells decreases the rate of firing of substantia nigra pars reticulata γ-aminobutyric acid output neurons. This obviously has an important bearing on the consequences of basal ganglion function [see Figure 14.26 (Section 14.5.5)]. The implication is that DA release is physiologically active both from dendrites and from nerve endings. There are no dense aggregations of typical storage vesicles in the dendrites of substantia nigra neurons (Hattori *et al.*, 1979), but endoplasmic reticulum can be seen near the dendritic surface. Histofluorescent studies following reserpine administration indicate that DA-binding protein is present in the dendrites. DA released from this location might not be acting in the close-contact environment seen at a typical axo-dendritic synapse, but rather at a distance, as has been shown for the pituitary. Studies of the axons of all monoaminergic neurons indicate that they have a varicose structure, and it is quite possible that DA is released from these varicosities, which do not have synapses associated with them.

We now turn to the effects of DA on its receptors. DA neurons have their own autoreceptors in addition to receptors that exist on postsynaptic neurons. Several classes of receptors have been reported, leading to a somewhat confusing literature. We shall concentrate on the two main receptor subtypes, D_1 and D_2. The characteristics of these two receptor subtypes are given in Table 9.3.

The principal difference between the two types is the manner of their linkage to

TABLE 9.3
Some Characteristics of Dopamine Receptors[a]

Characteristic	D_1	D_2
Effect on adenylate cyclase	Stimulates	None or inhibits
Guanine nucleotide sensitivity	+	+
Location of prototype receptor	Bovine parathyroid	Anterior pituitary
Striatal location	Intrinsic neurons	Intrinsic neurons; nigrostriatal and corticostriatal afferents
Dopamine action	Agonist (micromolar potency)	Agonist (nanomolar potency)
Selective agonist	SKF 38393[b]	LY 141865[c]
Selective antagonist	SCH 23390	Sulpiride
Apomorphine action	Partial agonist or antagonist	Agonist (nanomolar potency)
Dopaminergic ergots		
Bromocriptine	Antagonist	Agonist
Pergolide	Agonist	Agonist
Thioxanthine affinity	Nanomolar	Nanomolar
Phenothiazine affinity	Nanomolar	Nanomolar
Butyrophenone affinity	Nanomolar	Nanomolar

[a]From Creese *et al.* (1984) and Kebabian *et al.* (1984).
[b]2,3,4,5-Tetrahydro-7,8-dihydroxy-1-phenyl-1H-benzazepine.
[c]N-Propyl(pyrazolo-3′,4′)-6,7-hexahydroquinoline.

adenylate cyclase. Stimulation of D_1-receptors increases cyclic AMP (cAMP) formation, while stimulation of D_2-receptors either inhibits production of cAMP or has no effect (Kebabian and Calne, 1979; Kebabian et al., 1984). Quantitatively, the D_1-receptor is more common. Beef parathyroid is a gland that contains almost pure D_1-receptors, while the mammotropic region of the anterior pituitary contains almost pure D_2-receptors (Kebabian and Calne, 1979). DA has agonist potency at D_1-receptors in the micromolar range and at D_2-receptors, at least in the pituitary, in the nanomolar range. In the striatum, D_2-receptors apparently exist in a high-affinity and a low-affinity state (Grigoriadis and Seeman, 1984). Selective agonists for D_1- and D_2-receptors are, respectively, SKF 38393 (Arnt, 1985) and LY 141865. Selective antagonists for D_1- and D_2-receptors are, respectively, SCH 23390 (Schulz et al., 1985) and sulpiride.

A classic preparation for investigating DA-sensitive adenylate cyclase is the corpus striatum, even though it also has D_2-receptors. Kebabian et al. (1972) found that cAMP production was stimulated by concentrations of DA in the 1–10 µM range. The actual mechanism by which the agonist produces its postsynaptic effects, at least for the adrenergic receptor, was described in Section 5.4.1 and illustrated in Figure 5.4. Low concentrations of the DA agonist apomorphine, but not the β-adrenergic agonist isoproterenol, also caused stimulation. On the other hand, many known dopaminergic blockers, including the phenothiazines and butyrophenones, were found to block DA activation of adenylate cyclase.

Figure 9.11 illustrates these effects. The adenylate cyclase activity of a subcellular fraction of rat caudate nucleus, rich in synaptic membranes, was measured in the presence and absence of DA as well as fluphenazine. Fluphenazine is a relatively potent antipsychotic drug and a receptor antagonist for both D_1- and D_2-receptors. At about 40 µM, DA caused a maximal increase in enzyme activity, which was competitively inhibited by fluphenazine. With the concentration of fluphenazine at 10 µM, as in Figure 9.11, a

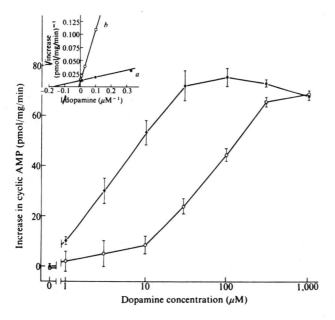

Figure 9.11. Effect of various concentrations of dopamine in the absence (●) or the presence (○) of 1×10^{-7} M fluphenazine on adenylate cyclase activity in a particulate fraction of rat caudate nucleus rich in synaptic membranes. Dopamine caused a sharp increase in cAMP formation, which was substantially blocked by the low concentration of fluphenazine. The inset shows a double reciprocal plot of the cAMP increase as a function of dopamine concentration from 3 to 300 µM. a, Control; b, 1×10^{-7} M fluphenazine. From Clement-Cormier et al. (1975).

considerably greater concentration of DA was required to produce a comparable stimulation of cAMP formation.

The adenylate cyclase receptor system is linked with GTP [see Figure 5.4 (Section 5.4.1)], so it would be anticipated that agonist binding, but not antagonist binding, would be affected by the guanine nucleotide. Guanine nucleotides reduce the binding affinity of DA or the D_1-agonist SKF 38393. The guanine nucleotides do not inhibit binding of the antagonist *cis*-flupenthixol. The most potent binding agent to D_1-receptors so far reported is the antagonist R-(+)-8-chloro-2,3,4,5-tetrahydro-3-methyl-5-phenyl-1H-3-benzazepine 7-ol (SCH 23390). It is at least 50-fold more potent at inhibiting DA-sensitive adenylate cyclase than at binding to any other known biochemical site (Schulz *et al.*, 1985). It has a K_i in the subnanomolar range for this inhibition, but of only 1 μM for inhibiting spiperone binding.

DA-sensitive adenylate cyclase has also been found in other structures of the CNS known to have high concentrations of dopaminergic terminals. These areas include the olfactory tubercle, the nucleus accumbens, the amygdala, and the cerebral cortex. The retina, which has dopaminergic activity, probably from interneurons, is also active. In contrast, the cerebellum, which does not appear to receive significant dopaminergic innervation, has no detectable DA-sensitive adenylate cyclase activity (Horn *et al.*, 1974).

Proof that DA-sensitive adenylate cyclase is attached to the postsynaptic neuron comes from lesion experiments. When the DA-containing nerve terminals of the basal ganglia are destroyed by surgical means, or chemically by 6-OHDA treatment, the DA-sensitive adenylate cyclase remains in homogenates of the denervated striatal tissue, showing that the enzyme is not localized in dopaminergic nerve terminals. When the intrinsic neurons of the rat basal ganglia are destroyed by means of local injections of kainic acid (KA), then the adenylate cyclase drops to 15–20% of normal, and that part responding to DA stimulation drops to zero. Since glial cells are preserved by this KA treatment, the DA-sensitive adenylate cyclase must be attached to the types of neurons destroyed by KA (E. G. McGeer *et al.*, 1976).

Another well-studied system in which D_1-receptors play a role is the superior cervical ganglion. Presynaptic fibers to this ganglion are cholinergic, and the majority of neurons are noradrenergic. However, SIF dopaminergic cells are scattered in the matrix.

DA-sensitive adenylate cyclase is present in homogenates of the ganglion. Electrical stimulation of the preganglionic cholinergic fibers causes an increase in the cAMP content. Similar results are obtained on application of exogenous DA to ganglionic slices. Histochemical studies show that this cAMP occurs primarily or exclusively in the postganglionic neuron (Kebabian *et al.*, 1975). Applied to the ganglion, exogenous cAMP causes hyperpolarization of the postganglionic neurons, mimicking the hyperpolarization seen on either preganglionic stimulation or application of DA. Phosphodiesterase inhibitors applied to the ganglion greatly increase the elevation of cAMP seen on preganglionic stimulation or on application of DA and also increase the size and duration of the DA-induced postsynaptic electrophysiological response.

Libet *et al.* (1975) found this electrophysiological property of DA in the superior cervical ganglion, in which DA produces a specific and enduring enhancement of postsynaptic responses to ACh. Liebet and colleagues have argued that this system is an adequate model for the memory trace, a concept that is discussed in considerably more detail in Chapter 16.

The D_2-receptor has different properties. The tuberoinfundibular DA system provides a model for its study. DA released at this location inhibits PRL release both *in vivo* and *in vitro*. This effect can be mimicked by apomorphine and other broad-spectrum DA agonists and blocked by broad-spectrum DA antagonists (Kebabian and Calne, 1979). However, this action is not linked to stimulation of adenylate cyclase activity. For example, cholera toxin, which is a nonspecific activator of adenylate cyclase, fails to inhibit PRL release.

In the intermediate lobe of the pituitary, stimulation of the D_2-receptor reduces adenylate cyclase activity, especially if it has just been stimulated by an agent such as cholera toxin. GTP is required. In the striatum, D_2-agonists inhibit the efflux and presumably the synthesis of cAMP from intact tissue (Kebabian *et al.*, 1984). Thus, in these cases at least, stimulation of D_2-receptors would seem to be producing a result similar to that shown in Figure 5.4 (Section 5.4.1) for inhibitory coupling to an adenylate cyclase system. In the anterior pituitary, D_2-receptors may be linked to an inositol phospholipid second messenger system (Simmonds and Strange, 1985), as shown in Figure 5.7 (Section 5.4.3).

The pharmacological actions of numerous DA agonists and antagonists can be explained on the basis of these different receptors. For example, the antipsychotic drugs molindone and sulpiride, as well as the antiemetic drug metoclopramide, are DA antagonists when tested against the anterior pituitary, but are either extremely weak or inactive at blocking the stimulation of striatal DA-sensitive adenylate cyclase. Agents that affect both receptors do so at different concentrations, according to the type of DA receptor. For example, nanomolar concentrations of DA elicit a response from the anterior pituitary, while micromolar concentrations are required to stimulate striatal adenylate cyclase (Kebabian and Calne, 1979) (Figure 9.11). Bromocriptine is active at a concentration of 2.9 nM in the anterior pituitary, but requires a 33,000-fold higher concentration to bring about an increase of cAMP in the retina.

Important evidence for distinct D_1- and D_2-receptors has come from a study of antipsychotic compounds. Many phenothiazine antipsychotic compounds block striatal adenylate cyclase activity, but butyrophenones, which are clinically more potent, do so only weakly. Thus, blockade of D_1-receptors does not correlate with clinical potency. If only binding to D_2-receptors is considered, then the strengths of binding of the phenothiazine, thioxanthine, thiothixine, and butyrophenone families of antipsychotics all show a similar correlation with clinical potency. Thus, it is believed that the antipsychotic action is one attached to blockade of D_2-receptors.

Bromocriptine, an agonist for the D_2- but not for the D_1-receptors, is capable of inducing psychotic symptoms. Its clinical usefulness is not only in Parkinson's disease, but also for arresting lactation in postpartum women and for treating women with secondary amenorrhea and infertility due to hyperprolactinemia. Serum PRL is reduced by bromocriptine with restoration of menstruation and fertility, presumably due to stimulation of pituitary D_2-receptors.

In addition to these postsynaptic DA receptors, there are DA autoreceptors distributed throughout the CNS (R. H. Roth, 1984). They are of the D_2 type, but appear to be more sensitive to DA than are postsynaptic D_2-receptors. They are particularly important in the nigrostriatal and mesolimbic systems. They have been located on DA nerve endings, cell bodies, and dendrites. In general, DA agonists at these receptors inhibit, and DA antagonists facilitate, the release of DA. As mentioned in Section 9.3.5, they may also be coupled to the control of catecholamine synthesis, with agonists inhibiting synthesis and antagonists reversing the effect, by interacting with calcium and the phosphorylation of TH. A selective agonist of these receptors is the (+)-enantiomer of 3(3-hydroxyphenyl)-*N-n*-propylpiperidine [(+)3-PPP]. (−)3-PPP appears to be agonistic, antagonistic, or both, depending on the experimental conditions (D. Clark *et al.*, 1985).

There is both pharmacological and lesion evidence suggesting that there may also be D_2-receptors on glutamate terminals of the corticostriatal pathway (Schwarcz *et al.*, 1978b; Garau *et al.*, 1978; Reubi and Cuenod, 1979). D_2-receptors appear to occur in both high- and low-affinity states, and other DA receptor types or subtypes have been described (Grigoriadis and Seeman, 1985; P. Sokoloff *et al.*, 1980), so it must be recognized that the story told so far is only a partial one, with many significant principles still to be elucidated.

9.5.2. Pharmacology of Dopamine

We can now turn to the effects of pharmacological agents on dopaminergic systems. The various classes include selective neurotoxins, receptor antagonists and agonists, vesicle storage depletors, false transmitters, pump inhibitors, synthesis inhibitors, DA precursors, and inhibitors of the metabolizing enzymes. They represent some of the most widely used agents in clinical medicine as well as some of the most important pharmacological tools in neuroscience research. Some representatives are listed in Table 9.3, and their sites of action are shown in Figure 9.12.

Toxic Agents.   These have become tools of inestimable value in elucidating the physiological effects of catecholamine neurons. The two most widely studied are 6-hydroxydopamine (6-OHDA) (Johnsson *et al.*, 1975) and 1-methyl-4-phenyl-1,2,3,6-tetrahydropyridine (MPTP) (Langston *et al.*, 1983). Both owe their selectivity to the

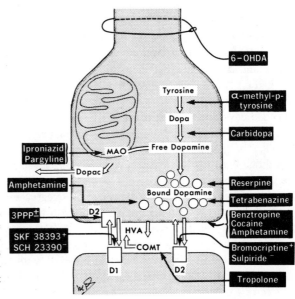

Figure 9.12. Site of action of agents modifying dopaminergic action at the synapse. Normal metabolic route is indicated with white arrows, action of drugs with black arrows. Agonist action at receptor site is indicated by a +, antagonist by a −.

selective catecholamine uptake processes and their toxicity to oxidation to unstable and toxic products.

As far as 6-OHDA is concerned, a high intraneuronal concentration must be reached for toxicity. Selectivity of the effect therefore depends on the availability of 6-OHDA, as well as on a strong pumping capacity to raise its intraneuronal concentration. Two hypotheses exist as to how destruction takes place at the molecular level. In one, either hydrogen peroxide or a superoxide radical (or both) generated by the *in vivo* oxidation of 6-OHDA is the causative agent. In the other, rapid nucleophilic reactions of the *p*-quinone of 6-OHDA or other quinone intermediates are responsible. Selective degeneration of dopaminergic vs. noradrenergic neurons depends on the conditions and site of administration. Intraventricularly administered 6-OHDA will attack most noradrenergic and dopaminergic neuronal systems. Direct injections into specific sites, such as the substantia nigra, will cause selective degeneration.

The nigrostriatal dopaminergic tract can thus be destroyed selectively by direct injections of 6-OHDA into the nucleus. The result of such bilateral injections is to produce animals with a profound motor deficit similar to that seen in Parkinson's disease. In the rat, decreases in self-stimulation, eating, drinking, and other behaviors that depend on motor capability occur. Protection is conferred by administration of agents that compete with the catecholamine neurons for uptake of the 6-OHDA. Combining 6-OHDA with a reuptake inhibitor of noradrenaline, such as desipramine, will give relative protection to noradrenergic pathways.

MPTP is a more interesting but dangerous tool, since peripheral administration will result in selective attack on nigrostriatal dopaminergic neurons. The story of its discovery is like a bizarre mystery thriller in which neurologists and neuroscientists became temporary detectives in the California illicit-drug scene. The saga commenced in 1982 when a drug addict and his sister entered a hospital in San Jose within a week of each other. They suffered from a disorder that would have been taken for Parkinson's disease by Ballard and Langston, the two neurologists who examined them, except that the patients were much too young. Soon after, Ballard met another neurologist, Tetrud, on a social occasion. Tetrud had just seen in nearby Watsonville two brothers with a similar condition who were also drug abusers. All four patients had been injecting a newly available "synthetic heroin" of then-unknown composition. Searching for a toxic compound in the impure material, analytical chemists were baffled at first. The key was provided by a case that had previously been reported in an obscure paper in a psychiatric journal after having first been rejected by two well-known medical journals. It described a graduate student who had developed a parkinsonian-like condition after self-administration of meperidine that he had synthesized himself. Unfortunately, the product was contaminated with MPTP. The graduate student later died from an overdose of cocaine, and autopsy established that there had been a loss of substantia nigra cells (G. C. Davis *et al.,* 1979). The medical journals had not recognized the significance of this finding, and the self-administration of a synthesized drug did not seem to carry with it the implications of an epidemic disease. Fortunately, Weingarten, a toxicologist who had been given the California "synthetic heroin," recalled seeing the paper. A telephone call to Irwin, the analytical chemist, who was trying to make sense of chemical fragmentation data, enabled him to solve the structural puzzle quickly. The California "synthetic heroin" was a sloppily prepared meperidine with MPTP as a significant and toxic contaminant.

In researching the literature on the by-products of meperidine synthesis, Langston

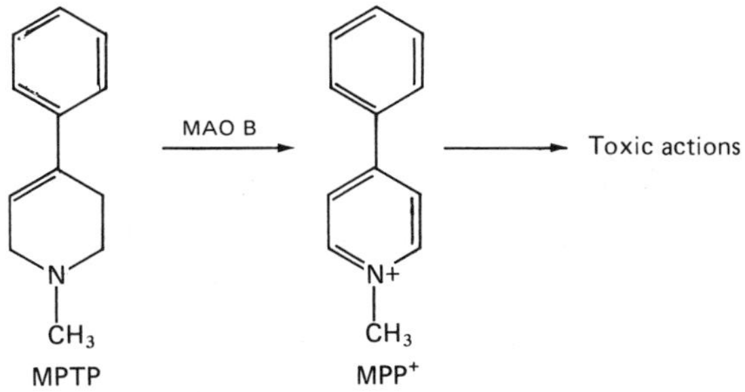

Figure 9.13. Conversion of MPTP to the neurotoxic MPP+ by MAO-B.

apparently stumbled across the tracks of the individual responsible for synthesizing the street drug. Someone had used a razor blade to cut the pages from the journal in the Stanford University library that referred to the method of synthesis, apparently in an attempt to prevent others in the area from making use of the information. A complication in the story, which probably caused the failure to identify MPTP as the culprit in the initial report of Davis and his NIH colleagues, is that MPTP is relatively nontoxic in rats. However, the toxicity of MPTP to monkeys and to mice was soon demonstrated in follow-up studies, although the species selectivity suggested that a metabolite rather than MPTP itself was responsible.

Burns *et al.* (1983) first reported on MPP+ as the *in vivo* metabolite of MPTP responsible for toxicity in monkeys. Chiba *et al.* (1984) next demonstrated that MPTP is converted to MPP+ by the action of MAO-B (Figure 9.13). Heikkila *et al.* (1984) then did the obvious but critical experiment. They administered MAO inhibitors to mice prior to administration of MPTP to determine whether there was a protective effect. Broad-spectrum MAO inhibitors, pargyline, nialamide, and tranylcypromine, all conferred protection, as did the specific MAO-B inhibitor deprenyl. The specific MAO-A inhibitor clorgyline was ineffective (Heikkila *et al.,* 1984).

A still-puzzling aspect of the MPTP story is where the toxic material is produced in brain. It is unlikely to be in the periphery, because MPP+ will not cross the blood–brain barrier (BBB). Within the brain, catecholamine neurons seem to contain MAO-A rather than MAO-B. This would argue against MPTP being converted to MPP+ within the DA neurons themselves. Further evidence against this possibility is provided by Nakamura and Vincent (1986), who did histochemical staining for MAO in brain using MPTP as substrate. Positive histochemical results were obtained with noradrenergic and serotonergic neurons of the brain stem and histamine neurons of the caudal hypothalamus, but not with DA neurons. Since 5-HT and histamine neurons innervate the substantia nigra, the possibility exists that MPP+ comes from an oxidizing source outside the DA neurons.

This concept is supported by two further observations. The first is that DA uptake blockers such as benztropine, buproprion, and mazindol protect against MPTP toxicity in the mouse just as do MAO-B inhibitors (Ricaurte *et al.,* 1985). The second is that striatal synaptosomes have an active uptake system for MPP+, but not for MPTP. Autoradiography of brain slices shows dense accumulation of [^{3}H]-MPP+ in the neostriatum. The DA-uptake inhibitor mazindol blocks MPP+ uptake (Javitch *et al.,* 1985).

Using positron emission tomography (PET) scanning, Calne *et al.* (1985) have tested

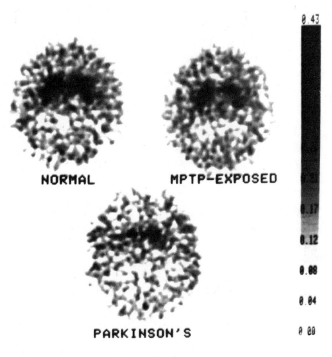

0.43

NORMAL   MPTP-EXPOSED

0.17

0.12

0.08

0.04

PARKINSON'S   0.00

Figure 9.14. Positron emission tomographs of normal, parkinsonian, and MPTP-exposed individuals. Scanning agent was 6-[18]fluorodopa. Brain scans shown are horizontal slices at the mid-basal ganglia level. Intensity of label accumulation is shown on the scale at the right. Note the very much decreased intensity in the parkinsonian patient and the somewhat decreased intensity in the MPTP-exposed individual. Photomicrograph courtesy of D. B. Calne and the UBC/TRIUMF PET team.

the ability of the basal ganglia to concentrate the DA analogue, 6-fluoro-DA, after peripheral administration of the L-dopa analogue, 6-fluorodopa, in normals, parkinsonian patients, and individuals exposed to MPTP. Figure 9.14 illustrates typical results. The basal ganglia stand out in such PET scans in normal individuals because of their active DA-metabolizing systems. In Parkinson's disease, damage to the nigrostriatal tract greatly reduces sites of DA formation and storage, and therefore the PET scan shows far less radioactivity in the basal ganglia. MPTP-exposed individuals, who do not show overt signs of parkinsonism, fall between normals and Parkinson patients. The drug abuser whose PET scan is shown in Figure 9.14 had acute, transient parkinsonian symptoms shortly after injecting the illicit drug contaminated with MPTP, but had clinical recovery. However, the impaired metabolism, as demonstrated by the PET scan, would suggest the existence of a pre-parkinsonian state. This might produce clinical symptoms in later years because of the dropout of nigral cells that is part of the normal aging phenomenon (P. L. McGeer *et al.,* 1977b).

The neuronal destruction following MPTP administration is not entirely restricted to the substantia nigra. There is some damage to the DA cells of the ventral tegmental area and to the NA cells of the locus coeruleus (Mitchell *et al.,* 1985a). The loss of non-dopaminergic cells in MPTP-induced parkinsonism may explain intellectual changes that have been observed (Y. Stern and Langston, 1985).

Precursors.   The precursors of DA are tyrosine and L-dopa. As mentioned in Section 9.2.1, tyrosine loads do not influence DA levels even in the presence of an MAO inhibitor, but L-dopa has a pronounced effect. Many books and articles now document the effectiveness of L-dopa in clinical medicine and describe the nuances of its use in Parkinson's disease (A. G. Gilman *et al.*, 1985). L-Dopa in doses of 1–8 g/day, accompanied by a peripheral DDC inhibitor, is now the treatment of choice in this disease. Comparable amounts (50–100 mg/kg) given in acute parenteral doses to rats will more than double brain levels of DA. It is not known, of course, what tissue levels of DA are achieved in humans, but HVA levels in the cerebrospinal fluid increase several-fold after therapeutic doses of L-dopa.

DA does not cross the BBB, but L-dopa does so with ease. However, many tissues contain DDC, including the endothelial cells lining brain capillaries. In addition, liver, erythrocytes, and other tissues possess the enzyme COMT. Thus, a substantial portion of every parenteral dose is metabolized before it can reach brain catecholamine cells. It is presumed that in Parkinson's disease, sufficient residual brain DDC exists to convert L-. dopa to DA, but other hypotheses exist, including decarboxylation by brain capillaries. Some L-dopa may be converted to DA by noradrenergic and serotonergic cells, and there is even the suggestion that L-dopa itself may be active at receptor sites without conversion.

L-Dopa in high doses produces stereotypy and choreiform movements, particularly of the head and neck. Nausea is almost universal at high doses. These two effects are usually the limiting factor in achieving therapeutic results in Parkinson's disease and are referable to dopaminergic stimulation of receptors in the striatum and chemoreceptor trigger zone, respectively.

In about 15% of cases, L-dopa will produce agitation, excitement, and even psychosis, presumably from stimulation of receptors in the cortex and limbic system. It will also stimulate the release of GH and inhibit lactation, as would be expected from the effects of stimulating the tuberoinfundibular system.

Synthesis Inhibitors. Since TH is the rate-limiting enzyme in catecholamine synthesis, it would be anticipated that inhibitors of the enzyme would greatly decrease catecholamine levels and produce physiological changes of practical benefit in clinical medicine. So far, no agent has emerged that meets the tests of safety and effectiveness. The most commonly used inhibitor in experimental situations is α-methyl-*p*-tyrosine (AMPT), but it is toxic to liver and kidney.

Many compounds have been screened as potential inhibitors, and they generally fall into the categories of analogues of tyrosine, inhibitors of the pteridine cofactor, catechols, and chelating agents. In addition to AMPT itself, its methyl ester and the compound 3-iodotyrosine have proved to be potent inhibitors.

In general, inhibitors that are successful in reducing catecholamine levels *in vivo*, such as AMPT, produce effects similar to those of reserpine and the phenothiazines even though 5-HT action is not affected. AMPT, for example, causes sedation, reduces motor activity, decreases blood pressure, and decreases intracranial self-stimulation.

A number of effective DDC inhibitors exist that are analogues of L-dopa. Hydrazine derivatives such as carbidopa have found great clinical usefulness as L-dopa adjuvants in the treatment of Parkinson's disease. They do not cross the BBB and thus inhibit the decarboxylation of L-dopa only in the periphery. The result is that the brain receives a much higher proportion of parenterally administered L-dopa than if the decarboxylase inhibitor had not been administered. The same therapeutic result can be achieved with one third to one quarter the dose of L-dopa, and the peripheral side effects can be substantially reduced. DDC inhibitors are not as toxic as TH inhibitors, though they are broader-spectrum in that they affect 5-HTP decarboxylation as well. Some hydrazine-type DDC inhibitors cross the BBB and have central as well as peripheral effects. At very high doses, they can completely inhibit catecholamine formation and, as described earlier (Section 9.2.4), can even be used to estimate rates of catecholamine synthesis. Generally speaking, however, these agents are not capable of performing the equivalent of a chemical sympathectomy because the amounts of DDC are so much higher than those of TH.

An unusual compound is α-methyldopa (Aldomet), which is a competitive inhibitor of DDC. It was originally thought that its sedative and antihypertensive properties were due to inhibition of this enzyme, but later it was realized that the decarboxylation products, α-methyl-DA and α-methyl-NA, could act as false transmitters. Thus, its action overlapped with those of a separate class of catecholaminergic drugs. α-Methyl-NA is a potent α_2-adrenergic agonist, and this effect is believed to be responsible for many of the actions of Aldomet.

False Transmitters.   False transmitters are agents that indirectly affect neurotransmitter action, substituting for the real neurotransmitter in mechanisms of storage or release or both. Although there is an implication that such molecules are themselves less effective at receptor sites and that failure of synaptic neurotransmission is inevitably associated with their accumulation, this is not necessarily required by the definition.

α-Methyldopa and α-methyl-*m*-tyrosine are converted into methylated catecholamines and methylated *m*-tyramines, respectively. A number of similar phenylethylamines can also be taken up into catecholamine storage granules for release on stimulation, including *p*-hydroxynorephedrine, which can be made *in vivo* from amphetamine. Amphetamine and tyramine probably replace neurotransmitters at the cytoplamic granules. Depending on the dose, time after administration, and strength of the false transmitter, the net effect may be excitatory or inhibitory to catecholaminergic action. Most of the properties of these agents are depressant and referable to the peripheral noradrenergic system. None is antipsychotic, and none is particularly useful in the treatment of Parkinson's disease. Amphetamine does not fall strictly within the definition of a false transmitter because it is not stored in granules *per se*.

Monoamine Oxidase Inhibitors.   MAO inhibitors have played an important role in elucidating the separate properties of the A- and B-type enzymes (see Section 9.2.2). These compounds also have clinical usefulness as antidepressants (see Section 15.5.2), in narcolepsy, and occasionally as adjuvants for Parkinson therapy. Most agents in clinical use are broad-spectrum, but deprenyl, a specific MAO-B inhibitor, avoids dangerous side effects associated with ingestion of L-dopa or tyramine-containing foods while peripheral MAO-A is inhibited. Structural features are shown in Figure 15.5B (Section 15.5).

Catechol-*O*-methyltransferase Inhibitors.   Pyrogallol and tropolone are the two best-known inhibitors of COMT. These agents have little physiological action, indicating the predominant role of MAO in metabolizing DA.

Dopamine Antagonists.   These have been extensively described in Section 9.5.1 (Table 9.3), and their chemical effects are further described in Section 15.5 (Figure 15.5 and Table 15.3). Structure–activity relationships of DA antagonists have been extensively studied, since potency is so strongly related to antipsychotic action. Several families of compounds have been identified that are clinically effective, each with numerous representatives. The families (and an example of each) include the phenothiazines (chlorpromazine), thioxanthines (chlorprothixene), butyrophenones (haloperidol), diphenylbutyl-piperidines (pimozide), indoles (molidone), dibenzoxazepines (loxapine), benzamides (sulpiride), and a number of compounds with miscellaneous structures. Potency in blocking D_2 receptors correlates with clinical efficacy, but it also correlates with inducing extrapyramidal reactions, particularly of the akinetic–rigid type seen in Parkinson's disease. However, these clinical properties are not indivisibly linked. It is possible to correct extrapyramidal reactions with anticholinergic agents and yet not disturb antipsychotic potential. While most of the clinically useful representatives also have some potency in blocking D_1 receptors, this effect is not quantitatively related either to antipsychotic action or to induction of extrapyramidal effects. As a consequence, the physiological significance of D_1-receptor action must at this stage be considered unknown. Some phenothiazines also have antihistaminic (e.g., promethazine) and anticholinergic (e.g., diethazine) activities. Structurally, the critical element seems to be the nature of the side chain attached to ring position 10. These properties reduce both antipsychotic activity and the tendency to induce extrapyramidal reactions.

Dopamine Agonists. The most widely studied groups are the aporphines (e.g., apomorphine) and the ergolines (e.g., bromocriptine, pergolide). Their most important clinical action is to overcome the effects of Parkinson's disease by stimulating dopaminergic receptor sites in the striatum. The search for long-acting, selective dopaminergic agonists is an active one because it is presumed that the neostriatal receptor sites in Parkinson's disease not only survive the disease process, but also develop supersensitivity due to DA deprivation.

DA agonists also stimulate receptor sites in the chemoreceptor trigger zone, inducing vomiting. This is hardly a clinical objective compatible with treating Parkinson's disease, although in the case of apomorphine, the classic dopaminergic agonist, this property has been capitalized on in its clinical employment as an emetic agent.

Vesicle Storage Inhibitors. Reserpine, which has already been mentioned several times in this chapter, is the classic intraneuronal storage inhibitor for DA, other catecholamines, and 5-HT. Even though reserpine is one of the most widely used drugs in basic neurological science, the mechanism by which it affects the vesicular storage process has still not been determined. The amounts required to displace amines are too low and too long-lasting for the inhibition to be competitive, and yet preloading of a nerve terminal with an amine precursor will protect against depletion. Studies using platelets have shown that reserpine sticks to the platelet membrane, suggesting that it exerts its effects at the membrane level of the intraneuronal vesicle. Reserpine has many physiological effects, but the only ones clearly attributable to DA are those of parkinsonian-like extrapyramidal reactions. DA probably also plays a prominent part in the antipsychotic and tranquilizing effects. Reserpine is further discussed in Section 15.5.1.

The action of tetrabenazine, the most useful of the benzoquinolizines, is comparable to that of reserpine, but of much shorter duration.

Pump Inhibitors. Since reuptake may be the primary method of terminating the action of DA and other neurotransmitters, inhibitors of this process can be expected to have prominent physiological effects in the direction of enhancing neurotransmitter function. Those inhibitors that appear to be preferential for DA, as opposed to NA or 5-HT, have widely varying properties because of their other actions. The tricyclic antidepressants, which inhibit DA uptake to some extent, are stronger NA and 5-HT pump inhibitors. Buproprion, an amphetamine-like compound, is considered an atypical antidepressant because it is a relatively selective inhibitor of dopamine uptake. All the antidepressants are discussed in Section 15.5.2.

Benztropine, amphetamine, and cocaine are three well-known DA pump inhibitors. Each has prominent but differing alternative actions that dominate their overall effects. Benztropine is a powerful muscarinic antagonist. Amphetamine is a releaser of DA and, to a lesser extent, of NA, as well as a competitive inhibitor for MAO. It is a stimulant, appetite suppressant, and, at high doses, a psychotogen (see Section 15.5.3). Cocaine is an NA pump inhibitor, a local anesthetic, and, due to its euphoriant action, a prominent drug of abuse. Amantidine, which delays neuronal DA reuptake, is a mildly effective antiparkinsonian agent. It also facilitates DA release. It was introduced as an antiviral agent.

9.6. Physiology and Pharmacology of Noradrenaline

9.6.1. Physiology of Noradrenaline

We learned in Section 9.4 that NA is quantitatively a less important catecholamine in brain than DA. It is the neurotransmitter for no more than one fifth as many neurons as is DA, and the NA neurons are all located caudal to the midbrain. Nevertheless, except for the limbic and extrapyramidal areas of brain, NA is the dominant catecholamine. It has a higher concentration than DA in the cortex, thalamus, hypothalamus, cerebellum, and hindbrain [see Table 9.2 (Section 9.4.1)]. Presumably, the levels of NA reported in the various brain areas listed in Table 9.2 reflect the density of terminals distributed from the A1-A7 cell groups of the pons and medulla [see Table 9.1 (Section 9.4.1)]. These cell groups all fall within the reticular system of the brain stem. Despite detailed anatomical knowledge, and an impressive armamentarium of drugs, there are as yet no satisfactory explanations of the physiological functions of the central noradrenergic system, although they are presumed to be somehow linked to the reticular activating system. There is no known disease entity that involves a clear-cut and selective loss of neurons of the noradrenergic complex or any part of it, so that significant clues have not come from human pathological studies.

The closest example that exists involves depression. NA uptake inhibitors have proved to be useful as antidepressants (see Section 15.5.2). However, even in this instance there is much confusion, since selective 5-HT uptake inhibitors also have clinical efficacy. Other diseases in which NA abnormalities may exist include Parkinson's disease, especially the variant from Guam that also involves dementia (Nakano and Hirano, 1983; Whitehouse *et al.*, 1983), and Alzheimer's disease (J. J. Mann *et al.*, 1981; P. L. McGeer, 1984). There is a loss of locus coeruleus neurons in both these diseases, but lesions more important to the symptoms occur in the dopaminergic system of substantia nigra in Parkinson's disease and the cholinergic system of the substantia innominata in Alzheimer's disease. There is a component of depression associated with these disorders, but it is questionable whether this is an integral part of the disease process.

To understand more about the physiological role of NA, we must turn to its action at receptor sites. Unfortunately, most of the meaningful physiological correlations have come from studies of peripheral systems. Centrally, it is not yet known how the receptors are organized.

The original concept of mutiple receptor sites for catecholamines was put forth by Ahlquist (1948), who proposed the designations "α-receptor" and "β-receptor" for sites on smooth muscle that produced, respectively, excitatory (i.e., constricting) and inhibitory (i.e., relaxing) responses. This original concept has long been superseded.

NA receptors are now divided into subtypes, α_1 and α_2 and β_1 and β_2, on the basis of the actions of various pharmacological agents and binding of ligands to the presumed receptor subtypes. All the receptor types have been found in brain, although properties of a given receptor may vary according to the site in which it is being tested. The following are some of the better-known peripheral tissues that have α-receptors: the radial muscle of the iris, most blood vessels, sphincter muscles of the stomach and intestine, the sphincter muscle of the urinary bladder, piloerector muscles and sweat glands of the skin, the salivary glands, and the capsule of the spleen. β-Receptors are found in the ciliary muscle of the eye, the muscles of the heart, some coronary and pulmonary vessels, the bronchial muscle of the lung, muscles having to do with motility in the stomach and intestine, and the detrusor muscle of the bladder (Table 9.4).

In brain, regional binding for α_1-receptors shows little variation, whereas for α_2-receptors there is considerable variation; β_1-receptors show marked regional variations in brain, while β_2-receptors are more constant. Destruction of noradrenergic neurons by 6-OHDA administration to neonatal rats caused a 65% decrease in β_1-receptor density in adult rat cerebral cortex, but no change in β_2-receptor density. The interpretation is that the β_1-receptors are the ones involved in neuronal function (Minneman et al., 1979b). It has been variously suggested that β_2-receptors are on glia, microvessels, and postsynaptic neurons.

The β_2-receptor has been purified from mammalian brain (Benovic et al., 1984). It has an apparent molecular weight of 64,000 and binds agonist and antagonist ligands with the expected selectivity and stereospecificity.

Langer (1980) has proposed that postsynaptic receptors are of the α_1, β_1, and β_2 types and that α_2 exists presynaptically. The α_2-agonist clonidine inhibits the release of NA evoked by nerve stimulation, while the α_2-antagonist yohimbine enhances such release. This could be a control mechanism for NA release, but C. Lee et al. (1983) have proposed an alternative regulatory mechanism based on reuptake of NA. In any case, not all α_2-receptors are presynaptic.

The α_1-receptor is differentiated from the α_2-receptor on the basis of agonism by clonidine and blockade by piperoxan. It is not known what effector the α-receptor is coupled to, but at some sites, α_1-agonists stimulate polyphosphoinositide turnover. The α_2-receptor is believed to be almost always coupled to inihibitory adenylate cyclase activity, at least in the periphery. It may exist in multiple affinity states (Bylund and U'Prichard, 1983).

The method for quantitating α_1- and α_2-receptors is by direct ligand binding using selective agents such as [^{3}H]-WB 4101 and [^{3}H]clonidine, respectively (Table 9.4). The concentrations of β_1- and β_2-adrenergic receptors are determined by labeling with a broad-spectrum β-antagonist such as [^{3}H]dihydroalprenol or [^{125}I]iodohydroxybenzyl-pindol and then displacing the binding with β_1- and β_2-selective drugs (Minneman et al., 1979a).

TABLE 9.4
Some Characteristics of Adrenergic Receptors

Characteristic	α_1	α_2	β_1	β_2
Effect on adenylate cyclase	?	Inhibits	Stimulates	Stimulates
Guanine nucleotide sensitivity	Insensitive	Decreases agonist binding	—	—
Selective agonists	Phenylephrine	Clonidine	Dobutamine	Terbutaline, salbutamol
Classic blockers	Phenoxybenzamine		Propranolol	
Selective antagonists	Prazosin, indoramin	Piperoxan, yohimbine	Practolol	Butoxamine
Binding ligands	[³H]-WB 4101[a]	[³H]Clonidine	[³H]Dihydroalprenol; [¹²⁵I]iodohydroxy-benzylpindol	
Physiological correlations				
Eye	Contracts iris	—	Relaxes ciliary muscle	
Heart	—	—	Increases heart rate	—
Lung	Decreases secretion	—	—	Relaxes bronchi
Veins	Constriction	—	—	Dilatation
Arterioles	Constriction		—	Dilatation
Stomach	Sphincter contraction		—	Decreases muscle tone
Gallbladder	—	—	—	Relaxes duct
Liver	—	—	—	Glycogenolysis
Bladder	Sphincter contraction		—	Relaxes detrusor
Sex organs	Ejaculation		—	—
Sweat glands	Localized secretion		—	—
Spleen capsule	Contraction		—	Relaxation
Skeletal muscle	—	—	—	Glycogenolysis
Pineal gland	—	—	Melatonin synthesis	—
Posterior pituitary	—	—	Vasopressin secretion	—

[a] 2-[(2',6'-Dimethoxy)phenoxyethanolamino]-methylbenzodioxan.

Some of the more important properties of these principal receptors are listed in Table 9.4. The β-adrenergic receptors stimulate a number of metabolic events by means of adenylate cyclase as shown in Figure 5.4 (Section 5.4.1). Cyclic AMP activates the catalytic unit of a protein kinase that then phosphorylates a number of proteins. Among these is phosphorylase kinase, which leads to the breakdown of glycogen. The result probably explains the effect of adrenergic stimulation on gluconeogenesis. At some locations, the activated protein kinase phosphorylates triglyceride lipase, which promotes hyperlipemia. The relationship of cAMP production to effects other than metabolic ones for β-adrenergic receptors is unclear, but generally stimulation of these receptors has a relaxant effect on smooth muscle.

Exposure of cells to a β-adrenergic receptor agonist, followed by a wash and reexposure, reduces the responsiveness of the cAMP-generating system. This is interpreted as being due to endocytosis or internalization of the receptor into vesicles within the cell (Pastan and Willingham, 1981; Glavin, 1985), which may have broad significance for receptors generally.

The cerebellum has been investigated as a model system (Hoffer *et al.*, 1973) for correlating electrophysiology with the biological actions of NA. NA slows the mean rate of discharge of Purkinje cells by interaction with β-receptors. NA hyperpolarizes the membrane of these cells, and this hyperpolarization is generally accompanied by increased membrane resistance, but never by increased membrane conductance. NA action is blocked by iontophoretic application of prostaglandins of the E series and is potentiated by phosphodiesterase inhibitors that would prolong the action of cAMP. Thus, it has been proposed that the synaptic action of NA is mediated by interaction with adenylate cyclase much as is the case for D_1 DA receptors. Hoffer and co-workers found that the effects of NA on Purkinje cells could be duplicated by stimulation of the locus coeruleus, indicating that endogenously released NA could duplicate the iontophoretic application.

The locus coeruleus is the principal source of noradrenergic innervation of most areas of brain. It receives information from a variety of sensory neuronal systems. In unanesthetized animals, its neurons discharge slowly; only painful stimuli will increase the rates. In awake animals, its neurons respond to many sensory stimuli with a pronounced response (Foote *et al.*, 1983) that correlates with cortical electroencephalographic desynchronization. Decreased locus coeruleus activity correlates with decreasing behavioral vigilance. The noradrenergic fibers from the nucleus carry impulses at the slow rate of 0.2–0.86 m/sec. Along their length are many varicosities that apparently release neurotransmitter even in the absence of synapses. Indeed, Descarries *et al.* (1977) estimate that as few as 5% of terminals labeled by uptake of [^{3}H]-NA make typical synaptic contacts. Maximal effects seem to correlate with short, high-frequency bursts of activity (e.g., 10 spikes at 50 Herz), which is analogous to the effects observed for substantia nigra DA cells [see Figure 9.11 (Section 9.5.1)].

Taken together, these data on noradrenergic mechanisms imply the existence of an amplifying system that originates in the brain stem, where there is ready access to a wide spectrum of sensory input. The output spreads diffusely to all regions of brain, where postsynaptic action of a varying type takes place according to the receptor subspecies.

9.6.2. Pharmacology of Noradrenaline

There are drugs that interact with all the sites previously mentioned for DA in Section 9.5.2. However, there is considerable overlap in their action, not only with DA, but also with adrenaline. The site of action of some of the prominent members of each class is shown in Figure 9.15.

Toxins. 6-OHDA can be used for selective action against central NA neurons through appropriate injection at sites remote from dopaminergic cells and pathways. A much more promising agent is *N*-(2-chloroethyl)-*N*-ethyl-2-bromobenzylamine (DSP-4), which is a selective noradrenergic neurotoxin (S. B. Ross, 1976). When administered parenterally, it produces a long-lasting depletion of central and peripheral NA levels, but leaves DA and 5-HT unaffected. Recovery can take place, especially in the periphery, indicating that terminals are more vulnerable than cell bodies. It is believed that the aziridinium derivative that is its oxidation product binds covalently to NA uptake sites and then exerts its toxic action in a manner analogous to that of MPP$^+$ (Zieher and Jaim-Etcheverry, 1980).

DSP-4 impairs the functioning of locus coeruleus neurons (Olpe *et al.*, 1983) and produces supersensitive α_2- and β-adrenergic receptors (Dooley *et al.*, 1983). It brings

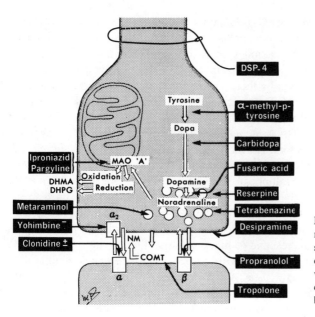

Figure 9.15. Site of action of agents modifying noradrenergic action at the synapse. Normal metabolic route is indicated with white arrows, action of drugs with black arrows. Agonist action at receptor site is indicated by a +, antagonist by a −.

about a number of changes in physiological responses of rats, such as impairment of a two-way avoidance response (Archer *et al.*, 1982) and potentiation of haloperidol-induced catalepsy (Asin *et al.*, 1982). This agent has not been as extensively utilized as 6-OHDA and MPTP, but it may be a key compound for future elucidation of noradrenergic mechanisms. Xylamine is a related agent with similar effects (Dudley *et al.*, 1981).

Precursors.   L-Dopa can act as a precursor for NA as well as DA. Most of the effects of L-dopa are not believed to be due to synthesis of NA, although it is conceivable that production of excess DA in NA cells could account for some of the effects. 3,4-Dihydroxyphenylserine (DOPS) can be decarboxylated directly to NA. Thus, it is a direct precursor, although again it cannot be stated with certainty that only NA in noradrenergic cells is formed after parenteral administration of DOPS. It is believed to cross the BBB poorly, and its effects have not been extensively studied.

Synthesis Inhibitors.   TH and DDC inhibitors influence NA as well as DA synthesis. Thus, the effects of such inhibitors are mixed as far as the catecholamines are concerned. Selective NA effects can come about only through the use of DBH inhibitors. DBH is inhibited by all copper chelators, especially FLA-63 (Anden and Fuxe, 1971) and disulfuram, which is also an inhibitor of alcohol dehydrogenase. Fusaric acid is an effective inhibitor *in vitro* and *in vivo* and also has antihypertensive action (Nagatsu *et al.*, 1970).

False Transmitters.   The administration of α-methyldopa leads to long-lasting depletion of NA from peripheral sympathetic nerves due to the accumulation of α-methyl-NA in these tissues. Following stimulation of sympathetic nerves, α-methyl-NA is released. Since α-methyl-NA is less active at receptor sites than the true neurotransmitter, hypotension results. A similar depletion is obtained following the administration of α-methyl-m-tyrosine, which is accompanied by an accumulation of β-hydroxy-α-methyl-m-tyramine or metaraminol. This produces a physiological response similar to that seen after α-methyldopa, i.e., decreased sympathetic action. Both drugs are employed clinically as antihypertensive agents (Iversen, 1967).

Storage Inhibitors.   The binding granules for NA are depleted of their neurotransmitter by reserpine and tetrabenazine in the same way as other catecholamine and 5-HT granules. Guanethidine is also an NA depleter, but its action is not clearly understood. It affects only peripheral stores, apparently because it does not cross the BBB. Given to newborn rats, it depletes NA from cells of the superior cervical ganglion, but not DA from SIF cells. Since guanethidine can itself be stored and released on stimulation, storage granules may be affected by the drug. It is used in the treatment of hypertension, partly because it is also a ganglionic blocking agent with local anesthetic action.

Noradrenaline Releasers.   Two potent releasers of NA are amphetamine and tyramine. These agents are taken up into catecholaminergic endings by active uptake processes. They promote release of NA by a replacement process that is not fully understood. The release cannot be primarily from granules into the intrasynaptosomal space, as is the case with reserpine, because NA, rather than dihydroxymandelic acid, reaches the extracellular fluid. Amphetamine and tyramine must be displacing NA from storage granules into the extracellular space. They therefore act indirectly, but their stimulant effects wear off as supplies of the endogenous neurotransmitter become depleted.

Monoamine Oxidase and Catechol-O-methyltransferase Inhibitors.   These agents were discussed in Sections 9.3 and 9.5.2. Representative compounds are shown in Figure 15.5 (Section 15.5). MAO inhibitors are used chiefly as antidepressants (see Section 15.5.2). There are as yet no clinical applications of COMT inhibitors.

Pump Inhibition.   NA pump inhibitors are clinical antidepressants and are discussed in Section 15.5.2. The best-known is desipramine, the demethylated derivative of imipramine, which is the most potent uptake inhibitor so far described. In the isolated rat heart, a concentration of 13 nM is sufficient to produce a 50% inhibition of NA uptake. Other well-known NA pump inhibitors are nortriptyline and amitriptyline. These tricyclic compounds differ only modestly in structure from the phenothiazines, which block the postsynaptic site rather than the presynaptic uptake site. The middle ring is seven-membered rather than six, which results in a molecule being bent out of its planar shape.

Cocaine also inhibits the uptake of NA. It produces a 50% inhibition at a concentration of 380 nM (Iversen, 1967). Cocaine inhibits the accumulation of exogenously administered NA in sympathetic nerve endings, and, following the administration of cocaine, there is a substantial increase in the amount of NA overflowing into the venous circulation following sympathetic nerve stimulation.

Agonists and Antagonists.   The actions of such agents have been described in a general way in Section 9.6.1. We present here some of the more prominent pharmacological actions. Phenylephrine is a peripheral α_1-agonist with little central action. It constricts arterioles, as does methoxamine, another α_1-agonist. It is used principally as a nasal decongestant.

Administered intracisternally, clonidine produces cardiodepressant and hypotensive effects that are blocked by centrally administered α_2-antagonists. Peripherally, it impairs adrenergic neurotransmission, apparently by inhibiting presynaptic α_2-receptors (Langer, 1980). In this way, it resembles α-methyldopa, which is metabolized to α-methyl-NA.

Dobutamine is a β_1-agonist that enhances the contractile action of the heart and thus is useful in congestive heart failure.

Terbutaline is an example of the β_2-agonist class of agents that are useful as bronchodilaters.

Prazosin is a powerfully selective antagonist of peripheral α_1-receptors. It is a hypotensive agent useful in the treatment of mild to moderate hypertension. Selective α_2-antagonists such as yohimbine do not have an established clinical role as yet, but should be effective in promoting NA release because of their presynaptic blocking action. Yohimbine crosses the BBB and produces a general picture of central excitation including sweating, nausea, and vomiting. It has an antidiuretic action and increases motor activity and irritability.

Practotol, a β_1-antagonist, has been used for the treatment of hypertension because of its cardioselectivity. Butoxamine is relatively selective for β_2-receptors and blocks chronotropic action on the heart. Most clinical agents, however, are broad-spectrum blockers such as the antihypertensive agent propranolol. CNS effects of β-blockade include sedation, depression, and sleep disturbances.

9.7. Physiology and Pharmacology of Adrenaline

The physiology and pharmacology of central adrenaline neurons are uncertain because of the very small concentration of adrenaline neurons (see Section 9.4), the heavy overlap with NA in terminal fields, and the lack of distinction between noradrenergic and adrenergic receptors. The picture is clearer in the periphery, where the physiology of the adrenal gland is better understood.

Coupland, an anatomist, first noticed that in species such as the rabbit, in which the adrenal cortex does not form a complete envelope around the medulla, little or no adrenaline is present in the portion of the medulla not adjacent to cortical cells. A similar situation exists in the dogfish. Wurtman, a physiologist, observed that patients suffering from pituitary disease had an unusual and prolonged sensitivity to insulin. These patients responded poorly to adrenocorticotropic hormone (ACTH), the pituitary hormone that stimulates adrenal cortical cells. Thus, a fast-acting agent, presumably adrenaline, was missing. Working with the biochemist Axelrod, Wurtman removed the pituitary in rats and noted that, within a week, PNMT activity fell to only 15–25% of normal. It could be restored by treating rats with ACTH. Activity could also be restored with very large doses of dexamethasone, but not by normal replacement doses of corticosteroids. Thus, it was evident that only the extremely high concentrations of corticosteroids found in the corticomedullary portal circulation, which are at least 100-fold greater than those present in systemic arterial blood, were adequate to induce PNMT. The effect could be blocked by concurrent administration of actinomycin D or puromycin, indicating that protein synthesis was required for the effect to be observed. The proposal was put forward that the mammalian adrenal medulla contains a single clone of cells, the members of which do or do not produce PNMT (and therefore adrenaline), depending on the extent to which they are stimulated by glucocorticoids (Wurtman *et al.*, 1972). In the frog, both adrenaline and PNMT activity exist in substantial quantities in the brain, heart, and spleen, as well as in the adrenals. After hypophysectomy in the frog, PNMT activity remains unchanged in all frog organs, indicating that the mechanism of PNMT induction is different in that species. In addition, mammalian pheochromocytomas are capable of synthesizing adrenaline in the absence of glucocorticoid stimulation, indicating that more than one mechanism may exist in mammals (Wurtman *et al.*, 1972).

Bohn (1983) has suggested, on the basis of developmental studies, that the phenotype for PNMT cells is derived independently of glucocorticoid stimulation during neurogenesis, but that the mRNA that codes for PNMT may be directly regulated by glucocorticoids in later life.

When released into the circulation, adrenaline increases the systolic pressure and the output of the heart. It stimulates gluconeogenesis and generally prepares the body for action and meeting stressful situations.

Presumably, central adrenergic pathways participate in a cooperative way with these activities.

Central adrenaline appears to stimulate the release of luteinizing hormone (LH) from the pituitary (Crowley and Terry, 1981), and Coen and Coombs (1983) have associated adrenaline with the preovulatory LH surge in female rats (see Section 15.2). Day *et al.* (1982) have suggested, on the basis of studies involving direct injections of adrenaline into the basal hypothalamus, that it may stimulate PRL release.

A role for central adrenaline in blood-pressure regulation has been suggested by an increased PNMT or adrenaline concentration in the hypothalamus and brain stem in genetically hypertensive rats (Saavedra *et al.*, 1979; Wijnen *et al.*, 1977, 1978) or rats with experimentally induced hypertension (Petty and Reid, 1979; Saavedra *et al.*, 1980).

Acute physical stress causes a marked increase in brain stem hypothalamic adrenaline levels (Kvetnansky *et al.*, 1978; Saavedra, 1980).

Taken together, these results imply central adrenergic functions that parallel peripheral functions in terms of endocrine, cardiovascular, and stress responses.

Drugs that interact exclusively with adrenaline but not NA or DA have not yet been reported. Thus, no separate discussion of adrenaline pharmacology will be attempted. It might be noted, however, that adrenaline is a much more powerful β-receptor agonist than NA. Therefore, agents that interact powerfully with β-receptors may be modifying the action of adrenaline more than that of NA. At this stage, such an interpretation is only a speculation, and a promising field for future catecholamine research will be to distinguish more clearly between NA and adrenaline physiology.

9.8. Summary

The catecholamines constitute an interesting family in that the biosynthetic sequence proceeds from dopamine (DA) through noradrenaline (NA) to adrenaline. Each catecholamine is represented by neurons centrally and peripherally, but DA and NA are also precursors and exist transiently in some neurons. This greatly complicates investigation of catecholamine physiology and pharmacology.

In the CNS, there are three principal dopaminergic pathways: the nigrostriatal, the mesolimbic–mesocortical, and the tuberoinfundibular. Interneurons and minor pathways are also present, as are neurons in the retina and sympathetic ganglia. In ruminant species, dopamine exists in many tissues as a constituent of mast cells.

The actions of dopaminergic neurons relate to extrapyramidal, behavioral, and endocrine function. The roles of interneurons are less certain.

Two major DA receptors have been identified and have been designated D_1 and D_2. Activation of D_1-receptors stimulates adenylate cyclase activity; activation of D_2-receptors does not. Both receptor types exist postsynaptically. A D_2-type of receptor also acts as an autoreceptor on dopaminergic nerve endings, providing negative feedback on DA release. D_2-type receptors may also exist on nerve endings of nondopaminergic pathways such as the corticostriatal glutamate neurons. Blockade of D_2-receptors correlates with antipsychotic action and the induction of extrapyramidal side effects. The phenothiazine, butyrophenone, and thioxanthine families of compounds, as well as other clinically useful antipsychotic agents, all show these two effects. The two responses probably involve different neuronal populations because anticholinergics counteract extrapyramidal side effects without canceling antipsychotic action. Most antipsychotics are also D_1-antagonists, but there is no correlation with clinical potency. D_2-receptor agonism is helpful in treating Parkinson's disease, particularly in smoothing out the peculiar on–off phenomenon associated with L-dopa therapy. L-Dopa is the agent of choice, presumably due to its ready conversion to dopamine in the brain. Excess L-dopa will induce choreiform extrapyramidal activity and, in some cases, psychotogenic symptoms.

MPTP is a relatively specific toxin for dopaminergic cells of the nigrostriatal tract when taken orally or administered parenterally. Its toxic action is due to conversion in brain to MPP^+ by MAO-B. Protection is afforded by specific MAO-B inhibitors such as deprenyl or by DA uptake inhibitors. Pump inhibitors of DA counteract drug-induced extrapyramidal reactions, but are not antidepressants. Amphetamine is a DA releaser, a stimulant, and, in high doses, an agent that produces paranoid psychosis.

NA neurons differ from DA neurons in possessing the enzyme DBH. This enzyme is associated with the noradrenergic storage granule. On stimulation of a noradrenergic nerve, both NA and DBH are released into the extracellular space. This is regarded as one confirmation of the theory of exocytosis for neurotransmitter release.

Noradrenergic cells are found in the reticular area of the brain stem. The principal cell group is in the locus coeruleus. Terminal fields of the ascending noradrenergic pathways heavily innervate the hypothalamus, with lesser innervation of the thalamus, limbic system, and cerebral cortex. A descending nonadrenergic pathway connects the more caudal cell groups with the dorsal, ventral, and lateral horns of the spinal cord.

The peripheral sympathetic nervous system is, of course, primarily noradrenergic. Exceptions are fibers to the salivary glands and adrenal medulla, but nerve fibers to the blood vessels, sweat glands, piloerector muscles of hair follicles, eye, heart, intestinal tract, spleen, bladder, lungs, and sex organs are noradrenergic.

Iontophoretically applied NA shows both excitatory and inhibitory characteristics. The action of NA at most of its receptor sites probably involves the participation of cAMP as a second messenger and possibly, at some sites, the prostaglandins as adjuvants.

Peripheral noradrenergic receptors are divided into α and β types. Classic α- and β-receptor blockers, such as phenoxybenzamine and propranolol, do not have antipsychotic action. There are now available a wide range of compounds that interact more specifically with NA receptors than do the classic agents. It is therefore possible to subdivide the receptors into α_1, α_2, β_1, and β_2 categories. All four types are found in brain, but their cellular locations are unclear. Receptors of the α_1 and β_2 types show little regional variation in brain, while α_2 and β_1 show marked variation. It has been suggested that many α_2-receptors are presynaptic, while β_1-receptors are on postsynaptic neurons and β_2-receptors are on microvessels and glia. The β-receptors are linked to an adenylate cyclase system that can be clearly differentiated from the one activated by DA at D_1-receptors. The clinical usefulness of noradrenergic receptor agonist and antagonist compounds is largely related to their action on peripheral smooth muscle; α_1-agonists are vasoconstrictors and decongestants, β_1-agonists stimulate the heart and are useful in congestive heart failure, and β_2-agonists are effective bronchodilators. As far as blockers are concerned, α_1-antagonists are antihypertensives, β_1-antagonists are also useful in hypertension, and β_2-antagonists block chronotropic actions of the heart.

DSP-4 is a selective toxin for NA nerve endings and, to a lesser extent, the cell bodies. It crosses the blood–brain barrier.

Uptake is the principal method for inactivating NA. NA pump inhibitors act as antidepressants, a property that distinguishes them from DA pump inhibitors. MAO inhibitors, which enhance the action of both DA and NA, are also antidepressants. COMT inhibitors, on the other hand, have little physiological action, suggesting that this enzyme plays only a minor role in catecholamine inactivation.

The adrenergic neuron is characterized by the presence of PNMT, which converts NA to adrenaline. Peripherally, adrenaline is contained in chromaffin cells of the adrenal medulla. PNMT can be induced in these cells by the high concentrations of glucocorticoids released into the adrenal portal circulation. Thus, the production of adrenaline is indirectly mediated in the periphery by ACTH from the pituitary, which acts on cells of the adrenal cortex.

There is no evidence that central adrenergic neurons are under similar control. Central neurons have been identified by immunohistochemistry using antibodies to PNMT. Cells in the brain stem have been located, with ascending and descending pathways. The descending pathways innervate areas of the spinal cord, while the ascending pathways terminate primarily in the hypothalamus but possibly also in more rostrally located structures.

There seems to be little pharmacology that is unique to adrenergic cells, although adrenaline is a more powerful activator of β-receptors than is NA. Central adrenaline pathways may participate in pituitary control, particularly the release of LH and PRL. They may also be involved in the central regulation of blood pressure and the response to stress.

10

Serotonin and Other Brain Indoles

10.1. Introduction

Despite an enormous amount of research, serotonin [5-hydroxytryptamine (5-HT)] remains a compound whose basic functions are not understood at either the cellular or the gross level. That it is a neurotransmitter is not in dispute. It is synthesized, stored, and released by central neurons and its synapses have morphology similar to that of other amine transmitters. Yet little can be said with regard to its precise action on postsynaptic neurons. The same state of affairs exists with respect to its overall actions. While hints can be obtained from the action of drugs, particularly ones that affect mood and behavior (see Chapter 15), many of these same agents also interact with the catecholamines, making it difficult to separate out a 5-HT component.

Perhaps it is fitting that 5-HT should first have been isolated from blood, in which it is present in high concentration in platelets but in which no clearly defined role has been identified. For years, Irvine Page of the Cleveland Clinic had been investigating materials in blood that caused vessel constriction. Rapport, an organic chemist, came to Page directly from graduate school, and Page assigned him the task of isolating the pharmacologically active material that had first been detected in 1868. After two years, Rapport got his first crystals—but they turned out to be the creatinine sulfate complex of 5-HT, which gave a formula analysis that was too complicated to decipher (Rapport *et al.*, 1948). It was not until 1949 that Rapport, by then working at Columbia University in New York, separated the complex into the picrates of 5-HT and creatinine. The structure was then readily deduced and rapidly confirmed by chemical synthesis (Rapport, 1949; Hamlin and Fisher, 1951).

Shortly afterward, another young investigator, Betty Twarog, arrived at the Cleveland Clinic. She was assigned the job of studying the distribution of the new amine. Using the clam heart as a bioassay system, she reported in 1953 an unusually high concentration of the active substance in brain (Twarog and Page, 1953). Even though brain has only a minor proportion of total body 5-HT (over 95% is in the gastrointestinal tract and blood platelets), this unexpected finding triggered a startling series of pharmacological investigations into mood and behavior.

Amin, Crawford and Gaddum (1954) followed up on Twarog's discovery by studying exactly where in the nervous system 5-HT is located. The highest concentrations were found to be in the hypothalamus and limbic systems. This tied in with the observation of Woolley and Shaw (1954) that antimetabolites of 5-HT, particularly lysergic acid diethylamide (LSD), seemed to disrupt behavioral patterns. They were so impressed by the consequence of tampering with this material that they proposed that either an excess or a deficiency of 5-HT is responsible for mental illness. Their hypothesis remains unrefuted to this day, with especial relevance to some forms of depression (see Chapter 15).

A few years later, Brodie, of the National Institutes of Health, inspired by the finding that reserpine caused depletion of 5-HT from body stores, speculated that 5-HT is concerned with trophotropic functions in brain and noradrenaline (NA) with ergotropic functions (Brodie et al., 1959).

Subsequent pharmacological investigations continued to link 5-HT, as well as the catecholamines, with mental function. The early groups of tranquilizers and antidepressants involved 5-HT in their spectrum of action along with the catecholamines. Although more specific agents have become available to help with the difficult task of sorting out the discrete functions of these substances, their specific roles in mental disease remain unproven (see Chapter 15).

The role of 5-HT as a neurotransmitter in brain was clearly established when it was discovered that the Falck technique for catecholamines also revealed 5-HT. Cell bodies that possessed a characteristic evanescent fluorescence were found to extend throughout the length of the raphe system, and terminals were detected in the diencephalic and telencephalic areas that had been previously shown to have high concentrations of 5-HT (Dahlstrom and Fuxe, 1964a).

Although the discovery of 5-HT in brain shifted the major focus of attention to the central nervous system (CNS), interest also continued to develop around its peripheral actions on smooth muscle. The Italian investigator Erspamer had detected "enteramine," a hormone secreted by argentaffin cells of the gut, and isolated it from the salivary glands of the octopus. It was only after the structure of 5-HT had been worked out by Rapport and the synthetic material had become available that Erspamer realized that the two substances were identical (Erspamer and Asero, 1952). Lembeck (1953a) next demonstrated the presence of large amounts of 5-HT in carcinoid tumors, which are derived from argentaffin cells. These tumors were well known to be associated with diarrhea and hot flushes, just as had been detected immediately following the administration of large doses of reserpine. Some aspects of the "carcinoid syndrome" were soon correctly attributed to release of large amounts of 5-HT.

In studies completely independent of those on 5-HT, Lerner was working with extracts of pineal gland that caused aggregation of melanin granules in frog skin, thus causing it to lighten. It took him some years to purify his material because he did not suspect its instability to light and kept losing activity. He named his hormone *melatonin* and finally established its chemical identity as 5-*O*-methyl-*N*-acetylserotonin (Lerner *et al.*, 1959).

Lerner's report immediately closed another gap. It explained the high concentration of 5-HT in the pineal gland and led Axelrod to conclude that the pineal must have an enzyme capable of methylating the 5-hydroxy group of 5-HT. Axelrod initiated a brilliant series of experiments on the functions of the pineal that contributed in part to his being awarded the Nobel Prize in 1969. Working with Wurtman, he established that there is a circadian rhythm in melatonin production in the rat, that it inhibits gonadal function, and that its production is inhibited by light (Axelrod and Wurtman, 1968). The functions of the pineal gland in man are still not known (for a comprehensive review, see Haymaker *et al.,* 1982).

While much progress has been made, the story of 5-HT has just begun. Many exciting discoveries lie ahead for imaginative investigators.

10.2. Chemistry and Metabolism of Serotonin

It is important to know how 5-HT is created and destroyed in order to understand the methods used for anatomical localization (see Section 10.4) and pharmacological manipulation (see Section 10.6) of serotonergic systems. The sequence of reactions is, in the main, closely analogous to that already described for the catecholamines (see Section 9.2). The dietary amino acid tryptophan is converted to 5-hydroxytryptophan (5-HTP) by the enzyme tryptophan hydroxylase (Tr-OH). 5-HTP decarboxylase then converts this intermediate amino acid to 5-HT. 5-HT is metabolized both pre- and postsynaptically by the enzyme monoamine oxidase (MAO), which produces the inactive metabolite 5-hydroxyindoleacetic acid (5-HIAA) (Figure 10.1). In this sequence, Tr-OH is the only enzyme unique to the 5-HT as compared to the catecholamine systems. 5-HT, however, is also converted in the pineal and Harderian glands, in the retina, and in some brain regions into *N*-acetylserotonin and, this may be further metabolized to melatonin [see Figure 10.3 (Section 10.2.3)]; analogous reactions have not been reported in the catecholamine systems.

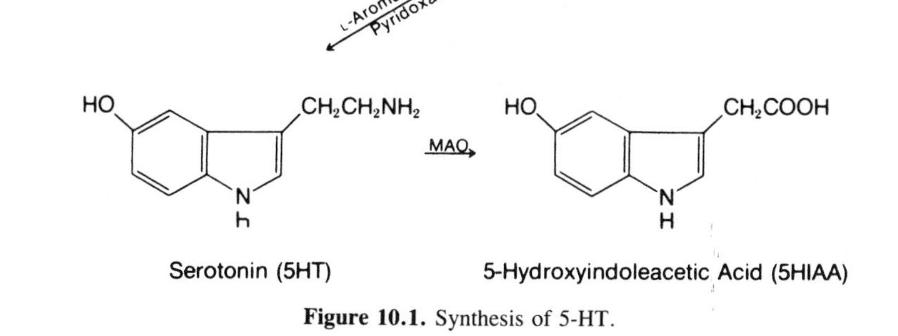

Figure 10.1. Synthesis of 5-HT.

TABLE 10.1

Levels of Serotonin and Tryptophan Hydroxylase and Turnover Rates for Serotonin in Various Brain Regions[a]

Brain region	5-HT (μg/g)	Tr-OH (nmoles/g-hr)	5-HT turnover (nmoles/g-hr)
Amygdala	2.10	12.1	ND
Hypothalamus	1.70	14.6	ND
Septal area	1.50	16.0	ND
Striatum	0.72	22.8	2.0
Hippocampus	0.65	4.3	1.1
Medulla	0.62	4.1	1.9
Pons	0.28	4.1	
Thalamus	0.57	8.4	ND
Diencephalon	ND	ND	0.6
Cerebral cortex	0.17	0.9	1.7
Telencephalon	ND	ND	2.0
Mesencephalon	ND	ND	2.8
Cerebellum	0.09	0.7	ND
Midbrain raphe	ND	17.3	ND

[a](ND) Not determined. 5-HT values are for the dog (Bogdanski *et al.*, 1957), Tr-OH levels for the cat (D. A. V. Peters *et al.*, 1968), and turnover rates for the rat (J. E. Smith *et al.*, 1978). Tr-OH activity in the pineal is reportedly at least 30-fold higher than in any brain area shown.

10.2.1. Synthesis of Serotonin

Tr-OH [tryptophan 5-monooxygenase (EC 1.16.4)] catalyzes the conversion of tryptophan to 5-HTP, the immediate precursor of 5-HT. It has a very limited distribution, being found only in cells that are specialized for synthesizing 5-HT. It is also the rate-limiting enzyme, which means that the control of 5-HT synthesis is exerted at this stage. Exactly how this control is achieved is not known, although both nerve stimulation and precursor availability seem to be important (see below and Section 10.3). It does seem certain that there is a considerable excess of enzyme capacity over normal synthetic needs. Presumably this is a reserve to meet peak demand. When the best possible conditions are chosen in test-tube experiments, including sufficient oxygen, tryptophan, and pteridine cofactor, the maximum rate of tryptophan hydroxylation by whole rat brain homogenates is more than 12 nmoles/g-hr. It has been shown by a variety of techniques that the amount converted *in vivo* is only 0.5–3.1 nmoles/g-hr (Carlsson, 1974; Gál and Whitacre, 1981). It would appear, therefore, that only about 10–25% of the available capacity is normally being used for synthesis at any given time. The Tr-OH activity and turnover rates, of course, vary from region to region of brain, although not as widely as do those of the catecholamines (Table 10.1).

Tr-OH has absolute requirements for molecular oxygen and tryptophan as substrates and 5,6,7,8-tetrahydrobiopterin (BH$_4$) as a cofactor (Tong and Kaufman, 1975; Gál, 1981). It is further stimulated by reduced pyridine nucleotide and dihydropteridine reductase. The reaction proceeds by a sequence comparable to that for tyrosine hydroxylase (TH) (see Section 9.2).

Though several scientific groups have purified Tr-OH, relatively little progress has been made in delineating its physical properties and kinetic mechanisms, apparently because of the great lability of the enzyme to standard purification procedures. It is particularly sensitive to oxygen. The catalytic activity *in vitro* appears to depend largely on the oxidation–reduction status of essential sulfhydryl and iron groups (D. M. Kuhn *et al.*, 1979). Initial purification attempts gave materials of low specific activities, but H. Nakata and Fujisawa (1982a) obtained an apparently homogeneous preparation from rat hindbrain with a specific activity about 5500 times that in brain homogenates. It appeared to be a tetramer composed of four, probably identical, subunits of molecular weight 59,000. The apparent K_m values for tryptophan and BH_4 were about 125 and 119 μM, respectively; Fe^{2+} was necessary for good activity (high V_{max}), although its presence or absence did not affect the $K_m(s)$. H. Nakata and Fujisawa (1982b) also purified from mouse mastocytoma cells a similar enzyme that had even higher specific activity (5280 nmoles/min) and seemed to be a tetramer composed of subunits of molecular weight about 53,000. Work with Tr-OH from different mouse strains suggests the existence of iso-enzymes that differ in their conformational flexibilities, state of *in vivo* activation, and, possibly, molecular weights (Knapp *et al.*, 1981).

A point of practical significance is whether there is adequate tryptophan, oxygen, and BH_4 in living brain to occupy the active centers on Tr-OH. The concentration of tryptophan required to half-saturate the enzyme in test-tube experiments has been esti-mated at 32–120 μM, while the overall tryptophan concentration in living rat brain has been estimated at 30–40 μM (D. A. V. Peters *et al.*, 1968; Knowles and Pogson, 1984). These data would suggest that the enzyme is not fully saturated *in vivo*, which explains why tryptophan loads will increase brain 5-HT levels while a tryptophan-low diet will lead to decreases. The brain levels of tryptophan are controlled not only by diet, and its effect on plasma tryptophan, but also by the concentration in plasma of other large neutral amino acids that may affect transport into brain, as well as by the activity in both liver and brain of the tryptophan pyrrolase (indoleamine 2,3-dioxygenase) that catalyzes the oxidative cleavage of tryptophan to kynurenine, the first step in nicotinic acid synthesis. One product of this pathway is quinolinic acid, which is found in brain and has been proposed as an excitatory amino acid neurotransmitter or endogenous excitotoxin or both (see Chapter 6).

Tryptophan pyrrolase is an inducible enzyme, so that chronic high levels of dietary tryptophan will lead to increased breakdown through this pathway; the enzyme activity is also modulated by corticosteroids. Changes in brain levels of tryptophan may therefore be of physiological importance in controlling 5-HT synthesis, but the regulation of brain (and plasma) levels is complex (S. N. Young, 1981; S. N. Young and Gauthier, 1981; Curzon, 1981).

Reduced 5-HT synthesis has also been observed under conditions of severe oxygen deprivation (J. N. Davis *et al.*, 1974), but this is a highly artificial circumstance, and it can be assumed that even at high altitudes, the enzyme is fully saturated with oxygen.

Experiments that inhibit BH_4 synthesis indicate that the normal pool is greatly in excess of that required for steady-state hydroxylations by TH and Tr-OH (Gál and Whitacre, 1981).

There are also considerations of substrate specificity. Strong similarities exist among phenylalanine hydroxylase, Tr-OH, and TH. Although each is localized to separate struc-tures, and is immunologically distinct, phenylalanine hydroxylase and Tr-OH show some capacity to accept each other's substrate. Liver phenylalanine hydroxylase has a significant capacity for hydroxylating tryptophan, and the apparently homogeneous rat brain Tr-OH catalyzed the hydroxylation of L-phenylalanine at approximately 39% the rate at which it catalyzed the hydroxylation of L-tryptophan, although it had no detectable activity toward L-tyrosine (H. Nakata and Fujisawa, 1982a). Under normal physiological conditions, the brain enzyme would be hydroxylating tryptophan almost exclusively. However, this may not be the case in a disease such as phenylketonuria in which the phenylalanine level is extremely high. There is deficient production of 5-HT in phenylketonuria, with reduced urinary excretion of 5-HIAA. The explanation may be simply that the enzyme Tr-OH has been sidetracked into hydroxylating phenylalanine (Curtius *et al.*, 1981). It is quite possible that future investigations may uncover other situations in which an imbalance of amino acid substrates could lead to a reduction in the synthesis rate of the proper neurotransmitter, or even the production of a false transmitter (Tong and Kaufman, 1975).

Tr-OH and the ability to synthesize 5-HT are not limited entirely to neurons; they occur in melatonin-producing cells (see Section 10.2.3) and in enterochromaffin cells of the gastric mucosa and other peripheral organs that account for the vast majority of the total body 5-HT. They have also been reported in cultured cerebrovascular endothelial cells, suggesting that these might be an extraneuronal source of 5-HT in the CNS (Maruki *et al.*, 1984).

5-Hydroxytryptophan decarboxylase [(L-aromatic amino acid decarboxylase; dopa decarboxylase (EC 4.1.1.28)] is a pyridoxal-dependent enzyme that catalyzes the conversion of 5-HTP to 5-HT. As the enzyme classification indicates, present evidence suggests that the enzyme is identical to the one that catalyzes the decarboxylation of L-3,4-dihydroxyphenylalanine (L-dopa; see Section 9.4). It is also capable of decarboxylating other aromatic amino acids such as phenylalanine and tryptophan. It is still possible, however, that highly similar isoenzymes with cross-catalytic and cross-immunological properties exist or that different aromatic amino acids have separate affinity sites on the same enzyme. The problem may be comparable to that with MAO; it was thought for many years that a single enzyme catalyzed the destruction of all aromatic monoamines, but it is now believed that at least two forms of the enzyme exist. Like MAO, the decarboxylase is not restricted to neurons that synthesize neurotransmitters. It is found, for example, in large quantities in the kidney, and it is this organ that was the source for most early studies on the enzyme (see Chapter 9).

The K_m value for 5-HT is of the order of 10^{-4} M, and the activity of 5-HTP decarboxylase is far in excess of that of Tr-OH both *in vivo* and *in vitro*, with most values in the literature being 50- to 100-fold greater. How much of this activity is attributable to serotonergic nerve endings and how much to other sources is still in doubt. Very small quantities of 5-HTP are found in brain tissue [e.g., 0.2 nmole/g vs. 7.0 nmoles/g for 5-HT (Aprison *et al.*, 1974)]. This means that almost all 5-HTP formed by the hydroxylation of tryptophan *in vivo* is immediately converted to 5-HT by virtue of the excess 5-HTP decarboxylase present in serotonergic nerve endings.

Inhibition of Tr-OH brings about a reduction in brain 5-HT levels, while moderate inhibition of 5-HTP decarboxylase is without effect. This is further evidence that Tr-OH is the rate-limiting step in 5-HT synthesis.

Although it is difficult to reduce 5-HT levels by inhibiting 5-HTP decarboxylase, the excess of this enzyme has been used to advantage in increasing 5-HT levels. 5-HTP crosses the blood–brain barrier (BBB) with ease, and, due to high levels of the decarboxylase, parenteral administration of this amino acid leads to sharp increases in brain 5-HT. These increases are accompanied by decreases in catecholamine levels and a rather strange behavioral syndrome in animals. For example, when 30 mg/kg of 5-HTP was administered to cats, most areas of brain showed more than a 30% increase in 5-HT, while dopamine (DA) dropped to about 25% of its initial level. Behavioral performance on conditioned escape or avoidance tests deteriorated markedly and the animals appeared to be detached from their surroundings (P. L. McGeer *et al.*, 1963). In view of the decline in catecholamines under these conditions, it is difficult to be certain that all the 5-HT formed was in 5-HT nerve endings. Some might have been produced in catecholamine nerve endings, acting as a false transmitter. Alternatively, 5-HTP might have interfered with uptake of catecholamine precursors, producing the catecholamine deficit. In any event, it cannot be assumed that the behavioral changes observed following administration of large doses of 5-HTP represent pure stimulation of 5-HT receptors.

10.2.2. Destruction of Serotonin

MAO (EC 1.4.3.4) has already been extensively discussed in Section 9.2.2. It deaminates aromatic amines in general (see Section 9.4) and, in particular, catalyzes the conversion of 5-HT to 5-hydroxyindoleacetaldehyde (Figure 10.2). Normally, the latter is rapidly oxidized to 5-HIAA. Under some circumstances, however, it can be reduced to 5-hydroxytryptophol (5-HTOH). Following the ingestion of alcohol, the metabolism of most aldehydes is retarded due to the formation of large quantities of acetyldehyde. As a result, the amount of 5-HTOH in the liver and urine is increased and 5-HIAA is decreased (Feldstein *et al.*, 1967). There is no indication that a similar shift takes place in brain, however, and only trace amounts of 5-HTOH are detectable (Diggory *et al.*, 1979). The tentative conclusion is that in brain the aldehyde intermediate is rapidly oxidized to 5-HIAA.

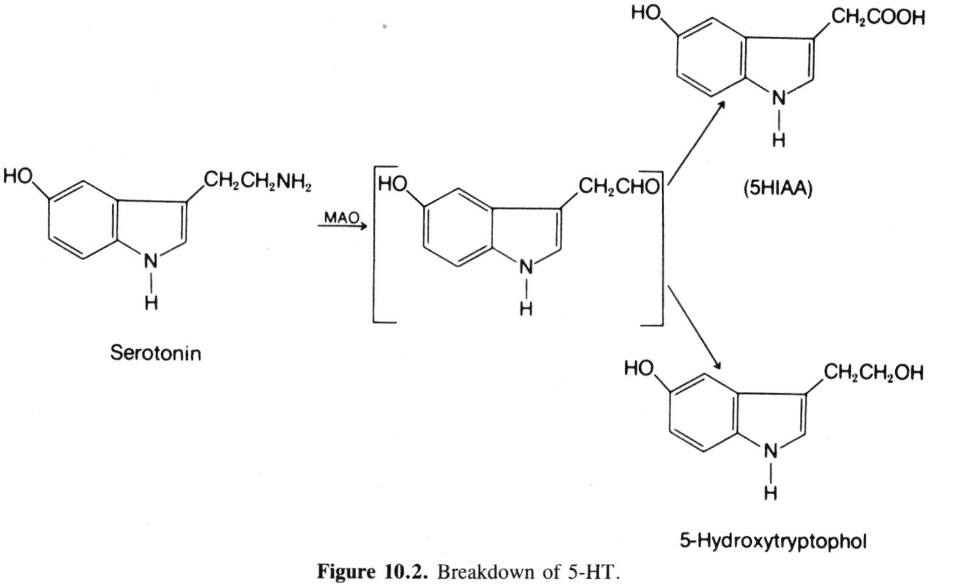

Figure 10.2. Breakdown of 5-HT.

Localization studies with monoclonal antibodies indicate that at least in the monkey, MAO-B is preferentially located in 5-HT cells (Westlund *et al.*, 1985) (see Section 9.2.2). However, in the rat, Green and Youdim (1975) measured the effects of inhibition of type A (with clorgyline), type B (with deprenyl), or combined (with tranylcypromine) MAO on hyperactivity and 5-HT levels after a tryptophan load. Combined inhibition was required to achieve a maximal effect. They concluded that 5-HT is metabolized by MAO-A preferentially but that, when MAO-A is inhibited, MAO-B takes over. Before the behavioral syndrome can be observed, both types must be almost totally inhibited. This finding is analogous to the findings of the same group with regard to DA (Green *et. al.*, 1977). In *in vitro* studies on rat brain homogenates, C. J. Fowler and Tipton (1982) found K_m values of 178 and 1170 μM and V_{max} values of 0.73 and 0.09 nmole/mg protein-min toward 5-HT for MAO-A and -B, respectively. These figures again indicate that MAO-A is much the more active form in 5-HT metabolism.

As indicated in Chapter 9, both MAO-A and MAO-B are found in both neurons and glia and are widely distributed in brain. Both types of MAO are also widely distributed in peripheral tissues.

10.2.3. Other Routes of Serotonin Metabolism

Another distinct route of 5-HT metabolism is shown in Figure 10.3; it involves initial N-acetylation of 5-HT followed by methylation of the hydroxy group to form *N*-acetyl--5-*O*-methylserotonin or melatonin. For many years, it was thought that this occurred only in the pineal gland, but the presence of both enzymes and of the product, melatonin, has been demonstrated in the retina and in some hypothalamic and midbrain regions (G. M. Brown *et al.*, 1983). The N-acetylating enzyme and the intermediate *N*-acetylserotonin have still more widespread distributions within the brain; they are particularly concentrated in cerebellar granular and Purkinje cells, but are also found in neuronal elements in other regions such as the hippocampus, locus coeruleus, and pontine reticular formation (Pulido *et al.*, 1981; G. M. Brown *et al.*, 1983).

A limited amount of work has been done on the enzymes. The *N*-acetyltransferase (EC 2.3.1.5) is found in many tissues and has been purified from both liver and brain. The K_m(s) toward 5-HT and acetyl-Co A are of the order of 200 and 60 μM, respectively (Schloot *et al.*, 1969). The hydroxyindole-*O*-methyltransferase (EC 2.2.2.4), as initially purified from beef pineal, had K_m(s) for *N*-acetylserotonin and *S*-adenosylmethionine of about 50 μM. *N*-Acetylserotonin was the best substrate of all the hydroxyindoles tested, but the enzyme would methylate (with about 10% relative activity) other hydroxyindoles such as 5-HIAA and 5-HT itself; hence, small amounts of the various methoxy derivatives may also be made *in vivo* (Axelrod and Weissbach, 1961).

The N-acetylation step is rate-limiting for melatonin synthesis, and the activity of the enzyme in the retina and pineal shows a circadian rhythm that can be entrained by environmental light cycles (see Section 10.7.1).

The primary site of melatonin catabolism is the liver, where it is converted to 6-hydroxymelatonin, which is excreted. In brain, some melatonin is converted to *N*-acetyl--5-methoxykynurenamine, which may also be of physiological importance (Figure 10.3).

The physiological actions of melatonin as a hormone are still not completely understood (see Section 10.7.1), and those of the acetylated derivatives in brain are obscure, but future research may attach some significance to this route.

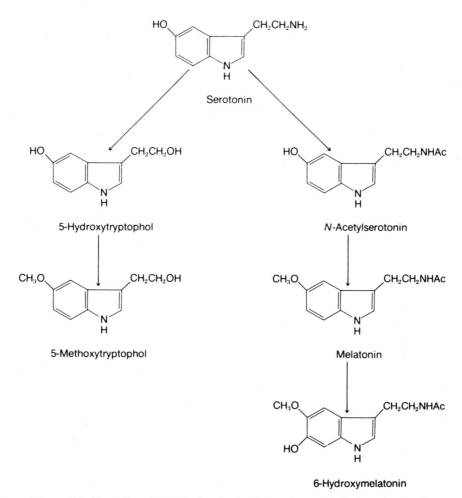

Figure 10.3. Metabolism of 5-HT in the pineal and formation and catabolism of melatonin.

A minor route of metabolism of brain 5-HT—as of brain DA—may be the formation of the O-sulfate derivative. Small amounts of 5-HT-O-sulfate have been isolated from brain following administration of an MAO inhibitor plus probenecid to prevent elimination of the metabolite into the peripheral circulation (Gál, 1972). It is unlikely that this compound is physiologically important. Sulfation is an important route of metabolism in the periphery; one third of the urinary metabolites of 5-HT are O-sulfates.

10.3. Storage, Release, and Turnover of Serotonin

10.3.1. Storage and Release of Serotonin

In common with other neurotransmitters, 5-HT is stored within vesicles in the nerve endings [see Figure 10.6 (Section 10.6.6)]. Some clues as to the general mechanisms and reasons for such packaging may be derived from studies on a specific 5-HT-binding protein (SBP), which is believed to play a key role in vesicular storage. SBP has been identified in serotonergic neurons in both the brain and the periphery. It is particularly concentrated in vesicles, undergoes fast axonal transport in serotonergic neurons, and is released with 5-HT during stimulated, Ca^{2+}-dependent exocytosis. The SBP in synaptosomes probably has a molecular weight of 45,000 and seems to be derived from a form of molecular weight 56,000 synthesized in the cell bodies. SBP binds 5-HT with extremely high affinity ($K_D = 10^{-8}-10^{-10}$ M) in the presence of Fe^{2+} and K^+ buffers, but loses its affinity for 5-HT in the presence of physiological concentrations of Na^+ and Ca^{2+}. As Gershon and Tamir (1984) explain:

> Within cells the dominant cation is K^+ and the concentrations of Na^+ and Ca^{++} are very low. In the extracellular fluid, however, these conditions are reversed; therefore, within synaptic vesicles, SBP would be expected to tightly bind 5-HT but the 5HT–SBP complex would be expected to dissociate when the vesicular interior becomes exposed to the extracellular fluid as occurs at the time of exocytosis.

Gershon and Tamir (1984) also suggest that one reason for such binding may be to reduce intravesicular osmotic pressure. The concentration of 5-HT within enteric serotonergic neurons has been estimated to be 7 mM, and the concentration within vesicles must be much higher. With such high concentrations of unbound small molecules, vesicular swelling would occur; this is presumably avoided by the binding of several molecules of 5-HT to each molecule of SBP.

SBP occurs only in serotonergic neurons. Other types of cells, such as platelets, enterochromaffin cells of the gastrointestinal mucosa, and mast cells (in some subprimate species), also contain 5-HT but have different, cell-type-specific, 5-HT-binding proteins. The material found in platelets, for example, is a glycoprotein called "serotonectin." A 5-HT-binding protein similar in properties to serotonectin is also found in cultured astrocytes (Hertz and Tamir, 1981). It has been suggested that such binding proteins play an important role in protecting against the possibility of a significant amount of free, active, circulating 5-HT. The danger arises because 5-HT is an extraordinarily active material that affects almost all tissues, including the endocrine, nervous, and cardiovascular systems. The phenomenon of lethal anaphylaxis in mice, characterized by vascular collapse and mediated by 5-HT, is only one example of the possible dangers (Gershon and Tamir, 1984).

10.3.2. Control of Serotonin Synthesis

While the synthetic and destructive pathways for 5-HT are well understood, the mechanisms by which the production of 5-HT is controlled are still not completely clear.

Several methods have been developed for determining the amount of 5-HT synthesized *in vivo* by brain. These include: (1) the administration of labeled L-tryptophan followed by the measurement of the specific activity of brain 5-HT; (2) the turnover in small regions of brain following the injection of labeled tryptophan intraventricularly or via an implanted cannula; (3) the immediate increase in brain 5-HT after MAO inhibition; (4) the accumulation of 5-HTP after 5-HTP decarboxylase inhibition; and (5) the accumulation of 5-HIAA following the administration of probenecid (Carlsson, 1974; Neckers and Meek, 1976).

These methods all give reasonably comparable values for the rate of synthesis of 5-HT *in vivo*, and in the rat they are in the range of 0.5–3.8 nmoles/hr-g of brain tissue, with the highest rates observed in the mesencephalon and the lowest in the diencephalon [see Table 10.1 (Section 10.2)]. It has already been mentioned that the capacity of brain to synthesize 5-HT is far in excess of this amount and that control must be exerted at the stage of tryptophan hydroxylation. As already described in Section 10.2.1, the supply of tryptophan to the brain appears to be one critical factor, and the regulation of this supply is complex.

Nerve stimulation is also an important factor. Excitation of 5-HT cell bodies in the raphe nucleus increases turnover in regions where the terminal fields are located (Aghajanian and Haigler, 1973; Boadle-Biber *et al.*, 1983). The nerve stimulation produces an increase in the activity of Tr-OH, probably through a calcium–calmodulin-dependent process that phosphorylates Tr-OH. A protein kinase (kinase II) with a specific distribution in brain that carries out this reaction has been identified (Boadle-Biber, 1982; D. M. Kuhn and Lovenberg, 1982; Yamauchi and Fujisawa, 1981, 1984). As already discussed in Section 10.2.1, the oxidation–reduction site of the enzyme may also play an important role.

5-HT itself does not appear to act directly as a feedback inhibitor of Tr-OH, although, as with the catecholamines, it may act on presynaptic autoreceptors to inhibit further release from synaptosomes (Gothert, 1982; Schlicker *et al.*, 1985) (see Section 10.5.1).

10.4. Anatomy of Serotonin Neurons

The original mapping of 5-HT neurons was done by Dahlstrom and Fuxe (1964a,b) in the rat using the Falck histofluorescence technique. Even though this method is relatively unsatisfactory due to the instability of the β-carboline yellow fluorescent product, it provided the main outlines of the 5-HT system in the CNS. Technical improvements in the histofluorescence method that have been applied with success to catecholamine mapping, such as the use of glyoxylic acid, have not proved useful for 5-HT mapping. Immunohistochemistry has been utilized for Tr-OH and for 5-HT, and the latter has been particularly valuable (see Figure 10.5).

10.4.1. Anatomical Distribution of Serotonin

In addition to the initial mapping of Dahlstrom and Fuxe, the distribution of 5-HT neurons has been determined in the rat using immunohistochemistry for Tr-OH (Pickel *et al.*, 1976). Immunohistochemistry for 5-HT has been used for mapping in the rat (Steinbusch, 1981, 1984; Lidof *et al.*, 1980), cat (Wiklund *et al.*, 1981; Jacobs *et al.*, 1984), and monkey (Felten and Sladek, 1983; Saavedra *et al.*, 1982; Takeuchi *et al.*, 1982). There is a high degree of concordance in the results, in terms of both species and techniques. Most descriptions have followed the original alphanumeric designation of Dahlstrom and Fuxe indicating cell groups B1–B9. While this method provides an easy framework, it does not relate to conventional anatomy and is hard to correlate with physiological systems. As with catecholamines, difficulties have arisen in fitting additions

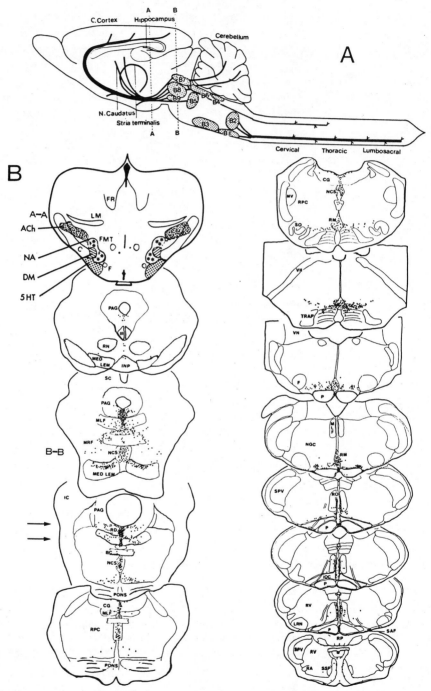

Figure 10.4. 5-HT cell groups and major pathways. (A, B) Schematic diagrams of the central serotonergic cell groups and projections from rat brain in sagittal (A) and cross-sectional (B) views. The detailed distribution of 5-HT-stained neurons in (B) is drawn from 25-μm tissue sections. Cell groups B₁–B₃ are shown in the right column, while the remainder are in the left column. Sections are in rostral (top) to caudal (bottom) order,

into the alphanumeric system. The approximate locations of the B1–B9 cell groups, as they appear in the rat, are shown in Figure 10.4 in horizontal (Figure 10.4A) and sagittal (Figure 10.4B) section (Fuxe and Johnson, 1974). Some of the major pathways are also indicated. More exact locations are given in a series of rat brain cross sections in Figure 10.4C.

The main cell groups are in the medulla (B1–B3), pons (B4–B6), and midbrain (B7–B9). In the medulla, the midline groups in the raphe obscurus (RO in Figure 10.4C) and raphe pallidus (RP) correspond to groups B1 and B2, while the raphe magnus (RM) group, which spreads laterally into the reticular formation, corresponds to group B3. In the pons, the cell group in the nucleus raphe pontis corresponds to the B5 group while a more or less continuous collection of cells in the central gray (CG), lying near the floor of the fourth ventricle and aqueduct, corresponds to the B4 and B6 cell groups. The largest group (B7) is found in the nucleus raphe dorsalis (Figure 10.5). Most of the neurons are in the dorsal medial and ventral medial mesencephalic part of the nucleus. In addition, 5-HT cells are found in the nucleus centralis superior [NCS (B8)] and the lemniscus medialis [MED LEM (B9)]. They are also distributed unevenly over the caudal aspect of the nucleus interpeduncularis (INP).

Most authors report that 5-HT cell clusters tend to spread laterally from the presumed anatomical outlines of the raphe nuclei to the surrounding reticular areas, and 5-HT cells by no means represent the only population of raphe neurons. Several authors have reported groups of neurons that are difficult to fit into the alphanumeric scheme, some of which have been revealed only after pretreatment with a tryptophan load and an MAO inhibitor.

5-HT neurons have been found in or near some catecholamine groups, particularly the substantia nigra. Wiklund et al. (1981), Steinbusch (1984), and Lidof and Molliver (1982) have all reported 5-HT-positive cell bodies in this area, although they have not been found by Felten and Sladek (1983) or Jacobs et al. (1984). Sladek and Walker (1977) and Jacobs et al. (1984) noted positive cells in the area of the locus coeruleus.

Steinbusch (1984) reported 5-HT cells in the area prostrema of the rat, although Jacobs et al. (1984) did not find these cells in the cat. Frankfurt et al. (1981) and Steinbusch (1984) report 5-HT-positive neurons in the dorsomedial hypothalamus.

In summary, it can be said that the numbers of 5-HT cell bodies in the brain are remarkably few, with the overwhelming majority being in the raphe and reticular systems. Newer techniques, especially the immunocytochemical technique for 5-HT, have confirmed the main outlines of the original distribution of 5-HT neurons described by

---

proceeding from the left column to the right. The section at level AA shows the distribution of rostrally projecting fiber bundles of various amines. The arrow points to the cross section that corresponds to Figure 10.5. Anatomy: (C) crus cerebri; (CG) central gray; (F) fornix; (FMT) fasciculus mamillothalamicus; (FR) fasciculus retroflexus; (Ic) inferior colliculus; (III) oculomotor nucleus; (INP) interpeduncular nucleus; (ioc) inferior olivary complex; (LM) lemniscus medialis; (LRN) lateral reticular nucleus; (MED LEM) medial lemniscus; (MLF) medial longitudinal fasciculus; (MRF) mesencephalic reticular formation; (MV) motor nucleus trigeminal complex; (NCS) nucleus centralis superior; (NGc) nucleus gigantocellularis; (P) tractus corticospinalis; (PAG) periaqueductal gray; (RA) nucleus retroambigualis; (RD) nucleus raphe dorsalis; (RM) nucleus raphe magnus; (RN) red nucleus; (RO) nucleus raphe obscurus; (RPC) nucleus reticularis pontis centralis; (SAF) superficial arcuate; (SC) superior colliculis; (SO) superior olive; (SPV) spinal nucleus trigeminal complex; (SSP) nucleus supraspinalis; (TRAP) trapezoid body; (VN) vestibular nucleus. After Bowker et al. (1982).

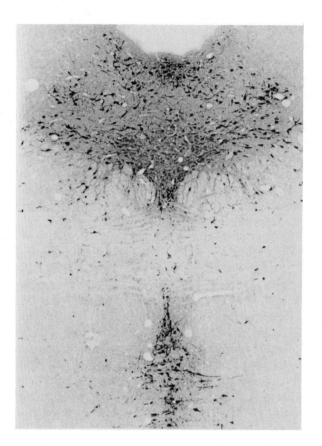

Figure 10.5. Immunohistochemical staining of 5-HT neurons in the rat dorsal raphe. The approximate level in the brain stem is indicated by the arrow in Figure 10.4C. Hyperimmune anti-serotonin rabbit IgG was diluted 1:50,000, indicating the strength of the antibody. Antiserum kindly supplied by Dr. H. Kimura (Shiga University, Japan). Photomicrograph courtesy of Dr. H. Tago (University of British Columbia).

Dahlstrom and Fuxe with a number of significant additions, particularly in terms of the spread of brain stem 5-HT groups laterally from the midline confines of the raphe nuclei. Some species differences may exist, which could account for minor discrepancies reported among the rat, cat, and monkey, such as in the area prostrema. Other areas of controversy—such as those that involve nuclei like the substantia nigra and hypothalamus that receive a dense 5-HT innervation and reveal cell bodies only after tryptophan loading and MAO inhibition—require further confirmation.

An intriguing and relatively unexplored area of 5-HT anatomy and function is that involving dual transmitters. Substance P (SP) (Hökfelt *et al.*, 1980b; Lovick and Hunt, 1983), the enkephalins (Charnay *et al.*, 1985), cholecystokinin, and thyrotropin-releasing hormone (Bowker *et al.*, 1983), and even the type II neurotransmitter γ-aminobutyric acid (Belin *et al.*, 1983), have been found to coexist with 5-HT in the raphe nuclei neurons [see Table 5.2 (Section 5.2)].

5-HT cells also occur outside the CNS. It has long been known that argentaffin cells of the gut are rich in 5-HT. In addition, 5-HT-immunoreactive cells have been reported in the superior cervical ganglion of the rat (Verhofstad *et al.*, 1981), in the guinea pig (Jaim-Etcheverry and Zieher, 1968) and rat (Koevary *et al.*, 1983) pancreas, and in the guinea pig celiac–superior mesenteric plexus (Ma *et al.*, 1985). SP (Lovick and Hunt, 1983) is said to coexist with 5-HT in intestinal enterochromaffin cells, and 5-HT has been colocalized with adrenaline in some adrenal medullary cells (Verhofstad and Jonsson, 1983) and with NA in nerve endings in the rat vas deferens and pineal (Jaim-Etcheverry and Zieher, 1980).

10.4.2. Serotonin Pathways and Terminals

While the numbers of 5-HT cell bodies in the CNS are considerably fewer than for the catecholamines, terminals are equally ubiquitous. Every area of brain and spinal cord receives 5-HT nerve endings. The more caudal 5-HT cells project particularly to the spinal cord, while the more rostral ones project to the diencephalon and forebrain. Projections are all overwhelmingly ipsilateral (see Figure 10.4B). Double-labeling experiments have generally established extensive overlap between cell body areas and terminal fields. Therefore, the main fiber bundles indicated diagrammatically in Figure 10.4 represent a mixture of neuronal groups of origin.

Pathways to the spinal cord are intermediate, ventral, and dorsal. The intermediate descending bulbospinal pathway arises mainly from the nuclei raphe pallidus and obscurus. It projects through the intermediate zone and terminates in the intermediate lateral cell column. The ventral descending bulbospinal pathway also arises from the nuclei raphe pallidus and obscurus and terminates in the region of anterior horn cells (R. F. Martin et al., 1978). The dorsal descending pathway travels down the dorsolateral funiculus to reach the dorsal horn. It contains only a minority of serotonergic fibers.

Bowker et al. (1981) have shown that medullary cell groups B1–B3, as well as B5, provide the major 5-HT input to the entire spinal cord, while cell groups B7 and B9 of the midbrain project to the cervical and probably the upper thoracic spinal cord. The densest innervation is to the substantia gelatinosa.

A pathway to the cerebellum traverses the pedunculus cerebellaris medius, terminating in the vicinity of Purkinje cells. The cells of origin include those in the nuclei raphe dorsalis, pontis, magnus, obscurus, and pallidus (Azmitia and Segal, 1978) and the nuclei gigantocellularis and paragigantocellularis (G. A. Bishop and Ho, 1985).

Ascending projections are dorsal, medial, and ventral. The dorsal ascending pathway arises primarily from the medial and rostral parts of the nucleus raphe dorsalis. It runs dorsolateral to the medial forebrain bundle and terminates predominantly in the neostriatum. The medial ascending pathway originates largely in the raphe dorsalis and projects primarily to the substantia nigra. The ventral ascending pathway is the major pathway. It innervates the diencephalon, limbic system, and cortex.

The distribution of 5-HT terminals from this major ascending pathway has been studied by histofluorescence (Dahlstrom and Fuxe, 1964b), autoradiography with [³H]-5-HT (Chan Palay, 1977; Descarries and Beaudet, 1978), and immunocytochemistry (Steinbusch, 1984). In the cortex, the density of 5-HT-containing axons exceeds that of noradrenergic fibers (Lidof et al., 1980). Layers V and VI receive a dense noradrenergic projection and a sparse 5-HT projection in the monkey primary visual cortex, whereas layer IV receives a dense 5-HT projection and is largely devoid of noradrenergic fibers (Morrison et al., 1982). These authors suggest that the raphe–cortical 5-HT projection preferentially innervates spiny stellate cells of layers IV, whereas the coerulial–cortical noradrenergic projection innervates primarily pyramidal cells.

Much more intense innervation of the limbic system and the hypothalamus is observed, particularly in the area of the medial forebrain bundle, islands of Calleja, some septal area nuclei, and the interstitial nucleus of the anterior commissure. Parts of the amygdala, thalamus, and hypothalamus also contain heavy innervation. In the midbrain, the substantia nigra pars reticularis and lateralis stain intensely, as do parts of the interpeduncular nucleus. Also receiving heavy innervation in the rhombencephalon are some lateral parts of the reticular formation near the superior cerebellar peduncles, the motor nucleus of the fifth nerve, the facial nerve, the central gray, the nucleus tractus solitarii, and the locus coeruleus.

There is a supraependymal plexus of nerve fibers that project through or lie on the cerebral ventricular ependyma. This appears in the aqueduct, the lateral ventricles, and the dorsal part of the third ventricle. In addition, there is serotonergic innervation of pial arteries and arterioles (Edvinsson *et al.*, 1983; Itakura *et al.*, 1985; Marco *et al.*, 1985), but the source is not entirely clear. The innervation is proposed to relate to cerebral vascular headaches.

10.5. Physiological Actions of Serotonin and Serotonin Receptors

5-HT-containing neurons in brain have been proposed to play a role in physiological mechanisms such as sleep, appetite and thermoregulation, pain perception, and control of pituitary secretions, as well as in the pathophysiology of migraine, myoclonus, and depressive illness. Much of the evidence is pharmacological, and the exact mechanisms are still not clear despite many hundreds of studies. In many of the functions traditionally associated with hypothalamic regions (e.g., appetite, sexual behavior, control of pituitary secretions) as well as in sleep, 5-HT and the catecholamines seem to have opposing effects, as was originally proposed by Brodie *et al.* (1959).

It has been difficult to sort out the contributions of the catecholamines and 5-HT, because many of the drugs commonly used have affected both systems and because of the intimate interrelationship of these systems in brain. Changes in 5-HT innervation are thought to modulate catecholamine release in some regions, as well as α_2- and β-adrenoreceptor sensitivity (D. S. Janowsky *et al.*, 1982). This has been postulated to be a factor in the action of antidepressant drugs (see Section 15.5.3). Many of the effects seen after pharmacological or surgical manipulations of 5-HT systems may also be indirectly mediated by changes in carbohydrate metabolism, in blood flow (see Section 10.5.1), or in the release of various pituitary hormones. Through innervation of other types of neurons in the hypothalamus (including DA and NA neurons) (C. O. Lynch *et al.*, 1984; Gudelsky *et al.*, 1984), 5-HT tends to facilitate release of adrenocorticotropic hormone, growth hormone, and prolactin, and to inhibit secretion of luteinizing hormone (LH), follicle-stimulating hormone, and thyroid-stimulating hormone. Depending on species, 5-HT innervation of the intermediate pituitary acts either to diminish or to increase the secretion of melanocyte-stimulating hormone (D. T. Krieger, 1978). It is such multiple actions and interactions that make it difficult to define the relationship of any brain functions to any single neurotransmitter system.

10.5.1. Serotonin Receptors

The actions of 5-HT on single neurons are variable and hard to interpret. The most general effect of iontophoretic application in mammalian CNS is a reduction in excitability, but a facilatory effect has also been found in many brain regions (F. E. Bloom *et al.*, 1973). In some situations, excitatory and inhibitory effects can even be observed in the same cell (M. H. Roberts and Straughan, 1967), illustrating, as with DA (see Chapter 9), the vagaries of this experimental approach. The responses observed are often of relatively slow onset and long duration, suggesting metabotropic actions. The variability in action and in response to pharmacological manipulation suggested the existence of multiple types of receptors, and binding studies have provided additional strong support for at least three, and possibly more, receptor subtypes. These are usually defined as:

5-HT$_1$ (or S$_1$): defined by high affinity (nanomolar) for 5-HT
    Subdivided into: 5-HT$_1$A with high affinity (2–13 nM) for spiperone
                    5-HT$_1$B with low affinity (35 μM) for spiperone
5-HT$_2$ (or S$_2$): defined by high affinity (nanomolar) for spiperone but low affinity (micromolar) for 5-HT; ketanserin (K_D = 0.4 nM) and cyprohep-tadine are highly selective ligands

8-Hydroxy-2-(di-*n*-propyl)aminotetralin (PAT) (Berge *et al.*, 1985; Marcinkiewicz *et al.*, 1984) and 1-[2-(3-bromoacetaminophenyl)ethyl]-4-(3-trifluoromethylphenyl)piperazine (Ransom *et al.*, 1985) have been said to be selective ligands for the 5-HT$_1$A as compared to the 5-HT$_1$B or 5-HT$_2$ sites. LSD labels both 5-HT$_1$ and 5-HT$_2$ binding sites.

A possible fourth type of binding site has been described in human, rat, and bovine cortex and has been designated the 5-HT$_3$ (or S$_3$) type; it has a 50–250 nM affinity for 5-HT but micromolar affinities for both spiperone and ketanserin (Todd and Ciaranello, 1985).

The general pattern of distribution of 5-HT$_1$ and 5-HT$_2$ sites in rat brain is indicated in Table 10.2. The 5-HT$_2$ sites are particularly concentrated in cortical and hippocampal areas and in the striatum, while the 5-HT$_1$ sites are more widely distributed. Of course, when [^{3}H]spiperone is used as the ligand for 5-HT$_2$ sites, DA binding sites are also labeled, but these can be distinguished by the use of selective displacing agents. DA sites appear to account for only about 15% of the total spiperone binding in rat cortex, but for almost the total binding in the striatum (Peroutka and Snyder, 1981). Particularly high concentrations of 5-HT$_1$ sites are found in the dorsal raphe, substantia nigra, interpeduncular nucleus, and certain subregions of the hippocampus (Biegon *et al.*, 1982). 5-HT$_1$A sites (labeled by PAT) occur to only a small extent in the choroid plexus, striatum, subiculum, substantia nigra, and lateral preoptic area, although these regions contain many 5-HT$_1$B sites (labeled by 5-HT but not by PAT) (Marcinkiewicz *et al.*, 1984).

The relationship of the various receptors to the multiple physiological actions of 5-HT is not yet clear. Drug specificity studies are said to indicate that synaptic excitation may be associated with 5-HT$_2$ and inhibition with 5-HT$_1$ receptors (Snyder and Peroutka, 1980). Some have suggested that the pharmacology of the 5-HT$_2$ sites corresponds with central activity of 5-HT antagonists and with the pharmacology of receptors in peripheral vasculature, while the pharmacology of the 5-HT$_1$ sites corresponds with that of central presynaptic receptors and those that modulate contraction of central vasculature (Deshmukh *et al.*, 1983; Peroutka *et al.*, 1981). In humans and dogs, 5-HT$_1$ (but not 5-HT$_2$) binding sites could be localized to the medial layer of the vascular wall in the basilar artery (Peroutka and Kuhar, 1984). Lesion studies suggest that many 5-HT$_1$ sites in the striatum are presynaptic, but those in the hippocampus seem to be on postsynaptic neuronal membranes (Gozlan *et al.*, 1983).

TABLE 10.2
**Regional Distribution of Putative 5-HT$_1$ and 5-HT$_2$ Binding Sites
in Rat Brain**

Brain region	Distribution (pmoles/g wet wt.)[a]				
	5-HT$_1$		5-HT$_2$		
	(1)	(2)	(2)	(3)	(4)
Frontal cortex	45.4	28	39	31	5.4
Temporal cortex		—	—	19	
Olfactory tubercle	—	—	—	21	—
Striatum	6.6	—	—	15	3.7
Hippocampus/amygdala	7.6	—	—	2.5–3	1.6
Midbrain (thalamus)	3.5	30	ND	(2.3–3)	1.6
Hypothalamus	2.4	39	ND	0.7	0.7
Medulla	2.5	13	4	2.3	0.4
Cerebellum	0.5	—	ND	ND	0.2
Spinal cord	—	8	ND	ND	—

[a](ND) Not detectable; (—) not determined. Studies: (1) [^{3}H]-5-HT binding with LSD as displacer (Peroutka and Snyder, 1981); (2) [^{3}H]-LSD binding with various displacers (Blackshear *et al.*, 1981); (3) [^{3}H]ketanserin binding (Leysen *et al.*, 1982); (4) [^{3}H]spiperone binding with cinanserin as displacer (Peroutka and Snyder, 1981).

The number of 5-HT receptors seems to be very responsive to pharmacological manipulation, with agents that increase or decrease 5-HT action causing, respectively, decreases or increases in receptor number (Fuller, 1984). Such modulation may be more effective for the postsynaptic as compared to the presynaptic receptors, and the down-regulation of 5-HT$_2$ sites has been postulated to play a key role in the clinical efficacy of antidepressants (Snyder and Peroutka, 1980; Blier and de Montigny, 1980) (cf. Chapter 15).

5-HT$_1$ sites, but not 5-HT$_2$ sites, are modulated by guanine nucleotides and are therefore probably linked to an adenylate cyclase, a possibility that is also supported by a correlation between the effects of various drugs on 5-HT$_1$ binding and 5-HT-stimulated adenylate cyclase (Snyder and Peroutka, 1980). In *Aplysia,* it has been shown that activation of 5-HT-sensitive adenylate cyclase causes a complete and prolonged closure of a particular type of K$^+$ channel, leading to increases in the duration of the action potential, Ca^{2+} influx, and neurotransmitter release (Siegelbaum *et al.*, 1982).

Linkage of 5-HT$_2$ receptors with a phosphoinositide second messenger system (see Section 5.4.3) is suggested by the report that the 5-HT-stimulated phosphoinositide hydrolysis seen in rat cortical slices is blocked by the selective 5-HT$_2$ antagonists ketanserin and pizotifen (Conn and Sanders-Bush, 1984).

Still other types of 5-HT receptors may be of physiological importance. 5-HT has been shown to stimulate glycogen hydrolysis in slices of mouse cortex, and antagonist studies suggest that a novel type of receptor (neither 5-HT$_1$ nor 5-HT$_2$) may be involved. The effect is apparently initiated by translocation of Ca^{2+} ions, which causes an activation of phosphorylase leading to glycogen breakdown. This suggests the possibility that 5-HT, like NA, histamine, and vasoactive intestinal polypeptide, may be involved in regulation of carbohydrate metabolism in the CNS (Quach *et al.*, 1982).

The probability that the highly divergent serotonergic systems in brain play a role in the control of carbohydrate metabolism as well as that of central vasculature may help to explain why this neurotransmitter has been reported to play some role in so many diverse physiological functions (see Section 10.8).

Gaddum and Picarelli (1957) suggested the presence in the periphery of at least two distinct types of 5-HT receptors, since they showed that 5-HT contracts the guinea pig ileum by two distinct mechanisms: (1) a direct action on smooth muscle that is blocked by dibenzyline and is therefore attributed to a "D"-type excitatory receptor on the muscle and (2) an indirect action mediated by the release of acetylcholine (ACh) from parasympathetic nerves that is blocked by morphine and is therefore attributed to an "M"-type receptor on the parasympathetic nerve endings. A comparison of the binding activities of various 5-HT analogues to cerebral membranes with their agonist actions at the M and D receptors indicates that both the latter differ from the 5-HT_1 site, but that the peripheral D receptors may be similar to the central 5-HT_2 sites (Richardson *et al.*, 1985).

10.5.2. Serotonin and Sleep

The cat normally spends about two thirds of its time sleeping. The sleep is uneven, with 20- to 30-min periods of light sleep alternating with 5- to 10-min periods of deeper sleep. Light sleep is characterized by slow, synchronized cortical EEG activity and by muscle tone not greatly reduced from the waking state. It is thus referred to as "slow-wave" sleep. Deep sleep is characterized by a profound relaxation of peripheral muscle accompanied by rapid eye movements (REM) and a paradoxical activation of the cortical EEG. It is thus referred to as "REM" or "paradoxical" sleep. In man, it has been established that this REM stage is when dreaming occurs. Jouvet and his colleagues reasoned that sleep was not just a passive process. Using the cat as an experimental model, they commenced searching for an area of brain that actively promotes sleep in the same way that the reticular activating system promotes wakefulness. They discovered that electrolytic lesions of the raphe system turned cats into insomniacs. There was a reduction in the time spent sleeping from 85 to 20% (Jouvet, 1973).

Because of the known relationship of the raphe to 5-HT systems, Jouvet next studied the possible role of 5-HT in this phenomenon. In studies extending more than a decade, Jouvet and his colleagues showed that inhibition of 5-HT biosynthesis by *p*-chlorophenylalanine (PCPA) (see Section 10.6.2) induces in the cat total insomnia that is accompanied by a permanent discharge of pontogeniculooccipital (PGO) activity. Small doses of 5-HTP immediately suppress PGO activity and restore both slow-wave and REM sleep. Insomnia can also be produced in the cat by treatment with 5-HT antagonists such as methysergide, LSD, methiothepin, and metergoline (Sallanon *et al.*, 1982).

Jouvet also investigated the role of catecholamines. Administration of L-dopa produced a long-lasting arousal in the cat. Inhibition of catecholamine synthesis by the administration of α-methyl-*p*-tyrosine decreased waking in normal cats and totally suppressed the behavioral and EEG arousal that normally follows the injection of amphetamine. Lesion experiments suggested that it was the noradrenergic neurons of the locus coeruleus that played an important role in cortical arousal (Jouvet, 1973).

Unfortunately for ease in interpretation, there appear to be some species differences, especially in the effects on slow-wave sleep. In rats, for example, experiments with the highly selective Tr-OH inhibitor 6-fluorotryptophan indicate that a decrease in 5-HT leads to a decrease in REM sleep and either no change (Sugden and Fletcher, 1981) or a decrease (Nicholson and Wright, 1981) in slow-wave sleep. In man, PCPA decreased REM sleep without affecting slow-wave sleep. 5-HTP not only reversed the PCPA effect; it actually elevated the time spent in REM sleep in normals. L-Dopa, on the other hand, reduced the duration of REM sleep, and its discontinuation was associated with a marked rebound effect (R. J. Wyatt *et al.*, 1970). The difficulties of interpreting experiments on the roles of the aromatic amines in REM sleep have been reviewed, with particular emphasis on the probable importance of modulation by 5-HT of α_2-adrenoceptors (Gaillard, 1983).

Despite the difficulties in interpretation, there is general agreement that 5-HT systems are somehow involved in the promotion of sleep, while catecholamine systems are involved in the promotion of wakefulness.

10.5.3. Serotonin and Sex

Serotonergic neurons in the raphe discharge at a slow rhythmic rate, suggesting that they serve a pacemaker function in the CNS (Aghajanian and Vandermaelen, 1982). The circadian changes in 5-HT levels, originally found by Quay (1965), and the heavy 5-HT innervation of the suprachiasmatic nucleus of the hypothalamus suggest that 5-HT may be important in controlling cyclic events (see Section 10.7.1), particularly cyclic changes in the reproductive neuroendocrine axis involving LH release. Most evidence suggests that 5-HT is inhibitory in female and, to a lesser extent, in male (McIntosh and Barfield, 1984) reproductive behavior, while NA is facilatory; it has been suggested that the decrease in the ratio of NA to 5-HT activity with aging may cause the age-related shutdown of reproductive function (R. F. Walker, 1984). A 5-HT antagonist, metergoline, has shown some promise as a therapeutic agent to restore menses in women with idiopathic normoprolactinemia secondary amenorrhea (Fadini *et al.*, 1984). 5,7-Dihydroxytryptamine (5,7-DHT) lesions in female rats produce facilitated sexual behavior. Fetal raphe cells transplanted into the hypothalamus reverse this effect (see Section 10.6.1). Reinnervation of the ventromedial nucleus of the hypothalamus by the transplants is associated with restoration of normal sexual function (Luine *et al.*, 1984) (Chapter 13).

10.5.4. Serotonin and Pain

There is a very large and growing literature on the role of 5-HT in neural mechanisms related to nociception. It has been proposed that analgesia produced by morphine or by stimulation of the periaqueductal gray involves a descending path with a synaptic relay in the nucleus raphe magnus. It is not known whether a 5-HT system is directly involved in this descending pathway or whether 5-HT neurons merely modulate the activity indirectly, but lesions of the 5-HT neurons or treatment of animals with PCPA or a 5-HT antagonist interferes with both stimulation- and morphine-induced analgesia. Oddly enough, in contrast to the central effects, 5-HT is an algesic agent when given peripherally, since it excites many nociceptive afferents (Willis, 1981).

10.5.5. Serotonin, Mood, and Mental Illness

The most fascinating aspect of 5-HT physiology is its possible involvement in mood and behavior. Ever since the original hypothesis of Woolley and Shaw (1954) that 5-HT

TABLE 10.3
Some Drugs That Affect Serotonin Systems[a]

Drugs and presumed mechanisms of action	Physiological effects
Toxins for serotonergic neurons	
5,6-Dihydroxytryptamine (5,6-DHT)	Not precisely determined
5,7-Dihydroxytryptamine (5,7-DHT)	Not precisely determined
p-Chloroamphetamine (PCA)	Not precisely determined
Inhibitors of tryptophan hydroxylase	
p-Chlorophenylalanine (PCPA)	Antisleep?
6-Fluorotryptophan	Antisleep?
N,N-Diethyl-3,4-dihydroxyphenylpropionamide	Convulsant?
Inhibitor of 5-HTP decarboxylase	
Carbidopa	None by itself
Serotonin-uptake inhibitors	
Chlorimipramine, imipramine, nortryptyline, clomipramine, fluoxetine, panuramine	Antidepressants (see Chapter 15)
Serotonin agonists	
8-Hydroxy-2-(di-n-propylamino)tetralin (PAT)	Not precisely determined
N,N-Dimethyl-5-methoxytryptamine	Hallucinogen
Quipazine	Not precisely determined
Serotonin antagonists	
Methysergid	Prophylaxis of migraine
Lysergic acid diethylamide (LSD)	Hallucinogen, antisleep
Ketanserin, cyproheptadine, pizotifen	Reduce intestinal motility
Metergoline	May promote menses

[a]For MAO inhibitors, vesicle storage inhibitors, and broad-spectrum receptor antagonists that affect both the catecholamine and the 5-HT systems, see the text and Section 9.6.

metabolism might be deranged in mental illness, there has been speculation that various psychopharmacologically active agents owe their efficacy to a 5-HT interaction. This applies to hallucinogens as well as to antidepressant drugs. There have also been suggestions that the 5-HT derivative melatonin may be involved in mental illness. These aspects are discussed in more detail in Chapter 15.

10.6. Pharmacology of Serotonin

There exists a wide variety of pharmacological agents that specifically interact with serotonergic systems. The spectrum is broadened where there is a high degree of similarity between 5-HT and catecholamine mechanisms. This is the case for the decarboxylase and MAO enzymes, the vesicular storage mechanism, and, to some extent, the reuptake pump system. Table 10.3 is a list of the more prominent compounds. Figure 10.6 is a semidiagrammatic representation of how they may act at the synaptic level.

10.6.1. Agents That Are Toxic to Serotonin Neurons

Three agents that selectively damage 5-HT neurons have so far been discovered: 5,6-dihydroxytryptamine (5,6-DHT), 5,7-DHT (Baumgarten and Lachenmayer, 1972), and p-chloroamphetamine (PCA). These neurotoxic agents are accumulated within 5-HT neu-

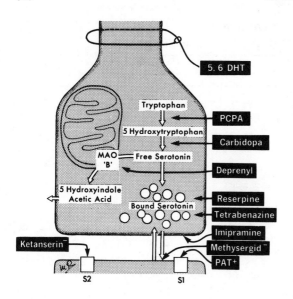

Figure 10.6. Mechanism of drug action at serotonergic synapse represented as a schematic diagram of a 5-HT nerve ending. The black on white printing shows the site of action of endogenous serotonergic elements. The white on black printing indicates various drugs that interact with 5-HT and their proposed sites of action (see Table 10.3).

rons by action of the membrane pump. The destructive effect of 5,6- and 5,7-DHT is probably due to the ease with which they are oxidized to materials that poison the respiratory enzymes of the cell. They share this characteristic with 6-hydroxydopamine (6-OHDA), which is pumped into catecholaminergic cells and is oxidized to a toxic derivative.

Uptake studies using labeled 5,6-DHT have shown that it competes with 5-HT for uptake and is accumulated by brain slices preferentially *in vitro*. It does not compete with NA or DA uptake. 5,7-DHT is 6 times less potent than 5,6-DHT in inhibiting 5-HT uptake, which may explain why it is a less potent neurotoxin (Horn *et al.*, 1973).

After a single intraventricular injection of 50–75 μg 5,6-DHT in the rat, damage to unmyelinated axons can be observed as early as 2 hr after the injection. About half the 5-HT in the forebrain disappears after 10 days. Due to its general toxicity, it has so far not been possible to administer high enough doses of 5,6-DHT intraventricularly to bring about total depletion of 5-HT in the brain, but well-placed local injections into ascending 5-HT tracts can produce a 90–95% reduction in the terminal areas (Björklund *et al.*, 1974).

With sublethal doses of 5,6-DHT, some regeneration of 5-HT neurons seems to take place over a period of weeks. The terminals apparently grow back into appropriate receptor areas, because the supersensitivity to 5-HT disappears as reinnervation takes place. This discovery of sprouting of 5-HT and catecholamine neurons was one of the first events leading to reevaluation of the classical concept that CNS neurons are incapable of regeneration (Fuxe and Johnsson, 1974) (see Chapter 13).

PCA is also toxic to serotonergic neurons. 5-HT levels of brain are reduced to less than half the normal values a few days following the injection of a single intraperitoneal dose of approximately 0.1 mmole/kg. It is a selective inhibitor of MAO. It does not inhibit 5-HT synthesis *in vitro*. The manner in which the selectively accumulated PCA damages 5-HT neurons is a mystery, because it is not readily oxidized to potentially toxic metabolites as are 6-OHDA and 5,6-DHT (Sanders-Bush *et al.*, 1974).

10.6.2. Inhibitors of Tryptophan Hydroxylase

The classical inhibitor of Tr-OH is PCPA. Koe and Weissman (1966) discovered the effects of PCPA by examining its ability to inhibit liver P-OH so that its specificity is not absolute. It has a complicated mechanism of action *in vivo* because it produces a depletion of 5-HT that persists long after the drug has been metabolized. A short-term effect on the enzyme has been documented *in vitro,* where PCPA is a competitive inhibitor of tryptophan with a K_i (the concentration required to produce 50% inhibition) of about 300 μM, but another explanation is needed to account for its long-term effect *in vivo.* Gál and Whitacre (1982) have shown that PCPA is incorporated as a tyrosyl residue into the enzyme protein in such a way as to render the catalytic site inactive, although it still reacted with antibodies to purified active enzyme. This is a good demonstration of the principle that the catalytic and immunological sites of a protein may be completely different. Gál and Whitacre (1982) suggest that PCPA is activated to form a PCPA~tRNA^{Phe} complex that is then hydrolyzed to L-tyrosyl~tRNA^{Phe}. Consequently, tyrosine would be on the wrong anticodon and would be incorporated at or near the active site [see Figure 1.12 (Section 1.2.3)]. This occurs in both phenylalanine hydroxylase and Tr-OH, again suggesting similarities between the catalytic sites of the two enzymes. Incorporation into TH also occurs, but does not lead to an inactive enzyme. The inhibition of Tr-OH is not reversed *in vivo* until new enzyme molecules are synthesized, a process that takes several days. The effect is nevertheless reversible, because PCPA is not toxic to the neurons themselves.

The catechol derivative *N,N*-diethyl-3,4-dihydroxyphenylpropionamide is a competitive inhibitor of Tr-OH that is believed to owe its audiogenic seizure-inducing properties to its selective effects on 5-HT systems (Stine and Kellogg, 1982).

Reversible and more selective inhibition is produced by 6-halotryptophan derivatives, particularly by 6-fluorotryptophan (E. G. McGeer *et al.,* 1968).

10.6.3. Inhibitors of 5-Hydroxytryptophan Decarboxylase

There is a variety of agents that inhibit the decarboxylation of 5-HTP, but they are nonselective, because the same enzyme acts on many aromatic amino acids, and are ineffective at lowering brain 5-HT, because the decarboxylase is present in so much higher concentration than Tr-OH (see Section 10.2.1). The better-known inhibitors have been discussed in Chapter 9 on catecholamines. Those that do not cross the BBB are not only of practical usage in the treatment of Parkinson's disease but also have been used in conjunction with 5-HTP administration to reduce the amount of peripheral decarboxylation and thus increase the amount of 5-HTP reaching the brain. The combination of carbidopa and 5-HTP, for example, has proven extremely useful in the treatment of myoclonus, particularly of postanoxic origin (Barbeau, 1978).

10.6.4. Inhibitors of Monoamine Oxidase

The properties of MAO and the MAO inhibitors (MAOIs) were extensively discussed in Section 9.4 (see also Section 15.5.3). Broad-spectrum MAOIs, affecting both type A and type B MAO, are mostly of the irreversible hydrazine type exemplified by phenylzine, but some, such as tranylcypromine, are reversible. Narrow-spectrum MAOIs are also available. Of these, clorgyline is the most widely used for MAO-A. Cimoxatone, 3-[4-(3-cyanophenylmethoxy)phenyl]-5-methoxymethyl-2-oxazolidinone, is another, very potent, fully reversible, and selective inhibitor of type A (C. J. Fowler and Benedetti, 1983). Other, less potent, type A inhibitors are harmaline, harmine, and α-ethyltryptamine, all of which are short-term competitive-type inhibitors. Deprenyl is the most commonly used material that is relatively selective for MAO-B.

MAOIs were the first type of "psychic energizer" used in the treatment of depression (see Section 15.5.3). They were gradually displaced by the tricyclic antidepressants (see Sections 10.6.5 and 15.5.3), largely because of the risk of hypertensive crises precipitated by ingestion of foods containing amines such as tyramine and tryptamine (the "beer and cheese syndrome"). Nevertheless, MAOIs are still used to some extent in depression.

So far, it is not known whether selective type A or type B inhibitors will produce more specific clinical effects than the nonselective MAO inhibitors that have been generally used.

10.6.5. Inhibitors of Serotonin Uptake

High-affinity uptake (pump) processes (cf. Chapters 5, 6, 7, and 9) are considered to be the major route for inactivation of most amino acid and amine neurotransmitters, with the major exceptions being histamine and ACh, and inhibition of the specific pump(s) prolongs neurotransmitter action at the receptor site. The tricyclic antidepressants are effective at inhibiting either NA or 5-HT uptake, although it is now questioned whether this is the key mechanism of antidepressant activity (see Section 15.5.3). Examples of compounds with relatively selective action on 5-HT, as opposed to NA, uptake systems are chloroimipramine, imipramine, nortryptyline, chlomipramine, and fluoxetine. The most widely studied of these compounds is imipramine. Quantitative autoradiographic work has shown that the distribution of binding sites for [^{3}H]imipramine parallels that of 5-HT nerve terminals (Biegon and Rainbow, 1983) and has given evidence for axonal transport of imipramine binding sites in ascending 5-HT axons of the medial forebrain bundle (T. M. Dawson et al., 1985).

Panuramine has recently been reported to be a novel, very potent, noncompetitive inhibitor of 5-HT uptake that, with an IC$_{50}$ of 22 nM, is as potent as, or more potent than, the tricyclics, and more selective in that the IC$_{50}$ for NA uptake is 850 nM (Blurton et al., 1984).

In some studies, particularly of psychiatric patients, uptake of 5-HT by blood platelets has been used as a "model" for neuronal uptake. The processes are pharmacologically different, however, and in fact DA and 5-HT seem to share a common uptake system in human platelets (Omenn and Smith, 1978).

Evidence from tissue culture work suggests that mammalian astrocytes can take up 5-HT by an Na$^+$-dependent, high-affinity system, sensitive to chlorimipramine and fluoxetine. This property had been considered to reside exclusively in 5-HT neurons (Kimelberg and Katz, 1985). If such uptake proves important in vivo, it may become desirable to seek selective antagonists of the glial and synaptosomal pump systems.

10.6.6. Inhibitors of Storage

The rauwolfia alkaloids and tetrabenazine are classical depletors, not just of 5-HT, but also of the catecholamines and, to some extent, of histamine. The mechanisms that bind these amines to granules are so similar that selective agents acting at this particular level are unknown. For many years, experiments were undertaken to try to sort out which of the clinical effects of reserpine were due to 5-HT and which to catecholamine depletion. Despite thousands of investigations, the question is not settled even today. The time frame for replenishing 5-HT stores following a single dose of reserpine closely parallels that for the catecholamines. It corresponds to the time taken to synthesize new vesicle-binding proteins in the cell bodies and transport them to the nerve endings. The minimum time is 18–24 hr, but it can take up to several days for complete replenishment.

Reserpine sedation can be temporarily overcome by an adequate dose of L-dopa but not of L-5-HTP. This indicates that the major tranquilizing effects are due to depletion of the catecholamines. On the other hand, the diarrhea and hot flush, which are found shortly after reserpine administration, parallel the symptoms of the carcinoid syndrome. Here there is overproduction of 5-HT by malignant enterochromaffin or related cells, which arise in the gastrointestinal tract. Studies with 5-HT antagonists, however, indicate that only the gastrointestinal symptoms are due to 5-HT; the flushing is probably due to histamine (see Chapter 11).

Tetrabenazine, unlike reserpine, does not bind irreversibly to the protein of storage vesicles. Thus, while its mechanism of action is similar, its effects are transient, persisting for 24–36 hr in man.

10.6.7. Receptor Site Agonists and Antagonists

5-HT receptor sites are blocked by many of the same broad-spectrum compounds that block catecholamine receptor sites. This is particularly true of the phenothiazine derivatives. As described in Section 15.6, however, there is strong evidence that blockade of dopamine is primarily responsible for the antipsychotic action.

Many of the drugs with more selective 5-HT, as opposed to catecholamine, antagonistic action are useful primarily because of peripheral effects. An example is methysergid, a congener of LSD that, unlike LSD, has comparatively little action on the CNS; it has been used for the prophylactic treatment of migraine and other vascular headaches, as well as in combating diarrhea and malabsorption in the carcinoid syndrome (see Section 10.6.6). Metergoline is another ergot alkaloid with somewhat similar properties.

Attention has shifted to antagonists selective for specific 5-HT receptor subtypes (see Section 10.5). Of these, ketanserin is the most widely used. This 5-HT_2 antagonist has been reported effective in the treatment of hypertension (Hedner et al., 1984), heart failure, and the intestinal symptoms of the carcinoid syndrome. The results may be due in part to an effect on α_1-receptors (Fuller, 1984).

Cyproheptadine is another 5-HT_2 antagonist used in intestinal conditions involving overproduction of 5-HT; it also has antihistaminic activity, which accounts for its anti-allergic actions (W. W. Douglas, 1980). The group of cyproheptadine derivatives with 5-HT-antagonist properties also includes pizotifen, mianserin, and others (Leysen et al., 1982).

A number of 5-HT agonists are commonly used in experimental studies in animals; they include PAT (see Section 10.5) (Hamon et al., 1984), N,N-dimethyl-5-methoxytryptamine, quipazine, and various other piperazine derivatives. They have not been used clinically, although fenfluoramine, an indirect agonist that releases 5-HT, has been used as an appetite suppressant in the treatment of obesity (Fuller, 1984).

10.7. Melatonin and Other Indoles

10.7.1. Melatonin and the Pineal Gland

For general references, the reader is referred to Haymaker *et al.* (1982), Preslock (1984), and Tamarkin *et al.* (1985).

Melatonin is the only indoleamine derivative other than 5-HT that is known to have a physiological function. It is produced principally, but not exclusively (see Section 10.2.3), by the pineal gland; the retina is another major source, and melatonin is also found in the Harderian gland, the intestines, and some regions of the brain.

The pineal gland is a legendary structure situated between the cerebral hemispheres just above the habenula. It has fascinated physiologists down through the centuries since it was first described by Herophilus more than 2300 years ago, perhaps because it is the only unpaired structure within the cranium. The French philosopher Descartes made the most spectacular suggestion, proposing that it was the seat of the soul. But it was not until Lerner's identification of melatonin in 1958, and Axelrod's recognition of its route of synthesis in 1961, with his later discovery of the effect of light, that any real understanding of its role began to emerge.

The pineal gland, although attached to the roof of the third ventricle by a vestigial and nonfunctional stalk, is actually outside the nervous system, being innervated by noradrenergic fibers of the superior cervical ganglion. It is made up of pinealocytes that, in all vertebrates so far investigated, contain the enzymes for producing 5-HT as well as for converting it to melatonin (see Section 10.2.3). The melatonin is released into both blood and cerebrospinal fluid (CSF) for delivery to specific target sites.

So far, the full range of physiological actions of melatonin is not known. In frogs, it causes dispersion of melanocytes and a lightening of the skin, a property that was critical to tracing its chemical identification but that apparently has no relevance to mammals. Its major mammalian effect seems to be that of modulating gonadal function, generally in an inhibitory fashion. One of the functions of the pineal, at least in small mammals, is to act as a neurochemical transducer, converting information about environmental lighting into chemical activity. To this end, the synthesis of melatonin in the pineal is controlled indirectly from the retina by a polysynaptic path (Figure 10.7). A key role is believed to be played by the suprachiasmatic nucleus of the hypothalamus, and the heavy serotonergic innervation of this nucleus from the raphe cells—which also show a circadian rhythm—is believed to have an important modulatory influence. The final relay is the sympathetic (noradrenergic) nervous system activity that inhibits the biosynthesis of melatonin. Melatonin production is greatly reduced during periods of light, and serum and CSF levels of melatonin and of 5-HT show a very marked nocturnal elevation in monkeys (Garrick *et al.,* 1983) and man, as well as lower mammals.

The pineal synthesis of melatonin may also be modulated by gonadal steroid feedback, particularly by estradiol and progesterone. In the rat, the effect of light is exerted mainly on the *N*-acetyltransferase, which is rate-limiting, and the effect of hormones on the hydroxyindole-*O*-methyltransferase.

In humans, work on the role of melatonin in gonadal function has focused primarily on its role in puberty. Although there are indications from some measurements that a fall in melatonin synthesis may play a part in human gonadal maturation, conflicting results have also been obtained, and there is no conclusive determination as yet (Preslock, 1984).

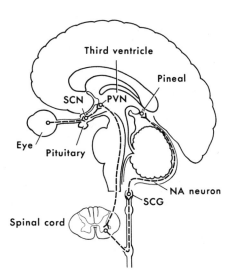

Figure 10.7. Diagram of the human brain (midsagittal section), showing the neural pathway (- - - -) from the eye to the pineal gland. Abbreviations: (SCN) suprachiasmatic nuclei; (PVN) paraventricular nuclei; (SCG) superior cervical ganglia. Modified from Tamarkin *et al.* (1985).

However, parenchymal (secreting) tumors tend to produce hypogenitalism, while non-parenchymal (destructive) tumors can produce sexual precocity in boys. This is consistent with animal studies that show an inhibitory action of melatonin on gonadal function.

The physiological sites and mechanism of action of melatonin are not well defined. It has been supposed that the major sites of action are on the hypothalamic–pituitary axis and the target organs. Specific cytoplasmic binding sites have been reported in the ovary, testis, uterus, brain, and some other organs of rodents (M. Cohen *et al.*, 1978). In the brain, surprisingly high binding is found in the midbrain, with much lower amounts in the hippocampus, hypothalamus, and striatum (Niles *et al.*, 1979). It has been proposed that the action of melatonin—at least on the pituitary—may be mediated through an inhibition of prostaglandin synthesis (Cardinali *et al.*, 1980), and there are indications that *N*-acetyl-5-methoxykynurenamine, a brain metabolite of melatonin (see Section 10.2.3), is an even more potent inhibitor than melatonin itself (Kelly *et al.*, 1984). Much remains to be learned about the functions of these products of the N-acetylation of 5-HT.

Small amounts of a number of other indoles such as 5-methoxytryptamine, 5-methoxytryptophol, and its *O*-acetyl derivative have been reported in mammalian pineal and hypothalamus, but there is no firm evidence of a physiological role, and some even suggest that the 5-methoxytryptamine may be an artifact of the isolation procedure (Narasimhachari *et al.*, 1980).

10.7.2. Tryptamine in the Brain

Tryptamine has frequently been proposed as a neurotransmitter candidate. It is a trace amine, with the amounts reported in rat brain being 0.05–5% of that of 5-HT. In the first edition of this book, we reviewed the available evidence bearing on a neurotransmitter role and concluded that the appearance of tryptamine in mammalian brain was likely to be a fortuitous result of the nonspecificity of L-aromatic amino acid decarboxylase (see Section 10.2.1), rather than of normal physiological importance (P. L. McGeer *et al.*,

1978a). Little has appeared in the last decade to change that view. There has been another report on the existence of specific binding sites for tryptamine in rat brain (Kellar and Cascio, 1982), but Dyck (1984) has found no evidence for either a high-affinity uptake process or K^+-stimulated release. In iontophoretic studies, very small amounts of co-released tryptamine have been found to enhance profoundly the depressant effects of 5-HT (R. S. G. Jones and Boulton, 1980), and Boulton (1979) has proposed that the trace amines may be modulators that may be particularly important in patients treated with psychoactive drugs that modify aromatic amine metabolism.

10.8. Summary

5-HT cell bodies in the CNS are confined to a remarkably limited anatomical area. Seven of the main nine groups that have so far been identified lie within the raphe system of the brain stem, while two groups (B3 and B9) are in the reticular formation. The axons of the more caudal neurons descend to the spinal cord to terminate mainly in the lateral horn, but terminals are also found in other spinal regions. The axons of the more rostral groups innervate the diencephalon, the limbic system, and, to a lesser extent, the striatum and cortex. The divergence of terminals and the pattern of firing of the cells suggest that the role of 5-HT is to exert a generalized tonic effect rather than to have a highly selective action on a limited group of neurons.

The enzyme Tr-OH, which converts tryptophan to 5-HTP, is contained exclusively in 5-HT-producing cells. It controls the synthesis of 5-HT by mechanisms not yet completely understood. 5-HTP decarboxylase, which converts 5-HTP to 5-HT, is present in excess and is probably the same enzyme that decarboxylates L-dopa and other aromatic amines. 5-HT is converted both pre- and postsynaptically to the inactive metabolite 5-HIAA by MAO. This identical pre- and postsynaptic enzymatic action has made it impossible to determine how much of the 5-HT that is synthesized is wasted intraneuronally and how much is released extraneuronally to act on receptor sites.

A major mechanism for removal of 5-HT from the synaptic cleft is by high-affinity uptake rather than metabolism.

Several pharmacologically distinct receptor subtypes for 5-HT have been described, and at some, if not all of these, 5-HT acts metabotropically through a second messenger that may be an adenylate cyclase at the $5HT_1$ or S_1 subtype and a phosphoinositide at the $5HT_2$ or S_2. Pharmacological agents are known that modify the synthesis, binding, uptake, metabolism, and receptor action of 5-HT, and many are of clinical importance.

The physiological actions of 5-HT are not really known, although they seem to follow the pattern of emphasizing trophotropic function. 5-HT promotes sleep, hyperthermia, and, at least peripherally, gut motility. It may have a mild depressing influence on sexual behavior and pain sensitivity, and it may also have a mild antidepressant effect. The possibility that 5-HT may be involved in schizophrenia remains a speculation, based largely on the psychotomimetic action of LSD and other indolealkylamine derivatives of close structural similarity to 5-HT (see Chapter 15).

While 5-HT and the catecholamines seem to be antagonistic in many of their physiological effects, it has nevertheless been extremely difficult to separate their actions through drug studies. Highly similar processes are involved in decarboxylation of the precursors, 5-HTP and L-dopa, in oxidative destruction of the amines, in their intraneuronal storage, in reuptake systems, and even in receptor activity. Thus, such important psychopharmacological agents as the phenothiazines, the rauwolfia alkaloids, the MAOIs, and many of the tricyclic antidepressants all act on 5-HT as well as catecholamine mechanisms (see Chapter 15).

Melatonin is the only other indoleamine known to have a physiological function. It is produced by the pineal, shows a circadian rhythm, and has an inhibitory action on the gonads.

11

Other Heterocyclic Putative Neurotransmitters: Histamine and Purines

11.1. Introduction

In previous chapters, we have dealt with the chemistry, anatomy, pharmacology, and physiology of a number of established nonpeptidergic neurotransmitters and some related neurotransmitter candidates. In this chapter, we will briefly consider histamine and the purines, which are the remaining nonpeptidergic putative neurotransmitters. The existence of histaminergic neurons in brain is well established.

The history of histamine shows several close parallels with that of acetylcholine (ACh). Like ACh, it was first synthesized as a chemical curiosity (Winders and Vogt, 1907), and its physiological role was suspected because of its powerful action on smooth muscle. It was soon recognized as a uterine stimulant in extracts of ergot by Barger and Dale (1910), who immediately undertook intensive pharmacological studies showing that it stimulated a host of smooth muscles (Dale and Laidlaw, 1910). They identified it chemically in extracts of intestinal mucosa and with characteristic acumen drew attention to the fact that it produced signs similar to those shown by animals sensitized to foreign proteins. This observation anticipated by more than 15 years the demonstration by T. Lewis that histamine is liberated by injurious stimulation of skin cells, including the union of antigen and antibody (T. Lewis and Grant, 1927).

Although much of the attention since has been given to histamine in peripheral tissues, knowledge of its presence in brain traces back as far as, or farther than, that of other biogenic amines (Abel and Kubota, 1919). In 1943, Kwiatkowski (1943) found that histamine is concentrated more in gray than in white matter of brain, and Harris *et al.*

(1952) determined that the hypothalamus contains considerably higher concentrations than any other brain region. Local synthesis of histamine in the brain from radioactive histidine was first shown by T. White (1959). By the early 1960s, it had been postulated that the histamine might be in histaminergic neurons, partly because of the central effects of antihistamine drugs.

Interest in brain histamine was rekindled about 1970 by the development of specific and much more sensitive assays for both histamine and its synthetic enzyme, L-histidine decarboxylase, as well as development of binding techniques to study presumed receptor sites. This permitted a rapid accumulation of evidence, mostly of a biochemical nature, on a possible neurotransmitter role (for reviews, see Mazurkiewicz-Kwilecki, 1984; J. C. Schwartz et al., 1980), followed by immunohistochemical evidence of the existence of histaminergic neurons.

The presence of high concentrations of histamine in mast cells was a stumbling block to early work aimed at establishing a neurotransmitter role for the amine (P. L. McGeer et al., 1978a); this obstacle has now been largely circumvented by a focus on other aspects of the histamine system that are more specific to neurons and by the discovery that mast cell histamine differs from neuronal histamine in a number of properties such as turnover rate, subcellular localization, and release by various agents.

11.2. Chemistry of Histamine

11.2.1. Synthesis of Histamine

Experiments performed in various animal species have shown that labeled histamine does not penetrate the blood-brain barrier readily except in newborn animals. Therefore, the brain depends on local biosynthesis, which takes place in specialized neurons. Histamine formation is a simple, one-step process catalyzed by L-histidine decarboxylase [HDC (EC 4.1.1.1)] (Figure 11.1) for which L-histidine (and possibly L-tele-methylhistidine) seem to be the only endogenous substrates.

HDC is clearly a different enzyme from 5-hydroxytryptophan or dopa decarboxylase (DDC) (see Chapters 9 and 10). It is inhibited by α-hydrazinohistidine, bromcresine, and α-methylhistidine, but is unaffected by DDC blockers such as α-methyldopa and Ro-4-4602. It is unaffected by 6-hydroxydopamine, which destroys dopaminergic neurons and reduces DDC activity (Garbarg et al., 1974) (see Chapter 9). Administration of HDC inhibitors, but not DDC blockers, brings about a rapid loss of endogenous histamine (J. C. Schwartz, 1975). The most effective of these histamine depleters is the irreversible, "suicide" inhibitor, α-fluoromethylhistidine, which has no effect on glutamic acid decarboxylase (GAD), DDC, or histamine-N-methyltransferase (HNMT) (cf. Garbarg et al., 1980, 1981; H. Wada et al., 1984). From work with this specific suicide inhibitor, the half-life of the enzyme in mouse brain has been calculated as 53 hr (Keeling et al., 1984). Pyridoxal phosphate antagonists also inhibit HDC, as they do many decarboxylases that require this cofactor.

HDC has been purified from a variety of sources such as rat basophilic leukemia cells (Grzanna, 1984), murine mastocytoma (L. Hammar and Hjerten, 1980), and fetal rat liver (H. Wada et al., 1984). In all cases studied, the mammalian enzyme is reported to have a

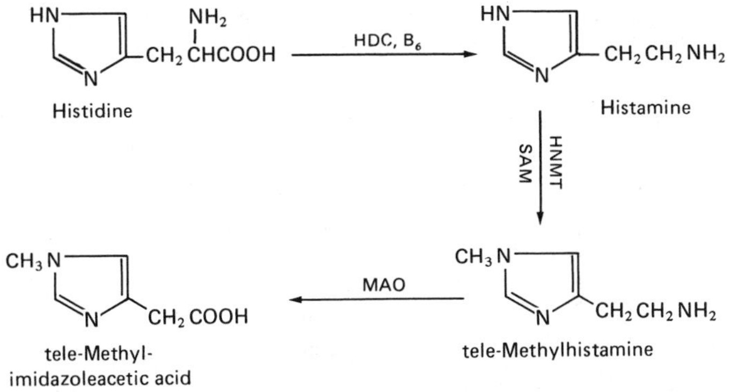

Figure 11.1. Synthesis and breakdown of histamine. Abbreviations: (HDC) L-Histidine decarboxylase; (HNMT) histamine-N-methyltransferase; (SAM) S-adenosylmethionine; (MAO) monoamine oxidase. The nomenclature follows Black and Ganellin (1974).

dimeric structure with a subunit molecular weight of 54,000–62,000. HDC partially purified from rat brain has a slightly higher pI value than the enzyme purified from rat fetuses (Yamada *et al.*, 1984), but the monoclonal antibodies raised against HDC from rat gastric mucosa, fetal rat liver, or other sources recognize brain HDC, indicating similarities in structure (Pollard *et al.*, 1985; H. Wada *et al.*, 1984). Bacterial HDC has been partially sequenced, but this bacterial enzyme differs in structure from the mammalian one (Huynh *et al.*, 1984). The gene complex for mouse kidney HDC has been localized on mouse chromosome 2 (S. A. M. Martin and Bulfield, 1984).

11.2.2. Destruction of Histamine

These are two well-known major pathways of histamine catabolism in peripheral tissue: (1) direct oxidative deamination to imidazoleacetic acid by diamine oxidase and (2) methylation to tele-methylhistamine followed by oxidative deamination by monoamine oxidase type B (MAO-B) (Waldmeier *et al.*, 1977; Suzuki *et al.*, 1979) to tele-methylimidazoleacetic acid (Figure 11.1). Treatment of rats with the MAO inhibitor pargyline reduces the levels of tele-methylimidazoleacetic acid in all brain regions (Khandelwal *et al.*, 1984). The relative contributions of these pathways vary among species and also from one organ to another. It seems probable that in mammalian species, however, the methylation pathway (Schayer and Reilly, 1973) is the only catabolic route of any significance in brain (Van Balgooy *et al.*, 1972). HNMT purified from guinea pig brain has a molecular weight of 29,000 and K_m(s) for histamine and S-adenosylmethionine (SAM) of 13.6 and 6.1 μM, respectively (Matuzewka and Borchardt, 1983). Excess SAM inhibits HNMT as it does other methyltransferases such as catechol-O-methyltransferase (J. C. Schwartz *et al.*, 1980). HNMT occurs in almost all organs and shows by far the highest activity in kidney; antibodies to purified rat renal enzyme (molecular weight 33,400) also immunoprecipitate the brain enzyme completely (Bowsher *et al.*, 1983).

Compartmentation of HNMT activity in brain is not definitely known. Subcellular distribution work indicated a glial location according to some authors (Snyder *et al.*, 1974) and a neuronal location according to others. The presence of HNMT in cultured glial cells (Garbarg *et al.*, 1975) supports a glial location, while the sharp increases in HNMT during postnatal development (Tuomisto, 1977) suggest a neuronal location.

11.2.3. Storage, Release, and Turnover of Neuronal and Mast Cell Histamine

Neuronal histamine is stored, like other neurotransmitters, in synaptic vesicles (Snyder *et al.*, 1974), from which it is presumably released on nerve stimulation.

The mechanism of control of neuronal histamine synthesis is unknown, although plasma and brain levels of histidine probably play a role. There is high-affinity (K_m = 35 μM) uptake of histidine into rat brain synaptosomes, with the density of the high-affinity site correlating with the density of proposed histaminergic innervation. The uptake, however, is K^+-dependent, rather than Na^+-dependent, and is not altered by prior depolarization of the synaptosomes. Such evidence is against the uptake mechanism playing a role in regulating histamine synthesis. Since hypothalamic HDC is inactivated under phosphorylating conditions *in vitro*, it is suggested that the enzyme activity *in vivo* is regulated through a cyclic AMP (cAMP)-dependent protein kinase (Huszti and Magyar, 1984). An analogous phosphorylation of tryptophan hydroxylase is believed to play a role in controlling serotonin [5-hydroxytryptamine (5-HT)] synthesis (see Section 10.2).

Histamine levels in brain increase following L-histidine loads (J. C. Schwartz *et al.*, 1972). Since the K_m of HDC for histidine is about 10^{-4} M (Palacios *et al.*, 1976) and the concentration of free histidine in brain tissue is about 6×10^{-5} M, the enzyme may not be saturated *in vivo*, thus explaining the response to histidine loads. In histidinemia, an inborn error of metabolism in which there is a deficiency of the liver enzyme histidase (which metabolizes histidine by a ring-opening reaction), increased levels of blood and urinary histidine are observed (Bulfield and Keser, 1975). It has been hypothesized that the mental defects seen in children with this condition are related to disturbances in brain histamine levels (J. C. Schwartz *et al.*, 1971).

So far, no specific uptake process for histamine has been demonstrated. It seems probable, therefore, that enzymatic breakdown is responsible for the inactivation of the amine and termination of its effects.

The regional distribution of HDC in rat brain is shown in Table 11.1 in comparison with the distributions of histamine, its metabolites (tele-methylhistamine and tele-methyl-imidazoleacetic acid), and the catabolizing enzyme, HNMT. Data on the half-life of histamine in various brain regions are also given. There is not always a correlation between histamine levels and those of the synthesizing enzyme; this inconsistency reflects the existence of the amine in mast cells as well as in neurons in brain. Thus, for example, in the rat brain, mast cells contribute up to 90% of the histamine in the thalamus (Goldschmidt *et al.*, 1985), but have little HDC activity (see Table 11.1). Some differences in properties of mast cell histamine and neuronal histamine are given in Table 11.2.

Numerous authors have estimated turnover rates for neuronal histamine. Again, there are marked regional differences, with hypothalamic histamine having the slowest turnover times. Most estimates are in the range of 10–30 min. Snyder *et al.* (1974) proposed that, although levels of HDC and histamine are generally lower than those of other putative nonpeptide neurotransmitters, the number of molecules of histamine synthesized per unit time may be greater than that of noradrenaline (NA), dopamine (DA), or 5-HT. This hypothesis suggests that there is relatively little excess HDC activity in brain.

Various other histidine and histamine derivatives such as carnosine, homocarnosine,

TABLE 11.1

Regional Rat Brain Levels of Histamine (HA), tele-Methylhistamine (tMHA), tele-Methylimidazoleacetic acid (tMImAA), Histidine Decarboxylase (HDC), and Histamine-N-methyltransferase (HNMT) Activity and Histamine Half-Life[a]

Brain region	HA (nm/g) (1)	tMHA (nm/g) (1)	tMImAA (nm/g) (2)	HDC (nm/g-hr) (3)	HNMT (nm/g-hr) (3)	HA half-life (min) (4)
Hypothalamus	4.5 (1.3)	1.5	2.2	6.1	1.7	57 (40)
Thalamus	1.9 (0.34)	0.66	1.5	1.5	1.8	42 (17)
Midbrain	0.35 (0.17)	0.18	1.1			16 (23)
Spinal cord	0.77 (—)	0.06	—	0.57	0.57	ND[b] (—)
Cortex	0.34 (0.12)	0.39	53	1.8	1.8	23 (11)
Amygdala	0.33 (—)	0.46	—	—	—	19 (—)
Striatum	0.29 (0.21)	0.21	0.94	1.4	1.9	8 (11)
Pons–medulla	0.19 (0.12)	0.10	0.46	0.35	1.4	25 (44)
Hippocampus	0.14 (0.14)	0.05	0.43	1.2	1.5	14 (19)
Cerebellum	0.04 (0.10)	0.03	0.15	0.11	0.47	10 (140)

[a](—) Not determined; (ND) not detectable. Studies: (1) Oishi *et al.* (1984) [Hegstrand and Hine (1985)]; (2) Khandelwal *et al.* (1984); (3) J. C. Schwartz *et al.* (1980); (4) Oishi *et al.* (1984) [Hough *et al.* (1984)].
[b]There was no measurable turnover in the spinal cord.

ergothionine, γ-glutamylhistamine, and N-acetylhistamine are found in mammalian brain, but there is little or no evidence for a neurotransmitter role for any of these except the dipeptide carnosine, which is believed to be a neurotransmitter in primary olfactory neurons. Formation of the γ-glutamylhistamine derivative seems to be a major route in the catabolism of histamine in the central nervous system (CNS) of *Aplysia* (Weinreich, 1979) and may play a role in mammalian brain when HNMT is inhibited. Although it has been suggested that the formation and deacetylation of histamine might play a role in regulating neuronal pools of the amine, N-acetylhistamine deacetylase has a relatively uniform distribution in rat brain, which argues against such a supposition (Hegstrand and Kalinke, 1985).

TABLE 11.2

Tentative Synopsis of Differential Properties of Neuronal and Mast Cell Stores of Histamine in Rat Brain

Property	Histamine	
	Neuronal	Mast-cell
Subcellular localization	Synaptosomes	Crude nuclear fraction
Appearance during development	At least 3 weeks postnatally	Before birth
Associated HDC activity	High	Low
Half-life	A few minutes	Several days
Effect of mast cell degranulaters, e.g., Cpd. 48/80	Not released	Released
Effect of K^+-induced depolarization	Released	Not released
Effect of reserpine	Released	Not released

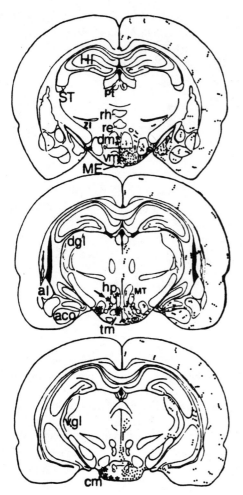

Figure 11.2. Distribution of HDC-containing cell bodies and terminals in cross sections of rat brain. Abbreviations: (aco) nucleus amygdaloideus corticalis; (al) nucleus amygdaloideus lateralis; (cm) caudal magnocellular nucleus; (dgl) nucleus dorsalis corporis geniculi lateralis; (dm) nucleus dorsomedialis hypothalami; (HI) hippocampus; (hp) nucleus hypothalamicus posterior; (ME) median eminence; (MT) fasciculus mammillothalamicus; (pt) nucleus periventricularis thalami; (re) nucleus reuniens; (rh) nucleus rhomboideus; (ST) stria terminalis; (tm) tuberomagnocellular nucleus; (vgl) nucleus ventralis corporis geniculi lateralis; (vm) nucleus ventromedialis hypothalami; (zi) zona incerta. After Watanabe *et al.* (1984).

11.3. Anatomy of Histamine Neurons

Histamine-containing neurons have been detected by immunocytochemistry using antibodies against HDC (Tran and Snyder, 1981; Watanabe *et al.*, 1983, 1984; H. Hayashi *et al.*, 1984; Pollard *et al.*, 1985) or against histamine conjugated to bovine serum albumin or hemocyanin (Panula *et al.*, 1984; Steinbusch and Mulder, 1984; Wilcox and Seybold, 1982). Histamine neurons have so far been observed only in the hypothalamus, although lesion studies have suggested that there may also be more caudally located groups. In the hypothalamus, the area surrounding the nucleus premammillaris ventralis (Figure 11.2) has been found to contain positive cells by immunohistochemistry for both HDC and histamine. This localization is clearly different from that of histamine-containing mast cells that can be observed at the basal surface of the hypothalamus. HDC-positive cells have also been observed in the dorsal hypothalamus (Figure 11.2).

An extensive system of fibers and terminals has also been observed by immunohistochemistry. The highest density of histamine-positive fibers is in the hypothalamus,
particularly in the median eminence and adjoining basal hypothalamic areas. The fibers in
the infundibulum can be traced to the pars nervosa of the pituitary. Heavy innervation is
found in the suprachiasmatic nucleus, the mammillary nuclei, and the posterior hypothalamic nucleus. Fibers are found in the nucleus of the diagonal band of Broca and the nucleus
accumbens. The subfornical organ is richly innervated, and some innervation occurs in
the amygdala and habenula nuclei. A few fibers can be seen throughout the cerebral
cortex, particularly in lamina 1, and some in the striatum and hippocampus. Caudally
located fibers can be seen in the central gray matter of the midbrain and pons, auditory
system, n. vestibularis medialis, n. originis nervi facialis, n. parabrachialis, n. com-
missuralis, n. tractus solitarii, and n. raphe dorsalis.

These anatomical results are generally in agreement with the lesion data that originally forecast the presence of histamine neurons in brain. Garbarg *et al.* (1974, 1976) first
found that lesions of the medial forebrain bundle caused decreases of HDC in all ipsilateral telencephalic areas. A descending tract from the hypothalamus to the brain stem,
especially the periventricular gray and ventral areas, was also predicted on the basis of
lesion studies (Barbin *et al.*, 1977; Pollard *et al.*, 1978).

The magnocellular neurons of the caudal mammillary region are known to have
diffuse rostral projections. The concept (Schwartz *et al.*, 1980) that ascending histaminergic neurons link the posterior hypothalamus directly to widespread telencephalic
areas and provide them with limbic, emotional, and visceral information now seems
reinforced by these anatomical data. A possible discrepancy is that immunohistochemistry
has not yet revealed any histaminergic neurons in the mesencephalic reticular formation
such as were suggested by Barbin *et al.* (1976, 1977).

There is a suggestion that γ-aminobutyric acid (GABA) may coexist with histamine
in the magnocellular neurons of the caudal hypothalamus (Panula *et al.*, 1984) region,
since similarly located neurons of identical morphology stain positively for GAD (Vincent
et al., 1983a) and GABA-α-oxoglutarate transaminase (Nagai *et al.*, 1983a). Such coexistence would obviously have important physiological implications.

Some histamine neurons are also found in the gastrointestinal tract (Tran and Snyder,
1981; Hakanson *et al.*, 1983; Panula *et al.*, 1984), where there is also a rapidly turning-
over pool of the amine, consistent with a neuronal localization (D. E. Duggan *et al.*,
1984). In the periphery, nonneuronal histamine is found in many organs, including the
gastrointestinal tract, and occurs not only in mast cells but also in enterochromaffin cells,
basophils, and platelets. It also occurs in NA-secreting cells of the adrenal medulla
(Happola *et al.*, 1985).

11.4. Physiological Actions of Histamine and Histamine Receptors

Histamine is known primarily for its actions on peripheral smooth muscle and for its
relationship to allergy. Many different substances are released from damaged tissues
following an allergic reaction, including histamine, ACh, 5-HT, heparin, choline, adenosine, and at least one proteolytic enzyme. But it is histamine, released from mast cells,
that is thought to be responsible for the serious reactions.

About 1966, two types of histamine receptors were demonstrated in peripheral tissue:

H_1 and H_2. The allergic reaction, which involves capillary dilatation and venous constriction, is mediated by H_1-receptors. Gastric secretion is mediated by H_2-receptors. H_1-receptors are blocked by an almost bewildering variety of conventional antihistamines, including various ethylenediamines, alkylamines, piperazines, and even certain phenothiazines. H_2-receptors are not blocked by these agents, but are blocked by metiamide and butamide (see Table 11.4) (Section 11.5)] (Ash and Schild, 1966).

These peripheral actions emphasize the general principle that many substances that have powerful effects on peripheral smooth muscle are also transmitters for central neurons (e.g., the catecholamines, ACh, 5-HT, and many peptides). The common central and peripheral mobilization of these substances in response to emotionally stressful situations might be behind many aspects of psychosomatic medicine.

Physiological studies on histaminergic neurons in the CNS may be considered to be of two kinds: release experiments (of which few have been done) and receptor studies, with information believed relevant to receptors coming from both physiological and chemical work.

Electrically and K^+-induced, Ca^{2+}-dependent release of histamine from brain tissue slices, which is typical of neurotransmitter release from nerve endings, has been demonstrated in a few studies (Subramanian and Mulder, 1976; Subramanian, 1982). Electrically stimulated release has not yet been demonstrated *in vivo*.

The iontophoretic application of histamine or its analogues into the immediate vicinity of single neurons has been reported to alter resting membrane potentials as well as firing rates, but the effects, whether excitatory or inhibitory, seem to depend, as with many other putative, metabotropic neurotransmitters, on the location of the neurons and the dose. In most regions, notably the cortex, small doses of histamine were generally depressant, but in the hypothalamus, they were excitatory in some nuclei and depressant in others (Geller *et al.*, 1984). In cultured hypothalamic neurons, such different responses have been attributed to the existence of both H_1 (excitatory) and H_2 (inhibitory) receptors (Geller, 1981). Much of the variation, however, may depend on the fact that histamine is a metabotropic neurotransmitter that modulates the membrane response to other inputs. In hippocampal slices, histamine has been shown to potentiate profoundly excitatory signals by blocking the Ca^{2+}-activated potassium conductance [$gk(Ca)$] that normally restricts strong excitations; it is suggested that histamine accelerates the removal of intracellular Ca^{2+} (Haas, 1984).

Both pharmacophysiological and binding studies have supported the existence of both H_1- and H_2-receptors in the CNS. An autoreceptor that modulates histamine release from rat cortical slices has been said to be a third, pharmacologically distinct type (Arrang *et al.*, 1983), but the evidence for this type is not yet strong.

Binding studies have been particularly useful in studying the regional distribution of H_1 binding sites (Table 11.3). A variety of ligands, such as [³H]mepyramine, [³H]pyrilamine, and [³H]doxepin, have been successfully used. There are marked differences in the regional distributions reported, which seem to depend on species differences rather than the nature of the ligand used (R. S. L. Chang *et al.*, 1979; Kanba and Richelson, 1984); similar species differences in the distribution of H_1- and H_2-receptors are seen in the heart (McNeill, 1984) and possibly in other organs as well. It has also been suggested that there may be two subtypes of H_1-receptors in rat brain, since the B_{max} for [³H]doxepin at the high-affinity site is only about 10% of the B_{max} found for [³H]mepyramine (J. E. Taylor and Richelson, 1982).

TABLE 11.3
Levels of H_1 and H_2 Binding Sites Reported in Various Regions of Rat, Guinea Pig, and Human Brain[a]

Brain region	H_1 Binding sites				H_2 Binding sites	
	Guinea pig (3)	Rat (2)	Human (1)	Human (4)	Guinea pig (3)	Rat (1)
Hypothalamus	—	100%	100%	100%	—	100%
Cortex	96	70%	200–450%	>188–273%	141	55%
Amygdala	—	—	310%	299%	—	—
Hippocampus	126	50%	155%	94%	97	70%
Striatum	95	50%	66–145%	61%	212	68%
Thalamus	—	67%	66–100%	61%	—	76%
Midbrain	—	73%	66–110%	28–57%	—	—
Pons–medulla	73	87%	22–110%	16–41%	ND	69%
Cerebellum	220	39%	11–45%	10–23%	ND	68%

[a](—) Not determined; (ND) not detectable. Studies: (1) R. S. L. Chang *et al.* (1979), using [³H]mepyramine—values are expressed as percentages of binding in hypothalamus; (2) Kendall *et al.* (1980), using [³H]cimetidine in the presence of Cu^{2+}, which greatly increased binding—values are expressed as percentages of binding in hypothalamus; (3) Norris *et al.* (1984), using [³H]mepyramine for H_1 and [³H]tiotidine for H_2—values are expressed in fmoles/mg protein; (4) Kanba and Richelson (1984), using [³H]doxepin—values are expressed as percentages of binding in hypothalamus.

H_2 sites have been much more difficult to label satisfactorily. Results with various antagonists ([³H]cimetidine, [³H]rantidine, [³H]metiamine, and [³H]tiotidine), agonists ([³H]impromidine), and [³H]histamine itself have not given results that satisfy the criteria of high-affinity binding sites resembling the physiological H_2-receptor (Gajtkowski *et al.*, 1983; Garbarg *et al.*, 1983).

According to both binding and physiological studies, some cultured astrocytes have H_1- and H_2-receptors (Hosli and Hosli, 1984); the subpopulation of glia carrying such receptors seems less than the subpopulation carrying α- and β-adrenoceptors.

Many papers have suggested that histamine acts at both H_1- and H_2-receptors to stimulate a histamine-specific adenylate cyclase, but variable results have been obtained depending on the species and type of preparation used. It appears that the H_2-receptor is directly coupled to a specific adenylate cyclase, with the coupling as usual depending on guanyl nucleotides (see Chapter 5), while the H_1-receptors mediate an indirect stimulatory effect on adenylate cyclases. Recent evidence suggests that the second messenger for histamine at the H_1 site is of the phosphoinositol type. Daum *et al.* (1983, 1984) say that the histamine-induced inositol phospholipid breakdown mirrors the H_1-receptor density in brain, and evidence has been provided that H_1-receptor activation stimulates phosphatidylinositol labeling in brain (Subramanian *et al.*, 1980), rabbit aorta (Villalobos-Molina and Garcia-Sainz, 1983), and rat mast cells (Ishizuka *et al.*, 1983). Figure 11.3 gives a hypothetical model of possible mechanisms by which the H_1 and H_2 agonists may act and interact.

Histamine has also been reported to stimulate glycogenolysis in mouse brain slices through H_1-receptors (Garbarg *et al.*, 1983) and prostaglandin synthesis in rat brain microvessels (Dux *et al.*, 1982).

The overall central actions of histamine are less well defined than its peripheral effects. There has been considerable speculation that central histamine systems play a role

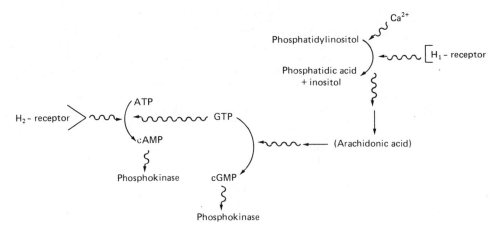

Figure 11.3. Simplified hypothetical model of the proposed second messengers and possible interactions of H_1- and H_2-receptor stimulation. More detailed models may be found in Snider *et al.* (1984) and McNeill (1984). (〰) Indicates activation. A mechanism similar to that proposed for H_1-receptors is proposed for muscarinic receptors (Snider *et al.*, 1984).

in arousal, mental disease, extrapyramidal function (P. L. McGeer *et al.*, 1961), central cardiovascular control, control of cerebral circulation (Karppanen, 1982), and such hypothalamic functions as the control of temperature, food intake, and hormone release (for reviews, see F. Roberts and Calcutt, 1983; Hough and Green, 1980). Some role in many physiological functions would not be suprising for any metabotropic neurotransmitter having such diffuse projections. These speculations are at present based largely on drug actions (see Section 11.5).

As with many neurotransmitters, there have been suggestions that histamine may play a trophic role during development. Histamine stimulates ornithine decarboxylase in brain regions of the developing rat, and this enzyme is associated with rapid tissue growth (G. Morris *et al.*, 1983).

11.5. Pharmacology of Histamine

A large number of drugs affect the histamine system. A few examples are listed in Table 11.4; they include H_1- and H_2-receptor agonists and antagonists, HDC inhibitors, histamine releasers, and HNMT inhibitors.

The clinical applications of drugs that affect primarily the histamine systems have been directed mainly toward the periphery. The use of H_1 antagonists in the treatment of allergies is well established. Classical examples are diphenhydramine and pyrilamine. The marked sedative effects of most such antihistamines used in therapeutics (W. W. Douglas, 1980) were an early clue that central histamine systems might be involved in arousal, but definitive evidence is still lacking. Some tranquilizers, such as chlorpromazine, are weak H_1 antagonists, a property that is believed to be irrelevant to their therapeutic effect. In contrast, almost all tricyclic antidepressants are potent, competitive H_1 antagonists, and this property has been suggested to be important to their therapeutic action (Kanof and

TABLE 11.4
Some Examples of Drugs That Affect the Histamine System

H$_1$ agonists	H$_2$ agonists
2-Thiazolylethylhistamine	Impromidine
2-Methylhistamine	Clonidine
2-Pyridylethylamine	Diamaprit
	Betazole
4-Methylhistamine	
H$_1$ antagonists[a]	H$_2$ antagonists
Mypramine	Metiamide
Diphenhydramine	Cimetidine
Chlorpheniramine	Ranitidine
Pyrilamine	Burimamide
Chlorpromazine	Tiotidine
Clozapine	Famotidine
Doxepin	
HDC inhibitors	HNMT inhibitors[a]
α-Fluoromethylhistidine	Metoprine
(+)Catechin	Quinacrine
Bromcresine	Amodiaquine
N$^{\alpha}$-Aminohistidine	
Naringenine	

[a]Most classic antihistamines and phenothiazines also have a concentration-dependent, biphasic action on HNMT.

Greengard, 1979). Hall and Ogren (1984), however, report that some of the newer antidepressants have only very limited action at the H$_1$-receptor, ruling out such antagonism as a mechanism to account for the action of all antidepressants. They found better correlation of the H$_1$-antagonist activity with sedative as opposed to antidepressant properties of the various drugs.

Some tricyclic antidepressants also appear to have H$_2$-antagonistic activity. Since others do not affect histamine-stimulated adenylate cyclase (Dam Trung Tuong *et al.*, 1980), this latter property is not believed to be a prerequisite for antidepressant properties.

The introduction of histamine H$_2$-receptor antagonists for the treatment of gastric and duodenal ulcers represents one of the more significant practical clinical advances of the past decade. Instead of being subjected to routine total gastrectomy, most cases can be treated successfully by a combination of an H$_2$ antagonist and an anticholinergic agent (Dobrilla, 1983). The agents generally used have been ranitidine, cimetidine, and pirenzepine, but famotidine is said to be preferable because of its greater potency and longer duration of action (Howard *et al.*, 1985).

Some specific HDC inhibitors among the bioflavinoids, such as (+)-catechin (3-hydroxy-5,7,3′,4′-tetrahydroflavan), its 3-*O*-methyl derivative, and the related naringenin, as well as α-fluoromethylhistidine, have also been shown to reduce gastric acid secretion in pylorus-ligated rats or cats with gastric fistulas (Parmar and Hennings, 1984; Parmar, 1983). In initial clinical trials, the catechin derivatives have been shown to reduce histamine levels in patients with duodenal ulcers (Wendt *et al.*, 1980), suggesting the possible application of HDC inhibitors, alone or in combination with H$_2$ blockers, in the treatment of such conditions.

Histamine agonists have not been widely used in clinical medicine, and nonspecific or H_1 agonists have well-known disadvantageous side effects such as bronchospasm, urticaria, and anaphylactic shock. Selective H_2 agonists, such as impromidine and dimaprit, may have clinical utility in post-acute myocardial infarction (Baumann *et al.*, 1984). 4-Methylhistamine has sometimes been cited as a selective H_2 agonist, but it is also active at H_1 sites in the guinea pig ileum (Barker and Hough, 1983).

No potent and selective inhibitor of HNMT is known (J. C. Schwartz *et al.*, 1980). A variety of antihistaminic drugs, such as diphenhydramine, are competitive inhibitors of HNMT in the presence of histamine concentrations below 10 μM, but enhance enzyme activity at higher histamine concentrations. In addition to agents used clinically as antihistamines, the phenothiazines (chlorpromazine and thioridazine), the tricyclic antidepressants (dimethylimipramine), and some butyrophenones (haloperidol) also show the biphasic effects on HNMT activity manifested by the antihistaminic drugs. The effects of most of these compounds on brain histamine levels are controversial.

11.6. Purines

For a general reference, the reader is referred to Stone (1981).

In 1954, Holton and Holton (1954) suggested, in connection with their work on antidromically elicited cutaneous vasodilation, that ATP might be a neurotransmitter. However, it was not until the 1970s that Burnstock proposed distinct populations of purinergic neurons that utilized either ATP or adenosine. It is still uncertain whether such distinctive purinergic neurons exist, but there is convincing evidence that adenosine plays an important role in modulating synaptic function by action at specific receptors in both the central and peripheral nervous systems. Adenosine depresses both spontaneous and evoked potentials in the CNS. It has potent effects on the synthesis of cAMP and inhibits the release of various neurotransmitters (Phillis and Wu, 1983). It has been suggested that caffeine may owe its stimulant effects to its blockade of adenosine receptors in the CNS (Phillis and Wu, 1981; Daly *et al.*, 1981).

11.6.1. Chemistry of Purines

The purine free bases we shall consider are adenine, hypoxanthine, and guanine. The corresponding ribosyl derivatives are called nucleo*sides* and are, in these cases, adenosine, inosine, and guanosine; the mono-, di-, and triphosphorylated derivatives are called nucleo*tides* and are named individually as derivatives of the corresponding nucleoside (Table 11.5). The structures are exemplified in Figure 11.4, which shows adenine and its derivatives.

The *de novo* biosynthesis (Figure 11.5) involves a multistep sequence of reactions starting with the reaction of glutamine with 5-phosphoribosyl-1-pyrophosphate (PRPP) to give the 5-phosphoribosyl-1-amine on which the skeleton of the purine ring is built. The first product in which the purine ring is complete is inosine monophosphate (IMP), which is converted to both adenosine monophosphate (AMP) and guanosine monophosphate (GMP); these are progressively phosphorylated to the diphosphate and triphosphate nucleotides. The nucleosides like adenosine are formed by hydrolysis of the nucleotides.

Nucleotides are the activated precursors of DNA and RNA, activated intermediates

TABLE 11.5
Some Purine Free Bases and the Corresponding
Ribonucleosides and Nucleotides

Free base	Nucleoside	Nucleotides
Adenine	Adenosine	AMP, ADP, ATP, cAMP
Guanine	Guanosine	GMP, GDP, GTP, cGMP
Hypoxanthine	Inosine	IMP, IDP, ITP

in many biosyntheses, components of several major coenzymes such as CoA, and, in the form of cAMP and cGMP, mediators of the action of many hormones and neurotransmitters. ATP is the universal currency of energy in biological systems. It is therefore not surprising that the purines are ubiquitous in the body and brain and that it is proving difficult to sort out a possible neurotransmitter pool of adenosine.

Nagata *et al.* (1984) have studied the distribution and properties of several enzymes in brain that might be capable of producing adenosine and suggest that the principal source is likely to be hydrolysis of AMP by the 5′-nucleotidase, which shows a high specific activity in brain membranes. It is not yet clear whether the adenosine that is active as a synaptic modulator is formed intra- or extracellularly.

The action of adenosine with its receptors seems to be terminated by its rapid uptake into surrounding cells. It differs from classical neurotransmitter uptake in being only par-

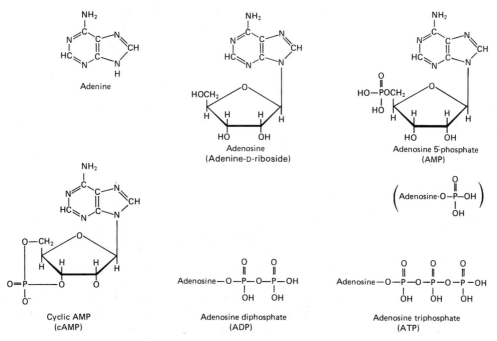

Figure 11.4. Structures of adenine, adenosine, and related nucleotides.

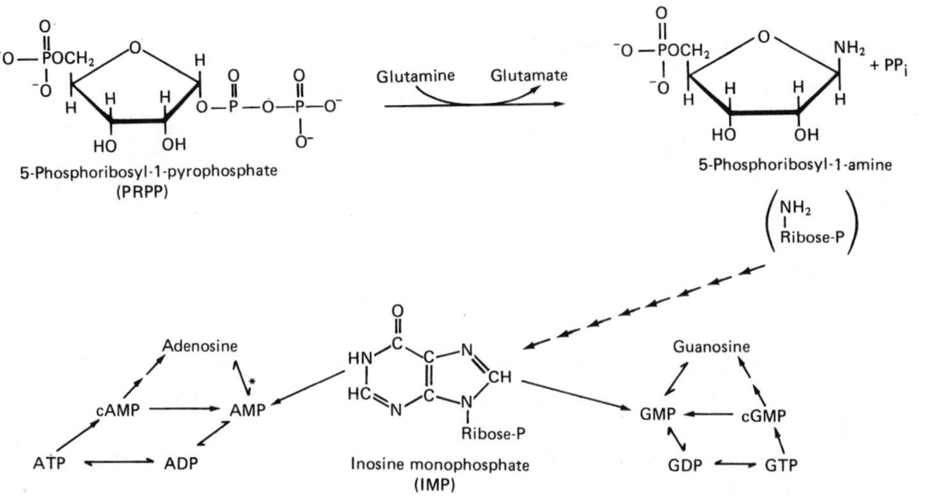

Figure 11.5. *De novo* route of synthesis of purines. There are 9 steps between phosphoribosyl-1-amine and IMP. *The enzyme that converts AMP to adenosine is 5'-nucleotidase; the enzyme that converts adenosine to AMP is adenosine kinase.

tially dependent on Na$^+$ and Ca^{2+} (Bender *et al.*, 1981a,b). The transport system shows a decided preference for adenosine as substrate [K_i = 3 μM (Geiger *et al.*, 1985)], but will act on any of the nucleosides; it exists in a wide variety of cells including erythrocytes, lymphocytes, endothelial cells, neurons, and glia (for a review, see Berne *et al.*, 1983b). Inhibitors of this adenosine uptake system potentiate the effects of adenosine in iontophoretic experiments (Phillis and Wu, 1983). Nitrobenzylthioinosine (NBI) is a potent and specific inhibitor of nucleoside transport, and radioactive NBI has been used to label the transport sites, which are clearly pharmacologically distinct from the "receptor" binding sites (Marangos, 1984) (see Section 11.6.3).

Some of the clinically active benzodiazepines (Phillis *et al.*, 1980) and antidepressants (Phillis, 1984) inhibit adenosine uptake, and it has been suggested that this may be important to their therapeutic effects. This seems unlikely in the case of the benzodiazepines, since many clinically effective ones are very weak inhibitors of [^{3}H]-NBI binding (Patel *et al.*, 1982a) and the inhibition of adenosine uptake is not antagonized by behaviorally effective antagonists (P. F. Morgan *et al.*, 1983). Similarly, some active antidepressants have little or no effect on the uptake system.

Adenosine deaminase [ADA (EC 3.5.4.4)] is the critical enzyme in a major metabolic pathway for adenosine. It is present in many peripheral tissues and has a very high specific activity in brain. This enzyme oxidatively deaminates adenosine to inosine, which is then readily hydrolyzed to hypoxanthine (Figure 11.6). Erythro-9-(1-hydroxy-3-nonyl)adenine is a competitive inhibitor of ADA. Parenteral administration results in a profound decrease in spontaneous motor activity in mice and rats and a marked reduction in electroencephalographically defined sleep (Mendelson *et al.*, 1983).

Alternatively, adenosine can be either rephosphorylated by adenosine kinase or hydrolyzed to adenine. The eventual degradation of purines is through hypoxanthine and xanthine to uric acid, which is excreted. A large part of the free purine bases are "sal-

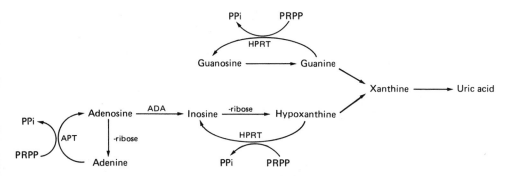

Figure 11.6. Routes of metabolism of adenosine and salvage of the purine free bases. *The enzyme converting adenine to adenosine is adenine pyrophoribosyl transferase (APT) and that converting hypoxanthine to inosine, or guanine to guanosine, is hypoxanthine–guanine phosphoribosyl transferase (HPRT). Other abbreviations: (ADA) adenosine deaminase; (PRPP) 5-phosphoribosyl-1-pyrophosphate; (PPi) inorganic phosphate.

vaged'' by special enzymes that convert them directly back to the nucleotides by reaction with hypoxanthine-guanine phosphoribosyltransferase (HPRT) (Figure 11.6). Children born with a congenital lack of one of these salvage enzymes (HPRT) suffer from one of the few neurological disorders (the Lesch–Nyhan syndrome) in which there is evidence that purines play a primary role. Plasma and urinary adenosine levels are low, and since HPRT is normally found in greatest quantities in the basal ganglia, an increase of transmitter release secondary to an adenosine deficiency (see Section 11.6.3) could account for the involuntary oral behavior and choreiform movements characteristic of the syndrome (Stone, 1981).

11.6.2. Distribution and Anatomy of Adenosine Systems

The amounts of adenosine in rat brain have been reported to range from 0.34–2.2 nmoles/g (M. E. Newman, 1983) to about 7.5–18.5 nmoles/g (W. J. Wojcik and Neff, 1982). Only a fraction may be in the neuromodulatory pool. Considerable adenosine release is seen in hypoxia, ischemia, or tissue injury, so that adenosine values vary with the condition of sacrifice. No marked regional variations exist (Table 11.6). Distribution studies on some of the enzymes possibly involved in synthesizing and degrading adenosine (see Section 11.6.1 and Table 11.6) and on one adenosine receptor subtype (see Section 11.6.3 and Table 11.6) suggest no more than a 3- to 5-fold variation among large brain regions for these other indices of adenosine systems.

An ADA-rich population of neurons has been identified in the basal hypothalamus by immunohistochemistry. It has been suggested that these neurons may release adenosine as a neurotransmitter. They have been shown by retrograde tracing methods to project to the cortex and striatum, and ADA-rich fibers were seen in these and other possible target areas. A few other neurons that stained weakly for ADA were seen in the septal nuclei and superior colliculi (Nagy et al., 1984). ADA-positive neurons in the hypothalamus and superior colliculus correspond with those that show high densities of [3H]-NBI binding (uptake) sites (Nagy et al., 1985). This finding cannot be considered definite evidence of adenosinergic neurons, however, and it cannot be stated that there are distinctive "pu-

TABLE 11.6
Distribution of Various Indices of Adenosine Systems in Brain[a]

Brain region	Rat brain				Human brain		
	Binding of:		NBI binding (uptake sites)	Adenosine levels	Deaminase	Kinase	Nucleotidase
	[³H]-CHA (A₁ "receptors")	[³H]-CADO					
Cerebellum	100% [100%]	100%	100%	100%	100%	100%	100%
Hippocampus	157% [—]	118%	142%	40%	—	—	—
Cortex	107% [91%]	55%	135%	42–80%	121–157%	60–83%	193–286%
Striatum	95% [—]	116%	323%	50%	—	—	—
Hypothalamus	63% [—]	48%	261%	—	250%	120%	269%
Pons	— [—]	60%	171%	66%	121%	108%	220%
Medulla	54% [46%]	48%	—	—	27%	107%	262%
Spinal cord	43% [78%]	—	—	—	—	—	—
Pineal	17% [—]	—	—	—	—	—	—
Pituitary	6% [—]	—	—	—	—	—	—

[a] The data are expressed as percentages of absolute cerebellar values, which are as follows: [³H]cyclohexyladenosine (CHA) binding: 157 fmoles/mg protein (Patel et al., 1982b); [578 fmoles/mg protein (J. D. Geiger et al., 1984)]; [³H]-2-chloroadenosine (CADO) binding: 62 fmoles/mg protein (M. Williams and Risley, 1980); [³H]nitrobenzylthioinosine (NBI) binding: 47 fmoles/mg protein (Marangos et al., 1982); adenosine levels: 185 pmoles/mg protein (W. J. Wojcik and Neff, 1982); enzyme activities in nmoles/min·g tissue are 140 for adenosine deaminase, 16.2 for adenosine kinase, and 290 for the 5'-nucleotidase; cortical adenosine and associated enzyme levels vary according to the region measured (E. Phillips and Newsholme, 1979). —, Not determined.

rinergic'' neurons in the brain. Other evidence, discussed in the following section, seems to suggest that adenosine is released from a variety of neurons and glia by a process that is different from that characteristic of neurotransmitter release.

11.6.3. Physiology of Adenosine

Release of radioactive adenosine from hippocampal or cortical slices preloaded with [^{3}H]adenine, or from a synaptosomal fraction of rat cortex preloaded with [^{3}H]adenosine, has been shown to be increased by electrical stimulation or by a depolarizing agent such as K^+, veratridine, or glutamate. The enhanced release, however, is not decreased in the absence of Ca^{2+} and shows other characteristics that differentiate it from the release of neurotransmitters such as NA or GABA (Fredholm and Hedqvist, 1980; Jonson and Fredholm, 1985; P. H. Wu et al., 1984). Axon terminals may be one important site of release, but postsynaptic membranes, axons, and glial cells also seem to contribute (Fredholm and Hedqvist, 1980; Maire et al., 1984).

Iontophoretically, adenosine is generally depressant, as originally shown for cortical neurons by Phillis et al. (1974). The depressant action in the cortex, hippocampus, olfactory cortex, and possibly other areas seems to be due to a presynaptic action whereby adenosine inhibits the release of the excitatory neurotransmitters glutamate and aspartate (Corradetti et al., 1984; Motley and Collins, 1983). Adenosine also acts presynaptically to inhibit the release of other neurotransmitters such as DA, NA, ACh, and GABA. The mechanism seems to involve action on a specific presynaptic receptor to inhibit the K^+-evoked Ca^{2+} flux into the nerve terminal that is essential for neurotransmitter release (P. H. Wu et al., 1982) (see Section 4.8). Postsynaptic actions of adenosine, such as modulation of adrenergic-receptor activity, have also been reported.

At least three different and specific types of receptors for adenosine have been characterized in the brain or peripheral tissue or both (Table 11.7). The two extracellular receptors, A_1 (R_i) and A_2 (R_a), were originally characterized by their effect on adenylate cyclase activity. Thus, they were designated as R_i (inhibitory riboside) and R_a (activating riboside). The A_1 and A_2 classification arose independently and is now favored because of increasing evidence that not all effects of adenosine are cAMP-dependent (Dunwiddie and Fredholm, 1984; Reddington and Schubert, 1979). Of these two receptors, the A_1 site has been the more widely studied and is generally believed to be responsible for the inhibition of glutamate/aspartate release and the generally depressant iontophoretic effects of adenosine (Dolphin and Prestwich, 1985). Within the hippocampus, for example, variations in the density of A_1 binding sites, determined autoradiographically, correlate with the magnitude of the effects of adenosine on cell firing (K. S. Lee et al., 1983). Characterization of the A_2 site has lagged because of the lack of really selective agonists or antagonists of this activating receptor.

Adenosine ''receptors'' are not limited to neuronal membranes, but also occur on the external surfaces of glia, mast cells, lymphocytes, basophils, capillary microvessels, platelets, and many other types of cells.

In addition to these cell surface receptors, Londos and Wolff (1977) described a third type of adenosine recognition site, the P-site, which is intracellular and located on the catalytic subunit of adenylate cyclase. At this somewhat mysterious site, relatively high concentrations of adenosine ($>10^{-5}$ M) act to inhibit cAMP formation (Daly, 1982). The nomenclature of adenosine receptors is confusing; in particular, it is essential not to

TABLE 11.7

Receptor Types Characterized for Adenosine and ATP by Cellular Localization, Agonist and Antagonist Ligands, Affinity for Adenosine, and Presumed Biochemical Action[a]

Name	Location	Agonists	Antagonists	Affinity for ADO	Biochemical response
Adenosine receptors					
A_1 (R_i)	Extracellular	CPA > CHA = L-PIA > 2-CADO > ADO > NECA > D-PIA	IIPX = DPPX = DJB = PACPX > 8PT > T	10^{-9}	↓ cAMP
A_2 (R_a)	Extracellular	MECA > NECA > 2-CADO > ADO > PIA	Alkylxanthines	10^{-6}	↑ cAMP
P	Intracellular	DDA >> ADO	5'-MTA	10^{-5}	↓ cAMP
ATP receptor					
P-2	Extracellular	ATP > ADP > AMP > ADO	AAPAT	—	↑ PGs

[a]Much of this table was compiled from data given by M. Williams (1983) and L. P. Davies et al. (1984). Compounds: (ADO) adenosine; (AAPAT) arylazidoaminopropionyladenosine triphosphate (Sneddon et al., 1982); (CHA) cyclohexyladenosine; (CPA) cyclopentyladenosine; (2-CADO) 2-chloroadenosine; (DDA) 2,5-dideoxyadenosine; (DJB) 4,6-bis-α-carbamoylethylthio-1-phenylpyrazolo[3,4-d]pyrimidine; (DPPX) 1,3-dipropyl-8-phenylxanthine; (IIPX) 1-isoamyl-3-isobutyl-8-phenylxanthine; (MECA) 5'-N-methylcarboxamidoadenosine; (NECA) 5'-ethylcarboxamidoadenosine; (PACPX) 1,3-dipropyl-8-(2-amino-4-chloro)phenylxanthine; (PG) prostaglandin; (PIA) phenylisopropyladenosine; (8PT) 8-phenyltheophylline; (T) theophylline.

confuse this internal P-site with the P_1 and P_2 designations originally suggested by Burnstock *et al.* (1978) for, respectively, adenosine and ATP receptors. The A_1 and A_2 sites may be considered subdivisions of Burnstock's P_1 type.

Examples of the known agonists and antagonists are also indicated in Table 11.7. In behavioral tests, adenosine and the metabolically stable agonists have been shown to have sedative, hypnotic, and anticonvulsant activity, while antagonists such as caffeine are stimulants. Adenosine has many other receptor-mediated physiological effects, including inhibition of lipolysis and platelet aggregation and, most important, dilation of blood vessels. A variety of experiments indicate that adenosine probably plays an important role in the adjustment of blood flow to the metabolic requirements of the heart, brain, and skeletal muscles (Berne *et al.*, 1983a). Only future experiments will tell whether or not the behavioral effects of adenosine and its analogues are dependent more on neuronal or on cardiovascular effects.

11.6.4. Other Suggested Purine Neurotransmitters

Burnstock (1975) originally suggested that the mammalian gastrointestinal tract, bladder, and possibly other organs are innervated by neurons that release ATP as their transmitter. Possibly the strongest bit of supporting evidence is the report that arylazido-aminopropionyladenosine triphosphate, which is said to be a specific ATP antagonist, blocks the excitatory junction potential caused in guinea pig vas deferens by motor nerve stimulation (Sneddon *et al.*, 1982). Despite such physiological results, and evidence for depolarization-induced ATP release in various preparations (e.g., T. D. White and Leslie, 1982), there is no firm evidence that any peripheral nerves use ATP as their major neurotransmitter. Evidence with regard to the CNS is even less convincing. ATP is certainly bound in vesicles with some neurotransmitters such as the catecholamines and, in these circumstances, is co-released on stimulation (see Chapter 9). Through this mechanism, ATP could have some important neuromodulatory role in the CNS, but such a function has not yet been established.

Stimulation or depolarization of nerves or brain slices loaded with [^{3}H]adenine leads to the appearance of radioactive hypoxanthine and inosine, as well as adenosine (see Section 11.6.3), in the extracellular fluid (Maire *et al.*, 1984). These purines have some activity as displacers of benzodiazepine binding and have even been suggested as endogenous ligands for such receptors (see Chapter 7). The concentrations of the purines required, however, seem too high for such a role.

A possible neuromodulatory role for guanosine has been suggested by S. J. Berger *et al.* (1985) on the basis of the more than 50-fold variation in the activity of guanine deaminase in different brain regions of the mouse. Activity ranged from very high in parts of the telencephalon, particularly the amygdala and olfactory tubercle, to undetectable in the cerebellum and brain stem areas. Variations of this magnitude are often observed with enzymes involved in the metabolism of neurotransmitters or second messengers. Guanine deaminase converts guanine—which is rapidly interconvertible with guanosine by an enzyme widely distributed in brain—to xanthine, which is then oxidized to uric acid (Figure 11.6). There are isolated reports suggesting that guanosine may be released on depolarization and have an excitatory action on neurons in hippocampal slices, but exploration of its possible role as a neuromodulator in the CNS is just beginning.

11.7. Summary

The evidence for histamine as a neurotransmitter in mammalian brain is convincing. A significant fraction of the cerebral amine is synthesized at a high rate, seems to be held in specific neuronal tracts, can be released during depolarization, and affects specific receptors. Most important, immunohistochemical techniques have allowed the localization of both histamine and its specific synthetic enzyme, histidine decarboxylase, to a group of neurons in the hypothalamus that seem to send diffuse projections through many brain areas.

Histamine seems to function as a modulatory, metabotropic neurotransmitter that acts through at least two pharmacologically different types of receptors, H_1 and H_2, which were initially demonstrated in the periphery. The H_1-receptor is important for allergic reactions that involve capillary dilation and venous constriction and the H_2-receptor for the control of gastric secretion. In the CNS, the second messenger for the H_1-receptor appears to be of the phosphoinositol type, while that for the H_2-receptor is a histamine-sensitive adenylate cyclase. The two types of receptors show different regional distributions, but the distributions tend to vary somewhat from species to species. Antagonists for the H_1- and H_2-receptors are clinically useful because of their peripheral effects in controlling allergic reactions and excessive gastric secretion. Other drugs known to affect histamine systems include inhibitors of synthesis, storage, and catabolism, but these have not yet achieved clinical importance. Many tranquilizing agents and most antidepressants act as histamine antagonists and also have a biphasic effect on the activity of the catabolic enzyme, histamine-N-methyltransferase. These actions on the histamine system are not believed to be essential for the primary therapeutic action, but are probably related to the sedative side effects.

The overall central effects of histamine are less well defined than its peripheral effects. As with many other metabotropic transmitters, histamine has been suggested to play some role in a wide variety of central functions including arousal, mental disease, extrapyramidal function, central cardiovascular control, control of cerebral circulation, and such hypothalamic functions as control of temperature, food intake, and hormone release.

There is, on the other hand, little convincing evidence that any purine is a neurotransmitter in the sense that it is the major material released by a specific set of neurons for action on postsynaptic sites. Purines are ubiquitous in the nervous system and serve many functions, including major roles in energy metabolism, nucleic acid synthesis, and as second messengers. There is also very strong evidence, however, that adenosine may be a ubiquitous neuroregulator with many important central effects, including presynaptic inhibition of the release of various neurotransmitters, modulation of the sensitivity of some postsynaptic receptors, and the homeostatic regulation of cerebral blood flow. The stimulated release and uptake of adenosine have different characteristics than the release and uptake of classical neurotransmitters, but specific adenosine receptors have been found in the CNS. Adenosine and its analogues are behavioral depressants, while antagonists, including methylxanthine derivatives such as caffeine, are stimulants. The increasing availability of selective and metabolically stable agonists and antagonists should result in a rapid accumulation of knowledge on the neuroregulatory mechanisms of adenosine and its possible involvement in various physiological functions or pathological conditions.

12

The Prominent Peptides

12.1. Introduction

We consider now the Type III neurotransmitters that were introduced in Section 5.2. They differ from the Type I (amino acid) or Type II (amine) transmitters in the following ways:

1. They are peptides containing 3–100 amino acid residues.
2. They occur in brain in the femtomole to picomole per gram of tissue concentration range, compared with the much higher levels of amino acid and amine transmitters [see Table 5.1 (Section 5.2)].
3. They are formed by cleavage of larger, precursor peptides that are synthesized from messenger RNA in the cell body.
4. There is no reuptake mechanism.
5. They occur very frequently in association with other peptides or with category I or II transmitters (Table 5.2); they are thus often referred to as "cotransmitters."
6. They are released in many locations at a distance from their intended receptors; they are thus often referred to as "neurohormones."

The term "peptidergic neuron" was first applied to the hypothalamic neurons that synthesize oxytocin and vasopressin. These peptides were identified by du Vigneaud *et al.* (cf. Taylor *et al.*, 1953) as the physiologically active substances released by the posterior lobe of the pituitary. The cells that make these factors had previously been predicted by Ernst and Berta Scharrer (1940) to be neurosecretory neurons: Like any other nerve cell, they were controlled by synaptic connections and yet, on suitable command, they could secrete a hormone into the bloodstream. This hypothesis of neurosecretion was strengthened when it was observed that the axons of these neurons not only released hormones into the bloodstream but also possessed terminals that form specialized junctions with epithelial cells of the pars intermedia of the pituitary. It is now known that these same cells send projections to many regions of the brain, where they form typical synapses with other neurons (see Section 12.3).

The property of neurosecretion is not exclusive to peptidergic neurons. We learned in Chapter 9, for example, that noradrenaline (NA) and adrenaline are released from the adrenal medulla to act on remote targets, while dopamine (DA) is released into the hypophyseal portal circulation to act on anterior pituitary cells. All the amine transmitters possess varicose axons, and presumably much transmitter is released from these varicosities. The receptors for these transmitters seem to have similar properties at conventional synapses and at locations more distant from the site of release. Therefore, such differing sites of release may not indicate any fundamental difference in the mechanism of action. It may indicate a lower requirement for speed of action and concentration of the transmitter at the receptors. This view would therefore imply that peptidergic transmitters differ from Type II metabotropic neurotransmitters and especially Type I ionotropic transmitters in not being associated with rapid responses and being able to exert their effects on receptors at a much lower concentration than conventional transmitters.

Over the past decade, a large number of peptides have assumed increasing importance as probable transmitters in brain as well as in the peripheral nervous system. In the first edition of this book in 1978, we wrote about what we chose then to call the "promising peptides." Despite the relatively low concentrations of these materials in mammalian brain (in the picomoles/gram range), the many symposia, books, and journals appearing on their neurobiology suggest that they should now be called the "prominent peptides" or even, in view of the rapidly increasing number, the "bewildering peptides." Many have suggested that, whereas the last 25 years has been the age of monoamines in neuroscience, the next 25 years may be the age of peptides.

Table 12.1 lists many of the peptides with more than two amino acid residues that are putative neurotransmitters or neuromodulators in mammalian brain, along with the sources from which they were originally identified. The table reveals that the hypothalamus and gut have been the richest of these sources. However, such identifications should not be taken to mean that these are the most significant sites of action; the distribution of peptides is very broad throughout the central nervous system (CNS). The abbreviations used in this chapter for the various peptides are also listed in Table 12.1. Because of the dynamism of the field, the list is a tentative one that will undoubtedly be subject to much future revision.

The literature on central peptides is already vast and is growing rapidly, but is still inconclusive with regard to many fundamental points of chemistry, anatomy, physiology, and pharmacology. For the peptides in Table 12.1, the amount of information varies markedly from a very substantial amount on substance P and the enkephalins to a minimal amount on bombesin, motilin, glucagon, and some other little-known peptides. It is beyond the scope of this book to discuss each of these peptides in detail. Fortunately, the investigation of each one, and the characteristics that have so far been reported, follow similar patterns. Hence, we shall discuss such peptides as a group, illustrated by a few specific examples. For each peptide, we mention at least one review or other recent paper that will lead interested readers into the pertinent and more detailed literature.

The rapid explosion of knowledge that has taken place in the field is due largely to the powerful new techniques that have become available for characterizing and assaying peptides. A brief resume of how some of the peptides were identified will illustrate how technical advances have been vital to expansion of knowledge.

TABLE 12.1

Alphabetical List of Some Putative Neurotransmitter Peptides and Their Synonyms, Abbreviations, Initial Sources, and Related Materials[a]

Name	Ref. No.[b]	Abbreviation	Initial source	Family	Related materials
$ACTH_{1-24}$	—	—	Hypothalamus	Propiomelanocortin (POMC)	See Table 12.3
$ACTH_{4-10}$	—	—	Hypothalamus	POMC	See Table 12.3
Adipokinetic hormone	1	AKH	Locust	—	—
Adrenocorticotropic hormone	—	ACTH	Hypothalamus	POMC	See Table 12.3
Adrenorphin (see metorphamide)					
Amidorphin	2	AMD	Adrenals	Proenkephalin	See Table 12.3
Angiotensin II	7	AII	Blood	Renin	AIII = des-Arg¹-AII
Atrial natriuretic factor	12	ANF	Atrial cardiac muscle	Atriopeptides	—
7B2	3		Pituitary		—
Bombesin	—	BO	Frog skin/gut	BO/GRP	Alytesin, litorin, NMD-B, -C, RT
Bradykinin	—	BK	Blood	Kinin	Polisteskinin, ranakinins, etc.
Calcitonin gene-related peptide	17	CGRP	mRNA	—	Calcitonin, hCGRP, β-CGRP
C'-fragment	—	LPH_{61-87}	Precursor analysis	POMC	See Table 12.3
Cholecystokinin (pancreozymin)	4	CCK	Gut	Gastrin/CCK	Caerulein (frog skin); molluscan neuropeptide, ceruletide
CLIP (= $ACTH_{18-39}$)	—	CLIP	Precursor analysis	POMC	See Table 12.3 (also see p. 380)
Corticotropin-releasing factor; corticoliberin	11	CRF	Hypothalamus	—	Urotensin I (teleosts); thymosin (thymus), sauvagine
Diazepam binding inhibitor	—	DBI	Brain	—	See Section 7.4
Dynorphin	16	DYN	Brain	Prodynorphin	See Table 12.3
α-Endorphin	16	α-End	Brain	POMC	See Table 12.3
β-Endorphin (=C-fragment)	16	β-end	Precursor analysis	POMC	See Table 12.3
γ-Endorphin	16	γ-End	Brain	POMC	See Table 12.3
methionine-Enkephalin	16	met-Enk	Brain	Proenkephalin	See Table 12.3
leucine-Enkephalin	16	leu-Enk	Brain	Proenkephalin	See Table 12.3
met-Enk-Arg⁶	16	—	Precursor analysis	Proenkephalin	See Table 12.3
met-Enk-Arg⁶-Phe⁷	16	—	Precursor analysis	Proenkephalin	See Table 12.3

(continued)

TABLE 12.1 (*Continued*)

Name	Ref. No.[b]	Abbreviation	Initial source	Family	Related materials
FMRF-amide	5	—	Invertebrate	—	LPLRF-amide
GAP	—	—	Precursor analysis	Pro-LHRH	See Table 12.3
Gastrin	—	—	Gut/brain	Gastrin/CCK	—
Gastric inhibiting peptide	6	GIP	Gut	GRF/PHI	VIP, SEC, PHI
Gastrin-releasing peptide	—	GRP	Gut	BO/GRP	See Bombesin
Glicentin (= proGLU?)	6	GLI	Gut	GRF/PHI	See Glucagon
Glucagon	6	GLU	Gut	GRF/PHI	VIP, GIP, SEC
Growth hormone-releasing factor (= somatocrinin)	6	GRF, GHRH	Hypothalamus	GRF/PHI	VIP, SEC, GLU
Leumorphin	—	LUM	Precursor analysis	Dynorphin	See Table 12.3
β-Lipotropin	—	β-LPH	Hypothalamus	POMC	See Table 12.3
Luteinizing hormone-releasing hormone (= luliberin = gonadotropin-releasing factor)	—	LHRH, LRF, GnRH	Hypothalamus	—	—
Melanocyte-stimulating hormones	—	α-MSH (= ACTH₁₋₁₃)	Hypothalamus	POMC	See Table 12.3
		β-MSH (= LPH₄₁₋₅₈)	Hypothalamus	POMC	See Table 12.3
		γ-MSH	Precursor analysis	POMC	See Table 12.3
Metorphamide (= adrenorphin)	13	MO, PIM	Caudate	Proenkephalin	See Table 12.3
Motilin	13	MO, PIM	Gut	Tachykinin	See Substance P
Neuromedin B	8	NMD-B	Spinal cord	BO/GRP	See Bombesin
Neuromedin C	8	NMD-C	Spinal cord	BO/GRP	See Bombesin
Neuromedin K (= neurokinin β)	—	NMD-K, β-NK	Spinal cord	Tachykinin	See Substance P
NMD-L (= substance K = neuromedin α)	14	α-NK, SK	Spinal cord, precursor analysis	Tachykinin	See Substance P
Neuropeptide Y	15	NPY	Brain	NPY/PP	Pancreatic polypeptides, peptide YY

Neurotensin	9	NT	Gut	—	Gln⁴-NT (plasma), Lan-6 (chick), neuromedin M (spinal cord), xanopsin (frog)
Oxytocin (= pitocin = α-hypophamine)	—	OXY	Hypothalamus	—	Isotoxin, mesotocin, MSH release-inhibiting factor 1 (MIF-1)
Peptide E	16	PEP-E	Precursor analysis	Proenkephalin	See Table 12.4 (Section 12.2.2)
Porcine intestinal peptide	—	PHI	Gut	GRF/PHI	VIP, PHM-27, SEC, GIP, GRF
Ranatensin	—	RT	Frog skin/brain	BO/GRP	See Bombesin
Secretin	—	SEC	Gut	GRF/PHI	VIP, GIP, PHI, GRF
Somatocrinin (see Growth hormone-releasing factor)					
Somatostatin (= growth hormone release-inhibiting factor = panhibin)	10	SOM, SRIF, SS, GHRIF	Hypothalamus	—	SS_{1-12} (brain), urotensin II (teleosts)
Substance K (see Neuromedin L)		SK			
Substance P	18	SP	Gut/brain	Tachykinin	Physalemin, phyllomedudin, kassinin (frog), eledoisin (octopus), α- and β-NK
Thyrotropin-releasing hormone (= thyroliberin)	—	TRH, TRF	Hypothalamus	—	—
Vasoactive intestinal polypeptide	—	VIP	Gut	GRF/PHI	GLU, SEC, GIP, PHI, GRF
Vasopressin (= β-hypophamine = arginine vasopressin)	—	AVP	Hypothalamus	—	See Figure 12.2 (Section 12.2.1)

[a]Where more than one abbreviation is given, the one used in this chapter is underlined. All the peptides listed have been reported in brain neurons, although in a few cases the evidence is slight.

[b]References: (1) Sasek et al. (1985); (2) Liebisch et al. (1985); (3) Iguchi et al. (1985); (4) Rehfeld (1985), (1985); (5) Dockray (1985); (6) Polak and Bloom (1980), R. J. Miller (1984), Guillemin et al. (1984), Ling et al. (1985), Holst (1980), Conlon (1980), Arimura et al. (1984); (7) M. I. Phillips (1980); (8) Namba et al. (1985), Minamino et al. (1984); (9) (1982), Reinecke (1985); (10) (1984); (11) (1985); (12) Papka et al. (1985), Jacobowitz et al. (1985); (13) Beinfeld and Korchak (1985); (14) Buck et al. (1984); (15) Solomon (1985); (16) Akil et al. (1984), Millan and Herz (1985), Olson et al. (1984); (17) (1985); (18) Cesaro (1984).

It has been more than half a century since the first physiological action of a neuropeptide was identified, but at the time its peptide structure was not suspected. Substance P (SP) was given its oddly appropriate name by its discoverers, Von Euler and Gaddum (1931), who obtained it by adding sulfuric acid to alcoholic extracts of horse intestine and brain. The preciptitate they obtained had an acetylcholine (ACh)-like activity on smooth muscle that differed in not being blocked by atropine (Von Euler and Gaddum, 1931). Early bioassay techniques (Pernow, 1953; Amin *et al.*, 1954) showed that SP occurred in the CNS largely in gray matter. It was particularly concentrated in the hypothalamus, basal ganglia. thalamus, and dorsal roots of the spinal cord. This latter finding, plus its property of causing peripheral vasodilatation, inspired Lembeck (1953b) to propose that this still chemically unidentified SP was the primary sensory transmitter. It was an insight that would wait a generation for experimental verification. The SP field really lay fallow until it was reopened by a serendipitous discovery in 1970. Leeman and her co-workers at Brandeis University were involved in an attempt to isolate a corticotropin-releasing factor from bovine hypothalamic extracts. When the various fractions from one preparative column were screened for biological activity, Leeman noticed that one fraction stimulated salivary secretion when injected into anesthetized rats. This "sialogogic effect" was not inhibited by atropine. It proved to be a simple quantitative assay from which isolation and purification of the peptide could proceed. The Leeman group then realized that their purified "sialogen" had pharmacological actions and chemical properties identical to those previously reported for partially purified SP. They concluded that sialogen and SP were identical. Techniques that were available soon made it possible to define exactly the undecapeptide structure (M. M. Chang and Leeman, 1970; Leeman and Mroz, 1974) (Figure 12.1).

SP was then synthesized (Yajima and Kitagawa, 1973; G. H. Fisher *et al.*, 1974), and antibodies were raised to the synthetic peptide. The antibodies made it relatively easy to localize SP to particular neuronal systems. At the same time, its synaptosomal localization and conditions of release were being studied. Then, with the development of radioactive binding assays as a chemical method for studying probable receptor sites (see Section 5.3), it became possible to demonstrate the regional distribution of such receptors.

Such binding techniques were crucial to the discovery of the first of the opioid peptides. Ingoglia and Dole (1970) opened the field when they injected labeled D- and L-isomers of the synthetic morphine substitute methadone intraventricularly. They found a 10% greater retention of the active L-isomer compared with that of the inactive D-isomer. This suggested to them specific binding to an active opiate receptor.

A. Goldstein *et al.* (1971) then recommended a strategy for eliminating nonspecific binding to membranes so that these receptors could be identified. Many technical developments contributed by Simon *et al.* (1973) and C. B. Pert and Snyder (1973) permitted true opiate binding to be identified and the distribution of the presumed receptor in brain to be determined.

The question of what endogenous materials might be intended for the receptors remained, for the moment, unanswered. The search for the ligand commenced without there being the slightest idea of the type of compound that might be involved. Work by Terenius and Wahlstrom (1975), using cerebrospinal fluid (CSF), and Hughes *et al.* (1975a), using brain extracts, suggested that low-molecular-weight peptides were involved.

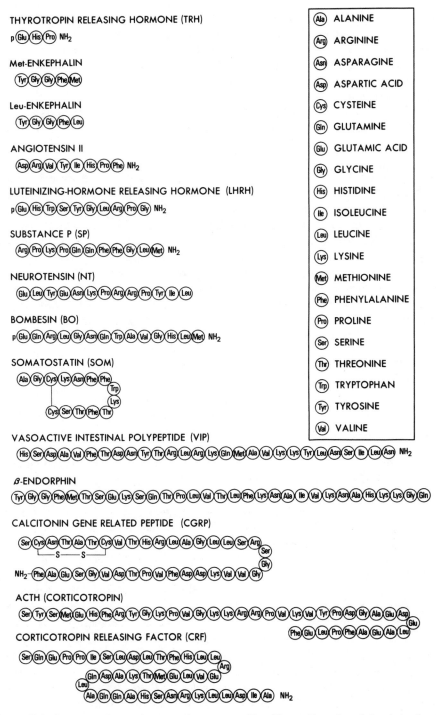

Figure 12.1. Structures of some representative neuropeptides. The configuration of the chains is arbitrary. (pGlu) Pyroglutamate.

It was only a few months later that Hughes *et al.* (1975b) announced the structures of two pentapeptides responsible for the activity, methionine enkephalin (met-Enk), and leucine enkephalin (leu-Enk) (Figure 12.1). It would be hard to overestimate the excitement generated among pharmacologists by this discovery. The structures were so simple that enkephalins were immediately synthesized and studies commenced in an amazing number of laboratories.

Antisera to these two peptides were prepared by linking them with large molecules such as polylysine or keyhole limpet hemocyanin. The antisera were then used for radioimmunoassays and immunohistochemical localization.

The discovery of many of the peptides listed in Table 12.1 followed a similar pattern: laborious isolation of an active peptide from brain, or other tissue, or even a nonmammalian source, followed by sequencing, synthesis, and preparation of antibodies to permit recognition of neuronal localization. The advent of molecular genetic techniques (see Sections 1.5 and 9.2.1) has been a significant addition that opened a whole new era of peptide identification. These techniques were initially applied to determining the sequence of complementary cDNA (cDNA) corresponding to messenger RNAs (mRNAs) that code for the large molecular precursors of known peptides. The amino acid sequence of the precursor could then be deduced and, given this sequence and the tendency toward cleavage at base pairs (see Section 12.2.2), the structure of other, possibly active peptides could be predicted. It was such work that led to the identification of β-endorphin (β-End), met-Enk-Arg[6]-Phe[7], and other products of the opionid precursors (see Section 12.2.2).

A later variant of this use of recombinant DNA technology was to examine the products of processing. This approach led to the identification of calcitonin gene-related peptide (CGRP). Its existence was predicted when the primary transcript of the rat calcitonin gene was found to be processed, in a tissue-specific manner, either into an mRNA that encoded a calcitonin precursor (in the thyroid) or, alternatively, into an mRNA that encoded a CGRP precursor (in the thyroid and in nervous tissue) (Amara *et al.*, 1982; Rosenfeld *et al.*, 1983). The carboxy-terminal fragment of the peptide predicted from the known structure of the CGRP mRNA was synthesized, and antibodies were raised against it. With these antibodies, CGRP-like immunoreactivity was demonstrated in the brain and throughout the entire nervous system of the rat, indicating the presence of the predicted peptide. CGRP is, in fact, the major product of the calcitonin gene. The human peptide (hCGRP) has a slightly different sequence but a very similar structure (Jonas *et al.*, 1985; H. R. Morris *et al.*, 1984). In man, a second calcitonin gene that encodes a second CGRP has been reported. The predicted sequence has three amino acid substitutions (Edbrooke *et al.*, 1985; Steenbergh *et al.*, 1985).

The speed at which recombinant DNA technology can provide the gene sequences that correspond to neuropeptide precursors of known, active peptides is remarkable. Biological evidence for a factor that causes the release of adrenocorticotropic hormone (ACTH) from the pituitary was provided as early as 1955. It took another 26 years to isolate the corticotropin-releasing factor (CRF) that was responsible and to work out the 41-amino-acid sequence. However, within 18 months, Furutani *et al.* (1983) had determined the sequence of the cDNA that codes for the CRF precursor.

Many of the peptides were initially named and have been subsequently classified according to the function by which they were first identified. This has led to the frequently used division of peptides into (1) gut–brain peptides, (2) hypothalamic factors, (3) opioid peptides, and (4) others. However, as Kastin *et al.* (1983) and others have pointed out, such naming and classification may lead to a restricted perspective on the true functions of these signaling agents. The gut–brain peptides, for example, are not localized exclusively to brain and gut, but are found in many tissues. The hypothalamic releasing factors do not act just on the pituitary, but also form neuron–neuron synapses in many areas of brain (see Section 12.3.2) and, like almost all neurotransmitters, are also found in the gut and many other body tissues. Most of the opioid peptides seem to be concerned with functions other than nociception (see Section 12.4).

A far better classification emerges with the recognition that the peptides belong to chemically and probably functionally related families, even though individual members of a particular family may have been discovered in unrelated circumstances. For example, SP is a member of a group of peptides known as the "tachykinins," which have a common C-terminal amino acid sequence. They share a similar spectrum of biological actions, including stimulation of salivary secretion, contraction of smooth muscle, and peripheral vasodilatation resulting in a lowering of blood pressure. Three members of this family, physalemin, eledoisin, and kassinin, were first isolated from invertebrate sources, and it was initially thought that SP might be the only tachykinin present in mammalian tissues. However, neuromedin K (β-NK), first isolated from spinal cord, proved to be a tachykinin, and physalemin-like immunoreactivity was subsequently identified in mammalian intestine and CNS (Lazarus *et al.*, 1980).

The structures of two SP precursors have been identified from an analysis of the nucleotide sequence of cloned cDNA from bovine brain (Nawa *et al.*, 1983). In addition to SP, one of these precursors also contains one copy of a peptide that is very similar in sequence to kassinin; this new peptide has been designated substance K. It is identical with neuromedin L from spinal cord, and evidence has been found for its existence in the mammalian CNS. This is only one of many similar examples of families of related peptides. Although we have given some indication of the traditional classification of the peptides in Table 12.1 by indicating the original source, we will stress in subsequent sections the family characteristics even though the relationships are just beginning to emerge. They can best be determined by seeking the common RNA transcripts and the genes that give rise to them.

12.2. Chemistry of the Neuropeptides

For a general reference, the reader is referred to J. D. White *et al.* (1985).

12.2.1. Structural Considerations

The number of amino acid residues in the chain and some other structural characteristics of many of the peptides are indicated in Table 12.2. It is evident from Table 12.2. that multiple forms of a given peptide can exist, as illustrated by cholecystokinin (CCK) and somatostatin (SOM). The predominant form of SOM in most brain regions appears to be the tetradecapeptide (SOM-14), but SOM-28 and even higher-molecular-weight forms have been reported to be released from the median eminence (Pierotti *et al.*, 1985). Most reports indicate that these various forms have similar physiological actions, but some differential effects of the SOM peptides in inhibiting the release of glucagon (SOM-14 > SOM-28) and insulin (SOM-28 > SOM-14) suggest that the larger peptides may be more than simple precursors of the smaller ones (Mandarino *et al.*, 1981).

TABLE 12.2
Chain Lengths and Some Structural Characteristics of Various Neuropeptides[a]

Name	Number of amino acid residues	C-terminal amide	N-terminal		O-SO$_4$	S-S cycle[b]
			PyroGlu	Acetyl		
ACTH$_{1-24}$	24	●	—	○	—	—
AKH	10	●	●	—	—	—
AVP	9	●	—	—	—	●
BK	9	—	—	—	—	—
BO	<u>14</u>, 27, 32	●	○	—	—	—
CCK	3, 4, <u>8</u>, 12, 33, 39	●	—	—	○	—
CGRP	37	○	—	—	—	●
CRF	<u>41</u>, 7	●	—	—	—	—
α-End	<u>15</u>, 16	—	—	—	—	—
β-End	31	●	—	○	—	—
γ-End	16, 17	—	—	○	—	—
FMRF-amide	4	●	—	—	—	—
GRF	44	●	—	—	—	—
GRP	27	●	—	—	—	—
LHRH	10	●	●	—	—	—
LPH	91	—	—	—	—	—
MO	≥22	—	—	—	—	—
MOA	8	●	—	—	—	—
α-MSH	13	—	—	○	—	—
γ-MSH	<u>11</u>, 12, 27	●	—	—	—	—
NPY	36	●	—	—	—	—
NT	13	—	●	—	—	—
OXY	9	●	—	—	—	●
RT	11	●	●	—	—	—
SOM	13, <u>14</u>, <u>28</u>, other	—	—	—	—	●
SP	11	●	—	—	—	—
TRH	3	●	●	—	—	—
VIP	<u>28</u>, other	●	—	—	—	—

[a] (●) Indicates a constant structural feature; (○) indicates a feature that may or may not appear in various forms of the neuropeptide. Where more than one chain length is indicated, that generally reported as the predominant one in mammalian brain is underlined. The peptide abbreviations are listed in Table 12.1 (Section 12.1). Chain lengths of some peptides not in this table are: ACTH (39), AII (8), AMD (26), ANF (126), GAP (56), GIP (43), GLI (100), GLU (≥29), β-MSH (18), β-NK (9), PHI (27), SEC (27), NMD-B (10), NMD-C (10), and SK (9).
[b] Indicates a cyclic structure due to a Cys–Cys linkage.

It is clear from Table 12.2 that activity can be conferred or enhanced by amidating the C-terminal or by cyclizing an N-terminal glutamate; these changes may also render the peptide more stable. Other chemical modifications are possible. CCK-8, for example, can exist in both the sulfated (on the tyrosyl residue) and unsulfated forms; the former is more potent because it is not cleaved by amidopeptidases.

Structural aspects that are not depicted in Table 12.2 are minor modifications in amino acid residues that occur either in different species or in closely related peptides in the same species. Vasopressin, as illustrated in Figure 12.2, is one example. Arginine-vasopressin (AVP) properly refers to the form that is found in all mammals except pigs; in pigs, the peptide is lysine-vasopressin. Figure 12.2 also illustrates the close structural similarities among AVP, oxytocin (OXY), and the corresponding materials from sub-

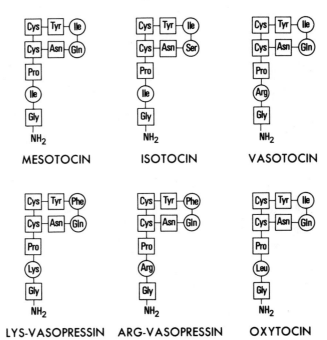

Figure 12.2. Structures of oxytocin, Lys-vasopressin, AVP, and some analogous peptides from nonmammalian species. Amino acids in squares are conserved; those in circles are variable. Lys-vasopressin is from pigs, AVP from other mammalian species, mesotocin from frogs, isotocin from fish, and vasotocin from other nonmammalian species.

mammalian species. It has been suggested that these similarities indicate that AVP and OXY may have arisen through mutations in duplicate copies of a single gene (F. E. Bloom, 1981).

12.2.2. Synthesis of the Neuropeptides

For general references, the reader is referred to Gainer *et al.* (1985), A. J. Turner (1984), and Loh *et al.* (1984).

Ever since the discovery of proinsulin about 19 years ago (Steiner and Oyer, 1967), it has been recognized that most secreted peptides and proteins are formed from larger precursor proteins by a variety of posttranslation processing steps. This seems to be true of all the putative peptide neurotransmitters of chain length greater than 2.

The biosynthetic process begins with the synthesis of the prepropeptide precursor, which is directed by the appropriate mRNA in the rough endoplasmic reticulum. The precursor is translocated to the Golgi apparatus (see Section 1.2.3). This prepropeptide, as synthesized, contains a "signal protein" as well as the active materials and other residues. An example is prepropressophysin, which consists of (Land *et al.*, 1982):

(19 AA signal peptide)-AVP-(Gly-Lys-Arg)-NpII-Arg-(39 AA glycoprotein)

The initial cleavage step removes the signal peptide. This step probably occurs during the translocation and is catalyzed by a microsomal endopeptidase. The remaining propeptide

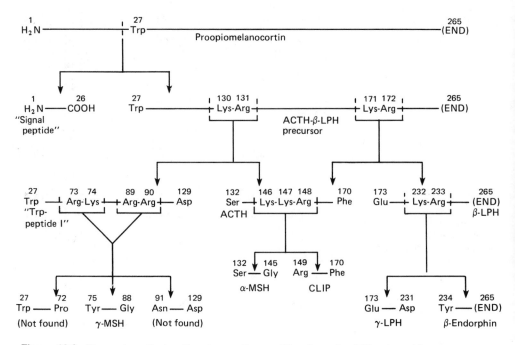

Figure 12.3. Proposed synthesis of various active peptides from the 265-amino-acid precursor, proopiomelanocortin. Cleavages at pairs or triplets of basic amino acids are indicated. Chain lengths are distorted and terminal NH₂ and COOH groups are only indicated for the signal peptide. (END) signifies last amino acid in precursor chain.

is thought to be packaged into secretory vesicles in the Golgi apparatus and to be further processed during transport down the axon. A good example of faulty processing due to a genetic defect involves the processing of prepressophysin in Brattleboro rats, which lack vasopressin and thus suffer from diabetes insipidus (Sections 13.2.5 and 12.4.2). In these rats there is a tiny error in the gene coding for the precursor: a single guanosine residue is missing in the mRNA in the sequence coding for the neurophysin section. As a result, the triplets coding for the amino acids become shifted with a drastic change in the structure predicted for the neurophysin section of the precursor. A stop codon is lost and an arginine residue is missed. Essentially, a nonsense protein is produced. The end result is failure to produce vasopressin itself, possibly either because of failure at the ribosomal translation stage due to the stop codon loss or because of failure to process the precursor completely because of terminus alterations (Schmale *et al.*, 1984).

Characteristically, but not universally, the sites of cleavage of prohormones are marked by a pair of basic residues (Lys or Arg) (Figure 12.3). Removal of these basic residues by the consecutive action of a trypsin-like and a carboxypeptidase-B-like enzyme generates the free peptide(s). The exact nature and specificity of the enzymes, which are frequently called "converting enzymes," and the factors that control differential processing of common precursors in different neurons, are not yet known, but further details may be found in the reviews cited for this section.

It should be noted that it is unsafe to assume that all cleavages will involve the removal of a base pair or that all base pairs will be cleaved. There are known examples in which cleavage occurs at a single base residue (as in the cleavage of the C-terminal glycopeptide from the AVP precursor) or in which uncleaved pairs of basic residues are retained within a neuropeptide [as in ACTH (see Figure 12.3)].

The cleavage process may proceed to different degrees in different tissues or different neurons, as evidenced by the existence of multiple forms of different chain lengths for a number of the neuropeptides (see Section 12.2.1). The multiple forms and functions of some peptides in brain and gut raise the question as to whether they stem from a single gene or from multiple genes. In the case of CCK, molecular genetic work has shown that it appears to be encoded by a single copy gene in the haploid human genome, suggesting that the same gene is expressed in both gut and brain (Y. Takahashi *et al.*, 1985).

Modifications of the peptide may also occur during processing. Many biologically active peptides have amide groups at the C-terminus [see Table 12.2 (Section 12.2.1)]. In all cases in which the precursor sequence has been established, the active peptide sequence is followed immediately by a Gly and then one or two basic residues. The amidation appears to be accomplished by an enzyme localized in secretory granules (Gomez *et al.*, 1984) that catalyzes a reaction such as the following:

$$\text{Peptide-Gly-Lys-Arg} + O_2 \rightarrow \text{Peptide-NH}_2 + \text{OHCCOOH} + H_2O + \text{Lys-Arg}$$

Ascorbic acid is required as a cofactor. The degree of amidation may vary from species to species. Thus, for example, rat CGRP has a free carboxy terminal, while human, cow, pig, and other CGRPs are amidated.

Posttranslational modifications, some of which were discussed in Section 12.2.1, are: cyclization of an N-terminal glutamate to pyroglutamate, as in thyrotropin-releasing hormone (TRH), luteinizing hormone-releasing hormone (LHRH), and neurotensin (NT); acetylation of the terminal amine, as in α-melanocyte-stimulating hormone (α-MSH) and β-End; and sulfation of tyrosyl hydroxyl groups, as in CCK. The specific enzymes involved in these processes have not generally been characterized.

Further information on some of the known precursor peptides and the products they yield is given in Table 12.3.

A given precursor may contain several copies of a single peptide [e.g., TRH and met-Enk (Table 12.3)] or a number of different peptides. Processing of the precursor in the latter case varies according to the particular type of neuron. The opioid family is an excellent example (Figure 12.3 and Tables 12.3 and 12.4). At first, it was considered that β-lipotropin (β-LPH) is the precursor from which both β-End and Met-Enk are derived and that dynorphins and neoendorphins are leu-Enk precursors. However, it became apparent that this was incorrect when the anatomical distributions of the putative precursors were shown to differ in many cases from those of the enkephalins that were supposed to be derived from them. Recombinant DNA techniques then revealed three distinct opioid precursors. The prohormone sequences were deduced from the isolated mRNAs (Table 12.3). Proopiomelanocortin (POMC), the common precursor for several neuropeptides including the endorphins, ACTH, and MSHs, was the first prohormone to be identified using this recombinant DNA approach (Nakanishi *et al.*, 1979).

A proenkephalin precursor in brain and adrenal tissues was then predicted on the basis of an isolated mRNA that contained the coding sequences for six copies of met-Enk

TABLE 12.3

**Neuropeptides and Related Materials Formed from Some of the Identified Higher-
Molecular-Weight Precursors**

Precursor			
Name	Number of amino acids	Products and copies/mole	Ref. No.[a]
Preprocholeocystokinin	115	CCK	1
Preprosomatostatin (PPS)	116	SOM-14, -28, -28_{1-12}; PPS_{25-100}	2
PreproTRH	255	TRH(5)	3
Propressophysin	166	AVP, NpII	4
Prooxyphysin	145	OXY, NpI	4
Angiotensinogen		AII, AIII	5
Prodynorphin (= preproenkephalin B)	256	α- and β-neoEnd; DYN_{1-17}, DYN_{1-8}, rimorphin; DYN B; leumorphin	6, 8
Preproenkephalin (= preproenkephalin A)	267	met-Enk (4), met-Enk-Arg^6-Phe^7, met-Enk-Arg^6-Gly^7-Leu^8; leu-ENK; peptides E and F; BAM-12P, -20P, -22P; amidorphin; MOA	7, 8
Proopiomelanocortin (POMC)	263	ACTH; CLIP; β- and γ-LPH; β-End; α-, β-, and γ-MSH; C'-fragment; β-End_{1-31}; glycylglutamine;	8
Pro-VIP		VIP; PHM-27 (very similar to PHI-27)	9
α-Preprotachykinin	112	SP	10
β-Preprotachykinin	130	SP, SK, and α-NK	10
GRF precursor	104–108	GRF	11
LHRH precursor	92	LHRH, GAP	12
ANF precursor	152	ANF	13
CGRP precursor	128–136	CGRP, calcitonin	14

[a]References: (1) Y. Takahashi et al. (1985), Deschenes et al. (1985), Kato et al. (1985), Kuwano et al. (1984); (2) Benoit et al. (1984), R. H. Goodman et al. (1983); (3) Lechan et al. (1986), Tavianini et al. (1984); (4) Land et al. (1982, 1983), Schmale et al. (1983); (5) Nakanishi et al. (1983), Ohkubo et al. (1983); (6) Kakidani et al. (1982); (7) Comb et al. (1982), Gubler et al. (1982), Noda et al. (1982a,b); (8) Dores et al. (1984); (9) N. Itoh et al. (1983) [the coding sequences for VIP and PHM-27 are located on two different (but close) exons of the human genome, a situation that may allow for alternative RNA processing (Bodner et al., 1985)]; (10) Nawa et al. (1983); (11) Mayo et al. (1985); (12) Seeburg and Adelman (1984); Nikolics et al. (1985); (13) Kangawa et al. (1984); (14) Amara et al. (1982); Nelkin et al. (1984); Rosenfeld et al. (1983).

and one of leu-Enk (Comb et al., 1982; Gubler et al., 1982; Noda et al., 1982a). The coding sequences were flanked by pairs of basic amino acid residues (Arg or Lys), suggesting that they were present as processing signals for trypsin-like enzymes. It was predicted that this processing would produce four copies of met-Enk and one each of leu-Enk, met-Enk-Arg^6-Phe^7, and met-Enk-Arg^6-Gly^7-Leu^8. The existence of these two carboxy terminal extended sequences of met-Enk in brain and adrenal chromaffin cells has been confirmed (A. S. Stern et al., 1980). The identification of enzymes capable of generating enkephalins from brain and adrenal precursors supports the assumption of opioid peptide biogenesis from proenkephalin.

The sequence of the precursor to the third opioid group, the dynorphins (DYNs), was determined from the nucleotide sequence of the porcine hypothalamic mRNA for proDYN (Kakidani et al., 1982). It was found to encode the sequences of DYN A (1–17), DYN A

TABLE 12.4
Amino Acid Sequences of Members of the Three Opioid Peptide Families

Precursor	Opioid peptides	Amino acid sequence[a]
Proopiomelanocortin	α-Endorphin	Tyr-Gly-Gly-Phe-Met-Thr-Ser-Glu-Lys-Ser-Gln-Thr-Pro-Leu-Val-Thr
	γ-Endorphin	Tyr-Gly-Gly-Phe-Met-Thr-Ser-Glu-Lys-Ser-Gln-Thr-Pro-Leu-Val-Thr-Leu
	β-Endorphin	Tyr-Gly-Gly-Phe-Met-Thr-Ser-Glu-Lys-Ser-Gln-Thr-Pro-Leu-Val-Thr-Leu-Phe-Lys-Asn-Ala-Ile-Lys-Asn-Ala-His-Lys-Lys-Gly-Gln (END)
Proenkephalin	met-Enkephalin	Tyr-Gly-Gly-Phe-Met
	leu-Enkephalin	Tyr-Gly-Gly-Phe-Leu
	met-Enk-Arg-Gly-Leu	Tyr-Gly-Gly-Phe-Met-Arg-Gly-Leu
	met-Enk-Arg-Phe	Tyr-Gly-Gly-Phe-Met-Arg-Phe (END)
	Peptide E	Tyr-Gly-Gly-Phe-Met-Arg-Arg-Val-Gly-Arg-Pro-Glu-Tyr-Trp-Met-Asn-Tyr-Gln-Lys-Arg-Tyr-Gly-Gly-Phe-Leu
Prodynorphin	Dynorphin A (1–17)	Tyr-Gly-Gly-Phe-Leu-Arg-Arg-Ile-Arg-Pro-Lys-Leu-Lys-Trp-Asp-Asn-Gln
	Dynorphin A (1–8)	Tyr-Gly-Gly-Phe-Leu-Arg-Arg-Ile
	α-Neoendorphin	Tyr-Gly-Gly-Phe-Leu-Arg-Lys-Tyr-Pro-Lys
	β-Neoendorphin	Tyr-Gly-Gly-Phe-Leu-Arg-Lys-Tyr-Pro
	Dynorphin B (= rimordin)	Tyr-Gly-Gly-Phe-Leu-Arg-Arg-Gln-Phe-Lys-Val-Val-Thr
	Leumorphin	Tyr-Gly-Gly-Phe-Leu-Arg-Arg-Gln-Phe-Lys-Val-Val-Thr-Arg-Ser-Gln-Gln-Asp-Pro-Asn-Ala-Tyr-Ser-Gly-Glu-Leu-Phe-Asp-Ala

[a]Tyr is the first amino acid of each peptide and forms the amino terminal; (END) signifies the final amino acid of the prohormone.

(1–8), the neoendorphins, and DYN B (rimorphin), all of which contain the leu-Enk sequence at the amino terminus.

Thus, these three precursors appear to account for all known opioid peptides discovered to date and form the basis for the present designation of three separate opioid families (Table 12.4). The processing of the precursors may show regional and species variations (Deftos and Catherwood, 1980; W. L. Miller *et al.*, 1980).

The high-molecular-weight precursors may also contain carrier molecules that are always associated with and co-released with a given peptide neurotransmitter even though they do not themselves appear to have neurotransmitter properties. The best-known examples are the cysteine-rich neurophysins (Np's) that coexist with OXY and AVP and are often used as markers of these neurons in immunohistochemical studies. In most mammals, the large propressophysin and prooxyphysin are common precursors of, respectively, AVP and its NpII and of OXY and its NpI (Russell *et al.*, 1981). Lipolytic peptide B is a material chemically related to the neurophysins that occurs in ACTH and α-MSH neurons of the hypothalamus (Loren *et al.*, 1980).

In summary, the synthetic processes for the peptides that have been outlined would clearly seem less efficient than the formation in the synapse of materials like γ-aminobutyric acid (GABA) and the amine neurotransmitters by the action of specific enzymes on low-molecular-weight precursors. In the latter case, each enzyme molecule transported from the cell body is capable of producing hundreds or thousands of copies of the neurotransmitter per minute. It is understandable, therefore, that evidence indicates that the number of molecules of peptides released per stimulation is considerably less than the number of molecules for low-molecular-weight neurotransmitters (see Section 12.4).

12.2.3. Destruction of the Neuropeptides

In no case has there been demonstrated for the peptides a high-affinity uptake system into either neurons or glia such as exists for many of the amine and amino acid neurotransmitters. Some uptake has occasionally been reported, but it would appear to be of the nonspecific type that can be observed in nerve endings with foreign materials such as horseradish peroxidase or dyes. Removal of the peptides from the synaptic cleft must therefore depend on inactivating enzymes, and rapid inactivation of injected peptides has indeed been repeatedly shown.

Examples of the types of cleavage reactions that have been demonstrated are shown in Figure 12.4. Thus, primary inactivating cleavages by peptidases present in highly purified rat brain synaptic membranes cleave NT at the 8–9, the 10–11, and the 11–12 positions, with further action being possible on some of the fragments (Checler *et al.*, 1985). CCK-8 seems to be cleaved first by a metalloendopeptidase to CCK-5, which is subsequently cleaved either to CCK-4 and glycine by an endopeptidase or to CCK-3 and the dipeptide glycyl-tryptophan by an amidopeptidase (Steardo *et al.*, 1985). Other examples are indicated in Figure 12.4.

There are apparently many different cleaving enzymes of the metalloendopeptidyl, dipeptidyl, and carboxypeptidyl types. There is no convincing evidence as yet of any peptidase that is specific for a single peptide, even though it has been suggested that the dipeptidyl carboxypeptidase named "enkephalinase A" is specific to the enkephalins (Matsas *et al.*, 1983). Another example of a probable misnomer is in the literature dealing with angiotensin converting enzyme (ACE) activity in the basal ganglia. It was once

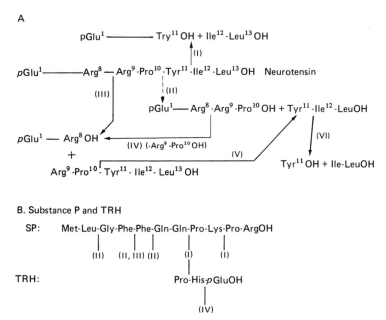

Figure 12.4. Hydrolysis of three neuropeptides by various peptidases. (A) Neurotensin: (I) endopeptidase 24–11 (= enkephalinase); (II) endopeptidase 24–11 + unidentified peptidase; (III) metalloendopeptidase; (IV) angiotensin converting enzyme (ACE); (V) postproline dipeptidyl aminopeptidase; (VI) bestatin-sensitive aminopeptidase. (B) Substance P and TRH: (I) a postproline cleaving enzyme (prolylendopeptidase) that is inhibited by bacitracin and also active, for example, on SOM and LHRH as well as TRH; (II) a peptidase or peptidases partially purified from human brain and also active on other peptides; (III) SP is cleaved at this position by ACE, pepsin, and cathepsin D as well as by the peptidase(s) from human brain; (IV) unspecified brain carboxypeptidases.

argued that ACE activity indicated the presence of angiotensin II (AII) in those brain regions in which it was concentrated. It was subsequently shown that the enzyme degrades SP, Enk, and probably other peptides (E. G. McGeer and Singh, 1979).

12.2.4. Biotransformation of the Neuropeptides

The metabolism of neuropeptides may produce materials with inherent biological activities distinct from the effects of the parent neuropeptide. This process has been termed *biotransformation,* and many examples have been reviewed by Griffiths and McDermott (1984). Only one or two instances will be cited here.

TRH is metabolized by a pyroglutamylaminopeptidase to yield histidylproline diketopiperazine [cyclo(His-Pro)]. This compound is found in brain and has a distribution different from that of the parent TRH. On iontophoresis, it stimulates cortical neurons (Stone, 1983), but it also often evokes inhibitory responses. Unlike TRH, it has no effect on the release of thyrotropin from the pituitary, but does inhibit prolactin (PRL) secretion. It opposes the action of TRH on body temperature. Its concentrations in brain regions other than the hypothalamus tend to be considerably higher than those of TRH. All these facts may indicate a separate physiological role.

Another example is the C-terminal tripeptide of OXY called PLG for its structure (Pro-Leu-Gly-NH$_2$) or MIF-I for its property of inhibiting the release of MSH from the pituitary. Binding sites for PLG have been characterised in many brain regions, and the compound has a wide range of behavioral effects. The fact that PLG enhances dopaminergic effects in animals led to a clinical trial in Parkinson's disease. The clinical benefits were limited, possibly because of rapid hydrolysis of the peptide (Kastin *et al.*, 1977; Mishra *et al.*, 1983).

Such biotransformations of neuropeptides into other active materials confer a further degree of flexibility on the physiological effects.

12.3. Distribution and Anatomy of Neuropeptide Systems

Much more is known about the neuronal distribution of the peptides than is known about their physiology or pharmacology. We will not attempt to present in this chapter all the immense anatomical detail already recorded in the literature. Rather, we will draw attention to some of the more significant anatomical features of certain key peptides as a means of providing clues regarding their possible functional role in brain. It must be kept in mind that a given peptide may be colocalized with other neurotransmitters [see Table 5.2 (Section 5.2)] and that the combinations do not follow any predictable pattern. Some ACh neurons, for example. contain SP as a cotransmitter, others vasoactive intestinal polypeptide (VIP), still others CGRP, and so forth. In many neurons, ACh appears to occur without a peptide cotransmitter.

We will first consider some of the special technical problems associated with peptidergic anatomical information and then describe some of the significant data.

12.3.1. Methodology

Almost all the evidence on the regional distribution and neuronal localization of peptides in the CNS and elsewhere is based on immunological techniques, i.e., radioimmunoassay and immunohistochemical methods. The usual problems with these techniques that lead to false-positive and false-negative results are particularly important as far as the peptides are concerned. It is quite possible, for example, that an antiserum produced against a particular peptide will cross-react with other peptides because they have an identical sequence that is serving as a common antigenic site. The problem of cross-reactivity may be particularly acute with monoclonal antibodies, since these antibodies recognize a single antigenic site, as opposed to polyclonal antibodies from immune sera that recognize a mixture of antigenic sites. The chemical similarity of many of the neuropeptides and neurohormones, which has already been mentioned (see Section 12.2.1 and Table 12.2), increases the probability of such cross-reactivity. Examples are the original erroneous identifications of CCK as gastrin and of neuropeptide Y (NPY) as avian pancreatic polypeptide in mammalian brain (DiMaggio *et al.*, 1985). Most investigators quite appropriately report their data as showing immunoreactivity that is Enk-like, CCK-like, SOM-like, etc. This is frequently indicated by Enki, CCKi, SOMi, etc.

Nature has further complicated the situation by making a single protein the precursor of a variety of peptide neurotransmitters and related molecules [see Table 12.3 (Section 12.2.2)]. As one consequence of nature's economy, it may often be unclear whether a particular peptide found in a neuron by immunohistochemistry is there as a precursor or as a neurotransmitter. This has been particularly true of the POMC family, and researchers are now beginning to report only that "end products of the POMC system" are found. An analogous case, of course, is that of DA, which occurs in both adrenergic and noradrenergic neurons only as a precursor, but is present in dopaminergic neurons as the neurotransmitter.

The increasing use of high-performance liquid chromatography (HPLC) separation techniques in combination with immunological techniques should increase the reliability of reported data.

The low concentration of most peptides may mean that, for visualization in the cell bodies, colchicine pretreatment is required (Sofroniew, 1985), as is the case for glutamic acid decarboxylase immunohistochemistry (see Section 7.3). Even such pretreatment may not be sufficient, because a number of the peptides exist in multiple forms (see Section 12.2.1), some of which may not be recognized by an otherwise satisfactory antibody.

Another problem in studying some of these peptide systems—notably LHRH and SOM—is that there seem to be considerable functional and probably seasonal effects on the immunohistochemical staining seen in animals, particularly in cortical and basal ganglia areas (Krisch, 1980). Pregnancy, estrus, and sex may have especially marked effects. Tsuruo et al. (1984), for example, have reported sex-dependent topographical differences in hypothalamic SP neurons and ultrastructural changes at different stages of the estrus cycle (see also Section 12.3.4).

Some reports are now appearing on the use of *in situ* hybridization of DNA probes with mRNA as a highly sensitive technique for the histochemical localization of cell bodies that express the genetic code for specific peptide precursors. An example is the demonstration of POMC neurons in the rat hypothalamus by *in situ* cDNA–mRNA hybridization (Gee *et al.*, 1983). Interpretation of the results may be complex because of the possible existence of the same precursor in a number of different peptide systems.

12.3.2. General Comments on Distribution

The relative concentrations and the presence or absence of cell bodies in various brain regions are indicated in Figure 12.5 for a number of the peptides. It is worthy of note that there may be extreme ranges of concentration in the different subnuclei in the large regions defined in Figure 12.5. The overall concentrations of the various peptides in rat brain range from a few hundred femtomoles to a few hundred picomoles per gram of tissue [see Table 5.1 (Section 5.2)]. Many of the peptides have most [ACTH, AVP, CRF, FMRF-amide, LHRH, TRH, bombesin (BO), motilin (MO), α-MSH] or all [bradykinin (BK), OXY] of their known cells in the hypothalamus. With few if any exceptions, the remainder have at least a few cells in that region. Some have their highest concentration in the neocortex (CCK, VIP) or the basal ganglia (SP, Enk), while others have cells in almost every brain region. One notable exception is the cerebellum, in which peptides appear to be relatively sparse.

Most peptides with all or almost all their cell bodies located in the hypothalamus tend to send diffuse projections to many brain regions. This is illustrated in Table 12.5, which lists many of the brain regions in which terminals of the AVP and OXY systems have been reported. The peptides with cell bodies in multiple areas tend, on the other hand, to have much shorter, less diffuse projection systems. The difference between long-axoned sys-

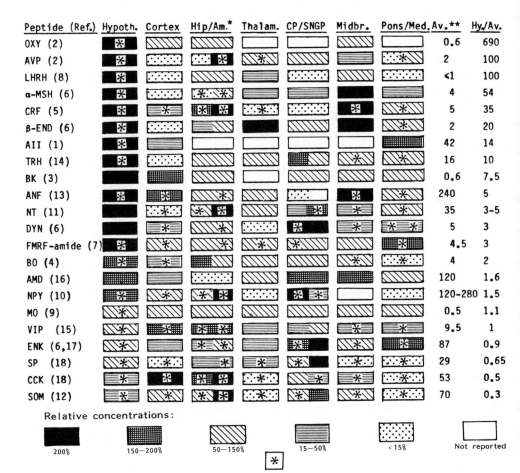

Figure 12.5. Relative regional concentrations and cell locations of some of the neuropeptides. The peptide abbreviations are listed in Table 12.1 (Section 12.1). *Cells in other parts of the limbic system are reported for AVP, CRF, ENK, FMRF-amide, LHRH, NPY, SP, SOM, CGRP, ANF, and DYN. CGRP is concentrated particularly in the brain stem and has cells not only there but also in the cortex, hippocampus, thalamus, and midbrain (18). GRF and GRP (19) have cells in the hypothalamus and pons–medulla. Particularly high levels of PHI are found in the cortex and hippocampus (20). **The average concentration shown (in pmoles/g) is the average of values reported for the regions shown, excluding the hypothalamus; it does not refer to pmoles/g whole brain. The levels are only indicative, since somewhat different results are reported from various species or for the same species by various authors. Data are reported for only one member of each of the opioid families, since the distributions of other family members tend to be parallel. References: (1) Lind *et al.* (1985); Brownfield *et al.* (1982), Hoffman *et al.* (1982), Ganten *et al.* (1983); (very variable concentrations are reported; the limited data cited are from experiments with HPLC confirmation of identity); (2) Sofroniew (1985), van Leeuwen *et al.* (1985), Brinton *et al.* (1983); (3) D. C. Perry and Snyder (1984); (4) Panula *et al.* (1982); Chronwall *et al.* (1985b) report RT⁺ cells in the pons and projections to the hippocampus; (5) Merchenthaler (1984), Merchenthaler *et al.* (1984), Fellmann *et al.* (1982), Vincent and Satoh (1984), Palkovits *et al.* (1985); (6) Dores *et al.* (1984), Eskay *et al.* (1979); (7) R. G. Williams and Dockray (1983), Dockray and Williams (1983); (8) King and Anthony (1984), Witkin and Silverman (1983), Jew *et al.* (1984); (9) Beinfeld and Bailey (1985), Beinfield and Korchak (1985); (10) Chronwall *et al.* (1985a), Card *et al.* (1983); (11) Emson *et al.* (1982), Manberg *et al.* (1982); (12) Krisch (1979), Ghatei *et al.* (1984); (13) Kawata *et al.* (1985); (14) Mori *et al.* (1982); (15) Obata-Tsuto *et al.* (1983), Ghatei *et al.* (1984), Kaneko *et al.* (1985); (16) Liebisch *et al.* (1985); (17) Schwartzberg and Nakane (1983); (18) Kawai *et al.* (1985); (19) Ling *et al.* (1985), R. J. Miller (1984), Bloch *et al.* (1983); (20) Polak and Bloom (1984).

TABLE 12.5

Extrahypothalamic Projections of Vasopressin and Oxytocin Fibers from the Suprachiasmatic and Paraventricular Nuclei[a]

Target area	AVP		OXY
	SCN	PVN	PVN
Forebrain			
Cortex	—	√	√
Medial septal area	—	√	√
Lateral septal area	√****	√	√+
Diagonal band	√**	√	—
Amygdala			
Medial	√***	√	√
Central, lateral, and basal	√+	√*	√*
Medial dorsal thalamus	√***	—	—
Lateral habenula	√****	—	—
Ventral hippocampus	—	√*	√*
Subcommissural organ	√	—	—
Brain stem			
Substantia nigra	—	√**	√****
Mesencephalic gray	√*	√+	√*
Dorsal raphe	√*	√+	√*
Interpeduncular nucleus	√+	—	—
Parabrachial	—	√*	√**
Locus coeruleus	√+	√+	√*
Raphe magnus	—	√*	√**
Tractus solitarius	√+	√**	√****
Dorsal motor vagus	√+	√**	√****
Ambiguus	—	√+	√+
Lateral reticular	—	√	√
Commissural	√+	√**	√****
Spinal cord			
Lamina I (dorsal horn)	—	√+	√*
Lamina X (central gray)	—	√*	√**
Intermediolateral gray	—	√+	√*

[a]Adapted from Sofroniew (1985) and E. A. Zimmerman *et al.* (1984). (SCN) Suprachiasmatic nucleus; (PVN) paraventricular nucleus. (√) Indicates reported staining of fibers in the region. (+, *, **, ***, ****) Indicate the density of fibers or terminals, or both, in ascending order; (+) indicates a few fibers and possibly some terminals.

tems that spread from a concentrated origin and short-axoned systems with multiple origins is well illustrated in Figure 12.6 by a comparison of schematic drawings of the sharply contrasting POMC and proenkephalin systems, which represent two of the three opioid peptide families.

As will be evident in the discussion of some particular regions in Sections 12.3.3–12.3.6, many peptides can be present in a particular small region of brain, and, as already discussed in Section 5.2, there is frequently coexistence of a peptide with either another neuropeptide or a nonpeptidergic neurotransmitter in the same neuron.

All the well-explored neurotransmitter peptides have been shown to be localized largely in the synaptic fraction of brain homogenates. There have been reports from electron microscopic studies that the immunoreactivities for LHRH, TRH, VIP, and SOM

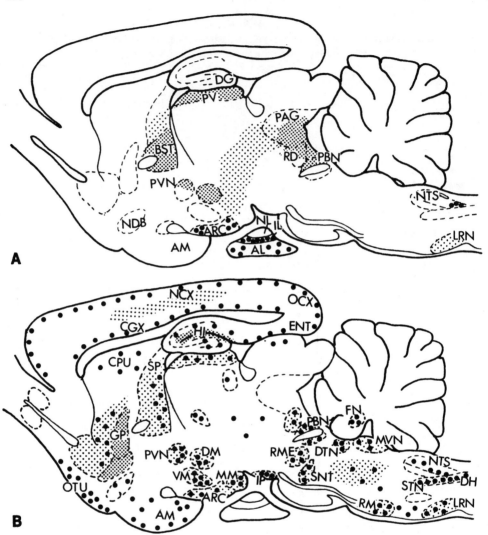

Figure 12.6. Distribution of cell bodies (•) and fibers (stippled areas) of the CNS opioid precursors POMC (A) and proEnk (B) in rat brain. Notice the extremely limited distribution of cell bodies of POMC (arcuate nucleus of hypothalamus, nucleus of the tractus solitarius, pars intermedia, and anterior lobe of the pituitary). ProEnk cell bodies appear at all levels of the brain, but not in the pituitary. Anatomy: (AL) anterior lobe of pituitary; (AM) amygdala; (ARC) arcuate nucleus; (BST) bed nucleus of stria terminalis; (CGX) cingulate cortex; (CPU) caudate–putamen; (DG) dentate gyrus; (DH) dorsal horn of spinal cord; (DM) dorsomedial nucleus of hypothalamus; (DTN) dorsal tegmental nucleus; (ENT) enthorhinal cortex; (FN) fastigial nucleus of cerebellum; (GP) globus pallidus; (HI) hippocampus; (IL) intermediate lobe of pituitary; (IP) interpeduncular nucleus; (LRN) lateral reticular nucleus; (MM) medial mammillary nucleus; (MVN) medial vestibular nucleus; (NCX) neocortex; (NDB) nucleus of diagonal band of Broca; (NL) neural lobe of pituitary; (NTS) nucleus tractus solitarius; (OCX) occipital cortex; (OTU) olfactory tubercle; (PAG) periaqueductal gray; (PBN) parabrachial nucleus; (PV) periventricular nucleus of thalamus; (PVN) paraventricular nucleus; (RD) nucleus raphe dorsalis; (RM) nucleus raphe magnus; (RME) nucleus raphe medialis; (SNT) sensory nucleus of trigeminal nerve; (SP) septal nuclei; (STN) spinal nucleus of trigeminal nerve; (VM) ventromedial nucleus of hypothalamus.

are associated with large (100-nm), dense vesicles and that of β-End with small (40 to 60-nm), clear vesicles. SP, the enkephalins, and MSH have been variously reported as associated either with large, dense or with small, clear vesicles (Pelletier and Dube, 1977; Guy et al., 1982; Pickel et al., 1979; Pelletier and Leclerc, 1979; Inagakai et al., 1986). Vesicular morphology therefore does not appear to be a good clue to the peptide or nonpeptide nature of the neurotransmitter.

A detailed description of the anatomical details of all or even a few of the many neuropeptides is beyond the scope of this book. Instead, we have chosen to describe some of the peptide systems in a few specific regions that may be associated with function (see also Section 14.4.2).

12.3.3. Hypothalamus

As indicated in the preceding section, the hypothalamus is extremely rich in many neuropeptides. The different subnuclei of the hypothalamus vary considerably in peptide content and type of axonal projection. We will describe briefly here some of the peptide neuroanatomy in the subnuclei involved with the neuroendocrine system.

Neuroendocrine System. The CNS exerts control over the periphery at three levels. The first is the musculoskeletal system. This system is the most highly developed, being voluntary and neocortical (see Section 14.3) with rapidly conducting fibers to striate muscles. It uses largely Type I (ionotropic) neurotransmitters, including the special case of ACh in the musculoskeletal system. The second is the autonomic system. This system is integrated at the subcortical level and involves a slower response because of the thin fibers that innervate smooth muscle (see Section 15.3). It uses largely Type II neurotransmitters. It is only partially under voluntary control. The third is the neuroendocrine system. This system is integrated by the hypothalamic–pituitary axis. This slowly responding system does not utilize peripheral nerve fibers and is not under voluntary control. Circulating hormones reach target cells in peripheral endocrine glands via the bloodstream. In turn, hormones released by these peripheral glands may return to the brain again via the bloodstream to provide a chemical feedback loop. It uses largely Type III neurotransmitters. In our discussion of this third system, we shall not be concerned with peripheral endocrine hormones. We shall discuss only the control of function by peptidergic hypothalamic neurons. We mention briefly the modulation of such control by other neuronal systems.

The pituitary gland is divided into two lobes—the anterior lobe and the posterior lobe—with the pars intermedia being embryologically derived from the anterior lobe but appearing as attached to the posterior lobe (Figure 12.7). Table 12.6 lists the pituitary hormones elaborated in each of these lobes, the target peripheral organ for each hormone, and the principal "hypothalamic releasing factors" that control release (McCann et al., 1984). It is worth noting that several of the "hormones" listed in Table 12.6 as being released by the pituitary (OXY, AVP, MSH, and ACTH) have also been found to serve as apparent neurotransmitters or neuromodulators in other brain regions [see Figure 12.5 (Section 12.3.2) and Section 12.4.1].

Many other neurotransmitters, not shown in Table 12.6, may also be involved in the regulation of pituitary hormonal release, at either the hypothalamic or the pituitary level. Jennes et al. (1983), for example, proposed a complex model to explain the multiple interactions of GABA and monoamine neurons with the effects of LHRH on the release of growth hormone (GH).

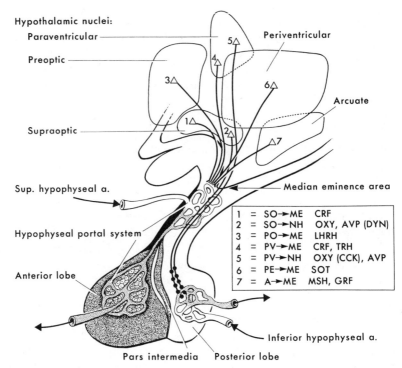

Figure 12.7. Diagrammatic representation of the human hypothalamus and pituitary indicating the nuclei of the hypothalamus containing peptidergic neurons that project to the pituitary. Pathways in most cases have been demonstrated in rat or other species. Anatomy of hypothalamic nuclei and pituitary areas: (SO) supraoptic; (PO) preoptic; (PV) paraventricular; (PE) periventricular; (A) arcuate; (ME) median eminence; (NH) neurohypophysis. The peptide abbreviations are listed in Table 12.1 (Section 12.1).

Figure 12.7 shows some of the nuclei of the hypothalamus that have major projecting fibers to the infundibulum, median eminence, and posterior lobe of the pituitary. Nerve endings are found in the posterior lobe and in the hypophyseal portal area.

The simplest arrangement is that for the posterior pituitary. Magnocellular OXY and AVP neurons of the supraoptic and paraventricular nuclei of the hypothalamus send axons through the infundibular stalk and the median eminence (Figure 12.7) to the neurohypophysis or posterior pituitary lobe, where they end adjacent to perivascular spaces. Suitable stimulation of these neurons causes release of AVP or OXY or both from the nerve terminals in the posterior pituitary and from there into the blood. Presumably, AVP or OXY are also released from the terminals in the other regions of the brain that receive projections from these neurons [see Table 12.5 (Section 12.3.2)].

Selective lesions of the median eminence area cause parallel decreases of DYN with AVP and of CCK with OXY, indicating the possibility of colocalization of DYN with AVP and of CCK with OXY (Palkovits, 1984).

MSH neurons in the arcuate nucleus project to the median eminence and release MSH, which is one of the main secretory elements of the intermediate lobe of the pituitary.

As shown in Figure 12.7, the principal hypothalamic releasing factors that control

TABLE 12.6
Pituitary Hormones and Some of the Hypothalamic Peptide Controlling Factors[a]

Lobe	Hormones	Target organs	Hypothalamic peptides	
			Releasing	Inhibiting
Posterior	Oxytocin	Mammary glands, uterus	—	—
	Vasopressin	Kidneys, vascular smooth muscle	—	—
Intermediate	Melanocyte-stimulating hormone (MSH)	Melanocytes	(CRF)	MIF-I
Anterior	Luteinizing hormone [= gonadotropin-releasing hormone (LH)]	Gonads	LHRH (NPY, GAP)	GIH[b]
	Follicle-stimulating hormone (FSH)	Gonads	FSH-RF[b] (NPY, GAP)	—
	Prolactin (PRL)	Mammary glands	PRF[b] (TRH, OXY, VIP, PHI)	PIF[b] (GAP, cyclo-His-Pro)
	ACTH[c]	Adrenal cortex	CRF (OXY, AII[d])	—
	Thyrotropin	Thyroid gland	TRH	(SOM)
	Growth hormone (GH)	Many tissues	GRF (MO, VIP, NPY)	SOM

[a]It is now recognized that release of pituitary hormones may be under the control of several peptide and nonpeptide factors, and the peptides shown in parentheses also have the activities indicated. DA and possibly GABA have PRL-inhibiting factor (PIF) activity, and DA also stimulates GH release and inhibits MSH release (see Section 9.5.1) (McCann et al. (1984). The peptide abbreviations are listed in Table 12.1 (Section 12.1).
[b]The peptide is not characterized, but there is evidence for a peptide factor other than those identified to date; the 56-amino-acid peptide GAP, predicted from the sequence of the precursor of LHRH, has potent PRL release-inhibiting activity (as well as some LH and FSH release-stimulating activity) and may be equivalent to PIF (Nikolics et al., 1985; H. S. Phillips et al., 1985).
[c]β-End is probably co-released (Risch et al., 1983).
[d]Sobel (1983).

anterior pituitary function arise in the medial preoptic (LHRH), paraventricular (CRF, TRH), periventricular (SOM), and arcuate [growth hormone-releasing factor (GRF)] nuclei of the hypothalamus and terminate in the median eminence. The peptides released from these terminals pass into the blood of the pituitary portal system to reach the anterior lobe.

The median eminence also contains fibers or terminals of numerous other neurotransmitters that may act directly or indirectly to modulate the release of the pituitary hormones. The tuberoinfundibular DA system was discussed in Section 9.5. A given neurotransmitter may mediate hormonal release at several levels. NPY, for example, has projections from cell bodies in the arcuate nucleus to both the periventricular and the paraventricular nuclei, as well as to the median eminence, so that its modulation of pituitary hormone release may be at both hypothalamic and pituitary levels. It is also a cotransmitter in many of the NA neurons that project from the brain stem to the hypothalamus (Sawchenko et al., 1985).

Relatively few neurons are involved in any one of these systems. For example, in the rat, the average numbers of OXY, AVP, and CRF cells have been estimated as about 3900, 4800, and 2000, respectively (Rhodes *et al.*, 1981; Sawchenko and Swanson, 1985).

Peripheral activities that are influenced by factors released from the hypothalamus are many and complex, and the reader is referred to many excellent reviews of the subject (McCann *et al.*, 1984; Palkovits, 1984; Blalock *et al.*, 1985). Among others, they influence sexual function, stress response, appetitive function, the immune system, antidiuretic and pressor responses, temperature regulation, and emotion.

Chemical feedback loops help regulate hypothalamic activity and pituitary hormone release. Corticosteroids, sex hormones, and other factors act directly on hypothalamic cells and on neurons with projections into the hypothalamus.

Suprachiasmiatic Nucleus.   The suprachiasmatic nucleus of the hypothalamus consists in the rat of 8000–10,000 small cells on either side of the midline. Its primary known function seems to be its involvement in the maintenance of various circadian rhythms and their entrainment to environmental light cycles [see Figure 10.7 (Section 10.7.1)]. This nucleus seems rich in many peptide neurotransmitters. Perikarya staining with antibodies against NT, ACTH, CGRP, CRF, met-Enk, DYN, glicentin, SP, BO, gastrin-releasing peptide, neurophysins, AVP, SOM, and VIP have been reported. Many of these types of cells appear to send axons to the contralateral suprachiasmatic nucleus, suggesting a high degree of commissural communication.

Besides the types of transmitters with intrinsic cell bodies, the nucleus also contains axons that react immunochemically in such ways as to suggest that serotonin [5-hydroxytryptamine (5-HT)], DA, NA, adrenaline, CCK, PRL, TRH, NPY, and ACh may all be important in the functions of the nucleus (van den Pol and Tsujimoto, 1985). An NPY projection from the lateral geniculate body to the suprachiasmatic nucleus has been demonstrated (Card and Moore, 1982), which would support a role related to circadian rhythms (see Figure 10.7). This is one of many similar examples in the literature of the incredibly complex diversity of putative neurotransmitters to be found within small nuclei.

12.3.4. Limbic System

Next to the hypothalamus, the richest source of peptides is the limbic system, and especially the amygdala. VIP-, SP-, SOM-, CCK-, met-Enk-, and NT-positive cells and fibers are all found in the amygdala with differential distributions among the various nuclei and complex intra- and extra-amygdaloid projections (G. W. Roberts *et al.*, 1982). Cells that stain for VIP are located mainly in the lateral nucleus, those for CCK in the cortical nucleus, those for NT and met-Enk in the central nucleus, and those for SP in the medial nucleus, while SOM cells are scattered throughout the amygdala. Ovoid NPY-containing cells are also disseminated in all the amygdaloid subnuclei (Nakagawa *et al.*, 1985), and many CRF cells (Fellmann *et al.*, 1982) and some DYN cells are found in the central nucleus. The VIP, SOM, and CCK cells in the basolateral amygdala nucleus form distinct groups, but all appear to be of the Golgi type II with few dendritic spines (McDonald, 1985b). Prominent peptide-containing amygdalofugal projections run from the central nucleus to the hypothalamus by way of the stria terminalis. They terminate as well in the bed nucleus of the stria terminalis, the nucleus accumbens, and the preoptic area (G. W. Roberts *et al.*, 1980). Among the peptides involved are enkephalin, NT, SOM, VIP, and NPY(Sakanaka *et al.*, 1981; Uhl *et al.*, 1978; Uhl and Snyder, 1979; Y. S. Allen *et al.*, 1984); many of the SP neurons are intrinsic to the amygdala (Emson *et al.*, 1978).

Another limbic area particularly rich in peptides is the bed nucleus of the stria terminalis. It not only receives peptide projections from the amygdala and other areas but also contains a profusion of peptidergic cells. These cells include some AVP-containing neurons (van Leeuwen *et al.*, 1985) and many NPY (Nakagawa *et al.*, 1985), NT (Biggins *et al.*, 1983), LHRH, CRF (Merchenthaler, 1984), met-Enk, DYN, VIP, SP, and SOM cells, as well as fibers that stain for these and other peptides (e.g., CCK) (Woodhams *et al.*, 1983). Many, such as VIP, show a heterogeneous distribution within the nucleus. The dorsal bed nucleus contains the highest concentration of VIP found in rat brain next to that of the suprachiasmatic nucleus, the concentration being 4-fold higher than in the frontal cortex. Lesion studies suggest VIP afferent inputs from the brain stem and one, or possibly two, projections that originate in the amygdala (Eiden *et al.*, 1985).

Woodhams *et al.* (1983) suggest that there are numerous local peptidergic connections within the bed nucleus; between the nucleus and such adjacent structures as the preoptic area, hypothalamus, and basal ganglia; and more distant projections to the amygdala and other forebrain brain areas. Like some of the hypothalamic nuclei previously described, this is a further example of a small region of brain that is extremely rich in various peptide neurotransmitters. It is also rich in GABA, NA, histamine, and DA.

AVP and OXY neurons and their associated neurophysins have been detected in the medial and lateral septum and the nucleus of the vertical limb of the diagonal band of Broca (Sofroniew, 1985). Among other peptides, SP-containing neurons are found in the lateral septum (Ljungdahl *et al.*, 1978), as are some met-Enk (Haber and Elde, 1982) and SOM (Vincent *et al.*, 1985) cells. The septum has one of the highest concentrations of atrial natriuretic factor (ANF) reported, with the level being about the same as in the hypothalamus (Kawata *et al.*, 1985).

It is notable that binding sites for many neuropeptides have also been found to be enriched in the amygdala and other structures associated with the limbic system. Thus, maps of the binding of opiates (La Motte *et al.*, 1978; Hiller *et al.*, 1973), SP (Rothman *et al.*, 1984), BO (A. Pert *et al.*, 1980), CCK (Gaudreau *et al.*, 1983), and NT (Quirion *et al.*, 1982) all showed enrichment in the amygdala and associated systems (C. B. Pert *et al.*, 1985).

Many of the neurons in the bed nucleus, like many in the peptide-rich hypothalamus, amygdala, and nucleus tractus solitarii, have been shown to concentrate estradiol and thus may be major sites of sex hormone–neuronal interactions (Stumpf and Jennes, 1984; Morrell *et al.*, 1984). The number of AVP-positive cells in the bed nucleus is greater in males than in females (van Leeuwen *et al.*, 1985), and the AVP innervation of the lateral septum is much denser in male than in female rats; this latter difference has been shown to be dependent on the presence of androgens about the 7th postnatal day (De Vries *et al.*, 1984) (see Section 13.2). TRH levels in human amygdala are about twice as high in males as in females (Biggins *et al.*, 1983). The peptide field may reveal a variety of interesting variations between the male and female brains. This is further discussed in Section 13.2.

Some of the highest levels of VIP in brain are in the hippocampus. The subiculum may carry efferent VIP fibers and those of CCK, NT, and SOM through the fornix and

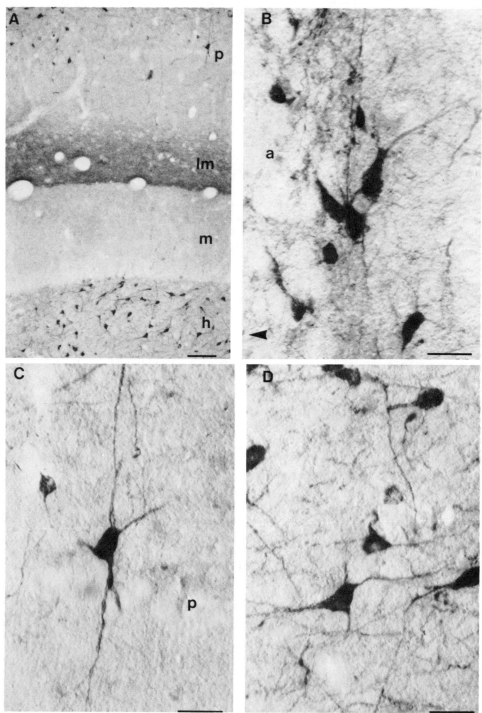

Figure 12.8. Immunoperoxidase micrographs illustrating SOM-immunoreactive structures in the rat hippocampus. (A) Many positive neurons are present in the hilus of the dentate gyrus (h) and above the pyramidal cell (p) layer in stratum oriens. The striatum lacunosum moleculare (lm) contains a dense terminal field, while the molecular layer (m) of the dentate has a sparser fiber network (B) Higher magnification of a cell in stratum oriens next to the alveus (a). (C) A cell in the pyramidal cell layer (p). (D) Cells in the dentate hilus. Scale bars: (A) 100 μM; (B–D) 25 μM. Courtesy of Dr. S. R. Vincent (University of British Columbia).

fimbria to other brain regions (G. W. Roberts et al., 1983; C. Kohler, 1982). Non-pyramidal, bipolar VIP neurons of the hippocampus receive commissural afferents (Leranth and Frotscher, 1983). The distribution of SOM cells is shown in Figure 12.8.

CCK perikarya are most dense in and around the stratum pyramidale, within the superficial layer of the subiculum, and in the polymorph zone of the hilus (Gall, 1984). CCK (and enkephalin) are probably in the mossy fibers and in the associational and commissural fibers (Stengaard-Pedersen et al., †983). α-MSH, NT, NPY, and SOM cells are also found in the hippocampus.

12.3.5. Cortex

CCK, VIP, SOM, and NPY all have numerous cells in the cerebral cortex, while some other peptides such as SP, BO, CRF, DYN, FMRF-amide, NT, and ANF have a few cells in that region.

The CCK cells in rat and monkey cortex are all small, nonpyramidal cells and are present in all layers, but are concentrated superficially in layers I–III. Most give rise to two to four long, radially ascending and descending processes, which branch at relatively few points and can extend through much of the thickness of the cortex. Labeled punctate structures that possibly correspond to positive axon terminals are located in several layers and are most densely distributed in layer VI. Ultrastructurally, the terminals form asymmetrical synapses on several different neuronal structures including dendrites and the somata of unlabeled nonpyramidal cells (S. H. C. Hendry et al., 1982). A. Peters et al. (1983) report that, in the rat cortex, CCK is not only in bipolar cells that form synapses that are asymmetrical and therefore probably excitatory, but also in bitufted and multipolar neurons that may be inhibitory. In adult rat visual cortex, CCK occurs mainly in bitufted and multipolar nonpyramidal cells of layers II and III (McDonald et al., 1982b). In human pre- and postcentral gyri, CCK is located mainly in the perikarya of nonpyramidal cells such as multipolar, bitufted, and bipolar cells and, to a lesser extent, in medium and small pyramidal cells (Sakamoto et al., 1984).

Because of the considerable mass of the cerebral cortex, the total brain content of CCK, about 1–2 mg in the human, is far greater than that of any other peptide, with the possible exception of the more recently discovered NPY (Crawley, 1985).

Like CCK, VIP is found in higher concentrations in the cerebral cortex than in any other large brain region [see Figure 12.5 (Section 12.3.2)]. In the rat visual cortex, the VIP cells are present predominantly in layers II and III; the majority are bipolar, but some multipolar forms are also present (McDonald et al., 1982a). At least 80% of the bipolar choline acetyltransferase-positive neurons in rat cortex (see Section 8.3) are also positive for VIP (Eckenstein and Baughman, 1984). Detailed morphological studies of VIP cells in rat cortex indicate that each is identified with a unique radial volume, supporting the idea that these cells play an important functional role within radially oriented columns (Morrison et al., 1984). While the cortical VIP cells in rats constitute a relatively homogeneous population of bipolar cells that probably serve as cortical interneurons, the hippocampal VIP cells are more heterogeneous (C. Kohler, 1982). The dense terminal innervation of cerebral blood cells (Hanko et al., 1977; Matsuyama et al., 1983) and the high VIP content of pial blood vessels (Edvinsson et al., 1980; Duckles and Said, 1982) suggest a role in control of cerebral circulation (see Section 12.4.2).

SOM perikarya occur in the cortex as in almost every other brain region. SOM cells

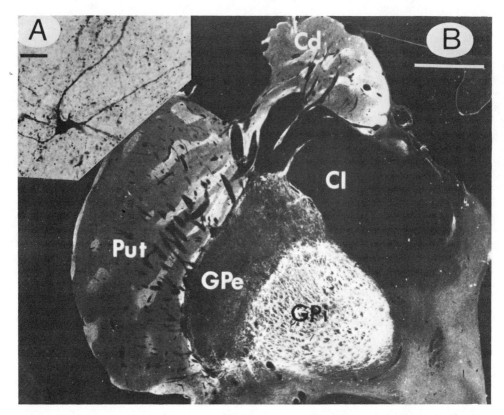

Figure 12.9. SP staining in the baboon basal ganglia. (A) Immunoperoxidase-stained neuron of the putamen. Scale bar: 30 μm. (B) Dark-field photograph of immunoperoxidase stain of basal ganglia slice at low power. Notice the positively staining light-colored patches in the caudate and putamen, the very intense staining of the internal segment, and the lack of staining in the external segment of the globus pallidus. (Cd) Caudate nucleus; (Put) putamen; (GPe) external segment of the globus pallidus; (GPi) internal segment of the globus pallidus; (CI) internal capsule. Scale bar: 3 mm. Courtesy of Dr. T. G. Beach (University of British Columbia).

are found in all layers of the cortex (Demeulemeester *et al.*, 1985; Vincent *et al.*, 1985). The morphological characteristics are varied and are consistent with long projection pathways as well as intrinsic neurons (Morrison *et al.*, 1983), although the majority appear to be interneurons. A proportion of these interneurons contain GABA as a cotransmitter. SOM projections from the insular cortex to the spinal cord have been shown by combined retrograde tracing and immunohistochemistry (Shimada *et al.*, 1985), and projections from various cortical regions to other subcortical regions have been suggested from lesion experiments.

In the cerebral cortex of the rat, NPY-positive cells are diffusely distributed throughout layers II–VI. The cells are about 15–20 μm in diameter and ovoid or bipolar in shape (Nakagawa *et al.*, 1985).

As mentioned above and as indicated in Figure 12.5 (Section 12.3.2), some other types of peptidergic cells have also been reported in the cortex, but these are less numerous than the systems described above and have so far received little study.

12.3.6. Basal Ganglia

The basal ganglia are extremely rich in many peptides, as they are in other, low-molecular-weight neurotransmitters. Some of the pathways of the known peptide systems in the basal ganglia are described in Section 14.5.3. These peptide systems must be important in the control of movement and other functions of the basal ganglia, but the physiological and pharmacological experiments necessary to establish their various roles have, by and large, not yet been done.

One of the most prominent peptides in the basal ganglia is SP. It originates in cell bodies associated with striosomes or islands in the caudate–putamen and projects preferentially to the internal segment of the globus pallidus and the pars reticulata of the substantia nigra (Figure 12.9). The SP neurons are separate from the GABA projecting cells of the striatum (Gale et al., 1977) and may represent an excitatory neostriatal output as opposed to the inhibitory GABA output. They are also separate from the enkephalin-containing cells, which also arise in patches (Graybiel et al., 1981) but project preferentially to the external pallidal segment (Haber and Elde, 1981). DYN distributions are similar to those of SP in the basal ganglia (Vincent et al., 1982b).

CCK is present in the caudate–putamen, but is associated primarily with nerve terminals of pathways that originate in the pyriform cortex and amygdala (D. K. Meyer et al., 1982). There is one report of CCK cells in the lateral part of the substantia nigra (Hökfelt et al., 1980a).

NPY and SOM have widely distributed cell bodies, including relatively high concentrations in the caudate–putamen, as indicated in Figure 12.5 (Section 12.3.2). Basal ganglial levels of a number of other peptides are also indicated in Figure 12.5.

12.3.7. Brain Stem and Nucleus Tractus Solitarius

In addition to the hypothalamus and limbic system, the midbrain and hindbrain also contain a rich variety of peptidergic neurons [see Figure 12.5 (Section 12.3.2)]. These neurons are distributed among many structures. We shall describe in detail only one, the nucleus tractus solitarius (NTS), which surrounds the tractus solitarius and is situated just lateral to the dorsal motor nucleus of the vagus nerve (Carpenter and Sutin, 1984). There are minor variations according to the species, but generally the dorsolateral aspect receives gustatory input; the ventrolateral aspect, respiratory input; and the medial and dorsomedial aspects, baroreceptor input.

SP, leu-Enk, CCK, NT, avian pancreatic polypeptide (NPY), and MSH fibers were all found concentrated in the medial and commissural regions, suggesting a role in blood pressure regulation (Yamazoe et al., 1984). SP and leu-Enk were more widely distributed. SP in particular may have a role in the respiratory part of the complex (Maley and Elde, 1982). SP, CCK, and NT neurons were found with the same preferential distributions as the fibers, as were the enkephalin neurons, although the detailed distributions had minor variations. NPY-positive cells were also distributed in these regions, but extended to the ventral and ventrolateral nuclei. MSH neurons were only in the commissural nucleus.

The NTS neurons project to the supraoptic and paraventricular nuclei of the hypothalamus and presumably are part of feedback loops that affect pituitary hormone release. Lesions of the NTS, for example, induce pressor responses due to enhanced AVP release (Kubo and Amano, 1986).

These data give some indication of the complexity of detail that is involved with peptide coordination of brain–peripheral autonomic and endocrine function. It will be some years before patterns emerge into which these data can be integrated.

Many instances of cotransmitter localization have been recorded for brain stem neuronal groups [see Table 5.2 (Section 5.2)], and these too are in the early stages of investigation. Again, the relationships are complex. For example, DA and CCK are colocalized in A10 and in substantia nigra pars lateralis neurons that project to the limbic system (Hökfelt *et al.*, 1980a), but not in substantia nigra neurons that project to the caudate–putamen. NA and NPY are colocalized in many neurons in the A1 group, but not in other noradrenergic groups. It can be anticipated that much new and significant information will emerge that will provide important insight into the detailed operation and need for such distinct chemical coding of these small but important neuronal subgroups.

12.3.8. Cerebellum

The cerebellum deserves mention because it seems to be unique among the large regions of brain in its relative scarcity of neuropeptides. SOM-like, MO-like, and CGRP-like immunoreactivities have been reported in some Purkinje cells (Chan-Palay *et al.*, 1981; Y. Kawai *et al.*, 1985) and Golgi cells (Vincent *et al.*, 1985). The human cerebellar cortex is said to have a particularly high level of CGRP-binding sites (Tschopp *et al.*, 1985). Positive staining for ACTH has been reported in some large neurons of cerebellar nuclei that project to the red nucleus (Leranth *et al.*, 1981). Although there are some TRH cells reported in the cerebellum, the cerebellar content of TRH in the monkey is very low, about equivalent to that of the cerebral cortex (Johansson and Hökfelt, 1980). Very few fibers in the cerebellum stain for any peptide, and by radioimmunoassay, the cerebellum seems to have extremely low or undetectable levels of most of the peptides listed in Figure 12.5. O'Donohue *et al.* (1981) have reported that the highest brain levels of secretin are in the cerebellum, pituitary, hippocampus, and thalamus, but its cellular localization is unknown.

12.3.9. Other Systems

Most of the neuropeptides listed in Table 12.1 (Section 12.1) have also been reported in retina and spinal cord. They also have a wide distribution in peripheral organs as well as in the CNS. They are particularly plentiful in the gastrointestinal tract (for a review, see M. Costa and Furness, 1982), but also appear in the pancreas, adrenals, lungs, heart, and reproductive organs. The occurrence of many peptides, but notably SP, in primary sensory afferents and other pathways involved in pain perception is mentioned in Section 12.4.2. Pearse and Takor Takor (1979) discuss 17 neuropeptides that have been identified in the so-called "amine precursor uptake and decarboxylation" (APUD) cells. These cells occur in brain and many peripheral locations, such as the pancreas, the gastrointestinal tract, the respiratory and urogenital tracts, the parathyroid and thyroid glands, the thymus, and the adrenal medulla. They also contain an aromatic amino acid decarboxylase (see Section 9.2.1).

VIP is an excellent example of a peptide that is very widely distributed in peripheral systems. VIP seems to occur in short-axoned neurons in sphincters throughout the body that may act to regulate local blood flow and secretion (Alumets *et al.*, 1979). A massive innervation of cerebral blood vessels by VIP neurons has also been noted (Edvinsson *et al.*, 1980).

SOM is another good example of a widely distributed peptide. It occurs not only in the spinal cord and retina, but also in many other organs. SOM has been found to be so widely distributed and to inhibit the secretion of so many materials, e.g., thyrotropin, glucagon, insulin, and gastrin, that the name "panhibin" has been suggested (McCann *et al.*, 1984).

The widespread occurrence of peptide systems in both brain and periphery makes it difficult to relate their general physiological effects to particular central neurons (see Section 12.4.2).

12.4. Physiology of the Neuropeptides

12.4.1. Iontophoretic Effects, Release, Binding Sites, and Second Messengers

The existence of a material in neurons or even its preferential synaptosomal localization within that neuron is not sufficient grounds on which to assign putative neurotransmitter status. Such status can be more reasonably assigned if the material is not only localized within neurons but also can be shown to be released on stimulation and to have a physiological action involving specific binding sites on postsynaptic membranes (see Chapter 5). An indication of how far these various criteria have been satisfied with peptides is given in Table 12.7.

Iontophoretic Effects.   Most of the peptides have been tested in iontophoretic experiments *in vivo* with quite variable results that often seem to depend on the particular test situation used. The type of data obtained is well illustrated in the reports of Phillis and Kirkpatrick (1980), who also reviewed much of the previous literature. Although there are usually more reports of an excitatory effect than of an inhibitory one, the same peptide may have both effects in a given brain area, and the percentage of cells that respond is variable. The amount of peptide released is difficult to assess. Gozlan *et al.* (1977) have suggested that the rapid metabolism may lead to erroneous conclusions in such iontophoretic experiments and might also help to explain some of the variable results obtained. Such problems have led to a search for a better model system, and investigators have turned to work on neurons in slice preparations or tissue culture. Studies on hippocampal pyramidal neurons in a slice have suggested that VIP, SOM, and CCK-8 may have effects in this preparation comparable to those of glutamate; i.e., each peptide caused a rapid depolarization accompanied by an increase in excitability. In contrast, SP had no effect in this preparation (J. S. Kelly and Dodd, 1981), but did act rapidly on cultured spinal cord neurons (McDonald and Norvak, 1981).

Elde and Hökfelt (1979) have suggested that the unconventional and somewhat variable effects of the peptides may be due, at least in part, to the relationship of peptidergic terminals to the receptors. In some cases, they postulate, the peptide is released into the extracellular spaces adjacent to fenestrated capillaries and is carried by the vascular system to its target, where it interacts with receptors. In other cases, the peptide receptor may be restricted to the postsynaptic specialization. In some cases, the peptide may act only to modulate the action of some more conventional neurotransmitter. There are many reports suggesting such modulation. For example, SP is said to depress ACh excitation of rat cortical cells (Lamour *et al.*, 1983). AVP is reported to potentiate the excitatory responses of septal area cells to both applied glutamate and fimbria–fornix

TABLE 12.7
Iontophoretic Effects, Release, Binding Sites, and Effects on Second Messengers for Some Putative Neurotransmitter Peptides[a]

Peptide	Ref. No.[b]	Iontophoretic effect[c]	Release[d]	Binding sites[e] Shown	K_m (nM)	B_{max} (fmoles/mg prot.)	Possible second messenger[f] cAMP	cAMP	PI
AII	1	E/I	—	√	0.1–3	3–20	—	—	√,
ACTH	2	—	—	—	—	—	√	√	√
AVP	3	E/I	√e	√	0.4, 13	6, 58	√	—	0
BK	4	I	—	√	1, 16	100, 1000	√, 0	√.	
CCK	5	E/I	√	√	≈1	7–115	—	—	√
CGRP	6	—	√	√	0.5	2–135	—	—	
CRF	7	E	√	√	—	—	√	—	
END	8	I/E	√e	√	—	—	—	—	
LHRH	9	E/I	√e	√	0.1, 6	—	0, √	—	
β-LPH	10	—	√	√	—	—	—	—	√
MIF-1	11	I	—	—	—	—	0	√	—
α-MSH	10	—	√	—	—	—	√	—	√
NT	12	E/I	√e	√	3	30	0	0	—
SOM	13	I/E/M	√e	√	0.4–5	95	0, √	0	—
SP	14	E/M	√e	√	0.3–3	30–100	0	0	√
TRH	15	I/E	√	√	2–200	3–60	√, 0	√, 0	√
VIP	16	I/E	√e	√	2, 125	2	√	—	

[a]No attempt has been made to cover the literature completely in this table, merely to give an indication of the type of data available. The peptide abbreviations are listed in Table 12.1 (Section 12.1).
[b]References: (1) Palovcik and Phillips (1984); Elliott et al. (1983), Bennett and Snyder (1980a,b), Israel et al. (1984), Speth et al. (1985); Gispen et al. (1985), Gispen (1982); (3) Yamane et al. (1984), Junig et al. (1985); (4) R. E. Lewis et al. (1985); (5) Pandol et al. (1985) Saito et al. (1981), S. E. Hays et al. (1981b); (6) Tschopp et al. (1985); (7) Suda et al. (1985); De Souza et al. (1984), Labrie et al. (19 Siggins et al. (1985); (8) Patey et al. (1985), Lindberg and Dahl (1981); (9) Liscovitch and Koch (1982), Leung et al. (1983); (10) Osborr al. (1979), Vermes et al. (1981); (11) Chiu et al. (1983); (12) Colmers et al. (1985), Uhl and Snyder (1981); (13) Pierotti et al. (1985); Hanley et al. (1980), Y. Nakata et al. (1978), D. M. White and Helme (1985); (15) Drummond (1985), Pazos et al. (1985), Funatsu et (1985); (16) D. P. Taylor and Pert (1979), Wang et al. (1985), Staun-Olsen et al. (1982, 1985), Kerwin et al. (1980), Albano et al. (19
[c](E) Excitation; (I) inhibition; (M) modulation. If either excitation or inhibition is reported to predominate, the letter is underlined.
[d](√) Demonstration of K⁺-stimulated, Ca²⁺-dependent release in vitro; (e) electrically stimulated release in vivo.
[e]The K_m and B_{max} data are only indicative, since considerable variation among regions and reports is found.
[f](√) Indicates that an effect on the particular system has been reported; (0) indicates the existence of negative reports.

stimulation (Joels and Urban, 1984), and, in the cat superior ganglion, VIP prolongs excitation produced by muscarinic but not by nicotinic agonists (Kawatani et al., 1985). Crawley et al. (1984) have reported that CCK potentiates the effects of DA in the nucleus accumbens but not in the caudate, and since CCK is believed to coexist with DA in the A10 projections to the nucleus accumbens but not in the nigrostriatal projections (see Chapter 9), such an observation may have particular significance.

The reverse type of modulation may also occur. Responses of dorsal root neurons to iontophoretically applied SP are said to be attenuated by iontophoretically applied 5-HT, even in cases in which 5-HT alone had no effect on the cell. 5-HT did not affect the responses to iontophoretically applied glutamate (J. E. Davies and Roberts, 1981). Such modulation suggests that there may be coupled receptor mechanisms, as indicated for the benzodiazepine–GABA receptors (see Chapter 7). As discussed in Chapter 7, the endogenous ligand(s) for the benzodiazepine site(s) may well be peptidergic in nature.

Release. K$^+$-stimulated, Ca^{2+}-dependent release has been demonstrated for most of the peptides from slices or synaptosomal fractions; in a few cases, electrical stimulation of selected pathways has also been shown to cause peptide release in putative projection areas (Table 12.7). As outlined in Chapter 5, such release is characteristic of neurotransmitters.

The amount of peptide released seems generally to be much less than that of low-molecular-weight neurotransmitters. For example, S. P. Wilson *et al.* (1980) indicated that the 1 : 60 molar opiate peptides/NA ratio in bovine splenic nerve was one of the higher peptide/nonpeptide ratios yet found. Moreover, at least in the case of AVP, a rapid decrease in the amount of neuropeptide secreted on continued electrical stimulation has been shown (Ingram *et al.*, 1982). These findings are not surprising, since the method of synthesis of neuropeptides (see Section 12.2.2) is such as to suggest that the reserve in the nerve endings may be small and slow to replace when depleted.

Binding Sites. Specific high-affinity binding to synaptic membrane fractions has been reported for most of the putative peptide neurotransmitters (Table 12.7), but, in the majority of cases, not enough data are available to ensure that the binding sites described are equivalent to the physiological receptors. The distribution in brain is usually consistent with immunohistochemical results for particular peptide systems, although some discrepancies are being found (Cascieri and Liang, 1983; Torrens *et al.*, 1983; Quirion *et al.*, 1983).

In many cases, binding data suggest the existence of multiple binding sites, and this would not be surprising in view of the existence of such multiple sites for many neurotransmitters [see Chapter 5 and Table 5.3]. The peptide example usually cited is that of the multiple opioid receptors (Table 5.3), for which the various enkephalins, endorphins, and synthetic analogues have differing affinities (Table 12.8). Like members of the opioid family, these binding sites show differential distributions in brain, and it is not yet entirely clear whether a single opioid peptide acts *in vivo* on a single type of site or on more than one. In another example, multiple sites, with different distributions, have been reported for SP in the brain and spinal cord. One appears to prefer SP and another β-NK. The latter type of site is preferentially labeled by [^{125}I]eledoisin and is found particularly in the cortex and hippocampus (Torrens *et al.*, 1984; Ninkovic *et al.*, 1985; Buck *et al.*, 1984; Ouirion and Dam, 1985).

The differences in relative potency of analogues of peptides such as SOM and CCK in various physiological tests are also indicative of multiple receptors (Guillemin, 1978).

Second Messengers. Since many of the peptides seem to have effects that are characteristic of metabotropic neurotransmitters, in that they are slow in onset and long-lasting and tend to vary between excitation and inhibition depending on the particular experimental conditions used, it is understandable that there have been a number of investigations into the effects of various peptides on proposed second messenger systems. As indicated in Table 12.7, positive results have been reported for many peptides, but the data in almost every case are controversial. The best evidence for interaction with a cyclic nucleotide system seems to be in the case of VIP, which has been repeatedly found to stimulate cAMP formation.

It has been hypothesized that opiate agonists act by decreasing intracellular cAMP levels. There is clear-cut evidence of inhibition in neuroblastoma cultures (A. Goldstein *et*

TABLE 12.8
Some Agonists and Antagonists for Some of the Multiple Opioid Binding Sites[a]

Subtype	Agonists		Ref. No.[b]	Antagonists	Ref. No.[b]
	Opioid family	Other			
μ/μ_1	POMC	Morphine	—	Naloxone	—
		Dihydromorphine	2	Naloxonazine	—
		Morphicetin	7	SKF-10047	8
		MOA	6	—	—
		Metkephamid	5	—	—
		Tyr-D-Ala-Gly-Phe-NH$_2$	1	—	—
κ	Dynorphin	Ethyl ketocyclazocine	—	—	—
δ	Enkephalin	Metkephamid	—	ICI 154129	4
		Tyr-D-Ser-Gly-Phe-Leu-Phe	3	—	—
		DADL	2	—	—
σ	POMC	N-Allylnorphenazocine	—	(Naloxone-insensitive)	—
		SKF-10047	8	—	—

[a]See also Table 5.3 (Section 5.3.4). The ε site has been omitted from this table and the μ and μ_1 sites considered together. The selectivity of indicated agonists and antagonists is not complete, and many others are reported in the literature.
[b]References: (1) Shimohigashi *et al.* (1982); (2) DADL = [D-Ala^2-D-Leu^5]enkephalin (Pfeiffer and Herz, 1982); (3) David *et al.* (1982); (4) ICI 154129 = N,N-biallyl-Tyr-Gly-Gly-ψ(CH$_2$S)-Phe-Leu-OH (Priestley *et al.*, 1985); (5) metkephamid = Tyr-D-Ala-Gly-Phe-(N-Me)Met-NH$_2$ (Frederickson *et al.*, 1981; Burkhardt *et al.*, 1982); (6) MOA = metaorphamid [see Table 12.1 (Section 12.1)] Zamir *et al.*, 1985); (7) morphicetin = Tyr-Pro-Phe-Pro-NH$_2$ (K. J. Chang, 1981); (8) Su (1981).

al., 1977), but the literature on opiate inhibition of brain adenylate cyclase has been equivocal, and some groups report stimulation or no effect, rather than inhibition. This may be because the effects on adenylate cyclase are dependent on the particular brain region and opiate used (Wolleman *et al.,* 1979) or because inhibition can be observed only in the presence of additives such as GTP and Na$^+$ (Law *et al.,* 1981).

Again, as with iontophoretic and binding studies, there is evidence in the literature for modulatory effects of the peptides. TRH, for example, is said to inhibit markedly the stimulation of adenylate cyclase by NA but to potentiate the DA stimulation (Tsang *et al.,* 1980).

A number of the peptides have been reported to affect polyphosphoinositide (PPI) metabolism (Table 12.7). A report that the rate of inositol phospholipid hydrolysis induced by SP in various areas of the CNS is proportional to the number of binding sites for [^{3}H]-SP (Mantyh *et al.,* 1984) is good evidence that the stimulation of such hydrolysis by SP, observed in both peripheral tissues and the hypothalamus (Hanley *et al.,* 1980; S. P. Watson and Downes, 1983), may correspond to the action of functional SP receptors.

Hypotheses have also been put forth that the effects of various peptides may be mediated through changes in prostaglandins, calmodulin compartmentation, or calcium transport, but the evidence is unconvincing. The mechanisms that mediate peptidergic synaptic actions—like the actions themselves—are so far unclear.

12.4.2. General Physiology of the Neuropeptides

Many of the putative peptide neurotransmitters were initially studied because of their effects on hormonal release or gut motility. Studies of the central effects of peptides are handicapped by their rapid degradation and, in most cases, lack of good pharmacological agonists and antagonists.

Most studies suggest that diffusion of most peptides across brain vesicular endothelium is severely restricted (Meisenberg and Simmons, 1983; Ermisch et al., 1985). Blood-borne peptides can act directly on receptors in the circumventricular organs, including the pineal gland, neurohypophysis, pituitary, and choroid plexus. Some peptides, including OXY, AVP, and MSH, can influence blood–brain barrier (BBB) permeability and thus alter the transport of essential substances to the brain (Meisenberg and Simmons, 1983; Ermisch et al., 1985). The BBB is a major obstacle for the development of pharmaceutically useful peptides. Analogues that cross the BBB and are resistant to peptidases will be required.

The potentiality for such agents has been suggested by studies involving direct injection into the ventricles or selected brain region. Many effects have been reported following such injections, including those on thermoregulation, blood pressure, heart rate, respiration, appetite, thirst, behavior, and pain perception. We will discuss briefly some of the literature on the possible importance of peptides in the regulation of appetite, thirst, and pain perception.

Appetite. For general references, the reader is referred to Levine and Yim (1984) and J. E. Morley et al. (1984). CCK has been called the satiety peptide because systemic administration to many species, including humans (Stacher, 1985), leads to a dose-related inhibition of food intake. The effect is counteracted by coadministration of the CCK antagonist proglumide (G. P. Smith and Gibbs, 1985). Part of the mechanism is certainly peripheral and involves activation of gastric vagal afferents to the nucleus tractus solitarius.

The existence and location of central sites of action of CCK in mediating satiety are more controversial. Administration of CCK into the lateral ventricle also leads to a dramatic reduction in food intake, even in species, such as sheep, that are not markedly affected by peripheral administration. Hypothalamic sites have been implicated but not proven to be involved in this phenomenon (Baile and Della-Fera, 1985). Other peptides, such as BO, glucagon, SOM, and GRF, also have marked satiating effects on food intake, with much of the action being at the peripheral level but some apparently in the hypothalamus (Gibbs and Bourne, 1985; Imaki et al., 1985).

The endogenous opioids have also received considerable attention as mediators of food intake. The anorexic effects of opioid antagonists are now well established and probably involve action at a μ-type receptor (G. K. W. Yim and Lowy, 1984). DYN, an endogenous ligand for κ-type opioid receptors, is, however, an appetite stimulant (J. E. Morley et al., 1984); so is NPY (Levine and Morley, 1984).

Much remains to be learned about the control of appetite, but it has already been suggested that agonists and antagonists of CCK and the opioids may be useful in the clinical treatment of obesity and anorexia.

Thirst. Much of the focus has been on AVP, which has a major role in regulating water intake. A dramatic demonstration of this is in Brattleboro rats with a hereditary diabetes insipidus; these rats lack hypothalamic AVP neurons. Alleviation of their excessive drinking by transplants of fetal hypothalamic cells to supply the missing AVP was one of the first demonstrations of normalization of function by such transplants (Gash and

Sladek, 1980) (see Section 13.2.5). In further confirmation of the importance of AVP, an autoimmune form of diabetes insipidus can be produced in normal rats by central injection of antibodies to AVP (Scherbaum and Bottazzo, 1983).

Other peptides may also be involved. Injections of AII into the brain parenchyma or lateral ventricle elicit vigorous drinking. The effect is counteracted by intracerebral administration of SP or other tachykinin; the tachykinins also inhibit drinking induced by water deprivation or a sodium chloride load (de Caro et al., 1979). The intracerebral injections of AII also stimulate the release of vasopressin, which would tend to have a counteracting effect. The action of AII is believed to be at the pituitary level or on the circumventricular organs or both; whether endogenous brain AII plays a role in the control of thirst remains uncertain (Scholkens et al., 1980; Share, 1979; Baertschi et al., 1981; Wright et al., 1985).

Pain.   Information perceived as pain reaches the CNS via peripheral nerves that terminate mainly in the substantia gelatinosa of the dorsal horn of the spinal cord. Internal processing which is not yet understood takes place within the dorsal horn prior to projection to the brain stem (see Section 17.2.3 and Figure 17.4). The periaqueductal gray (PAG) region of the brain stem seems to be particularly important. In the early 1970s, it was shown that electrical stimulation of this area produces analgesia in animals and humans. All these areas are rich in diverse peptides (Jessel, 1983) and peptide receptors (Pert et al., 1985).

The evidence for SP as a primary sensory neurotransmitter is convincing. It is released from the spinal cord after depolarization and into the CSF after peripheral nerve stimulation of nociceptor fibers (Jessel, 1981). It has been estimated that 20% of all ganglion cells in the cat and rat contain SP, and there is a correlation between the distribution of nociceptive and SP-responsive neurons in the rat (Salt et al., 1983). The SP neurons in the ganglion send extensive projections to the substantia gelatinosa, and a depletion of SP-containing axons in the substantia gelatinosa has been demonstrated in postmortem tissue from patients with familial dysautonomia who showed severe diminution of temperature and pain sensitivity (J. Pearson et al., 1982). This evidence is consistent with a major role for SP in such nociception.

Other peptides including CGRP, OXY, AVP, SOM, VIP, BO, and CCK have also been found in sensory neurons, with both CCK and CGRP being colocalized with SP in some (Kai-Kai et al., 1985; Uddman et al., 1985).

Evidence for the probable involvement of a number of peptides in the modulation of pain reception at the spinal cord level also comes from the effect of intrathecal injections. Intrathecal administration of SP produces naloxone-reversible antinociception (Doi and Jurna, 1982; A. G. Hayes and Tyres, 1979), but so does intrathecal injection of β-End and other opioids that are active at the μ-receptor (Tung and Yaksh, 1982; Quirion, 1984). Intrathecal injection of β-End in humans is effective in causing profound analgesia (Oyama et al., 1980). Intrathecal or intraventricular injections of SOM kept terminal cancer patients pain-free; the effect was not counteracted by naloxone. These results, as well as experiments in animals, suggest that SOM is a potent analgesic (Chrubasik et al., 1984). In accordance with its characterization in binding studies as a partial opiate agonist–antagonist, ACTH can produce elevations in pain thresholds yet antagonize β-End-induced antinociception. High doses of intrathecal AVP produced antinociception that was not antagonized by naloxone, while TRH had no effect by itself but, depending on dose, either attenuated or enhanced morphine-induced antinociception (L. R. Watkins *et al.*, 1986). A role for CCK as an antagonist of the analgesic action of opioids is suggested because, on intrathecal injection, CCK has no effect alone but antagonizes β-End-induced antinociception (Faris *et al.*, 1983). In rats immunized against CCK by subcutaneous administration of repeated doses of CCK conjugated to bovine serum albumin, morphine analgesia was both potentiated and prolonged (Faris *et al.*, 1984). A similar prevention of β-End-induced analgesia was noted in rats treated with the CCK antagonist proglumide (Katsuura and Itoh, 1985).

Other experiments suggest that a combination of peptides may be involved. Intrathecal administration of capsaicin (see Section 12.5.1) has an analgesic effect and profoundly reduces both SP and CCK staining in the spinal cord of the rat (Micevych *et al.*, 1983); it also lowers SOM levels in sensory neurons (Gamse *et al.*, 1981). On the other hand, no analgesia was seen after intrathecal administration of piperine, which depleted SP and SOM but not CCK, enkephalin, or 5-HT, or after intrathecal kainic acid, which did not alter any of the neuropeptides studied. This suggests a link between capsaicin-induced analgesia and the concomitant depletion of CCK, SOM, and SP (Micevych *et al.*, 1983).

Peptides also probably function in pain pathways at the level of the PAG. Here, much of the focus has been on the opioids (Basbaum and Fields, 1984). The analgesia produced by electrical stimulation of the PAG is partially blocked by naloxone and is associated with a rise in β-End levels in the CSF. During electrical stimulation for pain relief in humans, both β-End and α-MSH, another product of POMC, are released (Akil *et al.*, 1984). The PAG matter is rich in opiate receptors, and injections into the PAG of opioids with high activity at μ-receptors, such as β-End, have analgesic effects. The members of the DYN family of opioids, on the other hand, do not appear to have analgesic action (J. M. Walker *et al.*, 1982).

Again, however, β-End and related opioids are not the only peptides that modulate pain thresholds at the PAG level. Virtually every neuropeptide tested by microinjection into this region has significant effects. NT, BO, BK, α-MSH, VIP, CCK, and SP have all been shown to modulate pain thresholds in rodents on such injection, and the PAG contains receptors for all these peptides. SP, NT, and BK are of particular interest. The first two have been reported to be many times more potent than morphine as an analgesic following intracerebral injection (Malick and Goldstein, 1978; Nemeroff *et al.*, 1980). The effect of NT is not abolished by naloxone. BK has potent analgesic effects on injection into the PAG but, at the level of the skin, is the most potent pain-producing substance known in mammals; it is hypothesized to play a role in the primary mechanism of pain production following tissue damage (Inoki *et al.*, 1979). Other members of the kinin family are found in the venom of insects and snakes and play a major role in the pain produced by bee stings or snakebites.

Another peptide of some interest in connection with pain may be kyotorphin. This dipeptide (Tyr–Arg) occurs in brain in much higher concentrations than any of the opioid family of peptides. It is particularly concentrated in the midbrain, pons–medulla, and dorsal spinal cord, which are the most sensitive sites for the production of analgesia by microinjection of morphine or electrical stimulation. Kyotorphin does not appear to be derived from any of the opioid precursors, a neuronal localization has not been demonstrated, and it does not bind to opiate receptors. Nevertheless, on intracisternal administration to mice, it has a naloxone-sensitive analgesic effect that is about 4 times as great as that of met-Enk. Hence, it is speculated that this dipeptide may play some role in pain control in the CNS (Ueda *et al.*, 1980; Takagi *et al.*, 1979).

Another aspect of the pain story arises when one considers the innervation of the large cerebral blood vessels at the base of the brain, the electrical stimulation of which can elicit pain (H. G. Wolff, 1963). These vessels are innervated by nerve fibers that contain NA, ACh, 5-HT, VIP, SP, NT, NPY, CCK, and CGRP. NPY is a potent vasoconstrictor, and cerebral arteries are densely innervated by neurons that contain both NPY and NA. Projections to these large pial arteries of the circle of Willis come from the trigeminal ganglion, and some have been shown to contain CCK, SP, and CGRP, either independently or as cotransmitters (Liu-Chen *et al.*, 1985; Uddman *et al.*, 1985). CGRP is said to be the most potent vasoactive peptide yet discovered. It has been postulated that these trigeminal vascular projections may play a role in the pathophysiology of vascular headaches.

There has been considerable speculation that acupuncture might be effective in producing analgesic effects because it stimulates release of endogenous opioids. The questions of whether acupuncture is effective and whether it acts through endogenous opioids are still, however, unresolved (Wall and Woolf, 1980).

12.5. Pharmacology and Pathology of the Neuropeptides

A functional role of a neurotransmitter is often revealed by the identification of an abnormality that occurs in a neurological disease or by the interaction of the neurotransmitter with a pharmacological agent. Examples are the loss of DA in Parkinson's disease and the induction of extrapyramidal reactions with DA blockers (see Sections 9.5 and 14.5). Investigations along both these lines have been carried out on the peptides and will be briefly described here, although they have not yet yielded definitive information on any central function.

12.5.1. Pharmacology of the Neuropeptides

Agonists and Antagonists. Pharmacological investigations on the peptides have been hampered because of their multiple sites of action, their rapid degradation, and their poor penetration of the BBB. To get around the first two difficulties, investigators have prepared analogues that may be more selective and more stable. Frequently, stability is conferred by substitution of a D-amino acid for the normal L-form. A few such analogues have been reported for most of the peptides, but the most extensive work has probably been done on analogues of SOM, LHRH, and the opioid peptides. Here, we will first consider the SOM and LHRH analogues that have some clinical utility, although their actions are so far confined to the pituitary and peripheral organs. We will then discuss clinical trials of various other peptides and their analogues in neurological diseases, recognizing that all such trials have so far given controversial data and that central sites of action are questionable.

Extensive investigations of analogues have shown that SOM activity does not require the amino terminus and a 6-amino-acid residue can retain the full activity associated with the 14- to 28-amino-acid peptide. Linear and cyclic forms are equipotent. As previously described, SOM inhibits secretion from many glands. Therefore, an objective of any synthetic analogue program is to develop organ-selective agents. For example, analogues with relatively greater effects in suppressing insulin secretion from the pancreas than in suppressing GH secretion from the pituitary can be synthesized (Gottesman et al., 1982; Vale et al., 1978). A number of long-acting, potent preparations are available, and some, such as Pro-Phe-D-Trp-Lys-Thr-Phe and D-Phe-Cys-Phe-D-Try-Lys-Thr-Cys-Thr, are effective on oral administration (W. Bauer et al., 1982). SOM, or some such SOM analogue, offers a degree of promise as a temporary treatment for a variety of hyperfunctioning endocrine tumors such as GH-secreting adenomas, gastrinomas, and insulinomas. The most promising application, however, may be the use of SOM in the treatment of acute gastrointestinal hemorrhage, for which SOM seems somewhat superior to cimetidine (see Chapter 11), and in hemorrhagic pancreatitis (Reichlin, 1983). Although cortical SOM levels are decreased in Alzheimer's disease, Cutler et al. (1985) obtained no clinical improvement following treatment of 10 early cases with a potent SOM analogue.

Clinical use of LHRH and its analogues has been concentrated in reproductive medicine. The purpose is to enhance low endogenous levels of the peptide for induction of ovulation or spermatogenesis or to counteract delayed pubertal development. "Superagonist" LHRH analogues have been synthesized. Examples are Buserelin [(D-Ser[TBU]^6-des-Gly-NH$_2$^{10})–LHRH nonapeptide ethylamide] and LRF$_A$ [(D-Try^6-Pro^9-NE+)-LHRH], which are, respectively, 40 and 140 times more potent than the native compound (Ory, 1983).

The opioid peptides have caught both the popular and the scientific imagination. Their widespread distribution in brain [see Figure 12.6 (Section 12.3.2)] has suggested their possible involvement in diseases that affect the cortex, basal ganglia, limbic system, neuroendocrine system, and brain stem. Native peptides and their analogues, as well as such nonpeptidergic antagonists as naloxone and natrexone, have been tested in Alzheimer's disease (Reisberg et al., 1983; Panella and Blass, 1984), schizophrenia (Shah and Donald, 1982; Goldbloom, 1984), depression (Emrich, 1984), and various movement disorders such as the Tourette syndrome, Parkinson's disease, and tardive dyskinesia (Sandyk, 1985), as well as disorders that involve pain (see Section 12.4.2). Unfortunately, the literature on the clinical benefits in CNS disorders is controversial. Goldbloom (1984) summed up the general situation in his remarks describing some work on schizophrenia with one of the analogues:

> There is a sense of déjà vu in reviewing the clinical studies of des-tyr-γ-endorphin—the initial euphoria in studies where there are more investigators than patients and then the dénouement as replication of results does not occur.

Goldbloom's remarks apply equally to clinical studies involving other peptides in which a CNS effect was the primary objective.

CCK has aroused specific interest in schizophrenia research because of its selective coexistence with DA in mesolimbic neurons, which are considered to be the locus of antipsychotic drug action (see Section 15.6.1). As indicated in Table 12.9 (Section 12.5.2), decreases in CCK have been reported in the CSF, cortex, hippocampus, and amygdala of persons suffering from schizophrenia. Initial studies reported long-lasting antipsychotic effects in schizophrenic patients following single injections of CCK-8, CCK-33, or the analogue ceruletide (Moroji et al., 1982a,b, 1985; Nair et al., 1982, 1983), but subsequent double-blind, controlled studies with ceruletide showed no antipsychotic effect in patients on neuroleptics (Albus et al., 1984; Hommer and Pikar, 1985;

Hommer *et al.*, 1984; D. M. Bloom *et al.*, 1983; Chase *et al.*, 1985; Tamminga *et al.*, 1984) or in patients withdrawn from such drugs (Lotstra *et al.*, 1985). Infusions of ceruletide also had no effect on motor function in parkinsonians (Chase *et al.*, 1985). Intraventricular injections of CCK have neuroleptic-like effects and inhibit apomorphine-induced stereotypy as well as conditioned avoidance behavior (S. L. Cohen *et al.*, 1982), which might argue for continued investigation of CCK in schizophrenia.

Similar disappointments have been encountered with clinical trials of AVP analogues in memory disorders (Zager and Black, 1985). It has been widely claimed that the intracerebral administration of vasopressin improves certain types of cognitive behavior in animals (Burback and de Wied, 1981), but there have been failures to replicate the data (Saghal *et al.*, 1982). Nevertheless, the initial reports led to trials of the effects of Lys-vasopressin, AVP, or more stable analogues such as 1-desamino-8-D-Arg-vasopressin (DDAVP) on memory in humans. Some amnesic patients reportedly improved after vasopressin treatment (Oliveros *et al.*, 1978), while others did not (Jenkins *et al.*, 1982; Reding and DiPonte, 1983; N. L. Foster *et al.*, 1983b; Tinklenberg *et al.*, 1982); some memory tasks were affected significantly by vasopressin, while others were not (Legros *et al.*, 1978; Weingartner *et al.*, 1981; Beckwith *et al.*, 1982; Koch-Henriksen and Nielsen, 1981; Fehm-Wolfsdorf *et al.*, 1984).

$ACTH_{4-9}$, a peptide sequence found in both α-MSH and ACTH, has also been involved in controversial reports on attention and memory. In animals and in some human studies, it has been reported to have positive effects (Branconnier, 1981), but treatment of patients suffering from senile dementia with an analogue of this peptide, ORG 2766, has been reported to be of little benefit (J. C. Martin *et al.*, 1983).

Because of the well-known association of thyroid disturbances with mental function, there has been much interest in TRH pharmacology and psychiatric illness. There was originally a large number of reports that the thyrotropin response to TRH administration could be used in distinguishing mania from schizophrenia, as well as unipolar from dipolar depression, and that intravenous administration of TRH gave beneficial effects in both depression and schizophrenia. Subsequently, a number of adequately designed trials have refuted the claim of therapeutic benefit, although many of the studies did report a mild increase in energy (J. E. Morley, 1979; Sternbach *et al.*, 1982). TRH has also been tried in the treatment of the hyperkinetic syndrome, childhood autism, ataxia of spinocerebellar degeneration, and amyotrophic lateral sclerosis (Imoto *et al.*, 1984). Initial reports of benefit have generally been followed by double-blind studies in which the peptide showed little clinical effectiveness. The availability of analogues of TRH with more potent arousal actions in animal experiments and with enhanced biological half-life may offer better clinical promise (Metcalf, 1982).

Other Pharmacological Agents.   Relatively few agents except agonists and antagonists are known in the peptide pharmacological field.

Capsaicin, a neurotoxic material from the pepper plant, has aroused some interest, since it seems to cause a relatively specific release of peptides from sensory neurons and has been used particularly in the study of SP in such neurons. It has no apparent effect in the CNS. The SP-depleting action of capsaicin is antagonized by nerve growth factor (NGF), and doses of capsaicin that deplete SP also inhibit the retrograde axoplasmic transport of NGF. It is therefore suggested that NGF plays a role in the maintenance of normal peptide levels of sensory neurons and that the depleting effect of capsaicin is mediated through a decrease in availability of the NGF (M. S. Miller *et al.*, 1982; Otten *et al.*, 1983; Burks *et al.*, 1985).

Cysteamine selectively depletes SOM from the gut, pancreas, hypothalamus, retina, and spinal cord and throughout the CNS. The effect is dose-dependent and reversible, with SOM levels returning to normal within 1 week. Cysteamine does not affect hypothalamic levels of LHRH, TRH, SP, β-End, AVP, CCK, VIP, or enkephalin, but it does cause a marked decrease in PRL secretion (M. Brown *et al.*, 1985; McComb *et al.*, 1985; Millard *et al.*, 1985; Szabo and Reichlin, 1985). So far, little has been done with this compound in studying the SOM systems.

Despite the paucity of success of peptide pharmacology with CNS disturbances, the neuroscientist should not be discouraged. Much fundamental detail remains to be learned about these agents, and it may be that more basic work is required before practical benefits can be achieved.

12.5.2. Pathology of the Neuropeptides

Postmortem studies of peptide levels have indicated a surprising degree of stability (Sagar *et al.*, 1984). This stability may be related to vesicular storage, which confers protection against lysosomal peptidases. Valid pathological information may therefore come from postmortem human studies. Table 12.9 summarizes such data on a few brain areas in some movement disorders, dementia, and psychiatric disease. Also given in this table are some data on CSF measurements, but they must be viewed with particular caution, since studies on some peptides (SOM, SP, CCK) indicate that the level in the CSF probably does not reflect the activity of central neurons (Sørensen *et al.*, 1981; Nutt *et al.*, 1980; Tamminga *et al.*, 1985).

The data are clearly scattered, and there is no evidence as yet for a causal relationship of abnormalities of neuropeptides and the production of neurological symptoms. A few points about the data in Table 12.9, however, seem worthy of mention.

While losses in cortical SOM have been repeatedly confirmed in senile dementia, no loss has been found in the levels of other peptides measured in the cortex and hippocanmpus. This finding pertains, notably, to CCK, VIP, and NPY, which have many cells in the cortex. This is one more bit of evidence that senile dementia of the Alzheimer type is a specific neurodegenerative process and not a generalized cortical atrophy.

It is of interest that persons dying with Huntington's disease, who frequently show dementia, have no loss in cortical SOM. Most probably the loss of SOM in Alzheimer's disease is not fundamental to the dementia. An alternative explanation is that the dementia in the two diseases depends on separate mechanisms.

The loss of SP and enkephalin activity in the basal ganglia in Huntington's disease is to be expected, since there appears to be a general neuronal dropout in that nucleus. It is unusual that there should be an apparent sparing of the striatal SOM neurons.

The decreases in peptide levels in the basal ganglia of parkinsonian patients are also unexpected, since there is generally neuronal sparing of this area. They may be secondary to the prolonged loss of dopaminergic innervation, since the losses in peptides are quantitatively much less than the losses in DA. It is worth noting that Zamir *et al.* (1984) reported that, in MPTP-treated monkeys showing typical parkinsonian symptoms (see Section 9.5.2), there was no abnormality in SP in the basal ganglia and a decrease in enkephalin only in the substantia nigra.

TABLE 12.9
Representative Postmortem Data on Some Neuropeptide Systems in Several Brain Regions in Huntington's Disease, Parkinson's Disease, Alzheimer's Disease, Schizophrenia, and Depression[a,f]

Disease and peptides[b]	CP	GP	SN	N. acc	Cortex	Hipp.	Amyg.	CSF
Huntington's disease[c]								
CCK[13]	N	D	D	—	N	—	—	—
CCK binding[23,24]	D	D	D	—	D	—	—	—
ENK[30,37]	D	D	D	—	—	—	N	—
NT[14,15,28,32]	I/N	I	N	N	N	N	N	—
SOM[2,3,7,10,28,32]	I	I	I	I	N	N	N	D
SP[2,6,7,13,22,34]	D	D	D	—	—	—	—	N
TRH[32,41]	I	—	—	N	—	N	I	—
VIP[12]	N	N	N	—	N	—	—	—
Parkinson's disease[d]								
BO[5]	D	D	N	N	N	N	N	—
CCK[27,42]	N	—	D	N	N	—	—	D
ENK[6,13,26,35]	N/D	D	D	N	N	N	—	—
Opioid binding[26,36]	N–D	—	D	—	—	—	—	—
NT[5,15]	N	N	N	N/I	N	D/N	N	—
NT binding[44]	—	—	D	—	—	—	—	—
SOM[8,11,16]	—	—	—	—	N	—	—	D
SP[22,31]	—	D/N	D/N	—	—	—	—	—
TRH[25]	N	N	N	N	—	—	—	—
Alzheimer's disease[e]								
CCK[39]	N	—	—	N	N	—	—	—
NPY[1,19]	—	—	—	N	N	—	—	—
NT[15,32]	—	N	—	—	N	N	N	—
SOM[9,16,38]	—	—	—	D	D	—	—	—
SP[22,46]	N	N	N	—	N	N	—	—
TRH[4,46]	N	N	N	—	N	N	N	—
VIP[38,45]	—	—	—	—	N	N	—	N
Schizophrenia[f]								
CCK[18,20,21,27]	—	—	—	—	D	D	D	D/N
CCK binding[17]	—	—	—	—	D	D	—	—
NT[28,32]	N	—	—	N	N/I	—	N	D/N
SOM[12,20,33]	N	N	N	—	N–D	N/D	N	N
TRH[32]	N	—	—	D/N	N–D	N	N	—

[a]Entries indicate no change (N), a decrease (D), or an increase (I) compared to normal controls; (N–D) indicates different results in subnuclei; (N/D, N/I) indicate discrepant reports. Brain regions: (CP) caudate–putamen; (GP) globus pallidus; (SN) substantia nigra; (N. acc.) nucleus accumbens; (Hipp.) hippocampus; (Amyg) amygdala. The peptide abbreviations are listed in Table 12.1 (Section 12.1).

[b]References: (1) J. M. Allen et al. (1985); (2) Aronin et al. (1983); (3) Beal et al. (1984, 1985); (4) Biggins et al. (1983); (5) Bissette et al. (1985); (6) Buck et al. (1981); (7) Cramer et al. (1981); (8) Cramer et al. (1985); (9) Davies et al. (1980); (10) Dawbarn et al. (1985); (11) DuPont et al. (1982); (12) Emson et al. (1979); (13) Emson et al. (1980b); (14) Emson et al. (1980a); (15) Emson et al. (1985); (16) Epelbaum et al. (1983); (17) Farmery et al. (1985); (18) Ferrier et al. (1985); (19) Foster et al. (1986); (20) Gerner et al. (1985); (21) Gjerris et al. (1984); (22) Grafe et al. (1985); (23) S. E. Hays et al. (1981b); (24) S. E. Hays and Paul (1982); (25) Javoy-Agid et al. (1983); (26) Llorens-Cortez et al. (1984); (27) Lotstra et al. (1985); (28) Manberg et al. (1982); (29) Marshall and Landis (1985); (30) P. E. Marshall et al. (1983); (31) Mauborgne et al. (1983); (32) Nemeroff et al. (1983); (33) Nemeroff et al. (1984); (34) Nutt et al. (1980); (35) Ploska et al. (1982); (36) Reisine et al. (1979); (37) Rossor and Emson (1982); (38) Rossor et al., (1980); (39) Rossor et al. (1982a,b); (40) Sørensen et al. (1985); (41) Spindel et al. (1980); (42) Studler et al. (1982); (43) Sulkava et al. (1985); (44) Uhl et al. (1984); (45) Wikkelsø et al. (1985); (46) Yates et al. (1983).

[c]Also reported in Huntington's disease are normal levels of AVP in the SN[39] and CSF,[40] a decreased level of endorphin in the CSF (Kaiya et al., 1983), and an increased level of NPY in the CP.[10]

[d]Also reported in Parkinson's disease are normal levels of AVP in the SN and CSF,[39,40] of VIP in the cortex (Jegou et al., 1985), of CRF and GRF in the hypothalmus (Conte-Devolx et al., 1985), and of NPY in the cortex and hippocampus[1] and a decreased level of endorphin in the CSF (Nappi et al., 1985).

[e]Also reported in Alzheimer's disease are normal levels of CCK binding in the cortex,[24] of LHRH in the amygdala,[46] and of CRF in the CSF,[33] with decreased levels of AVP[40] and endorphin (Kaiya et al., 1983) in the CSF.

[f]Also reported in the CSF in schizophrenia are decreased levels of BO[20] and DYN$_{1-8}$ (Zhang et al., 1985) and normal levels of CRF,[33] VIP,[31] and gastrin.[21] In depression, the CSF is reported to show normal levels of AVP[40] and CCK [21,27] and increased levels of CRF[33] and SP (Rimon et al., 1984; Berrettini et al., 1985). CSF levels of VIP are reported decreased in nonendogenous depression, but normal in endogenous depression.[21]

The decrease in CCK in the substantia nigra with no change in the nucleus accumbens in Parkinson's disease is somewhat unexpected, since it is believed from animal studies that coexistence of the peptide with DA is far more frequent in neurons of the A10 area, which project to the nucleus accumbens, than in those of the substantia nigra. Since all DA neurons are adversely affected in Parkinson's disease, one might expect greater losses of CCK in the nucleus accumbens than in the nigra.

The loss of NT- and opioid-binding sites in the substantia nigra is quite marked in Parkinson's disease and suggests that many of these binding sites may be on dopaminergic neurons. This suggestion is consistent with data that indicate high levels of NT-binding sites in dopaminergic cell areas (Uhl and Kuhar, 1984) and the presence of opioid-binding sites on dopaminergic cell somata.

12.6. Summary

Few areas of neuroscience research are receiving as much attention as that involving the peptides. Powerful new techniques have become available for characterizing and assaying these materials. The ability to isolate, sequence, synthesize, and analyze proteins and peptides by relatively simple techniques has opened up an entirely new realm of neuroscience that is still in its infancy.

With respect to peptidergic neurons, there have emerged a number of principles that can act as a guide to neurobiologists in attempting to gain some perspective in dealing with the immense literature on the subject; they are as follows:

1. The CNS neurotransmitter peptides all have prominent distribution, and presumably function, outside the CNS.
2. Within the CNS, peptide neurotransmitters (Type III) all occur at much lower concentration than amino acid neurotransmitters (Type I) or amine transmitters (Type II) (see Section 5.2).
3. Peptide neurotransmitters are all synthesized by cleavage of large precursor peptides. Synthesis of these precursors takes place in the cell body, where RNA and ribosomes are located. Cleavage into active peptides can take place in the axon during transport or at the nerve ending.
4. Parent precursors often contain sequences of several active peptides.
5. Inactivation of peptide neurotransmitters takes place by further cleavage into smaller fragments. Reuptake mechanisms do not seem to exist. Some of the smaller fragments may have independent physiological actions.
6. The peptides frequently occur in neurons in combination with other peptides or other categories of neurotransmitters. The popular interpretation of Dale's law that only a single neurotransmitter may be released from each ending of a given neuron must be broadened to include the possibility of multiple neurotransmitters being released from each ending (see Chapter 5).
7. The peptides may be released into the synaptic cleft, but they may also be released directly into the bloodstream, either into some microcirculation or into the general circulation.
8. Multiple receptor sites appear to exist for the peptides, which implies that the same peptide may have different actions at different locations.

Principles (6)–(8) also apply to the amino acid and amine neurotransmitters, so that they are not unique to peptides.

While the neuronal distribution of peptides in brain is broad, the richest variety and highest concentrations are found in areas that coordinate autonomic function. This is particularly true of the hypothalamus and neurohypophysis. The amygdala and certain brain stem nuclei such as the nucleus tractus solitarius are other prominent locations.

The detailed functions of individual peptides are far better understood in the periphery than in the brain, and much about central action has been inferred from the peripheral results. At the iontophoretic level, peptidergic effects are often equivocal. This suggests that action in concert with other neurotransmitters may be an essential feature of their physiology.

CCK has been particularly identified with satiety, AVP with thirst and water intake, β-End and several other peptides with pain, and SP with sensory transmission.

Several peptides and their analogues have been tested as pharmacological agents. Some show promise for peripheral action, but none for the brain itself. Low penetration of the BBB and rapid inactivation by peptidases are significant practical problems for most agents.

The stability of many peptides to postmortem proteolysis suggests that valuable information regarding their role in disease may be obtained in the future. For example, SOM is selectively reduced in the cortex in Alzheimer's patients, but is selectively retained in the basal ganglia in Huntington's disease. CCK is reduced in the substantia nigra, but not in the nucleus accumbens, in Parkinson's disease.

This description of Type III neurons completes our story of how individual neuronal systems operate. We turn now to Part III of this volume, in which we consider the integrative aspects of brain function.

The Integrative Aspects of Brain Function

The detailed studies of anatomy and physiology in Part I and of neurochemistry in Part II have to be given functional meaning in the most important integrative performances carried out by the brain.

Before the four chapters on these themes, however, it is necessary to have an introductory Chapter 13, which gives an account of the mechanisms concerned in the embryological processes by which the brain is built. All neurons are generated in special mitosis of the germinal epithelium of the neuronal tube. There are extraordinary problems in developing hypotheses that attempt to account for the factors responsible for the migration of the neurons to their ultimate destinations and for the establishment of the correct synaptic connectivities. It has been proposed that chemical coding and sensing provide the key factors. The specific molecules involved appear to be a different class from the synaptic transmitters. They are assumed to be specific macromolecules, probably polypeptides associated with the surface membranes of the neurons and the glia (cf. Section 13.1.1).

The control of movement involves an immense range of neuronal performance, from the simplest spinal reflexes dealt with in Chapter 4 to the responses of progressively higher levels—the cerebellum, the basal ganglia, and ultimately the cerebral cortex. It is not yet possible to give a detailed and coherent account of all the operations of the neural machinery involved in carrying out even some simple movement. Much of our account in Chapter 14 will be impregnated with theories that go far beyond the present level of scientific investigation. But this procedure is justified because it reveals how far we are from fully understanding the control of movement, and also because it points the way to fruitful experimental investigations. As yet, our understanding is fragmentary at best. An important feature of this chapter concerns the complex neuronal systems that comprise what is often called the "extrapyramidal system." This system is of great clinical significance because its disorders are the cause of many important neurological diseases.

Consideration of the control of movement by all the skeletal musculature leads to the problems of behavior discussed in Chapter 15. Knowledge in this area is in an even more primitive state. Yet there is a tremendous overlap with movement, as is evidenced by the commonality of effects of drugs and disease on these two integrative aspects of brain function. This is where the interface between neurology and psychiatry exists in the clinical world and where psychology and pharmacology meet in the realm of basic science.

There is no more important function of the brain than learning and memory. Yet the essential mechanism is still obscure, despite the immense experimental efforts over the last decades. The reason for this discouraging situation is that there has been no well-defined hypothesis of the neural mechanisms concerned. In Chapter 16, an attempt is made to formulate such a hypothesis. It is encouraging that the same neurobiological principles seem to underlie the mechanism used by the cerebellum in motor learning and that used by the cerebrum in cognitive learning. There is even a suggestion that the basal ganglia may have an analogous mechanism. Many good experimental studies in molecular neurobiology can be developed to test the theoretical explanations of these structures.

It is appropriate that Chapter 17, the final chapter of this book, treats the ultimate problems of cerebral performance—perception, speech, and consciousness. It is possible to define only general principles in the enormous fields of perception and speech, but we believe it to be important for the molecular neurobiologist to be confronted with the problems that arise in higher brain functions. Finally, in the section on consciousness, there is an introduction to the brain–mind problem that goes beyond the realm of neuroscience but is nevertheless the ultimate scientific question involving brain function.

13

The Building of the Brain and Its Adaptive Capacity

13.1. Building of the Brain

13.1.1. Introduction

In the last few decades, there has been great progress in our understanding of the principal features of the most marvelous constructional operation—the building of brains (Gaze, 1970; Jacobson, 1970; Angevine, 1970; Sidman and Rakic, 1973; Rakic, 1977, 1981; Changeux, 1983; Cowan *et al.*, 1984; Kater and Letourneau, 1985). In its organized complexity, the human brain far exceeds that of other mammalian brains, and even more the brains of other animals, vertebrates and invertebrates. The foundations of neuroembryology were laid around the turn of the century by His, Held, Retzius, and Ramón y Cajal, and many of their basic discoveries with light microscopy have survived rigorous testing with powerful new techniques such as electron microscopy, radiolabeling with autoradiography, and DNA estimations of specific cell types. It is proposed to summarize in this chapter the early stages of neuroembryological development and then to treat in more detail the two best-understood developmental stories of the mammalian brain, that for the cerebral cortex and that for the cerebellar cortex.

In the earliest stage, the future central nervous system (CNS) can be recognized as an elongated neural plate already differentiated from the surrounding ectoderm by its thickening due to elongation of the constituent neuroepithelial cells that extend through the thickness of the plate. Soon, the rapid multiplication of these cells causes the inward longitudinal folding of the plate so that it eventually becomes the neural tube lying beneath the epidermis that has united over it. The neuroepithelial cells thus extend between an inner ventricular attachment and an outer attachment to mesodermal tissue that eventually becomes the pia mater. In the earlier stages, the neuroepithelial or germinal

417

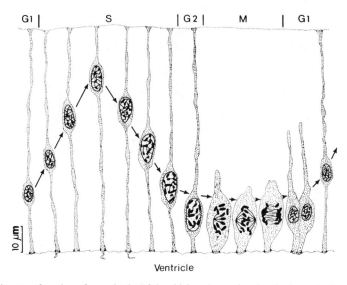

Figure 13.1. Diagram of section of neural tube of the chick embryo, showing the intermitotic migration of the nucleus of a single neuroepithelial germinal cell at approximately half-hour intervals throughout the mitotic cycle. From Jacobson (1970).

cells multiply in a simple clonal manner with no differentiation. Each mitosis is accompanied by a remarkable cycle of changes illustrated in Figure 13.1 in the sequential series for a single cell at approximately half-hour intervals (Sauer, 1935). Beginning at the left, the cell extends across the wall of the neural tube and the nucleus migrates outward and then inward during the stage of DNA replication by synthesis (S stage). There is often a short premitotic growth phase (G2), followed by retraction of the cell to the ventricular surface for the mitotic phase (M). The mitosis is transverse, so that the two daughter cells lie side by side in position to resume the extended form across the tube in phase G1.

The cycle then repeats for each of the daughter cells, the cycle time being no more than 6 hr for a chick embryo at an early stage. With a mouse fetus at 10 days, the cycle time is 8.5 hr, while at 15 days it has lengthened to 11 hr for that part of the neural tube forming the cerebral cortex. Since the mitotic cycles of the individual neuroepithelial cells are out of phase, in the postmitotic phase (G1 in Figure 13.1), there will be adjacent neuroepithelial cells stretching across the tube in phases S and G2 that can act as guides for the extension of the cytoplasmic processes of the daughter cells toward the external surface.

Soon there is an enormous population of these primitive germinal cells, and this simple clonal multiplication eventually ceases. In what is called a *differentiating mitosis,* a germinal cell divides to give rise to immature nerve cells. Alternatively, the differentiating mitosis may result in one immature nerve cell and one germinal cell (Rakic, 1973). These immature nerve cells eventually mature as nerve cells, and they never divide again. They have lost their mitotic competency, but they carry in their nuclei the genetic instructions to develop into their appointed roles in the nervous system. They set about growing first an axon and then dendrites, and these sprouts seem to know where they are going, as though guided by an intelligence and a knowledge of the ultimate design of the brain they are helping to build. It is a wonderful, self-organizing, self-developing process. This biological process of constructing a brain provides one of the most challenging scientific problems confronting us: How is it controlled? Some general statements can already be made:

1. At no stage in development are the neurons of the brain connected together as a random network.
2. In most regions, there is evidence of an initial excess of unused and unwanted neurons that rapidly die, as reviewed by Cowan *et al.* (1984). This is an important subtractive method for refining the design of neuronal systems. The factors responsible for cell death will be considered in relation to the special examples given below. No doubt there appears to be a redundancy in the enormous populations of neurons that survive to adulthood in many parts of the brain, but we may assume that this appearance derives from our deficient understanding of the modes of operation.
3. The organization of the brain is highly specific, not merely in terms of connections between particular neurons, but also in terms of the number and location of synaptic knobs on different parts of the same cell [cf. Figure 1.4B (Section 1.2.2)] and the precise distribution of synaptic knobs that arise from that cell.
4. The slow changes in performance of the brain during life, in particular learned performances, are due to subtle changes in the microstructure and microfunction that we shall consider later in this chapter and in Chapter 16.
5. Apart from these microchanges, the structure of the brain is precisely determined by genetic coding and the secondary instructions deriving therefrom, as well as from the medium in which the brain grows.

Besides forming neurons, the germinal cells also undergo differentiating mitoses to form glial cells. There is conflicting evidence with respect to the details of glial generation. By labeling techniques, it has been shown that there are separate lines of germinal cells for neurons and for glia (Levitt *et al.*, 1981). The relationship of astroglial and oligodendroglial production is still not known. It is now generally accepted, however, that microglia do not arise from neuroendothelial cells, but are phagocytic cells of mesenchymal origin that invade the brain from the pia mater. The immature glial cells or glioblasts assemble in the subventricular zone [S in Figure 13.2D (Section 13.1.2)], and from there may migrate laterally. Later in this chapter, we consider the probable glial role in the guidance of neuronal migration and in the regenerative replacement of degenerating synaptic terminals.

All types of glial cells differ from neurons in that they retain their mitotic competency. This generation of glial cells throughout life has one serious consequence, namely, the production of gliomas or brain tumors. Brain tumors rarely if ever have a neuronal origin.

Before the many complex features displayed by primitive neurons in their form and migration are considered in detail, it is important to present the recent fundamental investigations on cell adhesive molecules (CAMs), as reviewed by Edelman (1984). A special class of these sialoglycoproteins encrust about 1% of the surface membrane of embryonic nerve cells. These embryonic N-CAMs have molecular weights of 180,000–250,000 with a high concentration (about 30%) of sialic acid. By contrast, adult N-CAMs have about one third of the sialic acid and molecular weights of 140,000–180,000.

It was postulated by P. Weiss (1941) that in embryonic development, the outgrowing nerve fibers are attached to fascicles that guide them to their targets along tissue planes. We can now recognize these attachments as due to N-CAMs. More recently, Rakic (1972, 1973) has provided convincing evidence that in the developing cerebrum [see Figure 13.3 (Section 13.1.2)] and cerebellum [see Figure 13.7 (Section 13.1.3)], nerve cells are moved along by a guidance structure provided by a glial scaffolding. Edelman (1984) has attributed this growth to an identified adhesive molecule (molecular weight 135,000) that is a special type of CAM on embryonic neurons called Ng-CAM. The small ''g'' signifies that the CAM is on the neuronal surface, causing it to adhere to glial surfaces. Evidently, we have reached a new level of understanding of embryonic neuronal migration and association, as will be illustrated in later sections.

13.1.2. Building of the Cerebral Neocortex

For a general reference, the reader is referred to Rakic (1981).

Figure 13.2 gives diagrammatically the early stages in the development of the neocortex (Sidman and Rakic, 1973). Figure 13.2A shows the state of affairs depicted in Figure 13.1 (Section 13.1.1), but for a human fetus just less than 6 weeks of age. At 6–8 weeks, a cell-sparse marginal zone develops (M in Figure 13.2B), and some of the neuroepithelial cells undergo a differentiating mitosis and mature into fully developed nerve cells as they migrate up to form the intermediate zone (I in Figure 13.2C) at 8–10 weeks and sprout axons up into the marginal zone. The differentiating mitoses continue in zone V (Figure 13.2D at 8–10 weeks). The immature nerve cells so formed migrate more superficially to form the cortical plate (CP), which can be seen with its developing pyramidal cells (Figure 13.2E) at 10–12 weeks. An additional feature in Figure 13.2D and E is the subventricular zone (S) that is composed of subventricular cells migrating up from zone V and continuing their mitotic division there. The nerve cells so formed are not constrained by the vertical alignment of the cortical plate layer and may migrate laterally. Glia are also formed by differentiating mitoses in zone V.

An important feature is that the later differentiating neurons migrate to more superficial locations than the earlier neurons (Rakic, 1981). Thus, these later neurons have a

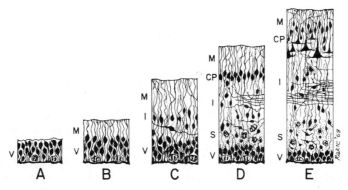

Figure 13.2. Semidiagrammatic drawing of the development of the basic embryonic zones and the cortical plate. Abbreviations: (CP) cortical plate; (I) intermediate zone; (M) marginal zone; (S) subventricular zone; (V) ventricular zone. From Sidman and Rakic (1973).

very long migration path across several millimeters of cortical thickness, whereas the earliest have a migration of only 100 μm. At the earlier stages, the migration of the cell appears to be accomplished simply by the movement of the nucleus up the cytoplasmic cylinder of the immature nerve cell that extends right across the cerebral wall, there being a subsequent retraction of this elongated cylinder (cf. J. Altman and Bayer, 1978). At later stages of development of the fetal monkey cerebrum, however, the thickness of the cerebral wall may be as much as 5 mm (Figure 13.3A). By radiolabeling, Rakic (1985) has shown that all neurons in primates are generated prenatally or very early postnatally.

A most attractive explanation was suggested by Rakic (1972). The glioblasts generated by the neuroepithelial cells in the S layer eventually grow into glial cells that extend right across the cerebral wall as thin fibers (RF in Figure 13.3B). Possibly this great elongation in parallel fascicles is to be explained by their early origin from neuroepithelial cells spanning the thin cerebral wall at early stages and being progressively stretched as the cerebral wall thickens. The small rectangle (*) in Figure 13.3B is drawn greatly enlarged in Figure 13.3C. The elongated glial filaments (RF) are shown cross-hatched. One neuron (A) with its nucleus (N) is shown closely attached to a glial filament. As it migrates upward, there is a leading process (LP) with several pseudopodia (PS) at its growing tip, one of which is already continuing along the filament. Below, there is a trailing process (TP) of the neuron that is eventually retracted. Thus, the upward migration of the neuron would seem to be guided by the glial filament, which may also aid in the movement. Two other neurons with associated glial filaments are also shown in Figure 13.3C. One (B) is moving out of the frame, the other (C) is coming out of the dense fiber mesh of the optic radiation (OR). That the glial filament provides a specific guidance is illustrated by the finding that the neuron follows all the bends of the glial filaments and ignores the multitude of alternative filamentous paths (nerve fibers and blood vessels) that it crosses on its upward migration. It can be conjectured that there must be some affinity between the surfaces of the glial filament and the neuron, which is in exact accord with the recent discoveries and hypotheses of Edelman (1984) about the Ng-CAM on the surfaces of embryonic nerve cells that gives adhesion to glia.

The radial guidance by glial cells can be demonstrated only at later stages of development, as in Figure 13.3. At earlier stages, glia may also be guiding the vertical migration of neurons. For example, by immunofluorescence, Antanitus et al. (1976) have discovered glial filaments organized for neuronal guidance as early as 10 weeks in the human fetus, which would be equivalent to the early stage of neuronal migration in the monkey at about the 60th embryonic day (Rakic, 1983). It is also possible that the vertical guidance by N-CAM may be shared by the vertically directed processes of the already migrated neurons that may be recognized in Figure 13.3C. After the glia have completed their task of vertical guidance, there is retraction of their long processes seen in Figure 13.3B with their transformation into astroglia.

A special problem in this vertical guidance relates to the inside-out positioning of the migrating neurons. As indicated in Figure 13.2E, the later migrating neurons take up a position external to the earlier, which have already been transformed into pyramidal cells. It is suggested by Sidman and Rakic (1973) that the final position of the neurons is dictated by interaction with processes of other neurons and even of afferent fibers that have already grown into the cortex, probably from the thalamus. The cerebral disorganization in the reeler mutant mouse provides evidence in support of this suggestion. The vertical migrations of the immature neurons to their final places in the cerebral cortex

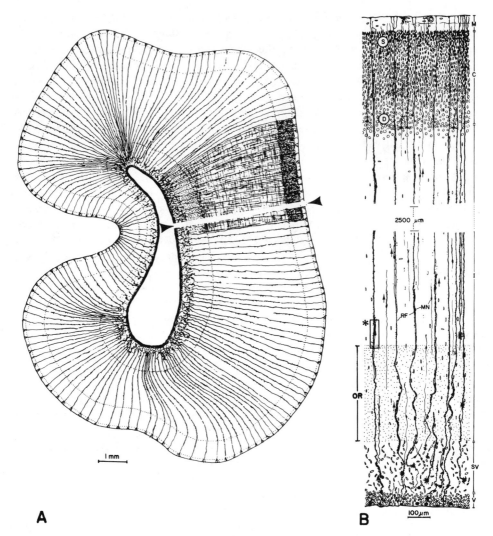

A   B

Figure 13.3. Development of the cerebral wall. (A) Cross section of cerebral wall as shown in Figure 13.2E.
From Rakic (1972). (B) Enlargement of the portion of cerebral wall indicated in (A). The middle 2500 μm of the
intermediate zone, also spanned by radial fibers, is omitted. The rectangle (*) shows the approximate position of
the three-dimensional reconstruction in (C). The 100-μm scale indicates the magnification. Abbreviations: (C)
cortical plate; (D) deep cortical cells; (I) intermediate zone; (M) marginal layer; (MN) migrating cell; (OR) optic
radiation; (RF) radial fiber; (S) superficial cortical cells; (SV) subventricular zone; (V) ventricular zone. From
Sidman and Rakic (1973). (C) Three-dimensional reconstruction of migrating neurons, based on electron
micrographs of semiserial sections. The reconstruction was made at the level of the intermediate zone indicated
by the rectangle (*) in (B). The subventricular zone lies some distance below the reconstructed area, whereas the
cortex is more than 2000 μm above it. The lower portion of the diagram contains uniform, parallel fibers of the
optic radiation (OR), and the remainder is occupied by a more variable and irregularly disposed fiber system.
Except at the lower portion of the figure, most of these fibers are deleted from the diagram to expose the radial

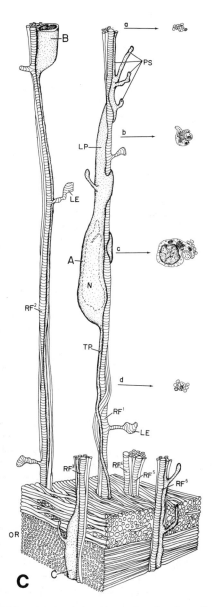

C

fibers (striped vertical shafts RF^1–RF^6) and their relationships to the migrating cells (A–C) and to other vertical processes. The soma of migrating cell A, with its nucleus (N) and voluminous leading process (LP), is within the reconstructed space, except for the terminal part of the attenuated trailing process (TP) and the tip of the vertical ascending pseudopodium. Cross sections of cell A in relation to the several vertical fibers in the fascicle are drawn at levels a–d at the right side of the figure. The perikaryon of cell B is cut off at the top of the reconstructed space, whereas the leading process of cell C is shown just penetrating between fibers of the OR on its way across the intermediate zone. (LE) Lamellate expansions; (PS) pseudopodia. From Rakic (1972).

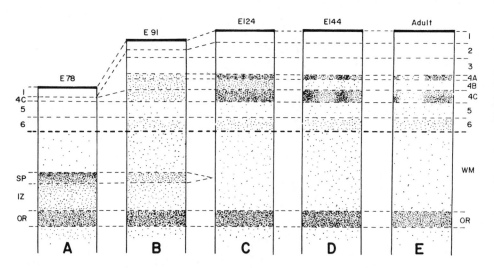

Figure 13.4. Semidiagrammatic summary of the development of geniculocortical connections and ocular dominance columns in the occipital lobe of the rhesus monkey from the end of the first half of pregnancy to adulthood. Each drawing (A–E) illustrates a portion of the lateral cerebral wall in the region of area 17 as seen in autoradiograms of animals that had received unilateral eye injections of a mixture of [³H]proline and [³H]fucose 14 days earlier. The age of animals at the time of sacrifice is indicated at the top of each drawing in embryonic days (E) and adulthood. Cortical layers 1–6 are delineated according to Brodmann's classification. Abbreviations: (IZ) intermediate zone; (OR) optic radiation; (SP) deep portion of subplate layer; (WM) white matter. From Rakic (1981).

form the developmental basis of the columnar organization of the cerebral cortex that is illustrated in Figure 17.2 (Section 17.1.1). The horizontal arrangements of dendrites and axonal branches develop at a much later stage.

The stages in the development of the synaptic connections into the cortical plate are illustrated in Figure 13.4 for cortical area 17 of the monkey cortex. The gestational period is about 160 days. The drawings of the cerebral cortex at the embryonic days indicated show the autoradiographic results on animals that had received unilateral eye injections of a mixture of radioactive proline and radioactive fucose 14 days earlier. In all diagrams, the optic radiation (OR) is highly active. At embryonic day 78, the geniculocortical fibers do not enter the cortical plate, but stay in a waiting compartment (SP) until the cortical neurons are ready for synaptic contact, which begins at embryonic day 91 in lamina 4 (B). At day 124 (C), this innervation of lamina 4 cells is well developed, especially in sublayers 4A and 4C. But only at a later stage, embryonic day 144, is there the development of the ocular dominance columns that can be seen in full clarity in the adult (E). Thus, Rakic (1981) proposed that there are three phases in the development of specific thalamic projections to cortical neurons:

1. The axonal terminals enter the correct cortical area, but stay in the waiting compartment.
2. There is a growth up to the correct cortical laminae.
3. There is restriction of axonal terminals to the ocular dominance columns.

There is evidence that Figure 13.4 also expresses in general the stages in cortical connectivity for callosal inputs into the frontal lobe and for the pyriform cortex. Thus, the transformation to the final specific synaptic pattern could be a general phenomenon in cortical development (Rakic, 1981).

After the immature neurons have taken up their final position, there is a relatively long period of dendritic development to the fully mature form. It is generally agreed that the sprouting of dendrites does not begin until the axonal process has grown. The direction of growth seems to be determined by external factors. The early stage of development of dendrites can be seen in Figure 13.2E at 11–13 weeks, but the full structural development of the human cerebral cortex is not achieved until after birth. The initial stages of dendritic outgrowth are determined by factors internal to the neuron, but in later stages the influence of the surrounding tissues is paramount. There is need of a far more intensive and systematic study of the details of cortical development than has yet been carried out.

Histologically identifiable synapses have been detected at the surprisingly early age of 8 weeks in the human fetus, and they increase very rapidly over the next few weeks (Molliver et al., 1973). These early synapses are located on dendrites both superficially and deep to the cortical plate (CP in Figures 13.2D and E). It is conjectured that the second phase of synaptogenesis, from 23 weeks onward (with synapses forming within the cortical plate), is due to the further development of dendrites and to the growth of afferent fibers into the cortical plate. The beautiful pictures of the Golgi-stained human neonatal cerebral cortex drawn by Ramón y Cajal (1911) reveal that it is in an advanced stage of development. By contrast, the rat cerebral cortex is very immature at birth. Recognizable synapses in the rat are rare even 14 days postnatally, but are up to the adult number by day 24 (Aghajanian and Bloom, 1967).

13.1.3. Building of the Cerebellum

The human cerebellum originates [see Figures 14.13 (Section 14.4.1) and 14.14 (Section 14.4.2)] in the same basic zones (V, I, and M) as does the cerebrum [see Figure 13.2A (Section 13.1.2)]. At 9–10 weeks, the Purkinje cells (cf. Figure 14.14) move outward much like the cortical plate cells (Figure 13.2D). They can be seen below the external granular (EG) layer in the right half of Figure 13.5A and as the P layer in all columns of Figure 13.5B. At 10–11 weeks, there is migration of subventricular cells out over the cerebellar surface from the rhombic lip (upper part of RL in Figure 13.5A) to form the external granular layer (EG in Figure 13.5A), in which there is a rapid clonal multiplication of germinal cells to give a layering 6–9 cells thick by 20–21 weeks, despite the great increase in cortical area that is evident in the drawings in Figure 13.5B (Sidman and Rakic, 1973). The maturation of the Purkinje cells begins in Figure 13.5B at 16 weeks and continues on to 25 weeks with profusely branching dendrites, some being from the soma, which is distinctively an embryonic feature (see Figure 13.6). Meanwhile, in the external granular layer, the simple clonal multiplication has been transformed by the onset of differentiating mitoses, immature granule cells being so formed in the external granule layer. By 16 weeks, these granule cells are migrating inward past the Purkinje cell bodies to form the internal granular layer (G in Figure 13.5B at 16 weeks and later). At term (Figure 13.5B at 40 weeks), the cerebellum is at an advanced stage of construction, but neurogenesis continues until 7 months postnatally (7 p.n.m. in Figure 13.5B), by which time the external granular layer has disappeared.

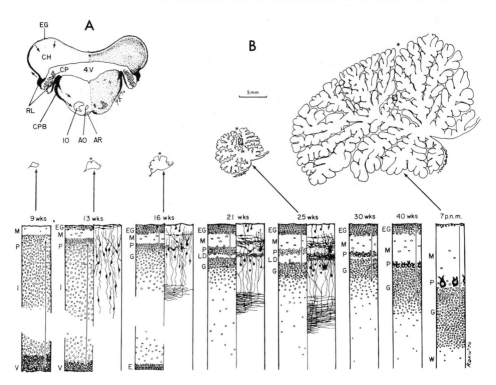

Figure 13.5. (A) Transverse section of the human rhombencephalon at the end of the 3rd lunar month of gestation. (→) Migration pathways (partly discussed in the text). Abbreviations: (AO) accessory olive; (AR) arcuate nucleus; (CH) cerebellar hemisphere; (CP) choroid plexus, (CPB) corpus pontobulbare; (EG) external granular layer; (IO) inferior olive; (RL) rhombic lip; (4V) fourth ventricle. (B) Semidiagrammatic summary of the main morphogenetic and histogenetic events during development of the human cerebellum from the 9th fetal week to the 7th postnatal month (7 p.n.m.). Above the columns are outlines of the cerebellum in the midsagittal plane at 9, 13, 16, and 25 fetal weeks and at 7 postnatal months, all at the same magnification as indicated by the 5-mm scale. (*) Primary fissure. The columns illustrate the histogenesis of the cerebellar cortex. At 9, 30, and 40 fetal weeks and the 7th postnatal month, only cresyl-violet-stained material was available, while at 13, 16, 21, and 25 weeks, both cresyl-violet- (left column) and silver-stained specimens (right column) were analyzed. Note the spread of Purkinje cell somata from multilayer to monolayer distribution as the cerebellar surface increases in area. Abbreviations: (E) ependyma; (EG) external granular layer; (G) granular layer; (I) intermediate layer; (LD) lamina dissecans; (M) molecular layer; (P) Purkinje layer; (V) ventricular zone; (W) white matter. From Sidman and Rakic (1973).

The monkey cerebellar development is rather earlier than the human, its maturation being almost completed at birth (Rakic, 1973). In the rat, all neuronal formation from the external granular layer occurs postnatally, and J. Altman (1969), in particular, has studied this process in more detail than has been possible with the primate cerebellum.

The valuable technique of pulse radiolabeling of developing cells depends on the facts that, before the stage of each mitotic division, there is a doubling of the DNA of the cell nucleus and that thymidine is one of the four constituent purines and pyrimidines used in the building of DNA. Thus, if [³H]thymidine is injected a few hours prior to this DNA synthesis, it will be incorporated into the DNA, thereby radiolabeling the two daughter cells. Since a single injection is effective in labeling for only a few hours, the time of mitotic origin—the "birthday"—of the labeled cell is specified. Furthermore, since neurons do not again divide, they carry the radiolabel on their nuclei throughout life. By contrast, after a brief period of labeling of germinal cells, the radiolabel is halved in each mitosis and so is rapidly diluted beyond recognition. Neurons formed earlier than the injection cannot, of course, acquire label. This radiolabeling establishes that the Purkinje cells and cerebellar nuclear cells are formed at almost the same time (11th–13th embryonic days of the mouse), and, since they are formed in the same place, the origin is probably from the same germinal cells in the ependyma of the rhombic lip (RL in Figure 13.5A) (Miale and Sidman, 1961). It should be mentioned that the Golgi cells of the cerebellar cortex [cf. Figures 14.14 (Section 14.4.2) and 14.15 (Section 14.4.3)] are also formed there about 2 days later, and like the Purkinje cells, they migrate up to the cortex.

Figure 13.6 is a composite of several drawings that Ramón y Cajal (1929) made of his Golgi-stained preparations [cf. Figure 1.2B (Section 1.2.1)] of the developing cerebellum, in which he brilliantly interpreted what was happening. The figure is constructed as a perspective drawing to show sections along the folium (left side) and across the folium (right side). It represents the situation at about the 10th postnatal day of the rat, corresponding approximately to Figure 13.5B at 21–25 weeks. On the surface (MU), there are about five layers of proliferating germinal cells. In the left frame, the small irregular dark cells (a) are the immature granule cells (GrN). They can be identified because they stain differently, and they are already sprouting little axons that rapidly grow out in both directions along the folium (b). These are the beginnings of the parallel fibers

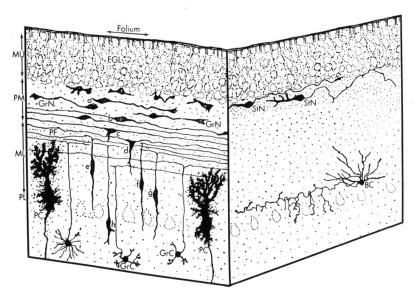

Figure 13.6. Neurogenesis of the external granular layer of the cerebellum. This montage perspective diagram was composed from several drawings by Ramón y Cajal to show the various stages of neurogenesis and morphogenesis from the cerebellar cortex, both along (left) and across a folium (right). It could represent approximately postnatal day 10 for the rat. Abbreviations: (BC) basket cell; (EGL) external granular layer; (GrC) granular cell; (GrN) immature granular cell; (ML) molecular layer; (MU) undifferentiated molecular layer; (PC) Purkinje cell; (PF) parallel fiber; (PL) Purkinje layer; (PM) plexiform molecular layer; (StN) stellate cell. From Eccles (1970a).

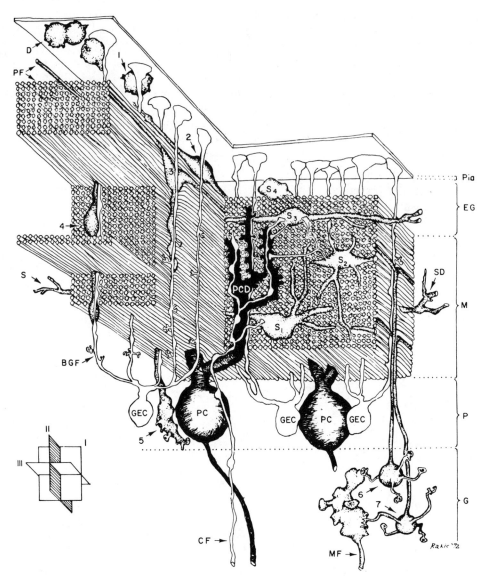

Figure 13.7. Four-dimensional (space–time) reconstruction of the developing cerebellar cortex in the rhesus monkey. The geometric figure in the lower left corner indicates the orientation of the planes: (I) transverse to the folium (sagittal); (II) longitudinal to the folium; (III) parallel to the pial surface. In the main figure, the thicknesses of the layers are drawn in their approximately true proportions for the 138-day monkey fetus, but the diameters of the cellular elements, particularly the parallel fibers, are exaggerated to make the reconstruction more explicit. Abbreviations: (BGF) Bergmann glial fiber; (CF) climbing fiber; (D) dividing external granule cell; (EG) external granule layer; (G) granular layer; (GEC) Golgi epithelial cell (Bergmann glia); (M) molecular layer; (MF) mossy fibers; (P) Purkinje layer; (PC) Purkinje cell; (PCD) Purkinje cell dendrite; (PF) parallel fiber; (S) stellate cell-fiber; (S$_{1-4}$) stellate cells; (SD) stellate cell dendrite; (1–7) granule cells. From Rakic (1973).

along the folium. They grow very rapidly and very nicely in parallel, as can be well seen in the figure.

What determines the direction of the growth? The cerebellar cortex has already become folded, and we suggest that the folding exercises a mechanical constraint on growth across the folium, thus favoring the longitudinal direction (Eccles, 1970a). Once a few of these fibers have grown, the others follow, presumably because of the CAMs on their surface. Hence, they all grow together by surface chemical identification (selective fasciculation) and, by growth alignment, make the dense bundles of parallel fibers that are so distinctive of the molecular layer of the adult cerebellar cortex [cf. Figure 14.14 (Section 14.4.2)]. It is important to realize that these immature granular cells are genetically coded to grow sprouts in two directions from the cell body so that a symmetrical structure with two oppositely directed primitive nerve fibers results.

The next stage in granule cell development is illustrated by cells c–h in Figure 13.6. When the two axonal sprouts are fully grown, pressure may rise in the soma because protein manufacture by the nucleus remains very active. So a sprout goes downward from the nuclear region, finding its way between the deeper parallel fibers. After a while, the nucleus follows down the sprout (cf. e–h in Figure 13.6), which meanwhile has grown down below the layer formed by the Purkinje cells.

Rakic (1973) has produced convincing evidence that the vertical orientation of the downward growth of granule cells is due to guidance by the fibers of the Bergmann glia, presumably being attributable to Ng-CAM on the surface of the neurons. As shown by Golgi pictures (J. Altman, 1975), Bergmann glial filaments are already in position as early as the 2nd postnatal day in the rat. They extend from the cell body at the Purkinje cell level up to the surface of the external granular layer. By day 8, the glial filaments are well formed for their function of guiding the downward migration of granular cells (Bignami and Dahl, 1973). In the elaborate drawing of Figure 13.7, Rakic (1973) shows how the Bergmann glial filaments (BGF) act as guides in the monkey. Granule cell 2 is poised ready to descend, 3 is already on the way, 4 is half down, and 5 has completed its descent and is beginning to grow dendrites, while 6 and 7 are mature granule cells with synapses from a mossy fiber (MF). There is an exact parallel with the stages in Figure 13.6.

We can surmise that the granule cells descend below the Purkinje cell layer to meet the mossy fiber sprouts that are growing into that location. Once below the Purkinje cell layer, the granule cell grows a lot of sprouts (i in Figure 13.6), but eventually only a few remain. The granule cells (GrC in Figure 13.6) are now fully formed. Meanwhile, the mossy fibers (MF) come in and form synapses on the granule cell dendrites (granule cells 6 and 7 in Figure 13.7). We can think that their synaptic "desire" has been consummated. It is a mutual desire. If the granule cells have been destroyed by virus or by X-rays, or are genetically absent as in weaver mutant mice, the mossy fibers go on looking for the granule cells and grow above the Purkinje cell layer, where they never normally go [see Figure 14.14 (Section 14.4.2)], and, apparently, finding no granule cells, they make aberrant synapses on dendrites of various Purkinje and basket cells there. We have to postulate that all cellular elements have the instructions adequate for the normal building operation. If you fool them by distorting the situation, as, for example, by massive depletion of some elements such as the granule cells, the remaining elements do the best they can, carrying on and searching for some means of satisfying their synaptic "desire" (cf. Section 13.3). Thus, there is displayed a quite wonderful living performance in the neurogenetic story.

We now will consider the right half of the perspective drawing of Figure 13.6 to see the simultaneous happenings in the generation of cells that are oriented across the folium. At the surface are the same layers of germinal cells, and below are two stellate cells (StN) with their dendrites branching from one pole and the axon from the other. The general orientation of growth is transverse both to the length of the folium and to the parallel fibers shown in the left half. Far down in the right half is a basket cell (BC) already in position. It developed much earlier, but it is essentially the same sort of cell as the stellate; both are inhibitory and both grow axons oriented transversely to the folium. You can see that the axon of the basket cell grows sprouts that are searching for the Purkinje cells, shown as faint dotted outlines. The stellate cells in the upper part of the right half have still to mature before they will synapse with the Purkinje cell dendrites that eventually will grow into the upper zone of the molecular layer. In Figure 13.7, S_2 and S_3 represent more mature stellate cells, while S_4 is a very immature stellate cell that has not yet formed any sprouts.

The stellate cells stay stacked in successive levels of the molecular layer according to the time of their development. This is also true of the parallel fibers that are stacked with a strict archeological layering. It can be conjectured that the stellate cells differ from granular cells in that they have no Ng-CAM for Bergmann glia, hence their inability to move vertically through the phalanx of stacked parallel fibers that are so effectively displayed in Figure 13.7. Instead, the basket and the stellate cells are seen in Figures 13.6 and 13.7 to grow transversely to the parallel fibers by what appears to be a series of avoidance reactions. Evidently, in their surface sensing, the immature basket and stellate cells display a pattern of desirability very different from that of immature granule cells. One can eventually expect an explanation in terms of N-CAMs. Despite their apparent avoidance of parallel fibers, the stellate neurons readily accept synapses from parallel fibers. On the output side, the basket and stellate cells make inhibitory synapses on Purkinje cells [see Figure 13.7 and Figure 14.14 (Section 14.4.2)].

There are many problems in the clonal mechanisms of the external granular layer. First, what is the mechanism whereby external germinal cells can form excitatory granule cells or inhibitory basket and stellate neurons? What sets the ratios of production of excitatory and inhibitory neurons? The germinal cells in Figure 13.6 make 80 times more granule cells than the combined basket and stellate cell populations. Why does the whole clonal process age? Germinal cells at the onset look as though they had an everlasting life of clonal multiplication, but at about the 4th postnatal day in the rat, they start differentiating into neurons, and this process occurs with increasing probability, so that eventually the whole neurogenesis comes to an end. All the germinal cells have been eliminated by mitotic transformation into neurons (7 p.n.m. in Figure 13.5), which never divide again. And why don't neurons ever divide again? These are all problems arising out of this early exploratory work on neurogenesis.

There are as yet no answers to these various questions, but one approach is to interfere with neurogenetic processes by radiation or by virus action (J. Altman, 1976; Sidman and Rakic, 1973). Mutations give all kinds of disorders and distortions of neurogenesis that lead to remarkable changes in the structure and function of the cerebellum (Rakic and Sidman, 1973). Fundamental problems thus arise in what we may term "molecular neurobiology."

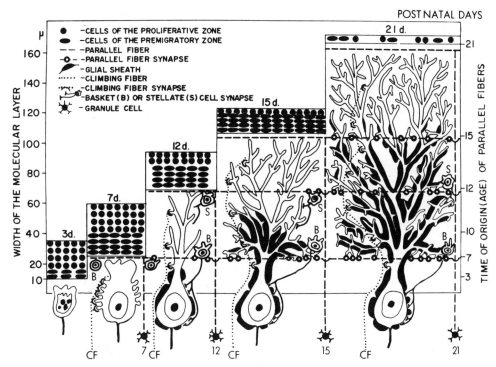

Figure 13.8. Diagram of maturation of a rat Purkinje cell and of the synaptic connections on it. Five stages of development are shown, at postnatal days 3, 7, 12, 15, and 21. Climbing fibers (CF), granule cells, and basket cells (B) are shown at days 7–21. Note that four granule cells are shown with their parallel fibers stacked at successive levels according to their "birthdays." Slightly modified from J. Altman (1972).

    In Figure 13.6, the two primitive Purkinje cells (PC) can be seen to have their axons passing downward. In fact, the axon trails behind as the body and dendrites migrate up to the cerebellar cortex, apparently in search of synaptic contacts. But the axon remains in position, ready to establish synaptic contacts with the deep nuclear cells [cf. Figure 14.14 (Section 14.4.2)]. We may call this the "axon-trailing" method of establishing synaptic contacts. The same axon-trailing process occurs with the granule cells that migrate, leaving their axons (the parallel fibers) in the molecular layer (see Figures 13.6 and 13.7).

    In Figure 13.8, J. Altman (1972) has summarized the maturation of the cerebellar cortex of the rat by drawings of the situations at significant periods: 3, 7, 12, 15, and 21 days postnatally. Reference has already been made to the numbers of cell layers in the proliferative and premigratory zones during the successive stages of development of the human cerebellum (see Figure 13.5). Figure 13.8 is particularly informative on the development of a Purkinje cell from the rudimentary form at 3 days to the almost mature form, with its profusely branching dendrites, at 21 days. A puzzling transitional relationship is shown by the three to four climbing fibers (CF) making synapses on the somatic spines of the Purkinje cell at 7 days (not shown in Figure 13.8) and the displacement of the single surviving climbing fiber up to the dendrites at 12 days (see Section 13.2.4). The somatic spines [cf. the two Purkinje cells (PC) in Figure 13.6] have meanwhile disappeared and been replaced by basket cell (B) synapses on the soma. Another interesting feature is the relatively late stage at which synapses are formed by parallel fibers on Purkinje cells. Granule cells formed at day 7 do not give parallel fiber synapses to Purkinje cells at day 12, but only at day 15. Similarly, 12-day and 15-day granule cells give parallel fiber synapses only at day 21.

There are many challenging problems in Purkinje cell maturation. What is the significance of the transitory somatic spines that are so striking in the drawing by Ramón y Cajal in Figure 13.6 (left side)? Is the displacement of the climbing fiber synapses from somatic spines related to the simultaneous development of basket cell synapses on the soma? Why are the parallel fiber synapses on Purkinje cells so delayed in formation relative to their synapses on basket and stellate cells? What is the significance of the massive development of glial coverage of the soma and main dendrites? What is the cause of the great proliferation of the Purkinje cell dendrites at 12 and 15 days, though no synapses are yet formed on them by parallel fibers? Apparently, there is as yet no recognition between the parallel fibers and these newly grown dendrites. Rakic and Sidman (1973) suggest that the maturation of Purkinje cells is influenced by early contact with climbing fibers, and their suggestion is supported by the experiments of Kawaguchi *et al.* (1975) and Bradley and Berry (1976), showing that climbing fibers induce the branching of dendrites and the growth of the dendritic spines on which the parallel fibers make synapses. A key role in the maturation of the Purkinje cell is played by the parallel fibers. When there is a deficiency of parallel fibers, the Purkinje cell dendrites are also deficient and deformed. It seems that the espaliered shape of Purkinje cells [see Figure 14.14 (Section 14.4.2)] is due to the sculpturing effect by parallel fibers by what has been called an "exclusion principle" (Eccles, 1970a; J. Altman, 1976).

13.1.4. Building of Brain Nuclei and the Hippocampus

In Figure 13.5A (Section 13.1.3), the lower rhombic lip (RL) line points to an intense clonal multiplication zone that gives rise to the corpus pontobulbare (CPB), from which there is migration of subventricular cells as shown by the two lower downward and medially pointing arrows. These immature neurons migrate along the two arrows to form deep collections of neurons such as the pontine nuclei, the lateral reticular nucleus, and the inferior olive (IO) and accessory olive (AO). On the right half of Figure 13.5A, the dots indicate neurons, and in the primitive cerebellum there can be seen a deep collection of neurons that will form the cerebellar nuclei; these arose from the dorsal RL at the same time as the Purkinje cells. In the human fetus, another structure analogous to the CPB can be seen bulging into the lateral ventricle dorsal to the CPB in Figure 13.5A. This is the ganglionic eminence, which is formed by rapidly mitosing subventricular cells, rather as in layer S of Figure 13.2E (Section 13.1.2). It appears before the 13th fetal week and persists until after birth. Sidman and Rakic (1973) have shown that the immature neurons so generated are destined to form the large nuclei of the diencephalon—the thalamus, the basal ganglia, and the pulvinar of the thalamus.

The hippocampus is formed by mitosing subventricular cells (layer S in Figure 13.2E), exactly as is the neocortex. Moreover, there is a glial scaffolding exactly as in Figures 13.3B and C (Section 13.1.2) on which the primitive neurons migrate to reach their adult position, presumably due to Ng-CAM molecules on the neuronal surfaces (Edelman, 1984). Recent investigations have had the advantage of precise identification of glial cells and fibers by immunocytochemical analysis. The difference from the neo-cortex is that the original fetal structure of the zone of the cortex destined to be trans-formed into the hippocampus is greatly distorted by folding during development (Eckenhoff and Rakic, 1984), so that the original dentate plate is separated by develop-mental stresses to assume its horseshoe shape and the hippocampus proper is also folded into its adult curvature. However, the radial glial fibers from the reticular surface to the pia (cf. Figure 13.3B) maintain their attachments and appropriately guide the neurons to form the hippocampus despite the developmental folding (Nowakowski and Rakic, 1981). E. R. Lewis and Cotman (1982) studied the septal lamination in the developing hippocam-pus by fiber outgrowths from septal implants. The situation is more complicated with the developing dentate gyrus, because the glial cells relinquish their attachments to the ventricular surface, though they keep a radial arrangement with respect to the pial surface infolded into the hippocampal fissure [cf. Figure 13.17 (Section 13.2.2)]. The primitive granule cells migrate along the glial framework (Eckenhoff and Rakic, 1984) to their adult position. Most astrocytes in the hippocampus seem to arise from the glial cells with their radial fibers exactly as occurs with the Bergmann glia of the cerebellum.

13.2. Principles of Neuronal Recognition and Connectivity

13.2.1. Introduction

So far, we have concentrated on neurogenesis in the mammalian brain, describing the genesis of the various cell types and the manner in which the immature neurons develop into the fully formed neurons in some of the principal regions of the mammalian brain. It has been a descriptive story built upon precise histological studies at all stages of embryonic development. Now we come to a fundamental question: What guides the growth of the axonal projections from the embryonic nerve cells so that eventually the connectivities of the adult brain are established? There must be a fantastic organizing process, or rather an assemblage of such processes. In the 1930s, it was believed that there was initially just neuronal disorder and that usage established the order. This has been disproved by the many investigations—e.g., of the retinocortical projections—showing that the connections are grown in almost their final form before they are used. These experimental findings indicate that in the developing brain, there are encoded extremely detailed instructions that control and guide the growing axons to the "correct" targets. This requirement led Sperry to formulate this most challenging hypothesis: that the precise building of the nervous system is due to an immense variety both in the chemical coding of neurons that are seeking to make synapses and in the complementary mechanisms for specific recognition by neuronal surfaces "ripe" for synapses. We quote from the elo-quent writings of Sperry (1971):

> The complicated nerve fiber circuits of the brain grow, assemble and organize themselves through the use of intricate chemical codes under genetic control. Early in development the nerve cells . . . acquire individual identification tags, molecular in nature, by which they can be recognized and distinguished one from another. . . . The outgrowing fibers in the developing brain are guided by a kind of probing chemical touch system. . . . By selective molecular preferences nerve fibers are guided correctly into their separate channels at each of the numerous forks or decision points which they encounter as they travel through what is essentially a three-dimensional multiple Y-maze of possible channel choices.

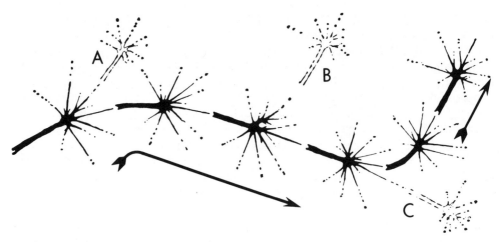

Figure 13.9. Sequential representation of steps in chemotactic guidance of a growing nerve fiber. From Sperry (1963).

Figure 13.9 illustrates Sperry's hypothesis by showing in diagrammatic form the sequences of growth of a nerve fiber that can be seen in time-lapse cinemicrography such as Pomerat so beautifully photographed in tissue cultures of nerve cells. It is easy to see that the fiber growing toward the right is sending out little searching probes. These several outgrowths at any one time are probing continuously, looking for the right kind of chemical surfaces with attractive molecular configurations. If they do not find them, they may regress, or they may grow on as in the main shaft of Figure 13.9, perhaps getting some encouragement from the general chemical environment. So the shoot grows on, at each stage sending out searching probes in all directions and then going on to follow up some attraction. In A, B, and C of Figure 13.9 can be seen "ghosts" of offshoots that finally failed after they had grown some distance. This diagram gives some kind of picture of what you can imagine happening to an incredible degree for the axon of every nerve cell finding its way to the neuronal surface on which it eventually forms synapses.

Letourneau (1983) has presented an excellent review of the present state of knowledge of the motile behavior of growth cones, which is now discussed at the level of adhesion molecules, of the intrinsic action of actin and myosin as a motive force, and of the microtubule growth in the elongating nerve fibers.

Sperry and his associates tested his chemoaffinity hypothesis particularly in the relatively simple retinotectal system of the frog, studying the regenerative processes that follow experimental lesions of the retina and optic nerve. Behavioral and histological studies seemed to indicate a remarkable ability of the severed optic nerve fibers to regenerate to the "correct" site in the optic tectum (Sperry, 1971). Gaze (1970, 1974), Jacobson (1970), and R. L. Meyer and Sperry (1974) have used electrophysiological field analysis to map the relationship between retina and tectum. Microelectrode recording has been carried out from a grid of tectal sites, and for each the corresponding retinal point has been determined by scanning the retina with a light spot. In general, there was confirmation of the regenerative chemoaffinity pattern assumed on the basis of the original behavioral experiments, but there was evidence that "incorrect" synapses were often formed initially, followed by a later reshuffling of synapses. Hence, Gaze (1970, 1974) suggested that optic nerve fibers grow to the tectum initially, not in a chemically specified manner, but with control by the order of embryonic development. It has long been evident that chemoaffinity guidance could be effective only in close proximity to the target neurons.

The crucial effect of order of development has been demonstrated by Rager and associates in work on the retinotectal system in the chick embryo (Rager, 1980a,b; Rager and von Oeynhausen, 1979; Rager and Rager, 1978). The first ganglion cells are developed in the central region of the retina (embryonic day 4), and their outgrowing axons are oriented in a crescent in the optic stalk (day 5) in a well-ordered array. This growth appears to be organized by contact guidance between the fibers as originally postulated by P. Weiss (1941) and in accord with the Ng-CAM guidance suggested by Edelman (1984). When arriving at the tectum, these earliest fibers grow over its surface until they reach the central region, where the tectal cells first mature with dendrites ready for receiving synaptic contacts. Thus, the initial connection from retina to tectum is governed by developmental priority. Subsequently maturing ganglion cells project as a series of crescents ordered by the initial crescent and so to the tectum for making synapses with the next maturing neurons. As Rager (1980b) states:

> . . . central retinal fibers continue to grow on the tectal surface until they meet dendrites mature enough to receive contacts. The synchrony between the arrival of central retinal axons and the maturation of central tectal neurons would be sufficient to determine the starting-place of the retino-tectal map. The mapping on to peripheral regions could be due to the temporo-spatial organization of the arriving fibre tract and to the fact that the maturation of tectal cells progresses continuously from the centre to the periphery.

The important role of fiber order in giving guidance for the developing visual pathways has been reviewed by C. Walsh and Guillery (1984), principally for mammals, but also for frogs and fish. In general, there is good agreement with the findings of Rager (1980b) with birds.

K. A. C. Martin and Perry (1983) present a comprehensive study of the role of fiber ordering in nerves and tracts in determining the precision of the terminal maps, emphasizing the "control of contact guidance mechanisms rather than chemospecificity cues." Possibly, chemospecificity is significant in the competitive synaptic operations between the optic fibers on the tectal cells. The fibers are produced in excess: 4 million on incubation days 10 and 11 are reduced to the final adult value of 2.4 million by day 18 (Rager and Rager, 1978). Similarly in the monkey, the optic nerve fiber counts show that the 2.85 million retinal axons in the fetal monkey are reduced to 1.2 million in the adult (Rakic and Riley, 1983). Presumably, death overtakes the fibers and retinal ganglion cells that are unable to secure an adequacy of synaptic contacts on tectal cells (see Section 13.2.4) (cf. Cowan et al., 1984). This cell death, and possible retraction of excess collateral branches made by optic nerve fibers, refines the retinotopic map.

It is generally agreed that there is a remarkable somatotopic relationship between the retina and the tectum in submammalian species. The chemospecificity hypothesis of Sperry has to be downplayed in favor of labeling by the temporal order of generation of neurons and targets and the contact guidance in the fasciculation manner proposed by P. Weiss (1941), with preservation of neighborhood relationships. It seems also that glial partitions in the region of the optic chiasma may be important (Rager, 1980b). Still unknown factors must be at work in generating channels in the retina to the optic stalk and in the guidance by structures in the optic stalk for the outgrowing growth cones. Nevertheless, the behavioral investigations of Sperry (1971) that supported his chemoaffinity hypothesis have yet to be satisfactorily explained by alternative hypotheses.

However, Bonhoeffer and Gierer (1984) have developed a new approach that could account for the experimental findings of Sperry (1971), Yoon (1975), and others on the apparent chemoaffinity of the growing nerve fibers for specific tectal neurons. *In vitro* experiments using fluorescent dye techniques measure the directional effects of growing nerve fibers from the retina in relation to competing targets that are set up as cell monolayers, e.g., of tectal vs. retinal cells. The former surface is preferred. Bonhoeffer and Gierer propose that the outgrowing nerve fibers read chemical gradients of specific molecules in a general way and not by the detailed specificity of the original Sperry hypothesis. Moreover, retinal fibers from different retinal locations grow along graded cues on the tectum. In addition, newly arriving fibers are guided primarily by the earlier fibers. Mechanical factors also play an important role, as do, of course, adhesion molecules. So we are presented comprehensively with a multifactor hypothesis that is still in a primitive form.

13.2.2. Neuronal Connectivity in Mammals

The complexity of the mammalian brain has made it difficult to investigate the development of pathways in the manner that has been done with the optic–tectal pathway in amphibia and fish (Sperry, 1971; Jacobson, 1970; Gaze, 1970, 1974). Sections 13.1.2–13.1.4 described the principal features in the building of the cerebral cortex, the cerebellum, and the hippocampus. Until recently, it was generally believed that all the embryonic know-how had been lost in the mammalian postnatal period, so that there was no regeneration of the mammalian brain and spinal cord after injury. But now we have well-designed experiments indicating that there is a considerable regenerative capacity in several regions of the mammalian brain, even in adults. Optimistically, one can predict that we are only at the beginning of an enterprise in which various surgical procedures, plus rehabilitation therapy and local administration of nerve growth factors, will be able to reduce some of the disabilities suffered by patients with lesions of the brain and spinal cord. We shall see in the subsequent sections that regeneration has already been demonstrated in many regions of the mammalian brain: the septal nuclei, the red nucleus, the hippocampus, the lateral geniculate nucleus, the superior colliculus, the cerebellothalamic tract, and the cerebral cortex. These regenerations differ from the regeneration of pathways in lower vertebrates in that regeneration usually occurs only for very short distances—perhaps for no more than 50 μm. An additional feature is that recovery from lesions can be aided by transplants, which is a technique still in its infancy.

Septal Nuclei. This work at higher levels of the CNS can best be introduced by considering the fine work of Raisman and his colleagues on neurons of the septal nuclei. As illustrated in Figure 13.10, the septal nuclei provide ideal conditions for the experimental investigations (Raisman, 1969, 1977, 1985; Raisman and Field, 1973). The two principal inputs to these nuclei are the fimbrial pathway from the hippocampus and the medial forebrain bundle (MFB). The former input forms synapses almost exclusively on the dendrites of septal cells, while the latter input ends on both dendrites and somata (Figure 13.10A).

After sectioning of either pathway in adult rats (3–6 months old), there was the usual

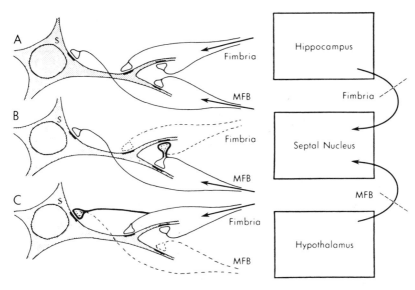

Figure 13.10. Schematic representation of synaptic regeneration of septal neurons. (A) In the normal situation, afferents from the medial forebrain bundle (MFB) terminate in boutons on the cell soma (S) and on dendrites, and the fimbrial fibers are restricted in termination to the dendrites. (B) Several weeks after a lesion of the fimbria, the MFB fiber terminals extend across from their own sites to occupy the vacated sites, thus forming double synapses (degenerated connections are shown with discontinuous lines, presumed plastic changes with a heavy black line). (C) Several weeks after a lesion of the MFB, the fimbrial fibers give rise to terminals occupying somatic sites, which are presumably those vacated as a result of the former lesion. Modified from Raisman (1969).

degeneration within a few days and disappearance of the synapses formed by the sectioned pathway. Electron microscopic observations revealed that the number of synapses was reduced almost to half. But some 30 days later, the full population was restored. Convincing evidence was presented that this restoration was due to sprouting of the fibers of the intact pathway, the sprouts growing to form new synapses that occupied the vacated synaptic sites, as illustrated in Figure 13.10B and C. In the electron micrographs, it could be seen that, some weeks after section of the MFB, the fimbrial fibers occupied a considerable number of synapses on the somata of the septal neurons, as illustrated in Figure 13.10C. Conversely, after section of the fimbrial pathway, there was evidence that the fibers of the MFB sprouted to form many new synapses, which often had the double configuration shown in Figure 13.10B. These new synapses would all be on the dendrites, which is the site of the vacated fimbrial synapses.

It appears that there has been a loss of the embryonic growth specificity, so that collaterals growing from the intact axons now heterotypically innervate synaptic sites originally reserved for and occupied by the sectioned pathway. Raisman (1969) regarded this heterotypic regeneration of synapses as being functionally meaningless. Nevertheless, their regeneration is of great interest because it shows that axonal sprouting and synaptic formation can occur in the adult rat, at least at microdistances. Raisman further suggested that this heterotypic regeneration may be aided by the chromatolytic reaction of the septal neurons that follows the inadvertent section of their axons in the initial operation (see

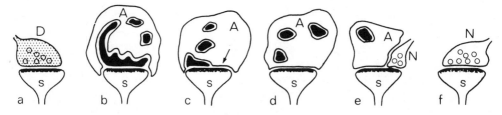

Figure 13.11. Proposed scheme for reinnervation of a deafferented dendritic spine. The figure is a schematic representation of some of the structures that have been found in apposition to dendritic spines in the septum during the period of the proposed collateral reinnervation. These configurations have been arranged in sequence from a to f to illustrate one possible series of changes that would result in reinnervation of the deafferented spine. (a) A degenerating terminal (D) lies in contact with the dendritic spine(s). (b) The degenerating terminal, now much darker and more shrunken, is surrounded by a swollen astrocytic process (A) that is indenting its surface and has partially engulfed two detached fragments of degenerating terminal cytoplasm. (c) The reactive astrocytic process has now partially displaced (←) the degenerating terminal from the region of the synaptic thickening. (d) The displacement of the degenerating terminal from the synaptic thickening is now completed ("vacated synaptic thickening"). (e) The reactive astrocytic process is partially displaced from the synaptic thickening by a nondegenerating axon terminal (N) ("shared synaptic thickening"). (f) Complete reinnervation of the synaptic site by the nondegenerating axon terminal. From Raisman and Field (1973).

Section 1.2.5). Since the axons of septal neurons are in both the fimbria and the MFB, many would be sectioned in the initial operation. Two important negative findings are illustrated in Figure 13.10B and C. First, the sectioned fibers make only abortive growths and do not regenerate pathways; second, there is no evidence for development of new synaptic sites on the septal cells, there being merely heterotypic reinnervation of old sites.

Raisman and Field (1973) have illustrated the manner in which the glial cells of the septal nucleus participate in the degenerative process. As depicted in Figure 13.11, the astroglia are specially concerned in the ingestion of the degenerating synapses, and at the same time, they move in to occupy the vacated sites, as described in the figure caption. The synaptic sites can be identified by the characteristic postsynaptic membrane densities that remain after the synaptic knobs disappear. Westrum and Black (1971) have reported a similar action of astroglia on degenerating synapses in the trigeminal nucleus.

Figures 13.10 and 13.11 are especially important because they lead to the formulation of questions relating to the problems of sprouting and reinnervation of the synaptic sites, of which we may ask two: How is the sprouting initiated? How is the sprout guided to the vacated synaptic sites?

As for the guidance, it has been suggested by Raisman and Field (1973) that glial cells may provide the guidelines for the newly growing fiber in the same way as Rakic (1972, 1973) has demonstrated for glia in the initial process of neurogenesis in the monkey cerebrum [see Figure 13.3C (Section 13.1.2)] and cerebellum [see Figure 13.7 (Section 13.1.3)]. As stated in Section 13.1.1, it can be presumed that the special adhesion molecules Ng-CAM (Edelman, 1984) are responsible for this glial guidance.

The second problem, however, concerns the initial process, which is the triggering of an axon to produce a branch. It is one thing for the glia to guide the branch home to the vacant synaptic site. Quite another problem is raised when one asks how the branch starts in the first place. Raisman and Field (1973) have shown by beautiful illustrations (Figure 13.11) how the astroglia ingest the degenerating synaptic terminal and eventually break it up and apparently digest it. It is suggested in their theory that because of this ingestion of the degenerating synapse, the astroglial cell develops a changed chemical constitution that affects the surface contacts it makes with presynaptic fibers so that it acts as a stimulant for their growth. Thus, we would have two separate functions for glia. One is the trigger function, whereby the glia through the ingested synaptic knobs stimulate the presynaptic fiber to branch; the second is the guiding role of glia, whereby the branch grows so as eventually to reach and occupy the vacated synaptic site.

It has been suggested that this synaptic regeneration, as disclosed in the septal nucleus, has no functional meaning because, as illustrated in Figure 13.10, synapses from one type of input regenerate to occupy synaptic sites of quite different inputs, the so-called "heterotypic regeneration." It could thus be maladaptive. It must be envisaged, however, that the experimental demonstration by Raisman and his colleagues could be obtained only if there were massive degeneration. If only a small fraction of the inputs from the fimbria or from the medial forebrain bundle were cut, then it would be impossible to discover whether there was any adaptive regeneration that could be selectively homotypic as described by Raisman (1985). We have to recognize that compared with possible naturally occurring random regenerations, recovery from the surgically induced degenerations as in Figure 13.10 must be maladaptive.

Red Nucleus. A most comprehensive study of synaptogenesis induced by synaptic degeneration has been carried out by Tsukahara and colleagues over the last decade, as reviewed by Tsukahara (1981, 1984). This study has been especially valuable because it has been built around the spatial locations of the synapses made by two types of well-defined inputs. The input from the interpositus nucleus is on the soma and adjacent proximal dendrites, whereas the corticorubral fibers make synapses on the more distal dendritic zones. The study's unique value also derives from the fact that regenerated synapses are tested by intracellular recording of the excitatory postsynaptic potentials (EPSPs) they generate.

The topography of the dendrites of the red nucleus (RN) neurons is the crucial factor in the analysis both of normal synaptic distribution and of the change brought about by new growths in response to lesions. Figure 13.12C shows the dendritic topography of an RN neuron with the typical tapering form over hundreds of microns. In Figure 13.12A, there is a plot of the dendritic diameter with distance from the soma for one dendrite, and in Figure 13.12B are the pooled measurements of 22 dendrites of several neurons. It is evident that the diameter of an observed dendrite provides an approximate measure of its distance from the soma.

Figure 13.13 illustrates the normal responses of an RN neuron as reported by Tsukahara *et al.* (1975). In (A), there is a monosynaptic EPSP of fast time–course in response to stimulation of the contralateral interpositus nucleus (c-IP). In contrast, stimulation of the ipsilateral cerebral peduncle (i-CP) evokes monosynaptically an EPSP of slow time–course (Figure 13.13C). The difference between these two types of EPSPs is well shown by histograms of the times to the peaks in Figure 13.13E and F. The explanation of this striking difference is diagrammed in Figure 13.13G, in which the cerebral peduncle projects to the distal dendrites on the same side and the interpositus nucleus to the soma (and proximal dendrites) of the contralateral RN neurons. The differences in the time–courses of the respective EPSPs are fully attributable to the electrotonic distortions that result in transmission from the distal generating sites to the intracellular recording

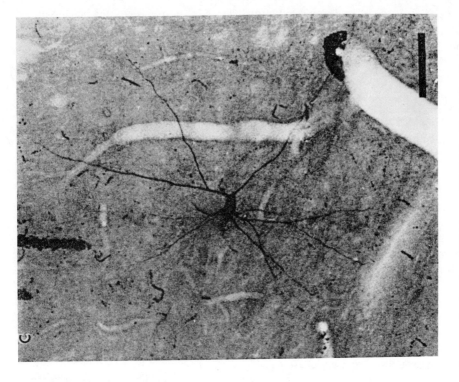

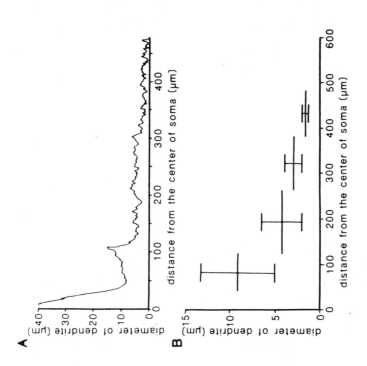

Figure 13.12. Relationship between diameter of dendrites and their distance from the soma of red nucleus (RN) neurons. (A) Relationship between diameter of a dendrite and the distance from the center of the soma of an RN neuron from a cat with nerve cross-innervation. Peaks that appear at 110 and 300 μm from the center of the soma are due to branching of the dendrite. (B) Relationship similar to that in (A), but for pooled data from 22 dendrites of cats with nerve cross-innervation. (C) Light micrograph of a horseradish peroxidase-filled rubrospinal neuron identified by antidromic invasion. Scale bar: 200 μm. From Murakami *et al.* (1984).

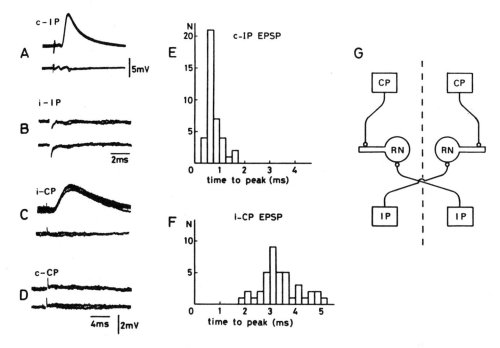

Figure 13.13. EPSPs produced in normal kitten red nucleus (RN) neurons. (A) EPSPs elicited in an RN cell by stimulating the contralateral nucleus interpositus (c-IP) of the cerebellum (upper trace). The lower traces of (A–D) are corresponding extracellular traces as recorded just outside the cells. (B) Records produced by stimulating the ipsilateral IP (i-IP). (C) EPSPs induced by stimulation of the ipsilateral cerebral peduncle (i-CP). (D) Records produced by stimulating the contralateral CP (c-CP). (E, F) Frequency distribution of the time-to-peak of the EPSPs elicited in 37- to 132-day-old kitten RN. (G) Diagram of the RN neuronal connections. The dashed line indicates the midline. From Tsukahara *et al.* (1983).

microelectrode in the soma of the RN neuron (Tsukahara *et al.*, 1975). The ipsilateral interpositus nucleus (B) and the contralateral cerebral peduncle (D) do not project to the RN.

When the interpositus nucleus in kittens or cats is destroyed, it is found that after 2–3 weeks, stimulation of the cerebral peduncle (CP) results in an EPSP (Figure 13.14C) with superposition of a newly developed fast "somatic" type of EPSP on the original slow dendritic type (Figure 13.14A). It was postulated that axonal sprouting of the CP fibers occupied the vacant sites left by the degenerating interpositus synapses, as indicated in Figure 13.14F. We can conjecture that, as with the septal nucleus, glia may have been responsible for the triggering of this growth to the vacant synaptic sites. Here again, we have a heterotypic regeneration just as with the septal nucleus (see Figure 13.10).

There has been a remarkable corroboration of the original hypothesis by a rigorous histological investigation (Murakami *et al.*, 1982). Horseradish peroxidase (HRP) was injected into an RN neuron identified by antidromic invasion, and the corticorubral synapses on it were identified by the degeneration resulting from a lesion of the sensorimotor cortex 2–6 days earlier. It was discovered that normally the degenerating synapses (CP) were located mostly on small diameter dendrites (Figure 13.15A), as would be predicted from the slow-rising EPSP (Figures 13.13C and F and 13.14A and D). When there had been a lesion of the contralateral interpositus nucleus 5–8 weeks previously,

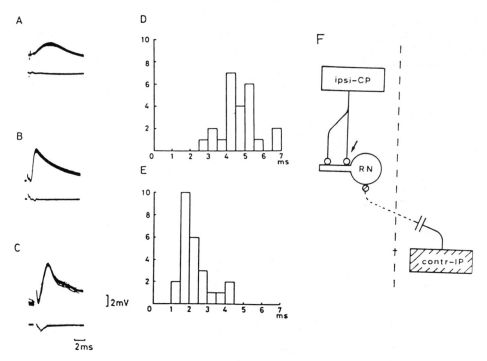

Figure 13.14. Intracellular responses of red nucleus (RN) neurons. (A–C) The upper traces are intracellular responses in RN neurons, while the lower traces show the corresponding field potentials recorded at a just-extracellular position. (A) A CP EPSP; (B) an IP EPSP; (C) a CP EPSP after IP destruction. Time and voltage calibrations for all intra- and extracellular responses are shown at (C). (D, E) Histograms in normal cats (D) and after IP lesion (E) illustrate the frequency distribution (number of cells on the ordinate) of the "time-to-peak" of CP EPSPs. (F) An illustrative diagram, with the arrow indicating a newly grown synapse on or close to the soma. From Tsukahara *et al.* (1983).

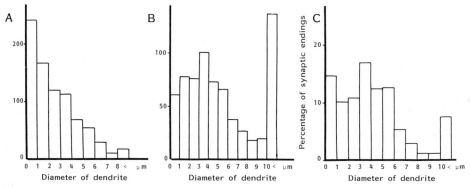

Figure 13.15. Histograms showing relationships between number of corticorubral degenerating terminals and the minor diameter of dendrites. (A) Summary of all degenerating terminals ($N = 837$) found in 15 Epon blocks from 5 normal cats. (B) Summary of all degenerating synaptic endings ($N = 612$) found in 15 Epon blocks from 7 cats with IP lesions. From Murakami *et al.* (1982). (C) Distribution of degenerating terminals in cats with nerve cross-union as described in Section 13.2.3. The total sample consists of 376 degenerating terminals found in the red nuclei of 9 animals. From Murakami *et al.* (1984).

many CP synapses were located on large dendrites, and even on the soma, as illustrated in the histogram of Figure 13.15B. This finding corroborates the predictions from the observed transformation of the EPSP produced on stimulation of the CP to a predominantly fast-rising type (Figure 13.14C and E).

This histological localization of synapses on dendrites either remote from the soma or in proximity to the soma is corroborated by the finding that, when a D.C. current hyperpolarizes the soma, there is a much larger increase in the fast-rising EPSPs than in the slow EPSPs (Tsukahara *et al.*, 1983).

It is important to realize that this relocation of the CP synapses to more proximal dendritic sites need not be on the same dendrites, with the synapses progressing along the dendrites as might be imagined from Figures 13.12C and 13.14F. All that is demonstrated in Figures 13.14 and 13.15 is that CP synapses have been developed on some proximal RN dendrites, which presumably are closely adjacent to the synapses of these same CP fibers on remote dendrites of other RN neurons.

Tsukahara *et al.* (1983) have also investigated the changed innervation of the RN neurons when there is a degeneration of the corticorubral input in kittens (<2 months after birth). The reactions evoked by this degeneration are more complicated than with IP destruction. As illustrated in Figure 13.16, there are three compensatory connectivities that are never observed in normal kittens. Figure 13.16A shows that stimulation of the contralateral CP can evoke the slow-rising EPSPs characteristic of the normal ipsilateral CP input. Figure 13.16B shows that stimulation of the ipsilateral interpositus nucleus (IP) also may evoke the slow type of EPSP. Finally, Figure 13.16C shows that stimulation of the contralateral IP evokes not only the normal fast EPSP (see Figure 13.14B), but also a later slow EPSP. Figure 13.16D gives a diagrammatic illustration of these three new types of synaptic input in contrast to the normal arrangement in Figure 13.13G. Thus, after the degeneration of the ipsilateral CP with its synapses on distal dendrites of RN neurons, there are developed the three compensatory inputs on the distal dendrites as indicated. The relative frequencies of the experimentally observed new synapses are shown in Figure 13.16E; those from the contralateral CP are the most frequent, followed by those from the contralateral IP. Most individual cells showed only one type of newly developed synaptic input as indicated in Figure 13.16E.

The probability of growth of such compensatory projections to the partially denervated RN neurons was as high as 100% with operations up to 20 days after birth, but declined rapidly to no more than 30% by 50 days, and such growth was never observed following operations on adult cats. There is no simple explanation of such compensatory synaptic growth that seems to occur over a considerable distance, which is in contrast to the small distances traversed by the regrowth after a lesion of the IP (see Figure 13.14F).

An important finding of Tsukahara and associates is that the synaptic replacements are functionally effective in generating EPSPs. Even a somatotopic relationship is preserved; axons from the forelimb cortical area sprout to make synapses with the RN neurons that project to the forelimb regions of the spinal cord, and correspondingly for the hindlimb (Tsukahara *et al.*, 1983).

Hippocampus. As indicated in the upper right diagram in Figure 13.17 (Cotman and Nieto-Sampedro, 1984), the granule cells of the hippocampal dentate gyrus have a complex regional input. The ipsilateral entorhinal cortex (EC) makes synapses on the thousands of spines on the outer two thirds of each granule cell, while the CA4 pyramidal

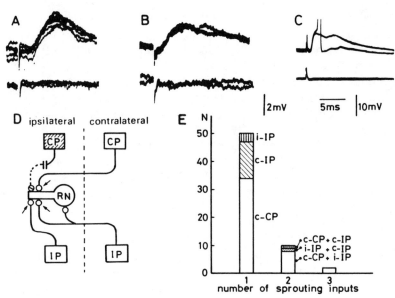

Figure 13.16. Newly appeared EPSPs in the kitten red nucleus (RN) after destruction of the ipsilateral cere-brorubral input. (A) Newly appeared EPSPs induced in an RN cell by stimulation of the contralateral cerebral peduncle (c-CP) (upper trace). Corresponding extracellular records are illustrated by the lower traces in (A–C). (B) EPSPs induced by stimulating the ipsilateral interpositus nucleus (i-IP) of the cerebellum. (C) Dual-peaked EPSPs induced by stimulation of the contralateral IP (c-IP). The voltage calibration in (B) also applies to (A). The time calibration in (C) also applies to (A) and (B). (D) Diagram illustrating the formation of new synapses in the kitten RN after destruction of the ipsilateral cerebral cortex. (→) Newly formed synapses. The dashed line shows the midline. (E) Frequency distribution of numbers of synaptogenesis sources on each RN cell. *Ordinate:* number of RN neurons. The left bar indicates RN cells for which only one source of synaptogenesis was found. (□) RN cells from which only c-CP EPSPs were recorded; (▨) RN cells from which only double-peaked c-IP EPSPs were recorded; (▥) RN cells from which only i-IP EPSPs were recorded. The middle bar shows RN cells in which two sources were simultaneously observed. (□) RN cells from which c-CP and i-IP EPSPs were recorded; (▨) RN cells from which i-IP and c-IP EPSPs were recorded; (▥) RN cells from which c-CP and c-IP EPSPs were recorded. The right bar shows RN cells from which all three new inputs were simultaneously recorded. From Tsukahara *et al.* (1983).

cells innervate the proximal third, there being also weaker innervation from the septal nucleus and a weak innervation (not shown) from the contralateral EC. When the EC on one side is excised, 90% of the synapses on the outer two thirds of the ipsilateral dendritic spines disappear within 1 day (Cotman and Nieto-Sampedro, 1984). However, recovery begins in 3 days, and in 60 days there is complete synaptic replacement with inputs indicated in the lower right diagram in Figure 13.17. The new synapses formed by the contralateral EC and CA4 are functional. The rate of replacement follows the rate of clearing of the dead synapses.

Synaptic replacement is selective, being restricted to fibers within or adjacent to the denervated zone (lower right diagram in Figure 13.17). The fibers most closely related to the degenerated synapses, those from the contralateral EC, are most favored. There is full replacement, but not hyperinnervation; the total synaptic surface of each dentate granule cell is conserved. An important feature is that the regeneration is restricted to fibers not more than 20–30 μm distant.

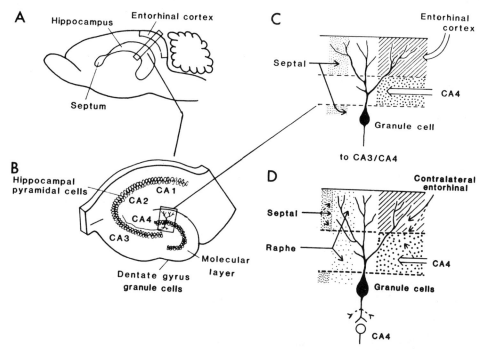

Figure 13.17. Two distinct neuronal populations in the hippocampal formation. (B) Pyramidal cells are the major cell type in subfields CA1–CA4 in the hippocampus proper. Granule cells are predominant in the dentate gyrus. The inputs to the dentate gyrus granule cells are strictly laminated along their dendritic tree. The entorhinal cortex collects input from about eight cortical and subcortical areas and projects to the outer two thirds of the dentate gyrus molecular layer. This projection is excitatory and provides about 60% of the total input to the granule cells. The other prominent extrinsic afferent originates in the septum, is partly cholinergic, and projects to both the dentate gyrus molecular layer (C) and the hippocampus proper. Minor inputs arrive from locus coeruleus, raphe, hypothalamus, and nucleus reuniens. The remaining innervation originates in the hippocampal formation itself. The dentate granule cells send their axons to hippocampal subfields CA3 and CA4 (B). Area CA4 cells, in their turn, project bilaterally back to the granule cells, innervating the inner third of the dendritic tree of the ipsilateral (C) (associational fibers) or the contralateral (commissural fibers) dentate gyrus (D). *Lower right:* After an entorhinal lesion, synapse replacement is achieved by selective growth of residual inputs. (- - -) Sprouting of undamaged inputs (lower right). From Cotman and Nieto-Sampedro (1984).

A study of the regenerated contralateral EC synapses shows that they have properties similar to those of the original ipsilateral EC synapses in respect of facilitation, depression, and long-term potentiation (see Section 16.2) (Steward, 1982). The small difference reported, that EPSPs during long-term potentiation are less effective in generating impulse discharges, may be accounted for either by the complications of inhibition or by the inadequacy of extracellular recording in providing precise information on the intensity of synaptic excitation.

In summary, these experiments display a remarkable ability for regeneration of the mammalian cerebral cortex. Just as with the septal nucleus and the red nucleus, the most probable explanation is that astroglia ingest the fragments of the degenerating synapses and form trophic substances that trigger the sprouting of nearby fibers and guide them to the vacant synaptic sites.

A

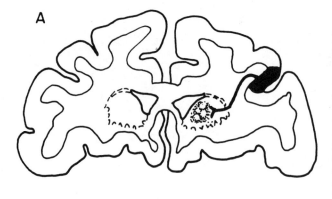

B

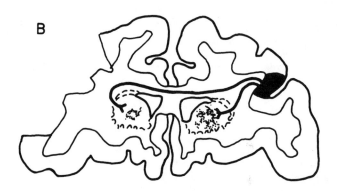

Figure 13.18. Diagrammatic illustration of the head of the caudate nuclei in the left and right hemispheres. (A) Normal monkey that received an injection of mixed [³H]proline and [³H]leucine in the right hemisphere at 5 days of age. The monkey was sacrificed 1 week later and its brain processed for autoradiography. The drawing shows the intricate and dense pattern of labeling in the right (ipsilateral) caudate nucleus. (B) Monkey in which the prospective dorsolateral prefrontal cortex in the left hemisphere was resected at E119, 6 weeks before birth. The fetus was returned to the uterus and subsequently delivered near term. On the 5th postnatal day, the right prefrontal cortex was injected with tritiated amino acids, and the animal was sacrificed 1 week later. Note the intricate pattern of grains in the right (ipsilateral) caudate nucleus, but also a distinct projection to the left (contralateral) caudate. Based on Goldman (1978).

Corticocaudate Projections. The dorsolateral prefrontal cortex in one hemisphere of a rhesus monkey was resected 6 weeks before birth and the fetus returned *in utero* for normal delivery (Goldman, 1978). A radiotracer was applied to the contralateral principal sulcus on the 5th postnatal day, and the monkey was sacrificed for autoradiography 1 week later. There was labeling of the caudate nucleus on the lesioned side, indicating the development of a much more powerful crossed corticocaudate pathway (Figure 13.18B) than is normally found (Figure 13.18A). These results have been confirmed in and extended to monkeys in which the dorsolateral cortex was resected as late as 2 months after birth (Goldman-Rakic, 1980).

Three possible explanations were proposed (Goldman, 1978). One is that there is a great expansion of the normal minimal crossed cortical innervation of the caudate nucleus, which is too weak to show in Figure 13.18A. A second and related suggestion is that there is already a strong crossed corticocaudate innervation in the fetus that normally atrophies but that survives when the competition from the ipsilateral cortex is removed by resection. A third possibility is that the normally existing callosal projection to the dorsolateral prefrontal cortex, having lost its target neurons, is attracted to the caudate nucleus denuded of its normal innervation, as can be imagined in Figure 13.18B. The second explanation can now be rejected in view of the success of the postnatal resection. The other two explanations are in accord with principles of regeneration already proposed following operative procedures on the septal nucleus, the red nucleus, and the hippocam-

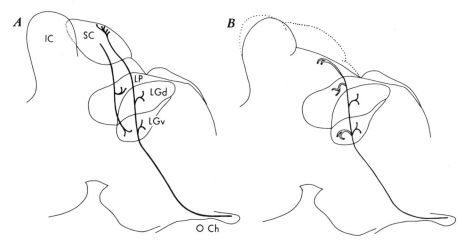

Figure 13.19. Formation of anomalous optic tract connections. (*A*) Lateral-view reconstruction of the rostral brainstem of the normal adult hamster. The heavy line depicts schematically the course of a group of optic-tract axons and some of their terminations. The tectothalamic pathway from the superior colliculus (SC) is shown in a similar manner. Abbreviations: (IC) inferior colliculus; (LGd) dorsal part of the lateral geniculate body; (LGv) ventral part of the lateral geniculate body; (LP) lateroposterior nucleus of thalamus; (OCh) optic chiasm. (B) Similar view of the brainstem of an adult hamster that had undergone destruction of the superficial layers of the SC as a neonate. Anomalous optic-tract connections are depicted by double lines. From Schneider (1973).

pus. However, special interest attaches to this regeneration that occurs over considerable distances and even in the young primate.

Superior Colliculus Resection. Schneider (1973, 1981) and Schneider *et al.* (1986) have described elegant investigations on the visual pathway of very young hamsters. In the visual pathway, the fibers decussate completely in the optic chiasma so that the left eye projects entirely to the right geniculate body and superior colliculus and vice versa for the right eye. The distribution of the axonal branches in these nuclei, as viewed from the right side, is shown in Figure 13.19*A*. From the optic chiasma (OCh), the fibers of the optic pathway traverse both components of the lateral geniculate body (LGv and LGd) and terminate in the superior colliculus (SC). When the right SC is removed from day 0 to day 5, the terminals of that part of the visual pathway that project to the SC are severed and there is a reconstitution of connections, as indicated in Figure 13.19*B*. The new growths have been studied histologically and are shown by double lines. Two explanations are offered by Schneider to account for the observed distribution of the new sprouts.

1. As shown in Figure 13.19*A*, the SC sends fibers to the lateroposterior (LP) and LGv nuclei. These will degenerate with removal of the SC, and this degeneration will induce sprouting in the optic fibers that are traversing these nuclei. It will be noted that there is no new formation of sprouts into the LGd nucleus, where there are no degenerating synapses from the SC. This explanation is in line with the explanations for the regenerations in the septal nucleus and the red nucleus as described above.

2. Schneider suggests that there is also another factor concerned in the sprouting of the severed terminals in the optic tract projection to the SC—namely, that these fibers

sprout from their severed ends and try to find neurons on which they can make synapses. This growth gives rise to a quite remarkable new tract that crosses the midline and ends on neurons on the medial side of the intact SC. If the eye on that side is removed, the crossed projection spreads to the whole colliculus because its ingrowth is not then limited by competition from the normal collicular input. Schneider suggests that this excessive sprouting of the severed fibers is related to the pruning effect obtained with plants. Certainly, with the eye intact, the formation of this tract to the contralateral SC cannot be attributed to the fibers growing to vacated synaptic sites. It would seem to be a growth, guided perhaps by glia, but in itself an exploration into new neuronal territory. Moreover, as would be expected, it is maladaptive. It must be pointed out that these results were obtained in very young hamsters, 0–5 days of age, and 14 days is the limiting age (G. E. Schneider, 1981).

Regeneration of the Cerebellothalamic Pathways. The pathways are shown by line drawings from the interpositus nucleus (IP) [see Figure 14.16 (Section 14.4.4)] and dentate (DE) [see Figure 14.17 (Section 14.4.4)] crossing the midline to go to the thalamus. Kawaguchi *et al.* (1979a,b) demonstrated that, after hemicerebellectomy in young kittens, there developed a pathway from the intact cerebellar nucleus (the dentate) to the ipsilateral ventrolateral (VL) thalamic nucleus that is never observed normally. Both electrophysiological and HRP tract-tracing methods demonstrated that the fibers of the intact cerebellar nuclei sprouted to innervate the denervated ipsilateral VL thalamus and that this innervation was powerful enough to evoke responses of the ipsilateral frontal motor cortex, which, however, were smaller than the contralateral. Also smaller than normal were responses in the ipsilateral parietal cortex. This sprouting probably occurred at the decussation, where the intact fibers would be closely admixed with the degenerating fibers, the sprouts then following the degenerated fibers to the VL thalamus, much as with the regenerations described above for the septal nuclei (Figure 13.10), the red nucleus (Figures 13.14 and 13.16), the hippocampus (Figure 13.17), the corticocaudate projection (Figure 13.18), and the optic nerve after superior colliculus resection (Figure 13.19).

More recently, Kawaguchi *et al.* (1981, 1985) have reported an extraordinary regeneration in kittens following complete transection of the decussation of the brachium conjunctivum, which is the pathway from the cerebellar nuclei to the thalamus, red nucleus (see Figures 14.16 and 14.17), inferior olive, and the pontine tegmental nucleus of the contralateral side. Extreme precautions were taken in the initial transection to ensure that it was complete, and the completeness was verified by postmortem histology. In 8 of these 82 transected kittens (ages 0 days to 3 months), there was regeneration across the lesion, the regenerating fibers mostly reinnervating the normal projection areas. Anterograde transport of HRP was used most effectively to trace the early stages of regeneration with growth cones and the eventual long course of the regenerated fibers, which resembled the normal, control course (cf. Figure 13.20A and B). This regeneration produced a dense mass of terminals in the thalamus by 19 days. Correspondingly, synaptic transmission from the cerebellar nuclei to the cerebral cortex was restored as tested electrophysiologically by recording from the frontal motor cortex and the parietal association cortex. A small proportion of the regenerated axons grew aberrantly. Many traveled to the ipsilateral areas corresponding to the normal contralateral distribution. Some of these aberrant fibers are indicated by the small arrow in Figure 13.20B.

This discovery of an effective long regenerating pathway in the brain of young mammals is quite unprecedented. However, there is the problem of why the transected

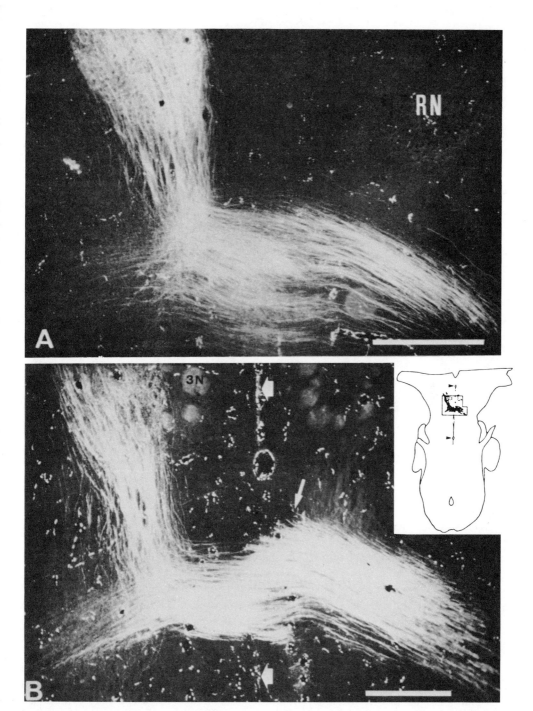

Figure 13.20. (A) Dark-field photomicrographs of the decussation of the superior cerebellar peduncle in a horizontal section of a 6-day-old kitten labeled with HRP injected into the cerebellar nuclei on the right side. (RN) Red nucleus. (B) As in (A), but showing regenerated fibers after complete transection of the decussation of the brachium conjunctivum in a kitten 6 days old and prepared 19 days later. Glial scars in the lesion rostral and caudal to the area of fiber crossing are shown by the thick arrows. The thin arrow indicates some fibers deviating to take an ipsilateral course. *Inset:* The two arrows show the two holes left by ventral extraction of the vertical arms of the cutting device before the histological preparation. (A, B) Scale bar: 500 μm. From Kawaguchi *et al.* (1986).

tracts regenerated, as shown in Figure 13.20B, in only about 10% of the animals. There was at most only marginal regeneration in the other animals. Kawaguchi *et al.* (1986) give evidence that dense gliosis prevented regeneration. Gliosis would seem to be a consequence of hemorrhage in the initial operation. These results can be regarded as opening up a new era in exploring regeneration of the mammalian brain. Hitherto, the evidence was histological or behavioral, but now there is precise labeling and electrophysiology. Even more surprising is the finding of Kawaguchi *et al.* (1982) that there is also regeneration in about 10% of young cats after section of the brachium conjunctivum by the same technique.

13.2.3. New Synaptic Connectivities Developed by Activation

In the preceding sections, the new connectivities developed in the mammalian brain can be regarded as responses to lesions. Here, we describe new synaptic connectivities arising with activation, in the absence of a brain lesion.

Red Nucleus.   A most surprising finding was the change in the cerebral peduncle (CP) synaptic endings on red nucleus (RN) neurons as an eventual consequence of a massive nerve cross-union of the forelimb nerves, the flexors and the extensors of one limb (Tsukahara *et al.*, 1982; Fujito *et al.*, 1982). It was found that the CP input to RN neurons on the contralateral side had been changed, with many synapses proximally on the dendrites, as indicated electrophysiologically (Figure 13.21B–E) and by histological identification [see Figure 13.15C (Section 13.2.2)]. In an attempt to understand this development of CP synapses in the absence of any brain lesion, it is important to recognize that, after the nerve cross-union had been fully established, the animals were using the limb in all manner of voluntary reaching actions and had recovered skills of movement, but the limb could not be normally integrated into walking actions (Tsukahara *et al.*, 1982). It would thus seem that the development of the corticorubral connectivity was related to the relearned skills, whereas the automatic movements of walking were subserved at lower levels of the CNS, where there was no such adaptive synaptic development. There is the enigma of how the relearned voluntary movements of the forelimb are related to the new synaptic growth in the RN whereby the corticorubral pathway is strengthened.

It is convenient to consider here the related phenomenon that classic conditioning is similarly effective in causing the CP input to RN neurons to develop synapses on the proximal dendrites (Tsukahara, 1984). In the experimental preparation, the corticofugal fibers below the RN input were severed and the conditioned stimulus (CS) was a pulse train to the CP, while the unconditioned stimulus (UCS) was applied to the skin of the forepaw and gave a flexion of the forelimb. The CSs (5 pulses at 500 Hz) were applied usually at 100 msec before the UCS. After 120 trials a day for 7 days, there was developed an elbow flexion to the CS alone, which all tests showed to be a classic conditioned response. Intracellular recording from RN neurons revealed that CP stimulation had developed an initial fast EPSP (Figure 13.21F, G, and I) just as in the nerve cross-union. Evidently, there had likewise been a sprouting of the CP fibers with growth to make new synapses on the proximal dendrites of the RN neurons.

In both these examples of synaptic new growth in the absence of any lesion of inputs

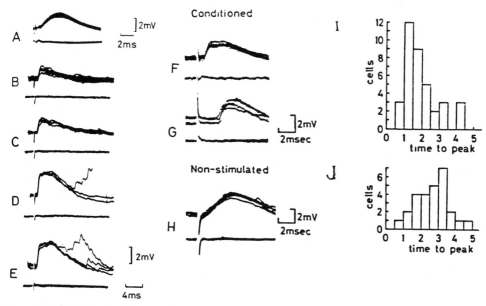

Figure 13.21. (A) Corticorubral EPSPs recorded intracellularly in a red nucleus (RN) neuron of a normal cat in response to a cerebral peduncle (CP) stimulus. (B–E) Similar recordings from a forelimb RN neuron with cross-innervation of forelimb nerves 147 days previously, there being increasing strength of CP stimulation from (B) to (E). The lower traces are extracellular records corresponding to upper traces. From Tsukahara *et al.* (1982). (F, G) Intracellular EPSPs of an RN neuron after establishment of a full conditioned reflex as described in the text, the stimuli being applied to the CP (F) and to the sensorimotor cortex (G). (H) stimulation of the CP in a cat with no conditioning, but with chronic section of the CP below the RN as was also done for (F) and (G). (I, J) Histograms for time-to-peak of the EPSPs in a large series of CP stimulations as in (F) and (H), respectively. From Tsukahara (1984).

to the RN, it has to be recognized that the corticorubral pathway showed itself capable of new growth designed to increase its effectiveness. Presumably, this is a consequence of the excess corticorubral activation, which in this way becomes more effective in driving the rubrospinal input to motoneurons and so to muscle contractions (cf. Section 16.3).

Hippocampus. A related phenomenon has been reported by K. S. Lee *et al.* (1980) in experiments designed to set up long-term potentiation in the hippocampus either *in vivo* or *in vitro* (cf. Chapter 16). A brief burst of high-frequency stimulation was applied to the Schaffer collaterals projecting to the CA1 pyramidal cells (100 Hz for 0.5 sec or 1.0 sec). About 10 min later, when the full long-term potentiation was developed [cf. Figure 16.3 (Section 16.3.1)], the tissue was prepared for electron microscopy. There was observed to be a significant increase of about 40–50% ($P < 0.02$) in the number of synapses on the dendritic shafts that could not be attributed to conversion of existing spine synapses to a dendritic location. Apparently, the activation had caused a new production of synapses, much as reported above for the red nucleus, but in an amazingly short time.

13.2.4. Regressive Events in Neurogenesis

For a general reference, the reader is referred to Cowan *et al.* (1984).

The heading of this section, which is taken from a recent review, sounds somewhat paradoxical. However, neurogenesis does not denote just neural growth, but rather selective neural growth. A simple analogy would be that the initial phase of neuron multiplication with the establishment of neural connectivities provides the crude structure that has to be fashioned by regression, either neuron death or fiber death, just as a sculptor fashions a block of marble by chiseling away to create the desired form. As stated by Cowan *et al.* (1984), regression by neuron death is now recognized as determining much of the refined pattern of neuronal connectivities. Neurons are created from germinal cells, as in Figure 13.2 (Section 13.1.2), by differentiating mitoses and thereby lose mitotic competency. It is therefore a "wise" biological strategy to create neurons in excess and discard the superfluous. The regression to about 50% in the optic nerve fibers of birds (Rager, 1980a,b) and monkeys (Rakic and Riley, 1983) was described in Section 13.2.1. The problem to be discussed herein is: What biological processes select neuronal deaths to achieve the exquisite neuronal designs of the brain? Several well-studied examples in mammalian brain will be presented before an attempt is made to formulate principles.

Climbing Fiber to Purkinje Cells.   The climbing fiber (CF) innervation of cerebellar Purkinje cells is extraordinarily simple numerically. In the adult mammal, it was discovered by Ramón y Cajal (1911) that every Purkinje cell is innervated by a single CF that twines around the branching dendrites to make a large number of synapses. In an early physiological study (Eccles *et al.*, 1966), it was found that as a rare exception, 2 cases in over 100, there was innervation by two CFs. A remarkable discovery (Crepel *et al.*, 1981; Mariani, 1983) was that in the neonatal rat at days 7–9, the initial CF innervation averaged 3.8 for all Purkinje cells, there being regression to the adult level of unity by day 15. It should be mentioned that the CF is the axon of a neuron of the inferior olivary nucleus [IO in Figure 14.17 (Section 14.4.4)]. A single axon branches to innervate about 15 Purkinje cells in the adult (M. Ito, 1984).

An important discovery was that severe depletion of the granule cell–parallel fiber system, which is the main synaptic input to a Purkinje cell, resulted in a continuation of the multiple CF innervation (average about 2.5) of most Purkinje cells (Crepel *et al.*, 1981; Mariani, 1983), rather than the normal pruning.

This depletion of granule cells could be produced by X-radiation just after birth, or it could be produced by a genetic defect in weaver- and reeler-type mice as reported by Mariani (1983). Hence, it has been proposed that the initial excess CF innervation of Purkinje cells in the first few days after birth undergoes regression only when there is the normal immense development of PF synapses on the Purkinje cell—up to 100,000 on each cell. However, since the initial phase of CF regression from day 6 to 8 occurs when granule cells are almost absent and also before there are, normally, any PF synapses [see Figure 13.8 (Section 13.1.3)], a simple competition of CFs and PFs for synaptic space is probably an untenable concept. We should concentrate on the extraordinary numerical fact that all adult Purkinje cells have innervation by just one CF, discounting the exceptional two of the initial investigation. Such a unitary relationship could come about only by a close interaction between the Purkinje cell and the CFs, and this before impulses are generated. The cell must give an intense trophic signal to one CF and its neuron in the inferior olive. Possibly the "unitary" choice is made when there is translocation of the initial CF synapses on the Purkinje cell soma to the dendrites (cf. Figure 13.8). When the Purkinje cell has lost most of its normal complement of PF synapses, it is not surprising that it continues to accept two or three CF synapses.

The postulate of an intense trophic action from the Purkinje cell to the inferior olive is in accord with the inferior olivary degeneration that occurs after ablation of the "corresponding" part of the cerebellar cortex. Brodal (1954) gives an account of his meticulous mapping of inferior olive projections to the cerebellum. For example, a cerebellar lesion in a 10-day kitten results, within 5 days, in complete degeneration of all neurons in a circumscribed zone of the inferior olive with replacement by glia. In adults, the retrograde changes are slower. Dow and Moruzzi [1958 (pp. 446–462)] have collected the numerous clinical reports of inferior olivary degeneration in cerebellar diseases. In Section 1.2.5, there is an account of the retrograde degeneration of neurons following section of their axons (chromatolysis). In Section 13.3.3, there is convincing evidence that this degeneration results from the loss of retrograde transport of specific substances (retrophins) from the synapses on the target organ. Such a strong trophic signal would provide an explanation of the close bonding of the Purkinje cell to its single CF (cf. Section 13.1.3). When there is massive destruction of the inferior olive, the axons of the few remaining neurons sprout to provide additional CFs to denuded cells, and so increase by 5–10 times the number of Purkinje cells innervated by a single inferior olive neuron (Benedetti et al., 1983). We can assume that the inferior olive neurons are able to utilize the increased retrograde transport of retrophins. However, there is as yet no physiological study of these surviving olivary neurons with their presumed surfeit of retrophins, nor is there any study of the adult olivary neurons that survive at least temporarily despite their complete loss of retrophins.

Plasticity of Ocular Dominance Columns. The elegant experiments of Wiesel and Hubel (1963) on kittens gave the first illustration of the complex factors concerned in the formation of neuronal connectivities in the mammalian cerebral cortex. It is some days after birth before kittens first open their eyes; nevertheless, at birth, before usage, the retina is connected with detailed topography to the visual cortex in the adult pattern—with ocular dominance columns as indicated diagrammatically in Figure 17.7 (Section 17.2.4). It has already been seen in Figure 13.4 (Section 13.1.2) that one eye projected diffusely to lamina 4 of area 17 of a monkey fetus at embryonic day 124. However, just before birth (embryonic day 144), there was already evidence of the adult pattern (Adult) of ocular dominance columns, the eyes projecting to alternate columns of lamina 4 to form a pattern that is zebra-like when it is seen from the surface (Hubel et al., 1977). The eyes project to alternate stripes in Brodmann area 17 to form this zebra-like pattern. This pattern is stabilized after birth by normal usage; if both eyes are kept closed for some weeks, the pattern becomes disorganized.

It was first discovered by Wiesel and Hubel (1963) that, if one eye is tightly closed by lid suture shortly after birth, then the open eye comes strongly to dominate almost all the cells of area 17. The zebra-like stripes belonging to the closed eye become progressively narrowed, but the repeat distance remains unaltered. If there is reversal of the open and closed eyes (reverse suturing) within a few days, the ocular columns show reversal of the eye dominance, and this is permanent (Garey et al., 1979).

In Figure 13.22, the dominance of the opened eye after 24 days is seen in the left

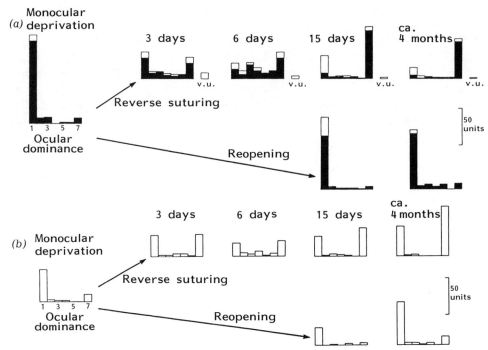

Figure 13.22. Histograms of the incidence of cells in the seven ocular dominance groups, showing pooled data for each animal. (*a*) Results for "nonlayer IVc" cells. (*b*) Results for nonoriented cells recorded within the defined boundaries of layer IVc. Since all recordings were from the right hemisphere, the left eye was always contralateral (thus, group 1 cells were monocularly driven by that eye). (■) Orientation-selective neurons; (□) nonoriented neurons; (v.u.) visually unresponsive cells. The vertical calibrations at right indicate the scale for all histograms. Each histogram on the extreme left is for the control animal, monocularly deprived in the right eye until 24 days. In (*a*) and (*b*), the top row shows, from left to right, data for 4 monkeys, all of which were reverse-sutured, for the lengths of time indicated; the bottom row shows results for 2 monkeys, one of which had had the right eye reopened for 15 days and the other for almost 4 months. From Blakemore *et al.* (1981).

histograms of (*a*) and (*b*). The convention of numbering 1–7 shows the population of cells according to their responses to inputs from one or the other eye. There is a strong dominance of the opened eye for cells of both laminae non-4C (*a*) and 4C (*b*). After 3 days of reverse suturing, there was approximate balance, and by 15 days, the ocular dominance had been fully reversed in (*a*) but less so in (*b*). There was stability up to 4 months. In an alternative experiment, the sutured eyelid was opened at 27 days, but with both eyes open, the previously closed eye failed to recover its quota of cells, even at 15 days or 4 months later.

The restriction of innervation by the closed eye was observed by four independent methods (Hubel *et al.*, 1977): physiological recording, transneuronal autoradiography, Nauta degeneration, and a reduced-silver stain for normal fibers. There must be retraction of the terminals of the geniculocortical fibers of the closed eye and expansion of the active fibers from the open eye into the vacated space, much as is observed in several other neuronal structures [see Figures 13.10, 13.14, 13.16, 13.17, and 13.19 (Section 13.2.2)]. It can be proposed that the normal zebra-like pattern is due to a competition for synaptic space with some collusion to give the stripe arrangement with an approximately standard repeat distance. The eye closure results in regression of the cortical terminals of the inactive geniculocortical fibers.

Further investigation has revealed that mere activity of the ocular cortical pathway is not enough to secure dominance. For example, Singer (1985) finds that dominance results only when the animal pays attention to the visual signals and uses them for control of behavior. There is now much evidence that diencephalic central core structures which provide an ascending arousal system are necessary for the ocular geniculate excitation to achieve dominance. In the anesthetized kitten, repetitive electrical stimulation was applied either to the mesencephalic reticular formation or to the intralaminar thalamic nucleus to provide a conditioning input into the visual cortex. Under such conditioning, significant modifications of ocular dominance were observed after 8–12 hr of monocular light stimulation.

Singer (1985) has developed the hypothesis that the conditioning stimulation is effective by conjunction with the visual input onto the same cortical neurons. It is proposed that there is an analogy with the cerebellar cortex investigations of M. Ito et al. (1982), in which the intense climbing fiber depolarization results in the opening of Ca^{2+} channels on the Purkinje cell dendrites. The increase in intracellular calcium is believed to cause long-term depression of the parallel fiber synapses [see Figure 16.17 (Section 16.7.2)]. This hypothesis has been tested by a systematic study of extracellular Ca^{2+} (H. Geiger and Singer, 1985) during the visual and conditioning stimulations alone and in conjunction. It was found that there was a remarkable collusive effect, both in the discharges from a visual cell and in the decline of extracellular Ca^{2+} in its environment. Usually, neither input alone caused a Ca^{2+} decrease, but together, at the right time sequence in which the conditioning led the visual by about 200 msec, a Ca^{2+} decrease was observed. With repetition of the conjunctive stimulation, the decrease in Ca^{2+} became larger and larger, and at the same time, there was an enhanced visual cell discharge. This potentiated effect lasted for hours after cessation of the stimulation.

In conclusion, there is good evidence that conjunctive stimulation is necessary for the development of the ocular dominance, and it is an attractive hypothesis that this effect is due to the increase in intracellular Ca^{2+}. The reader will find in Section 16.3.1 suggestive ideas about the manner of operation of the increased intracellular Ca^{2+} in provoking plastic responses (cf. Eccles, 1983).

Retinocollicular Projection.   For a general reference the reader is referred to Cowan et al. (1984).

In the first 2 weeks of the postnatal life of rats, more than 60% of the retinal ganglion cells and optic nerve fibers degenerate. At birth, there is a predominance of chiasmal decussation for the retinocollicular projection, but still a general distribution also to the ipsilateral colliculus. By 2 weeks, this ipsilateral projection is greatly reduced and is restricted to the rostromedial zone of the colliculus. By dye injection techniques, it was shown that death of these aberrant ipsilateral fibers was about 90%, which is several times higher than for the contralateral projection. Thus, it can be proposed that cell death is a program for reducing aberrancy. An interesting explanatory hypothesis has been formulated by Cowan et al. (1984) and is diagrammed in Figure 13.23, which shows two closely associated normal fibers and one aberrant fiber. The closely associated fibers, often firing in unison, are likely to be associated in evoking impulse discharges in the

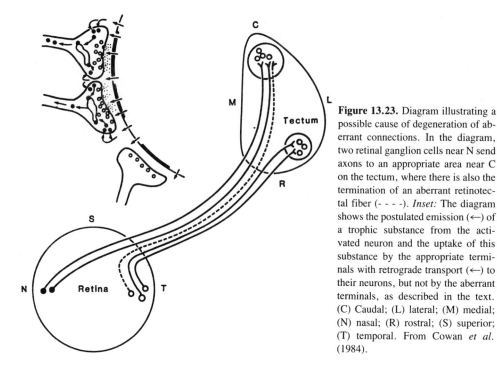

Figure 13.23. Diagram illustrating a possible cause of degeneration of aberrant connections. In the diagram, two retinal ganglion cells near N send axons to an appropriate area near C on the tectum, where there is also the termination of an aberrant retinotectal fiber (- - - -). *Inset:* The diagram shows the postulated emission (←) of a trophic substance from the activated neuron and the uptake of this substance by the appropriate terminals with retrograde transport (←) to their neurons, but not by the aberrant terminals, as described in the text. (C) Caudal; (L) lateral; (M) medial; (N) nasal; (R) rostral; (S) superior; (T) temporal. From Cowan *et al.* (1984).

target neuron. If these impulse discharges cause the liberation of a trophic substance into the synaptic clefts and if the exocytotic release of neurotransmitters is related to the endocytotic uptake of trophic substances, the normal fibers will be favorably placed for uptake of trophic substances, whereas the firing of the aberrant fiber will not be timed so as to benefit effectively from the trophic output, hence its poor survival chance. This hypothesis may be applicable to other examples of regression.

Commissural Projections in the Parietal Lobe of the Rat Brain. For a general reference, the reader is referred to Cowan *et al.* (1984).

These projections have been studied by the injection of dyes that are retrogradely transported. For a few days postnatally, there was a uniform commissural projection from the parietal cortex of one side to the other. After the 2nd postnatal week, dye injection showed that the commissural projection was virtually restricted to a few distinct patches. However, if the dye was injected early postnatally and the animal was allowed to survive for some weeks, the labeled cells were uniformly distributed in the contralateral hemisphere (O'Leary *et al.,* 1981). Thus, the developed patchiness was due, not to the death of cells with commisural axons, but rather to the death of collateral branches projecting commissurally. The regression of collateral connections may be a common occurrence in the restriction of the connectivity of surviving neurons (Cowan *et al.,* 1984), and it is suggested that both cell death and collateral elimination are attributable to competition for trophic factors that act in the same way as nerve growth factor (NGF) for sympathetic-ganglion cells (Campenot, 1977).

13.2.5. Transplantation

There is an enormous variety of possible transplantation experiments. The brain appears to be immunologically privileged, possibly because of the blood–brain barrier (BBB), so that survival of transplanted neurons occurs despite wide species and functional differences. The simplest type is that in which a selected piece of neural tissue from a young mammal is transplanted into the same tissue of an adult of the same species, i.e., a homologous transplant. Björklund and Stenevi (1984) and Sladek and Gash (1984) provide comprehensive surveys of the wide variety of intracerebral transplantation experiments in the last decade. This work is inspired largely by the hope that it will be possible in this way to substitute for losses in degenerative disease, e.g., to relieve parkinsonism by transplanting dopaminergic neurons (Schultzberg *et al.*, 1984). Transplants of tissue fragments from young animals survive for long periods, and embryonic neurons can survive when transplanted as a dissociated cell suspension. A general finding is that a transplant does better in a denervated region and that it is aided by local neuronal damage making a cavity for its reception. Presumably, these conditions provide the essential neuronotrophic substances that are considered in Section 13.5. Some special examples of transplantation will now be considered.

Monoaminergic and Cholinergic Transplants into the Hippocampus. Björklund and Stenevi (1979) have concentrated on these transplants because they can be easily monitored biochemically. Transplants into the hippocampus were from 16 to 17-day-old rat fetuses and were from the septal diagonal band (cholinergic), the locus coeruleus (noradrenergic), the raphe region (serotonergic), and the ventral mesencephalon (dopaminergic). It was found that fibers from the transplants grew well and closely mimicked the normal distribution, which is well known in the hippocampus. There is preferential growth into denervated hippocampal regions. It would seem that the new fiber growths are attracted to the filling of vacant fiber pathways and endings, as has been seen, for example, in Figures 13.10, 13.14, and 13.17 (Section 13.2.2). It may be assumed that since the ingrowing nerve terminals have the normal enzyme systems for transmitter manufacture, they are functionally effective. The transplants survive for over 1 year and provide a model for successful transplantation.

Hippocampal Transplants. Raisman and Ebner (1983, 1985) inserted fragments of the hippocampus of 1-day-old rats into the hippocampus of an adult rat. They employed the Timm stain as a specific histological marker for the axons of granule cells (the mossy fibers), which normally make large complex synapses on the basal dendrites of CA3 pyramidal cells [cf. Figure 13.17 and 16.3A (Section 16.3)]. The transplant–host relationships were examined usually at 30–38 days. At the early age of the transplant, very few granule cells would have been formed in the normal process of division of the stem granule cells. This generation occurs normally in the transplant, the axons of the newly formed granule cells projecting as normal mossy fibers to make the large complex synapses on the CA3 but not on the CA1 pyramidal cells. This selectivity was exhibited even if the transplant developed into a disorganized structure with clumps of the different neural components. In respect to the transplant–host relationship, it was surprising that, if the orientation was appropriate, the mossy fibers of the transplant made aberrant complex synapses on the proximal apical dendrites of the CA1 pyramidal cells of the host, but no synapses on the host's CA3 cells. In some cases, the host's mossy fibers made synapses with the CA3 cells of the transplant, but not the CA1. This was observed only if the transplant lacked dentate granule cells.

Raisman and Ebner interpret these findings by postulating that there is a relative preference of mossy fibers for CA3 cells over CA1, but that the transplant mossy fibers cannot form synapses with CA3 cells of the host because all synaptic space of the adult cells was already occupied. This study is of special interest because the Timm stain enables a unique marking of the mossy fibers in all their ramifications. It illustrates, furthermore, the deficient performance of transplants of normal synaptic systems relative to the neuron modulator systems of the preceding section.

Transplantation of Sympathetic Ganglion into Hippocampus. Very special interest attaches to this technique because for the first time there is a clear demonstration that a growth factor is released in the brain in quantities to cause very strong growth of a transplant. An essential prerequisite is that the septohippocampal (cholinergic) pathway be severed. A sympathetic ganglion transplanted into the hippocampus of an adult rat then hypertrophies, both in cell size and in fiber output, much as would occur peripherally under the influence of NGF. Furthermore, if there is axotomy of the locus coeruleus neurons innervating the hippocampus by intraventricular injection of a specific toxic agent, the regeneration of these adrenergic axons is greatly increased. Severing of the septohippocampal pathway can be delayed for as long as 2 months after the transplant and still be effective. It is postulated that the septal pathway normally suppresses the secretion of the neuronotrophic factors by the astroglia. There is evidence that astroglia in the hippocampus secrete neuronotrophic factors for both adrenergic and cholinergic nerve fibers (Varon and Bunge, 1978; Björklund and Stenevi, 1981). The special feature is that this secretion is assumed to be under inhibitory control by the septal input to the hippocampus. It leads to the hope that the poor regenerating ability of the mammalian brain may be attributable, in part, to as yet unknown controls of the secretion of neuronotrophic factors that facilitate regeneration.

Nerve Graft Technique. A special example of transplantation technique has been developed by Aguayo *et al.* (1983). In adult rats, a segment of a large peripheral nerve has been inserted into the brain stem or spinal cord, it being anticipated that the Schwann cells of the degenerating nerve may guide nerve sprouts from injured neurons of the CNS. When tested many weeks later, this invasion was demonstrated. The nerve trunk resembled a normal regenerating nerve with myelinated and unmyelinated nerve fibers ensheathed by Schwann cells. Retrograde labeling with HRP demonstrated that hundreds of neurons of a wide variety of brain stem nuclei or of the spinal cord had sprouted fibers for up to 30 mm along the nerve graft. Unfortunately, if the nerve graft was set up as a bridge, the growth of reentering fibers stopped within 2 mm. There seemed to be a glial block. What is lacking is a specific growth factor analogous to the NGF of the orthosympathetic system. If such a factor can be discovered, then this technique may be of value in restoring transmission in injured segments of the human CNS, particularly in the spinal cord.

Immunological Privilege and Cross-Species Transplantation. Transplantation studies have indicated a host tolerance for donor tissue in the CNS considerably broader than that observed in the periphery, where host rejection occurs in the absence of careful antigenic matching. The extent of brain immunological privilege in terms of strain and species crossover is unknown, but it may be fairly broad. Gash and Sladek (1980) first demonstrated a strain crossover at the functional level by transplanting the anterior hypothalamic area from fetal Wistar–Lewis rats into the third ventricle of Brattleboro rats. The Brattleboro rat is a mutant of the Long–Evans strain that has a congenital lack of vasopressin neurons resulting in severe diabetes insipidus. Following transplantation, affected animals showed a prompt reduction in the amount of water drunk and in the volume of urine excreted, as well as a return to normal of the osmolality of the urine. Grafted vasopressin neurons and their axonal and dendritic projections could be unequivocally identified in the host brain by immunohistochemical staining for vasopressin or vasopressin-specific neurophysin (Section 12.2.2). Thus, the transplants eliminated the polydipsia and polyuria in the same fashion that parenterally administered vasopressin would do. In addition to the survival of vasopressin and oxytocin neurons in such grafts, transplanted catecholamine neurons associated with the tuberoinfundibular system were also identified (Sladek and Gash, 1984).

Perlow *et al.* (1980) extended this principle considerably when they demonstrated that disbursed, cultured bovine adrenal chromaffin cells could be transplanted into the cerebral ventricles of neonatal and adult rats. No sign of rejection was evidenced when rats were sacrificed 2 months later. Chromaffin cells were identified by their strong fluorescent reaction indicating continued production of catecholamines. Transplantation of adrenal medullary cells obtained from adult rats into hosts of the same strain that had had unilateral nigrostriatal lesions induced by 6-hydroxydopamine (6-OHDA) produced some normalization of rotational behavior (Stromberg *et al.*, 1984). Kamo *et al.* (1985) have successfully transplanted human fetal adrenal chromaffin cells that had been maintained in culture for 3 weeks *in vitro* into the neostriatum of adult rats that had been unilaterally lesioned in the nigrostriatal tract with 6-OHDA. No signs of inflammation or rejection were observed up to 1 month later, the implanted area showed strong catecholamine histofluorescence, and rotational behavior was reduced (Kamo, personal communication).

The ability to transplant cultured cells and to have these cells extend over a broad cross-strain and cross-species donor–host spectrum opens up interesting therapeutic possibilities for the future. Cell lines that grow and multiply in culture are known to exist. With suitable addition of neuronotrophic factors, such cells might then be differentiated into specific neuronal lines that no longer divide. Transplantation studies utilizing such cell lines have not yet been reported, but they would be an obvious extension of these early experiments.

Cross-species transplants have also been demonstrated for cholinergic cells (Daniloff *et al.*, 1984). Cell suspensions from the septal region of mouse embryos were transplanted into the hippocampal dentate gyrus of young adult male rats; at the same time, the fornix–fimbrial cholinergic innervation of the hippocampus was lesioned. Choline acetyltransferase (ChAT) activity was partially restored in these rats, and F.H. Gage *et al.* (1984) have indicated that transplants that restore such activity also improve learning behavior in lesioned animals.

13.3. Dependence of Neurons on Trophic Factors

13.3.1. Introduction

This dependence is recognized by the reactions of neurons to lesions that interrupt connections with target organs, which can be either other neurons or muscles. There are two types of interruption: In one, there is severance of the axon of the neuron, which interrupts its efferent path and results in retrograde deprivation; in the other, there is severance of the presynaptic pathways onto the neuron, the so-called anterograde deprivation. Usually, it is not possible surgically to sever completely either the efferent output or the afferent inputs; hence, the experimental investigations are concentrated on neurons that provide the best opportunity for crucial study. However, the wide range of neuronal types that have been explored indicates that the specialized studies herein described can be regarded as providing a paradigm for mammalian neurons in general. The terms for the various neuronal and transneuronal deprivations and degenerations are based on the authoritative review by Cowan (1970).

13.3.2. Primary Retrograde Deprivation and Axonal Section

For general references, the reader is referred to Cragg (1970), Lieberman (1971), and Eccles (1986c).

It has been known for almost a century that section of motor axons leads to striking structural changes in motoneurons (chromatolysis) that have been described in Section 1.2.5 with appropriate references. The biochemical changes with intense protein anabolism can be regarded as being needed for the axon reaction of sprouting from the proximal end of the axon that eventually reinnervates the muscle. For this purpose, the neuron has to generate more than its own mass of axoplasm every day. Such is the biochemical stress on the neuron with nucleolar RNA synthesis, protein synthesis by the dispersed polyribosomes, and synthesis of other macromolecules that the surrounding glia react, the microglia pushing off boutons on the soma and proximal dendrites. This loss of excitatory synapses is compensated for by an increased excitability of the dendrites, which has been much investigated (cf. Eccles, 1986c). The chromatolytic reaction begins within 24 hr of the axon section and reaches a maximum in about 12–30 days. Recovery to normal occurs when reinnervation of the muscle is established in 8 weeks or more.

After section of their axons, sympathetic ganglion cells exhibited typical chromatolytic changes histologically (M. R. Matthews and Raisman, 1972; M. R. Matthews and Nelson, 1975). Intracellular recording showed halving of the EPSPs, matching the reduction of synaptic counts, and, in addition, evidence of increased dendritic excitability. It was of special interest that application of NGF (see Section 13.4) to the chromatolyzed ganglion greatly reduced the chromatolytic changes (Nja and Purves, 1978). This finding is in accord with the hypothesis of I. A. Hendry et al. (1974) and of Stockel et al. (1976) that the chromatolysis may be due to the cutting off of the normal retrograde transport of the trophic substance, NGF, from the target organ. In corroboration of the hypothesis, it was found that treating an intact sympathetic ganglion with antiserum to NGF produced a chromatolytic response.

Section of sensory nerves results in chromatolysis of dorsal root ganglion cells, but section of dorsal roots is without effect. Evidently, there is a retrograde flow of trophic substance from the peripherally directed axonal branch, but not from the central branch. Yet axonal transport occurs well along both branches (see Section 1.2.4) (Ochs, 1972). Section of axons of parasympathetic ganglion cells also evokes a chromatolytic reaction.

There has been no concentrated study of the CNS matching that of motoneurons, but the histological features of chromatolysis have been observed in a wide range of central neurons: pyramidal tract neurons, red nucleus neurons of the rubrospinal tract, reticular neurons of the reticulospinal tract, retinal ganglion cells, and thalamocortical neurons (Lieberman, 1971).

13.3.3. Hypothesis of Chromatolysis

It is proposed to formulate a general hypothesis based on the experimental evidence that chromatolysis of cells following axotomy is due to deprivation of a trophic factor that normally is retrogradely transported. It would be predicted that application to axons of substances that selectively depress axonal transport would cause chromatolytic changes. Such an effect was reported by Cull (1975) for hypoglossal motoneurons following 2 weeks of application of colchicine or vinblastin to the hypoglossal nerve of rats. Purves (1976) also was able to induce chromatolysis of sympathetic ganglion cells by blocking axon transport by colchicine.

Unfortunately, there is still a dearth of chemically identified candidate substances for this postulated retrograde trophic influence, even for motoneurons. I. A. Hendry *et al.* (1983) have proposed that the generic name ''retrophins'' be applied to all retrogradely acting substances. They have extracted from bovine heart a substance (not yet fully purified) that may be a retrophin for cholinergic parasympathetic ganglia (see Section 13.4.2).

13.3.4. Secondary and Tertiary Retrograde Transneuronal Degeneration

Cowan (1970) describes investigations that confirm findings from the last century that ablation of the limbic cortex (LiC) of very young rabbits results in degeneration, not only of the anterior thalamic nuclei (ATN), which would be a chromatolysis due to severance of the thalamocortical axons (see Section 13.3.2), but also of the medial mammillary nucleus (MMN) and the ventral tegmental nucleus (VTN). Between these nuclei, there is an almost exclusive chain of neuronal connectivity:

$$LiC \leftarrow ATN \leftarrow MMN \leftarrow VTN$$

With LiC ablation, there would be a primary chromatolysis of ATN neurons (Section 13.3.2), and this must affect the synaptic terminals of MMN neurons, which could lose their postsynaptic contacts in the manner described for chromatolyzed motoneurons and sympathetic ganglion cells (Section 13.3.2). Consequently, the MMN neurons would suffer chromatolytic degeneration from the failure of retrograde transport. This is a secondary transneuronal degeneration. As a further consequence, the synaptic terminals of VTN neurons on MMN neurons similarly suffer, with a consequent failure of retrograde transport and VTN chromatolysis, a tertiary retrograde transneuronal degeneration.

With limbic ablation, particularly of the cingulate gyrus, in newborn rabbits (day 0), there were severe retrograde degenerations with almost total neuronal loss in the MMN and VTN (Bleier, 1969). With increasing age at the time of the ablation, the degeneration was progressively less severe. Thus, with excision at days 7–15, the degeneration was milder, and by day 20, there were only occasional degenerated neurons. Cowan (1970) reported that similar retrograde transneuronal degenerations had been observed after limbic ablations in cats and monkeys.

As described above, this sequential retrograde transneuronal degeneration can be explained by an extension of the hypothesis for primary retrograde degeneration (see Section 13.3.3), i.e., by a loss of necessary trophic substances that normally are provided for a neuron by its target organ. In this way, there is a chain of retrograde reactions; ATN → MMN → VTN.

A comparable chain of retrograde transneuronal degeneration occurs after ablation of the visual cortex. Primary retrograde degeneration of the lateral geniculate cells leads to a secondary retrograde transneuronal degeneration of the retinal ganglion cells. As reported by Cowan (1970), Cowan and Hart made a precise study of part of the aforedescribed retrograde neuronal chain by observing the ventral tegmental nucleus after excision of the anterior thalamus. With a lesion in a 5-day rat, the cell loss increased from zero at 1 week to more than 20% at 2 weeks and to more than 35% by 7 weeks. The cell losses were greater the earlier the ablation and the longer the survival period.

An outstanding problem concerns the rapidly developing resistance to retrograde degeneration with age. As suggested by Bleier (1969), it may be attributable to the progressive growth of collateral branches with age. In a quantitative study of this suggestion, Fry and Cowan (1972) utilized very effectively the anatomical connections of the lateral mammillary nucleus (LMN) of adult cats. The axons of the principal mammillary tract (PMT) from this nucleus almost all bifurcate to form, on the one hand, the mammillothalamic tract (MThT) to the anterodorsal thalamus and, on the other hand, the mammillotegmental tract (MTgT). The effects of sections of one or other of these tracts on the neurons of the LMN were observed 9 months later. There was no effect after section of the MTgT and a small atrophy (about 15%) after section of the MThT. After the combined section, there was a severe atrophy; about 60% of the LMN cells died. As would be expected, there was also 60% loss after PMT section. On the basis of the retrograde trophic flow hypothesis (Section 13.3.3), it can be conjectured that the trophic flow into the LMN neurons was severely depleted by complete axon severance resulting from section of the PMT or of the MTgT + MThT. However, the trophic flow from one set of axon branches was adequate for the MThT set and almost adequate for the MTgT set.

13.3.5. Anterograde Transneuronal Deprivation

Anterograde Transneuronal Transport of Molecules. Grafstein (1971) first demonstrated the possible transneuronal transport of labeled macromolecules from the injection site in the retina across the synapses in the lateral geniculate body and so to the visual cortex. This work has been dramatically confirmed by Wiesel *et al.* (1974), who have shown by autoradiography that there is a highly specific transport from the injected eye to the zones of the visual cortex that receive the visual input from that eye. Thus, the path is selectively via the synapses of the lateral geniculate body [cf. Figure 17.7 (Section

17.2.4)] and not diffusely by nonspecific channels. Transneuronal transport has been demonstrated in several other sites, and Cowan (1970) and Graybiel (1975) envisage the possibility that transsynaptic transport of macromolecules may be a generalized phenomenon in the brain. We now report several examples of trophic deprivation.

Mammalian Visual System.   As reported by Cowan (1970), after enucleation of the eye of a monkey, the lateral geniculate neurons displayed two phases of atrophy. With the first, from 4 days to 2 weeks, there was a 25% shrinkage. With the second, which was a very slowly developing atrophic phase lasting for months, there was a 40% shrinkage, but there was little or no cell death. The prolonged atrophy may be correlated with the finding that there was a reduction of RNA to below half by the 5th day and thereafter to a maintained level as low as 30% of normal. The lateral geniculate nucleus of cats and rabbits had much less cell shrinkage after eye enucleation.

With rodents, eye enucleation resulted in severe atrophy of the superior colliculus in young animals with associated cell death, the effects being greatest with day 0 enucleation, less at day 13 and slight in adults. However, in the superior colliculus of carnivores and primates no effects were observed after eye enucleation.

Eye enucleation also resulted in atrophic changes in the visual cortex with loss of dendritic spines, but such changes could result from functional disuse and not from loss of trophic substances.

Other Sensory Systems.   For a general reference, the reader is referred to Cowan (1970).

Destruction of the cochlea resulted in degeneration comparable with that in the visual system. The atrophy of the cells of the ventral cochlear nucleus (20–30%) resembled that of the lateral geniculate nucleus and had a similar time–course. There was a similar secondary anterograde atrophy in the lateral superior olive and the medial trapezoid nucleus. However, there was no evidence of cell death in the auditory system.

In the olfactory system, there was comparable shrinkage of cells in the relay systems after excision of olfactory epithelium. After section of the olfactory tract, there was shrinkage of neurons of the prepyriform cortex with dendritic atrophy.

Other Neuronal Systems.   The most severe anterograde degenerations have been observed after lesions of the inputs to the inferior olivary nucleus and the pontine nuclei (Torvik, 1956). Within 4 days after lesions of the afferent inputs in young kittens, there was shrinkage of the cells that led in many cases to severe degenerative changes and cell death. The degenerative changes resembled those that follow axonal section; yet, at least in the case of the pontine nucleus, the lesions were so remote that axonal section of pontine cells could not have occurred.

A remarkable anterograde trophic influence was observed by Kawaguchi *et al.* (1975). When the inferior olive on one side was killed by thermocoagulation in kittens a few days old, there was degeneration of the climbing fibers to the Purkinje cells of the opposite side. These cells deprived of climbing fiber innervation failed to grow the elaborate branching dendritic tree [cf. Figure 14.14 (Section 14.4.2)] that provides optimum synaptic sites for the parallel fibers. Instead, the Purkinje cells had a single main dendrite running up through the molecular layer. The excitatory action of a mossy fiber input was correspondingly reduced. Bradley and Berry (1976) report quantitative investigations on Purkinje cells of rats. The climbing fiber input was destroyed by operation (cutting of the olivocerebellar decussation) in rats 3 days old. When examined at 30 days postpartum, the branching of the dendritic tree was found to be reduced to half, as also was the total length of the dendrites and the synaptic sites for parallel fibers. Thus, at this developmental level, there is good evidence of a trophic influence of the climbing fiber innervation on the receptivity for the parallel fiber input to the Purkinje cell. In Section 16.7.2, we shall see that this trophic influence has a further significance because it probably provides the basic mechanism concerned in the synaptic modifications that subserve learning.

General Observations on Anterograde Transneuronal Degeneration. For a general reference, the reader is referred to Cowan (1970).

As with retrograde degenerations, the younger the animal the more severe the degeneration, but the reason for this is still debatable. There are many other factors concerned in the degree of degeneration: the animal species, the location of the nuclei on the afferent pathway in both the visual and auditory systems, the completeness of the deafferentation, and the duration of the deprivation. It is generally believed that two main deprivation factors are concerned in anterograde degeneration: (1) functional disuse and (2) loss of an anterograde trophic influence from the presynaptic terminals. These influences have been recognized for motor nerve terminals to muscle fibers, but the relative significance of the two deprivations is controversial.

13.4. Nerve Growth Factor, Other Peptide Neuronotrophic Factors, and Gangliosides

13.4.1. Nerve Growth Factor

For general references, the reader is referred to Levi-Montalcini *et al.* (1980) and Perez-Polo (1985).

There is romance in the story of the simple way in which Levi-Montalcini discovered NGF 30 years ago and the tremendous significance that has been revealed in the ever-expanding stream of experiments right up to the present (see P. L. McGeer *et al.*, 1978a). NGF is the paradigm for other nerve growth factors, the neuronotrophic factors (NTFs), which, in contrast to NGF, have a strangely elusive existence.

NGF is a polypeptide that stimulates growth of sympathetic and dorsal root ganglia. It has been extensively characterized both structurally and physiologically. Snake venom and mouse salivary gland are the historic sources of material.

In the mouse, a large precursor molecule is synthesized from which the active β-subunit is derived. This consists of an N-terminal octapeptide followed by a defined 118-amino-acid sequence. Two units are held together to form a dimer of 26.5 kilodaltons; it is not known whether the active form is the mono.ner or the dimer (Calissano *et al.*, 1984).

The human gene for the β-subunit of NGF has been cloned and located on the proximal short arm of chromosome 1. It is hypothesized that it may belong to the insulin family of genes. A partial amino acid sequence for mouse NGF was used to synthesize an oligonucleotide, the code of which corresponded to this amino acid sequence. This oligonucleotide was used as a probe to identify the mouse β-NGF gene and to clone it. Complementary DNA to this mouse β-NGF was then used to hybridize with human chromosomal DNA. This cross-reactivity was possible because of the considerable homology between the human and mouse amino acid sequences (Francke *et al.*, 1983) (see Section 1.5).

By analogy with its action on PC12 cells (see below), it is postulated that NGF action on ganglion cells involves specific receptor sites for intracellular transport so that there can be action on organelles and on the nucleus. Reference was made in Section 13.3.3 to the role of NGF with its trophic influence on sympathetic ganglion cells. The highly restricted nature of this action has become established with the use of pure NGF. There is an amazing specificity in its action. After continuous systemic administration of NGF to rodent embryos, there is great hypertrophy of the sympathetic ganglia, but there is no change in the ciliary ganglion or in the noradrenergic secreting cells of the locus coeruleus. Yet the NGF penetrates into the CNS, because there is a remarkable growth of noradrenergic fibers from the paravertebral sympathetic ganglia into the spinal cord via the dorsal roots and up the spinal cord and brain stem as a prominent tract that, however, does not make any connection with the locus coeruleus. These abnormal growths atrophy on cessation of the NGF therapy. The NGF has no action on any parasympathetic ganglia, but it is effective in causing neurite growth on some nerve cells in tissue culture, dorsal root ganglion cells, and a cell line derived from a pheochromocytoma tumor referred to as PC12. A. P. Young *et al.* (1983) reported an analytical investigation of NGF on PC12 cells that gave insight into the cytological events produced by the NGF action. In view of the high specificity of NGF action, it is not surprising to find high-affinity receptor sites on the surface membrane, with cyclic AMP possibly playing a role in causing NGF internalization. The subsequent concentration of NGF in vesicles and lysosomes and on the nucleus may help to throw light on the manner in which it exerts its remarkable influence on the cell.

I. A. Hendry and Iversen (1973) proposed that retrograde transport of NGF from sympathetically innervated end organs exerts a trophic influence on sympathetic ganglion cells. Normally, this retrograde transport of NGF occurs by microtubules. Vinblastin destroys immature sympathetic nerve cells by disintegrating these tubules. NGF treatment protects against vinblastin, not only by preserving the microtubules, but also by its characteristic direct effect on the nerve cells causing ganglionic hypertrophy (see Section 13.3.2).

An interesting but unexplained effect of NGF is on central cholinergic systems (Gnahn *et al.*, 1983): Intracerebroventricular administration will produce prominent, dose-dependent increases in ChAT activity in striatal and medial basal forebrain cholinergic neurons (Mobley *et al.*, 1985).

It is appropriate to conclude this section with a quotation from Levi–Montalcini *et al.* (1980):

> Thus these *in vivo* and *in vitro* experiments provided strong evidence for the role played by chemical factors in directing the growing tip of nerve fibers. Since NGF is released in minute amounts from peripheral tissues which receive noradrenergic innervation, one can extrapolate from the above findings and see in this release, the tropic factor which directs the fibers toward their correct destination and provides for their subsistence once these connections are established.

13.4.2. Other Peptide Neuronotrophic Factors

In addition to the classical NGF, there is evidence for other NTFs, but the only one to have been significantly defined is that described by I. A. Hendry *et al.* (1983). By partially purifying a protein extract from bovine heart 1000-fold, they derived a material that exhibited properties resembling those of NGF, but for the parasympathetic system. Investigations have been limited by the very small amounts available. The molecular weight is about 50,000. This NTF selectively preserves parasympathetic ganglia in culture. When it is radiolabeled and injected into the eye, there is selective accumulation in the ipsilateral ciliary ganglion beginning at 4 hr, with a maximum at 18 hr. Autoradiography shows that the labeled NTF is within the ganglion cells, so it can be assumed that there is neurotubule transport of this parasympathetic NTF from nerve terminals in the iris to the ganglion cells.

Perez-Polo (1985) sums up the situation for putative NTFs other than the classical NGF:

> Except for the experiments of [I. A.] Hendry et al. (1983) with a partially purified fraction, little is known about the *in vivo* effects of these various NTFs. However, it is clear that biochemical, cell culture and immunologic tools equal to the task are available and it would seem likely that there will be a large array of NTFs available for *in vivo* experiments likely to provide new answers to developmental questions and perhaps a new understanding of neuronal regeneration.

13.4.3. Gangliosides

For a general reference, the reader is referred to Gorio (1984).

Although much of the attention has been focused on possible peptide NTFs, recent work suggests that other types of compounds may also be important. Gangliosides are sialic acid derivatives synthesized in the Golgi apparatus. They contain glycosphingolipids and are particularly abundant in neuronal membranes.

The initial indication that gangliosides might have a neuronotrophic role came from a study by Purpura and Suzuki (1976) of the neuronal morphology in ganglioside storage diseases in which aberrant secondary neurites and synapses were noted. Roisen *et al.* (1981) next showed that exogenous gangliosides promoted neuritogenesis in cultured neuroblastoma cells. A synergistic effect between NGF and gangliosides in promoting this growth has been seen (Katoh-Semba *et al.*, 1984). Antibodies to one ganglioside (GM_1) will block NGF-induced sprouting of chick embryo dorsal root ganglia (M. Schwartz and Spirman, 1982).

About the time of Purpura and Suzuki's initial work, Ceccarelli *et al.* (1976) reported that gangliosides accelerated the recovery of the cat nictitating membrane denervated pre- or postganglionically. Caccia *et al.* (1979) similarly showed that gangliosides promoted regeneration of denervated peripheral muscle.

It was not anticipated that peripherally administered gangliosides would promote central regeneration before Tettamanti *et al.* (1981) reported radiolabeled GM_1 crossing the BBB. Many studies since then have established that central regeneration involving cholinergic, dopaminergic, noradrenergic, serotonergic, and other neuronal pathways is stimulated by parenteral ganglioside administration (Table 13.1).

The mechanism of this ganglioside neurotrophism is unknown. It has been suggested that gangliosides enhance spontaneous sprouting and that they stabilize damaged neurons, thereby ameliorating the detrimental effects of the lesion.

TABLE 13.1
Ganglioside-Facilitated Recovery from Nerve Injury *in Vivo*

Site of injury	References
Peripheral neuronal systems	
Superior cervical ganglion to nictitating membrane	Ceccarelli *et al.* (1976)
Peroneal nerve to extensor digitorum longus	Caccia *et al.* (1979); Gorio *et al.* (1980)
Tail ventral nerves	Norido *et al.* (1981)
Sciatic nerve	Sparrow and Grafstein (1982); Kleinebeckel (1982); Verghese *et al.* (1982); Gorio *et al.* (1983, 1984)
Superior gluteal nerve to gluteus maximus muscle	Robb and Keynes (1984)
Central nervous system neuronal systems	
Goldfish optic nerve	Grafstein *et al.* (1982)
Entorhinal–hippocampal pathway	Karpiak (1983)
Septohippocampal cholinergic pathway	M. Wojcik *et al.* (1982)
Basal forebrain cortical cholinergic pathway	Casamenti *et al.* (1985)
Nigrostriatal dopaminergic pathway	Toffano *et al.* (1984); Agnati *et al.* (1985); Sabel *et al.* (1984)
Ascending noradrenergic pathways	H. Kojima *et al.* (1984)
Ascending serotonergic pathways	Jonsson *et al.* (1984)

13.5. Neurons and Neuronotrophic Factors

In the last two decades, there has been a transformation in our understanding of the plastic and trophic responses of the mammalian CNS. In contrast to the brains of amphibia and fish, the mammalian brain had been regarded as a structure oriented to degeneration with a negligible capacity for regeneration. This chapter has been devoted to a number of remarkable new discoveries that have altered this view. It would have been expected that brains of mammalian embryos would display regenerative features characteristic of the brains of more primitive vertebrates. To a large extent, the search for regenerative responses is still being conducted on the brains of young mammals as described in this chapter. It was a great surprise to discover that even adult brains exhibit remarkable regenerative capacities.

The vital dependence of a neuron on NTFs is most clearly demonstrated when the retrograde flow from the target organ is cut off by severance of its axon. Normally, a neuron enjoys an excess of NTF above minimal requirements. For example, the neurons of the lateral mammillary nucleus normally have a supply about double their needs (Fry and Cowan, 1972) (see Section 13.3.4). Neurons with a deficiency react by axon sprouting in an attempt to increase the input of NTF. There seems to be a continuous competition among axons for synaptic spaces. It occurs in normal development and appears in regeneration to vacant synaptic sites. However, vacated synaptic sites may not be filled, and alternatively there may be hyperinnervation of neurons that had no vacated synapses. The concept of "growth vigor" of axons (Schneider, 1973) relates to their normal synaptic complement. If this complement is depleted by section of some axon branches, there ensues "growth vigor" with sprouting and the establishment of new synapses that are not replacement synapses. The number of synapses for an axon is usually not at the maximum possible for that axon because of the competitive restraints; hence, it can regenerate to additional synaptic sites as described above.

A simple example of synaptic proliferation was provided by Benedetti *et al.* (1983) in their quantitative study of the growth of climbing fiber innervation to Purkinje cells, when only 1–2% of inferior olive cells survived a chemical poisoning. It was found that after some weeks, the few remaining inferior olive cells had expanded their climbing fiber input to 5–10 times more Purkinje cells than in the control. Complementarily, the survival of a neuron is at risk if its axon fails to innervate a critical minimum number of synapses. The growth vigor of an axon diminishes as its synapses approach the maximum possible. In the competition for "synaptic space," there is advantage in having a single concen-. trated end arbor. Schneider (1981) gives experimental evidence for all these "rules" of synaptic competition, which are mostly derived from his studies of regeneration after lesions of the visual system of the young hamster. It is shown that some of the new connections in the visual system are maladaptive [see Figure 13.19 (Section 13.2.2)], and this could be the case for other regenerations described above, e.g., in Figures 13.10, 13.16, and 13.17 (Section 13.2.2).

Another important result of the NTF economy is that, in the developing animal, neurons with a deficiency of NTF die [cf. Figure 13.23 (Section 13.2.4)], and so aberrant and redundant neurons are eliminated. The complementary effect is that the adequacy of NTF supply ensures the growth of neuronal axons and their conservation.

The other side of the coin is that neurons are generators, in some way not understood, of the NTFs that they donate to the synaptic boutons on their surfaces, thereby ensuring the maintenance of these boutons and the neurons to which they belong. There is evidence that neurons suffer from inadequacy of the special class(es) of NTFs provided by the boutons on their surface. They exhibit atrophy and even death when this anterograde trophic influence is cut off (see Section 13.3.5). However, the evidence for this is not fully satisfactory because the anterograde influence presumably might be due to functional disuse as well as to loss of NTF. The reciprocal exchange of NTFs across synapses for anterograde and retrograde transport is an intriguing problem for future research.

Cotman and Nieto-Sampedro (1984) formulated several generalizations relating to lesion-induced synaptic growth:

First, there is replacement of degenerated synapses with a time course over many days. This replacement is illustrated in Figures 13.10, 13.11, 13.14–13.17, and 13.19 (Section 13.2.2).

Second, there is selective axonal sprouting that is restricted to fiber populations near the degenerated zone; this is also indicated in the aforementioned figures.

Third, regeneration is biased for fibers similar to those degenerated. Figures 13.16 and 13.17 are examples of this selectivity. However, with transplanted hippocampal fragments (see Section 13.2.5), there is merely preference of mossy fibers for CA3 hippocampal pyramidal cells relative to CA1.

Fourth, no examples of permanent hyperinnervation have been reported.

Fifth, the total synaptic surface of the neuron is conserved. An example is Figure 13.19, in which the limited innervation of the contralateral superior colliculus was much increased if the eye projecting to that colliculus was excised.

Sixth, regeneration is generally effective only for very short distances, though regenerating central fibers can travel for millimeters along peripheral nerve grafts (Aguayo *et al.*, 1983). The regenerating neurites must find a continuous unbroken pathway to the desired target through a supporting milieu (Freed *et al.*, 1985). Figure 13.20 (Section 13.2.2) is a striking exception, as also is the innervation of the crossed superior colliculus in Figure 13.19. Moreover, some other regenerations, such as those of Figures 13.16 and 13.18B, would occur across spaces of hundreds of microns.

In Figures 13.15C and 13.21 (Section 3.2.3), the change of synaptic contacts of the corticorubral fibers from remote to more proximal dendrites is not explicable by replacement of degenerating synapses as in Figures 13.14, 13.16, and 13.19. It is proposed that the change results from increased synaptic activity, rather analogous to that reported for movement of stimulated synapses along the dendrites of CA1 pyramidal cells in the hippocampus (see Section 13.2.3).

13.6. Aging and Cell Death

Aging and cell death represent the other end of the spectrum from development. Much needs to be done in defining the particular cellular and chemical losses that occur with aging, but an even more fundamental question that needs to be addressed is why these losses should occur at all. We have been concerned in early parts of this chapter with the building of the brain and its ability to regenerate after injury. We have been less concerned with what maintains it in its dynamic state and what factors are lost when it begins to degenerate. Failure of trophic factors or other regenerative mechanisms may play a major role.

The inability of neurons to divide renders the nervous system the most susceptible of any organ to damage of all kinds. Neurons cannot be replaced, and a progressive dropout during life is suggested by the fact that the brain decreases somewhat in size with age and the ventricles enlarge. Figure 13.24 demonstrates dramatic losses of three types of neurons in three different areas of brain: the cholinergic neurons of the substantia innominata, the dopaminergic neurons of the substantia nigra pars compacta, and the noradrenergic neurons of the locus coeruleus. Each of these cell masses is relatively small, but is responsible for innervating a large and distant area of the brain. The substantia innominata contains 400,000–500,000 cholinergic neurons at birth (P. L. McGeer *et al.*, 1984b) and

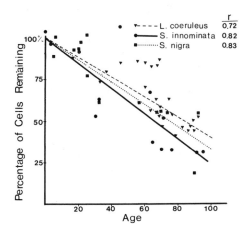

Figure 13.24. Percentage of neurons remaining vs. age in three areas of human brain. Locus coeruleus data from Vijayashankar and Brody (1979) (cf. also P. L. McGeer *et al.*, 1984b); substantia innominata data from P. L. McGeer *et al.* (1984b) substantia nigra data from P. L. McGeer *et al.* (1976).

is the principal source of cholinergic innervation of the neocortex [see Figure 8.5 (Section 8.3.1)]. The substantia nigra pars compacta has a similar complement of dopaminergic neurons that innervate the whole of the neostriatum and, less densely, other regions (see Chapter 9). The locus coeruleus has fewer than one tenth as many neurons (Vijayashankar and Brody, 1979), but provides the principal noradrenergic innervation for both the forebrain and the cerebellum (see Chapter 9).

As Figure 13.24 illustrates, the rate of dropout of cells from these three groups is strikingly similar. By age 70, only about half the initial complement remains. The figure suggests that if the individual lives long enough, all reserve capacity will eventually disappear. The minimum number of cells required to maintain normal function will no longer remain, and decompensation will occur. Disease produced by ''normal'' aging will appear. Groups such as these probably represent the weakest link in brain function. Investigating why there should be preferential degeneration of such cellular groups should be an important objective for future neurological research.

It could be anticipated that the neuronal groups shown in Figure 13.24 are disease-susceptible and that any pathological process that affects them should be sharply age-dependent. Alzheimer's disease affects neurons of the substantia innominata (see Figure 8.5) and locus coeruleus, Parkinson's disease affects neurons of the substantia nigra (see Chapter 14), and parkinsonian dementia affects neurons of all three areas (Nakano and Hirano, 1983, 1984; Whitehouse et al., 1983). These diseases are all sharply age-dependent.

Neuronal cell loss in other areas with aging in the absence of disease is less clear-cut. Brain weight decreases with age, and because the internal volume of the skull is fixed, cerebrospinal fluid (CSF) volume increases proportionately. Many computerized axial tomographic (CAT) studies have been carried out on aging brain to determine this volume change (for a review, see P. L. McGeer, 1986). Takeda and Matsuzawa (1984) examined the CAT scans of 980 subjects and found that the CSF space increased exponentially with age after the 30s. In males, the change was from an average of 39.9 ml in the 50 to 60-year-old group to 124.3 in the 80 to 89-year-old group. J. M. Anderson et al. (1983) found a loss of about 1% per annum of neocortical and hippocampal neurons in a 69- to 95-year-old group. Devaney and Johnson (1980) found the neuronal population of the macular projection area of the visual cortex to drop from 46 million/g at age 20 to 24 million/g at age 80. G. Henderson et al. (1980) examined 11 areas of cortex in individuals between 20 and 90 years of age and found the neuronal count to decrease from 12 to 43% for small neurons and from 36 to 60% for large neurons, depending on the area. While the quantitation in these studies is variable, taken together, they suggest a very significant dropout of cerebral cortical cells with aging.

It is well documented that there is an accumulation with age of pigments (lipofuscin pigments), which may be derived from incomplete degradation of damaged membranes, and an accumulation of abnormal neurofibrils in ''patchy'' localized regions of various parts of the brain, usually starting in the hippocampus. Samorajski and Rolsten (1973) report a fall in RNA and an alteration in the DNA/RNA ratio consistent with neuronal loss or glial proliferation in both aged human and aged monkey brain. It has therefore been postulated that the abnormal neurofibrils result from defective or inadequate translation mechanisms. It has also been hypothesized that the primary difficulty may be the accumulation of pigments that displace RNA and other normal neuronal components.

Another approach to neuronal changes with aging is to measure levels of the neurotransmitters or of their synthetic enzymes as indices of presynaptic neuronal integrity and of binding sites as indices of postsynaptic function. Much remains to be done, but present indications are that many—but not all—neuronal systems show some loss with "normal" aging, although others may even show an increase (E. G. McGeer and P. L. McGeer, 1982). In any consideration of the effect of age or disease, the particular region of the brain is as important as the particular neurotransmitter system. Dopamine systems, for example, seem particularly vulnerable in the striatum, γ-aminobutyric acid systems in thalamic regions and acetylcholine systems in cortical and hippocampal areas (P. L. McGeer and E. G. McGeer, 1976). The imbalances created by the apparent regional- and transmitter-specific nature of the age-related losses may have functional implications as important as those of the declines themselves.

It seems well to emphasize also that not all neurotransmitters and related enzymes show marked declines with age. For example, in the hypothalamus methionine-enkephalin levels are said to increase with age, and serotonin systems do not seem to be particularly affected. Many have postulated that a hypothalamic imbalance between serotonin and the catecholamines, resulting from their different vulnerabilities to the aging process, may be responsible for the hormonal and sleep changes that accompany aging (Kent, 1976; Cotzias et al., 1977; Meites et al., 1979). As D. S. Robinson et al. (1972) showed a number of years ago, enzymes such as monoamine oxidase may increase with age. This seems to be true of many enzymes that have glial as well as neuronal locations.

Losses of neurotransmitter indices are frequently greater than can be accounted for by cell death. They may be partly accounted for by decreased axonal transport of enzymes or other neuronal constituents essential for life of the divergent axons and boutons. In support of this possibility is the fact that losses in many neurotransmitter systems appear to be largest in regions that contain terminal fields of long-axoned neurons with high divergence numbers. There is some evidence from animal studies of a decrease in axonal transport in the septohippocampal system with aging (Geinisman et al., 1977). Degeneration of nerve terminals with consequent loss of neuronal indices may also occur as a secondary phenomenon consequent on degeneration of the postsynaptic dendrites or cell somata; the appearance of plaques and tangles in senile dementia and of atrophic dendrites in aged animals and humans (Geinisman et al., 1978; Scheibel et al., 1976) has been cited as an argument for this mechanism. Whether the initial event be loss of axonal transport, boutons, or cells, the reason for such loss is undetermined. The question may really be: Why do neurons live so much longer than other types of cells? Their life and death may well depend on the presence of absence of trophic factors (see Section 13.3), and research on such factors seems to be of as great importance to the aging problem as it does to development.

13.7. General Conclusions

We can now conjecture that the whole nervous system has communication not only by impulses that signal quickly in the manner described in Chapters 4, 14, 16, and 17, but also chemically by transport of specific proteins and other macromolecules. We can assume that this transport must go on in the most incredibly complex manner among the neurons of the brain, organizing all their interrelationships in a way that we still cannot even imagine, but with the life as well as the development of the neuron depending on it.

Apparently, a neuron does require the trophic transport from the synapses on it in order to stay alive, and this dependence goes on in turn from cell to cell. Every cell, as it were, is talking chemically to all the other cells that it is connected with and instructing them how to talk to the next ones, and so on. So this chemical manner of communication keeps the whole immense organized structure of the brain in some kind of dynamic state that we may call "trophic resonance" by this vast interlocking process of specific chemical communications. Here we are imagining far beyond experimental evidence, but such a vision is of the greatest value if it leads to the formulation of problems that can be experimentally attacked.

In the preceding and later chapters, we have dealt and will deal with elementary levels of brain operation, taking the brain and its constituent neurons as given. At all levels, the responses of the brain have been explained as being due to transmission by nerve impulses. The input from receptor organs via synaptic action and the many sequential synaptic relays lead eventually to the discharge of motoneurons, with the ensuing muscle contractions giving movement. At the higher levels of the brain, the neuronal pathways are of immense complexity, and in principle, it can be conceived that these pathways provide an adequate explanation of even the most complex and subtle human performances. This claim provides the theme for discussion in Chapter 17. Yet many problems remain, even if this program were successful in identifying all brain responses as being due to the operations of what we may call its constituent neuronal machinery.

First, as treated in this chapter, there is the construction of the brain with its detailed neuronal topography and lines of communication. But neurons are not isolated entities. They develop together with their chemical sensing, giving recognition of interconnected neurons, and there are many indications that this specific chemical communication continues throughout life. The neurons of the brain are linked not only by impulse communication but also by special chemical transmission, such as is displayed in the trophic reactions.

Second, there is plasticity in the connections and synaptic action at higher levels in the brain, as we shall see in Chapter 16. It is postulated that synaptic activity leads to their growth by its effect in causing the synthesis of RNA and so of the enzymes building proteins and other macromolecules.

Third, it should be noted that the brain is composed of glial cells as well as neurons. The glia have been relatively neglected in this account of the brain, which has concentrated on the neuronal mechanisms and impulse transmission. The investigations of Kuffler (1966) and his associates (Kuffler and Nicolls, 1976) have established that glia are not concerned in impulse transmission, but function in metabolic relationship with neurons, in guiding their growth, and possibly in limiting ion accumulation in the extracellular spaces. We can think of them as the "housemaids" and "nursemaids" of the brain, being particularly concerned in the transport of materials from the blood vessels to the neurons and vice versa. They fulfill an important role, not only in the nutrition of the brain but also in its protection against poisons by contributing to the so-called "blood-brain barrier" [see Figure 1.19 (Section 1.5.1)]. The growing evidence for chemical specialization of glia also emphasizes their importance in total brain function (see Section 1.3).

Finally, it seems probable that many of the histological and biochemical changes seen in the brain in disease, and even in so-called "normal" aging, are probably related to a failure of the neuronotrophic factors that protect the regenerative and plastic mechanisms of brain cells.

14

Control of Movement by the Brain

14.1. Introduction

In attempting an analysis of movement and its control by the brain, it is immediately evident that there are many hierarchical levels. This was appreciated by Sherrington (1906) in his great book *The Integrative Action of the Nervous System,* where in Chapter 9, "The Physiological Position and Dominance of the Brain," he recognized the simplest reflexes with a superimposition thereon of more and more complex controls at spinal, supraspinal, cerebellar, and cerebral levels. More recently, Granit (1970) has developed the same theme in his book *The Basis of Motor Control,* which in many ways can be considered an updating of Sherrington on the basis of the great advances of knowledge since 1904 when Sherrington delivered his classic lectures. Granit, however, was more concerned with lower levels of the hierarchy (cf. Chapter 4), particularly the spinal and supraspinal controls of automatic movements, and with the influence of inputs from muscle receptors on these controls. In this chapter, these levels will be treated as a base on which are superimposed the controls from higher levels—cerebellar and cerebral.

Our most complex muscle movements are carried out subconsciously and with consummate skill. The more subconscious we are in a golf stroke, the better it is, and the same with tennis, skiing, skating or any other skill. In all these performances, we do not have any appreciation of the complexity of muscle contractions and joint movements. All that we are voluntary conscious of is the general directive given by what we may call our voluntary command system. All the finesse and skill seem naturally and automatically to flow from that. Throughout life, particularly in the earlier years, we are engaged in an incessant learning program for motor performance (see Section 16.7). As a consequence, the brain can carry out all these remarkable tasks that we command it to do in the whole repertoire of our skill and movements.

What we will do in this chapter is describe the main influences on motor control, recognizing that much of our description must be speculative at this stage of our knowledge. It was once thought that a pyramidal system, including the motor cortex and the pyramidal tract fibers descending from it to the spinal cord, controlled voluntary or higher level movements. An extrapyramidal system, including the cerebellum and basal ganglia, was thought to control lower level, or automatic, movements. More recent anatomical and physiological studies make it clear, however, that the pyramidal and extrapyramidal systems are not separable.

The concept of an extrapyramidal system was introduced by the great neurologist S. A. Kinnier Wilson in 1912. Wilson never defined this system, and to this date it remains a vague entity. Wilson used the term to explain the fact that in hepatolenticular degeneration (later called "Wilson's disease"), a motor deficit of major proportions occurs in the absence of damage to the corticospinal tract. Today, the term *extrapyramidal* refers to the aspects of posture and motor activity that seem to be controlled by five interconnected subcortical nuclei that can be included in another loosely defined term, the basal ganglia: the caudate, putamen, globus pallidus, subthalamic nucleus, and substantia nigra. These nuclei are richly connected to some regions of the thalamus, cortex, and pontine reticular area. Cerebellar function is thought of separately. Thus, at the supraspinal level, there are really three systems playing on descending pathways to anterior horn cells in the spinal cord: the cerebral cortex, the basal ganglia, and the cerebellum. The principal descending pathway is the trunk line extending from pyramidal cells in the motor strip down the pyramidal tract to anterior horn cells. In addition, there is a series of auxiliary descending reticulospinal tracts that are interconnected with the three great nuclear systems.

Broad hints as to the function of individual parts of this incredibly complex array of motor influences come from diseases that affect various parts of the system. These influences are summarized in Figure 14.1. Muscular diseases (e.g., muscular dystrophy,

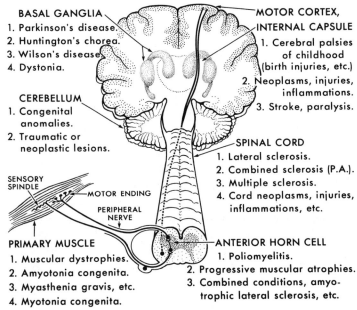

BASAL GANGLIA
1. Parkinson's disease.
2. Huntington's chorea.
3. Wilson's disease.
4. Dystonia.

CEREBELLUM
1. Congenital anomalies.
2. Traumatic or neoplastic lesions.

SENSORY SPINDLE
MOTOR ENDING
PERIPHERAL NERVE

PRIMARY MUSCLE
1. Muscular dystrophies.
2. Amyotonia congenita.
3. Myasthenia gravis, etc.
4. Myotonia congenita.

MOTOR CORTEX, INTERNAL CAPSULE
1. Cerebral palsies of childhood (birth injuries, etc.)
2. Neoplasms, injuries, inflammations.
3. Stroke, paralysis.

SPINAL CORD
1. Lateral sclerosis.
2. Combined sclerosis (P.A.).
3. Multiple sclerosis.
4. Cord neoplasms, injuries, inflammations, etc.

ANTERIOR HORN CELL
1. Poliomyelitis.
2. Progressive muscular atrophies.
3. Combined conditions, amyotrophic lateral sclerosis, etc.

Figure 14.1. Summary of some diseases that produce motor symptoms and the levels at which they attack the nervous system.

amyotonia congenita, myasthenia gravis) or diseases of anterior horn cells (e.g., poliomy-elitis, progressive muscular atrophy, amyotrophic lateral sclerosis) lead to flaccid paral-ysis of the limb. Spinal cord lesions (e.g., combined sclerosis of the cord from pernicious anemia, multiple sclerosis, cord neoplasms) lead to spastic paralysis. Diseases of the motor cortex or any portion of the pyramidal tract (e.g., cerebral palsies in childhood, neoplasms, injuries, multiple sclerosis, inflammations, strokes) also produce spastic pa-ralysis. Lesions of the cerebellum, or diseases of the cerebellum (e.g., congenital anoma-lies, traumatic or neoplastic lesions), produce tremor and uncertain movement, but not paralysis. There is an unsteady gait and a coarse tremor precipitated by movement. There is an inability to estimate the range of voluntary movement that is referred to as "past-pointing." Lesions of the basal ganglia (e.g., Parkinson's disease, Huntington's chorea, Wilson's disease) produce tremors at rest, rather than in motion. There may be difficulty in initiating movement (Parkinson's disease) or an inability to control it (Huntington's chorea).

To summarize, muscle cells, peripheral nerves, and motoneurons must be intact for there to be movement at all. There must be a pyramidal tract if there is to be good voluntary movement. The subcortical motor nuclei are concerned in voluntary movement, while the cerebellum is essential for its smooth execution. We shall now discuss each of these influences in turn, remembering that they all work in concert under normal circum-stances.

14.2. Motor Control from the Spinal Cord and Brain Stem

As already described [cf. Figure 3.1 (Section 3.2.1)], all movements are brought about by contractions induced in muscles by impulses that are discharged by motoneurons. The very simple reciprocal arrangement represented in Figure 4.1A (Sec-tion 4.1) can have functional meaning. When you are standing with slightly bent knees, your weight is stretching the knee extensor muscle (E) and the annulospiral (AS) stretch receptors are firing into the spinal cord, exciting the knee extensor motoneurons to fire impulses so that the extensor muscle contracts and holds your weight. If this muscle contraction is inadequate, the knee gives a little, stretching the extensor muscle more with more firing from its AS receptors [Figure 2.3A (Section 2.1)], giving an increased reflex discharge to the muscle, which in this way is nicely adjusted to give a steady posture. At the same time, the reciprocal inhibitory pathway prevents the antagonist motoneurons (F) from firing to give contraction of the antagonist flexor muscles (F). Such a contraction would oppose the extensors that are engaged in the essential task of supporting weight. This description of the mode of action of the pathways in Figure 4.1A illustrates a simple reflex performance with functional meaning. Superimposed on these simple pathways of Figure 4.1A are the γ-motoneurons with their biasing action on the AS endings as already described [Figure 2.3A (Section 2.1)].

Spinal motoneurons have an immense convergence of excitatory and inhibitory inputs that influence movement brought about from higher centers in ways beyond our present understanding. There was an initial description in Sections 4.7.1 and 4.7.2 of the important pathways from group Ia afferents (see Figure 4.1A) and the recurrent inhibition by Renshaw cells [see Figure 4.15A (Section 4.7.2)]. But the system is much more complex. For example, A. Lundberg [1979b (Figure 1)] diagrams a Ia inhibitory neuron that has many inputs besides the Ia excitatory input and the Renshaw inhibition shown in Figure 4.15A. Besides their propriospinal excitatory and inhibitory inputs, there are excitatory inputs from the pyramidal, vestibulospinal, rubrospinal, bulbospinal (reticulospinal), and tectospinal tracts. Evidently through these systems there is a wealth of convergent inputs to a motoneuron whose discharges are integrated from this diversity of excitatory information together with inputs from inhibitory interneurons. Despite the most sophisticated analysis, particularly by Lundberg and colleagues of the Göteborg school, we are still far from having a clear understanding of the manner in which this immense diversity of input is integrated by the motoneurons in the control of posture, rhythmic (breathing, stepping, running, swimming), automatic, and voluntary movements. Yet these movements are carried out with incredible precision and control.

14.3. Motor Control from the Cerebral Cortex

For a general reference, the reader is referred to Evarts (1984).

14.3.1. Motor Cortex

After this introduction, it is appropriate that we begin with motor control from the motor cortex, because it is preeminent in higher mammals and it is the control that has been most thoroughly investigated. Figure 14.2 shows the position of the left motor cortex as a band across the surface of the cerebral hemisphere. It lies just anterior to the central fissure (the fissure of Rolando), and many of its constituent nerve cells are pyramidal cells [cf. Figure 1.2B (Section 1.2.1)] the axons of which are in the nerve fibers that run down the pyramidal tract (C. G. Phillips and Porter, 1977). The motor cortex is a tremendously important structure, but it is not the prime initiator of a movement, such as a voluntary bending of your finger. It is only the final relay station of what has been going on in widely dispersed areas in the cerebral cortex, subcortical nuclei, and cerebellum, as is partially illustrated in Figure 14.27 (Section 14.6). The pyramidal cells of the motor cortex with their axons passing down the pyramidal tract are important because they provide a direct channel out from the brain to the motoneurons (Figure 14.3) that, in turn, cause the muscle contractions as described in earlier chapters and as illustrated in Figure 3.1C (Section 3.2.1).

When brief stimulating currents are passed through electrodes placed on the surface of the motor cortex, there are contractions of localized groups of muscles. In this way, it was discovered that all the various parts of the body can be shown on a strip map that is represented by the homunculus in Figure 14.3. This was done first with monkeys and anthropoid apes, but the map has now been completely established for humans, particularly by Penfield and his associates (Penfield and Jasper, 1954), because it is often important during brain surgery to locate an area on the motor map, using for this purpose the conventional stimulating technique. Labeled in the inset of Figure 14.3 along the strip of motor cortex are the general areas for toes, foot, leg, thigh, body, shoulder, arm, hand, fingers and thumb, neck, head, face, lips, and tongue, starting in the midline and progressing downward over the outer surface. There is a large representation for hands, fingers, and thumb and an even larger area for face and tongue as illustrated in the homunculus. The motor cortex is not uniformly spread in proportion to muscle size—far

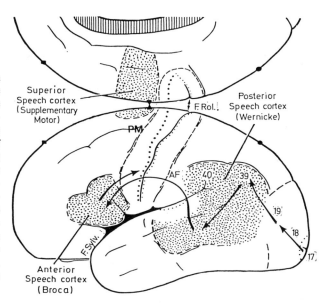

Figure 14.2. Left hemisphere from the lateral side with the frontal lobe to the left. The medial side of the hemisphere is shown as though reflected upward. Abbreviations: (AF) arcuate fasciculus; (F.Rol.) fissure of Rolando or central fissure; (F.Sylv.) fissure of Sylvius. The primary motor cortex is shown in the precentral cortex just anterior to the central sulcus and extending deeply into it. Anterior to the primary motor cortex is the premotor cortex (PM) with the supplementary motor area largely on the medial side of the hemisphere. Modified from Penfield and Roberts (1959).

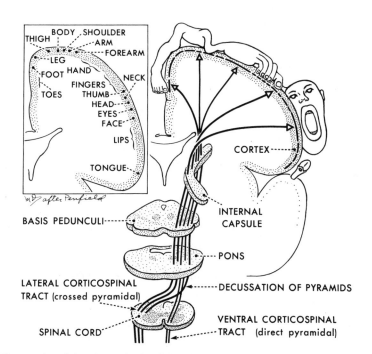

Figure 14.3. Homunculus of the motor strip, localizing the functions of large pyramidal cells. The descending tracts through the internal capsule and brain stem into the spinal cord are also shown. The tracts mostly decussate to descend in the dorsolateral column of the spinal cord on the opposite side. Adapted from Netter (1953).

from it. The muscles that control the thumb have a large representation, but then we use them in so many skilled actions; even more important are the areas for movement of the tongue, lips, and larynx, which are used in all the subtleties of expression in talking and singing. It is skill and finesse that are reflected in the representation.

Figure 14.3 also shows diagrammatically the course of the pyramidal tracts from the motor cortex to the spinal cord. After descending through the brain stem and giving off many branches (C. G. Phillips and Porter 1977), the pyramidal tracts cross or decussate in the medulla and so course down the spinal cord to terminate at various levels, in primates making strong monosynaptic connections on motoneurons. This very direct connection of the motor cortex with motoneurons is of the greatest importance in ensuring that the cerebral cortex, via the motor cortex, can bring about the desired movement very effectively and quickly. Nevertheless, two fundamental problems remain: How can one's willing of a muscle movement set in train neural events that lead to the discharge of pyramidal cells? How do the cerebellum and basal ganglia contribute to the finesse and skill of movement? The first question will be considered scientifically in this chapter and Chapter 16 (Section 16.7) and philosophically in Chapter 17; the second will be considered in the latter parts of this chapter.

14.3.2. Discharge of Motor Pyramidal Cells

The firing of the pyramidal cells in the motor cortex in relation to movement can be studied by a microelectrode implanted in the motor cortex so that impulse discharges from individual pyramidal cells are recorded. This requirement precludes human experimentation. Special reference will be made to the experiments performed by Evarts (1967, 1972, 1984) on monkeys. In an initial operation, the monkey in Figure 14.4A had an electrode implanted in its motor cortex in the right place for recording pyramidal cells concerned in an action it had been trained to do. During an experimental run it is seated comfortably in a cage, and it has to move the control bar from one stop to the other (shown in detail in Figure 14.4A), backward and forward in a time that had to be between 0.4 and 0.7 sec or else it was not rewarded by grape juice. As a variant, the amount of load on the movement could be changed.

In Figure 14.4B, the firing of the pyramidal cell is clearly related to the downward movement signaling flexion. Actually, there are two pyramidal cells: One is very close to the electrode and hence gives large action potentials, and the other is further away and gives smaller responses. We can assume that the size is only a matter of proximity and not of actual cell dimension. The activities of the two cells are nicely correlated with the downward movement. There is no doubt that these particular pyramidal cells are concerned with a particular movement, the flexion and not the extension. But if, as in the lowest trace, the monkey is carrying out some quite different movement, such as moving its shoulder, the two units are no longer correlated in their discharges, but go fast and slow in quite unrelated ways. The two pyramidal cells are correlated only when they are carrying out the specific actions that correspond to their locations on the motor map.

There are many more pyramidal cells in that location that would also be firing in effective relationship to the experimental movement.

This is a very important result, showing just what one would expect in the functional performance of the motor cortex when it is carrying out some specific action. There is a group or colony of pyramidal cells in close proximity (C. G. Phillips and Porter, 1977)

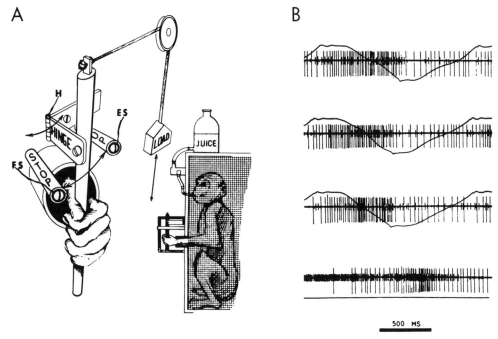

Figure 14.4. Pyramidal cell discharges during movement. From Evarts (1967).

that keep on firing and stopping and firing and stopping during a rhythmically repeated movement. And there would be other colonies specifically related to other movements, and so on for the whole extent of the motor cortex of Figure 14.3. It should be noticed in Figure 14.4B that the two cells are not silent during the reverse phase of the movement, but only slowed. They are, as it were, modulated in their frequency. As mentioned earlier, there is a general tendency for all cells to be firing all the time. Their responses are graded in frequency, coding intensity of action as frequency. Figure 14.4B is a beautiful example of this coding.

14.3.3. Cerebral Cortex Control of the Motor Cortex: Supplementary Motor Area

When we are carrying out some voluntary act, there is postulated to be an effective action of mental events (the intention) on the brain. Until recently, this action was ill-defined in its location, but in the last few years this unsatisfactory situation has been transformed by three independent investigations. In bringing about a voluntary movement, the site of the action of mental events on neural events can now be sharply delimited to a special area of the cerebral cortex that was initially investigated by Penfield and Welch (1951), who named it the *supplementary motor area* (SMA). It lies on the medial surface of each cerebral hemisphere anterior to the motor cortex representing the hindlimb and descends to the cingulate sulcus (Figure 14.2). Until recently, it was neglected because strong electrical stimulation was needed and the evoked movements were imprecise and apparently disorganized, adversive, in comparison with those evoked by the motor

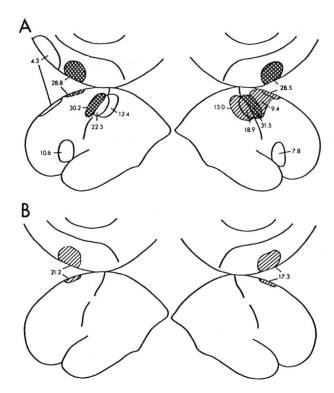

Figure 14.5. (A) Mean increase of the rCBF (%) during the motor sequence test performed with the contralateral hand. Values were corrected for diffuse increase of the blood flow. (▓) Indicates an increase of rCBF significant at the 0.0005 level; (▨) indicates an increase of rCBF significant at the 0.005 level; (□) indicates an increase of rCBF significant at the 0.05 level. *Left:* left hemisphere, 5 subjects. *Right:* right hemisphere, 10 subjects. (B) Mean increase of the rCBF (%) during internal programming of the motor sequence test. Values were corrected for diffuse increase of the blood flow. *Left:* left hemisphere, 3 subjects. *Right:* right hemisphere, 5 subjects. From Roland *et al.* (1980a).

cortex, whereby each movement of each joint is sharply represented in the well-known strip map. But now, the SMA seems cast for the master role in *voluntary movement* because of three converging lines of evidence:

1. The most remarkable studies have been made by Roland *et al.* (1980a,b), using an intracarotid injection of radioactive xenon to signal the regional cerebral blood flow (rCBF) of 254 small regions of the cerebral cortex on one side, so that a detailed cortical map is constructed. It is known that CBF is a reliable index of neuronal activity (cf. Reivich *et al.*, 1975, 1982; Frackowiak *et al.*, 1980b; Lebrun-Grandie *et al.*, 1983). The internal programming test (Roland *et al.*, 1980a) demonstrated that the *voluntary intention* to carry out a complex learned movement, the motor sequence test, specifically activates neurons of the SMA (Figure 14.5B). If the movement is actually carried out, there is a widely dispersed further excitation (Figure 14.5A), not only of the primary motor cortex and the associated sensory cortex, but also of the premotor cortex, the basal ganglia, the thalamus, and the cerebellum. This spreading action can be accounted for on the basis of the rich projections from the SMA (Kunzle, 1978; Wiesendanger *et al.*, 1973), not only to other cortical areas but also to subcortical structures including, for example, the basal ganglia, the cerebellum, and many thalamic nuclei as partly illustrated diagrammatically in Figure 14.17 (Section 14.4.4).

2. Complementary work by Brinkman and Porter (1979) has indicated that in the monkey, self-paced voluntary movements are initiated by increasing the firing rate of neurons in the SMA on both sides (Figure 14.6A and C). Some neurons appear to begin

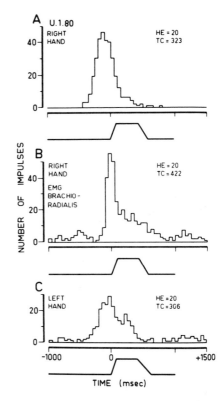

Figure 14.6. Illustration of the discharge patterns of a neuron associated with flexion of the elbow during the lever pull for both the right (A) and left hand (C). (B) Periresponse-time histogram demonstrating the electromyographic (EMG) activity of a representative elbow flexor, m. brachioradialis, in the right arm during the same 20 pulls as those in (A); the histogram shows that the neuron increased its discharge well before EMG activity increased. This was the case for the majority of neurons in which the discharge pattern could be compared with EMG changes. From Brinkman and Porter (1979).

firing before the motor pyramidal tract neurons that eventually would bring about the movements (Figure 14.6B). With complex movements, the SMA neurons displayed a wide variety of discharges over a considerable range of latencies, some even being inhibited. These findings would be anticipated for the diverse muscle contractions and relaxations that are concerned in the complex movement.

3. The primacy of the SMA is also indicated by Kristeva *et al.* (1979) on the readiness potentials recorded over the scalp during simple voluntary movements in humans. The readiness potential begins first and is maximum over the SMA (Figure 14.7). With the severe akinesia of bilateral parkinsonism, the readiness potential remains well developed over the SMA, but is greatly depressed over the motor cortex (Figure 14.7B) (Deecke and Kornhuber, 1978). The voluntary intention is fully expressed in the SMA response, the akinesia being due to failure of communication from the motor cortex.

If all complex voluntary movements are initiated by mental intentions acting on the SMA, there has to be in the SMA an immense inventory of motor programs, which are the operational components of all skilled movements, as defined by Brooks (1979). It is not necessary that the neural constituents of these motor programs be in the SMA, which surely would have inadequate storage facilities. It is enough for the SMA to function like a library card file with an inventory of all the skilled learned movements and the addresses of the storage sites, which could be elsewhere in the cerebral cortex, particularly the premotor cortex [PM in Figure 14.2 (Section 14.3.1)], and in subcortical sites, such as the basal ganglia, thalamus, and cerebellum [see Figure 14.27 (Section 14.6)]. The surface

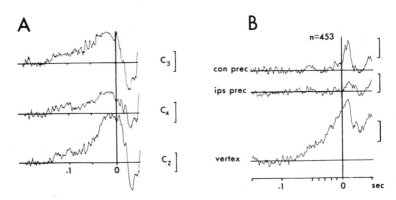

Figure 14.7. (A) Readiness potentials in response to brief voluntary flexions of the right hand (negative being upward). Zero time signals the onset of the electromyogram (EMG) of the contracting muscles (-1 being 1 sec before the EMG onset). C_3 and C_4 are, respectively, over the left and right motor cortical areas for hand movement; C_z is a midline lead over the SMA. From Kristeva *et al.* (1979). (B) A similar series of readiness potentials from a patient with bilateral parkinsonism. (con prec, ips prec) denote contralateral and ipsilateral precentral (motor cortex) locations of the recording electrode. Records are composed of 453 summed traces. The records display the averages of 800 recordings timed with respect to the EMG of the muscle contraction. From Deecke and Kornhuber (1978).

area of the human SMA is approximately 1% of the entire neocortex, so the neuronal population for the two SMAs would be about 100 million. Assuming a modular structure like that in other areas of the association cortex (Szentágothai, 1978a), there would be about 30,000 modules, 15,000 for each SMA. If in the SMA there is an inventory of motor programs or subroutines of movements, then one would expect it to have a fine grain of functional areas, each with its appropriate connections.

Unfortunately, most techniques of investigation do not have a sufficiently fine grain to differentiate the SMA into many discrete areas. For example, the injection techniques with autoradiographic analysis (Kunzle, 1978) displayed the wealth of connectivities of the SMA, but the injections involved the greater part of the SMA. Likewise, investigations of rCBF are most important in revealing the neuronal activity of the SMA under defined conditions but, as yet, do not have a sufficiently fine grain for topographic analysis. The activity of the SMA is seen globally (Roland *et al.*, 1980a,b). There is a similar deficiency in spatial analysis by techniques of investigation that involve antidromic testing from two or more sites in the brain stem and spinal cord (MacPherson *et al.*, 1983; Palmer *et al.*, 1981). Retrograde labeling by horseradish peroxidase (HRP) (MacPherson *et al.*, 1982) gave results similar to those from the antidromic investigation. There was intermingling throughout the SMA of neurons projecting to cervical and lumbar regions of the spinal cord. Retrograde labeling from the motor cortex (area 4) demonstrated somatotopic connections from the SMA with face, forelimb, and hindlimb in a rostrocaudal sequence, but no finer discrimination was attempted.

There is a long history of electrical stimulation of the primary motor cortex and of the precise somatotopic maps derived in this way [see Figure 14.3 (Section 14.3.1)]. With the SMA, the situation has been confused by conflicting reports. With SMA stimulation in awake man, Penfield and Welch (1951) found no somatotopy, but rather diffuse synergistic movements were evoked from the whole SMA. By contrast, in anesthetized monkeys, Woolsey *et al.* (1952) reported a discrete somatotopical organization, which formed the basis of a motor siminculus for the SMA resembling that for the motor cortex (Figure 14.3). Microstimulation (MacPherson *et al.*, 1983) has given unambiguous evidence of somatotopic organization of the SMAs with separate small sites for proximal and distal muscles of the forelimb and different sites also for the hindlimb; although there is some intermingling of these sites.

In a later study, Brinkman and Porter (1983) found a small focus of neurons in the premotor cortex that resembled the SMA neurons in many respects, but there were significant differences, so it is not sure if this premotor focus shares with the SMA in the initiation of voluntary movements. Wiesendanger (1981) gives an authoritative account of the secondary motor areas. In their recordings of single neurons of the SMA, Tanji *et al.* (1980) and Tanji and Kurata (1982) reported somatotopic arrangements of neurons carrying out instructional tasks. There was a remarkably sharp separation between SMA neurons oriented to forelimb and hindlimb movements, but somatotopic specificity for movements within a limb zone was not sought for in detail. However, SMA neurons often exhibited preferential responses to visual, auditory, or tactile stimulation or to any combination of two of these sensory inputs.

In summary, the whole range of experimental findings is certainly consistent with a somatotopic organization of the SMA, but still at a relatively crude level. According to the hypothetical postulate of an inventory of all motor programs in the SMA, there would be immensely complex somatotopic patterns of the modules of the SMA. The technique of single neuron recording offers the most hope for progress in the understanding of the functional structure of the SMA, which is symbolically the inventory of all skilled movements with addresses to their storage sites. Brinkman and Porter (1983) have concluded that the SMA is involved in the preprogramming of all skilled learned movements, and Roland *et al.* (1980b) state that the SMA elaborates programs for the motor subroutines necessary in skilled voluntary action.

It is undeniable that the planning of a movement is a mental event. On the hypothesis that there is activity of the SMA in all planned actions, different motor programs or motor subroutines would be brought into action for each movement. It would be predicted that different populations of SMA neurons would be excited, but testing of this prediction is beyond present techniques. It is not simply a matter of a particular neuron being activated or not; rather, one would expect there to be spatiotemporal patterns of neuronal activation. The findings of Brinkman and Porter (1979) certainly indicate activities of neurons that could be participating in complex spatiotemporal patterns in the SMA.

The criticism might be made that the discriminative role of the SMA is being exaggerated and that the execution of movements can be left to the primary motor cortex with control by feedback loops from the periphery with guidance by inputs from muscle, joint, skin, and visual receptors. However, well-planned, complex, skilled actions can be very fast, ballistic, with no time for feedback from the periphery. There is further reference to the feedback from receptors in Section 14.3.7.

14.3.4. Arrangement of Pyramidal Cells in Colonies

The arrangement of pyramidal cells in colonies having similar actions has long been postulated by C. G. Phillips (1966) because it has been recognized that the detailed cortical map defined by stimulation [see Figure 14.3 (Section 14.3.1)] could only thus be explained. The geometry of these colonies has now been investigated by Asanuma and Rosén (1972) and by Rosén and Asanuma (1972), using microstimulation of the monkey

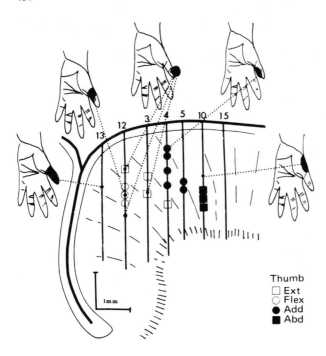

Figure 14.8. Correlation of cortical cell location and movement depicted in a reconstruction of electrode tracts and cell locations in a section through the right motor cortex of a monkey. Electrode penetrations, indicated by numbers, passed through efferent zones that project to various thumb muscles. The peripheral motor effects on the thumb produced by intracortical microstimulations of 5-mA strength at various locations are indicated by the symbols identified in the legend. (•) Positions of cells encountered along the track; (····) joining of these points to the drawings of the monkey's hand on which are indicated the receptive fields as determined by adequate stimulation. From Asanuma and Rosén (1972) and Rosén and Asanuma (1972).

Thumb
□ Ext
○ Flex
● Add
■ Abd

motor cortex as illustrated in Figure 14.8. The microelectrode was inserted in seven numbered tracks, and various sites at which one or another thumb movement was evoked by weak electrical stimulation were noted. The effective sites are indicated by the symbols in Figure 14.8. It can be seen that, as revealed by this stimulation, the actual cells performing one action or another tend to lie in columns the boundaries of which are indicated by the curved lines, orthogonal to the surface of the cortex. In Figure 14.8, this arrangement is particularly well shown for the movements of thumb extension (□) and adduction (●) that are produced along more than one track. There would be many hundreds of pyramidal tract cells in one such column, all presumably having the same action. It is all part of the motor map of Figure 14.3. In Figures 17.1 and 17.2 (Section 17.1), there is an account of the modular organization of the cerebral cortex.

14.3.5. α-Motoneurons and γ-Motoneurons and the γ-Loop

In primates, the pyramidal tract fibers directly excite a motoneuron such as that of Figure 3.1 (Section 3.2.1) to fire impulses and so to make the muscle contract. In Figure 3.1B, there is a transverse section of a muscle nerve containing only motor nerve fibers, all afferent fibers having been degenerated by dorsal root ganglia excision some 3 weeks earlier. It can readily be seen that, besides the large fibers that are concerned in the muscle contraction, there are also small fibers. In fact, there are two quite distinct populations. The small motor fibers provide the motor innervation of the muscle fibers in the muscle spindles that can be seen in the diagram of Figure 14.9 to be in parallel with the muscle fibers responsible for the contraction, the so-called "extrafusal fibers." The small motor nerve fibers come from small γ-motoneurons that lie interspersed with the large α-

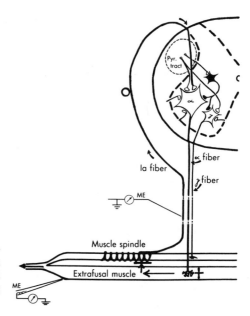

Figure 14.9. α-Innervation and γ-innervation in a diagrammatic representation of nerve pathways to and from the spinal cord showing the essential features of α- and γ-motoneuron action and interaction. (ME) measuring electrodes.

motoneurons for the same muscle, and the γ-axons (γ-fibers) exclusively innervate the muscle spindles, which are in parallel to the extrafusal fibers that are exclusively α-innervated. This α-innervation of extrafusal fibers was the theme of much of Chapter 3. The enormous literature on muscle spindles is excellently reviewed by P. B. C. Matthews (1981).

In the simplified diagram of Figure 14.9, it can be seen that an annulospiral ending around a muscle fiber of the spindle discharges along a group Ia afferent fiber, as already described in Section 4.1 (Figure 4.1A). The several intrafusal fibers bundled together in a spindle form two distinct species, and there are also secondary endings with smaller afferent fibers (group 2) than the large Ia afferent fibers of the annulospiral ending. Nevertheless, for our present purpose, the simplified diagram of Figure 14.9 is adequate. It is the Ia fiber that gives the monosynaptic innervation of motoneurons that was the theme of much of Chapter 4.

A pull on the tendon of the muscle, as shown by the leftmost arrow in Figure 14.9, excites the spindles to discharge impulses up the Ia fiber [cf. Figure 2.3A (Section 2.1)]. If the intrafusal fibers are excited to contract by γ-motor impulses, there is powerful excitation of the annulospiral endings (as illustrated in c to b to a in Figure 2.3A), and so there is an intensification of the monosynaptic activation of the α-motoneurons. If, on the other hand, a motoneuron discharge causes the extrafusal muscle fibers to contract, tension is taken off the muscle spindle that is in parallel with it, and the annulospiral ending will discharge less or not at all. But if the γ-motoneurons are fired at the same time as the α-motoneurons, the muscle spindle will contract and so will not be slackened. This arrangement gives a nice servomechanism performance. The more the α-motoneurons are firing in response in part to the Ia input, the stronger the extrafusal contraction and the less the Ia activation of the α-motoneurons by the so-called ''γ-loop.'' But the action of that loop can be biased over a wide range of levels by the discharge of γ-motoneurons. There is thus an adjustable servoloop control of muscle contraction in accord with the biasing by γ-motoneuron discharge (cf. Granit, 1970).

14.3.6. Pyramidal Tract Innervation of α-Motoneurons and γ-Motoneurons

In Figure 14.9, it is shown that the fast pyramidal tract fibers monosynaptically excite both α- and γ-motoneurons. There would also be inhibitory action via interneurons. When performing a voluntary movement, both α- and γ-motoneurons are caused to discharge by the pyramidal tract.

It has been shown in beautiful experiments by Hagbarth and Vallbo (cf. Vallbo, 1971) that in voluntary movement, the pyramidal tract impulses excite both α- and γ-motoneurons (coactivation), the whole α–γ complex being put into action in an approximately synchronous manner. The α-motoneurons are excited to discharge impulses and so bring about the muscle contraction. At about the same time, the γ-motoneurons discharge, thus exciting the muscle spindles and setting the γ-loop in operation. Because of the time involved in traversing the γ-loop, the α-motoneuron discharge so generated by the Ia impulses does not occur until after the muscle has started to contract. It thus occurs at just the right time for the onset of the servomechanism control.

These experiments of Vallbo (1971) have shown quite clearly that voluntary movements are not initiated by a prior activation of the γ-motoneurons with follow-up by γ-loop activation of the α-motoneurons. On the contrary, they show that there is a close follow-up of the γ-loop control after the initial α-contraction. In considering the meaning of this physiological arrangement, it is important to recognize that, in most voluntary movements, the load opposing the movement can be anticipated only approximately, hence the necessity for adjustment up or down with the utmost expedition. The servomechanism that operates via the γ-loop is thus of importance in giving an automatic adjustment at the spinal level, though controls at a higher level are also of great importance, as is discussed below.

14.3.7. Projection of Ia Fibers to the Cerebral Cortex

P. B. C. Matthews (1981) and McCloskey et al. (1983) have reviewed a remarkable series of experiments on conscious sensations arising from muscle receptors. It had long been thought that sensations of movement and force of muscle contraction were due to receptors in joints, fascia, and skin. It has now been demonstrated that a sense of the movement and the force of muscle contraction is recognized by subjects in which all other sensory inputs are blocked. The Ia afferent fibers would signal the movement as detected by the length of the muscle spindles, while the force of contraction would be from the tendon receptors with their Ib afferent fibers.

C. G. Phillips (1969) set the stage on which has been performed the wealth of investigations into the projection of the group Ia fibers of the muscle spindle receptors (Figure 14.9 (Section 14.3.5)] up to the cerebral cortex. There had long been a recognition of the myotactic reflex [cf. Figures 4.1 (Section 4.1) and 14.9], which is a monosynaptic segmental response of the spinal cord. The question raised is: Has there been in evolution a significant transfer of group Ia control of motor responses from segmental to suprasegmental levels with the attendant advantage of providing essential information to the overriding high-level control? This proposed suprasegmental path is known as a "long-loop reflex."

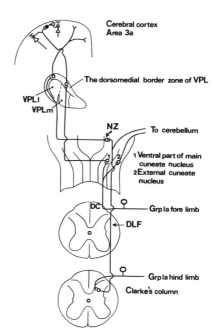

Figure 14.10. Projection paths from the primary muscle spindle afferents to area 3a of the cerebral cortex of the cat. Abbreviations: (DC) dorsal column; (DLF) dorsolateral fascicle; (NZ) nucleus Z of cuneate; (VPL) ventroposterolateral thalamic nucleus, with lateral (l) and medial (m) components. From Lándgrén *et al.* (1984).

Already in the *cat,* it had been discovered that the group Ia fibers of forelimb muscle project to the sensorimotor cortex by a very fast (latency 4.3– to 5.3-msec) and strong pathway (Oscarsson and Rosén, 1963; Oscarsson *et al.,* 1966). The details of the pathway are shown in Figure 14.10 for both the fore- and hindlimb, and the experimental investigations are fully reviewed by Landgren *et al.* (1984). The return pathway down the spinal cord could be either from area 3a of the sensory cortex or via a looping connection of 3a to 4 and so down the pyramidal tract.

With the *primate,* there are general similarities, but also important differences, as appear in the detailed reviews by E. G. Jones and Porter (1980), Wiesendanger (1981), and E. G. Jones (1983). The pathway from the forelimb resembles that of Figure 14.10 with the difference that the dorsal column pathway relays in the external cuneate nucleus and thence by the medial lemniscus to a special "shell" part of the VPLc component of the thalamus, as shown in Figure 14.11, and from there to two areas of the sensory cortex, 3a and 2. This distribution has been well shown by radiotracer techniques (D. O. Friedman and Jones, 1981). Physiologically, the main projection is to area 3a, just as in the cat (C. G. Phillips *et al.,* 1971; Maendly *et al.,* 1981), but two physiological studies also find a Ia input into area 2 (Burchfield and Duffy, 1972; Schwarz *et al.,* 1973). It is still an unsolved problem how the dominant Ia input into area 3a can activate a fast return circuit back to motoneurons. In the primate, area 3a itself has little or no pyramidal projection down the spinal cord, and careful radiotracer and HRP investigations (D. O. Friedman and Jones, 1981) failed to disclose a direct pathway from area 3a to area 4, yet activation of area 4 is essential for an effective return circuit to the motoneurons. E. G. Jones (1983) makes the tentative suggestion that the pathway to area 4 may be via area 2, as indicated in Figure 14.11. Thus, as in Figure 14.11, VPLc projects to area 2, which has a strong pathway to area 4 (Kosar *et al.,* 1985). It is doubtful, however, whether the VPLc path to

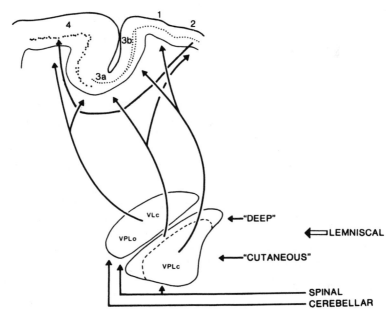

Figure 14.11. Schematic figure showing the distribution in sagittal section of inputs and outputs to components of the ventral nuclear thalamic group: VPLo (ventral posterolateral nucleus, oral part); VPLc (ventral posterolateral nucleus, caudal part); VLc (ventral lateral nucleus, caudal part). In the VPLc, an anterodorsal shell region that receives inputs from deep receptors including muscle afferents, traveling in the medial lemniscus, projects to both areas 3a and 2. The central core of the VPLc, receiving lemniscal inputs from cutaneous receptors, projects to a similar central core of the SI cortex consisting of areas 3b and 1. Cerebellar inputs relay in the common VPLo–VLc nucleus with no overlap into the VPLc. The common nucleus then projects to area 4, with little or no overlap into area 3a. Spinal inputs, about which far less is known, relay in both the VPLc and the VPLo. In accord with the text, a path is drawn from area 2 to area 4. From D. O. Friedman and Jones (1981).

area 2 is powerful enough to qualify as the link in the proposed Ia path: external cuneate → medial lemniscus → VPLc (deep) → area 2 → area 4 → pyramidal tract. There has been little investigation on the pathway for projection of Ia inputs from the hindlimb of primates. It is believed to resemble that of the cat as shown in Figure 14.10.

From Figure 14.11, it could be proposed that an alternative route for group Ia input to the motor cortex (area 4) could be by VPLo. The spinal path is by the spinothalamic tract, but that is excluded for two reasons: It is too slow for the very fast transmission of group Ia to the motor cortex; and this transmission is interrupted by section of the dorsal columns (Brinkman et al., 1978). Two investigations are urgently needed: further rigorous testing of the claim that area 3a does not project to area 4 in the monkey, and further physiological study of the projection of group Ia to area 2.

These investigations on animals are important in relation to the ingenious studies of Marsden et al. (1983a,b) on the reflex responses evoked in the flexor pollicis longus muscle in human subjects. This muscle is unique in that it is solely responsible for a joint movement, namely, flexion of the terminal joint of the thumb. When a steady flexor contraction is suddenly perturbed by an applied extension (Figure 14.12A), the muscle exhibits a series of reflex responses (Figure 14.12B). The earliest, at a latency of about 25

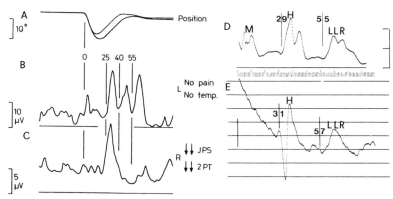

Figure 14.12. (A–C) Response to a fast brief stretch of the long thumb flexor in a patient with a lesion in the right brain stem, causing loss of pain and temperature sensation in the left arm, and loss of appreciation of joint position, vibration, and tactile discrimination in the right arm, but no motor deficit. (A) Records showing the angular position of the right and left thumbs. (B, C) Full-wave rectified electromyogram (EMG) recorded from the flexor pollicis longus of the left (B) and right hands (C). The subject held the thumb stationary against a standing force of 2 N. At time 0, indicated by the vertical marker, a force of 30 N was applied for 3 msec. Each trace is the average of 24 trials. In the EMG record from the left long thumb flexor, there are clear responses at spinal monosynaptic latency (25 msec), and later responses at 40 and 55 msec. In the record for the right long thumb flexor, the spinal monosynaptic response is still evident, but the LLRs are not apparent. From Marsden *et al.* (1978b). (D) Stimulus to the peroneal nerve and recording from the tibialis anterior muscle in the human. The stimulus was above threshold for motor nerve fibers as shown by the initial M-response, then followed the H-reflex (latency 29 msec) and an LLR (latency 55 msec). The LLR was dependent on the level of voluntary background and was zero at rest. From Iles (1977). (E) Stimulus to the tibial nerve just at threshold for motor fibers. The H-reflex with initial positivity had a latency of 31 msec and the LLR 57 msec. The LLR was produced when the subject was standing on toes and leaning forward so as to activate the motor cortex for the ankle extensor. The stimulus frequency was 3 Hz, with averaging of 50 responses. Deletis, Halter, and Táboříková, unpublished results (1985).

msec, is the ordinary tendon jerk produced by a circuit such as that of Figure 4.1A (Section 4.1). Two later responses, with mean latencies of 40 and 55 msec, are also automatic and not due to any voluntary response. There is much clinical evidence (Marsden *et al.*, 1978a,b) that these 40- and 55-msec automatic responses are due to a long-loop reflex (LLR) from the higher levels of the brain, in particular from the motor cortex. They are abolished or depressed by lesions that involve the pathways to and from the cerebral cortex (Figure 14.12C). However, cerebellar nuclei (interpositus and dentatus) may also be involved in the 55-msec latency response. It is not until much later that there is a conscious reaction to the sudden extension. The latencies of this reaction, which may be either a powerful resistance or a relaxation, average 126 msec, which is briefer than the conventional reaction times. With special efforts, the conscious reaction time of the flexor pollicis longus to all types of stretch can be reduced to a mean value of 99 msec. Similar LLRs have been observed in a comparable investigation by R. G. Lee *et al.* (1983) on the human wrist flexor suddenly stretched.

The clearest evidence for LLRs is displayed by responses to a very weak stimulus to muscle nerves, such as in the well-known H-reflex studies. Such a stimulus would selectively activate group Ia fibers, being even too weak to excite the motor fibers. In the resting situation, there is merely the sharp monosynaptic H-reflex. However, with a background of a steady voluntary contraction, there is a later sharp response with a latency just sufficient for the time of transmission up to and down from the motor cortex, i.e., for an LLR. With stimulation of the peroneal nerve and recording from the tibialis anterior muscle, Iles (1977) found a latency of about 29 msec for the H-reflex and 55-msec for the later response (Figure 14.12D). The size of this LLR was dependent on the intensity of the background voluntary contraction, being zero at rest. Similarly, with averaging techniques for the conventional H-reflex to soleus, it was found (Deletis, Halter, and Táboříková, unpublished findings, 1985) that when there was a steady voluntary contraction of the ankle extensors, the H-reflex evoked by a stimulus just at threshold for the motor fibers had a latency of 31 msec, and there was a sharp later response half the size of the H-reflex and at a latency of 57 msec (Figure 14.12E). This LLR was again dependent on the voluntary background, being negligible with zero background. Furthermore, this LLR was quite powerful, being largest with stimulation at 3 Hz and but little reduced at 5 Hz.

These automatic LLRs demonstrate that the Ia fibers excited in a voluntary contraction can activate a positive feedback loop to the motor cortex, which would be of value in sustaining a voluntarily initiated movement. In this context, C. W. Y. Chan (1983) has demonstrated that, with a sudden maintained stretch of biceps, the initial tendon jerk is followed by a silent period, and then discharges begin at the LLR latency and continue throughout the prolonged stretch.

This finding leads to a consideration of the functional role of the LLR from Ia fibers up to the motor cortex. As well stated by Wiesendanger and Miles (1982, p. 1256):

> The principle of a servomechanism is that the actual performance of a movement is continuously measured and is compared with the intended or desired performance. Any discrepancy between actual and intended movement results in a proportional error signal that in turn elicits a command signal to automatically compensate for the error. In the postulated transcortical position servomechanism, a perturbation occurring during a limb movement would activate muscle stretch receptors with an intensity proportional to the stimulus intensity. This error signal is transmitted to the motoneurons. The resulting change in muscular contraction, at M_2 latency (long loop), would then compensate for the initial load perturbation.

In carrying out a movement, there is an initial judgment of the requisite strength of the muscle contractions. Any error in this estimate will result in the activation of the muscle spindle receptors, including Ia. An error in the estimate from a hand movement will result in a corrective LLR, which causes an appropriate change in the discharge from the motor cortex, giving a corrective response of the hand movement with a latency of less than 50 msec. The corrective long-loop latency would be about 70 msec for ankle extensors. This corrective compensation is automatic and unconscious. If, as suggested by E. G. Jones (1983), the Ia path at the cortical level is via area 2 to the motor cortex, it would seem to be a good design. As shown by T. P. S. Powell and Mountcastle (1959), all deep sensory inputs from joints and fascia converge onto area 2 (see Figure 14.11). It would be appropriate if this input could be integrated with the Ia input, so that the area 2 projection to area 4 (the motor cortex) would carry essential information for the effective operation of the servoloop via the motor cortex. It is further suggested (cf. Figure 14.11) that the VPLc projection to area 3a is for the purpose of giving perception of the muscle contraction, while that to area 2 is relayed to area 4 for the servocontrol mechanism.

After a penetrating analysis of these problems of feedback to the cerebral cortex, Granit (1972) concludes that "coactivated spindles, which are the only end organs reflecting demand and execution, play a most essential role in our judgments about muscular exertion, difficult though it be to formulate the proprioceptive experience in the way we describe things seen and heard.'' Furthermore, Granit (1972) has considered the kinesthetic illusions that occur in relation to postural aftercontractions or the contractions of a muscle produced by vibratory stimulation of its tendon. The suggested explanation is that the illusion results from a perceived mismatch between the willed movement and the feedback to the cortex of group Ia afferent information. Only the unexpected is perceived with the vividness to give an illusion. On this basis, Granit has been able to explain many kinesthetic illusions that are described in the classic literature.

It is important to realize that much information going to the cerebral cortex from receptor organs does not immediately give a special sensation, as with hearing or vision, but rather something so much less vivid that it is not ordinarily recognized. For example, the vestibular receptors in the inner ear give us a feeling of the rightness of the situation. The world is as it should be. You move your head around and the world remains fixed. If this vestibular sense goes wrong, you experience vertigo in which the world is moving when your head is fixed. This demonstrates that the vestibular receptors do in fact signal to consciousness. It is the same with the Ia receptors of muscle. Quite a lot of the Ia input to the cerebrum is not appreciated normally, but when it is disordered, when there is some mismatch, as in the illusions, then you know that it is getting through to consciousness.

14.4. Motor Control from the Cerebellum

For general references, the reader is referred to Eccles *et al.* (1967), Eccles (1973), Brooks and Thach (1981, and M. Ito (1984).

14.4.1. Introduction

In Figure 14.13, three main components are seen in medial-to-lateral sequence, the vermis, the pars intermedia, and the hemispheres, together with their nuclei.

In Figure 1.1 (Section 1.1), it can be seen that, as the brain grew, the cerebellum grew commensurately through all the vertebrate evolution. We have to think of it as a part of the brain designed to function as a computer in handling all the complex inputs from receptors or from other parts of the brain. It is a computer that was originally built in relationship to the swimming of fish, using data from receptors in their lateral line organs and their vestibular mechanisms. But in the evolutionary process, it turned out that the cerebellum could be used for a wide variety of data computations, particularly in all the complex mechanisms involved in skilled movements, as, for example, in bird flight. It has long been known that the cerebellum is concerned very importantly indeed in the control of all complex and subtle movements, as, for example, in the playing of musical instruments. This relationship of the cerebellum to motor control was already appreciated in the last century, as a result of surgical lesions in animals and clinical lesions in man. Clumsiness and all kinds of badly disordered movements resulted from such lesions. With the great development of the cerebellum in the evolution of humans (cf. Figure 1.1) came all our motor skills. We are accustomed to think of these skills as being particularly exemplitied by tool manufacture and usage that eventually developed into our technology and civilization. But an even more important motor skill is that of being able to speak. It was this development of the cerebellum with the cerebrum, together as a linked evolutionary process, that gave the human species its immense superiority in survival.

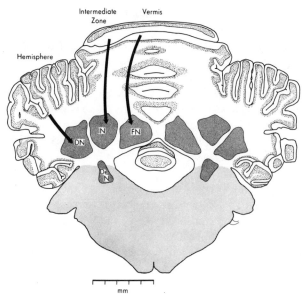

Figure 14.13. Drawing of a transverse section through the cerebellum and the cerebellar nuclei showing the lines of projection from the vermis, intermediate zone, and hemisphere to the respective nuclei: (FN) fastigal nucleus; (IN) interpositus nucleus; (DN) dentate nucleus. Deiters' nucleus (DeN) is also shown. Drawing kindly provided by P. Scheid.

The mammalian cerebellum has three principal components that are distinguished by their connectivities (Figure 14.13). With the evolutionary development of the primates, there has been a progressive increase in the cerebellar hemispheres. They account for 88% of the human cerebellum, with the pars intermedia and the vermis sharing the remaining 12%. These three components will be considered in turn. It is of the greatest importance to investigate the primate cerebellum because such studies relate particularly to the human cerebellum and provide a basis for understanding clinical disorders of the cerebellum (Gilman *et al.*, 1981). Furthermore, such studies open the way to discriminative investigations on the extremely complex interactions occurring in the cerebral control of learned movement. One can expect that the greater the evolutionary differentiation of function, the greater will be the opportunities for discriminative investigations into fundamental problems raised by cerebellar–cerebral interactions.

14.4.2. Neuronal Structure

In attempting to understand how the cerebellum carries out its amazing action as a computer, we are fortunate to have a precise picture of its neuronal arrangement, which we owe in the first place to the genius of Ramón y Cajal (1911). It is a surprisingly uniform, stereotyped structure that is laid out, as you might expect for a computer, with geometric precision. It is a rectangular laminated lattice as illustrated in Figure 14.14, in which a small fragment of a folium is shown in a perspective drawing. The principal neurons are the Purkinje cells (PCs), which provide the only output lines, their axons ending in the cerebellar nuclei (CN); i.e., these axons convey all the computational messages from the cerebellar cortex. The information that flows into the cerebellar cortex is conveyed entirely by two types of afferent fibers: the climbing fibers (CFs) that twine around the PC dendrites and the mossy fibers (MFs) that branch enormously and synapse

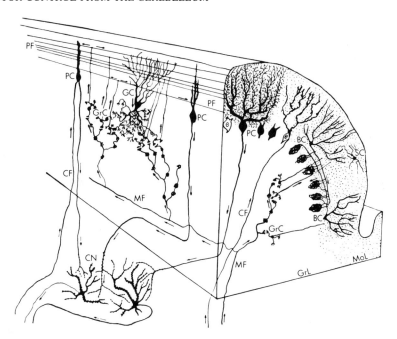

Figure 14.14. Schematic drawing of a segment of a cerebellar folium. Abbreviations: (BC) basket cell; (CF) climbing fiber; (CN) cerebellar nuclei; (GC) Golgi cell; (GrC) granule cell; (GrL) granular layer; (MF) mossy fiber; (MoL) molecular layer; (PC) Purkinje cell; (PF) parallel fiber; (SC) stellate cell. From Fox (1962).

on the little granule cells (GrCs) in the granular layer (GrL), the axons of which pass up to the molecular layer (MoL) to bifurcate and form the parallel fibers (PFs) that run along the folium for about 5 mm. Thus, they are orthogonal to the espalier-like dendritic trees of the PCs, basket cells (BCs), and stellate cells (SCs), with which they make numerous synapses, the so-called "crossing-over synapses." The BC axons are also transverse, traveling about 0.6 mm in either direction to form synapses on the PC somata. Finally, there are the Golgi cells (GCs), one being shown in Figure 14.14, which also receive synapses from the parallel fibers and have profusely branched axons that end on the GrC dendrites (Fox *et al.,* 1966).

The neuronal numbers are enormous. There are, for example, about 30 billion granule cells, 30 million Purkinje cells, and 200 million basket and stellate cells in the human cerebellum. Each Purkinje cell receives about 100,000 parallel fiber synapses on its dendritic spines, which are its mossy fiber input. In contrast, it receives only one climbing fiber, but this makes a massive series of synapses on the dendrites of the Purkinje cell.

This extraordinary double innervation has been maintained through all the exigences of evolution from primitive cerebella to the great efflorescence in mammals and birds. It is particularly remarkable that, when the cerebellar hemispheres were developed in step with the cerebral hemispheres, the inferior olive also hypertrophied. The cerebral efferents had to travel down to the medulla oblongata to excite the newly developed inferior olivary neurons for the essential climbing fiber input to the cerebellar hemispheres.

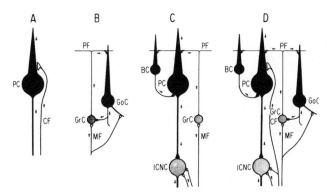

Figure 14.15. Synaptic connections of the cerebellar cortex. (A–C) Component circuits assembled in (D). The arrows show the lines of operation. Inhibitory cells are shown in black. Abbreviations: (BC) basket cell; (CF) climbing fiber; (GoC) Golgi cell; (GrC) granule cell; (ICNC) intracellular nuclear cell; (MF) mossy fiber; (PC) Purkinje cell; (PF) parallel fiber.

14.4.3. Neuronal Functions

From 1963 onward, there has been intensive investigation of the modes of action of the many synaptic connections in the cerebellar cortex; the results are summarized in very simplified form in Figure 14.15, to indicate the excitatory or inhibitory function of the various elements, which are drawn merely as single units (Eccles *et al.*, 1967; Eccles, 1973). All inhibitory neurons are shown in black. First, in Figure 14.15A, the climbing fiber (CF) has been shown to make an enormously powerful excitatory synapse with the Purkinje cell (PC) that fires several times for each CF impulse, with a subsequent prolonged inhibition (Strata, 1984). By contrast, the mossy fiber (MF) is very diversified, each fiber branching to contribute excitation [cf. Figures 14.14 (Section 14.4.2) and 14.15B] to about 400 granule cells (GrCs), and 100,000 GrCs via their parallel fibers (PFs) contribute to the excitation of each PC. The basket cells (BCs) are also excited by PFs (Figures 14.14C and 14.15C), and their axons go transversely to end as inhibitory synapses on the PC somata, which are in this way encased in a basket-like structure (cf. Figure 14.14). The Golgi cells (GoCs) are excited by parallel fibers (Figures 14.14 and 14.15B) and also by MFs, and their inhibitory synapses on GrCs complete a simple negative feedback loop [cf. Figure 4.17A (Section 4.7.4)]. Figure 14.15 is an ensemble of Figures 14.13 and 14.14.

We now have some basic information on which to build ideas of how the neuronal machinery of the cerebellar cortex can perform a computation. Two lines come in: (1) The mossy fiber is dispersed and powerful and gives both excitation and inhibition; (2) the climbing fiber gives a direct monosynaptic excitation with subsequent inhibition. Furthermore, the sole output from the cerebellar cortex is by Purkinje cells, which, as indicated in Figure 14.15D, inhibit the intracerebellar nuclear cells (ICNCs). In fact, all neurons of the cerebellar cortex are inhibitory except the granule cells. Nowhere else in the brain do we know of such dominance of inhibition.

In Chapter 16, there is reference to a hypothesis of learning according to which, as in a perceptron, the climbing fiber acts as a teacher in respect to the potency of the mossy fiber–parallel fiber synapses on a Purkinje cell. The dual innervation of the Purkinje cells is thus a further clever piece of evolutionary design (Eccles, 1977).

14.4.4. Cerebrocerebellar Pathways

For general references, the reader is referred to Evarts and Thach (1969), Eccles (1973), G. I. Allen and Tsukahara (1974), Massion and Sasaki (1979), Brooks and Thach (1981), and M. Ito (1984).

Closed Loop via the Pars Intermedia of the Cerebellar Cortex. Figure 14.16 is a greatly simplified diagram to show how the cerebrum and the cerebellum are linked together. All the neuronal pathways drawn in Figure 14.16 are securely based on anatomical and physiological investigations. Simplification is achieved by having the individual

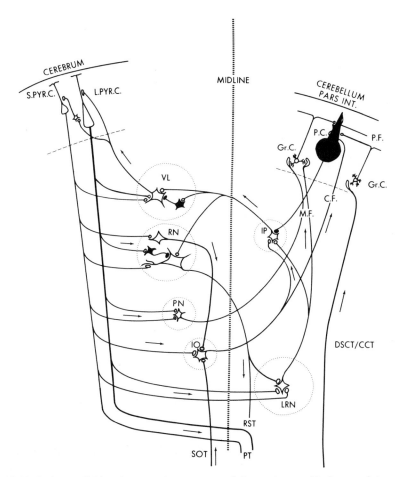

Figure 14.16. Pathways linking the sensorimotor areas of the cerebrum with the pars intermedia of the cerebellum. Abbreviations: (CF) climbing fiber; (DSCT/CCT) dorsal spinocerebellar tract/cervico-cerebellar tract; (Gr.C.) granule cell; (IO) inferior olive; (IP) interpositus nucleus; (L.PYR.C) large pyramidal cell; (LRN) lateral reticular nucleus; (M.F.) mossy fiber; (PARS INT.) pars intermedia; (P.C.) Purkinje cell; (P.F.) parallel fiber; (PN) nuclei pontis; (PT) pyramidal tract; (RN) red nucleus; (RST) rubrospinal tract; (SOT) spino-olivary tract; (S.PYR.C.) small pyramidal cell; (VL) ventrolateral thalamus.

species of cells reduced to just one example. Each species shown, however, has been recorded as individual cells, and the connectivities shown have all been checked by the timing of their discharges in each of the pathways. This is a fairly complete diagram so far as it goes, but of course it misses all the operational features that derive from the enormous number of cells in parallel with the wealth of convergence and divergence that has already been discussed to some extent. Thus, Figure 14.16 should be regarded as merely a skeleton drawing.

Let us start the operative sequences of Figure 14.16 by the firing of a large pyramidal cell (L.PYR.C.) of the motor cortex. These cells are, of course, the principal cells of origin of the pyramidal tract (PT) [cf. Figures 14.3 and 14.4 (Sections 14.3.1 and 14.3.2)]. Also shown is one small pyramidal cell (S.PYR.C.). Axons of these cells form small fibers of the PT but it is not known how effective these fibers are in the spinal cord. In Figure 14.16, the large fiber sends off branches that, after synaptic relay in the nuclei pontis (PN) and the lateral reticular nucleus (LRN), give mossy fiber (M.F.) inputs to the cerebellar cortex [pars intermedia (PARS INT.)] Thus, impulses fired down the PT to begin a movement (as we have seen in the monkey in Figure 14.4 making to-and-fro movements) will at the same time go to the cerebellar cortex on the opposite side from the cerebral cortex and on the same side as the movement. So there is an extremely fast and reliable input from the cerebrum to the cerebellum. The cerebrum cannot begin instituting any action without the cerebellum knowing about it immediately. There is little doubt that the cerebrum is the command center, but all instructions it fires to the motor machinery of the spinal cord are immediately fired into all the computational machinery of the cerebellar cortex via these two M.F. pathways, one via the PN and the other via the LRN. In addition, impulses fired from the small pyramidal cells go to these two nuclei, presumably aiding in their responses. More importantly, only small pyramidal cell discharges go to the inferior olive (IO), which is the exclusive source for the climbing fiber (C.F.) input to the cerebellar cortex (Desclin, 1974).

After the stage of interaction in the cerebellar cortex, there is the final stage of the return circuit to the motor cortex of the cerebrum. In Figure 14.16, the Purkinje cell (P.C.) inhibits the nuclear cell in the interpositus nucleus (IP), which is also excited by collaterals from M.F.s and C.F.s, as shown for the LRN and IO pathways, respectively. Here, then, is a further site for computation in the clash of excitatory and inhibitory actions on the IP cells. Thence, the pathway is very fast and direct, there being only one synaptic relay in the ventrolateral thalamus (VLc) on the way to the pyramidal cells of the motor cortex (Figure 14.11). Another pathway for action is also shown in Figure 14.16, namely, from the IP to the red nucleus (RN) and so via the rubrospinal tract (RST) to the motoneurons of the spinal cord. The cerebro-cerebellar interactions in the RN are important in displaying trophic reactions and in simple learning (see Sections 13.2.2 and 13.2.3).

We can thus appreciate the authority of the cerebellar influence on the course of all movements that are initiated by the motor cortex. As indicated in Figures 14.16 and 16.22 (Section 16.7.3), there is a very rapid and complete signaling to the cerebellar cortex of the whole array of impulse discharges down the pyramidal tract. We can assume that the input is computed in the cerebellar cortex with utilization of its memory stores (cf. Chapter 16), and, after a further computation in the cerebellar nuclei (IP), it is returned to the same motor area of the cerebrum, much as occurred from the periphery in Figure 14.8

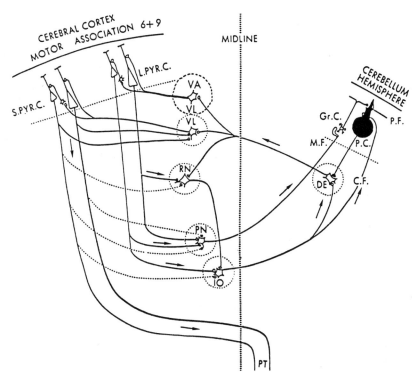

Figure 14.17. Cerebrocerebellar pathways linking association and motor cortices with the cerebellar hemisphere. Only a small cell is shown in the red nucleus (RN). Abbreviations: (C.F.) climbing fiber; (DE) dentate nucleus; (Gr.C.) granule cell; (IO) inferior olive; (L.PYR.C.) large pyramidal cell; (M.F.) mossy fiber; (P.C.) Purkinje cell; (P.F.) parallel fiber; (PN) nuclei pontis; (PT) pyramidal tract; (RN) red nucleus; (S.PYR.C.) small pyramidal cell; (VA) ventroanterior thalamus; (VL) ventrolateral thalamus. From G. I. Allen and Tsukahara (1974).

(Section 14.3.4). In the cat, the circuit time for this complete loop would be less than a hundredth of a second. In man, it would be longer, about a fiftieth of a second. With respect to the motor cortex, this system operates in a closed-loop manner.

Open-Loop System in the Cerebellar Hemispheres. The cerebellar hemispheres comprise almost 90% of the human cerebellum [cf. Figure 14.13 (Section 14.4.1)]; the principal cerebrocerebellar circuits are shown in Figure 14.17. In contrast to the pars intermedia, the cerebellar hemispheres receive most of their cerebral inputs from extensive areas of the cortex, such as the motor association cortex that is anterior to the motor cortex in Figure 14.2 (Section 14.3.1), and less from the motor cortex via collaterals of pyramidal tract fibers (dotted lines in Figure 14.17). This distinctive circuitry is well developed in primates and is preeminent in humans, in whom widespread zones of one cerebral hemisphere provide 20 million fibers passing to lower levels, as against only 500,000 pyramidal tract fibers (G.I. Allen and Tsukahara, 1974). In Figure 14.17, impulses discharged from the pyramidal cells of the association cortex pass to the contralateral cerebellar hemisphere via relays in the pontine nuclei (PN) and the inferior olive

(IO). After computation in the cerebellar hemisphere, the return circuit is via the ventrolateral (VLc) thalamus to the motor cortex (Figure 14.11) and so down the PT to effect the movement. Thus, the cerebrocerebellar circuit is essentially an open-loop system.

It has been shown in primates that wide areas of the association cortex project to the cerebellar hemisphere and the dentate nucleus (DE) of the opposite side. The strongest inputs to the DE come from Brodmann area 6 [the premotor and the supplementary motor area (SMA)], with additional inputs from the frontal, cingulate, and temporal cortices (G.I. Allen et al., 1978). In a very comprehensive study of cerebellar cortical responses, Sasaki et al. (1977) carefully differentiated the mossy fiber from the climbing fiber responses. These responses were evoked by stimulation of the frontal association cortex (areas 8 and 9), the premotor area 6, and the parietal association area 5. Areas 9 and 6 were particularly potent in evoking climbing and mossy fiber responses. However, it was also found that the forelimb zone of the motor cortex area 4 effectively projected to the cerebellar hemisphere, so the dotted lines suggesting these projections in Figure 14.17 could be redrawn as solid lines.

The converse projection from the DE to the contralateral cerebral cortex has been found to go not only to the motor cortex (area 4) but also strongly to premotor area 6 (including the SMA) and also to the frontal association area 9 (Sasaki, 1979; Sasaki et al., 1979). There is direct evidence that these projections are mediated by the VL and ventroanterior (VA) thalamic nuclei as indicated in Figure 14.17.

There are in these areas immense and challenging opportunities for investigation. In particular, the experiments of Sasaki et al. (1976, 1977, 1979) should be developed by recording the responses of single neurons in the cerebral and cerebellar cortices. This has already been done for the DE (G. I. Allen et al., 1978), in which the neurons displayed a complicated sequence of inhibitory and excitatory responses, as would be expected from interaction of impulses in the inhibitory Purkinje axons and in the excitatory axon collaterals from both the pontocerebellar fibers (not shown in Figure 14.17) and the olivocerebellar fibers.

This analytical work will be of great value in attempting to understand and to develop further the investigations of Thach (1975, 1978) on the responses of cerebellar nuclear cells and motor cortical pyramidal cells during voluntary movements of a trained monkey (cf. Brooks, 1985). There is good evidence in general that dentate neurons discharge before the motor pyramidal cells, which would be in accord with the circuits of Figures 14.16 and 14.17.

Dynamic Operation of the Cerebrocerebellar Circuits.   When pyramidal cells of the motor cortex (area 4) are firing impulses down the pyramidal tract (PT) to bring about a voluntary movement (a motor command), the patterns of this discharge in all details are transmitted to the cerebellum (pars intermedia) by virtue of the collateral branches of the PT fibers (G. I. Allen and Tsukahara, 1974). Computation occurs in the cerebellar cortex, and the resulting output is returned to the motor cortex so that there is an ongoing "comment" from the cerebellum within 10–20 msec of every motor command. We may regard this "comment" as being in the nature of an ongoing correction continuously provided by the cerebellum and being immediately incorporated into the modified motor commands issued by the motor cortex. Figure 14.16 also illustrates a longer feedback loop that operates through the same region of the cerebellum. When the motor command brings about a movement, this evolving movement excites a wide variety of peripheral receptors, e.g., in muscles, skin, joints, and these signal back to the same regions of the cerebellar cortex (upward pointing arrow) that were concerned in the more direct loop. A computation of the two sets of input forms the basis of the cerebellar response. Thus, the motor command centers are provided with an ongoing cerebellar comment synthesized from these two loops. In addition, the pars intermedia has a more direct path for influencing the spinal centers via the red nucleus (RN) and the rubrospinal tract (RST), as indicated in Figure 14.16.

In summary, we can regard the pars intermedia of the cerebellum as acting like the controlling system on a target-seeking missile. It does not give a single message for correction of a movement that is off-target. Instead, it provides sequences of correcting messages, thus giving a continuous updating control by closed dynamic loops.

In the primate, there is no effective peripheral input to the cerebellar hemispheres such as that for the pars intermedia (G. I. Allen and Tsukahara, 1974). Hence the only feedback information into this open-loop system is via the sensory pathways to the cerebral cortex from peripheral receptors in muscle, skin, joints, and elsewhere that project to the somesthetic areas, 3, 1, and 2 (cf. Figure 14.11).

Experimental investigation of unitary neuronal responses at all stages of the circuits suggests that, in principle, they give a satisfactory explanation of the performance of motor control (Eccles, 1973; G. I. Allen and Tsukahara, 1974). But much more experimental investigation is required, particularly on the mapping of related cerebral and cerebellar areas. We have to envisage that there is some sort of topographic design of the cerebellar hemispheres and pars intermedia somewhat on the lines of the cortical motor map of Figure 14.3 (Section 14.3.1). We know, however, that the maps are congruent only in part. The task of the cerebellum is to blend into a harmonious whole the movements of different joints of a limb, for example, so it would be anticipated that there would be convergence from these diverse motor areas to specific integrational areas of the cerebellar cortex. Tests of this conjecture have yet to be carried out on the primate cerebellum. The cerebellum is importantly concerned in motor learning and memory (see Section 16.7).

14.4.5. Spinocerebellar Connectivities

For general references, the reader is referred to Oscarsson (1973, 1976) and Eccles (1973).

These connections have been studied far more intensively than the cerebrocerebellar connections so far considered. They are especially concerned in walking, standing, reacting, and balancing and in all the postural adjustments that follow active movements—the stabilizing positions attained thereby. It can be seen in Figure 14.18 that essentially the same general circuits from the spinal cord act on the cerebellar cortex as those diagrammed in Figure 14.16 (Section 14.4.4). There are, first, the two main tracts up the spinal cord (SpC) that end as mossy fibers (MFs), the fast dorsal spinocerebellar tract (DSCT) and the slower tract (bVFRT) up to the lateral reticular nucleus (LRN). Second, there are the slow tracts to the inferior olive (IO) and so to the cerebellum as climbing fibers (CFs). These terminate in a remarkable series of parasagittal bands (Oscarsson, 1973, 1976) that must eventually be explained in terms of a subtle theory of information processing. The inhibitory outputs of the Purkinje cells go initially to the fastigial nucleus (FN) or to Deiters' nucleus (DN) and secondarily, via the FN, either to DN or to the reticular nucleus

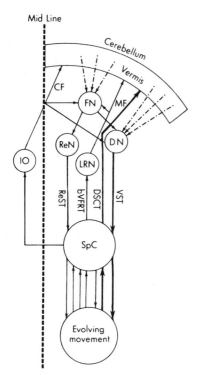

Figure 14.18. Pathways linking the cerebellar vermis with the spinal centers and so to the evolving movement. Abbreviations: (bVFRT) b fiber ventral reticular tract (CF) climbing fiber; (DCST) dorsal spinocerebellar tract; (DN) Deiters' nucleus; (FN) fastigial nucleus; (IO) inferior olive; (LRN) lateral reticular nucleus; (MF) mossy fiber; (ReN) reticular nucleus; (ReST) reticulospinal tract; (SpC) spinal cord; (VST) vestibulospinal tract.

(ReN) and so down the spinal cord to motoneurons via the vestibulospinal tract (VST) or reticulospinal tract (ReST), respectively.

The important feature of these connections of the cerebellar vermis is that there is only one loop in the dynamic control system. The evolving movement results in the discharge of various kinds of receptors that project via the diverse ascending pathways to the cerebellar cortex, thus modifying the Purkinje cell output to the cerebellar nuclei and so, via the ReST and VST, to the motoneurons. In this way, the cerebellum is able to control posture and movements by the simplified version of dynamic-loop operation. Investigations of the stages in the neuronal system (Eccles, 1973) showed that the various species of neurons give the responses that would be expected for inputs from receptor organs of skin and muscle via the DSCT and bVFRT.

In Figure 14.15C and D (Section 14.4.3), the intracellular nuclear cell (ICNC) is shown with excitatory inputs that provide the background against which is pitted the inhibitory action of the Purkinje cell. This excitatory background to the fastigial nucleus (FN) is provided by collaterals from two pathways to the cerebellum, climbing fibers (CFs) from the inferior olive (IO) and mossy fibers (MFs) from the lateral reticular nucleus (LRN). It was found experimentally that the fast MF pathway (DSCT) had almost no excitatory action on the FN (Eccles, 1973). Figure 14.16 (Section 14.4.4) shows a similar state of affairs for the collateral excitation of the interpositus nucleus (IP). These arrangements have been fully substantiated by stimulating the IO and LRN nuclei and observing the monosynaptic excitation of the cerebellar nuclei, FN and IP. They are also substantiated by anatomical studies employing radiotracer and HRP techniques.

Afferents from joint receptors require more study. Also to be integrated into this cerebellar control of posture and simple automatic movements are the inputs from the vestibular and neck receptors with their pathways down the vestibulospinal tract (VST) and the reticulospinal tract (ReST). These inputs have been extensively studied by V. J. Wilson (1979) and B. W. Peterson (1979).

14.4.6. General Comments on the Cerebellum

An intensive study of about 1000 Purkinje and 1000 nuclear cells has revealed a wide variety of responses and a surprising individuality. This individuality of response results from the built-in connectivities in the pathways to these cells that are shown only in skeletal form in Figures 14.16, 14.17, and 14.18 (Sections 14.4.4 and 14.4.5). We can assume that these connectivities are a result in part of the initial genetic coding and in part of superimposed "learned" connections that we will discuss in Chapter 16. The responses are all explicable in terms of these connectivities together with the basic integrated performance of the cerebellar cortex as illustrated in Figure 14.14 and 14.15 (Section 14.4.2).

Samplings of Purkinje cells by several microelectrode insertions into the cerebellum in any one experiment give but fragmentary glimpses of the immense number of colonies constituted by similarly behaving neurons. We had earlier made a comprehensive study of the responses evoked in single Purkinje cells by afferent volleys from many nerves (Eccles, 1973), both cutaneous and muscular, and had found remarkable examples of convergence. To preserve and develop the integration of information occurring in colonies that have somewhat similar inputs, it is postulated that such colonies of Purkinje cells would tend to project onto a common target of nuclear neurons. This arrangement would give averaging of inputs from many Purkinje cells, reducing noise, as already described, and it would give opportunity for further integration of different subsets of the total input to the cerebellum. If there were randomized projection, there would be a loss of all specificity of information. It is the diversified convergence that enhances the pattern-generating capability of the neuronal machinery of the cerebellar cortex. This postulate of organized projections from Purkinje cells to the nuclear cells has already been investigated by a systematic study of both nuclear cell responses and the neurons to which these nuclear cells project (Eccles, 1973; M. Ito 1984).

The relative simplicity of the neuronal design of the cerebellum together with its well-defined action in control of movement provide a most enticing challenge, both experimentally and theoretically. It is our belief that the cerebellum will be the first part of the brain to be "understood" in its total performance. For this reason, we have given a rather detailed account of the linkage between structure and function in specific cerebellar performances and of theoretical developments from these investigations. Much of this theory is speculative, going beyond what has been demonstrated. But this is the essence of scientific advance, to have theory leading and guiding experiments that essentially are tests of theoretical predictions.

Perhaps the most puzzling feature of the cerebellar design is the existence of the two quite different inputs, mossy and climbing fibers, that convey much the same information. These two inputs have been preserved from the most primitive cerebella. We now have a theory of cerebellar learning (Ito, 1984) that accounts satisfactorily for this linked input. There is further reference to this problem in Chapter 16.

Since, as we have seen, there is good reason to believe that the cerebellum functions as a special type of computer (Eccles, 1973), there have been many attempts at computer modeling of the cerebellum. Such modeling fails, we think, because it is premature. We do not yet have sufficient "hard data" as a basis for computer modeling. There has to be much more experimental investigation using the recordings of single cells. In such investigations, we are, as it were, listening in on the communication of data as coded in the cell discharges, in an attempt to understand these codes. At this level, computer modeling can be valuable in elemental coding problems. Gradually then, piece by piece, we are building a coherent story of cerebellar performance. In this way, we will gradually amass the requisite "hard data" for effective computer modeling. But there is still much to accomplish, and all too few experiments are at an adequate level.

14.5. Motor Control from the Basal Ganglia

For general references, the reader is referred to Carpenter and Sutin (1983); Evarts and Wise (1984); and E. G. McGeer et al. (1984).

14.5.1. Introduction

We now turn to the last group of nuclei that exert control over motor function. These are the subcortical masses loosely referred to as the basal ganglia. We shall include in the definition of the basal ganglia the caudate nucleus, the putamen, the globus pallidus, the subthalamic nucleus, and the substantia nigra. Although not directly concerned in our story of motor function, two other structures of the basal ganglia, the nucleus accumbens and the amygdala, should also be mentioned because they provide important anatomical links and play a prominent role in the associated behavioral phenomena that are discussed in Chapter 15.

Information on the physiological functioning of the basal ganglia is still extremely scanty. It cannot be said, for example, how these nuclei eventually affect peripheral musculature or how postural signals from muscle spindles influence their internal activity. Too little is known about the inflow and outflow routes. There must be integration with the major motor systems of the cerebellum and cortex, but how this integration is achieved is still a matter of speculation.

Much of the information regarding the functions of the basal ganglia has come from investigations that are very different from the precise neuroanatomy and elegant neurophysiology on which knowledge of cerebellar function has been built. Instead, information has been developed largely as a result of motor deficits produced by lesions, the effects of various drugs, and the consequences of certain diseases.

Pathological signs associated with disorders of the basal ganglia are of two kinds; dyskinesia (disorder of movement) and dystonia (disorder of muscle tone). They vary between two extremes: akinesia–rigidity and chorea–athetosis. It should be noted, however, that akinesia can occur without rigidity and that the term dystonia is applied to a particular clinical syndrome that is described in Section 14.5.4. Circumstances that will result in the appearance of these opposite signs are summarized in Table 14.1. Lesions, drugs, chemical poisons, altered neurotransmitter concentration, and a number of diseases of unknown etiology are all capable of producing the classical signs. The two extremes can be thought of as a postural rigidity on one hand, leading to poverty of movement, and a postural hyperactivity on the other, leading to incessant, involuntary activity. The

TABLE 14.1
Functional Output of the Basal Ganglia

Overactivity (Chorea–hypotonia)	Underactivity (Akinesia–rigidity)
	Diseases
Huntington's chorea	Parkinson's disease
Sydenham's chorea	Manganese poisoning
	Carbon monoxide poisoning
	Carbon disulfide poisoning
	Neurotransmitter functional activity
Excess DA	Excess ACh
Lowered ACh	Lowered DA
	Drug action
Cholinergic blockade	Dopaminergic blockade
(e.g., atropine, artane)	(butyrophenones, phenothiazines)
L-Dopa	Dopamine depletion
	(rauwolfia alkaloids)
	Lesions
Caudate	Substantia nigra
Putamen	

system is like a teeter-totter, with normal activity occurring only when these two opposing forces are in balance. Tremors at rest are frequently seen when the system is out of balance, indicating that rhythmic activity is involved in normal functioning and is probably a feature of the integration of this system with cerebellar, cortical, and peripheral motor loops.

Table 14.1 shows that there are many causes of akinesia–rigidity: Parkinson's disease, certain types of poisoning, reduced dopamine (DA) concentration or blockade of dopaminergic receptors, excess acetylcholine (ACh), or lesions to the substantia nigra.

Chorea–athetosis occurs in Huntington's and Sydenham's chorea, with excess DA, with cholinergic blockade, and with lesions of the caudate and putamen or blockade of their γ-aminobutyric acid (GABA) projections. While the true picture is much more complicated than indicated in Table 14.1, the table nevertheless serves as a starting point to show that definite correlations exist among neurotransmitter amines, subcortical motor nuclei, and disease states of unknown etiology.

14.5.2. Anatomical Interconnections of the Basal Ganglia

A simplified way of regarding the organization of the basal ganglia is to divide it into input, processing, and output neuronal arrangements. The major pathways are indicated in Figure 14.19.

The principal processing areas of the basal ganglia are the caudate and putamen. These twin nuclei, divided in primates by the internal capsule but not so divided in small rodents like the rat, are enormous assemblages of neurons of uncertain internal organization. Together, they are referred to as the *neostriatum*. Afferent input comes from the cortex, the thalamus, and the midbrain (Figure 14.19A).

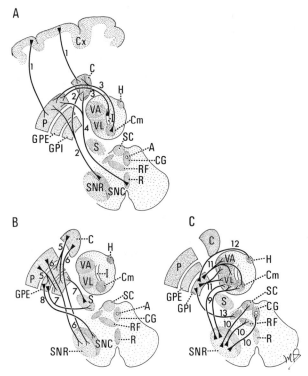

Figure 14.19. Pathways of the basal ganglia. (A) Input circuits to the basal ganglia. (B) Processing circuits within the basal ganglia. (C) Output circuits of the basal ganglia. Pathways: (1) corticostriatal; (2) nigrostriatal; (3) thalamostriatal; (4) raphe–striatal; (5) striatopallidal; (6) striatonigral; (7) pallidal–subthalamic–pallidal; (8) pallidonigral; (9) nigrothalamic; (10) nigrobrain stem; (11) pallidothalamic; (12) pallidohabenula; (13) pallidotegmental. Abbreviations: (A) aqueduct of Sylvius; (C) caudate nucleus; (CG) central gray; (Cm) centrum medianum thalamus; (Cx) cerebral cortex; (GPE) external globus pallidus; (GPI) internal globus pallidus; (H) hypothalamus; (I) intralaminar thalamic nucleus; (P) putamen; (R) raphe system; (RF) reticular formation; (S) subthalamic nucleus; (SC) superior colliculus; (SNC) substantia nigra, pars compacta; (SNR) substantia nigra, pars reticulata; (VA) ventroanterior thalamic nucleus; (VL) ventrolateral thalamic nucleus.

The cortical projection on the neostriatum is massive. Nearly all cortical regions participate in this corticostriatal pathway, which is organized more or less topographically (Kemp and Powell, 1971). The more frontal regions project to the caudate, the pre- and postcentral gyri project to the putamen, and so on. Some cortical fibers cross over to project to the contralateral side. The thalamus provides a heavy but less massive input from the intralaminar nuclei including the parafascicular nucleus and the centrum medianum (T. P. S. Powell and Cowan, 1956; Mehler and Nauta, 1974). Midbrain input comes from the zona compacta of the substantia nigra and the rostral raphe.

Processing systems within the basal ganglia are shown in Figure 14.19B. The striped appearance implied in the Latin term *corpus striatum* is due to bundles of striatal fugal fibers that converge radially on the globus pallidus. The majority terminate in the globus pallidus, but some course through the external and internal pallidal segments, "comb" through the cerebral peduncle, and terminate chiefly in the pars reticulata of the substantia nigra. Thus, the principal output of the caudate and putamen is to the globus pallidus and to a lesser extent the substantia nigra, particularly the pars reticulata. Considerable convergence is involved because the globus pallidus is much smaller than the neostriatum, and the substantia nigra pars reticulata is much smaller than the globus pallidus. The globus pallidus itself (the paleostriatum) is divided into two segments, the external and internal divisions. The internal division is equivalent to the entopeduncular nucleus in smaller mammals. The external globus pallidus projects primarily to the subthalamic nucleus, a small body embryologically related to the thalamus, but also to the pars compacta of the substantia nigra (SNC). The subthalamic nucleus projects back to the internal segment of the globus pallidus and thus can be thought of as a closed-loop control in the globus pallidus.

Figure 14.19C shows the main outflow circuits from the basal ganglia. The most important of these is from the internal segment of the globus pallidus to the thalamus. While the processing system within the basal ganglia are highly convergent from the caudate and putamen to the internal globus pallidus and substantia nigra, the output from these structures again becomes highly divergent. The fibers from the internal globus pallidus follow one of two routes: They traverse the internal capsule as the fasciculus lenticularis or loop around the internal capsule as the ansa lenticularis. Terminations are in the ventrolateral (VL) and ventroanterior (VA) nuclei of the thalamus, and to a lesser extent in the centrum medianum. Here they meet output pathways from the cerebellum. There is also a pallidotegmental tract terminating in the pedunculopontine nucleus (Carpenter and Sutin, 1983).

The substantia nigra has two main efferent projections emanating from different laminae of the pars reticulata (SNR). The intermediate lamina, lying just beneath the zona compacta, projects mainly to the thalamus (Faull and Mehler, 1976). It distributes partly to the VL and VA thalamic nuclei, but also to the dorsomedial nucleus, again overlapping with cerebellar projections to the thalamus. The most ventral cells of the pars reticulata project to the superior colliculus (SC) (Graybiel, 1978; Faull and Mehler, 1976) and to a wide medial zone of the ipsilateral midbrain tegmentum, including the anterior parabrachial and intrabrachial regions (Domesick, 1976) and the central gray (CG) (Hopkins and Niessen, 1976). Some fibers cross over. The nigrostriatal tract from the pars compacta (SNC) can be thought of as an internal loop, analogous to that from the subthalamic nucleus, but operating on the caudate (C) and putamen (P) instead of on the globus pallidus.

14.5.3. Neurotransmitters in the Basal Ganglia

The basal ganglia contain a rich variety of neurotransmitters, as discussed in Chapters 6–12. That information is now integrated in Figures 14.20–14.25, which also provide in block diagrams a somewhat more detailed picture of the anatomical pathways of these nuclei, as well as some of their interconnections with the limbic system.

Afferents to the Caudate–Putamen (Figure 14.20). The massive projection from the cortex to the caudate–putamen (CP) appears to use glutamate (Glu) as the neurotransmitter (see Chapter 6). This excitatory system innervates most of the neurons of the CP. There is some contribution from the contralateral cortex; this is one of the levels at which some integration between the hemispheres may occur (Leichnetz, 1981; J. E. Walker, 1983). Since many commissural pathways also probably use glutamate or aspartate (Chapter 6), such integration is highly dependent on these amino acid neurotransmitters.

The well-known DA system from the substantia nigra (SN) to the CP gives off some collaterals in the globus pallidus (GP). There may be at least one other neurotransmitter

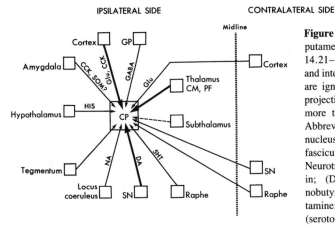

IPSILATERAL SIDE   CONTRALATERAL SIDE

Figure 14.20. Afferents to the caudate–putamen (CP). In this figure and in Figures 14.21–14.25, the relative sizes, locations, and internal structures of the various nuclei are ignored, as is the topography of the projections. A given tract may involve more than the neurotransmitters shown. Abbreviations: (CM) centrum medianum nucleus; (GP) globus pallidus; (PF) parafascicular nucleus; (SN) substantia nigra. Neurotransmitters: (CCK) cholecystokinin; (DA) dopamine; (GABA) γ-aminobutyric acid; (Glu) glutamate; (HIS) histamine; (5HT) 5-hydroxytryptamine (serotonin); (NA) noradrenaline; (SOM) somatostatin. For references to the pathways shown in Figures 14.20–14.25, see Carpenter and Sutin (1983) and E. G. McGeer *et al.* (1984).

involved in this tract, although it could account for no more than 5–15% of the fibers (Connor, 1975; Deniau *et al.*, 1978c; Fibiger *et al.*, 1972; Hedreen, 1978; van der Kooy *et al.*, 1981b). A peptide worth consideration for a minor role in this tract is neurotensin. Neurotensin-immunoreactive cells are found in the SN (Jennes *et al.*, 1982; Uhl *et al.*, 1979), there is a dense mosaic of positive fibers and terminals in the striatum (GP > caudate > putamen) (Goedert *et al.*, 1983), and neurotensin facilitates DA release in *vitro* from striatal slices (de Quidt and Emson, 1983). The levels are generally reported as relatively high in both the CP and the SN, although there is some variability in the available data (Goedert and Emson, 1983; Emson *et al.*, 1982; Manberg *et al.*, 1982; Uhl and Snyder, 1976).

A major unsolved question is the biochemical nature of the thalamostriatal tract. Despite some reports, it appears highly unlikely that this tract is served largely by glutamate, aspartate, or ACh (P. L. McGeer *et al.*, 1977a, 1982).

The relatively minor projections from the dorsal raphe and locus coeruleus are parts of the ascending serotonergic and noradrenergic systems of the brain. A second neurotransmitter (GABA) may be involved in the raphe–striatal tract (Belin *et al.*, 1983).

Lesion evidence indicating histamine (HIS) innervation of the CP is consistent with its origination in hypothalamic cell bodies (see Chapter 11).

The majority of the cholecystokinin (CCK) innervation of the CP seems to come from the amygdala (30%) and from the pyriform cortex or claustrum or both (70%) (D. K. Meyer *et al.*, 1982). The projection from the amygdala to the CP may also contain somatostatin (SOM). Other suggested sources of SOM innervation of the CP are the cortex, the periventricular nucleus of the hypothalamus, and the tegmental area. About half the SOM in the CP appears to be in intrinsic neurons (DiFiglia and Aronin, 1982; Finley *et al.*, 1978; Graybiel *et al.*, 1981). It has been estimated that 4–5% of striatal neurons in a rat contain SOM, and these neurons are distinct from the fewer than 5% that are intrinsic cholinergic neurons (Vincent *et al.*, 1983d); some of the SOM cells also contain neuropeptide Y [see Table 5.2 (Section 5.2)].

The pallidostriatal tract appears to be GABAergic and to land largely on the SOM neurons. There are also, of course, many GABA interneurons in the CP, as well as some thyrotropin-releasing hormone (TRH) perikarya (Spindel *et al.*, 1980). Relatively little is known about the TRH system.

Projections from the Caudate–Putamen (Figure 14.21). Five neurotransmitters have so far been implicated in these projections. There is excellent evidence for enkephalin (Enk) as a neurotransmitter in projections from the CP to the GP (Brann and Emson, 1980; Del Fiacco *et al.*, 1982; Somogyi *et al.*, 1982b; cf. Haber and Elde, 1981). The CP contains a dense network of met-Enk fibers and terminals (Yang *et al.*, 1983), so there may also be Enk interneurons (see Chapter 12).

Substance P (SP) projections arise throughout the CP and seem to project, probably by collaterals, to all three other nuclei, i.e., to the GP, the entopeduncular nucleus (EP), and the SN (Brownstein *et al.*, 1977; Jessel *et al.*, 1978; Staines *et al.*, 1980). In primates, the external segment (GP) shows relatively little staining for SP, with the staining being progressively much denser in the internal segment (EP) and in the SN (Beach and McGeer, 1984; Haber and Elde, 1981). Dynorphin (DY) distributions are similar to those of SP (Vincent *et al.*, 1982a), and this endorphin may occur in similar projections to all three nuclei; that to the SN is the best established (Vincent *et al.*, 1982b).

GABA pathways project from the CP and GP to the SN. Cells projecting to the SN are arranged in shells on the lateral border of the CP and the lateral border of the GP (Araki *et al.*, 1985). There are also GABA projections from the CP to the GP and from the CP to the EP. The GABA projections to the GP may be partly by collaterals of those to the SN (see Chapter 7).

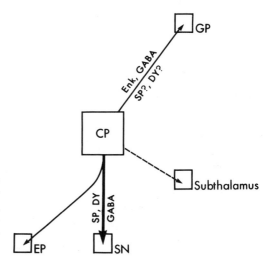

Figure 14.21. Projections from the caudate–putamen (CP). Abbreviations: (EP) entopeduncular nucleus; (GP) globus pallidus; (SN) substantia nigra. Neurotransmitters: (DY) dynorphin; (Enk) enkephalin; (GABA) γ-aminobutyric acid; (SP) substance P.

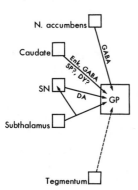

Figure 14.22. Afferents to the globus pallidus (GP). Abbreviations: (SN) substantia nigra. (DA) dopamine; (DY) dynorphin; (Enk) enkephalin; (GABA) γ-aminobutyric acid; (SP) substance P.

CCK may also be present in a projection to the SN, since injections of kainic acid (KA) into the rat CP lead to losses of CCK in the SN as well as the CP (Emson *et al.*, 1980b).

Afferents to the Globus Pallidus (Figure 14.22). Two of the main afferents have been previously described and comprise inputs from the CP and the SN. The input from the CP contains Enk and probably GABA; some SP and DY fibers may also occur. The projection from the SN certainly contains DA and may contain other substances.

Projections to the GP also arise in the subthalamus, the nucleus accumbens, and possibly the tegmental area. The transmitter that serves the tract from the subthalamus is almost certainly glutamate, despite some early reports that it might be GABA. The projection is excitatory; it is believed to have collaterals to the SN and EP (Deniau *et al.*, 1978b; van der Kooy and Hattori, 1980). The projection from the nucleus accumbens—which may be considered a ventral extension of the CP—is to the ventral pallidum rather than the GP proper and seems to use GABA (see Figure 14.24).

Projections from the Globus Pallidus (Figure 14.23). The GP projects mainly to the subthalamus (R. Kim *et al.*, 1976), but also to the EP, the SN, the CP, the dorsal

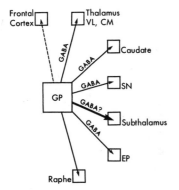

Figure 14.23. Projections from the globus pallidus (GP). Abbreviations: (CM) centrum medianum nucleus; (EP) entopeduncular nucleus; (SN) substantia nigra; (VL) ventrolateral thalamic nucleus; (GABA) γ-aminobutyric acid.

raphe, and, rather sparsely, to the reticular and medial dorsal nuclei of the thalamus and to the frontal cortex (H. J. W. Nauta, 1979).

GABA is the only neurotransmitter that has been definitively identified in projections of the GP. The GP, like the EP, probably contains a very high concentration of GABAergic neurons (see Chapter 7).

Afferents to the Substantia Nigra (Figure 14.24). The SN, although one of the smallest of the nuclei of the extrapyramidal system, seems nevertheless to have one of the most complicated wiring diagrams. These involve a number of connections with the contralateral side that provide, with the bilateral cortical innervation of the striatum and the commissural glutamate–aspartate tracts, an anatomical opportunity for hemispheric integration. Major afferents to the SNR and SNC that come from, respectively, the CP and the GP have already been discussed. There is also a projection from the nucleus accumbens to the SNC and from the midbrain tegmental area, most probably the pedunculopontine nucleus, to the SNR and SNC. The latter projection is bilateral, although the ipsilateral component is much heavier than the contralateral component. The SN also receives projections from the cortex, the lateral habenula, the parafascicular nucleus of the thalamus, the subthalamus, the amygdala, the raphe, and the hypothalamus. All these projections seem to be exclusively ipsilateral except for the hypothalamic afferents, which are bilateral. However, those from the ipsilateral hypothalamus come from the anterior portion, while those from the contralateral come from the posterior regions. As Figure 14.24 indicates, most of the transmitters are unknown, although those from the CP, GP, and raphe have been discussed.

Lesion studies indicate that at least part of the ipsilateral projection from the pedunculopontine nuclei (parabrachial area) may be cholinergic (E. G. McGeer and P. L. McGeer, 1984). Nicotinic binding sites are reportedly high in the SN (see Chapter 8).

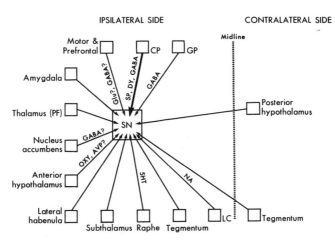

Figure 14.24. Afferents to the substantia nigra (SN). Abbreviations: (CP) caudate–putamen; (GP) globus pallidus; (LC) locus coeruleus; (PF) parafascicular nucleus; (AVP) arginine-vasopressin; (DY) dynorphin; (GABA) γ-aminobutyric acid; (Glu) glutamate; (5HT) 5-hydroxytryptamine (serotonin); (NA) noradrenaline; (OXY) oxytocin; (SP) substance P.

Almost certainly, other neurotransmitters are also involved in afferents to the SN. Histamine concentrations, histidine decarboxylase activity, and high-affinity glycine uptake are all relatively high in the SN, suggesting the possibility of afferents or internal neurons that use these neurotransmitters. Pollard *et al.* (1978) indicated that there were most probably histamine afferents to the SN. They presumably originate in the hypothalamic histamine cells (see Chapter 10). Some peptide binding sites, notably for neurotensin, also show high levels in the SN, suggesting the presence of the peptides in afferent systems. Reduction of neurotensin binding sites in the SN in parkinsonian brains is consistent with innervation of the DA cells of the pars compacta by this peptide.

Projections from the Substantia Nigra (Figure 14.25). Only two neurotransmitters, DA from the pars compacta and GABA from the pars reticulata, have so far been identified in projections from the SN. There are dopaminergic tracts to the CP, GP, amygdala (Meibach and Katzman, 1981), subthalamus (L. L. Brown *et al.*, 1979; Meibach and Katzman, 1979), olfactory bulb, and spinal cord. This part of the SN therefore provides not only an internal loop control to the CP, but also rich interconnections with the limbic system.

GABA is a neurotransmitter in the projections to the superior colliculus and the thalamus, some of which are collaterals (Bentivoglio *et al.*, 1979; M. E. Anderson and Yoshida, 1980). GABA is also contained in the projections to the tegmentum, and one group has estimated from physiological studies that some 40% of pars reticulata neurons send inhibitory fibers to the tegmentum, with about half these fibers having collaterals to the thalamus, the superior colliculus or both (Niijima and Yoshida, 1982; Beckstead and Frankfurter, 1982). Collaterals of these projection neurons within the SN have also been suggested (Grofova and Fonnum, 1982).

Internal Organization of the Nuclei.   These pathways may also provide an internal organization of the various motor nuclei. The neostriatum appears to be organized into a mosaic with preferential distribution of cells and fibers and thus preferential neu-

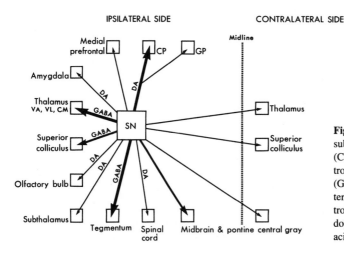

Figure 14.25. Projections from the substantia nigra (SN). Abbreviations: (CP) caudate–putamen; (CM) centromedian nucleus of the thalamus; (GP) globus pallidus; (VA) ventroanterior thalamic nucleus; (VL) ventrolateral thalamic nucleus; (DA) dopamine; (GABA) γ-aminobutyric acid.

rochemistry. Two compartments have so far been defined: islands, or striosomes, and a matrix surround. It has been suggested that these compartments represent two parallel but separate processing systems within the neostriatum (Gerfen, 1984). In the monkey, the islands consist of aggregates of 1500–15,000 densely packed neurons in ellipses about 300–600 μm in dimension (Goldman-Rakic, 1982). In the human caudate and putamen, they are 0.3–1 mm in diameter and about 3–7 mm long (Beach and McGeer, 1984). They have dense opiate receptor binding (C. B. Pert *et al.*, 1976) and are enriched in Enk and SP (Graybiel *et al.*, 1981; Beach and McGeer, 1984). The matrix, on the other hand, has a high acetylcholinesterase activity (Graybiel *et al.*, 1981) and dense SOM-like fibers and receives preferential innervation from the prefrontal cortex (Goldman-Rakic, 1982). Islands receive preferential input from the medial frontal cortex, which is also called the "prelimbic cortex" because it is heavily innervated from the amygdala and hippocampus (Gerfen, 1984).

14.5.4. Physiology, Pharmacology, and Pathology of the Basal Ganglia

We can now begin to integrate the clinical information in Section 14.5.1 with the information on the biochemical neuroanatomy in Section 14.5.3. To do so, we will need to provide additional information from pharmacological, lesion, and pathological investigations of extrapyramidal diseases.

The extremes of extrapyramidal dysfunction have already been described: akinetic rigidity and chorea–athetosis. We shall consider these extremes separately. Parkinson's disease is the classic example of akinetic rigidity. In advanced stages, such simple tasks as rising from a chair can require the kind of concentration that goes with executing a battle plan. Instituting simultaneous motor activities such as rising from a chair and extending the arm to shake hands becomes impossible (Schwab *et al.*, 1954). The embarrassed patient sinks back into the chair, failing in both motor activities. Marsden (1982) describes this as a deficit in motor planning. The parkinsonian patient is additionally handicapped by an inability to execute ballistic movements, so he must creep up on his objective in small movements. It must not be concluded from these fundamental deficits that the basal ganglia themselves are responsible for initiating movement. Muscles frequently contract in advance of neuronal discharge from the pallidal and nigral output zones (Iansek and Porter, 1980; DeLong *et al.*, 1984). The difficulty in Parkinson's disease is a subtle one of processing of motor information within the basal ganglia so that appropriate output information is available to integrate with other motor circuitry.

Parkinson's disease involves a loss of DA neurons from the SNC and a degeneration of the nigrostriatal tract. A human model for the disease exists among drug abusers who have inadvertently ingested 1-methyl-4-phenyl-1,2,3,6-tetrahydropyridine (MPTP). Positron emission tomography (PET) scans employing 6-fluorodopa demonstrate that such individuals have a deficit similar to that of idiopathic parkinsonians in metabolizing this analogue of the DA precursor (see Chapter 9) (Calne *et al.*, 1985). Animal models of parkinsonism can be created in some circumstances with MPTP, with the catecholamine neurotoxin 6-hydroxydopamine locally injected into the SN area, with DA-receptor blockers, and with the amine depletor reserpine. All these pharmacological and lesion models of Parkinson's disease consistently point to the nigrostriatal dopaminergic loss as being the lesion primarily responsible for the motor deficits. The nigrostriatal tract is part of a closed feedback loop and is thus concerned with internal processing of neuronal information in the CP and, if dendritic release of DA (P. L. McGeer *et al.*, 1979; Ruffieux and Schultz, 1980; Cheramy *et al.*, 1981) is of physiological significance, with SN output neurons as well. What could this internal processing involve? There is no information on this point as yet, but pharmacological data suggest that it must also involve the cholinergic cells that are scattered throughout the neostriatum.

One of the most fundamental principles of the pharmacology of the basal ganglia is the antagonism between dopaminergic and cholinergic tone (P. L. McGeer *et al.*, 1961; P. L. McGeer, 1963). It has been known from the time of the great French physician Charcot (1825–1893) that parkinsonian patients respond to the anticholinergic belladonna derivatives, and this balance theory preceded knowledge of either dopaminergic or cholinergic brain neurons.

Drugs that block or deplete DA cause large decreases in steady-state striatal ACh levels, while drugs that enhance DA cause substantial increases (P. L. McGeer *et al.*, 1974b; Sethy and Van Woert, 1974; Agid *et al.*, 1975). This would be expected if dopaminergic nerve endings were exerting a tonic inhibition on cholinergic interneurons. The cholinergic system is an internal one to the neostriatum, so that effects of perturbing this balance must be transmitted to efferent neurons, many of which are known to be GABAergic or to use SP or DYN.

The opposite clinical syndrome is chorea–athetosis. Huntington's disease is the classic example, but the syndrome also occurs in Sydenham's chorea. Huntington's disease involves pan-neuronal loss in the caudate and putamen and, to a much lesser extent, the globus pallidus. Experimentally, KA lesions of the neostriatum (see Chapter 6), L-3,4-dihydroxyphenylalanine (L-dopa) loads, or cholinergic blockade will induce the effect. Hemiballismus, the most bizarre of the movement disorders, can be thought of as an extreme form of chorea–athetosis. It involves violent, flinging movements of the proximal muscles of the limbs and is always associated with a unilateral lesion of the subthalamic nucleus or its connections.

In monkeys or baboons, discrete blockade of GABA receptors will reproduce these syndromes (Mitchell *et al.*, 1985b). Local infusion of bicuculline into the ventrolateral border of the lateral GP will produce chorea; infusions into the subthalamic nucleus will produce hemiballismus. Thus, impairment of the GABA systems of the GP, or interruptions of the subthalamic closed-loop control of the GP, will produce these effects. Both pharmacological lesions produced hypoactivity in the medial pallidal segment as measured by 2-deoxyglucose autoradiography (Mitchell *et al.*, 1985b).

Yet another extrapyramidal disorder is dystonia, in which there is discoordinated contraction of agonist and antagonist muscles even without abnormal posture. There may be a superimposed dystonic tremor of repetitive movement constituting dystonic spasm. Good animal and drug models of this disease do not exist; also unknown are the causes of dystonia musculorum deformans and its variants, writer's cramp and torticollis. However, infarctions of the putamen will produce this syndrome (Burton *et al.*, 1984), again emphasizing the relationship of extrapyramidal disorders to interruption of neuronal processing within the basal ganglia.

14.5.5. Summary of Motor Control by the Basal Ganglia

We can now synthesize the information of Sections 14.5.1–14.5.4 into a proposed scheme for how the basal ganglia integrate their actions to contribute to motor control (Figure 14.26).

Action in the basal ganglia is initiated when a command for movement comes from the cortex via the excitatory glutamate pathway. The DA input to the CP is a "hormonal" one that acts as a closed feedback loop, providing a continual damping of "static" so that the precise topography of the excitatory command can be appreciated. Parkinson's disease interferes with this loop. It results in hypermetabolism of glucose in the GP as measured by the PET–fluorodeoxyglucose method (W. R. W. Martin *et al.*, 1984). It is significant that in the monkey, the substantia nigra DA neurons discharge *in vivo* after a behavioral cue, but before and during the movement that follows.

The inputs from the thalamus and many other regions are seen as being primarily a means of keeping the CP continually informed about the status of all the external systems involved in movement. The CP has an integrator role and feeds its results via both excitatory and inhibitory pathways to the SNP/GP where they are translated into excitation or inhibition of GABA pathways that carry the detailed orders contributing to the desired movement. The GP has its own further processing system, which includes an internal loop from the external GP through the subthalamic nucleus to the internal GP. Chorea and athetosis may result, and hemiballismus does result, from a failure of this loop.

The connections of the SNR are also important. In a number of laboratories, it has been shown that injections of various GABA or glycine antagonists into the SN produce turning behavior even in animals in which the connection to the striatum has been severed (e.g., Andrews and Woodruff, 1982; Papadopoulos and Huston, 1980). Motamedi and York (1980) provided physiological evidence that the SNR-to-tegmental and -tectal projections influence cervical cord spinal interneurons and may mediate the head-turning

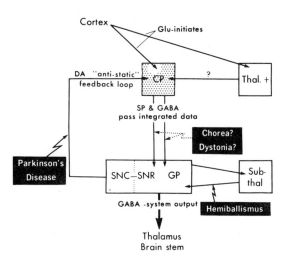

Figure 14.26. Hypothetical diagram of information flow in the basal ganglia. Parkinson's disease results from interruption of the dopaminergic feedback loop to the caudate–putamen (CP) and hemiballismus from the feedback loop to the globus pallidus (GP). (SNC,SNR) Substantia nigra, pars compacta and reticulata. Chorea and dystonia may result from failure of the SP- and GABA-integrated data flow.

behavior seen after basal ganglia activation. SNR neurons discharge mainly in relation to movement (Schultz, 1986b).

In addition to GP efferents, there is a high degree of convergence of SNR and cerebellar projections. Common target areas include the superior colliculus, the dorsal midbrain tegmentum, and the parafascicular and ventromedial nuclei of the thalamus. These overlapping inputs may allow for the coordination of basal ganglia and cerebellar influences on motor control.

Finally, there are many points of convergence between the extrapyramidal system and the limbic system (Mogenson *et al.*, 1980; E. G. McGeer *et al.*, 1984). The close relationship between these systems is discussed further in Chapter 15. The nucleus accumbens, the adjacent ventral caudate and putamen, and the olfactory tubercule are often referred to as the *ventral striatum*. There are close connections between these areas and the amygdala and allocortex. The SNC also projects to the amygdala and accumbens. Histamine provides an apparent connection with the hypothalamus, but there are also other tracts that complete a rich anatomical array of interactions between the two systems.

14.6. Synthesis of Various Neuronal Mechanisms Concerned with the Control of Voluntary Movement

We can now recapitulate the discussion by attempting to synthesize the contributions of cortex, basal ganglia, and cerebellum to voluntary movement. To do so, a proposed sequence of events is illustrated in Figure 14.27. Initially, there is the plan or program of a movement that results in the intention to move. This intention or idea achieves expression in patterns of excitation in the neurons of the SMA [see Figures 14.5 and 14.6 (Section 14.3.3)] which are projected to the association cortex and the motor cortex. The whole ensemble produces the readiness potential in diffuse scalp recording [see Figure 14.7 (Section 14.3.3)]. Meanwhile, the SMA and associated cortex activate both the basal ganglia [see Figure 14.26 (Section 14.5.5)] and the lateral cerebellum (the cerebellar hemispheres), initiating the open and closed dynamic loop sequences [see Figures 14.17 (Section 14.4.4) and 14.27] that fashion and guide the movement.

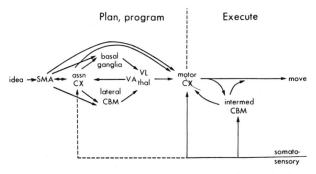

Figure 14.27. Diagrammatic representation of pathways concerned with the execution and control of voluntary movement. Abbreviations: (assn CX) association cortex; (lateral CBM) cerebellar hemisphere; (intermed CBM) pars intermedia of cerebellum; (motor CX) motor cortex; (SMA) supplementary motor area; (VA thal) ventroanterior thalamic nucleus; (VL thal) ventrolateral thalamic nucleus. The arrows represent neuronal pathways composed of hundreds of thousands of nerve fibers. To simplify the diagram, the intermed CBM is shown projecting directly to the motor CX and not via the VL thal as in Figure 14.16 (Section 14.4.4). Modified from G. I. Allen and Tsukahara (1974).

Loops playing through the basal ganglia (Figure 14.26) are required for the initiation of the movement. This is indicated by observations that in Parkinson's disease, there is failure or delay in executing a willed action (Figures 14.7B and 14.26). Loops playing through the cerebellum presumably would be able to incorporate stored motor programs (Brooks and Thach, 1981) [see Figure 16.22 (Section 16.7.3)] and project back to the motor cortex via the VL thalamus (Figure 14.17). There would also be projections back to the association cortex via the VA and VL thalamic nuclei, involving further dynamic loop circuitry (Figure 14.27), and giving the opportunity for further utilization of stored motor programs in both the cerebellar hemispheres and the association cortex (Figure 16.22). The synthesis of all these loop circuits provides what we may call "preprogrammed information" for the motor cortex. As a consequence of this preprogramming, the appropriate discharges are generated down the pyramidal tract as the final motor command for bringing about the desired movement. This story has been given a penetrating and vivid expression by Mountcastle (1975).

At the stage of motor discharge by means of the closed loop illustrated in Figure 14.16 (Section 14.4.4), the pars intermedia of the cerebellum makes an important contribution by updating the movement that is based on the sensory description of the limb position and velocity on which the intended movement is to be superimposed. This somatosensory input is indicated by the upward-pointing arrow to intermed CBM in Figure 14.27. The closed-loop operation is a kind of short-range planning as opposed to the long-range planning of the association cortex and the cerebellar hemispheres. Certainly, both cerebellar zones must cooperate in the performance of every skilled movement (Allen and Tsukahara, 1974). Superimposed on these inputs to the motor cortex is the servocontrol loop by the Ia afferents from the contracting muscle [see Figures 14.10–14.12 (Section 14.3.7)] that is indicated by the upward-pointing somatosensory-input arrow to the motor cortex in Figure 14.27.

The role of the basal ganglia can be envisaged by considering the consequences of the misfiring of its neurons. With damage to the neostriatum, as in Huntington's chorea, or with excess DA from L-dopa therapy, preprogrammed circuits play back in exaggerated manner to area 4 of the cortex even in the absence of cortical command. The result is the dancing movement from which chorea gains its name; the "volume control" of voluntary movement is turned too high and every movement is overdone. With damage to the nigrostriatal tract, as in Parkinson's disease, preprogrammed circuits can be played only with intense concentration. Even then, the "volume control" is too low and the movement is poverty-stricken. Background "static" produces a tremor at rest, which can sometimes be cured by lesions of the VL thalamic nucleus. The same poverty of movement can be produced in monkeys by local cooling of the globus pallidus (Hore and Vilis, 1976).

The role of the cerebellum, presumably the pars intermedia, in untrained exploratory movements is attested to by the clumsiness and slowness with which they are performed after cerebellar lesions [as illustrated in Figure 16.20 (Section 16.7.3)] (Gilman et al., 1981). The cerebrum and basal ganglia have to function in both preprogramming and updating without cerebellar cooperation. If only the cerebellar hemisphere with the circuits of Figure 14.17 is put out of action, a tremor often results, because the movement is so poorly preprogrammed that the pars intermedia can only ineffectively perform its normal function, which is updating a movement that is already a good guess. Presumably, the pars intermedia carries an immense store of information coded in its specific neuronal connectivities (Figure 14.16), so that in response to any pattern of pyramidal tract input, computation by the integrational machinery leads to an output to the cerebellum that appropriately corrects the pyramidal tract discharges.

In Brooks and Thach (1981), there is a review of a decade of investigation on the effect of localized cooling in the region of the cerebellar nuclei on trained movements of monkeys. Cooling of the dentate nucleus [DN in Figure 14.13 (Section 14.4.1)] results in slower, poorly controlled, even tremulous movements (dysmetria), which indicates failure of preprogramming. By contrast, cooling of the interpositus nucleus (IN in Figure 14.13) seemed merely to slow movements.

It is important to realize that no part of the cerebellar cortex or basal ganglia knows in general what other parts are doing. All the integration of input is going on in little subsets of machinery. Each small zone of the cerebellar cortex is working with some subsets of the total information input—e.g., from a limb—and the integration from the whole limb is accomplished in a piecemeal manner. It will be recognized that one could not integrate all the complex input coming from the movement of a limb in a single center. As would be expected, we find that the total information is broken down into subsets so that there are many centers for many different kinds of integration. These "computing centers," if we may so call them, are not sharply demarcated. On the contrary, they are diffusely organized, with much overlap and replication. But all this diversity of operation is integrated in the end in the smooth control of the evolving movement. It will be appreciated that it is this end result that is important, not all the confusing diversity of operation that may appear in the impulse discharges in single units at the various stages of the neuronal circuits. In summary, we can regard the basal ganglia and cerebellum as acting like the gross and fine control system on a target-seeking missile. They act similarly in that initially they aid in programming the movement, but they do not give a single message for correction of a movement that is off-target. Instead, they provide continuous sequences of corrections.

Here it is important to recognize the role of the cerebral cortex in evaluating the movement and guiding the learning experience that goes with each execution [cf. Figure 16.22 (Section 16.7.3)]. Athletes refer to this process as "concentration" and recognize the critical part it plays in their performance. The readout of the preprogrammed motion must be perfect. The tiniest failure in execution must be carefully analyzed and the readout improved in the next trial. Champion golfers, for example, go through this process hundreds of times each day.

When we come to the question of the reliability of basal ganglia and cerebellar action in control of movement, we have to consider not only the operational character of the continuous dynamic-loop controls, but also, and most importantly, the enormous numbers of cells that are involved, in contrast to the single units of Figures 14.16 and 14.17. At all stages of the synaptic relays, there is a convergence of many fibers onto the cell next in sequence, with the sole exception of the single climbing fiber input onto a Purkinje cell. Thus, there is averaging of many individual inputs, so that the unreliabilities of the units are minimized. It may be questioned whether this immense redundancy is worthwhile. The answer is that if reliability can be secured by having hundreds of units in parallel instead of one, then it certainly is a biological advantage. Moreover, we can assume that the multiplication of units gives in addition more subtlety and flexibility of performance in a way that we still do not fully understand. Anyway, this is what has happened in the evolutionary process, and the marvelous control of movements exhibited by birds and mammals shows that in neocortical, basal ganglia, and cerebellar design, it has paid off handsomely to go into numbers to give reliability, despite the counterweight of all the inherent complexity. We still cannot account for the relative numbers of the cerebellar cells, particularly for the enormous numbers of the granule cells—an estimated 30 billion in the human cerebellum. It would seem that in order to secure reliability, there is this prodigious neural cost. But in reality, the neural cost is low because the metabolism of single cells is so minute. The total brain consumes only 20 W for a total of tens of thousands of millions of nerve cells. Hence, in evolutionary terms, the animal could metabolically afford the prodigious neural cost because of its low metabolic expense.

15

Basic Behavioral Patterns

15.1. Introduction

Our discussion of the integrated action of the brain in Chapter 14 dealt only with the control of the musculoskeletal system. An attempt was made to account for how a willed movement could be smoothly executed, but no effort was made to deal with what may have motivated the movement. We need to consider *feelings*. They penetrate our consciousness, *motivate* our behavior, and produce noticeable changes in the functioning of our internal organs.

Feelings and motivation are *emotional* characteristics of the nervous system, but, like all other aspects of brain function, they must have an anatomical base and a physiological mode of expression. To date, relatively little is known about these aspects of brain function, but the practical value of such knowledge would be difficult to overestimate.

In this chapter, we shall discuss some basic behavioral patterns and, where possible, point out anatomical and biochemical correlations. Although we are concerned with the brain, it will be necessary for us to make some reference to visceral innervation and circulating hormones because they play such an important part in the overall behavioral response. To account for the actions of these hormones, as well as those of other agents that must work from within the brain itself, we introduce an entirely new signaling concept: *genotropic transmission*. Genotropic transmission refers to agents such as steroid hormones that enter the target cell and then act on nuclear receptors to initiate transcription of selective genes. This transcription process results in newly formed messenger RNA (mRNA) leaving the nucleus for the cytoplasm. The mRNA initiates translation into active protein products such as neurotransmitter receptors and other molecules that regulate cellular function. Genotropic effects differ from ionotropic or metabotropic effects in that they can by blocked by RNA synthesis inhibitors such as actinomycin D.

Attention will also be concentrated in this chapter on the operating characteristics of the parts of the brain that have been identified with behavior. We shall draw together relevant information from the actions of the individual neurotransmitters presented in Part II to discuss from a biochemical point of view the mystery of mental illness and the actions of psychopharmacological drugs. Throughout this chapter, a recurring theme will be the participation of certain metabotropic transmitters, particularly the catecholamines and serotonin. One of the amazing characteristics of the brain is that the relatively few neurons that use these transmitters (probably fewer than 0.01% of the total in the human) have apparently such a profound influence over so many others. We shall find that psychopharmacological agents and drugs of abuse interact as well with many other neurotransmitters.

15.2. Genotropic Action

At this stage in our knowledge of neuroscience, we can only speculate about the scope of genotropic action. It must be involved to a considerable extent in the building of the brain described in Chapter 13. But, as we shall see, there is much more to genotropic action than laying down the simple "hard-wiring" of brain. Involved as well is the chemical "burning-in" of some preformed circuits and their later response to signaling agents. The practical outcome is a permanent modification of behavioral patterns. Our model is the rat and its response to sex hormones as described principally by McEwen (1981) and his colleagues. Steroid hormones have been found to act on the nuclei of neurons of the hypothalamic–pituitary axis and limbic system to initiate transcription and to produce profound alterations in behavior.

During embryonic development, the gonads undergo differentiation into either ovaries or testes according to the genetically determined sex of the animals. Testicular secretions in the rat determine differentiation of the male reproductive tract. They also modify the brain to produce matching male patterns that differentiate the male brain from a female base pattern. Withdrawal of such male secretions through castration on the day of birth will result in a "feminized" male whose behavior is radically different from a male castrated as an adult because the base female brain pattern was never modified. Administration of testosterone to female rats shortly after birth, on the other hand, will modify the brain according to the male pattern, and the result will be a "defeminized" rat that will later show anovulatory sterility. If ovariectomized in adult life, the defeminized animal will show more masculine behavior following androgen administration and less feminine behavior following estrogen administration than ovariectomized normal female rats. These changes reflect permanent modifications in behaviorally oriented circuitry in the brain, particularly in the limbic system and hypothalamic–pituitary axis. Because of alterations in brain circuitry at this early stage, different transcriptional, or genotropic, responses to behavioral or hormonal stimulation will occur in later life.

The female brain retains a base pattern because it is never exposed to circulating sex hormones in the neonatal period. Newborn females have no testosterone, but they do have a neonatal estrogen-binding protein that complexes with circulating estradiol and other estrogens, thereby preventing entry of these female hormones into the brain.

The important question is what happens to testosterone when it reaches infant as well as adult rat brain tissue? There are two routes of metabolism which occur in different neuronal types: in one, testosterone is reduced and in the other it is aromatized (Figure

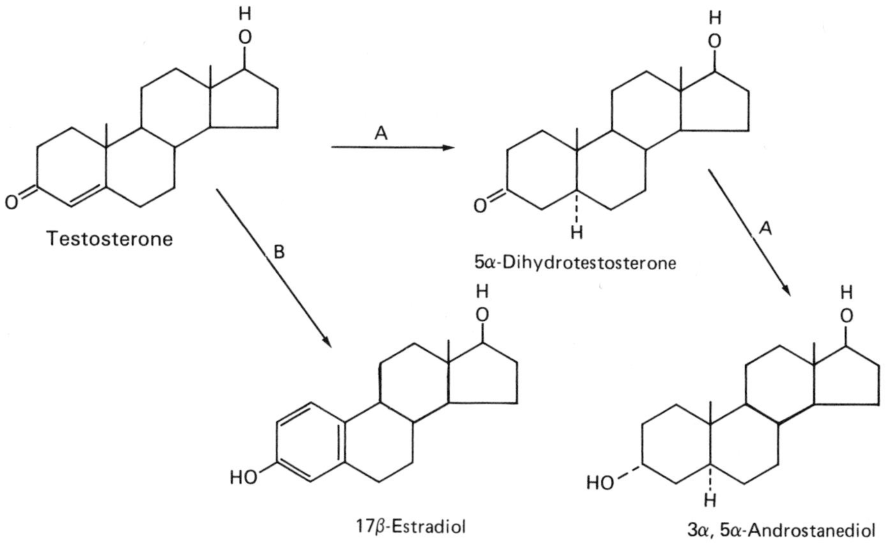

Figure 15.1. Testosterone reduction (A) and aromatization (B) reactions carried out by brain tissue.

15.1). The reduction route (A in Figure 15.1), where testosterone is converted to 5α-dihydrotestosterone, is a male or androgen-type chemical reaction. Such androgen-type reactions occur preferentially in the pituitary, midbrain, brain stem, and, to a lesser extent, in the hypothalamus. In the aromatization route, testosterone is converted to the female sex hormone 17β-estradiol (B in Figure 15.1). This conversion occurs preferentially in the hypothalamus and limbic structures. The continuing functional role of this neuronally produced female sex hormone is indicated by the fact that male sexual behavior can be facilitated by administration of estradiol to adults (P. G. Davis and Barfield, 1979) or blocked by inhibitors of this steroidal aromatization.

The "burn-in" of circuitry that takes place in infant male rats occurs at brain estrogen receptors. They are left unoccupied in neonatal females, and normal adult female sexual responses result.

These experiments demonstrate that a "female brain" can be mismatched with a male body and a "male brain" with a female body. The testosterone levels that reach the brains of neonatal rats determine the type of match.

We must now consider how these separate circuits in male and female brains come into play during adult sexual behavior. Let us begin by considering the estrous cycle of normal adult female rats. It is 4 days in duration. On day 1 of the cycle, estrogen and luteinizing hormone (LH) are low. At this time, a female rat will run away from, or fight off, a male. On day 2, estrogen levels start to rise. These levels peak on day 3, a few hours before a surge of LH that is the direct stimulus for ovulation. The circulating estrogen and progesterone prepare the reproductive tract and ovary to receive and maintain a fertilized egg. The behavior now changes. The female rat shows increased locomotor activity, dragging of the hindquarters to leave a scent, facilitation of the lordotic mating posture, an estrous scent, and decreased food intake. In short, efforts are made to attract a male. This behavior wanes on day 4 and does not commence again until the next cycle.

What is taking place in brain cells is revealed by studies of receptor binding (cf. Chapter 5) using radiolabeled steroids. [³H]Estradiol binds to brain tissue as shown in Figure 15.2. In the hypothalamus, the binding capacity corresponds to roughly 2000–3000 molecules per cell, but the figure is probably several times higher for those cells that are specifically binding the hormone (McEwen, 1981). The estradiol attaches to its protein binding site in the cytosol as shown in Figure 15.3. In cat uterus, the binding site is believed to be a tetrameric protein made up of 61,000-dalton subunits. Following binding, this tetramer is apparently hydrolyzed to a monomeric form through proteolytic activity. The estradiol–monomer complex is then translocated to the cell nucleus. The same process is presumed to take place in brain. [³H]Estradiol bound to the cell nucleus can be extracted using KCl in concentrations of 0.3 M or greater, and it is solubilized as a protein complex.

The estradiol–protein complex interacts with the genome to produce transcriptional activity. This is evidenced by a transient activation of hypothalamic cell nuclear RNA polymerase II activity following estradiol administration (Kelner et al., 1980). The RNA synthesis inhibitor actinomycin D blocks this transcription and blocks the estrogenic induction of female sexual behavior (Terkel et al., 1973), including the preovulatory LH surge. A similar blockade is produced by either of the protein synthesis inhibitors cyclo-heximide or anisomycin. Normally, estradiol induces feminine sexual behavior after a lag period of 18–24 hr, during which appropriate protein synthesis is taking place.

One of the most important of these synthetic actions is the production of progesterone receptors. Progesterone administration itself will stimulate sexual behavior and induce an LH surge with a lag time of no more than an hour. Progesterone facilitation of female sexual behavior is also blocked by the protein synthesis inhibitor anisomycin, indicating that it too is having a genotropic effect, although one that is much more rapid. Thus, the sequence of events that trigger sexual behavior would appear to be: first, transcriptional induction by estradiol to produce, after many hours, progesterone receptors in the cytosol; second, reaction of these receptors with progesterone to induce a new and more rapid genotropic transcriptional sequence. This latter sequence correlates with the LH pre-ovulatory surge.

Direct implants of estradiol have suggested that the basal medial hypothalamus is the region of brain that is primarily involved. Lesions of this region abolish lordosis, while electrical stimulation facilitates the response (McEwen et al., 1982). These changes are best seen when the normal female estrous cycle has been interrupted by ovariectomizing the female rat. In "feminized" males that have been castrated at birth, appropriately timed administration of estradiol and progesterone will produce lordotic behavior.

As previously mentioned, in male rats, testosterone acts in a dual fashion in separate classes of cells in the brain. In some cells, testosterone is converted to estradiol, inducing transcriptional changes that reinforce male behavior. In other cells, it undergoes reduction to dihydrotestosterone (see Figure 15.1), which also reinforces male sexual behavior.

Variations in response to sex hormones have also been noted in humans. Gladue et al. (1984) found no LH increase after an estrogen challenge in heterosexual men, a significant LH increase in homosexual men, and a greater increase in heterosexual women.

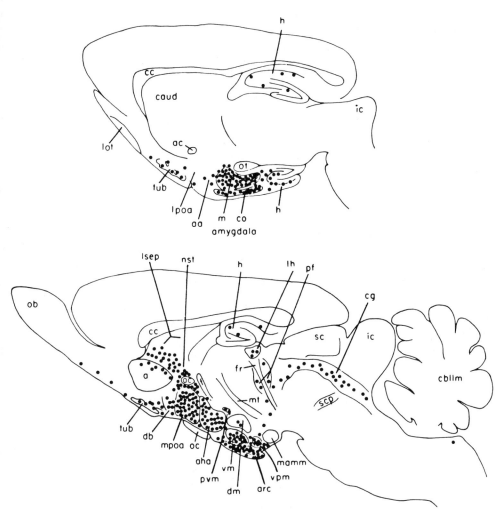

Figure 15.2. Uptake pattern of estradiol by rat brain represented schematically as the distribution of estrogen-concentrating neurons in the brain of the female rat in two sagittal sections. Estradiol-concentrating neurons are represented by black dots. Most labeled neurons are found in the amygdala (upper section) and hypothalamic, preoptic, and septal areas (lower section). Labeled cells are also found in the olfactory tubercle. Abbreviations: (a) nucleus accumbens; (aa) anterior amygdaloid area; (ac) anterior commissure; (aha) anterior hypothalamic area; (arc) arcuate nucleus; (caud) caudate; (cbllm) cerebellum; (cc) corpus callosum; (cg) central gray; (co) cortical nucleus of the amygdala; (db) diagonal band of Broca; (dm) nucleus dorsomedialis hypothalami; (f) fornix; (fr) fasciculus retroflexus; (h) hippocampus; (ic) inferior colliculus; (lh) lateral habenula; (lot) lateral olfactory tract; (lpoa) lateral preoptic area; (lsep) lateral septum; (m) medial nucleus of the amygdala; (mamm) mammillary body; (mpoa) medial preoptic area; (mt) mammillothalamic tract; (nst) bed nucleus of the stria terminalis; (ob) olfactory bulb; (oc) optic chiasm; (ot) optic tract; (pf) nucleus parafascicularis; (pvm) para-ventricular nucleus; (sc) superior colliculus; (scp) superior cerebellar peduncle; (tub) olfactory tubercle; (vm) ventromedial nucleus; (vpm) ventral premammillary nucleus. From Pfaff and Keiner (1973).

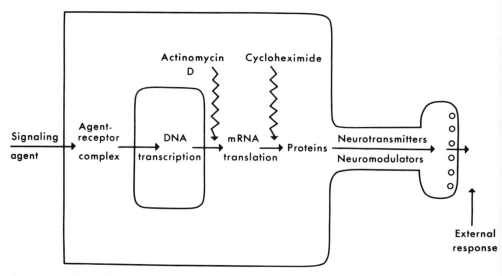

Figure 15.3. Proposed mechanism of action of a genotropic transmitter. The transmitter forms a complex with its cytosolic receptor. The complex is then translocated to the nucleus to initiate mRNA transcription. The mRNA synthesis inhibitor actinomycin D can arrest the process. The mRNA is translated into proteins. The protein synthesis inhibitor cycloheximide can also interrupt the process. The newly synthesized proteins give rise to neurotransmitter or neuromodulator response molecules.

The transcriptional changes that have been described are probably only a small part of a cascade of transcriptional events that are triggered by these particular hormonal agents. For example, other neurotransmitter receptors are affected. In estradiol-treated rats, binding increases to α_1-noradrenergic receptors in the ventromedial hypothalamus (A. E. Johnson *et al.*, 1985) and decreases to γ-aminobutyric acid type A (GABA$_A$) sites in the frontal cortex (O'Connor *et al.*, 1985). Changes in serotonin and α_2-noradrenergic receptors have also been reported (McEwen *et al.*, 1982). The lordotic response of females to progesterone is antagonized by cholinergic blockers (L. S. Kaufman *et al.*, 1985). Previously, there has been no connecting principle to explain the effects of various neurotransmitters on sexual behavior. A promising approach is revealed by these studies of transcriptionally induced modifications in levels of a spectrum of receptor molecules.

Glucocorticoids and mineralocorticoids also have direct action on selective neuronal populations. The mechanism appears to be similar to that described for the female sex hormones, although far less information is yet available. These compounds seem to interact with a protein receptor in the cytosol that then undergoes selective proteolysis. A subunit binding the steroid is then translocated to the nucleus, where it interacts with DNA to induce transcriptional events.

Table 15.1 summarizes available information regarding the classes of steroids that

TABLE 15.1
Central Nervous System Steroid Receptors[a]

Steroid class	Receptor concentration[b]	Intracellular binding protein[c]	
		Cytosol	Nucleus
Estrogen	P, H, A, S, Pr	8 S	4–5 S
Progestin	P, H, Pr, C	6–9 S	5 S
Androgen	P, H, A, S, Pr	≈7 S	3–4 S
Glucocorticoid	P, A, S, Hi	≈7 S	U
Mineralocorticoid	P, A, S, Hi	U	U

[a]After McEwen *et al.* (1982).
[b](P) Pituitary; (H) hypothalamus; (A) amygdala; (S) septum; (Pr) preoptic area; (C) cortex; (Hi) hippocampus.
[c](S) Svedberg sedimentation units; (U) unknown.

have been demonstrated to interact with proteins in the cytosol of neurons. The approximate size of protein that is then translocated to the nucleus is also given. The target areas of brain always include the limbic system, hypothalamus, and pituitary, although other areas can apparently be involved. It has been suggested, for example, that estrogens and progesterone may also act on motor areas because of their effects on tardive dyskinesia and chorea and that they act on limbic areas because of their influence on psychiatric disorders such as depression and schizophrenia (Maggi and Perez, 1985). An important approach is to measure high-affinity binding of various steroids to specific receptors in selective regions of brain as indicated in Table 15.1. For example, [^{3}H]corticosterone, which is the predominant adrenoglucocorticoid in the rat, binds to cell nuclei in the hippocampus at a level of about 16,000 molecules of corticosterone per cell nucleus (Sapolsky *et al.*, 1983). The corticomedial nucleus of the amygdala, the septum, and the induseum griseum are other areas that strongly bind this glucocorticoid. Proteins capable of binding [^{3}H]deoxycorticosterone and [^{3}H]aldosterone have also been reported for rat brain, but less is yet known about the details of this class of steroids.

These other steroid hormones also have a profound influence on development and behavior. Cushing's syndrome, which is caused by an overproduction of steroid hormones, is very frequently associated with depression or frank psychosis. Addison's disease, which involves a failure to produce such hormones, results in anorexia, apathy, irritability, and depression.

Thyroid hormones are also bound to cell nuclear receptors, and once again development is involved in early life, followed by metabolic and behavioral responses in later life. Excess triiodothyronine (T_3) in the neonatal period decreases the total number of brain cells (Balasz, 1971), while deficiency decreases the metabolic rate of brain tissue and can permanently impair body temperature regulation. T_3 receptor sites in cell nuclei increase markedly in the neonatal period in the rat and decrease gradually in the first postnatal month to adult levels. A distinctive T_3-labeling pattern in adult rat brain has been found (Dratman *et al.*, 1982). In adult life, hyperthyroidism in humans is associated with nervousness, irritability, and general emotional instability, while hypothyroidism produces apathy and depression.

It is tempting to speculate more generally on the possible implications of genotropism for behavior. The instinctive behavior of various animals is a never-ending source of fascination to professional scientists and the lay public. Environmental cues will often set off highly predictable responses from individual animals not previously exposed to such a stimulus.

The extent to which such patterned behavior might apply to humans is largely unexplored. However, some glimpses of the possible scope have come from observations of monozygotic twins separated at birth and raised in quite different social circumstances. It is known they have identical appearance. Can this be accompanied by identical behavior, even though the environmental upbringing is very different? Lykken (1982) described one pair of such separately raised monozygotic twin females who, when reunited in adult life, discovered each was in the habit of wearing seven rings. Another pair discovered that each had planned and built a circumarboreal bench in their backyard. Still another pair found that each had a fear of the water and that each had formed the habit of wading into the water backward and only to the knees. Furthermore, each had claustrophobia and each had separately and independently agreed to enter the psychiatrist's experimental chamber only if the door was tied open. Another pair were discovered to have developed the practical joke of sneezing in crowded elevators to enjoy the discomfort of other passengers. H. H. Newman *et al.* (1937) described monozygotic males reared by separate childless foster parents who had become repairmen in branches of the same telephone company. Each had obtained a fox terrier and named it Trixie. These are among many well-documented examples that have emerged from the limited number of studies that have been done on monozygotic twins separately raised and unaware of the other's existence until adult life.

We now consider briefly some of the biochemical aspects of learning and memory that involve a genotropic dimension. A more thorough neurophysiological presentation is given in Chapter 16. Following exposure to a stimulus, memory is converted from a labile to a permanent form over a period that typically extends for hours. Inhibitors of protein synthesis such as puromycin, anisomycin, and acetoxycycloheximide block long-term memory formation, but do not affect short-term acquisition. Since each of these three agents blocks a different step in protein synthesis, it has been concluded that inhibition of protein synthesis *per se* underlies the phenomenon (Vazquez, 1979). Similarly, blockers of RNA synthesis such as actinomycin D, 8-azaguanine, camptothecin, and amanitin produce a similar result (Agranoff, 1974, 1981). Evidence that this is a specific interference with genotropic action of neurons rather than a general interference with peripheral function comes from the fact that arabinosyl cytosine, a DNA blocker, does not affect learning or memory in the goldfish (Agranoff, 1974). These effects are entirely central and must involve the consolidation of information acquired through the stimulation of central neurons by sensory input. Circulating hormones are not critical to the process. Genotropic action is being stimulated in a fashion that is yet unknown, but that must involve central neurotransmitters rather than circulating hormones.

In summary, we have introduced the concept of genotropic transmission to account for (1) certain aspects of behavior that become "wired in" during early stages of brain development, and (2) a variety of neuronal responses to signaling agents that can be blocked by RNA or protein synthesis inhibitors. In the case of steroid hormones, the process involves binding to specific receptors in the cytosol, translocation of a hormone–protein complex to the nucleus, and induction of the transcription.

15.3. Central Coordination of Behavior

We consider now other methods of investigating the behavioral aspects of brain function. Electrical stimulation and ablation studies have particularly revealed the anatomically sensitive areas of brain. This information has in turn paved the way for detailed neurochemical investigation of these same areas. Amalgamation of anatomical and biochemical information has provided insight into the action of many psychopharmacological agents and paved the way for numerous theories regarding the causes of mental illness.

Hess was the first investigator to undertake a systematic exploration of the central integration of behavior commencing in the 1930s. He stimulated electrically various regions of the brain in conscious unanesthetized cats to observe their reactions. Hess recognized that such stimulation led to automatic responses amazingly similar to spontaneous behavior. Observations were recorded with moving pictures. The area was then lesioned with diathermy, often producing behavior opposite to the effects of stimulation.

Hess found that lesions in the caudal region of the hypothalamus produced functional defects opposite to those produced by lesions in the rostral region. He conceived of an ergotropic system designed to integrate sympathetic and somatomotor activities to produce behavioral patterns that prepared the body for positive action. The overall effects of ergotropic predominance would be arousal, increased sympathetic activity, enhanced skeletal muscle tone, and an activated psychic state, characterized by outgoing, exploratory behavior. He also conceived of a trophotropic system designed to integrate parasympathetic and somatomotor activities to produce rest and recuperation. With trophotropic predominance, there would be drowsiness and sleep, increased parasympathetic activity, decreased skeletal muscle tone and activity, withdrawal from external stimuli, increased appetite, enhanced digestive function, and elimination of bodily wastes. The systems would be in rhythmic opposition, resulting in alternating times of trophotropic predominance (sleep, relaxation, or apathy) and ergotropic predominance (alertness, exploration, or aggressive activity) (Hess, 1964).

He concluded that the integrating center for ergotropic function is the posterior hypothalamus, while that for trophotropic function is the anterior hypothalamus. Hess and many other investigators, however, have recorded highly similar behavioral sequences following stimulation or ablation of a wide variety of subcortical structures, indicating that the systems are distributed fairly broadly.

The types of behavior he induced ran the emotional range from happiness to rage and included many that were associated with vicarious sensations from the viscera. A desire to eat, or a desire to drink, was created by brain stimulation in the absence of any physiological need. Similarly, a desire for sex was created in the absence of appropriate stimulation, as was an urgency to urinate or defecate.

Table 15.2 summarizes some of the dominant behavioral patterns that various investigators have observed after stimulation or ablation of areas in the limbic system, diencephalon, and brain stem. These regions seem to be the ones most critically involved.

TABLE 15.2
Behavioral Responses Observed in Selected Regions of the Brain

Brain region	Response following:	
	Stimulation	Ablation
Cingulate gyrus	Tameness or aggression	Tameness, lack of anxiety
Septum	Defecation, micturition, tameness, hypersexuality	Irritability and attack
Amygdala	Aggression	Tameness, hypersexuality
Thalamus		
Paramedian	Somnolence, relaxation	—
Ventrolateral	Laughter	Apathy
Midline	Attack	—
Anterior	—	Docility
Dorsomedial	—	Rage
Hypothalamus		
Lateral	Eating and drinking	Adipsia and aphagia
Medial	Fight	Hyperphagia
Dorsomedial	Flight	—
Anterior	—	Trophotropic deficits
Posterior	—	Ergotropic deficits
Midbrain		
Central gray	Fight and flight	Hyperphagia

This table by no means covers all the phenomena observed and does not attempt to deal with the complicated effects seen when more than one of these areas is stimulated simultaneously (Hess, 1964; Mogenson and Huang, 1973; MacLean, 1958).

Apart from the hypothalamus, the structures that produce the most dramatic behavioral changes following stimulation or ablation are the amygdala and septum. Kluver and Bucy (1939) first noted that bilateral surgical excisions of the amygdala and its overlying pyriform cortex brought about drastic alterations in behavior. Monkeys, whose normally preferred diet is fruit, would eat raw meat or fish after ablation of this region, and wild animals lost their sense of fear and became tame. Animals also showed bizarre hypersexuality, as observed, for example, by Schreiner and Kling (1953). In their experiments, lesioned cats attempted to copulate with chickens.

Stimulation of the amygdala will produce behaviors that may be considered opposite. Mark and Ervin (1970) investigated a series of patients with epileptiform abnormalities of the temporal lobe whose problem was uncontrollable violence. In one of these patients, amygdaloid stimulation produced feelings ranging from elated and floating, through warm and pleasant, to unpleasant, according to the exact location of the electrodes. Stimulation of the medial part of the amygdala was dangerous, provoking the patient to lose control. This medial region of another patient with episodes of violence was stimulated while she was playing a guitar during a therapeutic session with her psychiatrist. As seizure activity took hold, the patient swung her guitar at the astonished psychiatrist, narrowly missing his head and smashing the instrument against the wall. Relief of episodes of rage was obtained in both these patients and several others, not by psychotherapy, but by surgical extirpation of the amygdala.

The septal and amygdala areas seem to have a partially reciprocal relationship.

Lesions in the septal area will produce the same effects as stimulation of the medial amygdala, i.e., irritability and a tendency to attack at the slightest provocation. Stimulation, on the other hand, may produce penile erection and other evidence of sexual arousal. It may also stimulate the desire to defecate. Thus, the suggestion has been made that one of the consequences of amygdala extirpation is to permit septal dominance to emerge.

Only mild effects are produced with stimulation of the cingulate gyrus. Both tameness and aggression have been reported. The cerebral cortex may play an inhibitory role. Bard and Mountcastle (1948) reported that removal of the neocortex produces placidity unless the amygdala is subsequently lesioned, in which case ferocity is observed, as in classic "sham rage."

Since the thalamus is the main sensory relay station of the brain, it would be reasonable to expect that stimulation or ablation of appropriate thalamic areas would greatly modify behavioral responses. Destruction of the dorsomedial nucleus or stimulation of midline nuclei increases irritability and attack reactions, while destruction of the anterior nucleus produces opposite behaviors. These ergotropic patterns are balanced by trophotropic ones in other areas. Stimulation in the paramedian fields of the thalamus in the region of the massa intermedia (the so-called "somnogenous zone" in the diencephalon) produces a decreased response to external stimuli that would normally provide a response, such as a noise of greater volume being required to disturb a cat from sleep. During stereotaxic operations for parkinsonism, Hassler and Reichert (1961) observed that stimulation of the inner margin of the ventral oral nucleus of the thalamus frequently produced smiles or laughter in awake patients, a response that could not be suppressed even when the surgeon asked for such suppression prior to stimulating the area. The patient seemed to find everything funny or to recall some amusing situation.

There is little doubt that the most prominent area for observing behavioral responses following stimulation or lesioning is the hypothalamus. In the cat, stimulation of the most medial sites results in considerable sympathetic arousal, with hissing, flattened ears, wide eyes, and a tendency to strike out at random with unsheathed claws. If the hypothalamus is stimulated more laterally and ventrally, the animal stalks quietly, but will attack a rat with lethal effect (fight). In this same perifornicular region, stimulation will also induce feeding and drinking behaviors, while medial stimulation will inhibit these behaviors. Thus, the concept has emerged that facilitatory systems for ingestive behavior are represented in the lateral hypothalamus, while inhibitory ones are in the ventromedial hypothalamus. Lesions of this ventromedial hypothalamus will cause rats or cats to eat incessantly until they become grotesquely obese. The role of the hypothalamus in feeding and drinking has been the subject of many reviews (see Epstein, 1971; Fitzsimons, 1972; Mogenson and Huang, 1973; J. C. Brown et al., 1983; Stricker and Zigmond, 1984; Rolls, 1984).

Lesioning of the central gray area of the midbrain will blunt rage reactions produced by hypothalamic stimulation, while midbrain stimulation will facilitate such attack. It is not known whether ascending or descending pathways or both are involved.

To summarize the results of these diverse experiments, it can be said that complex behavior can be triggered by external stimulation of widely separated subcortical areas of the brain, but especially the hypothalamus, amygdala, and septal areas.

Olds and Milner (1954) were responsible for an exciting yet remarkably simple advance over the external electrical stimulating techniques of Hess. When investigating subcortical centers in rats that might provoke an avoidance response, they found, to their surprise, that in certain brain areas the rats seemed to enjoy the stimulation. As Olds describes a typical rat he had prepared:

> He begins to search and pursue, eagerly after his very first stimulation. . . . Sometimes while the animal is self-stimulating, the circuit is cut off by the experimenter, so that stepping on the pedal will not produce any brain shock; in this case an animal that is self-stimulating will give a series of forceful ("frustrated-looking") responses and turn away finally from the pedal to groom himself for sleep; but he will go back from time to time and press the pedal (as though to make sure he is not missing anything).

Old and Milner designated as "reward centers" those areas of the brain in which such self-stimulation could be elicited and as "punishment centers," the areas in which no self-stimulation could be elicited and the rat would work to prevent being stimulated.

Olds concluded that about 35% of the cells of the rat brain lay in reward centers, about 5% in punishment centers, and 60% in neutral areas. Neocortical areas were neutral. The paleocortex showed rates of approximately 200 stimulations per hour, which increased up to 3000 per hour in certain subcortical nuclei, particularly in the septal area, up to 5000 times per hour in the hypothalamus, and up to 7000 times per hour in certain areas of the brain stem connected with the medial forebrain bundle. Rats were taught to run a maze for the privilege of self-stimulation. At times, they would run faster for electrical stimulation than for food, even when they were starved. Given the choice between food or electrical stimulation at the end of the maze, the rats would choose electrical stimulation.

In other experiments, a rat was obliged to run across a grid which gave successively more unpleasant shocks in order to reach the pedal for stimulation. The shock delivered by the grid could be raised to the point where the rat would forego the pleasure of brain stimulation and no longer cross the box. But the animal would accept electric shocks roughly twice as powerful for self-stimulation as for food, even if it had been starved. The implication, then, is that electrical stimulation in these areas satisfied the equivalent of hunger centers in the brain and that the electrical stimulation was more pleasant than the sensation that the brain received following food intake (Olds, 1962).

Since these historic experiments, literally thousands of rat self-stimulation studies have been undertaken. Self-stimulation is obtained in virtually all areas of the limbic system and, to a lesser extent, in areas of the extrapyramidal system. The medial forebrain bundle is the most preferred region of all. Comparable studies have also been undertaken in animals with more developed brains such as cats, monkeys, and dolphins.

Opportunities to make observations in conscious humans during subcortical electrical stimulation have also been presented by patients suffering from epilepsy, brain tumors, intractable pain, and mental illness. The pattern previously described for animals has so far been generally corroborated. Reward centers have been found in the human associated with intense, but nonspecific, feelings of well-being, with pleasant sensations ascribed to distinct parts of the body, and with sexual arousal. Punishment areas have also been found, the stimulation of which can elicit terror, anger, or pain (Delgado, 1976; Sem-Jacobsen, 1976). The implication of all these studies is that much learning and decision-making, which presumably take place chiefly in the cortex, are directed toward stimulating pleasure centers and away from stimulating punishment centers. In other words, the cells that are "neutral" for reward or punishment are marshaled to direct the animals' behavior in such a way as to provide a sensory input that is "just right" for pleasure and punishment cells. That many of the functions associated with pleasure and punishment

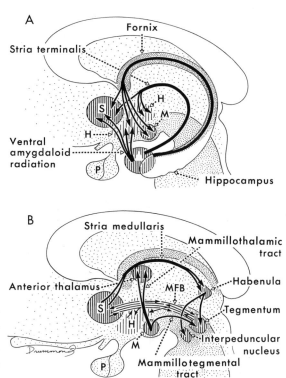

Figure 15.4. Some anatomical interconnections between the limbic system and brain stem. Abreviations: (A) amygdala; (H) hypothalamus; (M) mammillary body; (MFB) medial forebrain bundle; (P) pituitary; (S) septal area.

should have survival value for self and species is, no doubt, of more than casual consequence to our own existence.

Figure 15.4 indicates a number of the better-known connections of the midbrain, diencephalon, and limbic system. These may provide some insights into the functional interrelationships suggested by self-stimulation and the other behavioral phenomena described in Table 15.2. A key pathway is the medial forebrain bundle (MFB). It is shown in Figure 15.4 extending from the midbrain to the septal area(s) with prominent connections along the way to the hypothalamus (H). It serves as the main two-way-traffic highway between the brain stem, diencephalon, and limbic lobe. The septum has prominent connections with the amygdala (A) via the stria terminalis and the diagonal band of Broca. The septum also gives and receives fibers from the hippocampus via the fornix. The majority of fibers in the fornix, however, are traveling from the hippocampus to the mammillary bodies (M), with some being given off to other hypothalamic areas. From there, the mammillothalamic tract takes fibers to the anterior thalamus, then to the cingulate gyrus, and finally back to the hippocampus. The septum also sends fibers via the stria medullaris to the medial habenula, which sends its projection to the midbrain, particularly the interpeduncular nucleus. The hypothalamus also receives fibers from the amygdala and thalamus.

Close conjunctions are formed with the basal ganglia at a number of locations (see Figures 14.20, 14.22, 14.24 and Section 14.5.5). The nucleus accumbens forms an intermediate zone between the head of the caudate and the septal area. The tail of the caudate merges with the amygdala. Fibers of the mesolimbic and nigrostriatal dopaminergic tracts intermingle and lie adjacent to MFB fibers, and the internal globus pallidus sends fibers to the lateral habenula. This close relationship between behavior and extrapyramidal function was also mentioned in Chapter 9.

15.4. Central Amines and Behavior

Up to this point, we have considered only anatomical correlates of behavior. Are there chemical correlates as well? Do the behavioral pathways [Figure 15.4 (Section 15.3)] have neurotransmitters that are relatively specific to them and not used by other pathways with different functions? Are there drugs that specifically interact with, these neurotransmitters? And do they bring about chemical modification of behavior? The answer to all these questions is a partial yes. But the evidence is at best fragmentary; only the barest beginning has been made in exploring behavioral pathways. Most of the neurotransmitters for the pathways shown in Figure 15.4 are not known, and the mechanism of action of many major psychoactive agents is still doubtful.

The individual chapters of Part II made some reference to behavioral pathways and drugs that interact with them. The catecholamines and 5-HT are the neurotransmitters most involved in our behavioral story, no doubt because they have their richest innervation in the anatomical areas displayed in Figure 15.4. For example, noradrenaline (NA) [see Figure 9.8 (Section 9.4.3)], adrenaline [see Figure 9.9 (Section 9.4.4)], and 5-HT [see Figure 10.4 (Section 10.4.1)] all send rostral projections through the medial forebrain bundle (Figure 15.4) to serve the hypothalamus, septal area, and structures beyond. Just laterally, dopaminergic [see Figure 9.6 (Section 9.4.2)] fibers project to the cortex, accumbens, neostriatum, and amygdala. Dopaminergic fibers also project from the arcuate nucleus of the hypothalamus to the infundibulum. Histamine [see Figure 11.3 (Section 11.4)] has cell bodies in the hypothalamus and a rich plexus of terminals in most limbic system structures. It would appear that the overwhelming majority of all central pathways involving the catecholamines, 5-HT, and histamine are involved in behavioral and extrapyramidal circuits. Peripherally, of course, NA and adrenaline are the effectors of the sympathetic system, and dopamine (DA) modifies transmission in sympathetic ganglia. 5-HT occurs in argentaffin cells in the gut, and, while release is not directly stimulated by nerve fibers, it profoundly affects intestinal motility. Histamine cells are found in the gut as well.

Can we relate any of these amines and their pathways directly to behavioral response? Elmadjian (1959) was the first to do so. He measured urinary NA and adrenaline in athletes and psychiatric patients under various conditions. His experiments demonstrated that the "fight" sympathetic response was primarily associated with NA excretion, while the "flight" response was primarily associated with adrenaline excretion.

Elmadjian took pre- and postgame urine samples of professional hockey players. Typically, there was about a 6-fold increase in NA excretion, but some exceptions were noted. Two players who sat on the bench and were worried about their injuries had no increase in NA, but an appreciable increase in adrenaline. One player became involved in a fight and was ejected from the game. He showed a 9-fold increase in NA excretion and a 20-fold increase in adrenaline secretion.

Comparable results were obtained with measurements from boxers and basketball and baseball players (Elmadjian, 1959) and astronauts in training (Goodall, 1962). NA excretion was high whenever aggressive action was required without associated anxiety. If there was anxiety, there was an elevated excretion of adrenaline.

Public speaking caused a much sharper increase in plasma adrenaline than in plasma NA (Dimsdale and Moss, 1980), while the reverse was true of mental arithmetic (LeBlanc *et al.*, 1979). Under stress, coronary-prone individuals had a greater rise in plasma adrenaline than individuals not prone to coronaries (Glass *et al.*, 1980), indicating that plasma levels parallel urine levels.

The physiological response to intravenous infusion of these two catecholamines is also somewhat different. NA causes an increase in systolic blood pressure due to vasoconstriction, but no change in diastolic pressure. Adrenaline causes a greatly increased stroke volume of the heart, an increased systolic but decreased diastolic pressure, a mobilization of blood glucose, and a net vasodilation of blood vessels. Adrenaline clearly prepares the body for the more severe environmental stress of the "flight" reaction.

5-HT is associated with more passive, introverted activities related to comfort and relaxation. It is particularly involved with sleep (see Section 10.5.2). DA, of course, is primarily concerned with the execution of movement (Chapters 9 and 14), but it is also intimately involved with the actions of psychoactive drugs (see Sections 9.5 and 15.5). It also has an important governing influence on the endocrine system, particularly in promoting the release of growth hormone (GH) and inhibiting the release of prolactin (PRL) (see Chapters 9 and 12).

Other transmitters are suspected of playing a role in behavior, but the evidence is not as strong. These include acetylcholine (ACh) (see Chapter 8), GABA (see Chapter 7), and several of the peptides (see Chapter 12). Given the intimate association of neurons served by the various neurotransmitters and the wide distribution of many of them in limbic areas, it can be assumed that most will have at least some modifying influence on various behaviors.

By far the most convincing evidence as to the possible role of the aromatic amines in behavior has come from clinical observations on drug effects combined with basic studies on their molecular mode of action. The subtlety of action of any drug in a behavioral sense can never be appreciated through animal experiments. This is why detailed clinical observation is so important to establishing correlations between behavior and the anatomy and chemistry of the brain. For these reasons, we shall give brief descriptions of the behavioral effects of major psychoactive drugs, correlating, wherever possible, their spectrum of action with neurotransmitter amines.

15.5. Psychoactive Drugs

A classification of the most widely used psychopharmacological agents is outlined in Table 15.3, and the structures of representatives are shown in Figure 15.5. Detailed pharmacological and clinical information can be found in many useful textbooks on the subject (e.g., A. G. Gilman *et al.*, 1985).

15.5.1. Neuroleptics

Reserpine. Reserpine (Figure 15.5A) was the first of the neuroleptics. It is only of historical interest clinically, but it is still an extremely valuable research tool. It had an

TABLE 15.3
Classification and Properties of Some Psychoactive Drugs

Category	Typical examples	Broad pharmacological effects	Clinical
Neuroleptics	Phenothiazines Butyrophenones Thioxanthines Reserpine Benzoquinolizines	Decreased sympathetic activity; decreased motor activity; decreased conditioned avoidance reflex; no anesthesia	Antipsychotic effect; tranquilization; indifference to environment; in high doses, extrapyramidal symptoms produced; calming action
Antidepressants	Reuptake inhibitors MAO inhibitors Other	Reduced autonomic activity, especially of cholinergic systems; central sympathetic activation	Antidepressant action; stimulation
Anxiolytics	Benzodiazepines	Mild sedation, muscle relaxation	Tranquilization, but without any antipsychotic effect; sedation
	Barbiturates	As for other minor tranquilizers; also, there is some blockade of parasympathetic functions; in higher doses, ataxia and anesthesia	Tranquilization; hypnotic effect
Stimulants	Amphetamine	Increased motor activity; alerting behavior; reduction of appetite and sleeping time	Decreased fatigue; anxiety already present may be increased; psychotic symptoms, if already present, may be increased
Hallocinogenic agents	Lysergide Mescaline Psilocybin	Increased motor activity; variable autonomic effects; catatonia	Marked action on affect, perception, thought processes; psychotic disturbance may be aggravated

important impact in establishing the amine hypothesis of schizophrenia, and its side effects of parkinsonian-like extrapyramidal reactions and depression also did much to provide biochemical insight into Parkinson's disease and affective disorders.

When an animal is given a large dose of reserpine, the membranes of the amine intraneuronal storage vesicles are affected in a process that is Mg^{2+}- and ATP-dependent. 5-HT, DA, NA, adrenaline, and histamine are all depleted. The clinical result is tranquilization. There is indifference to environmental stimuli and a tendency to sleep, but arousal is nevertheless possible. In the initial phases, as amines are being released, there are paradoxical effects, including excitement and hypertension. There is also a characteristic flushing and diarrhea.

Reserpine has a prolonged action, apparently because it binds irreversibly to storage vesicles. Fluorescence histochemical studies have demonstrated that amines will begin to reappear within 12 hr after the administration of a single dose. The recovery of fluorescence commences in the cell body and only gradually extends to the nerve endings, so that full restoration of function lags behind. This delay is in keeping with the formation and axonal transport of new binding proteins to replace those inactivated by the drug.

It seems fairly certain that the parkinsonian-like and endocrine side effects of reserpine are due to depletion of DA and the antihypertensive effects to peripheral depletion of

A. Major Tranquilizers

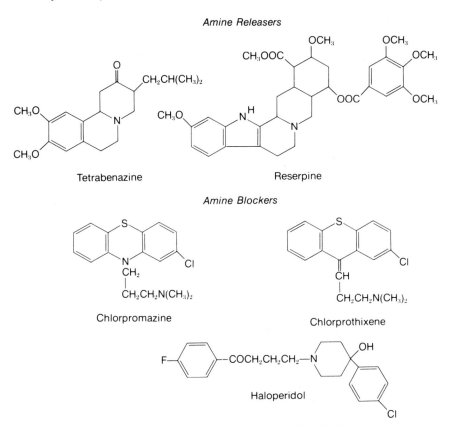

Figure 15.5. Structural formulas of some psychoactive drugs.

NA. But other actions, particularly those of tranquilization and depression, are still somewhat conjectural.

L-3,4-Dhydroxyphenylalanine (L-dopa) will temporarily reverse the sedation caused by reserpine in some species, but 5-hydroxytryptophan (5-HTP) will not. This has led to the conclusion that a catecholaminergic rather than a serotonergic system is involved in the tranquilization. DA is the most likely candidate, since the reversal takes place before significant amounts of NA are formed.

Further evidence that NA is not responsible is that 3,4-dihydroxyphenylserine, a synthetic β-hydroxy derivative of L-dopa that on decarboxylation produces NA directly, has proved to be relatively ineffective at reversing the reserpine tranquilization. It crosses the blood–brain barrier poorly, however, and may not have access comparable to that of L-dopa in reaching neurons.

The flushing and diarrhea experienced on initial dosing are probably due to 5-HT in addition to histamine release, because highly similar effects are seen in the carcinoid syndrome, in which rapid release of large stores of 5-HT takes place. It has been speculated that reserpine-induced depression is due to 5-HT depletion largely because *p*-chlo-

B. Antidepressants

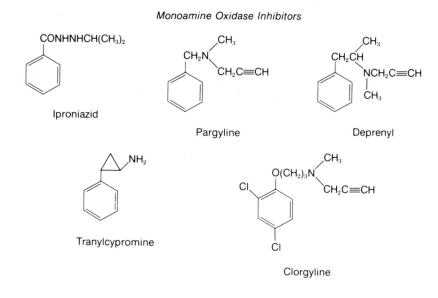

Monoamine Oxidase Inhibitors

Iproniazid

Pargyline

Deprenyl

Tranylcypromine

Clorgyline

Tricyclic Uptake Inhibitors

Atypical

Imipramine

Amitriptyline

Bupropion

Figure 15.5. (*continued*)

rophenylalanine (PCPA), which is a tryptophan hydroxylase inhibitor, and thus a selective 5-HT depletor, also produces depression in susceptible individuals.

Reserpine does not completely extinguish any of the many behavioral patterns it influences. Thus, none of them is absolutely dependent on the amines it depletes. Rather, the results argue for a modulatory effect of the amines on pathways the primary function of which may be governed by other neurotransmitters.

In addition to the rauwolfia alkaloids, some benzoquinolizines have a reserpine-like action, both pharmacologically and on the storage of monoamines. The best known of these is tetrabenazine (Figure 15.5A), which behaves qualitatively like reserpine, although it has a much shorter time of action.

C. Stimulants

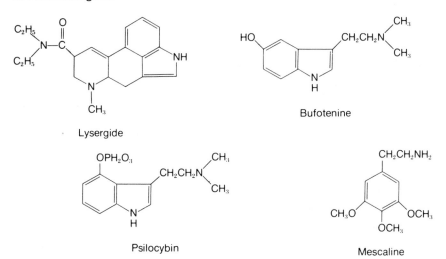

Amphetamine   Methamphetamine

D. Anxiolytics

Phenobarbital   Diazepam (Valium)

E. Hallucinogens

Lysergide   Bufotenine

Psilocybin   Mescaline

Figure 15.5. (*continued*)

Phenothiazines.   Phenothiazines were introduced into medicine for the treatment of intestinal worms. They were then utilized as antihistaminics and basal anesthetics. It was not until 1952 that Delay and Deniker (1952) of France first reported on the psychiatric effects of the most famous phenothiazine, chlorpromazine (Figure 15.5A). They were convinced that their results did not represent any simple relief of anxiety, but reflected a genuine antipsychotic action. For some years, skeptical psychiatrists thought chlorpromazine no more effective than placebo, but by the mid-1960s, over 50 million mental patients had been treated with a variety of phenothiazines and more than 10,000 articles had testified to their usefulness.

Delay and Deniker gave this classic description of the behavioral effects of chlorpromazine, on which their term ''neuroleptic'' was based:

> He is usually aware of improvement induced by the treatment but does not show euphoria. The apparent indifference or the showing of responses to external stimuli, the diminution of initiative and of anxiety without a change in the state of waking and consciousness or of intellectual faculties constitute the psychological syndrome attributable to the drug.

Phenothiazines have a prominent antiemetic effect that has been of important clinical benefit in such conditions as gastroenteritis, uremia, motion sickness, the nausea of pregnancy, carcinomatosis, and so on. They also strongly influence the endocrine system. Like reserpine, they block ovulation and interfere with menstrual cycles. They induce lactation, decrease testicular weight, and suppress growth. The latter can be overcome by injections of GH.

Phenothiazines block receptors for DA, NA, and 5-HT and therefore interfere with all the central and peripheral pathways described in Chapters 9 and 10. But many of the physiological actions are explicable, to a large extent, on the basis of DA blockade, including the antipsychotic, antiemetic, extrapyramidal, and endocrine effects. The antipsychotic effects are thought to be due even more specifically to blockade of D_2 receptors, presumably of the mesolimbic mesocortical pathways. Since phenothiazines prevent apomorphine-induced emesis, the antinauseant effect is probably due to the prevention of DA stimulation of receptors in the chemoreceptor trigger zone. The parkinsonian side effects are undoubtedly due to blockade of the nigrostriatal dopaminergic tract. Effects on the endocrine system, particularly suppression of growth and induction of lactation, could be due to blockade of hypothalamic–infundibular DA receptors, since DA promotes the release of GH and inhibits the release of PRL. The peripheral vascular effects are probably due to NA blockage. Phenothiazines can be classified into high-potency (e.g. fluphenazine) and low-potency (e.g., promazine) categories on the basis of the strength of their dopaminergic blockade. This potency varies inversely with anticholinergic and antihistaminergic action. Since control of movement in the basal ganglia appears to depend on a balance between ACh/histamine and DA systems (see Chapter 14), the potent tranquilizers with weak anticholinergic–antihistaminergic action have a great tendency to produce extrapyramidal side effects.

Butyrophenones.   Haloperidol (Figure 15.5A), the most important butyrophenone, was introduced in Europe in 1958 and in the United States in 1967. Although butyrophenones are highly important clinically, they do not provide a significantly different spectrum from that of some phenothiazines. There is also a considerable degree of structural similarity between butyrophenones and piperazine-substituted phenothiazines (Figure 15.5). Butyrophenones are virtually devoid of peripheral autonomic effects, but centrally they have been shown to block both DA and NA. They are potent antipsychotic and antiemetic agents and tend to produce parkinsonian side effects.

Although haloperidol blocks DA-sensitive adenylate cyclase, which is coupled to the D_1 receptor, its activity in this model system is far below what would be anticipated from its clinical potency (see Chapter 9). This was the principal reason for abandoning the concept that blockade of an adenylate cyclase-linked DA system (D_1 receptors) is responsible for antipsychotic action. However, the clinical potency of haloperidol does parallel its strong binding to D_2 sites. There is a good correlation between D_2 binding and clinical potency for both phenothiazines and butyrophenones.

Thioxanthenes. This family represents only a minor structural variation from the phenothiazines (Figure 15.5A), the nitrogen in the central ring being replaced by a carbon atom. Chlorprothixene (Figure 15.5A) is the main clinical representative, but others are in use as well. The spectrum of action of the thioxanthenes is almost identical to that of phenothiazines and butyrophenones. They are potent blockers of both D_1 and D_2 receptors. Only the *cis* isomers are active clinically or in binding assays.

15.5.2. Antidepressants

Until the late 1950s, the only pharmacological treatment for depression was stimulant drugs. Psychotherapy and electroshock treatment were therefore used almost exclusively. But the euphoriant properties of iproniazid (Figure 15.5B) were noted when it was being tested as an antitubercular agent in comparison with its analogue, isoniazid. Kline and colleagues (Loomer *et al.*, 1957) introduced iproniazid into psychiatry, coining the term "psychic energizer" to describe its action.

Monoamine Oxidase Inhibitors. As described in Chapter 9, there are two forms of MAO: A and B. MAO inhibitors that have been used clinically as antidepressants all inhibit both forms of the enzyme, although it has been suggested that only inhibition of MAO A is important to the clinical effect (Sandler, 1981). MAO inhibitors are now of largely historical interest because more effective and safer agents are available, but some clinicians still advocate their use in intractable depressive illness. Their danger is in toxicity to the liver, brain, and cardiovascular system. One particularly bizarre side effect can occur in MAO-inhibited individuals following the ingestion of strongly flavored cheese such as Camembert or Stilton. Violent headaches, hypertensive crises, and, in a few instances, cerebral hemorrhage and death have been brought on by the high content of tyramine in these sharp cheeses. If tyramine escapes destruction by liver MAO, it can act as a false transmitter for NA, causing the lethal increase in blood pressure (Asatoor *et al.*, 1963). Nonhydrazine MAO inhibitors such as tranylcypromine (Figure 15.5B) and pargyline (Figure 15.5B) are less toxic, but are not as effective as the tricyclic antidepressants (see below).

Mood elevation is the principal psychic effect, but the clinical response takes several weeks. In humans, the effects of overdosage include insomnia, agitation, hallucinations, hyperpyrexia, and convulsions. In some cases, treatment may bring about mania or hypomanic reactions.

A paradoxical effect of MAO inhibitors on blood pressure has never been adequately explained. Instead of the expected hypertensive effect due to enhanced NA, a hypotensive effect is often observed. Deprenyl (Figure 15.5B), which is selective for type B MAO inhibition, is useful in the treatment of Parkinson's disease and is said to have some antidepressant action despite the belief that inhibition of MAO A is more important.

Tricyclic Antidepressants. The tricyclic antidepressants include dibenzazepines, such as imipramine (Figure 15.5B) and dibenzocycloheptadienes such as amitriptyline (Figure 15.5B). Their three-ring structure is basically similar to that of the phenothiazines except that the sulfur atom of the central ring is replaced by an ethylene bridge. The result is to bend the molecule out of its planar shape and to alter substantially its spectrum of action. Modifications of the middle ring have been introduced, such as with maprotiline, which has a bridged structure.

The principal pharmacological action of the tricyclics is to interfere with the pump systems that are responsible for the reuptake of monoamines from the synaptic cleft back into the nerve ending, thus prolonging their postsynaptic action. Some, such as desipramine, nortryptiline, and protryptiline, are effective NA uptake inhibitors. Others, such as imipramine, chlorimipramine, and amitryptiline, preferentially inhibit 5-HT reuptake.

The first tricyclic to be investigated clinically, imipramine, was found not to be effective in schizophrenia as had been anticipated, but to be remarkably beneficial in depression (Kuhn, 1958). From that point on, many congeners have been synthesized and tested. They were found to be relatively safe and thus came to replace MAO inhibitors in many situations. Nonetheless, overdose can lead to complications, particularly with quinidine-like cardiac arrhythmias.

The fact that MAO inhibitors and tricyclics both act to enhance amine activity has considerably reinforced the notion that their clinical efficacy depends on this property and that a deficiency in NA, 5-HT, or both must underlie the depressive syndrome. There are two difficulties with this theory. The first is the delayed onset of clinical improvement. In animal studies, the effect of enhancing amine activity is extremely rapid, so that if a correlation existed, a clinical response within hours would be anticipated. The second difficulty is that new, atypical antidepressant agents have been found that have a non-tricyclic structure and that may not inhibit amine reuptake powerfully.

Atypical Antidepressants.   Trazodone is a heterocyclic triazole that is not an MAO inhibitor and falls in the atypical category. It inhibits 5-HT uptake, but also down-regulates β-adrenergic receptor binding sites (Tyrer and Marsden, 1985). Bupropion, a butylaminopropiophenone antidepressant with a structure analogous to that of amphetamine (Figure 15.5B), has less than 2% of the potency of imipramine in inhibiting NA uptake, but protects against 6-hydroxydopamine toxicity and is also thought to interact with the DA system. Mianserin, a bridged-ring derivative, is not a significant inhibitor of either NA or 5-HT reuptake, but blocks α_2-adrenoreceptors.

The failure to find a common structural or pharmacological profile for these diverse compounds has prompted Tyrer and Marsden (1985) to conclude that down-regulation of β-adrenergic receptors may be primarily responsible for the clinical efficacy of many antidepressants. This would account for the time delay in obtaining clinical improvement. The implication is that central adrenergic receptors are supersensitive in depression (Segawa et al., 1979). The theory does not explain why selective 5-HT reuptake inhibitors should be effective in many cases and is only one of a number of hypotheses suggesting that changes in various types of receptors are responsible for the clinical results. The possibility that depression may constitute a group of diseases with different underlying biochemical pathologies is strengthened by the clinical finding that individuals with similar symptoms respond differently to the different types of available drugs.

15.5.3. Stimulants

Amphetamine and methamphetamine (Figure 15.5C) and similar compounds are stimulants, but not antidepressants. In normals, they produce wakefulness, alertness, a decreased sense of fatigue, increased ability to concentrate or perform physical tasks, euphoria, and increased initiative and confidence. They can also produce headache, palpitation, dizziness, agitation, and, with the famous amphetamine "payback" principle, an overwhelming fatigue as the effect wears off. They suppress appetite and were at one time used to encourage weight reduction.

The clinical use of these stimulants is now largely restricted to narcolepsy and the treatment of hyperkinetic children. Illicit use, however, is common. The improvement in athletic performance they afford has been widely capitalized on by professional athletes with little evidence of serious harm. The same cannot be said for intravenous use by soft-drug addicts. The slogan "speed kills" can equally refer to driving or to "speed," the street name for amphetamine derivatives and the inevitable outcome of its promiscuous parenteral use for thrills. The drug produces an instant feeling of well-being on intravenous administration, but at high doses, this feeling is always replaced by a paranoia strikingly similar to the endogenous disease. The amphetamine psychosis is said to mimic schizophrenia more closely than any other form of drug-induced or toxic psychosis. It is this latter action that is of greatest behavioral interest, because it may provide a decisive clue to the etiology of schizophrenia, the most severe of the mental illnesses.

The effects of amphetamine, methamphetamine, and similar agents can probably be attributed exclusively to their interaction with catecholamines. This can take place in a number of ways. First of all, amphetamine acts as a releaser in the same way as mentioned previously for tyramine. It penetrates the nerve ending and releases the endogenous amine, producing an enhanced physiological result.

The extent to which the pharmacological effects of amphetamine reflect changes in dopaminergic systems, as opposed to noradrenergic or even adrenergic systems, is still controversial. The classic idea that the central stimulant and anorexic actions are mediated by NA whereas the stereotyped behaviors are mediated by DA has been challenged, with turnover rates suggesting that most of the effects may be due to DA (see Chapter 9).

A second important action of amphetamine is as a competitive inhibitor of MAO. Amphetamine, being a monoamine, is destroyed by MAO, and it competes successfully with catecholamines for this enzyme.

Amphetamine also inhibits reuptake of catecholamines, especially DA. Again, this appears to be a competitive phenomenon. In this manner, it acts similarly to cocaine or the tricyclic antidepressants. Amphetamine may also have some small influence at catecholamine receptor sites, but in view of the "payback" principle, which is reminiscent of tachyphylaxis, it is unlikely that this contributes in a significant way to its initial positive activity.

Whether the effects of amphetamine-like stimulants are due to DA or NA, they produce bizarre behavioral disturbances on overdosage. Thus, they lie on the borderline with the hallucinogenic agents (see Section 15.5.5).

15.5.4. Anxiolytic Agents

The classes of drugs we have considered so far, i.e., the antipsychotics, antidepressants, and stimulants, have all involved interactions with catecholamines and 5-HT systems. We now turn to another class, the anxiolytics, which are the most widely used of all prescription drugs. They are primarily sedatives and are of two chemical classes: the benzodiazepines and the barbiturates. They both appear to enhance GABA actions (see Chapter 7).

There are now at least 15 benzodiazepine derivatives in use. All the important ones have a 5-aryl-1,4-benzodiazepine ring structure as shown for diazepam in Figure 15.5D, and all have a similar pharmacological profile. Their effects are almost entirely on the central nervous system (CNS). As the dose increases, relaxation and sedation progress to hypnosis and then stupor. Anesthesia cannot be achieved without the use of other CNS drugs. They can cause anterograde amnesia, indicating an interference with the processes described in Chapter 16. They have muscle-relaxant and anticonvulsant properties and are therefore valuable in the treatment of disorders other than anxiety and tension. They are regarded as remarkably safe, although the depressant effects can be dangerous when enhanced with alcohol and other drugs.

As discussed in Sections 7.4 and 7.5 (see Figure 7.9), benzodiazepines affect GABA binding, and it is believed that this mechanism may account for most, if not all, of their physiological effects.

Barbiturates are the classical sedative agents. They have now been replaced for many uses by the safer benzodiazepines, although they retain an overlapping pharmacological profile. They are all derivatives of the barbituric acid ring structure, as exemplified by phenobarbital in Figure 15.5D. Some, such as thiopental, have a sulfur substituting for oxygen in the ring 2 position. In contrast to the benzodiazepines, barbiturates can induce anesthetic levels of CNS depression, although they are not used clinically for this purpose. They are still widely used as sedatives, and phenobarbitol continues to be one of the most valuable drugs for the treatment of epilepsy. Their dangers are CNS depression from overdose and habituation, both of which are much more severe than with the benzodiazepines.

Barbiturates are also thought to act by modifying GABA receptor mechanisms (see Section 7.5.1). They potentiate GABA-induced increases in Cl^- conductance. They do not potentiate this inhibitory action at glycinergic synapses. They inhibit glutamate-induced depolarization. At higher doses, they depress Ca^{2+}-dependent action potentials and reduce the Ca^{2+}-dependent release of neurotransmitters. Thus, the barbiturates appear to have a selective action in inhibiting channel operation, particularly the GABA-mediated Cl^- channel.

15.5.5. Hallucinogenic Agents

From a theoretical point of view, drugs that disturb normal behavior are as important as those that correct abnormal behavior. They are a major social problem instead of a significant benefit, but they may nevertheless be capable of yielding equivalent insights into the workings of the behavioral areas of the brain.

Hallucinogenic agents, also called "psychotogens," disturb the normal sensorium and thus affect mood. The terms are somewhat inappropriate, since they seldom cause either full-blown hallucinations or frank psychosis. They are a motley group of compounds with mechanisms of action that are not understood. We shall be concerned in this chapter with three main groups, which are thought to interact with the behavioral amines: indole derivatives, which may interact with 5-HT; phenylalkylamine derivatives, which may interact with catecholamines; and glycolic acid esters, which may interact with ACh.

Indolic Hallucinogens.   A number of these compounds have structures strikingly similar to that of 5-HT. The two closest are bufotenine [N,N-dimethyl-5-hydroxytryptamine (Figure 15.5E)] and psilocybin [N,N-dimethyl-4-phosphoryltryptamine (Figure

15.5E)]. The active principle of the latter compound is psilocin, the 4-hydroxy analogue of bufoteine, which is produced when the phosphoryl group is cleaved. Bufotenine is found both in plants and in the poison gland of the toad. Psilocybin is obtained from certain mushrooms found in the southern highlands of Mexico. N-Substituted tryptamines such as dimethyl- and diethyl-*N*-tryptamine are also hallucinogenic.

All these compounds produce disturbances in sensation and mood. They also produce thought disturbances. These latter effects have been pursued in relation to schizophrenia. Some investigators, for example, claim that bufoteine is a constituent of the urine of schizophrenic patients, normal subjects, or both. The implication is that schizophrenia may have as its fundamental cause the production of this hallucinogenic compound. Such results have not been generally corroborated, however. Although proof has been hard to obtain, it is thought that these materials interact with 5-HT receptor sites, with the N-methyl substituents making them poorer substrates than 5-HT for MAO.

By far the most dramatic hallucinogenic compound of all that have been discovered is D-lysergic acid diethylamide LSD) (Figure 15.5E). LSD is the most powerful known psychotomimetic compound. Its discovery by Hofman in 1943 was accidental. He had been synthesizing a series of ergot derivatives with a view to finding agents with oxytocic action. One day, he was unaccountably seized with peculiar sensations while at work. He went home to experience a period of exaggerated imagination coupled with, in his words, "fantastic pictures of extraordinary plasticity and intensive color." He retained enough insight to suspect that the effect might have come from the new compound LSD. He returned after a few days to ingest experimentally 250 μg, a dose he believed to be extremely minute. This is now known to be 5–10 times the dose of LSD required to cause disturbances in susceptible subjects, and the effect on poor Hofman was spectacular. From then on, there was no doubt that LSD was an extremely potent psychotogen (Hofman, 1963).

The social upheaval instigated by this compound is hard to credit even today. It not only revived theories of a toxic chemical etiology for schizophrenia, but also sparked a search for other psychotomimetic agents. An astonishing number of such agents were uncovered, including new synthetic compounds, as well as a variety of natural products that had been hidden in the legends and religions of primitive peoples. After a lag of almost 20 years, the use of LSD broke the bounds of scientific control and became the focus of a drug craze. Seldom has an agent unleashed such an unpredictable chain of events.

LSD is generally considered to be either a 5-HT agonist or antagonist. It has a powerful blocking action on peripheral smooth muscle preparations that respond to 5-HT. Centrally, there is considerable evidence that it stimulates rather than blocks some 5-HT receptor sites (see Chapter 10).

Phenylalkylamine Derivatives. The most famous representative of this group is mescaline, 3,4,5-trimethoxyphenylethylamine (Figure 15.5E). This constituent of the peyote button is famous for its use by Aldous Huxley and in religious rites of North American Indians. It produces vivid and striking visual hallucinations. There may be the appearance of brightly colored lights, vivid geometric patterns, animals, and people, all with a distortion of space and time. Several similar compounds such as 2,5-dimethoxy-4-methylamphetamine (DOM) and 2,5-dimethoxy-4-ethylamphetamine (DOET) are also hallucinogenic.

Exactly how mescaline and the amphetamine-like derivatives exert their hallucinogenic effects is not known. Their resemblance to the catecholamines, however, definitely suggests that it is through some interaction with them. In fact, the close structural resemblance of these compounds to DA has led to the suggestion that in schizophrenia, methylation of the ring hydroxyl groups of DA might lead to an endogenous toxin of structural similarity to mescaline. Unfortunately, there has been no solid evidence that significant amounts of this material are made in the human body, under either normal or abnormal circumstances.

Glycolic Acid Esters.   A series of compounds studied by Abood and Biel (1962), which are esters of a heterocyclic imino alcohol and a glycolic acid, proved to produce in humans delerium, hallucinations, and many abnormalities associated with psychotic-like behavior. Their effects are unpleasant, so that they have not become a problem on the illicit market. They have in common a strong anticholinergic action. The best-known representative is Ditran. Physostigmine has a temporary antidotal effect. Abnormal behavior is also noted with high doses of much milder anticholinergics such as atropine and hyoscine. Thus, ACh receptors are implicated in behavioral abnormalities along with 5-HT and catecholamine receptors.

Drugs of abuse continue to be a major social and medical problem, with new, so-called ''designer drugs'' continually adding to its scope (Langston and Langston, 1986).

15.5.6. Summary on Drugs

To summarize the results of this section on drugs, it can be said that the main interfaces of all these materials of widely differing structure are the catecholamines and 5-HT. It seems evident that perturbations in the metabolism of these amines can result in thought disorder and shifting of mood, as well as disturbances in movement, appetite, sleep, and functioning of the endocrine glands.

Yet it is equally clear that none of these functions is exclusively dependent on any of the aromatic amines, but involves many other types of neurotransmitters. The aromatic amines do, however, appear to modulate many functions. DA seems to be involved in modulation of movement, certain endocrine functions, and possibly also thought. NA seems to be involved in alertness, exploratory behavior for food, and similar behaviors. 5-HT, on the other hand, seems to be concerned with sleep, relaxation, and digestive function.

15.6. The Mystery of Mental Illness

We have now considered some anatomical pathways concerned with behavior and the neurotransmitters that govern their actions. We have classified psychopharmacological agents, drawing attention to the interaction of many of them with these same neurotransmitters. This leads to a consideration of the problems of mental illness and of whether the information developed so far leads to insights as to the etiology of this group of illnesses.

Mental disorders, by definition, are those in which physical damage to the brain cannot be detected postmortem by the usual pathological techniques. Nevertheless, it is clearly unacceptable to a large body of scientists that a deranged mind could exist in the presence of a normal brain. It is this conviction that spurs biological research into mental illness. It is strengthened not only by the effects of psychoactive drugs already described, but also by the knowledge that there is a genetic or familial nature to the major mental illnesses.

The genetic nature of schizophrenia seems to have been proven beyond question by the elegant studies of Kety and co-workers (Kety, 1974). Using the detailed birth records of the Danes, they were able to assemble statistically significant groups of offspring raised in foster homes to separate genetic from environmental factors.

Children with a schizophrenic genealogy raised by nonschizophrenic families had approximately the same attack rate as children with a similar genealogy raised by their own families. On the other hand, children with a clear genealogy raised by schizophrenic families did not experience a higher attack rate than the general population. This result confirmed the classic study of Kallman (1938), which had been criticized on the basis that the environment is different in schizophrenic compared with normal families. Kallman found that the incidence of schizophrenia in the general population was 0.85%. The concurrence rate for parent–offspring was 9%; for full siblings, 14%; for two-egg twins, 15%; and for one-egg twins, 86%.

Such impressive statistics have not been accumulated for manic–depressive psychosis, but the strong familial tendency has nevertheless been consistently noted in epidemiological studies. Thus, while it is recognized that environmental factors can precipitate a psychotic illness, there can be little doubt that there exists a constitutional vulnerability that must someday be revealed by appropriate physiological studies.

Another basis for belief in the theory of constitutional vulnerability is that psychoactive drugs frequently produce different responses in mental patients than in normals. For example, Gershon et al. (1975) administered L-dopa to psychiatric inpatients. In the nonpsychotic group, the behavioral effects were quite similar in quality and range to those seen in neurological patients such as parkinsonians. When L-dopa was administered to ten schizophrenic patients, however, a clear deterioration occurred in every instance. Therefore, L-dopa can induce paranoid psychosis in the occasional nonpsychotic patient, but dramatically aggravates psychopathology in schizophrenics. Similarly, when L-dopa was administered to manic–depressive patients, there was a tendency to precipitate hypermanic behavioral episodes that were strikingly similar to the natural episodes.

Amphetamine and methylphenidate are similar to L-dopa in that they cause dramatic exacerbation of symptoms in schizophrenics, but produce a reaction in normals only at very high doses.

Thus, in predisposed individuals, psychotic episodes can be precipitated by alterations in catecholamine metabolism. The effect is not restricted to catecholamines, however. In a well-known study conducted in 1961 at the U.S. National Institutes of Health, Pollin et al. (1961) administered various amino acids together with an MAO inhibitor to schizophrenic patients. Most of the amino acids brought about no changes, but L-tryptophan caused a feeling of euphoria, while methionine produced anxiety, a flood of associations, increased hallucinatory activity, and periods of disorientation.

In view of these observations, it is not surprising that biochemical theories of mental illness should emphasize disturbances in either catecholamine or 5-HT metabolism. Other neurotransmitters such as ACh and GABA have also been considered, and, as knowledge of the peptides accumulates, theories involving their action have been developed as well.

15.6.1. Hypotheses of Schizophrenia

Schizophrenia may represent a cluster of diseases with common symptomatology of thought disorder, delusions, and hallucinations. Crow (1980) has postulated two forms: type 1 corresponds to an acute disorder with florid symptoms but a good response to antipsychotic drugs; type 2 is a chronic illness with flattening of affect, poverty of speech, and loss of initiative. Such patients have a poor response to antipsychotic drugs and have been reported to have enlarged ventricles (Johnstone *et al.*, 1976). Crow suggests that type 1 schizophrenia is primarily a physiological abnormality, while type 2 may involve organic brain changes.

Dopamine Hypothesis. The most prominent theory of schizophrenia suggests that there is an abnormality in DA metabolism. The main evidence is as follows: (1) Nearly all antipsychotic drugs block DA D_2 receptor sites. Clinical dosage is in proportion to the strength of binding to these receptors. (2) Agents that will deplete DA, such as reserpine, or block its synthesis, such as α-methyl-p-tyrosine, or inhibit its release, as with low doses of apomorphine, will relieve schizophrenic symptoms (Snyder, 1982). (3) Excessive systemic doses of CNS stimulants, such as amphetamine, that are primarily false transmitters for catecholamines, will produce a drug-induced psychosis that is extremely difficult to distinguish from the disease itself. Low doses of amphetamine will exacerbate schizophrenic symptoms (Angrist *et al.*, 1980). (4) In about 15% of cases receiving high doses of L-dopa for other therapeutic purposes, a psychotic reaction is produced (Gershon *et al.*, 1975). Direct evidence to back this theory is scanty. Homovanillic acid (HVA) levels are not increased in the urine and cerebrospinal fluid (CSF), and DA and tyrosine hydroxylase levels are not elevated postmortem (P. L. McGeer *et al.*, 1978a; Rodnight, 1983). An exception may be the left amygdala, where substantially elevated DA levels have been reported (Reynolds, 1983). DA receptor binding sites of the D_2 type have been reproducibly found to be elevated postmortem in schizophrenic brain in the caudate, nucleus accumbens, and olfactory tubercle (Owen *et al.*, 1978; T. Lee *et al.*, 1978; Cross *et al.*, 1981). However, it is generally believed that this increase reflects only treatment with neuroleptics, since in animals it can be shown that such treatment significantly increases D_2 receptor numbers (Snyder, 1982). Owen *et al.* (1981) did find increased D_2 receptor numbers, though lesser in degree, in schizophrenic patients who were drug-free for 6 months prior to death. Moreover, Seeman *et al.* (1984) reported a bimodal pattern of D_2 receptor binding in schizophrenic basal ganglia that was independent of neuroleptic treatment. One subgroup had levels averaging 1.26 times normal and the other 2.3 times normal. Seeman and colleagues interpret the result as evidence for two distinct categories of schizophrenics. In support of this interpretation, there was no significant increase in D_2 receptors in a series of patients dying from Huntington's disease who had been treated with neuroleptics compared with other, nontreated Huntington's patients. K. L. Davis *et al.* (1985) have correlated the severity of schizophrenic symptoms with plasma HVA levels.

Noradrenaline Hypothesis. There is a close relationship between DA and NA metabolism and much overlap in the action of drugs on these catecholamines. It is therefore logical to suspect the possibility of some derangement in NA metabolism. Moreover, some neuroleptics, such as clozapine, are only weak DA blockers and more potent α-adrenergic antagonists. NA levels have been reported to be abnormally high (Farley *et al.*, 1978; R. J. Wyatt *et al.*, 1981) in schizophrenia. Significantly increased NA concentrations in the CSF have been reported (Gomes *et al.*, 1980; Sternberg *et al.*, 1981; Kemali *et al.*, 1982). Castellani *et al.* (1982) and Kemali *et al.* (1982) have also reported abnormal levels in the plasma of schizophrenic patients. Again, such findings need to be replicated and separated from the effects of drug treatment.

Serotonin Hypothesis. Woolley and Shaw (1954) put forward the original postulate that some form of defective 5-HT metabolism might be at the root of mental illness. The basis of speculation is the interaction of 5-HT with agents known to produce behavioral changes. LSD is a powerful blocker of some 5-HT receptor sites, and hallucinogenic agents such as psilocybin, bufotenine, and dimethyltryptamine are close structural relatives of 5-HT.

There are impressive overlaps between the LSD syndrome and schizophrenia. They include hallucinations, delusions, disturbances in mood and affect, and thinking disorders. However, LSD-induced hallucinations tend to be visual rather than auditory, and the individual usually retains an insight into his condition that is totally lacking in schizophrenia.

There is no direct evidence to support the 5-HT hypothesis. Many studies of the concentrations of 5-hydroxyindoleacetic acid (5-HIAA) in the lumbar CSF have failed to show any abnormality in schizophrenia. The concentration of tryptophan, 5-HIAA, and kynurenin are similar in postmortem brain tissue of schizophrenics and controls (Joseph *et al.*, 1979). An exception was the postmortem putamenal concentration of 5-HT in nine schizophrenic subjects, which again has been related to premortem neuroleptic medication. Equivocal studies have been conducted on 5-HT levels and the levels of its metabolites in peripheral tissues (for a review, see Rodnight 1983); again these studies have been inconclusive.

Since tryptophan is a precursor of 5-HT, the bizarre behavioral effects of tryptophan combined with an MAO inhibitor may support the hypothesis, but administration of the direct precursor 5-HTP does not produce such a result.

Hypotheses Involving Acetylcholine and γ-Aminobutyric Acid. Huntington's chorea involves a loss of ACh, GABA, and other neuronal types in the basal ganglia (see Chapter 14). It also results in a behavioral syndrome reminiscent of the early stages of schizophrenia. Both ACh and GABA cell types interact with DA pathways (see Chapter 14). It is logical to hypothesize, therefore, that an abnormality in one or both of these neurotransmitter types could underlie schizophrenia. In the case of ACh, it would be a deficiency in cells postsynaptic to ascending DA pathways. DA blockers, by releasing tonic DA inhibition, would boost the activity of cholinergic cells. Atropine, hyoscine, ditran-type cholinergic blockers, or cell destruction through Huntington's chorea would produce a cholinergic deficit and psychosis (cf. K. L. Davis *et al.*, 1978).

In the case of GABA, there would be a deficiency of GABA inhibition on dopaminergic cells of the midbrain and substantia nigra. The resulting disinhibition would result in overactivity of dopaminergic cells and would produce an amphetamine-like psychosis. However, no abnormality has been found in GABA or glutamic acid decarboxylase activity in schizophrenia (Cross *et al.*, 1979; Crow *et al.*, 1980), and the results of GABA levels in the CSF have been controversial (van Kammen *et al.*, 1982).

Peptide Hypotheses.   The large number of peptides now identified in brain tissue (see Chapter 12), their concentration in emotional parts of brain, and their apparent ability to modify other neurotransmitter functions has made this interesting family of compounds a prime source for generating new theories of psychiatric illness. The opioid peptides have been a source of particular attention because of initial reports that the opiate antagonist naloxone was capable of alleviating schizophrenia symptoms (Snyder, 1982). A report that the endorphin analogue des-tyrosine-γ-endorphin reduced schizophrenic symptoms was not confirmed. It can be anticipated, however, that much future attention will be focused on various peptides in this disorder.

Numerous other theories have been put forward over the years with respect to schizophrenia, only to be discredited. Some have been previously mentioned (P. L. McGeer *et al.*, 1978a), and they remain as historical curiosities and a testimony to the difficulty of experimentation in this field.

15.6.2. Hypotheses of Affective Psychoses

The affective psychoses may be bipolar (manic–depressive) or unipolar (depressive only). As with schizophrenia, there is a heavy genetic component to these illnesses. As well, psychopharmacological agents that are effective in treating these disorders interact primarily with the catecholamines and 5-HT. Not surprisingly, theories as to their causation primarily involve abnormalities in catecholamine and 5-HT metabolism, although cholinergic, GABAergic, peptidergic, and endocrine dysfunctions have all been suggested.

Noradrenaline Hypothesis.   NA has been preferred over DA in catecholamine theories of affective disorders because amphetamines, which are primarily DA releasers, are relatively ineffective at treating the depressive aspects, while many tricyclics that primarily enhance NA action are effective.

As discussed in Chapter 9, 3-methoxy-4-hydroxyphenylethylene glycol (MHPG) is the principal NA metabolite in brain. In the periphery, NA is converted primarily to vanillylmandelic acid (VMA). Between 25 and 60% of urinary MHPG is from brain, while nearly all VMA is from the periphery. Thus, some distinction can be made between central and peripheral NA metabolism. It has been reported that some patients with unipolar depression have low MHPG in urine, and it is these patients who respond well to maprotiline and the tricyclic antidepressants that are NA uptake inhibitors. It has been reported that patients with high concentrations of MHPG responded favorably to amitriptyline, a 5-HT uptake inhibitor. Some groups have confirmed this finding, while others have failed to do so (Maugh, 1981). Equally controversial reports have been obtained with respect to CSF MHPG.

A more subtle hypothesis might fit in with the action of lithium salts. This monovalent ion is considerably smaller in ionic radius than sodium and has the unusual property of being highly effective in the treatment of mania. Obviously, lithium could affect a tremendous variety of cellular activities because of its universal distribution, but it appears to have a somewhat selective action in reducing the availability of NA at receptor sites.

Serotonin Hypothesis.   Support for the hypothesis that deficient 5-HT production is connected with depression comes primarily from drug studies. Both PCPA and reserpine are noted for inducing depression in susceptible individuals. On the other hand, in patients being treated with 5-HTP for certain forms of epilepsy, euphoria is one of the symptoms of overdose. Isolated reports of successful treatment of depression with tryptophan and 5-HTP have appeared, but several groups have published negative clinical results with these agents after careful trials.

Three groups of workers have reported a decrease in the concentration of either 5-HT or 5-HIAA in the brains of suicide victims. Five of eight groups that studied 5-HIAA in the CSF drawn from the spinal canal of depressed patients found lower than normal levels, while the other three found no change. Trazodone, fluoxetine, and zimelidine are selective 5-HT reuptake inhibitors that are clinically effective antidepressants.

β-Noradrenergic Receptor Supersensitivity Hypothesis. This hypothesis is based on animal studies that show that the most common pharmacological property of clinically effective antidepressant drugs is the delayed down-regulation of β-adrenergic receptors. A pathologically high response to NA release in depressive patients is thus restored to normal. Drugs such as mianserin and zimeldine, which do not bring about such down-regulation, nevertheless decrease the ability of catecholamines to stimulate adenylate cyclase. This hypothesis does not completely explain why selective 5-HT reuptake inhibitors should be effective, although it has been suggested that 5-HT is somehow involved in β-adrenergic sensitivity. An intact 5-HT system is essential for β-receptor down-regulation (Tyrer and Marsden, 1985).

Dopamine Hypothesis. This hypothesis is based largely on studies of drug effects in which DA enhancers elevate mood and DA blockers depress mood. Depression frequently accompanies Parkinson's disease, which selectively damages brain stem DA cells. Studies of the content of HVA, the principal DA metabolite, in the CSF of depressed patients have been equivocal (for a review, see Willner, 1983).

Cholinergic Hypothesis. The cholinergic hypothesis is again largely based on drug effects. It suggests that if cholinergic activity predominates, depression results, and if catecholamine activity predominates, mania results. Most antidepressants have some anticholinergic activity, and anticholinergic agents that are effective in Parkinson's disease have mood-elevating properties (D. S. Janowsky et al., 1972; D. S. Janowsky and Risch, 1984; K. L. Davis et al., 1978).

Summary. There are numerous current theories regarding the biochemical etiology of depression. All have some support from the action of psychopharmacological agents, but none is supported by strong experimental evidence of altered neurotransmitter function. Moreover, the newer antidepressant drugs are revealing a much more diverse spectrum of action than had previously been thought permissible for clinical efficacy. This, combined with the lag time of days to weeks between the initiation of treatment and clinical response, strongly suggests that these agents act indirectly, influencing by slow readjustment some other primary process. Alternatively, the affective psychoses could represent a spectrum of diseases with common symptomatology. In such a case, one would anticipate that certain subgroups of depressives would respond selectively to certain biochemical classes of antidepressant medication. While such response has been suggested (Maugh, 1981), further confirmation is required.

15.7. Summary

In this chapter, we have dealt with the anatomy, physiology, neurochemistry, and neuropharmacology of behavior and emotion, recognizing that information is at a very primitive stage of development. We have introduced a new concept, genotropic transmission, to explain the manner in which signaling agents affect the development of brain. Circuits are thus prepared for appropriate behavioral responses in later life. At the molecular level, genotropic transmission involves interaction of the signaling agent with a receptor in the cytosol, translocation of the complex to the neuronal nucleus, and then initiation of appropriate gene transcription. The resulting mRNA produces proteins that then govern the output response of the neuron. Genotropic action can be blocked by inhibitors of RNA or protein synthesis and, in this respect, can be differentiated from ionotropic or metabotropic transmitters. The model used is sexual differentiation in neonatal rats and adult response to sex hormones. Selective cells in the limbic system and hypothalamic–pituitary axis are involved.

The generality of this mechanism is as yet unknown. At the molecular level, it seems to apply to other steroid hormones. It may also apply to mechanisms involved in the consolidation of memory, because RNA and protein synthesis inhibitors disrupt this process.

At a more speculative level, genotropic mechanisms might lie behind remarkable similarities in behavior that have been reported in monozygotic twins separated from birth and raised in different environments. Such mechanisms also help to provide a theoretical basis for the genetic aspects of schizophenia and affective disorders.

At the neurophysiological level, stimulation of appropriate subcortical centers in emotional areas of brain will precipitate complex behavioral patterns that, without knowledge that such stimulation had taken place, might be mistaken for voluntary but inappropriate action. Stimulation of areas in the limbic system, diencephalon, or brain stem can provoke evasive or aggressive reactions in the absence of any external provocation. Sexual initiative, feeding, or drinking in the absence of appropriate stimulus may be obtained. Ablation of selected areas in the same regions can bring about similar results, or, in some cases, reciprocal effects can be noted.

It is apparent from these stimulation and ablation studies that selected pathways in the brain, operating at the subcortical level, are responsible for the task of coordinating psychological and somatotropic action to prepare the body externally and internally to meet appropriate life situations.

Through the technique of self-stimulation, reward and punishment centers have been defined in rat brain. Reward centers are spread throughout the limbic, the diencephalic, and, to a lesser extent, the extrapyramidal system. Punishment zones are confined to a much smaller region in the periventricular area of the midbrain and diencephalon.

The areas involved, both central and peripheral, are richly served by pathways that use the neurotransmitters DA, NA, adrenaline, and 5-HT. ACh and GABA are also involved to a lesser degree. The use of selective metabolic inhibitors of these neurotransmitters as well as agents that interact with their receptors has permitted some sorting out of the probable functions that are modulated. For example, DA is most concerned with extrapyramidal movement and coordination of some thought processes. NA is particularly concerned with some ergotropic functions of brain and seems to modulate cell stimulation. 5-HT, on the other hand, is concerned with sleep, possibly digestive function, and increased temperature and comfort. None of these functions seems to depend absolutely on any of the amines. Their actions may be coordinated or modulated with a variety of peptides that occur throughout the nervous system (see Chapter 12), the individual functions of which are not yet understood even at a primitive level.

Much of the information regarding the individual roles of these amines has come from the discovery of psychoactive drugs. These drugs are of tremendous practical benefit in the treatment of mental illness. Antipsychotic agents act primarily on the DA system and its D_2-receptor sites. Antidepressants have a much broader spectrum of action. Some enhance amine levels, others inhibit their uptake, and still others block or down-regulate their receptors.

Out of these drugs studies have emerged a number of theories with regard to schizophrenia and affective psychosis. The theories provide an attractive basis on which to base further investigation, but as yet they offer no firm conclusions as to the causation of any form of mental illness.

The amines we have discussed in this chapter both modulate central neurotransmission and profoundly affect the action of peripheral smooth muscle and the endocrine glands; thus, they can be said to provide links between the psyche and the soma. Psychosomatic medicine is an important branch that, in the past, has relied more on art than on science but may, in the future, find a firmer physiological base.

This chapter on instinctive behavior makes no attempt to deal with two features of brain activity. One is the process of learning and memory, which is discussed in Chapter 16. The other is the phenomenon of consciousness itself and the realm of original, creative thought, which is the concern of Chapter 17.

16

Neuronal Mechanisms Involved in Learning and Memory

16.1. Introduction

Since this book is devoted to molecular neurobiology, attention will be focused in this chapter on the structural and functional changes in the brain that form the basis of learning. Learning is essentially a process of storage in the brain and memory the retrieval from the storage or, in modern terminology, from the data banks in the brain. There are two quite distinct types of learning and memory, though in many situations they may be employed in conjunction: First, there is *motor* learning and memory, which is the learning of all skilled movements. The repertoire is immense: the playing of all musical instruments, the playing of all games, the learning of all arts and crafts and of all technical procedures. Furthermore, there are all the expressive movements, as in speech, dance, song, and writing. Second, there is what we may call *cognitive* learning and memory. At the simplest level, there is the ability to recall some perceptual experience, but all levels can be involved, e.g., the remembrance of faces, names, scenes, events, pictures, musical themes. Then, at a higher level, there is the learning of language, and of stories, and of the contents of disciplines from the simplest technologies to the most refined academic studies in the humanities and the sciences.

It is a familiar observation that there may be an enduring cognitive memory of some single highly emotional experience. On the other hand, motor memories require reinforcement by continual practice if they are to be retained at a high level of skill. Quite distinct parts of the brain are concerned in these two types of memory. Nevertheless, it appears likely that the same kind of neural mechanism is concerned. We shall start with a brief resume of the various types of neural mechanism that have been proposed and of the related experimental evidence.

16.2. Structural and Functional Changes Possibly Related to Learning and Memory

For a general reference, the reader is referred to Eruklar (1983).

There have been theories of long-term memory based on a supposed analogy to genetic or immunological memory. For example, it has been conjectured that long-term memory is encoded in specific macromolecules, in particular RNA (Hyden, 1965), or that it is analogous to immunological memory. These theories fail for various reasons (cf. Eccles, 1970b; Szentágothai, 1971) and need not be further discussed. We therefore turn to the alternative, the *synaptic potentiation theory of learning* in the central nervous system (CNS). In general terms, following Ramón y Cajal (1911), Sherrington (1940), D. O. Hebb (1949), Lashley (1950), Eccles (1953), Kandel and Spencer (1968), and Szentágothai (1971), it is supposed that long-term memories are somehow encoded in the synaptic connectivities of the brain. It should be possible to secure the necessary changes in neuronal connectivity by means of microstructural changes in synapses (cf. Chapter 13). For example, they may be hypertrophied, or they may bud additional synapses, or, alternatively, they may regress. Since it would be expected that the increased synaptic efficacy would arise because of a strong conditioning synaptic activation, experiments such as those illustrated in Figure 16.1 have been carried out on many types of synapses.

Figure 16.1B is remarkable in showing that repetitive stimulation results in a large increase (up to 6 times) in the excitatory postsynaptic potentials (EPSPs) monosynaptically produced in an α-motoneuron by pyramidal tract fibers [cf. Figure 14.3 (Section 14.3.1)]. By contrast, in Figure 16.1A, the EPSPs generated monosynaptically in that same motoneuron by Ia fibers from muscle spindles [cf. Figure 4.2 (Section 4.2)] were not potentiated. Evidently, the pyramidal tract synapses display an extreme range of modifiability by what we may call *frequency potentiation*. This was described in Section 3.2 for neuromuscular transmission and attributed to the buildup of residual Ca^{2+} within the nerve terminal (Katz and Miledi, 1968). The emission of a synaptic vesicle is dependent on the Ca^{2+} ions that enter the depolarized nerve terminal and probably become effective by the activation of calmodulin (De Lorenzo, 1981). Rapid repetitive stimulation would build up an excess of Ca^{2+} ions, hence the frequency potentiation. Many types of synapses in the CNS have this ability to build up operationally during intense activation.

Although the synaptic potentiation during stimulation is of interest in relationship to the synaptic modification required for learning, any potentiation that persisted after the intense synaptic activation would be nearer to a possible paradigm of the learning process. This *posttetanic potentiation* (PTP) has been very fully investigated for monosynaptic action on spinal motoneurons (cf. Figure 4.2). In Figure 16.1C and D, the intracellularly recorded EPSPs were almost doubled in size after severe synaptic activations (640 per sec for 5 sec and 640 per sec for 30 sec), as may be seen in the inset records of the EPSPs at the height of potentiation and of the control (con.) EPSPs. It may be presumed that there was a large excess of intracellular Ca^{2+} ions (Katz and Miledi, 1968). However, the plotted points in Figure 16.1C and D show that all traces of potentiation had disappeared in a few minutes. Evidently, the synapses in the spinal cord exhibit no indefinitely prolonged increase in efficacy that could be a model of the learning process. This failure was not unexpected for the spinal cord, for at this level of the CNS, there is no good evidence for memory. As could have been predicted, the situation was transformed when

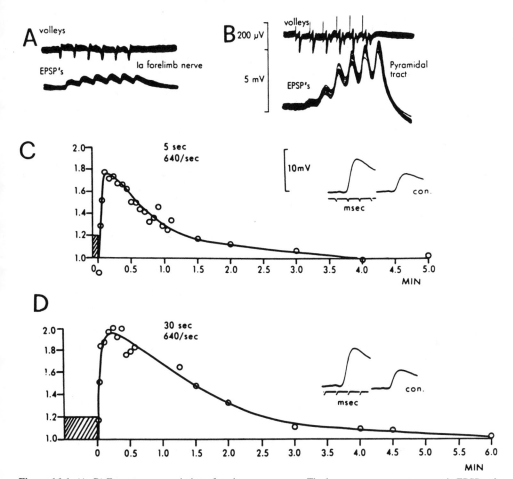

Figure 16.1. (A, B) Frequency potentiation of excitatory synapses. The lower traces are monosynaptic EPSPs of the same motoneuron of the cervical enlargement of the baboon spinal cord, there being in each case 6 stimuli at 200 per second to the Ia afferent pathway (A) and to the pyramidal tract (B) (Landgren *et al.*, 1962). (C, D) Posttetanic potentiation of EPSPs generated by Ia synapses on motoneurons. The conditioning tetani (▨) are specified. (○) Ratio of the potentiated EPSPs to the initial control at the indicated intervals after the end of the tetanus. For further description, see the text. From Curtis and Eccles (1959).

the effects of repetitive stimulation were studied in a high level of the brain, the hippocampus, a primitive cerebral cortex.

16.3. Hippocampus as a Model for Memory

In the hippocampus, there are three synaptic mechanisms that lend themselves to analytical investigations because of the simplicity of their monosynaptic connections (see Figure 16.3A). First, there are the fibers from the entorhinal cortex (ento) via the perforant pathways (pp) that make synapses on the apical dendrites of granule cells of the

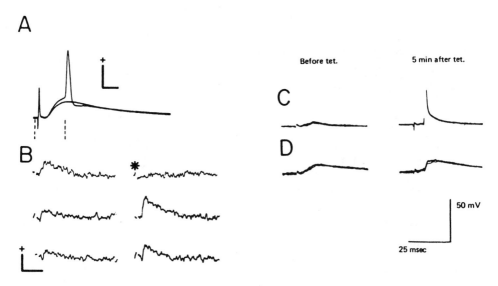

Figure 16.2. Responses of dentate granule cells to stimulation of perforant path fibers (cf. Figure 16.3A). (A) Just-threshold EPSP and a just-suprathreshold EPSP recorded intracellularly. (B) A series of 6 intracellular EPSPs showing fluctuations in the EPSP with very low intensity stimulation. Note the apparent response failure (*). Calibration: (A) 30 mV, 5 msec; (B) 1 mV, 10 msec. From McNaughton *et al.* (1981). (C, D) Intracellular recording from a CA1 pyramidal cell in response to stimulation of Schaffer collaterals (Sch in Figure 16.3A) before, and 5 min after, a tentanization (tet.) of 10 bursts of 50 Hz for 0.4 sec. (D) A hyperpolarizing pulse of 0.4 nA was applied just before the testing stimulation to reveal the long-term potentiation (LTP) increase in EPSP uncomplicated by the initiation of the spike potential seen in (C). From Andersen *et al.* (1980).

dentate area. Second, the axons of granule cells make synapses as mossy fibers (mf) on the apical dendrites of CA3 pyramidal cells. Third, the Schaffer axon collaterals (Sch) of the CA3 cells make synapses on the dendrites of CA1 pyramidal cells. Initially, the investigations were usually carried out *in vivo* on rabbits, but, in the last 10 years, the hippocampal slice technique has become increasingly used because the lamellar structure of the main hippocampal connectivities can be incorporated in a single slice (see Figure 16.3A), and the slice preparations exhibit responses similar to those found by the much more laborious *in vivo* techniques.

In developing a hypothesis of memory, the experimental observation on the perforant pathway from the entorhinal cortex to granule cells is of particular significance because of its inherent simplicity. It is necessary to give a brief review of the synaptic activation of granule cells before dealing with the special features on which a model for memory can be built.

The excitatory synaptic responses generated by these various monosynaptic connectivities (Figure 16.2A) exhibit the conventional characteristics (Bliss and Lømo, 1973; W. B. Levy and Steward, 1979). The EPSP is built up of the unitary potentials of the many convergent fibers as described in Sections 4.2 and 4.3. It is estimated (McNaughton *et al.*, 1981) that 400 perforant path fibers are necessary to discharge a granule cell, each fiber delivering a transmitter input that is close to a quantal content of 1. The EPSP from a single afferent fiber is about 0.1 mV (Figure 16.2B), and the EPSP for initiating a cell discharge is about 24 mV (Figure 16.2A).

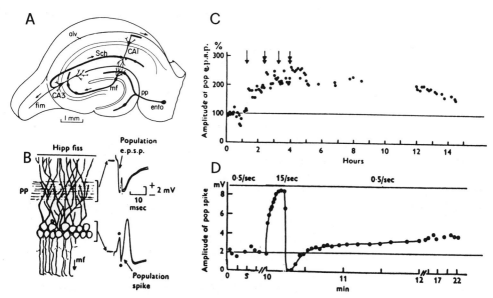

Figure 16.3. (A) Section of the hippocampus as described in the text. Abbreviations: (alv) alveus; (ento) entorhinal cortex; (fim) fimbria; (mf) mossy fiber; (pp) perforant pathway; (Sch) Schaffer axon collateral. (B) Drawing of granule cells with their bodies, dendrites, and axons that form the mossy fibers (mf). The fibers of the perforant pathway (pp) are shown traversing the dendrites on which they make excitatory synapses. Recording the field potentials at the level of the pp synapses (B) results in a large and prolonged negative potential (Population e.p.s.p.). When the recording electrode is advanced to the level of the cell bodies, the sharp negative spike (Population spike) signals the generation of impulses in the cell bodies. (C) Relative amplitudes of the population e.p.s.p. are plotted at up to 10 hr after four conditioning trains of stimulation indicated by the arrows. At the single-headed arrows, there was stimulation at 15 per sec for 15 sec; at the double-headed arrows, 100 per sec for 3 sec. The 100% line is drawn through the prestimulation responses at 0.5 per second; after the conditioning tetanus, the same low rate was resumed. (D) Time course of poststimulation potentiation as in (C), but for the population spike. There was only a single conditioning tetanus of 15 per second for 15 sec. From Bliss and Lømo (1973).

Other conventional features are synaptic facilitation with subsequent depression, frequency potentiation during the tetanus [see Figure 16.1B and D (Section 16.2)], and PTP (cf. Figure 16.1C and D).

There have been many anatomical and physiological studies designed to discover the synaptic properties on which a model for memory could be built. The first acceptable model was discovered by Bliss and Lømo (1973) in the synapses that the perforating pathway (pp) makes directly on granule cells of the dentate area of the hippocampus (Figure 16.3A and B). In (B), there is an illustration of the extracellular potentials recorded either at the level of the synapses, and hence largely a negative wave, a population EPSP, or at the level of the granule cell bodies, and hence a favorable site for recording the cell discharges, the negative population spike.

In Figure 16.3D, the population spike greatly increased during a brief tetanus (15 per second for 15 sec). Following this frequency potentiation (cf. Figure 16.1B), there was a brief depression and then a gradual increase that was still continuing 22 min later. More impressive is the effect of four bursts of conditioning stimulation on the population EPSP, which increased to more than double and was still large at 10 hr after the last conditioning tetanus (Figure 16.3C). Prolonged observations of the potentiation were made using implanted electrodes (Bliss and Gardner-Medwin, 1973). After a series of brief conditioning tetani, as in Figure 16.3C and D, the population spike was increased 4-fold. In 24 hr, it had declined to rather more than 2-fold, though it was still significant (2-fold) after 16 weeks, when the experiment was terminated. Intense interest was aroused in this long-term potentiation (LTP) because of its qualification for being a synaptic mechanism underlying memory. It is of great value that because of the extremely long duration of LTP, it can be fully investigated after the disappearance of the posttetanic effects described above, the PTP [cf. Figure 16.1C and D (Section 16.2)] and the subsequent depression. The LTP can be about 10,000 times longer than the PTP.

To demonstrate an LTP, there has to be strong repetitive presynaptic stimulation (McNaughton *et al.,* 1978) (ento or pp in Figure 16.3A) at a frequency above 5 per second and with many stimuli. The optimal conditions have been empirically determined to be bursts of about 8–10 stimuli at 400 per second and repeated every 10 sec perhaps 10 times, i.e., about 100 stimuli in just over a minute (R. M. Douglas, 1977). There has to be a convergence of many perforant path impulses on a granule cell to induce an LTP, a finding named "cooperativity" (McNaughton *et al.,* 1978). There are about 10,000 spines on a granule cell, and, as stated above, activation of about 400 is required for generation of an impulse discharge (McNaughton *et al.,* 1981), and possibly even more for producing an LTP. In LTP, the synaptic potency, as measured by the extracellular EPSP (Figure 16.3C), is rarely over double (Bliss and Lømo, 1973; W. B. Levy and Steward, 1979; McNaughton *et al.,* 1978), but it persists for days. After 12 daily reinforcements by stimulation through a chronically implanted electrode, the LTP declined with a time constant of 37 days (Barnes and McNaughton, 1980).

In investigations on the contralateral input into the proximal third of the dendrites of dentate granule cells [see Figure 13.17 (Section 13.2.2)], W. B. Levy and Steward (1979) found that stimulation of the contralateral entorhinal cortex evoked small EPSPs in the granule cells, but there was no LTP after a conditioning tetanus (Figure 16.4*B*). This is a further example that a large synaptic input is required to induce an LTP. There was a large LTP when the conditioning and testing were to the ipsilateral entorhinal cortex (EC) (Figure 16.4*A*) in the usual manner. However, when the *conditioning* stimulation was applied simultaneously to the ipsilateral and contralateral entorhinal cortices, *testing* by contralateral stimulation alone revealed a large potentiation, even more than double (Figure 16.4*C*). This observation is another example of cooperativity. The contralateral input to the granule cells was below threshold for evoking the LTP reaction, but when combined with the strong ipsilateral input, it shared in the LTP responses. This important finding is of great interest for the theory of memory. It was surprising that LTP of the contralateral entorhinal cortex occurred only if the burst stimulation of the contralateral entorhinal cortex was simultaneous with that to the ipsilateral entorhinal cortex or before it by up to 20 msec. If the contralateral entorhinal cortex followed the ipsilateral entorhinal cortex, it was depressed (Levy and Steward, 1983). As Levy and Steward (1983) point out, the weak contralateral input seems to open a window to the effect of the strong ipsilateral input for the brief time of about 20 msec.

Heterosynaptic potentiation was also demonstrated by Gustafsson and Wigstrom (1986), utilizing two separate inputs by Schaffer collaterals onto the CA1 pyramidal cells [see Figure 16.3A (Sch to CA1)]. In the picrotoxin-treated preparation (see Section

Figure 16.4. Effect of entorhinal conditioning stimulation, S_1 or S_2, on testing responses evoked by a testing entorhinal cortex (EC) stimulation (S). Extracellular recording from dentate granule cells gives population EPSPs as in Figure 16.3B. (A) S_1 stimulation of the EC excites the perforant pathway to the ipsilateral granule cells for both testing and conditioning stimulation. The conditioning was by 8 trains of 8–10 pulses at 400 Hz delivered every 10 sec. The traces show the population EPSP recorded by microelectrode R_1 before conditioning (a) and about 2 min after the conditioning tetanus (b). A second conditioning tetanus was applied at 16 min after the first, and the population EPSP about 2 min later (c) reveals a slight further increase in LTP. (B) Same EC stimulation procedures by S_1, but the recording by R_2 was for the population EPSP of the contralateral side. (b, c) Note the absence of LTP. (C) Recording by microelectrode R_2 as in (B), but conditioning by a conjoint stimulation of both entorhinal cortices, S_1

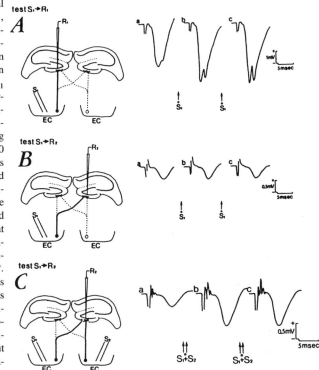

and S_2, as indicated. As a consequence there was a large LTP of the testing contralateral response S_1 alone (b, c), in great contrast to the absence of LTP in (B). From W. B. Levy and Steward (1979).

16.3.1), the optimum timing for the weak testing input was during the early part of the very brief conditioning tetanus (2 or 5 or 15 impulses at 50 Hz). The heterosynaptic LTP was then about half the homosynaptic LTP. Heterosynaptic LTP was also observed when the testing volley *preceded* the conditioning tetanus by as long as 40 msec; i.e., the "window opening" was about twice as long as that observed by Levy and Steward (1983). It is of great significance that the LTP is restricted to the synapses that are activated in temporal conjunction with the conditioning tetanus. Heterosynaptic LTP is observed even when the conditioning and testing inputs are remote from each other, as, for example on, respectively, the basal dendrites and apical dendrites of the CA1 pyramidal cells (Gustafsson and Wigstrom, 1986).

The pioneering experiments on LTP in the *in vivo* preparation have been greatly extended by analytical experiments on hippocampal slice preparations. In appropriate media, these slices give reliable performances for many hours and have the singular advantages of greatly simplifying the recording procedures and of allowing investigations on the effects of variations in the composition of the bathing medium.

That LTP has independence from the presynaptic events referred to in the introduction is indicated by the action of L-amino-phosphonobutyric acid (AP4) (Dunwiddie *et al.*, 1978), which blocks the neurotransmitter [glutamate (see Chapter 6)] action on the postsynaptic receptors, but has no presynaptic action. Strong repetitive presynaptic stimulation can be completely blocked by AP4 through this postsynaptic blockade. After removal of AP4, there was relief from the block, but no LTP was observed, which is contrary to what would be expected if LTP were presynaptically generated by an increased neurotransmitter output. This finding clearly indicates that LTP results from the stronger transsynaptic action, i.e., that it is initiated postsynaptically. There is a consideration of the alternative hypothesis of presynaptic generation in Section 16.3.3.

16.3.1. Calcium and Long-Term Potentiation

For general references, the reader is referred to Eccles (1983) and Gustafsson and Wigström (1986).

If the calcium of the bathing medium was reduced from 2.5 to 1 mM, synaptic transmission and PTP were well maintained, but LTP failed (Dunwiddie and Lynch, 1979). A special role for Ca^{2+} in LTP was also suggested by experiments in which Ca^{2+}-free bathing medium was substituted after LTP was fully developed (Dunwiddie and Lynch, 1978). All transmission was temporarily blocked during this episode, but on restoration of the normal Ca^{2+}-containing medium, the recovered synaptic transmission exhibited the LTP exactly as before the episode. This indicates that LTP is an enduring process that is fully set up in a few minutes by tetanization in the presence of Ca^{2+} and thereafter survives at full strength for many minutes in the absence of Ca^{2+}. A most important recent finding is that injection of the calcium chelator EGTA into the postsynaptic target cells blocks the development of LTP in those cells (G. Lynch *et al.*, 1983b).

The hypothesis is proposed that LTP is primarily postsynaptic and that the influx of Ca^{2+} ions into the synaptic spines (Figure 16.5) sets in train a series of processes that result in the enduring LTP. It is proposed that an essential prerequisite is the large and prolonged membrane depolarization that is set up by intense stimulation, the *cooperativity* of McNaughton *et al.* (1981). More recently it has been discovered by Wigström and Gustafsson (1983) that blockage of inhibition by picrotoxin in the slice preparation augments the LTP production. The conditioning tetanus need be only 10–15 volleys at 50 Hz. Apparently inhibitory postsynaptic potentials counteract the EPSPs so that stronger excitation is needed to give the depolarization that is large enough to open the Ca^{2+} gates. The electrochemical gradient across the membrane may be as high as -200 mV inside to outside, the inside Ca^{2+} concentration being as low as 10^{-8} M (Cheung, 1980). The empirical finding that LTP is most effectively induced by brief stimulus bursts repeated every 10 sec is attributable to the necessity for renewal of the extracellular Ca^{2+}. For example, after brief seizures of the pyriform cortex, the extracellular Ca^{2+} was reduced to half in a few seconds and recovery was far advanced 10 sec after the seizure (Galvan *et al.*, 1982).

Following Dingledine (1983) Gustafsson and Wigström (1986) have proposed that the spine synapses of the hippocampal neurons have two different sets of postsynaptic receptors for glutamate (cf. Section 6.4.1): the NMDA (*N*-methyl-D-aspartate) receptor channels that, when activated at resting potential, cause little EPSP but that, when strongly depolarized, carry a large Ca^{2+} influx; and the non-NMDA receptors that operate the sodium–potassium channels to give the main depolarizing EPSP.

The glutamate liberated by an impulse causes a slow activation of the NMDA receptors that lasts for about 60 msec. During that time, if the spine is sufficiently depolar-

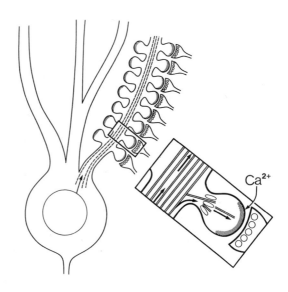

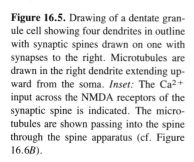

Figure 16.5. Drawing of a dentate granule cell showing four dendrites in outline with synaptic spines drawn on one with synapses to the right. Microtubules are drawn in the right dendrite extending upward from the soma. *Inset:* The Ca^{2+} input across the NMDA receptors of the synaptic spine is indicated. The microtubules are shown passing into the spine through the spine apparatus (cf. Figure 16.6*B*).

ized by its activated non-NMDA receptors together with the electrotonically spreading depolarization (the cooperativity), the NMDA Ca^{2+} channels are opened with a consequent Ca^{2+} influx and LTP.

Two hitherto inexplicable phenomena can be accounted for. First, the electrotonically spreading depolarization involves all spines, but only activated spines exhibit LTP. Evidently the Ca^{2+} input by the NMDA receptors on a spine is *private* to that spine in its LTP production. Second, the "window" effect described above is attributable to the long duration of NMDA activation. Even 40 msec after the testing stimulation there is lingering activity of the NMDA receptors with the consequence that their Ca^{2+} channels can be opened by the depolarization of the conditioning stimulation and LTP produced.

An important finding is that 2-amino-5-phosphonovalerate selectively blocks NMDA receptors and abolishes LTP. Much more investigation is required, but at least we have a model for heterosynaptic LTP that can be utilized in developing a hypothesis for long-term memory in the cerebral cortex.

We now confront the problem of how the increased intracellular Ca^{2+} can increase the effectiveness of the glutamate liberated into the synaptic cleft in close proximity to the postsynaptic density [Figure 16.6*B* (arrow)] that incorporates the glutamate receptors. There is now good evidence that, immediately after Ca^{2+} ions enter a nerve cell at a sufficient concentration, there is combination with calmodulin, a protein that has been very intensely studied (Cheung, 1980; Means *et al.*, 1982) [see Figure 5.4 (Section 5.4.1)]. Its four Ca^{2+} binding sites become fully occupied when the Ca^{2+} reaches a sufficient concentration. This complexed calmodulin (Ca^{2+}–calmodulin) becomes a powerful activator of enzyme systems and proteins. It is of special interest that trifluoperazine inhibits the action of calmodulin on enzyme systems by preventing its combination with Ca^{2+} (Cheung, 1982); this could cause the observed failure of LTP (G. Lynch *et al.*, 1983a).

The postsynaptic densities (PSDs) of the spine synapses of the cerebral cortex (including the hippocampus) are very thick structures [Figure 16.6*B* (arrow)] that have been

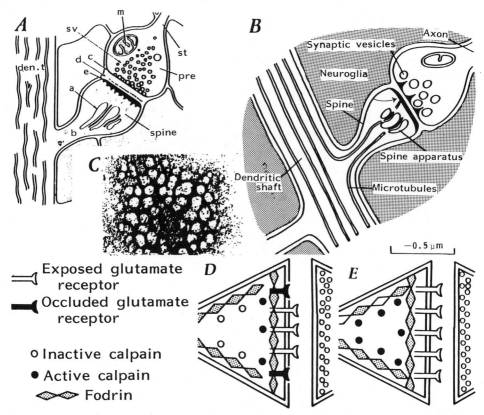

Figure 16.6. (*A*) Earliest diagram of a synaptic contact (pre) on a dendritic spine (spine) observed with the electron microscope after osmium tetroxide fixation. Abbreviations: (a) spine apparatus; (b) spine stalk; (c) presynaptic membrane with clustered vesicles; (d) synaptic cleft; (den.t) dendritic microtubules; (e) postsynaptic density; (m) mitochondria; (st) presynaptic fiber; (sv) synaptic vesicles. From Gray (1959). (*B*) Latest diagram illustrating in particular the microtubules going into the spine and the spine apparatus. From Gray (1982). (*C*) High-power electron microscopic picture of an isolated postsynaptic density to show a peripheral planar array of round subunits with a median diameter of 18 nm. They enclose a central fine granular material. The micrograph was made with a negative staining technique in which the subcellular particles are unstained and surrounded by an electron-dense background by uranyl acetate. From Matus (1981). (*D, E*) Illustrations of the hypothesis of G. Lynch and Baudry (1984). Influx of Ca^{2+} activates calpain, resulting in a breakdown of fodrin and exposure of the occluded glutamate receptors of the postsynaptic density, which is shown in two stages.

studied in detail by a variety of techniques (for a review see Matus, 1981). Several constituent proteins have been recognized—actin, tubulin, calmodulin, fodrin, and an unidentified protein of about 51,000 molecular weight (PSD 51) (Carlin *et al.*, 1983; Grab *et al.*, 1979, 1980; P. T. Kelly and Cotman, 1978; Matus, 1981)—that form a cytoskeletal framework (Figure 16.6*C*) (Matus, 1981), apparently for providing the attachment sites for glutamate receptors. The simplest explanation of LTP is that the Ca^{2+}-activated protein kinase calpain causes, presumably by means of Ca^{2+}–calmodulin activation, the uncovering of glutamate receptor sites, an explanation that has been suggested by the increased binding of glutamate (G. Lynch *et al.*, 1982, 1983a) and that is shown diagrammatically at two stages in Figure 16.6*D* and *E*.

The threshold and cooperativity exhibited for the LTP of granule cells (McNaughton *et al.*, 1978) and CA1 pyramidal cells (Gustafsson and Wigström, 1986) necessitate an explanation in terms of EPSPs generated by many activated spine synapses summating in producing dendritic depolarization (McNaughton *et al.*, 1981) that surpasses the threshold for opening Ca^{2+} gates of NMDA receptors. There could then be this sequence: influx of Ca^{2+}, production of the second messenger system, Ca^{2+}–calmodulin, and action of Ca^{2+}–calmodulin to produce the LTP.

The fastest production of LTP would be by Ca^{2+}–calmodulin-activated protein kinases, such as calpain, acting on the PSD (G. Lynch and Baudry, 1984), phosphorylating its proteins (Browning *et al.*, 1978; Schulman and Greengard, 1978) and activating or exposing the glutamate receptors (Figure 16.6D and E). The high concentration of calmodulin in the postsynaptic density, about 5–10%, would facilitate the Ca^{2+} action. G. Lynch and Baudry (1984) marshal evidence in support of the hypothesis that the conditioning stimulation leads to an increase in intracellular calcium that activates the proteinase (calpain), which then acts on its endogenous membrane substrate, the structural protein fodrin (molecular weight 230,000), which is an important constituent of PSDs (Carlin *et al.*, 1983). As a consequence of the breaking up of the protein network of the PSDs, there is an increase in the exposed glutamate receptors (Figure 16.6D and E), which is the cause of the LTP. Support for this hypothesis is provided by the finding that leupeptin, a selective inhibitor of calpain, antagonizes the effects of Ca^{2+} in exposing glutamate receptors (G. Lynch *et al.*, 1983a,b). As an addition to this hypothesis, it is here suggested that it is the Ca^{2+}–calmodulin complex that activates the calpain.

The long duration of LTP seems to require that the Ca^{2+}–calmodulin escapes from the spine of its origin. It could act on the protein transport up the dendrites by microtubules and into the spines (see Figure 16.5), this transport presumably being accelerated by Ca^{2+}–calmodulin just as with axonal microtubules (Iqbal and Ochs, 1980). Another relatively short latency effect would be by the Ca^{2+}–calmodulin-activated protein synthesis in the polyribosomes that are located in the dendrites, in proximity to the spine origin (Steward and Levy, 1982), with transport to the PSD of the adjacent spine. The most prolonged action would be by the Ca^{2+}–calmodulin system acting on the protein manufacture by the nucleus and perikaryal ribosomes (Section 1.2.3), with subsequent transport up the dendritic microtubules (Figures 16.5 and 16.6A and B). It would seem that this latter process is too slow to contribute to the onset of an LTP, which reaches 66% of its maximum in 26 sec after the initiating tetanus (McNaughton, 1983). Thus, it is envisaged that LTP is probably generated by a sequence of actions, as described above, but much more investigation is required. There is the crucial problem of how the specificity of the activated spines is preserved.

16.3.2. Synaptic Spine

The LTP of granule cells is associated with a hypertrophy of the spine heads by 30–40% (Fifkova *et al.*, 1982). This hypertrophy is prevented by a preceding injection of anisomycin, which induces a brief, less than 2-hr, suppression of the translation phase of protein synthesis (Fifkova *et al.*, 1982). A limited hypertrophy (about 20%) appears as the anisomycin suppression passes off. A related observation is that LTP in hippocampal slices is accompanied by protein synthesis (Duffy *et al.*, 1981).

The hypertrophy of the spine head in LTP is accompanied by an increase in the diameter of the spine stalks by up to 60%, which is significant at 4 min after the conditioning tetanus, but tends to decline at 90 min (Fifkova and Anderson, 1981). Rall (1970) first suggested that a decrease in the resistance of the spine stalk could provide a basic mechanism for learning, which is related to the observed increase in caliber of spine stalks. However, calculations by Jack et al. (1975) suggested that this mechanism for control of synaptic potency is inefficient. A doubling of the electrical length constant of the spine stalk would produce only a 10% change in the membrane voltage that a spine synapse induces in the parent dendrite. Koch and Poggio (1983) now give a more effective action in their calculation. However, there is much uncertainty in the parameters.

In summary, there is much evidence that LTP is induced postsynaptically by the following sequence of events: (1) Ca^{2+} influx into the activated synaptic spines, which are strongly depolarized by the bursting synaptic stimulation; (2) Ca^{2+} activation of calmodulin, with 4 Ca^{2+} ions to each calmodulin molecule; (3) Ca^{2+}–calmodulin acting as a second messenger system, producing a variety of actions on protein and enzyme systems that result in an increase in glutamate receptor sites on the PSDs and a swelling of the spines.

16.3.3. Possible Presynaptic Contributions to Long-Term Potentiation

In several publications, a presynaptic locus for LTP was supported by evidence that a conditioning repetitive stimulation produced for 1 hr an increase in the resting presynaptic release of a glutamate analogue (Skrede and Malthe-Sørenssen, 1981) or of labeled glutamate (Bliss and Dolphin, 1982; Dolphin et al., 1982). Such observations suggest that LTP may be induced in part by an increased presynaptic output of glutamate by the testing stimulus. The other evidence adduced in support of a presynaptic location for LTP could equally be attributed to a postsynaptic location as in the present hypothesis, e.g., the increase in uptake and retention of Ca^{2+} (R. W. Turner et al., 1982) that paralleled the LTP of CA1 pyramidal cells (Baimbridge and Miller, 1981). More significant evidence against the postsynaptic location of LTP is the finding by R. W. Turner et al. (1982) that there was no increased responsiveness to iontophoretically applied glutamate after a brief increase (5–10 min) in the Ca^{2+} of the bathing medium from 2 to 4 mM. There was, however, a prolonged increase, for at least 3 hr, in the response to presynaptic inputs. This resembled an LTP in both EPSP and population spike (R. W. Turner et al., 1982), but it could be presynaptic or postsynaptic or both.

Against this evidence for a presynaptic generation of LTP, there is the finding by Baudry et al. (1980) that stimulation of the hippocampal slice so as to induce an LTP was accompanied by a long-lasting (30-min) and highly significant increase (by about 20%) in the glutamate accumulation by the slice, the increase being restricted to the stimulated area. This finding indicates that there is an increase in the number of glutamate receptors during LTP. Furthermore, in isolated hippocampal membranes, the reversible glutamate binding was increased by 80% in the presence of 250 μM calcium (G. Lynch et al., 1983a). Presumably, this reversible binding was on the glutamate receptor sites of the PSD [see Figure 16.6D and E (Section 16.3.1)] (G. Lynch and Baudry, 1984).

16.3.4. Comprehensive Hypothesis for Long-Term Potentiation

Even Bliss and Dolphin (1982) admit that there is good evidence that LTP is in part postsynaptic. The question now arises: Is it possible to develop a comprehensive hypothesis that recognizes the primacy of the postsynaptic origin of LTP, but also accounts for such presynaptic effects as an increased transmitter output? It is important to comprehend the spine synapse (Figure 16.6*A, B, D,* and *E*) as a functional element, which we may term a *synaptic unit.* It is not simply a site for chemical transmission in the conventional sense, where the chemical transmitter is ejected from the presynaptic terminal into the synaptic cleft and diffuses to the postsynaptic receptor sites, where it binds and causes an opening of ionic gates across the postsynaptic membrane. There are more subtle trophic actions that are as yet rather ill defined (Varon and Bunge, 1978). For example, specific polypeptides (retrophins) are secreted postsynaptically and taken up presynaptically to be transported retrogradely to the presynaptic soma for necessary trophic actions (see Section 13.3). Failure of this transport, as after axonal section, results in chromatolysis, which may lead to cell death. Nerve growth factor (Levi-Montalcini *et al.,* 1980) is an example of a retrophin for sympathetic ganglion cells, and other examples are being discovered (see Sections 13.4 and 13.5).

Evidently there is an extraordinary degree of information flow across the synaptic cleft (cf. Eccles, 1984a). Hence, it is conceivable that when PSDs are transformed to give the increase in glutamate receptors during LTP (G. Lynch *et al.,* 1982), there can occur a complementary influence on the presynaptic terminal with increased probability of quantal output of neurotransmitter. Thus, though LTP may initially be an exclusive postsynaptic process, as proposed above, there could be secondary presynaptic changes such as those reported (Baimbridge and Miller, 1981; Bliss and Dolphin, 1982; Dolphin *et al.,* 1982; Skrede and Malthe-Sorenssen, 1981; R. W. Turner *et al.,* 1982).

16.4. Locations of Engrams

In a recent comprehensive review of the neuronal substrates of learning and memory, R. F. Thompson *et al.* (1983) raise the important issue of where the key memory changes or engrams occur in the brain, a problem proposed long ago by Lashley (1950) and D. O. Hebb (1949). It is generally agreed that the sensory and motor pathways are not the sites of the modifiable synapses that essentially are concerned in the "memory trace" of even the simplest conditioned responses, which typically have latencies of 100 msec or more. However, simple conditioned responses may be mediated subcortically, since they occur in the decorticate or even in the decerebrate animal (R. F. Thompson *et al.,* 1983). The elusive nature of the engram can be appreciated when it is recognized that recording of associated neuronal responses in a variety of pathways does not establish that the engrams are located in these responding structures. They may be merely on relay circuits that are instrumental in the input or output lines. Two techniques are used in attempting to locate engrams: selective lesions and recording of neural activity. R. F. Thompson *et al.* (1983) present a critical survey of attempts to localize engrams in the amygdala and hypothalamus, in the motor cortex, in the red nucleus, and in the cerebellum. As will be described in Section 16.7.2, extensive studies by M. Ito (1984) have shown the key role of the cerebellar flocculus in learning adaptations of the vestibuloocular reflex. However, the most convincing localization of an engram has been accomplished by Thompson and associates (T. W. Berger *et al.,* 1980; R. F. Thompson *et al.,* 1983) for the hippocampal role in the conditioned responses of the rabbit nictitating membrane. The unconditioned stimulus (UCS) was a brief (100-msec) air puff to the cornea of a rabbit that evoked a response of the nictitating membrane and response of a selected hippocampal cell. The

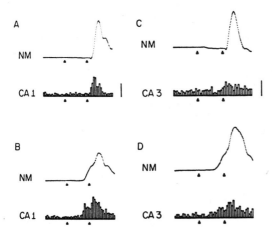

Figure 16.7. Hippocampal unit response from two paired conditioning animals. (NM) Average nictitating membrane response for one block of 8 trials. (CA1, CA3) Hippocampal unit poststimulus histogram (15-msec time bins) for one block of 8 trials. (A, C) First block of paired conditioning trials, day 1. (B, D) Last block of conditioning trials, day 1. The first cursor indicates tone onset; the second, air puff onset. The total trace length equals 750 msec in all histograms. (A) Vertical bar (for A and B) equals 37 unit counts per 15-msec time bin. (C) Vertical bar (for C and D) equals 35 counts. From T. W. Berger and Thompson (1978).

conditioned stimulus (CS) was a tone of 350-msec duration that began 250 msec before the UCS and that alone evoked no response of the nictitating membrane or hippocampus even when repeated many times. In Figure 16.7A, the CS was ineffective, but discharges of the CA1 pyramidal cell were evoked by the UCS shortly before the response of the nictitating membrane (NM). In the first day, there were 104 trials, and the responses of the last block of 8 trials (Figure 16.7B) show that there was now a conditioned response of the CA1 cell to the CS well before the UCS and also a corresponding early response of the nictitating membrane. Both responses, CA1 and nictitating membrane, appear in two phases, the CS preceding the UCS. Figure 16.7C and D shows a similar learning performance in a CA3 pyramidal cell of another animal. In a comprehensive study (T. W. Berger *et al.*, 1980), the time courses of the conditioned hippocampal responses correspond closely to the learned responses of the nictitating membrane. It was further shown that a significant part of the plasticity develops in the CA3–CA1 region and is not impressed on the hippocampus by the entorhinal input via the granule cells (R. F. Thompson *et al.*, 1983). It is suggested that LTP in the hippocampus is the probable basis of the learned response, which is then impressed on the limbic system and so to the nictitating membrane (T. W. Berger *et al.*, 1980). This study gives the best available demonstration of the role of the hippocampus in learning.

The conditioning tone of 350-msec duration was continuous during the whole air puff, so there could be conjunction of the subthreshold input from the tone with the later suprathreshold input from the air puff. It can be assumed that this conjunction of a weak hippocampal stimulus leading a strong stimulus is homologous to that of the hippocampal granule cell responses of Figure 16.4 (Section 16.3). It can be proposed that the conditioned learning of Figure 16.7B and D is an example of LTP resembling that of Figure 16.4 and that the engram has been located in the hippocampus. It must be appreciated that only a few of the hippocampal neurons responded in relation to the nictitating membrane response. It was a very selective response. Section 13.2.3 gives an account of another classical conditioning paradigm in which an increased effectiveness of the corticorubral pathway was brought about by the development of synapses close to the somata of the red nucleus neurons (Tsukahara, 1981).

16.5. Cognitive Learning and Memory

It will be recognized that cognitive learning and memory involve two distinct problems—storage and retrieval—or, in relation to the present theme, learning and remembering. It is proposed to deal with these problems on two levels.

First, they will be considered as a problem of neurobiology, namely, the structural and functional changes that form the neural basis of memory. We are all familiar with short-term memory of a few seconds' duration. For example, when we look up a telephone number, we have to "keep it in mind" by continual rehearsal, else it is lost beyond recall. A distraction during this time interrupts the rehearsal, and the short-term memory is lost. There is evidence that longer memories are of two kinds: intermediate, up to an hour or so, and truly long-term—for days, months, years. The necessity for keeping the remembered information "in mind" indicates that there is no special storage problem in short-term memory. As long as the information is carried encoded in the specific dynamic operations of the neuronal circuitry, it is available for readout. On the other hand, with memories enduring for minutes to years, it has to be discovered how the neuronal connectivities are changed so that there is stabilized some tendency for replay of the spatiotemporal patterns of neuronal activity that occurred in the initial experience and that have meanwhile subsided.

Second, the role of the self-conscious mind has to be considered. It will be proposed in Chapter 17 that a conscious experience arises when the self-conscious mind enters into an effective relationship with activated synapses. In the willed recall of a memory, the self-conscious mind must again be in relationship to a pattern of synaptic responses that resemble the original responses evoked by the event to be remembered, so that there is approximately the same experience. We will consider in Chapter 17 how the self-conscious mind is concerned in calling forth the neuronal events that give the remembered experience on demand, as it were. Furthermore, the self-conscious mind acts as an arbiter or assessor with respect to the correctness or relevance of the memory that is delivered on demand. For example, the name or number may be recognized as incorrect and a further recall process instituted, and so on. Thus, the recall of a memory involves two distinct processes in the self-conscious mind: first, that of recall from the "data banks" in the brain; second, the recognition memory that judges its correctness.

There are patients who suffer from a tragic loss of all but short-term memories. In the most severe and complete cases, there is an almost complete failure to establish new memories, a clinical condition known as the "amnestic" or "Korsakoff syndrome" (Milner, 1966, 1972; Victor et al., 1971; H. H. Kornhuber, 1973). The unfortunate victim is disoriented in time and place. Gaps in memory are covered up by confabulation: A bedridden patient, for example, may insist he has just taken a walk in the garden.

Many of these cases are not useful for our purpose of trying to discover the actual neural mechanism by which new memories are laid down for semipermanent storage, because the patients suffer from general cerebral degeneration, often from alcoholism (Victor et al., 1971). The mammillary bodies are involved. The mediodorsal and anterior thalamus plus the terminal portion of the fornices are also frequently affected. More exact information can be obtained by considering cases in which the lesion was sharply defined because it was due to an operative excision.

In particular, we shall begin with the case H.M., which has been the most extensively investigated. Because of an intractable epilepsy of hippocampal origin, there was a bilateral excision of the hippocampus and the adjacent part of the hippocampal gyrus. Since that time, this man has had an extremely severe loss of ability to lay down memory traces. There is an almost complete failure of memory for all happenings and experiences after the lesion; i.e., he has an anterograde amnesia. He lives entirely with the short-term memories of a few seconds' duration and the memories retained from before the operation.

He can keep current events in mind so long as he is not distracted. Distraction completely eliminates all trace of what he had been doing only a few seconds before. There are cited many remarkable examples of his failure to remember as soon as he is distracted. The only way in which this patient can hold onto new information is by constant verbal rehearsal. Forgetting occurs as soon as this rehearsal is prevented by some new activity claiming his attention. Marlen-Wilson and Teuber (1975) showed, by a testing procedure of prompting, that a minimal storage of information occurs even for experiences after the operation, but it is of no use to the patient.

In the operation of bilateral hippocampectomy on H.M. discussed by Milner (1972) other adjacent medial temporal structures, the uncus, the amygdala, and the hippocampal gyrus, were also removed bilaterally, so there have been claims (Horel, 1978) that damage to these other structures of the medial temporal lobe is the key factor in causing the amnesia, not the hippocampectomy. However, this thesis has now been disproved by a study of monkeys with lesions restricted to the temporal stem in comparison with monkeys with a hippocampal–amygdala lesion (Squire, 1982; Zola-Morgan et al., 1981). The monkeys lesioned in the temporal stem displayed no amnesia, in contrast to those with the hippocampal–amygdala lesions (Squire, 1982).

There are three other recorded cases in which a comparable severe anterograde amnesia resulted from destruction of both hippocampi (Milner, 1966). There was almost no recovery even after 11 years. The variable retrograde amnesia, i.e., the memory of events preceding the hippocampal destruction, however, showed a continued recovery. There are two other reported cases in which unilateral hippocampectomy resulted in a comparable anterograde amnesia, but there was evidence that the surviving hippocampus was severely damaged. We can conclude that the severe anterograde amnesia occurs only with grave bilateral hippocampal deficiency.

It is important to recognize that the hippocampus is not the seat of the memory traces. Memories from years before the hippocampectomy are as well recalled as in normal controls. The hippocampus is merely the instrument responsible for laying down the memory trace or engram, which presumably is very largely located in the cerebral cortex in the appropriate areas. There is no obvious impairment of intellect or personality in these subjects despite the acute failure of memory. In fact, they live either in the immediate present or with remembered experiences from before the time of the operation. There is one small relieving feature, namely, that they still have ability to learn motor acts, which indicates the distinctiveness of the neuronal mechanisms concerned in motor memory, as described above. Thus, the subject can build up skills in motor performances, such as drawing a line in the narrow space between line drawings of one five-pointed star and another surrounding star, using only the guidance provided by the view in a mirror of his hand and the two stars, but he has no memory of how he learned the skill.

Mishkin (1978, 1982) has called into question the significance of the hippocampus in memory by claiming that the memory of monkeys is severely impaired only when the hippocampectomy is linked with amygdalectomy, which of course was the case in the Scoville operation on H. M. Experimental testing certainly shows that the hippocampal lesion alone gives a memory deficit for both visual and tactual inputs (Mahut *et al.*, 1981; Squire, 1982). The addition of an amygdalectomy may add to the amnesia (Mishkin, 1978), but this is not yet certain (Mahut *et al.*, 1981; Squire, 1982, 1983).

The conclusion certainly is that the hippocampus is most importantly concerned in memory.

Partial amnestic syndromes have been observed in patients with a variety of lesions in structures related to the hippocampus: the cingulate gyrus, the fornix, the anterior and mediodorsal nuclei of the thalamus (Victor *et al.*, 1971), and the prefrontal lobe. We are now in a position to consider the neural pathways concerned in the laying down of memory traces in the neocortex.

16.5.1. Neural Pathways Involved in the Laying Down of Cognitive Memories According to the Instruction–Selection Hypothesis

The hypothesis here proposed is developed from the theory of H. H. Kornhuber (1973), which is illustrated in Figure 16.8. The sensory association areas play a key role, being on the input pathways to the hippocampus and the frontal cortex. Thus, the frontal cortex receives a "selection input" from the limbic system. It is to be noted that the hippocampus is given a dominant role in the two limbic circuits. One circuit is the so-called "Papez loop": hippocampus, mammillary body, anterior thalamic nucleus, cingulate gyrus, parahippocampus, hippocampus (cf. Figure 16.8). The other circuit is of special interest because it leads from the association cortices to the hippocampus via the cingulate gyrus and thence via the mediodorsal (MD) thalamus to the prefrontal lobe. H. H. Kornhuber (1973) conjectures that with special neurons of the sensory association areas: ". . . the synapses of afferents coming (directly or indirectly) from the limbic system are essential for forming long-term memory, while other synapses on the same neurons are essential for information processing and for recall." He even conjectures that "long-term memory could involve coincidence of thalamic and corticocortical afferents at

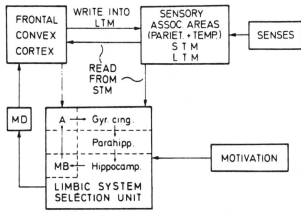

Figure 16.8. Scheme of anatomical structures involved in selection of information between short-term memory (STM) and long-term memory (LTM). (A) anterior thalamic nucleus; (MB) mammillary body; (MD) mediodorsal thalamic nucleus. From H. H. Kornhuber (1973).

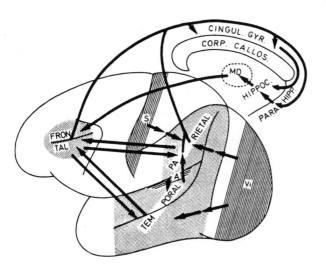

Figure 16.9. Scheme of pathways in the monkey brain involved in the flow of information from primary sensory areas via the sensory association areas of the temporal and parietal lobes and the cortex of the frontal convexity to the limbic system and then the loop back via the mediodorsal nucleus of the thalamus (MD) and the frontal cortex to the temporal and parietal areas for long-term storage. Medial areas have been translocated to the top. Primary sensory areas: (A) auditory; (S) somatosensory (the vestibular area is in the lower part of S); (Vi) visual. From H. H. Kornhuber (1973); based in part on data of Pandya and Kuypers (1969) and Pandya *et al.* (1971).

a given cortical neuron or cell column.'' These theoretical developments by Kornhuber provide the basis for the further developments here described.

A general scheme of the connectivities linking sensory association areas, the limbic system, and the prefrontal lobe is illustrated in Figure 16.9 for the monkey brain (H. H. Kornhuber, 1973). This diagram shows very well the inputs from frontal, temporal, and parietal cortical association areas to the hippocampus via the cingulate gyrus. The output from the hippocampus is shown via the mediodorsal thalamus (MD) to the convexity of the frontal lobe. The reciprocal connections of the frontal convexity to the parietal and temporal lobes are in accord with the diagram in Figure 16.8. Actually, both these figures greatly simplify the connectivities, particularly in the limbic system. For the present, we may note the important operation of motivation in Figure 16.8 and also the labeling of the limbic system as a selection unit. According to the present theory, the hippocampal output does indeed act to select, but this is done in the association cortex.

16.5.2. A Model for Cognitive Memory Built on Long-Term Potentiation

The proposed postsynaptic origin of LTP is of great significance in respect of the hypothesis, developed by Marr (1970), that long-term memory is encoded in synaptic potentiations that are set up by a conjunction process and that have an indefinitely long duration. An essential feature is the interaction between a strong conditioning synaptic input to a neuron and a weak synaptic input to that same neuron at about the same time. For example, in the hypothesis of cerebral learning (Marr, 1970), it was proposed that, when there was conjunction of the strong excitation of the apical dendrites of a pyramidal cell by the climbing fibers, the cartridge synapse of Szentágothai (1978a) (Figure 16.10) [cf. Figure 17.2 (Section 17.1)], with the weak excitation of the terminal branches of that same dendrite by horizontal fibers in lamina I, there would be induced a prolonged potentiation of the horizontal fiber synapses (Eccles, 1981a). The deficiency of the hypothesis was that there was no specification of the mode of interaction in either its nature or its timing.

Figure 16.10. Simplified diagram of connectivities in the neocortex showing pathways and synapses in the proposed theory of cerebral learning. The diagram shows three modules (A–C) that are vertical functional elements of the neocortex, each with about 4000 neurons. In lamina 1 and 2, there are horizontal fibers arising as bifurcating axons of commissural (COM) and association (ASS) fibers and also of Martinotti axons (MA) from module C. The horizontal fibers make synapses with the apical dendrites of the stellate pyramidal cells in module C and of pyramidal cells in modules A and B. Deeper, there is shown a spiny stellate cell (Sst) with axon (ax) making cartridge synapses with the shafts of apical dendrites of pyramidal cells (Py). Due to conjunction hypertrophy, the association fiber from module C has enlarged synapses on the apical dendrites of the pyramidal cell in module A.

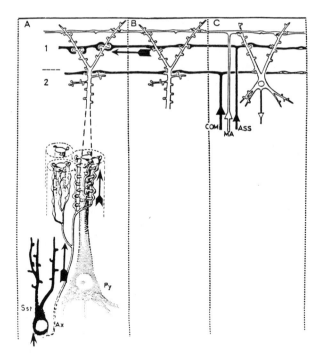

Figure 16.10 gives a diagram of the conjunction between the mediodorsal (MD) thalamic input by spiny stellate cells (Sst) to the cartridge synapses on the apical dendrites of pyramidal cells and the crossing-over synapses made by horizontal fibers that come from the sensory inputs of Figures 16.8 and 16.9 via the association (ASS) and commissural (COM) fibers and the Martinotti cell axons (MA). It was assumed by Marr (1970) that there was a selective potentiation of the synapses on the pyramidal cell dendrites and that it happened only in those with an approximate conjunction in time. Hitherto, this conjecture has had no experimental support. In fact, a rather similar conjunction in the cerebellar Purkinje cells results in a prolonged depression of the parallel fiber synapses (Albus, 1971; M. Ito *et al.*, 1982) (cf. Section 16.7).

The situation has now been transformed by the discovery of heterosynaptic LTP as described in Section 16.3.

The hypothesis of the postsynaptic origin of LTP based on increased intracellular Ca^{2+} can therefore be converted into a more developed variant of the Marr hypothesis. This is justified because K. S. Lee (1983) has found that slice preparations of several areas of the cerebral cortex exhibit an LTP matching that for the hippocampus. Thus, cerebral learning can be attributed to the calcium-induced changes in the postsynaptic densities on the spine synapses of the pyramidal cells, particularly for synapses made by horizontal fibers in lamina I, as outlined in a recent commentary (Eccles, 1983). There is urgent need for the experimental testing of this comprehensive hypothesis. From the unitary effect illustrated in Figure 16.10, there has been developed a hypothesis of the manner in which cognitive memories can be stored and retrieved in the cerebral cortex [Figures 7 and 8 in Eccles (1981a)].

16.5.3. Duration of Long-Term Potentiation and Cognitive Memory

As we inquire into the applicability of the postsynaptic calcium model for providing an explanation of cerebral memory, we have to consider the duration of the LTP. At this level of inquiry, we have to consider the question of the long duration of the LTP, particularly after repeated reinforcement.

In the extraordinary case of H.M., he had a retrograde amnesia for events occurring 1–3 years before the hippocampectomy (Squire, 1982, 1983). There is a transiently experienced retrograde amnesia for a similar duration of events after bilateral electroshock therapy (Squire, 1982, 1983). The period of sensitivity to disruption correlates with the observations on the normal course of forgetting. Thus, there would seem to be a period of 1–3 years involved in the process of consolidation (V. Bloch, 1970) of a long-term memory so that it is no longer susceptible to loss in the process of forgetting or in the process of memory disruption by bilateral hippocampectomy or electroshock therapy. Squire (1983) suggests that "the medial temporal region (hippocampus?) directly brings about those changes in memory storage whereby memory becomes gradually resistant to disruption." This suggestion should be incorporated into the theory of the hippocampally induced consolidation. Hitherto, this role was assumed to be a relatively transient intervention as in the LTP of Figure 16.4 (Section 16.3), though later relays of the memory experience were assumed to add to the stability of the patterned synaptic hypertrophy. We have now to envisage that to effect a "permanent" consolidation of a memory, the hippocampal input to the neocortex must be replayed much as in the initial experience in what we may name "recall episodes" for 1–3 years. Failure of this replay results in the ordinary process of forgetting. After 3 years, the memory codes in the cerebral cortex are much more securely established and apparently require no further reinforcing hippocampal inputs; hence, they are not lost in the disruption of bilateral hippocampectomy or electroconvulsive therapy.

The Papez loop illustrated in Figure 16.8 (Section 16.5.1) would have the properties of a reverberatory circuit and so could be the basis of repetitive burst discharges from the hippocampus to the mediodorsal thalamus and thence, via the spiny stellate cells, to the cartridge synapses on the pyramidal cell dendrites [Figure 16.10 (Section 16.5.2)]. It could thus resemble the repetitive burst discharges that are most effective in setting up LTPs. So it can be envisaged that, in the interaction illustrated in Figure 16.10, the synapses of the horizontal fibers are normally weak, but are potentiated by the conjunction (Figure 16.4C). However, this occurs only for synapses that, when activated, cause window-opening at the precise time.

Repeated replays of the conjunctional interaction would serve to prolong the potentiation of the horizontal fiber synapses on the pyramidal cell (cf. Figure 16.10), just as has been observed by Barnes and McNaughton (1980) for the LTP of dentate granule cells; hence, there is a plausible explanation of memory consolidation in the cerebral cortex.

16.5.4. Recall of Memory

At one stage of consideration, the recall of a memory can be explained by the replay of neocortical circuits that were consolidated by the learning process. Thus, the replayed circuits would closely resemble those that gave the original experience; hence, the remembered experience would be recognized as genuine. But there are many problems overlooked in this simple story. No special difficulty would attend an explanation of a memory triggered by some related present experience or a sequence of memories in some train of thought, but we recognize that we can attempt at will to recall a memory. Evidently, we are now involved with the mind–brain problem as illustrated diagrammatically in Figure 17.13 (Section 17.5.1). Even more challenging is the experience that we can assess the validity of a recalled memory, recognizing, for example, that the telephone number or ZIP code or name is almost correct, so that a new demand is made on the memory data banks for a delivery that can be recognized as correct. Thus, we have to entertain the idea of two kinds of long-term memory: (1) a data bank memory that is stored in the neocortex and (2) a recognition memory that is in the conscious mind, located in the inner sense column of Figure 17.13.

In retrieval of a memory, we have further to conjecture that the self-conscious mind is continuously searching to recover memories, e.g., words, phrases, sentences, ideas, events, pictures, melodies, aromas, by active scanning through the modular array and that, by its action on the preferred "open" modules, it tries to evoke the full neural patterned operation that it can read out as a memory. This searching and trying to evoke memories could be a trial-and-error process. We are all familiar with the ease or difficulty of recall of one or another memory and of the strategies we discover in our efforts to recover memories of names that for some unknown reason are refractory to recall. We can imagine that our self-conscious mind is under a continual challenge to recall the desired memory by discovering the appropriate entry into module operation that would, by development, give the appropriate patterned array of modules.

Penfield and Perot (1963) gave a most illuminating account of the experiential responses evoked in 53 patients by stimulation of the cerebral hemispheres during operations performed under local anesthesia. These responses differed from those produced by stimulation of the primary sensory areas, which were merely flashes of light or touches, and paresthesia, in that the patients had experiences that resembled dreams, the so-called "dreamy states." During the continued gentle electrical stimulation of sites on the exposed surface of their brains, the patients reported experiences that they often recognized as being recalls of long-forgotten memories. As Penfield pointed out, it is as though a past stream of consciousness is recovered during this electrical stimulation. The most common experiences were either visual or auditory, but there were also many cases of combined visual and auditory recollections. The recall of music and song provided very striking experiences for both the patient and the neurosurgeon. All these results were obtained from brains of patients with a history of epileptic seizures.

It can be concluded that the stimulation acts as a mode of recall of past experiences. We may regard this as an instrumental means for recovery of memories. It can be suggested that the storage of these memories is likely to be in cerebral areas close to the effective stimulation sites. It is important to recognize, however, that the experiential recall is evoked from areas in the region of the disordered cerebral function that is displayed by the epileptic seizures. Conceivably, the effective sites are abnormal zones that are thereby able to act by association pathways to the much wider areas of the cerebral cortex that are the actual storage sites for memories.

16.5.5. Final Comments on Cognitive Memory

The molecular neurobiology of learning has been much studied in the simpler nervous systems of invertebrates, which lend themselves readily to investigations of the learning process in single neurons. Special reference should be made to the refined studies in *Aplysia* by Kandel and associates (Kandel and Schwartz, 1982) into the molecular mechanisms involved in both short- and long-term memory that seem to grade into one another, which is in contrast to PTP and LTP in the mammalian brain. Associative learning is observed, so there is the suggestion that the studies in *Aplysia* may give important insights into learning processes in the mammalian brain. The evidence is that learning in *Aplysia* is a presynaptic mechanism, there being no postsynaptic events such as those that characterize the LTP of the mammalian hippocampus and presumably the rest of the cerebral cortex. However, it must be remembered that the memory processes may differ in other regions of the mammalian brain. There is, for example, the long-term depression of the mammalian cerebellum (M. Ito *et al.*, 1982) (see Section 16.7.2), and there is a presynaptic component in the hippocampal LTP, as described in Section 16.3. Thus, the vertebrate and invertebrate nervous systems may exhibit interesting similarities in learning systems. Nevertheless, it is important to recognize that the $Ca^{2+} \rightarrow$ calpain $\rightarrow$ fodrin mechanism for uncovering glutamate receptors [see Figure 16.6D and E (Section 16.3.1)] is probably unique for high brain levels in mammals. Preliminary tests have failed to disclose it even in lower vertebrates (G. Lynch and Baudry, 1984).

The manner in which memories can be laid down in the modular structure of the cerebral neocortex on the basis of the Marr (1970) hypothesis is exemplified at the first stage in Figure 16.10 (Section 16.5.2). Selectivity is given by the very short duration of 20 msec (W. B. Levy and Steward, 1983) or 40 msec (Gustafsson and Wigström, 1986) for the window-opening in the conjunction process as described in Section 16.3. It is important to recognize that the potentiation of synapses that are already effective will give merely a further consolidation of neural pathways and their associated mental events. What is needed is the development and consolidation of neural pathways not previously functional, as with the novel task of learning a name for a new thing. Classical conditioned reflexes display this learned conjunction at a primitive level, e.g., the recognition that the sound of a bell signals the presentation of food. From the conjunction potentiation of Figure 16.10, there has been developed a diagrammatic representation of the laying down and retrieval of associative memories [Figures 7 and 8 in Eccles (1981a)].

The conjunction interaction that gives LTP of previously unresponsive synapses on hippocampal cells is an ideal model for building a theory of cognitive learning of the cerebral cortex. It is important to recognize that in their conjunction studies, W. B. Levy and Steward (1979, 1983) take advantage of two easily discriminated inputs to a granule cell, the ipsi- and contra-entorhinal cortex pathways, but such a conjunction would also occur for the LTPs of two separate inputs from the ipsi-entorhinal cortex (McNaughton *et al.*, 1978), and from the distinctive inputs to CA1 pyramidal cells (Gustafsson and Wigström, 1986), as discussed above. It will be evident that the early studies of hippocampal LTP are opening up great fields of investigation in neurobiology and neurochemistry. Already, we can regard the LTP of the hippocampus as providing an excellent model for neural events in the neocortex that are concerned in the learning process. However, we are only at the beginning of this great enterprise of understanding how the brain can give us this wonderful ability to store and retrieve memories.

16.6. Neurochemistry of Learning

With regard to this topic, see also Sections 15.2, 16.3, and 16.5.

In Section 15.2, we developed the concept of genotropic action of neurotransmitters, specifically referring to processes that were arrested by inhibitors of RNA or protein synthesis as depending on this type of message transmission. Long-term memory was briefly mentioned as a specific example.

In neurochemistry and neuropharmacology, there have now been many fine studies by Barondes (1970), Agranoff (1981), and others that reveal that long-term learning (beyond 3 hr) does not occur when either cerebral protein synthesis or RNA synthesis is greatly depressed by poisoning of the specific enzymes by cycloheximide or puromycin. It is conjectured that in the process of learning, synaptic activation of neurons leads first to specific RNA synthesis and this in turn to protein synthesis, and so finally to unique structural and functional changes involved in the synaptic growth and coding of memory. In Section 16.4, we developed the hypothesis that, with the depolarization of intense synaptic activation, Ca^{2+} ions move into the dendrites through the opened Ca^{2+} channels and there combine with calmodulin to produce a second messenger system that increases protein synthesis. It is known (Barondes, 1970) that protein synthesis in the brain is increased in the learning process (cf. Section 16.5), and in some circumstances, it can be effective in laying down memory traces within minutes. Apparently, long-term memory can be established only if there is an intact protein-synthesizing capacity, an appropriate "state of arousal," and availability of the information in a short-term memory store (Barondes, 1970). Experiments performed by Agranoff (1976) illustrate the principles. He injected puromycin, which blocks protein synthesis by about 80%, directly into the brains of awake goldfish. The goldfish were trained in a shuttle box to swim over a barrier upon a light signal in order to avoid a punishing electrical shock administered through the water. If puromycin was given immediately after the training session, the effects of training were abolished when the response was measured several days later. The training was not abolished, however, when that same amount of puromycin was injected into the fish after they had been first returned to their home tanks for about 1 hr. When puromycin was injected before training, there was also no effect on subsequent performance. If puromycin was injected at various times after training, there was a slow gradient of abolition of the training effect. Thus, puromycin does not immediately obliterate memory, but rather interferes with some ongoing process that is initiated by the training process. Acetoxycycloheximide and cycloheximide, both inhibitors of protein synthesis, produce the same effect.

These experiments parallel those of a number of other workers indicating that inhibition of protein synthesis abolishes the consolidation of memory. They do not affect short-term memory.

Whether or not a critical protein or series of proteins is task-specific for memory is a matter of conjecture. If enzyme induction is critical to some aspect of the formation of long-term memory, a block at the transcriptional level of RNA formation [cf. anisomycin (Sections 15.2 and 16.3.2)] would be expected to produce a similar effect. Such agents are often toxic, making the interpretation of results difficult. Nevertheless, inhibitors of nucleic acid synthesis such as 8-azaguanine, actinomycin D, amanitin, and camptothecin all interfere with the consolidation of memory.

If RNA synthesis is truly involved in the memory consolidation process, then increased turnover would be anticipated during the period when consolidation is taking place. Glassman (1969) has reported increased [³H]uridine incorporation into macromolecular material, presumably RNA, in the diencephalon of mice during the consolidation period. Although such a change could be due to responses other than just the learning procedure, restrained control animals subjected to the same amount of training experience but not permitted to execute the trained response did not show labeling changes. Overtrained animals permitted to make the trained response also failed to show increased labeling. It appears, then, that some changes in RNA metabolism take place in the training situation. A difficulty is that concomitant alterations in protein labeling have not been found in these situations. Such changes should be detectable, since RNA subserves protein synthesis. As the theory presented in Sections 16.3–16.5 makes clear, however, the enhanced protein synthesis need only be associated with those selective synapses involved in the learning pathways.

16.7. Learning in the Motor System

16.7.1. Introduction

It is proposed that three distinct classes of phenomena are included in the general theme of motor learning. Examples of each will be considered in turn.

First, there is the learning of automatic movements in which the whole process is subconscious from the beginning. A much investigated learned movement of this type is the learning of a corrective vestibuloocular reflex. Similarly, the movements of a conditioned reflex are automatic from the start.

Second, there is the learning of motor skills by animals. There is an enormous literature on the training of animals in motor skills by operant conditioning techniques. However, attention is focussed on animal studies of the neuronal responses of the cerebrum and cerebellum. From mere behavioral descriptions, we shall never come to understand how the brain is effective in the learning of motor skills.

Third, there is the learning of human skills. In this learning, it is essential to have mental concentration with a planned strategy of action and a subsequent evaluation and correction of errors in successive attempts. When well learned, the skills can be accomplished without voluntary attention, as in walking, swimming, cycling, skiing, and the like; the performance of the skilled action has become automatic. Today, the list of learned motor skills is enormous. There is much more interest in motor skills as exhibited in all the sports than in intellectual and artistic performance with the exception of music and ballet. No doubt TV is responsible in part for this bias. It would be generally agreed that the cerebral cortex is primarily involved in human motor learning, but the cerebellum is most importantly involved, and also the basal ganglia and presumably other brain stem nuclei.

16.7.2. Learning of Automatic Movements

There is a very ancient part of the cerebellum, the flocculus, that controls the eye movements when the head is turned. One can think that the function of this part of the cerebellum is to maintain, as far as possible, a fixed position of the visual image on the retina. The movements of the head are sensed by the semicircular canals of the vestibular system. For our present purpose, we have only to consider the horizontal canal because, in the experiments of M. Ito et al. (1974), the head of the rabbit was rotated about a vertical axis. The pathway diagram of Figure 16.11 shows the direct projection of the vestibular nerve to the vestibular nucleus (VN) that in turn projects directly to the oculomotor

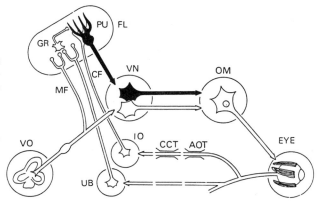

Figure 16.11. Construction of the flocculovestibuloocular system. Abbreviations: (AOT) accessory optic tract; (CCT) central tegmental tract; (CF) climbing fiber; (FL) flocculus; (GR) granule cell; (IO) inferior olive; (MF) mossy fiber; (OM) oculomotor neuron; (PU) Purkinje cell; (UB) unknown brain stem nucleus; (VN) vestibular nucleus; (VO) vestibular organ. Modified from M. Ito (1984).

nucleus (OM) for the eye muscles. By itself, this pathway can achieve some stabilization of the retinal image, but the control is much more regular in operation when it is aided by the pathway through the cerebellar flocculus. In Figure 16.11, there are two mossy fiber (MF) inputs, one directly from the vestibular nerve and the other a feedback from the visual system, which probably acts as a controlling device on the vestibuloocular reflex, improving the stability of the visual field during head movements. So far, we have neglected the role of the climbing fiber (CF) input, which is predominantly activated by the visual input via the pathway indicated to the inferior olive via the accessory optic tract (AOT) in Figure 16.11.

In Figure 16.12, the head of the rabbit is subjected to a sine wave horizontal rotation with a standard amplitude of 10° and a frequency of 0.15 Hz. The plotted left eye movements (Figure 16.12A) show that the compensation for the head rotation was much better when there was a fixed vertical strip of light than in darkness, but it was well below the 10° requirement for complete compensation. It was of great interest that, when the head rotation was continued for many hours (Figures 16.12B), there was a progressive increase in the compensation when tested with the illuminated slit (○). After several hours, it became almost perfect for eliminating retinal image slip. Evidently, there had been learning of the adaptive change, and this was very much less with the continual rotation when the measurements were carried out in darkness (●).

These observations provide clear evidence of an automatic learning process that, over some hours, tends to improve the stability of the retinal image. Two experiments define the neural pathways concerned in this learning. In the first experiment, after ablation of the flocculus, there was no learning. The second experiment showed that learning did not occur, or occurred rarely in a minor later form, when the inferior olive was destroyed. Thus, we have good evidence that the climbing fiber input to the cerebellar Purkinje cells of the flocculus (Figure 16.11) is effective in causing a learned cerebellar modification that gives improvement in the stabilization of the retinal image over and above that achieved by the vestibuloocular reflex. A remarkable feature is that in Figure 16.12, about 1300 trials were required for the good adaptive response.

Gonshor and Melvill Jones (1976a,b) performed similar experiments on human vestibuloocular reflexes that were disturbed by a horizontal visual inversion by dove

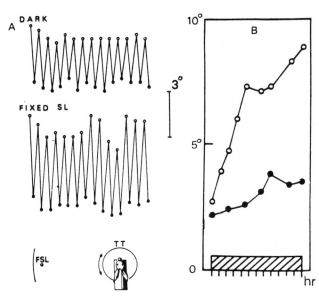

Figure 16.12. Horizontal eye movement induced by sinusoidal head rotation and its modification during presentation of slit lights in the visual field. (A) Most nasal (○) and most temporal (•) positions of the cornea mark on the left eye during each cycle of 10 head rotations. Plottings on the top indicate the eye movement in darkness and provide the control for the immediately succeeding measurement with slit light (SL) to the left eye, as plotted below. The diagram at the bottom is a dorsal view of the turntable (TT) mounting the rabbit and the fixed-slit light (FSL). (B) Changes of rabbit's horizontal vestibuloocular reflex during sustained head rotation with the fixed-slit light presented. Normal rabbit. *Ordinate:* mean angular amplitudes of the horizontal eye movement during 10° head rotation. *Abscissa:* time. (○) Eye movements measured with the fixed-slit light shown; (•) eye movements measured in temporary darkness. From M. Ito (1984).

prisms. Again, there was learning that minimized retinal image slip and that presumably is similarly explained by the responses of the flocculus (cf. D. A. Robinson, 1976).

Marr (1969) developed a most challenging theory of the neuronal events involved in cerebellar learning. He proposed that the single climbing fiber input into a Purkinje cell played the role of a teacher for the enormous number of parallel fiber synapses (about 100,000) on the dendrites of that same cell [see Figure 14.14 (Section 14.4.2)]. This enormous excitatory synaptic input is dominant in evoking the discharges of that Purkinje cell, which give simple spike (SS) discharges, usually at a frequency of about 50 per second. In contrast, the single climbing fiber impulse exerts a powerful excitation that generates a brief burst of impulses, which is called a "complex spike" (CS). According to the Marr hypothesis, learning is effected because the CS response (the teacher) increases the potency of the parallel fiber synapses that are activated at about the same time, which ensures a very sharp selective effect for the learned potentiation. A little later, Albus (1971), on the basis of perceptron theory, proposed a diametrically opposite plastic influence, namely, that there is a prolonged depression of the parallel fiber synapses that are activated at about the same time as the CS response. CS responses are infrequent, usually at 1–2 Hz.

With recording from the individual Purkinje cells of the flocculus during the head rotation, Ghelarducci *et al.* (1975) studied the phase relationship of SSs to CSs. Their evidence suggested a depressant influence of CSs on SSs, which is in accord with the Albus hypothesis. However, we now turn to a more direct testing procedure devised by M. Ito.

The plastic influence of a climbing fiber impulse on the excitatory synapses made by parallel fibers on the dendrites of that same Purkinje cell was first demonstrated (M. Ito *et al.*, 1982) by utilizing a vestibular nerve stimulus to set up a mossy fiber input into the

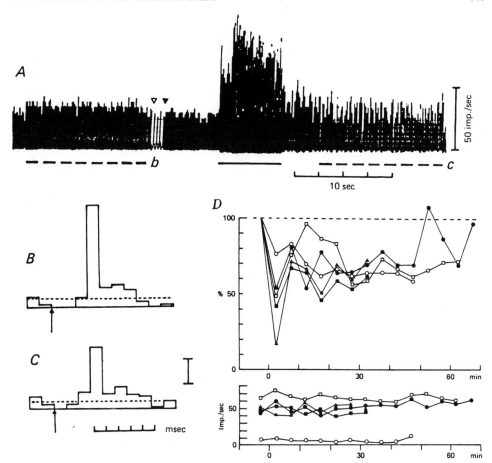

Figure 16.13. Effect of conjunctive stimulation of a vestibular nerve and the inferior olive. (*A*) Strip chart record of simple spike discharge from a Purkinje cell. Bin width: 0.5 sec. (∇ and ▼) mark a 3-min gap. (- - - -) Period of 2/sec stimulation of the ipsilateral vestibular nerve; (———) period of 4/sec conjunctive stimulation of the inferior olive and 20/sec stimulation of the ipsilateral vestibular nerve. (*B, C*) Peristimulus histograms constructed during periods (*b*) and (*c*), respectively, in (*A*). Calibration: 10 impulses/bin per 100 sweeps. (*D*) Ordinate: values of the firing index measured from peaks of the peristimulus histograms examplified in (*B*) and (*C*). The average values of two to three trials performed during each 5-min period are plotted. Note that plotted values of the firing index were normalized by control values before conjunctive stimulation. (- - - -) 100%. The bottom graph is the spontaneous discharge rate, averaged for each 5-min period. Measurements are from five different Purkinje cells. From M. Ito *et al.* (1982).

flocculus that activated the Purkinje cells via granule cells and parallel fibers (cf. Figure 16.11). The interacting climbing fiber impulse was set up by stimulation of the inferior olive. Conjunction was effected by simultaneous stimulation for 25 sec of the inferior olive at 4 Hz and the vestibular nerve at 25 Hz, frequencies that are in the normal physiological ranges. Any change in the effectiveness of parallel fiber synapses on the Purkinje cells was evaluated by observing the impulses generated by a test vestibular nerve stimulus at 2 per sec in comparison with the number before the conjunction (Figure 16.13*A*). The "firing index," as it is called, diminished sharply for some minutes (Figure

16.13 *B* and *C*), then partly recovered, to be followed by a depression for over 1 hr (Figure 16.13*D*). This was a clear demonstration of the Albus version of the conjunction phenomenon, namely, a prolonged synaptic depression of the parallel fiber synapses on the Purkinje cell dendrites. It is of interest that a similar conjunctive stimulation did not depress the synapses on basket cells.

The defect of this pioneering investigation was that the vestibular nerve stimulation was remote from the actual site of the presumed interaction on the Purkinje cell dendrites (Figure 16.11). M. Ito and Kano (1982) therefore set up the parallel fiber volley by direct stimulation and recorded the field potentials that the parallel fiber volley generated synaptically in the Purkinje cells at the presumed site of interaction. Again, conjunction often resulted in a significant synaptic depression that lasted more than an hour.

Figure 16.14A illustrates a still further refinement by Ekerot and Kano (1983). Both climbing fibers and parallel fibers were directly activated, the parallel fibers by two microelectrodes that set up beams projecting to the Purkinje cell from which impulse discharges were being selectively recorded. Conjunction was effected by stimulation of the climbing fiber (Cf) and one of the parallel fibers (Pf$_1$ or Pf$_2$) at 4 Hz for 1 min. The responses by the other parallel fiber electrode served as a control. The peristimulus time histograms in Figure 16.14B were constructed from the responses of the Purkinje cell to the parallel fiber stimulation, 60 stimuli at 2 Hz. The postconjunctional response was less than half the control shown above, whereas to the right there was no depression of the response evoked from the parallel fiber beam not activated in the conjunction. The importance of this restriction of the climbing fiber depressant action on parallel fiber synapses cannot be overestimated. It is the key to the selectivity that is an essential requisite for an effective learning mechanism. It is also observed in the LTP in the hippocampus (see Section 16.3). The time course of the depression is shown in Figure 16.14C for two conjunctions with the inactive control and then with 4-Hz conjunctive stimulation for 1 min of the control parallel fiber beam alone. In Figure 16.14D, a more prolonged depression was exhibited after conjunction in two other Purkinje cells. Depressions of parallel fiber synapses have been recorded for up to 2 hr. An interesting additional observation is that the amount of depression declined progressively, with conjunction intervals of climbing fiber leading parallel fiber by 20 msec up to 375 msec, the conjunction stimulation being then at 2 Hz. Thus, the climbing fiber stimulation opens, for a limited time of no more than 0.5 sec, a window for its depressant influence on a presently activated parallel fiber synapse.

It has been demonstrated by M. Ito *et al.* (1982) and M. Ito (1984) that the synaptic depression is due to a lowered sensitivity of the postsynaptic membrane to the synaptic neurotransmitter glutamate. Moreover, an hour-long desensitization to applied glutamate followed a conjunction of climbing fiber stimulation at 4 Hz with a prolonged (25- to 50-sec) glutamate iontophoresis (Figure 16.15A–C). It is of special interest that there was no depression of the much lower aspartate sensitivity. These iontophoretic experiments are closely analogous to the conjunction tests of Figures 16.13 and 16.14.

We now confront this intriguing question: How is the climbing fiber impulse instrumental in producing a depression of the parallel fiber synapses with which it is in conjunction? There is a spatial problem in the interaction because the synapses of the climbing fibers are restricted to the proximal two thirds of the Purkinje dendrites, while the parallel fiber synapses are at all levels (Eccles *et al.*, 1967). Ekerot and Oscarsson (1981) discovered a solution to this problem. Intracellular recording from a proximal

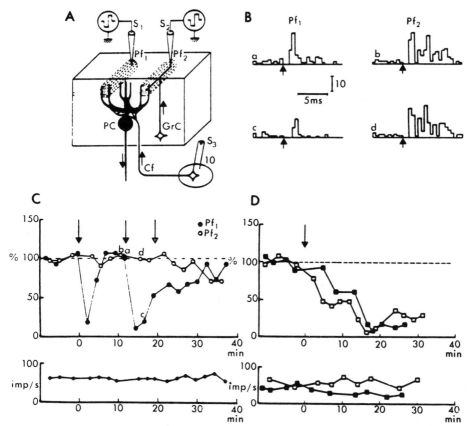

Figure 16.14. Effect of conjunctive activation of parallel fibers and climbing fibers on the parallel fiber–Purkinje cell transmission. (A) Experimental arrangement for stimulating two beams of parallel fibers (Pf_1, Pf_2) in the molecular layer and climbing fibers (Cf) in the inferior olive. Not shown is the extracellular recording from the Purkinje cell (PC). (GrC) Granule cell. (S_1, S_2, and S_3) Microelectrodes for stimulation. (B) Peristimulus histogram for discharge from the Purkinje cell. Upper histograms: Control responses of Purkinje cell for Pf_1 (a) and Pf_2 (b) activation. (↑) Time of activation. Lower histograms: (c) Responses of Purkinje cell after conjunctive activation of the beam Pf_1 with climbing fibers. Note that with such conjunctive activation of Pf_1, the response is lower than without such conjunctive activation (a). (d) Pf_2 was stimulated without conjunctive activation of the climbing fiber, and there was no change from (b) above. (C) Firing indices at the peaks of the peristimulus histograms. Solid arrowheads indicate the moments of conjunctive stimulation with Pf_1; open arrowhead indicates the moment of repetitive stimulation of Pf_2 alone. The lower histogram is the spontaneous firing level. (D) Similar to (C), but for another two Purkinje cells separately tested with only one beam of parallel fibers. From Ekerot and Kano (1983).

dendrite of the Purkinje cell revealed that the response to a climbing fiber impulse is composed of an initial CS response followed by a plateau depolarization for about 100 msec (Figure 16.16A). Prolonged depolarization of more distal dendrites was also produced by the climbing fiber impulse, as indicated by the negative potential recorded extracellularly (Figure 16.16B), but it was more variable in duration, from a few to several hundreds of milliseconds. Ekerot and Oscarsson suggested that a prolonged increase in Ca^{2+} conductance could be responsible for the depolarization, citing the finding of Støckle and ten Bruggencate (1978) that climbing fiber activation decreased the extra-

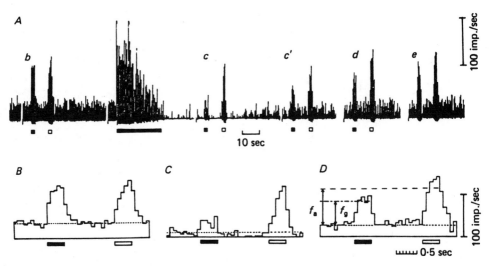

Figure 16.15. Responses of a Purkinje cell to amino acids and changes due to conjunctive application of glutamate and stimulation of the inferior olive. (A) Strip chart record showing simple spike discharges of a Purkinje cell. Bin widths; 0.5 sec. (■) Application of glutamate with a 9-nA current; (□) application of aspartate with a 69-nA current. Solid bar: Glutamate iontophoresis at 9 nA plus stimulation of the inferior olive at 4/sec. The record is interrupted between (c') and (d) and between (d) and (e); (d) 5 min after onset of conjunctive stimulation; (e) 30 min after onset. (B–D) Spike-density histograms expanding records (b), (c), and (d) in (A), respectively. The methods of measuring f_g and f_a are indicated in (D). (- - - -) Spontaneous discharge rates calculated from the initial 10 bins. (B, C). (D) Average discharge rates during application of glutamate (- - - -) and aspartate (- · - ·). From M. Ito et al. (1982).

cellular calcium. Llinás and Sugimori (1982) reported the prolonged depolarization induced by climbing fiber activation in *in vitro* studies of Purkinje cells, in both intrasomatic and intradendritic recording. The depolarization was shown to be due to Ca^{2+} input, since it was reduced and abolished by progressive $CdCl_2$ addition to the bath. The more complex bursts of spike responses are not seen *in vivo*, which raises the question of the effects of possible dendritic injury by the slicing technique. Ekerot and Oscarsson (1981) suggested that the increased intracellular Ca^{2+} could depress the receptor sensitivity in the manner described by Miledi (1980) for the acetylcholine receptors at motor endplates.

M. Ito (1984) illustrated this Ca^{2+} hypothesis (Figure 16.17). Depolarization by the climbing fiber impulse opens Ca^{2+} channels across the postsynaptic membrane, with the consequence that there is a partial self-regenerative depolarization that keeps the Ca^{2+} channels open for 100 msec or longer. The increased intradendritic Ca^{2+} sets in train

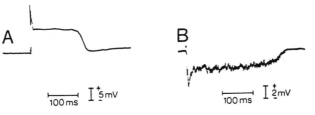

Figure 16.16. Plateau potentials evoked in Purkinje cell dendrites by impulses in climbing fibers. (A) Intracellular recording from proximal dendrite. (B) Extracellular recording from distal dendrites. From Ekerot and Oscarsson (1981).

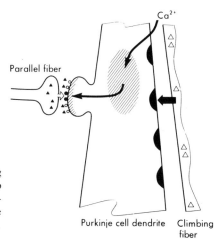

Figure 16.17. Diagram showing proposed action of a climbing fiber impulse in depolarizing Purkinje cell dendrites and so opening channels for Ca^{2+} entry. This Ca^{2+} activates a second-messenger system that depresses sensitivity of the spine synapse to the transmitter released by the parallel fiber impulse. From M. Ito (1984).

Parallel fiber

Purkinje cell dendrite   Climbing fiber

various metabolic reactions, including generating a second messenger system that then acts on the postsynaptic receptor sites of the parallel fiber synapses, reducing their sensitivity to the neurotransmitter (glutamate) and so depressing the synaptic effectiveness of the parallel fibers in the manner described by M. Ito *et al.* (1982) and Ekerot and Oscarsson (1981). However, there is the crucial problem of accounting for the selectivity of the depression, namely, that only the activated parallel fibers are depressed by the conjunction, yet all the dendritic synapses would be subjected to increased Ca^{2+}.

The hypothetical diagram (Figure 16.17) can account for the finding that, although the climbing fiber synapses are only on the proximal two thirds of the dendrites, all the activated parallel fiber synapses are depressed. By electrotonic transmission, the Purkinje dendrites would be depolarized through their whole length, and hence there is an increased Ca^{2+} input that would depress the sensitivity of even the most distal synapses. It also accounts for the finding of N. C. Campbell *et al.* (1983a) that, if by stimulation of parallel fibers there is produced an intense depolarization, there is opening of the Ca^{2+} channels with the consequent plateau afterdepolarization. Furthermore, if the parallel fiber volley activates inhibitory cells (basket or stellate) and so diminishes the initial excitatory depolarization by the climbing fiber impulse or by the massed parallel fibers, then there is depression or inhibition of the prolonged plateau, presumably because there is inadequate depolarization to open the Ca^{2+} channels (N. C. Campbell *et al.*, 1983b).

Summary.   The responses of the vestibuloocular reflex exhibit an adaptive learning response over many hours, and there is good evidence that the role of the climbing fiber input into the flocculus is to depress the parallel fiber synapses with which it is in conjunction, which is the Albus hypothesis. In the sine-wave head rotation, climbing fiber input into Purkinje cells tends to be out of phase with mossy fiber input (Ghelarducci *et al.*, 1975), and possibly the adaptation results from the conjunction depression. This depression is caused by a depression of the glutamate sensitivity of the parallel fiber synapses. Furthermore, there is evidence that this depression is brought about by the increase in dendritic Ca^{2+} consequent on the opening of the Ca^{2+} channels by strong depolarization. This complex hypothetical explanation of the automatic learning involved in the adaptive vestibuloocular reflex needs much more experimental testing, yet it will be

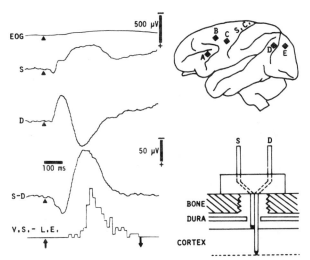

Figure 16.18. Specimen potentials recorded on the surface (S) and in the depth (D) (2.5–3.0 mm below the surface) of the forelimb motor cortex contralateral to the moving hand. (S–D) Surface potential minus depth potential, which is the same as the record of IV–C in Figure 16.19. (EOG) Simultaneously recorded electrooculogram (at the rostrolateral edge of the frontal bone above the orbita on the left side). Cortical potentials and EOG were led against the indifferent electrode in the bone just behind the ear and averaged 100 times with the time pulse of the onset of the light stimulus. Reaction times measured from the stimulus onset to the movement of lever elevation (V.S. - L.E) are plotted for the same 100 samples in the histogram at every 16 msec. Calibrations are 500 μV for EOG, 50 μV for cortical potentials, and 100 msec for all traces. (▲, ↑) Stimulus onset; (↓) end of the stimulus. Five sites for recording in the left hemisphere are illustrated in the upper right diagram (A–E) and correspond to those in Figure 16.19. (S.C.) sulcus centralis. The lower right schema illustrates a chronic recording electrode implanted on the surface (S) and in the depth (D) of the cortex. From Sasaki and Gemba (1981).

helpful in attempting to understand motor learning in the two higher classes. Since, by all evidence (anatomical, physiological, and pharmacological), all parts of the cerebellar cortex are similar, it can be assumed that the vermis, the pars intermedia, and the hemispheres have plastic properties similar to those demonstrated in Figures 16.13–16.15 for the flocculus. This conclusion will be most significant in the subsequent two sections.

16.7.3. Learning of Motor Skills by Animals

Highly significant experimental observations on the learning of a motor skill have been described by Sasaki and Gemba. The learning procedure is a very simple operant conditioning (Sasaki and Gemba, 1981, 1982). There is presented to the monkey a visual stimulus by a small green electric bulb that is turned on for 900 msec at random time intervals of 2.5–6 sec. The monkey has to learn that, if it lifts a lever by wrist extension during the light stimulus, there is a small juice reward. Initially, the monkey lifts the lever at random with an occasional reward, but gradually, after some weeks of training with about a thousand trials a week, there is motor learning, and eventually the monkey lifts the lever on every light stimulus and at a progressively diminishing latency, which reaches a uniform short time. There is still full success when the light stimulus is shortened to 500 msec.

Before the onset of the operant conditioning and under anesthesia, there had been implanted in the cerebral cortex of the monkey several bipolar recording electrodes, one on the pial surface, the other 2.5–3 mm deeper (Figure 16.18). Each bipolar electrode thus leads from a localized area of the cortex an electrical potential with recording of surface to depth, which is derived by subtraction as indicated in Figure 16.18 (S–D).

In Figure 16.19A–E are bipolar recordings from the sites indicated on the left

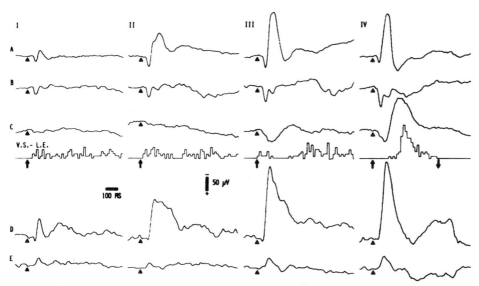

Figure 16.19. Specimen records from the five sites (A–E in Figure 16.18) in the cortex contralateral to the moving hand in the same monkey. (I–IV) Four different stages in the learning process of visually initiated hand movements, respectively 3, 21, 24, and 59 days after the commencement of training. Only S–D records (cf. Figure 16.18) are presented with the average of 100 times. (V.S.–L.E.) The histogram gives reaction times of the same 100 samples, but some parts were curtailed in I–III, because the light stimulus was 900 msec in I–III and the end was out of the trace presented. Calibrations are 50 μV for all potentials and 100 msec for all traces. From Sasaki and Gemba (1981).

cerebral cortex in Figure 16.18. In columns I–IV are the responses at various stages of the learning procedure, after approximately 500, 3,000, 4000, and 9000 trials. Row V. S. – L. E. gives the reaction times measured from the light-stimulus onset to the movement of the lever. In stages I and II, the lever movement is random relative to the light stimulus, but by stage III, there is a clear bunching of the motor responses with a latency of 250–600 msec, and in stage IV, the monkey was performing well, with a latency as brief as 200 msec and a mean latency of about 250 msec. It is surprising that many thousands of trials were necessary for this effective learning of a simple motor act, but eventually the monkey performs the lifting of the lever in a skilled manner.

The responses of the cerebral cortex develop with the learning. It was surprising to find that the prefrontal cortex (A in Figure 16.19) exhibited an increased response early in the learning procedure and that very large responses developed in area 18 (Figure 16.19D), whereas in area 17 (Figure 16.19E) there was a much smaller response. However, this could depend on the unfavorable location of electrode E [cf. Figure 16.18 (upper right diagram)]. Attention should be focused on the responses of the forelimb motor area (row C in Figure 16.19). Only at the stage of an effective motor reaction (III) does a delayed negative wave appear, and it becomes large in IV, preceding by about 50 msec the averaged reaction time (V.S. – L. E.). In contrast to the bilateral responses at the recording sites in the association cortex, the response in the forelimb motor cortex was strictly contralateral to the activated forelimb; this was also observed for the somatosensory cortex. Evidently there was much cerebral activity in various cortical regions (rows A

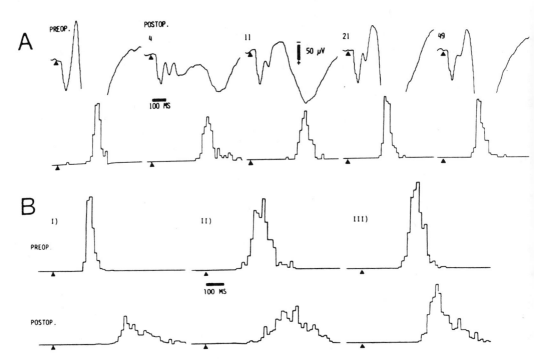

Figure 16.20. (A) Visually initiated premovement potentials (S–D) in the forelimb motor cortex (top row) and histograms of reaction times (bottom row) as influenced by cerebellar hemispherectomy when the interpositus was spared. (PREOP.) Before; (POSTOP.) at indicated days later. The calibration was 50 μV for potentials and 100-msec scale for potentials and histograms. There were 100 samples for all potentials and histograms. (B) Influence of cerebellar hemispherectomy on reaction times of visually initiated movements. Control data before operation (PREOP.) and those after (POSTOP.) are presented for three monkeys (I–III). Histograms of 300 movements (lever elevation) after onset of the visual stimulus (▲) were calculated for 16-msec time bins and 100-msec scale for all records. From Sasaki *et al.* (1982).

and D of Figure 16.19) before this learned response developed, the forelimb motor cortex responding only when the conditioned activity developed to a high level.

Sasaki and Gemba recognized that the late surface negative wave of the motor cortex resembled the response to projection from the neocerebellum via the ventrolateral thalamic nucleus (Sasaki *et al.*, 1979). This identification was tested by excising the opposite cerebellar hemisphere (Sasaki *et al.*, 1982). As shown in Figure 16.20A, the large negative potential of the motor cortex (PREOP.) was completely eliminated (POSTOP.), and at the same time the conditioned response was greatly delayed. Instead of a quick skilled lifting of the lever, there was a late disordered response that is also seen in Figure 16.20B for three other hemispherectomies. In Figure 16.20A, the interpositus nucleus was not excised, so a recovery of cerebello-thalamo-cerebral circuity was possible with further training, as is seen in the responses at 11, 21, and 49 postoperative days. After complete operative removal of the cerebellar hemisphere and intermediate lobe, there was no recovery.

It can be concluded that in the fast skilled lever-lifting of the trained monkey, the contralateral neocerebellum plays a key role in instructing the motor cortex. In support of this conclusion is the finding (Sasaki and Gemba, 1982) that in two monkeys hemi-spherectomized before the training procedures, there was failure to develop the negative wave over the contralateral motor cortex. The visually initiated hand movement required a much longer training procedure, and the learned movement was irregular, with a long latency and frequent failure.

In learning this simple skill in a visually initiated hand movement, the visual input to the cerebral cortex resulted in responses in several areas of the association cortex (rows A and D in Figure 16.19) that appear eventually to bring the contralateral neocerebellum into the response. It is postulated that, with continued training, the cerebro-cerebello-thalamic pathway grows in effectiveness so that the hand movement becomes more skillful and at a progressively shorter latency (row V.S. – L.E. in Figure 16.19).

It is an attractive hypothesis that this increased cerebellar influence arises because of learning in the cerebellum; such learning has already been established in the flocculus by direct experimental investigation [see Figures 16.13 and 16.14 (Section 16.7.2)]. In these experiments, it was seen that by conjunction of the climbing fiber and mossy fiber inputs into a Purkinje cell, there was initiated a prolonged depression (>1 hr) of the activated parallel fiber synapses on the Purkinje cell. Since the sole output by Purkinje cells is to inhibit the nuclear cells, depression of the mossy fiber excitation of Purkinje cells results in a disinhibition of the nuclear cells, which is an excitation. Thus, there would be an increase in the excitatory dentatus–interpositus input into the ventrolateral thalamus. This would result in a thalamocortical input into the motor cortex that is postulated to produce the negative wave in Figure 16.19 (row C-IV) and in Figure 16.20A (PREOP.).

In accord with this dentate $\rightarrow$ motor cortex proposal, Thach (1975) had already found that in fast wrist movements in trained monkeys, the activity of the dentate neurons tended to be earlier than the motor cortex activity, the mean values being respectively 70 and 54 msec before the muscle contraction. When the dentate nucleus was cooled (Meyer-Lohmann et al., 1977), there was a delay of 0.05–0.15 sec in the movement, which was in accord with a proposed chain of command: cerebellum $\rightarrow$ dentatus nucleus $\rightarrow$ thalamus $\rightarrow$ motor cortex. In this experiment, however, an earlier stage must be added. The GO signal must initially activate the cerebral cortex, the association areas projecting the command to the cerebellum, as proposed by Sasaki and Gemba (1981) on the basis of Figure 16.19.

A very different motor learning task from that of Sasaki and Gemba was investigated by P. F. C. Gilbert and Thach (1977). The right hand of the monkey grasped a handle that alternately gave to the wrist a flexor and extensor force with a variable duration of 1.5–4.0 sec, and the task of the monkey was to return the handle to the midposition, success being indicated by a signal light. After training, this return was accomplished skilfully in about 0.5 sec. Then the paradigm was altered by an increase or decrease in either the flexor or the extensor force, the other being unaltered. The smooth learned response was disturbed by the changed force, but a good compensation was learned in 12–100 trials, whereas at the same time there was little disturbance in the response to the unchanged force in the opposite direction. Recording of single Purkinje cell responses was simul-taneously performed through an electrode assembly that was previously implanted in the ipsilateral cerebellar cortex, usually in the pars intermedia of the anterior lobe. With adjustment of the recording electrode, the complex spikes (CSs) due to climbing fiber input could be clearly distinguished from the simple spikes (SSs) due to the mossy

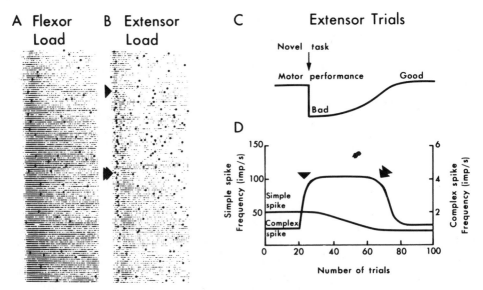

Figure 16.21. Cerebellum in adaptive motor control. (A, B) complex (CS) and simple (SS) spike frequency changes for Purkinje cell after change in load (as described in the text). Each dot represents a spike potential (SSs, small dots; CSs, large dots); each row of dots represents the discharge during a trial, beginning at the change in the direction of load. Successive trials are represented top to bottom; each flexor trace was followed by an extensor trace. At the single arrowheads shown in (B) and (D), the known extensor load of 300 g was changed to a novel 450-g load while the known flexor load of 310 g was kept constant. Before the load change (above the single arrowhead), there was a low frequency of related CS activity at about 100 msec after the start of extensor trials. After the load change [below the single arrowhead in (B)], CS frequency increased greatly and persisted for about 50 trials. This returned to the initially learned level [double arrowheads in (B) and (D)] as the newly learned adaptation was mastered. Associated with these transient increases in CS frequency, there were decreases in SS frequency as the new motor task was being learned. (C) Relationship of motor performance to number of trials. Performance deteriorated after the load change at 20 trials, but improved after about 50 more trials. (D) Change in CS and SS frequencies over multiple trials. SS frequency declined while the new motor task was being learned and then leveled off. CS frequency increased during the period of the roughly 50 trials (trials 20–70) required to learn good performance with the new load. From Brooks and Thach (1981).

fiber/parallel fiber inputs to that same Purkinje cell (Figure 16.21). Because of the changed force, some Purkinje cells were subjected to a climbing fiber input of short latency, often with a later prolonged increase. In some cells, it was the increased extensor load that triggered the increased climbing fiber input (Figure 16.21B). In other cells, a decreased extensor load was effective, and similarly for other Purkinje cells that were responsive to changes in flexor loads. The brief increase in climbing fiber responses corresponded to the learning time for the changed force, and there was an associated decrease in SSs, which continued indefinitely during the increased or decreased forces, the climbing fiber input meanwhile having ceased (Figure 16.21B–D).

P. F. C. Gilbert and Thach (1977) concluded that the increase in CSs in conjunction with parallel fiber inputs probably was responsible for the depression of the Purkinje cell responses (SSs) to parallel fiber inputs as proposed in the Albus (1971) version of cerebellar learning. Subsequently (cf. Section 16.7.2), this hypothesis has been tested by Ito and Ekerot with results in accord with prediction. Moreover, the experiments of

Gilbert and Thach on the cerebellar responses have been complemented by the studies of Sasaki and Gemba on the responses of the cerebral cortex (see Figures 16.18 and 16.19). Figure 16.21 beautifully illustrates the role of the climbing fiber input to the cerebellum in *error correction* by a learning process in the cerebellar cortex.

The learning of motor reactions to a changed force was quite fast (P. F. C. Gilbert and Thach, 1977), in great contrast to the learning of the visually initiated response (Figure 16.19). Changed forces in rhythmic animal movements would be a common experience, as, for example, in walking on a terrain with changing slopes; hence, rapid learning is essential. The learning of lever movement to irregularly presented light flashes is a much more esoteric experience for the animal, hence, presumably, the extreme slowness of that learning. This slowness was advantageous to Sasaki and Gemba in their study of the stages of learning. In further investigations, they are recording the cerebral potentials from many more areas, including the supplementary motor area (SMA). Brinkman and Porter (1983) suggest that the premotor area may be specially involved in motor learning, but the premotor responses were not large and augmented but little in the Sasaki and Gemba experiments (see row B in Figure 16.19).

Figure 16.22 represents diagrammatically the cerebrocerebellar contributions to the responses described in this section when an animal is learning a motor task in an operant conditioning. It is the basis of a hypothesis of motor learning. All training of animals by operant conditioning involves some sensory input–visual, auditory, somatosensory—so there will be, as shown, an involvement initially of primary sensory cortex and secondarily of various association areas. With a visual input, Sasaki and Gemba found the peristriate cortex strongly excited (see row D in Figure 16.19), but for other sensory inputs, there presumably would be other association areas. However, a prefrontal involvement (see row A in Figure 16.19) could be common to all. There is progressive augmentation through learning stages I–IV (see rows A and D in Figure 16.19). It is proposed that this learning response of the association cortices could be an example of the Marr (1970) model for conjunction potentiation in the cerebral cortex [cf. Figure 16.10 (Section 16.5.2)] (Eccles 1981a, 1983). At the next stage, there is projection from these association cortices to the motor cortex, giving a rather crude learned response, as is seen after cerebellectomy (see Figure 16.20). For a skilled learned response, the cerebellum seems to be necessary. As shown in Figure 16.22, there would be projection to the contralateral hemisphere and pars intermedia by both mossy fibers and climbing fibers (Sasaki *et al.*, 1977). It is proposed that there is a learned response (P. F. C. Gilbert and Thach, 1977; M. Ito, 1984) from their interaction in the Albus manner of conjunction, thus producing a prolonged depression of the response to mossy fiber inputs. At a further stage of the cerebral–cerebellar–cerebrum input circuit (see Figure 16.22), there is activation by disinhibition of the neurons of the dentate and interpositus nuclei (G. I. Allen *et al.*, 1977b, 1978), which project, in turn, to the ventrolateral thalamus and so to the motor cortex for a skilled response (see Figures 16.18–16.20).

It is thus proposed in Figure 16.22 that motor learning is accomplished in two stages: the cerebral, possibly on the Marr (1970) conjunction model, and the cerebellar, on the Albus (1971) model. This hypothesis is presented as a challenge for rigorous testing. For example, it is essential to have much more investigation in the manner of Sasaki and Gemba of the responses of the association cortices and, particularly, of the premotor area and the supplementary motor cortex. Again, the cerebellar cortex and nuclei must be studied in various learning procedures in the manner of Gilbert and Thach. Further

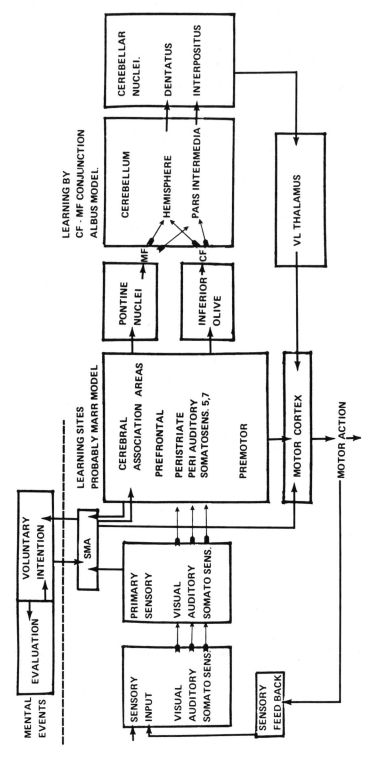

Figure 16.22. Diagrammatic representation of the proposed mental and neural events in learning a skilled movement. Mental events for human motor learning are shown above the dashed line that separates them from the neural events. The role of mental events in animal motor learning is left undefined. The arrows indicate directions of main pathways of action. See the text for a full description. All pathways are excitatory except for the inhibitory action of the cerebellum on the cerebellar nuclei.

outstanding problems relate to the responses of the thalamic nuclei. However, the most significant deficiency derives from the as yet largely unknown learning responses of the basal ganglia (cf. DeLong and Georgopoulos, 1983).

16.7.4. Learning of Motor Skills by Humans

The two preceding sections on motor learning, culminating in the diagram of Figure 16.22, provide a basis for our inquiry into the learning of the much more complex human motor skills, involving as they must a predominantly neocortical location. This inquiry will provide some important insight, but because of inadequate knowledge, the emerging story will be more fragmentary than in the two preceding sections.

In learning a motor skill, it is essential to concentrate on the planned action, so it can be postulated that the supplementary motor area (SMA) is initially activated by the intention (see Section 14.3.3), with subsequent excitation of other neocortical areas, particularly the premotor, with their stored motor programs. There is then excitation of the motor cortex to generate a discharge down the pyramidal tract that evokes the first "naïve" movement. Feedback by vision and somatosensory systems would be the basis of further intentions initiated by the SMA.

Meanwhile, on the basis of the discoveries of Sasaki and Gemba, it is postulated that the various cortical association areas are activated in a progressively enhanced manner [see rows A and D in Figure 16.19 (Section 16.7.3)] and in turn project to the neocerebellum and pars intermedia as in Figure 16.22 with an interaction of climbing fiber and mossy fiber inputs in the Albus model. Just as with the animal studies in Section 16.7.3, it can be expected that there is initiated from the Purkinje cell–nuclear cell interaction a discharge up to the ventrolateral thalamus and so by relay to the motor cortex (Figure 16.22). Thus, as learning proceeds, the motor cortex will be provided with a patterned input incorporating more and more the learned information that results in a pyramidal tract discharge giving the desired movement. The responses in row C of Figure 16.19 display the animal equivalent for this proposed cerebellar projection via the ventrolateral thalamus to the motor cortex.

The learning of an animal skill was postulated to be due to two distinct learning phenomena, in the cerebrum and in the cerebellum, as illustrated in Figure 16.22. Unfortunately, there is little evidence on the neuronal machinery involved in human motor learning. It is not possible to carry out precise electrical recording in the manner of P. F. C. Gilbert and Thach (1977) (Figure 16.21) or of Sasaki and Gemba (Figures 16.18–16.20). However, there are studies on clinical lesions, especially of the cerebellum, which result in clumsy and inefficient movements as originally described by Gordon Holmes. Brooks (1979) has given a comprehensive review of Holmes' descriptions of the motor disabilities resulting from cerebellar lesions and of clinical investigations since that time. Necessarily, the clinical material cannot be investigated by precise neuronal studies, so one has to rely on monkey experiments as partly described in Section 16.7.3.

It can be suggested that the readiness potential of Deecke and Kornhuber (1977, 1978) gives evidence of a cerebellar input to the motor cortex resembling that of Figure 16.19, row C, stage IV. After being symmetrical over the ipsi- and contralateral motor cortices in the early stages, asymmetry develops in the later stages [Figure 14.7A (Section 14.3.3)]. The readiness potential over the motor cortex that is involved in the movements becomes considerably increased, and this leads on to the motor potential. This developing asymmetry may be due, at least in part, to the cerebellothalamocortical input that is seen in row C of Figure 16.19, stages III and IV, and in the PREOP. response of Figure 16.20.

Studies by radio-emission tomography have disclosed areas of the cerebrum and deeper nuclei that are involved in voluntary movement as distinct from automatic movement (Roland *et al.*, 1982; Roland, 1984). All these movements had to be learned, but the neuronal processes involved in that learning are unknown. However, all voluntary movements seemed to be initiated in the SMA, with perhaps some help from the premotor cortex (Eccles, 1982a,b). Brinkman and Porter (1983) concluded from their studies on conscious monkeys that the premotor cortex may be especially involved in learning new motor programs. The visual input would seem to be of special importance for the premotor area.

It can therefore be assumed that, in the learning of some movement, the SMA is primarily involved by the mental intention (Figure 16.22), with perhaps some premotor assistance, and further that there is a continuous conscious judgment or appraisal (EVALUATION in Figure 16.22) of the successive trials. This appraisal is based on the sensory inputs (Figure 16.22) and results in intentions to carry out the movement more and more skilfully by action through the SMA. Thus, in the learning of human skills, mind–brain interaction is a key factor. Such mental activities as critical evaluation, redesigning of the intended movement, further critical evaluation of the new movement, and so on are experienced by all who try to learn a new skill or to improve an existing one. This conscious activity is diagrammed in the upper box of Figure 16.22 (above the dashed line) together with its proposed action on the SMA. The SMA is also shown to be influenced by the primary sensory areas in accord with the findings of Tanji and Kurata (1982) and many other investigators.

In contrast to the poverty of data on cerebral and cerebellar involvement in human motor learning, there is a large literature on behavioral studies with mechanical and electromyographic recordings. However, there has been no coherent attempt to understand the learning at the neuronal and synaptic level. For human motor learning, the diagram of Figure 16.22, derived from studies on monkeys, can be accepted provisionally. It is evident that the learning of motor skills provides an immense challenge for both monkey and human investigation.

It is probable that the learned responses of the association cortex with their progressive augmentation (see rows A and D in Figure 16.19) are dependent on the same neuronal factors as are cognitive memories (Section 16.3), namely, the Marr (1970) type of conjunction potentiation, as indicated in Figure 16.22. However, in contrast to cognitive learning, motor learning is not dependent on the hippocampus. The well-known subject of bilateral hippocampectomy (H.M.) was able to learn a complex motor skill despite an extreme disability in cognitive learning. For example, he learned within a normal time (3 days) to draw a line between the parallel lines of a double-lined star, the whole process being learned through vision in a mirror. In subsequent days, he had no recollection of learning the task; nevertheless, he had retained the motor skill (Milner, 1966). The hippocampus does not appear in Figure 16.22.

The hypothesis illustrated in Figure 16.22, with the addition of mental events for humans, is offered not as some final story, but provisionally. It needs testing at every level, but at least it is a coherent story of motor learning, whereas previously there were unrelated conjectures. It is interesting that an increase of intradendritic calcium is proposed as a key factor for both cerebellar learning (Figure 16.17) and hippocampal and cerebral learning (Sections 16.3.1 and 16.5.2). The difference is that Ca^{2+} is responsible for conjunction depression in the cerebellum, the Albus (1971) model, and for conjunction potentiation in the long-term potentiation of the hippocampus and for the Marr (1970) model in the neocerebrum. This difference in synaptic design presumably relates to the inhibitory action of the cerebellar Purkinje cells and to the excitatory action of all cerebral pyramidal cells.

17

Perception, Speech, and Consciousness

The most challenging problems for molecular neurobiology arise in an attempt to account for the amazing performance of the human cerebral cortex. These problems can be regarded as the ultimate goal, but all we can present in this chapter are the relatively crude discoveries concerning the neural machinery of the cerebral cortex—the constituent neurons and their connectivities. Somehow, the mental world comprised in the title of this chapter has to be related to the neural events of the cerebral cortex. We present a new hypothesis based on quantum mechanics. We must not suppose that the ultimate goal of cortical neuroscience is a final reduction of the whole range of mental events to the neural events as studied in neuroscience. Nevertheless, these neural events are coherent with the mental events, which exist in a different world. In a broad sense, these concepts are the basis of this chapter.

17.1. Cerebral Cortex

For general references, the reader is referred to Szentágothai (1978a, 1983).

This chapter is oriented to activities of the highest level of the brain, the mammalian cerebral cortex, so there will be a brief introduction to the structure and mode of action of this immense sheet of tissue that in the human has an area of about 2500 cm^2 and a complement of some 10,000 million nerve cells. A useful classification is to consider the great cerebral mantle of neocortex as being primarily concerned either with the sensory and motor projections into and out of the cortex, the koniocortex, or with the communications within the cortex, the association cortex. On the basis of the variations in structure, the human cerebral cortex has been subdivided into many areas, the best known being the almost 50 areas of Brodmann. The motor cortex has been briefly considered in Chapter 14

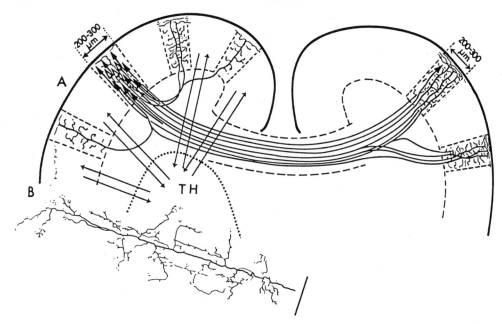

Figure 17.1. (A) General principle of corticocortical connectivity shown diagrammatically in a nonconvoluted brain. The connections are established in highly specific patterns between vertical columns with a diameter of 200–300 μm in both hemispheres. Ipsilateral connections are derived mainly from cells located in layer III (cells shown at left in outline), while contralateral connections (cells shown in solid black) derive from layers II–VI. (TH) Thalamus. The diagram does not try to show the convergence from afferents originating from different parts of the cortex to the same columns. (B) Golgi-stained branching of a single corticocortical afferent, oriented in relationship to the module with a single afferent in (A), but at several times higher magnification. It illustrates the profuse branching in all laminae. Scale bar: 100 μm. From Szentágothai (1978a).

[see Figures 14.2 and 14.3 (Section 14.3.1) and 14.8 (Section 14.3.4)], and the various sensory cortices follow in this chapter. Our present concern is with the association cortex, which forms almost 95% of the human neocortex.

In Figure 17.1A, Szentágothai (1978a) gives a diagrammatic representation of the extraordinary finding of Goldman and Nauta (1977) that radiotracer techniques reveal subdivision of the immense sheet of the association cortex into a mosaic of quasi-discrete space units. We will develop the thesis that these space units are the modules that form the basic anatomical elements in the functional design of the neocortex. At the left in Figure 17.1A are shown 12 closely packed pyramidal cells in one such module or column, which has a width of about 350 μm. The axons from these pyramidal cells project to 3 other modules of that same hemisphere and, after traversing the corpus callosum, to 2 modules of the other hemisphere. Thus, there is simply displayed both the association and the callosal projections of the pyramidal cells of a module.

There are several important features of Figure 17.1A: (1) Several pyramidal cells of a module project in a *completely overlapping manner* to other modules that are so defined and that are again about 350 μm across. In Figure 17.1A, this dimension is already represented by the branches of a single association fiber for 2 modules, and the overlapping distribution is illustrated for 2 association fibers and for 4 and 5 callosal fibers. (2) The callosal projection is mostly but not entirely to symmetrical modules on the con-

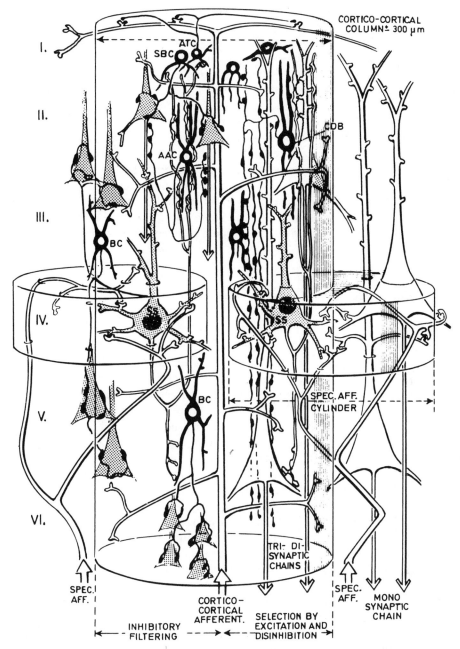

Figure 17.2. Neuron connectivity in a corticocortical column or module, the vertical cylindrical space of about 300 μm in the center. The module is sharing part of its space with two flat disks in lamina IV, in which specific afferents (SPEC. AFF.) arborize. The corticocortical afferent (indicated at bottom) terminates all over the corticocortical module, though with different densities of terminals. In lamina I, the tangential spread of the corticocortical fibers extends far beyond the module. The selection of pyramidal cells for output is envisaged in the right half of the diagram over excitatory interneurons [spiny stellates (SS)] or over disinhibitory interneurons, the *cellules à double bouquet* (CDB) of Ramón y Cajal, which are inhibitory interneurons that act specifically on inhibitory interneurons shown in solid black. The left side of the diagram explains the action of inhibitory interneurons as some kind of "filter" keeping out of action some of the pyramidal cells (stippled). Interneurons that can be defined as inhibitory with a considerable amount of confidence are indicated in solid black: the basket cell (BC) in the deeper laminae, the small basket cells (SBC) in lamina II, the axonal tuft cells (ATC), and a very specific axoaxonic cell (AAC), acting on the initial segments of pyramidal cell axons. Modified from Szentágothai (1979).

tralateral side. (3) There is reciprocity of callosal connections between symmetrical modules. In Figure 17.1B, there is shown at higher magnification the faint arborization of a single corticocortical afferent fiber, as revealed by Golgi staining. It is seen to traverse all laminae of the cortex and to branch extensively so that it is distributed to a column about 200 μm across, as is drawn for many association fibers in Figure 17.1A.

Figure 17.2 provides a diagrammatic summary of the essential components of the neocortical module together with the excitatory and inhibitory neurons, the former shown in outline in the right half, the latter in black in the left half. Centrally, there is the corticocortical afferent that forms the module (cf. Figure 17.1B) by its branches at all laminae. On either side are the two disks formed by the specific afferents from the thalamus with their branches in lamina IV. From these disks, there extend the vertical connectivities of the relay neurons, which are not in register with the module defined by the corticocortical afferents, of which there would be many hundreds instead of the one shown.

17.1.1. Modular Operation of the Neocortex

For general references, the reader is referred to Szentágothai (1978a, 1983).

Figure 17.1A illustrates in very simple form the distribution of the output from a module. It can be recognized that the module is a unit because other modules of the neocortex project onto it in a completely overlapping manner, the branches of any one fiber extending throughout the whole module. The simplest hypothesis is that this convergent input has no selectivity in its distribution; there is merely a pooling of the total input of information by the convergent association and callosal fibers. The complex integrative machinery (cf. Figure 17.2) of the whole module then operates to determine the output from the module from moment to moment. At any one instant, the output will have an intensity compounded of the number of pyramidal cells firing and the frequencies of these firings [cf. Figure 14.4 (Section 14.3.2)]. It will also have a temporal pattern that is given by the time course of the integrated frequency. An additional action of the module is to inhibit adjacent modules by the lateral distribution of inhibitory neurons (Marin-Padilla, 1969), and reciprocally it will receive inhibition from these surround modules. Not shown in Figure 17.2 are the axon collaterals of pyramidal cells that give widely distributed excitation to neurons of that module or adjacent modules in a selective manner (C. D. Gilbert, 1985).

To gain some insight into the manner of operation of this neuronal machine, we have to recognize that Figure 17.1A is greatly simplified. In the first place, there are about 4000 neurons in a module 350 μm across, at least 1500 of which are pyramidal cells. On the basis of the fiber count for the corpus callosum (200 million), there would be about 80 pyramidal cells projecting from one module to make callosal afferents to the other side, instead of the 6 in Figure 17.1A. On an estimate of 10 times as many association fibers as callosal fibers, there would be 800 pyramidal cells as association projectors to a module instead of the 3 shown in Figure 17.1A. Thus, there would be a much more intense convergence of inputs from any one module than is represented and also probably dispersion to many more modules. The projection from module to module is not random; on the contrary, it is very highly structured in accord with specific distributions, as is indicated for the sequential relays of the somatosensory and visual systems. We can think of the modular connectivities of Figure 17.1A as giving the grain of the large association connectivities. Szentágothai (1978a) suggests convergent and divergent numbers for modular interaction as high as 50.

But even if this massive addition were made to Figure 17.1A, the diagram would give the situation of connectivities for only a moment of time. If we assume a strong excitation by the convergent fibers with a powerful burst of impulse discharges from the constituent pyramidal cells, there will be within a few milliseconds strong excitatory inputs into many other modules by the association and callosal projections, as is shown in Figure 17.1A. Some of these secondary modules in turn will be excited sufficiently to discharge to tertiary modules and these in turn to quaternary. Some idea of the fast transmission from module to module is given by the observation of J. D. Newman (1976) that, in response to a brief click, some neurons of the dorsolateral prefrontal cortex respond with a latency as brief as 20 msec. At least two cortical relays by modular responses would be involved in this brief latency, as well as the transmission time to the primary auditory cortex. In a small fraction of a second, we have, in effect, a spreading pattern of excitation that is not random, but strictly specified by the sequential projections from the secondary, tertiary, and further-removed modules in this particular instance. There are approximately 3 million modules in the human neocortex, so there are immense possibilities for developing spatiotemporal patterns, even on the simple assumption that each module operates as a unit in its reception and projection. However, there is certainly a gradation of the responses of the pyramidal cells in a module, which is evidenced by the wide range in their firing frequencies, as reported by many investigators [cf. Figure 14.4 (Section 14.3.2)].

The special design features of a cortical module are: (1) the internal pattern of connectivities of its constituent neurons (Figure 17.2); (2) the inhibitory surround that is built up by some of its constituent inhibitory neurons and that sharpens its boundaries (Szentágothai, 1978a); (3) the output, which is graded both by number of pyramidal cells that are excited to fire impulses and by the frequencies of this firing (Figure 14.4). The simplest hypothesis is that a module acts as a fundamental unit in the performance of the neocortex and that fractionation of a module probably has no functional significance because there is complete overlap of the projections of its constituent pyramidal cells (cf. Figure 17.1A). Each module acts as a unit, processing the many convergent inputs from the other modules and in turn projecting divergently to many other modules. It thus embodies the Sherringtonian principles of convergence and divergence.

17.2. Perception

For general references, the reader is referred to Mountcastle (1975, 1978) and Edelman and Finkel (1984).

17.2.1. Introduction

It would be out of place in this book to present the immense field of perception. There are certain principles relating to the neural events that lead to perceptions of the various sensory experiences (Adrian, 1947; Mountcastle, 1975). Touch and vision have been most thoroughly investigated, but there is good reason to believe that all other sensory experiences are dependent on similar neuronal mechanisms. Of necessity, the crucial experimental investigation of perceptual experiences must be carried out on conscious human subjects, but both the design and interpretation of these experiments are dependent on the wonderful successes that have attended investigations on animal, and particularly monkey, sensory systems in the last few decades. The powerful techniques designed for precision and selectivity of stimulation have been matched by microelectrode recording from single neurons. But just as important, there has been success in defining the neural pathways from receptor organs to cerebral cortex and within the cerebral cortex by precise anatomical investigations.

There is a large variety of these receptor organs with built-in properties that enable them to encode in a highly selective manner some environmental change into a discharge of nerve impulses. In general, it can be stated that intensity of stimulus is encoded as frequency of discharge of impulses (cf. Chapter 2). In this way, there are transmitted from receptor organs to the higher levels of the central nervous system signals that result in the conscious experiences of vision, hearing, pain, and touch, for example. An introduction to the problem of conscious perception is best given in relation to cutaneous sensing. In the skin, there are receptor organs specialized for converting some mechanical stimulus such as touch or tap into impulse discharges in nerve fibers [see Figure 2.2A (Section 2.1)].

The pathways from receptor organs to the brain are never direct. There are always synaptic linkages from neuron to neuron at each of several relay stations. Each of these stages gives an opportunity for modifying the coding of the "messages" from the sensory receptors. Even the simplest stimulus, such as a flash of light or a tap on the skin, is signaled to the appropriate primary receiving area of the cerebral cortex in the form of a code of nerve impulses in various temporal sequences and in many fibers in parallel.

Our special interest is focused on the neural events that are necessary for giving a conscious experience. It is now generally agreed that a conscious experience does not light up as soon as impulses in some sensory pathway reach the primary sensory areas in the cerebral hemisphere. In response to some brief peripheral stimulus, the initial response is a sharp potential change, the evoked response in the appropriate primary cortical area. Immediately afterward, there is a change in the background frequency of firing of numerous neurons in this area—an increase or a decrease, or some complex temporal sequence thereof. Our present problem is to gain some insight into the neural events that have a necessary relationship to the conscious experience.

17.2.2. Cutaneous Perception (Somesthesis)

For general references, the reader is referred to Mountcastle (1978) and Libet (1973).

Figure 17.3 is a diagram of the simplest pathway from receptor organs in the skin up to the cerebral cortex. For example, a touch on the skin causes a receptor to fire impulses [see Figure 2.2A (Section 2.1)]. These travel up the dorsal columns of the spinal cord (the cuneate tract for the hand and arm), and then, after one synaptic relay in the cuneate nucleus and another in the thalamus, the pathway reaches the cerebral cortex. There are only two synapses on the way—and, you might say, why have any at all? Why not have a direct line? The point is that each of these relays gives an opportunity for an inhibitory action that sharpens the neuronal signals by eliminating all the weaker excitatory actions, such as would occur when the skin touches an ill-defined edge. In this way, a much more sharply defined signal eventually comes up to the cortex, and there again there would be the same inhibitory sculpturing of the signal by modular interaction [cf. Figures 17.1 and 17.2 (Section 7.1)]. As a consequence, touch stimuli can be more precisely located and evaluated. In fact, because of this inhibition, a strong cutaneous stimulus is often surrounded by a cutaneous area that has reduced sensitivity.

Also shown in Figure 17.3 are the pathways down from the cerebral cortex to both these relays on the cutaneous pathway. In this way, by exerting presynaptic and postsynaptic inhibition, the cerebral cortex is able to block these synapses and so protect itself from being bothered by cutaneous stimuli that can be ignored. This is what happens, of course, when you are very intensely occupied, e.g., in carrying out some action, or in experiencing, or in thinking. Under such situations, you can be oblivious even of severe stimulation. For example, in the heat of combat, severe injuries may be ignored.

At a less severe level, it has long been a practice to give counterirritation to relieve pain. Presumably, in this way, inhibitory suppression of the pain pathway to the brain is produced. Thus, we can account, at least in part, for the apparent anesthesias of hypnosis or of yoga or of acupuncture by the cerebral pathways' inhibition of the cutaneous pathways to the brain (cf. Section 17.2.3). In all these cases, discharges from the cerebral cortex down the pyramidal tract and other pathways will exert an inhibitory block at the relays in the spinocortical pathways such as those diagrammed in Figure 17.3. This ability of the cerebral cortex is important, because it is undesirable to have all receptor organ discharges from your body pouring into your brain all the time. The design pattern of successive synaptic relays, each with various central and peripheral inhibitory inputs, gives opportunity for turning off inputs according to the exigencies of situations.

Conventional studies on animals and man have defined the area of cortex that is primarily involved in responding to cutaneous sense, the somesthetic area [see Figures 14.2 and 14.3 (Section 14.3.1)]. The principal area is laid out as a long strip map in the postcentral gyrus with locations matching those of the motor cortex. All areas of the body surface, from the extreme caudal to the extreme rostral, lie in linear sequence along the postcentral gyrus from its dorsomedial end over the convex surface of the cerebral hemisphere. There is also a subsidiary somesthetic area that is partly concerned in the perception of pain. It can be seen in Figure 14.3 that cortical areas are apportioned in relationship to the fineness of motor discrimination and similarly for the cutaneous areas. The apportioning relates to sensitivity and not to size. This map has been explored in detail by two main procedures: recording in nonhuman primates of the cortical responses evoked by exploratory stimulation applied systematically to the whole surface of the body, limbs, neck and head; and electrical stimulation of the sensory cortex in conscious human subjects who report the skin areas to which the evoked sensations are referred (Penfield and Jasper, 1954).

Usually, the subjects report abnormal sensory experiences—paresthesia such as tingling, numbness, or "pins and needles"—though there are also reports of normal sensations—touch, tap, and pressure. The paresthesias are plausibly explained by the outrage that the applied stimulation perpetrates on the highly organized neuronal machinery of the cerebral cortex. Even the weakest electrical stimulus will excite in a manner dependent on the relationship of the immense neuronal assemblage to the applied electrical current. As a consequence, there will be a "neuronal shock wave" having little resemblance to the pattern of neuronal activation generated by a natural input from the receptor organs, hence the paresthesia, just as when the ulnar nerve at the elbow joint is bumped—the so-called "funny bone."

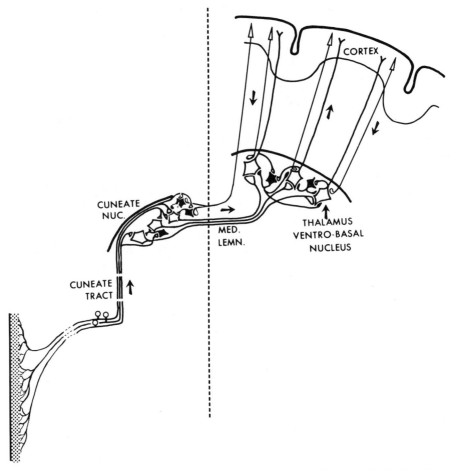

Figure 17.3. Pathways to and from the sensorimotor cortex for cutaneous fibers from the forelimb. Note the inhibitory cells shown in black in both the cuneate nucleus and the ventrobasal nucleus of the thalamus. The inhibitory pathway in the cuneate nucleus is of the feed-forward type, and in the thalamus it is the feedback type. Also shown is one presynaptic inhibitory pathway to an excitatory synapse of a cuneate tract fiber. Efferent pathways from the sensorimotor cortex are shown exciting the thalamocortical relay cells and exciting both postsynaptic and presynaptic inhibitory neurons in the cuneate nucleus.

The thalamocortical fibers in Figure 17.3 would have the distribution in the cortex illustrated for specific afferents in Figure 17.2 (Section 17.1), exciting many stellate cells—the excitatory ones that amplify and distribute the message in that column or module and the inhibitory ones that inhibit that module or adjacent modules. Figure 17.2 is a greatly simplified picture of the real state of affairs. An important design feature of the cerebral cortex is that the whole neuronal machinery, activated by many similar fibers of some specific cutaneous input, is arranged in a column that is vertically oriented to the cortical surface. So, as a first approximation, we can say that all these neurons in Figure 17.2 are engaged in an integrated operation of excitation and inhibition. The same kind of cutaneous sense (modality) from the cutaneous area provides the input to a column by many different fibers that give mutual help in neuronal activation (Mountcastle, 1978). You might ask, what is the point of this columnar arrangement? Isn't it enough for each afferent fiber to register its own ongoing activity in isolation? The point is that impulses in one fiber and discharges from one neuron or a few neurons will be virtually ineffective at the next stage of the synaptic relay. There must be many parallel lines because the input from one afferent fiber is lost in all the incessant "noise" of background firing, which is the actual state of the neurons of a "resting" cerebral cortex. But, if you have 100 or even perhaps only 20 fibers coming in with approximately synchronized bursts of impulses, the whole ensemble of cells in the column would be stirred up in a highly significant manner. This, in turn, would result in a spreading of meaningful signals widely and selectively in the cerebral cortex, a process that could lead eventually to a conscious experience.

It is important to realize that there is a considerable "incubation period" between the arrival of impulses at the primary sensory area in Figure 17.3 and the experience of a conscious sensation [Popper and Eccles, 1977 (Chapter E2)]. By accurately timed experiments with repetitive stimulation of the somesthetic cortex of conscious subjects, Libet (1973) and Libet et al. (1979) have shown that the "incubation period" is as long as 0.5 sec for weak cortical stimuli presented in a repetitive train of shocks at about 50 per second. Evidently there is opportunity for a great elaboration of neuronal activity in complex spatiotemporal patterns before there is a conscious experience, as has been investigated by Libet et al. (1979).

17.2.3. Pain

For general references, the reader is referred to Fields and Basbaum (1978), Kerr and Wilson (1978), Willis (1982), and Eccles (1984b).

Our present interest relates to the problem of how activities in the neuronal machinery of the brain can generate this extraordinary range of disagreeable affects that we refer to as "pain." Although affects are privately experienced, by intersubjective communication the objective reality of pain is ensured during the whole range of our experiences from babyhood onward. By this means, pain can be investigated in conscious human subjects, and it is most important in clinical investigations both as a symptom and as a sign. It is desirable first to consider the extraordinary nature of pain as a perception. All other perceptions [cf. Figure 17.13 (Section 17.5.1)] have some counterpart in the material world. Nevertheless, there is the same basic similarity that all are dependent on the stimulation of specific sense organs that cause the discharge of impulses carrying coded information along nerve fibers going to the brain, where the perceptual experiences arise as a result of processes but dimly understood. The perceptual experiences are not in the brain, where there is the coded information, but in the conscious mind (World 2) as indicated in Figure 17.13. Literally, we make our own pains!

This general philosophical introduction leads on to the scientific investigation of pain. Since the actual neural mechanism concerned in the generation of the experience of pain is not understood, the word *nociceptive* is used for pathways that are presumed to carry the information leading to the experience of pain.

The pathways for nociception in the human spinal cord and brain are shown in the greatly simplified diagram of Figure 17.4. There are, of course, tens of thousands of fibers and neurons instead of the units depicted. A painful stimulus on the hand excites three types of afferent fibers: The largest fiber is for touch, and the bold black arrow from

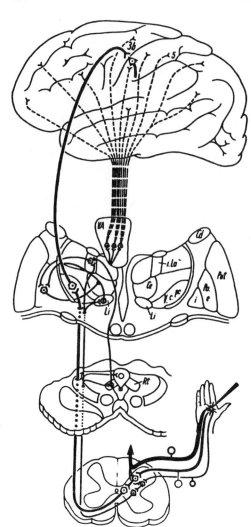

Figure 17.4. Diagrammatic representation of an injurious contact on the hand leading to transmission in three pathways. The one shown in bold black is for touch and ascends in the dorsal column as shown by the arrow. The other two are the nociceptive pathways. The branches of these two pathways in the brain stem are shown as described in the text for the human subject. Note the pathway from the external pallidum (*Pa.e*) diving under the nucleus centrum medianum (*Ce*) and intralaminar nucleus (*i.La*) to project to the ventroanterior thalamic nucleus (*VA*). Other abbreviations: (*Cd*) caudate; (*Ce*) nucleus centrum medianum; (*i.La*) intralaminar nucleus; (*Li*) nucleus limitans; (*Pa.e* and *i*) external and internal pallidum; (*Put*) putamen; (*Rt*) reticular formation; (*V.c.pc*) parvicellular ventrocaudal nucleus. From Hassler (1975) (derived from his extensive human experience).

the hand shows the beginning of the ascent to the brain (Figure 17.4). Our present concern is with the other two types. The medium black fiber (A fiber) signals pricking pain and the thin black fiber (C fiber) aching or burning pain. Both are shown traversing two neurons in the dorsal horn, with the second of these being the neuron of origin of the spinothalamic tract. The larger input fiber for nociception projects up the neospinothalamic tract to the thalamus [largely the parvicellular ventrocaudal nucleus (V.C.pc)] and thence to the somatosensory cortex (*3b*), where there is topographic representation. Thus, the pricking nociception can be sharply localized. The C fiber for nociception ascends via the paleospinothalamic tract and projects largely to two thalamic nuclei, limitans (*Li*) and intralaminar (*i.La*), whence it relays to the external pallidum (*Pa.e*) of the basal ganglia, which, in turn, relays to the ventroanterior thalamic nucleus (*VA*) and thence widely to the neocortex (- - -). In addition, this C fiber path gives collaterals to the reticular nucleus (*Rt*) throughout the whole length of the *Rt* in the brain stem. From both the *Rt* and the neocortex, there is projection to the limbic system.

It is conjectured that the painful affect results from the limbic projection, particularly to the amygdala. However, it is suggested that the conscious experience of pain is derived from some unique patterned performance of the cerebral cortex. Radioactive xenon studies (Lassen *et al.*, 1978) reveal that even a moderate pain causes an increased circulation over the whole frontal lobe, which corresponds to the distribution of nociceptive inputs in Figure 17.4. One can anticipate that, with severe pain, there would be an intense activation of the cerebral cortex matching our experience that in agonizing pain all thinking is suppressed. It seems that we have to envisage intensely acting reverberatory circuits to the neocortex involving the medial and anterior thalamus, the limbic system, and the hypothalamus. In particular, there is a specific reciprocal relationship between each cortical area with one or two thalamic sites [TH in Figure 17.1 (Section 17.1)] (E. G. Jones and Powell, 1968). An intense cerebral excitation would also be brought about by the reticular activating system that is excited by the paleospinothalamic tract (*Rt* in Figure 17.4).

In conclusion, we would suggest that three factors are concerned in the mental experience of pain: (1) a widely dispersed reverberatory activity of modules of the prefrontal and parietal cortices; (2) reverberatory circuits involving, on one hand, aversive limbic structures such as the amygdala and associated hypothalamic areas and, on the other, the cerebral cortex; (3) strong reinforcement from the reticular activating system. These suggestions are in line with the general belief that pain arises from most complex cerebral operations, which greatly complicates the difficulties attending the surgical relief of pain by a stereotaxic lesion. These neuronal operations must have a special property of urgency, perhaps because of the intensity of the modular operations activated by reverberatory circuits from the limbic system [see Figure 15.4 (Section 15.3)].

The synaptic transmissions on the pain pathway (cf. Eccles, 1982c, 1984b) are of great interest. For the initial synaptic relay in the dorsal horn, one transmitter is substance P (SP), as described in Section 12.4.2 (Nicoll *et al.*, 1980). Neurons with methionine-enkephalin as transmitter inhibit the SP synapses by presynaptic action [cf. Figure 4.18 (Section 4.8)]. There is a descending pathway from the raphe nuclei that uses serotonin as neurotransmitter and inhibits the synaptic relay on the cells of the spinothalamic tract (see Section 10.4.2) (Fields and Basbaum, 1978; Willis *et al.*, 1977; Willis, 1982).

17.2.4. Visual Perception

For general references, the reader is referred to Hubel (1982), Wiesel (1982), and Sections 13.1.2 and 13.2.4.

Highly complicated and exquisitely designed structures are involved in all steps of the visual pathways. The optical system of the human eye gives an image on the retina, which is a sheet of closely packed receptors, some 10^7 cones and 10^8 rods, that feed into the complexly organized neuronal systems of the retina. Thus, the first stage in visual perception is a radical fragmentation of the retinal picture into the independent responses of a myriad of punctate elements, the rods and cones. In some quite mysterious way, the retinal picture appears in conscious perception, but nowhere in the brain can there be found neurons that respond specifically with the refined detail of the perceived picture. The neuronal machinery of the visual system of the brain has been shown to accomplish a very inadequate reconstitution that can be traced in many sequences (cf. Kuffler, 1973; Gross *et al.*, 1985).

The initial stage of reconstitution of the picture occurs in the complex nervous system of the retina. As a consequence of this retinal synthetic mechanism, the output in the million or so nerve fibers in each optic nerve is not a simple translation of the retinal image into a corresponding pattern of impulse discharges that travel to the primary visual center of brain area 17 [see Figure 14.2 (Section 14.3.1)]. Already in the nervous system of the retina, there has begun the abstraction from the richly patterned retinal mosaic into elements of pattern, which we may call "features," and this abstraction continues in the many successive stages that have now been recognized in the visual centers of the brain and are shown diagrammatically by Popper and Eccles [1977 (Figures E1.7 and E1.8B)].

The complex interactions in the retinal nervous system are eventually expressed by the retinal ganglion cells that discharge impulses along the optic nerve fibers and so to the brain. These cells respond particularly to spatial and temporal changes of luminosity of the retinal image by two neuronal subsystems signaling brightness and darkness, respectively. The brightness contrasts of the retinal image are converted into contoured outlines by several neuronal stages of information processing. One type of ganglion cell is excited by a spot of light applied to the retina over it and is inhibited by light on the surrounding retina. The other type gives the reverse response—inhibition by light shone into the center and excitation by the surround. The combined responses of these two neuronal subsystems result in a contoured abstraction of the retinal image presented to the visual cortex. Hence, what the eye tells the brain by the million fibers of the optic nerve is an abstraction of brightness and color contrasts.

As illustrated in Figures 17.5 and 17.11 (Section 17.4.1), the optic nerves from each

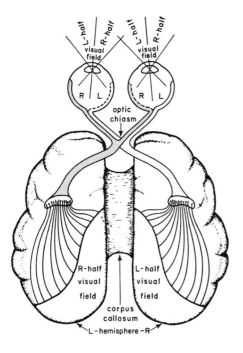

Figure 17.5. Diagram of visual pathways showing the L-half and R-half visual fields with the retinal images and the partial crossing in the optic chiasm so that the R-half of the visual field of each eye goes to the left visual cortex, after relay in the lateral geniculate body and correspondingly for the L-half visual field to the right visual cortex.

eye meet in the optic chiasm, where there is a partial crossing. The hemiretinas of both eyes (nasal of right and temporal of left) that receive the image from the right visual field have their optic nerve projections rearranged in the chiasm so that they coalesce to form the pathway to the left visual cortex, and vice versa for the left visual field projecting to the right visual cortex. Thus, with the exception of a narrow vertical (meridional) strip of the visual field that is directly in the line of vision, the visual imagery of the right and left fields comes to the left and right visual cortices, respectively, to form an ordered map, much as the cortical map for cutaneous sensation is formed. There is, of course, topographic distortion. The fine visual sensing of the center of the visual field results from a much more amplified cortical projection area than for the retina concerned with peripheral vision.

The center-surround orientations of the retina are preserved in the relay in the lateral geniculate body (LGB) (Figure 17.5) and also in the first relay of the geniculate fibers in the lower levels of lamina IV of the primary visual cortex of the monkey. These cells in lamina IV C are as strictly monocular as the lateral geniculate cells that project to them, but they in turn project to more superficial laminae, where there is convergence with inputs from IV C of the other eye (see Figure 17.7); hence, these cells may have dominance from one or the other eye or even equality. Various methods demonstrate the ocular dominance columns as the characteristic zebra-like pattern (see Section 13.2.4) (Hubel *et al.,* 1977; Hubel, 1982).

Microelectrodes can be used to record the impulses discharged from single nerve cells, for example, as has been done with great success by Hubel and Wiesel (1962) in the primary visual cortex. We use one aspect of their work to illustrate the very subtle and well-controlled experiments that are being carried out in many laboratories. In Figure 17.6A, there is a single cell firing impulses, having been "found" by a microelectrode that has been inserted into the primary visual cortex of the cat. The track of insertion is shown in Figure 17.6B as a sloping line with short transverse lines indicating the locations of many nerve cells along that track. With the microelectrode it is possible to record extracellularly the impulse discharges of a single cell if it is positioned carefully [cf. Figure 14.4 (Section 14.3.2)]. The cell has a slow background discharge (upper trace of Figure 17.6A) but, if the retina is swept with a band of light, as illustrated in the diagram to the left, there is an intense discharge of that cell when light sweeps across a certain zone of the retina and there is immediate cessation of the discharge as the light band leaves the zone (lowest trace of Figure 17.6A). If you rotate the direction of sweep, the cell discharges just a little, as in the middle trace. Finally, if the sweep is at right angles to the most favorable direction, it has no effect whatever (uppermost trace). It is a sign that this particular cell is most sensitive for movements of the light strip in one orientation and is quite insensitive for movements at right angles thereto.

The direction of the lines across the microelectrode track in Figure 17.6B indicates their orientational sensitivity. The same orientational sensitivity is found when the track runs along a column of cells that is orthogonal to the surface, as in the upper group of 12 cells. In Figure 17.6B, however, the track continues on across the central white matter and then proceeds to pass through three groups of cells with quite different orientation sensitivities. Evidently, the track crosses several columns with different orientational sensitivities in the way illustrated by the dotted sectors. This columnar arrangement for mutual reinforcement of similar receiving cells has already been illustrated in Figure 14.8 (Sections 14.3.4) and Figure 17.2 (Section 17.1).

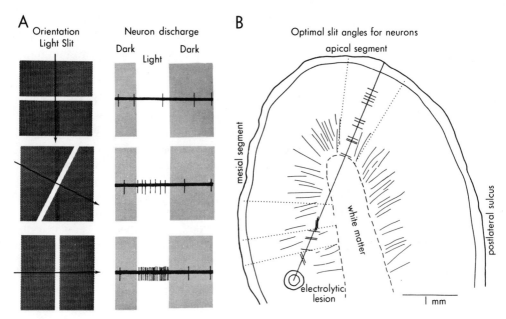

Figure 17.6. Orientational responses of neurons in primary visual cortex of the cat. From Hubel and Wiesel (1963).

In the visual cortex, neurons with similar orientation sensitivity tend to be arranged in columns that run orthogonally from the cortical surface. Thus, it can be envisaged that in the large area of the human primary visual cortex, the population of about 400 million neurons is arranged as a mosaic of columns, each with some thousands of neurons that have the same orientation sensitivity (Hubel and Wiesel, 1963).

This arrangement can be regarded as the first stage of reconstitution of the retinal image. It will be recognized, of course, that this orientation map is superimposed on the retinal field map, each zone of this field being composed of columns that collectively represent all orientations of bright lines or of edges between light and dark.

It has already been shown in Figure 17.5 that both the ipsilateral eye and the contralateral eye project to the LGB on the way to the visual cortex. However, in the primate, these projections are relayed in separate laminae, three for the ipsilateral (2i, 3i, 5i,) and three for the contralateral eye (1c, 4c, 6c) (Figure 17.7). The projection to the columns of area 17 is illustrated in highly diagrammatic form in this figure (Hubel and Wiesel, 1974). The ipsilateral and contralateral laminae of the LGB are shown projecting to alternating columns, the ocular dominance columns [cf. Figure 13.4 (Section 13.1.2) and Figure 13.22 (Section 13.2.4)]. Orthogonally, the columns are defined by the orientation specificities as indicated in Figure 17.6, and these can be seen to have a rotational sequence in Figure 17.7 on the upper surface of the cortex. The actual columnar elements are, of course, much less strictly arranged than is shown in this diagram for the monkey cortex.

There is a histological account of the development of ocular dominance columns in Section 13.1.2 (Figure 13.4) and of the manner of their definition by neuronal discharges in Section 13.2.4 to give the zebra-like pattern for left eye–right eye dominance over the

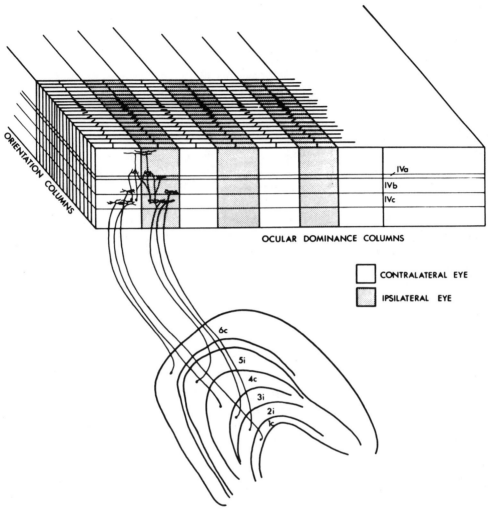

Figure 17.7. Idealized diagram showing for the monkey the projection from the lateral geniculate body (LGB) to the visual cortex (area 17). The six layers of the LGB are labeled accordingly as they are associated with the ipsilateral (i) or contralateral (c) eye. These (i) and (c) layers project to specific areas, thereby forming the ocular-dominance columns for the ipsilateral and contralateral eyes. The stacked slablike columns of the visual cortex are defined by the criteria of ocular dominance in one direction and orientation (shown on the upper surface). From Hubel and Wiesel (1974).

primary visual cortex (Hubel, 1982). Ocular dominance columns are excellently illustrated by the metabolic testing procedure of autoradiographic [^{14}C]deoxyglucose (C. Kennedy *et al.*, 1976). One eye is closed, and after an hour or so of normal visual usage by the open eye, the animal is sacrificed, and the projection of the open eye to the primary visual cortex is displayed autoradiographically. Cytochrome oxidase staining and neurotransmitter immunocytochemistry have given amazing displays of compartmentalization in the monkey visual cortex (Hubel, 1982; Hendrickson, 1985), but unfortunately the functional meaning is obscure.

In the upper part of lamina IV are the *simple cells* that are strictly monocular and that simply respond to lines or edges oriented as in Figure 17.6. At the next stage of image reconstitution are neurons at other levels in area 17 and in the surrounding secondary and tertiary visual areas [Brodmann areas 18 and 19 in Figure 14.2 (Section 14.3.1)]. In these areas, there are neurons that are specially sensitive to the length and thickness of bright or dark lines as well as to their orientation and even to two lines meeting at an angle. These so-called "complex" and "hypercomplex" neurons (Hubel and Wiesel, 1963, 1965) constitute a further stage of feature recognition. It is believed that these complex and hypercomplex neurons acquire their specific properties by means of synthesis of the neuronal circuits that are activated by "simple" cells, these circuits containing inhibitory as well as excitatory components as in Figure 17.2 (Section 17.1) (Hubel and Wiesel, 1965). In Figure 17.7, there are examples of two complex cells in the upper lamina that each receive inputs from two simple cells of different ocular dominance columns.

It has been demonstrated by Jung, Bishop, and others that retinal zones adjacent to those that give excitation have an inhibitory action on the firing of a neuron such as that of Figure 17.6A. For example, if the cell in Figure 17.6A had a relatively high background discharge, this discharge would be depressed by illumination of adjacent areas of the retina. This can be explained by the diagram in Figure 17.2, in which excitation of one area of the cerebral cortex results in inhibition of adjacent areas, much as with basket cell inhibition in the cerebellum. This observation has great interest because it explains the so-called "Mach bands" of perceptual physiology. At the edge between fields of uniform bright and dull illumination, there is perceived to be a narrow light–dark zone, the Mach band (Ratliff, 1972), which is explicable by this lateral inhibition.

So far, we have a relatively clear story, and there is identification in the visual cortex of neurons requisite for the various integrational tasks. This account is, of course, over-simplified. For example, we have neglected the neural events responsible for the various contrast phenomena and for dark recognition that form the basis of many visual illusions. Color recognition is dependent on coding by a three-color process in the retina, beginning with red, green, and blue cones that feed into relatively independent lines to the primary visual cortex (De Valois, 1973). At this stage, there are various synthetic mechanisms, but we are far from understanding the neuronal mechanisms involved in color recognition (Zeki, 1980).

Since the complex and hypercomplex cells receive their inputs from various assemblages of simple cells, they would be expected to have inputs from a more extensive visual field. This is indeed the case, but the loss of field specificity is more than would be expected. It prompts the as yet unanswered question: how can the field specificity be recovered in the further stages of reconstitution of the visual field?

One further stage of synthesis of visual information has recently been studied physiologically (Gross, 1973; Gross *et al.*, 1985). The main projection from visual areas 17, 18, and 19 is to areas 20 and 21 in the inferotemporal lobe. Many neurons in areas 20 and 21 have more exacting stimulus requirements than the lines and angles that were adequate for the complex and hypercomplex neurons of areas 17, 18, and 19. For example, neurons may be fired by rectangles in the visual field and not by disks, or by stars and not by circles. Evidently, some of the neurons have a remarkable feature-recognition propensity. In these neurons of areas 20 and 21, visual mapping is sacrificed to feature recognition even more than in neurons of areas 18 and 19. Large areas of the visual field can effectively influence one neuron, and the topography for each "feature-recognition neuron'' always includes the center of vision. Again, it can be envisaged that this specific response to geometric forms, such as squares, rectangles, triangles, and stars, is dependent on the ordered projection onto these feature-recognition neurons from complex and hypercomplex neurons sensitive to bright or dark lines or edges of a particular orientation and length and meeting at particular angles. For example, recognition of a triangle as a feature would be the property of a neuron receiving inputs from neurons in the extrastriate visual cortex that are responsive to the angles and line orientations that compose a triangle.

Weiskrantz (1974) has demonstrated the manner in which monkeys can build up a remembered three-dimensional model of an object that is repeatedly examined from only one angle. This ability is lost following lesions of the inferotemporal lobe (areas 20 and 21). Hence, Weiskrantz postulates that this lobe is concerned in building models and categories and is therefore importantly involved in visual thinking and imagination.

Each stage of the processing of visual information from the retina to cortical areas 20 and 21 can be regarded as having a hierarchical order with features in sequential array:

1. The visual field becomes progressively less specific. This increasing generalization results in a foveal representation for all neurons of areas 20 and 21. Furthermore, at this stage all neurons receive from both visual half-fields, including the fovea, through inputs to both occipital lobes via the splenium of the corpus callosum.
2. There is an increasing specificity of the adequate stimulus from a spot to a bright line or edge of particular orientation, then to lines of specified width and length and often with specificity for direction of movement, and finally to the more complex feature detection of some neurons of areas 20 and 21.
3. There is also evidence that neurons of areas 20 and 21 have an additional response feature, namely, the significance of the response to the animal, exactly as has been discovered for neurons of areas 5 and 7 of the somesthetic system (Mountcastle, 1975; Mountcastle et al., 1975).

In addition to the first visual system that is indicated in Figure 17.2 (Section 7.1), there is a second visual system, the residuum from an earlier visual system (Schneider, 1969). As indicated in Figure 17.8, the optic nerve pathway bifurcates, one projecting to the lateral geniculate nucleus as in Figure 17.5, the other to the superior colliculus on the same side. There are then many pathways, the simplest being indicated in Figure 17.8. The pulvinar is the principal further projection site, both ipsilateral and contralateral, and from thence there is projection to the association visual cortices on both sides. In normal subjects, there is no evidence for the projection of the left visual field of the right eye to the contralateral visual cortex. When the corpus callosum (Figure 17.5) is sectioned, however, preventing information transfer from the ipsilateral visual cortex (black and white striped arrows in Figure 17.8), information from the left peripheral visual field of the right eye can be received by the conscious subject that is in liaison only with the left hemisphere as indicated in Figure 17.11 (Section 17.4). Trevarthen and Sperry (1973) suggest that the decussating pathways (Figure 17.8) of the second visual system are responsible. This system does not give sharp visual information, but may be important in signaling brightness and some kinds of pattern.

Not shown in Figure 17.8 are the projections from lamina V pyramidal cells of the primary visual cortex to the superior colliculus of the same side and from lamina VI

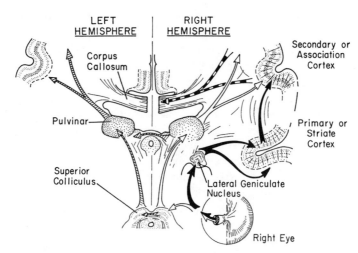

Figure 17.8. Anatomical pathways of a second visual system whereby, despite commissurotomy, objects far out in the visual field can be projected to the visual cortices of both hemispheres via the superior colliculus and the pulvinar. The first visual system (cf. Figure 17.5) is shown by bold black arrows from the right eye to lateral geniculate to striate (visual) cortex. From Trevarthen and Sperry (1973).

pyramidal cells to the lateral geniculate body in a highly specific fashion (Hubel and Wiesel, 1979). It will be evident that we are far from understanding the operation of all the complex neural machinery with the associated reverberatory circuits.

17.2.5. Auditory Perception

For general references, the reader is referred to Imig and Morel (1983) and Merzenich *et al.,* 1984).

There is a highly specialized transduction mechanism in the cochlea where, by a beautifully designed resonance mechanism, there is a frequency analysis of the complex patterns of sound waves and conversion into the discharges of neurons that project into the brain. After traversing several synaptic relays, the coded information reaches the primary auditory area (Heschl's gyrus) in the superior temporal gyrus (cf. Figure 17.9). The right cochlea projects mostly to the left primary auditory area and the left cochlea vice versa. There is a linear somatotopic distribution, the highest auditory frequencies being most medial in Heschl's gyrus (Figure 17.9) and the lowest most lateral. There is apparently no attempt at even a fragmentary reconstitution of the initiating stimulus such as occurs in the visual centers. It remains quite mysterious how a sequence of tones gives rise to a new synthesis, a melody. Nevertheless, there are parallels between the connections in cascade [Figures E1.7I–L in Popper and Eccles (1977)] and those for somesthesis and vision.

The primary auditory cortex has a columnar organization [cf. Figures 17.1 and 17.2 (Section 17.1)], as shown particularly by Merzenich *et al.* (1975). The detailed studies of the auditory pathway with the four cortical areas have been reviewed by Imig and Morel (1983) along with the strict tonotopic thalamocortical reciprocal connections.

As in the visual system, there is a crossed connection from the ear to the primary auditory sensory area, i.e., to Heschl's gyrus in the temporal lobe (Figure 17.9). The

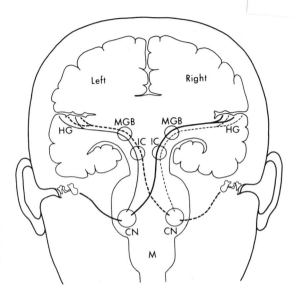

Figure 17.9. Schematic drawing of the auditory pathways to Heschl's gyrus (HG) on each side, showing the dominance of the crossed connections. Abbreviations: (CN) Cochlear nucleus; (IC) inferior colliculus; (M) medulla oblongata; (MGB) medial geniculate body.

situation differs from that in vision, however, where the connections of one visual field to the respective primary visual cortex are entirely crossed [see Figure 17.5 (Section 17.2.4)]. There exist ipsilateral connections from one ear to Heschl's gyrus of the same side. This ipsilateral connection is much weaker than the contralateral; furthermore, the ipsilateral pathways are suppressed by the contralateral during dichotic presentation (Milner *et al.*, 1968), presumably by inhibition in the cerebral cortex. Thus, we would attribute the right ear advantage to the fact that the right ear has a more direct access to that hemisphere in which the encoded auditory input is decoded into recognizable words, viz., the left hemisphere, the speech hemisphere.

17.2.6. Olfactory Perception

For a general reference, the reader is referred to Shepherd (1983).

In most lower mammals, olfaction (smell) is the dominant sensory input into the forebrain, but in the evolution of primates to man, olfaction became subordinated to vision and hearing, and even to somesthesis, particularly when touch became vital in manual skills. Chemical sensing in the olfactory mucosa is by receptor cells that are specialized neurons with axons that pass to the olfactory bulb, where there is a processing of information by a complex nervous system [see Figure 4.13 (Section 4.6)], much as in the retina. From the olfactory bulb (OLB), the lateral olfactory tract (LOT) passes to the brain (cf. Figure 17.10), where it has a complex distribution. The principal termination is in the pyriform cortex (PC), a primitive cerebral cortex. Thence there are connections to many structures of the limbic lobe. Connection to the neocortex (orbitofrontal cortex) is effected only after several relays in the limbic system and is in part via the mediodorsal (MD) thalamus (Tanabe *et al.*, 1975). Thus, the olfactory connections are quite different from those of the somesthetic, visual, and auditory systems, in which the connections are first to the neocortex and reach the limbic system only after several relays [Figure 16.9 (Section 16.5.1)].

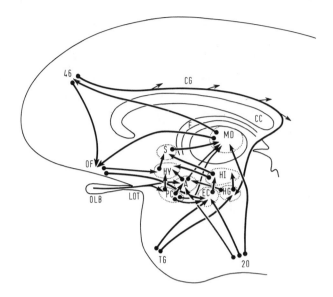

Figure 17.10. Schematic drawing of the medial surface of the right cerebral hemisphere to show connectivities from the neocortex to and from the mediodorsal thalamus (*MD*) and the limbic system. Abbreviations: (*20, 46*) Brodmann areas; (*A*) amygdala; (*CC*) corpus callosum; (*CG*) cingulate gyrus; (*EC*) entorhinal cortex; (*F*) fornix; (*HG*) hippocampal gyrus; (*HI*) hippocampus; (*HY*) hypothalamus; (*LOT*) lateral olfactory tract; (*MD*) mediodorsal thalamus; (*OF*) orbital surface of prefrontal cortex; (*OLB*) olfactory bulb; (*PC*) pyriform cortex; (*S*) septum; (*TG*) temporal pole.

17.2.7. Emotional Coloring of Conscious Perceptions

More discussion of this topic is to be found in Sections 15.3–15.5.

It is a common experience that the conscious perception derived from some sensory input is greatly modified by emotions, feelings, and appetitive drives. For example, when one is hungry, the sight of food gives an experience deeply colored by an appetitive drive! W. J. H. Nauta (1971) conjectures that the state of the organism's internal milieu (hunger, thirst, sex, fear, rage, pleasure) is signaled to the prefrontal lobes from the hypothalamus, the septal nuclei, and various components of the limbic system such as the hippocampus and the amygdala [see Figure 15.4 (Section 15.3)]. The pathways would be mainly through the MD thalamus to the prefrontal lobes [see Figure 16.9 (Section 16.5.1) and Figure 17.10 (Section 17.2.6)]. Thus, by their projections to the prefrontal lobes, the hypothalamus and the limbic system modify and color with emotion the conscious perceptions derived from sensory inputs and superimpose on them motivational drives. No other part of the neocortex has this intimate relationship with the hypothalamus.

Figure 16.9 shows for the somesthetic, visual, and auditory system the many projections to the prefrontal lobes from the primary sensory and the related association areas of the parietal, occipital, and temporal lobes. Simultaneously, these areas project to the limbic system, and, as shown in Figure 17.10, there are also projections from the prefrontal lobe (areas *46* and *OF*) to the limbic system. Thus, there are pathways for complicated circuitry from the various sensory inputs to the limbic system and back to the prefrontal lobe, with further circuits from that lobe to the limbic system and back again (W. J. H. Nauta, 1971). From the connectivities of Figure 17.10, it can be seen that the prefrontal and limbic systems are in reciprocal relationship and have the potentiality for continuously looping interaction. Thus, by means of the prefrontal cortex, the subject may be able to exercise a controlling influence on the emotions generated by the limbic system. An additional sensory input (olfaction) comes directly into the limbic system for cross-modal transfer to the other senses and thus contributes to the richness and variety of the perceptual experience. For example, the neocortical sensory systems project indirectly via areas *46, OF,* and *20* (Figure 17.10) to the hypothalamus, to the entorhinal cortex, and to the hippocampal gyrus, and so to the hippocampus, to the septal nuclei, and to the MD thalamus. After relay in the pyriform cortex and amygdala, the olfactory input also goes to the hypothalamus, septal nuclei, and MD thalamus. The MD nucleus is the receiving station for all inputs and in turn projects widely to the cortex of the prefrontal lobe. Thus, one can think of the prefrontal cortex as being the area in which all emotive information is synthesized with somesthetic, visual, and auditory to give conscious experiences to the subject and guidance to appropriate behavior.

Necessarily, physiological investigations of perception remain at the levels of neural events: neuronal activities with impulse discharges and synaptic transmissions. We have no idea how these events in specific areas of the cerebral cortex can give such diverse perceptual experiences as, for example, light, color, sound, touch, and pain. We return to this problem in Section 17.5.

17.3. Language Centers of the Human Brain

For general references, the reader is referred to Penfield and Roberts (1959), Geschwind (1970), Eccles (1981b), Ingvar (1983), and Damasio and Geschwind (1984).

17.3.1. Introduction

The representation of language in the cerebral cortex has been investigated by four methods: (1) the study of linguistic disorders arising from cerebral lesions; (2) the effects of stimulation of the exposed brain of conscious subjects and of the transient aphasias resulting from the exposure; (3) the effects of intracarotid injections of sodium amytal (a neural depressant); and (4) the dichotic listening test.

17.3.2. Aphasia

For over a century, disorders of speech (aphasia) have been associated with lesions of the left cerebral hemisphere [see Figure 14.2 (Section 14.3.1)]. There was first the motor aphasia described by Broca in 1861 as arising from lesions of an area that we now call the "anterior speech center of Broca." The patient had lost the ability to speak, although he could understand spoken language. Broca's area lies just in front of the cortical areas that control the speech muscles; nevertheless, motor aphasia is due not to paralysis of the vocal musculature, but to disorders in their usage. Much more important, however, is the large speech area that lies more posteriorly in the left hemisphere. On the basis of evidence from lesions, it was originally thought by Wernicke in 1874 to be only in the superior temporal convolution, but now it is recognized as having a much more extensive representation on the parietotemporal lobes (Figure 14.2). This posterior speech center of Wernicke is specially associated with the ideational aspect of speech. Lesions there lead to an aphasia characterized by failure to understand speech—either written or spoken. Although the patient can speak with normal speed and rhythm, his speech is remarkably devoid of content, being a kind of nonsense jargon. In the great majority of patients, lesions anywhere in the right hemisphere do not result in serious disorders of speech.

Aphasia itself has been subjected to most detailed and diverse descriptions and classifications, Areas specialized for reading and writing, for example, have been recognized by the alexia or agraphia resulting from their destruction. It is essential to recognize the incredible complexity of the encoding and decoding in speech. Computerized tomography has provided much more precise information on the cerebral lesions in aphasia (Damasio and Geschwind, 1984).

As an illustration, we can consider the neural events concerned in some simple linguistic performance. For example, in reading aloud, black symbols on white paper are projected from the retina to the brain, in the encoded form of impulse frequencies in the optic nerve fibers, and so eventually to the primary visual cortex [cf. Figures 14.2 and 17.5 (Section 17.2.4)]. In the next stage, the encoded visual information is transmitted to the visual association areas (Figure 14.2), where there is a further stage of reconstitution of the visual image. As described above, this reconstitution is still most inadequate. Neurons specifically respond to simple geometric forms, the so-called ''feature-recognition'' neurons. At the next stage, however, lesions of the posterior part of Wernicke's area [angular gyrus (39 in Figure 14.2)] result in dyslexia, suggesting that the relay from the visual association neurons provides information that is converted into word patterns and that the latter, in turn, are interpreted as meaningful sentences in the semantic process of conscious recognition of meaning that occurs in the Wernicke area.

The further stage in the process of reading aloud is via the arcuate fasciculus (AF in Figure 14.2) to the motor speech area (Broca's area). Lesions of the arcuate fasciculus result in conduction aphasia (Geschwind, 1970). There is comprehension of spoken language but a gross defect in its repetition and in normal speaking. At the terminal stage, appropriate patterns of neural activity in Broca's area lead to the motor areas for vocalization and so to the coordinated contraction of the speech muscles that give spoken language. A similarly complex chain of encoding and decoding is involved in writing out language that is heard.

As a general summary, it can be stated that great difficulties arise in formulating a strict classification of aphasia because of the irregular destructive action of clinical lesions. For our present purpose, it is not necessary to go into all the detailed disputation between the various experts on the many types of aphasia or on the causative cerebral lesions. The remarkable discovery is that the enormous proportion of aphasics have lesions in their left cerebral hemisphere. Only very rarely is a right cerebral lesion associated with aphasia. There was originally a general belief that right-handed patients had their speech centers on the left side and left-handed patients on the right. This has proved to be untrue. The great majority of left-handed subjects also have their speech centers in the left cerebral hemisphere (Penfield and Roberts, 1959).

17.3.3. Experiments on Exposed Brains

In the hands of Penfield and his associates, stimulation of the cerebral cortex has yielded quite remarkable discoveries relating to the localization of speech centers. Stimulation in either hemisphere of the motor areas [see Figure 14.2 (Section 14.3.1)] that innervate structures concerned in sound production, such as tongue and larynx, causes the patients to produce a variety of calls and cries (vocalization), but not recognizable words. These are the motor areas of voice control and are bilateral. Only rarely does a similar stimulation in animals give vocal responses. On the other hand, stimulation of the speech areas (Figure 14.2) results in interference with or arrest of speech. For example, if the subject is engaged in some speech performance such as the counting of numbers, his voice may be slurred or distorted, or the same number may be repeated. Often, the application of a gentle stimulating current to the speech areas causes a cessation of speech that is resumed as soon as the stimulation stops, or there is a temporary inability to name objects during the stimulation. One can imagine that the stimulus has caused widespread interference with the specific spatiotemporal patterns of neuronal activity that are responsible for speech. In this way, Penfield and his associates were able to delimit the two speech areas that have been recognized from clinical studies of aphasia, namely, the anterior and posterior speech areas, and also a subsidiary third area, the superior speech cortex (Figure 14.2).

Inadvertent sequelae to operative procedures have been important in demonstrating the cerebral hemisphere that is responsible for speech—whether it is the subject's right or left hemisphere. It has been observed that, after a cerebral operation involving exposure of one cerebral hemisphere, a transient aphasia often develops some days after the operation and continues for 2–3 weeks. This is attributed to the neuroparalytic edema resulting from brain exposure. A systematic study of the neuroparalytic aphasia of patients by Penfield and Roberts (1959) showed that it developed in over 70% of patients with a left hemisphere operation, regardless of whether they were right- or left-handed. By contrast, with operations on the right hemisphere, aphasia was very rare. These observations indicated the very strong dominance of speech representation (98%) in the left hemisphere. Other investigators using various techniques are in general agreement with these results, the dominance being about 95%, but left-handed patients have right hemisphere representation of speech more frequently, though still not as frequently as left hemisphere representation (Piercy, 1967).

17.3.4. Intracarotid Injections of Sodium Amytal

Another method of determining speech representation and relating it to handedness was developed by J. Wada and Rasmussen (1960) with the injection of sodium amytal into the common or internal carotid arteries of subjects in which it was important preoperatively to identify the speech hemisphere. There was likewise the overwhelming dominance of the left hemisphere representation of speech for right-handed subjects and a considerable dominance also for left-handed subjects. Speech in very young children is bilateral, the left hemisphere gradually assuming dominance over the first few years of life. By 4–5 years of age, speech has become fully lateralized (D. Kimura, 1967). Damage to the left hemisphere in infancy may result in the development of speech areas in the right hemisphere (Milner, 1974). There appears to be considerable neural plasticity at this early age.

17.3.5. Dichotic Listening Test

Through headphones, the subject receives simultaneously two different auditory stimuli, one to the right ear, the other to the left. The test was first tried on word recognition. Three pairs of digits (say 2, 5; then 3, 4; then 9, 7) were presented dichotically to normal subjects in rapid succession, after which the subject was asked to repeat in any order as many of the digits as he could. It was surprising to find that the digits presented to the right ear were more accurately reported than those presented to the left ear, although it was shown that there were no differences in the respective sensory auditory channels. The asymmetry in normal auditory recognition of words is explained by the dominance of crossed transmission in the neural pathways for hearing [cf. Figure 17.9 (Section 17.2.5)]. At the level of word recognition, the left hemisphere displays its linguistic superiority, but it has been shown that the auditory areas of both hemispheres perform equally well at the earlier stage of auditory pattern analysis.

17.3.6. Anatomical Substrates of Speech Mechanisms

For a general reference, the reader is referred to Damasio and Geschwind (1984).

The unique association of speech and consciousness with the dominant (left) hemisphere gives rise to the question: Is there some special anatomical structure in the dominant hemisphere that is not matched in the minor hemisphere? In general, the two hemispheres have been regarded as being mirror images at a crude anatomical level, but recently it has been discovered that, in about 65% of human brains, there is hypertrophy of a part, the planum temporale, of the left superior temporal gyrus in the region of the posterior speech area of Wernicke (Geschwind and Levitsky, 1968). J. A. Wada *et al.* (1975) found the left–right asymmetries of the planum temporale not only in infants who died at birth, but also in a 29-week-old fetus. Thus, speech localization appears to be genetically determined, the speech centers being built in preparation for their eventual usage after birth. On the other hand, handedness would appear to be much more flexible and to be determined at least in part by environmental habits. There will be further reference to the anatomical substrates of speech mechanisms in the next section, particularly in relation to the effects produced by global lesions of the cerebrum—commissurotomy and hemispherectomy.

Only brief reference need be made to the controversial topic of the linguistic performance of anthropoid apes. In their book *Speaking of Apes,* Sebeok and Umiker-Sebeok (1980) have assembled a wide range of critical articles by the authorities. It can be concluded that anthropoid apes use sign language pragmatically, in requesting things for themselves, and not mathetically in relation to objective enquiries about the world around them (Eccles, 1981b).

17.4. Language and Self-Consciousness

17.4.1. Effects of Global Cerebral Lesions

For general references, the reader is referred to Eccles (1980), Sperry (1982), and Sperry *et al.* (1979).

Commissurotomy.   The corpus callosum is a tract of about 200 million fibers that provides an immense commissural linkage in an approximately mirror-image manner between almost all regions of the cerebral hemispheres. The intense impulse traffic in the corpus callosum keeps the two hemispheres of the brain working together. The corpus callosum has been completely severed in a number of human subjects who were suffering from almost incessant epileptic seizures that could not be controlled, even by heavy

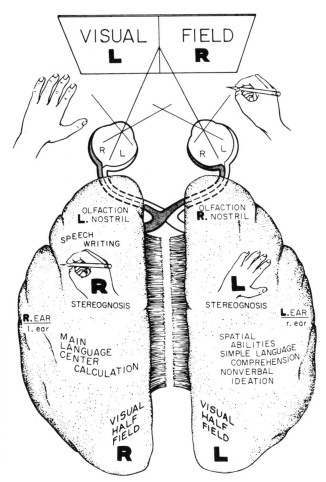

Figure 17.11. Schema showing the way in which the left and right visual fields are projected onto the right and left visual cortices, respectively, due to the partial decussation in the optic chiasm. The schema also shows other sensory inputs from right limbs to the left hemisphere and from left limbs to the right hemisphere. Similarly, hearing is largely crossed in its input, but olfaction is ipsilateral. From Sperry (1974).

medication. It was surmised that seizures developed in one cerebral hemisphere and then excited the other via the corpus callosum so that the seizure rapidly became generalized. Hence, it was proposed that section of the corpus callosum would, at least, keep one hemisphere free from the seizures. It turns out that the operation does better than predicted. There is a remarkable diminution of seizures in both hemispheres.

Sperry and associates (Sperry, 1974; Sperry *et al.*, 1979) have developed testing procedures in which information can be fed into one or the other hemisphere of these "split-brain" patients and in which the responses of either hemisphere can be observed independently. Essentially, the information is fed into the brain from the right or left visual field (cf. Figure 17.11). The subject fixates a central point and the signal, e.g., a word, is flashed on for only 0.1 sec to eliminate changes in the visual field by eye movements. All the seven subjects investigated had the speech areas in the left hemisphere [cf. Figure 14.2 (Section 14.3.1)], which was thus always the dominant hemisphere. The most remarkable discovery of these experiments was that all the neural activities in the right hemisphere (the so-called "minor hemisphere") are unknown to the speaking subject, who is only in liaison with the neuronal activities in the left hemisphere, the dominant hemisphere. Only through the dominant hemisphere can the subject communicate with language. Furthermore, in liaison with this hemisphere is the conscious being or self that is recognizably the same person as before the operation.

Figure 17.11 is a diagram drawn by Sperry several years ago. It is still valuable, however, as a basis of discussion of the whole split-brain story. The diagram illustrates the right and left visual fields with their highly selective projections to the crossed visual cortices, as indicated by the letters R and L [cf. Figure 17.5 (Section 17.2.4)]. Also shown in the diagram is the strictly unilateral projection of smell and the predominantly crossed projection of hearing [cf. Figure 17.9 (Section 17.2.5)].

The crossed representations of both motor and sensory innervation of the hands are indicated, as is the further finding that arithmetical calculation is predominant in the left hemisphere. Only very simple additions can be carried out by the right hemisphere.

We can say that the right hemisphere is a very highly developed brain except that it cannot express itself in language, so it is not able to disclose any experience of consciousness that we can recognize. Sperry (1982) postulates that there is another consciousness in the right hemisphere, but that its existence is obscured by its lack of expressive language. On the other hand, the left hemisphere has a normal linguistic performance, so it can be recognized as being associated with the prior existence of the ego or the self with all the memories of the past before the commissural section. According to this view, there has been split off from the talking mind a nontalking mind that cannot communicate by language, so it is there, but mute, or aphasic.

In general, the dominant hemisphere is specialized in respect to fine imaginative details in all descriptions and reactions; i.e., it is analytical and sequential. Also, it can add and subtract and multiply and carry out other computer-like operations. But, of course, its dominance derives from its verbal and ideational abilities and liaison to self-consciousness. Because of its deficiencies in these respects, the minor hemisphere deserves its title, but in many important properties it is preeminent, particularly in respect to its spatial abilities with a strongly developed pictorial and pattern sense. For example, the minor hemisphere programming the left hand is greatly superior in all kinds of geometric and perspective drawings. This superiority is also evidenced by the ability to assemble colored blocks so as to match a mosaic picture. The dominant hemisphere is unable to carry out even simple tasks of this kind and is almost illiterate in respect to pictorial and pattern sense, at least as displayed by its copying disability. It is an arithmetic hemisphere, but not a geometric hemisphere [see Figure 17.12 (Section 17.4.2)].

All the very fine work with flash testing has been superseded by another technique (Sperry et al., 1979), in which a contact lens is placed in the right eye with an optical device that limits the input into that eye to the left visual field no matter how the eye is moved. In this way, there can be up to 2 hr of continuous investigation of the subject, which gives opportunity for much more sophisticated testing procedures than with flash testing. The tests have been concerned with the ability of the right (minor) hemisphere to understand complex visual imagery, as shown by appropriate reactions with the left hand. It must be recognized that the left eye of the subject is covered and that the optical device mounted on the contact lens on the right eye allows an input to the retina only from the left visual field. Thus, there is elimination of all input to the subject from the right visual field; i.e., the subject is quite blind so far as any conscious visual experiences are concerned.

Examples of the pictorial understanding of the right hemisphere are given by testing procedures in which there is a picture of, say, a cat and below it the words "cat" and "dog." The subject can correctly point with the left hand to the appropriate word. Similarly, if there are two pictures, a cup and a knife, and below them only one word, "cup," the subject will point to the cup with the left hand. An even more sophisticated test of picture identification is provided by a drawing of landscapes below which are a correct and an incorrect name. For example, below the picture were the names "summer" and "winter," and the subject was able to point to the word "winter" rather than "summer" and correctly identify the picture. Despite this intelligent performance with pictorial and verbal presentation to the minor hemisphere, this hemisphere is totally unable to complete sentences, even the simplest. Most evidence from language testing of chimpanzees indicates likewise that they are unable to complete sentences, though some dubious claims have been made. This, of course, arises from the fact that neither the minor hemisphere nor the chimpanzee brain has a Wernicke area that provides the necessary semantic ability.

These more rigorous testing procedures (Sperry *et al.*, 1979) have shown that, after commissurotomy, the right hemisphere has access to a considerable auditory vocabulary, being able to recognize commands and to relate words presented by hearing or vision to pictorial representations. It is particularly effective in recognition of pictorial representations that occur in common experiential situations. It was also surprising that the right hemisphere responded to verbs as effectively as to action names. Response to verbal commands was not recognized by the flash technique. Despite all this display of language comprehension, the right hemisphere is extremely deficient in expression in speech or in writing, which is effectively zero. It is also incapable of understanding instructions that include many items that have to be remembered in correct order. The highly significant finding is the large difference between comprehension and expression in the performance of the right hemisphere.

Sperry *et al.* (1979) report investigations on two commissurotomy patients that were designed to test for aspects of self-consciousness and general social awareness in the right hemisphere. In these tests, a wide variety of pictures of persons as well as of familiar objects and scenes was presented in assembled arrays to the left visual field of the patient and hence *exclusively* to the right hemisphere (cf. Figure 17.11). The subject could always identify the familiar photograph in the ensemble of pictures, but there were difficulties in specifying what it was, and the investigators had to adopt a rather informative prompting system before the right hemisphere identification could be expressed in language, presumably by the left hemisphere. Their dramatic conclusions of approximate equality of the two hemispheres in identification may be criticized as being derived from a rather optimistic over-interpretation of the subject's responses, as illustrated in the experimental protocols. Nevertheless, there is remarkable evidence in favor of a limited self-consciousness of the right hemisphere.

These tests for the existence of mind and of self-conscious mind are at a relatively simple pictorial and emotional level. We can still doubt whether the right hemisphere has a full self-conscious existence. For example, does it plan and worry about the future, does it make decisions and judgments based on some value system? These are essential qualifications for personhood as ordinarily understood [Strawson, 1959; Popper and Eccles, 1977 (Sections P31 and P33)]. Let us now consider the bearing of these findings on the unity of the consciously experiencing self.

We think that in the light of these recent investigations by Sperry *et al.* (1979), there is some self-consciousness in the right hemisphere, but it is of a limited kind and would not qualify the right hemisphere to have personhood by the criteria mentioned above. Thus, the commissurotomy has split a fragment off from the self-conscious mind, but the person remains apparently unscathed with mental unity intact in its now exclusive left hemisphere association. However, it would be agreed that emotional reactions stemming from the right hemisphere can involve the left hemisphere via the unsplit limbic system; hence, the person also remains emotionally linked to the right hemisphere.

Hemispherectomy.   The investigations on patients subjected to commissurotomy have provided far more definitive and challenging information than have those on patients with other cerebral lesions. Nevertheless, the conclusions derived from the commissurotomy investigations can be evaluated in the light of testing procedures applied to patients with global or circumscribed cerebral lesions. The most radical are the hemispherectomized patients, who have had either the minor or the dominant hemisphere radically removed in the treatment of a gross cerebral tumor. Observations on these patients supplement the more definitive observations on the commissurotomy patients and are in general agreement therewith (Sperry *et al.*, 1979).

Excisions of the minor hemisphere result in symptoms that, except for the hemiplegia, are not appreciably different from those described in detail by Sperry in his study of patients with commissurotomy. Thus, after minor hemispherectomy, consciousness is derived from neural activities in the dominant hemisphere. One of the two cases of minor hemispherectomy reported by Gott (1973a) is remarkable because the operation was performed on a young woman who was a music major and an accomplished pianist. After the operation, there was a tragic loss of her musical ability. She could not play or carry a simple tune, but could still repeat correctly the words of familiar songs.

Excisions of left (dominant) hemisphere in the adult have much more serious sequelae. In the four cases that have been reported, there are traces of residual self-consciousness and some slight recovery in very primitive linguistic ability. But the patients were very difficult to study, since they were almost completely aphasic. A. J. Smith (1966) reported that his patient could use expletives and simple words in a song that he used to know. He had extreme restriction in language usage. Nevertheless, the isolated minor hemisphere had more linguistic ability than occurs in the minor hemisphere of Sperry's patients, in whom it is overshadowed by the dominant hemisphere. One wonders how much transfer of dominance had occurred in this patient before the operation, because there had been a severe lesion of the dominant hemisphere for at least 2 years, from age 45 to 47, before the time of the operation.

Hillier (1954) gave a more encouraging account of dominant hemispherectomy in a boy of 14 who survived about 2 years. This boy achieved a good recovery in general performance, but linguistically was very handicapped. Again, one suspects that there was some transfer of dominant hemisphere function before the operation, and the youthfulness of the patient could have helped in his recovery. Despite the rather optimistic tone of this report, it can be recognized that there was a tragic linguistic disability, a sequel that is to be expected after excision of the Wernicke speech area.

A better recovery was reported by Gott (1973b) in a girl who had complete hemispherectomy (dominant) at age 10. At age 8, there had been an excision of a tumor from that hemisphere, the final operation 2 years later being for a recurrence in the parietal area. At age 12, the patient had a linguistic ability that was gravely reduced but surpassed those in the cases of dominant hemispherectomy considered above. It is suggested by Gott that there was a better recovery because, at the age of 7–8, language may already have been in the process of transfer from the damaged dominant hemisphere. It was remarkable that despite her very limited speaking ability, the patient could sing well and liked to do so, usually with the correct words! Despite the grave linguistic handicap, it cannot be doubted that this girl had retained a self-conscious mind after the dominant hemispherectomy.

Infants present a much more encouraging situation. There is good evidence of a remarkable plasticity, the functions of the dominant hemisphere being effectively transferred up to 5 years of age (Milner, 1974). There is bilateral speech representation at birth. Then, over the first few years of life, cerebral dominance is established with regression of the linguistic ability of the minor hemisphere. When speech is transferred to the minor hemisphere, it is always defective, and there is in addition a deterioration of the normal functions of the minor hemisphere. Thus, there is a limit to the capabilities of the remaining hemisphere, so that it is deficient in mediating its normal functions and in accepting the functions from the other hemisphere.

Years after hemispherectomy performed in the first 5 years of life, tests reveal that the remaining hemisphere has taken over linguistic functions and hence assumed a "dominant" status, at least to a partial extent. Possibly a small transfer of this kind occurs even in adults because of the destruction of large areas of the dominant hemisphere in the years preceding the operation.

Zaidel (1976, 1977) has formulated a most interesting hypothesis. Up till the age of 4 or 5 years, both hemispheres develop together in linguistic competence, but the great increase in linguistic ability and skill coming on at that age demands fine motor control to give well-formed speech. It is at this stage that one hemisphere, usually the left, becomes dominant in linguistic ability because of its superior neurological endowment. Meanwhile, the other hemisphere, usually the right, regresses in respect to speech, but retains its limited competence in understanding. This understanding is particularly valuable when there are gestalt concepts to be interpreted. We suggest also that the right hemisphere is important for expressiveness and rhythm in speech, particularly in song, which is well preserved after dominant hemispherectomy and lost after minor hemispherectomy. This hypothesis accounts well for the transfer of speech to the minor hemisphere when there is severe damage to the linguistic areas of the dominant hemisphere before the age of 5 and for the progressive limitation of transfer at later ages.

17.4.2. Dominant and Minor Hemispheres

For general references, the reader is referred to Dimond and Beaumont (1974), J. Levy (1978), Sperry (1982), and Damasio and Geschwind (1984).

Figure 17.12 shows that the two hemispheres are complementary in their properties: The minor is coherent and the dominant is detailed. Furthermore, not only is the minor hemisphere pictorial, but also there is much recent evidence that it is musical. Music is essentially coherent and synthetic, being dependent on a sequential input of sounds. A coherent, synthetic, sequential imagery is made for us in some holistic manner by our musical sense. Furthermore, there is accumulating evidence by Milner (1974) that excision of the right temporal lobe does, in fact, seriously limit musical ability, as displayed in the Seashore tests. Testing by regional cerebral blood flow (rCBF) (see Section 14.3.3)

DOMINANT HEMISPHERE   MINOR HEMISPHERE

Liaison to self consciousness
Verbal
Linguistic description
Ideational Conceptual similarities
Analysis over time
Analysis of detail
Arithmetical and computer-like

No such liaison
Almost nonverbal
Musical
Pictorial and Pattern sense Visual similarities
Synthesis over time
Holistic — Images
Geometrical and Spatial

Figure 17.12. Various specific performances of the dominant and minor hemispheres as suggested by the conceptual developments of Levy-Agresti and Sperry (1968) and J. Levy (1978). There are some additions to their original list.

displays the dominance of the right superior temporal cortex in auditory discrimination (Roland *et al.*, 1981).

The distinctiveness of the functions of the two hemispheres listed in Figure 17.12 is also indicated by the results of the dichotic listening test described in Section 17.3.5. The test provides essentially a study of the subject's response to signals of a given modality that are applied simultaneously so as to give competitive inputs into the two hemispheres. This psychological technique of interhemispheric challenge has the great advantage that it can be applied to normal subjects, but, on the other hand, the results are not as discriminative as those of studies of the effects of hemispheral lesions, both global and circumscribed.

It is an attractive hypothesis of J. Levy (1973, 1978) and Sperry (1982) that the two hemispheres have complementary functions, which is an efficient arrangement because each can independently exercise its own peculiar abilities in developing and fashioning the pattern of neuronal responses to its new inputs. Then, as indicated in Figure 17.11 (Section 17.4.1), the two complementary performances can be combined by commissural transfer and integrated in the ideational, linguistic, and liaison areas.

It is our thesis that the philosophical problem of brain and mind has been transformed by these investigations of the functions of the separated dominant and minor hemispheres in split-brain subjects. The corpus callosum is an extremely strong communication system, so that all happenings in the minor hemisphere can very quickly and effectively be transmitted to the liaison brain of the dominant hemisphere, and so to the conscious self. This occurs for the contribution of the minor hemisphere for all perceptions, all experiences, all memories, and in fact for the whole content of consciousness. Removal of the immense interhemispheral communication by commissurotomy would deplete the performance of the minor hemisphere, though some indirect liaison with the self-conscious mind may survive, particularly in emotional responses.

17.5. Relationship of Brain to Mind

For general references, the reader is referred to Popper and Eccles (1977), Sperry (1983), and Eccles (1986b).

17.5.1. Introduction

In general terms, there are two theories about the way in which the behavior of an animal (and a man) can be organized into the effective unity that it so obviously is:

1. There is the explanation inherent in monist materialism including all the varieties of parallelism, panpsychism, epiphenomenalism, and the now-fashionable identity theory. As succinctly expressed by Feigl [1967 (pp. 79 and 90)]:

> The identity thesis which I wish to clarify and to defend asserts that the states of direct experience which conscious human beings "live through," and those which we confidently ascribe to some of the higher animals, are identical with certain (presumably configurational) aspects of the neural processes in those organisms. . . . processes in the central nervous system, perhaps especially in the cerebral cortex. . . . The neurophysiological concepts refer to complicated highly ramified patterns of neuron discharges.

The thesis implies the closedness of the physical world. In the identity hypothesis, the existence of mental events is not denied, but they are postulated to be "identical" with neural events of a special kind.

In current neurological theory, the diverse inputs into the brain interact on the basis of all the structural and functional connectivities, including the memory stores, to give some integrated output of motor performance. The aim of the neurosciences is to provide a more and more coherent and complete account of the manner in which the total performance of an animal or man is explicable on those terms, as has been described in Chapters 14–16 and in Sections 17.1–17.4 of this chapter. Without making too dogmatic a claim, we can state that the goal of the neurosciences is to formulate a theory that can, in principle, provide a complete explanation of all behavior of animals and man, including man's verbal behavior and conscious experiences. With some important reservations, we share this goal in our own scientific work and believe that it is acceptable for all automatic and subconscious movements, even of the most complex kind. We believe, however, that this reductionist strategy will fail in the attempt to account for the special brain activities that are coherent with conscious states or mental events.

2. There is the dualist–interactionist explanation that has been specially developed for the self-conscious mind and human brain. It is proposed that, superimposed on the neural machinery in all its performance, there are at certain sites of the cerebral hemispheres (the so-called "liaison areas") effective interactions with the self-conscious mind, both in receiving and in giving (Figure 17.13).

Before discussing the experimental testing of these two theories, it is useful to make brief reference to the fundamental classification of Popper (1972) in which everything in existence and experience is subsumed in one or another of the categories enumerated under Worlds 1, 2, and 3 (Figure 17.14). According to the theory of monist materialism or physicalism, the mental events of World 2 in Figures 17.13 and 17.14 have no independent existence, but are identical with specific neural events (Feigl, 1967), and so are in World 1. Thus, as Popper [Popper and Eccles, 1977 (p. 51)] points out, all materialist theories of the mind:

> assert that the physical world (World 1) is self-contained or *closed*. . . . This physicalist principle of the closedness of the physical World I is of decisive importance. . . . as the characteristic principle of physicalism or materialism.

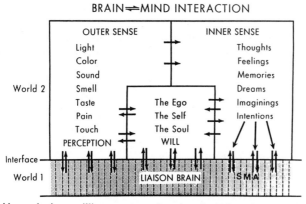

BRAIN⇌MIND INTERACTION

Figure 17.13. Information flow diagram for brain–mind interaction. The three components of World 2—outer sense, inner sense, and the psyche or self—are diagrammed with their connectivities. Also shown by reciprocal arrows are the lines of communication across the frontier between World 1 and World 2, i.e., from the liaison brain to and from these World 2 components. The liaison brain has the columnar arrangement indicated. It must be imagined that the area of the liaison brain is enormous, with open modules probably numbering a million or more and not just the 40 here depicted. Note that memories lie in the inner sense of World 2 as well as in the data banks of the modules of the liaison brain. (SMA) Supplementary motor area.

17.5.2. Testing of Mind–Brain Theories

We can raise the question whether there could be experimental testing of predictions from the dualist–interactionist hypothesis (Figure 17.13), on one hand, and the identity hypothesis, on the other. A simple diagram (Figure 17.15A) embodies the essential features of the identity hypothesis. In accord with Feigl (1967), mental–neural identity occurs only for neurons or neuron systems at a high level of the brain, especially in the

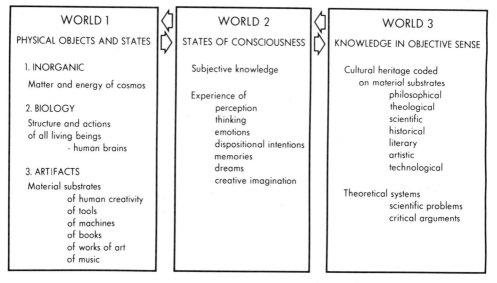

Figure 17.14. Tabular representation of the three worlds that comprise all existence and all experiences as defined by Popper and Eccles (1977).

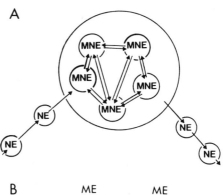

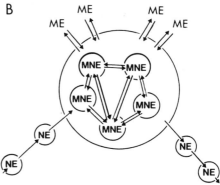

Figure 17.15. Diagrams of mind-brain theories. (A) Identity theory. (B) Dualist–interactionism. Assemblages of neurons are shown by circles. (NE) Neural event neurons, which are the conventional neurons that respond only to neural events. (MNE) Mental–neural event neurons, which are associated with both mental and neural events and are grouped in a larger circle representing the higher nervous system. In (B) ME arrows represent mental influences acting on the neural population that is associated with both mental and neural events. All other arrows in (A) and (B) represent the ordinary lines of neural communication which are shown in reciprocal action.

cerebral cortex. These neurons can be called mental–neural event (MNE) neurons, whereas other neurons in the brain, and in particular neurons on the input and output pathways, would be no more than simple neural event (NE) neurons, as in Figure 17.15. It would be predicted from the identity hypothesis that MNE neurons would be distinctive because in special circumstances their firing would be in unison (the identity) with mental events. But of course this firing would be in response to inputs from other neurons, MNE or NE, and is in no way *determined* by or *modified by* the mental events. This is the *closedness of the physical world* referred to above (Popper and Eccles, 1977).

It is remarkable that neurons or neuron systems in the cerebral cortex have been discovered experimentally that would seem to be MNE neurons, being distinctively related to intentional or attentional mental states [see Figure 14.5B (Section 14.3.3) and Figure 17.16].

Figure 17.16 illustrates the remarkable finding of Roland (1981) that, when the human subject was attending to a finger on which just detectable touch stimuli were to be applied, there was an increase in the rCBF over the finger-touch area of the postcentral gyrus of the cerebral cortex as well as in the mid-prefrontal area. These increases must have resulted from the mental attention, because no touch was applied during the recording. Thus, Figure 17.16A is a clear demonstration that the mental act of attention can activate appropriate regions of the cerebral cortex. A similar finding occurs with attention to the lips in expectation of a touch, but of course the activated somatosensory area is now that for the lips.

The effect of attention in causing an increased cerebral electrical response to finger

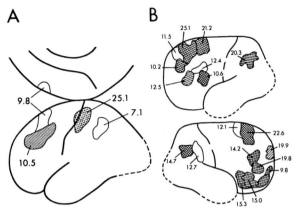

Figure 17.16. (A) Mean percentage increases of regional cerebral blood flow (rCBF) during pure selective somatosensory attention, i.e., somatosensory latent sensing without peripheral stimulation. The size and location of each focus shown are the geometric averages of the individual focus. Each individual focus has been transferred to a brain map of standard dimensions with a proportional stereotaxic system. The cross-hatched areas have an increase of rCBF significant at the 0.0005 level (Student's t test, one-sided significance level). For the other areas shown the rCBF increase is significant at the 0.05 level. There were eight subjects. From Roland (1981). (B) Mean percentage increases of rCBF and their average distribution in the cerebral cortex during silent arithmetic with successive subtraction of 3's from 50. Left hemisphere: 6 subjects, right hemisphere: 5 subjects. Cross-hatched areas have rCBF increases significant at the 0.005 level. With hatched areas $P < 0.01$ and with outlined areas $P < 0.05$. From Roland and Friberg (1985).

touch has been demonstrated by Desmedt and Robertson (1977). In a very ingenious investigation, they discovered that, with touch to the attended finger, there was a large increase in the late N 140 and P 500 evoked potentials relative to controls with touch to unattended fingers. This may be correlated with the increased rCBF that attention produced in the finger area of Figure 17.16A. In both these investigations, the mental event of attention is effecting selective neuronal responses.

A related finding is that, when the subject was attending to simple counting or other arithmetical mental activities during complete relaxation with eyes and ears closed, there was an increased rCBF in many cortical areas, but not in the primary sensory or motor areas (Roland and Friberg, 1985). As illustrated in Figure 17.16B for both the left and the right hemisphere, with the silent mental arithmetic of successive subtractions of 3 starting with 50, there was an increased rCBF in a medial strip of the frontal cortex anterior to the supplementary motor area (SMA) and also in other areas of the prefrontal cortex on both sides, as well as in the supramarginal and angular gyri of both parietal lobes. The patterns are more complex than for the silent thinking of a motor movement in Figure 14.5B (Section 14.3.3). Still more complex patterns were revealed both with a memory sequence based on a nonsense word sequence and with the visual imagery of route-finding (Roland and Friberg, 1985).

It can be predicted that the immense range of silent thinking of which we are capable will be found to initiate activity in such a wide variety of specific regions of the cerebral cortex that the greater part of the neocortex will be found to be under the mental influence of thinking (Ingvar, 1985). Of course, there is as yet no criterion for demonstrating a direct influence of silent thinking on a particular area. The areas of direct activation can immediately influence other areas, as occurred with the SMA (Figure 14.5B) activating the motor cortex (Figure 14.5A).

In Figure 17.15A, the identity theorists would have to postulate that the firing of the MNE neurons would be entirely neurally generated and explicable as responses either to NE inputs or to other MNE neurons of the higher brain. What, then, would be the position of the dualist–interactionist? It would be the crucial difference that the MNE neurons would have, in addition to the MNE and NE inputs, an input from mental events (ME) *per se,* as shown by the additional arrows in Figure 17.15B. The firing of MNE neurons would exhibit a response that is different from what it would be in the absence of the mental events of intention, attention, or silent thinking as shown in Figures 14.5 and 17.16A and B.

17.5.3. A New Hypothesis of Mind–Brain Interaction

The materialist critics argue that insuperable difficulties are encountered in the hypothesis that immaterial mental events can act in any way on material structures such as neurons, as diagrammed in Figures 17.13 (Section 17.5.1) and 17.15B (Section 17.5.2). Such a presumed action is alleged to be incompatible with the conservation laws of physics, in particular the first law of thermodynamics. This objection would certainly be sustained by 19th-century physicists and by neuroscientists and philosophers who are still idealogically in the physics of the 19th century, not recognizing the revolution wrought by quantum physicists in the 20th century. Unfortunately, it is rare for a quantum physicist to dare an intrusion into the brain–mind problem, but in a recent book, the quantum physicist Margenau (1984) makes a fundamental contribution. It is a remarkable transformation from 19th-century physics to be told that (p. 22):

> . . . some fields, such as the probability field of quantum mechanics, carry neither energy nor matter.

He goes on to state (p. 96):

> In very complicated physical systems such as the brain, the neurons and sense organs, whose constituents are small enough to be governed by probabilistic quantum laws, the physical organ is always poised for a multitude of possible changes, each with a definite probability; if one change takes place that requires energy, or more or less energy than another, the intricate organism furnishes it automatically. Hence, even if the mind has anything to do with the change, that is if there is a mind–body interaction, the mind would not be called upon to furnish energy.

In summary, Margenau states that (p. 97):

> The mind may be regarded as a field in the accepted physical sense of the term. But it is a nonmaterial field, its closest analogue is perhaps a probability field. It cannot be compared with the simpler nonmaterial fields that require the presence of matter. . . . Nor does it necessarily have a definite position in space. And so far as present evidence goes it is not an energy field in any physical sense, nor is it required to contain energy in order to account for all known phenomena in which mind interacts with brain.

These considerations should eventually convince the materialists or physicalists who are still imbued with 19th-century physics.

In formulating more precisely the dualist hypothesis of mind–brain interaction, the initial statement is that the whole world of mental events (World 2) has an existence as autonomous as the world of matter–energy (World 1) [see Figure 17.14 (Section 17.5.1)]. The present interactionist hypothesis does not relate to these ontological problems, but merely to the mode of action of mental events on neural events, i.e., events of the nature of the downward-pointing arrows across the frontier in Figures 17.13 and 17.15B. Following Margenau (1984), the hypothesis is that mind–brain interaction is analogous to a probability field of quantum mechanics that has neither mass nor energy, yet can cause effective action at microsites. More specifically, it is proposed that intentions (Figure 17.13) or attentions of the nonmaterial mind can cause neural events by a process analogous to the probability fields of quantum mechanics.

We can ask: What neural events could be appropriate recipients for mental fields that are analogous to quantal probability fields? We may already have the answer. In Figure 1.7 (Section 1.2.2), synaptic vesicles are organized in the paracrystalline presynaptic vesicular grid, there being one such grid with about 30–80 vesicles for each bouton. By precise analysis of the postsynaptic potentials generated when a presynaptic impulse activates a single bouton [see Figure 4.3 (Section 4.2)], it has been shown that the vesicular emission from a bouton (exocytosis) is probabilistic and below unity—the usual probability being 0.5 or less (Jack et al., 1981a; Korn and Faber, 1986). This probability can be varied up or down according to circumstances (Section 4.2). The presynaptic vesicular grid must have some subtle functional organization in controlling exocytosis of the embedded vesicles. It is proposed that mental events such as intention [see Figures 14.5 and 14.6 (Section 14.3.3) and Figures 17.13 and 17.15B], attention [see Figure 17.16A (Section 17.5.2)], or silent thinking (Figure 17.16B), acting in a manner analogous to a probability field of quantum mechanics, change the probability of synaptic vesicular emission in the manner suggested by Margenau [1984 (pp. 96–97)]. Vesicular opening [Figure 3.8 (Section 3.2.3)] involves displacement of about 10^{-18} g. It is not postulated that the firing of neurons is the target of the mental events, but merely that this firing is modified by alterations of the probabilities of quantal emission of the synapses that are engaged in actively exciting them. This is an important limitation in the target for mental events such as intention and attention.

By the hypothesis, it would be predicted that the interaction with mental events would be reduced to zero when the presynaptic firing background was reduced to zero. Lost of consciousness would occur and would be irreversible unless there were revival to a considerable degree of the impulse discharges in the cerebral cortex. An example is the "vigil coma" that supervenes when injury to the midbrain turns off the reticular activating system [Hassler, 1978; Eccles, 1980 (p. 160)]. In fact, the principal role of the reticular activating system may be to provide a background of excitatory synaptic actions on neurons of the cerebral cortex with an immense array of probabilistic vesicular emissions that are targets for the quantal probabilistic fields of mental influences.

Unconsciousness also occurs at the opposite extreme, the intense neuronal activity of an epileptic seizure [Eccles, 1980 (p. 157)]. It is to be expected that the postulated subtle probability fields would be ineffective on the cerebral cortex in such extreme states. However, they could still be effective on areas of the cerebral cortex not subjected to the seizure. The unconsciousness of sleep, strangely enough, is still controversial with respect to the levels and patterns of cortical activity that are responsible for the various types of sleep [cf. Eccles, 1980 (pp. 153–154)].

A general criticism of the hypothesis could be that a change in probability of vesicular emission of one synaptic bouton would be orders of magnitude too small for effective action on the patterns of neuronal activity in large areas of the brain. For example, in Figure 14.6 A and C, the intention of the monkey to pull the lever resulted in a great increase in the frequency of firing of that SMA neuron and also of many similar neurons in other recordings. However, there are many thousands of similar boutons on any one pyramidal cell of the supplementary motor area (SMA). The hypothesis is that the probability field of the mental intention is widely distributed not only to the boutons on that cell, but also to the boutons of a multitude of other cells with similar actions. Such a wide-spread excitatory influence is exhibited by the increased rCBF of the SMA in Figure 14.5A and B. Likewise, in mental attention (see Section 17.2.2), there must be an enormous increase in neuronal discharges, which may be attributed to a widespread distribution of the probability fields of mental influences in concentrated attention with consequent changes in the quantal probability of emissions from an immense number of boutons. Thus, by this massive integration, does the probability of units become the reliability of the ensemble.

We can therefore assume that the mental events achieve, in a global manner, interaction with the neural events of spatiotemporal patterns of activity (Eccles, 1982b) of the awake cerebral cortex. Even in one cortical module with its 4000 or so neurons, there must be an ongoing intense dynamic activity of unimaginable complexity. We now know the outlines of the neuronal structure of a module (Szentágothai, 1978a, 1983), and the physiological performance is now being systematically studied with intracellular recording by Oshima and associates (Ezure et al., 1985). All that we can surmise is that mental events acting as a field in the manner postulated by Margenau could effect changes in the spatiotemporal activity of a module by changing the probability of emission in many thousands of active synapses. There need be no violation of conservation laws.

One can ask how the monkey sets up the immense synaptic barrage that results in the neuronal firings in Figure 14.6A and C (Section 14.3.3), and that, through the well-known complex pathways shown schematically in Figures 14.3 (Section 14.3.1), 14.16, and 14.17 (Section 14.4.4), results in the desired motor action. The only answer is that this performance is at the end of a long line of training sessions. Motor learning is essential for all skilled actions that devolve from the cerebral cortex (Eccles, 1986d). Memory of some kind is required for all conscious experience and actions (see Sections 15.3 and 16.7).

It may be objected that the quantal emission of synapses has been rigorously investigated only at lower levels of the central nervous system, the motoneurons of the cat spinal cord and the fish Mauthner cells. However, investigations on the mammalian hippocampus, which is a primitive cerebral cortex, show quantal responses to single impulses resembling those of Figure 4.3C (Section 4.2) and with a similar probability of vesicular emission (McNaughton et al., 1981; Per Andersen, personal communication). Moreover, electron microscopy demonstrates that the synapses of the mammalian cerebral cortex also have presynaptic vesicular grids as in Figure 1.7 (Section 1.2.2) (Gray, 1982; Akert et al., 1969, 1975).

A general observation is that all hypotheses attempting to give some explanation of how conscious experiences derive from or relate to neural events in the active cerebral cortex have been of a nebulous nature. An example is that of Feigl (1967) described in Section 7.5.1. Sperry (1976) proposed that mental events are holistic configurational properties of the brain process. Mountcastle (1978) developed the concept of distributed systems that are ''composed of large numbers of modular elements linked together in echeloned parallel and serial arrangements'' and are thought to provide an objective mechanism of conscious awareness. Edelman (1978) suggested that ''the brain processes sensory signals and its own stored information upon this selective base in a phasic (cyclic) and re-entrant manner that is capable of generating the necessary conditions for conscious states.'' Szentágothai (1978b) suggested that ''dynamic patterns'' offer ''superstructures'' and might be helpful in giving a scientific explanation of the higher functions of the brain including even consciousness. Eccles (1982b) suggested that "the mental influence is exerted on an extremely complex dynamic system of interacting neurons."

The extreme alternative to these "nebular" hypotheses is now proposed, namely, that the essential locus of the action of nonmaterial mental events in the brain is at individual microsites, the presynaptic vesicular grids of the boutons, each of which operates in a probabilistic manner in releasing a single vesicle in response to a presynaptic impulse. It is this probability that is assumed to be modified by a mental influence acting analogously to a quantal probability field in the manner described above. The manner in which effective action at microsites becomes amplified by conventional neurocircuitry will be dependent on the complex circuits envisioned, for example, by Feigl (1967), Sperry (1976), Mountcastle (1978), Edelman (1978), Szentágothai (1978b), and Eccles (1982b). The *microsite hypothesis* can be proposed as a tentative beginning of a scientific study of the reflective loop proposed by Creutzfeldt (1979) as opening up the independent symbolic world of the mind, which is the World 2 of Popper and Eccles (1977). In contrast to the "nebular" hypotheses, it offers a unique challenge to molecular neurobiology.

Epilogue

We have told our story. Our objective in this monograph has been to build concepts of how the incredibly complex machinery of the mammalian brain operates. Topics of increasing subtlety have been treated as the book progressed, and we hope that the foundations laid at each stage are adequate for the next. It must be appreciated that we have not attempted to deal with every aspect of mammalian neurobiology. To do so would have been too ambitious for a single book and beyond the expertise of authors whose own horizons are limited. Rather, we have tried to select some of the areas that are fundamental to all branches of mammalian neurobiology and to explain in simple terms the underlying principles. To do so, we have selected for description only a few of the many critical experiments that have led to their establishment. Where possible, these principles have been presented in terms of the underlying hypotheses that have been formulated and the investigations that were undertaken to test them.

As these pages have attempted to illustrate, the whole brain, with its amazing powers of integration, is immeasurably greater than the sum of its parts. That is why a broad knowledge is more important in neurobiology than in any other branch of science. It is not always easy to obtain and assimilate the background information necessary to approach the now massive literature on neurobiology. We hope this book will assist in the endeavor.

In the early chapters, we described in much detail the ionotropic methods of signaling in the nervous system: impulse generation, impulse transmission along pathways, and synaptic action at terminals. But it is clear that the neurons participate as well in such other methods of signaling as metabotropic and genotropic. These methods can be described, as yet, only in their barest outline. There is the immense complexity of the building of the brain, which requires the correct establishment of all the lines of communication and their detailed topographic relationships. We have seen in Chapter 13 that this turns out to be a problem of genetic instruction that achieves expression as specific molecular labels on individual neurons. The neurons are born, in a sense, with the postal addresses of certain other neurons built into them. With the help of glia, their axon terminals are delivered on time, thence to begin the process of giving and receiving impulses and other, more subtle signals.

But the brain, once built, is not a fixed, unchanging structure. At the higher levels of the brain, modifiability is the essence of its performance, as is evidenced by learning and memory. In Chapter 16, we have seen that this could be explained by microstructural changes. It is postulated that synaptic activity leads to terminal growth by its effect in causing the synthesis of RNA and, through this, synthesis of the enzymes that build proteins and other macromolecules. There are many indications that not only this synaptic growth but also neuronal life itself depends on specific interneuronal chemical signals of a type that has yet barely been explored.

Our hypothesis of memory and learning implies that activity is as important to the brain as exercise is to the muscles. The DNA template is there from the earliest embryonic age at which a cell develops neuronal competency. It can, at least in theory, provide transcribable information from that moment throughout the life of the cell. But whether or not transcription occurs may depend on that cell's receiving appropriate genotropic signals from external sources. Although we do not know the nature of the signals, we are aware of the consequences of their disappearance. A neuron deprived of its nerve endings rapidly undergoes retrograde degeneration. It may attempt to sprout, as does a sectioned nigrostriatal neuron, but, if appropriate postsynaptic contacts are not made, it will give up and degenerate. It needs some signal from its postsynaptic neuron to survive. Similarly, a neuron deprived of its presynaptic input can also wither and die by the process referred to as "transneuronal degeneration."

What manner of signals are involved in these life-sustaining communications, without which the competency of the DNA template is of no value? What signals are involved in the plastic processes of cognitive and motor learning, which can be sustained only through reinforcement? By what means do they so dramatically alter subsequent responses to the familiar types of impulse transmission? They emanate from teaching cells, but what are the methods of instruction (Chapter 16)?

Is it possible that ways of substituting for these signals, or ways of altering their strength, can be found? Do there exist in the central nervous system macromolecules with properties that can alter the structure and function of the brain during life and help to sustain its machinery in the face of disease and degenerative processes?

We are only at the beginning of our understanding of signaling in the nervous system. In each chapter of this book, great gaps in our knowledge are evident. While these gaps present difficulties both for the authors and for the reader, they nevertheless point to areas wherein the most significant advances can be made. We have therefore felt a particular satisfaction in being able to present this book to students who represent a new generation of neuroscientists. We consider that no other field of science today offers as much challenge or opportunity to young investigators as does the brain.

Glossary

α-*adrenergic receptor:* a site at which the action of noradrenaline is blocked or mimicked by any of certain drugs that are known as, respectively, α-adrenergic *antagonists* or *agonists* [see Table 9.4 (Section 9.6.1)].

β-*adrenergic receptor:* a site at which the action of noradrenaline is blocked or mimicked by any of certain drugs that are known as, respectively, β-adrenergic *antagonists* or *agonists* [see Table 9.4 (Section 9.6.1)].

afferent: an axon that conveys impulses toward the specified area.

agonist: a substance that mimics the action of a neurotransmitter at the receptor.

allele: any of various forms of a gene that occupies a fixed locus on the chromosome (see *polymorphism*).

alveus: the thin sheet of pyramidal cell axons that covers the ventricular surface of the hippocampus.

amacrine cell: a special inhibitory cell in the retina.

amnestic syndrome: loss of all memory except short-term memory.

amplification: a term that in molecular genetics refers to the production of additional copies of a given length of DNA that typically carries with it coding instructions.

anaerobic: lacking oxygen.

anion: a negative ion such as Cl^-.

anode: positive pole, to which an anion is attracted.

anoxia: reduction of oxygen below physiological levels.

antagonist: a substance that tends to nullify the action of a neurotransmitter at the receptor.

anterograde transneuronal degeneration: degeneration of a neuron subsequent to destruction of afferents.

antibody: a circulating protein of the immunoglobulin class that interacts only with the antigen that induced its synthesis or (in cross-reactions) with molecules closely related to that antigen.

anticodon: a triplet of nucleotides found in transfer RNA that is complementary to a codon of mRNA.

antidromic: descriptive of an impulse that travels from axon terminals toward the neuronal soma, i.e., opposite to the usual (*orthodromic*) direction.

antigen: any substance (usually a protein) that is capable of inducing the formation of antibodies and of reacting specifically with the antibodies so induced.

antimitotic: inhibiting cell division.

aphasia: inability to express oneself properly through speech, or loss of verbal comprehension.

archicortex: the ancient part of the cerebral cortex including the hippocampus.

argentaffin cell: a type of cell in the gastrointestinal tract that is easily stained with silver salts and contains high
concentrations of serotonin (see *carcinoid syndrome*).

ataxia: motor incoordination of an irregular type during purposive movement.

atony: lack of normal strength or tone.

atrophy: a wasting away.

autoradiograph: a photograph picture made by exposing a photographic emulsion to the radiation emitted from
radioactive substances contained in tissue or other material.

autosome: any chromosome except the sex chromosome.

axolemma: a delicate membrane between an axon and the surrounding myelin sheath.

axon collateral: a branch from the main axon.

axoplasm: intracellular fluid within an axon.

bacteriophage: a virus that infects bacteria.

base pair (bp): a partnership of adenine with thymine or of cytosine with guanine in a DNA double helix.

bipolar cell: a neuron with two major processes arising from the cell body.

cable transmission: transmission of a potential charge along a nerve fiber in the manner of an electric current
through a cable and not as an impulse.

cap: structure at the 5' end of *mRNA* that is introduced after transcription.

carcinoid syndrome: symptoms characteristic of a person who has a tumor of the *argentaffin cells* of the
gastrointestinal tract.

carotid body: a nucleus at the bifurcation of each carotid artery with receptors sensitive to changes in oxygen
tension in the blood; impulses arising in these receptors help in reflex control of respiration.

cathode: negative pole, to which a *cation* is attracted.

cation: a positively charged ion such as Na^+, K^+, or Ca^{2+}.

caudal: toward the tail or posterior end (in humans the term is *inferior*).

causalgia: a burning pain, often accompanied by trophic skin changes, due to injury of a peripheral nerve.

cDNA: single-stranded DNA complementary to mRNA or a portion thereof that is synthesized from it by *in vitro*
reverse transcription.

cDNA library: a mixture of complementary DNAs made from an initial mixture of mRNAs using the enzyme
reverse transcriptase.

central nervous system (CNS): brain and spinal cord.

cerebellar nuclei: collections of nerve cells deep in the cerebellum that mediate the output from the cerebellar
Purkinje cells [see Figure 14.13 (Section 14.4.1)].

cerebral ventricle: a cavity (lumen) in the brain filled with *cerebrospinal fluid (CSF)*.

cerebrospinal fluid (CSF): clear liquid filling the ventricles of the brain and spaces between meninges, arach-
noid, and pia.

chemical specification: specific chemical structures on the surface of nerve fibers and nerve cells that control the
development of connectivities.

cholinergic: descriptive of neurons that release acetylcholine as a neurotransmitter.

cholinoceptive: descriptive of neurons or other structures that possess receptors for acetylcholine.

cholinomimetic: an agent that mimics the action of acetylcholine at cholinergic receptors (an acetylcholine
agonist).

chorea: involuntary jerky movements of skeletal muscles.

choroid plexus: folded processes rich in blood vessels that project into ventricles of the brain and secrete *cerebrospinal fluid.*

chromaffin cell: a cell that stains strongly with chromium salts due to catecholamine content; such cells are found in the adrenal gland, along sympathetic nerves, and in certain organs.

chromatin: complex of DNA and protein in the nucleus.

chromatography (paper, column, or gas): separation technique in which the materials to be separated are carried through an adsorbent (or stationary phase) by a liquid or gas as the moving phase.

clone: an assemblage of cells made by successive subdivisions of an initial parent cell.

cloning vector: any plasmid or bacteriophage into which a foreign DNA may be inserted to be cloned.

coding strand: length of DNA that is translated into *mRNA.*

codon: Triplet of nucleotides that represents an amino acid or a termination signal.

cognitive experience: a conscious thought or experience.

complex cell, hypercomplex cell: names given to cells in the visual cortex surrounding the primary visual cortex that exhibit sensitivity to special spatial properties of the retinal stimulation.

conditioned reflex: a reflex response that is developed in special training procedures.

conductance (g): reciprocal of electrical resistance and thus a measure of the ability of a circuit to conduct electricity; in excitable cells, a useful measure of permeability for an ion or ions.

contralateral: relating to the opposite side of the body.

coronal section: vertical section through the skull at right angles to the front–back (*sagittal*) axis, i.e., a plane parallel to the face.

corpus callosum: the tract of nerve fibers that connects the cerebral hemispheres.

CSF: see *cerebrospinal fluid.*

cytoplasm: protoplasm of a cell exclusive of the nucleus.

dalton: a unit of mass; equal to $1/16$th the mass of the oxygen atom, or approximately $1.65 \times 10^{-24}/g$.

deafferentation: elimination or interruption of afferent nerves.

decussation: intercrossing of similar structures.

denaturation (of DNA or RNA): conversion from a double-stranded state into a single-stranded state, usually accomplished by heating.

depolarization: reduction of membrane potential from the resting value toward zero.

desmosome: a thickening that forms the site of attachment between cells and consists of local differentiations of the opposing cell membranes.

dielectric coefficient: a measure of the insulating capacity of a substance.

diencephalon: the "in-between brain" consisting principally of the hypothalamus and thalamus.

differentiating mitosis: cell division in which the daughter cells have an effectively different genetic coding from that of the parents.

distal: away from the center of the body.

DNA probe: a length of labeled and identified DNA that can hybridize with unidentified DNA, thus marking complementary sequences.

dorsal: pertaining to or situated near the back of an animal.

dysgenesis: defective development.

dyslexia: reading disability.

dyspnea: difficult or labored breathing.

dystonia: a disordered tone of muscle.

ectoderm, ectodermal: outer, investing cellular membrane of multicellular animals; applies especially to the outer germ layer of embryos.

edema, cerebral: presence of abnormally large amounts of fluid in the intercellular spaces of the brain.

efferent: an axon that conveys impulses away from the named nucleus or area.

electrochemical gradient: value that represents the voltage across a membrane for a particular charged particle; it is compounded of the electrochemical potential for that particle and the voltage gradient across the membrane.

EC enzyme classification: numbers assigned to various enzymes, e.g.. monoamine oxidase (EC 1.4.3.4); taken in this book from *Enzyme Nomenclature (1972),* Elsevier Scientific Publishing Co., Amsterdam.

electrochemical potential: voltage difference between two solutions insulated from each other; it is derived by the Nernst equation from the logarithmic relationship of the concentrations of the charged particles on either side of the insulating membrane.

Electrophorus: a genus of cyprinoid fish that includes the electric eel.

endogenous: arising from within the body.

endonuclease: an enzyme that cleaves bonds at a specific site in a nucleic acid chain: endonucleases may be specific for single- or double-stranded DNA or for RNA.

endothelial cell: a cell of the type that composes the endothelium, the inner lining of the blood vessels and other closed cavities.

endplate: postsynaptic area of vertebrate skeletal muscle fiber.

engram: a patterned arrangement of neuronal connections via synapses that have been given increased effectiveness in learning.

ependyma: a layer of cells lining the cerebral ventricles and the central canal of the spinal cord.

EPSP: excitatory postsynaptic potential.

equilibrium potential: membrane potential predicted from the Nernst equation on the basis of concentration differences for an ion to which the membrane is permeable.

ergastoplasm: granular endoplasmic reticulum.

eukaryotic: possessing a nucleus (cf. *prokaryotic organism*).

exocrine gland: a gland that discharges its secretion through a duct opening on an internal or external surface of the body, as a lacrimal gland.

exocytosis: discharge from a cell of particles that are too large to diffuse through the wall.

exogenous: originating outside the organism.

exon: any segment of a gene that is translated into *mRNA.*

expectancy wave: a potential change that is recorded from the surface of the brain and arises in anticipation of a conditioned movement.

extensor muscle: a muscle that acts to straighten a joint.

external germinal layer: a layer of cells (germinal cells) that is formed on the surface of the developing cerebellum and generates many cell types of the cerebellum.

extrafusal: descriptive of muscle fibers that make up the mass of a skeletal muscle (i.e., not within the sensory muscle spindles).

extrasynaptic: descriptive of surface membrane that is not covered by synapses.

extravesicular: lying outside the vesicles.

facilitation: increased effectiveness of synaptic transmission by successive presynaptic impulses.

fasciculus: a bundle of nerve fibers.

fenestrated: pierced with one or more openings.

festinating: tending to take involuntary, short accelerating steps in walking.

fissure: any cleft or groove, normal or abnormal; especially a deep fold in the cerebral cortex that involves the entire thickness of the brain wall.

folium: any of the leaf-like subdivisions of the cerebellar cortex.

γ-loop: the operative path employing γ-fibers and γ-motoneurons in movement control: γ-motoneurons, muscle spindles, Ia fibers, α-motoneurons, extrafusal muscle contraction.

γ-motoneuron: a small motoneuron that innervates intrafusal muscle.

ganglion: collection of neurons that send and receive impulses; many in the autonomic system relay synaptically to the viscera.

gemmule: a bud-like excrescence of a nerve process, e.g., such as found in the olfactory bulb or in Purkinje cells of the cerebellum.

genome: the operative DNA of a cell that expresses genetic characteristics.

genomic DNA: native nuclear host DNA.

genomic DNA library: mixture of DNAs obtained from nuclear DNA material.

genotropic: descriptive of interaction between a neurotransmitter or other signaling agent and its receptor to produce an effect on the nucleus of the receptor cell that initiates transcription of selective genes.

genu: a general term used to designate any anatomical structure bent like the knee, e.g., the most rostral part of the corpus callosum.

germinal cell: an embryonic cell.

Golgi apparatus: a secretory organelle in the neuronal cytoplasm that is easily recognized in electron microscopic studies.

Golgi type II neuron: a short-axoned neuron with an axon that arborizes in dendrite-like fashion in the neighborhood of the cell body.

gonadotropin: a hormone that stimulates the gonads.

gray matter: part of the central nervous system that appears gray and consists largely of neuropile and cell soma.

heterogeneous nuclear (hn) RNA: a transcript of genomic nuclear DNA made by RNA polymerase II.

heterotypic regeneration: regeneration in which the vacated synaptic sites are occupied by presynaptic fibers different from the original fibers.

hippocampus: a special part of the cerebral cortex (see *archicortex*).

histone: a simple protein derived from cell nuclei that contains many basic groups and is thought to be involved in suppressing transcription.

horizontal cell: a special inhibitory cell in the retina.

hybridoma: a cell line, produced by fusing a myeloma with a lymphocyte, that can be grown in culture and continues indefinitely to express immunoglobulins derived from the parent lines.

hydrophilic: descriptive of a material that attracts water.

hypertonic: descriptive of a solution that has greater osmotic pressure than normal extracellular fluids when bathing body cells.

hypothermia: a state of low temperature of the body.

hypotonic: descriptive of solution that has less osmotic pressure than normal extracellular fluids.

hypoxia: low oxygen content or tension; deficiency of oxygen in the inspired air.

infundibulum: the funnel-shaped stalk of the pituitary gland.

inhibition: effect of one neuron on another that tends to prevent the affected neuron from initiating impulses: *postsynaptic inhibition* is mediated through a permeability change in the postsynaptic cell, holding the membrane potential away from threshold; *presynaptic inhibition* is mediated by an inhibitory fiber on an excitatory terminal, reducing the release of neurotransmitter; *electrical inhibition* is mediated by currents in the presynaptic fibers that hyperpolarize the postsynaptic cell and do not involve the secretion of a chemical neurotransmitter.

interneuron: a short-axoned excitatory or inhibitory nerve cell in the *central nervous system* (see *Golgi type II neuron*).

internode: myelinated portion of a nerve axon that lies between two nodes of Ranvier.

interstitial: pertaining to spaces between essential parts of organs or tissues.

intrafusal fiber: a fiber within a sensory motor spindle.

intron: segment of genomic DNA that has been transcribed into RNA but is removed from within the transcript by splicing together sequences (*exons*) and therefore never leaves the nucleus.

in vitro hybridization: in vitro association of an added single-stranded RNA or DNA (usually labeled in some fashion) with host DNA that has first been made single-stranded by denaturation.

ionotropic: descriptive of interaction of a neurotransmitter or other signaling agent with its receptor to produce an opening of ionic channels.

iontophoresis: Application of ions by passing current through a micropipette; used for applying charged molecules with a high degree of temporal and spatial resolution.

ipsilateral: on the same side of the body.

IPSP: inhibitory postsynaptic potential.

isotonic: descriptive of a physiological solution that has the same osmotic pressure as normal extracellular fluids.

karyoplasm: nucleoplasm, or protoplasm, of the nucleus of a cell.

K_i: an equilibrium constant that expresses the concentration of inhibitor required to reduce the product of a reaction by 50%.

kilobase (kb): an abbreviation for 1000 *base pairs* of DNA or 1000 bases of RNA.

kinase: see *protein kinase.*

myasthenia gravis: A neurological disease associated with partial paralysis due to a defect in nerve–muscle transmission and with extreme fatigability.

myelin: fused membranes of Schwann cells or glial cells forming a high-resistance sheath around an axon.

myoneural junction: a junction that involves both muscle and nerve; applied to nerve terminations in muscles (see *neuromuscular junction*).

myosin: the contractile protein of muscle fibers.

neocortex: the most recently developed part of the cerebral cortex, composing the cerebral hemispheres.

nerve fiber: an axon (the principal branch from a nerve cell) that may extend for long distances.

neuroblast: a special cell that is the precursor of and develops directly into a nerve cell.

neuroblastoma: a tumor that arises from *neuroblasts*.

neuroepithelium: epithelium composed of epithelial cells that receive terminations of sensory nerves such as the specific cells of the taste buds, olfactory mucosa, and the cochlear and vestibular apparatuses.

neurofibril: a fine fiber that runs along the interior of a nerve axon.

neurogenesis: the process whereby nerve cells are made by the sequence of germinal cells to neuroblasts to fully fashioned nerve cells.

neuroglia: nonneuron satellite cells associated with neurons; in the mammalian *central nervous system,* the main groupings are astrocytes and oligodendrocytes; in peripheral nerves, the satellite cells are called "Schwann cells."

neuroleptic: a neuropharmacological agent that has antipsychotic action that principally affects psychomotor activity and is generally without hypnotic effects; as a tranquilizer, it may produce extrapyramidal syndrome.

neuromuscular junction: a junction that involves both muscle and nerve (see *myoneural junction*).

neuron (nerve cell): the biological unit of the brain and of the remainder of the nervous system [see Figures 1.2 and 1.3 (Section 1.2.1) and 1.8 (Section 1.2.3)].

neuron theory: the theory that the nervous system is composed of individual neurons that are biologically independent but communicate informationally by synapses.

neuropil: a network of axons, dendrites, and synapses.

neurosecretory cell: a secreting nerve cell.

neurotoxic: poisonous or destructive to nerve tissue.

neurotubule: a very fine tubule that runs along the interior of a nerve axon.

nick translation: process by which the DNA polymerase I of *Escherichia coli* nicks a strand of double-stranded DNA, thereby causing degradation of a short length that can be replaced by resynthesis with labeled precursors.

Nissl body: an intracytoplasmic, basophilic mass that ramifies loosely through the cytoplasm; also known as *rough endoplasmic reticulum.*

Northern blotting: a technique for transferring RNA from a gel to a nitrocellulose filter on which it can be hybridized with a complementary DNA (cf. *Southern blotting*).

nucleolus: a dense, basophilic structure within the nucleus that contains RNA as well as DNA; females have a small nucleolar satellite known as a "Barr body."

nucleus: a large basophilic mass, usually centrally placed within the cell body; contains the DNA that provides the genetic instruction for the cell.

orthodromic: descriptive of a wave of neuronal excitation passing in the physiological direction from cell body to axonal ending.

orthostatic hypotension: a sudden drop in blood pressure on assuming a vertical posture due to gravitational pooling of blood.

parenchyma: a general term that designates the functional elements of an organ as distinguished from its framework, or stroma.

perikaryon: the body of the neuron.

perinuclear: situated or occurring around a nucleus.

phagocyte: a glial or other cell that ingests and destroys waste or foreign materials including cellular debris.

phospholipid: a molecule composed of fatty acid chains and a phosphorylated polar end that forms the basic structure of cell membranes.

pia mater: the innermost of the three membranes (*meninges*) that cover the brain and spinal cord.

pinocytosis: inhibition of liquids by cells, especially the phemonenon in which minute incuppings or invaginations are formed in the surface of cells, which close to form fluid-filled vacuoles.

plasmid: an autonomous, self-replicating extrachromosomal circular DNA that can grow independently in bacteria such as *Escherichia coli.*

polyadenylation: addition of a sequence of polyadenylic acid to the 3' end of *mRNA* after its transcription.

polymorphism: allelic variations in genomic DNA that occur in a population, which can result in variable patterns of DNA fragments being formed following digestion with the same restriction enzymes (see *allele*).

polysomal: descriptive of a structure that possesses many bodies, as in a complex of individual ribosomes.

positive feedback control: control by which the output is fed back to give intensification of input.

posterior: at or toward the hind end of the body; in a tailward or caudal direction.

postsynaptic membrane: the nerve cell or other receptor cell membrane immediately related to the synapse formed by presynaptic fibers ending on it.

postsynaptic potentiation: the increased synaptic action that follows intensive synaptic stimulation.

presynaptic fibers: the terminal branches of nerve fibers that end as synaptic knobs.

prokaryotic organism: an organism that lacks nuclei, as bacteria (cf. *eukaryotic*).

prolactin: one of the hormones of the anterior pituitary gland that stimulates and sustains lactation in postpartum mammals, the mammary gland having been prepared by other hormones, including estrogens, progesterone, growth hormone, and corticosteroids.

proliferative zone: that part of the developing cerebellar cortex in which the germinal cells are dividing.

promoter: a region of DNA involved in binding RNA polymerase to initiate the process of transcription.

prostaglandin: any of a group of naturally occurring, chemically related, long-chain hydroxy fatty acids that stimulate contractility of the uterine and other smooth muscle and have the ability to lower blood pressure and to affect the action of certain hormones.

prosthetic group: a chemical group on an enzyme molecule that is essential for the molecule's activity.

protein kinase: any protein that catalyzes the transfer of a phosphate group from ATP to form a phosphoprotein.

punctate: resembling or marked with points or dots.

pyriform cortex: refers to the anterior parahippocampal gyrus (see *limbic lobe*).

ramus: a branch; a general term for a smaller structure given off by a larger one, or into which the larger structure, such as a blood vessel or nerve, divides.

recombinant DNA: DNA that has been reconstituted following splicing techniques in which new DNA has been introduced.

resting potential: the steady electrical potential across the membrane in the quiescent state.

restriction enzyme: an enzyme obtained from bacteria that recognizes a specific short sequence of DNA and therefore cleaves the DNA at positions at which the sequence occurs.

reticular activating system: a system emanating from the reticular formation of the brainstem that, when stimulated, has a powerful recruiting influence on the cortex and other neuronal systems.

retrograde axonal transport: transport of substances along nerve axons toward the cell body (cf. *orthodromic*).

reverse transcription: synthesis of DNA from a template of RNA that is accomplished with the help of the enzyme reverse transcriptase.

rostral: toward the rostrum or nose; thus, in the anterior direction in the *central nervous system*.

rough endoplasmic reticulum: see *Nissl body*.

sacral: situated near or pertaining to the sacrus (the triangular base just below the lumbar vertebrae).

sagittal section: a section taken parallel to the anterior-posterior plane.

sarcoplasm: the interfibrillary matter of the striated muscles; the substance in which the fibrillae of the muscle fiber are embedded.

secondary ending: a receptor ending on the muscle spindle in addition to the primary ending or the annulospiral ending.

selective fasciculation: growth of nerve fibers along pioneering paths of earlier similar fibers so that nerve bundles or fascicles are built.

sensorium: the seat of sensation, located in the brain; often used to designate the condition of a subject relative to his consciousness or mental clarity.

sensory modality: the distinguishing names of all the various sensations arising from diverse inputs with their specific receptor organs.

servoloop control: automatic control by feedback pathways such as given by γ-loops.

servomechanism: a mechanism designed for feedback-control operation.

Southern blotting: a technique for transferring denatured DNA from agarose gel to a nitrocellulose filter on which it can be hybridized with a complementary nucleic acid (cf. *Northern blotting*).

stereotypy: persistent repetition of senseless acts or words.

stretch receptor: stretch-sensitive nerve endings embedded in skeletal muscle (muscle spindle) and tendon (Golgi tendon organ) that play a part in position sense and postural reflexes.

synaptic plasticity: the property of synapses whereby they are changed in functional efficiency, probably by virtue of changes in size.

syncitium: a multinucleate mass of protoplasm produced by the merging of cells.

tachyphylaxis: sudden diminution in response to repeated injections of the same material.

tegmentum: that part of the midbrain dorsal to the substantia nigra and ventral to the cerebral aqueduct.

tetraploid: denotes an individual or cell having four sets of chromosomes.

tonic: characterized by continuous tension or producing and restoring normal tone.

transcription: synthesis of RNA on a DNA template.

transducer: a device for converting one form of energy into another.

translation: synthesis of protein on the *mRNA* template.

trophic influence: an action from one part of a cell to another, or from one cell to another, that is concerned with the growth, maintenance, and metabolism of the cell.

trophic transport: postulated mechanism of transport of specific macromolecules within a cell and among cells that is concerned in exerting trophic influences.

ventral: toward the under or belly side of the body.

ventricle: see *cerebral ventricle*.

vertigo: an illusion of movement.

voltage-clamp: a technique for displacing membrane potential abruptly to a desired value and keeping the potential constant while measuring currents across the cell membrane; devised by Cole and Marmont.

white matter: part of the *central nervous system* that appears white; consists largely of myelinated fiber tracts.

References

The number(s) in brackets at the end of each reference indicates the page(s) on which it is cited.

Abel, J. J., and Kubota, S., 1919, On the presence of histamine (β-imidazolylethylamine) in the hypophysis cerebri and other tissues of the body and its occurrence among the hydrolytic decomposition products of proteins, *J. Pharmacol. Exp. Ther.* **13**:243–300. [349]

Abood, L. G., and Biel, J. H., 1962, Anticholinergic psychotomimetic agents, *Int. Rev. Neurobiol.* **4**:217–273. [544]

Adams, J. C., and Wenthold, R. J., 1979, Distribution of putative amino acid transmitters, choline acetyltransferase and glutamate decarboxylase in the inferior colliculus, *Neuroscience* **4**:1947–1951. [212]

Adams, P. R., Constanti, A., Brown, D. A., and Clark, R. B., 1982, Intracellular Ca^{2+} activates a fast voltage-sensitive K^+ current in vertebrate sympathetic neurones, *Nature (London)* **296**:746–749. [75]

Adams, R. N., and Marsden, C. A., 1982, Electrochemical detection methods for monoamine measurements *in vitro* and *in vivo*, *Handb. Psychopharmacol.* **15**:1–74. [296]

Adrian, E. D., 1947, *The Physical Background of Perception*, Clarendon Press, Oxford. [599]

Agamanolis, D. P., Potter, J. L., Herrick, M. K., and Sternberger, N. H., 1982, The neuropathology of glycine encephalopathy: A report of five cases with immunohistochemical and ultrastructural observations, *Neurology* **32**:975–985. [231]

Aghajanian, G. K., and Bloom, F. E., 1967, The formation of synaptic junctions in developing rat brain: A quantitative electron microscopic study, *Brain Res.* **6**:716–727. [425]

Aghajanian, G. K., and Haigler, H. J., 1973, Direct and indirect actions of LSD, serotonin and related compounds on serotonin-containing neurons, in: *Serotonin and Behavior* (J. Barchas and E. Usdin, eds.), pp. 263–266, Academic Press, New York and London. [329]

Aghajanian, G. K., and Vandermaelen, C. P., 1982, Intracellular recordings from serotonergic dorsal raphe neurons: Pacemaker potentials and the effect of LSD, *Brain Res.* **238**:463–469. [338]

Agid, Y., Guyenet, P., Glowinski, J., Beaujouan, J. C., and Javoy, F., 1975, Inhibitory influence of the nigrostriatal dopamine system on the striatal cholinergic neurons in the rat, *Brain Res.* **86**:488–492. [512]

Agnati, L. F., Fuxe, K., Benfanati, F., Zini, I., and Hökfelt, T., 1983, On the functional role of coexistence of 5-HT and substance P in bulbospinal 5-HT neurons: Substance P reduces affinity and increases density of ^3H-5-HT binding sites, *Acta Physiol. Scand.* **117**:299–301. [157]

Agnati, L. F., Fuxe, K., Zoli, M., *et al.* 1985, Effects of neurotoxic and mechanical lesions of the mesostriatal dopamine pathway on striatal polyamine levels in the rat: Modulation by chronic ganglioside GM_1 treatment, *Neurosci. Lett.* **61**:339–344. [467]

Agranoff, B. W., 1974, Biochemical concomitants of the storage of behavioral information, in: *Biochemistry of Sensory Functions: 25. Mosbacher Colloquium der Gesellschaft für Biologische Chemie* (L. Jaenicke, ed.), pp. 597–623, Springer-Verlag, Berlin. [526]

Agranoff, B. W., 1976, Learning and memory: Approaches to correlating behavioral and biochemical events, in: *Basic Neurochemistry* (G. J. Siegel, R. W. Albers, R. Katzman, and B. W. Agranoff, eds.), pp. 765–786. Little, Brown, Boston. [575]

Agranoff, B. W., 1981, Learning and memory: Biochemical approaches, in: *Basic Neurochemistry* (G. J. Siegel, R. W. Albers, B. W. Agranoff, and R. Katzman, eds.), pp. 801–820, Little, Brown, Boston. [526, 575]

Aguayo, A. J., Benfey, M., and David, S., 1983, A potential for axonal regeneration in neurons of the adult mammalian nervous system, in: *Nervous System Regeneration* (B. Haber, J. R. Perez-Polo, G. A. Hashim, and A. M. Guiffrida Stella, eds.), pp. 327–340, Alan R. Liss, New York. [458, 468]

Ahlquist, R. P., 1948, A study of adrenotropic receptors, *Am. J. Physiol.* **153**:586–600. [309]

Aickin, C. C., Deisz, R. A., and Lux, H. D., 1982, Ammonium action on postsynaptic inhibition in crayfish neurones: Implications for the mechanism of chloride expulsion, *J. Physiol. (London)* **329**:319–339. [131]

Airaksinen, E. M., Sihvola, P., Airaksinen, M. M., Sihvola, M., and Tuovinan, E., 1979, Uptake of taurine by platelets in retinitis pigmentosa, *Lancet* **1**:474–475. [232]

Akert, K., and Peper, K., 1975, Ultrastructure of chemical synapses: A comparison between presynaptic membrane complexes of the motor end plate and the synaptic junction of the central nervous system, in: *Golgi Centennial Symposium Proceedings* (M. Santini, ed.), pp. 521–527, Raven Press, New York. [13, 85]

Akert, K., Moor, H., Pfenninger, K., and Sandri, C., 1969, Contributions of new impregnation methods and freeze etching to the problems of synaptic fine structure, *Prog. Brain Res.* **31**:223–240. [12, 13, 631]

Akert, K., Peper, K., and Sandri, C., 1975, Structural organization of motor end plate and central synapses, in: *Cholinergic Mechanisms* (P. G. Waser, ed.), pp. 43–57, Raven Press, New York. [13, 631]

Akil, H., Watson, S. J., Young, E., Lewis, M. E., Khachaturian, H., and Walker, J. M., 1984, Endogenous opioids: Biology and function, *Annu. Rev. Neurosci.* **7**:223–255. [373, 407]

Albano, J., Bhoola, K. D., Kerwin, R. W., Pay, S., and Pycock, C. J., 1980, Regional distribution of rat brain vasoactive intestinal polypeptide sensitive adenylate cyclase: Effect of neurotoxic lesions on hypothalamic enzyme activity, *Br. J. Pharmacol.* **69**:283–284. [402]

Alberici, M., de Lores Arnaiz, G. R., and De Robertis, E., 1969, Glutamic acid decarboxylase inhibition and ultrastructural changes by the convulsant drug allylglycine, *Biochem. Pharmacol.* **18**:137–143. [223]

Albert, V. R., and Joh, T. H., 1985, Identification of DNA complementary to DOPA-decarboxylase mRNA, *Soc. Neurosci. Abstr.* **11**:1114. [46]

Albus, J. S., 1971, A theory of cerebellar function, *Math. Biosci.* **10**:25–61. [571, 588, 589, 593]

Albus, M., Ackenheil, M., Munch, U., and Naber, D., 1984, Ceruletide: A new drug for the treatment of schizophrenic patients?, *Arch. Gen. Psychiatry* **41**:528. [409]

Alger, B. E., and Nicoll, R. A., 1983, Ammonia does not selectively block IPSPs in rat hippocampal pyramidal cells, *J. Neurophysiol.* **49**:1381–1391. [131]

Alho, H., Costa, E., Ferrero, P., Fujimoto, M., Cosenza-Murphy, D., and Guidotti, A., 1985, Diazepam-binding inhibitor: A neuropeptide located in selected neuronal populations of rat brain, *Science* **229**:179–182. [219]

Allen, G. I., and Tsukahara, N., 1974, Cerebrocerebellar communication systems, *Physiol. Rev.* **54**:957–1006. [495, 497–499, 514, 515]

Allen, G. I., Eccles, J. C., Nicoll, R. A., Oshima, R., and Rubia, F. J., 1977a, The ionic mechanisms concerned in generating the IPSP of hippocampal pyramidal cells, *Proc. R. Soc. London Ser. B* **198**:363–384. [130, 131]

Allen, G. I., Gilbert, P. F. C., Marini, R., Schultz, W., and Yin, T. C. T., 1977b, Integration of cerebral and peripheral inputs by interpositus neurons in monkey, *Exp. Brain Res.* **27**:81–99. [589]

Allen, G. I., Gilbert, P. F., and Yin, T. C., 1978, Convergence of cerebral inputs onto dentate neurons in monkey, *Exp. Brain Res.* **32**:151–170. [498, 589]

Allen, J. M., Cross, A. J., Crow, T. J., Javoy-Agid, F., Agid, Y., and Bloom, S. R., 1985, Dissociation of neuropeptide Y and somatostatin in Parkinson's disease, *Brain Res.* **337**:197–200. [412]

Allen, S. J., Benton, J. S., Goodhardt, M. J., *et al.,* 1983, Biochemical evidence of selective nerve cell changes in the normal ageing human and rat brain, *J. Neurochem.* **41**:256–265. [261]

Allen, Y. S., Roberts, G. W., Bloom, S. R., Crow, T. J., and Polak, J. M., 1984, Neuropeptide Y in the stria terminalis: Evidence for an amygdalofugal projection, *Brain Res.* **321**:357–362.   [395]

Altman, H., ten Bruggencate, G., Pickelmann, P., and Steinberg, R., 1976, Effects of GABA, glycine, picrotoxin and bicuculline methiodide on rubrospinal neurones in cats, *Brain Res.* **111**:337–345.   [212]

Altman, J., 1969, Autoradiographic and histological studies of postnatal neurogenesis. III. Dating the time of production and onset of differentiation of cerebellar microneurons in rats, *J. Comp. Neurol.* **136**:269–294.   [426]

Altman, J., 1972, Postnatal development of the cerebellar cortex in the rat. II. Phases in the maturation of Purkinje cells and of the molecular layer, *J. Comp. Neurol.* **145**:399–464.   [431]

Altman, J., 1975, Postnatal development of the cerebellar cortex in the rat. IV. Spatial organization of bipolar cells, parallel fibers, and glial pallisades, *J. Comp. Neurol.* **163**:427–448.   [429]

Altman, J., 1976, Experimental reorganization of the cerebellar cortex. VII. Effects of late X-radiation schedules that interfere with cell acquisition after stellate cells are formed, *J. Comp. Neurol.* **165**:65–76.   [430, 432]

Altman, J., and Bayer, S. A., 1978, Prenatal development of the cerebral cortex in the rat. I. Histogenesis and cytogenesis, *J. Comp. Neurol.* **179**:23–48.   [421]

Altschuler, R. A., Neises, G. R., Harmison, G. G., Wenthold, R. J., and Fex, J., 1981, Immunocytochemical localization of aspartate aminotransferase immunoreactivity in cochlear nucleus of the guinea pig, *Proc. Natl. Acad. Sci. U.S.A.* **78**:6553–6557.   [178]

Altschuler, R. A., Mosinger, J. L., Harmison, G. G., Parakkal, M. H., and Wenthold, R. J., 1982, Aspartate aminotransferase-like immunoreactivity as a marker for aspartate/glutamate in guinea pig photoreceptors, *Nature (London)* **298**:657–659.   [178]

Altschuler, R. A., Wenthold, R. J., Schwartz, A. M., *et al.* 1984, Immunocytochemical localization of glutaminase-like immunoreactivity in the auditory nerve, *Brain Res.* **291**:173–178.   [179]

Altschuler, R. A., Monaghan, D. T., Haser, W. G., Wenthold, R. J., Curthoys, N. P., and Cotman, C. W., 1985, Immunocytochemical localization of glutaminase-like and aspartate aminotransferase-like immunoreactivities in the rat and guinea pig hippocampus, *Brain Res.* **330**:225–233.   [178, 179, 182]

Alumets, J., Fahrenkrug, J., Hakanson, R., de Muckadell, O. S., Sundler, F., and Uddman, R., 1979, A rich VIP nerve supply is characteristic of sphincters, *Nature (London)* **280**:155–156.   [400]

Amara, S. G., Jonas, V., Rosenfeld, M. G., Ong, E. S., and Evans, R. M., 1982, Alternative RNA processing in calcitonin gene expression generates mRNAs encoding different polypeptide products, *Nature (London)* **298**:240–244.   [376, 382]

Amin, A. H., Crawford, T. B., and Gaddum, J. H., 1954, The distribution of substance P and 5-hydroxytryptamine in the central nervous system of the dog, *J. Physiol. (London)* **126**:596–618.   [320, 374]

Anden, N. E., and Fuxe, K., 1971, A new dopamine-β-hydroxylase inhibitor: Effects on the noradrenaline concentration after L-dopa in the spinal cord, *Br. J. Pharmacol.* **43**:747–756.   [312]

Andersen, P., and Lømo, T., 1966, Mode of activation of hippocampal pyramidal cells by excitatory synapses on dendrites, *Exp. Brain Res.* **2**:247–260.   [72]

Andersen, P., Eccles, J. C., and Loyning, Y., 1964, Location of postsynaptic inhibitory synapses on hippocampal pyramids, *J. Neurophysiol.* **27**:592–607.   [139, 212, 214]

Andersen, P., Holmqvist, B., and Voorhoeve, P. E., 1966, Entorhinal activation of dentate granule cells, *Acta Physiol. Scand.* **66**:448–460.   [72, 140]

Andersen, P., Sundberg, S. H., Svenn, O., Swann, J. N., and Wigstrom, H., 1980, Possible mechanisms for long-lasting potentiation of synaptic transmission in hippocampal slices from guinea-pigs, *J. Physiol. (London)* **302**:463–482.   [556]

Anderson, J. M., Hubbard, B. M., Coghill, G. R., and Slidders, W., 1983, The effect of advanced old age on the neurone content of the cerebral cortex, *J. Neurol. Sci.* **58**:233–244.   [470]

Anderson, M., and Yoshida, M., 1977, Electrophysiological evidence for branching nigral projection to the thalamus and the superior colliculus, *Brain Res.* **137**:361–364.   [212]

Anderson, M. E., and Yoshida, M., 1980, Axonal branching patterns and location of nigrothalamic and nigrocollicular neurons in the cat, *J. Neurophysiol.* **43**:883–895.   [510]

Anderson, W. F., 1984, Prospects for human gene therapy, *Science* **226**:401–409.   [39]

Andrews, C. D., and Woodruff, G. N., 1982, Turning behaviour following nigral injections of dopamine agonists and glycine, *Eur. J. Pharmacol.* **84**:169–175.   [513]

Angevine, J. B., Jr., 1970, Critical cellular events in the shaping of neural centers, in: *The Neurosciences: Second Study Program* (F. O. Schmitt, ed.), pp. 62–72, Rockefeller University Press, New York.   [417]

Angrist, B., Rotrosen, J., and Gershon, S., 1980, Responses to apomorphine, amphetamine and neuroleptics in schizophrenic subjects, *Psychopharmacology* **67**:31–38.   [546]

Antanitus, D. B., Choi, B. H., and Lapham, L. W., 1976, The demonstration of glial fibrillary acidic protein in the cerebrum of the human fetus by indirect immunofluorescence, *Brain Res.* **103**:613–616.   [421]

Aprison, M. H., and Werman, R., 1965, The distribution of glycine in cat spinal cord and roots, *Life Sci.* **4**:2075–2083.   [198]

Aprison, M. H., Davidoff, R. A., and Werman, R., 1970, Glycine: Its metabolic and possible roles in nervous tissue, in: *Handbook of Neurochemistry,* Vol. 3 (A. Lajtha, ed.), pp. 381–397, Plenum Press, New York.   [228, 230]

Aprison, M. H., Tachiki, K. H., Smith, J. E., Lane, J. D., and McBride, W. J., 1974, Current status of 5-hydroxytryptophan in brain, *Adv. Biochem. Psychopharmacol.* **11**:31–41.   [324]

Araki, T., Ito, M., and Oscarsson, O., 1961, Anion permeability of the synaptic and non-synaptic motoneuron membrane, *J. Physiol. (London)* **159**:410–435.   [126, 128]

Araki, M., McGeer, P. L., and McGeer, E. G., 1984a, Presumptive γ-aminobutyric acid pathways from the midbrain to the superior colliculus studied by a combined horseradish peroxidase–γ-aminobutyric acid transaminase pharmacohistochemical method, *Neuroscience* **13**:433–439.   [211, 212, 215]

Araki, M., McGeer, P. L., and McGeer, E. G., 1984b, Retrograde HRP tracing combined with a pharmacohistochemical method for GABA transaminase for the identification of presumptive GABAergic projections to the habenula, *Brain Res.* **304**:271–277.   [211, 212, 215]

Araki, M., McGeer, P. L., and McGeer, E. G., 1985, Striatonigral and pallidonigral pathways studied by a combination of retrograde horseradish peroxidase tracing and a pharmacohistochemical method for γ-aminobutyric acid transaminase, *Brain Res.* **331**:17–24.   [211, 212, 215, 507]

Archer, T., Ogren, S. O., Johansson, G., and Ross, S. B., 1982, DSP4-induced two-way active avoidance impairment in rats: Involvement of central and not peripheral noradrenaline depletion, *Psychopharmacology (Berlin)* **76**:303–309.   [312]

Arendt, T., Bigl, V., Tennstedt, A., and Arendt, A., 1985, Neuronal loss in different parts of the nucleus basalis is related to neuritic plaque formation in cortical target areas in Alzheimer's disease, *Neuroscience* **14**:1–14.   [261]

Ariano, M. A., and Kenny, S. L., 1985, Peptide coincidence in rat superior cervical ganglion, *Brain Res.* **340**:181–185.   [157]

Arimura, A., Culler, M. D., Matsumoto, K., *et al.* 1984, Growth hormone releasing factors in the brain and the gut: Chemistry, actions and localization, *Peptides* **5**(Suppl. 1):41–44.   [373]

Armstrong, C. M., and Bezanilla, F., 1974, Charge movement associated with the opening and closing of the activation gates of the Na channels, *J. Gen. Physiol.* **63**:533–552.   [57]

Armstrong, D. M., Saper, C. B., Levey, A. I., Wainer, B. H., and Terry, R. D., 1982, Immunocytochemical localization of choline acetyltransferase in the rat brain, *Soc. Neurosci. Abstr.* **8**:662.   [250, 251]

Armstrong, D. M., Saper, C. B., Levey, A. I., Wainer, B. H., and Terry, R. D., 1983, Distribution of cholinergic neurons in rat brain: Demonstration by the immunocytochemical localization of choline acetyltransferase, *J. Comp. Neurol.* **216**:53–68.   [245, 250, 251]

Armstrong, M. D., McMillan, A., and Shaw, K. N. F., 1957, 3-Methoxy-4-D-mandelic acid, a urinary metabolite of norepinephrine, *Biochim. Biophys. Acta* **25**:422–425.   [278]

Arnt, J., 1985, Hyperactivity induced by stimulation of separate dopamine D-1 and D-2 receptors in rats with bilateral 6-OHDA lesions, *Life Sci.* **37**:717–723.   [298]

Aronin, N., Cooper, P. E., Lorenz, L. J., *et al.* 1983, Somatostatin is increased in the basal ganglia in Huntington disease, *Ann. Neurol.* **13**:519–526.   [412]

Aronin, N., Difiglia, M., Graveland, G. A., Schwartz, W. J., and Wu, J.-Y., 1984, Localization of immunoreactive enkephalins in GABA synthesizing neurons of the rat neostriatum, *Brain Res.* **300**:376–380.   [157]

Arrang, J. M., Garbarg, M., and Schwartz, J.-C., 1983, Auto-inhibition of brain histamine release mediated by a novel class (H_3) of histamine receptor, *Nature (London)* **302**:832–837.   [356]

Asano, T., Sakakibara, J., and Ogasawara, N., 1983, Molecular sizes of photolabeled GABA and benzodiazepine receptor proteins are identical, *FEBS Lett.* **151**:277–280.   [219]

Asanuma, H., and Rosen, I., 1972, Topographical organization of cortical efferent zones projecting to distal forelimb muscles in the monkey, *Exp. Brain Res.* **14**:243–256.   [483, 484]

Asatoor, A. M., Levi, A. J., and Milne, M. D. 1963, Tranylcypromine and cheese, *Lancet* **2**:733–734.   [539]

Ash, A. S. F., and Schild, H. O., 1966, Receptors mediating some actions of histamine, *Br. J. Pharmacol.* **27:**427–443. [356]

Asin, K. E., Wirtshafter, D., and Fibiger, H. C., 1982, Alterations in drug induced catalepsy and post-decapitation convulsions following brain and spinal cord depletion of norepinephrine by the neurotoxin, DSP-4, *Life Sci.* **30:**1531–1536. [312]

Awapara, J., Landua, A. J., Fuerst, R., and Seale, B., 1950, Free gamma-aminobutyric acid in brain, *J. Biol. Chem.* **187:**35–39. [231]

Axelrod, J., 1971, Noradrenaline: Fate and control of its biosynthesis, *Science* **173:**598–606. [281]

Axelrod, J., 1972, Dopamine-β-hydroxylase: Regulation of its synthesis and release from nerve terminals, *Pharmacol. Rev.* **24:**233–243. [280]

Axelrod, J., and Tomchek, R., 1958, Enzymatic O-methylation of epinephrine and other catechols, *J. Biol. Chem.* **233:**702–705. [278]

Axelrod, J., and Weissbach, H., 1961, Purification and properties of hydroxyindole-O-methyltransferase, *J. Biol. Chem.* **236:**211–213. [326]

Axelrod, J., and Wurtman, R. J., 1968, Photic and neural control of indoleamine metabolism in the rat pineal gland, *Adv. Pharmacol.* **6:**157–166. [321]

Azmitia, E. C., and Ségal, M., 1978, An autoradiographic analysis of the differential ascending projections of the dorsal and median raphe nuclei in the rat, *J. Comp. Neurol.* **179:**641–667. [333]

Babu, Y. S., Sack, J. S., Greenhough, T. J., Bugg, C. E., Means, A. R., and Cook, W. J., 1985, Three-dimensional structure of calmodulin, *Nature (London)* **315:**37–40. [169]

Bacq, Z. M., 1934, La pharmacologie du système nerveux autonome, et particulièrement du sympathique, d'après la theorie neurohumorale, *Ann. Physiol. Physiochim. Biol.* **10:**467–528. [266]

Baertschi, A. J., Zingg, H. H., and Dreifuss, J. J., 1981, Enkephalins, substance P, bradykinin and angiotensin II: Differential sites of action on the hypothalamo-neurohypophysial system, *Brain Res.* **220:**107–119. [406]

Baile, C. A., and Della-Fera, M. A., 1985, Central nervous system cholecystokinin and the control of feeding, *Ann. N. Y. Acad. Sci.* **448:**424–430. [405]

Baimbridge, K. G., and Miller, J. J., 1981, Calcium uptake and retention during long-term potentiation of neuronal activity in the rat hippocampal slice preparation, *Brain Res.* **221:**299–305. [564, 565]

Baker, B. R., and Gibson, R. E., 1971, Irreversible enzyme inhibition: Inhibition of brain choline acetyltransferase by derivatives of 4-stilbazole, *J. Med. Chem.* **14:**315–322. [260]

Baker, P. F., Hodgkin, A. L., and Shaw, T. I., 1962, Replacement of the axoplasm of giant nerve fibres with artificial solutions, *J. Physiol. (London)* **164:**330–354. [53]

Baker, P. F., Hodgkin, A. L., and Ridgway, E. P., 1970, Two phases of calcium entry during the action potential in giant axons of *Loligo*, *J. Physiol. (London)* **208:**80–82. [107]

Baker, P. F., Hodgkin, A. L., and Ridgway, E. B., 1971, Depolarization and calcium entry in squid giant axons, *J. Physiol. (London)* **218:**709–755. [59, 96, 107]

Balasz, R., 1971, Effects of hormones on the biochemical maturation of the brain, in: *Influence of Hormones on the Nervous System* (D. H. Ford, ed.), pp. 150–164, S. Karger, Basel. [525]

Baldissera, F., and Gustafsson, B., 1974, Firing behavior of a neurone model based on the afterhyperpolarization conductance time course and algebraical summation: Adaptation and steady state firing, *Acta Physiol. Scand.* **92:**27–47. [77]

Bannon, M. J., and Roth, R. H., 1983, Pharmacology of mesocortical dopamine neurons, *Pharmacol. Rev.* **35:**53–68. [296]

Barbeau, A., 1978, The last ten years of progress in clinical pharmacology of extrapyramidal symptoms, in: *Psychopharmacology: A Generation of Progress* (M. A. Lipton, A. DiMascio, and K. F. Killam, eds.), pp. 771–776, Raven Press, New York. [341]

Barber, R. P., Vaughn, J. E., Saito, K., McLaughlin, B. J., and Roberts, E., 1978, GABAergic terminals are presynaptic to primary afferent terminals in the substantia gelatinosa of the rat spinal cord, *Brain Res.* **141:**35–55. [203, 206, 217]

Barbin, G., Garbarg, M., Schwartz, J. C., and Storm-Mathisen, J., 1976, Histamine-synthesizing afferents to the hoppocampal region, *J. Neurochem.* **26:**259–264. [355]

Barbin, G., Garbarg, M., Schartz, J. C., and Storm-Mathisen, J., 1977, Neuronal and non-neuronal histamine in rat brain: Tentative localization of the nerve cell bodies, *Agents Actions* **7:**104–105. [355]

Barchi, R. L., 1983, Protein components of the purified sodium channel from rat muscle sarcolemma, *J. Neurochem.* **40:**1377–1385. [58]

Bard, P., and Mountcastle, V. B., 1948, Some forebrain mechanisms involved in expression of rage with special reference to suppression of angry behavior, *Res. Publ. Assoc. Res. Nerv. Ment. Dis.* **27**:362–404. [529]

Barger, G., and Dale, H. H., 1910, The presence in ergot and physiological activity of β-imidazolylethylamine, *J. Physiol. (London)* **40**:38–40. [349]

Barker, L. A., and Hough, L. B., 1983, Selectivity of 4-methylhistamine at H_1- and H_2-receptors in the guinea pig isolated ileum, *Br. J. Pharmacol.* **80**:65–71. [360]

Barnes, C. A., and McNaughton, B. L., 1980, Spatial memory and hippocampal synaptic plasticity in senescent and middle aged rats, in: *Psychology of Aging* (D. Stein, ed.), pp. 253–272, Elsevier/North-Holland, Amsterdam. [558, 572]

Barnett, K. C., 1980, Taurine deficiency retinopathy in the cat, *J. Small Anim. Pract.* **21**:521–534. [232]

Barondes, S. H., 1970, Multiple steps in the biology of memory, in: *The Neurosciences,* Vol. 2 (F. O. Schmitt, ed.), pp. 272–278, Rockefeller University Press, New York. [575]

Bartus, R. T., 1979, Physostigmine and recent memory: Effects in young and aged nonhuman primates, *Science* **206**:1087–1089. [261]

Bartus, R. T., and Johnson, H. R., 1976, Short-term memory in the rhesus monkey: Disruption from the anticholinergic scopolamine, *Pharm. Biol. B* **5**:39–46. [261]

Bartus, R. T., Dean, R. L., Beer, B., and Lippa, A. S., 1982, The cholinergic hypothesis of geriatric memory dysfunction, *Science* **217**:408–417. [261]

Basbaum, A. I., and Fields, H. L., 1984, Endogenous pain control systems: Brainstem spinal pathways and endorphin circuitry, *Annu. Rev. Neurosci.* **7**:309–338. [407]

Baudry, M., Oliver, M., Creager, R., Wieraszko, A., and Lynch, G., 1980, Increase in glutamate receptors following repetitive electrical stimulation in hippocampal slices, *Life Sci.* **27**:325–330. [564]

Bauer, B., and Ehinger, B., 1977, Stimulated release of [^{3}H]beta-alanine from the rabbit retina, *Brain Res.* **120**:447–457. [233]

Bauer, W., Briner, U., Doepfner, R., *et al.* 1982, SMS 201-995: A very potent and selective octapeptide analogue of somatostatin with prolonged action, *Life Sci.* **31**:1133–1140. [409]

Baughman, R. W., and Gilbert, C. D., 1981, Aspartate and glutamate as possible neurotransmitters in the visual cortex, *J. Neurosci.* **1**:427–439. [182]

Baumann, G., Permanetter, B., and Wirtzfeld, A., 1984, Possible value of H_2-receptor agonists for treatment of catecholamine-insensitive congestive heart failure, *Pharmacol. Ther.* **24**:165–77. [359]

Baumgarten, H. G., and Lachenmayer, L., 1972, Chemical lesioning of central monoamine axons by means of 5,6-dihydroxytryptamine and 5,7-dihydroxytryptamine, *Adv. Biochem. Pharmacol.* **10**:13–33. [339]

Baxter, C. F., 1976, Some recent advances in studies of GABA metabolism and compartmentation, in: *GABA in Nervous System Function* (E. Roberts, T. N. Chase, and D. B. Tower, eds.), pp. 61–87, Raven Press, New York. [202]

Bazemore, A. W., Elliott, K. A. C., and Florey, E., 1957, Isolation of factor I, *J. Neurochem.* **1**:334–339. [197]

Beach, T. G., and McGeer, E. G., 1984, The distribution of substance P in the primate basal ganglia: An immunohistochemical study of baboon and human brain, *Neuroscience* **13**:29–52. [507, 511]

Beal, M. F., Bird, E. D., Langlais, P. J., and Martin, J. B., 1984, Somatostatin is increased in the nucleus accumbens in Huntington's disease, *Neurology* **34**:663–666. [412]

Beal, M. F., Benoit, R., Bird, E. D., and Martin, J. B., 1985, Immunoreactive somatostatin-28_{1-12} is increased in Huntington's disease, *Neurosci. Lett.* **56**:377–380. [412]

Beale, R., and Osborne, N. N., 1983, Selective uptake of tritiated glycine, GABA and D-aspartate by retinal cells in culture: A study using autoradiography and simultaneous immunofluorescence, *Dev. Brain Res.* **7**:107–120. [205, 229]

Beauvillain, J. C., Tramu, G., and Dubois, M. P., 1981, Ultrastructural immunocytochemical evidence of the presence of a peptide related to ACTH in granules of LHRH nerve terminals in the median eminence of the guinea pig, *Cell Tissue Res.* **218**:1–6. [157]

Beckstead, R. M., and Frankfurter, A., 1982, The distribution and some morphological features of substantia nigra neurons that project to the thalamus, superior colliculus and pedunculopontine nucleus in the monkey, *Neuroscience* **7**:2377–2388. [510]

Beckstead, R. M., Domesick, V. B., and Nauta, W. J. H., 1979, Efferent connections of the substantia nigra and ventral tegmental area in the rat, *Brain Res.* **175**:191–217. [251]

Beckwith, B. E., Petros, T., Kanaan-Beckwith, S., Couk, D. I., and Haug, R. J., 1982, Vasopressin analog (DDAVP) facilitates concept learning in human males, *Peptides* **3:**627–630. [410]

Beinfeld, M. C., and Bailey, G. J., 1985, The distribution of motilin-like peptides in rhesus monkey brain as determined by radioimmunoassay, *Neurosci. Lett.* **54:**345–350. [388]

Beinfeld, M. C., and Korchak, D. M., 1985, The regional distribution and the chemical, chromatographic, and immunologic characterization of motilin brain peptides: The evidence for a difference between brain and intestinal motilin-immunoreactive peptides, *J. Neurosci.* **5:**2502–2509. [373, 388]

Beitz, A. J., Larson, A. A., Madl, J. E., Altschuler, R. A., and Mullett, M. M., 1985, Immunohistochemical localization of glutamate and glutaminase in neurons of the pontine nuclei and aspartate and AATase in neurons of the inferior olive, *Soc. Neurosci. Abstr.* **11:**865. [182]

Belcher, G., Ryall, R. W., and Schaffner, R., 1978, The differential effects of 5-hydroxytryptamine, noradrenaline and raphe stimulation on nociceptive and non-nociceptive dorsal horn interneurons in the cat, *Brain Res.* **151:**307–321. [229]

Belin, M. F., Aguera, M., Tappaz, M., McRae-Degueurce, A., Bobillier, P., and Pujol, J. F., 1979, GABA accumulating neurons in the nucleus raphe dorsalis and periaqueductal gray in the rat: A biochemical and radioautographic study, *Brain Res.* **170:**279–297. [205, 216]

Belin, M. F., Gamrani, H., Agnera, M., Calas, A., and Pujol, J. F., 1980, Selective uptake of [³H]gamma-aminobutyrate by rat supra- and subependymal nerve fibers: Histological and high resolution radioautographic studies, *Neuroscience* **5:**241–254. [202]

Berlin, M. F., Nanopoulos, D., Didier, M., *et al.* 1983, Immunohistochemical evidence for the presence of γ-aminobutyric acid and serotonin in one nerve cell: A study on the raphe nuclei of the rat using antibodies to glutamate decarboxylase and serotonin, *Brain Res.* **275:**329–339. [157, 215, 332, 506]

Ben-Ari, Y., Dingledine, R., Kanazawa, I., and Kelly, J. J., 1976, Acetylcholine evoked inhibition of thalamic neurones, *J. Physiol. (London)* **256:**112P–113P. [255]

Ben-Ari, Y., Lagowska, J., Tremblay, E., and Le Gal La Salle, G., 1979a, A new model of focal status epilepticus: Intra-amygdaloid application of kainic acid elicits repetitive secondarily generalized convulsive seizures, *Brain Res.* **163:**176–179. [194]

Ben-Ari, Y., Tremblay, E., Ottersen, O. P., and Naquet, R., 1979b, Evidence suggesting secondary epileptogenic lesions after kainic acid: Pretreatment with diazepam reduces distant but not local brain damage, *Brain Res.* **165:**362–365. [194]

Benavides, J., Rumigny, J. F., Bourguignon, J. J., Wermuth, C. G., Mandl, P., and Maitre, M., 1982, A high-affinity Na⁺-dependent uptake system for γ-hydroxybutyrate in membrane vesicles prepared from rat brain, *J. Neurochem.* **38:**1570–1575. [202]

Bender, A. S., Wu, P. H., and Phillis, J. W., 1981a, The rapid uptake and release of [³H]adenosine by rat cerebral cortical synaptosomes, *J. Neurochem.* **36:**651–660. [362]

Bender, A. S., Wu, P. H., and Phillis, J. W., 1981b, Some biochemical properties of the rapid adenosine uptake system in rat brain synaptosomes, *J. Neurochem.* **37:**1282–1290. [362]

Benedetti, F., Montarolo, P. G., Strata, P., and Tosi, L., 1983, Collateral reinnervation in the olivocerebellar pathway in the rat, in: *Nervous System Regeneration* (B. Haber, J. R. Perez-Polo, G. A. Hashim, and A. M. Guiffrida Stella, eds.), pp. 461–464, Alan R. Liss, New York. [453, 468]

Bennett, J. P., Jr., and Snyder, S. H., 1980a, Regulation of receptor binding interaction of ¹²⁵I-angiotensin II and ¹²⁵I-[sarcosine¹,leucine⁸] angiotensin II, and angiotensin antagonist by sodium ion, *Eur. J. Pharmacol.* **67:**1–9. [402]

Bennett, J. P., Jr., and Snyder, S. H., 1980b, Receptor binding interactions of angiotensin II antagonist, ¹²⁵I-[sarcosine¹,leucine⁸]angiotensin II, with mammalian brain and peripheral tissues, *Eur. J. Pharmacol.* **67:**11–25. [402]

Bennett, J. P., Jr., and Yamamura, H. I., 1985, Neurotransmitter, hormone, or drug receptor binding methods, in: *Neurotransmitter Receptor Binding* (H. I. Yamamura, S. J. Enna, and M. J. Kuhar, eds.), pp. 61–89, Raven Press, New York. [160, 161]

Benoit, R., Bohlen, P., Esch, F., and Ling, N., 1984, Neuropeptides derived from prosomatostatin that do not contain the somatostatin-14 sequence, *Brain Res.* **311:**23–29. [382]

Benovic, J. L., Shorr, R. G. L., Caron, M. G., and Lefkowitz, R. J., 1984, The mammalian β₂-adrenergic receptor: Purification and characterization, *Biochemistry* **23:**4510–4518. [309]

Bentivoglio, M., Van der Kooy, D., and Kuypers, H. G. J. M., 1979, The organization of the efferent projections of the substantia nigra in the rat: A retrograde fluorescent double labelling study, *Brain Res.* **174:**1–17. [510]

Berge, O.-G., Fasmer, O. B., Ogren, S. O., and Hole, K., 1985, The putative serotonin receptor agonist 8-hydroxy-2-(di-α-propylamino)tetralin antagonizes the antinociceptive effect of morphine, *Neurosci. Lett.* **54:**71–75. [335]

Berger, S. J., Carter, J. G., and Lowry, O. H., 1977, The distribution of glycine, GABA, glutamate and aspartate in rabbit spinal cord, cerebellum and hippocampus, *J. Neurochem.* **28:**149–158. [228. 229]

Berger, S. J., Carter, J. G., and Lowry, O. H., 1985, Distribution of guanine deaminase in mouse brain, *J. Neurochem.* **44:**1736–1740. [367]

Berger, T. W., and Thompson, R. F., 1978, Neuronal plasticity in the limbic system during classical conditioning of the rabbit nictating membrane response. 1. The hippocampus, *Brain Res.* **145:**323–346. [566]

Berger, T. W., Leham, R. I., and Thompson, R. F., 1980, Hippocampal unit behavior correlations during classical conditioning, *Brain Res.* **193:**229–248. [565, 566]

Berne, R. M., Knabb, R. M., Ely, S. W., and Rubio, R., 1983a, Adenosine in the local regulation of blood flow: A brief overview, *Fed. Proc. Fed. Am. Soc. Exp. Biol.* **42:**3136–3142. [367]

Berne, R. M., Rall, T. W., and Rubio, R. (eds.), 1983b, *Regulatory Functions of Adenosine,* Martinus Nijhoff, The Hague. [362]

Berrettini, W. H., Rubinow, D. R., Nurnberger, J. I., Jr., Simmons-Alling, S., Post, R. M., and Gershon, E. S., 1985, CSF substance P immunoreactivity in affective disorders, *Biol. Psychiatry* **20:**965–970. [412]

Berridge, M. J., 1985, The molecular basis of communication within the cell, *Sci. Am.* **253:**142–152. [167, 170]

Bertler, A., Falck, B., Hillarp, N. A., Rosengren, E., and Torp, A., 1959, Dopamine and chromaffin granules, *Acta Physiol. Scand.* **47:**251–258. [291]

Beutler, B. A., Noronha, A. B. C., Poon, M. M., and Arnason, B. G. W., 1981, The absence of unique kainic acid-like molecules in urine, serum, and CSF from Huntington's disease patients, *J. Neurol. Sci.* **5:**355–360. [195]

Bezanilla, F., and Armstrong, C. M., 1975, Kinetic properties and inactivation of the gating currents of sodium channels in squid axon, *Philos. Trans. R. Soc. London Ser. B* **270:**449–458. [56]

Biegon, A., and Rainbow, T. C., 1983, Distribution of imipramine binding sites in the rat brain studied by quantitative autoradiography, *Neurosci. Lett.* **37:**209–214. [342]

Biegon, A., Rainbow, T. C., and McEwen, B. S., 1982, Quantitative autoradiography of serotonin receptors in the rat brain, *Brain Res.* **242:**197–204. [335]

Biggins, J. A., Perry, E. K., McDermott, J. R., Smith, A. I., Perry, R. H., and Edwardson, J. A., 1983, Postmortem levels of thyrotropin-releasing hormone and neurotensin in the amygdala in Alzheimer's disease, schizophrenia and depression, *J. Neurol. Sci.* **58:**117–122. [395, 412]

Bigl, V., Woolf, N. J., and Butcher, L. L., 1982, Cholinergic projections from the basal forebrain to frontal, parietal, temporal, occipital, and cingulate cortices: A combined fluorescent tracer and acetylcholinesterase analysis, *Brain Res. Bull.* **8:**727–749. [247]

Bignami, A., and Dahl, D., 1973, Differentiation of astrocytes in the cerebellar cortex and the pyramidal tracts of the newborn rat: An immunofluorescence study with antibodies to a protein specific to astrocytes, *Brain Res.* **49:**393–402. [429]

Birkmayer, W., and Hornykiewicz, O., 1961, The effect of (3,4-dihydroxy-phenyl)-L-alanine (=dopa) on akinesia in Parkinson's disease, *Wien. Klin. Wochenschr.* **73:**787–788. [267]

Birks, R., and MacIntosh, F. C., 1961, Acetylcholine metabolism of a sympathetic ganglion, *Can. J. Biochem. Physiol.* **39:**788–827. [243]

Birks, R., Huxley, H. E., and Katz, B., 1960, The fine structure of the neuromuscular junction of the frog, *J. Physiol. (London)* **150:**134–144. [85]

Biscoe, T. J., and Curtis, D. R., 1966, Noradrenaline and inhibition of Renshaw cells, *Science* **151:**1230–1231. [228]

Biscoe, T. J., Davies, J., Dray, A., Evans, R. H., Martin, M. R., and Watkins, J. C., 1977, D-α-Aminoadipate, α,ε-diaminopimelic acid and HA-966 as antagonists of amino acid-induced and synaptic excitation of mammalian spinal neurones *in vivo, Brain Res.* **148:**543–548. [182]

Bishop, G. A., and Ho, R. H., 1985, The distribution and origin of serotonin immunoreactivity in the rat cerebellum, *Brain Res.* **331:**195–207. [333]

Bishop, G. H., 1958, The dendrite: receptive pole of the neurone, *Electroencephalogr. Clin. Neurophysiol.* **10:**12–21. [72]

Bissette, G., Nemeroff, C. H., Decker, M. W., Kizer, J. S., Agid, Y., and Javoy-Agid, F., 1985, Alterations in regional brain concentrations of neurotensin and bombesin in Parkinson's disease, *Ann. Neurol.* **17**:324–328. [412]

Biziere, K., and Coyle, J. T., 1978, Influence of corticostriatal afferents on striatal kainic acid neurotoxicity, *Neurosci. Lett.* **8**:303–310. [192]

Björklund, A., and Lindvall, O., 1984, Dopamine-containing systems in the CNS, in: *Handbook of Chemical Neuroanatomy*, Vol 2, *Classical Transmitters in the CNS, Part I* (A. Björklund and T. Hökfelt, eds.), pp. 55–122, Elsevier, Amsterdam. [286, 289, 295]

Björklund, A., and Stenevi, U., 1979, Regeneration of monaminergic and cholinergic neurons in the mammalian central nervous system, *Physiol. Rev.* **59**:62–100. [457]

Björklund, A., and Stenevi, U., 1981, *In vivo* evidence for a hippocampal adrenergic neuronotrophic factor specifically released on septal deafferentation, *Brain Res.* **229**:403–428. [458]

Björklund, A., and Stenevi, U., 1984, Intracellular neural implants: Neuronal replacement and reconstruction of damaged circuitries, *Annu. Rev. Neurosci.* **7**:279–308. [457]

Björklund, A., Baumgarten, H. G., and Nobin, A., 1974, Chemical lesioning of central monoamine axons by means of 5,6-dihydroxytryptamine and 5,7-dihydroxytryptamine, *Adv. Biochem. Pharmacol.* **10**:13–33. [340]

Björklund, A., Emson, P. C., Gilbert, R. F., and Skagerberg, G., 1979, Further evidence for the possible coexistence of 5-hydroxytryptamine and substance P in medullary raphe neurones of rat brain, *Br. J. Pharmacol.* **66**:112P–113P. [157]

Black, J. W., and Ganellin, C. R., 1974, Naming of substituted histamines, *Experientia* **30**:111–113. [351]

Blackshear, M. A., Steranka, L. R., and Sanders-Bush, E., 1981, Multiple serotonin receptors: Regional distribution and effect of raphe lesions, *Eur. J. Pharmacol.* **76**:325–334. [336]

Blackwood, W., and Corsellis, J. A. N., eds., 1976, *Greenfield's Neuropathology*, Arnold, London. [194]

Blakemore, C., Vital-Durand, F., and Garey, J., 1981, Recovery from monocular deprivation in the monkey. 1. Reversal of physiological effects in the visual cortex, *Proc. R. Soc. London Ser. B* **213**:399–423. [454]

Blalock, J. E., Harbour-McMenamin, D., and Smith, E. M., 1985, Peptide hormones shared by the neuroendocrine and immunologic systems, *J. Immunol.* **135**:858s–861s. [394]

Blaschko, H., 1939, The specific action of L-dopa decarboxylase, *J. Physiol. (London)* **96**:50P–51P. [266]

Bleier, R., 1969, Retrograde transsynaptic cellular degeneration in mammillary and ventral tegmental nuclei following limbic decortication in rabbits of various ages, *Brain Res.* **15**:365–393. [462]

Blier, P., and de Montigny, C., 1980, Effect of chronic tricyclic antidepressant treatment on the serotoninergic autoreceptor: A microiontophoretic study in the rat, *Naunyn-Schmiedeberg's Arch. Pharmacol.* **314**:123–128. [336]

Blinzinger, K., and Kreutzberg, G., 1968, Displacement of synaptic terminals from regenerating motoneurons by microglial cells, *Z. Zellforsch. Mikrosk. Anat.* **85**:145–157. [35]

Bliss, T. V. P., and Dolphin, A. C., 1982, What is the mechanism of long-term potentiation in the hippocampus?, *TINS* **5**:289–290. [564, 565]

Bliss, T. V. P., and Gardner-Medwin, A. R., 1973, Long-lasting potentiation of synaptic transmission in the dentate area of the unanesthetized rabbit following stimulation of the perforant path, *J. Physiol. (London)* **232**:357–374. [558]

Bliss, T. V. P., and Lømo, T., 1973, Long-lasting potentiation of synaptic transmission in the dentate area of the anesthetized rabbit following stimulation of the perforant path, *J. Physiol. (London)* **232**:331–356. [556, 557, 558]

Bloch, B., Brazeau, P., Bloom, F., and Ling, N., 1983, Topographical study of the neurons containing hpGRF immunoreactivity in monkey hypothlamus, *Neurosci. Lett.* **37**:23–28. [388]

Bloch, V. 1970, Facts and hypotheses concerning memory consolidation processes, *Brain Res.* **24**:561–575. [572]

Block, G. A., and Billier, R. B., 1981, Properties and regional distribution of nicotinic cholinergic receptors in the rat hypothalamus, *Brain Res.* **212**:152–158. [253]

Bloom, D. M., Nair, N. P. V., and Schwartz, G., 1983, CCK-8 in the treatment of chronic schizophrenia, *Psychopharmacol. Bull.* **19**:361–363. [410]

Bloom, F. E., 1981, Neuropeptides, *Sci. Am.* **245**:148–168. [379]

Bloom, F. E., and Iversen, L. L., 1971, Localizing ^{3}H GABA in nerve terminals of rat cerebral cortex by electron microscopic autoradiography, *Nature (London)* **229**:628–630. [205]

Bloom, F. E., Hoffer, B. J., Nelson, C., Shen, Y.-S., and Siggins, G. R., 1973, The physiology and pharmacology of serotonin mediated synapses, in: *Serotonin and Behavior* (J. Barchas and E. Usden, eds.), pp. 249–261, Academic Press, New York and London. [335]

Blurton, P. A., Broadhurst, A. M., Cross, J. A., Ennis, C., Wood, M. D., and Wyllie, M. G., 1984, Panuramine, a selective inhibitor of uptake of 5-hydroxytryptamine in the brain of the rat, *Neuropharmacology* **23**:1049–1052. [342]

Boadle-Biber, M. C., 1982, Further studies on the role of calcium in the depolarisation-induced activation of tryptophan hydroxylase, *Biochem. Pharmacol.* **31**:2495–2503. [329]

Boadle-Biber, M. C., Johannessen, J. N., Narasimhachari, N., and Phan, T. H., 1983, Activation of tryptophan hydroxylation by stimulation of central serotonergic neurons, *Biochem. Pharmacol.* **32**:185–188. [329]

Bodian, D., 1972, Poliomyelitis, in: *Pathology of the Nervous System* (J. Minckler, ed.), pp. 2323–2344, McGraw-Hill, New York. [262]

Bodner, M., Fridkin, M., and Gozes, I., 1985, Coding sequences for vasoactive intestinal peptide and PHM-27 peptide are located on two adjacent exons in the human genome, *Proc. Natl. Acad. Sci. U.S.A.* **82**:3548–3551. [382]

Bogdanski, D. F., Weissbach, H., and Udenfriend, S., 1957, The distribution of serotonin, 5-hydroxytryptophan decarboxylase and monoamine oxidase in brain, *J. Neurochem.* **1**:227–228. [322]

Bohn, M. C., 1983, Role of glucocorticoids in expression and development of phenylethanolamine *N*-methyltransferase (PNMT) in cells derived from the neural crest: A review, *Psychoneuroendocrinology* **8**:381–390. [315]

Bolam, J. P., Clarke, D. J., Smith, A. D., and Tomogyi, P., 1983, A type of aspiny neuron in the rat neostriatum accumulates [³H]gamma-aminobutyric acid: Combination of Golgi-staining, autoradiography, and electron microscopy, *J. Comp. Neurol.* **213**:121–134. [205]

Bolger, G. T., Weissman, B. A., Lueddens, H., *et al.* 1985, Late evolutionary appearance of "peripheral-type" binding sites for benzodiazepines, *Brain Res.* **338**:366–370. [222]

Bondareff, W., Mountjoy, C. O., and Roth, M., 1982, Loss of neurons of origin of the adrenergic projection to cerebral cortex (nucleus locus coeruleus) in senile dementia, *Neurology* **32**:167–168. [262]

Bonhoeffer, F., and Gierer, A., 1984, How do retinal axons find their targets on the tectum?, *TINS* **7**:378–381. [436]

Borbe, H. O., Muller, W. E., and Wolert, U., 1981, Specific [³H]strychnine binding associated with glycine receptors in bovine retina, *Brain Res.* **205**:131–139. [229]

Border, B. G., and Mihailoff, G. A., 1985, GAD-immunoreactive neural elements in the basilar pontine nuclei and nucleus reticularis tegmenti pontis of the rat. I. Light microscopic studies, *Exp. Brain Res.* **59**:600–614. [205]

Borman, J., Sakmann, B., and Seifert, W., 1983, Isolation of GABA-activated single-channel Cl currents in the soma membrane of rat hippocampal neurones, *J. Physiol. (London)* **341**:9P–10P. [119, 128]

Bostock, H., and Sears, T. A., 1978, The internodal axon membrane: Electrical excitability and continuous conduction in segmental demyelination, *J. Physiol. (London)* **280**:273–301. [66]

Bostock, H., Sears, T. A., and Sherratt, R. M., 1981, The effects of 4-aminopyridine and tetraethylammonium ions on normal and demyelinated mammalian nerve fibers, *J. Physiol. (London)* **313**:301–315. [59, 67]

Boulton, A. A., 1979, Trace amines in the central nervous system, *Int. Rev. Biochem.* **26**:179–206. [346]

Bowen, D. M., Smith, C. B., White, P., and Davison, A. N., 1976, Neurotransmitter-related enzymes and indices of hypoxia in senile dementia and other abiotrophies, *Brain* **99**:459–495. [261]

Bowen, D. M., Spillane, J. A., Curzon, G., Meir-Rose, W., White, P., and Goodhardt, M. J., 1979, Accelerated aging or selective neuronal loss as an important cause of dementia, *Lancet* **1**:11–14. [262]

Bowery, N. G., Dobloe, A., Hill, D. R., Hudson, A. L., Shaw, J., and Turnbull, M. J., 1980, A novel GABA receptor on central neurones, *Scott. Med. J.* **25**:S3–S11. [145, 219]

Bowker, R. M., Westlund, K. N., and Coulter, J. D., 1981, Origins of serotonergic projections to the spinal cord in rat: An immunocytochemical–retrograde transport study, *Brain Res.* **226**:187–199. [333]

Bowker, R. M., Westlund, K. N., Sullivan, M. C., and Coulter, J. D., 1982, A combined retrograde transport and immunocytochemical staining method for demonstrating the origins of serotonergic projections, *J. Histochem. Cytochem.* **30**:805–810. [331]

Bowker, R. M., Westlund, K. N., Sullivan, M. C., Wilber, J. F., and Coulter, J. C., 1983, Descending serotonergic, peptidergic and cholinergic pathways from the raphe nuclei: A multiple transmitter complex, *Brain Res.* **288**:33–48. [157, 332]

Bowsher, R. R., Verburg, K. M., and Henry, D. P., 1983, Rat histamine-*N*-methyl-transferase, *J. Biol. Chem.* **258**:12,215–12,220. [351]

Bradford, H. F., and Dodd, P. R., 1977, Convulsions and activation of epileptic foci induced by monosodium glutamate and related compounds, *Biochem. Pharmacol.* **26**:253–254. [194]

Bradford, H. F., and Richards, C. D., 1976, Specific release of endogenous glutamate from pyriform cortex stimulated *in vitro, Brain Res.* **105**:168–172. [182, 186]

Bradford, H. F., and Ward, H. K., 1976, On glutaminase in mammalian synaptosomes, *Brain Res.* **110**:115–125. [180]

Bradley, P., and Berry, M., 1976, The effects of reduced climbing and parallel fibre input on Purkinje cell dendritic growth, *Brain Res.* **109**:133–151. [432. 463]

Bradwejn, J., and de Montigny, C., 1984, Benzodiazepines antagonize cholecystokinin-induced activation of rat hippocampal neurones, *Nature (London)* **312**:363–364. [219]

Brady, S. T., 1985, A novel brain ATPase with properties expected for the fast axonal transport motor, *Nature (London)* **317**:73–75. [32]

Branconnier, R. J., 1981, The human behavioral pharmacology of the common core heptapeptides, *Pharmacol. Ther.* **14**:161–175. [410]

Brandon, C., 1985, Retinal GABA neurons: Localization in vertebrate species using an antiserum to rabbit brain glutamate decarboxylase, *Brain Res.* **344**:286–295. [205, 218]

Brandon, C., Lam, D. M. K., and Wu, J.-Y., 1979, Gamma-aminobutyric acid system in rabbit retina: Localization by immunocytochemistry and autoradiography, *Proc. Natl. Acad. Sci. U.S.A.* **76**:3557–3561. [205]

Brann, M. R., and Emson, P. C., 1980, Microiontophoretic injection of fluorescent tracer combined with simultaneous immunofluorescent histochemistry for the demonstration of efferents from the caudate putamen projecting to the globus pallidus, *Neurosci. Lett.* **16**:61–66. [507]

Brennan, M. J. W., and Cantrill, R. C., 1979, The effect of delta-aminolaevulinic acid on the uptake and efflux of amino acid neurotransmitters in rat brain synaptosomes, *J. Neurochem.* **33**:721–725. [225]

Brennan, M. J., Cantrill, R. C., and Krogsgaard-Larsen, P., 1981, GABA autoreceptors: Structure–activity relationships for agonists, *Adv. Biochem. Psychopharmacol.* **26**:157–167. [219]

Brigati, D. J., Myerson, D., Leary, J. J., *et al.* 1983, Detection of viral genomes in cultured cells and paraffin-embedded tissue sections using biotin-labeled hybridization probes, *Virology* **126**:32–50. [43]

Brightman, M. W., and Reese, T. S., 1975, Membrane specialization of ependymal cells and astrocytes, in: *The Nervous System,* Vol. 1, *The Basic Neurosciences* (D. B. Tower and R. O. Brady, eds.), pp. 267–277, Raven Press, New York. [37]

Brimijoin, S., 1983, Molecular forms of acetylcholinesterase in brain, nerve and muscle: Nature, localization and dynamics, *Prog. Neurobiol.* **21**:291–322. [240]

Brinkman, C., and Porter, R., 1979, Supplementary motor area in the monkey: Activity of neurons during performance of a learned motor task, *J. Neurophysiol.* **42**:681–709. [480, 481, 483]

Brinkman, C., and Porter, R., 1983, Supplementary motor area and premotor areas of the cerebral cortex in the monkey: Activity of neurons during performance of a learned movement task, in: *Motor Control Mechanisms in Man* (J. E. Desmedt, ed.), pp. 393–420, Raven Press, New York. [483, 589, 592]

Brinkman, C., Bush, B. M., and Porter, R., 1978, Deficient influences of peripheral stimuli on precentral neurones in monkeys with dorsal column lesions, *J. Physiol. (London)* **276**:27–48. [488]

Brinton, R. E., Deshmukh, P. P., Chen, A., Davis, T. P., Hsiao, S., and Yamamura, H. I., 1983, A non-equilibrium 24-hour vasopressin radioimmunoassay: Development and basal levels in the rat brain, *Brain Res.* **266**:344–347. [388]

Brisson, A., and Unwin, P. N. T., 1985, Quaternary structure of the acetylcholine receptor, *Nature (London)* **315**:474–480. [256]

Brock, L. G., Coombs, J. S., and Eccles, J. C., 1952, The recording of potentials from motoneuron with an intracellular electrode, *J. Physiol. (London)* **117**:431–460. [112, 114, 121]

Brodal, A., 1954, Afferent cerebellar connection, in: *Aspects of Cerebellar Anatomy* (J. Jansen and A. Brodal, eds.), pp. 82–188, Johan Grundt Tanum Forlag, Oslo. [453]

Brodie, B. B., Spector, S., and Shore, P. A., 1959, Interaction of drugs with norepinephrine in the brain, *Pharmacol. Rev.* **11**:548–564. [267, 320, 334]

Brooks, V. B., 1979, Control of intended limb movements by the lateral and intermediate cerebellum, in: *Integration in the Nervous System* (H. Asanuma and V. J. Wilson, eds.), pp. 321–356, Igaku-Shoin, Tokyo and New York. [481, 591]

Brooks, V. B., 1985, How are "move" and "hold" programs matched?, in: *Cerebellar Functions* (J. R. Bloedel, J. Dichgans, and W. Precht, eds.), pp. 1–23, Springer-Verlag, New York. [498]

Brooks, V. B., and Thach, W. T., 1981, Cerebellar control of posture and movement in the nervous system. *Handbook of Physiology,* pp. 877–946. [491, 495, 515, 516, 588]

Brothers, L. A., and Finch, D. M., 1985, Physiological evidence for an excitatory pathway from entorhinal cortex to amygdala in the rat, *Brain Res.* **359:**10–20. [182]

Brown, A. G., 1981, *Organization in the Spinal Cord: The Anatomy and Physiology of Identified Neurones,* Springer-Verlag, Berlin, Heidelberg and New York, 238 pp. [112–116]

Brown, G. M., Pulido, O., Niles, L. P., *et al.* 1983, Differential localization of melatonin and *N*-acetyl-serotonin in brain, in: *The Pineal Gland and Its Endocrine Role* (J. Axelrod, F. Fraschini, and G. P. Velo, eds.), pp. 257–276, Plenum Press, New York. [326]

Brown, J. C., McIntosh, C. H. S., and Pederson, R. A., 1983, The gastrointestinal peptides and nutrition, *Can. J. Physiol. Pharmacol.* **61:**282–289. [529]

Brown, J. H., and Brown, S. L., 1984, Agonists differentiate muscarinic receptors that inhibit cyclic AMP formation from those that stimulate phosphoinositide metabolism, *J. Biol. Chem.* **259:**3777–3781. [255]

Brown, L. L., Makman, M. H., Wolfson, L. I., Dvorkin, B., Warner, C., and Katzman, R., 1979, A direct role of dopamine in the rat subthalamic nucleus and adjacent interpeduncular area, *Science* **206:**1416–1418. [510]

Brown, M., Fisher, L., Mason, R. T., Rivier, J., and Vale, W., 1985, Neurobiological actions of cysteamine, *Fed. Proc. Fed. Am. Soc. Exp. Biol.* **44:**2556–2560. [411]

Brownfield, M. S., Reid, I. A., Ganten, D., and Ganong, W. F., 1982, Differential distribution of immunoreactive angiotensin and angiotensin-converting enzyme in rat brain, *Neuroscience* **7:**1759–1769. [388]

Browning, M., Dunwiddie, T., Bennett, W. W., Gispen, W., and Lynch, G., 1978, Synaptic phosphoproteins: Specific changes after repetitive stimulation of the hippocampal slice, *Science* **203:**60–62. [563]

Brownstein, M. J., Mroz, E. H., Tappaz, N. L., and Leeman, S. E., 1977, On the origin of substance P and glutamic acid decarboxylase (GAD) in the substantia nigra, *Brain Res.* **135:**315–323. [507]

Bruce, G., Wainer, B. H., and Hersh, L. B., 1985, Immunoaffinity purification of human choline acetyltransferase: Comparison of the brain and placental enzymes, *J. Neurochem.* **45:**611–620. [239]

Buck, S. H., Burks, T. F., Brown, M. R., and Yamamura, H. I., 1981, Reduction in basal ganglia and substantia nigra of substance P levels in Huntington's disease, *Brain Res.* **209:**464–469. [412]

Buck, S. H., Burcher, E., Shults, C. W., Lovenberg, W., and O'Donohue, T. L., 1984, Novel pharmacology of substance K-binding sites: A third type of tachykinin receptor, *Science* **226:**987–989. [373, 403]

Bulfield, C., and Keser, H., 1975, Histamine and histidine levels in the brain of the histadinaemic mutant mouse, *J. Neurochem.* **24:**403–405. [352]

Bunge, M. B., Bunge, R. P., and Ris, H., 1961, Ultrastructural study of remyelination in an experimental lesion in adult cat spinal cord, *J. Biophys. Biochem. Cytol.* **10:**67–94. [36]

Burbach, J. P. H., and deWied, D., 1981, Memory effects and brain proteolysis of neurohypophyseal hormones, *Dev. Endocrinol.* **13:**69–73. [410]

Burchfield, J. L., and Duffy, F. H., 1972, Muscle afferent input to single cells in primate somatosensory cortex, *Brain Res.* **45:**241–246. [487]

Burke, B. E., and De Lorenzo, R. J., 1982, Ca^{2+} and calmodulin-dependent phosphorylation of endogenous synaptic vesicle tubulin by a vesicle-bound calmodulin kinanse system, *J. Neurochem.* **38:**1205–1218. [169]

Burke, R. E., 1967, Composite nature of the monosynaptic excitatory postsynaptic potential, *J. Neurophysiol.* **30:**1114–1137. [114]

Burke, R. E., Walmsley, B., and Hodgson, J. A., 1979, HRP anatomy of group Ia afferent contacts on alpha motoneurones, *Brain Res.* **160:**347–352. [114]

Burkhardt, C., Frederickson, R. C. A., and Pasternak, G. W., 1982, Metkephamid (Tyr-D-Ala-Gly-Phe-*N*(Me)Met-NH$_2$), a potent opioid peptide: Receptor binding and analgesic properties, *Peptides* **3:**869–871. [404]

Burks, T. F., Buck, S. H., and Miller, M. S., 1985, Mechanisms of depletion of substance P by capsaicin, *Fed. Proc. Fed. Am. Soc. Exp. Biol.* **44:**2531–2534. [411]

Burlet, A., Tonon, M.-C., Tankosic, P., Coy, D., and Vaudry, H., 1983, Comparative immunocytochemical localization of corticotropin releasing factor (CRF-41) and neurohypophyseal peptides in the brain of Brattleboro and Long–Evans rats, *Neuroendocrinology* **37:**64–72. [157]

Burns, R. S., Chiueh, C. C., Markey, S. P., Ebert, M. H., Jacobowitz, D. M., and Kopin, I. J., 1983, A primate model of parkinsonism: Selective destruction of the pars compacta of the substantia nigra by *N*-methyl-4-phenyl-1,2,3,6-tetrahydropyridine, *Proc. Natl. Acad. Sci. U.S.A.* **80:**4546–4550. [303]

Burnstock, G., 1975, Purinergic transmission, in: *Handbook of Psychopharmacology*, Vol. 5 (L. L. Iversen, S. D. Iversen, and S. H. Snyder, eds.), pp. 131–194, Plenum Press, New York. [367]

Burnstock, G., Cocks, T., and Crowe, R., 1978, Evidence for purinergic innervation of the anococcygeus muscle, *Br. J. Pharmacol.* **64**:13–20. [367]

Burton, K., Farrell, K., Li, D., and Calne, D. B., 1984, Lesions of the putamen and dystonia: CT and magnetic resonance imaging, *Neurology* **34**:962–965. [512]

Butcher, L. L., Talbot, K., and Bilezikjian, L., 1975, Acetylcholinesterase neurons in dopamine-containing regions of the brain, *J. Neural Transm.* **37**:127–153. [240, 250]

Butcher, S. P., Roberts, P. J., and Collins, J. F., 1985, The subcellular distribution of DL-[^{3}H]2-amino-4-phosphonobutyrate binding sites in the rat brain, *Neurosci. Lett.* **61**:249–253. [189]

Butterworth, R. F., and Giguere, J. F., 1982, Glutamic acid in spinal-cord gray matter in Friedreich's ataxia, *N. Engl. J. Med.* **307**:897. [195]

Buzsaki, G., and Czeh, G., 1981, Commissural and perforant path interactions in the rat hippocampus, field potentials and unitary activity, *Exp. Brain. Res.* **43**:429–438. [212, 214]

Bylund, D. B., 1980, *Receptor Binding Techniques*, Society for Neuroscience, Washington, D.C. [162]

Bylund, D. B., and U'Prichard, D. C., 1983, Characterization of α_1 and α_2 adrenergic receptors, *Int. Rev. Neurobiol.* **24**:343–431. [309]

Bymaster, F. P., Perry, K. W., and Wong, D. T., 1985, IV. Measurement of acetylcholine and choline in brain by HPLC with electrochemical detection, *Life Sci.* **37**:1775–1781. [236]

Caccia, M. R., Meola, G., Cerri, C., Frattola, L., Scarlato, G., and Aporti, F., 1979, Treatment of denervated muscle by gangliosides, *Muscle Nerve* **2**:382–389. [466, 467]

Caffe, A. R., van Leeuwen, F. W., Buijs, R. M., de Vries, G. J., and Geffard, M., 1985, Coexistence of vasopressin, neurophysin and noradrenaline immunoreactivity in medium-sized cells of the locus coeruleus and subcoeruleus in the rat, *Brain Res.* **338**:160–164. [157]

Calissano, P., Cattaneo, A., Aloe, L., and Levi-Montalcini, R., 1984, The nerve growth factor (NGF), in: *Hormonal Proteins and Peptides*, Vol. XII (C. H. Li, ed.), pp. 1–56, Academic Press, New York. [464]

Calne, D. B., Langston, J. W., Martin, W. R. W. *et al.* 1985, Positron emission tomography after MPTP: Observations relating to the cause of Parkinson's disease, *Nature (London)* **317**:246–248. [303, 511]

Calvin, W. H., and Schwindt, P. C., 1972, Steps in production of motoneuron spikes during firing, *J. Neurophysiol.* **35**:297–310. [77]

Campbell, N. C., Ekerot, C. F., Hesslow, G., and Oscarsson, O., 1983a, Dendritic plateau potentials evoked in Purkinje cells by parallel fiber volleys in the cat, *J. Physiol. (London)* **340**:209–223. [583]

Campbell, N. C., Ekerot, C. F., and Hesslow, G., 1983b, Interaction between responses in Purkinje cells evoked by climbing fiber impulses and parallel fiber volleys in the cat, *J. Physiol. (London)* **340**:225–238. [583]

Campbell, W. C., Fisher, M. H., Stapley, E. O., Albers-Schonberg, G., and Jacob, T. A., 1983, Ivermectin: A potent new antiparasitic agent, *Science* **221**:823–828. [222]

Campenot, R. B., 1977, Local control of neurite development by nerve growth factor, *Proc. Natl. Acad. Sci. U.S.A.* **74**:4516–4519. [456]

Cannon, W. B., and Uridil, J. E., 1921, Studies on the conditions of activity in endocrine glands. VIII. Some effects on the denervated heart of stimulating the nerves of the liver, *Am. J. Physiol.* **58**:353–354. [266]

Canzek, V., and Reubi, J. E., 1980, The effect of cochlear nerve lesion on the release of glutamate, aspartate and GABA from cat cochlear nucleus *in vitro, Exp. Brain Res.* **38**:437–441. [180, 182, 184]

Card, J. P., and Moore, R. Y., 1982, Ventral lateral geniculate nucleus efferents to the rat suprachiasmiatic nucleus exhibit avian pancreatic polypeptide-like immunoreactivity, *J. Comp. Neurol.* **206**:390–396. [394]

Card, J. P., Brecha, N., and Moore, R. Y., 1983, Immunohistochemical localization of avian pancreatic polypeptide-like immunoreactivity in the rat hypothalamus, *J. Comp. Neurol.* **117**:123–136. [388]

Cardinali, D. P., Ritta, M. N., Fuentes, A. M., Gimeno, M. F., and Gimeno, A. L., 1980, Prostaglandin release by rat medial basal hypothalamus *in vitro:* Inhibition by melatonin at submicromolar concentrations, *Eur. J. Pharmacol.* **67**:151–153. [345]

Carlin, R. K., Bartelt, D. C., and Siekevitz, P., 1983, Identification of fodrin as a major calmodulin-binding protein in postsynaptic density preparations, *J. Cell Biol.* **96**:443–448. [562, 563]

Carlsson, A., 1959, Occurrence, distribution and physiological role of catecholamines in the nervous system, *Pharmacol. Rev.* **11**:300–304. [267]

Carlsson, A., 1974, Measurements of monoamine synthesis and turnover with special reference to 5-hydroxytryptamine, *Adv. Biochem. Psychopharmacol.* **10**:75–81. [322, 328]

Carpenter, M. B., and Sutin, J., 1983, *Human Neuroanatomy,* 8th ed., Williams & Wilkins, Baltimore. [399, 502, 505, 506]

Carroll, P. J., and Aspry, J. A. M., 1980, Subcellular origin of ACh released from mouse brain, *Science* **210**:641–642. [245]

Casamenti, F., Bracco, L., Bartolini, L., and Pepeu, G., 1985, Effects of ganglioside treatment in rats with a lesion of the cholinergic forebrain nuclei, *Brain Res.* **338**:45–52. [467]

Cascieri, M. A., and Liang, T., 1983, Characterization of the substance P receptor in rat brain cortex membranes and the inhibition of radioligand binding by guanine nucleotides, *J. Biol. Chem.* **258**:5158–5164. [403]

Castellani, S., Ziegler, M. G., van Kammen, D. P., Alexander, P. E., Sims, S. G., and Lake, C. R., 1982, Plasma norepinephrine and dopamine-β-hydroxylase activity in schizophrenia, *Arch. Gen. Psychiatry* **39**:1145–1149. [547]

Cawthon, R. M., Pintar, J. E., Haseltine, F. P., and Breakefield, X. O., 1981, Differences in the structure of A and B forms of human monoamine oxidase, *J. Neurochem.* **37**:363–372. [277]

Ceccarelli, B., Aporti, F., and Finesso, M., 1976, Effects of brain gangliosides on functional recovery in experimental regeneration and reinnervation, in: *Ganglioside Function* (G. Porcellati, B. Ceccarelli, and G. Tettamanti, eds.), pp. 275–293, Plenum Press, New York. [466, 467]

Cerione, R. A., Strulovivi, B., Benovic, J. L., Lefkowitz, R. J., and Caron, M. G., 1983, A role for N_1 in the hormonal stimulation of adenylate cyclase, *Nature (London)* **306**:562–566. [165]

Cesaro, P., 1984, Substance P, *Rev. Neurol. (Paris)* **140**:465–478. [373]

Chaminade, M., Foutz, A. S., and Rossier, J., 1983, Co-release of enkephalins and precursors with catecholamines by the perfused cat adrenal *in situ, Life Sci.* **33**:21–24. [157]

Chan, C. W. Y., 1983, Segmental versus suprasegmental contributions to long-latency stretch responses in man, *Adv. Neurol.* **39**:467–487. [490]

Chandler, W. K., and Meeves, H., 1965, Voltage clamp experiments on internally perfused giant axons, *J. Physiol. (London)* **180**:788–820. [106]

Chang, K. J., 1981, Morphiceptin (NH_4-Tyr-Pro-Phe-Pro-$CONH_2$): A potent and specific agonist for morphine (μ) receptors, *Science* **211**:75–77. [404]

Chang, M. M., and Leeman, S. E., 1970, Isolation of a sialogic peptide from bovine hypothalamus tissue and its characterization as substance P, *J. Biol. Chem.* **245**:4784–4790. [374]

Chang, R. S. L., Tran, V. T., and Snyder, S. H., 1979, Heterogeneity of histamine H_1-receptors: Species variations in [^{3}H]mepyramine binding of brain membranes, *J. Neurochem.* **32**:1653–1663. [356, 357]

Changeux, J. P., 1983, Concluding remarks: On the "singularity" of nerve cells and its ontogenesis, *Prog. Brain Res.* **58**:465–478. [417]

Chan-Palay, V., 1977, Indoleamine neurons and their processes in the normal rat brain and in chronic diet-induced thiamine deficiency demonstrated by uptake of ^{3}H-serotonin, *J. Comp. Neurol.* **176**:467–494. [333]

Chan-Palay, V., and Palay, S. L., 1979, Immunocytochemical localization of cyclic GMP: Light and electron microscopic evidence for involvement of neuroglia, *Proc. Natl. Acad. Sci. U.S.A.* **76**:1485–1488. [168]

Chan-Palay, V., Palay, S. L., and Wu, J.-Y., 1979, γ-Aminobutyric acid pathways in the cerebellum studied by retrograde and anterograde transport of glutamic acid decarboxylase antibody after *in vivo* injections, *Anat. Embryol. (Berlin)* **157**:1–14. [212]

Chan-Palay, V., Nilaver, G., Palay, S. L., *et al.* 1981, Chemical heterogeneity in cerebellar Purkinje cells: Existence and coexistence of glutamic acid decarboxylase and motilin-like immunoreactivities, *Proc. Natl. Acad. Sci. U.S.A.* **78**:7787–7791. [157, 400]

Chao, L. P., 1980, Choline acetyltransferase; purification and characterization, *J. Neurosci. Res.* **5**:88–115. [239]

Charnay, Y., Leger, L., Rossier, J., *et al.* 1985, Evidence for synenkephalin-like immunoreactivity in pontobulbar monoaminergic neurons of the cat, *Brain Res.* **335**:160–164. [157, 332]

Chase, T. N., Barone, P., Bruno, G., *et al.* 1985, Cholecystokinin-mediated synaptic function and the treatment of neuropsychiatric disease, *Ann. N. Y. Acad. Sci.* **448**:553–561. [410]

Checler, F., Vincent, J. P., and Kitabgi, P., 1985, Inactivation of neurotensin by rat brain synaptic membranes partly occurs through cleavage at the Arg^8–Arg^9 peptide bond by a metalloendopeptidase, *J. Neurochem.* **45**:1509–1513. [384]

Cheney, D. L., Panula, P., Revuelta, A. V., Thompson, H. K., Wu, J.-Y., and Costa, E., 1982, Immu-
    nohistochemical localization of glutamate decarboxylase and met-enkephalin-like immunoreactivity in the
    septal complex of the rat, *Soc. Neurosci. Abstr.* **8:**582.   [205]
Cheramy, A., Leviel, V., and Glowinski, J., 1981, Dendritic release of dopamine in the substantia nigra,
    Nature (London) **289:**537–542.   [296, 297, 512]
Cheung, W. Y., 1980, Calmodulin plays a pivotal role in cellular regulation, *Science* **207:**19–27.   [59, 96,
    560, 561]
Cheung, W. Y., 1982, Calmodulin, *Sci. Am.* **246:**62–70.   [96, 561]
Chevalier, G., Deniau, J. M., Thierry, A. M., and Feger, J., 1981, The nigro-tectal pathway: An elec-
    trophysiological reinvestigation in the rat, *Brain Res.* **213:**253–263.   [212]
Chiba, K., Trevor, A., and Castagnoli, N., Jr., 1984, Metabolism of the neurotoxic tertiary amine, MPTP, by
    brain monoamine oxidase, *Biochem. Biophys. Res. Commun.* **120:**574–578.   [303]
Childs, J. A., and Gale, K., 1983, Neurochemical evidence for a nigrotegmental GABAergic projection, *Brain
    Res.* **258:**109–114.   [212]
Chiodo, L. A., and Benjamin, S. B., 1983, Typical and atypical neuroleptics: Differential effects of chronic
    administration on the activity of A9 and A10 midbrain dopaminergic neurons, *J. Neurosci.* **3:**1607–
    1619.   [296]
Chiu, S. Y., Ritchie, J. M., Rogart, R. B., and Stagg, D., 1979, A quantitative description of membrane
    currents in rabbit myelinated nerve, *J. Physiol. (London)* **292:**149–166.   [59]
Chiu, S., Wong, Y. W., Ferris, J. A., Johnson, R. G., and Mishra, R. K., 1983, Binding studies of L-prolyl-L-
    leucyl-glycinamide (PLG), a novel antiparkinson agent, in normal human brain, *Pharmacol. Res. Com-
    mun.* **15:**41–51.   [402]
Christenson, J. G., Dairman, W. D., and Udenfriend, S., 1972, On the identity of dopa decarboxylase and 5-
    hydroxytryptophan decarboxylase, *Proc. Natl. Acad. Sci. U.S.A.* **69:**343–347.   [274]
Christie, M. J., Bridge, S., James, L. B., and Beart, P. M., 1985a, Excitotoxic lesions suggest an aspartatergic
    projection from rat medial prefrontal cortex to ventral tegmental area, *Brain Res.* **333:**169–172.
    [182]
Christie, M. J., James, L. B., and Beart, P. M., 1985b, An excitant amino acid projection from the medial
    prefrontal cortex to the anterior part of the nucleus accumbens in the rat, *J. Neurochem.* **45:**477–
    482. [182]
Chronwall, B. M., Chase, T. N., and O'Donohue, T. L., 1984, Coexistence of neuropeptide Y and somatostatin
    in rat and human cortical and rat hypothalamic neurons, *Neurosci. Lett.* **52:**213–217.   [157]
Chronwall, B. M., DiMaggio, D. A., Massari, V. J., Pickel, V. M., Ruggiero, D. A., and O'Donohue, T. L.,
    1985a, The anatomy of neuropeptide-Y-containing neurons in rat brain, *Neuroscience* **15:**1159–
    1181.   [388]
Chronwall, B. M., Pisano, J. J., Bishop, J. F., Moody, T. W., and O'Donohue, T. L., 1985b, Biochemical and
    histochemical characterization of ranatensin immunoreactive peptides in rat brain: Lack of coexistence with
    bombesin/GRP, *Brain Res.* **338:**97–113.   [388]
Chrubasik, J., Meynadier, J., and Bond, S., 1984, Somatostatin, a potent analgesic, *Lancet* **2:**1208–
    1209.   [406]
Chubb, I. E., Ranier, E., White, G. H., and Hodgson, A. J., 1983, The enkephalins are amongst the peptides
    hydrolyzed by purified acetylcholinesterase, *Neuroscience* **10:**1369–1377.   [240]
Churchich, J. E., and Moses, U., 1981, γ-Aminobutyrate aminotransferase: The presence of nonequivalent
    binding sites, *J. Biol. Chem.* **256:**1101–1104.   [201]
Cicero, T. J., Sharpe, L. G., Robins, E. P., and Grote, S. S., 1972, On the identity of dopa decarboxylase and
    5-hydroxytryptophan decarboxylase, *Proc. Natl. Acad. Sci. U.S.A.* **69:**343–347.   [283]
Clark, D., Hjorth, S., and Carlsson, A., 1985, Dopamine-receptor agonists: Mechanisms underlying autorecep-
    tor selectivity, *J. Neural Transm.* **62:**1–52.   [301]
Clark, W. C., Pass, P. S., Vankataraman, B., and Hodgetts, R. B., 1978, Dopa decarboxylase from *Drosophila
    melanogaster, Mol. Gen Genet.* **162:**287–297.   [273, 274]
Clarke, P. B. S., Pert, C. B., and Pert, A., 1984, Autoradiographic distribution of nicotine receptors in rat
    brain, *Brain Res.* **323:**390–395.   [254]
Claudio, T., Ballivet, M., Patrick, J., and Heinemann, S., 1983, Nucleotide and deduced amino acid sequences
    of *Torpedo californica* acetylcholine receptor γ-subunit, *Proc. Natl. Acad. Sci. U.S.A.* **80:**1111–
    1115.   [255]
Clement-Cormier, Y. C., Parrish, R. G., Petzold, G. L., Kebabian, J. W., and Greengard, P., 1975, Charac-

terization of a dopamine-sensitive adenylate cyclase in the rat caudate nucleus, *J. Neurochem.* **25:**143–150. [298]

Coen, C. W., and Coombs, M. C., 1983, Effects of manipulating catecholamines on the incidence of the preovulatory surge of luteinizing hormone and ovulation in the rat: Evidence for a necessary involvement of adrenaline in the normal or "midnight" occurrence of the surge, *Neuroscience* **10:**187–206. [315]

Cohen, A. I., McDaniel, M., and Orr, H., 1973, Absolute levels of some free amino acids in normal and biologically fractionated retinas, *Invest. Ophthalmol.* **12:**686–693. [229]

Cohen, M., Roselle, D., Chabner, B., Schmidt, T. J., and Lippman, M., 1978, Evidence for a cytoplasmic melatonin receptor, *Nature (London)* **272:**894–495. [345]

Cohen, S. L., Knight, M., Tamminga, C. A., and Chase, T. N., 1982, Cholecystokinin-octapeptide effects on conditioned avoidance behavior, stereotypy and catalepsy, *Eur. J. Pharmacol.* **83:**213–222. [410]

Cole, A. E., and Shinnick-Gallagher, P., 1984, Muscarinic inhibitory transmission in mammalian sympathetic ganglia mediated by increased potassium conductance, *Nature (London)* **307:**270–271. [255]

Collingridge, G. L., Thompson, P. A., Davies, J., and Melanby, J., 1981, *In vitro* effect of tetanus toxin on GABA release from rat hippocampal slices, *J. Neurochem.* **37:**1039–1041. [223]

Collins, G. G. S., 1972, GABA-T, GAD and half life of GABA in different areas of rat brain, *Biochem. Pharmacol.* **21:**2849–2858. [202]

Collins, G. G., 1979, Effect of chronic bulbectomy on the depth distribution of amino acid transmitter candidates in rat olfactory cortex, *Brain Res.* **171:**552–555. [182]

Collins, G. G., 1982, Some effects of excitatory amino acid receptor antagonists on synaptic transmission in the rat olfactory cortex slice, *Brain Res.* **244:**311–318. [182]

Collins, G. G., and Probett, G. A., 1981, Aspartate and not glutamate is the likely transmitter of the rat lateral olfactory tract fibres, *Brain Res.* **23:**231–234. [182, 186]

Collins, G. G. S., Massey, S. C., and Neal, M. J., 1979, The release of amino acids and [³H]ACh from the rabbit retina *in vivo*, *Br. J. Pharmacol.* **66:**110P. [212, 229]

Colmers, W. F., Lukowiak, K., and Pittman, O. J., 1985, Neuropeptide Y reduces orthodromically evoked population spike in rat hippocampal CA1 by a possibly presynaptic mechanism, *Brain Res.* **346:**404–408. [402]

Colquhoun, D., and Sakmann, B., 1981, Fluctuations in the microsecond time range of the current through single acetylcholine receptor ion channels, *Nature (London)* **294:**464–466. [100]

Comb, M., Seeburg, P. H., Adelman, J., Eiden, L., and Herbert, E., 1982, Primary structure of the human Met- and Leu-enkephalin precursor and its mRNA, *Nature (London)* **295:**663–666. [382]

Conlon, J. M., 1980, The glucagon-like polypeptides—order out of chaos?, *Diabetologia* **18:**85–88. [373]

Conn, P. J., and Sanders-Bush, E., 1984, Selective 5HT-2 antagonists inhibit serotonin stimulated phosphatidylinositol metabolism in cerebral cortex, *Neuropharmacology* **23:**993–996. [336]

Connett, R. J., and Kirshner, N., 1970, Purification and properties of bovine phenylethanolamine-N-methyltransferase, *J. Biol. Chem.* **245:**329–334. [275]

Connor, J. D., 1975, Electrophysiology of the nigro-caudate dopamine pathway, *Pharmacol. Ther. B* **1:**357–370. [506]

Conradi, S., 1969, Ultrastructure of dorsal root boutons on lumbosacral motoneurons of the adult cat, as revealed by dorsal root section, *Acta Physiol. Scand. (Suppl.)* **332:**85–115. [11, 142]

Conrath-Verrier, M., Dietl, M., and Tramu, G., 1984, Cholecystokinin-like immunoreactivity in the dorsal horn of the spinal cord of the rat: Light and electron microscopic study, *Neuroscience* **13:**871–878. [157]

Constantinides, P., 1974, *Functional Electronic Histology*, Elsevier, Amsterdam and New York. [15]

Conte-Devolx, B., Grino, M., Nieoullon, F., *et al.* 1985, Corticoliberin, somatocrinin and amine contents in normal and parkinsonian human hypothalamus, *Neurosci. Lett.* **56:**217–222. [412]

Contestible, A., and Fonnum, F., 1983, Cholinergic and GABAergic forebrain projections to the habenula and nucleus interpeduncularis: Surgical and kainic acid lesions, *Brain Res.* **275:**287–297. [212, 250]

Conti, F., Rustioni, A., and Petrusz, P., 1985, Morphology and laminar distribution of neurons with glutamate-like immunoreactivity in the rat somatosensory cortex, *Soc. Neurosci. Abstr.* **11:**755. [185]

Coombs, J. S., Eccles, J. C., and Fatt, P., 1955a, The electrical properties of the motoneuron membrane, *J. Physiol. (London)* **130:**291–325. [71, 74, 75]

Coombs, J. S., Eccles, J. C., and Fatt, P., 1955b, The specific ionic conductances and the ionic movements across the motoneuronal membrane that produce the inhibitory post-synaptic potential, *J. Physiol. (London)* **130:**326–373. [122, 129]

Coombs, J. S., Eccles, J. C., and Fatt, P., 1955c, Excitatory synaptic action in motoneurons, *J. Physiol. (London)* **130:**374–395. [117, 118]

Coombs, J. S., Curtis, D. R., and Eccles, J. C., 1957a, The interpretation of spike potentials of motoneurons, *J. Physiol. (London)* **139:**198–231. [63]

Coombs, J. S., Curtis, D. R., and Eccles, J. C., 1957b, The generation of impulses in motoneurons, *J. Physiol. (London)* **139:**232–249. [120]

Cooper, J. R., and Mayer, E. M., 1984, Possible mechanisms involved in the release and modulation of release of neuroactive agents, *Neurochem. Int.* **6:**419–423. [170]

Cooper, J. R., Bloom, F. E., and Roth, R. H., 1974, *The Biochemical Basis of Neuropharmacology,* 2nd ed., Oxford University Press, New York, p. 127. [12]

Corda, M. G., and Guidotti, A., 1983, Modulation of GABA receptor binding by Ca^{2+}, *J. Neurochem.* **41:**277–280. [218]

Corey, D. P., 1983, Patch clamp: Current excitement in membrane physiology, *Neurosci. Commen.* **1:**99–110. [99, 101]

Corradetti, R., Lo Conte, G., Moroni, F., Passani, M. B., and Pepeu, G., 1984, Adenosine decreases aspartate and glutamate release from rat hippocampal slices, *Eur. J. Pharmacol.* **104:**19–26. [365]

Costa, E., 1970, Simple neuronal models to estimate turnover rate of noradrenergic transmitters *in vivo, Adv. Biochem. Psychopharmacol.* **2:**169–216. [283]

Costa, M., and Furness, J. B., 1982, Neuronal peptides in the intestines, *Br. Med. Bull.* **38:**247–252. [400]

Costa, M., and Furness, J. B., 1984, Somatostatin is present in a subpopulation of noradrenergic nerve fibres supplying the intestine, *Neuroscience* **13:**911–919. [157]

Cotman, C. W., and Nieto-Sampedro, M., 1984, Cell biology of synaptic plasticity, *Science* **225:**1287–1294. [443–445, 468]

Cotzias, G. C., Papavasilion, P. S., and Gellene, R., 1969, Modification of parkinsonism—chronic treatment with L-dopa, *N. Engl. J. Med.* **280:**337–345. [267]

Cotzias, G. C., Miller, S. T., Tang, L. C., and Papavasiliou, P. S., 1977, Levodopa, fertility and longevity, *Science* **196:**549–551. [471]

Coull, B. M., and Cutler, R. W., 1978, Light-evoked release of endogenous glycine into the perfused vitreous of the intact rat eye, *Invest. Ophthalmol. Vis. Sci.* **17:**682–684. [229]

Couteaux, R., 1958, Morphological and cytochemical observations on the postsynaptic membrane at motor end-plates and ganglionic synapses, *Exp. Cell Res. Suppl.* **5:**294–322. [85]

Cowan, W. M., 1970, Anterograde and retrograde transneuronal degeneration in the central and peripheral nervous system, in: *Contemporary Research Methods in Neuroanatomy* (W. J. H. Nauta and S. O. E. Ebbesson, eds.), pp. 217–251, Springer-Verlag, Berlin, Heidelberg, and New York. [35, 460–464]

Cowan, W. M., Fawcett, J. W., O'Leary, D. D. M., and Stanfield, B. B., 1984, Regressive effects in neurogeneses, *Science* **225:**1258–1265. [417, 419, 435, 452, 455, 456]

Coyle, J. T., and Schwarcz, R., 1976, Lesion of striatal neurons with kainic acid provides a model for Huntington's chorea, *Nature (London)* **263:**244–246. [195]

Cragg, B. B., 1970, What is the signal for chromatolysis?, *Brain Res.* **23:**1–21. [33, 460]

Cragg, B. G., 1975, The density of synapses and neurons in normal, mentally defective and aging human brains, *Brain* **98:**81–90. [9]

Cramer, H., Kohler, J., Oepen, G., Schomburg, G., and Schroter, E., 1981, Huntington's chorea—measurements of somatostatin, substance P and cyclic nucleotides in the cerebrospinal fluid, *J. Neurol.* **225:**183–187. [412]

Cramer, H., Wolf, A., Rissler, K., Weigel, K., and Osterag, C., 1985, Ventricular somatostatin-like immunoreactivity in patients with basal ganglia disease, *J. Neurol.* **232:**219–222. [412]

Crawley, J. N., 1985, Comparative distribution of cholecystokinin and other neuropeptides: Why is this peptide different from all other peptides?, *Ann. N. Y. Acad. Sci.* **448:**1–8. [397]

Crawley, J. N., Hommer, D. W., and Skirboll, L. R., 1984, Behavioral and neurophysiological evidence for a facilatory interaction between co-existing transmitters: Cholecystokinin and dopamine, *Neurochem. Int.* **6:**755–760. [402]

Crawley, J. N., Hommer, D. W., and Skirboll, L. R., 1985, Topographical analysis of nucleus accumbens sites at which cholecystokinin potentiates dopamine-induced hyperlocomotion in the rat, *Brain Res.* **335:**337–341. [157]

Creed, R. S., Denny-Brown, D., Eccles, J. C., Liddell, E. G. T., and Sherrington, C. S., 1932, *Reflex Activity in the Spinal Cord,* Oxford University Press, London. [110]

Creese, I., Sibley, D. R., and Leff, S. E., 1984, Agonist interactions with dopamine receptors: Focus on radioligand-binding studies, *Fed. Proc. Fed. Am. Soc. Exp. Biol.* **43**:2779–2784. [297]

Crepel, F., Delhaye-Bouchaud, N., and Dupont, J. L., 1981, Fate of the multiple innervation of cerebellar Purkinje cells by climbing fibers in immature control, X-radiated and hypothyroid rats, *Dev. Brain Res.* **1**:59–71. [452]

Creutzfeldt, O. D., 1979, Neurophysiological mechanisms and consciousness, Ciba Foundation Series 69, pp. 217–233. [632]

Cross, A. J., Crow, T. J., and Owen, F., 1979, Gamma-aminobutyric acid in the brain in schizophrenia, *Lancet* **1**:560–561. [547]

Cross, A. J., Crow, T. J., and Owen, F., 1981, ^{3}H-Flupenthixol binding in post-mortem brains of schizophrenics: Evidence for a selective increase in D_2 receptors, *Psychopharmacology* **74**:122–124. [546]

Crossman, A. R., Walker, R. J., and Woodruff, G. N., 1973, Picrotoxin antagonism of gamma-aminobutyric acid inhibitory responses and synaptic inhibition in the rat substantia nigra, *Br. J. Pharmacol.* **49**:696–698. [212, 215]

Crow, T. J., 1980, Molecular pathology of schizophrenia: More than one disease process?, *Br. Med. J.* **280**:66–68. [546]

Crow, T. J., and Grove-White, I. G., 1973, An analysis of the learning deficits following hyoscine administration, *Br. J. Pharmacol.* **49**:322–327. [261]

Crow, T. J., Owen, F., Cross, A. J., *et al.* 1980, The dopamine receptor as the site of the primary disturbance in the Type I syndrome of schizophrenia, in: *Enzymes and Neurotransmitters in Mental Disease* (E. Usdin, T. L. Sourkes, and M. B. H. Youdin, eds.), pp. 559–572, John Wiley, Chichester. [547]

Crowley, W. R., and Terry, L. C., 1981, Effect of an epinephrine synthesis inhibitor, SKF 64139, on the secretion of luteinizing-hormone in ovariectomized female rats, *Brain Res.* **204**:231–235. [315]

Cuenod, M., Beaudet, A., Canzek, V., Streit, P., and Reubi, J. C., 1981, Glutamatergic pathways in the pigeon and the rat brain, *Adv. Biochem. Psychopharmacol.* **27**:57–68. [182]

Cull, R. E., 1975, Role of axonal transport in maintaining central synaptic connections, *Exp. Brain Res.* **24**:97–101. [461]

Cunningham, V. T., and Cremer, J. E., 1985, Current assumptions behind the use of PET scanning for measuring glucose utilization in brain, *TINS* **8**:96–99. [69]

Curtis, D. R., 1965, The actions of amino acids upon mammalian neurones, in: *Studies in Physiology* (D. R. Curtis and A. K. MacIntyre, eds.), pp. 34–42, Springer-Verlag, New York. [182]

Curtis, D. R., 1978, Pre- and postsynaptic action of GABA in the mammalian spinal cord, in: *Advances in Pharmacology and Therapeutics,* Vol. 2 (P. Simon, ed.), pp. 281–295, Pergamon Press, Oxford. [144]

Curtis, D. R., and Eccles, J. C., 1959, The time courses of excitatory and inhibitory synaptic actions, *J. Physiol. (London)* **145**:529–546. [122, 555]

Curtis, D. R., and Johnston, G. A. R., 1974, Amino acid transmitters in the mammalian central nervous system, *Ergeb. Physiol.* **69**:97–188. [116, 141]

Curtis, D. R., and Lodge, D., 1982, The depolarization of feline ventral horn group Ia spinal afferent terminals by GABA, *Exp. Brain Res.* **46**:215–233. [144]

Curtis, D. R., and Watkins, J. C., 1960, The excitation and depression of spinal neurones by structurally-related amino acids, *J. Neurochem.* **6**:117–141. [175, 231]

Curtis, D. R., and Watkins, J. C., 1963, Acidic amino acids with strong excitatory actions on mammalian neurones, *J. Physiol. (London)* **166**:1–14. [175]

Curtis, D. R., Hosli, L., and Johnston, G. A. R., 1967, Inhibition of spinal neurons by glycine, *Nature (London)* **215**:1502–1503. [230]

Curtis, D. R., Duggan, A. W., Felix, D., and Johnston, G. A. R., 1970, Bicuculline and GABA receptors, *Nature (London)* **228**:676–677. [224]

Curtis, D. R., Felix, D., Game, C. J. A., and McCullock, R. M., 1973, Tetanus toxin and the synaptic release of GABA, *Brain Res.* **51**:362–385. [223]

Curtius, H.-C., Niederwieser, A., Viscontini, M., *et al.* 1981, Serotonin and dopamine synthesis in phenylketonuria, *Adv. Exp. Med. Biol.* **133**:277–291. [324]

Curzon, G., 1981, Influence of plasma tryptophan on brain 5HT synthesis and serotonergic activity, *Adv. Exp. Med. Biol.* **133**:207–219. [323]

Cutler, N. R., James, M. D., Haxby, J. V., *et al.* 1985, Evaluation of an analogue of somatostatin (L363,586) in Alzheimer's disease, *N. Engl. J. Med.* **312**:725. [409]

Dahlstrom, A., and Fuxe, K., 1964a, Evidence for the existence of monoamine-containing neurons in the central nervous system, *Acta Physiol. Scand.* **62** *(Suppl. 232)*. [268, 285, 320, 329]

Dahlstrom, A., and Fuxe, K., 1964b, A method for the demonstration of monoamine-containing fibers in the central nervous system, *Acta Physiol. Scand.* **60**:293–295. [268, 285, 320, 329, 333]

Dale, H. H., 1914, The action of certain esters and ethers of choline, and their relation to muscarine, *J. Pharmacol.* **6**:147–190. [80]

Dale, H. H., 1935, Pharmacology and nerve endings, *Proc. R. Soc. Med.* **28**:319–332. [82, 151]

Dale, H. H., 1938, Acetylcholine as a chemical transmitter substance of the effects of nerve impulses: The William Henry Welch Lectures, 1937, *J. Mt. Sinai Hosp.* **4**:401–429. [80, 82]

Dale, H. H., and Dudley, H. W., 1929, The presence of histamine and acetylcholine in the spleen of the ox and the horse, *J. Physiol. (London)* **68**:97–123. [81]

Dale, H. H., and Laidlaw, P. P., 1910, The physiological action of β-imidazlylethylamine, *J. Physiol. (London)* **41**:318–344. [349]

Dale, H. H., Feldberg, W., and Vogt, M., 1936, Release of acetylcholine at voluntary motor nerve endings, *J. Physiol. (London)* **86**:353–380. [81]

Dalsgaard, C. T., Vincent, S. R., Hökfelt, T., *et al.* 1982, Coexistence of cholecystokinin- and substance P-like peptides in neurons of the dorsal root ganglia of the rat, *Neurosci. Lett.* **33**:159–163. [157]

Daly, J. W., 1982, Adenosine receptors: Targets for future drugs, *J. Med. Chem.* **25**:197–207. [365]

Daly, J. W., Bruns, R. F., and Snyder, S. H., 1981, Adenosine receptors in the central nervous system: Relationship to the central action of methylxanthines, *Life Sci.* **28**:2083–2097. [360]

Dam, M., Gram, L., Philbert, A., *et al.* 1983, Progabide: A control trial in partial epilepsy, *Epilepsia* **24**:127–134. [225]

Dam Trung Tuong, M., Garbarg, M., and Schwartz, J. C., 1980, Pharmacological specificity of brain histamine H_2-receptors differs in intact cells and cell-free preparations, *Nature (London)* **287**:548–551. [359]

Damasio, A. R., and Geschwind, N., 1984, The neural basis of language, *Annu. Rev. Neurosci.* **7**:127–147. [615, 616, 618, 623]

Damsma, G., Westerink, B. H. C., and Horn, A. S., 1985, A simple, sensitive, and economic assay for choline and acetylcholine using HPLC, an enzyme reactor, and an electrochemical detector, *J. Neurochem.* **45**:1649–1652. [236]

Daniloff, J. K., Wells, J., and Ellis, J., 1984, Cross-species septal transplants: Recovery of choline acetyltransferase activity, *Brain Res.* **324**:151–154. [459]

Dantry-Yarsat, A., and Lodish, H. F., 1983, The Golgi complex and the sorting of membranes secreting proteins, *TINS* **6**:484–490. [23]

Dasgupta, P., and Narayanaswami, A., 1981, Glyoxylate aminotransferase activity in vertebrate retina, *J. Neurochem.* **37**:1603–1606. [227]

Daum, P. R., Downes, C. P., and Young, J. M., 1983, Histamine stimulation of inositol phospholipid breakdown mirrors H_1-receptor density in brain, *Eur. J. Pharmacol.* **87**:497–498. [357]

Daum, P. R., Downes, C. P., and Young, J. M., 1984, Histamine stimulation of inositol 1-phosphate accumulation in lithium-treated slices from regions of guinea pig brain, *J. Neurochem.* **43**:25–32. [357]

David, M., Moisand, C., Mennier, J. C., Morgat, J. L., Gacel, G., and Roques, B. P., 1982, [^{3}H]Tyr-D-Ser-Gly-Phe-Leu-Thr: A specific probe for the delta-opiate receptor subtype in brain membranes, *Eur. J. Pharmacol.* **78**:385–387. [404]

Davidoff, R. A., Graham, L. T., Jr., Shank, R. P., Werman, R., and Aprison, M. H., 1967, Changes in amino acid concentrations associated with loss of spinal interneurons, *J. Neurochem.* **14**:1025–1031. [182]

Davies, J., and Watkins, J. C., 1979, Selective antagonism of amino-acid-induced and synaptic excitation in the cat spinal cord, *J. Physiol. (London)* **297**:621–635. [182]

Davies, J., and Watkins, J. C., 1982, Actions of D and L forms of 2-amino-5-phosphonovalerate and 2-amino-4-phosphonobutyrate in the cat spinal cord, *Brain Res.* **235**:378–386. [182]

Davies, J., and Watkins, J. C., 1985, Depressant actions of γ-D-glutamylaminomethyl sulfonate (GAMS) on amino acid-induced and synaptic excitation in the cat spinal cord, *Brain Res.* **327**:113–120. [188]

Davies, J., Francis, A. A., Jones, A. W., and Watkins, J. C., 1980, 2-Amino-5-phosphonovalerate (APV), a potent and selective antagonist of amino acid-induced and synaptic excitation, *Neurosci. Lett.* **21**:77–81. [182]

Davies, J. E., and Roberts, M. H. T., 1981, 5-Hydroxytryptamine reduces substance P responses on dorsal horn interneurones: A possible interaction of neurotransmitters, *Brain Res.* **217**:399–404. [402]

Davies, L. P., Chen Chow, S., Skerritt, J. H., Brown, D. J., and Johnston, G. A. R., 1984, Pyrazolo-[3,4-d]pyrimidines as adenosine antagonists, *Life Sci.* **34:**2117–2128. [366]

Davies, P., 1979, Neurotransmitter-related enzymes in senile dementia of the Alzheimer type, *Brain Res.* **171:**319–327. [261]

Davies, P., and Maloney, A. J. R., 1976, Selective loss of central cholinergic neurons in Alzheimer's disease, *Lancet* **2:**1403. [261]

Davies, P., and Verth, A. H., 1977, Regional distribution of muscarinic acetylcholine receptor in normal and Alzheimer's type dementia brains, *Brain Res.* **138:**385–392. [253, 261]

Davies, P., Katzman, R., and Terry, R. D., 1980, Reduced somatostatin-like immunoreactivity in cerebral cortex from cases of Alzheimer disease and Alzheimer senile dementia, *Nature (London)* **288:**279–280. [412]

Davis, G. C., Williams, A. C., Markey, S. P., *et al.* 1979, Chronic parkinsonism secondary to intravenous injection of meperidine analogues, *Psychiatr. Res.* **1:**249–254. [302]

Davis, J. N., Carlson, A., MacMillan, V., and Siesjo, B. K., 1974, Brain tryptophan hydroxylation: Dependence on arterial oxygen tension, *Science* **182:**72–73. [323]

Davis, K. L., Berger, P. A., Hollister, L. E., and Barchas, J. D., 1978, Cholinergic involvement in mental disorders, *Life Sci.* **22:**1865–1872. [547, 549]

Davis, K. L., Mohs, R. C., and Tinklenberg, J. R., 1979, Enhancement of memory by physostigmine, *N. Engl. J. Med.* **301:**946. [261]

Davis, K. L., Davidson, M., Mohs, R. C., *et al.* 1985, Plasma homovanillic acid concentration and the severity of schizophrenic illness, *Science* **227:**1601–1602. [546]

Davis, P. G., and Barfield, R. J., 1979, Activation of masculine sexual behavior by intracranial estradiol benzoate implants in male rats, *Neuroendocrinology* **28:**217–227. [521]

Davis, W. E., 1977, GABAergic innervation of the mammalian cochlear nucleus, in: *Inner Ear Biology* (M. Portmann and J. M. Aran, eds.), pp. 155–164, Inserm, Paris. [212]

Davson, H., 1976, The blood–brain barrier, *J. Physiol. (London)* **255:**1–28. [37, 38]

Dawbarn, D., de Quidt, M. E., and Emson, P. C., 1985, Survival of basal ganglia neuropeptide Y–somatostatin neurones in Huntington's disease, *Brain Res.* **340:**251–260. [412]

Dawson, R. M., and Jarrott, B., 1980, Regional distribution of the muscarinic cholinoceptor and acetylcholinesterase in guinea pig brain, *Neurochem. Res.* **5:**809–815. [253]

Dawson, T. D., Gehlert, D. R., Snowhill, E. W., and Wamsley, J. K., 1985, Quantitative autoradiographic evidence for axonal transport of imipramine receptors in the central nervous system of the rat, *Neurosci. Lett.* **55:**261–266. [342]

Day, T. A., Jervois, P. M., Menadue, M. F., and Willoughby, J. O., 1982, Catecholamine mechanisms in medio-basal hypothalamus influence prolactin but not growth hormone secretion, *Brain Res.* **253:**213–219. [315]

De Belleroche, D. S., and Gardiner, I. M., 1982, Cholinergic action in the nucleus accumbens: Modulation of dopamine and acetylcholine release, *Br. J. Pharmacol.* **75:**359–365. [241]

De Caro, G., Massi, M., and Micossi, L. G., 1978, Antidipsogenic effect of intracranial injections of substance P in rats, *J. Physiol. (London)* **279:**133–140. [406]

Deecke, L., and Kornhuber, H. H., 1977, Cerebral potentials and the initiation of voluntary movement, *Prog. Clin. Neurophysiol.* **1:**132–150. [591]

Deecke, L., and Kornhuber, H. H., 1978, An electrical sign of participation of the mesial "supplementary" motor cortex in human voluntary finger movement, *Brain Res.* **159:**473–476. [481, 482, 591]

De Feudis, F. V., 1978, Can binding of GABA, glycine and beta-alanine to synaptic receptors be determined in presence of physiological concentration of Na^+?, *Experientia* **34:**1314–1315. [225]

Deftos, L. J., and Catherwood, B. D., 1980, Dissociation between ACTH and β-endorphin immunoreactivity in cells of the rat pituitary gland, *Life Sci.* **27:**223–228. [384]

Delay, J., and Deniker, P., 1952, Trente-huit cas de psychoses traitées par la cure prolongée et continuée de 4560 R.P. Les Congrés des Alet Neurol de Langue Fr., in: *Compte Rendu du Congres,* Masson, Paris. [537]

Del Castillo, J., and Katz, B., 1954, Quantal components of the end-plate potential, *J. Physiol. (London)* **124:**560–573. [93]

Del Castillo, J., and Katz, B., 1956, Biophysical aspects of neuromuscular transmission, *Prog. Biophys. Mol. Biol.* **6:**121–170. [236]

Del Fiacco, M., Paxinos, G., and Cuello, A. C., 1982, Neostriatal enkephalin-immunoreactive neurons project to the globus pallidus, *Brain Res.* **231**:1–17. [507]

Delgado, J. M. P., 1976, New orientations in brain stimulation in man, in: *Brain-Stimulation Reward* (A. Wanquier and E. T. Rolls, eds.), pp. 481–503, Elsevier, Amsterdam. [530]

DeLong, M. R., and Georgopoulos, A. P., 1983, Motor functions of the basal ganglia, *Am. Physiol. Soc. Handb.* **21**:1017–1061. [591]

DeLong, M. R., Alexander, G. E., Georgopoulos, A. P., Crutcher, M. D., Mitchell, S. J., and Richardson, R. T., 1984, Role of basal ganglia in limb movements, *Hum. Neurobiol.* **2**:235–244. [511]

De Lorenzo, R. J., 1981, The calmodulin hypothesis of neurotransmission, *Cell Calcium* **2**:365–385. [96, 97, 101, 554]

De Lorenzo, R. J., 1982, Calmodulin modulation of the calcium signal in synaptic transmission, in: *Neurotransmitter Interaction and Compartmentation* (H. F. Bradford, ed.), pp. 101–120, Plenum Press, New York. [169, 170]

Demeulemeester, H., Vandesande, F., and Orban, G. A., 1985, Immunocytochemical localization of somatostatin and cholecystokinin in the cat visual cortex, *Brain Res.* **332**:361–364. [398]

De Montes, G., Beaumont, K., Javoy-Agid, F., *et al.* 1982, Glycine receptors on the human substantia nigra as defined by [³H]strychnine binding, *J. Neurochem.* **38**:718–724. [229]

Deniau, J. M., Chevalier, G., and Feger, J., 1978a, Electrophysiological study of the nigrotectal pathway in the rat, *Neurosci. Lett.* **10**:215–220. [212]

Deniau, J. M., Hammond, C., Chevalier, G., and Feyer, J., 1978b, Evidence for branched subthalamic nucleus projections to substantia nigra, entopeduncular nucleus and globus pallidus, *Neurosci. Lett.* **9**:117–121. [508, 212]

Deniau, J. M., Hammond, C., Riszk, A., and Feger, J., 1978c, Electrophysiological properties of identified output neurons of the rat substantia nigra (pars compacta and pars reticulata): Evidence for the existence of branched neurons, *Exp. Brain Res.* **32**:409–422. [506]

Denis-Donini, S., Glowinski, J., and Prochiantz, A., 1984, Glial heterogeneity may define the three-dimensional shape of mouse mesencephalic dopaminergic neurons, *Nature (London)* **307**:641–664. [37]

Denny, R. M., Fritz, R. R., Patel, N. T., and Abell, C. W., 1982, Human liver MAO-A and MAO-B separated by immunoaffinity chromatography with MAO-B-specific monoclonal antibody, *Science* **215**:1400–1403. [277]

De Quidt, M. E., and Emson, P. E., 1983, Neurotensin facilitates dopamine release *in vitro* from rat striatal slices, *Brain Res.* **274**:376–380. [506]

De Robertis, E., and Bennett, H. S., 1955, Some features of the submicroscopic morphology of synapses in frog and earthworm, *J. Biophys. Biochem. Cytol.* **1**:47–58. [236]

De Robertis, E., de Iraldi, A. P., De Lores Arnaiz, G. R., and Salganicoff, L., 1962, Cholinergic and non-cholinergic nerve endings in rat brain. I. Isolation and subcellular distribution of acetylcholine and acetylcholinesterase, *J. Neurochem.* **9**:23–35. [236]

De Robertis, E., De Lores Arnaiz, G. R., Salganicoff, L., Peregrino de Iraldi, A., and Zieher, L. M., 1963, Isolation of synaptic vesicles and structural organization of the acetylcholine system within brain nerve endings, *J. Neurochem.* **10**:225–235. [236]

Descarries, L., and Beaudet, A., 1978, The serotonin innervation of adult rat hypothalamus, in: *Cell Biology of Hypothalamic Neurosecretion* (J. D. Vincent and C. Korolon, eds.), *Colloq. Int. C.N.R.S.* **280**:135–153. [333]

Descarries, L., Watkins, K. C., and Lapierre, Y., 1977, Noradrenergic axon terminals in the cerebral cortex of the rat. III. Topometric ultrastructural analysis, *Brain Res.* **133**:197–222. [311]

Deschenes, R. J., Haun, R. S., Sunkel, D., Roos, B. A., and Dixon, J. E., 1985, Modulation of cholecystokinin gene expression, *Ann. N.Y. Acad. Sci.* **448**:53–60. [382]

Desclin, J. C., 1974, Histological evidence supporting the inferior olive as the major source of cerebellar climbing fibers in the rat, *Brain Res.* **77**:365–384. [496]

Deshmukh, P. P., Yamamura, H. I., Woods, L., and Nelson, D. L., 1983, Computer-assisted autoradiographic localization of subtypes of serotonin receptors in rat brain, *Brain Res.* **288**:338–343. [335]

Desmedt, J. E., and Robertson, D., 1977, Differential enhancement of early and late components of the cerebral somatosensory evoked potentials during forced–paced cognitive tasks in man, *J. Physiol. (London)* **271**:761–782. [628]

De Souza, E. B., Perrin, M. H., Insel, T. R., Rivier, J., Vale, W. W., and Kuhar, M. J., 1984, Corticotropin-releasing factor receptors in rat forebrain: Autoradiographic identification, *Science* **224**:1449–1451. [402]

De Valois, R. L., 1973, Central mechanisms of color vision, in: *Handbook of Sensory Physiology,* Vol. VII/3A, (R. Jung, ed.), pp. 209–253, Springer-Verlag, Berlin, Heidelberg, and New York. [610]

Devanandan, M. S., Eccles, R. M., and Westerman, R. A., 1965, Single motor units of mammalian muscle, *J. Physiol. (London)* **178**:359–367. [83, 84]

Devaney, K. O., and Johnson, H. A., 1980, Neuron loss in the aging visual cortex of man, *J. Gerontol.* **35**:836–841. [470]

Devillers-Thiery, A., Giraudat, J., Bentaboulet, M., and Changeux, J.-P., 1983, Complete mRNA coding sequence of the acetylcholine binding alpha-subunit of *Torpedo marmorata* acetylcholine receptor: A model for the transmembrane organization of the polypeptide chain, *Proc. Natl. Acad. Sci. U.S.A.* **80**:2067–2071. [255]

De Vries, G. J., Buijs, R. M., and Sluiter, A. A., 1984, Gonadal hormone actions on the morphology of the vasopressinergic innervation of the adult rat brain, *Brain Res.* **298**:121–145. [395]

Dhondt, J.-L., 1984, Tetrahydrobiopterin deficiencies: Preliminary analysis from an international survey, *J. Pediatr.* **104**:501–508. [273]

Di Chiara, G., Porceddu, M. L., Morelli, M., Mulas, M. L., and Gessa, G. L., 1979, Evidence for a GABAergic projection from the substantia nigra to the ventromedial thalamus and to the superior colliculus of the rat, *Brain Res.* **176**:273–284. [212]

DiFiglia, M., and Aronin, N., 1982, Ultrastructural features of immunoreactive somatostatin neurons in the rat caudate nucleus, *J. Neurosci.* **2**:1267–1274. [506]

Diggory, G. L., Ceasar, P. M., Hazelby, D., and Taylor, K. T., 1979, Endogenous 5-hydroxytryptophol in mouse brain, *J. Neurochem.* **32**:1323–1325. [325]

Di Lauro, A., Schmid, R. W., and Meek, J. L., 1981, Is aspartic acid the neurotransmitter of the perforant pathway?, *Brain Res.* **207**:476–480. [212]

DiMaggio, D. A., Chronwall, B. M., Buchanan, K., and O'Donohue, T. L., 1985, Pancreatic polypeptide immunoreactivity in rat brain is actually neuropeptide Y, *Neuroscience* **15**:1149–1157. [386]

Dimond, S. J., and Beaumont, J. G. (eds.), 1974, *Hemisphere Function in the Human Brain,* John Wiley, New York. [623]

Dimsdale, J. C., and Moss, J., 1980, Plasma catecholamines in stress and exercise, *J. Am. Med. Assoc.* **243**:340–342. [533]

Dingledine, R., 1983, *N*-Methylaspartate activates voltage-dependent calcium conductance in rat hippocampal pyramidal cells, *J. Physiol. (London)* **343**:385–405. [561]

Divac, I., 1975, Magnocellular nuclei of the basal forebrain project to neocortex, brain stem, and olfactory bulb: Review of some functional correlates, *Brain Res.* **93**:385–398. [247]

Divac, I., Fonnum, F., and Storm-Mathiesen, J., 1977, High affinity uptake of glutamate in terminals of cortico striatal axons, *Nature (London)* **266**:377–378. [182, 183]

Dixon, W. E., 1906, Vagus inhibition, *Br. Med. J.* **2**:1807. [80]

Dobrilla, G., 1983, Placebo in the evaluation of antiulcer drugs, *Int. J. Tissue React.* **5**:329–337. [359]

Dockray, G. J., 1985, Characterization of FMRF amide-like immunoreactivity in rat spinal cord by region-specific antibodies in radiimminoassay and HPLC, *J. Neurochem.* **45**:152–158. [373]

Dockray, G. J., and Williams, R. G., 1983, FMRF amide-like immunoreactivity in rat brain: Development of a radioimmunoassay and its application in studies of distribution and chromatographic properties, *Brain Res.* **266**:295–303. [388]

Dodge, F. A., and Rahaminoff, R., 1967, Cooperative action of calcium ions in transmitter release at the neuromuscular junction, *J. Physiol. (London)* **193**:419–432. [96]

Dodge, F. A., Jr., Miledi, R., and Rahamimoff, R., 1969, Strontium and quantal release of transmitter at the neuromuscular junction, *J. Physiol. (London)* **200**:267–283. [106]

Doi, T., and Jurna, I., 1982, Intrathecal substance P depresses spinal motor and sensory responses to stimulation of nociceptive afferents—antagonism by naloxone, *Naunyn-Schmiedeberg's Arch. Pharmacol.* **319**:154–160. [406]

Dolphin, A. C., and Prestwich, S. A., 1985, Pertussis toxin reverses adenosine inhibition of neuronal glutamate release, *Nature (London)* **316**:148–150. [365]

Dolphin, A. C., Errington, M. L., and Bliss, T. V. P., 1982, Long-term potentiation of the perforant path *in vivo* associated with increased glutamate release, *Nature (London)* **297**:496–498. [564, 565]

Dolphin, A. C., Detre, J. A., Schlichter, D. J., *et al.*, 1983, Cyclic nucleotide-dependent protein kinases and some major substrates in the rat cerebellum after neonatal X-irradiation, *J. Neurochem.* **40:**577–581. [168]

Domesick, V. B., 1976, Projections of the nucleus of the diagonal band of Broca in the rat, *Anat. Rec.* **184:**391–392. [505]

Dooley, D. J., Bittiger, H., Hauser, K. L., Bischoff, S. F., and Waldmeier, P. C., 1983, Alteration of central alpha 2- and beta-adrenergic receptors in the rat after DSP-4, a selective noradrenergic neurotoxin, *Neuroscience* **9:**889–898. [311]

Dores, R. M., Akil, H., and Watson, S. J., 1984, Strategies for studying opioid peptide regulation at the gene, message and protein levels, *Peptides* **5**(Suppl. 1):9–17. [382, 388]

Douglas, R. M., 1977, Long lasting synaptic potentiation in the rat dentate gyrus following brief high frequency stimulation, *Brain Res.* **126:**361–365. [558]

Douglas, W. W., 1980, Histamine and 5-hydroxytryptamine (serotonin) and their antagonists, in: *Goodman and Gilman's The Pharmacological Basis of Therapeutics,* 6th ed., pp. 608–646, Macmillan, New York. [343, 358]

Dow, R. S., and Moruzzi, G., 1958, *The Physiology and Pathology of the Cerebellum,* University of Minnesota Press, Minneapolis, 675 pp. [453]

Dowling, J. E., and Boycott, B. B., 1966, Neural connections of the retina: Fine structure of the inner plexiform layer, *Cold Spring Harbor Symp. Quant. Biol.* **30:**393–402. [134]

Downing, A. C., Gerard, R. W., and Hill, A. V., 1926, The heat production of nerve, *Proc. R. Soc. London Ser. B* **100:**223–251. [69]

Drachman, D. A., 1977, Memory and cognitive function in man: Does the cholinergic system have a specific role?, *Neurology* **27:**783–790. [261]

Drachman, D. B., 1978, Myasthenia gravis, *N. Engl. J. Med.* **298:**136–142 and 186–298. [260]

Drachman, D. A., and Leavitt, J., 1974, Human memory and the cholinergic system: A relationship to aging?, *Arch. Neurol.* **30:**113–121. [261]

Dratman, M. B., Futaesaku, Y., Crutchfield, F. L., *et al.* 1982, Iodine-125-labeled triiodothyronine in rat brain: Evidence for localization in discrete neural systems, *Science* **215:**309–312. [526]

Dray, A., and Oakley, N. R., 1978, Projections from nucleus accumbens to globus pallidus and substantia nigra in the rat, *Experientia* **34:**68–70. [212]

Drayna, D., and White, R., 1985, The genetic linkage map of the human X chromosome, *Science* **230:**753–758. [45]

Droz, B., 1975, Synaptic machinery and axoplasmic transport: Maintenance of neuronal connectivity, in: *The Nervous System,* Vol. 1, *The Basic Neurosciences* (D. B. Tower and R. O. Brady, eds.), pp. 111–128, Raven Press, New York. [32]

Druce, D., Peterson, D., De Belleroche, J., and Bradford, H. F., 1982, Differential amino acid neurotransmitter release in rat neostriatum following lesioning of the cortico-striatal pathway, *Brain Res.* **247:**303–307. [182]

Drummond, A. H., 1985, Bidirectional control of cytosolic free calcium by thyrotropin-releasing hormone in pituitary cells, *Nature (London)* **315:**752–755. [402]

Dubois, B., Mayo, W., Agid, Y., Le Moal, M., and Simon, H., 1985, Profound disturbance of spontaneous and learned behaviors following lesions of the nucleus basalis magnocellularis in the rat, *Brain Res.* **338:**249–258. [261]

Duckles, S. P., and Said, S. I., 1982, Vasoactive intestinal peptide as a neurotransmitter in the cerebral circulation, *Eur. J. Pharmacol.* **78:**371–374. [397]

Dudley, M. W., Butcher, L. L., Kammerer, R. C., and Cho, A. K., 1981, The actions of xylamine on central adrenergic neurons, *J. Pharmacol. Exp. Ther.* **217:**834–840. [312]

Duffy, C. J., Teyler, T. J., and Shashona, V. E., 1981, Long-term potentiation in the hippocampal slice: Evidence for stimulated secretion of newly synthesized proteins, *Science* **212:**1148–1151. [563]

Duggan, A. W., and Johnston, G. A. R., 1970, Glutamate and related amino acids in cat spinal roots, dorsal root ganglia and peripheral nerves, *J. Neurochem.* **17:**1205–1208. [182]

Duggan, A. W., Lodge, D., and Biscoe, T. J., 1973, The inhibition of hypoglossal motoneurons by impulses in the glossopharyngeal nerve of the rat, *Exp. Brain Res.* **17:**261–270. [230]

Duggan, D. E., Hooke, K. F., and Maycock, A. L., 1984, Inhibition of histamine synthesis *in vitro* and *in vivo* by *S*-α-fluoromethylhistidine, *Biochem. Pharmacol.* **33:**4003–4009. [355]

Dundee, J. W., and Pandit, D. K., 1972, Anterograde amnesic effects of pethidine, hyoscine and diazepam in adults, *Br. J. Pharmacol.* **44:**140–144. [261]

Dunwiddie, T. V., and Fredholm, B. B., 1984, Adenosine receptors mediating inhibitory electrophysiological responses in rat hippocampus are different from receptors mediating cyclic AMP accumulation, *Naunyn-Schmiedeberg's Arch. Pharmacol.* **326:**294–301. [365]

Dunwiddie, T., and Lynch, G., 1978, Long-term potentiation and depression of synaptic responses in the rat hippocampus: Localization and frequency dependency, *J. Physiol. (London)* **276:**353–367. [560]

Dunwiddie, T. V., and Lynch, G., 1979, The relationship between extracellular calcium concentrations and the induction of hippocampal long-term potentiation, *Brain Res.* **169:**103–110. [560]

Dunwiddie, T., Madison, D., and Lynch, G., 1978, Synaptic transmission is required for initiation of long-term potentiation, *Brain Res.* **150:**413–417. [559]

Dupont, E., Christensen, S. E., Hansen, A., de Fine Olivarius, B., and Ørskov, H., 1982, Low cerebrospinal fluid somatostatin in Parkinson disease: An irreversible abnormality, *Neurology* **32:**312–314. [412]

Dustin, P., 1980, Microtubules, *Sci. Am.* **243:**66–76. [25]

Duvoisin, R. C., Chokroverty, S., Lepore, F., and Nicklas, W., 1983, Glutamate dehydrogenase deficiency in patients with olivopontocerebellar atrophy, *Neurology* **33:**1322–1326. [179]

Dux, E., Joo, F., Geese, A., *et al.* (1982, Histamine-stimulated prostaglandin synthesis in rat brain microvessels, *Agents Actions* **12:**146–148. [357]

Dyck, L. E., 1984, Tryptamine transport in rat brain slices: A comparison with 5-hydroxytryptamine, *Neurochem. Res.* **9:**617–628. [346]

Eccles, J. C., 1953, *The Neurophysiological Basis of Mind,* Clarendon Press, Oxford. [61, 554]

Eccles, J. C., 1957, *The Physiology of Nerve Cells,* Johns Hopkins Press, Baltimore. [29, 70, 71, 74, 75, 77, 112]

Eccles, J. C., 1960, The properties of dendrites, in: *Structure and Function of the Cerebral Cortex* (D. B. Tower and J. P. Schade, eds.), pp. 192–202, Elsevier, Amsterdam. [72]

Eccles, J. C., 1961, Membrane time constants of cat motoneurons and time courses of synaptic action, *Exp. Neurol.* **4:**1–22. [118]

Eccles, J. C., 1964a, Excitatory responses of spinal neurons, *Prog. Brain Res.* **12:**1–34. [70, 71]

Eccles, J. C., 1964b, *The Physiology of Synapses,* Springer-Verlag, Heidelberg. [8, 106, 112, 119, 121, 126, 128, 131, 144]

Eccles, J. C., 1964c, Presynaptic inhibition in the spinal cord, *Prog. Brain Res.* **12:**65–91. [142–145]

Eccles, J. C., 1969, *The Inhibitory Pathways of the Central Nervous System (Sherrington Lecture),* Liverpool University Press, Liverpool. [112, 141]

Eccles, J. C., 1970a, Neurogenesis and morphogenesis in the cerebellar cortex, *Proc. Natl. Acad. Sci. U.S.A.* **66:**294–301. [427, 429, 432]

Eccles, J. C., 1970b, *Facing Reality: Philosophical Adventures by a Brain Scientist,* Springer-Verlag, New York. [554]

Eccles, J. C., 1973, The cerebellum as a computer: Patterns in space and time, *J. Physiol. (London)* **229:**1–32. [491, 494, 495, 499–502]

Eccles, J. C., 1976, The plasticity of the mammalian central nervous system with special reference to new growths in response to lesions, *Naturwissenschaften* **63:**8–15. [37]

Eccles, J. C., 1977, An instruction–selection theory of learning in the cerebellar cortex, *Brain Res.* **127:**327–352. [494]

Eccles, J. C., 1980, *The Human Psyche,* Springer-Verlag, Berlin, Heidelberg, and New York. [69, 618, 630]

Eccles, J. C., 1981a, The modular operation of the cerebral neocortex considered as the material basis of mental events, *Neuroscience* **6:**1839–1856. [570, 571, 574, 578, 589]

Eccles, J. C., 1981b, Language, thought, and brain, *Epistemologia (Special Issue)* **4:**97–126. [615, 618]

Eccles, J. C., 1982a, The initiation of voluntary movements by the supplementary motor area, *Arch. Psychiatr. Nervenkr.* **231:**423–441. [592]

Eccles, J. C., 1982b, How the self acts on the brain, *Psychoneuroendocrinology* **7:**271–283. [592, 631, 632]

Eccles, J. C., 1982c, Cerebral mechanisms in pain perception, *Panminerva Med.* **24:**1–13. [605]

Eccles, J. C., 1982d, The synapse: From electrical to chemical transmission, *Annu. Rev. Neurosci.* **5:**325–339. [82, 110]

Eccles, J. C., 1983, Calcium in long-term potentiation as a model for memory, *Neuroscience* **10:**1071–1081. [560, 561, 571, 589, 455]

Eccles, J. C., 1984a, The form and function of synapses, *Exp. Brain Res. Suppl.* **9:**301–304. [565]

Eccles, J. C., 1984b, Physiological and pharmacological investigations on pain control, *Schweiz. Mschr. Zahnmed.* **94:**1004–1013. [603]

Eccles, J. C., 1986a, Chemical transmission and Dale's principle, in: *Neuropeptides in the Central Nervous System* (T. Hökfelt, K. Fuxe, and A. Björklund, eds.), Elsevier, Amsterdam. [151]

Eccles, J. C., 1986b, Do mental events cause neural events analogously to the probability fields of quantum mechanics?, *Proc. R. Soc. London Ser. B* **227**:411–428. [624, 632]

Eccles, J. C., 1986c, Chromatolysis of neurons after axon section, in: *Recent Achievements in Restorative Neurology: Restorative Procedures in Progressive Neuromuscular Diseases* (M. R. Dimitrijevic, B. Kakulas, and G. Vrbova, eds.), pp. 318–331, S. Karger, Basel. [460]

Eccles, J. C., 1986d, Learning in the motor system, in: *Oculomotor and Skeletal Motor System* (J. Noth, ed.), *Progr. Brain Res.* **64**:3–18. [631]

Eccles, J. C., and McGeer, P. L., 1979, Ionotropic and metabotropic neurotransmission, *TINS* **2**:39–40. [155]

Eccles, J. C., and Sherrington, C. S., 1930, Numbers and contraction values of individual motor units examined in some muscles of the limb, *Proc. R. Soc. London Ser. B* **106**:326–357. [65, 83]

Eccles, J. C., Fatt, P., and Koketsu, K., 1954, Cholinergic and inhibitory synapses in a pathway from motor-axon collaterals to motoneurons, *J. Physiol. (London)* **126**:524–562. [137, 138]

Eccles, J. C., Eccles, R. M., and Lundberg, A., 1957, Synaptic actions on motoneurons in relation to the two components of the group I muscle afferent volley, *J. Physiol. (London)* **136**:527–546. [113]

Eccles, J. C., Eccles, R. M., and Lundberg, A., 1958a, The action potentials of the alpha motoneurones supplying fast and slow muscles, *J. Physiol. (London)* **142**:275–291. [74]

Eccles, J. C., Libet, B., and Young, R. R., 1958b, The behavior of chromatolyzed motoneurons studied by intracellular recording, *J. Physiol. (London)* **143**:11–40. [73]

Eccles, J. C., Schmidt, R. F., and Willis, W. D., 1963, Inhibition of discharges into the dorsal and ventral spinocerebellar tracts, *J. Neurophysiol.* **26**:635–645. [212]

Eccles, J. C., Llinás, R., and Sasaki, K., 1966, The excitatory synaptic action of climbing fibers on the Purkinje cells of the cerebellum, *J. Physiol. (London)* **182**:268–296. [452]

Eccles, J. C., Ito, M., and Szentagothai, J., 1967, *The Cerebellum as a Neuronal Machine*, Springer-Verlag, Heidelberg. [8, 141, 142, 491, 494, 580]

Eccles, J. C., Nicoll, R. A., Oshima, T., and Rubia, F. J., 1977, The anionic permeability of the inhibitory postsynaptic membrane of hippocampal pyramidal cells, *Proc. R. Soc. London Ser. B.* **198**:345–361. [128]

Eckenhoff, M. F., and Rakic, P., 1984, Radial organization of the hippocampal dentate gyrus: A Golgi, ultrastructural, and immunocytochemical analysis in the developing rhesus monkey, *J. Comp. Neurol.* **223**:1–21. [433]

Eckenstein, F., and Baughman, R. W., 1984, Two types of cholinergic innervation in cortex, one co-localized with vasoactive intestinal polypeptide, *Nature (London)* **309**:153–155. [397]

Eckenstein, F., and Thoenen, H., 1983, Cholinergic neurons in the rat cerebral cortex demonstrated by immunohistochemical localization of choline acetyltransferase, *Neurosci. Lett.* **36**:211–215. [252]

Eckernas, S.-A., Sahlstrom, L., and Aquilonius, S.-M., 1980, Effects of atropine treatment on cortical, striatal and hippocamal high-affinity uptake of choline in the mouse, *Life Sci.* **27**:641–645. [243]

Edbrooke, M. R., Parker, D., McVey, J. H., et al. 1985, Expression of the human calcitonin/CGRP gene in lung and thyroid carcinoma, *Evr. Mol. Biol. Org. J.* **4**:715–719. [376]

Edelman, G. M., 1978, Group selection and phasic reentrant signalling: A theory of higher brain function, in: *The Mindful Brain* (F. O. Schmitt, ed.), pp. 51–100, MIT Press, Cambridge, Massachusetts. [631, 632]

Edelman, G. M., 1984, Modulation of cell adhesion during induction, histogenesis, and prenatal development of the nervous system, *Annu. Rev. Neurosci.* **7**:339–377. [26, 419–421, 432, 435, 438]

Edelman, G. M., and Finkel, L. H., 1984, Neuronal group selection in the cerebral cortex, in: *Dynamic Aspects of Neocortical Function* (G. M. Edelman, W. E. Gall, and W. M. Cowan, eds.), pp. 653–695, John Wiley, New York. [599]

Edvinsson, L., Fahrenkrug, J., Hanko, J., Owman, C., Sundler, F., and Uddman, R., 1980, VIP (vasoactive intestinal polypeptide)-containing nerves of intracranial arteries in mammals, *Cell Tissue Res.* **208**:135–142. [397, 400]

Edvinsson, L., Degueurce, A., Duverger, D., MacKenzie, E. T., and Scatton, B., 1983, Central serotonergic nerves project to the pial vessels of the brain, *Nature (London)* **306**:55–57. [334]

Edwards, F. R., Redman, S. J., and Walmsley, B., 1976, The effect of polarizing currents on unitary Ia excitatory post-synaptic potentials evoked in spinal motoneurones, *J. Physiol. (London)* **259**:705–723. [118]

Edwardson, J. M., Phillips, N. I., Kirby, N., and Fowler, L. J., 1985, A monoclonal antibody to rabbit brain GABA transaminase, *J. Neurochem.* **44:**1679–1684. [201]

Egan, T. M., and North, R. A., 1986, Acetylcholine hyperpolarizes central neurones by acting on an M_2 muscarinic receptor, *Nature (London)* **319:**405–407. [255]

Ehinger, B., 1970, Autoradiographic evidence of GABA uptake in rabbit retinal neurons, *Experientia* **26:**1063–1064. [205]

Ehinger, B., and Lindberg, B., 1974, Light-evoked release of glycine from the retina, *Nature (London)* **251:**727–728. [229]

Eide, E., Jurna, I., and Lundberg, A., 1968, Conductance measurements from motoneurons during presynaptic inhibition, in: *Structure and Function of Inhibitory Neuronal Mechanisms* (C. von Euler, S. Skoglund, and U. Soderberg, eds.), pp. 215–219, Pergamon Press, Oxford. [143, 144]

Eiden, L. E., Hökfelt, T., Brownstein, M. J., and Palkovits, M., 1985, Vasoactive intestinal polypeptide afferents to the bed nucleus of the stria terminalis in the rat: An immunohistochemical and biochemical study, *Neuroscience* **15:**999–1013. [395]

Ekerot, C. F., and Kano, M., 1983, Climbing fiber induced depression on Purkinje cell responses to parallel fiber stimulation, *Proc. IUPS* **15:**393–470. [580, 581]

Ekerot, C. F., and Oscarsson, O., 1975, Inhibitory spinal paths to the lateral reticular nucleus, *Brain Res.* **99:**157–161. [141]

Ekerot, C. F., and Oscarsson, O., 1981, Prolonged depolarization elicited in Purkinje cell dendrites by climbing fibre impulses in cat, *J. Physiol. (London)* **318:**207–221. [60, 72, 580, 582, 583]

Elde, R., and Hökfelt, T., 1979, Localization of hypophysiotropic peptides and other biologically active peptides within the brain, *Annu. Rev. Physiol.* **41:**587–602. [401]

Elliott, M. E., Farese, R. V., and Goodfriend, T. L., 1983, Effects of angiotensin II and dibutyryl cyclic adenosine monophosphate on phosphatidylinositol metabolism, $^{45}Ca^{2+}$ fluxes, and aldosterone synthesis in bovine adrenal glomerulosa cells, *Life Sci.* **33:**1771–1778. [402]

Elmadjian, F., 1959, Excretion and metabolism of epinephrine, *Pharmacol. Rev.* **11:**409–415. [532]

Emrich, H. M., 1984, Endorphins in psychiatry, *Psychiatr. Dev.* **2:**97–114. [409]

Emson, P. C., and Lindvall, O., 1986, Neuroanatomical aspects of neurotransmitters affected in Alzheimer's disease, *Br. Med. Bull.* **42:**57–62. [262]

Emson, P. C., Jessell, T., Paxinos, G., and Cuello, A. C., 1978, Substance P in the amygdaloid complex, bed nucleus and stria terminalis of the rat, *Brain Res.* **149:**97–105. [395]

Emson, P. C., Fahrenkrug, J., and Spokes, E. G., 1979, Vasoactive intestinal polypeptide (VIP): Distribution in normal human brain and in Huntington's disease, *Brain Res.* **173:**174–178. [412]

Emson, P. C., Arregui, A., Clement-Jones, V., Sanberg, B. E. B., and Rossor, M., 1980a, Regional distribution of methionine-enkephalin and substance P-like immunoreactivity in normal human brain and in Huntington's disease, *Brain Res.* **199:**147–160. [412]

Emson, P. C., Rehfeld, J. F., Langevin, H., and Rossor, M., 1980b, Reduction in cholecystokinin-like immunoreactivity in the basal ganglia in Huntington's disease, *Brain Res.* **198:**497–500. [412, 508]

Emson, P. C., Geodert, M., Horsfield, P., Rioux, F., and St-Pierre, S., 1982, The regional distribution and chromatographic characterization of neurotensin-like immunoreactivity in the rat central nervous system, *J. Neurochem.* **38:**992–999. [388, 506]

Emson, P. C., Horsfield, P. M., Goedert, M., Rossor, M. N., and Hawkes, C. H., 1985, Neurotensin in human brain: Regional distribution and effects of neurological illness, *Brain Res.* **347:**239–244. [412]

Engberg, I., and Marshall, K. C., 1979, Reversal potential for Ia excitatory post synaptic potentials in spinal motoneurones of cats, *Neuroscience* **4:**1583–1591. [118, 182]

Epelbaum, J., Ruberg, M., Moyse, E., Javoy-Agid, F., Dubois, B., and Agid, Y., 1983, Somatostatin and dementia in Parkinson's disease, *Brain Res.* **278:**376–379. [412]

Epstein, A. N., 1971, The lateral hypothalamic syndrome: Its implications for the physiological psychology of hunger and thirst, in: *Progress in Physiological Psychology,* Vol. 4 (J. M. Sprague, ed.), pp. 263–317, Academic Press, New York. [529]

Eranko, O., 1955, Histochemistry of noradrenaline in the adrenal medulla of rats and mice, *Endocrinology* **57:**363–368. [267]

Erdo, S. L., Rosdy, B., and Szporny, L., 1982, Higher GABA concentrations in fallopian tube than in brain of the rat, *J. Neurochem.* **38:**1174–1176. [218]

Ermisch, A., Ruhle, H.-J., Landgraf, R., and Hess, J., 1985, Blood–brain barrier and peptides, *J. Cereb. Blood Flow Metab.* **5:**350–357. [405]

Erspamer, V., and Asero, B., 1952, Identification of enteramine, the specific hormone of the enterochromaffin cell system, as 5-hydroxytryptamine, *Nature (London)* **169**:800–801. [320]

Eruklar, S. D., 1983, The modulation of neurotransmitter release at synaptic junctions, *Rev. Physiol. Biochem. Pharmacol.* **98**:63–175. [554]

Eskay, R. L., Brownstein, M. J., and Long, R. T., 1979, α-Melanocyte-stimulating hormone: Reduction in adult rat brain after monosodium glutamate treatment of neonates, *Science* **205**:827–829. [388]

Evans, E. F., and Nelson, P. G., 1973, On the functional relationship between the dorsal and ventral divisions of the cochlear nucleus of the cat, *Exp. Brain Res.* **17**:428–442. [212]

Evans, R. H., Francis, A. A., Hunt, K., Oakes, D. J., and Watkins, J. C., 1979, Antagonism of excitatory amino acid-induced responses and of synaptic excitation in the isolated spinal cord of the frog, *Br. J. Pharmacol.* **67**:591–603. [182]

Evans, R. H., Francis, A. A., Jones, A. W., Smith, D. A. S., and Watkins, J. C., 1982, The effects of a series of ω-phosphonic α-carboxylic amino acids on electrically evoked and excitant amino acid-induced responses in isolated spinal cord preparations, *Br. J. Pharmacol.* **75**:65–75. [182]

Evans, W. O., 1968, The psychopharmacology of the normal human: Trends in research strategy, in: *Psychopharmacology* (D. H. Efron, ed.), pp. 1003–1011, U.S. Government Printing Office, Washington, D.C. [194]

Evarts, E. V., 1961, Effects of sleep and waking on activity of single units in the unrestrained cat, in: *The Nature of Sleep* (G. E. W. Wolstenholme and M. O'Connor, eds.), pp. 171–182, J. & A. Churchill, London. [69]

Evarts, E. V., 1967, Representation of movements and muscles by pyramidal tract neurons of the precentral motor cortex, in: *Neurophysiological Basis of Normal and Abnormal Motor Activity* (M. D. Yahr and D. P. Purpura, eds.), pp. 215–253, Raven Press, New York. [478, 479]

Evarts, E. V., 1972, Contrasts between activity of precentral and postcentral neurons of cerebral cortex during movement in the monkey, *Brain Res.* **40**:25–31. [478]

Evarts, E. V., 1984, Hierarchies and emergent features in motor control, in: *Dynamic Aspects of Neocortical Function* (G. Edelman, W. E. Gall, and W. M. Cowan, eds.), pp. 557–579, John Wiley, New York. [476, 478]

Evarts, E. V., and Thach, W. T., 1969, Motor mechanisms of the CNS, cerebrocerebellar interrelations, *Annu. Rev. Physiol.* **31**:451–498. [495]

Evarts, E. V., and Wise, S. P., 1984, Basal ganglia outputs and motor control, *Ciba Found. Symp.* **107**:83–107. [502]

Ewing, G., Bigelow, J. C., and Wightman, R. M., 1983, Direct *in vivo* monitoring of dopamine released from two striatal compartments, *Science* **221**:169–171. [296]

Ezure, K., Oguri, M., and Oshima, T., 1985, Vertical spread of neuronal activity within the cat motor cortex investigated with epicortical stimulation and intracellular recording, *Jpn. J. Physiol.* **35**:193–221. [631]

Fadini, R., Meschia, M., Crosignani, P. G., *et al.* 1984, Metergoline in the management of normoprolactinemic secondary amenorrhea, *Gynecol. Obstet. Invest.* **17**:47–53. [338]

Fagg, G. E., and Foster, A. C., 1983, Amino acid neurotransmitters and their pathways in the mammalian central nervous system, *Neuroscience* **9**:701–719. [186]

Fagg, G. E., Jordan, C. C., and Webster, R. A., 1978, Descending fibre-mediated release of endogenous glutamate and glycine from the perfused spinal cord *in vivo*, *Brain Res.* **158**:158–170. [182, 228]

Fahn, S., 1983, High dosage anticholinergic therapy in dystonia, *Adv. Neurol.* **37**:177–188. [258]

Falck, B., Hillarp, N. A., Thieme, G., and Torp, A., 1962, Fluorescence of catecholamines and related compounds condensed with formaldehyde, *J. Histochem. Cytochem.* **10**:348–354. [267]

Fallon, J. F., Wang, C., Kim, Y., Canepa, N., Loughlin, J., and Seroogy, K., 1983, Dopamine- and cholecystokinin-containing neurons of the crossed neostriatal projection, *Neurosci. Lett.* **40**:233–238. [157]

Fallon, J. R., Nitkin, R. M., Reist, N. E., Wallace, B. G., and McMahan, U. J., 1985, Acetylcholine receptor-aggregating factor is similar to molecules concentrated at neuromuscular junction, *Nature (London)* **315**:571–574. [257]

Fan, S. G., Wuseman, M., and Iversen, L. L., 1981, 3-Mercaptopropionic acid inhibits GABA release from rat brain slices *in vitro*, *Brain Res.* **229**:371–377. [223]

Faris, P. L., Komosaruk, B. R., Watkins, L. R., and Mayer, D. J., 1983, Evidence for the neuropeptide cholecystokinin as an antagonist of opiate analgesia, *Science* **219**:310–312. [407]

Faris, P. L., McLaughlin, C. L., Baile, C. A., and Olney, J. W., 1984, Morphine analgesia potentiated but tolerance not affected by active immunization against cholecystokinin, *Science* **226:**1215–1217. [407]

Farley, I. J., Price, K. S., McCullough, E., Deck, J. N., Hordynski, W., and Hornykiewicz, O., 1978, Norepinephrine in chronic paranoid schizophrenia: Above normal levels in limbic forebrain, *Science* **200:**465–468. [547]

Farmery, S. M., Owen, F., Poulter, M., and Crow, T. J., 1985, Reduced high affinity cholecystokinin binding in hippocampus and frontal cortex of schizophrenic patients, *Life Sci.* **36:**473–477. [412]

Fatt, P., 1957, Electric potentials occurring around a neurone during its antidromic activation, *J. Neurophysiol.* **20:**27–60. [72]

Fatt, P., and Katz, B., 1951, An analysis of the end-plate potential recorded with an intra-cellular electrode, *J. Physiol. (London)* **115:**320–370. [86, 87, 112]

Fatt, P., and Katz, B., 1952, Spontaneous subthreshold activity at motor nerve endings, *J. Physiol. (London)* **117:**109–128. [90, 91, 99]

Faull, R. L. M., and Mehler, W. R., 1976, Subdivision of the ventral tier nuclei in the rat thalamus based on their afferent fiber connections, *Anat. Rec.* **184:**400. [505]

Fehm-Wolfsdorf, G., Born, J., Voight, K.-L., and Fehm, H. L., 1984, Human memory and neurohypophyseal hormones: Opposite effects of vasopressin and oxytocin, *Psychoneuroendocrinology* **9:**285–292. [410]

Feigl, H., 1967, *The "Mental" and the "Physical,"* University of Minnesota Press, Minneapolis, 179 pp. [625, 626, 631, 632]

Feldberg, W., and Gaddum, J. H., 1934, The chemical transmitter at synapses in a sympathetic ganglion, *J. Physiol. (London)* **81:**305–319. [81]

Feldberg, W., and Krayer, O., 1933, Das Auftreten eines acetylcholinartigen Stoffes im Herzvenenblut bei Reizung der Nervi vagi, *Arch. Exp. Pathol.* **172:**170–193. [81]

Feldstein, A., Hoagland, H., Freeman, H., and Williamson, O., 1967, Effect of ethanol ingestion on serotonin-C^{14} metabolism in man, *Life Sci.* **6:**53–61. [325]

Fellmann, D., Bugnon, C., and Gouget, A., 1982, Immunocytochemical demonstration of corticoliberin-like immunoreactive (CLI) in neurones of the rat amygdala central nucleus (ACN), *Neurosci. Lett.* **34:**253–258. [388, 394]

Felten, D. L., and Sladek, J. R., Jr., 1983, Monoamine distribution in primate brain v. monoaminergic nuclei: Anatomy, pathways and local organization, *Brain Res. Bull.* **10:**171–284. [329, 331]

Fenstermacher, J. D., 1975, Mechanisms of ion distribution between blood and brain, in: *The Nervous System,* Vol. 1, *The Basic Neurosciences* (D. B. Tower and R. O. Brady, eds.), pp. 299–311, Raven Press, New York. [37, 38]

Ferrero, P., Guidotti, A., Conti-Tronconi, B., and Costa, E., 1984, A brain octadecaneuropeptide generated by tryptic digestion of DBI (diazepam binding inhibitor) functions as a proconflict ligand of benzodiazepine recognition sites, *Neuropharmacology* **23:**1359–1362. [220]

Ferrier, I. N., Crow, T. J., Farmery, S. M., *et al.* 1985, Reduced cholecystokinin levels in the limbic lobe in schizophrenia, *Ann. N. Y. Acad. Sci.* **448:**495–506. [412]

Fex, J., Altschuler, R. A., Wenthold, R. J., and Parakkal, M. H., 1982, Aspartate aminotransferase immunoreactivity in cochlea of guinea pig, *Hear. Res.* **7:**149–160. [178]

Ffrench-Mullen, J. M. H., Koller, K., Zaczek, R., Coyle, J. T., Hori, N., and Carpenter, D. O., 1985, *N*-Acetylaspartylglutamate: Possible role as the neurotransmitter of the lateral olfactory tract, *Proc. Natl. Acad. Sci. U.S.A.* **82:**3897–3900. [186]

Fibiger, H. C., Pudritz, R. E., McGeer, P. L., and McGeer, E. G., 1972, Axonal transport in nigro-striatal and nigro-thalamic neurons: Effects of medial forebrain bundle lesions and 6-hydroxydopamine, *J. Neurochem.* **19:**1697–1708. [506]

Fields, H. L., and Basbaum, A. I., 1978, Brainstem control of spinal pain-transmission neurons, *Annu. Rev. Physiol.* **40:**217–248. [603, 605]

Fifkova, E., and Anderson, C. L., 1981, Stimulation-induced changes in dimensions of stalks of dendritic spines in dentate molecular layer, *Exp. Neurol.* **74:**621–627. [564]

Fifkova, E., Anderson, C. L., Yong, J. J., and van Harreveld, A., 1982, Effect of anisomycin on stimulation induced changes in dendritic spines of dentate granule cells, *J. Neurocytol.* **11:**183–210. [563]

Finkel, A. S., and Redman, S. J., 1983, The synaptic current evoked in cat spinal motoneurones by impulses in single group Ia axons, *J. Physiol. (London)* **342:**615–632. [118]

Finley, J. C. W., Grossman, G. H., Dimeo, P., and Petrusz, P., 1978, Somatostatin containing neurons in the rat brain: Widespread distribution revealed by immunocytochemistry after pretreatment with pronase, *Am. J. Anat.* **153**:483–488. [506]

Fisher, G. H., Humphries, J., Folkers, K., Pernow, B., and Bowers, C. Y., 1974, Synthesis and some biological activities of substance P, *J. Med. Chem.* **17**:843–846. [374]

Fisher, S. K., Klinger, P. D., and Agranoff, B. W., 1983, Muscarinic agonist binding and phospholipid turnover in brain, *J. Biol. Chem.* **258**:7358–7363. [255]

Fitzpatrick, D., Penny, G. R., and Schmechel, D. E., 1984, Glutamic acid decarboxylase-immunoreactive neurons and terminals in the lateral geniculate nucleus of the cat, *J. Neurosci.* **4**:1809–1829. [205]

Fitzsimons, J. T., 1972, Thirst, *Physiol. Rev.* **52**:468–561. [529]

Flicker, C., Dean, R. L., Watkins, D. L., Fisher, S. K., and Bartus, R. T., 1983, Behavioral and neurochemical effects following neurotoxic lesions of a major cholinergic input to the cerebral cortex in the rat, *Pharmacol. Biochem. Behav.* **18**:973–982. [261]

Flint, R. S., Rea, M. A., and McBride, W. J., 1981, *In vitro* release of endogenous amino acids from granule cell-, stellate cell- and climbing fiber-deficient cerebella, *J. Neurochem.* **37**:1425–1430. [178, 182, 212]

Florey, E., 1953, Uber einen nervösen Hemmungsfaktor im Gehirn und Rückenmark, *Naturwissenschaften* **40**:295–296. [197]

Fonnum, F., 1968, The distribution of glutamate decarboxylase and aspartate transaminase in subcellular fractions of rat and guinea-pig brain, *Biochem. J.* **106**:401–412. [178]

Fonnum, F., 1984, Glutamate: A neurotransmitter in mammalian brain, *J. Neurochem.* **42**:1–11. [177, 186]

Fonnum, F., and Walaas, I., 1978, The effect of intrahippocampal kainic acid injections and surgical lesions on neurotransmitters in hippocampus and septum, *J. Neurochem.* **31**:1173–1181. [182, 212]

Fonnum, F., Storm-Mathisen, J., and Walberg, F., 1970, Glutamate decarboxylase in inhibitory neurons: A study of the enzyme in Purkinje cell axons and boutons in the cat, *Brain Res.* **20**:259–275. [212, 213]

Fonnum, F., Walaas, I., and Iversen, E., 1977, Localization of GABAergic, cholinergic and aminergic structures in the mesolimbic system, *J. Neurochem.* **29**:221–230. [212]

Fonnum, F., Gottesfield, Z., and Grofova, I., 1978a, Distribution of glutamate decarboxylase, choline acetyltransferase and aromatic amino acid decarboxylase in the basal ganglia of normal and operated rats: Evidence for striatopallidal, striatoentopeduncular and striatonigral GABAergic fibers, *Brain Res.* **143**:125–128. [212]

Fonnum, F., Grofova, I., and Rinvik, E., 1978b, Origin and distribution of glutamate decarboxylase in the nucleus subthalamicus of the cat, *Brain Res.* **153**:370–374. [212]

Fonnum, F., Storm-Mathisen, J., and Divac, I., 1981a, Biochemical evidence for glutamate as neurotransmitter in corticostriatal and corticothalamic fibres in rat brain, *Neuroscience* **6**:863–873. [182, 184]

Fonnum, F., Søreide, A., Kvale, I., Walker, J., and Walaas, I., 1981b, Glutamate in cortical fibers, in: *Glutamate as a Neurotransmitter* (G. DiChiara and G. L. Gessa, eds.), pp. 29–41, Raven Press, New York. [182, 184]

Foote, S. L., Bloom, F. E., and Aston-Jones, G., 1983, The nucleus locus coeruleus: New evidence of anatomical and physiological specificity, *Physiol. Rev.* **63**:844–914. [311]

Foster, A. C., and Fagg, G. E., 1984, Acidic amino acid binding sites in mammalian neuronal membranes: Their characteristics and relationship to synaptic receptors, *Brain Res. Rev.* **7**:103–164. [187–189]

Foster, A. C., and Roberts, P. J., 1980, Morphological and biochemical changes in the cerebellum induced by kainic acid *in vivo*, *J. Neurochem.* **34**:1191–1200. [212]

Foster, A. C., and Schwarcz, R., 1985, Characterization of quinolinic acid phosphoribosyltransferase in human blood and observations in Huntington's disease, *J. Neurochem.* **45**:199–205. [195]

Foster, A. C., Zinkand, W. C., and Schwarcz, R., 1985a, Quinolinic acid phosphoribosyltransferase in rat brain, *J. Neurochem.* **44**:446–454. [192]

Foster, A. C., Whetsell, W. O., Jr., Bird, E. D., and Schwarcz, R., 1985b, Quinolinic acid phosphoribosyltransferase in human and rat brain: Activity in Huntington's disease and quinolinate-lesioned rat striatum, *Brain Res.* **336**:207–214. [192, 195]

Foster, N. L., Chase, T. N., Fedio, P., Patronas, N. J., Brooks, R. A., and Di Chiro, G., 1983a, Alzheimer's disease: Focal cortical changes shown by positron emission tomography, *Neurology* **33**:961–965. [262]

Foster, N. L., Cox, C., Harlan, A. L., Fedio, P., and Chase, T. N., 1983b, Vasopressin in Alzheimer's disease, *Neurology* **33**:1635. [410]

Foster, N. L., Chase, T. N., Mansi, L., *et al.* 1984, Cortical abnormalities in Alzheimer's disease, *Ann. Neurol.* **16**:649–654. [69]

Foster, N. L., Tamminga, C. A., O'Donohue, T. L., Tanimoto, K., Bird, E. D., and Chase, T. N., 1986, Brain choline acetyltransferase activity and neuropeptide Y concentrations in Alzheimer's disease, *Neurosci. Lett.* **63**:71–75. [412]

Fowler, C. J., and Benedetti, M. S., 1983, Cimoxatone is a reversible tight-binding inhibitor of the A form of rat brain monoamine oxidase, *J. Neurochem.* **40**:510–513. [342]

Fowler, C. J., and Tipton, K. F., 1982, Deamination of 5-hydroxytryptamine by both forms of monoamine oxidase in the rat brain, *J. Neurochem.* **38**:733–736. [326]

Fowler, L. J., Lovell, D. H., and John, R. A., 1983, Reaction of muscimol with 4-aminobutyrate aminotransferase, *J. Neurochem.* **41**:1751–1754. [201]

Fox, C. A., 1962, The structure of the cerebellar cortex, in: *Correlative Anatomy of the Nervous System* (E. C. Crosby, T. H. Humphrey, and E. W. Lauer, eds.), pp. 193–198, Macmillan, New York. [8, 493]

Fox, C. A., Hillman, D. E., Siegesmund, K. A., and Dutta, C. R., 1966, The primate cerebellar cortex: A Golgi and electron microscope study, *Prog. Brain Res.* **25**:174–225. [493]

Frackowiak, R. S. J., Jones, T., Lenzi, G.-L., and Heather, S. D., 1980a, Regional cerebral oxygen utilization and blood flow in normal man using oxygen-15 and positron emission tomography, *Acta Neurol. Scand.* **62**:336–344. [69]

Frackowiak, R. S. J., Lenzi, G.-L., Jones, T., and Heather, J. D., 1980b, Quantitative measurement of regional cerebral bood flow and oxygen metabolism in man using ^{15}O and positron emission tomography: Theory, procedure and normal values, *J. Comput. Assist. Tomogr.* **4**:727–736. [480]

Francis, A. A., Jones, A. W., and Watkins, J. C., 1980, Dipeptide antagonists of amino acid-induced and synaptic excitation in the frog spinal cord, *J. Neurochem.* **35**:1458–1460. [182]

Francke, U., De Martinville, B., Coussens, L., and Ullrich, A., 1983, The human gene for the β subunit of nerve growth factor is located on the proximal short arm of chromosome 1, *Science* **222**:1248–1250. [465]

Frank, K., and Fuortes, M. G. F., 1957, Presynaptic and postsynaptic inhibition of monosynaptic reflexes, *Fed. Proc. Fed. Am. Soc. Exp. Biol.* **16**:39–40. [142, 144]

Frankfurt, M., Lauder, J. M., and Azmitia, E. C., 1981, The immunocytochemical localization of serotonergic neurons in the rat hypothalamus, *Neurosci. Lett.* **24**:227–232. [331]

Frederickson, R. C. A., Smithwick, E. L., Shuman, R., and Bemis, K. G., 1981, Metkephamid, a systematically active analog of methionine enkephalin with potent opioid δ-receptor activity, *Science* **211**:603–605. [404]

Fredholm, B. B., and Hedqvist, P., 1980, Modulation of neurotransmitters by purine nucleosides and nucleotides, *Biochem. Pharmacol.* **29**:1635–1643. [365]

Freed, W. J., de Medinaceki, L., and Wyatt, R. J., 1985, Promoting functional plasticity in the damaged nervous system, *Science* **227**:1544–1552. [469]

Frey, H. H., and Loscher, W., 1980, Acetyl GABA: Effect on convulsant threshold in mice and acute toxicity, *Neuropharmacology* **19**:217–220. [225]

Friedman, D. O., and Jones, E. G., 1981, Thalamic input to areas 3a and 2 in monkeys, *J. Neurophysiol.* **45**:59–85. [487, 488]

Friedman, S., and Kaufman, S., 1965, 3,4-Dihydroxyphenylethylamine β-hydroxylase: Physical properties, copper content and the role of copper in the catalytic activity, *J. Biol. Chem.* **240**:4763–4773. [274]

Frueh, B. R., Felt, D. P., Wojno, T. H., and Musch, D. C., 1984, Treatment of blepharospasm with botulinum toxin, *Arch. Ophthalmol.* **102**:1464–1468. [259]

Fry, F. H., and Cowan, W. M., 1972, A study of retrograde cell degeneration in the lateral mammillary nucleus of the cat with special reference to the role of axonal branching in the preservation of the cell, *J. Comp. Neurol.* **144**:1–24. [462, 467]

Fujito, Y., Tsukahara, N., Oda, Y., and Yoshida, M., 1982, Formation of functional synapses in the adult cat red nucleus from the cerebrum following cross-innervation of forelimb flexor and extensor nerves. II. Analysis of newly-appeared synaptic potentials, *Exp. Brain Res.* **45**:13–18. [450]

Fuller, R. W., 1984, Serotonin receptors, *Monogr. Neural Sci.* **10**:158–181. [336, 343]

Funatsu, K., Teshima, S., and Inanaga, K., 1985, Various types of thyrotropin-releasing hormone receptors in discrete brain regions and the pituitary of the rat, *J. Neurochem.* **45**:390–397. [402]

Funk, C., 1911, Syntheses of *dl*-3:4-dihydroxyphenylalanine, *J. Chem. Soc.* **99**:554–557. [266]

Furutani, Y., Morimoto, Y., Shibahara, S., *et al.* 1983, Cloning and sequence analysis of cDNA for ovine corticotropin-releasing factor precursor, *Nature (London)* **301**:537–540. [376]

Fuxe, K., and Johnsson, G., 1974, Further mapping of central 5-hydroxytryptamine neurons: Studies with the neurotoxic dihydroxytryptamines, *Adv. Biochem. Psychopharmacol.* **10**:1–12. [331, 340]

Fuxe, K., Agnati, L. F., Ganten, D., *et al.* 1982, Morphometric evaluation of the coexistence of renin-like and oxytocin-like immunoreactivity in nerve cells of the paraventricular hypothalamic nucleus of the rat, *Neurosci. Lett.* **33**:19–24. [157]

Fuxe, K., Agnati, L. F., McDonald, T., *et al.* 1983, Immunohistochemical indications of gastrin releasing peptide–bombesin-like immunoreactivity in the nervous system of the rat: Codistribution with substance P-like immunoreactive nerve terminal systems and coexistence with substance P-like immunoreactivity in dorsal root ganglion cell bodies, *Neurosci. Lett.* **37**:17–22. [157]

Gabbott, P. L. A., Somogyi, J., Stewart, M. G., and Hamori, J., 1985, GABA-immunoreactive neurons in the rat dorsal lateral geniculate nucleus: Light microscopical observations, *Brain Res.* **346**:171–175. [205]

Gaddum, J. H., and Picarelli, Z. P., 1957, Two kinds of tryptamine receptor, *Br. J. Pharmacol.* **12**:323–328. [337]

Gage, F. H., Björklund, A., Stenevi, U., Dunnett, S. B., and Kelly, P. A. T., 1984, Intrahippocampal septal grafts ameliorate learning impairments in aged rats, *Science* **225**:533–536. [459]

Gage, P. W., and Moore, J. W., 1969, Synaptic current at the squid giant synapse, *Science* **160**:510–512. [107, 119]

Gahwiler, B. H., 1984, Slice cultures of cerebellar, hippocampal and hypothalamic tissue, *Experientia* **40**:235–243. [119]

Gaillard, J. M., 1983, Biochemical pharmacology of paradoxical sleep, *Br. J. Clin. Pharmacol.* **16**:205S–230S. [338]

Gainer, H., Russell, J. T., and Loh, P., 1985, The enzymology and intracellular organization of peptide precursor processing: The secretory vesicle hypothesis, *Neuroendocrinology* **40**:171–184. [379]

Gajtkowski, G. A., Norris, D. P., Rising, T. J., and Wood, T. P., 1983, Specific binding of ^{3}H-tiotidine to histamine H_2 receptors in guinea pig cerebral cortex, *Nature (London)* **304**:65–67. [357]

Gál, E. M., 1972, 5-Hydroxytryptamine-O-sulfate: An alternative route of serotonin inactivation in brain, *Brain Res.* **44**:309–312. [327]

Gál, E. M., 1981, Synthesis and quantitative aspects of dihydrobiopterin control of cerebral serotonin levels, *Adv. Exp. Med. Biol.* **133**:197–206. [322]

Gál, E. M., and Whitacre, D. H., 1981, Biopterin. VII. Inhibition of synthesis of reduced biopterins and its bearing on the function of cerebral tryptophan-5-hydroxylase *in vivo*, *Neurochem. Res.* **6**:233–241. [322, 323]

Gál, E. M., and Whitacre, D. H., 1982, Mechanism of irreversible inactivation of phenylalanine-4 and tryptophan-5-hydroxylases by [4-^{36}Cl,2-^{14}C]-*p*-chlorophenylalanine, *Neurochem. Res.* **7**:13–24. [341]

Gale, K., Hong, J. S., and Guidotti, A., 1977, Presence of substance P and GABA in separate striatonigral neurones, *Brain Res.* **136**:371–375. [399]

Galindo, A., Krnjevic, K., and Schwartz, S., 1967, Microiontophoretic studies on neurons in the cuneate nucleus, *J. Physiol. (London)* **192**:359–377. [230]

Gall, C., 1984, The distribution of cholecystokinin-like immunoreactivity in the hippocampal formation of the guinea pig: Localization in the mossy fibers, *Brain Res.* **306**:73–83. [397]

Gallager, D. W., Mallorga, P., Oertel, W., Henneberry, R., and Tallman, J., 1981, [^{3}H]Diazepam binding in mammalian central nervous system: A pharmacological characterization, *J. Neurosci.* **1**:218–225. [220]

Gallagher, J. P., Higashi, H., and Nishi, S., 1978, Characterization and ionic basis of GABA-induced depolarizations recorded *in vitro* from cat primary afferent neurones, *J. Physiol. (London)* **275**:263–282. [144]

Galvan, M., Grafe, P., and ten Bruggencate, G., 1982, Convulsive actions of 4-aminopyridine on neurons and extracellular K^+ and Ca^{2+} activities in guinea pig olfactory cortex slices, in: *Physiology and Pharmacology of Epileptogenic Phenomena* (M. R. Klee, H. D. Lux, and E. J. Speckman, eds.), pp. 353–360, Raven Press, New York. [560]

Galzigna, L., Garbin, L., Bianchi, M., and Marzotto, A., 1978, Properties of two derivatives of γ-aminobutyric acid (GABA) capable of abolishing cardiazol- and bicuculline-induced convulsion in the rat, *Arch. Int. Pharmacodyn. Ther.* **235**:73–85. [225]

Galzigna, L., Bianchi, M., Bertazzon, A., Barthez, A., Quadro, G., and Coletti-Previero, M. A., 1984, An N-protected γ-aminobutyric acid dipeptide with anticonvulsant action, *J. Neurochem.* **42**:1762–1766. [225]

Gamrani, C. A. H., Belin, M. F., Aguera, M., and Pujol, J. F., 1979, High resolution radioautographic identification of [³H]GABA labeled neurons in the rat nucleus raphe dorsalis, *Neurosci. Lett.* **15**:43–48. [205]

Gamse, R., Leeman, S. E., Holzer, P., and Lembeck, F., 1981, Differential effects of capsaicin on the content of somatostatin, substance P and neurotensin in the nervous system of the rat, *Naunyn-Schmiedeberg's Arch. Pharmacol.* **317**:140–148. [407]

Ganten, D., Hermann, K., Bayer, C., Unger, T. H., and Lang, R. E., 1983, Angiotensin synthesis in the brain and increased turnover in hypertensive rats, *Science* **221**:869–871. [388]

Garau, L., Govoni, S., Stefani, E., Trabucchi, M., and Spano, P. F., 1978, Dopamine receptors: Pharmacological and anatomical evidence indicates two distinct dopamine receptor populations are present in striatum, *Life Sci.* **23**:1745–1750. [301]

Garbarg, M., Barbin, G., Bischoff, S., Pollard, H., and Schwartz, J. C., 1974, Evidence for a specific decarboxylase involved in histamine synthesis in an ascending pathway in rat brain, *Agents Actions* **4**:181. [350, 355]

Garbarg, M., Baudry, M., Benda, P., *et al.* 1975, Simultaneous presence of HMNT and COMT in neuronal and glial cells in culture, *Brain Res.* **83**:538–541. [352]

Garbarg, M., Barbin, G., Bischoff, S., Pollard, H., and Schwartz, J. C., 1976, Dual localization of histamine in an ascending neuronal pathway and in non-neuronal cells evidenced by lesions in the lateral hypothalamic area, *Brain Res.* **106**:333–348. [355]

Garbarg, M., Barbin, G., Rodergas, E., and Schwartz, J. C., 1980, Inhibition of histamine synthesis in brain by α-fluoromethylhistidine, a new irreversible inhibitor: *In vitro* and *in vivo* studies, *J. Neurochem.* **35**:1045–1052. [350]

Garbarg, M., Barbin, G., Rodergas, E., and Schwartz, J. C., 1981, Biochemical studies on histaminergic systems in mammalian brains, *Agents Actions* **11**:144–146. [350]

Garbarg, M., Arrang, J. M., Duchemin, A. M., Quach, T. T., Rose, C., and Schwartz, J. C., 1983, Histamine-induced responses in mammalian brain slices: Characterization of H₁ and H₂ receptors, *Adv. Biochem. Psychopharmacol.* **37**:421–428. [357]

Garey, L. J., Blakemore, C., and Vital Durand, F., 1979, Visual deprivation in monkeys: Its effects and its reversal, *Prog. Brain Res.* **51**:445–456. [453]

Garrick, N. A., Tamarkin, L., Taylor, P. L., Markey, S. R., and Murphy, D. L., 1983, Light and propranolol suppress the nocturnal elevation of serotonin in the cerebrospinal fluid of rhesus monkeys, *Science* **221**:474–476. [344]

Garvey, J. M., Rossor, M., and Iversen, L. L., 1984, Evidence for multiple muscarinic receptor subtypes in human brain, *J. Neurochem.* **43**:299–302. [254]

Gash, D., and Sladek, J. R., Jr., 1980, Vasopressin neurons grafted into Battleboro rats: Viability and activity, *Peptides* **1**:11–14. [405, 459]

Gaudreau, P., Quirion, R., St-Pierre, S., and Pert, C. B., 1983, Characterization and visualization of cholecystokinin receptors in rat brain using [³H]pentagastrin, *Peptides* **4**:755. [395]

Gaze, R. M., 1970, *The Formation of Nerve Connections,* Academic Press, New York. [417, 434–436]

Gaze, R. M., 1974, Neuronal specificity, *Br. Med. Bull.* **30**:116–121. [434–436]

Gee, C. E., Chen, C.-L. C., Roberts, J. L., Thompson, R., and Watson, S. J., 1983, Identification of POMC neurons in the rat hypothalamus by *in situ* hybridization, *Nature (London)* **306**:374–376. [387]

Gee, K. W., Ehlert, F. J., and Yamamura, H. I., 1983, Differential effect of gamma-aminobutyric acid on benzodiazepine receptor subtypes labeled by [³H]propyl β-carboxylate in rat brain, *J. Pharmacol. Exp. Ther.* **225**:132–137. [220]

Geiger, H., and Singer, W., 1986, Ca⁺⁺-currents correlated with developmental plasticity in the kitten visual cortex, *J. Physiol. (London)* (in press). [455]

Geiger, J. D., LaBella, F. S., and Nagy, J. I., 1984, Ontogenesis of adenosine receptors in the central nervous system of the rat, *Dev. Brain Res.* **13**:97–104. [364]

Geiger, J. D., LaBella, F. S., and Nagy, J. I., 1985, Characterization of nitrobenzylthioinosine binding to nucleoside transport sites selective for adenosine in rat brain, *J. Neurosci.* **5**:735–740. [362]

Geinisman, Y., Bondariff, W., and Telser, A., 1977, Transport of [³H] glucose labeled glycoproteins in the septohippocampal pathway of young adult and senescent rats, *Brain Res.* **125**:182–186. [471]

Geinisman, Y., Bondareff, D., and Dodge, J. T., 1978. Dendritic atrophy in the dentate gyrus of the senescent rat, *Am. J. Anat.* **152**:321–330. [471]

Geller, H. M., 1981, Histamine actions on activity of cultured hypothalamic neurons: Evidence for mediation by H₁- and H₂-histamine receptors, *Dev. Brain Res.* **1**:89–101. [356]

Geller, H. M., Springfield, S. A., and Tiberio, A. R., 1984, Electrophysiological actions of histamine, *Can. J. Physiol. Pharmacol.* **62:**715–719. [356]

Gerfen, C. R., 1984, The neostriatal mosaic: Compartmentalization of corticostriatal input and striatonigral output systems, *Nature (London)* **311:**461–464. [511]

German, D. C., Bruce, G., and Hersh, L. B., 1985, Immunohistochemical staining of cholinergic neurons in the human brain using a polyclonal antibody to human choline acetyltransferase, *Neurosci. Lett.* **61:**1–5. [245]

Gerner, R. H., van Kammen, D. P., and Ninan, P. T., 1985, Cerebrospinal fluid cholecystokinin, bombesin and somatostatin in schizophrenia and normals, *Prog. Neuro-Psychopharmacol. Biol. Psychiatry* **9:**73–82. [412]

Gerschenfeld, H. M., and Paupardin-Tritsch, D., 1974, Ionic mechanisms and receptor properties underlying the responses of molluscan neurons to 5-hydroxytryptamine, *J. Physiol. (London)* **243:**427–456. [131]

Gershon, M. D., and Tamir, H., 1984, Serotonectin and the family of proteins that bind serotonin, *Biochem. Pharmacol.* **33:**3115–3118. [328]

Gershon, M. D., Dreyfus, C. F., Pickel, V. M., Joh, T. H., and Reis, D. J., 1975, Drugs, diagnosis and disease, in: *Biology of the Major Psychoses* (D. X. Freedman, ed.), pp. 85–96, Raven Press, New York. [545, 546]

Geschwind, N., 1970, The organization of language and the brain, *Science* **170:**940–944. [615, 616]

Geschwind, N., and Levitsky, W., 1968, Human brain: Left–right asymmetries in temporal speech region, *Science* **161:**186–187. [618]

Ghatei, M. A., Bloom, S. R., Langevin, H., *et al.* 1984, Regional distribution of bombesin and seven other regulatory peptides in the human brain, *Brain Res.* **293:**101–109. [388]

Ghelarducci, B., Ito, M., and Yagi, N., 1975, Impulse discharges from flocculus Purkinje cells of alert rabbits during visual stimulation combined with horizontal head rotation, *Brain Res.* **87:**66–72. [578, 583]

Giacobini, E., 1983, Imino acids of the brain, in: *Handbook of Neurochemistry*, 2nd ed., Vol. 3 (A. Lajtha, ed.), pp. 583–605, Plenum Press, New York. [232]

Gibbins, I. L., Furness, J. B., Costa, M., MacIntyre, J., Hillyard, C. J., and Girgis, S., 1985, Co-localization of calcitonin gene-related peptide-like immunoreactivity with substance P in cutaneous, vascular and visceral sensory neurons of guinea pigs, *Neurosci. Lett.* **57:**125–130. [157]

Gibbs, J., and Bourne, E. W., 1985, Effect of bombesin on feeding behavior, *Life Sci.* **37:**147–153. [405]

Gibson, G. E., and Blass, J. P., 1976, Impaired synthesis of acetylcholine in brain accompanying mild hypoxia and hypoglycemia, *J. Neurochem.* **27:**37–42. [239]

Gibson, K. M., 1983, Succinic semialdehyde dehydrogenase deficiency: An inborn error of gamma-aminobutyric acid metabolism, *Clin. Chim. Acta* **133:**33–42. [225]

Gilad, G. M., and Reis, D. J., 1979, Transneuronal effects of olfactory bulb removal on choline acetyltransferase and glutamic acid decarboxylase activities in the olfactory tubercle, *Brain Res.* **178:**185–190. [212]

Gilbert, C. D., 1985, Horizontal integration in the neocortex, *TINS* **8:**160–165. [598]

Gilbert, P. F. C., and Thach, W. T., 1977, Purkinje cell activity during motor learning, *Brain Res.* **128:**309–328. [587–589, 591]

Gilles, C., Lotstra, F., and Vandenhaeghen, J. J., 1983, CCK nerve terminals in the rat striatal and limbic areas originate partly in telencephalic structures, *Life Sci.* **32:**1683–1690. [157]

Gilman, A. G., Goodman, L. S., Rall, T. W., and Murad, F. (eds.), 1985, *Goodman and Gilman's The Pharmacological Basis of Therapeutics*, 7th ed., Macmillan, New York. [258, 305, 533]

Gilman, S., Bloedel, J. R., and Lechtenberg, R., 1981, *Disorders of the Cerebellum*, F. A. Davis, Philadelphia. [492, 515]

Gispen, W. H., 1982, ACTH and brain membrane phosphorylation: A model for modulation by neuropeptides, *Acta Biol. Med. Ger.* **41:**279–288. [402]

Gispen, W. H., Leunissen, J. L. M., Oestreicher, A. B., Verkleij, A. J., and Zwiers, H., 1985, Presynaptic localization of B-50 phosphoprotein: The (ACTH)-sensitive protein kinase substrate involved in rat brain polyphosphoinositide metabolism, *Brain Res.* **328:**381–385. [402]

Gjerris, A., Rafaelson, O. J., Vendsborg, P., Fahrenkrug, J., and Rehfeld, J. F., 1984, Vasoactive intestinal peptide decreases in cerebrospinal fluid (CSF) in atypical depression, *J. Affective Disorders* **7:**325–337. [412]

Gladue, B. A., Green, R., and Hellman, R. E., 1984, Neuroendocrine response to estrogen and sexual orientation, *Science* **225:**1496–1499. [522]

Glass, D. C., Krakoff, L. R., Contrada, R., *et al.* 1980, Effect of harassment and competition upon cardiovascular and plasma catecholamine responses in Type A and Type B individuals, *Psychophysiology* **17:**453–463. [533]

Glassman, E., 1969, Some considerations of the effects of short term learning on the incorporation of uridine into RNA and polysomes of mouse brain, in: *The Future of the Brain Sciences* (S. Bogoch, ed.), pp. 281–291, Plenum Press, New York. [575]

Glavin, G. B., 1985, Stress and brain noradrenaline: A review, *Neurosci. Biobehav. Rev.* **9:**233–243. [310]

Gnahn, H., Hefti, F., Heumann, R., Schwab, M. E., and Thoenen, H., 1983, MGF-mediated increase of choline acetyltransferase (ChAT) in the neonatal rat forebrain: Evidence for a physiological role of NGF in the brain?, *Dev. Brain Res.* **9:**45–52. [465]

Godfrey, D. A., Ross, C. D., Carter, J. A., Lowry, O. H., and Matschinsky, F. M., 1980, Effect of intervening lesions on amino acids distributions in rat olfactory cortex and olfactory bulb, *J. Histochem. Cytochem.* **28:**1157–1169. [212]

Godukhin, O. V., Zharikova, A. D., and Novoselov, V. I., 1980, The release of labeled L-glutamic acid from rat neostriatum *in vivo* following stimulation of frontal cortex, *Neuroscience* **5:**2151–2154. [182]

Goedert, M., and Emson, P. C., 1983, The regional distribution of neurotensin-like immuoreactivity in central and peripheral tissues of the cat, *Brain Res.* **272:**291–297. [506]

Goedert, M., Mantyh, P. W., Hunt, S. P., and Emson, P. C., 1983, Mosaic distribution of neurotensin-like immunoreactivity in the cat striatum, *Brain Res.* **274:**176–179. [506]

Gold, B. I., and Huger, F. P., 1982, Calcium stimulated glutamate decarboxylase activity in brain slice, *Biochem. Pharmacol.* **31:**882–884. [201]

Goldberg, M. E., Satama, A. I., and Blum, S. W., 1971, Inhibition of choline acetyltransferase and hexobaritone-metabolizing enzymes by naphthyl vinyl pyridine and analogues, *J. Pharm. Pharmacol.* **23:**384–385. [260]

Goldbloom, D. S., 1984, Endogenous opiates and schizophrenia: Directions in clinical research, *Can. J. Psychiatry* **29:**355–360. [409]

Goldenring, J. R., McGuire, J. S., Jr., and DeLorenzo, R. J., 1984, Identification of the major postsynaptic density protein as homologous with the major calmodulin-binding subunit of a calmodulin-dependent protein kinase, *J. Neurochem.* **42:**1077–1084. [169]

Goldman, P. S., 1978, Neuronal plasticity in primate telencephalon: Anomalous crossed cortico-caudate projections induced by prenatal removal of frontal association cortex, *Science* **202:**765–770. [446]

Goldman, P. S., and Nauta, W. J. H., 1977, An intricately patterned prefrontocaudate projection in the rhesus monkey, *J. Comp. Neurol.* **171:**369–386. [596]

Goldman-Rakic, P. S., 1980, Development and plasticity of primate frontal association cortex, in: *The Organization of the Cerebral Cortex* (F. O. Schmitt, F. G. Worden, G. Adelman, and S. G. Dennis, eds.), pp. 69–97, MIT Press, Cambridge, Massachusetts. [446]

Goldman-Rakic, P. S., 1982, Cytoarchitectonic heterogeneity of the primate neostriatum: Subdivision into island and matrix cellular compartments, *J. Comp. Neurol.* **205:**398–413. [511]

Goldowitz, D., Vincent, S. R., Wu, J.-Y., and Hökfelt, T., 1982, Successful immunohistochemical demonstration of plasticity in GABA neurons of the adult rat dentate gyrus, *Brain Res.* **238:**413–420. [205]

Goldschmidt, R. C., Hough, L. B., and Glick, S. D., 1985, Rat brain mast cells: Contribution to brain histamine levels, *J. Neurochem.* **44:**1943–1947. [352]

Goldstein, A., Lowney, L. I., and Pal, B. K., 1971, Stereospecific and nonspecific interactions of the morphine congener levorphanol in subcellular fractions of mouse brain, *Proc. Natl. Acad. Sci. U.S.A.* **68:**1742–1747. [374]

Goldstein, A., Cox, B. M., Klee, W. A., and Nemeroff, M., 1977, Endorphin from pituitary inhibits cyclic AMP formation in homogenates of neuroblastoma × glioma hybrid cells, *Nature (London)* **265:**362–363. [403]

Goldstein, M., 1984, Regulator mechanisms of dopamine biosynthesis at the tyrosine hydroxylase step, *Ann. N. Y. Acad. Sci.* **430:**1–5. [283]

Gomes, U. C. R., Shanley, B. C., Potgieter, L., and Roux, J. T., 1980, Noradrenergic overactivity in chronic schizophrenia: Evidence based on cerebrospinal fluid noradrenaline and cyclic nucleotide concentrations, *Br. J. Psychiatry* **137:**346–351. [547]

Gomez, S., diBello, C., Lam, T. H., *et al.* 1984, C-terminal amidation of neuropeptides: Gly-Lys-Arg extension an efficient precursor of C-terminal amide, *FEBS Lett.* **167:**160–164. [381]

Gonce, M., Schoenen, J., Charlier, M., and Delwaide, P. J., 1983, Successful treatment of hemiballismus with progabide, a new GABA-mimetic agent, *J. Neurol.* **22:**121–124. [225]

Gonon, F. G., and Buda, M. J., 1985, Regulation of dopamine release by impulse flow and by autoreceptors as studied by *in vivo* voltammetry in the rat striatum, *Neuroscience* **14**:765–774. [296]

Gonshor, A., and Melvill Jones, G., 1976a, Short-term adaptive changes in the human vestibulo-ocular reflex arc, *J. Physiol. (London)* **256**:361–379. [577]

Gonshor, A., and Melvill Jones, G., 1976b, Extreme vestibulo-ocular adaptation induced by prolonged optical reversal of vision, *J. Physiol. (London)* **256**:381–414. [577]

Goodall, M., 1951, Studies of adrenaline and noradrenaline in mammalian heart and suprarenals, *Acta Physiol. Scand.* **24**(Suppl.):85. [267]

Goodall, M., 1962, Sympathoadrenal response to gravitational stress, *J. Clin. Invest.* **41**:197–202. [532]

Goodman, R. H., Aron, D. C., and Roos, B. A., 1983, Rat pre-prosomatostatin structure and processing by microsomal membranes, *J. Biol. Chem.* **258**:5570–5573. [382]

Gorio, A. (ed.), 1984, Neurobiology of gangliosides, *J. Neurosci. Res.* **12**:147–509. [466]

Gorio, A., Carmignoto, G., Facci, L., and Finesso, M., 1980, Motor nerve sprouting induced by ganglioside treatment: Possible implications for gangliosides in neuronal growth, *Brain Res.* **197**:236–241. [467]

Gorio, A., Marini, P., and Zanoni, R., 1983, Muscle reinnervation. III. Motoneuron sprouting capacity, enhancement by exogenous gangliosides, *Neuroscience* **8**:417–429. [467]

Gorio, A., Aporti, F., Di Gregorio, F., Schiavinato, A., Siliprandi, R., and Vitadello, M., 1984, Ganglioside treatment of genetic and alloxan-induced diabetic neuropathy, in: *Ganglioside Structure, Function, and Biomedical Potential* (R. W. Ledeen, R. K. Yu, M. M. Rapport, and K. Suzuki, eds.), pp. 549–564, Plenum Press, New York. [467]

Goswell, M. J., and Sedgwell, E. M., 1971, Inhibition in the substantia nigra following stimulation of the caudate nucleus, *J. Physiol. (London)* **218**:84P–85P. [212, 215]

Gothert, M., 1982, Modulation of serotonin release in the brain via presynaptic receptors, *Trends Pharmacol. Sci.* **3**:437–440. [329]

Goto, M., Inomata, N., Ono, H., Saito, K., and Fukuda, H., 1981, Changes of electroretinogram and neurochemical aspects of GABAergic neurons of retina after intraocular injection of kainic acid in rats, *Brain Res.* **211**:305–314. [212]

Gott, P. S., 1973a, Cognitive abilities following right and left hemispherectomy, *Cortex* **9**:266–274. [622]

Gott, P. S., 1973b, Language after dominant hemispherectomy, *J. Neurol. Neurosurg. Psychiatry* **36**:1082–1088. [622]

Gottesfeld, Z., and Jacobowitz, D. M., 1978a, Further evidence for GABAergic afferents to the lateral habenula, *Brain Res.* **152**:609–613. [212]

Gottesfeld, Z., and Jacobowitz, D. M., 1978b, Cholinergic projection of the diagnonal band to the interpeduncular nucleus of the rat brain, *Brain Res.* **156**:329–332. [250]

Gottesfeld, Z., and Jacobowitz, D. M., 1979, Cholinergic projections from the septal–diagonal band area to the habenular nuclei, *Brain Res.* **176**:391–394. [250]

Gottesfeld, Z., Hoover, D. B., Muth, E. A., and Jacobowitz, D. M., 1978, Lack of biochemical evidence for a direct habenulo-raphe GABAergic pathway, *Brain Res.* **141**:353–356. [216]

Gottesman, I. S., Mandarino, L. J., and Gerich, J. E., 1982, Somatostatin, in: *Special Topics in Endocrinology and Metabolism*, Vol. 4 (M. Cohen and P. Foa, eds.), pp. 177–243, Alan R. Liss, New York. [409]

Gozlan, H., Le Gal La Salle, G., Michelot, R., and Ben-Ari, Y., 1977, Rapid degradation of substance P and related peptides during microiontophoretic experiments, *Neurosci. Lett.* **6**:27–33. [401]

Gozlan, H., El Mestikawy, S., Pichat, L., Glowinski, J., and Hamon, M., 1983, Identification of presynaptic serotonin autoreceptors using a new ligand: ^{3}H-PAT, *Nature (London)* **305**:140–142. [335]

Grab, D. J., Berzins, K., Cohen, R. S., and Siekevitz, P., 1979, Presence of calmodulin in postsynaptic densities isolated from canine cerebral cortex, *J. Biol. Chem.* **254**:8690–8696. [562]

Grab, D. J., Carlin, R. K., and Siekevitz, P., 1980, The presence and functions of calmodulin in the postsynaptic density, *Ann. N. Y. Acad. Sci.* **356**:55–72. [562]

Grafe, M. R., Lysia, D., Forno, S., and Eng, L. F., 1985, Immunocytochemical studies of substance P and met-enkephalin in the basal ganglia and substantia nigra in Huntington's and Alzheimer's disease, *J. Neuropathol. Exp. Neurol.* **44**:47–59. [412]

Grafstein, B., 1971, Transneuronal transfer of radioactivity in the central nervous system, *Science* **172**:177–179. [462]

Grafstein, B., Yip, H. K., and Meiri, H., 1982, Techniques for improving axonal regeneration: Assay in goldfish optic nerve, in: *Nervous System Regeneration* (A. M. Giuffrida-Stella, B. Haber, G. Hashim, and J. R. Perez-Polo, eds.), pp. 105–118, Alan R. Liss, New York. [467]

Graham, L. T., 1972, Intraretinal distribution of GABA content and GAD activity, *Brain Res.* **36**:476–479. [216]

Graham, L. T., Shank, R. P., Werman, R., and Aprison, M. H., 1967, Distribution of some synaptic transmitter suspects in cat spinal cord: Glutamic acid, aspartic acid, glycine and glutamine, *J. Neurochem.* **14**:465–472. [182]

Granit, R., 1970, *The Basis of Motor Control,* Academic Press, New York. [473, 486]

Granit, R., 1972, Constant errors in the execution and appreciation of movement, *Brain* **95**:649–660. [490, 491]

Granit, R., Kernell, D., and Shortess, G. K., 1963, Quantitative aspects of repetitive firing of mammalian motoneurones, caused by injected currents, *J. Physiol. (London)* **168**:911–931. [76]

Gray, E. G., 1959, Axo-somatic and axo-dendritic synapses of the cerebral cortex: An electron microscope study, *J. Anat.* **93**:420–423. [11, 562]

Gray, E. G., 1962, A morphological basis for pre-synaptic inhibition?, *Nature (London)* **193**:82–83. [142]

Gray, E. G., 1964, Tissue of the central nervous system, in: *Electron Microscopic Anatomy*, pp. 369–417, Academic Press, New York. [12, 13]

Gray, E. G., 1970, The fine structure of nerve, *Comp. Biochem. Physiol.* **36**:419–448. [10]

Gray, E. G., 1982, Rehabilitating the dendritic spine, *TINS* **5**:5–6. [562, 631]

Gray, E. G., 1983, Neurotransmitter release mechanisms and microtubules, *Proc. R. Soc. London Ser. B* **218**:253–258. [11, 97]

Gray, E. G., Gordon-Weeks, P. R., and Burgoyne, R. D., 1982, Nerve-terminal ultrastructure: A role for neurotubules, in: *Neurotransmitter Interaction and Compartmentation* (H. F. Bradford, ed.), pp. 1–13, Plenum Press, New York. [98]

Graybiel, A. M., 1975, Wallerian degeneration and anterograde tracer methods, in: *The Use of Axonal Transport for Studies of Neuronal Connectivity* (W. M. Cowan and M. Cuenod, eds.), pp. 173–216, Elsevier, Amsterdam. [463]

Graybiel, A. M., 1978, Organization of the nigrotectal connections: An experimental tracer study in the cat, *Brain Res.* **143**:339–348. [505]

Graybiel, A. M., Ragsdale, C. W., Jr., Yoneoka, E. S., and Elde, R. P., 1981, An immunohistochemical study on enkephalins and other neuropeptides in the striatum of the cat with evidence that the opiate peptides are arranged to form mosaic patterns in register with the striasomal compartments visible by acetylcholinesterase staining, *Neuroscience* **6**:377–397. [399, 506, 511]

Green, A. R., and Youdim, M. B. H., 1975, Effects of monoamine oxidase inhibition by clorgyline, deprenyl or tranylcypromine on 5-hydroxytryptamine concentrations in rat brain and hyperactivity following subsequent tryptophan administration, *Br. J. Pharmacol.* **55**:415–422. [326]

Green, A. R., Mitchell, B. D., Tordoff, A. F. C., and Youdim, M. B. H., 1977, Evidence for dopamine deamination by both type A and type B monoamine oxidase in rat brain *in vivo* and for the degree of inhibition of enzyme necessary for increased functional activity of dopamine and 5-hydroxytryptamine, *Br. J. Pharmacol.* **60**:343–349. [326]

Greenmyre, J. T., Penney, J. B., Young, A. B., D'Amato, C. J., Hicks, S. P., and Shoulson, I., 1985, Alterations in L-glutamate binding in Alzheimer's and Huntington's disease, *Science* **227**:1496–1499. [195]

Griffiths, E. C., and McDermott, J. R., 1984, Biotransformation of neuropeptides, *Neuroendocrinology* **39**:573–581. [385]

Grigoriadis, D., and Seeman, P., 1984, The dopamine/neuroleptic receptor, *Can. J. Neurol. Sci.* **11**:108–113. [298]

Grigoriadis, D., and Seeman, P., 1985, Complete conversion of brain D_2 dopamine receptors from the high- to low-affinity state for dopamine agonists, using sodium ions and guanine nucleotide, *J. Neurochem.* **44**:1925–1935. [301]

Grima, B., Lamouroux, A., Blanot, F., Biguet, N. F., and Mallet, J., 1985, Complete coding sequence of rat tyrosine hydroxylase mRNA, *Proc. Natl. Acad. Sci. U.S.A.* **82**:617–621. [271]

Grofova, I., and Fonnum, F., 1982, Extrinsic and intrinsic origin of GAD in pars compacta of the rat substantia nigra, *Soc. Neurosci. Abstr.* **8**:961. [510]

Gross, C. G., 1973, Visual functions of inferotemporal cortex, in: *Handbook of Sensory Physiology*, Vol. VII/3B (R. Jung, ed.), pp. 451–482, Springer-Verlag, Berlin, Heidelberg and New York. [610]

Gross, C. G., Desimone, R., Albright, T. D., and Schwartz, E. L., 1985, Inferior temporal cortex and pattern recognition in: *Pattern Recognition Mechanisms* (C. Chagas, R. Gattass, and C. Gross, eds.), pp. 179–201, Pontificia Academius Scientiarum, Scripta Varia, Città dello Stato del Vaticano. [606, 610]

Grzanna, R., 1984, Histidine decarboxylase: Isolation and molecular characteristics, *Neurochem. Res.* **9**:993–1009. [350]

Guarneri, P., Corda, M. G., Concas, A., and Biggio, G., 1981, Kainic acid-induced lesion of rat retina: Differential effect on cyclic GMP and benzodiazepine and GABA receptors, *Brain Res.* **209**:216–220. [212]

Gubler, U., Seeburg, P., Hoffman, B. J., Gage, L. P., and Udenfriend, S., 1982, Molecular cloning establishes proenkephalin as precursor of enkephalin-containing peptides, *Nature (London)* **295**:206–208. [382]

Gudelsky, G. A., Simonovic, M., and Meltzer, H. Y., 1984, Dopaminergic and serotonergic control of neuroendocrine function, *Monogr. Neural Sci.* **10**:85–102. [334]

Guertzenstein, P. G., and Silver, A., 1974, Fall in blood pressure produced from discrete regions of the ventral surface of the medulla by glycine and lesions, *J. Physiol. (London)* **242**:489–503. [229]

Guidotti, A., Badiani, G., and Pepeu, G., 1972, Taurine distribution in cat brain, *J. Neurochem.* **19**:431–435. [232]

Guillemin, R., 1978, Peptides in the brain: The new endocrinology of the neuron, *Science* **202**:390–402. [403]

Guillemin, R., Brazeau, P., Bohlen, P., Esch, F., *et al.*, 1984, Somatocrinin, the growth hormone releasing factor, *Recent Prog. Horm. Res.* **40**:233–299. [373]

Guillery, R. W., 1970, Light- and electron-microscopical studies of normal and degenerating axons, in: *Contemporary Research Methods in Neuroanatomy* (W. J. H. Nauta and S. O. E. Ebbesson, eds.), pp. 77–105, Springer-Verlag, Heidelberg. [33]

Guldin, W. O., and Markowitsch, H. J., 1981, No detectable remote lesions following massive intrastriatal injections of ibotenic acid, *Brain Res.* **225**:446–451. [192]

Gundlach, A. L., and Beart, P. M., 1981, [³H]Strychnine binding suggests glycine receptors in the ventral tegmental area of rat brain, *Neurosci. Lett.* **22**:289–294. [229]

Gundlach, A. L., and Beart, P. M., 1982, Neurochemical studies of the mesolimbic dopaminergic pathway: Glycinergic mechanisms and glycinergic–dopaminergic interactions in the rat ventral tegmentum, *J. Neurochem.* **38**:574–581. [229]

Gusella, J. F., Wexler, N. S., Conneally, P. M., *et al.*, 1983, A polymorphic DNA marker genetically linked to Huntington's disease, *Nature (London)* **306**:234–238. [44]

Gusella, J. F., Tanzi, R. E., Bader, P. I., *et al.*, 1985, Deletion of Huntington's disease-linked G8 (D4S10) locus in Wolf-Hirschhorn syndrome, *Nature (London)* **318**:75–78. [45]

Gustafsson, B., 1974, Afterhyperpolarization and the control of repetitive firing in spinal neurones of the cat, *Acta Physiol. Scand. (Suppl.)* **416**:1–47. [74, 75, 77]

Gustafsson, B., and Wigström, H., 1986, Hippocampal long-lasting potentiation produced by pairing single volleys and brief conditioning tetani evoked in separate afferents, *J. Neurosci.* **6**:1575–1582. [558–561, 563, 574]

Gutierrez, M. D. C., and Giacobini, E., 1985, Identification and characterization of pipecolic acid binding sites in mouse brain, *Neurochem. Res.* **10**:691–702. [233]

Guy, J., Vaudry, H., and Pelletier, G., 1982, Further studies on the identification of neurons containing immunoreactive alpha-melanocyte-stimulating hormone (α-MSH) in the rat brain, *Brain Res.* **239**:265–270. [391]

Guyenet, P. G., and Crane, J. K., 1981, Non-dopaminergic nigrostriatal pathway, *Brain Res.* **213**:291–305. [286]

Haas, H. L., 1984, Histamine potentiates neuronal excitation by blocking a calcium-dependent potassium conductance, *Agents Actions* **14**:534–537. [356]

Haber, S., and Elde, R., 1981, Correlation between met-enkephalin and substance P immunoreactivity in the primate globus pallidus, *Neuroscience* **6**:1291–1297. [399, 507]

Haber, S., and Elde, R., 1982, The distribution of enkephalin immunoreactive fibres and terminals in the monkey central nervous system: An immunohistochemical study, *Neuroscience* **7**:1049–1095. [395]

Hablitz, J. J., and Langmoen, I. A., 1982, Excitation of hippocampal pyramidal cells by glutamate in the guinea pig and rat, *J. Physiol. (London)* **325**:317–331. [182]

Hagiwara, S., and Byerly, L., 1981, Calcium channel, *Annu. Rev. Neurosci.* **4**:69–125. [59]

Hagiwara, S., and Tasaki, K., 1958, A study of the mechanism of impulse transmission across the giant synapse of the squid, *J. Physiol. (London)* **143**:114–137. [103]

Hakanson, R., Wahlestedt, C., Westlin, L., Vallgren, S., and Sundler, F., 1983, Neuronal histamine in the gut wall releasable by gastrin and cholecystokinin, *Neurosci. Lett.* **42:**305–310. [355]

Halász, N., Ljungdahl, Å, and Hökfelt, T., 1979, Transmitter histochemistry of the rat olfactory bulb. III. Autoradiographic localization of [^{3}H]GABA, *Brain Res.* **167:**221–240. [205]

Halász, N., Parry, D. M., Blackett, N. M., Ljungdahl, A., and Hökfelt, T., 1981, [^{3}H]Gamma-aminobutyrate autoradiography of the rat olfactory bulb: Hypothetical grain analysis of the distribution of silver grains, *Neuroscience* **6:**473–479. [205]

Haldeman, S., Huffman, R. D., Marshall, K. C., and McLennan, H., 1972, The antagonism of the glutamate-induced and synaptic excitation of thalamic neurones, *Brain Res.* **39:**419–425. [182]

Hall, H., and Ogren, S. O., 1984, Effects of antidepressant drugs on histamine H_1-receptors in the brain, *Life Sci.* **34:**597–605. [359]

Hamberger, A. C., Chiang, G. H., Nylen, E. S., Scheff, S. W., and Cotman, C. W., 1979, Glutamate as a CNS transmitter. I. Evaluation of glucose and glutamine as precursors for the synthesis of preferentially-released glutamate, *Brain Res.* **168:**513–530. [178, 182]

Hamill, O. P., Bormann, J., and Sakmann, B., 1983, Activation of multiple-conductance state chloride channels in spinal neurones by glycine and GABA, *Nature (London)* **305:**805–808. [128]

Hamlin, K. E., and Fisher, F. E., 1951, The synthesis of 5-hydroxytryptamine, *J. Am. Chem. Soc.* **73:**5007–5008. [319]

Hamlyn, L. H., 1962, An electron microscope study of pyramidal neurons in the Ammon's horn of the rabbit, *J. Anat.* **97:**189–201. [9]

Hammar, L., and Hjerten, S., 1980, Purification and immunochemical analysis of histidine decarboxylase from murine mastocytoma, *Agents Actions* **10:**92–93. [350]

Hammer, R., Berrie, C. P., Birdsall, N. J. M., Burgen, A. S. V., and Hulme, E. C., 1980, Pirenzepine distinguishes between different sub-classes of muscarinic receptors, *Nature (London)* **283:**90–92. [254]

Hammerschlag, R., and Stone, G. C., 1982, Membrane delivery by fast axoplasmic transport, *TINS* **5:**12–16. [33]

Hamon, M., Bourgoin, S., Gozlan, H., *et al.* 1984, Biochemical evidence for the 5-HT agonist properties of PAT [8-hydroxy-2-(di-*n*-propyl-)aminotetralin] in the rat brain, *Eur. J. Pharmacol.* **100:**263–276. [343]

Hámori, J., Pasik, T., Pasik, P., and Szentágothai, J., 1974, Triadic synaptic arrangements and their possible significance in the lateral geniculate nucleus of the monkey, *Brain Res.* **80:**379–393. [134–136]

Hamos, J. E., Davis, T. L., and Sterling, P., 1983, Four types of neuron in layer IVab of cat cortical area 17 accumulate ^{3}H-GABA, *J. Comp. Neurol.* **217:**449–457. [205]

Hampton, C. K., and Redburn, D. A., 1983, Autoradiographic analysis of ^{3}H-glutamate, ^{3}H-dopamine and ^{3}H-GABA accumulation in rabbit retina after kainic acid treatment, *J. Neurosci. Res.* **9:**239–251. [205]

Hanko, J., Edvinsson, L., Fahrenkrug, J., *et al.* 1977, Immunohistochemical demonstration of vasodilatory peptidergic nerves in brain, *Acta Physiol. Scand. (Suppl.)* **452:**61–63. [397]

Hanley, M. R., Sandberg, B. E. B., Lee, C. M., Iversen, L. L., Brundish, D. E., and Wade, R., 1980, Specific binding of ^{3}H-substance P to rat brain membranes, *Nature (London)* **286:**810–812. [402, 404]

Happola, O., Soinila, S., Paivarinta, H., Joh, T. H., and Panula, P., 1985, Histamine-immunoreactive endocrine cells in the adrenal medulla of the rat, *Brain Res.* **339:**393–396. [355]

Harandi, M., Nieoullon, A., and Calas, A., 1983, High resolution radioautographic investigation of [^{3}H] GABA accumulating neurons in cat sensorimotor cortical areas, *Brain Res.* **260:**306–312. [205]

Harbaugh, R. E., Roberts, D. W., Coombs, D. W., Saunders, R. L., and Reeder, T. M., 1984, Preliminary report: Intracranial cholinergic drug infusion in patients with Alzheimer's disease, *Neurosurgery* **15:**514–518. [258]

Hare, M. L. C., 1928, Tyramine oxidase. I. A new enzyme system in liver, *Biochem. J.* **22:**968–970. [275]

Harris, G. W., Jacobson, D., and Kahlson, G., 1952, The occurrence of histamine in the cerebral regions related to the hypophysis, *Ciba Found. Colloq. Endocrinol. Proc.* **4:**186–194. [349]

Hartline, H. K., 1934, Intensity and duration of the excitation of single photoreceptors, *J. Cell. Comp. Physiol.* **5:**229–247. [51]

Harvey, J. A., Scholfield, C. N., Graham, L. T., and Aprison, M. H., 1975, Putative transmitters in denervated olfactory cortex, *J. Neurochem.* **24:**445–449. [182, 186]

Hassler, R., 1975, Central interactions of the systems of rapidly and slowly conducted pain, *Adv. Neurosurg.* **3:**143–150. [604]

Hassler, R., 1978, Interaction of reticular activating system for vigilance and the truncothalamic and pallidal systems for directing awareness and attention under striatal control, in: *Cerebral Correlates of Conscious Experience* (P. A. Busher and A. Rougeul-Buser, eds.), pp. 111–129, Elsevier, Amsterdam. [630]

Hassler, R., and Reichert, T., 1961, Wirkungen der Reizungen und Koagulationen in den Stammganglien bei stereotaktischen Hirnoperation, *Nervenarzt* **32**:97–109. [529]

Hassler, R., Haugh, P., Nitsch, C., Kim, J. S., and Paik, K., 1982, Effect of motor and premotor cortex ablation on concentrations of amino acids, monoamines, and acetylcholine and on the ultrastructure in rat striatum: A confirmation of glutamate as the specific corticostriatal transmitter, *J. Neurochem.* **38**:1087–1098. [182, 184]

Hattori, T., McGeer, P. L., Fibiger, H. C., and McGeer, E. G., 1973, On the source of GABA-containing terminals in the substantia nigra: Electron microscopic, autoradiographic and biochemical studies, *Brain Res.* **54**:103–114. [205, 212]

Hattori, T., Singh, V. K., McGeer, E. G., and McGeer, P. L., 1976, Immunohistochemical localization of choline acetyltransferase containing neostriatal neurons and their relationship with dopaminergic synapses, *Brain Res.* **102**:164–173. [238, 250]

Hattori, T., McGeer, E. G., Singh, V. K., and McGeer, P. L., 1977, Cholinergic synapse of the interpeduncular nucleus, *Exp. Neurol.* **55**:102–111. [238, 250]

Hattori, T., McGeer, P. L., and McGeer, E. G., 1979, Dendroaxonic neurotransmission. II. Morphological sites for the synthesis, binding and release of neurotransmitters in dopaminergic dendrites in the substantia nigra and cholinergic dendrites in the neostriatum, *Brain Res.* **170**:71–83. [238, 297]

Haubrich, D. R., and Pflueger, A. B., 1979, Choline administration—central effect mediated by stimulation of acetylcholine synthesis, *Life Sci.* **24**:1083–1090. [260]

Hayashi, H., Suga, M., Satake, M., and Tsubaki, T., 1981, Reduced glycine receptors in the spinal cord in amyotrophic lateral sclerosis, *Ann. Neurol.* **9**:292–294. [228]

Hayashi, H., Takagi, H., Takeda, N., et al. 1984, Fine structure of histaminergic neurons in the caudal magnocellular nucleus of the rat as demonstrated by immunohistochemistry using histidine decarboxylase as a marker, *J. Comp. Neurol.* **229**:233–241. [354]

Hayashi, T., 1952, A physiological study of epileptic seizures following cortical stimulation in animals and its application of human clinic, *Jpn. J. Physiol.* **3**:46–64. [175]

Hayden, M. R., and Nichols, J. L., 1984, Molecular genetic approaches to the study of the nervous system, *Dev. Neurosci.* **6**:189–214. [39]

Hayes, A. G., and Tyres, N. B., 1979, Effects of intrathecal and intracerebroventricular injections of substance P on nociception in the rat and mouse, *Br. J. Pharmacol.* **66**:488. [406]

Haymaker, W., Lis, L., Vogel, F. S., Johnson, J. E., Jr., Adams, R. D., and Scharenberg, K., 1982, The pineal gland, in: *Histology and Pathology of the Nervous System* (W. Haymaker and R. D. Adams, eds.), pp. 1801–2023, Charles C. Thomas, Springfield, Illinois. [321, 344]

Hays, S. E., and Paul, S. M., 1982, CCK receptors and human neurological disease, *Life Sci.* **31**:319–322. [412]

Hays, S. E., Goodwin, F. K., and Paul, S. M., 1981a, Cholecystokinin receptors are decreased in basal ganglia and cerebral cortex of Huntington's disease, *Brain Res.* **225**:452–456. [412]

Hays, S. E., Houston, S., Beinfeld, M., and Paul, S., 1981b, Postnatal ontogeny of cholecystokinin receptors in rat brain, *Brain Res.* **213**:237–241. [402]

Hebb, C. O., and Waites, G., 1956, Choline acetylase in antero and retrograde degeneration of cholinergic nerve, *J. Physiol. (London)* **132**:667–671. [240]

Hebb, D. O., 1949, *The Organization of Behavior*, John Wiley, New York. [554, 565]

Hedner, T., Persson, B., and Berglund, G., 1984, Experience with ketanserin, a serotonin (S$_2$) antagonist, in longterm treatment of essential hypertension, *Clin. Exp. Hypertens—Theory Pract.* **A6**:743–751. [343]

Hedreen, J., 1978, Nondopaminergic and dopaminergic nigrostriatal pathways in rats, *Soc. Neurosci. Abstr.* **4**:45. [506]

Hedreen, J. C., and Chalmers, J. P., 1972, Neuronal degeneration in rat brain induced by 6-hydroxydopamine: A histological and biochemical study, *Brain Res.* **47**:1–36. [286]

Hedreen, J. C., Bacon, S. J., Cork, L. C., et al. 1983, Immunocytochemical identification of cholinergic neurons in the monkey central nervous system using monoclonal antibodies against choline acetyltransferase, *Neurosci. Lett.* **43**:173–177. [245, 247]

Hegstrand, L. R., and Hine, R. J., 1985, Measurement of brain histamine: A reappraisal, *Neurochem. Res.* **10**:307–314. [353]

Hegstrand, L. R., and Kalinke, T. H., 1985, Properties of *N*-acetylhistamine deacetylase in mammalian brain, *J. Neurochem.* **45**:300–307. [353]

Heidmann, T., and Changeux, J. P., 1978, Structural and functional properties of the acetylcholine receptor protein in its purified and membrane-bound states, *Annu. Rev. Biochem.* **47**:317–357. [165]

Heikkila, R. E., Manzino, L., Cabbat, F. S., and Duvoisin, R. C., 1984, Protection against the dopaminergic neurotoxicity of 1-methyl-4-phenyl-1,2,5,6-tetrahydropyridine by monoamine oxidase inhibitors, *Nature (London)* **311**:467–469. [303]

Heimer, L., and Robards, M. J., 1981, *Neuroanatomical Tracing Methods,* Plenum Press, New York. [33]

Heistad, D. D. (ed.), 1984, The blood–brain barrier, *Fed. Proc. Fed. Am. Soc. Exp. Biol.* **43**:185–219. [37]

Henderson, A. S., 1986, The epidemiology of Alzheimer's disease, *Br. Med. Bull.* **42**:3–10. [261]

Henderson, G., Tomlinson, B. E., and Gibson, P. H., 1980, Cell counts in cerebral cortex in normal adults throughout life using an image analysing computer, *J. Neurol. Sci.* **46**:113–136. [470]

Hendrickson, A. E., 1985, Dots, stripes and columns in monkey visual cortex, *TINS* **8**:406–410. [609]

Hendrickson, A. E., Ogren, M. P., Vaughn, J. E., Barber, R. P., and Wu, J.-Y., 1983, Light and electron microscopic immunocytochemical localization of glutamic acid decarboxylase in monkey geniculate complex: Evidence for GABAergic neurons and synapses, *J. Neurosci.* **3**:1245–1262. [205]

Hendry, I. A., and Iversen, L. L., 1973, Changes in tissue and plasma concentrations of nerve growth factor following removal of the submaxillary glands in adult mice and their effects on the sympathetic nervous system, *Nature (London)* **243**:500–504. [465]

Hendry, I. A., Stockel, K., Thoenen, H., and Iversen, L. L., 1974, The retrograde axonal transport of nerve growth factor, *Brain Res.* **68**:103–121. [460]

Hendry, I. A., Bonyhady, R. E., and Hill, C. E., 1983, The role of target tissues in development and regeneration—retrophins, in: *Nervous System Regeneration* (B. Haber, J. R. Perez-Polo, G. A. Hashima, and A. M. Guiffrida Stella, eds.), pp. 397–405, Alan R. Liss, New York. [461, 465]

Hendry, S. H. C., and Jones, E. G., 1981, Sizes and distribution of intrinsic neurons incorporating tritiated GABA in monkey sensory-motor cortex, *J. Neurosci.* **1**:390–408. [205]

Hendry, S. H. C., Valentino, K. L., Jones, E. G., and Beinfeld, M. C., 1982, An investigation of neurons displaying cholecystokinin-like immunoreactivity in the cerebral cortex of the monkey and rat, *Neurosci. Abstr.* **8**:585. [397]

Hendry, S. H. C., Houser, C. R., Jones, E. G., and Vaughn, J. E., 1983, Synaptic organization of immunocytochemically identified GABA neurons in the monkey sensory-motor cortex, *J. Neurocytol.* **12**:639–660. [205]

Henn, F. A., 1976, Neurotransmission and glial cells: A functional relationship?, *J. Neurosci. Res.* **2**:271–282. [36]

Hertz, L., 1982, Astrocytes, in: *Handbook of Neurochemistry,* 2nd ed., Vol. 1 (A. Lajtha, ed.), pp. 319–355, Plenum Press, New York. [36, 37]

Hertz, L., and Tamir, H., 1981, Some properties of an astrocytic protein fraction that binds serotonin, *J. Neurochem.* **37**:1331–1334. [328]

Hess, W. R., 1964, *The Biology of Mind,* University of Chicago Press, Chicago. [527, 528]

Heydorn, W. E., Creed, G. J., Wada, H., and Jacobowitz, D. M., 1985, Immunological evidence for existence of two subforms of soluble glutamic oxaloacetic transaminase (sGOT) in human and rat brain, *Neurochem. Int.* **7**:833–841. [178]

Heyes, M. P., Garnett, E. S., and Brown, R. R., 1985, Normal excretion of quinolinic acid in Huntington's disease, *Life Sci.* **37**:1811–1816. [195]

Hicks, T. P., and McLennan, H., 1979, Amino acids and the synaptic pharmacology of granule cells in the dentate gyrus of the rat, *Can. J. Physiol. Pharmacol.* **57**:973–978. [182]

Hildebrandt, J. D., Hanoune, J., and Birnbaumer, L., 1982, Guanine nucleotide inhibition of cyc-S49 mouse lymphoma cell membrane adenyl cyclase, *J. Biol. Chem.* **257**:14,723–14,725. [167]

Hill, A. V., 1960, The heat production of nerve, in: *Molecular Biology* (D. Nachmansohn, ed.), pp. 153–162, Academic Press, New York. [69]

Hill, D. R., and Bowery, N. G., 1981, ^{3}H-Baclofen and ^{3}H-GABA bind to bicuculline-insensitive GABA$_B$ sites in rat brain, *Nature (London)* **290**:149–152. [218]

Hill, D. R., Bowery, N. G., and Hudson, A. L., 1984, Inhibition of GABA$_B$ receptor binding by guanyl nucleotides, *J. Neurochem.* **42**:652–657. [218]

Hille, B., 1975, The receptor for tetrodotoxin and saxitoxin: A structural hypothesis, *Biophys. J.* **15:**615–619. [57]

Hille, B., 1976, Gating in sodium channels of nerve, *Annu. Rev. Physiol.* **38:**139–142. [56]

Hille, B., 1977, Local anaesthetics: Hydrophilic and hydrophobic pathways for the drug receptor interaction, *J. Gen. Physiol.* **69:**497–515. [59]

Hiller, J. M., Pearson, J., and Simon, E. J., 1973, Distribution of stereospecific binding of the potent narcotic analgesic etorphine in the human brain: Predominance in the limbic system, *Res. Commun. Chem. Pathol. Pharmacol.* **6:**1052. [395]

Hillier, W. F., Jr., 1954, Total left cerebral hemispherectomy for malignant glioma, *Neurology* **4:**718–721. [622]

Hirsch, J. D., and Margolis, F. L., 1980, Influence of unilateral olfactory bulbectomy on opiate and other binding sites in the contralateral bulb, *Brain Res.* **199:**39–47. [212]

Hirst, G. D. S., Redman, S. J., and Wong, K., 1981, Post-tetanic potentiation and facilitation of synaptic potentials evoked in cat spinal motoneurones, *J. Physiol. (London)* **321:**97–109. [114, 116]

Hodgkin, A. L., 1964, *The Conduction of the Nervous Impulse,* Liverpool University Press, Liverpool. [51, 52, 54, 64, 106]

Hodgkin, A. L., and Huxley, A. F., 1952, A quantitative description of membrane current and its application to conduction and excitation in nerve, *J. Physiol. (London)* **117:**500–544. [54–56]

Hoffer, B. J., Jiggins, G. R., Oliver, A. P., and Bloom, F. E., 1973, Activation of the pathway from the locus coeruleus to rat cerebellar Purkinje neurons: Pharmacological evidence of noradrenergic central inhibition, *J. Pharmacol. Exp. Ther.* **184:**553–569. [311]

Hoffman, D. L., Krupp, L., Schrag, D., *et al.* 1982, Angiotensin immunoreactivity in vasopressin cells in rat hypothalamus and its relative deficiency in homozygous Brattleboro rats, *Ann. N. Y. Acad. Sci.* **394:**135–141. [157, 388]

Hofman, A., 1963, Psychotomimetic substances, *Indian J. Pharm.* **25:**245–256. [543]

Hökfelt, F., and Ljungdahl, A., 1972, Autoradiographic identification of cerebral and cerebellar cortical neurons accumulating labeled gamma-aminobutyric acid (^{3}H-GABA), *Exp. Brain Res.* **14:**331–353. [205, 210]

Hökfelt, T., Fuxe, K., Goldstein, M., and Johansson, O., 1974, Immunohistochemical evidence for the existence of adrenaline neurons in the rat brain, *Brain Res.* **66:**235–251. [268, 294]

Hökfelt, T., Skirboll, L., Rehfeld, J. F., Goldstein, M., Markey, K., and Dann, O., 1980a, A subpopulation of mesencephalic dopamine neurons projecting to limbic areas contains a choleocystokinin-like peptide: Evidence from immunohistochemistry combined with retrograde tracing, *Neuroscience* **5:**2093–2124. [157, 399, 400]

Hökfelt, T., Lundberg, J. M., Schultzberg, M., Johansson, O., Ljungdahl, A., and Rehfeld, J., 1980b, Coexistence of peptides and putative transmitters in neurons, *Adv. Biochem. Psychopharmacol.* **22:**1–23. [157, 332]

Hökfelt, T., Vincent, S., Dalsgaard, C. J., *et al.* 1982, Distribution of substance P in brain and periphery and its possible role as a cotransmitter, *Ciba Found. Symp.* **91:**84–106. [157]

Hökfelt, T., Lundberg, J. M., Lagercrantz, H., *et al.* 1983, Occurrence of neuropeptide Y (NPY)-like immunoreactivity in catecholamine neurons in the human medulla oblongata, *Neurosci. Lett.* **36:**217–222. [157]

Hökfelt, T., Johansson, O., and Goldstein, M., 1984, Central catecholamine neurons as revealed by immunohistochemistry with special reference to adrenaline neurons, in: *Handbook of Chemical Neuroanatomy,* Vol. 2, *Classical Transmitters in the CNS, Part I* (A. Björklund and T. Hökfelt, eds.), pp. 157–276, Elsevier, Amsterdam. [284, 289, 293, 294]

Hollister, L. E., 1985, Alzheimer's disease: Is it worth treating? *Drugs* **29:**483–488. [260]

Holst, J. J., 1980, Evidence that glicentin contains the entire sequence of glucagon, *Biochem. J.* **187:**337–343. [373]

Holton, F. A., and Holton, P., 1954, The capillary dilator substances in dry powders of spinal roots: A possible role of ATP in chemical transmission from nerve endings, *J. Physiol. (London)* **126:**124–140. [360]

Holtz, P., 1939, Dopa decarboxylase, *Naturwissenschaften* **27:**724. [266]

Holtz, P., Credner, K., and Kroneberg, G., 1947, Uber des sympathicomimetische pressorische Princip des Harn ("Urosympathin"), *Arch. Exp. Pathol. Pharmakol.* **204:**228–243. [266]

Homma, S., Suzuki, T., Murayama, S., and Otsuka, M., 1979, Amino acid and substance P contents in spinal cord of cats with experimental hind limb rigidity produced by occlusion of spinal cord blood supply, *J. Neurochem.* **32**:691–698. [182, 212]

Hommer, D. W., and Pikar, D., 1985, The effects of cholecystokinin-like peptides in schizophrenics and normal human subjects, *Ann. N. Y. Acad. Sci.* **448**:542–552. [409]

Hommer, D. W., Pikar, D., Roy, A., Ninan, P., Boronow, J., and Paul, S. M., 1984, The effect of ceruletide in schizophrenia, *Arch. Gen. Psychiatry* **41**:528. [410]

Hopkins, D. A., and Niessen, L. W., 1976, Substantia nigra projections to the reticular formation, superior colliculus and central gray in the rat, *Neurosci. Lett.* **2**:253–259. [505]

Hore, J., and Vilis, T., 1976, Initiation of monkey arm movements during globus pallidus cooling, *Soc. Neurosci. Abstr.* **2**:63. [515]

Horel, J. A., 1978, The neuroanatomy of amnesia: A critique of the hippocampal memory hypothesis, *Brain* **101**:403–445. [568]

Hori, N., Auker, C. R., Braitman, D. J., and Carpenter, D. O., 1981, Lateral olfactory tract transmitter: Glutamate, aspartate, or neither?, *Cell. Mol. Neurobiol.* **1**:115–120. [182]

Horn, A. S., Baumgarten, H. G., and Schlossberger, H. G., 1973, Inhibition of uptake of 5-hydroxytryptamine, noradrenaline and dopamine into rat brain homogenates by various hydroxylated tryptamines, *J. Neurochem.* **21**:232–236. [340]

Horn, A. S., Cuello, A. C., and Miller, R. J., 1974, Dopamine in the mesolimbic system of the rat brain: Endogenous levels and the effects of drugs on the uptake mechanism and stimulation of adenylate cyclase activity, *J. Neurochem.* **22**:265–270. [299]

Hosli, E., and Hosli, L., 1984, Autoradiographic localization of binding sites for [^{3}H]histamine and H_1- and H_2-antagonists on cultured neurones and glial cells, *Neuroscience* **13**:863–870. [357]

Hosli, E., Mohler, H., Richards, J. G., and Hosli, L., 1980, Autoradiographic localization of binding sites for [^{3}H]gamma-aminobutyrate, [^{3}H]muscimol, (+) [^{3}H]bicuculline methiodide and [^{3}H]flunitrazepam in cultures of rat cerebellum and spinal cord, *Neuroscience* **5**:1657–1665. [218]

Hough, L. B., and Green, J. P., 1980, Possible functions of brain histamine, *Psychopharmacol. Bull.* **16**:42–44. [358]

Hough, L. B., Khandelwal, J. K., and Green, J. P., 1984, Histamine turnover in regions of rat brain, *Brain Res.* **291**:103–109. [353]

Houseman, D., and Gusella, J. F., 1982, Molecular genetic approaches to neural degenerative disorders, in: *Molecular Genetic Neuroscience* (F. O. Schmitt, S. J. Bird, and F. E. Bloom, eds.), pp. 415–422, Raven Press, New York. [44]

Houser, C. R., Vaughn, J. E., Barber, R. P., and Roberts, E., 1980, GABA neurons are the major cell type of the nucleus reticularis thalami, *Brain Res.* **200**:341–354. [205]

Houser, C. R., Hendry, S. H., Jones, E. G., and Vaughn, J. E., 1983a, Morphological diversity of immunocytochemically identified GABA neurons in the monkey sensory-motor cortex, *J. Neurocytol.* **12**:617–638. [205]

Houser, C. R., Lee, M., and Vaughn, J. E., 1983b, Immunocytochemical localization of glutamic acid decarboxylase in normal and deafferented superior colliculus: Evidence for reorganization of gamma-aminobutyric acid synapses, *J. Neurosci.* **3**:2030–2042. [205]

Houser, C. R., Crawford, G. D., Barber, R. P., Salvaterra, P. M., and Vaughn, J. E., 1983c, Organization and morphological characteristics of cholinergic neurons: An immunocytochemical study with a monoclonal antibody to choline acetyltransferase, *Brain Res.* **266**:97–119. [245, 250, 252]

Howard, J. M., Chremos, A. N., Collen, M. J., *et al.* 1985, Famotidine, a new, potent, long lasting histamine H_2-receptor antagonist: Comparison with cimetidine and ranitidine in the treatment of Zollinger–Ellison syndrome, *Gastroenterology* **88**:1026–1033. [359]

Hrbek, J., Komenda, S., and Macakova, J., 1974, The effect of scopolamine (0.6 mg) and physostigmine (1.0 mg) on higher nervous system activity in man, *Acta Nerv. Super. (Prague)* **16**:213–215. [261]

Hubbard, J. I., 1973, Microphysiology of vertebrate neuromuscular transmission, *Physiol. Rev.* **53**:674–723. [94, 97, 99, 101, 102]

Hubel, D. H., 1982, Exploration of the primary visual cortex, 1955–1978, *Nature (London)* **299**:515–524. [605, 607, 609]

Hubel, D. H., and Wiesel, T. N., 1962, Receptive fields, binocular interaction and functional architecture in the cat's visual cortex, *J. Physiol. (London)* **160**:106–154. [607]

Hubel, D. H., and Wiesel, T. N., 1963, Shape and arrangement of columns in the cat's striate cortex, *J. Physiol. (London)* **165**:559–568. [608, 610]

Hubel, D. H., and Wiesel, T. N., 1965, Receptive fields and functional architecture in two nonstriate visual areas (18 and 19) of the cat, *J. Neurophysiol.* **28:**229–289. [610]

Hubel, D. H., and Wiesel, T. N., 1974, Sequence regularity and geometry of orientation columns in the monkey striate cortex, *J. Comp. Neurol.* **158:**267–294. [608, 609]

Hubel, D. H., and Wiesel, T. N., 1979, Brain mechanisms of vision, *Sci. Am.* **241:**130–144. [612]

Hubel, D. H., Wiesel, T. N., and LeVay, S., 1977, Plasticity of ocular dominance columns in monkey striate cortex, *Philos. Trans. R. Soc. London Ser. B* **278:**377–409. [453, 454, 607]

Hudson, D. B., Valcana, T., Bean, G., and Timiras, P. S., 1976, Glutamic acid: A strong candidate as the neurotransmitter of the cerebellar granule cells, *Neurochem. Res.* **1:**73–81. [184]

Hughes, J., Smith, J. W., Kosterliz, H. W., Fothergill, L. A., Morgan, B. A., and Morris, H. R., 1975a, Identification of methionine-enkephalin structure, *Nature (London)* **258:**577–579. [374]

Hughes, J., Smith, T., Morgan, B., and Fothergill, L., 1975b, Purification and properties of enkephalin—the possible endogenous ligand for the morphine receptor, *Life Sci.* **16:**1753–1758. [376]

Hultborn, H., 1976, Transmission of the pathway of reciprocal Ia inhibition to motoneurones and its control during the tonic stretch reflex, *Prog. Brain Res.* **44:**235–255. [137, 138]

Hultborn, H., and Pierrot-Deseilligny, E., 1979, Changes in recurrent inhibition during voluntary soleus contractions in man studied by an H-reflex technique, *J. Physiol. (London)* **297:**229–251. [138]

Hunt, R., and Taveau, R. de M., 1906, On the physiological action of certain choline derivatives and new methods for detecting choline, *Br. Med. J.* **2:**1788–1791. [80]

Hunt, S. P., Gilbert, E. R., Goldstein, M., and Kimmell, J. R., 1981a, Presence of avian pancreatic polypeptide-like immunoreactivity in catecholamine and methionine-enkephalin containing neurones within the central nervous system, *Neurosci. Lett.* **21:**125–130. [157]

Hunt, S. P., Kelly, J. S., Emson, P. C., Kimmel, J. R., Miller, R. J., and Wu, J.-Y., 1981b, An immunohistochemical study of neuronal populations containing neuropeptides or gamma-aminobutyrate within the superficial layers of the rat dorsal horn, *Neuroscience* **6:**1883–1893. [205]

Huszti, Z., and Magyar, K., 1984, Regulation of histidine decarboxylase activity in rat hypothalamus *in vitro* by ATP and cyclic AMP: Enzyme inactivation under phosphorylating conditions, *Agents Actions* **14:**546–549. [352]

Huxley, A. F., 1959, Ion movements during nerve activity, *Ann. N. Y. Acad. Sci.* **81:**221–246. [52]

Huxley, A. F., and Stampfli, R., 1949, Evidence for saltatory conduction in peripheral myelinated nerve fiber, *J. Physiol. (London)* **108:**315–339. [66]

Huxtable, R., and Barbeau, A. (eds.), 1976, *Taurine,* Raven Press, New York. [232]

Huxtable, R., Azari, J., Reisine, T., Johnson, P., Yamamura, H., and Barbeau, A., 1979, Regional distribution of amino acids in Friedreich's ataxia brains, *Can. J. Neurol. Sci.* **6:**255–258. [195]

Huynh, O. K., Recsei, P. A., Vaaler, G. L., and Snell, E. E., 1984, Histidine decarboxylase of *Lactobacillus* 30a, *J. Biol. Chem.* **259:**2833–2839. [351]

Hwang, B. H., and Wu, J.-Y., 1984, Ultrastructural studies on catecholaminergic terminals and GABAergic neurons in nucleus tractus solitarius of the rat medulla oblongata, *Brain Res.* **302:**57–67. [205]

Hydén, H., 1967, Biochemical changes accompanying learning, in: *The Neurosciences* (G. C. Ouarton, T. Melnechuk, and F. O. Schmitt, eds.), pp. 765–771, Rockefeller University Press, New York. [554]

Iansek, R., and Porter, R., 1980, The monkey globus pallidus: Neuronal discharge properties in relation to movement, *J. Physiol. (London)* **301:**439–455. [511]

Iansek, R., and Redman, S. J., 1973, The amplitude, time course and charge of unitary post-synaptic potentials evoked in spinal motoneurone dendrites, *J. Physiol. (London)* **234:**665–688. [116]

Iguchi, H., Chan, J. S. D., Dennis, M., Seidah, N. G., and Chrétien, M., 1985, Regional distribution of a novel pituitary protein (7B2) in the rat brain, *Brain Res.* **338:**91–96. [373]

Iles, J. F., 1977, Responses in human pretibial muscles to sudden stretch and to nerve stimulation, *Exp. Brain Res.* **30:**541–570. [489, 490]

Imaki, T., Shibasaki, T., Hotta, M., *et al.* 1985, The satiety effect of growth hormone-releasing factor in rats, *Brain Res.* **340:**186–188. [405]

Imig, T. J., and Morel, A., 1983, Organization of the thalamocortical auditory system in the cat, *Annu. Rev. Neurosci.* **6:**90–120. [612]

Imoto, K., Saida, K., Iwamura, K., Saida, T., and Nishitani, H., 1984, Amyotrophic lateral sclerosis: A double-blind crossover trial of thyrotropin releasing hormone, *J. Neurol. Neurosurg. Psychiatry* **47:**1332–1334. [410]

Inagaki, S., Kubota, Y., and Kito, S., 1986, Ultrastructural localization of enkephalin immunoreactivity in the substantia nigra of the monkey. *Brain Res.* **362**:171–174. [391]

Ingoglia, N. A., and Dole, V. P., 1970, Localization of D- and L-methadone after intraventricular injection into rat brains, *J. Pharmacol. Exp. Ther.* **175**:84–87. [374]

Ingram, C. D., Bicknell, R. J., Brown, D., and Leng, G., 1982, Rapid fatigue of neuropeptide secretion during continual electrical stimulation, *Neuroendocrinology* **35**:424–428. [403]

Ingvar, D. H., 1983, Serial aspects of language and speech related to prefrontal cortical activity: A selective review, *Hum. Neurobiol.* **2**:177–189. [69, 615]

Ingvar, D. G., 1985, Memory of the future: An essay on the temporal organization of conscious awareness, *Hum. Neurobiol.* **4**:127–136. [628]

Inoki, R., Matsumoto, K., Kudo, T., Kotani, Y., and Oka, M., 1979, Bradykinin as an algesic (pain producing) substance in the pulp, *Naunyn-Schmiedeberg's Arch. Pharmacol.* **306**:29–36. [407]

Iqbal, Z., and Ochs, S., 1980, Calmodulin in mammalian nerve, *J. Neurobiol.* **11**:311–318. [563]

Ishizuka, Y., Imai, A., Nakashima, J., and Nozawa, Y., 1983, Evidence for *de novo* synthesis of phosphatidylinositol coupled with histamine release in activated rat mast cells, *Biochem. Biophys. Res. Commun.* **111**:581–587. [357]

Israel, A., Correa, F. M. A., Niwa, M., and Saavedra, J., 1984, Quantitative determination of angiotensin II binding sites in rat brain and pituitary gland by autoradiography, *Brain Res.* **322**:341–345. [402]

Itakura, T., Yokote, H., Kimura, H., *et al.* 1985, 5-Hydroxytryptamine innervation of vessels in the rat cerebral cortex, *J. Neurosurg.* **62**:42–47. [334]

Ito, M., 1984, *The Cerebellum and Neural Control,* Raven Press, New York. [452, 491, 495, 501, 565, 577, 578, 580, 582, 583, 589]

Ito, M., and Kano, M., 1982, Long-lasting depression of parallel fiber–Purkinje cell transmission induced by conjunctive stimulation of parallel fibers and climbing fibers in the cerebellar cortex, *Neurosci. Lett.* **33**:253–258. [580]

Ito, M., and Oshima, T., 1964, The extrusion of sodium from cat spinal motoneurones, *Proc. R. Soc. London Ser. B.* **161**:109–131. [74]

Itoh, M., and Uchimura, H., 1981, Regional differences in cofactor saturation of glutamate decarboxylase (GAD) in discrete brain nuclei of the rat. Effect of repeated administration of haloperidol on GAD activity in the substantia nigra, *Neurochem. Res.* **6**:1283–1290. [200]

Ito, M., and Yoshida, M., 1964, The cerebellar-evoked monosynaptic inhibition of Deiters' neurons, *Experientia* **20**:515–516. [212, 216]

Ito, M., Kostyuk, P. G., and Oshima, T., 1962, Further study on anion permeability of inhibitory post-synaptic membrane of cat motoneurons, *J. Physiol. (London)* **164**:150–156. [127]

Ito, M., Udo, M., and Mano, N., 1970, Long inhibitory and excitatory pathways coverging onto cat reticular and Deiters' neurons and their relevance to reticulofugal axons, *J. Neurophysiol.* **33**:210–226. [141]

Ito, M., Shiida, T., Yagi, N., and Yamamoto, M., 1974, The cerebellar modification of rabbits' horizontal vestibulo-ocular reflex induced by sustained head rotation combined with visual stimulation, *Proc. Jpn. Acad.* **50**:85–89. [576]

Ito, M., Sakurai, M., and Tongroach, P., 1982, Climbing fiber induced depression of both mossy fiber responsiveness and glutamate sensitivity of cerebellar Purkinje cell, *J. Physiol. (London)* **324**:113–134. [455, 571, 574, 578–580, 582, 583]

Itoh, M., Ebadi, M., and Swanson, S., 1983, The presence of zinc-binding protein in brain, *J. Neurochem.* **41**:823–829. [201]

Itoh, N., Obata, K.-I., Yanaihara, N., and Okamoto, H., 1983, Human preprovasoactive intestinal polypeptide contains a novel PHI-27-like peptide, PHM-27, *Nature (London)* **304**:547–549. [382]

Iversen, L. L., 1967, *The Uptake and Storage of Noradrenaline in Sympathetic Nerves,* Cambridge University Press, Cambridge. [280, 313]

Iversen, L. L., 1973, Neuronal and extraneuronal catecholamine uptake mechanisms, in: *Frontiers in Catecholamine Research* (E. Usdin and S. H. Snyder, eds.), pp. 403–407, Pergamon Press, New York. [281]

Iversen, L. L., and Bloom, F. E., 1972, Studies of the uptake of ^{3}H-glycine in slices and homogenates of rat brain and spinal cord by electron microscopic autoradiography, *Brain Res.* **41**:131–143. [205, 228]

Iversen, L. L., and Glowinski, J., 1966, Regional studies of catecholamines in various brain regions, *J. Neurochem.* **13**:671–682. [283]

Iversen, L. L., and Schon, F. E., 1973, The use of autoradiographic techniques for the identification and mapping of transmitter specific neurons in CNS, in: *New Concepts in Neurotransmitter Mechanisms* (A. J. Mandell, ed.), pp. 153–193, Plenum Press, New York. [205]

Iwata, H., Nakayama, K., Matsuda, T., and Baba, A., 1984, Effect of taurine on a benzodiazepine–GABA–chloride ionophore receptor complex in rat brain membranes, *Neurochem. Res.* **9**:535–544. [231]

Izumi, K., Butterworth, R. F., and Barbeau, A., 1977, Effect of taurine on Ca^{++} binding to microsomes isolated from cerebral cortex, *Life Sci.* **20**:943–950. [231]

Jack, J. J. B., Noble, D., and Tsien, R. W., 1975, *Electric Current Flow in Excitable Cells*, Clarendon Press, Oxford. [564]

Jack, J. J. B., Redman, S. J., and Wong, K., 1981a, The components of synaptic potentials evoked in cat spinal motoneurones by impulses in single group Ia afferents, *J. Physiol. (London)* **321**:65–96. [114–116, 630]

Jack, J. J. B., Redman, S. J., and Wong, K., 1981b, Modifications to synaptic transmission at group Ia synapses on cat spinal motoneurones by 4-aminopyridine, *J. Physiol. (London)* **321**:111–126. [114, 116]

Jacobowitz, D. M., Skofitsch, G., Eskay, R. L., Keiser, H. R., and Zamir, N., 1985, Evidence for the existence of atrial natriuretic factor-containing neurons in the rat brain, *Neuroendocrinology* **40**:92–94. [373]

Jacobs, B. L., Gannon, P. J., and Azmitia, E., 1984, Atlas of serotonergic cell bodies in the cat brainstem: An immunocytochemical analysis, *Brain Res. Bull.* **13**:1–31. [329–331]

Jacobson, M., 1970, *Developmental Neurobiology*, Holt, Rinehard & Winston, New York. [417, 418, 434, 436]

Jacoby, J. H., Lytle, L. D. (eds.), 1978, Serotonin neurotoxins, *Ann. N. Y. Acad. Sci.* **305**:1–702. [339]

Jaeger, C. B., Ruggiero, D. A., Albert, V. R., Park, D. H., Joh, T. H., and Reis, D. J., 1984, Aromatic L-amino acid decarboxylase in the rat brain: Immunocytochemical localization in neurons of the brain stem, *Neuroscience* **11**:691–713. [273, 284]

Jaffe, E. H., Cuello, A. C., and Priestly, J. V., 1983, Localization of ^{3}H-GABA in the rat olfactory bulb: An *in vivo* and *in vitro* autoradiographic study, *Exp. Brain Res.* **50**:100–106. [205]

Jaim-Etcheverry, G., and Zieher, M. Z., 1968, Electron microscopic cytochemistry of 5-hydroxytryptamine (5-HT) in the beta cells of guinea pig endocrine pancreas, *Endocrinology* **83**:917–923. [332]

Jaim-Etcheverry, G., and Zieher, L. M., 1980, Stimulation–depletion of serotonin and noradrenaline from vesicles of sympathetic nerves in the pineal gland of the rat, *Cell Tissue Res.* **207**:13–20. [333]

Jaim-Etcheverry, G., and Zieher, L. M., 1982, Coexistence of monoamines in peripheral adrenergic neurons, in: *Co-transmission* (A. C. Cuello, ed.), pp. 189–206, Macmillan, New York. [157]

James, T. A., and Starr, M. S., 1979, Is glycine an inhibitory synaptic transmitter in the substantia nigra?, *Eur. J. Pharmacol.* **57**:115–125. [229]

Jankowska, E., 1979, New observations on neuronal organization of reflexes evoked from muscle spindle afferents, *Prog. Brain. Res.* **50**:29–36. [137]

Jankowska, E., and Lindstrom, S., 1971, Morphological identification of Renshaw cells, *Acta Physiol. Scand.* **81**:428–430. [138]

Janowsky, A., Okada, F., Manier, D. H., Applegate, C. D., Sulser, F., and Steranka, L. R., 1982, Role of serotonergic input in the regulation of the β-adrenergic receptor-coupled adenylate cyclase system, *Science* **218**:900–901. [334]

Janowsky, D. S., and Risch, S. C., 1984, Cholinomimetic and anticholinergic drugs used to investigate an acetylcholine hypothesis of affective disorders and stress, *Drug Dev. Res.* **4**:125–142. [549]

Janowsky, D. S., El-Yousef, M. K., Davis, J. M., and Sekerke, H. J., 1972, A cholinergic–adrenergic hypothesis of mania and depression, *Lancet* **2**:632–635. [549]

Jansen, J. K. S., and Matthews, P. B. C., 1962, The central control of the dynamic response of muscle spindles, *J. Physiol. (London)* **61**:357–378. [52]

Jasper, H. H., and Koyama, I., 1969, Rate of release of amino acids from the cerebral cortex of the cat as affected by brain stem and thalamic stimulation, *Can. J. Physiol. Pharmacol.* **47**:889–905. [231]

Javitch, J. A., D'Amato, R. J., Strittmatter, S. M., and Snyder, S. H., 1985, Parkinsonism-induced neurotoxin, N-methyl-4-phenyl-1,2,3,6-tetrahydropyridine: Uptake of the metabolite N-methyl-4-phenylpyridine by dopamine neurons explains selective toxicity, *Proc. Natl. Acad. Sci. U.S.A.* **82**:2173–2177. [303]

Javoy-Agid, F., Grouselle, D., Tixier-Vidal, A., and Agid, Y., 1983, Thyrotropin releasing hormone content is unchanged in brains of patients with Parkinson's disease, *Neuropeptides* **3**:405–410. [412]

Jeffery, D. R., and Roth, J. A., 1985, Purification and kinetic mechanism of human brain soluble catechol-O-methyltransferase, *J. Neurochem.* **44**:881–885. [278]

Jefferys, J. G. R., 1979, Initiation and spread of action potentials in granule cells maintained *in vitro* in slices of guinea-pig hippocampus, *J. Physiol. (London)* **289**:375–388. [60, 72]

Jegou, S., Javoy-Agid, F., Delbende, C., Ruberg, M., Vaudry, H., and Agid, Y., 1985, Cortical vasoactive intestinal peptide in relation to dementia in Parkinson's disease, *J. Neurol. Neurosurg. Psychiatry* **48**:842–845. [412]

Jenden, D. J., 1977, Estimation of acetylcholine and the dynamics of its metabolism, in: *Cholinergic Mechanisms and Psychopharmacology* (D. J. Jenden, ed.), pp. 138–162, Plenum Press, New York. [236]

Jenkins, J. S., Mather, H. M., and Coughlan, A. K., 1982, Effect of desmopressin on normal and impaired memory, *J. Neurol. Neurosurg. Psychiatry* **45**:830–831. [410]

Jennes, L., Stumpf, W. E., and Kalivas, P. W., 1982, Neurotensin: Topographical distribution in rat brain by immunohistochemistry, *J. Comp. Neurol.* **210**:211–224. [506]

Jennes, L., Stumpf, W. E., and Tappaz, M. L., 1983, Anatomical relationships of dopaminergic and GABAergic systems with the GnRH-systems in the septohypothalamic area, *Exp. Brain Res.* **50**:91–99. [391]

Jessel, T. M., 1981, The role of substance P in sensory transmission and pain perception, *Adv. Biochem. Psychopharmacol.* **28**:189–198. [406]

Jessel, T. M., 1983, Neuropeptide function in neurons that transmit and regulate nociceptive information, in: *Brain Peptides* (D. T. Krieger, J. D. Martin, and M. J. Brownstein, eds.), pp. 315–332, Wiley, New York. [406]

Jessel, T. M., Emson, P. C., Paxinos, G., and Cuello, A. C., 1978, Topographic projections of substance P and GABA pathways in the striato- and pallidonigral system: A biochemical and immunohistochemical study, *Brain Res.* **152**:487–498. [507]

Jessen, K. R., 1981, GABA and the enteric nervous system: A neurotransmitter function?, *Mol. Cell. Biochem.* **38**:69–76. [218]

Jessen, K. R., Mirsky, R., Dennison, M. E., and Burnstock, G., 1979, GABA may be a neurotransmitter in the vertebrate peripheral nervous system, *Nature (London)* **281**:71–74. [218]

Jew, J. Y., Léranth, C., Arimura, A., and Palkovits, M., 1984, LH-RH and somatostatin in the rat median eminence: An experimental light and electron microscopic immunocytochemical study, *Neuroendocrinology* **38**:169–175. [388]

Joels, M., and Urban, I. J. A., 1984, Arginine8-vasopressin enhances the responses of lateral septal neurons in the rat to excitatory amino acids and fimbria–fornix stimuli, *Brain Res.* **311**:201–209. [402]

Joh, T. H., and Goldstein, M., 1973, Isolation and characterization of multiple forms of phenylethanolamine-*N*-methyltransferase, *Mol. Pharmacol.* **9**:117–129. [275]

Joh, T. H., Baetge, E. E., Ross, M. E., and Reis, D. J., 1984, Biochemistry and molecular biology of catecholamine neurons: A single gene or gene family hypothesis, *Clin. Exp. Hypertens.—Theory Pract.* **A6**:11–21. [268]

Johansson, O., and Hökfelt, T., 1980, Thyrotropin releasing hormone, somatostatin, and enkephalin: Distribution studies using immunohistochemical techniques, *J. Histochem. Cytochem.* **28**:364–366. [400]

Johansson, O., Hökfelt, T., Pernow, B., *et al.* 1981, Immunohistochemical support for three putative transmitters in one neuron: Coexistence of 5-hydroxytryptamine, substance P- and thyrotropin releasing hormone-like immunoreactivity in medullary neurons projecting to the spinal cord, *Neuroscience* **6**:1857–1881. [157]

Johnson, A. E., Nock, B., McEwen, B. S., and Feder, H. H., 1985, Sex difference in α_1-noradrenergic receptor binding following estradiol treatment in the guinea pig assessed by quantitative autoradiography, *Soc. Neurosci. Abstr.* **11**:772. [524]

Johnson, I. P., Pullen, A. H., and Sears, T. A., 1985, Target dependence of Nissl body ultrastructure in cat thoracic motoneurones, *Neurosci. Lett.* **61**:201–205. [33]

Johnson, J. L., and Aprison, M. H., 1970, The distribution of glutamic acid, a transmitter candidate, and other amino acids in the dorsal sensory neuron of the cat, *Brain Res.* **24**:285–292. [182]

Johnsson, G., Malmfors, T., and Sachs, C., 1975, *6-Hydroxydopamine as a Denervation Tool in Catecholamine Research,* North-Holland, Amsterdam. [301]

Johnston, G. A. R., 1968, The intraspinal distribution of some depressant amino acids, *J. Neurochem.* **15**:1013–1017. [182]

Johnston, G. A., 1977, Effects of calcium on potassium-stimulated release of radioactive β-alanine and gamma-aminobutyric acid from slices of rat cerebral cortex and spinal cord, *Brain Res.* **121**:179–181. [233]

Johnston, J. P., 1968, Some observations upon a new inhibitor of monoamine oxidase in brain tissue, *Biochem. Pharmacol.* **17**:1285–1297. [275]

Johnston, M. V., and Coyle, J. T., 1980, Ontogeny of neurochemical markers for noradrenergic, GABAergic, and cholinergic neurons in neocortex lesioned with methylazoxymethanol acetate, *J. Neurochem.* **34**:1429–1441. [212]

Johnston, M. V., McKinney, M., and Coyle, J. T., 1979, Evidence for a cholinergic projection to the neocortex from neurons in basal forebrain, *Proc. Natl. Acad. Sci. U.S.A.* **76**:5392–5396. [247]

Johnstone, E. C., Crow, T. J., Frith, C. D., Husband, J., and Kreel, L., 1976, Cerebral ventricular size and cognitive impairment in chronic schizophrenia, *Lancet* **2**:924–926. [546]

Jolles, J., van Dungen, C. J., ten Haaf, J., and Gispen, W. H., 1982, Polyphosphoinositide metabolism in rat brain: Effects of neuropeptides, neurotransmitters and cyclic nucleotides, *Peptides* **3**:709–714. [402]

Jonas, V., Lin, C. R., Kawashima, E., *et al.* 1985, Alternative RNA processing events in human calcitonin/calcitonin gene-related peptide gene expression, *Proc. Natl. Acad. Sci. U.S.A.* **82**:1994–1998. [376]

Jones, A. W., Croucher, M. J., Meldrum, B. S., and Watkins, J. C., 1984, Suppression of audiogenic seizures in DBA/2 mice by two new dipeptide NMDA receptor antagonists, *Neurosci. Lett.* **45**:157–161. [187]

Jones, D. L., and Mogenson, G. J., 1980, Nucleus accumbens to globus pallidus GABA projection: Electrophysiological and iontophoretic investigations, *Brain Res.* **188**:93–105. [212]

Jones, E. G., 1983, The nature of the afferent pathways conveying short-latency inputs to primate motor cortex, *Adv. Neurol.* **39**:263–285. [83, 487, 490]

Jones, E. G., and Porter, R., 1980, What is area 3a?, *Brain Res.* **203**:1–43. [487]

Jones, E. G., and Powell, T. P. S., 1968, The projection of the somatic sensor cortex upon the thalamus of the cat, *Brain Res.* **10**:369–391. [605]

Jones, E. G., and Powell, T. P. S., 1969, Electron microscopy of synaptic glomeruli in the thalamic relay nuclei of the cat, *Proc. R. Soc. London Ser. B* **172**:153–171. [136]

Jones, E. G., Burton, H., Saper, C. B., and Swanson, L. W., 1976, Midbrain, diencephalic and cortical relationships of the basal nucleus of Meynert and associated structures in primates, *J. Comp. Neurol.* **167**:385–420. [247]

Jones, I. M., Jordan, C. C., Morton, I. K. M., Stagg, C. J., and Webster, R. A., 1974, The effect of chronic dorsal root section on the concentration of free amino acids in the rabbit spinal cord, *J. Neurochem.* **23**:1239–1244. [182]

Jones, R. S. G., and Boulton, A. A., 1980, Tryptamine and 5-hydroxytryptamine: Actions and interactions on cortical neurones in the rat, *Life Sci.* **27**:1849–1856. [346]

Jonson, B., and Fredholm, B. B., 1985, Release of purines, noradrenaline and GABA from rat hippocampal slices by field stimulation, *J. Neurochem.* **44**:217–224. [365]

Jonsson, G., Gorio, A., Hallman, H., Jangro, D., Kojima, H., and Zanoni, R., 1984, Effect of GM_1 ganglioside on neonatally neurotoxin induced degeneration of serotonin neurons in the rat brain, *Dev. Brain Res.* **16**:171–180. [467]

Jope, R. S., 1979, High affinity choline transport and acetyl CoA production in brain and their roles in the regulation of acetylcholine synthesis, *Brain Res.* **180**:313–344. [238, 239, 242, 245]

Jope, R. S., 1981, Acetylcholine turnover and compartmentation in rat brain synaptosomes, *J. Neurochem.* **36**:1712–1721. [245]

Joseph, M. H., Baker, H. F., Crow, T. J., Riley, G. L., and Risby, D., 1979, Tryptophan metabolism in schizophrenia: A post mortem study of the serotonin and kynurenine pathways in schizophrenic and control subjects, *Psychopharmacology* **62**:279–285. [547]

Jouvet, M., 1973, Serotonin and sleep in the cat, in: *Serotonin and Behavior* (J. Barchas and E. Usdin, eds.), pp. 385–400, Academic Press, New York. [337]

Junig, J. T., Abood, L. G., and Skrobala, A. M., 1985, Two classes of arginine vasopressin binding sites on rat brain membranes, *Neurochem. Res.* **10**:1187–1202. [402]

Kai-Kai, M. A., Swann, R. W., and Keen, P., 1985, Localization of chromatographically characterized oxytocin and arginine-vasopressin to sensory neurones in the rat, *Neurosci. Lett.* **55**:83–88. [406]

Kaiya, H., Tanaka, T., Takeuchi, K., *et al.* 1983, Decreased level of β-endorphin-like immunoreactivity in cerebrospinal fluid of patients with senile dementia of Alzheimer type, *Life Sci.* **33**:1039–1043. [412]

Kakidani, H., Furutani, Y., Takahashi, H., *et al.* 1982, Cloning and sequence analysis of cDNA for porcine β-neo-endorphin/dynorphin precursor, *Nature (London)* **298:**245–249.   [382]

Kallman, F. J., 1938, *The Genetics of Schizophrenia,* Augustin, New York.   [545]

Kamo, H., Kim, S. U., McGeer, P. L., and Shin, D. H., 1985, Transplantation of cultured fetal human adrenal chromaffin cells to rat brain, *Neurosci. Lett.* **57:**43–48.   [459]

Kanba, S., and Richelson, E., 1984, Histamine H₁ receptors in human brain labeled with [³H]doxepin, *Brain Res.* **304:**1–7.   [356, 357]

Kandel, E. R., and Schwartz, J. H., 1982, Molecular biology of learning: Modulation of transmitter release, *Science* **218:**433–443.   [574]

Kandel, E. R., and Spencer, W. A., 1968, Cellular neurophysiological approaches in the study of learning, *Physiol. Rev.* **48:**65–134.   [554]

Kandel, E. R., Spencer, W. A., and Brinley, F. J., 1961, Electrophysiology of hippocampal neurons. 1. Sequential invasion and synaptic organization, *J. Neurophysiol.* **24:**225–265.   [139]

Kaneko, T., Tashiro, K., Sugimoto, T., Konishi, A., and Mizuno, N., 1985, Identification of the thalamic neurons with vasoactive intestinal polypeptide-like immunoreactivity in the rat, *Brain Res.* **347:**390–393.   [388]

Kangawa, K., Tawaragi, Y., Oikawa, S., *et al.* 1984, Identification of rat γ atrial natriuretic polypeptide and characterization of the cDNA encoding its precursor, *Nature (London)* **312:**152–155.   [382]

Kanof, P. D., and Greengard, P., 1979, Pharmacological properties of histamine-sensitive adenylate cyclase from mammalian brain, *J. Pharmacol. Exp. Ther.* **209:**87–96.   [358]

Karczmar, A. G., 1981, Basic phenomena underlying novel use of cholinergic agents, anticholinesterases and precursors in neurological including peripheral and psychiatric disease, *Adv. Behav. Biol.* **25:**853–869.   [258]

Karlin, A., 1983, The anatomy of a receptor, *Neurosci. Commun.* **1:**111–123.   [101]

Karlsson, A., Fonnum, F., Malthe-Sorenssen, D., and Storm-Mathisen, J., 1974, Effect of the convulsive agent 3-mercaptopropionic acid on the levels of GABA, other amino acids and glutamate decarboxylase in different regions of the rat brain, *Biochem. Pharmacol.* **23:**3052–3061.   [223]

Karpiak, S. E., 1983, Ganglioside treatment improves recovery of alternation behavior after unilateral entorhinal cortex lesion, *Exp. Neurol.* **81:**330–339.   [467]

Karppanen, H., 1982, Control of circulation by cerebral catecholaminergic and histaminergic mechanisms, *Acta Med. Scand. (Suppl.)* **660:**40–48.   [358]

Kase, Y., Takahama, K., Hashimoto, T., Kaisaku, J., Okano, Y., and Miyata, T., 1980, Electrophoretic study of pipecolic acid, a biogenic imino acid, in the mammalian brain, *Brain Res.* **193:**608–613.   [233]

Kastin, A. J., Barbeau, A., Plotnikoff, N. P., Schally, A. V., and Ehrensing, R. H., 1977, MIF-I: Actions in man, in: *Clinical Neuroendocrinology* (L. Martini and G. M. Besser, eds.), pp. 393–400, Academic Press, New York, San Francisco, and London.   [386]

Kastin, A. J., Banks, W. A., Zadina, J. E., and Graf, M., 1983, Brain peptides: The dangers of constricted nomenclatures, *Life Sci.* **32:**295–301.   [376]

Kataoka, K., Nakamura, Y., and Hassler, R., 1973, Habenulo-interpeduncular tract: A possible cholinergic neuron in rat brain, *Brain Res.* **62:**264–267.   [250]

Kataoka, K., Sorimachi, M., Okuno, S., and Mizuno, N., 1977, Cholinergic and GABAergic fibers in the stria medullaris of the rabbit, *Brain Res. Bull.* **2:**461–464.   [212]

Kater, S., and Letourneau, P. (eds.), 1985, Biology of the growth cone, *J. Neurosci. Res.* **13:**1–335.   [417]

Kato, K., Takahashi, Y., and Matsubara, K., 1985, Molecular cloning of the human cholecystokinin gene, *Ann. N. Y. Acad. Sci.* **448:**613–615.   [382]

Katoh-Semba, R., Skaper, S. D., and Varon, S., 1984, Interaction of GM₁ ganglioside with PC12 pheochromocytoma cells: Serum- and NGF-dependent effects on neuritic growth (and proliferation), *J. Neurosci. Res.* **12:**299–310.   [466]

Katsuura, G., and Itoh, S., 1985, Potentiation of β-endorphin effects by proglumide in rats, *Eur. J. Pharmacol.* **107:**363–366.   [407]

Katz, B., 1958, Microphysiology of the neuromuscular junction: A physiological "quantum of action" at the myoneural junction, *Bull. Johns Hopkins Hosp.* **102:**275–295.   [93]

Katz, B., 1962, The transmission of impulses from nerve to nerve and the subcellular unit of synaptic action, *Proc. R. Soc. London Ser. B* **155:**455–479.   [93]

Katz, B., 1966, *Nerve, Muscle and Synapse,* McGraw-Hill, New York.   [53, 86, 92, 94]

Katz, B., 1969, *The Release of Neural Transmitter Substances,* Liverpool University Press, Liverpool. [86, 91, 94]

Katz, B., and Miledi, R., 1965a, Propagation of electric activity in motor nerve terminals, *Proc. R. Soc. London Ser. B* **161**:453–482. [94, 107]

Katz, B., and Miledi, R., 1965b, Measurement of synaptic delay and time course of acetylcholine release at neuromuscular junction, *Proc. R. Soc. London Ser. B.* **161**:483–495. [92, 93]

Katz, B., and Miledi, R., 1965c, The effect of calcium on acetylcholine release from motor nerve terminals, *Proc. R. Soc. London Ser. B* **161**:496–503. [94]

Katz, B., and Miledi, R., 1966, Input–output relation of a single synapse, *Nature (London)* **212**:1242–1245. [60, 103]

Katz, B., and Miledi, R., 1967a, A study of synaptic transmission in the absence of nerve impulses, *J. Physiol. (London)* **192**:407–436. [95, 105]

Katz, B., and Miledi, R., 1967b, Release of acetylcholine from nerve endings by graded electric pulses, *Proc. R. Soc. London Ser. B* **167**:23–38. [94]

Katz, B., and Miledi, R., 1968, The role of calcium in neuromuscular facilitation, *J. Physiol. (London)* **195**:481–492. [97, 554]

Katz, B., and Miledi, R., 1969a, Tetrodotoxin-resistant electric activity in presynaptic terminals, *J. Physiol. (London)* **203**:459–487. [59, 106, 107]

Katz, B., and Miledi, R., 1969b, Spontaneous and evoked activity of motor nerve endings in calcium Ringer, *J. Physiol. (London)* **203**:689–706. [106, 107]

Katz, B., and Miledi, R., 1972, The statistical nature of the acetylcholine potential and its molecular components, *J. Physiol. (London)* **224**:665–699. [97]

Katz, B., and Miledi, R., 1973, The binding of acetylcholine to receptors and its removal from the synaptic cleft, *J. Physiol. (London)* **231**:549–574. [97]

Katz, B., and Miledi, R., 1977, Suppression of transmitter release at the neuromuscular junction, *Proc. R. Soc. London Ser. B* **196**:465–469. [105]

Katzman, R., 1978, Dementias, *Postgrad. Med.* **64**:119–125. [25, 201]

Kaufman, D. L., McGinnis, J. F., Wood, T. L., and Tobin, A. J., 1985, Brain glutamate decarboxylase (GAD) cDNA cloned on the basis of its antigenicity and ability to produce GABA, *Soc. Neurosci. Abstr.* **11**:1115. [46]

Kaufman, L. S., Pfaff, D. W., and McEwen, B. S., 1985, Effects of intracranial application of various cholinergic drugs and estrogen on lordotic behavior in Long Evans rats, *Soc. Neurosci. Abstr.* **11**:1894. [524]

Kawaguchi, S., and Ono, T., 1973, Bicuculline and picrotoxin sensitive inhibition in interpositus neurones of cat, *Brain Res.* **58**:260–265. [213]

Kawaguchi, S., Yamamoto, Y., Mizuno, N., and Iwahori, N., 1975, The role of climbing fibers in the development of Purkinje cell dendrites, *Neurosci. Lett.* **1**:301–304. [432, 463]

Kawaguchi, S., Yamamoto, T., Samejima, A., Itoh, K., and Mizuno, N., 1979a, Morphological evidence for axonal sprouting of cerebello-thalamic neurons in kittens after neonatal hemicerebellectomy, *Exp. Brain Res.* **35**:511–518. [448]

Kawaguchi, S., Yamamoto, T., and Samejima, A., 1979b, Electrophysiological evidence for axonal sprouting of cerebello-thalamic neurons in kittens after neonatal hemicerebellectomy, *Exp. Brain Res.* **36**:21–39. [448]

Kawaguchi, S., Miyata, H., Kawamura, M., and Harada, Y., 1981, Morphological and electrophysiological evidence for axonal regeneration of axotomized cerebello-thalamic neurons in kittens, *Neurosci. Lett.* **25**:13–18. [448]

Kawaguchi, S., Miyata, H., and Kato, N., 1982, Axonal regeneration of axotomized cerebello-thalamic projection neurones in adult cats, *J. Physiol. Soc. Jpn.* **44**:383. [450]

Kawaguchi, S., Miyata, H., and Kato, N., 1986, Regeneration of the cerebellofugal projection after transection of the superior cerebellar peduncle in kittens: Morphological and electrophysiological studies, *J. Comp. Neurol.* **245**:258–273. [448–450]

Kawai, N., Yamagishi, S., Saito, M., and Furuya, K., 1983, Blockade of synaptic transmission in the squid giant synapse by a spider toxin (JSTX), *Brain Res.* **278**:346–349. [104, 187]

Kawai, Y., Takami, K., Shiosaka, S., *et al.* 1985, Topographical localization of calcitonin gene-related peptide in the rat brain: An immunohistochemical analysis, *Neuroscience* **15**:747–763. [388, 400]

Kawata, M., Kakao, K., Morii, N., *et al.* 1985, Atrial natriuretic polypeptide: Topographical distribution in the rat brain by radioimmunoassay and immunohistochemistry, *Neuroscience* **16**:521–546. [373, 388, 395]

Kawatani, M., Rutigliano, M., and De Groat, W. C., 1985, Selective facilitatory effect of vasoactive intestinal polypeptide (VIP) on muscarinic firing in vesical ganglia of the cat, *Brain Res.* **336**:223–234. [402]

Kay, D. W. K., 1986, The genetics of Alzheimer's disease, *Br. Med. Bull.* **42**:19–23. [261]

Kebabian, J. W., and Calne, D. B., 1979, Multiple receptors for dopamine, *Nature (London)* **277**:93–96. [298, 300]

Kebabian, J. W., Petzold, G. L., and Greengard, P., 1972, Dopamine-sensitive adenyl cyclase in caudate nucleus of rat brain and its similarity to the "dopamine receptor," *Proc. Natl. Acad. Sci. U.S.A.* **69**:2145–2149. [298]

Kebabian, J. W., Steiner, A. L., and Greengard, P., 1975, Muscarinic cholinergic regulation of cyclic guanosine 3′,5′-monophosphate in autonomic ganglia: Possible role in synaptic transmission, *J. Pharmacol. Exp. Ther.* **193**:474–488. [299]

Kebabian, J. W., Beaulieu, M., and Itoh, Y., 1984, Pharmacological and biochemical evidence for the existence of two categories of dopamine receptor, *Can. J. Neurol. Sci.* **11**:114–117. [297, 298, 300]

Keeling, D. J., Smith, I. R., and Tipton, K. F., 1984, A coupled assay for histidine decarboxylase: *In vivo* turnover of the enzyme in mouse brain, *Naunyn-Schmiedeberg's Arch. Pharmacol.* **326**:215–221. [350]

Kehoe, J., 1972, Ionic mechanisms of a two-component cholinergic inhibition in *Aplysia* neurones, *J. Physiol. (London)* **225**:85–114. [131]

Keilhauer, G., Faissner, A., and Schachner, M., 1985, Differential inhibition of neurone–neurone, neurone–astrocyte and astrocyte–astrocyte adhesion by L1, L2 and N-CAM antibodies, *Nature (London)* **316**:728–730. [37]

Kellar, K. J., and Cascio, C. S., 1982, [³H]Tryptamine: High affinity binding sites in rat brain, *Eur. J. Pharmacol.* **78**:475–478. [346]

Keller, F., Rimvall, K., and Waser, P. G., 1984, Slice cultures confirm the presence of cholinergic neurons in the rat habenula, *Neurosci. Lett.* **52**:299–304. [250]

Kelly, J. S., and Dodd, J., 1981, Cholecystokinin and gastrin as neurotransmitters in the mammalian nervous system, in: *Neurosecretion and Brain Peptides: Implications for Brain Functions and Neurological Disease* (J. B. Martin, S. Reichlin, and K. L. Bick, eds.), pp. 133–144, Raven Press, New York. [401]

Kelly, J. S., Dick, F., and Schon, F., 1975, The autoradiographic localization of the GABA-releasing nerve terminals in cerebellar glomeruli, *Brain Res.* **85**:255–259. [205]

Kelly, P. T., and Cotman, C. W., 1978, Synaptic proteins: Characterization of tubulin and actin and identification of a distinct postsynaptic density protein, *J. Cell Biol.* **79**:173–183. [562]

Kelly, R. B., Deutsch, J. W., Carlson, S. S., and Wagner, J., 1979, Biochemistry of neurotransmitter release, *Annu. Rev. Neurosci.* **2**:399–446. [96]

Kelly, R. W., Amato, F., and Seamark, R. F., 1984, *N*-Acetyl-5-methoxykynurenamine, a brain metabolite of melatonin, is a potent inhibitor of prostaglandin biosynthesis, *Biochem. Biophys. Res. Commun.* **121**:372–379. [345]

Kelner, K. L., Miller, A. L., and Peck, E. J., Jr., 1980, Estrogens and the hypothalamus: Nuclear receptor and RNA polymerase activation, *J. Recept. Res.* **1**:215–237. [522]

Kemali, D., Del Vecchio, M., and Maj, M., 1982, Increased noradrenaline levels in CSF and plasma of schizophrenic patients, *Biol. Psychiatry* **17**:711–717. [547]

Kemp, J. M., and Powell, T. P. S., 1971, The structure of the caudate nucleus of the cat: Light and electron microscopy, *Philos. Trans. R. Soc. London Ser. B* **262**:383–401. [250, 504]

Kendall, D. A., Ferkany, J. W., and Enna, S. J., 1980, Properties of ³H-cimetidine binding in rat brain membrane fractions, *Life Sci.* **26**:1293–1302. [357]

Kennedy, C., Des Rosiers, M. H., Sukurada, O., *et al.* 1976, Metabolic mapping in primary visual system of the monkey by means of the autoradiographic [C¹⁴]deoxyglucose technique, *Proc. Natl. Acad. Sci. U.S.A.* **73**:4230–4234. [609]

Kennedy, M. B., 1983, Experimental approaches to understanding the role of protein phosphorylation in the regulation of neuronal function, *Annu. Rev. Neurosci.* **6**:493–525. [97]

Kent, S., 1976, Neurotransmitters may be weak link in the aging brain's communication network. *Geriatrics* **31**:105–111. [471]

Kerr, F. W. L., and Wilson, P. R., 1978, Pain, *Annu. Rev. Neurosci.* **1**:83–102. [603]

Kerwin, R. W., and Pycock, C. J., 1979a, The effect of some putative neurotransmitters on the release of 5-

hydroxytryptamine and gamma-aminobutyric acid from slices of the rat midbrain raphe area, *Neuroscience* **4:**1359–1365. [228]

Kerwin, R. W., and Pycock, C. J., 1979b, Specific stimulating effect of glycine on ³H-dopamine efflux from substantia nigra slices of the rat, *Eur. J. Pharmacol.* **54:**93–98. [229]

Kerwin, R. W., Pay, S., Bhoola, K. D., and Pycock, C. J., 1980, Vasoactive intestinal polypeptide (VIP)-sensitive adenylate cyclase in rat brain: Regional distribution and localization on hypothalamic neurons, *J. Pharm. Pharmacol.* **32:**561–566. [402]

Kety, S. S., 1961, Sleep and the energy metabolism of the brain, in: *The Nature of Sleep* (G. E. W. Wolstenholme and M. O'Connor, eds.), pp. 375–385, J. A. Churchill, London. [69]

Kety, S. S., 1974, From rationalization to reason, *Am. J. Psychiatry* **131:**957–963. [545]

Keynes, R. D., 1975, Organization of the ionic channels in nerve membranes, in: *The Basic Neurosciences*, Vol. 1 (D. B. Tower, ed.), pp. 165–175, Raven Press, New York. [57, 68]

Keynes, R. D., 1983, Voltage-gated ion channels in the nerve membrane, *Proc. R. Soc. London Ser. B* **220:**1–30. [53, 56–59]

Khandelwal, J. K., Hough, L. R., and Green, J. P., 1984, Regional distribution of the histamine metabolite tele-methylimidazoleacetic acid in rat brain: Effects of pargyline and probenecid, *J. Neurochem.* **42:**519–522. [351, 353]

Kievit, J., and Kuypers, H. G., 1975, Rasal forebrain and hypothalamic connections to frontal and parietal cortex in the rhesus monkey, *Science* **187:**660–662. [247]

Kilcoyne, M. M., Hoffman, D. L., and Zimmerman, E. A., 1980, Immunocytochemical localization of angiotensin II and vasopressin in rat hypothalamus: Evidence for production in the same neuron, *Clin. Sci.* **59**(Suppl. 6):57s–60s. [157]

Kilpatrick, I. C., and Starr, M. S., 1981, The nucleus tegmenti pedunculopontinus and circling behavior in the rat, *Neurosci. Lett.* **26:**11–16. [212]

Kilpatrick, I. C., Starr, M. S., Fletcher, A., James, T. A., and MacLeod, N. K., 1980, Evidence for a GABAergic nigrothalamic pathway in the rat. I. Behavioural and biochemical studies, *Exp. Brain Res.* **40:**45–54. [212]

Kim, J. S., and Giacobini, E., 1984, Quantitative determination and regional distribution of pipecolic acid in rodent brain, *Neurochem. Res.* **9:**1559–1569. [233]

Kim, J. S., Hassler, R., Hang, P., and Paik, K.-S., 1977, Effect of frontal cortex ablation on striatal glutamic acid level in rat, *Brain Res.* **132:**370–374. [182]

Kim, R., Nakano, K., Hayaramon, A., and Carpenter, M. B., 1976, Projections of the globus pallidus and adjacent structures: An autoradiographic study in the monkey, *J. Comp. Neurol.* **169:**263–290. [508]

Kimelberg, H. K., and Katz, D. M., 1985, High affinity uptake of serotonin into immunocytochemically identified astrocytes, *Science* **228:**889–890. [342]

Kimura, D., 1967, Functional asymmetry of the brain in dichotic listening, *Cortex* **3:**163–178. [617]

Kimura, H., McGeer, P. L., Peng, J. H., and McGeer, E. G., 1980, Choline acetyltransferase containing neurons in rodent brain by immunohistochemistry, *Science* **208:**1057–1059. [85, 247, 250]

Kimura, H., McGeer, P. L., Peng, J. H., and McGeer, E. G., 1981, The central cholinergic system studied by choline acetyltransferase immunohistochemistry in the cat, *J. Comp. Neurol.* **200:**151–201. [236, 245, 247, 250–252]

Kimura, H., McGeer, E. G., Peng, F., and McGeer, P. L., 1983, Cholinergic systems in the cat cortex studied by choline acetyltransferase immunohistochemistry, in: *Structure and Function of Peptidergic and Aminergic Neurons* (Y. Sano, Y. Ibata, and E. Z. Zimmerman, eds.), pp. 263–274, Japan Science Society Press, Tokyo. [247, 250]

Kimura, H., McGeer, P. L., Peng, J. H., and McGeer, E. G., 1984, Choline acetyltransferase-containing neurons in the rat brain, in: *Handbook of Chemical Neuroanatomy* (A. Björklund, J. Hökfelt, and M. J. Kuhar, eds.), pp. 51–67, Elsevier, Amsterdam. [245, 250–252]

King, J. C., and Anthony, E. L. P., 1984, LHRH neurons and their projections in humans and other mammals: Species comparisons, *Peptides* **5**(Suppl. 1):195–207. [388]

Kita, H., 1978, The inhibition and the possible transmitter substance from the frontal cortex to the lateral hypothalamic area in the rat, *Futuoka Igaku Zasshi* **69:**223–250. [229]

Kita, H., and Oomura, Y., 1982, Evidence for a glycinergic cortico-lateral hypothalamic inhibitory pathway in the rat, *Brain Res.* **235:**131–136. [229]

Klee, C. B., and Haiech, J., 1980, Concerted role of calmodulin and calcineurin in calcium regulation, *Ann. N. Y. Acad. Sci.* **256:**43–54. [169]

Klee, C. B., Crouch, T. H., and Richman, P. G., 1980, Calmodulin, *Annu. Rev. Biochem.* **49:**489–515. [169]

Kleinebeckel, D., 1982, Acceleration of muscle re-innervation in rats by ganglioside treatment: An electromyographic study, *Eur. J. Pharmacol.* **80:**243–245. [467]

Klinman, J. P., Krueger, M., Brenner, M., and Edmondson, D. E., 1984, Evidence for two copper atoms/subunit in dopamine β-monooxygenase catalysis, *J. Biol. Chem.* **259:**3399–3402. [274]

Kluver, H., and Bucy, P., 1939, Preliminary analysis of functions of the temporal lobes in monkeys, *Arch. Neurol. Psychiatry* **42:**979–1000. [528]

Knapp, S., Mandell, A. J., Russo, P. V., Vitto, A., and Stewart, D. K., 1981, Strain differences in kinetic and thermal stability of two mouse brain tryptophan hydroxylase activities, *Brain Res.* **230:**317–336. [323]

Knowles, R. G., and Pogson, C. I., 1984, Tryptophan uptake and hydroxylation in rat forebrain synaptosomes, *J. Neurochem.* **42:**677–684. [323]

Kobayashi, R. M., Palkovits, M., Hruska, R. E., Rothschild, R., and Yamamura, H. I., 1978, Regional distribution of muscarinic cholinergic receptors in rat brain, *Brain Res.* **154:**13–23. [253]

Koch, C., and Poggio, T., 1983, Electrical properties of dendritic spines, *TINS* **6:**80–83. [564]

Koch-Henriksen, N., and Nielsen, H., 1981, Vasopressin in post-traumatic amnesia, *Lancet* **1:**38–39. [410]

Koe, B. K., and Weissman, A., 1966, p-Chlorophenylalanine: A specific depletor of brain serotonin, *J. Pharmacol. Exp. Ther.* **154:**499–516. [341]

Koelle, G. B., 1970, Anticholinesterase agents, in: *The Pharmacological Basis of Therapeutics* (L. S. Goodman and A. Gilman, eds.), pp. 442–465, Macmillan, London. [259]

Koerner, J. F., and Cotman, C. W., 1981, Micromolar L-2-amino-4-phosphonobutyric acid selectively inhibits perforant path synapses from lateral entorhinal cortex, *Brain Res.* **216:**192–198. [182]

Koevary, S. B., McEvoy, R. C., and Azmitia, E. C., 1983, Evidence for the presence of serotonergic perikarya in the fetal rat pancreas as demonstrated by high affinity uptake of [3H]5-HT, *Brain Res.* **280:**368–372. [332]

Kohler, C., 1982, Distribution and morphology of vasoactive intestinal peptide-like immunoreactive neurons in regio superior of the rat hippocampal formation, *Neurosci. Lett.* **33:**265–270. [397]

Kohler, C., and Chan-Palay, V., 1983a, Distribution of gamma aminobutyric acid containing neurons and terminals in the septal area: An immunohistochemical study using antibodies to glutamic acid decarboxylase in the rat brain, *Anat. Embryol. (Berlin)* **167:**53–65. [205]

Kohler, C., and Chan-Palay, V., 1983b, Gamma-aminobutyric acid interneurons in the rat hippocampal region studied by retrograde transport of glutamic acid decarboxylase antibody after *in vivo* injections, *Anat. Embryol. (Berlin)* **166:**53–66. [205, 212]

Kohler, C., and Schwarcz, R., 1983, Comparison of ibotenate and kainate neurotoxicity in rat brain: A histological study, *Neuroscience* **8:**819–836. [192]

Köhler, G., 1986, Derivation and diversification of monoclonal antibodies, *Science* **233:**1281–1286. [48]

Köhler, G., and Milstein, C., 1975, Continuous cultures of fused cells secreting antibody of predefined specificity, *Nature (London)* **256:**495–497. [46]

Koike, H., Okada, Y., Oshima, T., and Takashi, K., 1968, Accommodative behavior of cat pyramidal tract cells investigated with intracellular injection of currents, *Exp. Brain Res.* **5:**173–188. [120]

Kojima, H., Gorio, A., Janigro, D., and Jonsson, G., 1984, GM₁ ganglioside enhances regrowth of noradrenaline nerve terminals in rat cerebral cortex lesioned by the neurotoxin 6-hydroxydopamine, *Neuroscience* **13:**1011–1022. [467]

Kojima, T., Saito, K., and Kakimi, S., 1975, *An Electron Microscopic Atlas of Neurons,* University of Tokyo Press, Tokyo. [142]

Koller, K. J., Zaczek, R., and Coyle, J. T., 1984, N-Acetyl-aspartyl-glutamate: Regional levels in rat brain and the effects of brain lesions as determined by a new HPLC method, *J. Neurochem.* **43:**1136–1142. [186]

Kondo, H., Kuramato, H., Wainer, B. H., and Yanaihara, N., 1985, Discrete distribution of cholinergic and vasoactive intestinal polypeptidergic amacrine cells in the rat retina, *Neurosci. Lett.* **54:**213–218. [157]

Kondo, Y., and Iwatsubo, K., 1978, Increased release of preloaded [3H]GABA from substantia nigra *in vivo* following stimulation of caudate nucleus and globus pallidus, *Brain Res.* **154:**395–400. [212, 215]

Korn, H., and Faber, D. S., 1986, Regulation and significance of probabilistic release mechanisms at central synapses, in: *New Insights into Synaptic Function* (G. M. Edelman, W. E. Gall, and W. M. Cowan, eds.), Neurosciences Research Foundation, John Wiley, New York (in press). [116, 630]

Korn, H., Mallet, A., Triller, A., and Faber, D. S., 1982, Transmission at a central inhibitory synapse. II.

Quantal description of release, with a physical correlate for binomial n., *J. Neurophysiol.* **48:**679–707. [116]

Kornhuber, H. H., 1973, Neural control of input into long-term memory: Limbic system and amnestic syndrome in man, in: *Memory and Transfer of Information* (H. P. Zippel, ed.), pp. 1–22, Plenum Press, New York. [567, 569, 570]

Kornhuber, J., Kim, J.-S., Kornhuber, M. E., and Kornhuber, H. H., 1984, The cortico-nigral projection: Reduced glutamate content in the substantia nigra following frontal cortex ablation in the rat, *Brain Res.* **322:**124–126. [182]

Kosaka, T., Hataguchi, Y., Hama, K., Nagatsu, I., and Wu, J.-Y., 1985, Coexistence of immunoreactivities for glutamate decarboxylase and tyrosine hydroxylase in some neurons in the periglomerular region of the rat main olfactory bulb: Possible coexistence of gamma-aminobutyric acid (GABA) and dopamine, *Brain Res.* **343:**166–171. [157]

Kosar, E., Waters, A. S., Tsukahara, N., and Asanuma, H., 1985, Anatomical and physiological properties of the projection from the sensory cortex to the motor cortex in normal cats: The difference between corticocortical and thalamocortical projections, *Brain Res.* **345:**68–78. [487]

Koski, G., Streaty, R. A., and Klee, W. A., 1982, Modulation of sodium-sensitive GTPase by partial opiate agonists: An explanation for the dual requirement for Na^+ and GTP in inhibitory regulation of adenylate cyclase, *J. Biol. Chem.* **257:**14,035–14,040. [167]

Kraus, J. P., and Rosenberg, L. P., 1982, Purification of low abundance messenger RNAs from rat liver by polysome immunoadsorption, *Proc. Natl. Acad. Sci. U.S.A.* **79:**4015–4019. [46]

Krieger, D. T., 1978, Endocrine processes and serotonin, in: *The Central Nervous System,* Vol. 3, *Serotonin in Health and Disease* (E. W. B. Essman, ed.), pp. 51–67, Spectrum, New York. [334]

Krieger, N. R., Megill, J. R., and Sterling, P., 1983, Granule cells in the rat olfactory tubercle accumulate ^3H-gamma-aminobutyric acid, *J. Comp. Neurol.* **215:**465–471. [205]

Krisch, B., 1979, Immunohistochemical results on the distribution of somatostatin in the hypothalamus and in limbic structures of the rat, *J. Histochem. Cytochem.* **27:**1389–1390. [388]

Krisch, B., 1980, Two types of luliberin-immunoreactive perikarya in the preoptic area of the rat, *Cell Tissue Res.* **212:**443–455. [387]

Kristeva, R., Keller, E., Deeke, L., and Kornhuber, H. H., 1979, Cerebral potentials preceding unilateral and simultaneous bilateral finger movements, *Electroencephalogr. Clin. Neurophysiol.* **47:**229–238. [481, 482]

Kristt, D. A., McGowan, R. A., Jr., Martin-MacKinnon, N., and Solomon, J., 1985, Basal forebrain innervation of rodent neocortex: Studies using acetylcholinesterase histochemistry, Golgi and lesion strategies, *Brain Res.* **337:**19–39. [247]

Krnjevic, J. F., and Cotman, C. W., 1981, Micromolar L-2-amino-4-phosphonobutyric acid selectively inhibits perforant path synapse from lateral entorhinal cortex, *Brain Res.* **216:**192–198. [182]

Krnjevic, K., 1974, Chemical nature of synaptic transmission in vertebrates, *Physiol Rev.* **54:**418–540. [79, 186, 254, 255]

Krnjevic, K., and Miledi, R., 1958, Acetylcholine in mammalian neuromuscular transmission, *Nature (London)* **182:**805–806. [89]

Krnjevic, K., and Phillis, J. W., 1963, Iontophoretic studies of neurons in the mammalian cerebral cortex, *J. Physiol. (London)* **249:**167–182. [175]

Krnjevic, K., and Schwartz, S., 1966, Is γ-aminobutyric acid an inhibitory transmitter?, *Nature (London)* **211:**1372–1374. [197, 212]

Krogsgaard-Larsen, P., Hjeds, H., Curtis, D. R., Leah, J. D., Peet, M. J., 1982, Glycine antagonists structurally related to muscimol, THIP, or isoguvacine, *J. Neurochem.* **39:**1319–1324. [230]

Kubo, T., and Amano, H., 1986, Vasopressin-induced pressor responses in rats to bilateral electrolytic lesioning of the caudal portion of the nucleus tractus solitarii, *Brain Res.* **363:**183–187. [399]

Kuffler, S. W., 1942, Electric potential changes at an isolated nerve–muscle junction, *J. Neurophysiol.* **5:**18–26. [87, 88]

Kuffler, S. W., 1966, Physiological properties of vertebrate and invertebrate neuroglial cells and the movement of substances through the nervous system, *Proc. R. Soc. London Ser. B* **168:**1–21. [472]

Kuffler, S. W., 1973, The single-cell approach in the visual system and the study of receptive fields, *Invest. Ophthalmol.* **12:**794–813. [606]

Kuffler, S. W., and Edwards, C., 1958, Mechanism of gamma-aminobutyric acid (GABA) and its relation to synaptic inhibition, *J. Neurophysiol.* **21:**589–610. [197]

Kuffler, S. W., and Nicholls, J. G., 1976, *From Neuron to Brain,* Sinauer Associates, Sunderland, Massachusetts. [472]

Kuffler, S. W., and Yoshikami, D., 1975a, The distribution of acetylcholine sensitivity at the postsynaptic membrane of vertebrate skeletal twitch muscles: Iontophoretic mapping in the micron range, *J. Physiol. (London)* **244:**703–730. [90, 99]

Kuffler, S. W., and Yoshikami, D., 1975b, The number of transmitter molecules in a quantum: An estimate from iontophoretic application of acetylcholine at the neuronmuscular synapse, *J. Physiol. (London)* **251:**465–482. [90, 94, 99]

Kuhar, M., and Zarbin, M. A., 1984, Axonal-transport of muscarinic cholinergic receptors and its implications, *Trends Pharm.* **5:**53–54. [254]

Kuhar, M. J., Sethy, V. H., Roth, R. H., and Aghajanian, C. K., 1973, Choline: Selective accumulation by central cholinergic neurons, *J. Neurochem.* **20:**581–593. [243]

Kuhl, D. E., Metter, E. J., Riege, W. H., and Phelps, M. E., 1982, Effects of human aging on patterns of local cerebral glucose utilization determined by the [^{18}F] fluorodeoxyglucose method, *J. Cereb. Blood Flow Metab.* **2:**163–171. [69]

Kuhn, D. M., and Lovenberg, W., 1982, Role of calmodulin in the activation of tryptophan hydroxylase, *Fed. Proc. Fed. Am. Soc. Exp. Biol.* **41:**2258–2264. [329]

Kuhn, D. M., Ruskin, B., and Lovenberg, W., 1979, Studies of the oxygen sensitivity of tryptophan hydroxylase, *Adv. Exp. Med. Biol.* **133:**253–264. [323]

Kuhn, R., 1958, The treatment of depressive states with G 22355 (imipramine hydrochloride), *Am. J. Psychiatry* **115:**459–464. [540]

Kuno, M., and Llinás, R., 1970, Enhancement of synaptic transmission by dendritic potentials in chromatolysed motoneurones of the cat, *J. Physiol. (London)* **210:**807–821. [72]

Kunzle, H., 1978, An autoradiographic analysis of the efferent connections from premotor and adjacent prefrontal regions (areas 6 and 9) in *Macaca* fascicularis, *Brain Behav. Evol.* **15:**185–234. [480, 482]

Kurihara, E., Kuriyama, K., and Yoneda, Y., 1980, Interconnection of GABAergic neurons in rat extrapyramidal tract: Analysis using intracerebral microinjection of kainic acid, *Exp. Neurol.* **68:**12–26. [212]

Kuriyama, K., 1980, Taurine as a neuromodulator, *Fed. Proc. Fed. Am. Soc. Exp. Biol.* **39:**2680–2684. [231]

Kuwano, R., Araki, K., Usui, H., et al. 1984, Molecular cloning and nucleotide sequence of cDNA coding for rat brain cholecystokinin precursor, *J. Biochem. (Tokyo)* **96:**923–926. [382]

Kvetnansky, R., Kopin, I. J., and Saavedra, J. M., 1978, Changes in epinephrine in individual hypothalamic nuclei after immobilization stress, *Brain Res.* **155:**387–390. [315]

Kwiatkowski, H., 1943, Histamine in nervous tissue, *J. Physiol. (London)* **102:**32–41. [349]

Kwok, R. H. M., 1968, Chinese restaurant syndrome, *N. Engl. J. Med.* **278:**769. [194]

Labrie, F., Gagne, B., and Lefevre, G., 1982, Corticotropin-releasing factor stimulates adenylate cyclase activity in the anterior pituitary gland, *Life Sci.* **31:**1117–1121. [402]

Laduron, P., 1980, Axoplasmic transport of muscarinic receptors, *Nature (London)* **286:**287–288. [254]

Lake, J. A., 1981, The ribosome, *Sci. Am.* **245:**84–97. [20]

Lam, D. M. K., Su, Y. Y. T., Swain, L., Marc, R. E., Brandon, C., and Wu, J.-Y., 1979, Immunocytochemical localization of L-glutamic acid decarboxylase in the goldfish retina, *Nature (London)* **278:**565–567. [205]

Lam, D., Funk, S. C., and Kond, Y. C., 1980, Postnatal development of GABA-ergic neurons in the rabbit retina, *J. Comp. Neurol.* **193:**89–102. [218]

La Motte, C. C., Snowman, A., Pert, C. B., and Snyder, S. H., 1978, Opiate receptor binding in rhesus monkey brain: Association with limbic structures, *Brain Res.* **155:**374–379. [395]

Lamour, Y., Dutar, P., and Jobert, A., 1983, Effects of neuropeptides on rat cortical neurons: Laminar distribution and interaction with the effects of acetylcholine, *Neuroscience* **10:**107–117. [401]

Lamouroux, A., Bignet, N. F., Samolyk, D., *et al.,* 1982, Identification of cDNA clones coding for rat tyrosine hydroxylase antigen, *Proc. Natl. Acad. Sci. U.S.A.* **79:**3881–3885. [46, 270, 271]

Land, H., Schutz, G., Schmale, H., and Richter, D., 1982, Nucleotide sequence of cloned cDNA encoding bovine arginine vasopressin–neurophysin II precursor, *Nature (London)* **295:**299–303. [379, 382]

Land, H., Grez, M., Ruppert, S., *et al.,* 1983, Deduced amino acid sequence from the bovine oxytocin–neurophysin I precursor cDNA, *Nature (London)* **302:**342–344. [382]

Landgren, S., Phillips, C. G., and Porter, R., 1962, Cortical fields of origin of the monosynaptic pyramidal pathways to some alpha motoneurones of the baboon's hand and forearm, *J. Physiol. (London)* **161:**112–125. [555]

Landgren, S., Silfvenius, H., and Olsson, K. A., 1984, The sensorimotor integration in area 3a of the cat, *Exp. Brain Res. Suppl.* **9:**359–375. [487]

Lang, W., and Henke, H., 1983, Cholinergic receptor binding and autoradiography in brains of non-neurological and senile dementia of Alzheimer-type patients, *Brain Res.* **267:**271–280. [262]

Langer, S. Z., 1980, Presynaptic regulation of the release of catecholamines, *Pharmacol. Rev.* **32:**337–362. [309, 314]

Langley, J. N., 1905, On the reaction of cells and nerve-endings to certain poisons, chiefly as regards the reaction of striated muscle to nicotine and to curare, *J. Physiol. (London)* **33:**374–413. [27]

Langston, J. W., and Langston, E. B., 1986, Neurologic complications of drugs of abuse, in: *Diseases of the Nervous System* (A. Astbury, G. W. McKhann, and W. I. McDonald, eds.), W. B. Saunders, Philadelphia (in press). [544]

Langston, J. W., Ballard, P., Tetrud, J. W., and Irwin, I., 1983, Chronic parkinsonism in humans due to a product of meperidine-analog synthesis, *Science* **219:**979–980. [301]

Lanoir, J., Soghomonian, J. J., and Cadenel, G., 1982, Radioautographic study of ³H-GABA uptake in the oculomotor nucleus of the cat, *Exp. Brain Res.* **48:**137–143. [205]

Lashley, K. S., 1950, In search of the engram, *Symp. Soc. Exp. Biol.* **4:**454–482. [554, 565]

Lassen, N. A., Ingvar, D. H., and Skinhoj, E., 1978, Patterns of activity in the human cerebral cortex graphically displayed, *Sci. Am.* **239:**50–53. [69, 605]

Law, P., Wu, J., Koehler, J., and Loh, H., 1981, Demonstration and characterization of opiate inhibition of the striatal adenylate cyclase, *J. Neurochem.* **36:**1834–1846. [404]

Lazarus, L. H., Linnoila, R. I., Hernandez, O., and Diaustine, R. P., 1980, A neuropeptide in mammalian tissue with physalaemin-like immunoreactivity, *Nature (London)* **287:**555–558. [377]

LeBlanc, A., Cote, J., Jobin, J., and Labrie, F., 1979, Plasma catecholamine and cardiovascular response to cold and mental activity, *J. Appl. Physiol.* **47:**1207–1211. [533]

Lebrun-Grandié, P., Baron, J.-C., Soussaline, F., Loch'h, C., Sastre, J., and Bousser, M.-G., 1983, Coupling between regional blood flow and oxygen utilization in the normal human brain: A study with positron emission tomography and oxygen 15, *Arch. Neurol.* **40:**230–236. [480]

Lechan, R. M., Wu, P., Jackson, I. M. D., *et al.*, 1986, Thyrotropin-releasing hormone precursor: Characterization in rat brain, *Science* **231:**159–161. [382]

Lee, C., Javitch, J., and Snyder, S. H., 1983, Recognition sites for norepinephrine uptake: Regulation by neurotransmitter, *Science* **220:**626–629. [309]

Lee, K. S., 1983, Sustained modification of neuronal activity in the hippocampus and cerebral cortex, in: *Molecular, Cellular and Behavioral Neurobiology of the Hippocampus* (W. Seifert, ed.), pp. 265–272, Academic Press, New York. [571]

Lee, K. S., Schoppler, F., Oliver, M., and Lynch, G., 1980, Brief bursts of high-frequency stimulation produce two types of structural change in rat hippocampus, *J. Neurophysiol.* **44:**247–258. [451]

Lee, K. S., Schubert, P., Reddington, M., and Kreutzberg, G. W., 1983, Adenosine receptor density and the depression of evoked neuronal activity in the rat hippocampus *in vitro, Neurosci. Lett.* **37:**81–85. [365]

Lee, R. G., Murphy, J. T., and Tatton, W. G., 1983, Long-latency myotatic reflexes in man: Mechanisms, functional significance and changes in patients with Parkinson's disease or hemiplegia, *Adv. Neurol.* **39:**489–508. [489]

Lee, T., Seeman, P., Tourtelotte, W. W., Farley, I. J., and Hornykiewicz, O., 1978, Binding of ³H-neuroleptics and ³H-apomorphine in schizophrenic brains, *Nature (London)* **274:**897–900. [546]

Lee, Y., Kawai, Y., Shiosaka, S., *et al.*, 1985, Coexistence of calcitonin gene-related peptide and substance P-like peptide in single cells of the trigeminal ganglion of the rat: Immunohistochemical analysis, *Brain Res.* **330:**194–196. [157]

Leeman, S. E., and Mroz, E. A., 1974, Substance P, *Life Sci.* **15:**2033–2044. [374]

Leeming, R. J., Pheasant, A. E., and Blair, J. A., 1981, The role of tetrahydrobiopterin in neurological disease: A review, *J. Ment. Defic. Res.* **25:**231–241. [273]

Le Gal La Salle, G., Paxinos, G., Emson, P., and Ben-Ari, Y., 1978, Neurochemical mapping of GABAergic systems in the amygdaloid complex and bed nucleus of the stria terminalis, *Brain Res.* **155:**397–403. [212]

Legros, J. J., Gilot, P., Seron, X., *et al.* 1978, Influence of vasopressin on learning and memory, *Lancet* **1:**41–42. [410]

Lehmann, J., and Fibiger, H. C., 1979, Acetylcholinesterase and the cholinergic neuron, *Life Sci.* **25**:1939–1947. [250]

Lehmann, J. C., Nagy, J. I., Atmadja, S., and Fibiger, H. C., 1980, The nucleus basalis magnocellularis: The origin of a cholinergic projection to the neocortex of the rat, *Neuroscience* **5**:1161–1174. [247]

Leichnetz, G. R., 1981, The median subcallosal fasciculus in the monkey: A unique prefrontal corticostriate and corticocortical pathway revealed by anterogradely transported horseradish peroxidase, *Neurosci. Lett.* **21**:137–142. [505]

Lembeck, F., 1953a, 5-Hydroxytryptamine in a carcinoid tumor, *Nature (London)* **172**:910–911. [320]

Lembeck, F., 1953b, Zur Frage der zentralen Ubertragung afferenter Impulse, *Arch. Exp. Pathol. Pharmakol.* **219**:197–213. [374]

Leonardelli, J., and Tramu, G., 1979, Immunoreactivity for beta-endorphin in LH-RH neurons of the fetal human hypothalamus, *Cell Tissue Res.* **203**:201–207. [157]

Léranth, C., and Frotscher, M., 1983, Commissural afferents to the rat hippocampus terminate on vasoactive intestinal polypeptide-like immunoreactive non-pyramidal neurons: An EM immunocytochemical degeneration study, *Brain Res.* **276**:357–361. [397]

Léranth, C., Williams, T. H., Hámori, J., and Chrétien, M., 1981, Light and electron microscopic immunocytochemical localization of adrenocorticotropin-like activity in rat cerebellar and red nuclei, *Neuroscience* **6**:481–487. [400]

Léranth, C., Sakamoto, H., MacLusky, N. J., Shanabrough, M., and Naftolin, F., 1985, Estrogen responsive cells in the arcuate nucleus of the rat contain glutamic acid decarboxylase (GAD): An electron microscopic immunocytochemical study, *Brain Res.* **331**:376–381. [205, 216]

Lerer, B., and Friedman, E., 1982, Neocortical cholinergic deficit and behavioral impairment produced by subcortical neurotoxic lesions, *Soc. Neurosci. Abstr.* **8**:838. [261]

Lerner, A. B., Chase, J. D., and Heinzelman, R. V., 1959, Structure of melatonin, *J. Am. Chem. Soc.* **81**:6084–6085. [320]

Lester, B. R., and Peck, E. J., 1979, Kinetic and pharmacologic characterization of gamma-aminobutyric acid receptive sites from mammalian brain, *Brain Res.* **161**:79–97. [225]

Letourneau, P. C., 1983, Axonal growth and guidance, *TINS* **6**:451–455. [434]

Leung, P. C., Raymond, V., and Labrie, F., 1983, Stimulation of phosphatidic acid and phosphatidylinositol labeling in luteal cells by LHRH, *Endocrinology* **112**:1138–1140. [402]

LeVay, S., and Sherk, H., 1983, Retrograde transport of [³H] proline: A widespread phenomenon in the central nervous system, *Brain Res.* **271**:131–134. [233]

Levey, A. I., Rye, D., and Wainer, B. H., 1982, Immunochemical studies of bovine and human choline-O-acetyltransferase using monoclonal antibodies, *J. Neurochem.* **39**:1652–1659. [239]

Leviel, V., Cheramy, A., Nieoullon, A., and Glowinski, J., 1979, Symmetrical bilateral changes in dopamine release from the caudate nuclei of the cat induced by unilateral nigral application of glycine and GABA-related compounds, *Brain Res.* **175**:259–270. [229]

Levi-Montalcini, R., Aloe, L., Chen, M. G., and Chen, J. S., 1980, New features of the nerve growth factor–target cells interaction, in: *Cellules Nerveuses, Transmetteurs et Comportement* (R. Levi-Montalcini, ed.), pp. 11–39, Pontificia Academia Scientiarum, Città dello Stato del Vaticano. [464, 465, 565]

Levine, A. S., and Morley, J. E., 1984, Neuropeptide Y: A potent inducer of consummatory behavior in rats, *Peptides* **5**(Suppl. 1):1025–1029. [405]

Levine, A. S., and Yim, G. K. W., 1984, Neuropeptidergic regulation of food intake, *Fed. Proc. Fed. Am. Soc. Exp. Biol.* **43**:2888. [405]

Levitt, P., Cooper, M. L., and Rakic, P., 1981, Coexistence of neuronal and glial precursor cells in the cerebral ventricular zone of the fetal monkey: An ultrastructural immunoperoxidase analysis, *J. Neurosci.* **1**:27–39. [419]

Levy, J., 1973, Lateral specialization of the human brain: Behavioral manifestations and possible evolutionary basis, in: *The Biology of Behavior* (J. A. Kiger, Jr., ed.), pp. 159–180, Oregon State University Press, Corvallis, Oregon. [624]

Levy, J., 1978, Lateral differences in the human brain in cognition and behavioural control in: *Cerebral Correlates of Conscious Experience* (P. Buser and A. Rougeul-Buser, eds.), pp. 285–298, Elsevier, Amsterdam. [623, 624]

Levy, W. B., and Steward, O., 1979, Synapses as associative elements in the hippocampal formation, *Brain Res.* **175**:233–245. [556, 558, 559, 574]

Levy, W. B., and Steward, O., 1983, Temporal contiguity requirements for long-term associative potentiation/depression in the hippocampus, *Neuroscience* **8**:791–797. [558, 559, 574]

Levy-Agresti, J., and Sperry, R. W., 1968, Differential perceptual capacities in major and minor hemispheres, *Proc. Natl. Acad. Sci. U.S.A.* **61**:1151. [624]

Lewin, B., 1985, *Genes,* John Wiley, New York. [39]

Lewis, E. R., and Cotman, C. W., 1982, Mechanisms of septal lamination in the developing hippocampus revealed by outgrowth of fibers from septal implants. III. Competitive interactions, *Brain Res.* **233**:29–44. [433]

Lewis, R. E., Childers, S. R., and Phillips, M. I., 1985, [^{125}I]Tyr-bradykinin binding in primary rat brain cultures, *Brain Res.* **346**:263–272. [402]

Lewis, T., and Grant, R. T., 1927, *The Blood Vessels of the Human Skin and Their Responses,* Shaw & Sons, London. [349]

Leysen, J. E., Niemegeers, C. J. E., Van Nueten, J. M., and Laduron, P. M., 1982, [^{3}H]Ketanserin (R 41 468), a selective ^{3}H-ligand for serotonin$_2$ receptor binding sites, *Mol. Pharmacol.* **21**:301–314. [336, 343]

Li, P. P., Warsh, J. J., and Godse, D. D., 1984, Formation and clearance of norepinephrine glycol metabolites in mouse brain, *J. Neurochem.* **43**:1425–1433. [278]

Libet, B., 1973, Electrical stimulation of cortex in human subjects and conscious sensory aspects, in: *Handbook of Sensory Physiology,* Vol. 2 (A. Iggo, ed.), pp. 743–790, Springer-Verlag, New York. [600, 603]

Libet, B., Kobayashi, H., and Tanaka, T., 1975, Synaptic coupling with the production and storage of a neuronal memory trace, *Nature (London)* **258**:155–157. [299]

Libet, B., Wright, E. W., Feinstein, B., and Pearl, D. K., 1979, Subjective referral of the timing for a conscious experience: A functional role for the somatosensory specific projection system in man, *Brain* **102**:191–222. [603]

Lichtensteiger, W., 1979, The neuroendocrinology of dopamine systems, in: *The Neurobiology of Dopamine* (A. S. Horn, J. Korf, and B. H. C. Westerink, eds.), pp. 491–521, Academic Press, London, New York, and San Francisco. [295]

Lidof, H. G. W., and Molliver, M. E., 1982, Immunocytochemical study of the development of serotonergic neurons in the rat CNS, *Brain Res. Bull.* **9**:559–604. [331]

Lidof, H. G. W., Grzanna, R., and Molliver, M. E., 1980, The serotonin innervation of the cerebral cortex in the rat—an immunocytochemical analysis, *Neuroscience* **5**:207–227. [329, 333]

Lieberman, A. R., 1971, The axon reaction: A review of the principal features of perikaryal responses to axon-injury, *Int. Rev. Neurobiol.* **14**:49–124. [14, 33, 460, 461]

Liebisch, D. C., Seizinger, B. R., Michael, G., and Herz, A., 1985, Novel opioid peptide amidorphin: Characterization and distribution of amidorphin-like immunoreactivity in bovine, ovine, and porcine brain, pituitary, and adrenal medulla, *J. Neurochem.* **45**:1495–1503. [373, 388]

Liley, A. W., 1956, The quantal components of the mammalian end-plate potential, *J. Physiol. (London)* **133**:571–587. [92, 93]

Lin, C.-T., Li, H.-Z., and Wu, J.-Y., 1983, Immunocytochemical localization of L-glutamate decarboxylase, gamma-aminobutyric acid transaminase, cysteine sulfinic acid decarboxylase, aspartate aminotransferase and somatostatin in rat retina, *Brain Res.* **270**:273–283. [205]

Lin, C.-T., Song, G.-X., and Wu, J.-Y., 1985, Ultrastructural demonstration of L-glutamate decarboxylase and cysteine sulfinic acid decarboxylase in rat retina by immunocytochemistry, *Brain Res.* **331**:71–80. [205, 232]

Lind, R. W., Swanson, L. W., Bruhn, T. O., and Ganten, D., 1985, The distribution of angiotensin II-immunoreactive cells and fibers in the paraventriculo-hypophysial system of the rat, *Brain Res.* **338**:81–89. [388]

Lindberg, I., and Dahl, J. L., 1981, Characterization of enkephalin release from rat striatum, *J. Neurochem.* **36**:506–512. [402]

Lindvall, O., Björklund, A., Hökfelt, T., and Ljungdahl, A., 1973, Application of the glyoxylic acid method to Vibratome sections for improved visualization of central catecholamine neurons, *Histochemie* **35**:31–38. [284]

Ling, N., Zeytin, F., Bohlen, P., *et al.,* 1985, Growth hormone releasing factors, *Annu. Rev. Biochem.* **54**:403–423. [373, 388]

Liscovitch, M., and Koch, Y., 1982, Characterization and subcellular localization of GnRH analog binding in rat brain, *Peptides* **3**:55–60. [402]

Liu-Chen, L.-Y., Norregaard, T. V., and Moskowitz, M. A., 1985, Some cholecystokinin-8 immunoreactive fibers in large pial arteries originate from trigeminal ganglion, *Brain Res.* **359:**166–176. [408]

Ljungdahl, A., and Hökfelt, T., 1973, Autoradiographic uptake patterns of [^{3}H]-GABA and [^{3}H]glycine in central nervous tissues with special reference to the cat spinal cord, *Brain Res.* **69:**587–595. [228]

Ljungdahl, A., Hökfelt, T., and Nilsson, G., 1978, Distribution of substance P-like immunoreactivity in the central nervous system of the rat. I. Cell bodies and nerve terminals, *Neuroscience* **3:**861–943. [395]

Llinás, R., and Nicholson, C., 1971, Electrophysiological properties of dendrites and somata in alligator Purkinje cells, *J. Neurophysiol.* **34:**532–551. [72]

Llinás, R., and Sugimori, M., 1980, Electrophysiological properties of *in vitro* Purkinje cell dendrites in mammalian cerebellar slices, *J. Physiol. (London)* **305:**197–213. [60, 72]

Llinás, R., and Sugimori, M., 1982, Functional significance of the climbing fiber input to Purkinje cells: An *in vitro* study in mammalian cerebellar slices, in: *The Cerebellum: New Vistas* (S. L. Palay, and V. Chan-Palay, eds.), pp. 402–411, Springer-Verlag, Berlin, Heidelberg, and New York. [582]

Llorens-Cortes, C., Javoy-Agid, F., Taquet, H., and Schwartz, J. C., 1984, Enkephalinergic markers in substantia nigra and caudate nucleus from parkinsonian subjects, *J. Neurochem.* **43:**874–877. [412]

Lloyd, D. P. C., 1946, Facilitation and inhibition of spinal motoneurons, *J. Neurophysiol.* **9:**421–438. [112, 121]

Lo, M. M. S., Niehoff, D. L., Kuhar, M. J., and Snyder, S. H., 1983, Autoradiographic differentiation of multiple benzodiazepine receptors by detergent solubilization and pharmacologic specificity, *Neurosci. Lett.* **39:**37–44. [220]

Lodge, D., Headley, P. M., and Curtis, D. R., 1978, Selective antagonism by D-α-aminoadipate of amino acid and synaptic excitation of cat spinal neurons, *Brain Res.* **152:**603–608. [182]

Loewi, O., 1960, An autobiographic sketch, *Perspect. Biol. Med.* **4:**3–25. [181]

Loh, Y. P., Brownstein, M. J., and Gainer, H., 1984, Proteolysis in neuropeptide processing and other neural functions, *Annu. Rev. Neurosci.* **7:**189–222. [379]

Loiseau, P., 1983, Double-blind crossover trial of progabide versus placebo in severe epilepsies, *Epilepsia* **24:**703–715. [225]

Lombard, J. P., Gilbert, J. G., and Donofrio, A. F., 1955, The effects of glutamic acid upon intelligence, social maturity and adjustment of a group of mentally retarded children, *Am. J. Ment. Defic.* **60:**127–132. [194]

Londos, C., and Wolff, J., 1977, Two distinct adenosine-sensitive sites on adenylate cyclase, *Proc. Natl. Acad. Sci. U.S.A.* **74:**5482–5486. [365]

Loomer, H. P., Saunders, J. C., and Kline, N. S., 1957, A clinical and pharmacodynamic evaluation of iproniazid as a psychic energizer, *Psychiatr. Res. Rep. Wash.* **8:**129–141. [267, 539]

López-Colomé, A., Tapia, R., Salceda, R., and Pasantes-Morales, H., 1978a, Potassium-stimulated release of GABA, glycine, and taurine from the chick retina, *Neurochem. Res.* **3:**431–442. [229]

López-Colomé, A., Tapia, R., Salceda, R., and Pasantes-Morales, H., 1978b, K$^+$-stimulated release of labeled γ-aminobutyrate, glycine and taurine in slices of several regions of rat central nervous system, *Neuroscience* **3:**1069–1074. [228]

Lopez-Lahoya, J., Garcia, M. L., Benavides, J., and Ugarte, M., 1981, Inhibition by methylmalonate of glycine uptake by synaptosomes from rat spinal cord, *J. Neurochem.* **36:**325–327. [230]

Lorén, I., Schwandt, P., Alumets, J., *et al.*, 1980, Evidence that lipolytic peptide B occurs in the ACTH/MSH-cells of the pituitary and in the brain: An immunocytochemical study of its distribution in correlation to the distribution of neurophysin, *Cell Tissue Res.* **205:**349–359. [384]

Lorenz, R. G., Saper, C., Wong, D. L., *et al.*, 1985, Co-localization of substance P- and phenylethanolamine-N-methyltransferase-like immunoreactivity in neurons of ventrolateral medulla that project to the spinal cord: Potential role in control of vasomotor tone, *Neurosci. Lett.* **55:**255–260. [157]

Lotstra, F., Verbanck, P. M. P., Gilles, C., Mendlewicz, J., and Vanderhaeghen, J.-J., 1985, Reduced cholecystokinin levels in cerebrospinal fluid of parkinsonian and schizophrenic patients, *Ann. N. Y. Acad. Sci.* **448:**507–517. [410, 412]

Lovick, T. A., and Hunt, S. P., 1983, Substance P-immunoreactive and serotonin-containing neurones in the ventral brainstem of the cat, *Neurosci. Lett.* **36:**223–228. [157, 332]

Luabeya, M. K., Maloteaux, J.-M., and Laduron, P. M., 1984, Regional and cortical laminar distributions of serotonin S$_2$, benzodiazepine, muscarinic and dopamine D$_2$ receptors in human brain, *J. Neurochem.* **43:**1068–1071. [253]

Luine, V. N., Renner, K. J., Frankfurt, M., and Azmitia, E. C., 1984, Facilitated sexual behavior reversed and serotonin restored by raphe nuclei transplanted into denervated hypothalamus, *Science* **226**:1436–1438. [338]

Luini, A., Tal, N., Goldberg, O., and Teichberg, V. I., 1984, An evaluation of selected brain constituents as putative excitatory neurotransmitters, *Brain Res.* **324**:271–277. [186]

Lund, R. D., 1969, Synaptic patterns of the superficial layer of the superior colliculus of the rat, *J. Comp. Neurol.* **135**:179–208. [136]

Lundberg, A., 1979a, Multisensory control of spinal reflex pathways, *Prog. Brain Res.* **50**:11–26. [137]

Lundberg, A., 1979b, Integration in a propriospinal motor centre controlling the forelimb in the cat, in: *Integration in the Nervous System* (H. Asanuma and V. J. Wilson, eds.), pp. 47–64, Igaku-Shoin, Tokyo and New York. [475]

Lundberg, J. M., Hökfelt, T., Schultzberg, M., Uvnas-Wallenstein, K., Kohler, C., and Said, S. I., 1979, Occurrence of vasoactive intestinal polypeptide (VIP)-like immunoreactivity in certain cholinergic neurons of the cat: Evidence from combined immunohistochemistry and acetylcholinesterase staining, *Neuroscience* **4**:1539–1560. [157]

Lundberg, J. M., Terenius, L., Hökfelt, T., and Goldstein, M., 1983, High levels of neuropeptide Y in peripheral noradrenergic neurons in various mammals including man, *Neurosci. Lett.* **42**:167–172. [157]

Lundberg, J. M., Fakrenkrug, J., Larsson, O., and Anggard, A., 1984, Corelease of vasoactive intestinal polypeptide and peptide histidine isoleucine in relation to atropine-resistant vasodilation in cat submandibular salivary gland, *Neurosci. Lett.* **52**:37–42. [157]

Lund-Karlsen, R., 1978, The toxic effect of sodium glutamate and DL-α-aminoadipic acid on rat retina: Changes in high affinity uptake of putative transmitters, *J. Neurochem.* **31**:1055–1061. [212, 229]

Lund-Karlsen, R., and Fonnum, F., 1978, Evidence for glutamate as a neurotransmitter in the corticofugal fibres to the dorsal lateral geniculate body and the superior colliculus in rats, *Brain Res.* **151**:457–467. [182, 184]

Lux, H. D., 1971, Ammonium and chloride extrusion: Hyperpolarizing synaptic inhibition in spinal motoneurones, *Science* **173**:555–557. [129]

Lux, H. D., Loracher, C., and Neher, E., 1970, Action of ammonium on postsynaptic inhibition of cat spinal motoneurones, *Exp. Brain Res.* **11**:431–447. [129]

Lykken, D. T., 1982, Presidential address 1981: Research with twins: The concept of emergenesis, *Psychophysiology* **19**:361–373. [526]

Lynch, C. O., Jr., Johnson, M. D., and Crowley, W. R., 1984, Effects of the serotonin agonist, quipazine, on luteinizing hormone and prolactin release: Evidence for serotonin–catecholamine interaction, *Life Sci.* **35**:1481–1487. [334]

Lynch, G., and Baudry, M., 1984, The biochemical intermediates in memory formation: A new and specific hypothesis, *Science* **224**:1057–1063. [562–564, 574]

Lynch, G., Halpain, S., and Baudry, M., 1982, Effects of high frequency synaptic stimulation on glutamate binding studied with a modified *in vitro* hippocampal slice preparation, *Brain Res.* **244**:101–111. [562, 565]

Lynch, G., Halpain, S., and Baudry, M., 1983a, Structural and biochemical effects of high frequency stimulation in hippocampus, in: *Molecular, Cellular and Behavioral Neurobiology of the Hippocampus* (W. Seifert, ed.), pp. 253–262, Academic Press, New York. [561–564]

Lynch, G., Larson, J., Kelso, S., Barrionirevo, G., and Schottler, F., 1983b, Intracellular injections of EGTA block induction of hippocampal long-term potentiation, *Nature (London)* **305**:719–721. [560, 563]

Ma, R. C., Horwitz, J., Kiraly, M., Perlman, R. L., and Dun, N. J., 1985, Immunohistochemical and biochemical detection of serotonin in the guinea pig celiac-superior mesenteric plexus, *Neurosci. Lett.* **56**:107–112. [332]

Macdonald, R. L., Pun, R. Y. K., Neale, E. A., and Nelson, P. G., 1983, Synaptic interactions between mammalian central neurons in cell culture. I. Reversal potential for excitatory postsynaptic potentials, *J. Neurophysiol.* **49**:1428–1441. [119]

MacLean, P. D., 1958, Contrasting functions of limbic and neocortical systems of the brain and their relevance to psycho-physiological aspects of medicine, *Am. J. Med.* **25**:611–626. [528]

MacLeod, N. K., James, T. A., Kilpatrick, I. C., and Starr, M. S., 1980, Evidence for a GABAergic nigrothalamic pathway in the rat. II. Electrophysiological studies, *Exp. Brain Res.* **40**:55–61. [212]

MacPherson, J., Marangoz, C., Miles, T. S., and Wiesendanger, M., 1982, Micro-stimulation of the supplementary motor area (SMA) in the awake monkey, *Exp. Brain Res.* **45**:410–416. [482]

MacPherson, J., Wiesendanger, M., Marangoz, C., and Miles, T. S., 1983, Corticospinal neurons of the supplementary motor area of monkeys, *Exp. Brain Res.* **48:**81–88. [482, 483]

Madarasz, M., Somogyi, G., Somogyi, J., and Hámori, J., 1985, Numerical estimation of γ-aminobutyric acid (GABA)-containing neurons in three thalamic nuclei of the cat: Direct GABA immunocytochemistry, *Neurosci. Lett.* **61:**73–78. [205]

Maendly, R., Ruegg, D., Wiesendanger, M., Wiesendanger, R., Lagouska, J., and Hess, B., 1981, Thalamic relay for group I muscle afferents of forelimb nerves in the monkey, *J. Neurophysiol.* **46:**901–917. [487]

Maggi, A., and Perez, J., 1985, Role of female gonadal hormones in the CNS: Clinical and experimental aspects, *Life Sci.* **37:**893–906. [525]

Magnusson, K. R., Beitz, A. J., Larson, A. A., Madl, J. E., and Altschuler, R., 1985, Immunohistochemical localization of glutamate, glutaminase and aspartyl aminotransferase neurons in the spinal trigeminal nucleus of the rat, *Soc. Neurosci. Abstr.* **11:**579. [182]

Mahut, H., Moss, M., and Zola-Morgan, S., 1981, Retention deficits after combined amygdalo-hippocampal and selective hippocampal resections in the monkey, *Neuropsychologia* **19:**201–225. [569]

Main, A. R., 1976, Structure and inhibitors of cholinesterase, in: *Biology of Cholinergic Function* (A. M. Goldberg and I. Hanin, eds.), pp. 269–377, Raven Press, New York. [259]

Maire, J. C., Medilanski, J., and Straub, R. W., 1984, Release of adenosine, inosine and hypoxanthine from rabbit non-myelinated nerve fibres at rest and during activity, *J. Physiol. (London)* **257:**67–77. [365, 367]

Maitre, M., Rumigny, J. F., and Mandel, P., 1983, Positive cooperativity in high affinity binding sites for γ-hydroxybutyric acid in rat brain, *Neurochem. Res.* **8:**113–120. [202]

Makara, G. B., Rappay, G., and Stark, E., 1975, Autoradiographic localization of ^{3}H-gamma-aminobutyric acid in the medial hypothalamus, *Exp. Brain Res.* **22:**449–455. [212]

Maley, B., and Elde, R., 1982, Immunohistochemical localization of putative neurotransmitters within the feline nucleus tractus solitarii, *Neuroscience* **7:**2469–2490. [399]

Maley, B., and Newton, B. W., 1985, Immunohistochemistry of γ-aminobutyric acid in the cat nucleus tractus solitarius, *Brain Res.* **330:**364–368. [205]

Malick, J. B., and Goldstein, J. M., 1978, Analgesic activity of substance P following intracerebral administration in rats, *Life Sci.* **23:**835–844. [407]

Malthe-Sørenssen, D., Skrede, K. K., and Fonnum, F., 1979, Calcium-dependent release of D[^{3}H]aspartate evoked by selective electrical stimulation of excitatory afferent fibres to hippocampal pyramidal cells *in vitro*, *Neuroscience* **4:**1255–1263. [182]

Malthe-Sørenssen, D., Odden, E., and Walaas, I., 1980a, Selective destruction by kainic acid of neurons innervated by putative glutaminergic afferents in septum and diagonal band, *Brain Res.* **182:**461–465. [212]

Malthe-Sørenssen, D., Skrede, K. K., and Fonnum, F., 1980b, Release of D-[^{3}H]aspartate from the dorsolateral septum after electrical stimulation of excitatory afferent fibres to hippocampal pyramidal cells *in vitro*, *Neuroscience* **5:**127–133. [182]

Manberg, P. J., Youngblood, W. W., Nemeroff, C. B., et al. 1982, Regional distribution of neurotensin in human brain, *J. Neurochem.* **38:**1777–1780. [388, 412, 506]

Mandarino, L., Stenner, D., Blanchard, W., et al. 1981, Selective effects of somatostatin-14, -25 and -28 on *in vitro* insulin and glucagon secretion, *Nature (London)* **291:**76–77. [377]

Maneckjee, R., and Baylin, S. B., 1983, Use of radiolabeled monofluoromethyl-dopa to define the subunit structure of human L-dopa decarboxylase, *Biochemistry* **22:**6058–6063. [273, 274]

Mann, D. M. A., Lincoln, J., Yates, P. O., Stamp, J. E., and Toper, S., 1980, Changes in the monoamine containing neurons of the human central nervous system in senile dementia, *Br. J. Psychiatry* **136:**533–541. [262]

Mann, D., Yates, P., and Marcyniuk, B., 1984a, A comparison of changes in the nucleus basalis and locus coeruleus in Alzheimer's disease, *J. Neurol. Neurosurg. Psychiatry* **47:**201–203. [261]

Mann, D. M. A., Yates, P. O., and Marcyniuk, B., 1984b, Changes in nerve cells of the nucleus basalis of Meynert in Alzheimer's disease and their relationship to ageing and to the accumulation of lipofuscin pigment, *Mech. Ageing Dev.* **25:**189–204. [261]

Mann, J. J., Stanley, M., Neophytides, A., De Leon, M. J., Ferris, S. H., and Gershon, S., 1981, Central amine metabolism in Alzheimer's disease: *In vivo* relationship to cognitive deficit, *Neurobiol. Aging* **2:**57–60. [262, 308]

Mantyh, P. W., and Hunt, S. P., 1984, Neuropeptides are present in projection neurones at all levels in visceral and taste pathways: From periphery to sensory cortex, *Brain Res.* **299:**297–311. [157]

Mantyh, P. W., Pinnock, R. D., Downes, C. P., Goedert, M., and Hunt, S. P., 1984, Correlation between inositol phospholipid hydrolysis and substance P receptors in rat CNS, *Nature (London)* **309:**795–797. [404]

Marangos, P. J., 1984, Differentiating adenosine receptors and adenosine uptake sites in brain, *J. Recept. Res.* **4:**231–244. [362]

Marangos, P. J., Patel, J., Clark-Rosenberg, R., and Martino, A. M., 1982, [³H]Nitrobenzylthioinosine binding as a probe for the study of adenosine uptake sites in brain, *J. Neurochem.* **39:**184–191. [364]

Marc, R. E., Stell, W. K., Bok, D., and Lam, D. M., 1978, GABA-ergic pathways in the goldfish retina, *J. Comp. Neurol.* **182:**221–245. [205]

Marchbanks, R., 1975, Biochemistry of cholinergic neurons, in: *Handbook of Psychopharmacology*, Vol. 3, *Biochemistry of Biogenic Amines* (L. L. Iversen, S. D. Iversen, and S. H. Snyder, eds.), pp. 247–326, Plenum Press, New York. [237]

Marcinkiewicz, M., Vergé, D., Gozlan, H., Pichat, L., and Hamon, M., 1984, Autoradiographic evidence for the heterogeneity of 5-HT₁ sites in the rat brain, *Brain Res.* **291:**159–163. [335]

Marco, E. J., Balfagón, G., Salaices, M., Sánchez-Ferrer, C. F., and Marin, J., 1985, Serotonergic innervation of cat cerebral arteries, *Brain Res.* **338:**137–139. [334]

Margenau, H., 1984, *The Miracle of Existence*, Ox Bow Press, Woodbridge Connecticut. [629, 630]

Mariani, J., 1983, Elimination of synapses during the development of the central nervous system, *Prog. Brain Res.* **58:**383–392. [452]

Marin-Padilla, M., 1969, Origin of the pericellular baskets of the pyramidal cells of the human motor cortex: A Golgi study, *Brain Res.* **14:**633–646. [598]

Mark, V. H., and Ervin, F. R., 1970, *Violence and the Brain*, Harper and Row, New York. [528]

Marlen-Wilson, W. D., and Teuber, H. L., 1975, Memory for remote events in anterograde amnesia: Recognition of public figures from news photographs, *Neuropsychologia* **13:**353–364. [568]

Marr, D., 1969, A theory of cerebellar cortex, *J. Physiol. (London)* **202:**437–470. [578]

Marr, D., 1970, A theory for cerebral neocortex, *Proc. R. Soc. London Ser. B* **176:**161–234. [570, 571, 574, 589, 592, 593]

Marsden, C. D., 1982, The mysterious motor function of the basal ganglia: The Robert Wartenberg lecture, *Neurology* **32:**514–539. [511]

Marsden, C. D., Merton, P. A., Morton, H. B., Adam, J. E. R., and Hallett, M., 1978a, Automatic and voluntary responses to muscle stretch in man, *Prog. Clin. Neurophysiol.* **4:**167–177. [489]

Marsden, C. D., Merton, P. A., Morton, H. B., and Adam, J., 1978b, The effect of lesions of the central nervous system on long-latency stretch reflexes in the human thumb, *Prog. Clin. Neurophysiol.* **4:**334–341. [489]

Marsden, C. D., Bennett, G. W., Irons, J., Gilbert, R. F., and Emson, P. C., 1982, Localization and release of 5-hydroxytryptamine, thyrotropin releasing hormone and substance P in rat ventral spinal cord, *Comp. Biochem. Physiol. C* **72:**263–270. [157]

Marsden, C. D., Meadows, J. C., and Merton, P. A., 1983a, "Muscular wisdom" that minimizes fatigue during prolonged effort in man: Peak rates of motoneuron discharge and slowing of discharge during fatigue, *Adv. Neurol.* **39:**169–211. [488]

Marsden, C. D., Rothwell, J. C., and Day, B. L., 1983b, Long-latency automatic responses to muscle stretch in man: Origin and function, *Adv. Neurol.* **39:**509–539. [488]

Marshall, J., and Voaden, M., 1974, An investigation of the cells incorporating [³H]-GABA and [³H]glycine in the isolated retina of the rat, *Exp. Eye Res.* **18:**367–370. [205, 229]

Marshall, P. E., and Landis, D. M. D., 1985, Huntington's disease is accompanied by changes in the distribution of somatostatin-containing neuronal processes, *Brain Res.* **329:**71–82. [412]

Marshall, P. E., Landis, D. M. D., and Zalneraitis, E. L., 1983, Immunocytochemical studies of substance P and leucine-enkephalin in Huntington's disease, *Brain Res.* **289:**11–26. [412]

Martin, J. C., Ballinger, B. R., Cockram, L. L., McPherson, F. M., Pigache, R. M., and Tregaskis, D., 1983, Effect of a synthetic peptide, ORG 2766, in patients with severe senile dementia: A controlled clinical trial, *Acta Psychiatr. Scand.* **67:**205–207. [410]

Martin, K. A. C., and Perry, V. H., 1983, The role of fiber ordering and axon collateralization in the formation of topographic projections, *Prog. Brain Res.* **58:**321–337. [435]

Martin, M. R., 1980, The effects of iontophoretically-applied antagonists on auditory nerve and amino acid-evoked excitation of anteroventral cochlear nucleus neurons, *Neuropharmacology* **19**:519–528. [182]

Martin, M. R., and Adams, J. C., 1979, Effects of DL-α-aminoadipate on synaptically and chemically evoked excitation of anteroventral cochlear nucleus neurons in the cat, *Neuroscience* **4**:1097–1105. [178]

Martin, R., and Voigt, K. H., 1982, Leucine-enkephalin-like immunoreactivity in vasopressin terminals is enhanced by treatment with peptidases, *Life Sci.* **31**:1729–1732. [157]

Martin, R., Moll, U., and Voigt, K. H., 1983, An attempt to characterize by immunocytochemical methods the enkephalin-like material in oxytocin endings of the rat neurohypophysis, *Life Sci.* **33**(Suppl. 1):69–72. [157]

Martin, R. F., Jordan, L. M., and Willis, W. D., 1978, Differential projections of cat medullary raphe neurons demonstrated by retrograde labeling following spinal cord lesions, *J. Comp. Neurol.* **182**:77–88. [333]

Martin, S. A. M., and Bulfield, R. C., 1984, A structural gene (*Hdc-s*) for mouse kidney histidine decarboxylase, *Biochem. Genet.* **22**:645–656. [351]

Martin, W. R. W., Beckman, J. H., Calne, D. B., *et al.*, 1984, Cerebral glucose metabolism in Parkinson's disease, *Can. J. Neurol. Sci.* **11**:169–173. [513]

Martin del Rio, R., and Caballero, A. L., 1980, Presence of γ-aminobutyric acid in rat ovary, *J. Neurochem.* **34**:1584–1586. [218]

Martinez-Rodriguez, R., Fernandez, B., Cevallos, C., and Gonzalez, M., 1974, Histochemical location of glutamic dehydrogenase and aspartate aminotransferase in chicken cerebellum, *Brain Res.* **69**:31–40. [178]

Maruki, C., Spatz, M., Ueki, Y., Nagatsu, I., and Bembry, J., 1984, Cerebrovascular endothelial cell culture: Metabolism and synthesis of 5-hydroxytryptamine, *J. Neurochem.* **43**:316–319. [324]

Marx, J. L., 1985a, The polyphosphoinositides revisited, *Science* **228**:312–313. [170, 171]

Marx, J. L., 1985b, "Anxiety peptide" found in brain, *Science* **227**:934. [219]

Mash, D. C., Flynn, D. D., and Potter, L. T., 1985, Loss of M_2 muscarine receptors in the cerebral cortex in Alzheimer's disease and experimental cholinergic denervation, *Science* **228**:1115–1117. [262]

Massion, J., and Sasaki, K., 1979, *Cerebro-cerebellar Interactions,* Elsevier North-Holland, Amsterdam. [495]

Mata, M. M., Schrier, B. K., and Morre, R. Y., 1977, Interpeduncular nucleus: Differential effects of habenula lesions on choline acetyltransferase and glutamic acid decarboxylase, *Exp. Neurol.* **57**:913–921. [216]

Matalon, R., 1984, Current status of biopterin screening, *J. Pediatr.* **104**:579–581. [273]

Matsas, R., Fulcher, I. S., Kenney, A. J., and Turner, A. J., 1983, Substance P and (Leu)-enkephalin are hydrolyzed by an enzyme in pig caudate synaptic membranes that is identical with the endopeptidase of kidney microvilli, *Proc. Natl. Acad. Sci. U.S.A.,* **80**:3111–3115. [384]

Matsuyama, T., Shiosaka, S., Matsumoto, M., *et al.*, 1983, Overall distribution of vasoactive intestinal polypeptide-containing nerves on the wall of cerebral arteries: An immunohistochemical study using whole-mounts, *Neuroscience* **10**:89–96. [397]

Matthews, D. A., Salvaterra, P. M., Crawford, G. D., Houser, C. R., and Vaughn, J. E., 1983, Distribution of choline acetyltransferase positive neurons and terminals in hippocampus, *Soc. Neurosci. Abstr.* **9**:79. [252]

Matthews, M. R., and Nelson, V. H., 1975, Detachment of structurally intact nerve endings from chromatolytic neurones of rat superior cervical ganglion during the depression of synaptic transmission induced by post-ganglionic axotomy, *J. Physiol. (London)* **245**:91–135. [460]

Matthews, M. R., and Raisman, G., 1972, A light and electron microscopic study of the cellular response to axonal injury in the superior cervical ganglion of the rat, *Proc. Roy. Soc. London Ser. B* **181**:43–79. [460]

Matthews, P. B. C., 1981, Evolving views on the internal operation and functional role of the muscle spindle, *J. Physiol. (London)* **320**:1–30. [485, 486]

Matus, A., 1981, The postsynaptic density, *TINS* **4**:51–53. [562]

Matus, A. I., and Dennison, M. E., 1972, An autoradiographic study of uptake of exogenous glycine by vertebrate spinal cord slices *in vitro, J. Neurobiol.* **1**:27–34. [228]

Matus, A., Bernhardt, R., and Hugh-Jones, T., 1981, High molecular weight microtubule-associated proteins are preferentially associated with dendritic microtubules in brain, *Proc. Natl. Acad. Sci. U.S.A.,* **78**:3010–3014. [25]

Matute, C., and Martinez-Millan, L., 1985, Selective retrograde labeling in some afferents of the rabbit lateral geniculate nucleus following injections of tritiated neurotransmitter-related compounds, *Neurosci. Lett.* **53:**9–14. [184]

Matuzewka, B., and Borchardt, R. T., 1983, Guinea pig brain histamine *N*-methyltransferase: Purification and partial characterization, *J. Neurochem.* **41:**113–118. [351]

Mauborgne, A., Javoy-Agid, F., Legrand, J. C., Agid, Y., and Casselin, F., 1983, Decrease of substance P-like immunoreactivity in the substantia nigra and pallidum of parkinsonian brains, *Brain Res.* **268:**167–170. [412]

Maugh, T. H., II, 1981, Biochemical markers identify mental states, *Science* **214:**39–41. [548, 549]

Mayo, K. E., Cerelli, G. M., Rosenfeld, M. G., and Evans, R. M., 1985, Characterization of cDNA and genomic clones encoding the precursor to rat hypothalamic growth hormone-releasing factor, *Nature (London)* **314:**464–467. [382]

Mazurkiewicz-Kwilecki, I. M., 1984, Possible role of histamine in brain function: Neurochemical, physiological and pharmacological indications, *Can. J. Physiol. Pharmacol.* **62:**709–714. [350]

McBean, G. J., and Roberts, P. J., 1984, Chronic infusion of ʟ-glutamate causes neurotoxicity in rat striatum, *Brain Res.* **290:**372–375. [195]

McBride, W. J., Aprison, M. H., and Kusano, K., 1976a, Contents of several amino acids in the cerebellum, brain stem and cerebrum of the "staggerer," "weaver" and "nervous" neurologically mutant mice, *J. Neurochem.* **26:**867–870. [182, 184]

McBride, W. J., Nadi, N. S., Altman, J., and Aprison, M. H., 1976b, Effects of selective doses of X-irradiation on the levels of several amino acids in the cerebellum of the rat, *Neurochem. Res.* **1:**141–152. [182]

McCann, S. M., Lumpkin, M. D., Mizunuma, H., Khorram, O., and Samson, W. K., 1984, Recent studies on the role of brain peptides in control of anterior pituitary hormone secretion, *Peptides* **5**(Suppl. 1):3–7. [391, 393, 394, 401]

McClintock, B., 1984, The significance of responses of the genome to challenge, *Science* **226:**792–801. [39]

McCloskey, D. I., Gandivia, S., Potter, E. L., and Colebatch, G., 1983, Muscle sense and effort: Motor commands and judgments about muscular contractions, *Adv. Neurol.* **39:**151–167. [486]

McComb, D. J., Cairns, P. D., Kovacs, K., and Szabo, S., 1985, Effects of cysteamine on the hypothalamic–pituitary axis in the rat, *Fed. Proc. Fed. Am. Soc. Exp. Biol.* **44:**2551–2555. [411]

McCormick, D. A., and Prince, D. A., 1986, Acetylcholine induces firing in thalamic neurones by activating a potassium conductance, *Nature (London)* **319:**401–405. [255]

McDonald, A. J., 1985a, Immunohistochemical identification of γ-aminobutyric acid-containing neurons in the rat basolateral amygdala, *Neurosci. Lett.* **53:**203–207. [205]

McDonald, A. J., 1985b, Morphology of peptide-containing neurons in the rat basolateral amygdaloid nucleus, *Brain Res.* **338:**186–191. [394]

McDonald, J. K., Parnavelas, J. G., Karamanlidis, A. N., and Brecha, N., 1982a, The morphology and distribution of peptide-containing neurons in the adult and developing visual cortex of the rat. II. Vasoactive intestinal polypeptide, *J. Neurocytol.* **11:**825–837. [397]

McDonald, J. K., Parnavelas, J. G., Karamanlidis, A. N., Rosenquist, G., and Brecha, N., 1982b, The morphology and distribution of peptide-containing neurons in the adult and developing visual cortex of the rat. III. Cholecystokinin, *J. Neurocytol.* **11:**881–895. [397]

McDonald, R. L., and Nowak, L. M., 1981, Substance P and somatostatin actions on spinal cord neurons in primary dissociated cell culture, in: *Neurosecretion and Brain Peptides: Implications for Brain and Neurological Disease* (J. B. Martin, S. Reichlin, and K. L. Bick, eds.), pp. 159–174, Raven Press, New York. [401]

McEwen, B. S., 1981, Endocrine effects on the brain and their relationship to behavior, in: *Basic Neurochemistry* (G. J. Siegel, R. W. Albers, B. W. Agranoff, and R. Katzman, eds.), pp. 775–799, Little, Brown, Boston. [520, 522]

McEwen, B. S., Biegon, A., Davis, P. G., *et al.,* 1982, Steroid hormones: Humoral signals which alter brain cell properties and functions, *Recent Prog. Horm. Res.* **38:**41–92. [522, 524, 525]

McGeer, E. G., and McGeer, P. L., 1976, Duplication of biochemical changes of Huntington's chorea by intrastriatal injections of glutamic and kainic acids, *Nature (London)* **263:**517–519. [195]

McGeer, E. G., and McGeer, P. L., 1979, Localization of glutaminase in the rat neostriatum, *J. Neurochem.* **32:**1071–1975. [178]

McGeer, E. G., and McGeer, P. L., 1982, Neurotransmitters in normal aging, in: *Geriatrics I: Cardiology and*

Vascular System, Central Nervous System (D. Platt, ed.), pp. 263–282, Springer-Verlag, Berlin, Heidelberg, and New York. [471]

McGeer, E. G., and McGeer, P. L., 1984, Substantia nigra cell death from kainic acid or folic acid injections into the pontine tegmentum, *Brain Res.* **298:**339–342. [193, 509]

McGeer, E. G., and McGeer, P. L., 1985, Neurotoxin-induced animal models of human diseases, in: *Neurotoxicology* (K. Blum and L. Manzo, eds.), pp. 515–533, Mercel Dekker, New York and Basel. [193, 195, 251]

McGeer, E. G., and Singh, E. A., 1979, Inhibition of angiotensin converting enzyme by substance P, *Neurosci. Lett.* **14:**105–108. [385]

McGeer, E. G., and Singh, E. A., 1980, Possible cortico-hypothalamic glycinergic tract, *Neurosci. Lett.* **17:**85–87. [229]

McGeer, E. G., Ling, G. M., and McGeer, P. L., 1963, Conversion of tyrosine to catecholamines by cat brain *in vivo*, *Biochem. Biophys. Res. Commun.* **13:**291–296. [267]

McGeer, E. G., Peter, D. A. V., and McGeer, P. L., 1968, Inhibition of rat brain tryptophan hydroxylase by 6-halotryptophans, *Life Sci.* **7:**605–616. [341]

McGeer, E. G., Wada, J. A., Terao, A., and Jung, E., 1969, Amine synthesis in various brain regions with caudate or septal lesions, *Exp. Neurol.* **24:**277–284. [247]

McGeer, E. G., Innanen, V. T., and McGeer, P. L., 1976, Evidence on the cellular localization of adenyl cyclase in the neostriatum, *Brain Res.* **118:**356–358. [286, 299]

McGeer, E. G., McGeer, P. L., and Singh, K., 1978, Kainic acid-induced degeneration of neostriatal neurons: Dependency upon corticostriatal tract, *Brain Res.* **139:**381–383. [192]

McGeer, E. G., Scherer-Singler, U., and Singh, E. A., 1979, Confirmatory data on habenular projections, *Brain Res.* **168:**375–376. [212, 216]

McGeer, E. G., McGeer, P. L., and Vincent, S. R., 1982, GABA enzymes and pathways, in: *NATO Advanced Study Institutes Series, Series A: Life Sciences,* Vol. 48, *Neurotransmitter Interaction and Compartmentation* (H. F. Bradford, ed.), pp. 299–327, Plenum Press, New York. [212]

McGeer, E. G., McGeer, P. L., and Thompson, S., 1983, GABA and glutamate enzymes, in: *Glutamine, Glutamate and GABA in the Central Nervous System* (L. Hertz, E. Kvamme, E. G. McGeer, and A. Schousboe, eds.), pp. 3–17, Alan R. Liss, New York. [180]

McGeer, E. G., Staines, W. A., and McGeer, P. L., 1984, Neurotransmitters in the basal ganglia, *Can. J. Neurol. Sci.* **11:**89–99. [215, 502, 506, 514]

McGeer, P. L., 1963, Central amines and extrapyramidal functions, *J. Neuropsychiatry* **4:**247–250. [512]

McGeer, P. L., 1984, Aging, Alzheimer's disease, and the cholinergic system: The 12th J. A. F. Stevenson Memorial Lecture, *Can. J. Physiol. Pharmacol.* **62:**741–754. [308]

McGeer, P. L., 1986, Brain imaging in Alzheimer's disease, *Br. Med. J.* **42:**24–28. [70, 262, 470]

McGeer, P. L., and McGeer, E. G., 1964, Formation of adrenaline by brain tissue, *Biochem. Biophys. Res. Commun.* **17:**502–507. [268]

McGeer, P. L., and McGeer, E. G., 1975, Evidence for glutamic acid decarboxylase containing interneurons in the neostriatum, *Brain Res.* **91:**331–335. [212, 215]

McGeer, P. L., and McGeer, E. G., 1976, Enzymes associated with the metabolism of catecholamines, acetylcholine and GABA in human controls and patients with Parkinson's disease and Huntington's chorea, *J. Neurochem.* **26:**65–76. [261, 471]

McGeer, P. L., and McGeer, E. G., 1981a, Kainate as a selective lesioning agent, in: *Glutamate: Transmitter in the Central Nervous System* (P. J. Roberts, J. Storm-Mathisen, and G. A. R. Johnston, eds.), pp. 55–75, John Wiley, London. [192]

McGeer, P. L., and McGeer, E. G., 1981b, Amino acid neurotransmitters, in: *Basic Neurochemistry,* 3rd ed. (G. J. Siegel, R. W. Albers, B. W. Agranoff, and R. Katzman, eds.), pp. 233–253, Little Brown, Boston. [231]

McGeer, P. L., and McGeer, E. G., 1984, Cholinergic systems and cholinergic pathology, in: *Handbook of Neurochemistry,* Vol. 6 (A. Lajtha, ed.), pp. 379–410, Plenum Press, New York. [262]

McGeer, P. L., Boulding, J. C., Gibson, W. C., and Foulkes, R. G., 1961, Drug-induced extrapyramidal reactions, *J. Am. Med. Assoc.* **177:**665–670. [358, 512]

McGeer, P. L., McGeer, E. G., and Wada, J. A., 1963, Central aromatic amines and behaviour, *Arch. Neurol.* **9:**81–89. [325]

McGeer, P. L., McGeer, E. G., Fibiger, H. C., and Wickson, V., 1971, Neostriatal choline acetylase and cholinesterase following selective brain lesions, *Brain Res.* **35:**308–314. [250]

McGeer, P. L., McGeer, E. G., Singh, V. K., and Chase, W. H., 1974a, Choline acetyltransferase localization in the central nervous system by immunohistochemistry, *Brain Res.* **81**:373–379. [250]

McGeer, P. L., Grewaal, D. S., and McGeer, E. G., 1974b, Influence of noncholinergic drugs on rat striatal acetylcholine levels, *Brain Res.* **80**:211–217. [512]

McGeer, P. L., Hattori, T., and McGeer, E. G., 1975, Chemical and autoradiographic analysis of γ-aminobutyric acid transport in Purkinje cells of the cerebellum, *Exp. Neurol.* **47**:26–41. [210, 212, 213]

McGeer, P. L., McGeer, E. G., and Suzuki, J. S., 1976, Aging and extrapyramidal function, *Arch. Neurol.* **34**:33–35. [286, 469]

McGeer, P. L., McGeer, E. G., Scherer, U., and Singh, K., 1977a, A glutamatergic cortico-striatal path?, *Brain Res.* **128**:369–373. [182, 183, 506]

McGeer, P. L., McGeer, E. G., and Hattori, T., 1977b, Dopamine–acetylcholine–GABA neuronal linkages in the extrapyramidal and limbic systems, in: *Nonstriatal Dopaminergic Neurons* (E. Costa and G. L. Gessa, eds.), pp. 397–402, Raven Press, New York. [212, 304]

McGeer, P. L., Eccles, J. C., and McGeer, E. G., 1978a, *Molecular Neurobiology of the Mammalian Brain,* Plenum Press, New York. [233, 345, 350, 464, 546–548]

McGeer, P. L., McGeer, E. G., and Hattori, T., 1978b, Transmitters in the basal ganglia, in: *Amino Acids as Chemical Transmitters* (F. Fonnum, ed.), pp. 123–141, Raven Press, New York. [212]

McGeer, P. L., McGeer, E. G., and Innanen, V. T., 1979, Dendro axonic transmission. I. Evidence from receptor binding of dopaminergic and cholinergic agents, *Brain Res.* **169**:433–441. [512]

McGeer, P. L., Kimura, H., McGeer, E. G., and Peng, J. H., 1982, Cholinergic systems in the CNS, in: *Neurotransmitter Interaction and Compartmentation, NATO Advanced Study Institutes Series A,* Vol. 48 (H. F. Bradford, ed.), pp. 253–289, Plenum Press, New York. [506]

McGeer, P. L., McGeer, E. G., and Nagai, T., 1983, GABAergic and cholinergic indices in various regions of rat brain after intracerebral injections of folic acid, *Brain Res.* **260**:107–116. [193]

McGeer, P. L., McGeer, E. G., and Peng, J. H., 1984a, ChAT: Purification and immunohistochemical localization, *Life Sci.* **34**:2319–2338. [236, 239, 245]

McGeer, P. L., McGeer, E. G., Suzuki, J., Dolman, C. E., and Nagai, T., 1984b, Aging, Alzheimer's disease and the cholinergic system of the basal forebrain, *Neurology* **34**:741–745. [245, 261, 469]

McGeer, P. L., Kamo, H., Harrop, R., *et al.,* 1986, Positron emission tomography in patients with clinically diagnosed Alzheimer's disease, *Can. Med. Assoc. J.* **134**:597–607. [262]

McIntosh, T. K., and Barfield, R. J., 1984, Brain monoaminergic control of male reproductive behavior, *Behav. Brain Res.* **12**:255–281. [338]

McKusick, V. A., 1982, Genetic disorders of the human nervous system: A commentary, in: *Molecular Genetic Neuroscience* (F. O. Schmitt, S. J. Bird, and F. E. Bloom, eds.), pp. 401–405, Raven Press, New York. [18]

McLaughlin, B. J., Wood, J. G., Saito, K., *et al.,* 1974, The fine structural localization of glutamate decarboxylase in synaptic terminals of rodent cerebellum, *Brain Res.* **76**:377–391. [203, 213]

McMichael, A. J., and Fabre, J. W., 1982, *Monoclonal Antibodies in Clinical Medicine,* Academic Press, New York. [48]

McNaughton, B. L., 1983, Activity dependent modulation of hippocampal synaptic efficiency: Some implications for memory processes, in: *Molecular, Cellular and Behavioral Neurobiology of the Hippocampus* (W. Seifert, ed.), pp. 233–249, Academic Press, New York. [563]

McNaughton, B. L., Douglas, R. M., and Goddard, G. V., 1978, Synaptic enhancement in fascia dentata: Cooperativity among coactive afferents, *Brain Res.* **157**:277–293. [558, 563, 574]

McNaughton, B. L., Barnes, C., and Andersen, P., 1981, Synaptic efficiency and EPSP summation in granule cells of rat fascia dentata studied *in vitro, J. Neurophysiol.* **46**:952–966. [556, 558, 560, 563, 631]

McNeill, J. H., 1984, Histamine and the heart, *Can. J. Physiol. Pharmacol.* **62**:720–726. [356, 358]

Means, A. R., Tash, J. S., and Chafouleas, J. G., 1982, Physiological implications of the presence, distribution and regulation of calmodulin in eukaryotic cells, *Physiol. Rev.* **62**:1–39. [96, 561]

Medina, J. H., and De Robertis, E., 1984, Taurine modulation of the benzodiazepine–γ-aminobutyric acid receptor complex in brain membranes, *J. Neurochem.* **42**:1212–1217. [231]

Meeley, M. P., Ruggiero, D. A., Ishitsuka, T., and Reis, D. J., 1985, Intrinsic γ-aminobutyric acid neurons in the nucleus of the solitary tract and the rostral ventrolateral medulla of the rat: An immunocytochemical and biochemical study, *Neurosci. Lett.* **58**:83–89. [205, 212]

Meeves, H., 1975, Calcium currents in squid giant axon, *Philos. Trans. R. Soc. London Ser. B* **270**:377–387. [59]

Mefford, I. N., Foutz, A., Noyce, N., et al., 1982, Distribution of norepinephrine, epinephrine, dopamine, serotonin, 3,4-dihydroxyphenylacetic acid, homovanillic acid and 5-hydroxyindole-3-acetic acid in dog brain, Brain Res. 236:339–349. [286, 287]

Mehler, W. R., and Nauta, W. J., 1974, Connections of the basal ganglia and of the cerebellum, Confin. Neurol. 36:205–222. [504]

Meibach, R. C., and Katzman, R., 1979, Catecholaminergic innervation of the subthalamic nucleus: Evidence for a rostral continuation of the A9 (substantia nigra) dopaminergic cell group, Brain Res. 173:364–368. [510]

Meibach, R. C., and Katzman, R., 1981, Origin, course and termination of dopaminergic substantia nigra neurons projecting to the amygdaloid complex in the cat, Neuroscience 6:2159–2171. [510]

Meier, E., Hansen, G. H., and Schousboe, A., 1985, The trophic effect of GABA on cerebellar granule cells is mediated by GABA-receptors, Int. J. Dev. Neurosci. 3:401–407. [219]

Meisenberg, G., and Simmons, W. H., 1983, Peptides and the blood–brain barrier, Life Sci. 32:2611–2623. [37, 38, 405]

Meites, J., Simpkins, J. W., and Huang, H. H., 1979, The relation of hypothalamic biogenic amines to secretion of gonadotropin and prolactin in the aging rat, Aging 8:87–94. [471]

Melander, T., Staines, W. A., Hökfelt, T., et al., 1985, Galanin-like immunoreactivity in cholinergic neurons in the septal–basal forebrain complex projecting to the hippocampus of the rat, Brain Res. 360:130–138. [157]

Meldrum, B. S., Croucher, M. J., Badman, G., and Collins, J. F., 1983, Antiepileptic action of excitatory amino acid antagonists in the photosensitive baboon, Neurosci. Lett. 39:101–104. [194]

Mendell, L. M., and Henneman, E., 1971, Terminals of single Ia fibers: Location, density and distribution within a pool of 300 homogeneous motoneurons, J. Neurophysiol. 34:171–187. [114, 115]

Mendelson, W. B., Kuruvilla, A., Watlington, T., Goehi, K., Paul, S. M., and Skolnick, P., 1983, Sedative and electroencephalographic actions of erythro-9-(2-hydroxy-3-nonyl)adenine (EHNA): Relation to inhibition of adenosine deaminase, Psychopharmacology 79:126–129. [362]

Menini, C., Meldrum, B. S., Riche, D., Silva-Comte, C., and Stutzmann, J. M., 1980, Sustained limbic seizures induced by intraamygdaloid kainic acid in the baboon: Symptomatology and neuropathological consequences, Ann. Neurol. 8:501–509. [194]

Merchenthaler, I., 1984, Corticotropin releasing factor (CRF)-like immunoreactivity in the rat central nervous system: Extrahypothalamic distribution, Peptides 5(Suppl. 1):53–69. [388, 395]

Merchenthaler, I., Hynes, M. A., Vigh, S., Schally, A. V., and Petrusz, P., 1984, Corticotropin releasing factor (CRF): Origin and course of afferent pathways to the median eminence (ME) of the rat hypothalamus, Neuroendocrinology 39:296–306. [388]

Merzenich, M. M., Knight, P. L., and Roth, G. L., 1975, Representation of the cochlea within primary auditory cortex in the cat, J. Neurophysiol. 38:231–249. [612]

Merzenich, M. M., Jenkins, W. M., and Middlebrooks, J. C., 1984, Observations and hypotheses on special organizational features of the central auditory nervous system, in: Dynamic Aspects of Cortical Function (G. M. Edelman, W. E. Gall, and W. M. Cowan, eds.), pp. 397–424, John Wiley, New York. [612]

Mesulam, M.-M., Mufson, E. J., Levey, A. I., and Wainer, B. H., 1983a, Cholinergic innervation of cortex by basal forebrain: Cytochemistry and cortical connections of the septal area, diagonal band nuclei, nucleus basalis (substantia innominata) and hypothalamus in the rhesus monkey, J. Comp. Neurol. 214:170–197. [247]

Mesulam, M.-M., Mufson, E. J., Wainer. B. H., and Levey, A. I., 1983b, Central cholinergic pathways in the rat: An overview based on an alternative nomenclature (Ch1–Ch6), Neuroscience 10:1185–1201. [245]

Mesulam, M.-M., Mufson, E. J., Levey, A. I., and Wainer, B. H., 1984, Atlas of cholinergic neurons in the forebrain and upper brainstem of the macaque based on monoclonal choline acetyltransferase immunohistochemistry and acetylcholinesterase histochemistry, Neuroscience 12:669–686. [245, 247, 250, 251]

Metcalf, G., 1982, Regulatory peptides as a source of new drugs—the clinical prospects for analogues of TRH which are resistant to metabolic degradation, Brain Res. Rev. 4:389–408. [410]

Meyer, D. K., Beinfeld, M. C., Oertel, W. H., and Brownstein, M. J., 1982, Origin of the cholecystokinin-containing fibers in the caudatoputamen, Science 215:187–188. [399, 506]

Meyer, D. L., Krause, E., and Dubberstein, B., 1982, Secretory protein translocation across membranes—the role of the ''docking protein,'' Nature (London) 297:647–650. [22]

Meyer, E. M., and Cooper, J. R., 1981, Correlations between Na$^+$-K$^+$ ATPase activity and acetylcholine release in rat cortical synaptosomes, *J. Neurochem.* **36:**467–475. [241]

Meyer, H., and Lux, H. D., 1974, Action of ammonium on a chloride pump-removal of hyperpolarizing inhibition in an isolated neuron, *Pfluegers Arch. Gesamte Physiol.* **350:**185–195. [131]

Meyer, R. L., and Sperry, R. W., 1974, Explanatory models for neuroplasticity in retinotectal connections, in: *Plasticity and Recovery of Function in the Central Nervous System* (D. G. Stein, J. I. Rosen, and N. Butters, eds.), pp. 45–63, Academic Press, New York. [434]

Meyer-Lohmann, J., Hore, J., and Brooks, V. B., 1977, Cerebellar participation in generation of prompt arm movement, *J. Neurophysiol.* **38:**871–908. [587]

Meyers, D., Oertel, W., and Brownstein, M., 1980, Deafferentation studies on the glutamic acid decarboxylase content of the supraoptic nucleus of the rat, *Brain Res.* **200:**165–168. [212]

Miale, I. L., and Sidman, R. L., 1961, An autoradiographic analysis of histogenesis in the mouse cerebellum, *Exp. Neurol.* **4:**277–296. [427]

Micevych, P. E., Yaksh, T. L., and Szolcsányi, J., 1983, Effect of intrathecal capsaicin analogues on the immunofluorescence of peptides and serotonin in the dorsal horn of rats, *Neuroscience* **8:**123–131. [407]

Michael, A. C., Justice, J. B., Jr., and Neill, D. B., 1985, *In vivo* voltammetric determination of the kinetics of dopamine metabolism in the rat, *Neurosci. Lett.* **56:**365–369. [296]

Miledi, R., 1967, Spontaneous synaptic potentials and quantal release of transmitter in the stellate ganglion of the squid, *J. Physiol. (London)* **192:**379–406. [104]

Miledi, R., 1969, Transmitter action in the giant synapse of the squid, *Nature (London)* **223:**1284–1286. [104]

Miledi, R., 1973, Transmitter release induced by injection of calcium ions into nerve terminals, *Proc. R. Soc. London Ser. B* **183:**421–425. [104]

Miledi, R., 1980, Intracellular calcium and desensitization of acetylcholine receptors, *Proc. R. Soc. London Ser. B* **209:**447–452. [582]

Miledi, R., and Parker, I., 1981, Calcium transport recorded with arsenazo III in the presynaptic terminal of the squid giant synapse, *Proc. R. Soc. London Ser. B* **212:**197–211. [168]

Miledi, R., and Slater, C. R., 1966, The action of calcium on neuronal synapses in the squid, *J. Physiol. (London)* **184:**473–498. [105]

Millan, M. J., and Herz, A., 1985, The endocrinology of the opioids, *Int. Rev. Neurobiol.* **26:**1–83. [373]

Millar, T., Ishimoto, I., Johnson, C. D., Epstein, M. L., Chubb, I. W., and Morgan, I. G., 1985, Cholinergic and acetylcholinesterase-containing neurons of the chicken retina, *Neurosci. Lett.* **61:**311–316. [252]

Millard, W. J., Sagar, S. M., and Martin, J. B., 1985, Cysteamine-induced depletion of somatostatin and prolactin, *Fed. Proc. Fed. Am. Soc. Exp. Biol.* **44:**2546–2550. [411]

Miller, M. S., Buck, S. H., Sipes, I. G., Yamamura, H. I., and Burks, T. F., 1982, Regulation of substance P by nerve growth factor: Disruption by capsaicin, *Brain Res.* **250:**193–196. [411]

Miller, R. J., 1984, PHI and GRF: Two new members of the glucagon/secretin family, *Med. Biol.* **62:**159–162. [373, 388]

Miller, W. L., Johnson, L. K., Baxter, J. D., and Roberts, J. L., 1980, Processing of the precursor to corticotropin and beta-lipotropin in humans, *Proc. Natl. Acad. Sci. U.S.A.* **77:**5211–5215. [384]

Milner, B., 1966, Amnesia following operation on the temporal lobes, in: *Amnesia* (C. W. M. Whitty and O. L. Zangwill, eds.), pp. 109–133, Butterworths, London. [567, 568, 592]

Milner, B., 1972, Disorders of learning and memory after temporal-lobe lesions in man, *Clin. Neurosurg.* **19:**421–446. [567, 568]

Milner, B., 1974, Hemispheric specialization: Scope and limits, in: *The Neurosciences: Third Study Program* (F. O. Schmitt and F. G. Worden, eds.), pp. 75–89, MIT Press, Cambridge, Massachusetts. [617, 623]

Milner, B., Taylor, L., and Sperry, R. W., 1968, Lateralized suppression of dichotically presented digits after commissural section, *Science* **161:**184–186. [613]

Milstein, C., 1981, The Wellcome Foundation Lecture, 1980: Monoclonal antibodies from hybrid myelomas, *Proc. R. Soc. London Ser. B* **211:**393–412. [46]

Milstien, S., and Kaufman, S., 1983, Tetrahydro-sepiapterin is an intermediate in tetrahydrobiopterin biosynthesis, *Biochem. Biophys. Res. Commun.* **115:**888–893. [273]

Minamino, N., Kangawa, K., and Matsuo, H., 1984, Neuromedin B is a major bombesin-like peptide in rat brain: Regional distribution of neuromedin BJ and neuromedin C in rat brain, pituitary and spinal cord, *Biochem. Biophys. Res. Commun.* **124:**925–932. [373]

Minneman, K. P., Hegstrand, L. R., and Molinoff, P. B., 1979a, Simultaneous determination of β_1 and β_2 adrenergic receptors in tissues containing both receptor subtypes, *Mol. Pharmacol.* **16**:34–46. [309]

Minneman, K. P., Hegstrand, L. R., and Molinoff, P. B., 1979b, β_1 and β_2 adrenergic receptors in rat cerebral cortex are independently regulated, *Science* **204**:866–868. [309]

Mishina, M., Tobimatsu, T., Imoto, K., et al. 1985, Location of functional regions of acetylcholine receptor α-subunit by site-directed mutagenesis, *Nature (London)* **313**:364–369. [165, 257]

Mishkin, M., 1978, Memory in monkeys severely impaired by combined, but not by separate removal of amygdala and hippocampus, *Nature (London)* **273**:297–298. [568, 569]

Mishkin, M., 1982, A memory system in the monkey, *Philos. Trans. R. Soc. London Ser. B* **298**:85–95. [568]

Mishra, R. K., Chiu, J., Chiu, P., and Mishra, C. P., 1983, Pharmacology of L-prolyl-L-leucyl-glycinamide (PLG): A review, *Methods Find. Exp. Clin. Pharmacol.* **5**:203–233. [386]

Mitchell, I. J., Cross, A. J., Sambrook, M. A., and Crossman, A. R., 1985a, Sites of the neurotoxic action of 1-methyl-4-phenyl-1,2,3,6-tetrahydropyridine in the macaque monkey include the ventral tegmental area and the locus coeruleus, *Neurosci. Lett.* **61**:195–200. [304]

Mitchell, I. J., Jackson, A., Sambrook, M. A., and Crossman, A. R., 1985b, Common neural mechanisms in experimental chorea and hemiballismus in the monkey: Evidence from 2-deoxyglucose autoradiography, *Brain Res.* **339**:346–350. [512]

Miyata, T., Kamata, K., Noguicji, M., Okano, Y., and Kase, Y., 1973, Pharmacological studies on alicyclic amines, XV: Intracerebral administration of pipecolic acid, *Jpn. J. Pharmacol. (Suppl.)* **23**:81–85. [233]

Miyata, Y., and Otsuka, M., 1972, Distribution of γ-aminobutyric acid in cat spinal cord and the alteration produced by local ischemia, *J. Neurochem.* **19**:1833–1834. [212]

Miyata, Y., and Otsuka, M., 1975, Quantitative histochemistry of γ-aminobutyric acid in cat spinal cord with special reference to presynaptic inhibition, *J. Neurochem.* **25**:239–244. [212]

Miyoshi, R., Kito, S., Kishida, T., Itoga, E., and Ogawa, N., 1985, Immunohistochemical localization of neurotensin and β-endorphin in the rat anterior pituitary gland, *Brain Res.* **331**:386–388. [157]

Mize, R. R., Spencer, R. F., and Sterling, P., 1981, Neurons and glia in cat superior colliculus accumulate [3H] gamma-aminobutyric acid (GABA), *J. Comp. Neurol.* **202**:385–396. [205]

Mizukawa, K., McGeer, P. L., Tago, H., Peng, J. H., McGeer, E. G., and Kimura, H., 1986, The cholinergic system of the human hindbrain studied by choline acetyltransferase immunohistochemistry and acetylcholinesterase histochemistry, *Brain Res.* **379**:39–55. [245, 252]

Mobley, W. C., Rutkowski, J. L., Tennekoon, G. I., Buchanan, K., and Johnston, M. V., 1985, Choline acetyltransferase activity in striatum of neonatal rats increased by nerve growth factor, *Science* **229**:284–287. [465]

Mogenson, G. J., and Huang, Y. H., 1973, The neurobiology of motivated behavior, *Prog. Neurobiol.* **1**:53–83. [528, 529]

Mogenson, G. J., Jones, D. L., and Yim, C. Y., 1980, From motivation to action: Functional interface between the limbic system and the motor system, *Prog. Neurobiol.* **14**:69–97. [514]

Molineaux, C. J., and Cox, B. M., 1982, Subcellular localization of immunoreactive dynorphin and vasopressin in rat pituitary and hypothalamus, *Life Sci.* **31**:1765–1768. [157]

Molliver, M. E., Kostovic, I., and Van der Loos, H., 1973, The development of synapses in cerebral cortex of the human fetus, *Brain Res.* **50**:403–407. [425]

Monaghan, P. L., Beitz, A. J., Madl, J. E., and Larson, A. A., 1985, An analysis of monoclonal glutamate immunoreactive neurons in the deep cerebellar nuclei and their projections to the red nucleus and the thalamus, *Soc. Neurosci. Abstr.* **11**:690. [182]

Mondrup, K., and Pedersen, E., 1984a, The effect of the GABA-agonist progabide on stretch and flexor reflexes and on voluntary power in spastic patients, *Acta Neurol. Scand.* **69**:191–199. [225]

Mondrup, K., and Pedersen, E., 1984b, The clinical effect of the GABA-agonist progabide on spasticity, *Acta Neurol. Scand.* **69**:200–206. [225]

Moore, R. Y., and Bloom, F. E., 1978, Central catecholamine neuron systems: Anatomy and physiology of the dopamine system, *Annu. Rev. Neurosci.* **1**:129–169. [291]

Moore, R. Y., and Card, J. P., 1984, Noradrenaline containing neuron systems, in: *Handbook of Chemical Neuroanatomy,* Vol. 2, *Classical Transmitters in the CNS, Part I* (A. Björklund and T. Hökfelt, eds.), pp. 123–156, Elsevier, Amsterdam. [292]

Moos, F., and Richard, P., 1982, Excitatory effect of dopamine on oxytocin and vasopressin reflex release in the rat, *Brain Res.* **241**:249–260. [295]

Morest, D. K., 1971, Dendrodendritic synapses of cells that have axons: The fine structure of the Golgi type II cell in the medial geniculate body of the cat, *Z. Anat. Entwicklungsgesch.* **133:**216–246. [136]

Morgan, I. G., and Ingram, C. A., 1981, Kainic acid affects both plexiform layers of chicken retina, *Neurosci. Lett.* **21:**275–280. [212]

Morgan, P. F., Lloyd, H. G. E., and Stone, T. W., 1983, Inhibition of adenosine accumulation by a CNS benzodiazepine antagonist (Ro 15-1788) and a peripheral benzodiazepine receptor ligand (Ro 05-4864), *Neurosci. Lett.* **41:**183–188. [362]

Mori, M., Jayaraman, A., Prasad, C., Pegues, J., and Wilber, J. F., 1982, Distribution of histidyl-proline diketopiperazine [cyclo(His-Pro)] and thyrotropin-releasing hormone (TRH) in the primate central nervous system, *Brain Res.* **245:**183–186. [388]

Morley, B. J., Lorden, J. F., Brown, G. B., Kemp, G. E., and Bradley, R. J., 1977, Regional distribution of nicotinic acetycholine receptor in rat brain, *Brain Res.* **134:**161–166. [253]

Morley, J. E., 1979, Extrahypothalamic thyrotropin releasing hormone (TRH)—its distribution and its functions, *Life Sci.* **25:**1539–1550. [410]

Morley, J. E., Levine, A. S., Gosnell, B. A., and Billington, C. J., 1984, Neuropeptides and appetite: Contribution of neuropharmacological modeling, *Fed. Proc. Fed. Am. Soc. Exp. Biol.* **43:**2903–2907. [405]

Moroji, T., Watanabe, N., Aoki, N., and Itoh, S., 1982a, Antipsychotic effects of ceruletide (caerulein) on chronic schizophrenia, *Arch. Gen. Psychiatry* **39:**485–486. [409]

Moroji, T., Watanabe, N., Aoki, N., and Itoh, S., 1982b, Antipsychotic effects of caerulein, a decapeptide chemically related to cholecystokinin octapeptide, on schizophrenia, *Int. Pharmacopsychiatry* **17:**255–273. [409]

Moroji, T., Itoh, K., and Itoh, K., 1985, Antipsychotic effects of ceruletide in chronic schizophrenia, *Ann. N. Y. Acad. Sci.* **448:**518–534. [409]

Morrell, J. I., Schwanzel-Fukuda, M., Fahrbach, S. E., and Pfaff, D. W., 1984, Axonal projections and peptide content of steroid hormone concentrating neurons, *Peptides* **5**(Suppl. 1):227–239. [395]

Morris, G., Seidler, F. J., and Slotkin, T. A., 1983, Stimulation of ornithine decarboxylase by histamine or norepinephrine in brain regions of the developing rat: Evidence for biogenic amines as trophic agents in neonatal brain development, *Life Sci.* **32:**1565–1571. [358]

Morris, H. R., Panico, M., Etienne, T., Tippins, J., Girgis, S. I., and MacIntyre, I., 1984, Isolation and characterization of human calcitonin gene-related peptide, *Nature (London)* **308:**746–748. [376]

Morris, J. L., Gibbins, I. L., Furness, J. B., Costa, M., and Murphy, R., 1985, Co-localization of neuropeptide Y, vasoactive intestinal polypeptide and dynorphin in non-noradrenergic axons of the guinea pig uterine artery, *Neurosci. Lett.* **62:**31–37. [157]

Morrison, J. H., Foote, J. L., Molliver, M. E., Bloom, F. E., and Lidov, H. G., 1982, Noradrenergic and serotonergic fibers innervate complementary layers in monkey primary visual cortex: An immunohistochemical study, *Proc. Natl. Acad. Sci. U.S.A.* **79:**2401–2405. [333]

Morrison, J. H., Benoit, R., Magistretti, P. J., and Bloom, F. E., 1983, Immunohistochemical distribution of pro-somatostatin-related peptides in cerebral cortex, *Brain Res.* **262:**344–351. [398]

Morrison, J. H., Magistretti, P. J., Benoit, R., and Bloom, F. E., 1984, The distribution and morphological characteristics of the intracortical VIP-positive cell: An immunohistochemical analysis, *Brain Res.* **292:**269–282. [397]

Morselli, P. L., Fournier, V., Bossi, L., and Musch, B., 1985, Clinical activity of GABA agonists in neuroleptic and L-dopa-induced dyskinesia, *Psychopharmacology (Suppl.)* **2:**128–136. [225]

Motamedi, F., and York, D. H., 1980, Effects of a nigral descending pathway on cervical spinal cord afferent fibers and interneurons, *Exp. Neurol.* **68:**258–268. [513]

Motley, S. J., and Collins, G. G. S., 1983, Endogenous adenosine inhibits excitatory transmission in the rat olfactory cortex slice, *Neuropharmacology* **22:**1081–1086. [365]

Mountcastle, V. B., 1966, The neural replication of sensory events in the somatic afferent system, in: *Brain and Conscious Experience* (J. C. Eccles, ed.), pp. 85–115, Springer-Verlag, New York. [51]

Mountcastle, V. B., 1975, The view from within: Pathways to the study of perception, *Johns Hopkins Med. J.* **136:**109–131. [515, 599, 611]

Mountcastle, V. B., 1978, An organizing principle for cerebral function: The unit module and the distributed system, in: *The Mindful Brain* (F. O. Schmitt, ed.), pp. 7–50, MIT Press, Cambridge, Massachusetts. [599, 600, 603, 631, 632]

Mountcastle, V. B., Lynch, J. C., Georgopoulos, A., Sakata, H., and Acuna, C., 1975, Posterior parietal association cortex of the monkey: Command functions for operation within extrapersonal space, *J. Neurophysiol.* **38**:871–908. [611]

Mountjoy, C. Q., 1986, Correlations between neuropathological and neurochemical changes, *Br. Med. Bull.* **42**:81–85. [262]

Mountjoy, C. Q., Roth, M., Evans, N. J. R., and Evans, H. M., 1983, Cortical neuronal counts in normal elderly controls and demented patients, *Neurobiol. Aging* **4**:1–11. [262]

Mufson, E. J., Levey, A., Wainer, B., and Mesulam, M.-M., 1982, Cholinergic projections from the mesencephalic tegmentum to neocortex in rhesus monkey, *Soc. Neurosci. Abstr.* **8**:135. [250]

Müller, W. E., and Snyder, S. H., 1977, Delta-aminolevulinic acid: Influences on synaptic GABA receptor binding may explain CNS symptoms of porphyria, *Ann. Neurol.* **2**:340–342. [225]

Murakami, F., Katsumaru, H., Saito, K., and Tsukahara, N., 1982, A quantitative study of synaptic reorganization in red nucleus neuron after lesion of the nucleus interpositus of the cat: An electron microscope study involving intracellular injection of horseradish peroxidase, *Brain Res.* **242**:41–53. [441, 442]

Murakami, F., Katsumaru, H., Maeda, J., and Tsukahara, N., 1984, Reorganization of corticorubral synapses following cross-innervation of flexor and extensor nerves of adult cat: A quantitative electron microscope study, *Brain Res.* **306**:299–306. [440, 442]

Muramoto, O., Kanazawa, I., and Ando, K., 1981, Neurotransmitter abnormality in rolling mouse Nagoya, an ataxic mutant mouse, *Brain Res.* **215**:295–304. [229]

Murray, C. L., and Fibiger, H. C., 1985, Learning and memory deficits after lesions of the nucleus basalis magnocellularis: Reversal by physostigmine, *Neuroscience* **14**:1025–1032. [261]

Murray, J. M., Davies, K. E., Harper, P. S., Meredith, L., Mueller, C. R., and Williamson, R., 1982, Linkage relationship of cloned DNA sequence on the short arm of the X chromosome to Duchenne muscular dystrophy, *Nature (London)* **300**:69–71. [45]

Mushiake, S., Shosaku, A., and Kayama, Y., 1984, Inhibition of thalamic ventrobasal complex neurons by glutamate infusion into the thalamic reticular nucleus in rats, *J. Neurosci. Res.* **12**:93–100. [212]

Nadler, J. V., White, W. F., Vaca, K. W., Perry, B. W., and Cotman, C. W., 1978, Biochemical correlates of transmission mediated by glutamate and aspartate, *J. Neurochem.* **31**:147–155. [180, 182, 184, 186]

Nagai, T., Satoh, K., Imamoto, K., and Maeda, T., 1981, Divergent projections of catecholamine neurons of locus coeruleus as revealed by fluorescent double labeling technique, *Neurosci. Lett.* **23**:117–123. [293]

Nagai, T., Kimura, H., Maeda, T., McGeer, P. L., and McGeer, E. G., 1982, Cholinergic projections from the basal forebrain of rat to the amygdala, *J. Neurosci.* **2**:513–520. [247]

Nagai, T., McGeer, P. L., and McGeer, E. G., 1983a, Distribution of GABA-T intensive neurons in the rat forebrain and midbrain, *J. Comp. Neurol.* **218**:220–238. [206, 209, 213, 214, 216, 355]

Nagai, T., McGeer, P. L., Peng, J. H., McGeer, E. G., and Dolman, C. E., 1983b, Choline acetyltransferase immunohistochemistry in brains of Alzheimer's disease patients and controls, *Neurosci. Lett.* **36**:195–199. [245, 261, 262]

Nagai, T., Pearson, T., Peng, J. H., McGeer, E. G., and McGeer, P. L., 1983c, Immunohistochemical staining of the human forebrain with monoclonal antibody to human choline acetyltransferase, *Brain Res.* **265**:300–306. [245]

Nagai, T., Maeda, T., Imai, H., McGeer, P. L., and McGeer, E. G., 1985, Distribution of GABA-T-intensive neurons in the rat hindbrain, *J. Comp. Neurol.* **231**:260–269. [206, 209]

Nagata, H., Mimori, Y., Nakamura, S., and Kameyama, M., 1984, Regional and subcellular distribution in mammalian brain of the enzymes producing adenosine, *J. Neurochem.* **42**:1001–1007. [361]

Nagatsu, T., Levitt, M., and Udenfriend, S., 1964, Tyrosine hydroxylase: The initial step in norepinephrine biosynthesis, *J. Biol. Chem.* **239**:2910–2917. [269, 270]

Nagatsu, T., Hidaka, H., Kuzaya, H., and Takeya, K., 1970, Inhibition of dopamine-beta-hydroxylase by fuscaric acid (5-butylpicolinic acid) *in vitro* and *in vivo, Biochem. Pharmacol.* **19**:35–44. [312]

Nagy, J. I., and Fibiger, H. C., 1980, A striatal source of glutamic acid decarboxylase activity in the substantia nigra, *Brain Res.* **187**:237–242. [212]

Nagy, J. I., Carter, D. A., Lehmann, J., and Fibiger, H. C., 1978, Evidence for a GABA-containing projection from the entopeduncular nucleus to the lateral habenula in the rat, *Brain Res.* **145**:360–364. [212]

Nagy, J. I., LaBella, L. A., Buss, M., and Daddona, P. E., 1984, Immunohistochemistry of adenosine deaminase: Implications for adenosine neurotransmission, *Science* **224**:166–168. [363]

Nagy, J. I., Geiger, J. D., and Daddona, P. E., 1985, Adenosine uptake sites in rat brain: Identification using [³H]nitrobenzylthioinosine and co-localization with adenosine deaminase, *Neurosci. Lett.* **55**:47–53. [363]

Nair, N. P. V., Bloom, D. M., and Nestoros, J. N., 1982, Cholecystokinin appears to have antipsychotic properties, *Prog. Neuro-Psychopharmacol. Biol. Psychiatry* **6**:509–512. [409]

Nair, N. P. V., Bloom, D. M., Nestoros, J. N., and Schwartz, G., 1983, Therapeutic efficacy of cholecystokinin in neuroleptic-resistant schizophrenic subjects, *Psychopharmacol. Bull.* **19**:134–136. [409]

Nakagawa, Y., Shiosaka, S., Emson, P. C., and Tohyama, M., 1985, Distribution of neuropeptide Y in the forebrain and diencephalon: An immunohistochemical analysis, *Brain Res.* **361**:52–60. [394, 395, 398]

Nakamura, S., and Vincent, S. R., 1986, Histochemistry of MPTP oxidation in the rat brain: Sites of synthesis of the parkinsonism-inducing toxin MPP^+, *Neurosci. Lett.* **65**:321–325. [303]

Nakanishi, S., Inoue, A., Kita, T., *et al.*, 1979, Nucleotide sequence of cloned cDNA for bovine corticotropin–β-lipotropin precursor, *Nature (London)* **278**:423–427. [381]

Nakanishi, S., Ohkubo, H., Nawa, H., Kitamura, N., Kageyama, R., and Ujihara, M., 1983, Angiotensin and kininogen: Cloning and sequence analysis of the cDNAs, *Clin. Exp. Hypertens.* **5**:997–1003. [382]

Nakano, I., and Hirano, A., 1983, Neuron loss in the nucleus basalis of Meynert in parkinsonism–dementia complex of Guam, *Ann. Neurol.* **13**:87–91. [308, 470]

Nakano, I., and Hirano, A., 1984, Parkinson's disease: Neuron loss in the nucleus basalis without concomitant Alzheimer's disease, *Ann. Neurol.* **15**:415–418. [470]

Nakata, H., and Fujisawa, H., 1982a, Purification and properties of tryptophan 5-monooxygenase from rat brain-stem, *Eur. J. Biochem.* **122**:41–47. [323, 324]

Nakata, H., and Fujisawa, H., 1982b, Tryptophan 5-monooxygenase from mouse mastocytoma P815, *Eur. J. Biochem.* **124**:595–601. [323]

Nakata, Y., Kusaka, Y., Segawa, T., Yajima, H., and Kitagawa, K., 1978, Substance P: Regional distribution and specific binding to synaptic membranes in rabbit central nervous system, *Life Sci.* **22**:259–268. [402]

Namba, M., Ghatei, M. A., Gibson, S. J., Polak, J. M., and Bloom, S. R., 1985, Distribution and localization of neuromedin B-like immunoreactivity in pig, cat and rat spinal cord, *Neuroscience* **15**:1217–1226. [373]

Nanopoulos, D., Belin, M. F., Maitre, M., Vincendon, G., and Pujol, J. F., 1982, Immunocytochemical evidence for the existence of GABAergic neurons in the nucleus raphe dorsalis: Possible existence of neurons containing serotonin and GABA, *Brain Res.* **232**:375–379. [205, 215]

Nappi, G., Petraglia, F., Martignoni, E., Facchinetti, F., Bono, G., and Genazzani, A. R., 1985, β-Endorphin cerebrospinal fluid decrease in untreated parkinsonian patients, *Neurology* **35**:1371–1374. [412]

Narasimhachari, N., Kempster, E., and Anbar, M., 1980, 5-Methoxytryptamine in rat hypothalamus and human CSF: A fact or artifact?, *Biomed. Mass Spectrom* **7**:231–235. [345]

Nauta, H. J. W., 1979, Projections of the pallidal complex: An autoradiographic study in the cat, *Neuroscience* **4**:1853–1873. [509]

Nauta, W. J. H., 1971, The problem of the frontal lobe: A reinterpretation, *J. Psychiatr. Res.* **8**:167–187. [614]

Nawa, H., Hirose, T., Takashima, H., Inayama, S., and Nakanishi, S., 1983, Nucleotide sequences of cloned cDNAs for two types of bovine brain substance P precursor, *Nature (London)* **306**:32–36. [377, 382]

Neal, M. J., and Massey, S. C., 1980, Release of acetylcholine and amino acids from retina, *Neurochem. Int.* **1**:191–208. [229]

Neale, E. A., Nelson, P. G., Macdonald, R. L., Christian, C. N., and Bowers, L. M., 1983, Synaptic interactions between mammalian central neurons in cell culture. III. Morphophysiological correlates of quantal synaptic transmission, *J. Neurophysiol.* **49**:1459–1468. [116, 119]

Neckers, L. M., and Meek, J. L., 1976, Measurement of 5HT turnover rate in discrete nuclei of rat brain, *Life Sci.* **19**:1579–1584. [328]

Neff, N. H., Yang, H. T., Goridis, C., and Bialek, D., 1974, The metabolism of indolealkylamines by type A and B monoamine oxidase of brain, in: *Serotonin—New Vistas* (E. Costa, G. L. Gessa, and M. Sandler, eds.), Advances in Biochemical Psychopharmacology, Vol. 11, pp. 51–58, Raven Press, New York. [277]

Neher, E., and Sakmann, B., 1976, Single-channel currents recorded from membrane of denervated frog muscle fibers, *Nature (London)* **260**:779–802. [99]

Nelkin, B. D., Rosenfeld, K. I., de Bustros, A., Leong, S. S., Roos, B. A., and Baylin, S. B., 1984, Structure and expression of a gene encoding human calcitonin and calcitonin gene related peptide, *Biochem. Biophys. Res. Commun.* **123**:648–655. [382]

Nelson, P. G., Marshall, K. C., Pun, R. Y. K., *et al.*, 1983, Synaptic interactions between mammalian central neurons in cell culture. II. Quantal analysis of EPSPs, *J. Neurophysiol.* **49**:1442–1458. [116, 119]

Nemeroff, C. B., Luttinger, D., and Prange, A. J., Jr., 1980, Neurotensin: Central nervous system effects of a neuropeptide, *TINS* **3**:212–215. [407]

Nemeroff, C. B., Youngblood, W. W., Manberg, P. J., Prange, A. J., Jr., and Kizer, J. S., 1983, Regional brain concentrations of neuropeptides in Huntington's chorea and schizophrenia, *Science* **221**:972–974. [412]

Nemeroff, C. B., Widerlov, E., Bissette, G., *et al.* 1984, Elevated concentrations of CSF corticotropin-releasing factor-like immunoreactivity in depressed patients, *Science* **226**:1342–1344. [412]

Netter, F. H., 1953, *The Ciba Collection of Medical Illustrations*, Vol. 1, *Nervous System*, Case–Hoyt, Rochester, New York. [477]

Newman, H. H., Freeman, F. N., and Holzinger, K. J., 1937, *Twins: A Study of Heredity and Environment*, University of Chicago Press, Chicago. [526]

Newman, J. D., 1976, Single unit analysis of prefrontal cortex in primates, *Exp. Brain Res.* **1**:459–462. [599]

Newman, M. E., 1983, Adenosine binding sites in brain: Relation to endogenous levels of adenosine and its physiological and regulatory roles, *Neurochem. Int.* **5**:21–25. [363]

Nichol, C. A., Lee, C. L., Edelstein, M. P., Chao, J. Y., and Duch, D. S., 1983, Biosynthesis of tetrahydrobiopterin by *de novo* and salvage pathways in adrenal medulla extracts, mammalian cell cultures, and rat brain *in vivo*, *Proc. Natl. Acad. Sci. U.S.A.* **80**:1546–1550. [270, 273]

Nicholson, A. N., and Wright, C. M., 1981, (+)-6-Fluorotryptophan, an inhibitor of tryptophan hydroxylase: Sleep and wakefulness in the rat, *Neuropharmacology* **20**:335–339. [338]

Nicoll, R. A., 1969, Inhibitory mechanisms in the rabbit olfactory bulb: Dendro-dendritic mechanisms, *Brain Res.* **14**:157–172. [133, 134, 212]

Nicoll, R. A., 1975, The action of presumed blockers of chloride transport on synaptic and amino acid responses in the frog spinal cord, *Soc. Neurosci. Abstr.* **1**:377. [144]

Nicoll, R. A., and Alger, B. E., 1979, Presynaptic inhibition: Transmitter and ionic mechanisms, *Int. Rev. Neurobiol.* **21**:217–258. [142, 144]

Nicoll, R. A., Schenker, C., and Leeman, S. E., 1980, Substance P as a transmitter candidate, *Annu. Rev. Neurosci.* **3**:227–268. [605]

Nieoullon, A., and Dusticier, N., 1983, Glutamate uptake, glutamate decarboxylase and choline acetyltransferase in subcortical areas after sensorimotor cortical ablations in the cat, *Brain Res. Bull.* **10**:287–293. [182]

Niijima, K., and Yoshida, M., 1982, Electrophysiological evidence for branching nigral projections to pontine reticular formation, superior colliculus and thalamus, *Brain Res.* **239**:279–282. [212, 510]

Nikolics, K., Mason, A. J., Szonyi, E., Ramachandran, J., and Seeburg, P. H., 1985, A prolactin-inhibiting factor within the precursor for human gonadotropin-releasing hormone, *Nature (London)* **316**:511–517. [382, 393]

Niles, L. P., Wong, Y.-W., Mishra, R. K., and Brown, G. M., 1979, Melatonin receptors in brain, *Eur. J. Pharmacol.* **55**:219–220. [345]

Ninkovic, M., Beaujouan, J. C., Torrens, Y., Saffroy, M., Hall, M. D., and Glowinski, J., 1985, Differential localization of tachykinin receptors in rat spinal cord, *Eur. J. Pharmacol.* **106**:463–464. [403]

Nishi, S., Minota, S., and Karczmar, A. G., 1974, Primary afferent neurones: The ionic mechanism of GABA-mediated depolarization, *Neuropharmacology* **13**:215–219. [144]

Nishimura, Y., Schwartz, M. L., and Rakic, P., 1985, Localization of γ-aminobutyric acid and glutamic acid decarboxylase in rhesus monkey retina, *Brain Res.* **359**:351–355. [205]

Nishizuka, Y., 1984, Turnover of phospholipids and signal transduction, *Science* **225**:1365–1370. [170]

Nitsch, C., Kim, J. K., Shimada, C., and Okada, G., 1979, Effect of hippocampus extirpation in the rat on glutamate levels in target structures of hippocampal efferents, *Neurosci. Lett.* **11**:295–299. [182]

Nixon, J. C., Lee, C. L., Milstien, S., Kaufman, S., and Bartholome, K., 1980, Neopterin and biopterin levels in patients with atypical forms of phenylketonuria, *J. Neurochem.* **35**:898–904. [273]

Nja, A., and Purves, D., 1978, The effects of nerve growth factor and its antiserum on synapses in the superior cervical ganglion of the guinea-pig, *J. Physiol. (London)* **277**:53–75. [460]

Noble, M., Albrechtsen, M., Moller, C., *et al.*, 1985, Glial cells express N-CAM/D2-CAM-like polypeptides *in vitro*, *Nature (London)* **316**:725–728. [37]

Noda, M., Furutani, Y., Takahashi, H., *et al.*, 1982a, Cloning and sequence analysis of cDNA for bovine adrenal preproenkephalin, *Nature (London)* **295**:202–206. [382]

Noda, M., Teranishi, Y., Takahashi, H., *et al.*, 1982b, Isolation and structure organization of the human preproenkephalin gene, *Nature (London)* **297**:431–434. [382]

Noda, M., Shimitzu, S., Tanabe, T., *et al.*, 1984, Primary structure of the *Electrophorus electricus* sodium channel deduced from a cDNA sequence, *Nature (London)* **312:**121–127. [46, 58, 255]

Noell, W. K., 1959, The visual cell: Electric and metabolic manifestations of its life processes, *Am. J. Ophthalmol.* **48:**347–370. [212, 216]

Nomura, Y., Okuma, Y., Segawa, T., Schmidt-Glenewinkel, T., and Giacobini, E., 1979, A calcium-dependent high potassium-induced release of pipecolic acid from brain slices, *J. Neurochem.* **33:**803–805. [233]

Nomura, Y., Schmidt-Glenewinkel, T., and Giacobini, E., 1980, Uptake of piperidine and pipecolic acid by synaptosomes from mouse brain, *Neurochem. Res.* **5:**1163–1173. [233]

Nomura, Y., Schmidt-Glenewinkel, T., Giacobini, E., and Ortiz, J., 1983, Metabolism of cadaverine and pipecolic acid in brain and other organs of the mouse, *J. Neurosci. Res.* **9:**279–289. [233]

Nonner, W., Rojas, E., and Stämpfli, R., 1975, Gating currents in the node of Ranvier: Voltage and time dependence, *Philos. Trans. R. Soc. London Ser. B* **270:**483–492. [58]

Norido, F., Canella, R., and Aporti, F., 1981, Acceleration of nerve regeneration by gangliosides estimated by the somatosensory evoked potentials (SEP), *Experientia* **37:**301–302. [467]

Norris, D. B., Gajtkowski, G. A., and Rising, T. J., 1984, Histamine H_2-binding studies in guinea pig brain, *Agents Actions* **14:**541–545. [357]

Nowakowski, R. S., and Rakic, P., 1981, The site of origin and route and rate of migration of neurons to the hippocampal region of the rhesus monkey, *J. Comp. Neurol.* **196:**129–154. [433]

Nutt, J. G., Mroz, E. A., Leeman, S. E., Williams, A. C., Engel, W. K., and Chase, T. N., 1980, Substance P in human cerebrospinal fluid: Reductions in peripheral neuropathy and anatomic dysfunction, *Neurology* **30:**1280–1285. [411, 412]

Nyback, H., 1972, Effect of brain lesions and chlorpromazine on accumulation and disappearance of catecholamines formed *in vivo* from ^{14}C-tyrosine, *Acta Physiol. Scand.* **84:**54–64. [284]

Obata, K., and Takeda, K., 1969, Release of GABA into the fourth ventricle induced by stimulation of the cat cerebellum, *J. Neurochem.* **16:**1043–1047. [212, 213]

Obata, K., and Yoshida, M., 1973, Caudate-evoked inhibition and actions of GABA and other substances on cat pallidal neurons, *Brain Res.* **65:**455–459. [212, 215]

Obata, K., Ito, M., Ochi, R., and Sato, N., 1967, Pharmacological properties of the postsynaptic inhibition of Purkinje cell axons and the action of γ-aminobutyric acid on Deiters' neurons, *Exp. Brain Res.* **4:**43–57. [197]

Obata, K., Takeda, K., and Shinozaki, H., 1970, Further studies on pharmacological properties of the cerebellar-induced inhibition of Deiters' neurones, *Exp. Brain Res.* **11:**327–342. [230]

Obata-Tsuto, H. L., Okamura, H., Tsuto, T., *et al.*, 1983, Distribution of the VIP-like immunoreactive neurons in the cat central nervous system, *Brain Res. Bull.* **10:**653–660. [388]

Ochs, S., 1972, Fast transport of materials in mammalian nerve fibers, *Science* **176:**252–260. [30, 31, 461]

Ochs, S., 1981, Axoplasmic transport, in: *Basic Neurochemistry*, 3rd ed. (G. J. Siegel, R. W. Albers, B. W. Agranoff, and R. Katzmann, eds.), pp. 425–442, Little, Brown, Boston. [30, 32]

Ochs, S., 1982, *Axoplasmic Transport and Its Relation to Other Nerve Functions*, John Wiley, New York. [30, 32]

Ochs, S., 1983, Axoplasmic transport, in: *Handbook of Neurochemistry*, Vol. 5 (A. Lajtha, ed.), pp. 355–379, Plenum Press, New York. [31, 32]

Ochs, S., and Iqbal, Z., 1982, The role of calcium in axoplasmic transport in nerve, in: *Calcium and Cell Function*, Vol. III (W. Y. Cheung, ed.), pp. 325–355, Academic Press, New York. [30]

O'Connor, L. H., Nock, B., and McEwen, B. S., 1985, Quantitative autoradiography of $GABA_A$ receptors in rat forebrain: Receptor distribution and effects of estradiol, *Soc. Neurosci. Abstr.* **11:**772. [524]

O'Donohue, T. L., Charlton, C. G., Miller, R. L., Boden, G., and Jacobowitz, D. M., 1981, Identification, characterization, and distribution of secretin immunoreactivity in rat and pig brain, *Proc. Natl. Acad. Sci. U.S.A.* **78:**5221–5224. [400]

Oertel, W. H., Schmechel, E., Mugnaini, M. L., Tappaz, M. L., and Koplin, I. J., 1981, Immunocytochemical localization of glutamate decarboxylase in rat cerebellum with a new antiserum, *Neuroscience* **6:**2715–2735. [203, 206, 213]

Oertel, W. H., Mugnaini, E., Tappaz, M. L., *et al.*, 1982a, Central GABAergic innervation of neurointermediate pituitary lobe: Biochemical and immunocytochemical study in the rat, *Proc. Natl. Acad. Sci. U.S.A.* **79:**675–679. [206]

Oertel, W. H., Tappaz, M. L., Berod, A., and Mugnaini, E., 1982b, Two-color immunohistochemistry for dopamine and GABA neurons in rat substantia nigra and zona incerta, *Brain Res. Bull.* **9**:463–474. [206]

Oertel, W. H., Graybiel, A. M., Mugnaini, E., Elde, R. P., Schmechel, D. E., and Kopin, I. J., 1983a, Coexistence of glutamic acid decarboxylase- and somatostatin-like immunoreactivity in neurons of the feline nucleus reticularis thalami, *J. Neurosci.* **3**:1322–1332. [157]

Oertel, W. H., Riethmuller, G., Mugnaini, E., *et al.*, 1983b, Opioid peptide-like immunoreactivity localized in GABAergic neurons of rat neostriatum and central amygdaloid nucleus, *Life Sci.* **33**(Suppl. 1):73–76. [157]

Ohkubo, H., Kageyama, R., Ujihara, M., Hirose, T., Inayama, S., and Nakanishi, S., 1983, Cloning and sequence analysis of cDNA for rat angiotensinogen, *Proc. Natl. Acad. Sci. U.S.A.* **80**:2196–2200. [382]

Ohta, M., and Oomura, Y., 1979, Monosynaptic facilitatory pathway from the hypothalamic ventromedial nucleus to the frontal cortex in the rat, *Brain Res. Bull.* **4**:223–229. [229]

Oishi, R., Nishibori, M., and Saeki, K., 1984, Regional differences in the turnover of neuronal histamine in the rat brain, *Life Sci.* **34**:691–699. [353]

Oka, K., Kojima, K., and Nagatsu, T., 1983, Characterization of tyrosine hydroxylase from bovine adrenal medulla, *Biochem. Int.* **7**:387–393. [269, 270]

Okamoto, S., 1951, Epileptogenic action of glutamate directly applied into the brain of animals and inhibitory effect of proteins and tissue emulsions on its action, *J. Physiol. Soc. Jpn.* **13**:555–562. [175]

Okuno, S., and Fujisawa, H., 1982, Purification and some properties of tyrosine 3-monooxygenase from rat adrenal, *Eur. J. Biochem.* **122**:49–55. [269]

Okuno, S., and Fujisawa, H., 1984, Purification and characterization of rat dopamine β-monooxygenase and monoclonal antibodies to the enzyme, *Biochim. Biophys. Acta* **799**:260–269. [274]

O'Kusky, J., and Colonnier, M., 1982a, Postnatal changes in the number of astrocytes, oligodendrocytes and microglia in the visual cortex (area 17) of the macaque monkey: A stereological analysis in normal and monocularly deprived animals, *J. Comp. Neurol.* **210**:307–315. [8, 9, 35]

O'Kusky, J., and Colonnier, M., 1982b, A laminar analysis of the number of neurons, glia and synapses in the visual cortex (area 17) of the adult macaque monkey, *J. Comp. Neurol.* **210**:278–290. [8, 9]

Oldendorf, W. H., 1975, Permeability of the blood–brain barrier, in: *The Basic Neurosciences,* Vol. 1, *The Nervous System* (D. B. Tower and R. O. Brady, eds.), pp. 279–289, Raven Press, New York. [37, 38]

Olds, J., 1962, Hypothalamic substrates of reward, *Physiol. Rev.* **42**:554–604. [530]

Olds, J., and Milner, P., 1954, Positive reinforcement produced by electrical stimulation of septal area and other regions in rat brain, *J. Comp. Physiol.* **47**:419–427. [529]

O'Leary, D. D. M., Stanfield, B. B., and Cowan, W. M., 1981, Evidence that the early postnatal restriction of the cells of origin of the callosal projection is due to the elimination of axonal collaterals rather than to the death of neurons, *Dev. Brain Res.* **1**:607–617. [456]

Oliver, G., and Schafer, E. A., 1895, The physiological effects of extracts of the suprarenal capsules, *J. Physiol. (London)* **18**:230–276. [266]

Oliveros, J. C., Jandali, M. K., Timsit-Berthier, M., *et al.*, 1978, Vasopressin in amnesia, *Lancet* **1**:42. [410]

Olney, J. W., 1978, Neurotoxicity of excitatory amino acids, in: *Kainic Acid as a Tool in Neurobiology* (E. G. McGeer, J. W. Olney, and P. L. McGeer, eds.), pp. 95–122, Raven Press, New York. [190]

Olney, J. W., and Price, M. T., 1978, Excitotoxic amino acids as neuroendocrine probes, in: *Kainic Acid as a Tool in Neurobiology* (E. G. McGeer, J. W. Olney, and P. L. McGeer, eds.), pp. 239–264, Raven Press, New York. [193]

Olney, J. W., Fuller, T. A., de Gubareff, T., and Labruyere, J., 1981, Intrastriatal folic acid mimics the distant but not the local brain damage properties of kainic acid, *Neurosci. Lett.* **25**:185–191. [193]

Olpe, H. R., Laszlo, J., Dooley, D. J., Heid, J., and Steinmann, M. W., 1983, Increased activity of locus coeruleus neurons in the rat after DSP-4 treatment, *Neurosci. Lett.* **40**:81–84. [311]

Olsen, R. W., 1981, GABA–benzodiazepine–barbiturate receptor interactions, *J. Neurochem.* **37**:1–13. [222]

Olsen, R. W., and Leeb-Lundberg, F., 1981, Convulsant and anticonvulsant drug binding sites related to GABA-regulated chloride ion channels, *Adv. Biochem. Psychopharmacol.* **26**:93–102. [219]

Olson, G. A., Olson, R. D., and Kastin, A. J., 1984, Endogenous opiates: 1983, *Peptides* **5**(Suppl. 1):975–992. [373]

O'Malley, K., Mauron, A., Makk, G., *et al.*, 1983, Dopamine β-hydroxylase rat mRNA: Structure, regulation, and tissue localization, *Cold Spring Harbor Symp. Quant. Biol.* **48**:319–325. [274]

Omenn, G. S., and Smith, L. T., 1978, A common uptake system for serotonin and dopamine in human platelets, *J. Clin. Invest.* **62**:235–240. [342]

Ory, S. J., 1983, Clinical uses of luteinizing hormone-releasing hormone, *Fertil. Steril.* **39**:577–591. [409]

Osborne, H., Przewlocki, R., Hollt, V., and Herz, A., 1979, Release of β-endorphin from rat hypothalamus *in vitro, Eur. J. Pharmacol.* **55**:425–428. [402]

Oscarsson, O., 1973, Functional organization of spinocerebellar pathways, in: *Handbook of Sensory Physiology,* Vol. 2 (A. Iggo, ed.), pp. 339–380, Springer-Verlag, Berlin. [499]

Oscarsson, O., 1976, Spatial distribution of climbing and mossy fibre inputs into the cerebellar cortex, in: *Afferent and Intrinsic Organization of Laminated Structures in the Brain* (O. Creutzfeld, ed.), pp. 36–44, Springer-Verlag, Heidelberg. [499]

Oscarsson, O., and Rosén, I., 1963, Projection to cerebral cortex of large muscle-spindle afferents in forelimb nerves of the cat, *J. Physiol. (London)* **169**:924–945. [487]

Oscarsson, O., Rosén, I., and Sulg, I., 1966, Organization of neurones in the cat cerebral cortex that are influenced from group I muscle afferents, *J. Physiol. (London)* **183**:189–210. [487]

Oshima, T., 1969, Studies of pyramidal tract cells, in: *Basic Epilepsies* (R. O. Brady, H. H. Jasper, A. A. Ward, and A. Pope, eds.), pp. 253–261, Little, Brown, Boston. [71, 76]

Ostfeld, A. M., and Araguette, A., 1962, Central nervous system effects of hyoscine in man, *J. Pharmacol. Exp. Ther.* **137**:133–139. [261]

Otten, U., Lorez, H. P., and Businger, F., 1983, Nerve growth factor antagonizes the neurotoxic action of capsaicin on primary sensory neurones, *Nature (London)* **301**:515–517. [411]

Owen, F., Crow, T., Poulter, M., Cross, A. J., Longden, A., and Riley, G. J., 1978, Increased dopamine receptor sensitivity in schizophrenia, *Lancet* **2**:223–226. [546]

Owen, F., Cross, A. J., Crow, T. J., Poulter, M., and Waddington, J. L., 1981, Increased dopamine receptors in schizophrenia: Specificity and relationship to drugs and symptomatology, in: *Biological Psychiatry 1981* (C. Perris, G. Struwe, and B. Jansson, eds.), pp. 699–706, Elsevier/North-Holland, Amsterdam. [546]

Oyama, T., Jin, T., Yamaya, R., Ling, N., and Guillemin, R., 1980, Profound analgesic effects of β-endorphin in man, *Lancet* **1**:122–124. [406]

Pacitti, C., Fiadene, G., Gasbarri, A., Civitelli, D., and Scarnati, E., 1982, Electrophysiological evidence for an inhibitory accumbens–entopenduncular pathway in the rat, *Neurosci. Lett.* **33**:35–40. [212]

Palacios, J. M., Mengod, G., Picatoste, F., Grau, M., and Blanco, I., 1976, Properties of rat brain histidine decarboxylase, *J. Neurochem.* **27**:1455–1460. [352]

Palkovits, M., 1984, Neuropeptides in the hypothalamo-hypophyseal system: Lateral retrochiasmatic area as a common gate for neuronal fibers towards the median eminence, *Peptides* **5**(Suppl. 1):35–39. [392, 394]

Palkovits, M., Brownstein, M. J., and Vale, W., 1985, Distribution of corticotropin-releasing factor in rat brain, *Fed. Proc. Fed. Am. Soc. Exp. Biol.* **44**:215–219. [388]

Palmer, C., Schmidt, E. M., and McIntosh, J. S., 1981, Corticospinal and corticorubral projections from the supplementary motor area in the monkey, *Brain Res.* **209**:305–314. [482]

Palovcik, R. A., and Phillips, M. I., 1984, Saralasin increases activity of hippocampal neurons inhibited by angiotensin II, *Brain Res.* **323**:345–348. [402]

Pandol, S. J., Thomas, M. W., Schoeffield, M. S., Sachs, G., and Muallem, S., 1985, Role of calcium in cholecystokinin-stimulated phosphoinositide breakdown in exocrine pancreas, *Am. J. Physiol.* **248**:G551–560. [402]

Pandya, D. N., and Kuypers, H. G., 1969, Cortico-cortical connections in the rhesus monkey, *Brain Res.* **13**:13–36. [570]

Pandya, D. N., Dye, P., and Butters, N., 1971, Efferent cortico-cortical projections of the prefrontal cortex in the rhesus monkey, *Brain Res.* **31**:35–46. [570]

Panella, J. J., Jr., and Blass, J. P., 1984, Lack of clinical benefit from naloxone in a dementia day hospital, *Ann. Neurol.* **15**:308. [409]

Panula, P., Yang, H. Y., and Costa, E., 1982, Neuronal location of the bombesin-like immunoreactivity in the central nervous system of the rat, *Regulat. Peptides* **4**:275–283. [388]

Panula, P., Yang, H. Y. T., and Costa, E., 1984, Histamine containing neurons in the rat hypothalamus, *Proc. Natl. Acad. Sci. U.S.A.* **81**:2572–2576. [216, 354, 355]

Papadopoulos, G., and Huston, J. P., 1980, Removal of the telencephalon spares turning induced by injection of GABA agonists and antagonists into the substantia nigra, *Behav. Brain Res.* **1**:25–38. [513]

Papka, R. E., Traurig, H. H., and Wekstein, M., 1985, Localization of peptides in nerve terminals in the paracervical ganglion of the rat by light and electron microscopic immunohistochemistry: Enkephalin and atrial natriuretic factor, *Neurosci. Lett.* **61**:285–290. [373]

Pappas, C. D., 1975, Ultrastructural basis of synaptic transmission, in: *The Basic Neurosciences,* Vol. 1, *The Nervous System* (D. B. Tower and R. O. Brady, eds.), pp. 19–30, Raven Press, New York. [14]

Parmar, N. S., 1983, The gastric anti-ulcer activity of narinogen, a specific histidine decarboxylase inhibitor, *Int. J. Tissue React.* **5**:415–420. [359]

Parmar, N. S., and Hennings, G., 1984, The gastric antisecretory activity of 3-methoxy-5,7,3′,4′-tetrahydroxyflavan (ME)—a specific histidine decarboxylase inhibitor in rat, *Agents Actions* **15**:143–145. [359]

Pasantes-Morales, H., and Gamboa, A., 1980, Effect of taurine on $^{45}Ca^{2+}$ accumulation in rat brain synaptosomes, *J. Neurochem.* **34**:244–246. [231]

Pasantes-Morales, H., Ademe, R. M., and Lope-Colomé, A. M., 1979, Taurine effects on $^{45}Ca^{2+}$ transport in retinal subcellular fractions, *Brain Res.* **172**:131–138. [231]

Pasantes-Morales, H., Chatagner, F., and Mandel, P., 1980, Synthesis of taurine in rat liver and brain *in vivo,* *Neurochem. Res.* **5**:441–453. [231]

Pasantes-Morales, H., Quesada, O., and Carabez, A. J., 1981, Light-stimulated release of taurine from retinas of kainic acid-treated chicks, *J. Neurochem.* **36**:1583–1586. [229]

Pastan, I., and Willingham, M., 1981, Receptor mediated endocytosis of hormones in culture cells, *Annu. Rev. Physiol.* **43**:239–250. [310]

Patel, J., Marangos, P. J., Skolnick, P., Paul, S. M., and Martino, A. M., 1982a, Benzodiazepines are weak inhibitors of [³H]nitrobenzylthioinosine binding to adenosine uptake sites in brain, *Neurosci. Lett.* **29**:79–82. [362]

Patel, J., Marangos, P. J., Stivers, J., and Goodwin, F. K., 1982b, Characterization of adenosine receptors in brain using cyclohexyl[³H]adenosine, *Brain Res.* **237**:203–214. [364]

Patel, A. J., Weir, M. D., Hunt, A., Tahourdin, C. S. M., and Thomas, D. G. T., 1985, Distribution of glutamine synthetase and glial fibrillary acidic protein and correlation of glutamine synthetase with glutamate decarboxylase in different regions of the rat central nervous system, *Brain Res.* **331**:1–9. [200]

Patey, G., Cupo, A., Mazarguil, H., Morgat, J.-L., and Rossier, J., 1985, Release of proenkephalin-derived opioid peptides from rat striatum *in vitro* and their rapid degradation, *Neuroscience* **15**:1035–1044. [402]

Pazos, A., Cortés, R., and Palacios, J. M., 1985, Thyrotropin-releasing hormone receptor binding sites: Autoradiographic distribution in the rat and guinea pig brain, *J. Neurochem.* **45**:1448–1463. [402]

Pearse, A. G. E., and Takor Takor, T., 1979, Embryology of the diffuse neuroendocrine system and its relationship to the common peptides, *Fed. Proc. Fed. Am. Soc. Exp. Biol.* **38**:2288–2294. [273, 400]

Pearson, J., Brandeis, L., and Cuello, A. C., 1982, Depletion of substance P-containing axons in substantia gelatinosa of patients with diminished pain sensitivity, *Nature (London)* **295**:61–63. [406]

Pearson, R. C. A., 1983, The cortical relationships of certain basal ganglia and the cholinergic basal forebrain nuclei, *Brain Res.* **26**:327–330. [247]

Pelletier, G., and Dube, D., 1977, Electron microscopic immunohistochemical localization of alpha-MSH in the rat brain, *Am. J. Anat.* **150**:201–205. [391]

Pelletier, G., and Leclerc, R., 1979, Localization of Leu-enkephalin in dense core vesicles of axon terminals, *Neurosci. Lett.* **12**:159–163. [391]

Penfield, W., and Jasper, H., 1954, *Epilepsy and the Functional Anatomy of the Human Brain,* Little, Brown, Boston. [476, 601]

Penfield, W., and Perot, P., 1963, The brain's record of auditory and visual experience, *Brain* **86**:596–696. [573]

Penfield, W., and Roberts, L., 1959, *Speech and Brain-Mechanisms,* Princeton University Press, Princeton, New Jersey. [477, 615–617]

Penfield, W., and Welch, K., 1951, The supplementary motor area of the cerebral cortex, *Arch. Neurol. Psychiatry* **66**:289–317. [479, 482]

Peng, J. H., McGeer, P. L., and McGeer, E. G., 1983, Anti-human choline acetyltransferase fragments antigen binding (FAB)–sepharose chromatography for enzyme purification, *Neurochem. Res.* **8**:1481–1486. [239]

Peng, J. H., McGeer, P. L., and McGeer, E. G., 1986, Membrane-bound choline acetyltransferase from human brain: Purification and properties, *Neurochem. Res.* **11**:959–971. [238]

Penney, J. B., Jr., and Young, A. B., 1981, GABA as the pallidothalamic neurotransmitter: Implications for basal ganglia function, *Brain Res.* **207**:195–199. [212]

Peper, K., and McMahan, U. J., 1972, Distribution of acetylcholine receptors in the vicinity of nerve terminals on skeletal muscle of the frog, *Proc. R. Soc. London Ser. B* **181**:431–440. [89]

Perez-Polo, J. R., 1985, Neuronotropic factors, in: *Cell Culture in the Neurosciences* (J. Bottenstein and G. Sato, eds.), pp. 95–123, Plenum Press, New York. [464, 466]

Perkins, M. N., and Stone, T. W., 1983, Quinolinic acid: Regional variations in neuronal sensitivity, *Brain Res.* **259**:172–176. [193]

Perlow, M. J., Kumakura, K., and Guidotti, A., 1980, Prolonged survival of bovine adrenal chromaffin cells in rat cerebral ventricles, *Proc. Natl. Acad. Sci. U.S.A.* **77**:5278–5281. [459]

Pernow, B., 1953, Studies of substance P: Purification, occurrence and biological actions, *Acta Physiol. Scand.* **29**(Suppl. 105):1–90. [374]

Peroutka, S. J., and Kuhar, M. J., 1984, Autoradiographic localization of 5-HT$_1$ receptors to human and canine basilar arteries, *Brain Res.* **310**:193–196. [335]

Peroutka, S. J., and Snyder, S. H., 1981, Two distinct serotonin receptors: Regional variations in receptor binding in mammalian brain, *Brain Res.* **208**:339–347. [335, 336]

Peroutka, S. J., Lebovitz, R. M., and Snyder, S. H., 1981, Two distinct central serotonin receptors with different physiological functions, *Science* **212**:827–829. [335]

Perry, D. C., and Snyder, S. H., 1984, Identification of bradykinin in mammalian brain, *J. Neurochem.* **43**:1072–1080. [388]

Perry, E. K., 1986, The cholinergic hypothesis—ten years later, *Br. Med. Bull.* **42**:63–69. [261]

Perry, E. K., Gibson, P. H., Blessed, G., and Tomlinson, B. E., 1977a, Neurotransmitter enzyme abnormalities in senile dementia, *J. Neurol. Sci.* **34**:247–265. [261]

Perry, E. K., Perry, R. H., Blessed, G., and Tomlinson, B. E., 1977b, Necropsy evidence of central cholinergic deficits in senile dementia, *Lancet* **1**:189. [261]

Perry, E. K., Perry, R. H., Gibson, P. H., Blessed, G., and Tomlinson, B. E., 1977c, A cholinergic connection between normal aging and senile dementia in human hippocampus, *Neurosci. Lett.* **6**:85–89. [261, 262]

Perry, T. L., and Hansen, S., 1981, Amino acid abnormalities in epileptogenic foci, *Neurology* **31**:872–876. [194]

Perry, T. L., Berry, K., Diamond, S., and Mok, C., 1971, Regional distribution of amino acids in human brain obtained at autopsy, *J. Neurochem.* **18**:513–519. [181]

Perry, T. L., Hansen, S., Currier, R. D., and Berry, K., 1978, Abnormalities in neurotransmitter amino acids in dominantly inherited cerebellar disorders, *Adv. Neurol.* **21**:303–314. [195]

Persson, T., 1970, *Catecholamine Turnover in Central Nervous System,* Scandinavian University Books, Göteberg. [283]

Pert, A., Moody, T. W., Pert, C. B., Dewald, A., and Rivier, J., 1980, Bombesin: Receptor distribution in brain and effects on nociception and locomotor activity, *Brain Res.* **193**:209–220. [395]

Pert, C. B., and Snyder, S. H., 1973, Properties of opiate receptor binding in rat brain, *Proc. Natl. Acad. Sci. U.S.A.* **70**:2243–2247. [374]

Pert, C. B., Kuhar, M. J., and Snyder, S. H., 1976, The opiate receptor: Autoradiographic localization in rat brain, *Proc. Natl. Acad. Sci. U.S.A.* **73**:3729–3733. [511]

Pert, C. B., Ruff, M. R., Weber, R. J., and Herkenham, M., 1985, Neuropeptides and their receptors: A psychosomatic network, *J. Immunol.* **135**:820s–825s. [395, 406]

Peters, A., Palay, S. L., and Webster, H. de F., 1976, *The Fine Structure of the Nervous System,* W. B. Saunders, Philadelphia. [22, 23]

Peters, A., Miller, M., and Kimerer, L. M., 1983, Cholecystokinin-like immunoreactive neurons in rat cerebral cortex, *Neuroscience* **8**:431–448. [397]

Peters, D. A. V., McGeer, P. L., and McGeer, E. G., 1968, The distribution of tryptophan hydroxylase in cat brain, *J. Neurochem.* **15**:1431–1435. [322, 323]

Peterson, B. W., 1979, Reticulo-motor pathways: Their connections and possible roles in motor behavior, in: *Integration in the Nervous System* (H. Asanuma and V. J. Wilson, eds.), pp. 185–201, Igaku-Shoin, Tokyo. [501]

Peterson, D. W., Collins, J. F., and Bradford, H. F., 1983, The kindled amygdala model of epilepsy: Anticonvulsant action of amino acid antagonists, *Brain Res.* **275**:169–172. [194]

Peterson, D. W., Collins, J. F., and Bradford, H. F., 1984, Anticonvulsant action of amino acid antagonists against kindled hippocampal seizures, *Brain Res.* **311**:176–180. [194]

Petty, M. A., and Reid, J. L., 1979, Catecholamine synthesizing enzymes in brain stem and hypothalamus during the development of renovascular hypertension, *Brain Res.* **163**:277–288. [315]

Pfaff, D. W., and Keiner, M., 1973, Atlas of estradiol-concentrating cells in the central nervous system of the rat, *J. Comp. Neurol.* **151**:121–158. [523]

Pfeiffer, A., and Herz, A., 1982, Mixed type inhibition of [D-Ala2,D-Leu5]-enkephalin binding to mu-opiate binding sites by mu- but not by kappa-opiate ligands, *Eur. J. Pharmacol.* **77**:359–361. [404]

Phelps, M. E., and Mazziotta, J. C., 1985, Positron emission tomography: Human brain function and biochemistry, *Science* **228**:799–809. [69, 70]

Phillips, C. G., 1966, Changing concepts of the precentral motor area, in: *Brain and Conscious Experience* (J. C. Eccles, ed.), pp. 389–421, Springer-Verlag, New York. [483, 486]

Phillips, C. G., 1969, Motor apparatus of the baboon's hand, *Proc. R. Soc. London Ser. B* **173**:141–174. [486]

Phillips, C. G., and Porter, R., 1977, *Corticospinal Neurones: Their Role in Movement,* Academic Press, London and New York. [476, 478]

Phillips, C. G., Powell, T. P. S., and Wiesendanger, M., 1971, Projection from low threshold muscle afferents of hand and forearm to area 3a of baboon's cortex, *J. Physiol. (London)* **217**:419–446. [487]

Phillips, E., and Newsholme, E. A., 1979, Maximum activities, properties and distribution of 5′-nucleotidase, adenosine kinase and adenosine deaminase in rat and human brain, *J. Neurochem.* **33**:553–558. [364]

Phillips, H. S., Nikolic, K., Cranton, D., and Seeburg, P. H., 1985, Immunocytochemical localization in rat brain of a prolactin release-inhibiting sequence of gonadotropin-releasing hormone prohormone, *Nature (London)* **316**:542–545. [393]

Phillips, M. I., 1980, The central renin–agiotensin system, in: *The Role of Peptides in Neuronal Function* (S. L. Barker and T. G. Smith III, eds.), pp. 389–430, Marcel Dekker, New York. [373]

Phillis, J. W., 1970, *The Pharmacology of Synapses,* Pergamon Press, New York. [254, 255]

Phillis, J. W., 1984, Potentiation of the action of adenosine on cerebral cortical neurones by the tricyclic antidepressants, *Br. J. Pharmacol.* **83**:567–575. [362]

Phillis, J. W., and Kirkpatrick, J. R., 1980, The action of motilin, luteinizing hormone releasing hormine, cholescystokinin, somatostatin, vasoactive intestinal peptide, and other peptides on rat cerebral cortical neurons, *Can. J. Physiol. Pharmacol.* **58**:612–623. [401]

Phillis, J. W., and Wu, P. H., 1981, The role of adenosine and its nucleotides in central synaptic transmission, *Prog. Neurobiol.* **16**:187–239. [360]

Phillis, J. W., and Wu, P. H., 1983, The role of adenosine in central neuromodulation, in: *Regulatory Function of Adenosine* (R. M. Berne, T. W. Rall, and R. Rubio, eds.), pp. 419–437, Martinus Nijhoff, The Hague. [360, 362]

Phillis, J. W., Kostopoulos, G. K., and Limbacher, J. J., 1974, Depression of corticospinal cells by various purines and pyrimidines, *Can. J. Physiol. Pharmacol.* **52**:1226–1229. [365]

Phillis, J. W., Bender, A. S., and Wu, P. H., 1980, Benzodiazepines inhibit adenosine uptake into rat brain synaptosomes, *Brain Res.* **195**:494–498. [362]

Pickel, V. M., Joh, T. H., and Reis, D. J., 1976, Monoamine-synthesizing enzymes in central dopaminergic, noradrenergic and serotonergic neurons: Immunocytochemical localization by light and electron microscopy, *J. Histochem. Cytochem.* **24**:792–806. [329]

Pickel, V. M., Joh. T. H., Reis, D. J., Leeman, S. E., and Miller, R. J., 1979, Electron microscopic localization of substance P and enkephalin in axon terminals related to dendrites of catecholaminergic neurons, *Brain Res.* **160**:387–400. [391]

Piercy, M., 1967, Studies of the neurological basis of intellectual function, *Modern Trends Neurol.* **4**:106–124. [617]

Pierotti, A. R., Harmar, A. J., Tannahill, L. A., and Arbuthnott, G. W., 1985, Different patterns of molecular forms of somatostatin are released by the rat median eminence and hypothalamus, *Neurosci. Lett.* **57**:215–220. [377, 402]

Pinching, A. J., 1970, Synaptic connectior.s in the glomerular layer of the olfactory bulb, *J. Physiol. (London)* **210**:14P–15P. [14]

Pinching, A. J., and Powell, T. P., 1971, The neuropil of the glomeruli of the olfactory bulb, *J. Cell Sci.* **9**:347–377. [14]

Ploska, A., Jaquet, H., Javoy-Agid, F., *et al.,* 1982, Dopamine and methionine-enkephalin in human brain, *Neurosci. Lett.* **33**:191–196. [412]

Polak, J. M., and Bloom, S. R., 1984, Regulatory peptides—the distribution of two newly discovered peptides PHI and NPY, *Peptides* **5**(Suppl. 1):79–89. [373, 388]

Pollard, H., Llorens-Cortes, C., Barbin, G., Garbarg, M., and Schwartz, J. C., 1978, Histamine and histidine decarboxylase in brain stem nuclei: Distribution and decrease after lesions, *Brain Res.* **157**:178–181. [355, 510]

Pollard, H., Pachot, I., and Schwartz, J. C., 1985, Monoclonal antibody against L-histidine decarboxylase for localization of histaminergic cells, *Neurosci. Lett.* **54**:53–58. [351, 354]

Pollin, W., Cardon, P. V., and Kety, S. S., 1961, Effect of amino acid feeding in schizophrenic patients treated with iproniazid, *Science* **133**:104–105. [345]

Popper, K. R., 1972, *Objective Knowledge: An Evolutionary Approach,* The Clarendon Press, Oxford. [625]

Popper, K. R., and Eccles, J. C., 1977, *The Self and Its Brain,* Springer-Verlag, Heidelberg. [603, 606, 612, 621, 624–627, 632]

Poritsky, R., 1969, Two and three-dimensional ultrastructure of boutons and glial cells on the motoneural surface in the cat spinal cord, *J. Comp. Neurol.* **135**:423–452. [10]

Potter, I. T., Flynn, D. D., Hanchett, H. E., Kalinoski, D. L., Luber-Narod, J., and Mash, D. C., 1983, Subtypes of muscarinic receptors. *TIPS Suppl.* 23–31. [254]

Potter, P. E., Hadjiconstantinon, M., Meek, J. L., and Neff, N. H., 1984, Measurement of acetylcholine turnover rate in brain: An adjunct to a simple HPLC method for choline and acetylcholine, *J. Neurochem.* **43**:288–290. [236, 242]

Pourcho, R. G., 1980, Uptake of [³H] glycine and [³H] GABA by amacrine cells in the cat retina, *Brain Res.* **198**:333–346. [205, 229]

Pourcho, R. G., and Goebel, D. J., 1986, Immunocytochemical demonstration of glycine in retina. *Brain Res.* **348**:339–342. [227]

Powell, J. F., 1984, Genetic analysis of the relationship between the different forms of monoamine oxidase, *J. Pharm. Pharmacol.* **36**(Suppl.):1W. [277]

Powell, J. F., Boni, C., Lamouroux, A., Craig, I. W., and Mallet, J., 1984, Assignment of the human tyrosine hydroxylase gene to chromosome 11, *FEBS Lett.* **175**:37–40. [271]

Powell, T. P. S., and Cowan, W. M., 1956, A study of thalamo-striate relations in the monkey, *Brain* **75**:571–584. [504]

Powell, T. P. S., and Mountcastle, V. B., 1959, Some aspects of the functional organization of the cortex of the postcentral gyrus of the monkey: A correlation of findings obtained in a single unit analysis with cytoarchitecture, *Bull. Johns Hopkins Hosp.* **105**:133–162. [490]

Precht, W., Schwindt, P. C., and Baker, R., 1973, Removal of vestibular comissural inhibition by antagonists of GABA and glycine, *Brain Res.* **62**:222–226. [230]

Preslock, J. P., 1984, The pineal gland: Basic implications and clinical correlations, *Endocrinol. Rev.* **5**:282–308. [344]

Price, D. L., Stocks, A., Griffin, J. W., Young, A., and Peck, K., 1976, Clycine-specific synapses in rat spinal cord: Identification by electron microscope autoradiography, *J. Cell Biol.* **68**:389–395. [228]

Price, D. L., Whitehouse, P. J., Struble, R. G., *et al.,* 1982, Basal forebrain cholinergic systems in Alzheimer's disease and related dementias, *Neurosci. Commun.* **1**:84–92. [261]

Price, J. C., Waelsch, H., and Putnam, T., 1943, DL-Glutamic acid hydrochloride in treatment of petit mal and psychomotor seizures, *J. Am. Med. Assoc.* **122**:1153–1156. [194]

Priestley, T., Turnbull, M. J., and Wei, E., 1984, *In vivo* evidence for the selectivity of ICI 154129 for the delta-opioid receptor, *Neuropharmacology* **24**:107–110. [404]

Puil, E., 1981, S-Glutamate: Its interaction with spinal neurons, *Brain Res. Rev.* **3**:229–322. [187]

Pujol, J. F., Belin, M. F., Gamrani, H., Aguera, M., and Calas, A., 1981, Anatomical evidence for GABA–5HT interaction in serotonergic neurons, *Adv. Exp. Med. Biol.* **133**:67–79. [205]

Pulido, O. M., Brown, G. M., and Grota, L. J., 1981, Localization of N-acetylserotonin (NAS) in the rat hindbrain by immunohistology, *Prog. Neuropsychopharmacol.* **5**:573–576. [326]

Purpura, D. P., and Suzuki, K., 1976, Distortion of neuronal geometry and formation of aberrant synapses in neuronal storage disease, *Brain Res.* **116**:1–21. [466]

Purves, D., 1976, Functional and structural changes in mammalian synmpathetic neurones following colchicine application to post-ganglionic nerves, *J. Physiol. (London)* **259**:159–175. [461]

Quach, T. T., Rose, C., Duchemin, A. M., and Schwartz, J. C., 1982, Glycogenolysis induced by serotonin in brain: Identification of a new class of receptor, *Nature (London)* **298**:373–375. [336]

Quastel, J. H., Tennenbaum, M., and Wheatley, A. H. M., 1936, Choline ester formation in, and choline esterase activities of tissues *in vitro, Biochem. J.* **30**:1668–1681. [235, 238]

Quay, W. B., 1965, Retinal and pineal hydroxyindole-O-methyl transferase activity in vertebrates, *Life Sci.* **4**:983–991. [338]

Quirion, R., 1984, Pain, nociception and spinal opioid receptors, *Prog. Neuropsychopharmacol.* **8:**571–579. [406]

Quirion, R., and Dam, T.-V., 1985, Multiple tachykinin receptors in guinea pig brain: High densities of substance K (neurokinin A) binding sites in the substantia nigra, *Neuropeptides* **6:**191–204. [403]

Quirion, R., Gaudreau, P., St.-Pierre, S., Rioux, F., and Pert, C. B., 1982, Autoradiographic distribution of [³H]neurotensin receptors in rat brain: Visualization by tritium-sensitive film, *Peptides* **3:**757–763. [395]

Quirion, R., Shults, C. W., Moody, T. W., Pert, C. B., Chase, T. N., and O'Donohue, T. L., 1983, Autoradiographic distribution of substance P receptors in rat central nervous system, *Nature (London)* **303:**714–716. [403]

Raabe, W., and Gumnit, R. J., 1975, Disinhibition in cat motor cortex by ammonia, *J. Neurophysiol.* **28:**347–355. [131]

Rager, G., 1980a, Specificity of nerve connections by unspecific mechanisms, *TINS* **3:**43–44. [435, 452]

Rager, G. H., 1980b, *Development of the Retino-tectal Projection in the Chicken,* Springer-Verlag, Berlin, Heidelberg, and New York, 92 pp. [435, 452]

Rager, G. H., and Rager, V., 1978, Systems matching by degeneration. 1. A quantitative electron microscopic study of the generation and degeneration of retinal ganglion cells in the chicken, *Exp. Brain Res.* **33:**65–78. [435]

Rager, G. H., and von Oeynhausen, B., 1979, Ingrowth and ramification of retinal fibers in the developing optic tectum of the chick embryo, *Exp. Brain Res.* **35:**213–227. [435]

Raisman, G., 1969, Neuronal plasticity in the septal nuclei of the adult rat, *Brain Res.* **14:**25–48. [436, 437]

Raisman, G., 1977, Formation of synapses in the adult rat after injury: Similarities and differences between a peripheral and a central neurons site, *Philos. Trans. R. Soc. London Ser. B* **278:**349–359. [436]

Raisman, G., 1985, Synapse formation in the septal nucleus of adult rats, in: *Synaptic Plasticity* (C. W. Cotman, ed.), pp. 13–38, Guilford Press, New York. [436, 439]

Raisman, G., and Ebner, F. F., 1983, Mossy fibre projections into and out of hippocampal transplants, *Neuroscience* **9:**783–801. [457]

Raisman, G., and Ebner, F. F., 1986, Hippocampal transplants demonstrate the ability of the adult brain to receive and produce mossy fiber connections, in: *Recent Achievements in Restorative Neurology,* Vol. 1, *Upper Motor Neurone Functions and Dysfunctions* (J. C. Eccles and S. Dimitrijevic, eds.), Karger, Basel, pp. 280–290. [457]

Raisman, G., and Field, P. M., 1973, A quantitative investigation of the development of collateral reinnervation after partial deafferentiation of the septal nuclei, *Brain Res.* **50:**241–264. [37, 436, 438]

Rakic, P., 1971, Neuron–glia relationship during granule cell migration in developing cerebellar cortex: A Golgi and electron microscope study in *Macacus rhesus, J. Comp. Neurol.* **141:**283–312. [37]

Rakic, P., 1972, Mode of cell migration to the superficial layers of fetal monkey neocortex, *J. Comp. Neurol.* **145:**61–84. [37, 420–423, 438]

Rakic, P., 1973, Kinetics of proliferation and latency between final cell division and onset of differentiation of cerebellar, stellate and basket neurons, *J. Comp. Neurol.* **147:**523–540. [418, 420, 426, 428, 429, 438]

Rakic, P., 1977, Prenatal development of the visual system in rhesus monkey, *Philos. Trans. R. Soc. London Ser. B* **278:**245–260. [417]

Rakic, P., 1981, Developmental events leading to laminar and areal organization of the neocortex, in: *Organization of the Cerebral Cortex* (F. O. Schmitt, F. G. Worden, G. Edelmanm, and S. G. Dennis, eds.), pp. 7–28, MIT Press, Cambridge, Massachusetts. [417, 420, 424, 425]

Rakic, P., 1983, Geniculo-cortical connections in primates: Normal and experimentally altered development, *Prog. Brain Res.* **58:**393–404. [421]

Rakic, P., 1985, Limits of neurogenesis in primates, *Science* **227:**1054–1056. [421]

Rakic, P., and Riley, K. P., 1983, Overproduction and elimination of retinal axons in the fetal rhesus monkey, *Science* **219:**1441–1444. [435, 452]

Rakic, P., and Sidman, R. L., 1973, Sequence of developmental abnormalities leading to granule cell deficit in cerebellar cortex of weaver mutant mice, *J. Comp. Neurol.* **152:**103–132. [430, 432]

Rall, W., 1970, Dendritic neuron theory and dendrodendritic synapses in a simple cortical system, in: *The Neurosciences,* Vol. 2 (F. O. Schmitt, ed.), pp. 552–565, Rockefeller University Press, New York. [114, 564]

Rall, W., Shepherd, G. M., Reese, T. S., and Brightman, M. W., 1966, Dendrodendritic synaptic pathway for inhibition in the olfactory bulb, *Exp. Neurol.* **14:**44–56. [14, 133, 212]

Ralston, H. J., and Herman, M. M., 1969, The fine structure of neurons and synapses in the ventrobasal thalamus of the cat, *Brain Res.* **14:**77–97. [136]

Rambourg, A., and Droz, B., 1980, Smooth endoplasmic reticulum and axonal transport, *J. Neurochem.* **35:**16–25. [33]

Ramón y Cajal, S., 1909, *Histologie du Systéme Nerveux de l'Homme et des Vertébrés*, Vol. 1, Maloine, Paris. [6, 50]

Ramón y Cajal, S., 1911, *Histologie du Systéme Nerveux de l'Homme et des Vertébrés*, Vol. 2, Maloine, Paris. [6, 8, 133, 139, 425, 452, 492, 554]

Ramón y Cajal, S., 1929, *Studies on Vertebrate Neurogenesis* [translated from the Spanish by L. Guth (1959)], Charles C. Thomas, Springfield, Illinois. [427]

Ramón y Cajal, S., 1934, Les preuves objectives de l'unité anatomique des cellules nerveuses, *Trab. Lab. !nvest. Biol. Univ. Madrid* **29:**1–137. [6]

Ransom, R. W., Asarch, K. B., and Shih, J. C., 1985, A trifluoromethylphenyl piperazine derivative with high affinity for 5-hydroxytryptamine-1A sites in rat brain, *J. Neurochem.* **44:**875–880. [335]

Rapport, M. M., 1949, Serum vasoconstrictor (serotonin) v. the presence of creatinine in the complex: A proposed study of the vasoconstrictor principle, *J. Biol. Chem.* **180:**961–969. [319]

Rapport, M. M., Green, A. A., and Page, I. H., 1948, Crystalline serotonin, *Science* **108:**329–330. [319]

Raminsky, M., and Sears, T. A., 1971, Internodal conduction in normal and demyelinated mammalian single nerve fibres, *Proc. Physiol. Soc.* **217:**66P–67P. [67]

Ratliff, F., 1972, Contour and contrast, *Sci. Am.* **226:**91–101. [610]

Rea, M. A., McBride, W. J., and Rohde, B., 1980, Regional and synaptosomal levels of amino acid neurotransmitters in the 3-acetylpyridine deafferented rat cerebellum, *J. Neurochem.* **34:**1106–1108. [182, 186]

Recasens, M., and Delaunoy, J. P., 1981, Immunological properties and immunohistochemical localization of cysteine sulfinate or aspartate aminotransferase-isoenzymes in rat CNS, *Brain Res.* **205:**351–361. [178]

Reddington, M., and Schubert, P., 1979, Parallel investigation of the effects of adenosine on evoked potentials and cyclic AMP accumulation in hippocampus slices of the rat, *Neurosci. Lett.* **14:**37–42. [365]

Reding, M. J., and DiPonte, P., 1983, Vasopressin in Alzheimer's disease, *Neurology (N.Y.)* **33:**1634–1635. [410]

Redman, S. J., 1981, Mechanisms of transmitter release at Ia afferent terminations, in: *Regulatory Functions of the CNS: Principles of Motion and Organization* (J. Szentágothai, M. Palkovits, and J. Hámori, eds.), pp. 93–100, Akademiai Kiado, Budapest. [114, 115]

Rehfeld, J. F., 1985, Neuronal cholecystokinin: One or multiple transmitters?, *J. Neurochem.* **44:**1–10. [373]

Reichardt, L. F., and Kelly, R. B., 1983, A molecular description of nerve terminal function, *Annu. Rev. Biochem.* **52:**871–926. [169]

Reichlin, S., 1983, Somatostatin, *N. Engl. J. Med.* **309:**1495–1501 and 1556–1563. [409]

Reijnierse, G. L. A., Veldstra, H., and van den Berg, C. J., 1975, Subcellular localization of γ-aminobutyrate transaminase and glutamate dehydrogenase in adult rat brain, *Biochem. J.* **152:**469–475. [179]

Reinecke, M., 1985, Neurotensin: Immunohistochemical localization in central and peripheral nervous system and in endocrine cells and its functional role as neurotransmitter and endocrine hormone, *Prog. Histochem. Cytochem.* **16:**1–172. [373]

Reisberg, B., Ferris, S. H., Anand, R., *et al.*, 1983, Effects of naloxone in senile dementia: A double-blind trial, *N. Engl. J. Med.* **308:**1039–1043. [409]

Reisine, T. D., Yamamura, H. I., Bird, E. D., Spokes, E., and Enna, S. J., 1978, Pre and postsynaptic neurochemical alterations in Alzheimer's disease, *Brain Res.* **159:**477–481. [262]

Reisine, T. D., Rossor, M., Spokes, E., Iversen, L. L., and Yamamura, H. I., 1979, Alterations in brain opiate receptors in Parkinson's disease, *Brain Res.* **173:**378–382. [412]

Reivich, M., Sokoloff, L., Kennedy, C., and Des Rosiers, M., 1975, An autoradiographic method for the measurement of local glucose metabolism in the brain, in: *Brain Work* (D. Ingvar and N. A. Lassen, eds.), pp. 377–384, Munksgaard, Copenhagen. [480]

Reivich, M., Alavi, A., Wolf, A., *et al.*, 1982, Use of 2-deoxy-D(1–11C) glucose for the determination of local cerebral glucose metabolism in humans: Variation within and between subjects, *J. Cereb. Blood Flow Metab.* **2:**307–319. [480]

Remtulla, M. A., Katz, S., and Applegarth, D. A., 1979, Effect of taurine on passive ion transport in rat brain synaptosomes, *Life Sci.* **24:**1885–1892. [231]

Retz, K. C., and Coyle, J. T., 1982, Kainic acid lesion of mouse striatum: Effects on energy metabolites, *Life Sci.* **27:**2495–2500. [190]

Reubi, J. C., 1980, Comparative study of the release of glutamate and GABA, newly synthesized from glutamine, in various regions of CNS, *Neuroscience* **5:**2145–2150. [178]

Reubi, J. C., and Cuenod, M., 1979, Glutamate release *in vitro* from corticostriatal terminals, *Brain Res.* **176:**185–188. [182, 301]

Reubi, J. C., Toggenburger, G., and Cuenod, M., 1980, Asparagine as precursor for transmitter aspartate in corticostriatal fibres, *J. Neurochem.* **35:**1015–1017. [180]

Reuter, H., Stevens, C. F., Tsiern, R. W., and Yellen, G., 1982, Properties of single calcium channels in cardiac cell culture, *Nature (London)* **297:**501–504. [168]

Reynolds, G. P., 1983, Increased concentrations and lateral asymmetry of amygdala dopamine in schizophrenia, *Nature (London)* **305:**527–529. [546]

Rhodes, C. H., Morrell, J. I., and Pfaff, D. W., 1981, Immunohistochemical analysis of magnocellular elements in rat hypothalamus: Distribution and numbers of cells containing neurophysin, oxytocin, and vasopressin, *J. Comp. Neurol.* **198:**45–64. [394]

Ribak, C. E., 1978, Aspinous and sparsely-spinous stellate neurons in visual cortex of rats contain glutamic acid decarboxylase, *J. Neurocytol.* **7:**461–478. [205]

Ribak, C. E., Vaughn, J. E., Saito, K., Barber, R., and Roberts, E., 1977a, Glutamate decarboxylase localization in neurons of the olfactory bulb, *Brain Res.* **126:**1–18. [134, 206, 216]

Ribak, C. E., Vaughn, J. E., Saito, K., Barber, R., and Roberts, E., 1977b, Immunocytochemical localization of glutamate decarboxylase in rat substantia nigra, *Brain Res.* **116:**287–298. [204, 206, 215]

Ribak, C. E., Vaughn, J. E., and Saito, K., 1978, Immunocytochemical localization of glutamic acid decarboxylase in neuronal somata following cholchicine inhibition of axonal transport, *Brain Res.* **140:**315–332. [206, 213, 214]

Ribak, C. E., Vaughn, J. E., and Roberts, E., 1979, The GABA neurons and their axon terminals in rat corpus striatum as demonstrated by GAD immunocytochemistry, *J. Comp. Neurol.* **187:**261–283. [206, 214]

Ribak, C. E., Vaughn, J. E., and Roberts, E., 1980, GABAergic nerve terminals decrease in the substantia nigra following hemitransections of the striatonigral and pallidonigral pathways, *Brain Res.* **192:**413–420. [206]

Ribak, C. E., Vaughn, J. E., and Barber, R. P., 1981, Immunocytochemical localization of GABAergic neurones at the electron microscopical level, *Histochem. J.* **13:**555–582. [206]

Ribeiro Da-Silva, A., and Coimbra, A., 1980, Neuronal uptake of [³H]GABA and [³H]glycine in lamina II–III (substantia gelatinosa Rolandi) of the rat spinal cord: An autoradiographic study, *Brain Res.* **188:**449–464. [206, 228]

Ricaurte, G. A., Langston, J. W., DeLanney, L. E., Irwin, I., and Brooks, J. D., 1985, Dopamine uptake blockers protect against the dopamine depleting effect of 1-methyl-4-phenyl-1,2,3,6-tetrahydropyridine (MPTP) in the mouse striatum, *Neurosci. Lett.* **59:**259–264. [303]

Rice, M. E., Oke, A. F., Bradberry, C. W., and Adams, R. N., 1985, Simultaneous voltammetric and chemical monitoring of dopamine release *in situ*, *Brain Res.* **340:**151–155. [296]

Richardson, B. P., Engel, G., Donatsch, P., and Stadler, P. A., 1985, Identification of serotonin M-receptor subtypes and their specific blockade by a new class of drugs, *Nature (London)* **316:**126–131. [337]

Richman, L., Ribak, C. E., Issacs, S., Houser, C. R., and Fallon, J. H., 1980, Multiple neurotransmitter studies in the islands of Calleja complex of the basal forebrain, *Soc. Neurosci. Abstr.* **6:**7. [206]

Richter, D., 1937, Adrenaline and amine oxidase, *Biochem. J.* **31:**2022–2028. [275]

Rimon, R., Le Greves, P., Nyberg, F., Heikkila, L., and Salmela, L., 1984, Elevation of substance P-like peptides in the CSF of psychiatric patients, *Biol. Psychiatry* **19:**509–516. [412]

Ringertz, N. R., and Savage, R. E., 1976, *Cell Hybrids,* Academic Press, New York. [46]

Risch, S. C., Kalin, N. H., Janowsky, D. S., Cohen, R. M., Pickar, D., and Murphy, D. L., 1983, Co-release of ACTH and β-endorphin immunoreactivity in human subjects in response to central cholinergic stimulation, *Science* **222:**77. [393]

Ritchie, J. M., 1979, A pharmacological approach to the structure of sodium channels in myelinated axons, *Annu. Rev. Neurosci.* **2:**341–362. [59]

Riveros, N., and Orrego, F., 1984, A study of possible excitatory effects of *N*-acetylaspartylglutamate in different *in vivo* and *in vitro* brain preparations, *Brain Res.* **299:**393–395. [186]

Rizzoli, A. A., 1968, Distribution of glutamic acid, aspartic acid, γ-aminobutyric acid and glycine in six areas of cat spinal cord after transection, *Brain Res.* **11:**11–18. [182]

Robb, G. A., and Keynes, R. J., 1984, Stimulation of nodal and terminal sprouting of mouse motor nerves by gangliosides, *Brain Res.* **295:**368–371. [467]

Roberts, E., 1981, Strategies for identifying sources and sites of formation of GABA-precursor or transmitter glutamate in brain, *Adv. Biochem. Psychopharmacol.* **27**:91–102. [179]

Roberts, E., and Frankel, S., 1950, γ-Aminobutyric acid in brain: Its formation from glutamic acid, *J. Biol. Chem.* **187**:55–63. [197, 231]

Roberts, F., and Calcutt, C. R., 1983, Histamine and the hypothalamus, *Neuroscience* **9**:721–739. [358]

Roberts, G. W., Woodhams, P. L., Crow, T. J., and Polak, J. M., 1980, Loss of immunoreactive VIP in the bed nucleus following lesions of the stria terminalis, *Brain Res.* **195**:471–475. [395]

Roberts, G. W., Woodhams, P. L., Polak, J. M., and Crow, T. J., 1982, Distribution of neuropeptides in the limbic system of the rat: The amygdaloid complex, *Neuroscience* **7**:99–131. [394]

Roberts, G. W., Allen, Y., Crow, T. J., and Polak, J. M., 1983, Immunocytochemical localization on neuropeptides in the fornix of rat, monkey and man, *Brain Res.* **263**:151–155. [397]

Roberts, M. H., and Straughan, D. W., 1967, Excitation and depression of cortical neurones by 5-hydroxytryptamine, *J. Physiol. (London)* **193**:269–294. [335]

Roberts, P. J., and Keen, P., 1974, Effects of dorsal root section in amino acids of rat spinal cord, *Brain Res.* **74**:333–337. [182]

Roberts, P. J., and Mitchell, J. F., 1972, The release of amino acids from the hemisected spinal cord during stimulation, *J. Neurochem.* **19**:2473–2481. [182]

Roberts, R. C., Ribak, C. E., and Oertel, W. H., 1985, Increased numbers of GABAergic neurons occur in the inferior colliculus of an audiogenic model of genetic epilepsy, *Brain Res.* **361**:324–338. [205]

Robertson, J. D., 1958, Structural alterations in nerve fibers produced by hypnotic and hypertonic solutions, *J. Biophys. Biochem. Cytol.* **4**:349–364. [65]

Robinson, D. A., 1976, Adaptive gain control of the vestibulo-ocular reflex by the cerebellum, *J. Neurophysiol.* **39**:954–969. [578]

Robinson, D. S., Nies, A., Davis, J. N., *et al.*, 1972, Aging, monoamines and monoamine oxidase levels, *Lancet* **1**:290–291. [471]

Robson, J. A., and Martin-Elkins, C. L., 1985, The effects of monocular deprivation on the size of GAD+ neurons in the cat's dorsal lateral geniculate nucleus, *J. Comp. Neurol.* **239**:62–74. [205]

Rodnight, R., 1983, Schizophrenia: Some current neurochemical approaches, *J. Neurochem.* **41**:12–21. [546, 547]

Roffler-Tarlov, S., Beart, P. M., O'Gorman, S., and Sidman, R. L., 1979, Neurochemical and morphological consequences of axon terminal degeneration in cerebellar deep nuclei of mice with inherited Purkinje cell degeneration, *Brain Res.* **168**:75–95. [212]

Rogers, J. D., Brogan, D., and Mirra, S. S., 1985, The nucleus basalis of Meynert in neurological disease: A quantitative morphological study, *Ann. Neurol.* **17**:163–170. [261]

Rohde, B. H., Rea, M. A., Simon, J. R., and McBride, W. J., 1979, Effects of γ-irradiation induced loss of cerebellar granule cells on the synaptosomal levels and the high affinity uptake of amino acids, *J. Neurochem.* **32**:1431–1435. [182, 184]

Roisen, F. J., Bartfeld, H., Nagele, R., and Yorke, G., 1981, Ganglioside stimulation of axonal sprouting *in vitro*, *Science* **214**:577–578. [466]

Rojas, E., and Keynes, R. D., 1975, On the relation between displacement currents and activation of the sodium conductance in the squid giant neuron, *Philos. Trans. R. Soc. London Ser. B* **270**:459–482. [56]

Roland, P. E., 1981, Somatotopical tuning of postcentral gyrus during focal attention in man, *J. Neurophysiol.* **46**:744–754. [627]

Roland, P. E., 1984, Organization of motor control by normal human brain, *Hum. Neurobiol.* **2**:205–216. [592]

Roland, P. E., and Friberg, L., 1985, Localization of cortical areas by thinking, *J. Neurophysiol.* **53**:1219–1243. [628]

Roland, P. E., Larsen, B., Lassen, N. A., and Skinhoj, E., 1980a, Supplementary motor area and other cortical areas in organization of voluntary movements in man, *J. Neurophysiol.* **43**:118–136. [69, 480, 482]

Roland, P. E., Skinhoj, F., Lassen, N. A., and Larsen, B., 1980b, Different cortical areas in man in organization of voluntary movements in extrapersonal space, *J. Neurophysiol.* **43**:137–150. [480, 482, 483]

Roland, P. E., Skinhoj, E., and Lassen, N. A., 1981, Focal activation of human cerebral cortex during auditory discrimination, *J. Neurophysiol.* **45**:1139–1151. [624]

Roland, P. E., Meyer, E., Shibasaki, T., Yamamoto, L., and Thompson, C. J., 1982, Regional cerebral blood flow changes in cortex and basal ganglia during voluntary movements in normal human volunteers, *J. Neurophysiol.* **48**:467–480. [69, 592]

Rolls, E. T., 1984, The neurophysiology of feeding, *Int. J. Obesity* **8**(Suppl. I):139–150. [529]

Ronzio, R. A., Rowe, W. B., and Meister, A., 1969, Studies on the mechanism of inhibition of glutamine synthetase by methionine sulfoximine, *Biochemistry* **8**:1066–1075. [223]

Rosén, I., and Asanuma, H., 1972, Peripheral afferent inputs to the forelimb area of monkey motor cortex: Input–output relations, *Exp. Brain Res.* **14**:257–273. [483, 484]

Rosenfeld, M. G., Mermod, J.-J., Amara, S. G., *et al.*, 1983 Production of a novel neuropeptide encoded by the calcitonin gene via tissue-specific RNA processing, *Nature (London)* **304**:129–135. [376, 382]

Ross, C. A., Ruggiero, D. A., Meeley, M. P., Park, D. H., Joh, T. H., and Reis, D. J., 1984, A new group of neurons in hypothalamus containing phenylethanolamine *N*-methyltransferase (PNMT) but not tyrosine hydroxylase, *Brain Res.* **306**:349–353. [293]

Ross, S. B., 1976, Long term effects of *N*-2-chloroethyl-*N*-ethyl-2-bromo-benzylamine hydrochloride on noradrenergic neurons in the rat brain and heart, *Br. J. Pharmacol.* **58**:521–527. [311]

Rossor, M. N., and Emson, P. C., 1982, Neuropeptides in degenerative diseases of the central nervous system, *TINS* **5**:399–401. [412]

Rossor, M. N., Fahrenkrug, J., Emson, P. C., Mountjoy, C. O., Iversen, L. L., and Roth, M., 1980, Reduced cortical choline acetyltransferase activity in senile dementia of Alzheimer type is not accompanied by changes in vasoactive intestinal polypeptide, *Brain Res.* **201**:249–253. [412]

Rossor, M. N., Emson, P. C., Iversen, L. L., *et al.* 1982a, Neuropeptides and neurotransmitters in cerebral cortex in Alzheimer's disease, in: *Alzheimer's Disease: A Report of Progress* (S. Corkin, ed.), pp. 15–23, Raven Press, New York. [261, 412]

Rossor, M. N., Hunt, S. P., Iversen, L. L., *et al.*, 1982b, Extrahypothalamic vasopressin is unchanged in Parkinson's disease and Huntington's disease, *Brain Res.* **253**:341–343. [412]

Roth, K. A., Weber, E., and Barchas, J. C., 1982, Immunoreactive corticotropin releasing factor (CRF) and vasopressin are colocalized in a subpopulation of the immunoreactive vasopressin cells in the paraventricular nucleus of the hypothalamus, *Life Sci.* **31**:1857–1860. [157]

Roth, K. A., Weber, E., Barchas, J. D., Chang, D., and Chang, J. K., 1983, Immunoreactive dynorphin (1–8) and corticotropin releasing factor in a subpopulation of hypothalamic neurons, *Science* **219**:189–191. [157]

Roth, R. H., 1984, CNS dopamine autoreceptors: Distribution, pharmacology, and function, *Ann. N. Y. Acad. Sci.* **430**:27–53. [300]

Rothman, J. E., 1981, The Golgi apparatus—two organelles in tandem, *Science* **213**:1212–1219. [23]

Rothman, R. B., Herkenham, M., Pert, C. B., Liang, T., and Cascieri, M. A., 1984, Visualization of rat brain receptors for the neuropeptide substance P, *Brain Res.* **309**:47–54. [395]

Rouzaire-Dubois, B., Hammond, C., Hamon, B., and Feger, J., 1980, Pharmacological blockade of the GABA-induced inhibitory response of subthalamic cells in the rat, *Brain Res.* **200**:321–329. [212]

Rouzaire-Dubois, B., Scarnati, E., Hammond, C., Crossman, A. R., and Shibazaki, T., 1983, Microiontophoretic studies on the nature of the neurotransmitter in the subthalamo-entopeduncular pathway of the rat, *Brain Res.* **271**:11–20. [212]

Rowlands, G. J., and Roberts, P. J., 1980, Specific calcium-dependent release of endogenous glutamate from rat striatum is reduced by destruction of the corticostriatal tract, *Exp. Brain Res.* **39**:239–240. [182]

Ruffieux, A., and Schultz, W., 1980, Dopaminergic activation of reticulata neurones in the substantia nigra, *Nature (London)* **285**:240–241. [297, 512]

Ruggiero, D. A., Meeley, M. P., Ross, C. A., Park, D. H., Joh, T. H., and Reis, D. J., 1983, PNMT neurons in the central nervous system: New anatomical and biochemical findings, *Soc. Neurosci. Abstr.* **7**:343. [293]

Ruggiero, D. A., Meeley, M. P., Anwar, M., and Reis, D. J., 1985, Newly identified GABAergic neurons in regions of the ventrolateral medulla which regulate blood pressure, *Brain Res.* **339**:171–177. [205]

Rumigny, J. F., Maitre, M., Cash, C., and Mandel, P., 1981, Regional and subcellular localization in rat brain of the enzymes that can synthesize γ-hydroxybutyric acid, *J. Neurochem.* **36**:1433–1438. [202]

Russell, J. T., Brownstein, M. J., and Gainer, H., 1981, Biosynthesis of neurophypophyseal polypeptides: The order of peptide components in pro-pressophysin and pro-oxyphysin, *Neuropeptides* **2**:59–65. [384]

Rustioni, A., and Cuenod, M., 1982, Selective retrograde transport of D-aspartate in spinal interneurons and cortical neurons of rats, *Brain Res.* **236**:143–155. [182, 184]

Rustioni, A., Spreafico, R., Cheema, S., and Cuenod, M., 1982, Selective retrograde labelling of somatosensory cortical neurons after ³H-D-aspartate injections in the ventrobasal complex of rats and cats, *Neuroscience* **7**(Suppl.):S183. [182]

Rylett, R. J., and Colhoun, E. H., 1984, An evaluation of irreversible inhibition of synaptosomal high-affinity choline transport by choline mustard aziridinium ion, *J. Neurochem.* **43**:787–794. [260]

Saavedra, J. M., 1980, Brain stem adrenergic neurons participate in the regulation of the stress response and in genetic and experimental hypertension in: *Central Adrenaline Neurons: Basic Aspects and Their Role in Cardiovascular Functions* (K. Fuxe, M. Goldstein, B. Hökfelt, and T. Hökfelt, eds.), pp. 235–244, Pergamon Press, Oxford and New York. [315]

Saavedra, J. M., Kvetnansky, R., and Kopin, I. J., 1979, Adrenaline and dopamine levels in specific brain stem areas of acutely immobilized rats, *Brain Res.* **160**:271–280. [315]

Saavedra, J. M., Correa, F. M., and Iwai, J., 1980, Discrete changes in adrenaline-forming enzyme activity in brain stem areas of genetic salt-sensitive hypertensive (Dahl) rats, *Brain Res.* **193**:299–303. [315]

Saavedra, J. P., Pausek, T., and Pasik, P., 1982, *Immunochemistry of Serotonergic Neurons in the Central Nervous System of Monkeys, IBRO Monograph Series,* Vol. 10, pp. 81–96, Raven Press, New York. [329]

Sabel, B. A., Slavin, M. D., and Stein, D. G., 1984, GM$_1$ ganglioside treatment facilitates behavioral recovery from bilateral brain damage, *Science* **225**:340–342. [467]

Saffrey, M. J., Marcus, N., Jessen, K. R., and Burnstock, G., 1983, Distribution of neurons with high-affinity uptake sites for GABA in the myenteric plexus of the guinea-pig, rat and chicken, *Cell Tissue Res.* **234**:231–235. [218]

Sagar, S. M., Beal, M. F., Marshall, P. E., Landis, D. M. D., and Martin, J. B., 1984, Implications of neuropeptides in neurological diseases, *Peptides* **5**(Suppl. 1):255–262. [411]

Sahgal, A., Keitz, A. B., Wright, C., and Edwardson, J. A., 1982, Failure of vasopressin to enhance memory in a passive avoidance task, *Neurosci. Lett.* **28**:87–92. [410]

Saito, A., Goldfine, I. D., and Williams, J. A., 1981, Characterization of receptors for cholecystokinin and related peptides in mouse cerebral cortex, *J. Neurochem.* **37**:483–490. [402]

Saito, A., Wu, J.-Y., and Lee, J.-F., 1985, Evidence for the presence of cholinergic nerves in cerebral arteries: An immunohistochemical demonstration of choline acetyltransferase, *J. Cereb. Blood Flow Metab.* **5**:327–334. [242]

Saito, H., Saito, S., Ohuchi, T., Oka, M., Sano, T., and Hosoi, E., 1984, Co-storage and co-secretion of somatostatin and catecholamine in bovine adrenal medulla, *Neurosci. Lett.* **52**:43–47. [157]

Saito, K., Barber, R., Wu, J.-Y., Matswuda, T., Roberts, E., and Vaughn, J. E., 1974a, Immunohistochemical localization of glutamate decarboxylase in rat cerebellum, *Proc. Natl. Acad. Sci. U.S.A.* **71**:269–277. [203, 206, 213]

Saito, K., Schousboe, A., Wu, J.-Y., and Roberts, E., 1974b, Some immunochemical properties of and species specificity of GABA-alpha-ketoglutarate transaminase from mouse brain, *Brain Res.* **65**:287–296. [201]

Sakamoto, N., Takatsuji, K., Shiosaka, S., *et al.,* 1984, Cholecystokinin-8-like immunoreactivity in the pre- and post-central gyri of the human cerebral cortex, *Brain Res.* **307**:77–83. [397]

Sakanaka, M., Shiosaka, S., Takatsuki, K., *et al.,* 1981, Experimental immunohistochemical studies on the amygdalofugal peptidergic (substance P and somatostatin) fibers in the stria terminalis of the rat, *Brain Res.* **221**:231–242. [395]

Sakmann, B., Hamill, O., and Bormann, J., 1983, Patch-clamp measurements of elementary chloride currents activated by the putative inhibitory transmitters GABA and glycine in mammalian spinal neurons, *J. Neural Transm. Suppl.* **18**:83–95. [128, 129]

Sallanon, M., Buda, C., Janin, M., and Jouvet, M., 1982, 5-HT antagonists suppress sleep and delay its restoration after 5-HTP in p-chlorophenylalanine-pretreated cats, *Eur. J. Pharmacol.* **82**:29–35. [337]

Salt, T. E., Morris, R., and Hill, R. G., 1983, Distribution of substance P-responsive and nociceptive neurones in relation to substance P-immunoreactivity within the caudal trigeminal nucleus of the rat, *Brain Res.* **273**:217–228. [406]

Samorajski, T., and Rolsten, C., 1973, Age and regional differences in the chemical composition of brains of mice, monkeys and humans, *Prog. Brain Res.* **40**:251–265. [470]

Sandberg, M., Ward, H. K., and Bradford, H. F., 1985, Effect of cortico-striate pathway lesion on the activities of enzymes involved in synthesis and metabolism of amino acid neurotransmitters in the striatum, *J. Neurochem.* **44**:142–147. [178, 179]

Sanders-Bush, E., Gallaher, D. A., and Sulser, F., 1974, On the mechanism of brain 5-hydroxytryptamine depletion by p-chloroamphetamine and related drugs and the specificity of their reaction, *Adv. Biochem. Psychopharmacol.* **10**:185–194. [340]

Sandler, M., 1981, Monoamine oxidase inhibitor efficacy in depression and the "cheese effect," *Psychol. Med.* **11**:455–458. [539]

Sandner, G., Dessort, D., Schmitt, P., and Karli, P., 1981, Distribution of GABA in the periaqueductal gray matter: Effects of medial hypothalamic lesions, *Brain Res.* **224**:279–290. [212]

Sandyk, R., 1985, The endogenous opioid system in neurological disorders of the basal ganglia, *Life Sci.* **37**:1655–1663. [409]

Sanger, F., Nicklen, S., and Coulson, A. R., 1977, DNA sequencing with chain-terminating inhibitors, *Proc. Natl. Acad. Sci. U.S.A.* **74**:5463–5467. [271]

Saper, C. B., and Loewy, A. D., 1980, Efferent connections of the parabrachial nucleus in the rat, *Brain Res.* **197**:291–317. [251, 252]

Saper, C. B., and Loewy, A. D., 1982, Reciprocal parabrachial–cortical connections in the rat, *Brain Res.* **242**:33–40. [251, 252]

Saper, C. B., German, D. C., and White, C. L., 1985, Neuronal pathology in the nucleus basalis and associated cell groups in senile dementia of the Alzheimer's type: Possible role in cell loss, *Neurology* **35**:1089–1095. [261]

Sapolsky, R. M., McEwen, B. S., and Rainbow, T. C., 1983, Quantitative autoradiography of [^{3}H]corticosterone receptors in the rat brain, *Brain Res.* **271**:331–334. [525]

Sasaki, K., 1979, Cerebro-cerebellar interconnections in cats and monkeys, in: *Cerebro-cerebellar Interactions* (J. Massion and K. Sasaki, eds.), pp. 105–124, Elsevier/North-Holland, Amsterdam. [498]

Sasaki, K., and Gemba, H., 1981, Changes of premovement field potentials in the cerebral cortex during learning processes of visually initiated hand movements in the monkey, *Neurosci. Lett.* **27**:125–130. [584, 585, 587]

Sasaki, K., and Gemba, H., 1982, Development and change of cortical field potentials during learning processes of visually initiated hand movements in the monkey, *Exp. Brain Res.* **48**:429–437. [584, 587]

Sasaki, K., Kawaguchi, S., Oka, H., Sakai, M., and Mizuno, N., 1976, Electro-physiological studies on the cerebellocerebral projections in monkeys, *Exp. Brain Res.* **24**:495–507. [498]

Sasaki, K., Oka, H., Kawaguchi, S., Jinnai, K., and Yasuda, T., 1977, Mossy fiber and climbing fiber responses produced in the cerebellar cortex by stimulation of the cerebral cortex in monkeys, *Exp. Brain Res.* **29**:419–428. [498, 589]

Sasaki, K., Jinnai, K., Gemba, H., Hashimoto, S., and Mizuno, N., 1979, Projection of the cerebellar dentate nucleus onto the frontal associating cortex in monkeys, *Exp. Brain Res.* **37**:193–198. [498, 586]

Sasaki, K., Gemba, H., and Mizuno, N., 1982, Cortical field potentials preceding visually initiated hand movements and cerebellar actions in the monkey, *Exp. Brain Res.* **46**:29–36. [586]

Sasek, C. A., Schueler, P. A., Herman, W. S., and Elde, R. P., 1985, An antiserum to locust adipokinetic hormone reveals a novel peptidergic system in the rat central nervous system, *Brain Res.* **343**:172–175. [373]

Satoh, K., and Fibiger, H. C., 1985, Distribution of central cholinergic neurons in the baboon (*Papio papio*), *J. Comp. Neurol.* **236**:197–214. [245, 251]

Satoh, K., Armstrong, D. M., and Fibiger, H. C., 1983, A comparison of the distribution of central cholinergic neurons as demonstrated by acetylcholinesterase pharmacohistochemistry and choline acetyltransferase immunohistochemistry, *Brain Res. Bull.* **11**:693–720. [250, 251]

Sauer, F. C., 1935, Mitosis in the neural tube, *J. Comp. Neurol.* **62**:377–405. [418]

Sawchenko, P. E., and Swanson, L. W., 1985, Localization, colocalization and plasticity of corticotropin-releasing factor immunoreactivity in rat brain, *Fed. Proc. Fed. Am. Soc. Exp. Biol.* **44**:221–227. [394]

Sawchenko, P. E., Swanson, L. W., Grzanna, R., Howe, P. R. C., Bloom, S. R., and Polak, J. M., 1985, Colocalization of neuropeptide Y immunoreactivity in brainstem catecholaminergic neurons that project to the paraventricular nucleus of the hypothalamus, *J. Comp. Neurol.* **241**:138–153. [393]

Scarnati, E., Campana, E., and Pacitti, C., 1983, The functional role of the nucleus accumbens in the control of the substantia nigra: Electrophysiological investigations in intact and striatum–globus pallidus lesioned rats, *Brain Res.* **265**:249–257. [212]

Schachner, M., 1982, Glial antigens and the expression of glial phenotypes, *TINS* **5**:225–228. [37]

Scharrer, E., and Scharrer, B., 1940, Secretory cells within the hypothalamus, *Proc. Assoc. Res. Nerv. Ment. Dis.* **20P**:170–194. [369]

Schayer, R. W., and Reilly, M. A., 1973, Formation and fate of histamine in rat and mouse brain, *J. Pharmacol. Exp. Ther.* **184**:33–40. [351]

Scheibel, M. E., Lindsay, R. D., Tomiyasu, U., and Scheibel, A. B., 1976, Progressive dendritic changes in aging human limbic system, *Exp. Neurol.* **53**:420–430. [471]

Scherbaum, W. A., and Bottazzo, G. F., 1983, Autoantibodies to vasopressin cells in idiopathic diabetes insipidus: Evidence for an autoimmune variant, *Lancet* **1**:897–901. [406]

Schiller, P. H., 1970, The discharge characteristics of single units in the oculomotor and abducens nuclei of the unanesthetized monkey, *Exp. Brain Res.* **10**:347–362. [52]

Schlicker, E., Brandt, F., Classen, K., and Gothert, M., 1985, Serotonin release in human cerebral cortex and its modulation via serotonin receptors, *Brain Res.* **331**:337–341. [329]

Schloot, W., Tigges, F.-J., Blaesner, H., and Goedde, H. W., 1969, *N*-Acetyltransferase and serotonin metabolism in man and other species, *Hoppe-Seyler's Z. Physiol. Chem.* **350**:1353–1361. [326]

Schmale, H., Heinsohn, S., and Richter, D., 1983, Structural organization of the rat gene for the arginine vasopressin–neurophysin precursor, *Eur. Mol. Biol. Org. J.* **2**:763–767. [382]

Schmale, H., Ivell, R., Breindl, M., Darmer, D., and Richter, D., 1984, The mutant vasopressin gene from diabetes insipidus (Brattleboro) rats is transcribed but the message is not efficiently translated, *Eur. Mol. Biol. Org. J.* **3**:3289–3293. [380]

Schmechel, D. E., Vickrey, B. G., Fitzpatrick, D., and Elde, R. P., 1984, GABAergic neurons of mammalian cerebral cortex: Widespread subclass defined by somatostatin content, *Neurosci. Lett.* **47**:227–232. [157]

Schmid, R., Hong, J. S., Meek, J., and Costa, E., 1980, The effect of kainic acid on the hippocampal content of putative transmitter amino acids, *Brain Res.* **200**:355–362. [212]

Schmidt, M., Wassle, H., and Humphrey, M., 1985, Number and distribution of putative cholinergic neurons in the cat retina, *Neurosci. Lett.* **59**:235–240. [252]

Schmidt, R. F., 1971, Presynaptic inhibition in the vertebrate central nervous system, *Ergeb. Physiol.* **63**:21–108. [142, 144, 145]

Schmitt, F. O., Bird, S. J., and Bloom, F. E. (eds.), 1982, *Molecular Genetic Neuroscience,* Raven Press, New York. [39]

Schneider, G. E., 1969, Two visual systems, *Science* **163**:895–902. [611]

Schneider, G. E., 1973, Early lesions of superior colliculus: Factors affecting the formation of abnormal retinal projections, *Brain Behav. Evol.* **8**:73–109. [447, 467]

Schneider, G. E., 1981, Early lesions and abnormal neuronal connections, *TINS* **4**:187–192. [447, 448, 468]

Schneider, G. E., Thaver, S., Edwards, M. A., and So, K.-F., 1986, Regeneration, rerouting and redistribution of axons after early lesions—changes with age and functional impact, in: *Recent Achievements in Restorative Neurology,* Vol. 1, *Upper Motor Neurone Functions and Dysfunctions* (J. C. Eccles and M. R. Dimitrijevic, eds.), pp. 291–310, S. Karger, Basel. [447]

Schoemaker, H., Morelli, M., Desmukh, P., and Yamamura, H. I., 1982, [^{3}H]Ro5-4684 benzodiazepine binding in the kainate lesioned striatum and Huntington's diseased basal ganglia, *Brain Res.* **248**:396–401. [220]

Scholkens, B. A., Jung, W., Rascher, W., Schomig, A., and Ganten, D., 1980, Brain angiotensin II stimulates release of pituitary hormones, plasma catecholamines and increases blood pressure in dogs, *Clin. Sci.* **59**(Suppl. 6):53s–56s. [406]

Schousboe, A., Wu, J.-Y., and Roberts, E., 1973, Purification and characterization of the 4-aminobutyrate-2-ketoglutarate transaminase from mouse brain, *Biochemistry* **12**:2868–2873. [201]

Schousboe, A., Larsson, O. M., Hertz, L., and Krogsgaard-Larsen, P., 1981, Heterocyclic GABA analogues as new selective inhibitors of astroglial GABA transport, *Drug Dev. Res.* **1**:115–127. [224]

Schramm, M., and Selinger, Z., 1984, Message transmission: Receptor controlled adenylate cyclase system, *Science* **225**:1350–1356. [167]

Schreiner, L., and Kling, A., 1953, Behavioral changes following rhinencephalic injuries in cat, *J. Neurophysiol.* **16**:643–659. [528]

Schulman, H., and Greengard, P., 1978, Stimulation of brain membrane protein phosphorylation by calcium and an endogenous heat-stable protein, *Nature (London)* **271**:478–479. [563]

Schultz, W., 1986a, in: *Neurotransmitter Interactions in the Basal Ganglia* (C. Feuerstein, M. Sandler, and B. Scatton, eds.), Raven Press, New York (in press). [295]

Schultz, W., 1986b, The activity of pars reticulata neurons of the monkey substantia nigra in relation to motor, sensory and complex events, *J. Neurophysiol.* **55**:660–677. [295, 296, 514]

Schultzberg, M., Dunerett, S. B., Iversen, S., *et al.*, 1984, Dopamine and cholecystokinin immunoreactive neurones in mesencephalic grafts reinnervating the neostriatum: Evidence for selective growth regulation, *Neuroscience* **12**:17–32. [457]

Schulz, D. W., Stanford, E. J., Wyrick, S. W., and Mailman, R. B., 1985, Binding of [^{3}H]SCH23390 in rat brain: Regional distribution and effects of assay conditions and GTP suggest interactions at a D$_1$-like dopamine receptor, *J. Neurochem.* **45**:1601–1611. [298, 299]

Schwab, R. S., Chafetz, M. E., and Walker, S., 1954, Control of two simultaneous voluntary motor acts in normals and parkinsonism, *Arch. Neurol.* **72**:591–598. [511]

Schwarcz, R., and Kohler, C., 1983, Differential vulnerability of central neurons of the rat to quinolinic acid, *Neurosci. Lett.* **38**:85–90. [193]

Schwarcz, R., Zaczek, R., and Coyle, J. T., 1978a, Microinjection of kainic acid into the rat hippocampus, *Eur. J. Pharmacol.* **50**:209–220. [212, 214]

Schwarcz, R., Creese, I., Coyle, J. T., and Snyder, S. H., 1978b, Dopamine receptors localized on cerebral cortical afferents to rat corpus striatum, *Nature (London)* **271**:766–768. [301]

Schwarcz, R., Kohler, C., Fuxe, K., Hökfelt, T., and Goldstein, M., 1979, On the mechanism of selective neuronal degeneration in the rat brain: Studies with ibotenic acid, *Adv. Neurol.* **23**:655–668. [192]

Schwarcz, R., Collins, J. F., and Parks, D. A., 1982, α-Amino-ω-phosphono carboxylates block ibotentate but not kainate neurotoxicity in rat hippocampus, *Neurosci. Lett.* **33**:85–90. [192]

Schwarcz, R. C., Whetsell, W. D., Jr., and Mangano, R. M., 1983, Ouinolinic acid: An endogenous metabolite that produces axon-sparing lesions in rat brain, *Science* **219**:316–318. [193, 195]

Schwarcz, R., Foster, A. C., French, E. D., Whetsell, W. O., Jr., and Kohler, C., 1984, II. Excitotoxic models for neurodegenerative disorders, *Life Sci.* **35**:19–32. [193, 195]

Schwartz, I. R., 1981, The differential distribution of label following uptake of ^{3}H-labeled amino acids in the dorsal cochlear nucleus of the cat, *Exp. Neurol.* **73**:601–617. [206]

Schwartz, J. C., 1975, Histamine as a transmitter in brain, *Life Sci.* **17**:503–518. [350]

Schwartz, J. C., Lampart, C., Rose, C., Rehault, M. C., Bischoff, S., and Pollard, H., 1971, Histamine formation in rat brain during development, *J. Neurochem.* **18**:1787–1789. [352]

Schwartz, J. C., Lampart, C., and Rose, C., 1972, Histamine formation in rat brain *in vivo*: Effects of histidine loads, *J. Neurochem.* **19**:801–810. [352]

Schwartz, J. C., Barbin, G., Baudry, M., *et al.*, 1980, Metabolism and function of histamine in the brain, in: *Current Developments in Pharmacology*, Vol. 5 (W. B. Essman and L. Valzelli, eds.), pp. 173–261, Spectrum, New York. [350, 351, 353, 355, 360]

Schwartz, J. H., 1979, Axonal transport: Components, mechanisms and specificity, *Annu. Rev. Neurosci.* **2**:467–500. [30]

Schwartz, M., and Spirman, N., 1982, Sprouting from chicken embryo dorsal root ganglia induced by nerve growth factor is specifically inhibited by affinity-purified antiganglioside antibodies, *Proc. Natl. Acad. Sci. U.S.A.* **79**:6080–6083. [466]

Schwartz, R. D., McGe R. Jr., and Kellar, K. J., 1982, Nicotinic cholinergic receptors labeled by [^{3}H] acetylcholine in rat b...*Mol. Pharmacol.* **22**:56–62. [254]

Schwartzberg, D. G., and Nakane, P. K., 1983, ACTH-related peptide containing neurons within the medulla oblongata of the rat, *Brain Res.* **276**:351–356. [388]

Schwartzkroin, P. A., and Slawsky, M., 1977, Probable calcium spikes in hippocampal neurons, *Brain Res.* **135**:157–161. [60]

Schwarz, D. W., Deecke, L., and Fredrickson, J. M., 1973, Cortical projection of Group I muscle afferents to areas 2, 3a and the vestibular field in the rhesus monkey, *Exp. Brain Res.* **17**:516–526. [487]

Schwindt, P. C., and Calvin, W. H., 1972, Membrane-potential trajectories between spikes underlying motoneurone firing rate, *J. Neurophysiol.* **35**:311–325. [77]

Sears, T. (ed.), 1982, *Neuronal–Glial Cell Relationships: Ontogeny, Maintenance, Injury, Repair, Life Sciences Research Reports*, Vol. 20, Springer-Verlag, Berlin, Heidelberg, and New York. [35]

Sebeok, T. A., and Umiker-Sebeok, D. J., 1980, *Speaking of Apes*, Plenum Press, New York. [618]

Seeburg, P. H., and Adelman, J. P., 1984, Characterization of cDNA for precursor of human luteinizing hormone releasing hormone, *Nature (London)* **311**:666–668. [382]

Seeman, P., Ulpian, C., Bergeron, C., *et al.*, 1984, Bimodal distribution of dopamine receptor densities in brains of schizophrenics, *Science* **225**:728–731. [546]

Segal, D. S., Knapp, S., Kuczenski, R. T., and Mandell, A. J., 1971, Effects of long-term reserpine treatment on brain tyrosine hydroxylase activity and behavioral activity, *Science* **173**:847–849. [284]

Segal, M., 1981, The actions of glutamic acid on neurons in the rat hippocampal slice, *Adv. Biochem. Psychopharmacol.* **27**:217–225. [182]

Segal, M., Dudai, Y., and Amsterdam, A., 1978, Distribution of an alpha-bungarotoxin-binding cholinergic nicotinic receptor in rat brain, *Brain Res.* **148**:105–119. [253]

Segawa, T., Mizuta, T., and Nomura, Y., 1979, Modifications of central 5-hydroxytryptamine binding sites in synaptic membranes from rat brain after long-term administration of tricyclic antidepressants, *Eur. J. Pharmacol.* **58**:75–83. [540]

Sem-Jacobsen, C. W., 1976, Electrical stimulation and self-stimulation in man with chronic implanted electrodes, in: *Brain-Stimulation Reward* (A. Wanquier and E. T. Rolls, eds.), pp. 505–526, Elsevier, Amsterdam. [530]

Seress, L., and Ribak, C. E., 1983, GABAergic cells in the dentate gyrus appear to be local circuit and projection neurons, *Exp. Brain Res.* **50**:173–182. [206, 214]

Sethy, V. H., and Van Woert, M. H., 1974, Modification of striatal acetylcholine concentration by dopamine receptor agonists and antagonists, *Res. Commun. Chem. Pathol. Pharmacol.* **8**:13–28. [512]

Shah, N. S., and Donald, A. G. (eds.), 1982, *Endorphins and Opiate Antagonists in Psychiatric Research: Clinical Implications,* Plenum Press, New York. [409]

Share, L., 1979, Interrelations between vasopressin and the renin–angiotensin system, *Fed. Proc. Fed. Am. Soc. Exp. Biol.* **38**:2267–2271. [406]

Sharif, N. A., Zuhowski, E. G., and Burt, D. R., 1983, Benzodiazepines compete for thyrotropin-releasing hormone receptor binding: Micromolar potency in rat pituitary, retina and amygdala, *Neurosci. Lett.* **41**:301–306. [219]

Shepherd, G. M., 1974, *The Synaptic Organization of the Brain,* Oxford University Press, London. [132, 136]

Shepherd, G. M., 1983, *Neurobiology,* Oxford University Press, Oxford and New York, 611 pp. [613]

Sherrington, C. S., 1906, *The Integrative Action of the Nervous System,* Yale University Press, New Haven, Connecticut. [8, 11, 77, 109, 473]

Sherrington, C. S., 1940, *Man on his Nature,* Cambridge University Press, London. [554]

Shimada, S., Shiosaka, S., Takami, K., Yamano, M., and Tohyama, M., 1985, Somatostatinergic neurons in the insular cortex project to the spinal cord: Combined retrograde axonal transport and immunohistochemical study, *Brain Res.* **326**:197–200. [398]

Shimohama, S., Tsukahara, T., Taniguchi, T., and Fujiwara, M., 1985, The binding of ^{3}H-acetylcholine to cholinergic receptors in bovine cerebral arteries, *Life Sci.* **37**:1887–1893. [242]

Shimohigashi, Y., Costa, T., Chen, H.-C., and Rodbard, D., 1982, Dimeric tetrapeptide enkephalins display extraordinary selectivity for the δ opiate receptor, *Nature (London)* **297**:333–335. [404]

Shiraishi, T., Senba, E., Tohyama, M., Wu, J.-Y., Kubo, T., and Matsunaga, T., 1985, Distribution and fine structure of neuronal elements containing glutamate decarboxylase in the rat cochlear nucleus, *Brain Res.* **347**:183–187. [205]

Sholl, D. A., 1956, *The Organization of the Cerebral Cortex,* John Wiley, New York. [5]

Sidman, R. L., and Rakic, P., 1973, Neuronal migration with special reference to developing human brain: A review, *Brain Res.* **62**:1–35. [417, 420–422, 425, 426, 430, 432]

Siegelbaum, S. A., Camardo, J. S., and Kandel, E. R., 1982, Serotonin and cyclic AMP close single K$^+$ channels in *Aplysia* sensory neurones, *Nature (London)* **299**:413–417. [336]

Siggins, G. R., Gruol, D., Aldenhoff, J., and Pittman, Q., 1985, Electrophysiological actions of corticotropin-releasing factor in the central nervous system, *Fed. Proc. Fed. Am. Soc. Exp. Biol.* **44**:237–242. [402]

Silinsky, E. M., 1985, The biophysical pharmacology of calcium-dependent acetylcholine secretion, *Pharmacol. Rev.* **37**:81–132. [92]

Simmonds, S. H., and Strange, P. G., 1985, Inhibition of inositol phospholipid breakdown by D$_2$ dopamine receptors in dissociated bovine anterior pituitary, *Neurosci. Lett.* **60**:267–272. [300]

Simon, E. J., Hiller, J. M., and Edelman, I., 1973, Stereospecific binding of the potent narcotic analgesic [^{3}H]etorphine to rat brain homogenates, *Proc. Natl. Acad. Sci. U.S.A.* **70**:1947–1949. [374]

Singer, W., 1985, Central control of developmental plasticity in the mammalian visual cortex, *Vision Res.* **25**:389–396. [455]

Singh, V. K., and McGeer, P. L., 1977, Studies on choline acetyltransferase isolated from human brain, *Neurochem. Res.* **2**:281–291. [242]

Sitaram, N., Weingartner, H., and Gillin, J. C., 1978, Human serial learning: Enhancement with arecoline and impairment with scopolamine correlated with performance on placebo, *Science* **201**:274–276. [261]

Skagerberg, G., Björklund, A., Lindvall, O., and Schmidt, R. H., 1982, Origin and termination of the diencephalo-spinal dopamine system in the rat, *Brain Res. Bull.* **9**:237–244. [291]

Skirboll, L., Hökfelt, T., Dockray, G., Rehfeld, J., Brownstein, M., and Cuello, A. C., 1983, Evidence for periaqueductal choleocystokinin–substance P neurons projecting to the spinal cord, *J. Neurosci.* **3:**1151–1157. [157]

Skrede, K. K., and Malthe-Sørenssen, D., 1981, Increased resting and evoked release of transmitter following repetitive electrical tetanization in hippocampus: A biochemical correlate to long-lasting synaptic potentiation, *Brain Res.* **208:**436–441. [182, 564, 565]

Sladek, J. R., Jr., and Gash, D. M. (eds.), 1984, *Neural Transplants: Development and Function,* Plenum Press, New York. [457, 459]

Sladek, J. R., and Walker, P., 1977, Serotonin-containing neuronal perikarya in the primate locus coeruleus and subcoeruleus, *Brain Res.* **134:**359–366. [331]

Sloper, J. J., 1971, Dendro-dendritic synapses in the primate motor cortex, *Brain Res.* **34:**186–192. [136]

Smith, A. J., 1966, Speech and other functions after left (dominant) hemispherectomy, *J. Neurol. Neurosurg. Psychiatry* **29:**467–471. [622]

Smith, G. K., and Nichol, C. A., 1984, Two new tetrahydropterin intermediates in the adrenal medullary *de novo* biosynthesis of tetrahydrobiopterin, *Biochem. Biophys. Res. Commun.* **120:**761–766. [273]

Smith, G. P., and Gibbs, J., 1985, The satiety effect of cholecystokinin: Recent progress and current problems, *Ann. N. Y. Acad. Sci.* **448:**417–423. [405]

Smith, J. E., Co, C., and Lane, J. D., 1978, Turnover rates of serotonin, norepinephrine and dopamine concurrently measured in seven rat brain regions, *Prog. Neuro-Psychopharmacol.* **2:**359–367. [322]

Smith, L. F. P., and Pycock, C. J., 1982, Potassium stimulated release of radiolabeled taurine and glycine from the isolated rat retina, *J. Neurochem.* **39:**653–658. [229]

Sneddon, P., Westfall, D. P., and Fedan, J. S., 1982, Cotransmitters in the motor nerves of the guinea pig vas deferens: Electrophysiological evidence, *Science* **218:**693–695. [366, 367]

Snider, R. M., McKinney, M., Forray, C., and Richelson, E., 1984, Neurotransmitter receptors mediate cyclic GMP formation by involvement of arachidonic acid and lipoxygenase, *Proc. Natl. Acad. Sci. U.S.A.* **81:**3905–3909. [358]

Snyder, S. H., 1982, Schizophrenia, *Lancet* **2:**970–973. [546, 548]

Snyder, S. H., 1984, Drug and neurotransmitter receptors in the brain, *Science* **224:**22–31. [162]

Snyder, S. H., and Peroutka, S. J., 1980, Multiple neurotransmitter receptors: Two populations of serotonin receptors with different physiological functions in: *Psychopharmacology and Biochemistry of Neurotransmitter Receptors* (H. Yamamura, R. W. Olsen, and E. Usdin, eds.), pp. 313–324, Elsevier/North Holland, Amsterdam. [335, 336]

Snyder, S. H., Brown, R., and Kuhar, M. J., 1974, The subsynaptosomal localization of histamine, histidine decarboxylase and histamine methyltransferase in rat hypothalamus, *J. Neurochem.* **23:**37–45. [351, 352]

Sobel, D. O., 1983, Characterization of angiotensin mediated ACTH release, *Neuroendocrinology* **36:**249–253. [393]

Sofroniew, M. V., 1980, Projections from vasopressin, oxytocin, and neurophysin neurons to neural targets in the rat and human, *J. Histochem. Cytochem.* **28:**475–478. [389]

Sofroniew, M. V., 1985, Vasopressin- and neurophysin-immunoreactive neurons in the septal region, medial amygdala and locus coeruleus in colchicine-treated rats, *Neurosci. Lett.* **15:**347–358. [387–389, 395]

Sofroniew, M. V., Eckenstein, F., Thoenen, H., and Cuello, A. C., 1982, Topography of choline acetyltransferase-containing neurons in the forebrain of the rat, *Neurosci. Lett.* **33:**7–12. [245, 250]

Sokoloff, L., 1981, Localization of functional activity in the central nervous system by measurement of glucose utilization with radioactive deoxyglucose, *J. Cereb. Blood Flow Metab.* **1:**7–36. [69]

Sokoloff, P., Martres, M. P., and Schwartz, J. C., 1980, Three classes of dopamine receptor (D-2, D-3, D-4) identified by binding studies with ³H-apomorphine and ³H-domperidone, *Naunyn-Schmiedeberg's Arch. Pharmacol.* **315:**89–102. [301]

Sokolovsky, M., 1984, Muscarinic receptors in the central nervous system, *Int. Rev. Neurobiol.* **25:**139–183. [254]

Solomon, T. E., 1985, Pancreatic polypeptide, peptide YY, and neuropeptide Y family of regulatory peptides, *Gastroenterology* **88:**838–841. [373]

Sommerville, J., 1985, Organizing the nucleolus, *Nature (London)* **318:**410–411. [8]

Somogyi, P., Freund, T. F., Halasz, N., and Kisvarday, Z. F., 1981, Selectivity of neuronal [³H] GABA accummulation in the visual cortex as revealed by Golgi staining of the labeled neurons, *Brain Res.* **225:**431–436. [205]

Somogyi, P., Freund, T. F., and Cowey, A., 1982a, The axo-axonic interneuron in the cerebral cortex of the rat, cat and monkey, *Neuroscience* **7**:2577–2607. [14]

Somogyi, P., Priestley, J. V., Cuello, A. C., Smith, A. D., and Takagi, H., 1982b, Synaptic connections of nekephalin-immunoreactive nerve terminals in the neostriatum: A correlated light and electron microscopic study, *J. Neurocytol.* **11**:779–807. [507]

Somogyi, P., Cowey, A., Kisvárday, Z. F., Freund, T. F., and Szentágothai, J., 1983, Retrograde transport of γ-amino[3H]butyric acid reveals specific interlaminar connections in the striate cortex of monkey, *Proc. Natl. Acad. Sci. U.S.A.* **80**:2385–2389. [212]

Somogyi, P., Hodgson, A. J., DePotter, R. W., *et al.*, 1984a, Chromogranin immunoreactivity in the central nervous system, *Brain Res. Rev.* **8**:193–230. [279]

Somoygi, P., Hodgson, A. J., Smith, A. D., Nunzi, M. G., Gorio, A., and Wu, J.-Y., 1984b, Different populations of GABAergic neurons in the visual cortex and hippocampus of cat contain somatostatin or cholecystokinin immunoreactive material, *J. Neurosci.* **4**:2590–2603. [157]

Somogyi, P., Freund, T. F., Hodgson, A. J., Somogyi, J., Beroukas, D., and Chubb, I. W., 1985, Identified axo-axonic cells are immunoreactive for GABA in the hippocampus and visual cortex of the cat, *Brain Res.* **332**:143–149. [205]

Sorensen, K. V., Christensen, S. E., Hansen, A. P., Ingerslev, J., Pedersen, E., and Orskov, H., 1981, The origin of cerebrospinal fluid somatostatin: Hypothalamic or disperse central nervous system secretion, *Neuroendocrinology* **32**:335–338. [411]

Sørensen, P. S., Gjerris, A., and Hammer, M., 1985, Cerebrospinal fluid vasopressin in neurological and psychiatric disorders, *J. Neurol. Neurosurg. Psychiatry* **48**:50–57. [412]

Spano, P. F., and Neff, N. H., 1972, Metabolic fate of caudate nucleus dopamine, *Brain Res.* **42**:139–145. [283]

Sparrow, J. R., and Grafstein, B., 1982, Sciatic nerve regeneration in 13 ganglioside-treated rats, *Exp. Neurol.* **77**:230–235. [467]

Spehlmann, R., Norcross, K., and Grimmer, E. J., 1977, GABA in the caudate nucleus: A possible synaptic transmitter of interneurons, *Experientia* **33**:623–625. [212]

Spencer, H. J., 1976, Antagonism of cortical excitation of striatal neurons by glutamic acid diethyl ester: Evidence for glutamic acid as an excitatory transmitter in the rat striatum, *Brain Res.* **102**:91–101. [182, 188]

Spencer, W. A., and Kandel, E. R., 1961, Electrophysiology of hippocampal neurones, IV. Fast prepotentials, *J. Neurophysiol.* **24**:272–285. [72]

Sperry, R. W., 1963, Chemoaffinity in the orderly growth of neural circuits, *Growth* **10**:63–87. [434]

Sperry, R. W., 1971, How a developing brain gets itself properly wired for adaptive function, in: *Biopsychology of Development* (E. Tobach, R. Avonson, and E. Shaw, eds.), pp. 27–44, Academic Press, New York. [433, 434, 436]

Sperry, R. W., 1974, Lateral specialization in the surgically separated hemispheres, in: *The Neurosciences: Third Study Program* (F. O. Schmitt and F. G. Worden, eds.), pp. 5–19, MIT Press, Cambridge, Massachusetts. [619]

Sperry, R. W., 1976, Mental phenomena as causal determinants in brain function in: *Consciousness and the Brain* (G. G. Globus, G. Maxwell, and I. Savodik, eds.), pp. 163–177, Plenum Press, New York. [631, 632]

Sperry, R., 1982, Some effects of disconnecting the cerebral hemispheres, *Science* **217**:1223–1226. [618, 620, 623, 624]

Sperry, R., 1983, *Science and Moral Priority*, Columbia University Press, New York. [624]

Sperry, R. W., Zaidel, E., and Zaidel, D., 1979, Self recognition and social awareness in the deconnected minor hemisphere, *Neuropsychologia* **17**:153–166. [618–622]

Speth, R. C., Wamsley, J. K., Gehlert, D. R., Chernicky, C. L., Barnes, K. L., and Ferrario, C. M., 1985, Angiotensin II receptor localization in the canine CNS, *Brain Res.* **326**:137–143. [402]

Spindel, E. R., Wurtman, R. J., and Bird, E. D., 1980, Increased TRH content of the basal ganglia in Huntington's Disease, *N. Engl. J. Med.* **303**:1235–1236. [412, 507]

Squire, L. R., 1982, The neuropsychology of human memory, *Annu. Rev. Neurosci.* **5**:241–273. [568, 569, 572]

Squire, L. R., 1983, The hippocampus and the neuropsychology of memory, in: *Molecular, Cellular, and Behavioral Neurobiology of the Hippocampus* (W. Seifert, ed.), pp. 491–507, Academic Press, New York. [569, 572]

Stacher, G., 1985, Satiety effects of cholecystokinin and ceruletide in lean and obese man, *Ann. N. Y. Acad. Sci.* **448**:431–436. [405]

Staines, W. A., Nagy, J. I., Vincent, S. R., and Fibiger, H. C., 1980, Neurotransmitters contained in the efferents of the striatum, *Brain Res.* **194**:391–402. [212, 507]

Stamford, J. A., 1985, *In vivo* voltammetry: Promise and perspective, Brain Res. **357**:119–135. [296]

Starr, M., and Kilpatrick, I., 1981, Distribution of gamma-aminobutyrate in the rat thalamus: Specific decreases in thalamic gamma-aminobutyrate following lesions or electrical stimulation of the substantia nigra, *Neuroscience* **6**:1095–1104. [212]

Staun-Olsen, P., Ottersen, B., Bartels, P. D., Nielsen, M. N., Gummeltoft, S., and Fahrenkrug, J., 1982, Receptors for vasoactive intestinal peptide on isolated synaptosomes from rat cerebral cortex: Heterogeneity of binding and desensitization of receptors, *J. Neurochem.* **39**:1242–1251. [402]

Staun-Olsen, P., Ottersen, B., Gammeltoft, S., and Fahrenkrug, J., 1985, The regional distribution of receptors for vasoactive intestinal peptide (VIP) in the rat central nervous system, *Brain Res.* **330**:317–321. [402]

Steardo, L., Knight, M., Tamminga, C. A., Barone, P., Kask, A. M., and Chase, T. N., 1985, CCK_{26-33} degrading activity in brain and nonneural tissue: A metalloendopeptidase, *J. Neurochem.* **45**:784–790. [384]

Stedman, E., and Stedman, E., 1939, The mechanism of biological synthesis of acetylcholine, *Biochem. J.* **33**:811–821. [235]

Steenbergh, P. H., Hoppener, J. W. M., Zandberg, J., Lips, C. J. M., and Jansz, H. S., 1985, A second human calcitonin/CGRP gene, *FEBS Lett.* **183**:403–407. [376]

Steinbusch, H. W. M., 1981, Distribution of serotonin-immunoreactivity in the central nervous system of the rat cell bodies and terminals, *Neuroscience* **6**:557–618. [329]

Steinbusch, H. W. M., 1984, Serotonin-immunoreactive neurons and their projections in the CNS, in: *Handbook of Chemical Neuroanatomy*, Vol. 3, *Classical Transmitters and Transmitter Receptors in the CNS, Part II* (A. Björklund, T. Hökfelt, and M. J. Kuhar, eds.), pp. 68–125, Elsevier, Amsterdam. [331, 333]

Steinbusch, H. W. M., and Mulder, A. H., 1984, Immunohistochemical localization of histamine in mast cells and neurons in the rat brain, in: *Handbook of Chemical Neuroanatomy*, Vol. 3, *Classical Transmitters and Transmitter Receptors in the CNS, Part II* (A. Björklund, T., Hökfelt, T., and M. J. Kuhar, eds.), pp. 126–140, Elsevier, Amsterdam. [354]

Steiner, D. F., and Oyer, P. E., 1967, The biosynthesis of insulin by a human islet adenoma, *Proc. Natl. Acad. Sci. U.S.A.* **57**:473–480. [379]

Steiner, H. X., McBean, G. J., Kohler, C., Roberts, P. J., and Schwarcz, R., 1984, Ibotenate-induced neuronal degeneration in immature rat brain, *Brain Res.* **307**:117–124. [192]

Steinhardt, R. A., and Alderton, J. M., 1982, Calmodulin confers calcium sensitivity on secretory exocytosis, *Nature (London)* **295**:154–155. [169]

Stengaard-Pedersen, K., Fredens, K., and Larsson, L.-I., 1983, Comparative localization of enkephalin and cholecystokinin immunoreactivities and heavy metals in the hippocampus, *Brain Res.* **273**:81–96. [397]

Sterling, P., and Davis, T. L., 1980, Neurons in cat lateral geniculate nucleus that concentrate exogenous [^{3}H]-γ-aminobutyric acid (GABA), *J. Comp. Neurol.* **192**:737–749. [206]

Stern, A. S., Lewis, R. V., Kimura, S., Rossier, J., Stein, S., and Underfriend, S., 1980, Opioid hexapeptides and heptapeptides in adrenal medulla and brain: Possible implications on the biosynthesis of enkephalins, *Arch. Biochem. Biophys.* **205**:606–613. [382]

Stern, Y., and Langston, J. W., 1985, Intellectual changes in patients with MPTP-induced parkinsonism, *Neurology* **35**:1506–1509. [304]

Sternbach, H., Gerner, R. H., and Gwirtsman, H. E., 1982, The thyrotropin releasing hormone stimulation test: A review, *J. Clin. Psychiatry* **43**:4–6. [410]

Sternberg, D. E., van Kammen, D. P., Lake, C. R., Ballenger, J. C., Morder, S. R., and Bunney, W. E., Jr., 1981, The effect of pimozide on CSF norepinephrine in schizophrenia, *Am. J. Psychiatry* **138**:1045–1050. [547]

Stevens, C. F., 1985, AChR structure: A new twist in the story, *TINS* **8**:1–2. [101]

Steward, O., 1982, Assessing the functional significance of lesion-induced neuronal plasticity, *Int. Rev. Neurobiol.* **23**:197–254. [445]

Steward, O., and Levy, W. B., 1982, Preferential localization of polyribosomes under the base of dendritic spines in granule cells of the dentate gyrus, *J. Neurosci.* **2**:284–291. [563]

Stine, S., and Kellogg, C., 1982, Inhibition of tryptophan hydroxylase: Neurochemical action of a catecholamide seizure-inducing agent, *Neurochem. Res.* **7**:87–98. [341]

Stockel, K., Paravicini, U., and Thoenen, H., 1976, Specificity of the retrograde axonal transport of nerve growth factor, *Brain Res.* **76:**413–421. [460]

Støckle, H., and ten Bruggencate, G., 1978, Climbing fiber-mediated rhythmic modulations of potassium and calcium in cat cerebellar cortex, *Exp. Neurol.* **61:**226–230. [581]

Stolz, F., 1904, Uber Adrenaline and Alkylaminoacetobrenzcatechin. *Ber. Dtsch. Chem. Ges.* **37:**4149–4154. [226]

Stone, T. W., 1979a, Amino acids as neurotransmitters of corticofugal neurones in the rat: A comparison of glutamate and aspartate, *Br. J. Pharmacol.* **67:**545–551. [182, 187]

Stone, T. W., 1979b, Glutamate as the neurotransmitter of cerebellar cells in the rat: Electrophysiological evidence, *Br. J. Pharmacol.* **66:**291–296. [182]

Stone, T. W., 1981, Physiological roles for adenosine and adenosine 5'-triphosphate in the nervous system, *Neuroscience* **6:**523–555. [360, 363]

Stone, T. W., 1983, Actions of TRH and cyclo (His-Pro) on spontaneous and evoked activity of cortical neurones, *Eur. J. Pharmacol.* **92:**113–118. [385]

Stone, T. W., and Connick, J. H., 1985, Quinolinic acid and other kynurenines in the central nervous system, *Neuroscience* **15:**597–617. [191]

Storm-Mathisen, J., 1975, Accumulation of glutamic acid decarboxylase in the proximal parts of presumed GABA-ergic neurons after axotomy, *Brain Res.* **87:**107–109. [214]

Storm-Mathisen, J., 1977, Glutamic acid and excitatory nerve endings: Reduction of glutamic acid uptake after axotomy, *Brain Res.* **120:**379–386. [182, 184]

Storm-Mathisen, J., and Fonnum, F., 1971, Quantitative histochemistry of glutamate decarboxylase in the rat hippocampal region, *J. Neurochem.* **18:**1105–1111. [212, 214]

Storm-Mathisen, J., and Woxen-Opsahl, M., 1978, Aspartate and/or glutamate may be transmitters in hippocampal efferents to septum and hypothalamus, *Neurosci. Lett.* **9:**65–70. [182]

Storm-Mathisen, J., Leknes, A. K., Bore, A. T., *et al.*, 1983, First visualization of glutamate and GABA in neurones by immunocytochemistry, *Nature (London)* **301:**517–520. [185, 203]

Strahlendorf, H. K., and Barnes, C. D., 1983, Control of substantia nigra pars reticulata neurons by the nucleus accumbens, *Brain Res. Bull.* **11:**259–263. [212]

Straschell, M., and Perwein, J., 1969, The inhibition of retinal ganglion cells by catecholamines and γ-aminobutyric acid, *Pfluegers Arch.* **312:**45–54. [212, 216]

Strata, P., 1984, Recent aspects of the function of the inferior olive, *Exp. Brain Res. Suppl.* **9:**184–189. [494]

Strauss, W., Hilt, D., Puckett, C., 1985, Cloning of human choline acetyltransferase cDNA, *Soc. Neurosci. Abstr.* **11:**1114. [46, 239]

Strawson, P., 1959, *Individuals,* Methuen, London. [621]

Streit, P., 1980, Selective retrograde labeling indicating the transmitter of neuronal pathways, *J. Comp. Neurol.* **191:**429–463. [182]

Streit, P., Knecht, E., and Cuenod, M., 1979, Transmitter-specific retrograde labeling in the striatonigral and raphe–nigral pathways, *Science* **205:**306–308. [212]

Stricker, E. M., and Zigmond, M. J., 1984, Brain catecholamines and the central control of food intake, *Int. J. Obesity* **8**(Suppl. I):39–50. [529]

Strittmatter, W. J., Hirata, F., Axelrod, J., Mallorga, P., Tallman, J. F., and Henneberry, R. C., 1979, Benzodiazepine and β-adrenergic receptor ligands independently stimulate phospholipid methylation, *Nature (London)* **282:**857–859. [222]

Stromberg, I., Herrera-Marschitz, Hultgren, L., Ungerstedt, U., and Olson, L., 1984, Adrenal medullary implants in the dopamine-denervated rat striatum. I. Acute catecholamine levels in grafts and host caudate as determined by HPLC-electrochemistry and fluorescence histochemical image analysis, *Brain Res.* **297:**41–51. [459]

Stryer, L., 1975, *Biochemistry,* Freeman, San Francisco. [239]

Studler, J. M., Javoy-Agid, F., Casselin, F., Legrand, J. C., and Agid, Y., 1982, CCK-8-immunoreactivity distribution in human brain: Selective decrease in the substantia nigra from parkinsonian patients, *Brain Res.* **243:**176–179. [412]

Stumpf, W. E., and Jennes, L., 1984, The A-B-C (allocortex–brainstem–core) circuitry of endocrine–autonomic integration and regulation: A proposed hypothesis on the anatomical–functional relationships between estradiol sites of action and peptidergic–aminergic neuronal systems, *Peptides* **5**(Suppl. 1):221–226. [395]

Su, T. P., 1981, Psychotomimetic opioid binding: Specific binding of [3H]SKF-10047 to etorphine-inaccessible sites in guinea pig brain, *Eur. J. Pharmacol.* **75**:81–82. [404]

Subramanian, N., 1982, Electrically induced release of radiolabeled histamine from rat hippocampal slices: Opposing roles for H_1 and H_2 receptors, *Life Sci.* **31**:557–562. [356]

Subramanian, N., and Mulder, A. H., 1976, Potassium-induced release of tritiated histamine from rat brain tissue slices, *Eur. J. Pharmacol.* **35**:203–206. [356]

Subramanian, N., Whitmore, W. L., Seidler, F. J., and Slotkin, T. A., 1980, Histamine stimulates brain phospholipid turnover through a direct H-1 receptor-mediated mechanism, *Life Sci.* **27**:1315–1319. [357]

Suda, T., Yajima, F., Tomori, N., Demura, H., and Shizume, K., 1985, *In vitro* study of immunoreactive corticotropin-releasing factor release from the rat hypothalamus, *Life Sci.* **37**:1499–1505. [402]

Sugden, D., and Fletcher, A., 1981, Changes in the rat sleep–wake cycle produced by DL-6-fluorotryptophan, a competitive inhibitor of tryptophan hydroxylase, *Psychopharmacology* **74**:369–373. [338]

Sugimoto, T., and Hattori, T., 1983, The nucleus tegmenti pedunculopontinus pars compacta in the rat: Organization, efferent projections and cholinergic aspects, *Soc. Neurosci. Abstr.* **9**:283. [251]

Sugimoto, T., Itoh, K., Yasui, Y., Kaneko, T., and Mizuno, N., 1985, Coexistence of neuropeptides in projection neurons of the thalamus in the cat, *Brain Res.* **347**:381–384. [157]

Sulkava, R., Erkinjuntti, T., and Laatikainen, T., 1985, CSF β-endorphin and β-lipotropin in Alzheimer's disease and multi-infarct dementia, *Neurology* **35**:1057–1058. [412]

Sung, S. C., and Ruff, B. A., 1983, Molecular forms of sucrose extractable and particulate acetylcholinesterase in the developing and adult rat brain, *Neurochem. Res.* **8**:303–311. [240]

Sutcliffe, J. G., Milner, R. J., Shinnick, T. H., and Bloom, F. E., 1983, Identifying the protein products of brain-specific genes with antibodies to chemically synthesized peptides, *Cell* **33**:671–682. [43, 44, 46]

Suzuki, O., Katsumata, Y., and Oya, M., 1979, 1,4-Methylhistamine is a specific substrate for type B monoamine oxidase, *Life Sci.* **24**:2227–2230. [351]

Swanson, L. W., 1976, The locus coeruleus: A cytoarchitectonic Golgi and immunohistochemical study in the albino rat, *Brain Res.* **110**:39–56. [286]

Swanson, L. W., 1982, The projections of the ventral tegmental area and adjacent regions: A combined fluorescent retrograde tracer and immunofluorescent study in the rat, *Brain Res. Bull.* **9**:321–353. [286]

Swanson, L. W., and Hartman, B. K., 1975, The central adrenergic system: An immunofluorescence study of the location of cell bodies and their efferent connections in the rat utilizing dopamine-β-hydroxylase as a marker, *J. Comp. Neurol.* **163**:467–506. [286]

Szabo, S., and Reichlin, S., 1985, Somatostatin depletion by cysteamine: Mechanism and implication for duodenal ulceration, *Fed. Proc. Fed. Am. Soc. Exp. Biol.* **44**:2540–2545. [411]

Sze, P. Y., Alderson, R. F., and Hedrick, B. J., 1983, Two forms of striatal tyrosine hydroxylase from DEAE–cellulose chromatography, *Brain Res.* **268**:129–137. [283]

Szentágothai, J., 1969, Architecture of the cerebral cortex, in: *Basic Mechanisms of the Epilepsies* (H. H. Jasper, A. A. Ward, and A. Pope, eds.), pp. 13–28, Little, Brown, Boston. [8]

Szentágothai, J., 1971, Memory functions and the structural organization of the brain, in: *Biology of Memory* (G. Adam, ed.) *Symp. Biol. Hung.* **10**:21–25. [554]

Szentágothai, J., 1978a, The neuron network of the cerebral cortex: A functional interpretation, *Proc. R. Soc. London Ser. B* **201**:219–248. [482, 570, 595, 596, 598, 599, 631]

Szentágothai, J., 1978b, The local neuronal apparatus of the cerebral cortex, in: *Cerebral Correlates of Conscious Experience* (P. Buser and A. Rongeul-Buser, eds.), pp. 131–138, Elsevier/North-Holland, Amsterdam. [631, 632]

Szentágothai, J., 1979, Local neuron circuits of the neocortex, in: *The Neurosciences: Fourth Study Program* (F. O. Schmitt and F. G. Worden, eds.), pp. 399–415, MIT Press, Cambridge, Massachusetts. [597]

Szentágothai, J., 1983, The modular architectonic principle of neural centers, *Rev. Physiol. Biochem. Pharmacol.* **98**:11–61. [595, 598, 631]

Tagliavini, F., Pilleri, G., Bouras, C., and Constantinidis, J., 1984, The basal nucleus of Meynert in patients with progressive supranuclear palsy, *Neurosci. Lett.* **44**:37–42. [261]

Takagi, H., Shiomi, H., Ueda, H., and Amano, H., 1979, Morphine-like analgesia by a new dipeptide L-tyrosyl-L-arginine (kyotorphin) and its analogue, *Eur. J. Pharmacol.* **55**:109–111. [408]

Takahama, K., Miyata, T., Okano, Y., Kataoka, M., Hitoshi, T., and Kase, Y., 1982, Potentiation of phenobarbital-induced anticonvulsant activity by pipecolic acid, *Eur. J. Pharmacol.* **81**:327–331. [233]

Takahashi, T., and Otsuka, M., 1975, Regional distribution of substance P in the spinal cord and nerve roots of the cat and the effect of dorsal root section, *Brain Res.* **87**:1–11. [182]

Takahashi, Y., Kato, K., Hayashizaki, Y., *et al.*, 1985, Molecular cloning of the human cholecystokinin gene by use of a synthetic probe containing deoxyinosine, *Proc. Natl. Acad. Sci. U.S.A.* **82:**1931–1935. [381, 382]

Takai, T., Noda, M., Mishima, M., *et al.*, 1985, Cloning, sequencing and expression of cDNA for a novel subunit of acetylcholine receptor from calf muscle, *Nature (London)* **315:**761–764. [46, 165]

Takami, K., Kawai, Y., Shiosaka, S., *et al.*, 1985, Immunohistochemical evidence for the coexistence of calcitonin gene-related peptide- and choline acetyltransferase-like immunoreactivity in neurons of the rat hypoglossal, facial and ambiguus nuclei, *Brain Res.* **328:**386–389. [157]

Takeda, S., and Matsuzawa, T., 1984, Brain atrophy during aging: A quantitative study using computed tomography, *J. Am. Geriatr. Soc.* **32:**520–524. [470]

Takeuchi, A., and Takeuchi, N., 1960, On the permeability of the end-plate membrane during the action of transmitter, *J. Physiol. (London)* **154:**52–67. [107, 119, 125]

Takeuchi, A., and Takeuchi, N., 1962, Electrical changes in pre and postsynaptic axons of the giant synapse of *Loligo, J. Gen. Physiol.* **45:**1181–1193. [103, 187]

Takeuchi, Y., Kimura, H., Matsunra, T., and Sano, Y., 1982, Immunohistochemical demonstration of the organization of serotonin neurons in the brain of the monkey (*Macaca fuscata*), *Acta Anat. (Basel)* **114:**106–124. [329]

Tamarkin, L., Baird, C. J., and Almeida, O. F. X., 1985, Melatonin: A coordinating signal for mammalian reproduction?, *Science* **227:**714–720. [344, 345]

Tamminga, C. A., Lucignani, G., Porrino, L. J., Littman, B., Thaker, G. K., and Alphs, L., 1984, Cholecystokinin: Potential peptidergic antipsychotic, *Clin. Neuropharmacol.* **7**(Suppl. 1):556–557. [410]

Tamminga, C. A., LeWitt, P. A., and Chase, T. N., 1985, Cholecystokinin and neurotensin gradients in human CSF, *Arch. Neurol.* **42:**354–355. [411]

Tanabe, T., Yarita, H., Iino, M., Ooshima, Y., and Tahagi, S. F., 1975, An olfactory projection area in orbitofrontal cortex of the monkey, *J. Neurophysiol.* **38:**1269–1283. [613]

Taniyama, K., Miki, Y., and Tabaka, C., 1982a, Presence of γ-aminobutyric acid and glutamic acid decarboxylase in Auerbach's plexus of cat colon, *Neurosci. Lett.* **29:**53–56. [218]

Taniyama, K., Kusunoki, M., Taito, N., and Tanaka, C., 1982b, Release of γ-aminobutyric acid from cat colon, *Science* **217:**1039–1040. [218]

Tanji, J., and Kurata, K., 1982, Comparison of movement-related activity in two cortical motor areas of primates, *J. Neurophysiol.* **48:**633–653. [483, 592]

Tanji, J., Taniguchi, K., and Saga, T., 1980, Supplementary motor area: Neuronal response to motor instructions, *J. Neurophysiol.* **43:**60–68. [483]

Tappaz, M., Aguera, M., Belin, M. F., and Pujol, J. F., 1980, Autoradiography of GABA in the rat hypothalamic median eminence, *Brain Res.* **186:**379–391. [206]

Tappaz, M. L., Wassef, M., Oertel, W. H., Paut, L., and Pujol, J. F., 1983, Light- and electron-microscopic immunocytochemistry of glutamic acid decarboxylase (GAD) in the basal hypothalamus: Morphological evidence for neuroendocrine gamma-aminobutyrate (GABA), *Neuroscience* **9:**271–287. [206]

Tasaki, I., 1939, The electro-saltatory transmission of the nerve impulse and the effect of narcosis upon the nerve fiber, *Am. J. Physiol.* **127:**211–227. [66]

Tavianini, M. A., Hayes, T. E., Magazin, M. D., Minth, C. D., and Dixon, J. E., 1984, Isolation, characterization, and DNA sequence of the rat somatostatin gene, *J. Biol. Chem.* **259:**11,798–11,803. [382]

Taylor, D. P., and Pert, C. B., 1979, Vasoactive intestinal polypeptide: Specific binding to rat brain membranes, *Proc. Natl. Acad. Sci. U.S.A.* **76:**660–664. [402]

Taylor, J. E., and Richelson, E., 1982, High affinity binding of [^{3}H]doxepin to histamine H_1-receptors in rat brain: Possible identification of a subclass of histamine H_1-receptors, *Eur. J. Pharmacol.* **78:**279–285. [356]

Taylor, S. P., Jr., Du Vigneaud, V., and Kunkel, H. G., 1953, Electrophoretic studies of oxytocin and vasopressin, *J. Biol. Chem.* **205:**45–53. [369]

Terence, C. F., From, G. H., and Rousseau, M. S., 1983, Baclofen—Its effect on seizure frequency, *Arch. Neurol.* **40:**28–29. [225]

Terenius, L., and Wahlstrom, A., 1975, Search for an endogenous ligand for the opiate receptor, *Acta Physiol. Scand.* **94:**74–81. [374]

Terkel, A. S., Shryne, J., and Gorski, R. A., 1973, Inhibition of estrogen facilitation of sexual behavior by the intracerebral infusion of actinomycin-D, *Horm. Behav.* **4:**377–386. [522]

Terzuolo, C. A., and Araki, T., 1961, An analysis of intra- versus extra-cellular potential changes associated with activity of single spinal motoneurones, *Ann. N. Y. Acad. Sci.* **94**:547–558. [70–72]

Tettamanti, G., Venerando, B., Roberti, S., *et al.*, 1981, The fate of exogenously administered brain gangliosides, in: *Gangliosides in Neurological and Neuromuscular Function, Development and Repair* (M. M. Rapport and A. Gorio, eds.), pp. 225–240, Raven Press, New York. [466]

Thach, W. T., 1975, Timing of activity in cerebellar dentate nucleus and cerebral motor cortex during prompt volitional movement, *Brain Res.* **88**:233–241. [498, 587]

Thach, W. T., 1978, Single unit studies of long-loops involving the motor cortex and cerebellum during limb movement in monkeys, *Prog. Clin. Neurophysiol.* **4**:94–106. [498]

Thangnipon, W., and Storm-Mathiesen, J., 1981, K^+ induced, Ca^{2+} dependent release of [^{3}H]aspartate from terminals of the cortico-pontine pathway, *Neurosci. Lett.* **23**:181–187. [182]

Thangnipon, W., Taxt, T., Brodal, P., and Storm-Mathisen, J., 1983, The corticopontine projection: Axotomy-induced loss of high affinity L-glutamate and D-aspartate uptake, but not of gamma-aminobutyrate uptake, glutamate decarboxylase or choline acetyltransferase, in the pontine nuclei, *Neuroscience* **8**:490–497. [182]

Thompson, G. C., Cortez, A. M., and Lam, D. M.-K., 1985, Localization of GABA immunoreactivity in the auditory brainstem of guinea pigs, *Brain Res.* **339**:119–122. [206]

Thompson, R. F., Berger, T. W., and Madden, J., 1983, Cellular processes of learning and memory in the mammalian CNS, *Annu. Rev. Neurosci.* **6**:447–491. [565, 566]

Thompson, S. G., Wong, P. T.-H., Leong, S. F., and McGeer, E. G., 1985, Regional distribution in rat brain of 1-pyrroline-5-carboxylate dehydrogenase and its localization to specific glial cells, *J. Neurochem.* **48**:1791–1796. [180]

Tillakaratne, N. J. K., Huttner, S. L., and Tobin, A. J., 1985, NGF induction of glutamate decarboxylase (GAD) mRNA in pheochromocytoma cell lines, *Soc. Neurosci. Abstr.* **11**:355. [201]

Tinklenberg, J. R., Pigache, R., Berger, P. A., and Kopell, B. S., 1982, Desglycinamide-9-arginine-8-vasopressin (DGAVP, Organon 5667) in cognitively impaired patients, *Psychopharmacol. Bull.* **18**:202–204. [410]

Todd, R. D., and Ciaranello, R. D., 1985, Demonstration of inter- and intraspecies differences in serotonin binding sites by antibodies from an autistic child, *Proc. Natl. Acad. Sci. U.S.A.* **82**:612–616. [335]

Toffano, G., Savoini, G. E., Moroni, F., Lombardi, G., Calza, L., and Agnati, L. F., 1984, Chronic GM_1 ganglioside treatment reduces dopamine cell body degeneration in the substantia nigra after unilateral hemitransection in rat, *Brain Res.* **296**:233–239. [467]

Tong, J. H., and Kaufman, S., 1975, Tryptophan hydroxylase: Purification and some properties of the enzyme from rabbit hindbrain, *J. Biol. Chem.* **250**:4152–4158. [322, 324]

Torda, C., 1973, Depolarization by acetylcholine (acetylcholine activation of triphosphoinositide phosphomonoesterase), *Experientia* **29**:536–537. [255]

Torrens, Y., Beaujouan, J. C., Viger, A., and Glowinski, J., 1983, Properties of a ^{125}I-substance P derivative binding to synaptosomes from various brain structures and the spinal cord of the rat, *Naunyn-Schmiedebergs Arch. Pharmakol.* **324**:134–139. [403]

Torrens, Y., Lavielle, S., Chassing, G., Marquet, A., Glowinski, J., and Beaujouan, J. C., 1984, Neuromedin K, a tool in further distinguishing two central tachykinin binding sites, *Eur. J. Pharmacol.* **102**:381–382. [403]

Torvik, A., 1956, Transneuronal changes in the inferior olive and pontine nuclei in kittens, *J. Neuropathol. Exp. Neurol.* **15**:119–145. [463]

Tramu, G., Beauvillain, J. C., Croix, D., Pillez, A., and Garaud, J. C., 1984, Coexistence of hypothalamic factors with other neuropeptides: Demonstration in the median eminence of rats and guinea pigs, *Neurochem. Int.* **6**:721–730. [157]

Tran, V. T., and Snyder, S. H., 1981, Histidine decarboxylase, purification from fetal rat liver, immunologic properties and histochemical localization in brain and stomach, *J. Biol. Chem.* **256**:680–686. [354, 355]

Trauner, D. A., 1985, Olivopontocerebellar atrophy with dementia, blindness, and chorea, *Arch. Neurol.* **42**:757–758. [225]

Trebecis, A. K., 1973, Transmitters and reticulospinal neurons, *Exp. Neurol.* **40**:297–308. [228]

Trevarthen, C. B., and Sperry, R. W., 1973, Perceptual unity of the ambient visual field in human commissurotomy patients, *Brain* **96**:547–570. [611, 612]

Tsang, D., Lal, S., and Finlayson, M. H., 1980, Effect of TRH on cyclic AMP formation in human pineal gland homogenates, *Brain Res.* **188**:278–281. [404]

Tschopp, F. A., Henke, H., Petermann, J. B., *et al.*, 1985, Calcitonin gene-related peptide and its binding sites in the human central nervous system and pituitary, *Proc. Natl. Acad. Sci. U.S.A.* **82:**248–252. [400, 402]

Tsui, J. K., Eisen, A., Mak, E., Carruthers, J., Scott, A., and Calne, D. B., 1985, A pilot study of the use of botulinum toxin in spasmodic torticollis, *Can. J. Neurol. Sci.* **12:**314–316. [259]

Tsuji, M., Sobue, K., and Nakajima, T., 1978, Studies on the formation of γ-aminobutyric acid from putrescine, *Neurochem. Res.* **3:**678. [202]

Tsukahara, N., 1981, Synaptic plasticity in the mammalian central nervous system, *Annu. Rev. Neurosci.* **4:**351–379. [439, 566]

Tsukahara, N., 1984, Classical conditioning mediated by the red nucleus: An approach beginning at the cellular level, in: *Neurobiology of Learning and Memory* (G. Lynch, J. L. McGaugh, and N. M. Weinberger, eds.), pp. 166–180, Guilford Press, New York. [439, 450, 451]

Tsukahara, N., Hultborn, H., Murakami, F., and Fujito, Y., 1975, Electrophysiological study of formation of new synapses and collateral sprouting in red nucleus neurons after partial denervation, *J. Neurophysiol.* **38:**1359–1372. [439, 441]

Tsukahara, N., Fujito, Y., Oda, Y., and Maeda, J., 1982, Formation of functional synapses in the adult cat red nucleus from the cerebrum following cross-innervation of forelimb flexor and extensor nerves, *Exp. Brain Res.* **45:**1–12. [450, 451]

Tsukahara, N., Fujito, Y., and Kubota, M., 1983, Specificity of the newly formed cortico-rubral synapses in the kitten red nucleus, *Exp. Brain Res.* **51:**45–56. [441–444]

Tsuruo, Y., Hisano, S., Okamura, Y., Tsukamoto, N., and Daikoku, S., 1984, Hypothalamic substance P-containing neurons: Sex-dependent topographical differences and ultrastructural transformations associated with stages of the estrous cycle, *Brain Res.* **305:**331–341. [387]

Tung, A. S., and Yaksh, T. L., 1982, *In vivo* evidence for multiple opiate receptors mediating analgesia in the rat spinal cord, *Brain Res.* **247:**75–83. [406]

Tuomisto, L., 1977, Ontogenesis and regional distribution of histamine and histamine-*N*-methyltransferase in the guinea pig brain, *J. Neurochem.* **25:**271–276. [352]

Turner, A. J., 1984, Neuropeptide processing enzymes, *TINS* **7:**258–230. [379]

Turner, R. W., Baimbridge, K. G., and Miller, J. J., 1982, Calcium-induced long-term potentiation in the hippocampus, *Neuroscience* **7:**1411–1416. [564, 565]

Twarog, B. M., and Page, I. H., 1953, Serotonin content of some mammalian tissues and urine and a method for its determination, *Am. J. Physiol.* **175:**157–161. [319]

Tyrer, P., and Marsden, C., 1985, New antidepressant drugs: Is there anything new they tell us about depression?, *TINS* **8:**427–431. [540, 549]

Uchizono, K., 1967, Synaptic organization of the Purkinje cells in the cerebellum of the cat, *Exp. Brain Res.* **4:**97–113. [11]

Uddman, R., Edvinsson, L., Ekman, R., Kingman, T., and McCulloch, J., 1985, Innervation of the feline cerebral vasculature by nerve fibers containing calcitonin gene-related peptide: Trigeminal origin and coexistence with substance P, *Neurosci. Lett.* **62:**131–136. [406, 408]

Udenfriend, S., 1966, Tyrosine hydroxylase, *Pharmacol. Rev.* **18:**43–51. [267]

Ueda, H., Shiomi, H., and Takagi, H., 1980, Regional distribution of a novel analgesic dipeptide kyotorphin (Tyr-Arg) in the rat brain and spinal cord, *Brain Res.* **198:**460–464. [408]

Uhl, G. R., and Kuhar, M. J., 1984, Chronic neuroleptic treatment enhances neurotensin receptor binding in human and rat substantia nigra, *Nature (London)* **309:**350–352. [413]

Uhl, G. R., and Snyder, S. H., 1976, Regional and subcellular distributions of brain neurotensin, *Life Sci.* **19:**1827–1832. [506]

Uhl, G. R., and Snyder, S. H., 1979, Neurotensin: A neuronal pathway projecting from amygdala through stria terminalis, *Brain Res.* **161:**522–526. [395]

Uhl, G. R., and Snyder, S. H., 1981, Neurotensin, *Adv. Biochem. Psychopharmacol.* **28:**87–106. [402]

Uhl, G. R., Kuhar, M. J., and Snyder, S. H., 1978, Enkephalin-containing pathway: Amygdaloid efferents in the stria terminalis, *Brain Res.* **149:**223–228. [395]

Uhl, G. R., Goodman, R. R., and Snyder, S. H., 1979, Neurotensin-containing cell bodies, fibers and nerve terminals in the brain stem of the rat: Imunohistochemical mapping, *Brain Res.* **167:**77–91. [506]

Uhl, G. R., Whitehouse, P. J., Price, D. L., Tourtelotte, W. W., and Kuhar, M. J., 1984, Parkinson's disease: Depletion of substantia nigra neurotensin receptors, *Brain Res.* **308:**186–190. [412]

Ungerstedt, U., 1971, Stereotaxic mapping of the monoamine pathways in the rat brain, *Acta Physiol. Scand. Suppl.* **367**:1–48. [289, 292]

Usherwood, P. N. R., 1981, Glutamate synapses and receptors in insect muscle, in: *Glutamate as a Neurotransmitter* (G. Di Chiara and G. L. Gessal, eds.), pp. 183–192, Raven Press, New York. [187]

Vaccarino, F., Conti Tronconi, B. M., Panula, P., Guidotti, A., and Costa, E., 1985, GABA-modulin: A synaptosomal basic protein that differs from small myelin basic protein of rat brain, *J. Neurochem.* **44**:278–290. [220]

Vale, W., Rivier, J., Ling, N., and Brown, M., 1978, Biologic and immunologic activities and applications of somatostatin analogs, *Metabolism* **27**(Suppl. 1):1391–1401. [409]

Valentino, K. L., Winter, J., and Reichardt, L. F., 1985, Applications of monoclonal antibodies to neuroscience research, *Annu. Rev. Neurosci.* **8**:199–232. [46]

Vallbo, A. B., 1971, Muscle spindle response at the onset of isometric voluntary contractions in man: Time difference between fusimotor and skeletomotor effects, *J. Physiol. (London)* **318**:405–431. [486]

Van Balgooy, J. N. A., Marshall, F. D., and Roberts, E., 1972, Metabolism of intracerebrally administered histidine, histamine and imidazoleacetic acid in mice and frogs, *J. Neurochem.* **19**:2341–2353. [351]

Van den Pol, A. N., and Tsujimoto, K. L., 1985, Neurotransmitters of the hypothalamic suprachiasmatic nucleus: Immunocytochemical analysis of 25 neuronal antigens, *Neuroscience* **15**:1049–1086. [394]

Van der Heyden, J. A. M., Venema, K., and Korf, J., 1979, *In vivo* release of endogenous GABA from rat substantia nigra measured by a novel method, *J. Neurochem.* **32**:469–476. [212, 215]

Van der Kooy, D., and Hattori, T., 1980, Single subthalamic nucleus neurons project to both the globus pallidus and substantia nigra in rat, *J. Comp. Neurol.* **192**:751–768. [508]

Van der Kooy, D., Hattori, T., Shannak, K., and Hornykiewicz, O., 1981a, The pallido-subthalamic projection in rat: Anatomical and biochemical studies, *Brain Res.* **199**:466–472. [212]

Van der Kooy, D., Coscina, D. V., and Hattori, T., 1981b, Is there a non-dopaminergic nigrostriatal pathway?, *Neuroscience* **6**:345–357. [506]

Van Kammen, D. P., Sternberg, D. E., Hare, T. A., and Waters, R. N., 1982, CSF levels of γ-aminobutyric acid in schizophrenia, *Arch. Gen. Psychiatry* **39**:91–97. [547]

Van Leeuwen, F. W., Caffe, A. R., and de Vries, G. J., 1985, Vasopressin cells in the bed nucleus of the stria terminalis of the rat: Sex difference and the influence of androgens, *Brain Res.* **325**:391–394. [388, 395]

Varon, S. S., and Bunge, R. P., 1978, Trophic mechanisms in the peripheral nervous system, *Annu. Rev. Neurosci.* **1**:327–361. [458, 565]

Vazquez, D., 1979, Inhibitors of protein biosynthesis, *Mol. Biol. Biochem. Biophys.* **30**:1–312. [526]

Venter, J. C., Eddy, B., Schaber, J. S., Lilly, L., and Fraser, C. M., 1983, Purification and molecular characterization of neurotransmitter receptors, *Prog. Clin. Biol. Res.* **135**:183–206. [165]

Verghese, J. P., Bradley, W. G., Mitsumoto, H., and Chad, D., 1982, A blind controlled trial of adrenocorticotropin and cerebral gangliosides in nerve regeneration in the rat, *Exp. Neurol.* **77**:455–458. [467]

Verhofstad, A. A. J., and Jonsson, G., 1983, Immunohistochemical and neurochemical evidence for the presence of serotonin in the adrenal medulla of the rat, *Neuroscience* **10**:1443–1453. [157, 333]

Verhofstad, A. A., Steinbusch, H. W. M., Penke, B., Varga, J., and Joosten, H. W. J., 1981, Serotonin-immunoreactive cells in the superior cervical ganglion of the rat: Evidence for the existence of separate serotonin and catecholamine-containing small ganglionic cells, *Brain Res.* **212**:39–49. [332]

Vermes, I., Mulder, G. H., Berkenbosch, F., and Tilders, F. J. H., 1981, Release of β-lipotropin and β-endorphin from rat hypothalami *in vitro*, *Brain Res.* **211**:248–254. [402]

Vickroy, T. W., Roeske, W. R., Gehlert, D. R., Wamsley, J. K., and Yamamura, H. I., 1985, Quantitative light microscopic autoradiography of [^{3}H]hemicholinium-3 binding sites in the rat central nervous system: A novel biochemical marker for mapping the distribution of cholinergic nerve terminals, *Brain Res.* **329**:368–373. [260]

Victor, M., Adams, R. D., and Collins, G. H., 1971, *The Wernicke–Korsakoff-Syndrome*, Blackwell, Oxford. [567, 569]

Vijayashankar, N., and Brody, H. A., 1979, A quantitative study of the pigmented neurons in the nucleus locus coeruleus and subcoeruleus in man as related to aging, *J. Neuropathol. Exp. Neurol.* **38**:490–498. [469, 470]

Villalobos-Molina, R., and Garcia-Sainz, J. A., 1983, H_1-histaminergic activation stimulates phosphatidylinositol labeling in rabbit aorta, *Eur. J. Pharmacol.* **90**:457–459. [357]

Vincent, S. R., and Johansson, O., 1983, Striatal neurons containing both somatostatin- and avian pancreatic polypeptide (APP)-like immunoreactivities and NADPH-diaphorase activity: A light and electron microscopic study, *J. Comp. Neurol.* **217**:264–270. [157]

Vincent, S. R., and Satoh, K., 1984, Corticotropin-releasing factor (CRF) immunoreactivity in the dorsolateral pontine tegmentum: Further studies on the micturition reflex system, *Brain Res.* **308**:387–391. [388]

Vincent, S. R., Hattori, T., and McGeer, E. G., 1978, The nigrotectal projection: A biochemical and ultrastructural characterization, *Brain Res.* **151**:159–164. [212]

Vincent, S. R., Kimura, H., and McGeer, E. G., 1980, The pharmacohistochemical demonstration of GABA-transaminase, *Neurosci. Lett.* **16**:345–348. [213]

Vincent, S. R., Hökfelt, T., Christensson, I., and Terenius, L., 1982a, Dynorphin-immunoreactive neurons in the central nervous system of the rat, *Neurosci. Lett.* **33**:185–190. [507]

Vincent, S. R., Hökfelt, T., Christensson, I., and Terenius, L., 1982b, Immunohistochemical evidence for a dynorphin immunoreactive striatonigral pathway, *Eur. J. Pharmacol.* **85**:251–252. [399, 507]

Vincent, S. R., Hökfelt, T., and Wu, J.-Y., 1982c, GABA neuron systems in hypothalamus and the pituitary gland, *Neuroendocrinology* **34**:117–125. [206]

Vincent, S. R., Johansson, O., Hökfelt, T., *et al.*, 1982d, Neuropeptides: Coexistence in human cortical neurones, *Nature (London)* **298**:65–67. [157]

Vincent, S. R., Skirboll, L., Hökfelt, T., *et al.*, 1982e, Coexistence of somatostatin- and avian pancreatic polypeptide (APP)-like immunoreactivity in some forebrain neurons, *Neuroscience* **7**:439–446. [157]

Vincent, S. R., Hökfelt, T., Skirboll, L., and Wu, J.-Y., 1983a, Hypothalamic γ-aminobutyric acid neurons project to the cortex, *Science* **120**:1309–1310. [212, 355]

Vincent, S. R., Hökfelt, T., Wu, J.-Y., Elde, R. P., Morgan, L. M., and Kimmel, J. R., 1983b, Immunohistochemical studies of the GABA system in the pancreas, *Neuroendocrinology* **36**:197–204. [218]

Vincent, S. R., Satoh, K., Armstrong, D. M., and Fibiger, H. C., 1983c, Substance P in the ascending cholinergic reticular system, *Nature (London)* **306**:688–691. [157, 252]

Vincent, S. R., Staines, W. A., and Fibiger, H. C., 1983d, Histochemical demonstration of separate populations of somatostatin and cholinergic neurons in the rat striatum, *Neurosci. Lett.* **35**:111–114. [250, 507]

Vincent, S. R., McIntosh, C. H. S., Buchan, A. M. J., and Brown, J. C., 1985, Central somatostatin systems revealed with monoclonal antibodies, *J. Comp. Neurol.* **238**:169–186. [395, 398, 400]

Viveros, O. H., Lee, C.-L., Abou-Donia, M. M., Nixon, J. C., and Nichol, C. A., 1981, Biopterin cofactor biosynthesis: Independent regulation of GTP cyclohydrolase in adrenal medulla and cortex, *Science* **213**:349–350. [273]

Vogt, M., 1954, The concentration of sympathin in different parts of the central nervous system under normal conditions and after the administration of drugs, *J. Physiol. (London)* **123**:451–481. [267]

Von Euler, U. S., and Gaddum, J. H., 1931, An unidentified depressor substance in certain tissue extracts, *J. Physiol. (London)* **72**:74–87. [374]

Von Schwarzenfeld, I., 1979, Origin of transmitters released by electrical stimulation from a small metabolically very active vesicular pool of cholinergic synapses in guinea pig cerebral cortex, *Neuroscience* **4**:477–493. [245]

Von Wendt, L., Simila, S., Saukkonen, A. L., and Koivists, M., 1980, Failure of strychnine treatment during the neonatal period in three Finnish children with nonketotic hyperglycinemia, *Pediatrics* **65**:1166–1169. [231]

Vulliet, P. R., Woodgett, J. R., and Cohen, P., 1984, Phosphorylation of tyrosine hydroxylase by calmodulin-dependent multiprotein kinase, *J. Biol. Chem.* **259**:13,680–13,683. [283]

Wada, H., Watanabe, T., Maeyama, K., Taguchi, Y., and Hayashi, H., 1984, Mammalian histidine decarboxylase and its suicide substrate α-fluoromethylhistidine, in: *Chemical and Biological Aspects of Vitamin B_6 Catalysis: Part A* (A. E. Evangelopoulos, ed.), pp. 245–254, Alan R. Liss, New York. [350, 351]

Wada, J., and Rasmussen, T., 1960, Intracarotid injection of sodium amytal for the lateralization of cerebral speech dominance: Experimental and clinical observations, *J. Neurosurg.* **17**:266–282. [617]

Wada, J. A., Clark, R., and Hamm, A., 1975, Cerebral hemispheric asymmetry in humans, *Arch. Neurol.* **32**:239–246. [618]

Waddington, J. L., and Cross, A. J., 1978, Neurochemical changes following kainic acid lesions of nucleus accumbens—implications for a GABAergic accumbal–ventral tegmental pathway, *Life Sci.* **22**:1011–1014. [212]

Wadhwa, S., Hámori, J., and Bijlani, V., 1985, Immunohistochemical localization of GABAergic cells in the developing human dorsal lateral geniculate nucleus, *Neurosci. Lett.* **61**:97–101. [206]

Walaas, I., 1981, Biochemical evidence for overlapping neocortical and allocortical glutamate projections to the nucleus accumbens and rostral caudatoputamen in the rat brain, *Neuroscience* **6**:399–405. [182]

Walaas, I., and Fonnum, F., 1979a, The distribution and origin of glutamate decarboxylase and choline acetyltransferase in ventral pallidum and other basal forebrain regions, *Brain Res.* **177**:325–336. [212]

Walaas, I., and Fonnum, F., 1979b, Biochemical evidence for gamma-aminobutyrate containing fibres from the nucleus accumbens to the substantia nigra and ventral tegmental areas in the rat, *Neuroscience* **5**:63–72. [212]

Walaas, I., and Fonnum, F., 1979c, The effects of surgical and chemical lesions on neurotransmitter candidates in the nucleus accumbens of the rat, *Neuroscience* **4**:209–216. [184]

Walaas, I., and Fonnum, F., 1980, Biochemical evidence for glutamate as a transmitter in hippocampal efferents to the basal forebrain and hypothalamus in the rat brain, *Neuroscience* **5**:1691–1698. [182, 212]

Waldmeier, P. C., Feldtrauer, J. J., and Maitre, L., 1977, Methylhistamine: Evidence for selective deamination by MAO-B in the rat brain *in vivo*, *J. Neurochem.* **29**:785–790. [351]

Walker, J. E., 1983, Glutamate, GABA, and CNS disease: A review, *Neurochem. Res.* **8**:521–550. [505]

Walker, J. E., and Fonnum, F., 1983, Effect of regional cortical ablations on high-affinity D-aspartate uptake in striatum, olfactory tubercle, and pyriform cortex of the rat, *Brain Res.* **278**:283–286. [184]

Walker, J. M., Moises, H., Coy, D., Baldrighic, G., and Akil, H., 1982, Nonopiate effects of dynorphin and des-Tyr-dynorphin, *Science* **218**:1136–1138. [407]

Walker, R. F., 1984, Impact of age-related changes in serotonin and norepinephrine metabolism on reproductive function in female rats: An analytical review, *Neurobiol. Aging* **5**:121–139. [338]

Wall, P. D., and Woolf, C. J., 1980, What we don't know about pain, *Nature (London)* **287**:185–186. [408]

Walsh, C., and Guillery, R. W., 1984, Fiber order in the pathways from the eye to the brain, *TINS* **7**:208–211. [435]

Walsh, T. J., Tilson, H. A., DeHaven, D. L., Mailman, R. B., Fisher, A., and Hanin, I., 1984, AF64A, a cholinergic neurotoxin, selectively depletes acetylcholine in hippocampus and cortex, and produces long-term passive avoidance and radial-arm maze deficits in the rat, *Brain Res.* **321**:91–102. [260]

Walters, J. R., Bunney, B. S., and Roth, R. H., 1975, Piribedil and apomorphine: Pre- and post-synaptic effects on dopamine synthesis and neuronal activity, in: *Dopaminergic Mechanisms* (D. B. Calne, T. N. Chase, and A. Barbeau, eds.), pp. 273–284, Raven Press, New York. [296]

Wamsley, J. K., Zarbin, M. A., Birdsall, N. J. M., and Kuhar, M. J., 1980, Muscarinic cholinergic receptors: Autoradiographic localization of high and low affinity agonist binding sites, *Brain Res.* **200**:1–12. [254]

Wamsley, J. K., Zarbin, M. A., and Kuhar, M. J., 1981, Muscarinic cholinergic receptors flow in the sciatic nerve, *Brain Res.* **217**:155–161. [254]

Wang, J.-Y., Yaksh, T. L., and Go, Y. L. W., 1985, Studies on the *in vivo* release of vasoactive intestinal polypeptide (VIP) from the cerebral cortex: Effects of cortical, brainstem and somatic stimuli, *Brain Res.* **326**:317–334. [402]

Ward, H. W., Thanki, C. M., Peterson, D. W., and Bradford, H. F., 1982, Brain glutaminase activity in relation to transmitter glutamate biosynthesis, *Biochem. Soc. Trans.* **10**:369–370. [179]

Waszczak, B. L., and Walters, J. R., 1983, Dopamine modulation of the effects of γ-aminobutyric acid on substantia nigra pars reticulata neurons, *Science* **220**:218–221. [297]

Watanabe, T., Taguchi, Y., Hayashi, H., *et al.*, 1983, Evidence for the presence of a histaminergic neuron system in the rat brain: An immunohistochemical analysis, *Neurosci. Lett.* **39**:249–254. [354]

Watanabe, T., Taguchi, Y., Shiosaka, S., *et al.*, 1984, Distribution of the histaminergic neuron system in the central nervous system of rats: A fluorescent immunohistochemical analysis with histidine decarboxylase as a marker, *Brain Res.* **295**:13–25. [354]

Watkins, J. C., and Evans, R. H., 1981, Excitatory amino acid transmitters, *Annu. Rev. Pharmacol.* **21**:165–204. [182, 187]

Watkins, L. R., Suberg, S. N., Thurston, C. L., and Culhane, E. S., 1986, Role of spinal cord neuropeptides in pain sensitivity and analgesia: Thyrotropin releasing hormone and vasopressin, *Brain Res.* **362**:308–317. [407]

Watson, M., Roeske, W. R., and Yamamura, H. I., 1982, [3H] Pirenzepine selectively identifies a high-affinity population of muscarinic cholinergic receptors in the rat cerebral cortex, *Life Sci.* **31**:2019–2023. [254]

Watson, S. P., and Downes, C. P., 1983, Substance P induces hydrolysis of inositol phospholipids in guinea pig ileum and rat hypothalamus, *Eur. J. Pharmacol.* **93**:245–253. [404]

Weinberg, R. J., Bentivoglio, M., Phend, K., Schmechel, D. E., and Rustioni, A., 1985, A new double-labeling method demonstrates transmitter-specific projections, *Neurosci. Lett.* **55**:349–353. [206, 212]

Weingartner, H., Gold, P., Ballenger, J. C., *et al.,* 1981, Effects of vasopressin on human memory functions, *Science* **211**:601–603. [410]

Weinreich, D., 1979, γ-Glutamylhistidine: A major product of histamine metabolism in ganglia of the marine mollusk, *Aplysia Californica, J. Neurochem.* **32**:363–369. [353]

Weiskrantz, L., 1974, The interaction between occipital and temporal cortex in vision: An overview, in: *The Neurosciences: Third Study Program* (F. O. Schmitt and F. G. Worden, eds.), pp. 189–204, MIT Press, Cambridge, Massachusetts. [611]

Weiss, D. G., 1982, General properties of axoplasmic transport, in: *Axoplasmic Transport in Physiology and Pathology* (D. G. Weiss and A. Gorio, eds.), pp. 1–14, Springer-Verlag, Berlin, Heidelberg, and New York. [33]

Weiss, P., 1941, Nerve patterns: The mechanics of nerve growth, *Third Growth Symp.* **5**:163–203. [420, 435]

Weiss, P., and Hiscoe, H. B., 1948, Experiments on the mechanism of nerve growth, *J. Exp. Zool.* **107**:315–396. [30]

Wendt, P., Reimann, H. J., Swoboda, K., Hennings, G., and Blumel, G., 1980, The use of flavenoids as inhibitors of histidine decarboxylase in gastric disease: Experimental and clinical studies, *Naunyn-Schmiedebergs Arch. Pharmakol.* **313**(Suppl.):R60. [359]

Wenthold, R. J., 1979, Release of endogenous glutamic acid, aspartic acid and GABA from cochlear nucleus slices, *Brain Res.* **162**:338–343. [178, 182]

Wenthold, R. J., 1980, Glutaminase and aspartate amino transferase decrease in cochlear nucleus after lesion of the auditory nerve, *Brain Res.* **190**:283–297. [178]

Wenthold, R. J., and Gulley, R. L., 1978. Glutamic acid and aspartic acid in the cochlear nucleus of the waltzing guinea pig, *Brain Res.* **158**:295–302. [178, 182]

Werman, R., Davidoff, R. A., and Aprison, M. H., 1968, Inhibitory effect of glycine in spinal neurons in the cat, *J. Neurophysiol.* **31**:81–95. [230]

Westlund, K. N., Bowker, R. M., Ziegler, M. G., and Coulter, J. D., 1983, Origins and terminations of descending noradrenergic projections to the spinal cord of monkey, *Brain Res.* **292**:1–16. [293]

Westlund, K. N., Denney, R. M., Kochersperger, L. M., Rose, R. M., and Abell, C. W., 1985, Distinct monoamine oxidase A and B populations in primate brain, *Science* **230**:181–183. [278, 326]

Westman, J., Blomqvist, A., Kohler, C., and Wu, J.-Y., 1984, Light and electron microscopic localization of glutamic acid decarboxylase and substance P in the dorsal column nuclei of the cat, *Neurosci. Lett.* **51**:347–352. [157]

Westrum, L. E., and Black, R. G., 1971, Fine structural aspects of the synaptic organization of the spinal trigeminal nucleus (pars interpolaris) of the cat, *Brain Res.* **25**:265–287. [438]

Wheal, H. V., and Miller, J. J., 1980, Pharmacological identification of acetylcholine and glutamate excitatory systems in the dentate gyrus of the rat, *Brain Res.* **182**:145–155. [182]

White, D. M., and Helme, R. D., 1985, Release of substance P from peripheral nerve terminals following electrical stimulation of the sciatic nerve, *Brain Res.* **336**:27–31. [402]

White, J. D., Stewart, K. D., Krause, J. E., and McKelvy, J. F., 1985, Biochemistry of peptide-secreting neurons, *Physiol. Rev.* **65**:553–606. [377]

White, P., Goodhardt, M. J., Kent, J. P., *et al.,* 1977, Neocortical cholinergic neurons in elderly people, *Lancet* **1**:668–671. [262]

White, T., 1959, Formation and catabolism of histamine in brain tissue *in vitro, J. Physiol. (London)* **149**:34–42. [350]

White, T. D., and Leslie, R. A., 1982, Depolarization-induced release of adenosine 5′-triphosphate from isolated varicosities derived from the myenteric plexus of the guinea pig small intestine, *J. Neurosci.* **2**:206–215. [367]

White, W. F., Nadler, J. V., Hamberger, A., Cotman, C. W., and Cummins, J. T., 1977, Glutamate as transmitter of hippocampal perforant path, *Nature (London)* **270**:356–357. [182]

White, W. F., Nadler, J. V., and Cotman, C. W., 1979, The effect of acidic amino acid antagonists on synaptic transmission in the hippocampal formation *in vitro, Brain Res.* **164**:177–194. [182]

White, W. F., Snodgrass, S. R., and Dichter, M., 1980, Identification of GABA neurons in rat cortical cultures by GABA uptake autoradiography, *Brain Res.* **190**:139–152. [206]

Whitehouse, P. J., Price, D. L., Clark, A. W., Coyle, J. T., and DeLong, M. R., 1981, Alzheimer's disease: Evidence for selective loss of cholinergic neurons in the nucleus basalis, *Ann. Neurol.* **10**:122–126. [261]

Whitehouse, P. J., Price, D. L., Struble, R. G., Clark, A. W., Coyle, J. T., and DeLong, M. R., 1982,

Alzheimer's disease and senile dementia: Loss of neurons in the basal forebrain, *Science* **215**:1237–1239. [261]

Whitehouse, P. J., Hedreen, J. C., White, C. L., and Price, D. L., 1983, Basal forebrain neurons in the dementia of Parkinson disease, *Ann. Neurol.* **13**:243–248. [308, 470]

Whitnall, M. H., Gainer, H., Cox, B. M., and Molineaux, C. J., 1983, Dynorphin-A-(-8) is contained within vasopressin neurosecretory vesicles in rat pituitary, *Science* **222**:1137–1139. [157]

Whittaker, V. P., and Gray, E. G., 1962, The synapse: Biology and morphology, *Br. Med. Bull.* **18**:223–228. [11, 12]

Whittaker, V. P., Michaelson, I. A., and Kirkland, R. J. A., 1964, The separation of synaptic vesicles from nerve ending particles (''synaptosomes''), *Biochem. J.* **90**:293–303. [237]

Wieraszko, A., and Lynch, G. S., 1978, Stimulation-dependent release of possible transmitter substances from hippocampal slices with localized perfusion, *Brain Res.* **160**:372–376. [182]

Wiesel, T. N., 1982, The postnatal development of the visual cortex and the influence of environment (Nobel lecture), *Biosci. Rep.* **2**:351–377. [605]

Wiesel, T. N., and Hubel, D. H., 1963, Single-cell responses in striate cortex of kittens deprived of vision in one eye, *J. Neurophysiol.* **26**:1003–1017. [453]

Wiesel, T. N., Hubel, D. H., and Lam, D. M. K., 1974, Autoradiographic demonstrations of ocular-dominance columns in the monkey striate cortex by means of transneuronal transport, *Brain Res.* **79**:273–279. [462]

Wiesendanger, M., 1981, Organization of secondary motor areas of the cerebral cortex, in: *Handbook of Physiology: Neurophysiology*, Vol. VII, *Cerebral Control Mechanisms*, pp. 1121–1147, American Physiology Society, Bethesda, Maryland. [483, 487]

Wiesendanger, M., and Miles, J. S., 1982, Ascending pathways of low threshold muscle afferents to the cerebral cortex and its possible role in muscle control, *Physiol. Rev.* **62**:1234–1270. [490]

Wiesendanger, M., Séguin, J. J., and Künzle, H., 1973, The supplementary motor area—a control system for posture?, in: *Control of Posture and Locomotion* (R. B. Stein, K. C. Pearson, R. S. Smith, and J. B. Redford, eds.), pp. 331–346, Plenum Press, New York. [480]

Wiesenfeld-Hallin, Z., Hökfelt, T., Lundberg, J. M., 1984, Immunoreactive calicotinin gene-related peptide and substance P coexist in sensory neurons to the spinal cord and interact in spinal behavioral responses of the rat, *Neurosci. Lett.* **52**:199–204. [157]

Wigstrom, H., and Gustafsson, B., 1983, Heterosynaptic modulation of homosynaptic long-lasting potentiation in the hippocampal slice, *Acta Physiol. Scand.* **119**:455–458. [560]

Wijnen, H. J. L. M., Versteeg, D. H. G., Palkovits, M., and De Jong, W., 1977, Increased adrenaline content of individual nuclei of the hypothalamus and the medulla oblongata of genetically hypertensive rats, *Brain Res.* **135**:180–185. [315]

Wijnen, H. J. L. M., Palkovits, M., De Jong, W., and Versteeg, D. H. G., 1978, Elevated adrenaline content in nuclei of the medulla oblongata and the hypothalamus during the development of spontaneous hypertension, *Brain Res.* **157**:191–195. [315]

Wikkelsø, C., Fahrenkrug, J., Blomstrand, C., and Johansson, B. B., 1985, Dementia of different etiologies: Vasoactive intestinal polypeptide in CSF, *Neurology* **35**:592–595. [412]

Wiklund, L., Leger, L., and Persson, M., 1981, Monoamine cell distribution in the cat brain stem: A fluorescence histochemical study with quantification of indolaminergic and locus coeruleus groups, *J. Comp. Neurol.* **203**:613–647. [329, 331]

Wiklund, L., Toggenburger, G., and Cuenod, M., 1982, Aspartate: Possible neurotransmitter in cerebellar climbing fibers, *Science* **216**:78–80. [182, 186]

Wilcox, B. J., and Seybold, V. S., 1982, Localization of neuronal histamine in rat brain, *Neurosci. Lett.* **29**:105–110. [354]

Wilkin, G. P., Hudson, A. L., Hill, D. R., and Bowery, N. G., 1981, Autoradiographic localization of $GABA_B$ receptors in rat cerebellum, *Nature (London)* **294**:584–587. [229]

Williams, M., 1983, Adenosine receptors in the mammalian central nervous system, *Prog. Neuropsychopharmacol. Biol. Psychiatry* **7**:443–450. [366]

Williams, M., and Risley, E. A., 1980. Biochemical characterization of putative central purinergic receptors by using 2-chloro[^{3}H]adenosine, a stable analog of adenosine, *Proc. Natl. Acad. Sci. U.S.A.* **77**:6892–6896. [364]

Williams, R. G., and Dockray, G. J., 1983, Immunohistochemical studies of FMRF-amide-like immunoreactivity in rat brain, *Brain Res.* **276**:213–229. [388]

Williams, R. G., Gayton, R. J., Zhu, W., and Dockray, G. J., 1981, Changes in brain cholecystokinin octapeptide following lesions of the medial forebrain bundle, *Brain Res.* **213**:227–230. [157]

Willis, W. D., 1981, Role of serotonin in processing of nociceptive information, *Adv. Exp. Med. Biol.* **133**:101–104. [338]

Willis, W. D., 1982, *Control of Nociceptive Transmission in the Spinal Cord,* Springer-Verlag, Heidelberg and New York, 159 pp. [603, 605]

Willis, W. D., Haber, L. H., and Martin, R. F., 1977, Inhibition of spinothalamic tract cells and interneurons by brain stem stimulation in the monkey, *J. Neurophysiol.* **40**:968–981. [605]

Willner, P., 1983, Dopamine and depression: A review of recent evidence, *Brain Res.* **287**:211–246. [549]

Wilson, S. P., Klein, R. L., Chang, K.-J., Gasparis, M. S., Viveros, O. H., and Yang, W.-H., 1980, Are opioid peptides co-transmitters in noradrenergic vesicles of sympathetic nerves?, *Nature (London)* **288**:707–709. [157, 403]

Wilson, V. J., 1979, Electrophysiological and dynamic studies of vestibulospinal reflexes, in: *Integration in the Nervous System* (H. Asanuma and V. J. Wilson, eds.), pp. 167–184, Igaku-Shoin, Tokyo. [501]

Wilson, V. J., and Burgess, P. R., 1962, Disinhibition in the cat spinal cord, *J. Neurophysiol.* **25**:392–404. [138, 141]

Wilson, V. J., Yoshida, M., and Schor, R. H., 1970, Supraspinal monosynaptic excitation and inhibition of the thoracic back motoneurons, *Exp. Brain Res.* **11**:282–295. [141]

Winders, A., and Vogt, W., 1907, Synthese des Imidazolyläthylamins, *Ber. Dtsch. Chem. Ges.* **40**:3691–3695. [349]

Witkin, J. W., and Silverman, A. J., 1983, Luteinizing hormone-releasing hormone (LHRH) in rat olfactory systems, *J. Comp. Neurol.* **218**:426–432. [388]

Wofsey, A. R., Kuhar, M. J., and Snyder, S. H., 1971, A unique synaptosomal fraction which accumulates glutamic and aspartic acids in brain tissue, *Proc. Natl. Acad. Sci. U.S.A.* **68**:1102–1106. [176]

Wojcik, W. J., and Neff, N. H., 1982, Adenosine measurement by a rapid HPLC-fluorometric method: Induced changes of adenosine content in regions of rat brain, *J. Neurochem.* **39**:280–282. [363, 364]

Wojcik, M., Ulas, J., and Oderfeld-Nowak, B., 1982, The stimulating effect of ganglioside injections on the recovery of choline acetyltransferase and acetylcholinesterase activities in the hippocampus of the rat after septal lesions, *Neuroscience* **7**:495–499. [467]

Wolf, P., Olpe, H. R., Avrith, D., and Haas, H. L., 1978, GABAergic inhibition of neurons in the ventral tegmental area, *Experientia* **34**:73–74. [212]

Wolff, H. G., 1963, *Headache and Other Head Pain,* Oxford University Press, New York. [408]

Wolff, J. R., and Chronwall, B. M., 1982, Axosomatic synapses in the visual cortex of adult rat: A comparison between GABA-accumulating and other neurons, *J. Neurocytol.* **11**:409–425. [206]

Wolfson, B., Manning, R. W., Davis, L. G., Arentzen, R., and Baldino, F., Jr., 1985, Co-localization of corticotropin releasing factor and vasopressin mRNA in neurones after adrenalectomy, *Nature (London)* **315**:59–61. [157]

Wollemann, M., Szebeni, A., Bajusz, S., and Gráf, L., 1979, Effect of met-enkephalin and (d-met^2, pro^5)-enkephalinamide on the adenylate cyclase activity of rat brain, *Neurochem. Res.* **4**:627–631. [404]

Wong, P. T.-H., and McGeer, E. G., 1981, Postnatal changes of GABAergic and glutamatergic parameters, *Dev. Brain Res.* **1**:519–529. [179]

Wong, P. T.-H., McGeer, E. G., and McGeer, P. L., 1981, A sensitive radiometric assay for ornithine aminotransferase: Regional and subcellular distribution in rat brain, *J. Neurochem.* **36**:501–505. [179]

Wong, P. T.-H., McGeer, E. G., and McGeer, P. L., 1982a, Effects of kainic acid injection and cortical lesion on ornithine and aspartate aminotransferases in rat striatum, *J. Neurosci. Res.* **8**:643–650. [178]

Wong, P. T.-H., McGeer, P. L., Rossor, M., and McGeer, E. G., 1982b, Ornithine aminotransferase in Huntington's disease, *Brain Res.* **231**:466–471. [179]

Wong, P. T.-H., McGeer, E. G., Singh, K., and McGeer, P. L., 1983, Changes in muscarinic binding in distant brain regions showing neuronal losses after folic acid injections into the substantia innominata, *J. Neurochem.* **40**:1754–1757. [160, 161]

Wong, R. K. S., and Prince, D. A., 1978, Participation of calcium spikes during intrinsic burst firing in hippocampal neurons, *Brain Res.* **159**:385–390. [60, 560]

Wood, J. D., 1975, The role of gamma-aminobutyric acid in the mechanism of seizures, *Prog. Neurobiol.* **5**:77–95. [223]

Wood, J. G., McLaughlin, B. J., and Vaughn, J. E., 1976, Immunocytochemical localization of glutamate decarboxylase in electron microscopic preparations of rodent CNS, in: *GABA in Nervous System Function* (E. Roberts, T. N. Chase, and D. B. Tower, eds.), pp. 133–148, Raven Press, New York. [206]

Woodhams, P. L., Roberts, G. W., Polak, J. M., and Crow, T. J., 1983, Distribution of neuropeptides in the limbic system of the rat: The bed nucleus of the stria terminalis, septum and preoptic area, *Neuroscience* **8:**677–703. [395]

Woolley, D. W., and Shaw, E., 1954, A biochemical and pharmacological suggestion about certain mental disorders, *Science* **119:**587–588. [320, 338, 547]

Woolsey, C. N., Settlage, P. H., and Meyer, D. R., 1952, Patterns of localization in precentral and ''supplementary'' motor areas and their relation to the concept of a premotor area, *Res. Publ. Assoc. Res. Nerv. Ment. Dis.* **30:**238–264. [482]

Worton, R. G., Duff, C., Sylvester, J. C., *et al.*, 1984, Duchenne muscular dystrophy involving translocation of the DMD gene next to ribosome RNA genes, *Science* **224:**1447–1449. [45]

Wright, J. W., Sullivan, M. J., Petersen, E. P., and Harding, J. W., 1985, Brain angiotensin II and III binding and dipsogenicity in the rabbit, *Brain Res.* **358:**376–379. [406]

Wu, J.-Y., 1976, Purification, characterization and kinetic studies of GAD and GABA-T from mouse brain, in: *GABA in Nervous System Function* (E. Roberts, T. N. Chase, and D. B. Tower, eds.), pp. 7–60, Raven Press, New York. [200, 201]

Wu, J.-Y., Brandon, C., Su, G. G., and Lam, D. M., 1981, Immunocytochemical and autoradiographic localization of GABA system in the vertebrate retina, *Mol. Cell. Biochem.* **39:**229–238. [206]

Wu, P. H., Phillis, J. W., and Thierry, D. L., 1982, Adenosine receptor agonists inhibit K^+-evoked Ca^{2+} uptake by rat brain cortical synaptosomes, *J. Neurochem.* **39:**700–708. [365]

Wu, P. H., Moron, M., and Barraco, R., 1984, Organic calcium channel blockers enhance [^{3}H]purine release from rat brain cortical synaptosomes, *Neurochem. Res.* **9:**1019–1031. [365]

Wu, S. M., and Dowling, J. E., 1980, Effects of GABA and glycine on the distal cells of the cyprina retina, *Brain Res.* **199:**401–414. [218]

Wurtman, R. J., Pohorecky, L. A., and Baliga, B. S., 1972, Adrenocortical control of the biosynthesis of epinephrine and proteins in the adrenal medulla, *Pharmacol. Rev.* **24:**411–426. [315]

Wyatt, H. J., and Daw, N. W., 1976, Specific effects of neurotransmitter antagonists on ganglion cells in rabbit retina, *Science* **191:**204–205. [229]

Wyatt, R. J., Chase, T. N., Scott, J., and Snyder, F., 1970, Effect of L-dopa on the sleep of man, *Nature (London)* **228:**999–1000. [338]

Wyatt, R. J., Potkin, S. G., Kleinman, J. E., Weinberger, D. R., Luchins, D. J., and Jeste, D. V., 1981, The schizophrenia syndrome: Examples of biological tools for subclassification, *J. Nerv. Ment. Dis.* **169:**100–112. [547]

Yajima, H., and Kitagawa, K., 1973, Studies on peptides. XXXIV. Conventional synthesis of the undecapeptide amide corresponding to the entire amino acid sequence of bovine substance P, *Chem. Pharm. Bull. (Tokyo)* **21:**682–683. [374]

Yamada, M., Watanabe, T., Fukul, H., Taguchi, Y., and Wada, H., 1984, Comparison of histidine decarboxylase from rat stomach and brain with that from whole bodies of rat fetus, *Agents Actions* **14:**143–152. [351]

Yamamoto, C., and Matsui, S., 1976, Effect of stimulation of excitatory nerve tract on release of glutamic acid from olfactory cortex slices *in vitro, J. Neurochem.* **26:**487–491. [182]

Yamamura, H. I., and Snyder, S. H., 1973, High affinity transport of choline into synaptosomes of rat brain, *J. Neurochem.* **21:**1355–1374. [242, 243]

Yamamura, H. I., Kuhar, M. J., Greenberg, D., and Snyder, S. H., 1974, Muscarinic cholinergic receptor binding: Regional distribution on monkey brain, *Brain Res.* **66:**541–546. [244]

Yamamura, H. I., Enna, S. J., and Kuhar, M. J. (eds.), 1985, *Neurotransmitter Receptor Binding,* Raven Press, New York. [162]

Yamane, Y., Masatsugu, N., Yamamoto, J., Umeda, Y., and Ogino, K., 1984, Release of vasopressin by electrical stimulation of the intermediate portion of the nucleus of the tractus solitarius in rats with cervical spinal cordotomy and vagotomy, *Brain Res.* **324:**358–360. [402]

Yamauchi, T., and Fujisawa, H., 1981, A calmodulin-dependent protein kinase that is involved in the activation of tryptophan 5-monooxygenase is specifically distributed in brain tissues, *FEBS Lett.* **129:**117–119. [329]

Yamauchi, T., and Fujisawa, H., 1984, Calmodulin-dependent protein kinase (kinase 11) which is involved in the activation of tryptophan 5-monooxygenase catalyses phosphorylation of tubulin, *Arch. Biochem. Biophys.* **234**:89–96. [329]

Yamazoe, M., Shiosaka, S., Shibasaki, T., *et al.*, 1984, Distribution of six neuropeptides in the nucleus tractus solitarii of the rat: An immunohistochemical analysis, *Neuroscience* **13**:1243–1266. [399]

Yang, H.-Y. T., Panula, P., Tang, J., and Costa, E., 1983, Characterization and localization of Met[5]-enkephalin-Arg[6]-Phe[7] stored in various rat brain regions, *J. Neurochem.* **40**:969–976. [507]

Yates, C. M., Harmar, A. J., Rosie, R., *et al.*, 1983, Thyrotropin-releasing hormone, luteinizing hormone-releasing hormone and substance P immuno-reactivity in post-mortem brain from cases of Alzheimer-type dementia and Down's syndrome, *Brain Res.* **258**:45–52. [412]

Yim, C. Y., and Morgensen, G. J., 1980, Effects of picrotoxin and nipecotic acid on inhibitory response of dopaminergic neurons in the ventral tegmental area to stimulation of the nucleus accumbens, *Brain Res.* **199**:466–472. [212]

Yim, G. K. W., and Lowy, M. T., 1984, Opioids, feeding and anorexias, *Fed. Proc. Fed. Am. Soc. Exp. Biol.* **43**:2893–2897. [405]

Yoon, M. G., 1975, On topographic polarity of the optic tectum in the goldfish, *Cold Spring Harbor Symp. Quant. Biol.* **40**:503–519. [436]

Yoshida, M., and Omata, S., 1979, Blocking by picrotoxin of nigra-evoked inhibition of neurons of ventromedial nucleus of the thalamus, *Experientia* **35**:794–795. [212]

Young, A. B., Oster-Granite, M. L., Herndon, R. M., and Snyder, S. H., 1974, Glutamic acid: Selective depletion by viral-induced granule cell loss in hamster cerebellum, *Brain Res.* **73**:1–13. [182, 184]

Young, A. B., Bromberg, M. B., and Penney, J. B., 1981, Decreased glutamate uptake in subcortical areas deafferented by sensorimotor cortex ablation in the cat, *J. Neurosci.* **1**:241–249. [182]

Young, A. B., Penney, J. B., Dauth, G. W., Bromberg, M. B., and Gilman, S., 1983, Glutamate or aspartate as a possible neurotransmitter of cerebral corticofugal fibers in the monkey, *Neurology* **33**:1513–1516. [182, 184]

Young, A. P., Gunning, P. W., Landreth, G. E., Ignatius, M., and Shooter, E. M., 1983, The nerve growth factor induced differentiation of a pheochromocytoma PC 12 cell line, in: *Nervous System Regeneration* (B. Haber, J. R. Perez-Polo, G. A. Hashim, and A. M. Guiffrida Stella, eds.), pp. 33–46, Alan R. Liss, New York. [465]

Young, E. F., Ralston, E., Blake, J., Ramachandran, J., Hall, Z. W., and Stroud, R. M., 1985, Topological mapping of acetylcholine receptor: Evidence for a model with five transmembrane segments and a cytoplasmic COOH-terminal peptide, *Proc. Natl. Acad. Sci. U.S.A.* **82**:626–630. [255–257]

Young, S. N., 1981, Mechanism of decline in rat brain 5-hydroxytryptamine after induction of liver tryptophan pyrrolase by hydrocortisone: Roles of tryptophan catabolism and kynurenine synthesis, *Br. J. Pharmacol.* **74**:695–700. [323]

Young, S. N., and Gauthier, S., 1981, Tryptophan availability and the control of 5-hydroxytryptamine and tryptamine synthesis in human CNS, *Adv. Exp. Med. Biol.* **133**:221–230. [323]

Zaczek, R., Hedreen, J. C., and Coyle, J. T., 1979, Evidence for a hippocampal–septal glutamatergic pathway in the rat, *Exp. Neurol.* **65**:145–156. [182]

Zager, E. L., and Black, P. M., 1985, Neuropeptides in human memory and learning processes, *Neurosurgery* **17**:355–369. [410]

Zaidel, E., 1976, Auditory language comprehension in the right hemisphere following cerebral commisurotomy and hemispherectomy: A comparison with child language and aphasia, in: *The Acquisition and Breakdown of Language: Parallels and Divergencies* (A. Caramazza and E. B. Zurif, eds.), pp. 228–275, Johns Hopkins University Press, Baltimore. [623]

Zaidel, E., 1977, Unilateral auditory language comprehension on the Token Test following cerebral commissurotomy and hemispherectomy, *Neuropsychologia* **15**:1–18. [623]

Zakut, H., Matzkel, A., Schejter, E., Avni, A., and Soreq, H., 1985, Polymorphism of acetylcholinesterase in discrete regions of the developing human fetal brain, *J. Neurochem.* **45**:382–389. [240]

Zamir, N., Skofitsch, G., Bannon, M. J., Helke, C. J., Kopin, I. J., and Jacobowitz, D. M., 1984, Primate model of Parkinson's disease: Alterations in multiple opioid systems in the basal ganglia, *Brain Res.* **322**:356–360. [411]

Zamir, N., Weber, E., Palkovits, M., and Brownstein, M. J., 1985, Distribution of immunoreactive metorphamide (adrenorphin) in discrete regions of the rat brain: Comparison with Met-enkephalin-Arg[6]-Gly[7]-Leu[8], *Brain Res.* **361**:193–199. [404]

Zarbin, M. A., Wamsley, J. K., and Kuhar, M. J., 1981, Glycine receptor: Light microscopic autoradiographic localization with [³H] strychnine, *J. Neurosci.* **1**:532–547.   [230]

Zeki, S. M., 1980, The representation of colours in the cerebral cortex, *Nature (London)* **284**:412–418.   [610]

Zhang, A. Z., Zhou, G. Z., Xi, G. F., *et al.*, 1985, Lower CSF level of dynorphin(1–8) immunoreactivity in schizophrenic patients, *Neuropeptides* **5**:553–556.   [412]

Zieglgansberger, W., and Puil, E. A., 1973, Actions of glutamic acid on spinal neurones, *Exp. Brain Res.* **17**:35–49.   [182, 187]

Zieher, L. M., and Jaim-Etcheverry, G., 1980, Neurotoxicity of *N*-(2-chloroethyl)-*N*-ethyl-2-bromobenzylamine hydrochloride (DSP-4) on noradrenergic neurons is mimicked by its cyclic aziridinium derivative, *Eur. J. Pharmacol.* **65**:249–256.   [311]

Zigmond, R. E., Chalazonitis, A., and Joh, T., 1980, Preganglionic nerve stimulation increases the amount of tyrosine hydroxylase in the rat superior cervical ganglion, *Neurosci. Lett.* **20**:61–65.   [282]

Zimmerman, E. A., Nilaver, G., Hou-Yu, A., and Silverman, A. J., 1984, Vasopressinergic and oxytocinergic pathways in the central nervous system, *Fed. Proc. Fed. Am. Soc. Exp. Biol.* **43**:91–96.   [389]

Zimmerman, F. T., and Ross, S., 1944, Effect of glutamic acid and other amino acids on maze learning in the white rat, *Arch. Neurol. Psychiatry* **51**:446–451.   [194]

Zimmerman, F. T., Burgemeister, B. B., and Putnam, T. J., 1946, Effects of glutamic acid on mental functioning in children and in adolescents, *Arch. Neurol. Psychiatry* **56**:489–506.   [194]

Zlobina, G. P., Kondakova, L. I., and Khalansky, A. S., 1984, Specific binding of [³H] diazepam in mouse glioblastoma: The influence of clonazepam and Ro 5-4864 on [³H] diazepam binding, *Neurosci. Lett.* **52**:259–262.   [222]

Zola-Morgan, S., Mishkin, M., and Squire, L. R., 1981, The anatomy of amnesia: Hippocampus and amygdala vs. temporal stem, *Soc. Neurosci. Abstr.* **7**:236.   [568]

Index